Lecture Notes in Computer Science 9452

Commenced Publication in 1973
Founding and Former Series Editors:
Gerhard Goos, Juris Hartmanis, and Jan van Leeuwen

More information about this series at http://www.springer.com/series/7410

Tetsu Iwata · Jung Hee Cheon (Eds.)

Advances in Cryptology – ASIACRYPT 2015

21st International Conference on the Theory
and Application of Cryptology and Information Security
Auckland, New Zealand, November 29 – December 3, 2015
Proceedings, Part I

 Springer

Editors
Tetsu Iwata
Nagoya University
Nagoya
Japan

Jung Hee Cheon
Seoul National University
Seoul
Korea (Republic of)

ISSN 0302-9743 ISSN 1611-3349 (electronic)
Lecture Notes in Computer Science
ISBN 978-3-662-48796-9 ISBN 978-3-662-48797-6 (eBook)
DOI 10.1007/978-3-662-48797-6

Library of Congress Control Number: 2015953256

LNCS Sublibrary: SL4 – Security and Cryptology

Springer Heidelberg New York Dordrecht London

Printed on acid-free paper

Springer-Verlag GmbH Berlin Heidelberg is part of Springer Science+Business Media
(www.springer.com)

Preface

ASIACRYPT 2015, the 21st Annual International Conference on Theory and Application of Cryptology and Information Security, was held on the city campus of the University of Auckland, New Zealand, from November 29 to December 3, 2015. The conference focused on all technical aspects of cryptology, and was sponsored by the International Association for Cryptologic Research (IACR).

The conference received 251 submissions from all over the world. The program included 64 papers selected from these submissions by a Program Committee (PC) comprising 43 leading experts of the field. In order to accommodate as many high-quality submissions as possible, the conference ran in two parallel sessions, and these two-volume proceedings contain the revised versions of the papers that were selected. The revised versions were not reviewed again and the authors are responsible for their contents.

The selection of the papers was made through the usual double-blind review process. Each submission was assigned three reviewers and submissions by PC members were assigned five reviewers. The selection process was assisted by a total of 339 external reviewers. Following the individual review phase, the selection process involved an extensive discussion phase.

This year, the conference featured three invited talks. Phillip Rogaway gave the 2015 IACR Distinguished Lecture on "The Moral Character of Cryptographic Work," Gilles Barthe gave a talk on "Computer-Aided Cryptography: Status and Perspectives," and Masayuki Abe spoke on "Structure-Preserving Cryptography." The proceedings contain the abstracts of these talks. The conference also featured a traditional rump session that contained short presentations on the latest research results of the field.

The best paper award was decided based on a vote by the PC members, and it was given to "Improved Security Proofs in Lattice-Based Cryptography: Using the Rényi Divergence Rather than the Statistical Distance" by Shi Bai, Adeline Langlois, Tancrède Lepoint, Damien Stehlé, and Ron Steinfeld. Two more papers, "Key-Recovery Attacks on ASASA" by Brice Minaud, Patrick Derbez, Pierre-Alain Fouque, and Pierre Karpman, and "The Tower Number Field Sieve" by Razvan Barbulescu, Pierrick Gaudry, and Thorsten Kleinjung, were solicited to submit full versions to the *Journal of Cryptology*.

ASIACRYPT 2015 was made possible by the contributions of many people. We would like to thank the authors for submitting their research results to the conference. We are deeply grateful to all the PC members and all the external reviewers for their hard work to determine the program of the conference. We sincerely thank Steven Galbraith, the general chair of the conference, and the members of the local Organizing Committee for handling all the organizational work of the conference. We also thank Nigel Smart for organizing and chairing the rump session.

We thank Shai Halevi for setting up and letting us use the IACR conference management software. Springer published the two-volume proceedings and made these available at the conference. We thank Alfred Hofmann, Anna Kramer, and their colleagues for handling the editorial process. Last but not least, we thank the speakers, session chairs, and all the participants for coming to Auckland and contributing to ASIACRYPT 2015.

December 2015   Tetsu Iwata
                                                              Jung Hee Cheon

ASIACRYPT 2015

The 21st Annual International Conference on Theory and Application of Cryptology and Information Security

Sponsored by the International Association for Cryptologic Research (IACR)

November 29–December 3, 2015, Auckland, New Zealand

General Chair

Steven Galbraith   University of Auckland, New Zealand

Program Co-chairs

Tetsu Iwata   Nagoya University, Japan
Jung Hee Cheon   Seoul National University, Korea

Program Committee

Daniel J. Bernstein   University of Illinois at Chicago, USA and Technische
                          Universiteit Eindhoven, The Netherlands
Ignacio Cascudo   Aarhus University, Denmark
Chen-Mou Cheng   National Taiwan University, Taiwan
Sherman S.M. Chow   Chinese University of Hong Kong, Hong Kong, SAR China
Kai-Min Chung   Academia Sinica, Taiwan
Nico Döttling   Aarhus University, Denmark
Jens Groth   University College London, UK
Dawu Gu   Shanghai Jiaotong University, China
Dong-Guk Han   Kookmin University, Korea
Marc Joye   Technicolor, USA
Nathan Keller   Bar-Ilan University, Israel
Aggelos Kiayias   National and Kapodistrian University of Athens, Greece
Kaoru Kurosawa   Ibaraki University, Japan
Xuejia Lai   Shanghai Jiaotong University, China
Hyang-Sook Lee   Ewha Womans University, Korea
Jooyoung Lee   Sejong University, Korea
Soojoon Lee   Kyung Hee University, Korea
Arjen Lenstra   EPFL, Switzerland
Hemanta K. Maji   UCLA, USA
Alexander May   Ruhr University Bochum, Germany
Bart Mennink   KU Leuven, Belgium
Tatsuaki Okamoto   NTT Secure Platform Laboratories, Japan
Raphael C.-W. Phan   Multimedia University, Malaysia

Josef Pieprzyk	Queensland University of Technology, Australia
Bart Preneel	KU Leuven, Belgium
Damien Robert	Inria Bordeaux, France
Giovanni Russello	University of Auckland, New Zealand
Ahmad-Reza Sadeghi	TU Darmstadt, Germany
Rei Safavi-Naini	University of Calgary, Canada
Palash Sarkar	Indian Statistical Institute, India
Yu Sasaki	NTT Secure Platform Laboratories, Japan
Peter Schwabe	Radboud University, The Netherlands
Jae Hong Seo	Myongji University, Korea
Nigel Smart	University of Bristol, UK
Damien Stehlé	ENS de Lyon, France
Tsuyoshi Takagi	Kyushu University, Japan
Mehdi Tibouchi	NTT Secure Platform Laboratories, Japan
Dominique Unruh	University of Tartu, Estonia
Serge Vaudenay	EPFL, Switzerland
Vesselin Velichkov	University of Luxembourg, Luxembourg
Huaxiong Wang	Nanyang Technological University, Singapore
Hongjun Wu	Nanyang Technological University, Singapore
Vassilis Zikas	ETH Zurich, Switzerland

Additional Reviewers

Masayuki Abe
Divesh Aggarwal
Shashank Agrawal
Shweta Agrawal
Hyunjin Ahn
Janaka Alawatugoda
Martin Albrecht
Gergely Alpár
Joël Alwen
Prabhanjan Ananth
Elena Andreeva
Yoshinori Aono
Daniel Apon
Hassan Jameel Asghar
Tomer Ashur
Nuttapong Attrapadung
Maxime Augier
Jean-Philippe Aumasson
Christian Badertscher
Yoo-Jin Baek
Shi Bai

Foteini Baldimtsi
Razvan Barbulescu
Achiya Bar-On
Harry Bartlett
Lejla Batina
Aurélie Bauer
Carsten Baum
Anja Becker
Fabrice Benhamouda
Shivam Bhasin
Sanjay Bhattacherjee
Begül Bilgin
Gaëtan Bisson
Jonathan Bootle
Joppe W. Bos
Elette Boyle
Zvika Brakerski
Mark Bun
David Cash
Guilhem Castagnos
Andrea Cerulli

Pyrros Chaidos
Debrup Chakraborty
Donghoon Chang
Seunghwan Chang
Yun-An Chang
Chien-Ning Chen
Jie Chen
Ming-Shing Chen
Yu-Chi Chen
Dooho Choi
Seung Geol Choi
Ji Young Chun
Stelvio Cimato
Sandro Coretti
Jean-Marc Couveignes
Joan Daemen
Bernardo David
Angelo De Caro
Jeroen Delvaux
Gregory Demay
Patrick Derbez

Jintai Ding
Itai Dinur
Christophe Doche
Ming Duan
Léo Ducas
Alina Dudeanu
Orr Dunkelman
Keita Emura
Martianus Frederic
  Ezerman
Xiong Fan
Antonio Faonio
Pooya Farshim
Sebastian Faust
Marc Fischlin
Eiichiro Fujisaki
Philippe Gaborit
Martin Gagné
Steven Galbraith
Nicolas Gama
Wei Gao
Peter Gaži
Essam Ghadafi
Hossein Ghodosi
Irene Giacomelli
Benedikt Gierlichs
Zheng Gong
Dov Gordon
Robert Granger
Sylvain Guilley
Jian Guo
Qian Guo
Zheng Guo
Divya Gupta
Florian Göpfert
Jaecheol Ha
Xue Haiyang
Keisuke Hakuta
Shuai Han
Neil Hanley
Malin Md Mokammel
  Haque
Yasufumi Hashimoto
Gottfried Herold
Javier Herranz
Shoichi Hirose
Viet Tung Hoang

Dennis Hofheinz
Justin Holmgren
Deukjo Hong
Wei-Chih Hong
Tao Huang
Yun-Ju Huang
Pavel Hubáček
Michael Hutter
Andreas Hülsing
Jung Yeon Hwang
Laurent Imbert
Sorina Ionica
Zahra Jafargholi
Tibor Jager
Jérémy Jean
Ik Rae Jeong
Hyungrok Jo
Thomas Johansson
Antoine Joux
Handan Kılınç
Taewon Kim
Alexandre Karlov
Pierre Karpman
Kenji Kashiwabara
Aniket Kate
Marcel Keller
Carmen Kempka
Dmitry Khovratovich
Dakshita Khurana
Jinsu Kim
Jongsung Kim
Min Kyu Kim
Sungwook Kim
Tae Hyun Kim
Taechan Kim
Taewan Kim
Paul Kirchner
Elena Kirshanova
Susumu Kiyoshima
Thorsten Kleinjung
Jessica Koch
Markulf Kohlweiss
Ilan Komargodski
Venkata Koppula
Ranjit Kumaresan
Po-Chun Kuo
Stefan Kölbl

Pascal Lafourcade
Russell W.F. Lai
Adeline Langlois
Martin M. Lauridsen
Changhoon Lee
Changmin Lee
Eunjeong Lee
Hyung Tae Lee
Juhee Lee
Tancrède Lepoint
Wen-Ding Li
Yang Li
Benoît Libert
Seongan Lim
Changlu Lin
Fuchun Lin
Tingting Lin
Wei-Kai Lin
Feng-Hao Liu
Junrong Liu
Shengli Liu
Ya Liu
Zhen Liu
Zhenhua Liu
Zhiqiang Liu
Satya Lokam
Carl Löndahl
Yu Long
Steve Lu
Yiyuan Luo
Atul Luykx
Vadim Lyubashevsky
Alex J. Malozemoff
Avradip Mandal
Giorgia Azzurra Marson
Luke Mather
Takahiro Matsuda
Christian Matt
Peihan Miao
Daniele Micciancio
Andrea Miele
Eric Miles
Kazuhiko Minematsu
Marine Minier
Takaaki Mizuki
Ameer Mohammed
Paweł Morawiecki

Daisuke Moriyama
Kirill Morozov
Nicky Mouha
Nadia El Mrabet
Pratyay Mukherjee
Yusuke Naito
Chanathip Namprempre
Mridul Nandi
María Naya-Plasencia
Khoa Nguyen
Ruben Niederhagen
Jesper Buus Nielsen
Ivica Nikolić
Svetla Nikova
Tobias Nilges
Ryo Nishimaki
Wakaha Ogata
Go Ohtake
Claudio Orlandi
Ilya Ozerov
Jiaxin Pan
Giorgos Panagiotakos
Omkant Pandey
Kostas Papagiannopoulos
Cheol-Min Park
Bryan Parno
Anat Paskin-Cherniavsky
Chris Peikert
Bo-Yuan Peng
Clément Pernet
Léo Perrin
Giuseppe Persiano
Thomas Peters
Christophe Petit
Albrecht Petzoldt
Thomas Peyrin
Le Trieu Phong
Cécile Pierrot
Bertram Poettering
Joop van de Pol
Antigoni Polychroniadou
Carla Ràfols
Yogachandran
    Rahulamathavan
Sergio Rajsbaum
Somindu C. Ramanna
Samuel Ranellucci

Vanishree Rao
Christian Rechberger
Oded Regev
Michał Ren
Oscar Reparaz
Reza Reyhanitabar
Vincent Rijmen
Matthieu Rivain
Vladimir Rožić
Saeed Sadeghian
Yusuke Sakai
Subhabrata Samajder
Simona Samardjiska
Katerina Samari
Alessandra Scafuro
Jacob C.N. Schuldt
Karn Seth
Yannick Seurin
Setareh Sharifian
Ji Sun Shin
Bo-Yeon Sim
Siang Meng Sim
Leonie Simpson
Shashank Singh
Arkadii Slinko
Mate Soos
Pierre-Jean Spaenlehauer
Martijn Stam
Ron Steinfeld
Christoph Striecks
Le Su
Koutarou Suzuki
Alan Szepieniec
Björn Tackmann
Katsuyuki Takashima
Syh-Yuan Tan
Qiang Tang
Christophe Tartary
Sidharth Telang
Isamu Teranishi
Stefano Tessaro
Ivan Tjuawinata
Daniel Tschudi
Yiannis Tselekounis
Yu-Hsiu Tung
Himanshu Tyagi
Aleksei Udovenko

Praveen Vadnala
Srinivas Vivek Venkatesh
Frederik Vercauteren
Damien Vergnaud
Gilles Villard
Dhinakaran
    Vinayagamurthy
Vanessa Vitse
Damian Vizár
Lei Wang
Qingju Wang
Weijia Wang
Bogdan Warinschi
Hoeteck Wee
Benjamin Wesolowski
Carolyn Whitnall
Daniel Wichs
Xiaodi Wu
Hong Xu
Sen Xu
Shota Yamada
Naoto Yanai
Bo-Yin Yang
Guomin Yang
Shang-Yi Yang
Masaya Yasuda
Takanori Yasuda
Kazuki Yoneyama
Taek-Young Youn
Ching-Hua Yu
Shih-Chun Yu
Yu Yu
Aaram Yun
Thomas Zacharias
Mark Zhandry
Bingsheng Zhang
Hui Zhang
Jiang Zhang
Liang Feng Zhang
Liting Zhang
Tao Zhang
Ye Zhang
Yongjun Zhao
Bo Zhu
Jens Zumbrägel

Organizing Committee

Chair

Steven Galbraith   University of Auckland, New Zealand

Local Committee Members

Peter Gutmann   Computer Science, University of Auckland, New Zealand
Hinne Hettema   ITS, University of Auckland, New Zealand
Giovanni Russello   Computer Science, University of Auckland, New Zealand
Arkadii Slinko   Mathematics, University of Auckland, New Zealand
Clark Thomborson   Computer Science, University of Auckland, New Zealand

Sponsors

The University of Auckland
Microsoft Research
Intel
STRATUS
Centre for Discrete Mathematics and Theoretical Computer Science

Invited Talks

Structure-Preserving Cryptography

Masayuki Abe

NTT Secure Platform Laboratories, NTT Corporation, Tokyo, Japan
abe.masayuki@lab.ntt.co.jp

Bilinear groups has been a common ground for building cryptographic schemes since its introduction in seminal works [3, 5, 6]. Not just being useful for directly designing schemes for their rich mathematical structure, they aim to modular construction of complex schemes from simpler building blocks that work over the same bilienar groups. Namely, given a description of blinear groups, several building blocks exchange group elements each other, and the security of the resulting scheme is proven based on the security of the underlying building blocks. Unfortunately, things are not that easy in reality. Building blocks often require grues that bridge incompatible interfaces or they have to be modified to work together and the security has to be re-proved.

Structure-preserving cryptography [2] is a paradigm for designing cryptographic schemes over bilinear groups. A cryptographic scheme is called structure preserving if its all public inputs and outputs consist of group elements of bilinear groups and the functional correctness can be verified only by computing group operations, testing group membership and evaluating pairing product equations. Due to the regulated interface, structure-preserving schemes are highly inter-operable as desired in modular constructions. In particular, combination of structure-preserving signatures and noninteractive proof system of [4] yields numerous applications that protect signers' or receivers' privacy. The required properties on the other hand make some important primitives such as pseudo-random functions and collision resistant shrinking commitments unavailable in the world of structure-preserving cryptography. Interestingly, however, the constraints on the verification of correctness aim to argue non-trivial lower bounds in some aspects of efficiency such as signature size in the structure-preserving signature schemes.

Since the first use of the term "structure-preserving" in [1] in 2010, intensive research has been done for the area. In this talk, we overview state of the art on several structure-preserving schemes including commitments and signatures with a careful look about underlying assumptions, known bounds, and impossibility results. We also show open questions and discuss promising directions for further research.

References

1. Abe, M., Fuchsbauer, G., Groth, J., Haralambiev, K., Ohkubo, M.: Structure-preserving signatures and commitments to group elements. In: Advances in Cryptology - CRYPTO 2010, 30th Annual Cryptology Conference, Santa Barbara, CA, USA, 15–19 August 2010. Proceedings, pp. 209–236 (2010)

2. Abe, M., Fuchsbauer, G., Groth, J., Haralambiev, K., Ohkubo, M.: Structure-preserving signatures and commitments to group elements. J. Cryptol. (2015). doi:http://dx.doi.org/10.1007/s00145-014-9196-7.
3. Boneh, D., Boyen, X.: Efficient selective-id secure identity-based encryption without random oracles. In: Advances in Cryptology - EUROCRYPT 2004, International Conference on the Theory and Applications of Cryptographic Techniques, Interlaken, Switzerland, 2–6 May 2004. Proceedings, pp. 223–238 (2004)
4. Groth, J., Sahai, A.: Efficient noninteractive proof systems for bilinear groups. SIAM J. Comput. **41**(5), 1193–1232 (2012)
5. Menezes, A., Okamoto, T., Vanstone, S.A.: Reducing elliptic curve logarithms to logarithms in a finite field. IEEE Trans. Inf. Theory **39**(5), 1639–1646 (1993)
6. Sakai, R., Kasahara, M.: ID based cryptosystems with pairing on elliptic curve. IACR Cryptology ePrint Archive 2003, vol. 54 (2003)

Computer-Aided Cryptography:
Status and Perspectives

Gilles Barthe

IMDEA Software Institute, Madrid, Spain

Computer-aided cryptography is an emerging discipline which advocates the use of computer tools for building and mechanically verifying the security of cryptographic constructions. Computer-aided cryptography builds on the code-based game-based approach to cryptographic proofs, and adopts a program verification approach to justify common patterns of reasoning, such as equivalence up to bad, lazy sampling, or simply program equivalence. Technically, tools like EasyCrypt use a program verification method based on probabilistic couplings for reasoning about the relationship between two probabilistic programs, and standard tools to reason about the probability of events in a single probabilistic program. The combination of these tools, together with general mechanisms to instantiate or combine proofs, can be used to verify many examples from the literature.

Recent developments in computer-aided cryptography have explored two different directions. On the one hand, several groups have developed fully automated techniques to analyze cryptographic constructions in the standard model or hardness assumptions in the generic group model. In turn, these tools have been used for synthesizing new cryptographic constructions. *Transformational* synthesis tools take as input a cryptographic construction, for instance a signature in Type I setting and outputs another construction, for instance a batch signature or a signature in Type III setting. In contrast, *generative* synthesis tools take as input some size constraints and output a list of secure cryptographic constructions, for instance padding-based encryption schemes, modes of operations, or tweakable blockciphers, meeting the size constraints. On the other hand, several groups are working on carrying security proofs to (assembly-level) implementations, building on advances in programming languages, notably verified compilers. These works open the possibility to reason formally about mitigations used by cryptography implementers and to deliver strong mathematical guarantees, in the style of provable security, for cryptographic code against more realistic adversaries.

For further background information, please consult: www.easycrypt.info.

The Moral Character of Cryptographic Work

Phillip Rogaway[1]

Department of Computer Science
University of California, Davis, USA

Abstract. Cryptography rearranges power: it configures who can do what, from what. This makes cryptography an inherently *political* tool, and it confers on the field an intrinsically *moral* dimension. The Snowden revelations motivate a reassessment of the political and moral positioning of cryptography. They lead one to ask if our inability to effectively address mass surveillance constitutes a failure of our field. I believe that it does. I call for a community-wide effort to develop more effective means to resist mass surveillance. I plea for a reinvention of our disciplinary culture to attend not only to puzzles and math, but, also, to the societal implications of our work.

Keywords: Cryptography · Democracy · Ethics · Mass surveillance · Privacy · Snowden revelations · Social responsibility

[1] Work on the paper and talk associated to this abstract has been supported by NSF Grant CNS 1228828. Many thanks to the NSF for their continuing support.

Contents – Part I

ABE and IBE

Zero-Knowledge

Multiparty Computation II

Contents – Part II

Side-Channel Attacks

Design of Block Ciphers

Authenticated Encryption

Best Paper

Improved Security Proofs in Lattice-Based Cryptography: Using the Rényi Divergence Rather Than the Statistical Distance

Shi Bai[1]([✉]), Adeline Langlois[2,3], Tancrède Lepoint[4],
Damien Stehlé[1], and Ron Steinfeld[5]

[1] Laboratoire LIP (U. Lyon, CNRS, ENSL, INRIA, UCBL),
ENS de Lyon, Lyon, France
{shi.bai,damien.stehle}@ens-lyon.fr
http://perso.ens-lyon.fr/shi.bai, http://perso.ens-lyon.fr/damien.stehle
[2] EPFL, Lausanne, Switzerland
adeline.langlois@epfl.ch
http://lasec.epfl.ch/alangloi/
[3] CNRS/IRISA, Rennes, France
[4] CryptoExperts, Paris, France
tancrede.lepoint@cryptoexperts.com
https://www.cryptoexperts.com/tlepoint/
[5] Faculty of Information Technology, Monash University, Clayton, Australia
ron.steinfeld@monash.edu
http://users.monash.edu.au/rste/

Abstract. The Rényi divergence is a measure of closeness of two probability distributions. We show that it can often be used as an alternative to the statistical distance in security proofs for lattice-based cryptography. Using the Rényi divergence is particularly suited for security proofs of primitives in which the attacker is required to solve a search problem (e.g., forging a signature). We show that it may also be used in the case of distinguishing problems (e.g., semantic security of encryption schemes), when they enjoy a public sampleability property. The techniques lead to security proofs for schemes with smaller parameters, and sometimes to simpler security proofs than the existing ones.

1 Introduction

Let D_1 and D_2 be two non-vanishing probability distributions over a common measurable support X. Let $a \in (1, +\infty)$. The *Rényi divergence* [Rén61,EH12] (RD for short) $R_a(D_1 \| D_2)$ of order a between D_1 and D_2 is defined as the $((a-1)$th root of the) expected value of $(D_1(x)/D_2(x))^{a-1}$ over the randomness of x sampled from D_1. For notational convenience, our definition of the RD is the exponential of the classical definition [EH12]. The RD is an alternative to the statistical distance (SD for short) $\Delta(D_1, D_2) = \frac{1}{2} \sum_{x \in X} |D_1(x) - D_2(x)|$ as measure of distribution closeness, where we replace the difference in SD, by the ratio in RD. RD enjoys several properties that are analogous of those enjoyed

© International Association for Cryptologic Research 2015
T. Iwata and J.H. Cheon (Eds.): ASIACRYPT 2015, Part I, LNCS 9452, pp. 3–24, 2015.
DOI: 10.1007/978-3-662-48797-6_1

by SD, where addition in the property of SD is replaced by multiplication in the analogous property of RD (see Subsect. 2.3).

SD is ubiquitous in cryptographic security proofs. One of its most useful properties is the so-called *probability preservation property*: For any measurable event $E \subseteq X$, we have $D_2(E) \geq D_1(E) - \Delta(D_1, D_2)$. RD enjoys the analogous property $D_2(E) \geq D_1(E)^{\frac{a}{a-1}}/R_a(D_1 \| D_2)$. If the event E occurs with significant probability under D_1, and if the SD (resp. RD) is small, then the event E also occurs with significant probability under D_2. These properties are particularly handy when the success of an attacker against a given scheme can be described as an event whose probability should be negligible, e.g., the attacker outputs a new valid message-signature pair for a signature scheme. If in the attacker succeeds with good probability in the real scheme based on distribution D_1, then it also succeeds with good probability in the simulated scheme (of the security proof) based on distribution D_2.

To make the SD probability preservation property useful, it must be ensured that the SD $\Delta(D_1, D_2)$ is smaller than any $D_1(E)$ that the security proof must handle. Typically, the quantity $D_1(E)$ is assumed to be greater than some success probability lower bound ε, which is of the order of $1/\mathrm{poly}(\lambda)$ where λ refers to the security parameter, or even $2^{-o(\lambda)}$ if the proof handles attackers whose success probabilities can be sub-exponentially small (which we believe better reflects practical objectives). As a result, the SD $\Delta(D_1, D_2)$ must be $< \varepsilon$ for the SD probability preservation property to be relevant. Similarly, the RD probability preservation property is non-vacuous when the RD $R_a(D_1 \| D_2)$ is $\leq \mathrm{poly}(1/\varepsilon)$. In many cases, the latter seems less demanding than the former: in all our applications of RD, the RD between D_1 and D_2 is small while their SD is too large for the SD probability preservation to be applicable. In fact, as we will see in Subsect. 2.3, the RD becomes sufficiently small to be useful before the SD when $\sup_x D_1(x)/D_2(x)$ tends to 1. This explains the superiority of the RD in several of our applications.

Although RD seems more amenable than SD for search problems, it seems less so for distinguishing problems. A typical cryptographic example is semantic security of an encryption scheme. Semantic security requires an adversary $\mathcal{A}$ to distinguish between the encryption distributions of two plaintext messages of its choosing: the distinguishing advantage $\mathrm{Adv}_{\mathcal{A}}(D_1, D_2)$, defined as the difference of probabilities that $\mathcal{A}$ outputs 1 using D_1 or D_2, should be large. In security proofs, algorithm $\mathcal{A}$ is often called on distributions D_1' and D_2' that are close to D_1 and D_2 (respectively). If the SDs between D_1 and D_1' and D_2 and D_2' are both bounded from above by ε, then, by the SD probability preservation property (used twice), we have $\mathrm{Adv}_{\mathcal{A}}(D_1', D_2') \geq \mathrm{Adv}_{\mathcal{A}}(D_1, D_2) - 2\varepsilon$. As a result, SD can be used to distinguishing problems in a similar fashion as for search problems. The multiplicativity of the RD probability preservation property seems to prevent RD from being applicable to distinguishing problems.

We replace the statistical distance by the Rényi divergence in several security proofs for lattice-based cryptographic primitives. *Lattice-based cryptography* is a relatively recent cryptographic paradigm in which cryptographic primitives are shown at least as secure as it is hard to solve standard problems over lattices (see the survey [MR09]). Security proofs in lattice-based cryptography involve different types of distributions, often over infinite sets, such as continuous Gaussian distributions and Gaussian distributions with lattice supports. The RD seems particularly well suited to quantify the closeness of Gaussian distributions. Consider for example two continuous distributions over the reals, both with standard deviation 1, but one with center 0 and the other one with center c. Their SD is linear in c, so that c must remain extremely small for the SD probability preservation property to be useful. On the other hand, their RD of order $a = 2$ is bounded as $\exp(O(c^2))$ so that the RD preservation property remains useful event for slightly growing c.

RD was first used in lattice-based cryptography by [LPR13], in the decision to search reduction for the Ring Learning With Errors problem (which serves as a security foundation for many asymptotically fast primitives). It was then exploited in [LSS14] to decrease the parameters of the Garg et al. (approximation to) cryptographic multilinear maps [GGH13]. In the present work, we present a more extensive study of the power of RD in lattice-based cryptography, by showing several independent applications of RD. In some cases, it leads to security proofs allowing to take smaller parameters in the cryptographic schemes, hence leading to efficiency improvements. In other cases, this leads to alternative security proofs that are conceptually simpler.

Our applications of RD also include distinguishing problems. To circumvent the aforementioned a priori limitation of the RD probability preservation property for distinguishing problems, we propose an alternative approach that handles a class of distinguishing problems, enjoying a special property that we call *public sampleability*. This public sampleability allows to estimate success probabilities via Hoeffding's bound.

The applications we show in lattice-based cryptography are as follows:

- Smaller signatures for the Hash-and-Sign GPV signature scheme [GPV08].
- Smaller storage requirement for the Fiat-Shamir BLISS signature scheme [DDLL13, PDG14, Duc14].
- Alternative proof that the Learning With Errors (LWE) problem with noise chosen uniformly in an interval is no easier than the Learning With Errors problem with Gaussian noise [DMQ13]. Our reduction does not require the latter problem to be hard, and it is hence marginally more general as it also applies to distributions with smaller noises. Further, our reduction preserves the LWE dimension n, and is hence tighter than the one from [DMQ13] (the latter degrades the LWE dimension by a constant factor).[1]
- Smaller parameters in the dual-Regev encryption scheme from [GPV08].

---

[1] Note that LWE with uniform noise in a small interval is also investigated in [MP13], with a focus on the number of LWE samples. The reduction from [MP13] does not preserve the LWE reduction either.

We think RD is likely to have further applications in lattice-based cryptography, for search and for distinguishing problems.

Related Works. The framework for using RD in distinguishing problems was used in [LPSS14], in the context of the k-LWE problem (a variant of LWE in which the attacker is given extra information). In [PDG14], Pöpplemann, Ducas and Güneysu used the Kullback-Leibler divergence (which is the RD of order $a = 1$) to lower the storage requirement of [DDLL13]. Asymptotically, using the Kullback-Leibler divergence rather than SD only leads to a constant factor improvement. Our approach allows bigger savings in the case where the number of signature queries is limited, as explained in Sect. 3.

Very recently, Bogdanov *et al.* [BGM+15] adapted parts of our RD-based hardness proof for LWE with noise uniform in a small interval, to the Learning With Rounding problem. In particular, they obtained a substantial improvement over the hardness results of [BPR12, AKPW13].

Road-Map. In Sect. 2, we provide necessary background on lattice-based cryptography, and on the Rényi divergence. In Sect. 3, we use RD to improve some lattice-based signature scheme parameters. Section 4 contains the description of the framework in which we can use RD for distinguishing problems, which we apply to the dual-Regev encryption scheme. In Sect. 5, we describe an alternative hardness proof for LWE with noise uniformly chosen in an interval.

Notations. If x is a real number, we let $\lfloor x \rceil$ denote a closest integer to x. The notation ln refers to the natural logarithm and the notation log refers to the base 2 logarithm. We define $\mathbb{T} = ([0,1], +)$. For an integer q, we let $\mathbb{Z}_q$ denote the ring of integers modulo q. We let $\mathbb{T}_q$ denote the group $\mathbb{T}_q = \{i/q \mod 1 : i \in \mathbb{Z}\} \subseteq \mathbb{T}$. Vectors are denoted in bold. If $\boldsymbol{b}$ is a vector in $\mathbb{R}^d$, we let $\|\boldsymbol{b}\|$ denote its Euclidean norm. By default, all our vectors are column vectors.

If D is a probability distribution, we let $\mathrm{Supp}(D) = \{x : D(x) \neq 0\}$ denote its support. For a set X of finite weight, we let $U(X)$ denote the uniform distribution on X. The statistical distance between two distributions D_1 and D_2 over a countable support X is $\Delta(D_1, D_2) = \frac{1}{2}\sum_{x \in X} |D_1(x) - D_2(x)|$. This definition is extended in the natural way to continuous distributions. If $f : X \to \mathbb{R}$ takes nonnegative values, then for all countable $Y \subseteq X$, we define $f(Y) = \sum_{y \in Y} f(y) \in [0, +\infty]$. For any vector $\boldsymbol{c} \in \mathbb{R}^n$ and any real $s > 0$, the (spherical) Gaussian function with standard deviation parameter s and center $\boldsymbol{c}$ is defined as follows: $\forall \boldsymbol{x} \in \mathbb{R}^n, \rho_{s,\boldsymbol{c}}(\boldsymbol{x}) = \exp(-\pi \|\boldsymbol{x} - \boldsymbol{c}\|^2/s^2)$. The Gaussian distribution is $D_{s,\boldsymbol{c}} = \rho_{s,\boldsymbol{c}}/s^n$. When $\boldsymbol{c} = \boldsymbol{0}$, we may omit the subscript $\boldsymbol{c}$.

We use the usual Landau notations. A function $f(\lambda)$ is said negligible if it is $\lambda^{-\omega(1)}$. A probability $p(\lambda)$ is said overwhelming if it is $1 - \lambda^{-\omega(1)}$.

The distinguishing advantage of an algorithm $\mathcal{A}$ between two distributions D_0 and D_1 is defined as $\mathrm{Adv}_{\mathcal{A}}(D_0, D_1) = |\Pr_{x \hookleftarrow D_0}[\mathcal{A}(x) = 1] - \Pr_{x \hookleftarrow D_1}[\mathcal{A}(x) = 1]|$, where the probabilities are taken over the randomness of the input x and the internal randomness of $\mathcal{A}$. Algorithm $\mathcal{A}$ is said to be an (ε, T)-distinguisher if it runs in time $\leq T$ and if $\mathrm{Adv}_{\mathcal{A}}(D_0, D_1) \geq \varepsilon$.

2 Preliminaries

We assume the reader is familiar with standard cryptographic notions, as well as with lattices and lattice-based cryptography. We refer to [Reg09a, MR09] for introductions on the latter topic.

2.1 Lattices

A (full-rank) n-dimensional *Euclidean lattice* $\Lambda \subseteq \mathbb{R}^n$ is the set of all integer linear combinations $\sum_{i=1}^{n} x_i \boldsymbol{b}_i$ of some $\mathbb{R}$-basis $(\boldsymbol{b}_i)_{1 \leq i \leq n}$ of $\mathbb{R}^n$. In this setup, the tuple $(\boldsymbol{b}_i)_i$ is said to form a $\mathbb{Z}$-basis of Λ. For a lattice Λ and any $i \leq n$, the ith successive minimum $\lambda_i(\Lambda)$ is the smallest radius r such that Λ contains i linearly independent vectors of norm at most r. The dual Λ^* of a lattice Λ is defined as $\Lambda^* = \{\boldsymbol{y} \in \mathbb{R}^n : \boldsymbol{y}^t \Lambda \subseteq \mathbb{Z}^n\}$.

The (spherical) *discrete Gaussian distribution over a lattice* $\Lambda \subseteq \mathbb{R}^n$, with standard deviation parameter $s > 0$ and center $\boldsymbol{c}$ is defined as:

$$\forall \boldsymbol{x} \in \Lambda, D_{\Lambda,s,c} = \frac{\rho_{s,c}(\boldsymbol{x})}{\rho_{s,c}(\Lambda)}.$$

When the center is $\boldsymbol{0}$, we omit the subscript $\boldsymbol{c}$.

The *smoothing parameter* [MR07] of an n-dimensional lattice Λ with respect to $\varepsilon > 0$, denoted by $\eta_\varepsilon(\Lambda)$, is the smallest $s > 0$ such that $\rho_{1/s}(\Lambda^* \setminus \{0\}) \leq \varepsilon$. We use the following properties.

Lemma 2.1 ([MR07, Lemma 3.3]). *Let Λ be an n-dimensional lattice and $\varepsilon > 0$. Then*

$$\eta_\varepsilon(\Lambda) \leq \sqrt{\frac{\ln(2n(1 + 1/\varepsilon))}{\pi}} \cdot \lambda_n(\Lambda).$$

Lemma 2.2 (Adapted from [GPV08, Lemma 5.3]). *Let $m, n \geq 1$ and q a prime integer, with $m \geq 2n \ln q$. For $A \in \mathbb{Z}_q^{n \times m}$ we define $A^\perp$ as the lattice $\{\boldsymbol{x} \in \mathbb{Z}^m : A\boldsymbol{x} = \boldsymbol{0} \mod q\}$. Then,*

$$\forall \varepsilon < 1/2 : \Pr_{A \hookleftarrow U(\mathbb{Z}_q^{n \times m})} \left[\eta_\varepsilon(A^\perp) \geq 4\sqrt{\frac{\ln(4m/\varepsilon)}{\pi}} \right] \leq q^{-n}.$$

Lemma 2.3 (Adapted from [GPV08, Corollary 2.8]). *Let Λ, Λ' be n-dimensional lattices with $\Lambda' \subseteq \Lambda$ and $\varepsilon \in (0, 1/2)$. Then for any $\boldsymbol{c} \in \mathbb{R}^n$ and $s \geq \eta_\varepsilon(\Lambda')$ and any $x \in \Lambda/\Lambda'$ we have*

$$(D_{\Lambda,s,c} \mod \Lambda')(x) \in \left[\frac{1 - \varepsilon}{1 + \varepsilon}, \frac{1 + \varepsilon}{1 - \varepsilon} \right] \cdot \frac{\det(\Lambda)}{\det(\Lambda')}.$$

2.2   The SIS and LWE Problems

The Small Integer Solution (SIS) problem was introduced by Ajtai in [Ajt96]. It serves as a security foundation for numerous cryptographic primitives, including, among many others, hash functions [Ajt96] and signatures [GPV08,DDLL13].

Definition 2.4. *Let* $m \geq n \geq 1$ *and* $q \geq 2$ *be integers, and* β *a positive real. The* $\mathrm{SIS}_{n,m,q,\beta}$ *problem is as follows: given* $A \hookleftarrow U(\mathbb{Z}_q^{n \times m})$, *the goal is to find* $\boldsymbol{x} \in \mathbb{Z}^m$ *such that* $A\boldsymbol{x} = \boldsymbol{0} \mod q$ *and* $0 < \|\boldsymbol{x}\| \leq \beta$.

The SIS problem was proven by Ajtai [Ajt96] to be at least as hard as some standard worst-case problems over Euclidean lattices, under specific parameter constraints. We refer to [GPV08] for an improved (and simplified) reduction.

The Learning With Errors (LWE) problem was introduced in 2005 by Regev [Reg05,Reg09b]. LWE is also extensively used as a security foundation, for encryption schemes [Reg09b,GPV08], fully homomorphic encryption schemes [BV11], and pseudo-random functions [BPR12,AKPW13], among many others. Its definition involves the following distribution. Let χ be a distribution over $\mathbb{T}$, $q \geq 2$, $n \geq 1$ and $\boldsymbol{s} \in \mathbb{Z}_q^n$. A sample from $A_{\boldsymbol{s},\chi}$ is of the form $(\boldsymbol{a}, b) \in \mathbb{Z}_q^n \times \mathbb{T}$, with $\boldsymbol{a} \hookleftarrow U(\mathbb{Z}_q^n)$, $b = \frac{1}{q}\langle \boldsymbol{a}, \boldsymbol{s} \rangle + e$ and $e \hookleftarrow \chi$.

Definition 2.5. *Let* χ *be a distribution over* $\mathbb{T}$, $q \geq 2$, *and* $m \geq n \geq 1$. *The search variant* $\mathrm{sLWE}_{n,q,\chi,m}$ *of the* LWE *problem is as follows: given* m *samples from* $A_{\boldsymbol{s},\chi}$ *for some* $\boldsymbol{s} \in \mathbb{Z}_q^n$, *the goal is to find* $\boldsymbol{s}$. *The decision variant* $\mathrm{LWE}_{n,q,\chi,m}$ *consists in distinguishing between the distributions* $(A_{\boldsymbol{s},\chi})^m$ *and* $U(\mathbb{Z}_q^n \times \mathbb{T})^m$, *where* $\boldsymbol{s} \hookleftarrow U(\mathbb{Z}_q^n)$.

In some cases, it is convenient to use an error distribution χ whose support is $\mathbb{T}_q$. In these cases, the definition of LWE is adapted in that $U(\mathbb{Z}_q^n \times \mathbb{T})$ is replaced by $U(\mathbb{Z}_q^n \times \mathbb{T}_q)$. Note also that for a fixed number of samples m, we can represent the LWE samples using matrices. The $\boldsymbol{a}_i$'s form the rows of a matrix A uniform in $\mathbb{Z}_q^{m \times n}$, and the scalar product is represented by the product between A and $\boldsymbol{s}$.

Regev [Reg09b] gave a quantum reduction from standard worst-case problems over Euclidean lattices to sLWE and LWE, under specific parameter constraints. Classical (but weaker) reductions have later been obtained (see [Pei09, BLP+13]). We will use the following sample-preserving search to decision reduction for LWE.

Theorem 2.6 (Adapted from [MM11, Proposition 4.10]). *If* $q \leq$ poly (m, n) *is prime and the error distribution* χ *has support in* $\mathbb{T}_q$, *then there exists a reduction from* $\mathrm{sLWE}_{n,q,\chi,m}$ *to* $\mathrm{LWE}_{n,q,\chi,m}$ *that is polynomial in* n *and* m.

2.3   The Rényi Divergence

For any two discrete probability distributions P and Q such that $\mathrm{Supp}(P) \subseteq \mathrm{Supp}(Q)$ and $a \in (1, +\infty)$, we define the Rényi divergence of order a by

$$R_a(P\|Q) = \left(\sum_{x \in \text{Supp}(P)} \frac{P(x)^a}{Q(x)^{a-1}} \right)^{\frac{1}{a-1}}.$$

We omit the a subscript when $a = 2$. We define the Rényi divergences of orders 1 and $+\infty$ by

$$R_1(P\|Q) = \exp \left(\sum_{x \in \text{Supp}(P)} P(x) \log \frac{P(x)}{Q(x)} \right) \quad \text{and} \quad R_\infty(P\|Q) = \max_{x \in \text{Supp}(P)} \frac{P(x)}{Q(x)}.$$

The definitions are extended in the natural way to continuous distributions. The divergence R_1 is the (exponential of) the Kullback-Leibler divergence.

For any fixed P, Q, the function $a \mapsto R_a(P\|Q) \in (0, +\infty]$ is non-decreasing, continuous over $(1, +\infty)$, tends to $R_\infty(P\|Q)$ when a grows to infinity, and if $R_a(P\|Q)$ is finite for some a, then $R_a(P\|Q)$ tends to $R_1(P\|Q)$ when a tends to 1 (we refer to [EH12] for proofs). A direct consequence is that if $P(x)/Q(x) \le c$ for all $x \in \text{Supp}(P)$ and for some constant c, then $R_a(P\|Q) \le R_\infty(P\|Q) \le c$. In the same setup, we have $\Delta(P, Q) \le c/2$.

The following properties can be considered the multiplicative analogues of those of the SD. We refer to [EH12, LSS14] for proofs.

Lemma 2.7. *Let $a \in [1, +\infty]$. Let P and Q denote distributions with $\text{Supp}(P) \subseteq \text{Supp}(Q)$. Then the following properties hold:*

- **Log. Positivity:** $R_a(P\|Q) \ge R_a(P\|P) = 1$.
- **Data Processing Inequality:** $R_a(P^f\|Q^f) \le R_a(P\|Q)$ *for any function f, where P^f (resp. Q^f) denotes the distribution of $f(y)$ induced by sampling $y \hookleftarrow P$ (resp. $y \hookleftarrow Q$).*
- **Multiplicativity:** *Assume P and Q are two distributions of a pair of random variables (Y_1, Y_2). For $i \in \{1, 2\}$, let P_i (resp. Q_i) denote the marginal distribution of Y_i under P (resp. Q), and let $P_{2|1}(\cdot|y_1)$ (resp. $Q_{2|1}(\cdot|y_1)$) denote the conditional distribution of Y_2 given that $Y_1 = y_1$. Then we have:*
  - $R_a(P\|Q) = R_a(P_1\|Q_1) \cdot R_a(P_2\|Q_2)$ *if Y_1 and Y_2 are independent.*
  - $R_a(P\|Q) \le R_\infty(P_1\|Q_1) \cdot \max_{y_1 \in X} R_a(P_{2|1}(\cdot|y_1)\|Q_{2|1}(\cdot|y_1))$.
- **Probability Preservation:** *Let $A \subseteq \text{Supp}(Q)$ be an arbitrary event. If $a \in (1, +\infty)$, then $Q(A) \ge P(A)^{\frac{a}{a-1}}/R_a(P\|Q)$. Further, we have*

$$Q(A) \ge P(A)/R_\infty(P\|Q).$$

Let P_1, P_2, P_3 be three distributions with $\text{Supp}(P_1) \subseteq \text{Supp}(P_2) \subseteq \text{Supp}(P_3)$. Then we have:

- **Weak Triangle Inequality:**

$$R_a(P_1\|P_3) \le \begin{cases} R_a(P_1\|P_2) \cdot R_\infty(P_2\|P_3), \\ R_\infty(P_1\|P_2)^{\frac{a}{a-1}} \cdot R_a(P_2\|P_3) & \text{if } a \in (1, +\infty). \end{cases}$$

Getting back to the setup in which $P(x)/Q(x) \leq c$ for all $x \in \text{Supp}(P)$ and for some constant c, the RD probability preservation property above is relevant even for large c, whereas the analogous SD probability preservation property starts making sense only when $c < 2$.

Pinsker's inequality is the analogue of the probability preservation property for $a = 1$: for an arbitrary event $A \subseteq \text{Supp}(Q)$, we have $Q(A) \geq P(A) - \sqrt{\ln R_1(P\|Q)/2}$ (see [PDG14, Lemma 1] for a proof). Analogously to the statistical distance, this probability preservation property is useful for unlikely events A only if $\ln R_1(P\|Q)$ is very small. We refer to Subsect. 3.1 for additional comments on this property.

2.4   RD Bounds

We will use the following result, adapted from [LSS14].

Lemma 2.8. *For any n-dimensional lattice $\Lambda \subseteq \mathbb{R}^n$ and $s > 0$, let P be the distribution $D_{\Lambda,s,c}$ and Q be the distribution $D_{\Lambda,s,c'}$ for some fixed $c, c' \in \mathbb{R}^n$. If $c, c' \in \Lambda$, let $\varepsilon = 0$. Otherwise, fix $\varepsilon \in (0,1)$ and assume that $s \geq \eta_\varepsilon(\Lambda)$. Then, for any $a \in (1, +\infty)$:*

$$R_a(P\|Q) \in \left[\left(\frac{1-\varepsilon}{1+\varepsilon} \right)^{\frac{2}{a-1}}, \left(\frac{1+\varepsilon}{1-\varepsilon} \right)^{\frac{2}{a-1}} \right] \cdot \exp\left(a\pi \frac{\|c - c'\|^2}{s^2} \right).$$

It may be checked that also $R_1(P\|Q)$ is of the order of $\exp(\|c' - c\|^2/s^2)$, $R_\infty(P\|Q) = +\infty$ and $\Delta(P,Q)$ is of the order of $\|c' - c\|^2/s^2$. In that setup, the RD of order $a = \infty$ is useless, and the probability preservation properties of the SD and RD of order $a = 1$ lead to interesting bounds for events occurring only when $\|c' - c\|/s = o(\varepsilon)$. Oppositely, for any $a \in (1, +\infty)$, the probability preservation property for the RD of order $a \in (1, +\infty)$ may be used with $\|c' - c\|/s = O(\sqrt{\log(1/\varepsilon)})$ while still leading to probabilistic lower bounds of the order of $\varepsilon^{O(1)}$.

As we have already seen, if two distributions are close in a uniform sense, then their RD is small. We observe the following immediate consequence of Lemma 2.3, that allows replacing the SD with the RD in the context of smoothing arguments, in order to save on the required parameter s. In applications of Lemma 2.3, it is customary to use $s \geq \eta_\varepsilon(\Lambda')$ with $\varepsilon \leq 2^{-\lambda}$, in order to make the distribution $D_{\Lambda/\Lambda',s,c} = D_{\Lambda,s,c} \bmod \Lambda'$ within SD $2^{-\lambda}$ of the uniform distribution $U(\Lambda/\Lambda')$. This translates via Lemma 2.1 to use $s = \Omega(\sqrt{\lambda + \log n} \cdot \lambda_n(\Lambda'))$. Whereas if using an RD bound $R_\infty(D_{\Lambda/\Lambda',s,c}\|U_{\Lambda/\Lambda'}) = O(1)$ suffices for the application, one can take $\varepsilon = O(1)$ in the corollary below, which translates to just $s = \Omega(\sqrt{\log n} \cdot \lambda_n(\Lambda'))$, saving a factor $\Theta(\sqrt{\lambda})$.

Lemma 2.9. *Let Λ, Λ' be n-dimensional lattices with $\Lambda' \subseteq \Lambda$ and $\varepsilon \in (0, 1/2)$. Let $D_{\Lambda/\Lambda',s,c}$ denote the distribution on Λ/Λ' induced by sampling from $D_{\Lambda,s,c}$*

and reducing modulo Λ', and let $U_{\Lambda/\Lambda'}$ denote the uniform distribution on Λ/Λ'. Then for any $c \in \mathbb{R}^n$ and $s \geq \eta_\varepsilon(\Lambda')$ and any $x \in \Lambda/\Lambda'$ we have

$$R_\infty(D_{\Lambda/\Lambda',s,c}\|U_{\Lambda/\Lambda'}) \leq \frac{1+\varepsilon}{1-\varepsilon}.$$

3   Application to Lattice-Based Signature Schemes

In this section, we use the RD to improve the security proofs of the GPV and BLISS signature schemes [GPV08, DDLL13], allowing to take smaller parameters for any fixed security level.

3.1   Sampling Discrete Gaussians and the BLISS Signature Scheme

We show that the use of RD in place of SD leads to significant savings in the required precision of integers sampled according to a discrete Gaussian distribution in the security analysis of lattice-based signature schemes. These savings consequently lower the precomputed table storage for sampling discrete Gaussians with the method described in [DDLL13, PDG14]. In Table 1, we provide a numerical comparison of RD and SD based on an instantiation of BLISS-I.

Discrete Gaussian Sampling. In the BLISS signature scheme [DDLL13] (and similarly in earlier variants [Lyu12]), each signature requires the signing algorithm to sample $O(n)$ independent integers from the 1-dimensional discrete Gaussian distribution $D_{\mathbb{Z},s}$, where $s = O(m)$ is the standard deviation parameter (here the variable m denotes a parameter related to the underlying lattice dimension, and is typically in the order of several hundreds)[2].

In [DDLL13], a particularly efficient sampling algorithm for $D_{\mathbb{Z},s}$ is presented. To produce a sample from $D_{\mathbb{Z},s}$, this algorithm samples about $\ell = \lceil \log(0.22s^2(1 + 2\tau s)) \rceil$ Bernoulli random variables of the form $B_{\exp(-\pi 2^i/s^2)}$, where $i = 0, \ldots, \ell - 1$ and $\tau = O(\sqrt{\lambda})$ is the tail-cut factor for the Gaussian. To sample those Bernoulli random variables, the authors of [DDLL13] use a precomputed table of the probabilities $c_i = \exp(-\pi 2^i/s^2)$, for $i = 1, \ldots, \ell$. Since these probabilities are real numbers, they must be truncated to some bit precision p in the precomputed table, so that truncated values $\tilde{c}_i = c_i + \varepsilon_i$ are stored, where $|\varepsilon_i| \leq 2^{-p}c_i$ are the truncation errors.

In previous works, the precision was determined by an analysis either based on the statistical distance (SD) [DDLL13] or the Kullback-Leibler divergence (KLD) [PDG14]. In this section, we review and complete these methods, and we propose an RD-based analysis that leads to bigger savings, asymptotically and in practice (see Table 1). More precisely, we give sufficient lower bounds on the precision p to ensure security on the scheme implemented with truncated values against adversaries succeeding with probability $\geq \varepsilon$ and making $\leq q_s$ signing

---

[2] Note that [Lyu12, DDLL13] consider the unnormalized Gaussian function $\rho'_{\sigma,c}(x) = \exp(-\|x - c\|/(2\sigma^2))$ instead of $\rho_{s,c}$. We have $\rho_{s,c} = \rho'_{\sigma,c}$ when $\sigma = s/\sqrt{2\pi}$.

queries. For any adversary, the distributions Φ' and Φ denote the signatures in the view of the adversary in the untruncated (resp. truncated) cases.

SD-based analysis [DDLL13]. Any forging adversary $\mathcal{A}$ with success probability $\geq \varepsilon$ on the scheme implemented with truncated Gaussian has a success probability $\varepsilon' \geq \varepsilon - \Delta(\Phi, \Phi')$ against the scheme implemented with perfect Gaussian sampling. We select parameters to handle adversaries with success probabilities $\geq \varepsilon/2$ against the untruncated scheme; we can set the required precision p so that $\Delta(\Phi, \Phi') \leq \varepsilon/2$. Each signature requires $\ell \cdot m$ samples from the Bernoulli random variables $(B_{\tilde{c}_i})_i$. To ensure security against q_s signing queries, each of the truncated Bernoulli random variables $B_{\tilde{c}_i}$ should be within SD $\Delta(\Phi, \Phi')/(\ell \cdot m \cdot q_s)$ of the desired B_{c_i} (by the union bound). Using $\Delta(B_{\tilde{c}_i}, B_{c_i}) = |\varepsilon_i| \leq 2^{-p}c_i \leq 2^{-p-1}$ leads to a precision requirement

$$p \geq \log(\ell \cdot m \cdot q_s/\Delta(\Phi, \Phi')) \geq \log(\ell \cdot m \cdot q_s/\varepsilon). \tag{1}$$

The overall precomputed table is hence of bit-size $L_{\text{SD}} = p \cdot \ell \geq \log(\ell \cdot m \cdot q_s/\varepsilon) \cdot \ell$.

Note that in [DDLL13], the authors omitted the term $\ell \cdot m \cdot q_s$ in their analysis: they only ensured that $\Delta(B_{\tilde{c}_i}, B_{c_i}) \leq \varepsilon$, leading to the requirement that $p \geq \log(1/\varepsilon)$.

One may also set the precision p_i depending on i for $0 \leq i \leq \ell - 1$. It is sufficient to set

$$2^{-p_i}c_i = 2^{-p_i}\exp(-\pi 2^i/s^2) \leq (\varepsilon/2)/(\ell \cdot m \cdot q_s).$$

Hence the precision p_i is

$$p_i \geq \log\left(\frac{\ell \cdot m \cdot q_s}{\varepsilon} \cdot \exp(-\pi 2^i/s^2)\right) + 1. \tag{2}$$

The bit-size of the overall precomputed table can be computed as a sum of the above p_i's. Using the symmetry of the Bernoulli variable, we can further drop the bit-size of the precomputed table.

KLD-based analysis [PDG14]. In [PDG14], Pöppelman, Ducas and Güneysu replace the SD-based analysis by a KLD-based analysis (i.e., using the RD of order $a = 1$) to reduce the precision p needed in the precomputed table. They show that any forging adversary $\mathcal{A}$ with success probability ε on the scheme implemented with truncated Gaussian has a success probability $\varepsilon' \geq \varepsilon - \sqrt{\ln R_1(\Phi\|\Phi')/2}$ on the scheme implemented with perfect Gaussian (see remark at the end of Subsect. 2.3). By the multiplicative property of the RD over the $\ell \cdot m \cdot q_s$ independent samples needed for signing q_s times, we get that $R_1(\Phi\|\Phi') \leq (\max_{i=1,\dots,\ell} R_1(B_{\tilde{c}_i}\|B_{c_i}))^{\ell \cdot m \cdot q_s}$. Now, we have:

$$\ln R_1(B_{\tilde{c}_i}\|B_{c_i}) = (1 - c_i - \varepsilon_i)\ln\frac{1 - c_i - \varepsilon_i}{1 - c_i} + (c_i + \varepsilon_i)\ln\frac{c_i + \varepsilon_i}{c_i}$$

$$\leq -(1 - c_i - \varepsilon_i)\frac{\varepsilon_i}{1 - c_i} + (c_i + \varepsilon_i)\frac{\varepsilon_i}{c_i} = \frac{\varepsilon_i^2}{(1 - c_i)c_i}.$$

Using $|\varepsilon_i| \leq 2^{-p}c_i$ and $1 - c_i \geq 1/2$, we obtain $\ln R_1(B_{\tilde{c}_i}\|B_{c_i}) = 2^{-2p+1}\frac{c_i}{1-c_i} \leq 2^{-2p}$. Therefore, we obtain $\varepsilon' \geq \varepsilon - \sqrt{\ell \cdot m \cdot q_s \cdot 2^{-2p}}$. We can select parameters such that $\sqrt{\ell \cdot m \cdot q_s \cdot 2^{-2p+1}} \leq \varepsilon/2$. This leads to a precision requirement

$$p \geq \frac{1}{2}\log\left(\frac{\ell \cdot m \cdot q_s}{\varepsilon^2}\right) + \frac{1}{2}. \tag{3}$$

The overall precomputed table is hence of bit-size $L_{\text{KLD}} \geq (\log(\ell \cdot m \cdot q_s/\varepsilon^2)/2 + 1/2) \cdot \ell$. This KLD-based analysis may save some storage if ε is not too small.

Note that in [PDG14], the authors selected $\varepsilon = 1/2$ and $\ell \cdot m \cdot q_s = 2^\lambda$ where λ is the desired bit-security, and hence obtained $p \geq \lambda/2 + 1$.

One may also set the precision p_i depending on i. It is sufficient to set

$$p_i \geq \frac{1}{2}\log\left(\frac{\ell \cdot m \cdot q_s}{\varepsilon^2} \cdot \frac{c_i}{1-c_i}\right) + 1. \tag{4}$$

Using symmetry, we may assume $c_i \leq 1/2$.

R_∞-*based analysis.* The probability preservation property of the Rényi divergence from Lemma 2.7 is multiplicative for $a > 1$ (rather than additive for $a = 1$). Here we use the order $a = \infty$. This property gives that any forging adversary $\mathcal{A}$ having success probability ε on the scheme implemented with truncated Gaussian sampling has a success probability $\varepsilon' \geq \varepsilon/R_\infty(\Phi\|\Phi')$ on the scheme implemented with perfect Gaussian. If $R = R_\infty(\Phi\|\Phi') \leq O(1)$, then $\varepsilon' = \Omega(\varepsilon)$. By the multiplicative property of the RD (over the $\ell \cdot m \cdot q_s$ samples needed for signing q_s times), we have $R_\infty(\Phi\|\Phi') \leq R_\infty(B_{\tilde{c}_i}\|B_{c_i})^{\ell \cdot m \cdot q_s}$. By our assumption that $c_i \leq 1/2$, we have $R_\infty(B_{\tilde{c}_i}\|B_{c_i}) = 1 + |\varepsilon_i|/c_i \leq 1 + 2^{-p}$. Therefore, we get $\varepsilon' \geq \varepsilon/(1 + 2^{-p})^{\ell \cdot m \cdot q_s}$. We select parameters to get adversaries with success probabilities $\geq \varepsilon/2$ against the untruncated scheme and set the precision so that $(1 + 2^{-p})^{\ell \cdot m \cdot q_s} \leq 2$. This yields an approximated precision requirement

$$p \geq \log(\ell \cdot m \cdot q_s). \tag{5}$$

Note above estimate may not be accurate unless $\ell \cdot m \cdot q_s$ is much smaller than 2^p. Hence we may also require that $p \geq \log(\ell \cdot m \cdot q_s) + C$ for some constant C. This condition essentially eliminates the term ε from the precision needed by the SD-based and KLD-based analyses.[3] Overall, we get a precomputed table of bit-size $L_{\text{RD}} = \log(\ell \cdot m \cdot q_s) \cdot \ell$.

R_a-*based analysis.* We may also consider R_a-based analysis for general $a > 1$. It should noted that the reductions here are not tight: for R_a-based analysis with $a > 1$, the probability preservation shows $\varepsilon' > \varepsilon^{a/(a-1)}/R_a(\Phi\|\Phi')$. The Rényi

---

[3] Note that the resulting precision is not independent of ε. The parameters $m = m(\varepsilon)$ and $\ell = \ell(\varepsilon)$ are chosen in [DDLL13] so that any forging adversary has success probability at most ε on the scheme implemented with perfect Gaussian sampling.

divergence can be computed by

$$(R_a(\varPhi\|\varPhi'))^{a-1} = \frac{(1 - c_i - \varepsilon_i)^a}{(1 - c_i)^{a-1}} + \frac{(c_i + \varepsilon_i)^a}{c_i^{a-1}}$$

$$= (1 - c_i - \varepsilon_i)\left(1 - \frac{\varepsilon_i}{1 - c_i}\right)^{a-1} + (c_i + \varepsilon_i)\left(1 + \frac{\varepsilon_i}{c_i}\right)^{a-1}.$$

If a is much smaller than 2^p, we get

$$(R_a(\varPhi\|\varPhi'))^{a-1} \approx (1 - c_i - \varepsilon_i)\left(1 - \frac{(a-1)\varepsilon_i}{1 - c_i}\right) + (c_i + \varepsilon_i)\left(1 + \frac{(a-1)\varepsilon_i}{c_i}\right)$$

$$= 1 + \frac{\varepsilon_i^2(a-1)}{c_i(1 - c_i)} \leq 1 + 2^{-2p}(a-1)\frac{c_i}{1 - c_i} \leq 1 + 2^{-2p}(a-1).$$

For instance if we take $a = 2$, we have $R_2(\varPhi\|\varPhi') \leq 1 + 2^{-2p}$ and hence $\varepsilon' \geq \varepsilon^2/R_2(\varPhi\|\varPhi')$. To get a success probability lower bound $\varepsilon^2/2$, it is sufficient to set

$$p \geq \frac{1}{2}\log(\ell \cdot m \cdot q_s). \tag{6}$$

On the other hand, if a is much larger than 2^p, then we have

$$(R_a(\varPhi\|\varPhi'))^{a-1} = (1 - c_i - \varepsilon_i)\left(1 - \frac{\varepsilon_i}{1 - c_i}\right)^{a-1} + (c_i + \varepsilon_i)\left(1 + \frac{\varepsilon_i}{c_i}\right)^{a-1}$$

$$\approx (c_i + \varepsilon_i)\exp\left(\frac{(a-1)\varepsilon_i}{c_i}\right).$$

Hence the Rényi divergence

$$R_a(\varPhi\|\varPhi') \approx (c_i + \varepsilon_i)^{1/(a-1)}\exp\left(\frac{\varepsilon_i}{c_i}\right) \approx 1 + \frac{\varepsilon_i}{c_i}.$$

As $a \to \infty$, $R_a(\varPhi\|\varPhi') \to 1 + 2^{-p}$.

Thus if the tightness of the reduction is not a concern, using R_a with small a reduces the precision requirement. Furthermore, we can amplify the success probability of the forger on the truncated Gaussian from ε' to some $\varepsilon'' > \varepsilon'$.

Numerical Examples. In Table 1, we consider a numerical example which gives the lower bound on the precision for the scheme BLISS-I (with $\varepsilon = 2^{-128}$) when allowing up to $q_s = 2^{64}$ sign queries to the adversary. For the BLISS-I parameters, we use $m = 1024$, $\ell = 29$, $s = \lceil\sqrt{2\pi} \cdot 254 \cdot \sqrt{1/(2\ln 2)}\rceil = 541$ and $\tau = 13.4/\sqrt{2\pi} \approx 5.4$). The reductions in the table are tight, except for R_2 (as ε' in the reduction does not depend directly on ε but on ε^2), and we are a little bit loose for the R_∞ case.

When instantiating BLISS-I with the parameters of Table 1, the table bit-size can be reduced from about 6000 bits to about 1200 bits by using R_2 in place of SD. If the tightness of the reduction is concerned, we may use R_∞ instead, which leads to a table of about 2300 bits.

Table 1. Comparison of the precision to handle adversaries with success probability $\geq \varepsilon$ making $\leq q_s$ sign queries to BLISS-I. Our Rényi-based parameters are on the last two lines.

	Lower bound on the precision p	Example p	Table bit-size
SD (Eq. (1))	$p \geq \log(\ell \cdot m \cdot q_s / \varepsilon)$	$p \geq 207$	6003
SD (Eq. (2))	$p_i \geq \log(\ell \cdot m \cdot q_s \cdot e^{-\pi 2^i / s^2} / \varepsilon) + 1$	–	4598
KLD (Eq. (3))	$p \geq \log(\ell \cdot m \cdot q_s / \varepsilon^2)/2 + 1/2$	$p \geq 168$	4872
KLD (Eq. (4))	$p_i \geq \log(\ell \cdot m \cdot q_s / \varepsilon^2 \cdot c_i/(1 - c_i))/2 + 1$	–	3893
R_∞ (Eq. (5))	$p \geq \log(\ell \cdot m \cdot q_s)$	$p \geq 79$	2291
R_2 (Eq. (6))	$p \geq \log(\ell \cdot m \cdot q_s)/2$	$p \geq 40$	1160

3.2 GPV Signature Scheme

The RD can also be used to reduce the parameters obtained via the SD-based analysis of the GPV signature scheme in [GPV08].

In summary, the signature and the security proof from [GPV08] work as follows. The signature public key is a matrix $A \in \mathbb{Z}_q^{n \times m}$ with n linear in the security parameter λ, $q = \mathrm{poly}(n)$, and $m = O(n \log q)$. The private signing key is a short basis matrix T for the lattice $A^\perp = \{x \in \mathbb{Z}^m : A \cdot x = 0 \mod q\}$, whose last successive minimum satisfies $\lambda_m(A^\perp) \leq O(1)$ when $m = \Omega(n \log q)$ (see [GPV08]). A signature (σ, s) on a message M is a short vector $\sigma \in \mathbb{Z}^m$ and a random salt $s \in \{0,1\}^\lambda$, such that $A \cdot \sigma = H(M, s) \mod q$, where H is a random oracle hashing into $\mathbb{Z}_q^n$. The short vector σ is sampled by computing an arbitrary vector t satisfying $A \cdot t = H(M, s) \mod q$ and using T along with a Gaussian sampling algorithm (see [GPV08, BLP+13]) to produce a sample from $t + D_{A^\perp, r, -t}$.

The main idea in the security proof from the SIS problem [GPV08] is based on simulating signatures without T, by sampling σ from $D_{\mathbb{Z}^m, r}$ and then programming the random oracle H at (M, s) according to $H(M, s) = A \cdot \sigma \mod q$. As shown in [GPV08, Lemma 5.2], the conditional distribution of σ given $A \cdot \sigma \mod q$ is exactly the same in the simulation and in the real scheme. Therefore, the SD between the simulated signatures and the real signatures is bounded by the SD between the marginal distribution D_1 of $A \cdot \sigma \mod q$ for $\sigma \hookleftarrow D_{\mathbb{Z}^m, r}$ and $U(\mathbb{Z}_q^m)$. This SD for one signature is bounded by ε if $r \geq \eta_\varepsilon(A^\perp)$. This leads, over the q_s sign queries of the attacker, in the SD-based analysis of [GPV08], to take $\varepsilon = O(2^{-\lambda} q_s^{-1})$ and thus $r = \Omega(\sqrt{\lambda + \log q_s})$ (using Lemma 2.2), in order to handle attackers with success probability $2^{-o(\lambda)}$.

Now, by Lemma 2.9, we have that the RD $R_\infty(D_1 \| U)$ is bounded by $1 + c \cdot \varepsilon$ for one signature, for some constant c. By the multiplicativity property of Lemma 2.7, over q_s queries, it is bounded by $(1 + c\varepsilon)^{q_s}$. By taking $\varepsilon = O(q_s^{-1})$, we obtain overall an RD bounded as $O(1)$ between the view of the attacker in the real attack and simulation, leading to a security proof with respect to SIS but with a smaller $r = \Omega(\sqrt{\log \lambda + \log(nq_s)})$. When the number of sign queries q_s allowed to the adversary is much smaller than 2^λ, this leads to significant parameter savings, because SIS's β is reduced and hence n, m, q may be set smaller for the same security parameter λ.

4   Rényi Divergence and Distinguishing Problems

In this section, we prove Theorem 4.1 which allows to use the RD for distinguishing problems, and we show how to apply it to the dual-Regev encryption scheme.

4.1   Problems with Public Sampleability

A general setting one comes across in analyzing the security of cryptographic schemes has the following form. Let P denote a decision problem that asks to distinguish whether a given x was sampled from distribution X_0 or X_1, defined as follows:

$$X_0 = \{x : r \hookleftarrow \Phi, x \hookleftarrow D_0(r)\}, \quad X_1 = \{x : r \hookleftarrow \Phi, x \hookleftarrow D_1(r)\}.$$

Here r is some parameter that is sampled from the same distribution Φ in both X_0 and X_1. The parameter r then determines the conditional distributions $D_0(r)$ and $D_1(r)$ from which x is sampled in X_0 and X_1, respectively, given r. Now, let P' denote another decision problem that is defined similarly to P, except that in P' the parameter r is sampled from a different distribution Φ' (rather than Φ). Given r, the conditional distributions $D_0(r)$ and $D_1(r)$ are the same in P' as in P. Let X'_0 (resp. X'_1) denote the resulting marginal distributions of x in problem P'. Now, in the applications we have in mind, the distributions Φ' and Φ are "close" in some sense, and we wish to show that this implies an efficient reduction between problems P' and P, in the usual sense that every distinguisher with efficient run-time T and non-negligible advantage ε against P implies a distinguisher for P' with efficient run-time T' and non-negligible advantage ε'. In the classical situation, if the SD $\Delta(\Phi, \Phi')$ between Φ' and Φ is negligible, then the reduction is immediate. Indeed, for $b \in \{0, 1\}$, if p_b (resp. p'_b) denotes the probability that a distinguisher algorithm $\mathcal{A}$ outputs 1 on input distribution X_b (resp. X'_b), then we have, from the SD probability preservation property, that $|p'_b - p_b| \le \Delta(\Phi, \Phi')$. As a result, the advantage $\varepsilon' = |p'_1 - p'_0|$ of $\mathcal{A}$ against P' is bounded from below by $\varepsilon - 2\Delta(\Phi, \Phi')$ which is non-negligible (here $\varepsilon = |p_1 - p_0|$ is the assumed non-negligible advantage of $\mathcal{A}$ against P).

Unfortunately, for general decision problems P, P' of the above form, it seems difficult to obtain an RD-based analogue of the above SD-based argument, in the weaker setting when the SD $\Delta(\Phi, \Phi')$ is non-negligible, but the RD $R = R(\Phi\|\Phi')$ is small. Indeed, the probability preservation property of the RD in Lemma 2.7 does not seem immediately useful in the case of general decision problems P, P'. With the above notations, it can be used to conclude that $p'_b \ge p_b^2/R$ but this does not allow us to usefully relate the advantages $|p'_1 - p'_0|$ and $|p_1 - p_0|$.

Nevertheless, we now make explicit a special class of "publicly sampleable" problems P, P' for which such a reduction can be made. In such problems, it is possible to efficiently sample from both distributions $D_0(r)$ (resp. $D_1(r)$) given the single sample x from the unknown $D_b(r)$. This technique is implicit in the application of RD in the reductions of [LPR13], and we abstract it and make it explicit in the following.

Theorem 4.1. *Let* Φ, Φ' *denote two distributions with* $\mathrm{Supp}(\Phi) \subseteq \mathrm{Supp}(\Phi')$, *and* $D_0(r)$ *and* $D_1(r)$ *denote two distributions determined by some parameter* $r \in \mathrm{Supp}(\Phi')$. *Let* P, P' *be two decision problems defined as follows:*

- *Problem P: Distinguish whether input* x *is sampled from distribution* X_0 *or* X_1, *where*

$$X_0 = \{x : r \hookleftarrow \Phi, x \hookleftarrow D_0(r)\}, \quad X_1 = \{x : r \hookleftarrow \Phi, x \hookleftarrow D_1(r)\}.$$

- *Problem P': Distinguish whether input* x *is sampled from distribution* X_0' *or* X_1', *where*

$$X_0' = \{x : r \hookleftarrow \Phi', x \hookleftarrow D_0(r)\}, \quad X_1' = \{x : r \hookleftarrow \Phi', x \hookleftarrow D_1(r)\}.$$

Assume that $D_0(\cdot)$ *and* $D_1(\cdot)$ *satisfy the following* public sampleability *property: there exists a sampling algorithm* S *with run-time* T_S *such that for all* (r, b), *given any sample* x *from* $D_b(r)$:

- $\mathsf{S}(0, x)$ *outputs a fresh sample distributed as* $D_0(r)$ *over the randomness of* S,
- $\mathsf{S}(1, x)$ *outputs a fresh sample distributed as* $D_1(r)$ *over the randomness of* S.

Then, given a T-time distinguisher $\mathcal{A}$ *for problem* P *with advantage* ε, *we can construct a distinguisher* $\mathcal{A}'$ *for problem* P' *with run-time and distinguishing advantage respectively bounded from above and below by (for any* $a \in (1, +\infty]$):

$$O\left(\frac{1}{\varepsilon^2} \log\left(\frac{R_a(\Phi \| \Phi')}{\varepsilon^{a/(a-1)}}\right) \cdot (T_S + T)\right) \quad and \quad \frac{\varepsilon}{4 \cdot R_a(\Phi \| \Phi')} \cdot \left(\frac{\varepsilon}{2}\right)^{\frac{a}{a-1}}.$$

Proof. Distinguisher $\mathcal{A}'$ is given an input x sampled from $D_b(r)$ for some r sampled from Φ' and some unknown $b \in \{0, 1\}$. For an ε' to be determined later, it runs distinguisher $\mathcal{A}$ on $N = O(\varepsilon^{-2} \log(1/\varepsilon'))$ independent inputs sampled from $D_0(r)$ and $D_1(r)$ calling algorithm S on $(0, x)$ and $(1, x)$ to obtain estimates $\hat{p}_0$ and $\hat{p}_1$ for the acceptance probabilities $p_0(r)$ and $p_1(r)$ of $\mathcal{A}$ given as inputs samples from $D_0(r)$ and $D_1(r)$ (with the r fixed to the value used to sample the input x of $\mathcal{A}'$). By the choice of N and the Hoeffding bound, the estimation errors $|\hat{p}_0 - p_0|$ and $|\hat{p}_1 - p_1|$ are $< \varepsilon/8$ except with probability $< \varepsilon'$ over the randomness of S. Then, if $\hat{p}_1 - \hat{p}_0 > \varepsilon/4$, distinguisher $\mathcal{A}'$ runs $\mathcal{A}$ on input x and returns whatever $\mathcal{A}$ returns, else distinguisher $\mathcal{A}'$ returns a uniformly random bit. This completes the description of distinguisher $\mathcal{A}'$.

Let $\mathcal{S}_1$ denote the set of r's such that $p_1(r) - p_0(r) \geq \varepsilon/2$, $\mathcal{S}_2$ denote the set of r's that are not in $\mathcal{S}_1$ and such that $p_1(r) - p_0(r) \geq 0$, and $\mathcal{S}_3$ denote all the remaining r's. Then:

- If $r \in S_1$, then except with probability $< \varepsilon'$ over the randomness of S, we will have $\hat{p}_1 - \hat{p}_0 > \varepsilon/4$ and thus $\mathcal{A}'$ will output $\mathcal{A}(x)$. Thus, in the case $b = 1$, we have $\Pr[\mathcal{A}'(x) = 1 | r \in S_1] \geq p_1(r) - \varepsilon'$ and in the case $b = 0$, we have $\Pr[\mathcal{A}'(x) = 1 | r \in S_1] \leq p_0(r) + \varepsilon'$.
- Assume that $r \in S_2$. Let $u(r)$ be the probability over the randomness of S that $\hat{p}_1 - \hat{p}_0 > \varepsilon/4$. Then $\mathcal{A}'$ will output $\mathcal{A}(x)$ with probability $u(r)$ and a uniform bit with probability $1 - u(r)$. Thus, in the case $b = 1$, we have $\Pr[\mathcal{A}'(x) = 1 | r \in S_2] = u(r) \cdot p_1(r) + (1 - u(r))/2$, and in the case $b = 0$, we have $\Pr[\mathcal{A}'(x) = 1 | r \in S_2] = u(r) \cdot p_0(r) + (1 - u(r))/2$.
- If $r \in S_3$, except with probability $< \varepsilon'$ over the randomness of S, we have $\hat{p}_1 - \hat{p}_0 < \varepsilon/4$ and $\mathcal{A}'$ will output a uniform bit. Thus, in the case $b = 1$, we have $\Pr[\mathcal{A}'(x) = 1 | r \in S_3] \geq 1/2 - \varepsilon'$, and in the case $b = 0$, we have $\Pr[\mathcal{A}'(x) = 1 | r \in S_3] \leq 1/2 + \varepsilon'$.

Overall, the advantage of $\mathcal{A}'$ is bounded from below by:

$$\sum_{r \in S_1} \Phi'(r) \left(p_1(r) - p_0(r) - 2\varepsilon' \right) + \sum_{r \in S_2} \Phi'(r) u(r) \left(p_1(r) - p_0(r) \right) - \sum_{r \in S_3} \Phi'(r) 2\varepsilon'$$

$$\geq \Phi'(S_1) \cdot \frac{\varepsilon}{2} - 2\varepsilon'.$$

Without loss of generality, we may assume that the advantage of $\mathcal{A}$ is positive. By an averaging argument, the set S_1 has probability $\Phi(S_1) \geq \varepsilon/2$ under distribution Φ. Hence, by the RD probability preservation property (see Lemma 2.7), we have $\Phi'(S_1) \geq (\varepsilon/2)^{\frac{a}{a-1}} / R_a(\Phi \| \Phi')$. The proof may be completed by setting $\varepsilon' = (\varepsilon/8) \cdot (\varepsilon/2)^{\frac{a}{a-1}} / R_a(\Phi \| \Phi')$. □

4.2   Application to Dual-Regev Encryption

Let m, n, q, χ be as in Definition 2.5 and Φ denote a distribution over $\mathbb{Z}_q^{m \times n}$. We define the LWE variant $\text{LWE}_{n,q,\chi,m}(\Phi)$ as follows: Sample $A \hookleftarrow \Phi$, $s \hookleftarrow U(\mathbb{Z}_q^n)$, $e \hookleftarrow \chi^m$ and $u \hookleftarrow U(\mathbb{T}^m)$; The goal is to distinguish between the distributions $(A, \frac{1}{q} As + e)$ and (A, u) over $\mathbb{Z}_q^{m \times n} \times \mathbb{T}^m$. Note that standard LWE is obtained by taking $\Phi' = U(\mathbb{Z}_q^{m \times n})$.

As an application to Theorem 4.1, we show that LWE with non-uniform and possibly statistically correlated a_i's of the samples (a_i, b_i)'s (with b_i either independently sampled from $U(\mathbb{T})$ or close to $\langle a_i, s \rangle$ for a secret vector s) remains at least as hard as standard LWE, as long as the RD $R(\Phi \| U)$ remains small, where Φ is the joint distribution of the given a_i's and U denotes the uniform distribution.

To show this result, we first prove in Corollary 4.2 that there is a reduction from $\text{LWE}_{n,q,\chi,m}(\Phi')$ to $\text{LWE}_{n,q,\chi,m}(\Phi)$ using Theorem 4.1 if $R_a(\Phi \| \Phi')$ is small enough. We then describe in Corollary 4.3 how to use this first reduction to obtain smaller parameters for the dual-Regev encryption. This allows us to save an $\Omega(\sqrt{\lambda/\log \lambda})$ factor in the Gaussian deviation parameter r used for secret key generation in the dual-Regev encryption scheme [GPV08], where λ refers to the security parameter.

Corollary 4.2. *Let Φ and Φ' be two distributions over $\mathbb{Z}_q^{m \times n}$ with $\mathrm{Supp}(\Phi) \subseteq \mathrm{Supp}(\Phi')$. If there exists a distinguisher $\mathcal{A}$ against $\mathrm{LWE}_{n,q,\chi,m}(\Phi)$ with run-time T and advantage $\varepsilon = o(1)$, then there exists a distinguisher $\mathcal{A}'$ against $\mathrm{LWE}_{n,q,\chi,m}(\Phi')$ with run-time $T' = O(\varepsilon^{-2} \log \frac{R_a(\Phi \| \Phi')}{\varepsilon^{a/(a-1)}} \cdot (T + \mathrm{poly}(m, \log q)))$ and advantage $\Omega\left(\frac{\varepsilon^{1+a/(a-1)}}{R_a(\Phi \| \Phi')}\right)$, for any $a \in (1, +\infty]$.*

Proof. Apply Theorem 4.1 with $r = A \in \mathbb{Z}_q^m$, $x = (A, b) \in \mathbb{Z}_q^{m \times n} \times \mathbb{T}^m$, $D_0(r) = (A, A \cdot s + e)$ with $s \hookleftarrow U(\mathbb{Z}_q^n)$ and $e \hookleftarrow \chi^m$, and $D_1(r) = (A, u)$ with $u \hookleftarrow U(\mathbb{Z}_q^n)$. The sampling algorithm S is such that $\mathsf{S}(0, x)$ outputs $(A, A \cdot s' + e')$ for $s' \hookleftarrow U(\mathbb{Z}_q^n)$ and $e' \hookleftarrow \chi^m$, while $\mathsf{S}(1, x)$ outputs (A, u') with $u' \hookleftarrow U(\mathbb{Z}_q^m)$. $\quad\square$

We recall that the dual-Regev encryption scheme has a general public parameter $A \in \mathbb{Z}_q^{m \times n}$, a secret key of the form $sk = x$ with $x \hookleftarrow D_{\mathbb{Z}^m, r}$ and a public key of the form $u = A^t x \bmod q$. A ciphertext for a message $M \in \{0, 1\}$ is obtained as follows: Sample $s \hookleftarrow U(\mathbb{Z}_q^n)$, $e_1 \hookleftarrow \chi^m$ and $e_2 \hookleftarrow \chi$; return ciphertext $(c_1, c_2) = (\frac{1}{q} As + e_1, \frac{1}{q} \langle u, s \rangle + e_2 + \frac{M}{2}) \in \mathbb{T}^m \times \mathbb{T}$.

Corollary 4.3. *Suppose that q is prime, $m \geq 2n \log q$ and $r \geq 4\sqrt{\log(12m)/\pi}$. If there exists an adversary against the IND-CPA security of the dual-Regev encryption scheme with run-time T and advantage ε, then there exists a distinguishing algorithm for $\mathrm{LWE}_{n,q,\chi,m+1}$ with run-time $O((\varepsilon')^{-2} \log(\varepsilon')^{-1} \cdot (T + \mathrm{poly}(m)))$ and advantage $\Omega((\varepsilon')^2)$, where $\varepsilon' = \varepsilon - 2q^{-n}$.*

Proof. The IND-CPA security of the dual-Regev encryption scheme as described above is at least as hard as $\mathrm{LWE}_{n,q,\chi,m+1}(\Phi)$ where Φ is obtained by sampling $A \hookleftarrow U(\mathbb{Z}_q^{m \times n})$, $u \hookleftarrow A^t \cdot D_{\mathbb{Z}^m, r} \bmod q$ and returning the $(m+1) \times n$ matrix obtained by appending u^t at the bottom of A. We apply Corollary 4.2 with $\Phi' = U(\mathbb{Z}_q^{(m+1) \times n})$.

Since q is prime, if A is full rank, then the multiplication by A^t induces an isomorphism between the quotient group $\mathbb{Z}^m / A^\perp$ and $\mathbb{Z}_q^n$, where $A^\perp = \{x \in \mathbb{Z}^m : A^t \cdot x = 0 \bmod q\}$. By Lemma 2.2, we have $\eta_{1/3}(A^\perp) \leq 4\sqrt{\log(12m)/\pi} \leq r$, except for a fraction $\leq q^{-n}$ of the A's. Let BAD denote the union of such bad A's and the A's that are not full rank. We have $\Pr[BAD] \leq 2q^{-n}$.

By the multiplicativity property of Lemma 2.7, we have:

$$R_\infty(\Phi \| \Phi') \leq \max_{A \notin BAD} R_\infty(D_{\mathbb{Z}^m, r} \bmod A^\perp \| U_{\mathbb{Z}^m / A^\perp}).$$

Thanks to Lemma 2.9, we know that the latter is ≤ 2. The result now follows from Corollary 4.2. $\quad\square$

In all applications we are aware of, the parameters satisfy $m \leq \mathrm{poly}(\lambda)$ and $q^{-n} \leq 2^{-\lambda}$, where λ refers to the security parameter. The $r = \Omega(\sqrt{\log \lambda})$ bound of our Corollary 4.3, that results from using $\delta = 1/3$ in the condition $r \geq \eta_\delta(A^\perp)$ in the RD-based smoothing argument of the proof above, improves on the corresponding bound $r = \Omega(\sqrt{\lambda})$ that results from the requirement to use

$\delta = O(2^{-\lambda})$ in the condition $r \geq \eta_\delta(A^\perp)$ in the SD-based smoothing argument of the proof of [GPV08, Theorem 7.1], in order to handle adversaries with advantage $\varepsilon = 2^{-o(\lambda)}$ in both cases. Thus our RD-based analysis saves a factor $\Omega(\sqrt{\lambda/\log \lambda})$ in the choice of r, and consequently of a^{-1} and q. (The authors of [GPV08] specify a choice of $r = \omega(\sqrt{\log \lambda})$ for their scheme because they use in their analysis the classical "no polynomial attacks" security requirement, corresponding to assuming attacks with advantage $\varepsilon = \lambda^{-O(1)}$, rather than the stronger $\varepsilon = \omega(2^{-\lambda})$ but more realistic setting we take.)

5   Application to LWE with Uniform Noise

The LWE problem with noise uniform in a small interval was introduced in [DMQ13]. In that article, the authors exhibit a reduction from LWE with Gaussian noise, which relies on a new tool called *lossy codes*. The main proof ingredients are the construction of lossy codes for LWE (which are lossy for the uniform distribution in a small interval), and the fact that lossy codes are pseudorandom.

We note that the reduction from [DMQ13] needs the number of LWE samples to be bounded by $\mathrm{poly}(n)$ and that it degrades the LWE dimension by a constant factor. The parameter β (when the interval of the noise is $[-\beta, \beta]$) should be at least $mn^\sigma \alpha$ where α is the LWE Gaussian noise parameter and $\sigma \in (0,1)$ is an arbitrarily small constant.

We now provide an alternative reduction from the $\mathrm{LWE}_{n,q,D_\alpha,m}$ distinguishing problem to the $\mathrm{LWE}_{n,q,U([-\beta,\beta]),m}$ distinguishing problem, and analyze it using RD. Our reduction preserves the LWE dimension n, and is hence tighter than the one from [DMQ13]. We also require that $\beta = \Omega(m\alpha)$.

Theorem 5.1. *Let $m \geq n \geq 1$ and with $q \leq \mathrm{poly}(m,n)$ prime. Let $\alpha, \beta > 0$ be real numbers with $\beta = \Omega(m\alpha)$. Then there is a polynomial-time reduction from $\mathrm{LWE}_{n,q,D_\alpha,m}$ to $\mathrm{LWE}_{n,q,\phi,m}$, with $\phi = \frac{1}{q}\lfloor qU([-\beta,\beta])\rceil$.*

Proof. In the proof, we let U_β denote the distribution $U([-\beta, \beta])$, to ease notations. Our reduction relies on four steps:

- A reduction from $\mathrm{LWE}_{n,q,D_\alpha,m}$ to $\mathrm{LWE}_{n,q,\psi,m}$ with $\psi = D_\alpha + U_\beta$,
- A reduction from $\mathrm{LWE}_{n,q,\psi,m}$ to $\mathrm{sLWE}_{n,q,\psi,m}$,
- A reduction from $\mathrm{sLWE}_{n,q,\psi,m}$ to $\mathrm{sLWE}_{n,q,U_\beta,m}$,
- A reduction from $\mathrm{sLWE}_{n,q,U_\beta,m}$ to $\mathrm{LWE}_{n,q,U_\beta,m}$.

First step. The reduction is given m elements $(\boldsymbol{a}_i, b_i) \in \mathbb{Z}_q^n \times \mathbb{T}$, all drawn from $A_{\boldsymbol{s},D_\alpha}$ (for some $\boldsymbol{s}$), or all drawn from $U(\mathbb{Z}_q^n \times \mathbb{T})$. The reduction consists in adding independent samples from U_β to each b_i. The reduction maps the uniform distribution to itself, and $A_{\boldsymbol{s},D_\alpha}$ to $A_{\boldsymbol{s},\psi}$.

Second step. Reducing the distinguishing variant of LWE to its search variant is direct.

Third step. The reduction from sLWE$_{n,q,\psi,m}$ to sLWE$_{n,q,U_\beta,m}$ is vacuous: by using the RD (and in particular the probability preservation property of Lemma 2.7), we show that an oracle solving sLWE$_{n,q,U_\beta,m}$ also solves sLWE$_{n,q,\psi,m}$.

Lemma 5.2. *Let* α, β *be real numbers with* $\alpha \in (0, 1/e)$ *and* $\beta \geq \alpha$. *Let* $\psi = D_\alpha + U_\beta$. *Then*

$$R_2(U_\beta \| \psi) = \frac{\alpha}{\beta} \int_0^\beta \frac{1}{\int_{-\beta}^\beta e^{\frac{-\pi(x-y)^2}{\alpha^2}} dy} dx \leq 1 + 16 \frac{\alpha}{\beta} \sqrt{\ln(1/\alpha)/\pi}.$$

Proof. The density function of ψ is the convolution of the density functions of D_α and U_β:

$$f_\psi(x) = \frac{1}{2\alpha\beta} \int_{-\beta}^\beta e^{\frac{-\pi(x-y)^2}{\alpha^2}} dy.$$

Using Rényi of order 2, we have:

$$R_2(U_\beta \| \psi) = \int_{-\beta}^\beta \frac{\frac{1}{(2\beta)^2}}{\frac{1}{2\alpha\beta} \int_{-\beta}^\beta e^{\frac{-\pi(x-y)^2}{\alpha^2}} dy} dx = \frac{\alpha}{\beta} \int_0^\beta \frac{1}{\int_{-\beta}^\beta e^{\frac{-\pi(x-y)^2}{\alpha^2}} dy} dx.$$

The denominator in the integrand is a function for $x \in [0, \beta]$.

$$\phi(x) = \alpha - \int_{\beta+x}^\infty \exp(\frac{-\pi y^2}{\alpha^2}) \, dy - \int_{\beta-x}^\infty \exp(\frac{-\pi y^2}{\alpha^2}) \, dy.$$

For standard Gaussian, we use the following tail bound [CDS03]:

$$\frac{1}{\sqrt{2\pi}} \int_z^\infty e^{-x^2/2} dx \leq \frac{1}{2} e^{-z^2/2}.$$

Then we have

$$\phi(x) \geq \alpha \left(1 - \frac{1}{2} \exp\left(\frac{-\pi(\beta+x)^2}{\alpha^2} \right) - \frac{1}{2} \exp\left(\frac{-\pi(\beta-x)^2}{\alpha^2} \right) \right).$$

Taking the reciprocal of above, we use the first-order Taylor expansion. Note here

$$t(x) = \frac{1}{2} \exp\left(\frac{-\pi(\beta+x)^2}{\alpha^2} \right) + \frac{1}{2} \exp\left(\frac{-\pi(\beta-x)^2}{\alpha^2} \right).$$

We want to bound the function $t(x)$ by a constant $c \in (0, 1)$. Here $t(x)$ is not monotonic. We take the maximum of the first-half and the maximum of the second-half of $t(x)$. An upper bound ($\beta \geq \alpha$) is:

$$t(x) \leq \frac{1}{2} e^{-\pi\beta^2/\alpha^2} + \frac{1}{2} =: \sigma_{\alpha,\beta} + \frac{1}{2} < 1.$$

We then use the fact that $\frac{1}{1-t(x)} = 1 + \frac{1}{1-t(x)}t(x) \le 1 + \frac{1}{1-2\sigma_{\alpha,\beta}}t(x)$ to bound the Rényi divergence of order 2.

$$
\begin{aligned}
R_2(U_\beta\|\psi) &= \frac{\alpha}{\beta}\int_0^\beta \frac{1}{\phi(x)}\,\mathrm{d}x \\
&\le \frac{1}{\beta}\int_0^\beta \frac{1}{1 - \frac{1}{2}\exp\left(\frac{-\pi(\beta+x)^2}{\alpha^2}\right) - \frac{1}{2}\exp\left(\frac{-\pi(\beta-x)^2}{\alpha^2}\right)}\,\mathrm{d}x \\
&\le \frac{1}{\beta}\int_0^\beta \left(1 + \frac{1}{1-2\sigma_{\alpha,\beta}}\exp\left(\frac{-\pi(\beta+x)^2}{\alpha^2}\right)\right. \\
&\quad \left.+ \frac{1}{1-2\sigma_{\alpha,\beta}}\exp\left(\frac{-\pi(\beta-x)^2}{\alpha^2}\right)\right)\mathrm{d}x \\
&= 1 + \frac{1}{(1-2\sigma_{\alpha,\beta})\beta}\int_0^{2\beta}\exp\left(\frac{-\pi x^2}{\alpha^2}\right)\mathrm{d}x \\
&= 1 + \frac{1}{2(1-2\sigma_{\alpha,\beta})\beta}\int_{-2\beta}^{2\beta}\exp\left(\frac{-\pi x^2}{\alpha^2}\right)\mathrm{d}x \\
&= 1 + \frac{\alpha}{(1-2\sigma_{\alpha,\beta})\beta}(1 - 2D_\alpha(2\beta)) \le 1 + \frac{1}{1-2\sigma_{\alpha,\beta}}\frac{\alpha}{\beta}.
\end{aligned}
$$

Hence we have the bound,

$$
R_2(U_\beta\|\psi) \le 1 + \frac{1}{1-e^{-\pi\beta^2/\alpha^2}}\frac{\alpha}{\beta}. \qquad \square
$$

We use Lemma 5.2 with m samples and $\beta = \Omega(m\alpha)$ to ensure that the mth power of the RD is ≤ 2. The RD multiplicativity and probability preservation properties (see Lemma 2.7) imply that $\varepsilon' \ge \varepsilon^2/R_2^m(U_\beta\|\phi)$; hence if an oracle solves sLWE$_{n,q,U_\beta,m}$ with probability ε, then it also solves sLWE$_{n,q,\psi,m}$ with probability $\ge \varepsilon^2/2$.

Fourth step. We reduce sLWE$_{n,q,U_\beta,m}$ with continuous noise U_β to sLWE$_{n,q,\phi,m}$ with discrete noise $\phi = \frac{1}{q}\lfloor qU_\beta\rceil$ with support contained in $\mathbb{T}_q$, by rounding to the nearest multiple of $\frac{1}{q}$ any provided b_i (for $i \le m$). We reduce sLWE$_{n,q,\phi,m}$ to LWE$_{n,q,\phi,m}$ by invoking Theorem 2.6. $\qquad \square$

6   Open Problems

Our results show the utility of the Rényi divergence in several areas of lattice-based cryptography. However, they also suggest some natural open problems, whose resolution could open up further applications. In particular, can we extend the applicability of RD to more general distinguishing problems than those satisfying our 'public sampleability' requirement? This may extend our results further. For instance, can we use RD-based arguments to prove the hardness of LWE with uniform noise without using the search to decision reduction of [MM11]? This may allow the proof to apply also to Ring-LWE with uniform noise.

Acknowledgments. We thank Léo Ducas, Vadim Lyubashevsky and Fabrice Mouhartem for useful discussions. This work has been supported in part by ERC Starting Grant ERC-2013-StG-335086-LATTAC, an Australian Research Fellowship (ARF) from the Australian Research Council (ARC), and ARC Discovery Grants DP0987734, DP110100628 and DP150100285. This work has been supported in part by the European Union's H2020 Programme under grant agreement number ICT-644209.

References

[Ajt96] Ajtai, M.: Generating hard instances of lattice problems (extended abstract). In: Proceedings of of STOC, pp. 99–108. ACM (1996)

[AKPW13] Alwen, J., Krenn, S., Pietrzak, K., Wichs, D.: Learning with rounding, revisited. In: Canetti, R., Garay, J.A. (eds.) CRYPTO 2013, Part I. LNCS, vol. 8042, pp. 57–74. Springer, Heidelberg (2013)

[BGM+15] Bogdanov, A., Guo, S., Masny, D., Richelson, S., Rosen, A.: On the hardness of learning with rounding over small modulus. Cryptology ePrint Archive, Report 2015/769 (2015). http://eprint.iacr.org/

[BLP+13] Brakerski, Z., Langlois, A., Peikert, C., Regev, O., Stehlé, D.: Classical hardness of learning with errors. In: Procedings of STOC, pp. 575–584. ACM (2013)

[BPR12] Banerjee, A., Peikert, C., Rosen, A.: Pseudorandom functions and lattices. In: Pointcheval, D., Johansson, T. (eds.) EUROCRYPT 2012. LNCS, vol. 7237, pp. 719–737. Springer, Heidelberg (2012)

[BV11] Brakerski, Z., Vaikuntanathan, V.: Efficient fully homomorphic encryption from (standard) LWE, In: Proceedings of FOCS, pp. 97–106. IEEE Computer Society Press (2011)

[CDS03] Chiani, M., Dardari, D., Simon, M.K.: New exponential bounds and approximations for the computation of error probability in fading channels. IEEE Trans. Wireless. Comm. $2(4)$, 840–845 (2003)

[DDLL13] Ducas, L., Durmus, A., Lepoint, T., Lyubashevsky, V.: Lattice signatures and bimodal gaussians. In: Canetti, R., Garay, J.A. (eds.) CRYPTO 2013, Part I. LNCS, vol. 8042, pp. 40–56. Springer, Heidelberg (2013)

[DMQ13] Döttling, N., Müller-Quade, J.: Lossy codes and a new variant of the learning-with-errors problem. In: Johansson, T., Nguyen, P.Q. (eds.) EUROCRYPT 2013. LNCS, vol. 7881, pp. 18–34. Springer, Heidelberg (2013)

[Duc14] Ducas, L.: Accelerating Bliss: the geometry of ternary polynomials. Cryptology ePrint Archive, Report 2014/874 (2014). http://eprint.iacr.org/

[EH12] van Erven, T., Harremoës, P.: Rényi divergence and Kullback-Leibler divergence. CoRR, abs/1206.2459 (2012)

[GGH13] Garg, S., Gentry, C., Halevi, S.: Candidate multilinear maps from ideal lattices. In: Johansson, T., Nguyen, P.Q. (eds.) EUROCRYPT 2013. LNCS, vol. 7881, pp. 1–17. Springer, Heidelberg (2013)

[GPV08] Gentry, C., Peikert, C., Vaikuntanathan, V.: Trapdoors for hard lattices and new cryptographic constructions. In: Proceedings of STOC, pp. 197–206. ACM (2008)

[LPR13] Lyubashevsky, V., Peikert, C., Regev, O.: On ideal lattices and learning with errors over rings. J. ACM $60(6)$, 43 (2013)

[LPSS14] Ling, S., Phan, D.H., Stehlé, D., Steinfeld, R.: Hardness of k-LWE and applications in traitor tracing. In: Garay, J.A., Gennaro, R. (eds.) CRYPTO 2014, Part I. LNCS, vol. 8616, pp. 315–334. Springer, Heidelberg (2014)

[LSS14] Langlois, A., Stehlé, D., Steinfeld, R.: GGHLite: more efficient multilinear maps from ideal lattices. In: Nguyen, P.Q., Oswald, E. (eds.) EUROCRYPT 2014. LNCS, vol. 8441, pp. 239–256. Springer, Heidelberg (2014)

[Lyu12] Lyubashevsky, V.: Lattice signatures without trapdoors. In: Pointcheval, D., Johansson, T. (eds.) EUROCRYPT 2012. LNCS, vol. 7237, pp. 738–755. Springer, Heidelberg (2012)

[MM11] Micciancio, D., Mol, P.: Pseudorandom knapsacks and the sample complexity of LWE search-to-decision reductions. In: Rogaway, P. (ed.) CRYPTO 2011. LNCS, vol. 6841, pp. 465–484. Springer, Heidelberg (2011)

[MP13] Micciancio, D., Peikert, C.: Hardness of SIS and LWE with small parameters. In: Canetti, R., Garay, J.A. (eds.) CRYPTO 2013, Part I. LNCS, vol. 8042, pp. 21–39. Springer, Heidelberg (2013)

[MR07] Micciancio, D., Regev, O.: Worst-case to average-case reductions based on Gaussian measures. SIAM J. Comput. **37**(1), 267–302 (2007)

[MR09] Micciancio, D., Regev, O.: Lattice-based cryptography. In: Bernstein, D.J., Buchmann, J., Dahmen, E. (Eds), Post-Quantum Cryptography, pp. 147–191. Springer, Heidelberg (2009)

[PDG14] Pöppelmann, T., Ducas, L., Güneysu, T.: Enhanced lattice-based signatures on reconfigurable hardware. In: Batina, L., Robshaw, M. (eds.) CHES 2014. LNCS, vol. 8731, pp. 353–370. Springer, Heidelberg (2014)

[Pei09] Peikert, C.: Public-key cryptosystems from the worst-case shortest vector problem. In: Proceedings of STOC, pp. 333–342. ACM (2009)

[Reg05] Regev, O.: On lattices, learning with errors, random linear codes, and cryptography. In: Proceedings of STOC, pp. 84–93 (2005)

[Reg09a] Regev, O.: Lecture notes of lattices in computer science, taught at the Computer Science Tel Aviv University, (2009). http://www.cims.nyu.edu/regev

[Reg09b] Regev, O.: On lattices, learning with errors, random linear codes, and cryptography. J. ACM **56**(6), 1–40 (2009)

[Rén61] Rényi, A.: On measures of entropy and information. In: Proceedings of the Fourth Berkeley Symposium on Mathematical Statistics and Probability, vol. 1, pp. 547–561 (1961)

Indistinguishability Obfuscation

Multi-input Functional Encryption
for Unbounded Arity Functions

Saikrishna Badrinarayanan[1]([⊠]), Divya Gupta[1], Abhishek Jain[2],
and Amit Sahai[1]

[1] UCLA, Los Angeles, USA
{bsaikrishna7393,divyagupta.iitd,amitsahai}@gmail.com
[2] Johns Hopkins University, Baltimore, USA
abhishekjain.itbhu@gmail.com

Abstract. The notion of multi-input functional encryption (MI-FE) was recently introduced by Goldwasser et al. [EUROCRYPT'14] as a means to non-interactively compute aggregate information on the joint private data of multiple users. A fundamental limitation of their work, however, is that the total number of users (which corresponds to the arity of the functions supported by the MI-FE scheme) must be a priori *bounded* and fixed at the system setup time.

In this work, we overcome this limitation by introducing the notion of unbounded input MI-FE that supports the computation of functions with *unbounded arity*. We construct such an MI-FE scheme with indistinguishability security in the selective model based on the existence of public-coin differing-inputs obfuscation for turing machines and collision-resistant hash functions.

Our result enables several new exciting applications, including a new paradigm of *on-the-fly* secure multiparty computation where new users can join the system dynamically.

1 Introduction

Functional Encryption. Traditionally, encryption has been used as a tool for private end-to-end communication. The emergence of cloud computing has opened up a host of new application scenarios where more functionality is desired

---

A. Jain—Supported in part by a DARPA/ARL Safeware Grant W911NF-15-C-0213 and NSF CNS-1414023

A. Sahai—Research supported in part from a DARPA/ONR PROCEED award, a DARPA/ARL SAFEWARE award, NSF Frontier Award 1413955, NSF grants 1228984, 1136174, 1118096, and 1065276, a Xerox Faculty Research Award, a Google Faculty Research Award, an equipment grant from Intel, and an Okawa Foundation Research Grant. This material is based upon work supported by the Defense Advanced Research Projects Agency through the U.S. Office of Naval Research under Contract N00014-11-1-0389. The views expressed are those of the author and do not reflect the official policy or position of the Department of Defense, the National Science Foundation, or the U.S. Government.

© International Association for Cryptologic Research 2015
T. Iwata and J.H. Cheon (Eds.): ASIACRYPT 2015, Part I, LNCS 9452, pp. 27–51, 2015.
DOI: 10.1007/978-3-662-48797-6_2

from encryption beyond the traditional privacy guarantees. To address this challenge, the notion of functional encryption (FE) has been developed in a long sequence of works [3,4,13,16,17,19,21]. In an FE scheme for a family $\mathcal{F}$, it is possible to derive decryption keys K_f for any function $f \in \mathcal{F}$ from a master secret key. Given such a key K_f and an encryption of a message x, a user can compute $f(x)$. Intuitively, the security of FE says that an adversarial user should only learn $f(x)$ and "nothing else about x."

Multi-input Functional Encryption. Most of the prior work on FE focuses on the problem of computing a function over a *single* plaintext given its corresponding ciphertext. However, many applications require the computation of aggregate information from *multiple* data sources (that may correspond to different users). To address this issue, recently, Goldwasser et al. [10] introduced the notion of multi-input functional encryption (MI-FE). Let $\mathcal{F}$ be a family of n-ary functions where n is a polynomial in the security parameter. In an MI-FE scheme for $\mathcal{F}$, the owner of the master secret key (as in FE) can compute decryption keys K_f for any function $f \in \mathcal{F}$. The new feature in MI-FE is that K_f can be used to compute $f(x_1, \ldots, x_n)$ from n ciphertexts $CT_1, \ldots, CT_n$ of messages $x_1, \ldots, x_n$ respectively, where each CT_i is computed *independently*, possibly using a different encryption key (but w.r.t. the same master secret key).

As discussed in [10] (see also [11,12]), MI-FE enables several important applications such as computing on multiple encrypted databases, order-revealing and property-revealing encryption, multi-client delegation of computation, secure computation on the web [14] and so on. Furthermore, as shown in [10], MI-FE, in fact, implies program obfuscation [2,8].

A fundamental limitation of the work of Goldwasser et al. [10] is that it requires an a priori (polynomial) bound on the arity n of the function family $\mathcal{F}$. More concretely, the arity n of the function family must be fixed during system setup when the parameters of the scheme are generated. This automatically fixes the number of users in the scheme and therefore new users cannot join the system at a later point of time. Furthermore, the size of the system parameters and the complexity of the algorithms depends on n. This has an immediate adverse impact on the applications of MI-FE: for example, if we use the scheme of [10] to compute on multiple encrypted databases, then we must a priori fix the number of databases and use decryption keys of size proportional to the number of databases.

Our Question: Unbounded Arity MI-FE. In this work, we seek to overcome this limitation. Specifically, we study the problem of MI-FE for general functions $\mathcal{F}$ with *unbounded* arity. Note that this means that the combined length of all the inputs to any function $f \in \mathcal{F}$ is unbounded and hence we must work in the Turing machine model of computation (as opposed to circuits). In addition, we also allow for each individual input to f to be of unbounded length.

More concretely, we consider the setting where the owner of a master secret key can derive decryption keys K_M for a general Turing machine M. For any index $i \in 2^\lambda$ (where λ is the security parameter), the owner of the master secret key can (at any point in time) compute an encryption key EK_i. Finally, given a list of ciphertexts $CT_1, \ldots, CT_\ell$ for any arbitrary ℓ, where each CT_i is encryption of some message x_i w.r.t. EK_i, and a decryption key K_M, one should be able to learn $M(x_1, \ldots, x_\ell)$.

We formalize security via a natural generalization of the indistinguishability-based security framework for bounded arity MI-FE to the case of unbounded arity. We refer the reader to Sect. 3 for details but point out that similar to [10], we also focus on selective security where the adversary declares the challenge messages at the beginning of the game.

1.1   Our Results

Our main result is an MI-FE scheme for functions with unbounded arity assuming the existence of public-coin differing-inputs obfuscation (pc-diO) [20] for general Turing machines with unbounded input length and collision-resistant hash functions. We prove indistinguishability-based security of our scheme in the selective model.

Theorem 1 (Informal). *If public-coin differing-inputs obfuscation for general Turing machines and collision-resistant hash functions exist, then there exists an indistinguishably-secure MI-FE scheme for general functions with unbounded arity, in the selective model.*

Discussion. Recently, Pandey et al. [20] defined the notion of pc-diO as a weakening of differing-inputs obfuscation (diO) [1,2,5]. In the same work, they also give a construction of pc-diO for general Turing machines with unbounded input length based on pc-diO for general circuits and public-coin (weak) succinct non-interactive arguments of knowledge (SNARKs).[1] We note that while the existence of diO has recently come under scrutiny [9], no impossibility results are known for pc-diO.

On the Necessity of Obfuscation. It was shown by Goldwasser et al. [10] that MI-FE for bounded arity functions with indistinguishability-based security implies indistinguishability obfuscation for general circuits. A straightforward extension of their argument (in the case where at least one of the encryption keys is known to the adversary) shows that MI-FE for functions with unbounded arity implies indistinguishability obfuscation for Turing machines with unbounded input length.

Applications. We briefly highlight a few novel applications of our main result:

- *On-the-fly secure computation*: MI-FE for unbounded inputs naturally yields a new notion of *on-the-fly* secure multiparty computation in the correlated randomness model where new parties can join the system *dynamically* at any

---

[1] A recent work by [6] shows that SNARKs with privately generated auxiliary inputs are impossible assuming the existence of pc-diO for circuits. We stress, however, that [20] only assumes the existence of a much weaker notion of *public-coin* SNARKs for their positive result. Therefore, the impossibility result of [6] is *not* applicable to [20].

point in time. To the best of our knowledge, no prior solution for secure computation (even in the interactive setting) exhibits this property.

In order to further explain this result, we first recall an application of MI-FE for bounded inputs to secure computation on the web [14] (this is implicit in [10]): consider a group of n parties who wish to jointly compute a function f over their private inputs using a web server. Given an MI-FE scheme that supports f, each party can simply send an encryption of its input x_i w.r.t. to its own encryption key to the server. Upon receiving all the ciphertexts, the server can then use a decryption key K_f (which is given to it as part of a correlated randomness setup) to compute $f(x_1, \ldots, x_n)$. Note that unlike the traditional solutions for secure computation that require simultaneous participation from each player, this solution is completely non-interactive and asynchronous (during the computation phase), which is particularly appealing for applications over the web.[2]

Note that in the above application, since the number of inputs for the MI-FE scheme are a priori bounded, it means that the number of parties must also be bounded at the time of correlated randomness setup. In contrast, by plugging in our new MI-FE scheme for unbounded inputs in the above template, we now no longer need to fix the number of users in advance, and hence new users can join the system on "on-the-fly." In particular, the same decryption key K_f that was computed during the correlated randomness setup phase can still be used even when new users are dynamically added to the system.

- *Computing on encrypted databases of dynamic size*: In a similar vein, our MI-FE scheme enables arbitrary Turing machine computations on an encrypted database where the size of the database is not fixed a priori and can be increased dynamically.[3] Concretely, given a database of initial size n, we can start by encrypting each record separately. If the database owner wishes to later add new records to the database, then she can simply encrypt these records afresh and then add them to the existing encrypted database. Note that a decryption key K_M that was issued previously can still be used to compute on the updated database since we allow for Turing machines of unbounded input length.

We finally remark that this solution also facilitates "flexible" computations: suppose that a user is only interested in learning the output of M on a subset S of the records of size (say) $\ell \ll n$. Then, if we were to jointly compute on the entire encrypted database, the computation time would be proportional to n. In contrast, our scheme facilitates selective (joint) decryption of the encryptions of the records in S; as such, the running time of the resulting computation is only proportional to ℓ.

---

[2] One should note, however, that due to its non-interactive nature, this solution only achieves a weaker indistinguishability-based notion of security for secure computation where the adversary also gets access to the residual function $f(\boldsymbol{x}_H^b, \cdot)$. Here $(\boldsymbol{x}_H^0, \boldsymbol{x}_H^1)$ are vectors of inputs of the honest parties.

[3] The same idea can be naturally extended to multiple databases.

1.2   Technical Overview

In this work, we consider the indistinguishability-based selective security model for unbounded arity multi-input functional encryption[4]. The starting point for our construction is the MiFE scheme for bounded arity functions [10]. Similar to their work, in our construction, each ciphertext will consist of two ciphertexts under pk_1 and pk_2, and some other elements specific to the particular encryption key used. At a high level, a function key for a turing machine M will be an obfuscation of a machine which receives a collection of ciphertexts, decrypts them using sk_1, and returns the output of the turing machine on the decrypted messages. Before we decrypt the ciphertext with sk_1, we also need to have a some check that the given ciphertext is a valid encryption corresponding to a certain index. This check needs to be performed by the functional key for the turing machine M. Moreover, there is a distinct encryption key for each index and we do not have any a-priori bound on the number of inputs to our functions. Hence, the kinds of potential checks which need to be performed are unbounded in number. Dealing with unbounded number of encryption keys is the main technical challenge we face in designing an unbounded arity multi-input functional encryption scheme. We describe this in more detail below.

In the indistinguishability based security game of MiFE, the adversary can query for any polynomial number of encryption keys and is capable of encrypting under those. Finally, it provides the two challenge vectors. For the security proof to go through, we need to switch-off all encryption keys which are not asked by the adversary. The construction of [10] achieves this by having a separate "flag" value for each encryption key; this flag is part of the public parameters and also hardcoded in all the function keys that are given out. This approach obviously does not work in our case because we are dealing with unbounded number of encryption keys. This is one of the main technical difficulties which we face in extending the construction of MiFE for bounded arity to our case. We would like to point out that these problems can be solved easily using diO along with signatures, but we want our construction to only rely on pc-diO.

At a high level, we solve this issue of handling and blocking the above mentioned unbounded number of keys as follows: The public parameters of our scheme will consist of a pseudorandom string $u = G(z)$ and a random string α. An encryption key EK_i for index i will consist of a proof that either there exists a z such that $u = G(z)$ or there exists a string x such that $x[j] = i$ and $\alpha = h(x)$, where h is a collision resistant hash function. Our programs only contain u and α hardcoded and hence their size is independent of the number of keys we can handle. In our sequence of hybrids, we will change u to be a random string and $\alpha = h(I)$, where I denotes the indices of the keys given out to the adversary. The encryption keys (which are asked by the adversary) will now use a proof for the second part of the statement and we show that a valid proof for an encryption key which is not given out to the adversary leads to a collision in the hash function.

---

[4] It was shown in [10] that simulation-based security even for bounded arity MiFE implies the strong notion of black-box obfuscation. Hence, we do not consider that notion in this paper.

Another issue which occurs is relating to the challenge ciphertexts for the indices for which the encryption key is not given to the adversary. Consider the setting when there is some index, say i^*, in challenge vector such that EK_{i^*} is secret. In the security game of MiFE we are guaranteed that output of M on any subset of either of the challenge ciphertexts along with any collection of the ciphertexts which the adversary can generate, is identical for both the challenge vectors. As mentioned before, for security proof to go through we need to ensure that for i^*, there should only exist the encryption of $x_{i^*}^0$ and $x_{i^*}^1$ (which are the challenge messages) and nothing else. Otherwise, if the adversary is able to come up with a ciphertext of $y^* \neq x_{i^*}^b$, he might be able to distinguish trivially. This is because we do not have any output restriction corresponding to y^*. In other words, we do not want to rule out all ciphertexts under EK_{i^*}; we want to rule out everything except $x_{i^*}^0$ and $x_{i^*}^1$. In the MiFE for bounded inputs [10], this problem was solved by hardcoding these specific challenge ciphertexts in public parameters as well as function keys. In our case, this will clearly not work since there is no bound on length of challenge vectors. We again use ideas involving collision resistant hash functions to deal with these issues. In particular, we hash the challenge vector and include a commitment to this hash value as part of the public parameters as well as the function keys. Note that we can do this because we only need to prove the selective security of our scheme.

We note that since collision resistant hash-functions have no trapdoor secret information, they work well with pc-diO assumption. We will crucially rely on pc-diO property while changing the program from using sk_1 to sk_2. Note that there would exist inputs on which the programs would differ, but these inputs would be hard to find for any PPT adversary even given all the randomness used to sample the two programs.

MiFE with unbounded arity implies iO for turing machines with unbounded inputs. First we recall the proof for the fact that MiFE with bounded number of inputs implies iO for circuits. To construct an iO for circuit C with n inputs, consider an MiFE scheme which supports arity $n+1$. Under the first index EK_1, encrypt C and under keys $\{2, \ldots, n+1\}$ give out encryptions of both 0 and 1 under each index. Also, the secret key corresponding to universal circuit is given out. For our case, consider the setting of two encryption keys EK_1 and EK_2. We give out the encryption of the machine M under EK_1 and also the key EK_2. That is, we are in the partial public key setting. We also give out the secret key corresponding to a universal turing machine which accepts inputs of unbounded length. Now, the user can encrypt inputs of unbounded length under the key EK_2 by encrypting his input bit by bit. Note that our construction allows encryption of multiple inputs under the same key.

2   Preliminaries

In this section, we describe the primitives used in our construction. Let λ be the security parameter.

2.1 Public-Coin Differing-Inputs Obfuscation

The notion of public coin differing-inputs obfuscation (pc-diO) was recently introduced by Yuval Ishai, Omkant Pandey, and Amit Sahai [15].

Let $\mathbb{N}$ denote the set of all natural numbers. We denote by $\mathcal{M} = \{\mathcal{M}_\lambda\}_{\lambda \in \mathbb{N}}$, a parameterized collection of Turing machines (TM) such that $\mathcal{M}_\lambda$ is the set of all TMs of size at most λ which halt within polynomial number of steps on all inputs. For $x \in \{0,1\}^*$, if M halts on input x, we denote by $\mathsf{steps}(\mathsf{M}, x)$ the number of steps M takes to output $\mathsf{M}(x)$. We also adopt the convention that the output $\mathsf{M}(x)$ includes the number of steps M takes on x, in addition to the actual output. The following definitions are taken almost verbatim from [15].

Definition 1 (Public-Coin Differing-Inputs Sampler for TMs). *An efficient non-uniform sampling algorithm* $\mathsf{Sam} = \{\mathsf{Sam}_\lambda\}$ *is called a public-coin differing-inputs sampler for the parameterized collection of TMs* $\mathcal{M} = \{\mathcal{M}_\lambda\}$ *if the output of* Sam_λ *is always a pair of Turing Machines* $(\mathsf{M}_0, \mathsf{M}_1) \in \mathcal{M}_\lambda \times \mathcal{M}_\lambda$ *such that* $|\mathsf{M}_0| = |\mathsf{M}_1|$ *and for all efficient non-uniform adversaries* $\mathcal{A} = \{\mathcal{A}_\lambda\}$, *there exists a negligible function* ϵ *such that for all* $\lambda \in \mathbb{N}$:

$$\Pr_r \left[\begin{array}{c} \mathsf{M}_0(x) \neq \mathsf{M}_1(x) \wedge \\ \mathsf{steps}(\mathsf{M}_0, x) = \mathsf{steps}(\mathsf{M}_1, x) = t \end{array} \middle| \begin{array}{c} (\mathsf{M}_0, \mathsf{M}_1) \leftarrow \mathsf{Sam}_\lambda(r); \\ (x, 1^t) \leftarrow \mathcal{A}_\lambda(r) \end{array} \right] \leqslant \epsilon(\lambda)$$

By requiring $\mathcal{A}_\lambda$ *to output* 1^t, *we rule out all inputs* x *for which* $\mathsf{M}_0, \mathsf{M}_1$ *may take more than polynomial steps.*

Definition 2 (Public-Coin Differing-Inputs Obfuscator for TMs). *A uniform* PPT *algorithm* $\mathcal{O}$ *is called a public-coin differing-inputs obfuscator for the parameterized collection of TMs* $\mathcal{M} = \{\mathcal{M}_\lambda\}$ *if the following requirements hold:*

- **Correctness** : $\forall \lambda, \forall \mathsf{M} \in \mathcal{M}_\lambda, \forall x \in \{0,1\}^*$, *we have*
  $\Pr[\mathsf{M}'(x) = \mathsf{M}(x) : \mathsf{M}' \leftarrow \mathcal{O}(1^\lambda, \mathsf{M})] = 1$.
- **Security** : *For every public-coin differing-inputs sampler* $\mathsf{Sam} = \{\mathsf{Sam}_\lambda\}$ *for the collection* $\mathcal{M}$, *for every efficient non-uniform distinguishing algorithm* $\mathcal{D} = \{\mathcal{D}_\lambda\}$, *there exists a negligible function* ϵ *such that for all* λ :

$$\left| \begin{array}{c} \Pr[\mathcal{D}_\lambda(r, \mathsf{M}') = 1 : (\mathsf{M}_0, \mathsf{M}_1) \leftarrow \mathsf{Sam}_\lambda(r), \mathsf{M}' \leftarrow \mathcal{O}(1^\lambda, \mathsf{M}_0)] - \\ \Pr[\mathcal{D}_\lambda(r, \mathsf{M}') = 1 : (\mathsf{M}_0, \mathsf{M}_1) \leftarrow \mathsf{Sam}_\lambda(r), \mathsf{M}' \leftarrow \mathcal{O}(1^\lambda, \mathsf{M}_1)] \end{array} \right| \leq \epsilon(\lambda)$$

  where the probability is taken over r *and the coins of* $\mathcal{O}$.
- **Succinctness and input − specific running time** : *There exists a (global) polynomial* s' *such that for all* λ, *for all* $\mathsf{M} \in \mathcal{M}_\lambda$, *for all* $\mathsf{M}' \leftarrow \mathcal{O}(1^\lambda, \mathsf{M})$, *and for all* $x \in \{0,1\}^*$, $\mathsf{steps}(\mathsf{M}', x) \leqslant s'(\lambda, \mathsf{steps}(\mathsf{M}, x))$.

We note that the size of the obfuscated machine M' is always bounded by the running time of $\mathcal{O}$ which is polynomial in λ. More importantly, the size of M' is independent of the running time of M. This holds even if we consider TMs which always run in polynomial time. This is because the polynomial bounding the running time of $\mathcal{O}$ is independent of the collection $\mathcal{M}$ being obfuscated. It is easy to obtain a uniform formulation from our current definitions.

2.2 Non Interactive Proof Systems

We start with the syntax and formal definition of a non-interactive proof system. Then, we give the definition of non-interactive witness indistinguishable proofs (NIWI) and strong non-interactive witness indistinguishable proofs (sNIWI).

Syntax : Let R be an efficiently computable relation that consists of pairs (x, w), where x is called the statement and w is the witness. Let L denote the language consisting of statements in R. A non-interactive proof system for a language L consists of the following algorithms:

- **Setup** CRSGen(1^λ) is a PPT algorithm that takes as input the security parameter λ and outputs a common reference string crs.
- **Prover** Prove(crs, x, w) is a PPT algorithm that takes as input the common reference string crs, a statement x and a witness w. If $(x, w) \in R$, it produces a proof string π. Else, it outputs fail.
- **Verifier** Verify(crs, x, π) is a PPT algorithm that takes as input the common reference string crs and a statement x with a corresponding proof π. It outputs 1 if the proof is valid, and 0 otherwise.

Definition 3 (Non-interactive Proof System). *A non-interactive proof system* (CRSGen, Prove, Verify) *for a language L with a* PPT *relation R satisfies the following properties:*

- **PerfectCompleteness** : *For every* $(x, w) \in R$, *it holds that*

$$\Pr\left[\mathsf{Verify}(\mathsf{crs}, x, \mathsf{Prove}(\mathsf{crs}, x, w))\right] = 1$$

where crs $\xleftarrow{\$}$ CRSGen(1^λ), *and the probability is taken over the coins of* CRSGen, Prove *and* Verify.
- **StatisticalSoundness** : *For every adversary $\mathcal{A}$, it holds that*

$$\Pr\left[x \notin L \wedge \mathsf{Verify}(\mathsf{crs}, x, \pi) = 1 \;\middle|\; \mathsf{crs} \leftarrow \mathsf{CRSGen}(1^\lambda); (x, \pi) \leftarrow \mathcal{A}(\mathsf{crs})\right] \leqslant \mathsf{negl}(\lambda)$$

If the soundness property only holds against PPT adversaries, then we call it an argument system.

Definition 4. *(Strong Witness Indistinguishability* sNIWI*). Given a non-interactive proof system* (CRSGen, Prove, Verify) *for a language L with a* PPT *relation R, let $\mathcal{D}_0$ and $\mathcal{D}_1$ be distributions which output an instance-witness pair (x, w). We say that the proof system is strong witness-indistinguishable if for every adversary $\mathcal{A}$ and for all* PPT *distinguishers D', it holds that*

If $\left| \Pr[D'(x) = 1 | (x, w) \leftarrow \mathcal{D}_0(1^\lambda)] - \Pr[D'(x) = 1 | (x, w) \leftarrow \mathcal{D}_1(1^\lambda)] \right| \leqslant \mathsf{negl}(\lambda)$

Then $\begin{aligned} & | \Pr[\mathcal{A}(\mathsf{crs}, x, \mathsf{Prove}(\mathsf{crs}, x, w)) = 1 | (x, w) \leftarrow \mathcal{D}_0(1^\lambda)] - \\ & \Pr[\mathcal{A}(\mathsf{crs}, x, \mathsf{Prove}(\mathsf{crs}, x, w)) = 1 | (x, w) \leftarrow \mathcal{D}_1(1^\lambda)] | \leqslant \mathsf{negl}(\lambda) \end{aligned}$

The proof system of [7] is a strong non-interactive witness indistinguishable proof system.

2.3 Collision Resistent Hash Functions

In this section, we describe the collision resistant hash functions mapping arbitrary polynomial length strings to $\{0,1\}^\lambda$. We begin by defining a family of collision resistant hash functions mapping 2λ length strings to λ length strings.

Definition 5. *Consider a family of hash functions $\mathcal{H}'_\lambda$ such that every $h' \in \mathcal{H}'_\lambda$ maps $\{0,1\}^{2\lambda}$ to $\{0,1\}^\lambda$. $\mathcal{H}'_\lambda$ is said to be a collision resistant hash family if for every PPT adversary $\mathcal{A}$,*

$$\Pr\left[h' \xleftarrow{\$} \mathcal{H}'_\lambda; (x,y) \leftarrow \mathcal{A}(h'); h'(x) = h'(y)\right] \leqslant \mathsf{negl}(\lambda)$$

In our scheme, we will need hash functions which hash unbounded length strings to $\{0,1\}^\lambda$. We describe these next, followed by a simple construction using Merkle trees [18]. In our construction, each block will consists of λ bits. Note that it is sufficient to consider a hash family hashing 2^λ blocks to λ bits, i.e., hashing strings of length at most $\lambda 2^\lambda$ to λ bits.

Definition 6. *[Family of collision resistant hash functions for unbounded length strings] Consider a family of hash functions $\mathcal{H}_\lambda$ such that every $h \in \mathcal{H}_\lambda$ maps strings of length at most $\{0,1\}^{\lambda 2^\lambda}$ to $\{0,1\}^\lambda$. Additionally, it supports the following functions:*

- H.Open(h, x, i, y): *Given a hash function key h, a string $x \in \{0,1\}^*$ such that $|x| \leqslant \lambda 2^\lambda$, an index $i \in [|x|]$, and $y \in \{0,1\}^\lambda$, it outputs a short proof $\gamma \in \{0,1\}^{\lambda^2}$ that $x[i] = y$.*
- H.Verify(h, y, u, γ, i): *Given a hash function key h, a string $y \in \{0,1\}^\lambda$, a string $u \in \{0,1\}^\lambda$, a string $\gamma \in \{0,1\}^{\lambda^2}$ and an index $i \in [2^\lambda]$, it outputs either accept or reject. This algorithm essentially verifies that there exists a x such that $y = h(x)$ and $x[i] = u$.*

For security it is required to satisfy the following property of collision resistance.
CollisionResistance. *The hash function family $\mathcal{H}_\lambda$ is said to be collision resistant if for every PPT adversary $\mathcal{A}$,*

$$\Pr\left[h \xleftarrow{\$} \mathcal{H}_\lambda; (x, u, \gamma, i) \leftarrow \mathcal{A}(h) \text{ s.t. } h(x) = y; x[i] \neq u; \mathsf{H.Verify}(h, y, u, \gamma, i) = accept\right] \leqslant \mathsf{negl}(\lambda)$$

Construction: The above described scheme can be constructed by a merkle hash tree based construction on standard collision resistant hash functions of Definition 5.

3 Unbounded Arity Multi-input Functional Encryption

Multi-input functional encryption(MiFE) for bounded arity functions (or circuits) was first introduced in [11,12]. In other words, for any bound n on the number of inputs, they designed an encryption scheme such that the owner of the master secret key MSK, can generate function keys sk_f corresponding to functions f accepting n inputs. That is, sk_f computes on $\mathsf{CT}_1, \ldots, \mathsf{CT}_n$ to produce $f(x_1, \ldots, x_n)$ as output where CT_i is an encryption of x_i. In this work, we remove the a-priori bound n on the cardinality of the function.

In this work, we consider multi-input functional encryption for functions which accept unbounded number of inputs. That is, the input length is not bounded at the time of function key generation. Since we are dealing with FE for functions accepting unbounded number of inputs, in essence, we are dealing with TMs (with unbounded inputs) instead of circuits (with bounded inputs). Similar to MiFE with bounded inputs which allows for multi-party computation with bounded number of players, our scheme allows multiparty computation with a-priori unbounded number of parties. In other words, our scheme allows for more parties to join on-the-fly even after function keys have been given out. Moreover, similar to original MiFE, we want that each party is able to encrypt under different encryption keys, i.e., we want to support unbounded number of encryption keys. We want to achieve all this while keeping the size of the public parameters, master secret key as well as the function keys to be bounded by some fixed polynomial in the security parameter.

As mentioned before, we consider unbounded number of encryption keys, some of which may be made public, while rest are kept secret. When all the encryption keys corresponding to the challenge ciphertexts of the adversary are public, it represents the "public-key setting". On the other hand, when none of the keys are made public, it is called the "secret-key" setting. Our modeling allows us to capture the general setting when any polynomial number of keys can be made public. This can correspond to any subset of the keys associated with the challenge ciphertexts as well as any number of other keys. Note that we have (any) unbounded polynomial number of keys in our system unlike previous cases, where the only keys are the ones associated with challenge ciphertext.

As another level of generality, we allow that the turing machines or the functions can be invoked with ciphertexts corresponding to any subset of the encryption keys. Hence, if CT_j is an encryption of x_j under key EK_{i_j} then sk_M on $\mathsf{CT}_1, \ldots, \mathsf{CT}_n$ computes $\mathsf{M}((x_1, i_1), \ldots, (x_n, i_n))$. Here sk_M corresponds to the key for the turing machine M.

Now, we first present the syntax and correctness requirements for unbounded arity multi-input functional encryption in Sect. 3.1 and then present the security definition in Sect. 3.2.

3.1 Syntax

Let $\mathcal{X} = \{\mathcal{X}_\lambda\}_{\lambda \in \mathbb{N}}$, $\mathcal{Y} = \{\mathcal{Y}_\lambda\}_{\lambda \in \mathbb{N}}$ and $\mathcal{K} = \{\mathcal{K}_\lambda\}_{\lambda \in \mathbb{N}}$ be ensembles where each $\mathcal{X}_\lambda, \mathcal{Y}_\lambda, \mathcal{K}_\lambda \subseteq [2^\lambda]$. Let $\mathcal{M} = \{\mathcal{M}_\lambda\}_{\lambda \in \mathbb{N}}$ be an ensemble such that each $\mathsf{M} \in \mathcal{M}_\lambda$ is a turing machine accepting an (a-priori) unbounded polynomial (in λ) length of inputs. Each input string to a function $\mathsf{M} \in \mathcal{M}_\lambda$ is a tuple over $\mathcal{X}_\lambda \times \mathcal{K}_\lambda$. A turing machine $\mathsf{M} \in \mathcal{M}_\lambda$, on input a n length tuple $((x_1, i_1), (x_2, i_2), \ldots, (x_n, i_n))$ outputs $M((x_1, i_1), (x_2, i_2), \ldots, (x_n, i_n)) \in \mathcal{Y}_\lambda$, where $(x_j, i_j) \in \mathcal{X}_\lambda \times \mathcal{K}_\lambda$ for all $j \in [n]$ and $n(\lambda)$ is any arbitrary polynomial in λ.

An unbounded arity multi-input functional encryption scheme FE for $\mathcal{M}$ consists of five algorithms (FE.Setup, FE.EncKeyGen, FE.Enc, FE.FuncKeyGen, FE.Dec) described below.

- **Setup** FE.Setup(1^λ) is a PPT algorithm that takes as input the security parameter λ and outputs the public parameters PP and the master secret key MSK.
- **Encryption Key Generation** FE.EncKeyGen(PP, i, MSK) is a PPT algorithm that takes as input the public parameters PP, an index $i \in \mathcal{K}_\lambda$ and master secret key MSK, and outputs the encryption key EK_i corresponding to index i.
- **Encryption** FE.Enc(PP, EK_i, x) is a PPT algorithm that takes as input public parameters PP, an encryption key EK_i and an input message $x \in \mathcal{X}_\lambda$ and outputs a ciphertext CT encrypting (x, i). Note that the ciphertext also incorporates the index of the encryption key.
- **Function Key Generation** FE.FuncKeyGen(PP, MSK, M) is a PPT algorithm that takes as input public parameters PP, the master secret key MSK, a turing machine $\mathsf{M} \in \mathcal{M}_\lambda$ and outputs a corresponding secret key SK_M.
- **Decryption** FE.Dec(SK_M, CT_1, CT_2, ..., CT_n) is a deterministic algorithm that takes as input a secret key SK_M and a set of ciphertexts $\mathsf{CT}_1, \ldots, \mathsf{CT}_n$ as input and outputs a string $y \in \mathcal{Y}_\lambda$. Note that there is no a-priori bound on n.

Definition 7 (Correctness). *An unbounded arity multi-input functional encryption scheme* FE *for* $\mathcal{M}$ *is correct if* $\forall \mathsf{M} \in \mathcal{M}_\lambda$, $\forall n$ *s.t* $n = p(\lambda)$, *for some polynomial* p, *all* $(x_1, x_2, \ldots, x_n) \in \mathcal{X}_\lambda^n$ *and all* $\mathsf{I} = (i_1, \ldots, i_n) \in \mathcal{K}_\lambda^n$:

$$\Pr \left[\begin{array}{l} (\mathsf{PP}, \mathsf{MSK}) \leftarrow \mathsf{FE.Setup}(1^\lambda); \mathsf{EK}_\mathsf{I} \leftarrow \mathsf{FE.EncKeyGen}(\mathsf{PP}, \mathsf{I}, \mathsf{MSK}); \\ \mathsf{SK}_\mathsf{M} \leftarrow \mathsf{FE.FuncKeyGen}(\mathsf{PP}, \mathsf{MSK}, \mathsf{M}); \\ \mathsf{FE.Dec}(\mathsf{SK}_\mathsf{M}, \mathsf{FE.Enc}(\mathsf{PP}, \mathsf{EK}_{i_1}, x_1), \ldots, \mathsf{FE.Enc}(\mathsf{PP}, \mathsf{EK}_{i_n}, x_n)) \neq \\ \mathsf{M}((x_1, i_1), \ldots, (x_n, i_n)) \end{array} \right] \leqslant \mathsf{negl}(\lambda)$$

Here, EK_I denotes a set of encryption keys corresponding to the indices in the set I. For each $i \in \mathsf{I}$, we run FE.EncKeyGen(PP, i, MSK) and we denote that in short by FE.EncKeyGen(PP, I, MSK).

3.2   Security Definition

We consider indistinguishability based selective security (or IND-security, in short) for unbounded arity multi-input functional encryption. This notion will be defined very similar to the security definition in original MiFE papers [11, 12]. We begin by recalling this notion.

Let us consider the simple case of 2-ary functions $f(\cdot, \cdot)$ such that adversary requests the function key for f as well as the encryption key for the second index. Let the challenge ciphertext be (x^0, y^0) and (x^1, y^1). For the indistinguishability of challenge vectors, first condition required is that $f(x^0, y^0) = f(x^1, y^1)$. Moreover, since the adversary has the encryption key for the second index, he can encrypt any message corresponding to the second index. Hence, if there exists a y^* such that $f(x^0, y^*) \neq f(x^1, y^*)$, then distinguishing is easy! Hence, they additionally require that $f(x^0, \cdot) = f(x^1, \cdot)$ for all the function queries made by the adversary. That is, the function queries made have to be compatible with the encryption keys requested by the adversary; otherwise the task of distinguishing is trivial.

Similar to this notion, since in our case as well, the adversary can request any subset of the encryption keys, we require that the function key queries are compatible with encryption key queries. Since we allow the turing machine to be invoked with any subset of the key indices and potentially unbounded number of key indices, this condition is much more involved in our setting. At a high level, we require that the function outputs should be identical for any subset of the two challenge inputs combined with any vector of inputs for indices for which adversary has the encryption keys. More formally, we define the notion of I-compatibility as follows:

Definition 8 (I-Compatibility). *Let* {M} *be any set of turing machines such that every turing machine* M *in the set belongs to* $\mathcal{M}_\lambda$. *Let* $I \subseteq \mathcal{K}_\lambda$ *such that* $|I| = q(\lambda)$ *for some polynomial* q. *Let* $\boldsymbol{X}^0$ *and* $\boldsymbol{X}^1$ *be a pair of input vectors, where* $\boldsymbol{X}^b = \{(x_1^b, k_1), (x_2^b, k_2), \ldots, (x_n^b, k_n)\}$ *such that* $n = p(\lambda)$ *for some polynomial* p. *We say that* {M} *and* $(\boldsymbol{X}^0, \boldsymbol{X}^1)$ *are* I-compatible *if they satisfy the following property:*

- *For every* M $\in$ {M}, *every* $I' = \{i_1, \ldots, i_\alpha\} \subseteq I$, *every* $J = \{j_1, \ldots, j_\beta\} \subseteq [n]$, *and every* $y_1, \ldots, y_\alpha \in \mathcal{X}_\lambda$ *and every permutation* $\pi : [\alpha + \beta] \to [\alpha + \beta]$:

$$
\text{M}\Big(\pi\big((y_1, i_1), (y_2, i_2), \ldots, (y_\alpha, i_\alpha), (x_{j_1}^0, k_{j_1}), (x_{j_2}^0, k_{j_2}), \ldots, (x_{j_\beta}^0, k_{j_\beta})\big)\Big) =
$$
$$
\text{M}\Big(\pi\big((y_1, i_1), (y_2, i_2), \ldots, (y_\alpha, i_\alpha), (x_{j_1}^1, k_{j_1}), (x_{j_2}^1, k_{j_2}), \ldots, (x_{j_\beta}^1, k_{j_\beta})\big)\Big)
$$

Here, $\pi(a_1, a_2, \ldots, a_{\alpha+\beta})$ *denotes the permutation of the elements* $a_1, \ldots, a_{\alpha+\beta}$.

We now present our formal security definition for IND-secure unbounded arity multi-input functional encryption.

Selective IND-Secure MiFE. This is defined using the following game between the challenger and the adversary.

Definition 9 (Indistinguishability-Based Selective Security). *We say that an unbounded arity multi-input functional encryption scheme* FE *for* $\mathcal{M}$ *is* IND-*secure if for every* PPT *adversary* $\mathcal{A} = (\mathcal{A}_0, \mathcal{A}_1)$, *for all polynomials* p, q *and for all* $m = p(\lambda)$ *and for all* $n = q(\lambda)$, *the advantage of* $\mathcal{A}$ *defined as*

$$
\text{Adv}_{\mathcal{A}}^{\text{FE,IND}}(1^\lambda) = \left| \Pr\left[\text{IND}_{\mathcal{A}}^{\text{FE}}(1^\lambda) = 1\right] - \frac{1}{2} \right|
$$

is negl(λ) *where the experiment is defined below (Fig. 1).*

In the above experiment, we require :

- *Let* {M} *denote the entire set of function key queries made by* $\mathcal{A}_1$. *Then, the challenge message vectors* $\boldsymbol{X}^0$ *and* $\boldsymbol{X}^1$ *chosen by* $\mathcal{A}_1$ *must be* I-compatible *with* {M}.

Experiment $\mathsf{IND}^{\mathsf{FE}}_{\mathcal{A}}(1^\lambda)$:

$(\mathrm{I}, \boldsymbol{X}^0, \boldsymbol{X}^1, st_0) \leftarrow \mathcal{A}_0(1^\lambda)$ where $|\mathrm{I}| = m$; $\boldsymbol{X}^\ell = \{(x_1^\ell, k_1), (x_2^\ell, k_2), \ldots, (x_n^\ell, k_n)\}$
$(\mathsf{PP}, \mathsf{MSK}) \leftarrow \mathsf{FE.Setup}(1^\lambda)$
Compute $\mathsf{EK}_i \leftarrow \mathsf{FE.EncKeyGen}(\mathsf{PP}, i, \mathsf{MSK})$, $\forall i \in \mathrm{I}$. Let $\mathsf{EK}_{\mathrm{I}} = \{\mathsf{EK}_i\}_{i \in \mathrm{I}}$.
$b \stackrel{\$}{\leftarrow} \{0,1\}$; $\mathsf{CT}_i \leftarrow \mathsf{FE.Enc}(\mathsf{EK}_{k_i}, x_i^b)$, $\forall i \in [n]$. Let $\mathbf{CT} = \{\mathsf{CT}_1, \ldots, \mathsf{CT}_n\}$
$b' \leftarrow \mathcal{A}_1^{\mathsf{FE.FuncKeyGen}(\mathsf{PP}, \mathsf{MSK}, \cdot)}(st_0, \mathsf{PP}, \mathsf{EK}_{\mathrm{I}}, \mathbf{CT})$
Output: $(b = b')$

Fig. 1.

4   A Construction from Public-Coin Differing-Inputs Obfuscation

Notation : Without loss of generality, let's assume that every plaintext message and encryption key index is of length λ where λ denotes the security parameter of our scheme. Let $(\mathsf{CRSGen}, \mathsf{Prove}, \mathsf{Verify})$ be a statistically sound, non-interactive strong witness-indistinguishable proof system for NP, $\mathcal{O}$ denote a public coin differing-inputs obfuscator, $\mathsf{PKE} = (\mathsf{PKE.Setup}, \mathsf{PKE.Enc}, \mathsf{PKE.Dec})$ be a semantically secure public key encryption scheme, com be a statistically binding and computationally hiding commitment scheme and G be a pseudorandom generator from $\{0,1\}^\lambda$ to $\{0,1\}^{2\lambda}$. Without loss of generality, let's say com commits to a string bit-by-bit and uses randomness of length λ to commit to a single bit. Let $\{H_\lambda\}$ be a family of merkle hash functions such that every $h \in H_\lambda$ maps strings from $\{0,1\}^{\lambda 2^\lambda}$ to $\{0,1\}^\lambda$. That is, the merkle tree has depth λ.
We now describe our scheme $\mathsf{FE} = (\mathsf{FE.Setup}, \mathsf{FE.EncKeyGen}, \mathsf{FE.Enc}, \mathsf{FE.FuncKeyGen}, \mathsf{FE.Dec})$ as follows:

- **Setup $\mathsf{FE.Setup}(1^\lambda)$:**
  The setup algorithm first computes $\mathsf{crs} \leftarrow \mathsf{CRSGen}(1^\lambda)$. Next, it computes $(\mathsf{pk}_1, \mathsf{sk}_1) \leftarrow \mathsf{PKE.Setup}(1^\lambda)$, $(\mathsf{pk}_2, \mathsf{sk}_2) \leftarrow \mathsf{PKE.Setup}(1^\lambda)$, $(\mathsf{pk}_3, \mathsf{sk}_3) \leftarrow \mathsf{PKE.Setup}(1^\lambda)$ and $(\mathsf{pk}_4, \mathsf{sk}_4) \leftarrow \mathsf{PKE.Setup}(1^\lambda)$. Let $\alpha = \mathsf{com}(0^\lambda; u)$, $\beta_1 = \mathsf{com}(0^\lambda; u_1)$ and $\beta_2 = \mathsf{com}(0^\lambda; u_2)$ where u, u_1 and u_2 are random strings of length λ^2. Choose a hash function $h \leftarrow H_\lambda$. Choose $z \stackrel{\$}{\leftarrow} \{0,1\}^\lambda$ and compute $Z = G(z)$.
  The public parameters are $\mathsf{PP} = (\mathsf{crs}, \mathsf{pk}_1, \mathsf{pk}_2, \mathsf{pk}_3, \mathsf{pk}_4, h, \alpha, \beta_1, \beta_2, Z)$.
  The master secret key is $\mathsf{MSK} = (\mathsf{sk}_1, z, u, u_1, u_2)$.
- **Encryption Key Generation $\mathsf{FE.EncKeyGen}(\mathsf{PP}, i, \mathsf{MSK})$:**
  Given an index i, this algorithm first defines $b_i = z||0^\lambda||0^{\lambda^2}||0^{\lambda^2}||0^\lambda$. Then, it computes $d_i = \mathsf{PKE.Enc}(\mathsf{pk}_4, b_i; r)$ for some randomness r and $\sigma_i \leftarrow \mathsf{Prove}(\mathsf{crs}, st_i, w_i)$ for the statement that $st_i \in L_1$ using witness $w_i = (b_i, r)$ where $st_i = (d_i, i, \mathsf{pk}_4, \alpha, Z)$.
  L_1 is defined corresponding to the relation R_1 defined below.

Relation R_1:

Instance : $st_i = (d_i, i, \mathsf{pk}_4, \alpha, Z)$

Witness : $w = (b_i, r)$, where $b_i = z||\mathsf{hv}||\gamma||u||t$

$R_1(st_i, w) = 1$ if and only if the following conditions hold:

1. $d_i = \mathsf{PKE.Enc}(\mathsf{pk}_4, b_i; r)$ AND
2. The OR of the following statements must be true:
   - (a) $G(z) = Z$
   - (b) $\mathsf{H.Verify}(h, \mathsf{hv}, i, \gamma, t) = 1$ and $\mathsf{com}(\mathsf{hv}; u) = \alpha$

The output of the algorithm is the ith encryption key $\mathsf{EK}_i = (\sigma_i, d_i, i)$, where σ_i is computed using witness for statements 1 and 2(a) of R_1.

- **Encryption $\mathsf{FE.Enc}(\mathsf{PP}, \mathsf{EK}_i, x)$:**

To encrypt a message x with the ith encryption key EK_i, the encryption algorithm first computes $c_1 = \mathsf{PKE.Enc}(\mathsf{pk}_1, x||i; r_1)$ and $c_2 = \mathsf{PKE.Enc}(\mathsf{pk}_2, x||i; r_2)$. Define string $a = x||i||r_1||0^{\lambda^2}||0^{\lambda}||0^{\lambda^2}||x||i||r_2||0^{\lambda^2}||0^{\lambda}||0^{\lambda^2}||0^{\lambda}$ and compute $c_3 = \mathsf{PKE.Enc}(\mathsf{pk}_3, a; r_3)$. Next, it computes a proof $\pi \leftarrow \mathsf{Prove}(\mathsf{crs}, y, w)$ for the statement that $y \in L_2$ using witness w where:

$y = (c_1, c_2, c_3, \mathsf{pk}_1, \mathsf{pk}_2, \mathsf{pk}_3, \mathsf{pk}_4, \beta_1, \beta_2, i, d_i, \alpha, Z)$

$w = (a, r_3, \sigma_i)$

L_2 is defined corresponding to the relation R_2 defined below.

Relation R_2:

Instance : $y = (c_1, c_2, c_3, \mathsf{pk}_1, \mathsf{pk}_2, \mathsf{pk}_3, \mathsf{pk}_4, \beta_1, \beta_2, i, d_i, \alpha, Z)$

Witness : $w = (a, r_3, \sigma_i)$ where $a = x_1||i_1||r_1||u_1||\mathsf{hv}_1||\gamma_1||x_2||i_2||r_2||u_2||\mathsf{hv}_2||\gamma_2||t$

$R_2(y, w) = 1$ if and only if the following conditions hold:

1. $c_3 = \mathsf{PKE.Enc}(\mathsf{pk}_3, a; r_3)$ AND
2. The OR of the following two statements 2(a) and 2(b) is true:
   - (a) The OR of the following two statements is true:
     - i. $(c_1 = \mathsf{PKE.Enc}(\mathsf{pk}_1, (x_1||i_1); r_1)$ AND $c_2 = \mathsf{PKE.Enc}(\mathsf{pk}_2, (x_1||i_1); r_2)$
       AND $i_1 = i$ AND $\mathsf{Verify}(\mathsf{crs}, st_i, \sigma_i) = 1$ such that $st_i = (d_i, i, \mathsf{pk}_4, \alpha, Z) \in L_1)$; OR
     - ii. $(c_1 = \mathsf{PKE.Enc}(\mathsf{pk}_1, (x_2||i_2); r_1)$ AND $c_2 = \mathsf{PKE.Enc}(\mathsf{pk}_2, (x_2||i_2); r_2)$
       AND $i_2 = i$ AND $\mathsf{Verify}(\mathsf{crs}, st_i, \sigma_i) = 1$ such that $st_i = (d_i, i, \mathsf{pk}_4, \alpha, Z) \in L_1)$;
   - (b) c_1, c_2 encrypt $(x_1||i_1), (x_2||i_2)$ respectively, which may be different but then both β_1 and β_2 contain a hash of one of them (which may be different). That is,
     - i. $c_1 = \mathsf{PKE.Enc}(\mathsf{pk}_1, (x_1||i_1); r_1)$ AND $c_2 = \mathsf{PKE.Enc}(\mathsf{pk}_2, (x_2||i_2); r_2)$
     - ii. $\mathsf{H.Verify}(h, \mathsf{hv}_1, (x_1||i_1), \gamma_1, t) = 1$ AND $\beta_1 = \mathsf{com}(\mathsf{hv}_1; u_1)$ OR
       $\mathsf{H.Verify}(h, \mathsf{hv}_1, (x_2||i_2), \gamma_1, t) = 1$ AND $\beta_1 = \mathsf{com}(\mathsf{hv}_1; u_1)$
     - iii. $\mathsf{H.Verify}(h, \mathsf{hv}_2, (x_1||i_1), \gamma_2, t) = 1$ AND $\beta_2 = \mathsf{com}(\mathsf{hv}_2; u_2)$ OR
       $\mathsf{H.Verify}(h, \mathsf{hv}_2, (x_2||i_2), \gamma_2, t) = 1$ AND $\beta_2 = \mathsf{com}(\mathsf{hv}_2; u_2)$

Program G_M

Input : $CT_1, CT_n, \ldots, CT_n$
Constants : (sk_1, PP), i.e. $(sk_1, (crs, pk_1, pk_2, pk_3, pk_4, h, \alpha, \beta_1, \beta_2, Z))$
1. For every $i \in [n]$:
   (a) Parse $CT_i = (c_{i,1}, c_{i,2}, c_{i,3}, d_{k_i}, \pi_i, k_i)$
   (b) Let $y_i = (c_{i,1}, c_{i,2}, c_{i,3}, pk_1, pk_2, pk_3, pk_4, \beta_1, \beta_2, k_i, d_{k_i}, \alpha, Z)$ be the statement corresponding to the proof string π_i. If $\mathsf{Verify}(crs, y_i, \pi_i) = 0$, then stop and output $\bot$. Else, continue to the next step.
   (c) Compute $(x_i \| k_i) = \mathsf{PKE.Dec}(sk_1, c_{i,1})$
2. Output $M((x_1, k_1), (x_2, k_2), \ldots, (x_n, k_n))$

Fig. 2.

The output of the algorithm is the ciphertext $CT = (c_1, c_2, c_3, d_i, \pi, i)$. π is computed for the AND of statements 1 and 2(a)i of R_2.

– **Function Key Generation** $\mathsf{FE.FuncKeyGen}(PP, MSK, M)$: The algorithm computes $SK_M = \mathcal{O}(G_M)$ where the program G_M is defined as follows (Fig. 2):
– **Decryption** $\mathsf{FE.Dec}(SK_M, CT_1, \ldots, CT_n)$: It computes and outputs $SK_M(CT_1, \ldots, CT_n)$.

5   Security Proof

We now prove that the proposed scheme FE is selective IND-secure.

Theorem 2. *Let $\mathcal{M} = \{\mathcal{M}_\lambda\}_{\lambda \in \mathbb{N}}$ be a parameterized collection of Turing machines (TM) such that $\mathcal{M}_\lambda$ is the set of all TMs of size at most λ which halt within polynomial number of steps on all inputs. Then, assuming there exists a public-coin differing-inputs obfuscator for the class $\mathcal{M}$, a non-interactive strong witness indistinguishable proof system, a public key encryption scheme, a non-interactive perfectly binding computationally hiding commitment scheme, a pseudorandom generator and a family of merkle hash functions, the proposed scheme FE is a selective IND-secure MIFE scheme with unbounded arity for Turing machines in the class $\mathcal{M}$ according to Definition 9.*

We will prove the above theorem via a series of hybrid experiments $H_0, \ldots, H_{20}$ where H_0 corresponds to the real world experiment with challenge bit $b = 0$ and H_{20} corresponds to the real world experiment with challenge bit $b = 1$.

– **Hybrid H_0:** This is the real experiment with challenge bit $b = 0$. The public parameters are
$PP = (crs, pk_1, pk_2, pk_3, pk_4, h, \alpha, \beta_1, \beta_2, Z)$ such that $\alpha = com(0^\lambda; u)$, $\beta_1 = com(0^\lambda; u_1), \beta_2 = com(0^\lambda; u_2)$ and $Z = G(z)$, where $z \xleftarrow{\$} \{0,1\}^\lambda$.

- **Hybrid H_1:** This hybrid is identical to the previous hybrid except that β_1 and β_2 are computed differently. β_1 is computed as a commitment to hash of the string $s_1 = (x_1^0||k_1, \ldots, x_n^0||k_n)$ where $\{(x_1^0, k_1), \ldots, (x_n^0, k_n)\}$ is the challenge message vector $\boldsymbol{X}^0$. Similarly, β_2 is computed as a commitment to hash of the string $s_2 = (x_1^1||k_1, \ldots, x_n^1||k_n)$ where $\{(x_1^1, k_1), \ldots, (x_n^1, k_n)\}$ is the challenge message vector $\boldsymbol{X}^1$. That is, $\beta_1 = \mathsf{com}(h(s_1); u_1)$ and $\beta_2 = \mathsf{com}(h(s_2); u_2)$. There is no change in the way the challenge ciphertexts are computed.

  Note that s_1 and s_2 are padded with sufficient zeros to satisfy the input length constraint of the hash function.

- **Hybrid H_2:** This hybrid is identical to the previous hybrid except that we change the third component (c_3) in every challenge ciphertext. Let the ith challenge ciphertext be $\mathsf{CT}_i = (c_{i,1}, c_{i,2}, c_{i,3}, d_{k_i}, \pi_i, k_i)$ for all $i \in [n]$. Let $s_1 = (x_1^0||k_1, \ldots, x_n^0||k_n)$ and $s_2 = (x_1^1||k_1, \ldots, x_n^1||k_n)$. In the previous hybrid $c_{i,3}$ is an encryption of $a_i = x_i^0||k_i||r_1||0^{\lambda^2}||0^\lambda||0^{\lambda^2}||x_i^0||k_i||r_2||0^{\lambda^2}||0^\lambda||0^{\lambda^2}||0^\lambda$. Now, a_i is changed to $a_i = x_i^0||k_i||r_1||u_1||h(s_1)||\gamma_{1,i}||x_i^1||k_i||r_2||u_2||h(s_2)||\gamma_{2,i}||i$ where $\gamma_{1,i}, \gamma_{2,i}$ are the openings for $h(s_1)$ and $h(s_2)$ w.r.t. $x_i^0||k_i$ and $x_i^1||k_i$, respectively. That is, $\gamma_{1,i} = \mathsf{H.Open}(h, s_1, i, x_i^0||k_i)$ and $\gamma_{2,i} = \mathsf{H.Open}(h, s_2, i, x_i^1||k_i)$. Since a_i has changed, consequently, ciphertext $c_{i,3}$ which is an encryption of a_i, witness w_i for π_i and proof π_i change as well for all $i \in [n]$. Note that for all challenge ciphertexts, π still uses the witness for statement 1 and 2(a).

- **Hybrid H_3:** This hybrid is identical to the previous hybrid except that we change the second component in every challenge ciphertext. Let the ith challenge ciphertext be CT_i where $i \in [n]$. Let's parse $\mathsf{CT}_i = (c_{i,1}, c_{i,2}, c_{i,3}, d_{k_i}, \pi_i, k_i)$. We change $c_{i,2}$ to be an encryption of $x_i^1||k_i$. Further, π_i is now computed using the AND of statements 1 and 2(b) in the relation R_2.

- **Hybrid H_4:** This hybrid is identical to the previous hybrid except that α is computed as a commitment to hash of the string $s = (k_1, k_2, \ldots, k_m)$ where $\{k_1, \ldots, k_m\}$ is the set of indices I for which the adversary requests encryption keys. i.e. $\alpha = \mathsf{com}(h(s); u)$.

  Note that in this hybrid, for any encryption key EK_i, the proof σ_i is unchanged and is generated using the AND of statements 1 and 2(a).

- **Hybrid H_5:** This hybrid is identical to the previous hybrid except that we change the second component d_{k_i} for every encryption key EK_{k_i} that is given out to the adversary. First, let's denote $s = (k_1, \ldots, k_m)$ as in the previous hybrid. d_{k_i} is an encryption of $b_{k_i} = z||0^\lambda||0^{\lambda^2}||0^{\lambda^2}||0^\lambda$. Now, b_{k_i} is changed to $b_{k_i} = z||h(s)||\gamma_i||u_1||i$ where u_1 is the randomness used in the commitment of α and γ_i is the opening of the hash values in the merkle tree. That is, $\gamma_i = \mathsf{H.Open}(h, s, i, k_i)$. Consequently, d_{k_i} which is an encryption of b_{k_i} also changes. Since b_{k_i} has changed, the witness used in computing the proof σ_{k_i} has also changed. Note that σ_{k_i} still uses the witness for statements 1 and 2(a).

- **Hybrid H_6:** This hybrid is identical to the previous hybrid except that for every encryption key EK_{k_i} that is given out to the adversary, σ_{k_i} is now computed using the AND of statements 1 and 2(b) in the relation R_1.

- **Hybrid H_7:** This hybrid is identical to the previous hybrid except that in the public parameters Z is chosen to be a uniformly random string. Therefore, now $G(z) \neq Z$ except with negligible probability.
- **Hybrid H_8:** Same as the previous hybrid except that the challenger sets the master secret key to have sk_2 instead of sk_1 and for every function key query M, the corresponding secret key SK_M is computed as $SK_M \leftarrow \mathcal{O}(G'_M)$ where the program G'_M is the same as G_M except that :
  1. It has secret key sk_2 as a constant hardwired into it instead of sk_1.
  2. It decrypts the *second* component of each input ciphertext using sk_2. That is, in step 1(C), $x_i || k_i$ is computed as $x_i || k_i = PKE.Dec(sk_2, c_{i,2})$
- **Hybrid H_9:** This hybrid is identical to the previous hybrid except that in the public parameters Z is chosen to be the output of the pseudorandom generator applied on the seed z. That is, $Z = G(z)$.
- **Hybrid H_{10}:** This hybrid is identical to the previous hybrid except that for every encryption key EK_{k_i} that is given out to the adversary, we change σ_{k_i} to now be computed using the AND of statements 1 and 2(a) in the relation R_1.

Remark: Note that statement 2(b) is true as well for all EK_{k_i} but we choose to use 2(a) due to the following technical difficulty. Observe that at this point we need to somehow change each $c_{i,1}$ to be an encryption of $x_i^1 || k_i$ instead of $x_i^0 || k_i$. When we make this switch, the statement 2(b) in R_2 is no longer true. This is because β_1 will not be valid w.r.t. $c_{i,1}$ and $c_{i,2}$ since both are now encryptions of $x_i^1 || k_i$. So we need to make statement 2(a) true for all challenge ciphertexts including the ones under some EK_{k_j} such that $k_j \notin I$.

- **Hybrid H_{11}:** This hybrid is identical to the previous hybrid except that we change the first component in every challenge ciphertext. Let the ith challenge ciphertext be CT_i where $i \in [n]$. Let's parse $CT_i = (c_{i,1}, c_{i,2}, c_{i,3}, d_{k_i}, \pi_i, k_i)$. We change $c_{i,1}$ to be an encryption of $x_i^1 || k_i$. Then, we change the proof π_i to be computed using the AND of statements 1 and 2(a) in the relation R_2.
- **Hybrid H_{12}:** This hybrid is identical to the previous hybrid except that β_1 is computed differently. β_1 is computed as a commitment to hash of the string $s_2 = (x_1^1 || k_1, \ldots, x_n^1 || k_n)$ where $\{(x_1^1, k_1), \ldots, (x_n^1, k_n)\}$ is the challenge message vector $\boldsymbol{X}^1$. That is, $\beta_1 = com(h(s_2); u_1)$
  Note that s_2 is padded with sufficient zeros to satisfy the input length constraint of the hash function. There is no change in the way the challenge ciphertexts are computed.
- **Hybrid H_{13}:** This hybrid is identical to the previous hybrid except that we change the proof in every challenge ciphertext. Let the ith challenge ciphertext be CT_i where $i \in [n]$. Let's parse $CT_i = (c_{i,1}, c_{i,2}, c_{i,3}, d_{k_i}, \pi_i, k_i)$. We change π_i to now be computed using the AND of statements 1 and 2(b) in the relation R_2.
- **Hybrid H_{14}:** This hybrid is identical to the previous hybrid except that for every encryption key EK_{k_i} that is given out to the adversary, we change σ_{k_i} to now be computed using the AND of statements 1 and 2(b) in the relation R_1.
- **Hybrid H_{15}:** This hybrid is identical to the previous hybrid except that in the public parameters Z is chosen to be a uniformly random string.

- **Hybrid H_{16}:** This hybrid is identical to the previous hybrid except that the master secret key is set back to having sk_1 instead of sk_2 and for every function key query M, the corresponding secret key SK_M is computed using obfuscation of the original program G_M, i.e. $\mathsf{SK}_\mathsf{M} \leftarrow \mathcal{O}(\mathsf{G}_\mathsf{M})$.
- **Hybrid H_{17}:** This hybrid is identical to the previous hybrid except we change Z to be the output of the pseudorandom generator applied on the seed z. That is, $Z = G(z)$.
- **Hybrid H_{18}:** This hybrid is identical to the previous hybrid except that for every encryption key EK_{k_i} that is given out to the adversary, σ_{k_i} is now computed using the AND of statements 1 and 2(a) in the relation R_1.
- **Hybrid H_{19}:** This hybrid is identical to the previous hybrid except that we change the second component d_{k_i} for every encryption key EK_{k_i} that is given out to the adversary. We change b_{k_i} to be $b_{k_i} = z||0^\lambda||0^{\lambda^2}||0^{\lambda^2}||0^\lambda$ and consequently d_{k_i} also changes as it is the encryption of b_{k_i}. Since b_{k_i} has changed, the witness used in computing the proof σ_{k_i} has also changed. Note that σ_{k_i} still uses the witness for statements 1 and 2(a).
- **Hybrid H_{20}:** This hybrid is identical to the previous hybrid except that we change α to be a commitment to 0^λ. That is, $\alpha = \mathsf{com}(0^\lambda; u)$.
- **Hybrid H_{21}:** This hybrid is identical to the previous hybrid except that for every challenge ciphertext key CT_i that is given out to the adversary, π_i is now computed using the AND of statements 1 and 2(a) in the relation R_2.
- **Hybrid H_{22}:** This hybrid is identical to the previous hybrid except that we change the third component in every challenge ciphertext. Let the ith challenge ciphertext be CT_i where $i \in [n]$. Let's parse $\mathsf{CT}_i = (c_{i,1}, c_{i,2}, c_{i,3}, d_{k_i}, \pi_i, k_i)$ where $c_{i,3}$ is an encryption of a_i. Now, a_i is changed to $a_i = x_i^1||k_i||r_1||0^{\lambda^2}||0^\lambda||0^{\lambda^2}||x_i^1||k_i||r_2||0^{\lambda^2}||0^\lambda||0^{\lambda^2}||0^\lambda$. Consequently, ciphertext $c_{i,3}$ which is an encryption of a_i will also change. Note that for all challenge ciphertexts, π still uses the witness for statement 1 and 2(a).
- **Hybrid H_{23}:** This hybrid is identical to the previous hybrid except that β_1 and β_2 are both computed to be commitments of 0^λ. That is, $\beta_1 = \mathsf{com}(0^\lambda; u_1)$ and $\beta_2 = \mathsf{com}(0^\lambda; u_2)$. This is identical to the real experiment with challenge bit $b = 1$.

Below we will prove that $(H_0 \approx_c H_1)$, $(H_1 \approx_c H_2)$, and $(H_7 \approx_c H_8)$. The indistinguishability of other hybrids will follow along the same lines.

Lemma 1 ($H_0 \approx_c H_1$). *Assuming that* com *is a (computationally) hiding commitment scheme, the outputs of experiments* H_0 *and* H_1 *are computationally indistinguishable.*

Proof. The only difference between the two hybrids is the manner in which the commitments β_1 and β_2 are computed. Let's consider the following adversary $\mathcal{A}_{\mathsf{com}}$, which internally executes the hybrid H_0 except that it does not generate the commitments β_1 and β_2 on it's own. Instead, after receiving the challenge message vectors $\boldsymbol{X}^0$ and $\boldsymbol{X}^1$ from $\mathcal{A}$, it sends two sets of strings, namely $(0^\lambda, 0^\lambda)$ and $(h(s_1), h(s_2))$ to the outside challenger where s_1 and s_2 are defined the same way as in H_1. In return, $\mathcal{A}_{\mathsf{com}}$ receives two commitments β_1, β_2 corresponding to either the first or the second set of strings. It then gives these to $\mathcal{A}$. Now, whatever bit b that $\mathcal{A}$ guesses, $\mathcal{A}_{\mathsf{com}}$ forwards the guess to the outside challenger. Clearly, $\mathcal{A}_{\mathsf{com}}$ is a polynomial time algorithm and violates the hiding property of com unless $H_0 \approx_c H_1$.

Lemma 2 ($H_1 \approx_c H_2$). *Assuming the semantic security of* PKE *and the strong witness indistinguishability of the proof system, the outputs of experiments* H_1 *and* H_2 *are computationally indistinguishable.*

Proof. Recall that strong witness indistinguishability asserts the following: let $\mathcal{D}_0$ and $\mathcal{D}_1$ be distributions which output an instance-witness pair for an NP-relation R and suppose that the first components of these distributions are computationally indistinguishable, i.e., $\{y : (y,w) \leftarrow \mathcal{D}_0(1^\lambda)\} \approx_c \{y : (y,w) \leftarrow \mathcal{D}_1(1^\lambda)\}$; then $\mathcal{X}_0 \approx_c \mathcal{X}_1$ where $\mathcal{X}_b : \{(\mathsf{crs}, y, \pi) : \mathsf{crs} \leftarrow \mathsf{CRSGen}(1^\lambda); (y,w) \leftarrow \mathcal{D}_b(1^\lambda); \pi \leftarrow \mathsf{Prove}(\mathsf{crs}, y, w)\}$ for $b \in \{0,1\}$.

Suppose that H_1 and H_2 can be distinguished with noticeable advantage δ. Note that we can visualize Hybrid H_2 as a sequence of n hybrids $H_{1,0}, \ldots, H_{1,n}$ where in each hybrid, the only change from the previous hybrid happens in the ith challenge ciphertext CT_i. $H_{1,0}$ corresponds to H_1 and $H_{1,n}$ corresponds to H_2. Therefore, if H_1 and H_2 can be distinguished with advantage δ, then there exists i such that $H_{1,i-1}$ and $H_{1,i}$ can be distinguished with advantage δ/n where n is a polynomial in the security parameter λ. So, let's fix this i and work with these two hybrids $H_{1,i-1}$ and $H_{1,i}$.

Observe that both hybrids internally sample the following values in an identical manner: $\zeta = (\mathsf{pk}_1, \mathsf{pk}_2, \mathsf{pk}_3, \mathsf{pk}_4, h, \alpha, \beta_1, \beta_2, Z, c_{i,1}, c_{i,2}, d_{k_i}, k_i)$. This includes everything except $crs, c_{i,3}$ and π_i. By simple averaging, there is at least a $\delta/2n$ fraction of strings st such that the two hybrids can be distinguished with advantage at least $\delta/2n$ when $\zeta = st$. Call such a ζ to be good. Fix one such ζ, and denote the resulting hybrids by $H_{1,i-1}^\zeta$ and $H_{1,i}^\zeta$. Note that the hybrids have inbuilt into them all other values used to sample ζ namely : $\boldsymbol{X}^0, \boldsymbol{X}^1$ received from $\mathcal{A}$, randomness for generating the encryptions and the commitments, and the master secret key msk.

The first distribution $\mathcal{D}_0^{(\zeta)}$ is defined as follows: compute $c_{i,3} = \mathsf{PKE.Enc}$ $(\mathsf{pk}_3, a_i; r_{i,3})$ where $a_i = x_i^0||k_i||r_{i,1}||0^{\lambda^2}||0^\lambda||0^{\lambda^2}||x_i^0||k_i||r_{i,2}||0^{\lambda^2}||0^\lambda||0^{\lambda^2}||0^\lambda$ and let statement $y = (c_{i,1}, c_{i,2}, c_{i,3}, \mathsf{pk}_1, \mathsf{pk}_2, \mathsf{pk}_3, \mathsf{pk}_4, \beta_1, \beta_2, k_i, d_{k_i}, \alpha, Z)$, witness $w = (a_i, r_{i,3}, \sigma_{k_i})$. It outputs (y,w). Note that y is identical to ζ except that h has been removed and $c_{i,3}$ has been added. Define a second distribution $\mathcal{D}_1^{(\zeta)}$ identical to $\mathcal{D}_0^{(\zeta)}$ except that instead of a_i , it uses $a_i^* = x_i^0||k_i||r_1||u_1||h(s_1)||\gamma_{i,1}||x_i^1||k_i||r_2||u_2||h(s_2)||\gamma_{i,2}||i$. Here, $\gamma_{i,1}, \gamma_{i,2}$ are the openings of the hash values in the merkle tree. That is, $\gamma_{i,1} = \mathsf{H.Open}(h, s_1, i, x_i^0||k_i)$ and $\gamma_{i,2} = \mathsf{H.Open}(h, s_2, i, x_i^1||k_i)$ where $s_1 = (x_1^0||k_1, \ldots, x_n^0||k_n)$ and $s_2 = (x_1^1||k_1, \ldots, x_n^1||k_n)$. Then, it computes $c_{i,3}^* = \mathsf{PKE.Enc}(\mathsf{pk}_3, a_i^*; r_{i,3})$, $y^* = (c_{i,1}, c_{i,2}, c_{i,3}^*, \mathsf{pk}_1, \mathsf{pk}_2, \mathsf{pk}_3, \mathsf{pk}_4, \beta_1, \beta_2, k_i, d_{k_i}, \alpha, Z)$, and $w^* = (a_i^*, r_{i,3}, \sigma_i)$. It outputs (y^*, w^*). It follows from the security of the encryption scheme that the distribution of y sampled by $\mathcal{D}_0^{(\zeta)}$ is computationally indistinguishable from

y^* sampled by $\mathcal{D}_1^{(\zeta)}$, i.e., $y \approx_c y^*$. Therefore, we must have that $\mathcal{X}_0 \approx_c \mathcal{X}_1$ with respect to these distributions. We show that this is not the case unless $\mathsf{H}_{1,i-1}^\zeta \approx_c \mathsf{H}_{1,i}^\zeta$.

Consider an adversary $\mathcal{A}'$ for strong witness indistinguishability who incorporates $\mathcal{A}$ and ζ (along with sk_1 and all values for computing ζ described above), and receives a challenge (crs, y, π) distributed according to either $\mathcal{D}_0^{(\zeta)}$ or $D_1^{(\zeta)}$; here y has one component $c_{i,3}$ that is different from ζ. The adversary $\mathcal{A}'$ uses $\mathsf{crs}, \mathsf{sk}_1$ and other values used in defining ζ to completely define PP, answer encryption key queries, generate other challenge ciphertexts and answer the function key queries and feeds it to $\mathcal{A}$. Then, it uses $(c_{i,3}, \pi)$ to define the ith challenge ciphertext $\mathsf{CT}_i = (c_{i,1}, c_{i,2}, c_{i,3}, d_{k_i}, \pi, k_i)$. The adversary $\mathcal{A}'$ outputs whatever $\mathcal{A}$ outputs. We observe that the output of this adversary is distributed according to $\mathsf{H}_{1,i-1}^m$ (resp., $\mathsf{H}_{1,i}^m$) when it receives a tuple from distribution $\mathcal{X}_0$ (resp., $\mathcal{X}_1$). A randomly sampled m is good with probability at least $\delta/2n$, and therefore it follows that with probability at least $\frac{\delta^2}{4n^2}$, the strong witness indistinguishability property will be violated with non-negligible probability unless δ is negligible.

Lemma 3 ($\mathsf{H}_7 \approx_c \mathsf{H}_8$). *Assuming the correctness of* PKE, *that* $\mathcal{O}$ *is a public-coin differing-inputs obfuscator for for Turing machines in the class* $\mathcal{M}$, G *is a pseudorandom generator*, com *is a perfectly binding and (computationally) hiding commitment scheme and* H_λ *is a family of merkle hash functions, the outputs of experiments* H_7 *and* H_8 *are computationally indistinguishable.*

Proof. Suppose that the claim is false and $\mathcal{A}'s$ output in H_7 is noticeably different from its output in H_8. Suppose that $\mathcal{A}'s$ running time is bounded by a polynomial μ so that there are at most μ function key queries it can make. We consider a sequence of μ hybrid experiments between H_7 and H_8 such that hybrid $\mathsf{H}_{7,v}$ for $v \in [\mu]$ is as follows.

Hybrid $\mathsf{H}_{7,v}$. It is identical to H_7 except that it answers the function key queries as follows. For $j \in [\mu]$, if $j \leqslant v$, the function key corresponding to the jth query, denoted by M_j , is an obfuscation of program $\mathsf{G}_{\mathsf{M}_j}$. If $j > v$, it is an obfuscation of program $\mathsf{G}'_{\mathsf{M}_j}$. We define $\mathsf{H}_{7,0}$ to be H_7 and observe that $\mathsf{H}_{7,\mu}$ is the same as H_8.

We see that if $\mathcal{A}'s$ advantage in distinguishing between H_7 and H_8 is δ, then there exists a $v \in [\mu]$ such that $\mathcal{A}$'s advantage in distinguishing between $\mathsf{H}_{7,v-1}$ and $\mathsf{H}_{7,v}$ is at least δ/μ. We show that if δ is not negligible, then we can use $\mathcal{A}$ to violate the indistinguishability of the obfuscator $\mathcal{O}$. To do so, we define a sampling algorithm $\mathsf{Sam}_{\mathcal{A}}^v$ and a distinguishing algorithm $\mathcal{D}_{\mathcal{A}}^v$ and prove that $\mathsf{Sam}_{\mathcal{A}}^v$ is a public-coin differing inputs sampler outputting a pair of differing-input TMs yet $\mathcal{D}_{\mathcal{A}}^v$ can distinguish an obfuscation of left TM from that of right TM that is output by $\mathsf{Sam}_{\mathcal{A}}^v$. The description of these two algorithms is as follows:

Sampler $\mathsf{Sam}_{\mathcal{A}}^v(\rho)$:

1. Receive $(\boldsymbol{X}^0, \boldsymbol{X}^1, \mathrm{I})$ from $\mathcal{A}$.
2. Parse ρ as (crs, h, τ).

3. Proceed identically to H_7 using τ as randomness for all tasks except for sampling the hash function which is set to h, and the CRS, which is set to crs. This involves the following steps:

   (a) Parse $\tau = (\tau_1, \tau_2, \tau_3, \tau_4, r_{i,1}, r_{i,2}, r_{i,3}, r_\ell, u, u_1, u_2)$ for all $i \in [n]$ and for all $\ell \in [|I|]$.

   (b) Use τ_1 as randomness to generate $(\mathsf{pk}_1, \mathsf{sk}_1)$, τ_2 as randomness to generate $(\mathsf{pk}_2, \mathsf{sk}_2)$ τ_3 as randomness to generate $(\mathsf{pk}_3, \mathsf{sk}_3)$ τ_4 as randomness to generate $(\mathsf{pk}_4, \mathsf{sk}_4)$.

   (c) Use u as randomness to generate $\alpha = \mathsf{com}(h(s); u)$, where $s = (1||k_1, 2||k_2, \ldots, t||k_m)$ and $\{k_1, \ldots, k_m\} = I$.

   (d) Use u_1, u_2 as randomness to generate $\beta_1 = \mathsf{com}(h(s_1); u_1)$ and $\beta_2 = \mathsf{com}(h(s_2); u_2)$, where $s_1 = (1||x_1^0||k_1, \ldots, n||x_n^0||k_n)$ and $s_2 = (1||x_1^1||k_1, \ldots, n||x_n^1||k_n)$.

   (e) Define Z to be a uniform random string of length 2λ. Define the public parameters $\mathsf{PP} = (\mathsf{crs}, \mathsf{pk}_1, \mathsf{pk}_2, \mathsf{pk}_3, \mathsf{pk}_4, h, \alpha, \beta_1, \beta_2, Z)$. Send PP to $\mathcal{A}$.

   (f) For all $k_i \in I$, to generate the ith encryption key EK_{k_i}, compute $b_{k_i} = z||h(s)||\gamma_i||u_1||i$ and $d_{k_i} = \mathsf{PKE.Enc}(\mathsf{pk}_4, b_{k_i}; r_i)$. Using witness $w_{k_i} = (b_{k_i}, r_i)$, compute proof σ_{k_i} using the AND of statements 1 and 2(b) in the relation R_1.

   Send the encryption key EK_{k_i} for all $k_i \in I$ to $\mathcal{A}$.

   (g) For all $i \in [n]$, we generate the ith challenge ciphertext in the following manner. We use $r_{i,1}$ and $r_{i,2}$ as randomness to generate $c_{i,1} = \mathsf{PKE.Enc}(\mathsf{pk}_1, x_i^0||k_i; r_{i,1})$ and $c_{i,2} = \mathsf{PKE.Enc}(\mathsf{pk}_2, x_i^1||k_i; r_{i,2})$. Use $a_i = x_i^0||k_i||r_{i,1}||u_1||h(s_1)||\gamma_{i,1}||x_i^1||k_i||r_{i,2}||u_2||h(s_2)||\gamma_{i,2}||i$ where $\gamma_{i,1}, \gamma_{i,2}$ are the openings for $h(s_1)$ and $h(s_2)$ w.r.t. $x_i^0||k_i$ and $x_i^1||k_i$ respectively. That is, $\gamma_{i,1} = \mathsf{H.Open}(h, s_1, i, x_i^0||k_i)$ and $\gamma_{i,2} = \mathsf{H.Open}(h, s_2, i, x_i^1||k_i)$. Compute $c_{i,3} = \mathsf{PKE.Enc}(\mathsf{pk}_3, a_i; r_{i,3})$. Then, use witness $w_i = (a_i, r_{i,3}, \sigma_{k_i})$ to compute proof π_i using the AND of statements 1 and 2(b) in the relation R_2. The ith challenge ciphertext is $(c_{i,1}, c_{i,2}, c_{i,3}, d_{k_i}, \pi_i, k_i)$.

   Send all the challenge ciphertexts to $\mathcal{A}$.

   (h) Answer the function key queries of $\mathcal{A}$ as follows. For all queries M_j, until $j < v$, send an obfuscation of $\mathsf{G}_{\mathsf{M}_j}$.

   (i) Upon receiving the vth function key query M_v, output $(\tilde{\mathsf{M}}_0, \tilde{\mathsf{M}}_1)$ and halt, where :

   $$\tilde{\mathsf{M}}_0 = \mathsf{G}_{\mathsf{M}_v}, \qquad \tilde{\mathsf{M}}_1 = \mathsf{G}'_{\mathsf{M}_v}.$$

Distinguisher $\mathcal{D}_{\mathcal{A}}^v(\rho, \mathsf{M}')$: on input a random tape ρ and an obfuscated TM M', the distinguisher simply executes all steps of the sampler $\mathsf{Sam}_{\mathcal{A}}^v(\rho)$, answering function keys for all $j < v$ as described above. The distinguisher, however, does not halt when the vth query is sent, and continues the execution of $\mathcal{A}$ answering function key queries for M_j as follows :

- if $j = v$, send M' (which is an obfuscation of either $\tilde{\mathsf{M}}_0$ or $\tilde{\mathsf{M}}_1$).
- if $j > v$, send an obfuscation of $\mathsf{G}'_{\mathsf{M}_j}$.

The distinguisher outputs whatever $\mathcal{A}$ outputs.

We can see that if M' is an obfuscation of $\tilde{\mathsf{M}}_0$, the output of $\mathcal{D}^v_{\mathcal{A}}(\rho, \mathsf{M}')$ is identical to $\mathcal{A}'s$ output in $\mathsf{H}_{7,k-1}$ and if M' is an obfuscation of $\tilde{\mathsf{M}}_1$, it is identical to $\mathcal{A}'s$ output in $\mathsf{H}_{7,k}$. We have that $\mathcal{D}^v_{\mathcal{A}}(\rho, \mathsf{M}')$ distinguishes $\mathsf{H}_{7,k-1}$ and $\mathsf{H}_{7,k}$ with at least δ/μ advantage.

All that remains to prove now is that $\mathsf{Sam}^v_{\mathcal{A}}(\rho)$ is a public-coin differing-inputs sampler.

Theorem 3. $\mathsf{Sam}^v_{\mathcal{A}}(\rho)$ *is a public-coin differing inputs sampler.*

Proof. We show that if there exists an adversary B who can find differing-inputs to the pair of TMs sampled by $\mathsf{Sam}^v_{\mathcal{A}}(\rho)$ with noticeable probability, we can use B and $\mathsf{Sam}^v_{\mathcal{A}}(\rho)$ to construct an efficient algorithm $\mathsf{CollFinder}_{\mathsf{B},\mathsf{Sam}^v_{\mathcal{A}}(\rho)}$ which finds collisions in h with noticeable probability.

$\mathsf{CollFinder}_{\mathsf{B},\mathsf{Sam}^v_{\mathcal{A}}(\rho)}(h)$:
On input a random hash function $h \leftarrow H_\lambda$, the algorithm first samples uniformly random strings (crs, τ) to define a random tape $\rho = (\mathsf{crs}, h, \tau)$. Then, it samples $(\tilde{\mathsf{M}}_0, \tilde{\mathsf{M}}_1) \leftarrow \mathsf{Sam}^v_{\mathcal{A}}(\rho)$ and computes $e^* \leftarrow \mathsf{B}(\rho)$ e^* is the differing input and corresponds to a set of ciphertexts. Let $e^* = (e_1^*, \ldots, e_\ell^*)$ where each $e_j^* = (e_{j,1}^*, e_{j,2}^*, e_{j,3}^*, d_{k_j^*}^*, \pi_j^*, k_j^*)$ for $j \in [\ell]$. For each j, if π_j^* is a valid proof, compute $a_j^* = \mathsf{PKE.Dec}(\mathsf{sk}_3, e_{j,3}^*)$ and let $a_j^* = x_{j,1}^* \| k_{j,1}^* \| r_{j,1}^* \| u_1 \| \mathsf{hv}_1^* \| \gamma_{j,1}^* \| x_{j,2}^* \| k_{j,2}^* \| r_{j,2}^* \| u_2 \| \mathsf{hv}_2^* \| \gamma_{j,2}^* \| t^*$. Let $(\boldsymbol{X}^0, \boldsymbol{X}^1)$ be the challenge message vectors output by $\mathcal{A}$ initially. Let $\boldsymbol{X}^0 = \{(x_1^0, k_1), \ldots, (x_n^0, k_n)\}$ and $\boldsymbol{X}^1 = \{(x_1^1, k_1), \ldots, (x_n^1, k_n)\}$. Define $s_1 = (x_1^0 \| k_1, \ldots, x_n^0 \| k_n)$ and $s_2 = (x_1^1 \| k_1, \ldots, x_n^1 \| k_n)$ Let the encryption key queries be $I = \{k_1, \ldots, k_t\}$. Define $s = (k_1, \ldots, k_t)$. If $h(s_1) = h(s_2)$, output (s_1, s_2) as collisions to the hash function.

Claim. For all $j \in [\ell]$, π_j^* is a valid proof.

Proof. Since e^* is a differing input, $\tilde{\mathsf{M}}_0(e^*) \neq \tilde{\mathsf{M}}_1(e^*)$. Now, suppose for some $j \in [\ell]$, π_j^* was not a valid proof. Then, both $\tilde{\mathsf{M}}_0$ and $\tilde{\mathsf{M}}_1$ would output $\bot$ on input e^* which means that e^* is not a differing input.

Condition A : A ciphertext $C = (c_1, c_2, c_3, d_k, \pi, k)$ for which π is valid satisfies condition A with respect to challenge message vectors $(\boldsymbol{X}^0, \boldsymbol{X}^1)$ and encryption key queries I iff

1. c_1 and c_2 encrypt the same message and $k \in I$ (OR)
2. $\exists i \in [n]$ such that $\{(x_1 \| k_1), (x_2 \| k_2)\} = \{(x_i^0 \| k_i), (x_i^1 \| k_i)\}$, where $x_1 \| k_1 = \mathsf{PKE.Dec}(\mathsf{sk}_1, c_1)$ and $x_2 \| k_2 = \mathsf{PKE.Dec}(\mathsf{sk}_2, c_2)$.

Claim. For every $j \in [\ell]$, if e_j^* satisfies condition A, then e is not a differing input.

Proof. Suppose the above two conditions are true for every $j \in [\ell]$. Then, from the definition of I-compatibility of challenge message vectors $(\boldsymbol{X}^0, \boldsymbol{X}^1)$ and function query M_v, we see that $\tilde{\mathsf{M}}_0(e^*) = \tilde{\mathsf{M}}_1(e^*)$ which means that e^* is not a differing input.

Therefore, since we have assumed that e^* is a differing input, there exists $j \in [\ell]$ such that e_j^* does not satisfy condition A.

Claim. If there exists $j \in [\ell]$ such that e_j^* does not satisfy condition A, then we can find a collision in the hash function h.

Proof. Let's fix $j \in [\ell]$ such that e_j^* does not satisfy condition A. Since π_j^* is a valid proof, by the soundness of the strong witness indistinguishable proof system, one of the following two cases must hold:

- **case 1:** π_j^* was proved using statements 1 and 2(a) of relation R_2.
  Now, since e_j^* does not satisfy condition A, it doesn't satisfy condition A(1) as well. Therefore, either $e_{j,1}^*$ and $e_{j,2}^*$ encrypt different messages or $k_j^* \notin I$. If $e_{j,1}^*$ and $e_{j,2}^*$ encrypt different messages, statement 2(a) would clearly be false and π_j^* would not be valid. However, we already proved that π_j^* is valid. Therefore, it must be the case that $k_j^* \notin I$.
  Since 2(a) is true in R_2, we have $\mathsf{Verify}(\mathsf{crs}, st_{k_j^*}, \sigma_{k_j^*}) = 1$ where $st_{k_j^*} = (d_{k_j^*}, k_j^*, \mathsf{pk}_4, \alpha, Z)$ and $\sigma_{k_j^*}$ is a proof that $st_{k_j^*} \in L_1$. Further, since Z is a uniform random string, $Z \neq \mathsf{G}(z)$ for any z except with negligible probability. As a result, $\sigma_{k_j^*}$ must be proved using statements 1 and 2(b) in relation R_1. Therefore, there exists $\mathsf{hv}^*, \gamma^*, t^*$ such that $\mathsf{H.Verify}(h, \mathsf{hv}^*, k_j^*, \gamma^*, t^*) = 1$ and $\mathsf{com}(\mathsf{hv}^*; u) = \alpha$. Since the commitment scheme is perfectly binding, $\mathsf{hv}^* = h(s)$. We know that $s = (k_1, \dots, k_t)$. Therefore, $s[t^*] \neq k_j^*$. Thus, there exists γ^*, t^* such that $\mathsf{H.Verify}(h, h(s), k_j^*, \gamma^*, t^*) = 1$ and $s[t^*] \neq k_j^*$. By definition 6, we have found a collision in the hash function h.
- **case 2:** π_j^* was proved using statements 1 and 2(b) of relation R_2.
  Since e_j^* does not satisfy condition A, it doesn't satisfy condition A(2) as well. Therefore, $\forall i \in [n]$ $\{(x_{j,1}^*\|k_{j,1}^*), (x_{j,2}^*\|k_{j,2}^*)\} \neq \{(x_i^0\|k_i), (x_i^1\|k_i)\}$. Since π_j^* was proved using 2(b), $\exists \mathsf{hv}_1^*, \mathsf{hv}_2^*, \gamma_1^*, \gamma_2^*, t^*$ such that 2(b)(ii) and 2(b)(iii) are true. Without loss of generality, let's say that the first of the two conditions in 2(b)(ii) is true and the second of the two conditions in 2(b)(iii) is true. That is, $\mathsf{H.Verify}(h, \mathsf{hv}_1^*, x_{j,1}^*\|k_{j,1}^*, \gamma_1^*, t^*) = 1$, $\beta_1 = \mathsf{com}(\mathsf{hv}_1^*; u_1)$ and $\mathsf{H.Verify}(h, \mathsf{hv}_2^*, x_{j,2}^*\|k_{j,2}^*, \gamma_2^*, t^*) = 1$, $\beta_2 = \mathsf{com}(\mathsf{hv}_2^*; u_2)$. Since the commitment scheme is perfectly binding, $\mathsf{hv}_1^* = h(s_1)$ and $\mathsf{hv}_2^* = h(s_2)$. We know that $\{(x_{j,1}^*\|k_{j,1}^*), (x_{j,2}^*\|k_{j,2}^*)\} \neq \{(x_{t^*}^0\|k_{t^*}), (x_{t^*}^1\|k_{t^*})\}$. Without loss of generality, let's say $(x_{j,1}^*\|k_{j,1}^*) \neq (x_{t^*}^0\|k_{t^*})$. Since $s_1 = (x_1^0\|k_1, \dots, x_n^0\|k_n)$, we have $s_1[t^*] \neq (x_{j,1}^*\|k_{j,1}^*)$. Thus, there exists γ_1^*, t^* such that $s_1[t^*] \neq x_{j,1}^*\|k_{j,1}^*$ and $\mathsf{H.Verify}(h, h(s_1), x_{j,1}^*\|k_{j,1}^*, \gamma_1^*, t^*) = 1$. By Definition 6, we have found a collision in the hash function h.

References

1. Ananth, P., Boneh, D., Garg, S., Sahai, A., Zhandry, M.: Differing-inputs obfuscation and applications (2013)
2. Barak, B., Goldreich, O., Impagliazzo, R., Rudich, S., Sahai, A., Vadhan, S.P., Yang, K.: On the (im)possibility of obfuscating programs. In: Kilian, J. (ed.) CRYPTO 2001. LNCS, vol. 2139, pp. 1–18. Springer, Heidelberg (2001)

3. Boneh, D., Sahai, A., Waters, B.: Functional encryption: definitions and challenges. In: Ishai, Y. (ed.) TCC 2011. LNCS, vol. 6597, pp. 253–273. Springer, Heidelberg (2011)

4. Boneh, D., Waters, B.: Conjunctive, subset, and range queries on encrypted data. In: Vadhan, S.P. (ed.) TCC 2007. LNCS, vol. 4392, pp. 535–554. Springer, Heidelberg (2007)

5. Boyle, E., Chung, K.M., Pass, R.: On extractability (aka differing-inputs) obfuscation. In: TCC (2014)

6. Boyle, E., Pass, R.: Limits of extractability assumptions with distributional auxiliary input. IACR Cryptology ePrint Archive 2013/703 (2013). http://eprint.iacr.org/2013/703

7. Feige, U., Lapidot, D., Shamir, A.: Multiple noninteractive zero knowledge proofs under general assumptions. SIAM J. Comput. **29**(1), 1–28 (1999)

8. Garg, S., Gentry, C., Halevi, S., Raykova, M., Sahai, A., Waters, B.: Candidate indistinguishability obfuscation and functional encryption for all circuits. In: 2013 IEEE 54th Annual Symposium on Foundations of Computer Science (FOCS), pp. 40–49. IEEE (2013)

9. Garg, S., Gentry, C., Halevi, S., Wichs, D.: On the implausibility of differing-inputs obfuscation and extractable witness encryption with auxiliary input. In: Garay, J.A., Gennaro, R. (eds.) CRYPTO 2014, Part I. LNCS, vol. 8616, pp. 518–535. Springer, Heidelberg (2014)

10. Goldwasser, S., Gordon, S.D., Goyal, V., Jain, A., Katz, J., Liu, F.-H., Sahai, A., Shi, E., Zhou, H.-S.: Multi-input functional encryption. In: Nguyen, P.Q., Oswald, E. (eds.) EUROCRYPT 2014. LNCS, vol. 8441, pp. 578–602. Springer, Heidelberg (2014)

11. Goldwasser, S., Goyal, V., Jain, A., Sahai, A.: Multi-input functional encryption. Cryptology ePrint Archive, Report 2013/727 (2013)

12. Gordon, S.D., Katz, J., Liu, F.H., Shi, E., Zhou, H.S.: Multi-input functional encryption. IACR Cryptology ePrint Archive 2013/774 (2013)

13. Goyal, V., Pandey, O., Sahai, A., Waters, B.: Attribute-based encryption for fine-grained access control of encrypted data. In: Proceedings of the 13th ACM conference on Computer and communications security, pp. 89–98. ACM (2006)

14. Halevi, S., Lindell, Y., Pinkas, B.: Secure computation on the web: computing without simultaneous interaction. In: Rogaway, P. (ed.) CRYPTO 2011. LNCS, vol. 6841, pp. 132–150. Springer, Heidelberg (2011)

15. Ishai, Y., Pandey, O., Sahai, A.: Public-coin differing-inputs obfuscation and its applications. In: Dodis, Y., Nielsen, J.B. (eds.) TCC 2015, Part II. LNCS, vol. 9015, pp. 668–697. Springer, Heidelberg (2015)

16. Katz, J., Sahai, A., Waters, B.: Predicate encryption supporting disjunctions, polynomial equations, and inner products. In: Smart, N.P. (ed.) EUROCRYPT 2008. LNCS, vol. 4965, pp. 146–162. Springer, Heidelberg (2008)

17. Lewko, A., Okamoto, T., Sahai, A., Takashima, K., Waters, B.: Fully secure functional encryption: attribute-based encryption and (hierarchical) inner product encryption. In: Gilbert, H. (ed.) EUROCRYPT 2010. LNCS, vol. 6110, pp. 62–91. Springer, Heidelberg (2010)

18. Merkle, R.C.: A digital signature based on a conventional encryption function. In: Pomerance, C. (ed.) CRYPTO 1987. LNCS, vol. 293, pp. 369–378. Springer, Heidelberg (1988)

19. O'Neill, A.: Definitional issues in functional encryption. IACR Cryptology ePrint Archive 2010/556 (2010)

20. Pandey, O., Prabhakaran, M., Sahai, A.: Obfuscation-based non-black-box simulation and four message concurrent zero knowledge for np. IACR Cryptology ePrint Archive 2013/754 (2013)
21. Sahai, A., Waters, B.: Fuzzy identity-based encryption. In: Cramer, R. (ed.) EUROCRYPT 2005. LNCS, vol. 3494, pp. 457–473. Springer, Heidelberg (2005)

Multi-party Key Exchange for Unbounded Parties from Indistinguishability Obfuscation

Dakshita Khurana[1][(✉)], Vanishree Rao[2], and Amit Sahai[1]

[1] Department of Computer Science, Center for Encrypted Functionalities,
UCLA, Los Angeles, CA, USA
{dakshita,sahai}@cs.ucla.edu
[2] PARC, a Xerox Company, Palo Alto, CA, USA
Vanishree.Rao@parc.com

Abstract. Existing protocols for non-interactive multi-party key exchange either (1) support a bounded number of users, (2) require a trusted setup, or (3) rely on knowledge-type assumptions.

We construct the first non-interactive key exchange protocols which support an unbounded number of parties and have a security proof that does not rely on knowledge assumptions. Our non-interactive key-exchange protocol does not require a trusted setup and extends easily to the identity-based setting. Our protocols suffer only a polynomial loss to the underlying hardness assumptions.

1 Introduction

Non-interactive key exchange (NIKE) enables a group of parties to derive a shared secret key without any interaction. In a NIKE protocol, all parties simultaneously broadcast a message to all other parties. After this broadcast phase, each party should be able to locally compute a shared secret key for any group of which he is a member. All members of a group should generate an identical shared key, and the shared key for a group should look random to a non-member.

This notion was introduced by Diffie and Hellman [16], who also gave a protocol for non-interactive key exchange in the two-party setting. More than two decades later, Joux [32] constructed the first non-interactive key exchange protocol for three parties. Given a set of N parties (where N is a polynomial in

D. Khurana and A. Sahai—Research supported in part from a DARPA/ARL SAFE-WARE award, NSF Frontier Award 1413955, NSF grants 1228984, 1136174, 1118096, and 1065276, a Xerox Faculty Research Award, a Google Faculty Research Award, an equipment grant from Intel, and an Okawa Foundation Research Grant. This material is based upon work supported by the Defense Advanced Research Projects Agency through the ARL under Contract W911NF-15-C-0205. The views expressed are those of the authors and do not reflect the official policy or position of the Department of Defense, the National Science Foundation, or the U.S. Government.
V. Rao—Work done while studying at UCLA.

© International Association for Cryptologic Research 2015
T. Iwata and J.H. Cheon (Eds.): ASIACRYPT 2015, Part I, LNCS 9452, pp. 52–75, 2015.
DOI: 10.1007/978-3-662-48797-6_3

the security parameter), Boneh and Silverberg [4] obtained a multiparty NIKE protocol based on multilinear maps. The recent candidates for multilinear maps given by [14,15,22,27] can be used to instantiate the scheme of [4], assuming a trusted setup. After the advent of candidate constructions for indistinguishability obfuscation (iO) starting with the result of Garg et. al. [23], Boneh and Zhandry [6] demonstrated how to obtain static secure NIKE and ID-NIKE based on indistinguishability obfuscation, without relying on a trusted setup.

However, all these constructions require an a-priori bound on the number of parties. The only known protocols which can possibly handle an unbounded number of parties are the ones by Ananth et. al. and Abusalah et. al. [1,2], but their solutions rely on *differing-inputs obfuscation* (diO)[2,3,7]. Unfortunately, diO is a knowledge-type assumption, and recent work [9,25,31] demonstrates that it may suffer from implausibility results. In this paper, we address the following question:

Can we obtain NIKE supporting an a-priori unbounded number of parties and not requiring any setup, based on indistinguishability obfuscation?

We give a positive answer to this question, by demonstrating non-interactive key exchange protocols that achieve static security and support an a-priori unbounded number of parties, based on indistinguishability obfuscation.

1.1 Our Contributions

We consider a setting where an a-priori unbounded number of parties may broadcast or publish messages, such that later any party can derive a shared secret key for a group of which it is a member. In our setting, parameters do not grow with the number of parties. Our results can be summarized as follows.

Theorem 1. *Assuming indistinguishability obfuscation and fully homomorphic encryption, it is possible to obtain static-secure non-interactive multi-party key exchange for an a-priori unbounded number of parties, without any setup.*

Theorem 2. *Assuming indistinguishability obfuscation and fully homomorphic encryption, it is possible to obtain static-secure identity-based non-interactive multi-party key exchange for an a-priori unbounded number of parties.*

Fully homomorphic encryption was first constructed by Gentry [26] and subsequently Brakerski and Vaikuntanathan [10] constructed it under the learning with errors assumption. Alternatively, it can be constructed based on sub-exponentially secure iO for circuits and sub-exponential one-way functions [12].

1.2 Technical Overview

Bottlenecks in known constructions. Our starting point the static-secure NIKE protocol of Boneh and Zhandry [6] based on indistinguishability obfuscation, in the simplest case where parties have access to a trusted setup. The adversary fixes a set of parties to corrupt, independent of the system parameters. Then the setup generates public parameters, and each party broadcasts its public values. We require that the shared group key for all the honest parties should look indistinguishable from random, from the point of view of the adversary.

The basic Boneh-Zhandry construction uses a trusted setup to generate an obfuscated program with a secret PRF key. Parties pick a secret value uniformly at random, and publish the output of a length-doubling PRG applied to this value, as their public value. To derive the shared secret key for a group of users, a member of the group inputs public values of all users in the group (including himself) according to some fixed ordering, as well as his own secret value into this obfuscated program. The program checks if the PRG applied to the secret value corresponds to one of the public values in the group, and if the check passes, outputs a shared secret key by applying a PRF to the public value of all parties in the group.

To prove security, we note that the adversary never corrupts any party in the challenge set, so we never need to reveal the secret value for any party in this set. Thus, (by security of the PRG) we can set the public values of parties in the challenge set, to uniformly random values in the co-domain of the PRG. Then with overwhelming probability, there exists no pre-image for any of the public values in the challenge set, and therefore there exists no secret value for which the PRG check in the program would go through. This allows us to puncture the program at the challenge set, and then replace the shared secret key for this set with random. However, in this construction it is necessary to set an a-priori bound on the number of participants, since the size of inputs to the setup circuit must be bounded.

The construction of Ananth et. al. [2] works without this a-priori bound, by making use of differing-inputs obfuscation and collision resistant hashing. To obtain a shared group key, parties first hash down the group public values and then generate a short proof of membership in the group. The program takes the hashed value and proof, and outputs a shared secret key for the group if and only if the proof verifies. However, because the hash is compressing, there exist collisions and thus there exist false proofs of membership. The only guarantee is that these proofs are 'hard' to find in a computational sense. Unfortunately, proving security in such a situation requires the use of differing-inputs obfuscation, since we must argue that any non-member which distinguishes from random the shared key of a group, could actually have generated false proofs. Such an argument inherently involves an extractable assumption.

First attempt. Now, it seems that we may benefit from using an iO-friendly tool for hashing, such that the proof of membership is unconditionally sound for a select piece of the input to the hash. Moreover, the description of the hash should computationally hide which part of the input it is sound for. Then, like in the previous proof, we can set the public values of parties in the challenge set to uniformly random values in the co-domain of the PRG, and try to puncture the program at each position one by one. This resembles the "selective enforcement" techniques of Koppula et. al. [34] who construct accumulators (from iO) as objects enabling bounded commitments to an unbounded storage, which are unconditionally binding for a select piece of the storage.

At this point, it may seem that we can use the accumulator hash function and we should be done. Parties can hash down an unbounded number of public values to a bounded size input using the accumulator, and generate a short proof of correctness. On input the hashed value and proof, the program can be set to output a shared key if the proof verifies. To prove security, we could begin by generating the public values for parties in the challenge set, uniformly at random in the co-domain of the PRG. Then, it should be possible to make the accumulator binding at the first index, and consequently puncture out the first index from the program. Then we can continue across indices and finally puncture out the hash value at the challenge set for all indices.

Even though this proof seems straightforward, the afore-mentioned approach fails. This is because an accumulator can be made unconditionally sound at a particular index, *only conditioned on the previous indices being equal to an a-priori fixed sequence.* In the setting of obfuscating unbounded-storage Turing Machines, for which such accumulators were first introduced [34], there was indeed a well-defined "correct" path that was generated by the machine itself, and consequently there was a way to enforce correct behaviour on all previous indices. However, in our setting the adversary is allowed to hash completely arbitrary values for all indices. Very roughly, we require a tool that enables selective enforcing even when the adversary is allowed to behave arbitrarily on all other indices.

Our solution. At this point, we require a hash function which can enforce soundness at hidden indices, while allowing arbitrary behaviour on all other indices. Such a hash function was introduced recently in the beautiful work of Hubacek and Wichs [30], in the context of studying the communication complexity of secure function evaluation with long outputs. They call it a *somewhere statistically binding* (SSB) hash and give a construction based on fully homomorphic encryption. An SSB hash can be used like other hash functions, to hash an a-priori unbounded number of chunks of input, each of bounded size, onto a bounded space. Moreover, this hash can operate in various *modes*, where each mode is statistically binding on some fixed index i determined at setup; yet, the description of the hash function computationally hides this index.

Equipped with this tool, it is possible to argue security via a selective enforcing hybrid argument. As usual, we begin by generating the public values of all parties in the challenge set, uniformly at random in the co-domain of the PRG. With overwhelming probability, this ensures that there exist no secret values that could generate the public values in the challenge set. Now, we zoom into each index (of the hash) one by one, and make the hash function statistically binding at that particular index. Specifically, we know that a specific output of the hash h^* can only be achieved via a single fixed public value at the enforcing index (say i). Moreover, with overwhelming probability, this public value lies outside the range of the PRG, and thus there exist no false proofs for hash value h^* at the enforcing index i.

This allows us to alter the obfuscated circuit to always ignore the value h^* at index i. Once we have changed the program, we generate the hash function to be statistically binding at index $(i + 1)$, and repeat the argument. Note that once we are at this next index, there may exist false proofs for previous indices – however, at this point, we have already programmed the obfuscation to eliminate the value h^* for all previous indices.

In the identity-based NIKE setting, we generate secret keys for identities as PRF outputs on the identity (in a manner similar to [6]). In addition to using the enforce-and-move technique detailed above, we need to avoid simultaneously programming in an unbounded number of public values. We handle this using Sahai-Waters [37] punctured programming techniques(using PRGs) to puncture and then *un-puncture* the PRF keys before moving on to the next value.

1.3   Other Related Work

Cash, Kiltz, and Shoup [13] and Freire, Hofheinz, Kiltz, and Paterson [19] formalized various security models in the two-party NIKE setting. Bones and Zhandry [6] first resolved NIKE for bounded $N > 3$ parties without relying on a trusted setup, assuming indistinguishability obfuscation. However, their security proofs worked only for the static and semi-static scenarios. Hofheinz et. al. [29] realized adaptive secure bounded N-party NIKE in the random oracle model without setup, and Rao [36] realized bounded N-party NIKE with adaptive security and without setup based on assumptions over multilinear maps. A recent independent work of Yamakawa et. al. [39] gives multilinear maps where the multilinearity levels need not be bounded during setup, and the size of the representations of elements is independent of the level of multi-linearity. In the same paper, these maps are used to construct multiparty NIKE with unbounded parties, however, requiring a trusted setup. Furthermore, their scheme does not seem to extend directly to the identity-based NIKE setting.

In the identity-based NIKE setting, there is a trusted master party that generates secret values for identities using a master secret key. This (seemingly weaker) setting has been extensively studied both in the standard and the random oracle models, and under various static and adaptive notions of security [18,20,35,38], but again for an a-priori bounded number of parties.

Zhandry [40] uses somewhere statistically binding hash along with obfuscation to obtain adaptively secure broadcast encryption with small parameters.

2   Preliminaries

2.1   Indistinguishability Obfuscation and PRFs

Definition 1 (Indistinguishability Obfuscator (iO)). *A uniform PPT machine iO is called an indistinguishability obfuscator for circuits if the following conditions are satisfied:*

- *For all security parameters $\kappa \in \mathbb{N}$, for all circuits C, for all inputs x, we have that*

$$\Pr[C'(x) = C(x) : C' \leftarrow iO(\kappa, C)] = 1$$

- *For any (not necessarily uniform) PPT adversaries Samp, D, there exists a negligible function α such that the following holds: if $\Pr[|C_0| = |C_1|$ and $\forall x, C_0(x) = C_1(x) : (C_0, C_1, \sigma) \leftarrow Samp(1^\kappa)] > 1 - \alpha(\kappa)$, then we have:*

$$\left| \Pr\left[D(\sigma, iO(\kappa, C_0)) = 1 : (C_0, C_1, \sigma) \leftarrow Samp(1^\kappa) \right] \right.$$

$$\left. - \Pr\left[D(\sigma, iO(\kappa, C_1)) = 1 : (C_0, C_1, \sigma) \leftarrow Samp(1^\kappa) \right] \right| \leq \alpha(\kappa)$$

Such indistinguishability obfuscators for circuits were constructed under novel algebraic hardness assumptions in [24].

Definition 2. *A puncturable family of PRFs F is given by a triple of Turing Machines Key_F, $\mathrm{Puncture}_F$, and Eval_F, and a pair of computable functions $n(\cdot)$ and $m(\cdot)$, satisfying the following conditions:*

- *[Functionality preserved under puncturing] For every PPT adversary A such that $A(1^\kappa)$ outputs a set $S \subseteq \{0,1\}^{n(\kappa)}$, then for all $x \in \{0,1\}^{n(\kappa)}$ where $x \notin S$, we have that:*

$$\Pr\left[\mathrm{Eval}_F(K, x) = \mathrm{Eval}_F(K_S, x) : K \leftarrow \mathrm{Key}_F(1^\kappa), K_S = \mathrm{Puncture}_F(K, S) \right] = 1$$

- *[Pseudorandom at punctured points] For every PPT adversary (A_1, A_2) such that $A_1(1^\kappa)$ outputs a set $S \subseteq \{0,1\}^{n(\kappa)}$ and state σ, consider an experiment where $K \leftarrow \mathrm{Key}_F(1^\kappa)$ and $K_S = \mathrm{Puncture}_F(K, S)$. Then we have*

$$\left| \Pr\left[A_2(\sigma, K_S, S, \mathrm{Eval}_F(K, S)) = 1 \right] - \Pr\left[A_2(\sigma, K_S, S, U_{m(\kappa) \cdot |S|}) = 1 \right] \right| = negl(\kappa)$$

where $\mathrm{Eval}_F(K, S)$ denotes the concatenation of $\mathrm{Eval}_F(K, x_1)), \ldots,$ $\mathrm{Eval}_F(K, x_k))$ where $S = \{x_1, \ldots, x_k\}$ is the enumeration of the elements of S in lexicographic order, $negl(\cdot)$ is a negligible function, and U_ℓ denotes the uniform distribution over ℓ bits.

For ease of notation, we write $\mathsf{PRF}(K, x)$ to represent $\mathrm{Eval}_F(K, x)$. We also represent the punctured key $\mathrm{Puncture}_F(K, S)$ by $K\{S\}$.

The GGM tree-based construction of PRFs [28] from one-way functions are easily seen to yield puncturable PRFs, as recently observed by [5,8,33]. Thus,

Imported Theorem 1. *[5, 8, 28, 33] If one-way functions exist, then for all efficiently computable functions $n(\kappa)$ and $m(\kappa)$, there exists a puncturable PRF family that maps $n(\kappa)$ bits to $m(\kappa)$ bits.*

2.2 Somewhere Statistically Binding Hash

Definition 3. *We use the primitive somewhere statistically binding hash (SSB hash), constructed by Hubacek and Wichs [30]. Intuitively, these are a special type of collision resistant hash function that is binding on a hidden index, and can be used with indistinguishability obfuscation. An SSB hash is a tripe of algorithms* (Gen, Open, Ver) *where:*

- Gen(s, i) *takes as input two integers* s *and* i, *where* $s \leq 2^{\kappa}$ *denotes the number of blocks that will be hashed, and* $i \in [s]$ *indexes a particular block. The output is a function* $H : \Sigma^s \to \mathcal{Z}$. *The size of the description of* H *is independent of* s *and* i *(though it will depend on the security parameter).*
- Open$(H, x = \{x_\ell\}_{\ell \in [s]}, j)$ *for* $x_\ell \in \Sigma$ *and* $j \in [s]$ *produces an "opening"* π *that proves that the* j^{th} *element in* x *is* x_j.
- Ver$(H, h \in \mathcal{Z}, j \in [s], u \in \Sigma, \pi)$ *either accepts or rejects. The idea is that* Ver *should only accept when* $h = H(x)$ *where* $x_j = u$.
- *Correctness:* Ver$(H, H(x), j, x_j, \text{Open}(H, x, j))$ *accepts.*
- *Index hiding:* Gen(s, i_0) *is computationally indistinguishable from* Gen(s, i_1) *for any* i_0, i_1.
- *Somewhere Statistically Binding: If* $H \leftarrow \text{Gen}(s, i)$ *then if* Ver(H, h, i, u, π) *and* Ver(H, h, i, u', π') *accept, it must be that* $u = u'$.

Remark. Note that using SSB hash functions, one can efficiently hash down an a-priori unbounded (polynomial) number of values in the security parameter κ.

Imported Theorem 2. *Assuming the existence of FHE, there exists a somewhere statistically binding hash function family mapping unbounded polynomial size inputs to outputs of size* κ *bits (where* κ *denotes the security parameter), according to Definition 3.*

3 Definitions

Definition 4 (Multiparty Non-interactive Key Exchange). *An adaptive multiparty NIKE protocol has the following three algorithms:*

- Setup(1^{κ}) : *The setup algorithm takes a security parameter* κ, *and outputs public parameters* params.
- Publish$(1^{\kappa}, i)$: *Each party executes the publishing algorithm, which takes as input the index of the party, and outputs two values: a secret key* sv_i *and a public value* pv_i. *Party* P_i *keeps* sv_i *as his secret value, and publishes* pv_i *to the other parties.*
- KeyGen$(\text{params}, \mathcal{S}, (pv_i)_{i \in \mathcal{S}}, j, sv_j)$: *To derive the common key* $k_{\mathcal{S}}$ *for a subset* $\mathcal{S}$, *each party in* $\mathcal{S}$ *runs* KeyGen *with* params, *its secret value* sv_j *and the public values* $(pv_i)i \in \mathcal{S}$ *of the parties in* $\mathcal{S}$.

Then, these algorithms should satisfy the following properties:

- *Correctness: For all $\mathcal{S}, i, i' \in \mathcal{S}$,*

$$\mathsf{KeyGen}(\mathsf{params}, \mathcal{S}, (pv_j)_{j \in \mathcal{S}}, i, sv_i) = \mathsf{KeyGen}(\mathsf{params}, \mathcal{S}, (pv_j)_{j \in \mathcal{S}}, i', sv_{i'}).$$

- *Security: The adversary is allowed to (statically) corrupt any subset of users of his choice. More formally, for $b \in \{0, 1\}$, we denote by $\mathsf{expmt}(b)$ the following experiment, parameterized only by the security parameter κ and an adversary $\mathcal{A}$, and $\mathsf{params} \xleftarrow{\$} \mathsf{Setup}(1^\kappa)$ and $b' \leftarrow \mathcal{A}^{\mathsf{Reg}(\cdot), \mathsf{RegCor}(\cdot, \cdot), \mathsf{Ext}(\cdot), \mathsf{Rev}(\cdots), \mathsf{Test}(\cdots)}$ $(1^\kappa, \mathsf{params})$ where:*
  - *$\mathsf{Reg}(i \in [2^\kappa])$ registers an honest party P_i. It takes an index i, and runs $(sv_i, pv_i) \leftarrow \mathsf{Publish}(\mathsf{params}, i)$. The challenger then records the tuple $(i, sk_i, pv_i, \mathsf{honest})$ and sends pv_i to A.*
  - *$\mathsf{RegCor}(i \in [2^\kappa], pk_i)$ registers a corrupt party P_i^*. It takes an index i and a public value pv_i. The challenger records $(i, \bot, pv_i, \mathsf{corrupt})$. The adversary may make multiple queries for a particular identity, in which case the challenger only uses the most recent record.*
  - *$\mathsf{Ext}(i)$ extracts the secret key for an honest registered party. The challenger looks up the tuple $(i, sv_i, pv_i, \mathsf{honest})$ and returns sv_i to $\mathcal{A}$.*
  - *$\mathsf{Rev}(\mathcal{S}, i)$ reveals the shared secret for a group $\mathcal{S}$ of parties, as calculated by the ith party, where $i \in \mathcal{S}$. We require that party P_i was registered as honest. The challenger uses the secret key for party P_i to derive the shared secret key $k_\mathcal{S}$, which it returns to the adversary.*
  - *$\mathsf{Test}(\mathcal{S})$: Takes a set $\mathcal{S}$ of users, all of which were registered as honest. Next, if $b = 0$ the challenger runs KeyGen to determine the shared secret key (arbitrarily choosing which user to calculate the key), which it returns to the adversary. Else if $b = 1$, the challenger generates a random key k to return to the adversary.*

A static adversary $\mathcal{A}$ must have the following restrictions:

- *$\mathcal{A}$ commits to a set S^* before seeing the public parameters, and,*
- *$\mathcal{A}$ makes a single query to Test, and this query is on the set S^*.*

We require that all register queries and register-corrupt queries are for distinct i, and that $pv_i \neq pv_j$ for any $i \neq j$. For $b = 0$, let W_b be the event that $b' = 1$ in $\mathsf{expmt}(b)$ and we define $\mathsf{Adv}_{\mathsf{NIKE}}(\kappa) = |\Pr[W_0] - \Pr[W_1]|$.

Then, a multi-party key exchange protocol $(\mathsf{Setup}, \mathsf{Publish}, \mathsf{KeyGen})$ is statically secure if $\mathsf{Adv}_{\mathsf{NIKE}}(\kappa)$ is $\mathsf{negl}(\kappa)$ for any static PPT adversary $\mathcal{A}$.

4   Static Secure NIKE for Unbounded Parties

4.1   Construction

Let PRF denote a puncturable PRF mapping κ bits to κ bits, and PRG denote a length-doubling pseudorandom generator with inputs of size κ bits. Let $(\mathsf{Gen}, \mathsf{Open}, \mathsf{Ver})$ denote the algorithms of a somewhere statistically binding hash

---

NIKE- Public Parameters

Constants: H, PRF key K.
Input: h, i, pv, sv, π, t.

1. If $i > t$, output $\perp$ and abort.
2. Set $K_t = \mathsf{PRF}(K, t)$.
3. If $\mathsf{Ver}(H, h, i, pv, \pi) = 1$ and $\mathsf{PRG}(sv) = pv$, output $\mathsf{PRF}(K_t, h)$.
4. Else output $\perp$.

Fig. 1. Static Secure NIKE Parameters P_{KE}

scheme as per Definition 3. Then the static secure NIKE algorithms are constructed as follows.

Setup(1^κ): Pick puncturable PRF key $K \xleftarrow{\$} \{0,1\}^\kappa$. Run $\mathsf{Gen}(1^\kappa, 2^\kappa, 0)$ to obtain H. Obfuscate using iO the program P_{KE} in Fig. 1, padded to the appropriate length. Output $P_{\mathsf{iO}} = \mathsf{iO}(P_{KE})$ as public parameters.

Publish: Party i chooses a random seed $s_i \in \{0,1\}^\lambda$ as a secret value, and publishes $x_i = \mathsf{PRG}(s_i)$.

KeyGen($P_{KE}, i, s_i, S, \{pv_j\}_{j \in S}$): Compute $h = H(\{pv_j\}_{j \in S})$. Compute $\pi = \mathsf{Open}(H, \{pv_j\}_{j \in S}, i)$. Run program P_{KE} on input $(h, i, pv, sv, \pi, |S|)$ to obtain shared key K_S.

4.2 Security Game and Hybrids

Hybrid$_0$: This is the real world attack game, where $\mathcal{A}$ commits to a set $\hat{S}$. In response $\mathcal{A}$ gets the public parameters from the setup, and then makes the following queries.

- Register honest user queries: $\mathcal{A}$ submits an index i. The challenger chooses a random s_i, and sends $x_i = \mathsf{PRG}(s_i)$ to $\mathcal{A}$.
- Register corrupt user queries: $\mathcal{A}$ submits an index i such that $i \notin \hat{S}$, along with a string x_i as the public value for party i. We require that i was not, and will not be registered as honest.
- Extract queries: $\mathcal{A}$ submits an $i \in [N] \setminus \hat{S}$ that was previously registered as honest. The challenger responds with s_i.
- Reveal shared key queries: The adversary submits a subset $S \neq \hat{S}$ of users, of which at least one is honest. The challenger uses PRF to compute and send the group key.
- Finally, for set $\hat{S}$, the adversary receives either the correct group key (if $b = 0$) or a random key (if $b = 1$). The adversary outputs a bit b' and wins if $\Pr[b' = b] > \frac{1}{2} + 1/\mathsf{poly}(\kappa)$ for some polynomial $\mathsf{poly}(\cdot)$.

We now demonstrate a sequence of hybrids, via which we argue that the advantage of the adversary in guessing the bit b is $\mathsf{negl}(\kappa)$, where $\mathsf{negl}(\cdot)$ is a function that is asymptotically smaller than $1/\mathsf{poly}(\kappa)$ for all polynomials $\mathsf{poly}(\cdot)$. We give

---

NIKE- Public Parameters

Constants: H, PRF key K, i^*, h^*.
Input: h, i, pv, sv, π, t.

1. If $i > t$, output $\perp$ and abort.
2. Compute $K_t = \mathsf{PRF}(K, t)$.
3. If $i \leq i^*$, $h = h^*$, output $\perp$ and abort.
4. If $\mathsf{Ver}(H, h, i, pv, \pi) = 1$ and $\mathsf{PRG}(sv) = pv$, output $\mathsf{PRF}(K_t, h)$.
5. Else output $\perp$.

Fig. 2. Static Secure NIKE Parameters P_{KE,i^*}

short overviews of indistinguishability between the hybrids, with full proofs in Appendix A. We use <u>underline</u> changes between subsequent hybrids.

Hybrid_1: For each $i \in \hat{S}$, choose $x_i \xleftarrow{\$} \{0,1\}^{2\lambda}$. When answering register honest user queries for $i \in \hat{S}$, use these x_i values instead of generating them from PRG. Follow the rest of the game same as Hybrid_0. This hybrid is indistinguishable from Hybrid_0, by security of the PRG.

Let $t^* = |S|$. We start with $\mathsf{Hybrid}_{2,1,a}$, and go across hybrids in the following sequence: $\mathsf{Hybrid}_{2,1,a}, \mathsf{Hybrid}_{2,1,b}, \mathsf{Hybrid}_{2,2,a}, \mathsf{Hybrid}_{2,2,b} \ldots \mathsf{Hybrid}_{2,t^*,a},$ $\mathsf{Hybrid}_{2,t^*,b}$. The hybrids $\mathsf{Hybrid}_{2,i^*,a}$ and $\mathsf{Hybrid}_{2,i^*,b}$ are described below, for $i^* \in [t^*]$.

$\mathsf{Hybrid}_{2,1,a}$: <u>Generate hash function $H \xleftarrow{\$} \mathsf{Gen}(1^\kappa, 2^\kappa, 1)$.</u> Follow the rest of the game same as Hybrid_1. This is indistinguishable from Hybrid_1 because of indistinguishability of statistical binding index.

$\mathsf{Hybrid}_{2,i^*,a}$: <u>Generate hash function $H \xleftarrow{\$} \mathsf{Gen}(1^\kappa, 2^\kappa, i^*)$.</u> Follow the rest of the game same as $\mathsf{Hybrid}_{2,i^*-1,b}$. This is indistinguishable from $\mathsf{Hybrid}_{2,i^*-1,b}$ because of indistinguishability of statistical binding index.

$\mathsf{Hybrid}_{2,i^*,b}$: Generate hash function $H \xleftarrow{\$} \mathsf{Gen}(1^\kappa, 2^\kappa, i^*)$. Let $h^* = H(\{pv_j\}_{j \in S})$. <u>Set P_{KE} to C_{i^*}, an obfuscation of the circuit in Fig. 2.</u> This is indistinguishable from $\mathsf{Hybrid}_{2,i^*,a}$ because of iO between functionally equivalent circuits P_{KE,i^*} and P_{KE,i^*-1} when the hash is statistically binding at index i^*.

$\mathsf{Hybrid}_{2,t^*,b}$: <u>Generate hash function $H \xleftarrow{\$} \mathsf{Gen}(1^\kappa, 2^\kappa, t^*)$.</u> Let $h^* = H(\{pv_j\}_{j \in S})$. Set P_{KE} to C_i which is an obfuscation of the circuit in Fig. 3.

Hybrid_3 : Generate hash function $H \xleftarrow{\$} \mathsf{Gen}(1^\kappa, 2^\kappa, t^*)$. Let $h^* = H(\{pv_j\}_{j \in S})$. <u>Set $k^* = \mathsf{PRF}(K, t^*)$. Set P_{KE} to C_i which is an obfuscation of the circuit in Fig. 4, using punctured key $K\{t^*\}$.</u> Follow the rest of the game the same as in $\mathsf{Hybrid}_{2,t^*,b}$. This program is functionally equivalent to the program in Hybrid_3, and thus the hybrid is indistinguishable by iO.

Hybrid_4 : Generate hash function $H \xleftarrow{\$} \mathsf{Gen}(1^\kappa, 2^\kappa, t^*)$. Let $h^* = H(\{pv_j\}_{j \in S})$. <u>Set $k^* \xleftarrow{\$} \{0,1\}^\kappa$.</u> Follow the rest of the game honestly according to Hybrid_3, with

---

NIKE- Public Parameters

Constants: H, PRF key K, t^*, h^*.
Input: h, i, pv, sv, π, t.

1. If $i > t$, output $\perp$ and abort.
2. Compute $K_t = \mathsf{PRF}(K, t)$.
3. If $i \leq t^*$, $h = h^*$, output $\perp$ and abort.
4. If $\mathsf{Ver}(H, h, i, pv, \pi) = 1$ and $\mathsf{PRG}(sv) = pv$, output $\mathsf{PRF}(K_t, h)$.
5. Else output $\perp$.

Fig. 3. Static Secure NIKE Parameters P_{KE,t^*}

NIKE- Public Parameters

Constants: H, PRF key $K\{t^*\}, t^*, h^*, k^*$.
Input: h, i, pv, sv, π, t.

1. If $i > t$, output $\perp$ and abort.
2. If $t = t^*$, set $K_t = k^*$. Else compute $K_t = \mathsf{PRF}(K\{t^*\}, t)$.
3. If $i \leq t^*$, $h = h^*$, output $\perp$ and abort.
4. If $t = t^*$, $h = h^*$, output $\perp$ and abort.
5. If $\mathsf{Ver}(H, h, i, pv, \pi) = 1$ and $\mathsf{PRG}(sv) = pv$, output $\mathsf{PRF}(K_t, h)$.
6. Else output $\perp$.

Fig. 4. Static Secure NIKE Parameters $P_{KE'}$

this value of k^*. This hybrid is indistinguishable from Hybrid_3 because of security of the punctured PRF.

Hybrid_5 : Generate hash function $H \xleftarrow{\$} \mathsf{Gen}(1^\kappa, 2^\kappa, t^*)$. Set $h^* = H(\{pv_j\}_{j \in S})$. Set $k^* \xleftarrow{\$} \{0, 1\}^\kappa$. Set P_{KE} to C_i which is an obfuscation of the circuit in Fig. 5, using punctured key $k^*\{h^*\}$. This program is functionally equivalent to the program in Hybrid_4, thus the hybrids are indistinguishable by iO.

Finally, by security of punctured PRF, $\mathcal{A}$'s advantage in Hybrid_5 is $\mathsf{negl}(\kappa)$.

4.3   Removing the Setup

We note that in our protocol, the Publish algorithm is independent of the Setup algorithm. In such a scenario, [6] gave the following theorem, which can be used to remove setup from our scheme in the case of static corruptions.

Imported Theorem 3. *[6] Let* (Setup, Publish, KeyGen) *be a statically secure NIKE protocol where* Publish *does not depend on* params *output by* Setup, *but instead just takes as input* (λ, i). *Then there is a statically secure NIKE protocol* (Setup', Publish', KeyGen') *with no setup.*

NIKE- Public Parameters

Constants: H, PRF key $K\{t^*\}, t^*, h^*, k^*\{h^*\}$.
Input: h, i, pv, sv, π, t.

1. If $i > t$, output $\perp$ and abort.
2. If $t = t^*$, set $K_t = k^*\{h^*\}$. Else compute $K_t = \mathsf{PRF}(K, t)$.
3. If $i \leq t^*$, $h = h^*$, output $\perp$ and abort.
4. If $t = t^*$, $h = h^*$, output $\perp$ and abort.
5. If $\mathsf{Ver}(H, h, i, pv, \pi) = 1$ and $\mathsf{PRG}(sv) = pv$, output $\mathsf{PRF}(K_t, h)$.
6. Else output $\perp$.

Fig. 5. Static Secure NIKE Parameters P_{KE}

5 ID-NIKE for Unbounded Parties

In the identity-based NIKE setting, there is a trusted setup that outputs public parameters, and generates secret keys for parties based on their identity. These parties then run another key-generating algorithm on their secret keys and setup, to compute shared group keys.

Our NIKE scheme can be extended to obtain identity-based NIKE for unbounded parties with a polynomial reduction to the security of indistinguishability obfuscation and fully homomorphic encryption. In this section, we describe our protocol for ID-NIKE for a-priori unbounded parties, and give an overview of the hybrid arguments.

5.1 Construction

Let κ denote the security parameter, PRF denote a puncturable pseudo-random function family mapping κ bits to κ bits and PRG denote a length-doubling pseudo-random generator with inputs of size κ bits. ID-NIKE consists of the following algorithms.

- $\mathsf{Setup}(1^\kappa)$: Sample random PRF keys K_1 and K_2. Compute the program P_{IBKE} in Fig. 6 padded to the appropriate length, and compute $P_{\mathsf{iO}} = \mathsf{iO}(P_{\mathsf{IBKE}})$. Sample SSB hash $H \leftarrow \mathsf{Gen}(1^\kappa)$. Publish the public parameters params $= P_{\mathsf{iO}}, H$.
- $\mathsf{Extract}(K_2, \mathsf{id})$: Output $sk_{\mathsf{id}} = \mathsf{PRF}(K_2, \mathsf{id})$.
- $\mathsf{KeyGen}(\mathsf{params}, S, \mathsf{id}, s_{\mathsf{id}})$: To compute shared key k_S for lexicographically ordered set S, compute $h = H(S)$ and $\pi = \mathsf{Open}(h, S, i)$; where i denotes the index of id in sorted S. Then obtain output $k_S = P_{\mathsf{iO}}(h, i, \mathsf{id}, \pi, s_{\mathsf{id}}, |S|)$.

5.2 Security Game and Hybrids

Hybrid_0: This is the real world attack game, where $\mathcal{A}$ commits to a set $\hat{S}$. In response $\mathcal{A}$ gets the public parameters as hash function H and the obfuscation of P_{IBKE} and then makes the following queries.

```
┌─────────────────────────────────────────────────────────────────┐
│                    NIKE- Public Parameters                        │
│                                                                   │
│  Constants: H, PRF keys K₁, K₂.                                   │
│  Input: h, i, id, π, s_id, t.                                     │
│                                                                   │
│   1. If i > t, output ⊥ and abort.                                │
│   2. Compute Kₜ = PRF(K₁, t).                                     │
│   3. If Ver(H, h, i, id, π) = 0 or PRG(s_id) ≠ PRG(PRF(K₂, id)),  │
│      output ⊥ and abort.                                          │
│   4. Else output PRF(Kₜ, h).                                      │
└─────────────────────────────────────────────────────────────────┘
```

Fig. 6. Static Secure ID-NIKE Parameters P_{IBKE}

- Obtain secret keys: $\mathcal{A}$ submits an identity id such that id $\leq \mathsf{poly}(\kappa)$ and id $\notin \hat{S}$. The challenger outputs $\mathsf{Extract}(K_2, \mathsf{id})$ to $\mathcal{A}$.
- Reveal shared keys: The adversary submits a subset $\mathcal{S} \neq \hat{S}$ of users, of which at least one is honest. The challenger uses the public parameters to compute and send the group key.
- Finally, for set $\hat{S}$, the adversary receives either the correct group key (if $b = 0$) or a random key (if $b = 1$). The adversary outputs a bit b' and wins if $\Pr[b' = b] > \frac{1}{2} + 1/\mathsf{poly}(\kappa)$ for some polynomial $\mathsf{poly}(\cdot)$.

We now demonstrate a sequence of hybrids, via which we argue that the advantage of the adversary in guessing the bit b is $\mathsf{negl}(\kappa)$, where $\mathsf{negl}(\cdot)$ is a function that is asymptotically smaller than $1/\mathsf{poly}(\kappa)$ for all polynomials $\mathsf{poly}(\cdot)$. We give short arguments for indistinguishability between the hybrids, with complete proofs in the full version.

Let $\hat{S} = \{\mathsf{id}_1^*, \mathsf{id}_2^*, \mathsf{id}_3^*, \dots \mathsf{id}_{|\hat{S}|}^*\}$. Then, for $p \in [1, |\hat{S}|]$, we have the sequence of hybrids:

$\mathsf{Hybrid}_{p-1,j}, \mathsf{Hybrid}_{p,a}, \mathsf{Hybrid}_{p,b}, \ \dots \mathsf{Hybrid}_{p,j}, \mathsf{Hybrid}_{p+1,a}, \mathsf{Hybrid}_{p+1,b}, \dots$. Here, $\mathsf{Hybrid}_0 \equiv \mathsf{Hybrid}_{0,j}$. We now write out the experiments and demonstrate the sequence of changes between $\mathsf{Hybrid}_{p-1,j}$ and $\mathsf{Hybrid}_{p,j}$ for any $p \in [1, |\hat{S}|]$.

$\mathsf{Hybrid}_{p-1,j}$: This is the same as Hybrid_0 except that the hash is generated as $H \leftarrow \mathsf{Gen}(1^\kappa, 2^\kappa, p-1)$. The challenger computes $h^* = H(\hat{S})$ and outputs the program P_{IBKE} in Fig. 7 padded to the appropriate length. He publishes $P_{\mathsf{iO}} = \mathsf{iO}(P_{\mathsf{IBKE}})$.

$\mathsf{Hybrid}_{p,a}$: This is the same as $\mathsf{Hybrid}_{p-1,j}$ except that the challenger computes $r_p^* = \mathsf{PRF}(K_2, \mathsf{id}_p^*)$, $z_p^* = \mathsf{PRG}(r_p^*)$. He computes punctured PRF key $K_2\{\mathsf{id}_p^*\}$ and using the program P_{IBKE} in Fig. 8 padded to the appropriate length, computes $P_{\mathsf{iO}} = \mathsf{iO}(P_{\mathsf{IBKE}})$. This is indistinguishable from $\mathsf{Hybrid}_{p-1,j}$ because of iO between functionally equivalent circuits.

$\mathsf{Hybrid}_{p,b}$: This is the same as $\mathsf{Hybrid}_{p,a}$ except that the challenger picks $r_p^* \xleftarrow{\$} \{0,1\}^\kappa$ and sets $z_p^* = \mathsf{PRG}(r_p^*)$. This is indistinguishable from $\mathsf{Hybrid}_{p,a}$ by security of the puncturable PRF.

NIKE- Public Parameters

Constants: H, PRF keys K_1, K_2, h^*.
Input: $h, i, \mathsf{id}, \pi, s_{\mathsf{id}}, t$.

1. If $i > t$, output $\perp$ and abort.
2. Compute $K_t = \mathsf{PRF}(K_1, t)$.
3. If $\mathsf{Ver}(H, h, i, \mathsf{id}, \pi) = 0$, output $\perp$ and abort.
4. If $i \leq (p-1), h = h^*$, then output $\perp$ and abort.
5. Else if $\mathsf{PRG}(s_{\mathsf{id}}) \neq \mathsf{PRG}(\mathsf{PRF}(K_2, \mathsf{id}))$, output $\perp$ and abort.
6. Else output $\mathsf{PRF}(K_t, h)$.

Fig. 7. Static Secure ID-NIKE Parameters P_{IBKE}

NIKE- Public Parameters

Constants: H, PRF keys $K_1, K_2\{\mathsf{id}_p^*\}, \mathsf{id}_p^*, z_p^*, h^*$.
Input: $h, i, \mathsf{id}, \pi, s_{\mathsf{id}}, t$.

1. If $i > t$, output $\perp$ and abort.
2. Compute $K_t = \mathsf{PRF}(K_1, t)$.
3. If $\mathsf{Ver}(H, h, i, \mathsf{id}, \pi) = 0$, output $\perp$ and abort.
4. If $i \leq (p-1), h = h^*$, then output $\perp$ and abort.
5. If $\mathsf{id} = \mathsf{id}_p^*$, then if $\mathsf{PRG}(s_{\mathsf{id}}) \neq z_p^*$, output $\perp$ and abort.
6. If $\mathsf{id} \neq \mathsf{id}_p^*$, then if $\mathsf{PRG}(s_{\mathsf{id}}) \neq \mathsf{PRG}(\mathsf{PRF}(K_2\{\mathsf{id}_p^*\}, \mathsf{id}))$, output $\perp$ and abort.
7. Else output $\mathsf{PRF}(K_t, h)$.

Fig. 8. Static Secure ID-NIKE Parameters P_{IBKE}

$\mathsf{Hybrid}_{p,c}$: This is the same as $\mathsf{Hybrid}_{p,b}$ except that the challenger sets $z_p^* \xleftarrow{\$} \{0,1\}^{2\kappa}$. This is indistinguishable from $\mathsf{Hybrid}_{p,b}$ by security of the PRG.

$\mathsf{Hybrid}_{p,d}$: This is the same as $\mathsf{Hybrid}_{p,c}$ except that the challenger computes the program P_{IBKE} in Fig. 9 padded to the appropriate length, and publishes $P_{\mathsf{iO}} = \mathsf{iO}(P_{\mathsf{IBKE}})$.

With probability $1/2^\kappa$ over random choice of z_1^*, the value z_1^* does not lie in the co-domain of the length-doubling PRG. Then this hybrid is indistinguishable from $\mathsf{Hybrid}_{1,c}$ because of iO between functionally equivalent circuits.

$\mathsf{Hybrid}_{p,e}$: In this hybrid, the challenger generates $H \xleftarrow{\$} \mathsf{Gen}(1^\kappa, 2^\kappa, p)$ (such that it is statistically binding at index p). The rest of the game is same as $\mathsf{Hybrid}_{p,d}$. This hybrid is indistinguishable from $\mathsf{Hybrid}_{p,d}$ because of indistinguishability of statistical binding index.

$\mathsf{Hybrid}_{p,f}$: This is the same as $\mathsf{Hybrid}_{p,e}$ except that the challenger outputs the program P_{IBKE} in Fig. 10 padded to the appropriate length. He publishes $P_{\mathsf{iO}} = \mathsf{iO}(P_{\mathsf{IBKE}})$.

NIKE- Public Parameters

Constants: H, PRF keys $K_1, K_2\{\mathsf{id}_p^*\}, \mathsf{id}_p^*, h^*$.
Input: $h, i, \mathsf{id}, \pi, s_{\mathsf{id}}, t$.

1. If $i > t$, output $\perp$ and abort.
2. Compute $K_t = \mathsf{PRF}(K_1, t)$.
3. If $\mathsf{Ver}(H, h, i, \mathsf{id}, \pi) = 0$, output $\perp$ and abort.
4. If $i \le (p-1), h = h^*$, then output $\perp$ and abort.
5. If $\mathsf{id} = \mathsf{id}_p^*$, then output $\perp$ and abort.
6. If $\mathsf{id} \ne \mathsf{id}_p^*$, then if $\mathsf{PRG}(s_{\mathsf{id}}) \ne \mathsf{PRG}(\mathsf{PRF}(K_2\{\mathsf{id}_p^*\}, \mathsf{id}))$, output $\perp$ and abort.
7. Else output $\mathsf{PRF}(K_t, h)$.

Fig. 9. Static Secure ID-NIKE Parameters P_{IBKE}

NIKE- Public Parameters

Constants: H, PRF keys $K_1, K_2\{\mathsf{id}_p^*\}, h^*$.
Input: $h, i, \mathsf{id}, \pi, s_{\mathsf{id}}, t$.

1. If $i > t$, output $\perp$ and abort.
2. Compute $K_t = \mathsf{PRF}(K_1, t)$.
3. If $\mathsf{Ver}(H, h, i, \mathsf{id}, \pi) = 0$, output $\perp$ and abort.
4. If $i \le p, h = h^*$, then output $\perp$ and abort.
5. Else if $\mathsf{PRG}(s_{\mathsf{id}}) \ne \mathsf{PRG}(\mathsf{PRF}(K_2\{\mathsf{id}_p^*\}, \mathsf{id}))$, output $\perp$ and abort.
6. Else output $\mathsf{PRF}(K_t, h)$.

Fig. 10. Static Secure ID-NIKE Parameters P_{IBKE}

This is indistinguishable from $\mathsf{Hybrid}_{p,e}$ because of iO between functionally equivalent circuits. The circuits are functionally equivalent because the hash is statistically binding at index p.

$\mathsf{Hybrid}_{p,g}$: This is the same as $\mathsf{Hybrid}_{p,f}$, except that the challenger picks $z_p^* \xleftarrow{\$} \{0,1\}^{2\kappa}$, and then outputs the program P_{IBKE} in Fig. 11 padded to the appropriate length. He publishes $P_{\mathsf{iO}} = \mathsf{iO}(P_{\mathsf{IBKE}})$.

With probability $1/2^\kappa$ over the randomness of choice of z_p^*, the value z_p^* lies outside the co-domain of the PRG. Thus, with over whelming probability, the extra statement is never activated and the circuit is functionally equivalent to the one in $\mathsf{Hybrid}_{p,f}$. Then this is indistinguishable from $\mathsf{Hybrid}_{p,f}$ because of iO between functionally equivalent circuits.

$\mathsf{Hybrid}_{p,h}$: This is the same as $\mathsf{Hybrid}_{1,g}$ except that the challenger picks $r_p^* \xleftarrow{\$} \{0,1\}^\kappa$ and sets $z_p^* = \mathsf{PRG}(r_p^*)$. It follows the rest of the game same as $\mathsf{Hybrid}_{p,g}$ with this value of z_p^*. This hybrid is indistinguishable from $\mathsf{Hybrid}_{p,g}$ because of security of length-doubling PRGs.

NIKE- Public Parameters

Constants: H, PRF keys $K_1, K_2\{\mathsf{id}_p^*\}, h^*, \underline{z_p^*}$.
Input: $h, i, \mathsf{id}, \pi, s_{\mathsf{id}}, t$.

1. If $i > t$, output $\perp$ and abort.
2. Compute $K_t = \mathsf{PRF}(K_1, t)$.
3. If $\mathsf{Ver}(H, h, i, \mathsf{id}, \pi) = 0$, output $\perp$ and abort.
4. If $i \leq p, h = h^*$, then output $\perp$ and abort.
5. If $\mathsf{id} = \mathsf{id}_p^*$ then if $\mathsf{PRG}(s_{\mathsf{id}}) \neq z_p^*$ output $\perp$ and abort.
6. Else if $\mathsf{PRG}(s_{\mathsf{id}}) \neq \mathsf{PRG}(\mathsf{PRF}(K_2\{\mathsf{id}_p^*\}, \mathsf{id}))$, output $\perp$ and abort.
7. Else output $\mathsf{PRF}(K_t, h)$.

Fig. 11. Static Secure ID-NIKE Parameters P_{IBKE}

NIKE- Public Parameters

Constants: H, PRF keys K_1, K_2, h^*.
Input: $h, i, \mathsf{id}, \pi, s_{\mathsf{id}}, t$.

1. If $i > t$, output $\perp$ and abort.
2. Compute $K_t = \mathsf{PRF}(K_1, t)$.
3. If $\mathsf{Ver}(H, h, i, \mathsf{id}, \pi) = 0$, output $\perp$ and abort.
4. If $i \leq p, h = h^*$, then output $\perp$ and abort.
5. Else if $\mathsf{PRG}(s) \neq \mathsf{PRG}(\mathsf{PRF}(K_2, \mathsf{id}))$, output $\perp$ and abort.
6. Else output $\mathsf{PRF}(K_t, h)$.

Fig. 12. Static Secure ID-NIKE Parameters P_{IBKE}

$\mathsf{Hybrid}_{p,i}$: This is the same as $\mathsf{Hybrid}_{p,h}$ except that the challenger sets $r_p^* = \mathsf{PRF}(K_2, \mathsf{id}_p^*)$ and $z_p^* = \mathsf{PRG}(r_p^*)$. It follows the rest of the game same as $\mathsf{Hybrid}_{p,h}$ with this value of z_p^*. This hybrid is indistinguishable from $\mathsf{Hybrid}_{p,h}$ because of security of the puncturable PRF.

$\mathsf{Hybrid}_{p,j}$: This is the same as $\mathsf{Hybrid}_{p,i}$ except that the challenger outputs the program P_{IBKE} in Fig. 12 padded to the appropriate length. He publishes $P_{\mathsf{iO}} = \mathsf{iO}(P_{\mathsf{IBKE}})$.

This is indistinguishable from $\mathsf{Hybrid}_{p,i}$ by iO between functionally equivalent circuits. Note that at this stage, we have un-punctured the PRF at value id_p^*. This is crucial for our hybrid arguments to go through, because we will eventually have to program in an a-priori un-bounded number of identities.

$\mathsf{Hybrid}_{|\hat{S}|,j}$: This is the final hybrid in the sequence, where the hash is generated as $H \leftarrow \mathsf{Gen}(1^\kappa, 2^\kappa, |\hat{S}|)$. The challenger computes $h^* = H(\hat{S})$ and outputs the program P_{IBKE} in Fig. 7 padded to the appropriate length. He publishes $P_{\mathsf{iO}} = \mathsf{iO}(P_{\mathsf{IBKE}})$.

Finally, we have the following two hybrids

NIKE- Public Parameters

Constants: H, PRF keys K_1, K_2, h^*.
Input: $h, i, \mathsf{id}, \pi, s_{\mathsf{id}}, t$.

1. If $i > t$, output $\perp$ and abort.
2. Compute $K_t = \mathsf{PRF}(K_1, t)$.
3. If $\mathsf{Ver}(H, h, i, \mathsf{id}, \pi) = 0$, output $\perp$ and abort.
4. If $i \leq \hat{S}, h = h^*$, then output $\perp$ and abort.
5. Else if $\mathsf{PRG}(s) \neq \mathsf{PRG}(\mathsf{PRF}(K_2, \mathsf{id}))$, output $\perp$ and abort.
6. Else output $\mathsf{PRF}(K_t, h)$.

Fig. 13. Static Secure ID-NIKE Parameters P_{IBKE}

NIKE- Public Parameters

Constants: H, PRF keys $\underline{K_1\{t^*\}}, K_2, h^*, \underline{k^*}$.
Input: $h, i, \mathsf{id}, \pi, s_{\mathsf{id}}, t$.

1. If $i > t$, output $\perp$ and abort.
2. If $t = t^*$, set $K_t = k^*$, else compute $K_t = \mathsf{PRF}(K_1, t)$.
3. If $\mathsf{Ver}(H, h, i, \mathsf{id}, \pi) = 0$, output $\perp$ and abort.
4. If $i \leq |\hat{S}|, h = h^*$, then output $\perp$ and abort.
5. Else if $\mathsf{PRG}(s) \neq \mathsf{PRG}(\mathsf{PRF}(K_2, \mathsf{id}))$, output $\perp$ and abort.
6. Else output $\mathsf{PRF}(K_t, h)$.

Fig. 14. Static Secure ID-NIKE Parameters P_{IBKE}

Hybrid$_{\mathsf{ante-penultimate}}$: In this hybrid, the challenger generates the hash function as $H \leftarrow \mathsf{Gen}(1^\kappa, 2^\kappa, |\hat{S}|)$. The challenger computes $h^* = H(\hat{S})$ and sets $k^* = \mathsf{PRF}(K_1, t^*)$. He punctures the PRF key K_1 on input t^* to obtain $K\{t^*\}$. He outputs the program P_{IBKE} in Fig. 14 padded to the appropriate length. He publishes $P_{\mathsf{iO}} = \mathsf{iO}(P_{\mathsf{IBKE}})$.

This is indistinguishable from Hybrid$_{|\hat{S}|,j}$ because of indistinguishability between functionally equivalent circuits.

Hybrid$_{\mathsf{penultimate}}$: This is the same as Hybrid$_{\mathsf{ante-penultimate}}$, except that the challenger sets $k^* \xleftarrow{\$} \{0,1\}^\kappa$. This is indistinguishable from Hybrid$_{\mathsf{ante-penultimate}}$ because of security of the puncturable PRF.

Hybrid$_{\mathsf{ultimate}}$: This is the same as Hybrid$_{\mathsf{penultimate}}$, except that the challenger punctures PRF key k^* on value h^*. Then, he sets the program P_{IBKE} in Fig. 15 using punctured key $\underline{k^*\{h^*\}}$ padded to the appropriate length. He publishes $P_{\mathsf{iO}} = \mathsf{iO}(P_{\mathsf{IBKE}})$.

This hybrid is indistinguishable from Hybrid$_{\mathsf{penultimate}}$ because of iO between functionally equivalent programs. Finally, the distinguishing advantage of the adversary in this hybrid is at most $\mathsf{negl}(\kappa)$, by security of the puncturable PRF.

NIKE– Public Parameters

Constants: H, PRF keys $K_1\{t^*\}, K_2, h^*, k^*\{h^*\}$.
Input: $h, i, \mathsf{id}, \pi, s_{\mathsf{id}}, t$.

1. If $i > t$, output $\perp$ and abort.
2. If $t = t^*$, set $K_t = k^*\{h^*\}$, else compute $K_t = \mathsf{PRF}(K_1, t)$.
3. If $\mathsf{Ver}(H, h, i, \mathsf{id}, \pi) = 0$, output $\perp$ and abort.
4. If $i \leq |\hat{S}|, h = h^*$, output $\perp$ and abort.
5. If $t = t^*, h = h^*$, output $\perp$ and abort.
6. Else if $\mathsf{PRG}(s) \neq \mathsf{PRG}(\mathsf{PRF}(K_2, \mathsf{id}))$, output $\perp$ and abort.
7. Else output $\mathsf{PRF}(K_t, h)$.

Fig. 15. Static Secure ID-NIKE Parameters P_{IBKE}

6 Conclusion

We construct static-secure protocols that allow NIKE and ID-NIKE between an unbounded number of parties, relying on more feasible assumptions such as indistinguishability obfuscation and fully homomorphic encryption; as opposed to 'knowledge-type' assumptions such as differing-inputs obfuscation. It would be interesting to design protocols that tolerate more active attacks by adversaries, for an unbounded number of parties.

A NIKE: Proofs of Indistinguishability of the Hybrids

Lemma 1. *For all (non-uniform) PPT adversaries $\mathcal{D}$, $\mathcal{D}(\mathsf{Hybrid}_0) \approx_c \mathcal{D}(\mathsf{Hybrid}_1)$.*

Proof. We define a sub-sequence of polynomially many (concretely, $|\hat{S}|$) sub-hybrids $\mathsf{Hybrid}_{0,j}$ for $j \in |\hat{S}|$, where $\mathsf{Hybrid}_{0,j}$ is the same as Hybrid_0 except that the challenger samples the first j public values for parties in the set $\hat{S}$ at random in $\{0,1\}^{2\kappa}$ instead of generating them as the output of the PRG. Note that $\mathsf{Hybrid}_{0,0} \equiv \mathsf{Hybrid}_0$ and $\mathsf{Hybrid}_{0,|\hat{S}|} \equiv \mathsf{Hybrid}_1$.

We show that hybrids $\mathsf{Hybrid}_{0,j}$ and $\mathsf{Hybrid}_{0,j+1}$ are computationally indistinguishable. We will prove this by contradiction.

Suppose there exists a distinguisher $\mathcal{D}$ which distinguishes between $\mathsf{Hybrid}_{0,j}$ and $\mathsf{Hybrid}_{0,j+1}$ with advantage $1/\mathsf{poly}(\kappa)$ for some polynomial $\mathsf{poly}(\cdot)$. We construct a reduction that uses this distinguisher to break security of the underlying PRG. The reduction obtains a challenge value x, which may either be the output of the PRG on a uniform input, or may be chosen uniformly at random in $\{0,1\}^{2\kappa}$. It picks the first j public values for parties in the set $\hat{S}$ at random in $\{0,1\}^{2\kappa}$ instead of generating them as the output of the PRG. It sets the $(j+1)^{th}$ public value to the challenge value x. It samples the remaining public values for parties in the set $\hat{S}$ by picking a uniform $s_i \in \{0,1\}^{\kappa}$ and computing $x_i = \mathsf{PRG}(s_i)$.

If x is the output of a PRG this is the experiment in $\mathsf{Hybrid}_{0,j}$ else it is $\mathsf{Hybrid}_{0,j+1}$. Therefore, the reduction can mimic the output of the distinguisher between $\mathsf{Hybrid}_{0,j}$ and $\mathsf{Hybrid}_{0,j+1}$, thereby breaking the security of the PRG with advantage $1/\mathsf{poly}(\kappa)$.

Lemma 2. *For all (non-uniform) PPT adversaries* $\mathcal{D}$, $\mathcal{D}(\mathsf{Hybrid}_1) \approx_c$ $\mathcal{D}(\mathsf{Hybrid}_{2,1,a})$.

Proof. Suppose there exists a distinguisher which distinguishes between Hybrid_1 and $\mathsf{Hybrid}_{2,1,a}$ with advantage $1/\mathsf{poly}(\kappa)$ for some polynomial $\mathsf{poly}(\cdot)$. We construct a reduction that uses this distinguisher to break index-hiding security of the somewhere statistically hiding hash.

The reduction gives indices $\{0,1\}$ to the hash challenger. The challenger then generates hash $H = \mathsf{Gen}(s,b)$ for $b \xleftarrow{\$} \{0,1\}$ and sends them to the reduction. The reduction uses this function H as the hash (instead of generating the hash itself), and continues the game with the distinguisher. If $b = 0$, this corresponds to Hybrid_1, and if $b = 1$ the game corresponds to $\mathsf{Hybrid}_{2,1,a}$. Therefore, the reduction can mimic the output of the distinguisher between $\mathsf{Hybrid}_{2,i^*-1,b}$ and $\mathsf{Hybrid}_{2,i^*,a}$, thereby breaking index-hiding security of the hash function with advantage $1/\mathsf{poly}(\kappa)$.

Lemma 3. *For all (non-uniform) PPT adversaries* $\mathcal{D}$, $\mathcal{D}(\mathsf{Hybrid}_{2,i^*-1,b}) \approx_c$ $\mathcal{D}(\mathsf{Hybrid}_{2,i^*,a})$.

Proof. Suppose there exists a distinguisher which distinguishes between $\mathsf{Hybrid}_{2,i^*-1,b}$ and $\mathsf{Hybrid}_{2,i^*,a}$ with advantage $1/\mathsf{poly}(\kappa)$ for some polynomial $\mathsf{poly}(\cdot)$. We construct a reduction that uses this distinguisher to break index-hiding security of the somewhere statistically hiding hash.

The reduction gives indices $\{i^*-1, i^*\}$ to the hash challenger. The challenger then generates hash $H = \mathsf{Gen}(s,b)$ for $b \xleftarrow{\$} \{i^*-1, i^*\}$ and sends them to the reduction. The reduction uses this function H as the hash (instead of generating the hash itself), and continues the game with the distinguisher. If $b = i^* - 1$, this corresponds to $\mathsf{Hybrid}_{2,i^*-1,b}$, and if $b = i^*$ the game corresponds to $\mathsf{Hybrid}_{2,i^*,a}$. Therefore, the reduction can mimic the output of the distinguisher between Hybrid_1 and $\mathsf{Hybrid}_{2,1,a}$, thereby breaking index-hiding security of the hash function with advantage $1/\mathsf{poly}(\kappa)$.

Lemma 4. *For all (non-uniform) PPT adversaries* $\mathcal{D}$, $\mathcal{D}(\mathsf{Hybrid}_{2,i^*,a}) \approx_c$ $\mathcal{D}(\mathsf{Hybrid}_{2,i^*,b})$.

Proof. Note that in the experiment of $\mathsf{Hybrid}_{2,1,a/b}$, $H \leftarrow Gen(1^\kappa, 2^\kappa, i^*)$. Thus, by the statistical binding property of the hash function, if $\mathsf{Ver}(H, h, i^*, u, \pi)$ and $\mathsf{Ver}(H, h, i^*, u', \pi')$ accept, then it must be that $u = u'$.

Consider the programs P_{KE,i^*-1} and P_{KE,i^*}. Note that the only place where the two programs may differ is on inputs of the form $(h^*, i^*, pv, sv, \pi, t)$, where $h^* = H(\hat{S})$. Denote the public values $\{pv_j\}_{j \in S}$ by $pv_{x_1}, pv_{x_2}, \ldots pv_{x_{|S|}}$. In this case, P_{KE,i^*-1} (in $\mathsf{Hybrid}_{2,i^*,a}$) checks if $\mathsf{Ver}(H, h^*, i^*, pv, \pi) = 1$ and if $\mathsf{PRG}(sv) = pv$, then outputs $\mathsf{PRF}(K, h)$ else outputs $\perp$. On the other hand, P_{KE,i^*} (in $\mathsf{Hybrid}_{2,i^*,b}$) always outputs $\perp$. Because of the statistical binding property of the hash at index i, if $\mathsf{Ver}(H, h^*, i^*, pv, \pi)$ accepts for any value of (pv, π), then $pv = pv_{x_{i^*}}$. Moreover, since $pv_{x_{i^*}}$ is uniformly chosen in the range of the PRG, then with overwhelming probability, there does not exist any value sv such that $\mathsf{PRG}(sv) = pv_{x_{i^*}}$. Thus, the 'if' condition in P_{KE,i^*-1} in $\mathsf{Hybrid}_{2,i^*,a}$ will never be activated, and the two programs are functionally equivalent.

Therefore, the obfuscated circuits $iO(P_{KE,i^*-1})$ and $iO(P_{KE,i^*})$ must be indistinguishable by security of the iO. Suppose they are not, then consider a distinguisher $\mathcal{D}$ which distinguishes between these hybrids with non-negligible advantage. $\mathcal{D}$ can be used to break selective security of the indistinguishability obfuscation (according to Definition 1) via the following reduction to iO. The reduction acts as challenger in the experiment of $\mathsf{Hybrid}_{2,i^*,a}$.

The iO challenger $\mathsf{Samp}(1^\kappa)$ first activates the reduction, which samples the two circuits P_{KE,i^*-1}, P_{KE,i^*} and gives them to $\mathsf{Samp}(1^\kappa)$. The challenger then samples challenge circuit $C \xleftarrow{\$} \{P_{KE,i^*-1}, P_{KE,i^*}\}$, and sends $C' = iO(C)$ to the reduction. The reduction continues the game of $\mathsf{Hybrid}_{2,i^*,a}$ with C' in place of the obfuscation of program P_{KE,i^*-1}.

If $C = P_{KE,i^*-1}$, this corresponds to $\mathsf{Hybrid}_{2,i^*-1,b}$, and if $C = P_{KE,i^*}$ the game corresponds to $\mathsf{Hybrid}_{2,i^*,a}$. Therefore, the reduction can mimic the output of the distinguisher between $\mathsf{Hybrid}_{2,i^*,a}$ and $\mathsf{Hybrid}_{2,i^*,b}$, thereby breaking security of the iO with advantage $1/\mathsf{poly}(\kappa)$.

Lemma 5. *For all (non-uniform) PPT adversaries $\mathcal{D}$, $\mathcal{D}(\mathsf{Hybrid}_{2,t^*,b}) \approx_c \mathcal{D}(\mathsf{Hybrid}_3)$.*

Proof. Consider the programs P_{KE,t^*} ($\mathsf{Hybrid}_{2,t^*,b}$) and $P_{KE'}$ (Hybrid_3). Note that the only place where the two programs may differ is on inputs where $t = t^*$.

In this case, if $i \le t^*$, then for all $h = h^*$, both programs output $\perp$ and abort. If $i > t^*$ and $t = t^*$, then $i > t$ and both programs output $\perp$ and abort in Step 1. Moreover, for $t = t^*$, $k^* = \mathsf{PRF}(K, t^*)$ and thus the programs are functionally equivalent.

Therefore, the obfuscated circuits $iO(P_{KE,t^*})$ and $iO(P_{KE'})$ must be indistinguishable by security of the iO. Suppose they are not, then consider a distinguisher $\mathcal{D}$ which distinguishes between these hybrids with non-negligible advantage. $\mathcal{D}$ can be used to break security of the indistinguishability obfuscation (according to Definition 1) via the following reduction to iO. The reduction acts as challenger in the experiment of Hybrid_3.

The iO challenger $\mathsf{Samp}(1^\kappa)$ first activates the reduction, which samples the two circuits $P_{KE,t^*}, P_{KE'}$ and gives them to $\mathsf{Samp}(1^\kappa)$. The challenger then samples challenge circuit $C \xleftarrow{\$} \{P_{KE,t^*}, P_{KE'}\}$, and sends $C' = iO(C)$ to the reduction. The reduction continues the game of Hybrid_3 with C' in place of the obfuscation of program $P_{KE'}$.

If $C = P_{KE,t^*}$, this corresponds to $\mathsf{Hybrid}_{2,t^*,b}$, and if $C = P_{KE'}$ the game corresponds to Hybrid_3. Therefore, the reduction can mimic the output of the distinguisher between $\mathsf{Hybrid}_{2,t^*,b}$ and Hybrid_3, thereby breaking security of the iO with advantage $1/\mathsf{poly}(\kappa)$.

Lemma 6. *For all (non-uniform) PPT adversaries* $\mathcal{D}$, $\mathcal{D}(\mathsf{Hybrid}_3) \approx_c \mathcal{D}(\mathsf{Hybrid}_4)$.

Proof. Suppose there exists a distinguisher $\mathcal{D}$ which distinguishes between these hybrids with non-negligible advantage. $\mathcal{D}$ can be used to break selective security of the puncturable PRF via the following reduction. The reduction acts as challenger in the experiment of Hybrid_3.

It obtains challenge set $\hat{S}$ from the distinguisher, and computes $t^* = |\hat{S}|$. Then, it gives t^* to the PRF challenger, and obtains punctured PRF key $K\{t^*\}$ and a challenge a, which is either chosen uniformly at random or is the output of the PRF at t^*. Then, the reduction continues the experiment of Hybrid_3 as challenger, except that he sets $r^* = a$.

If $a = \mathsf{PRF}(K, t^*)$ then this is the experiment of Hybrid_3, and if a is chosen uniformly at random, then this is the experiment of Hybrid_4. Therefore, the reduction can mimic the output of the distinguisher between Hybrid_3 and Hybrid_4, thereby breaking security of the puncturable PRF with advantage $1/\mathsf{poly}(\kappa)$.

Lemma 7. *For all (non-uniform) PPT adversaries* $\mathcal{D}$, $\mathcal{D}(\mathsf{Hybrid}_4) \approx_c \mathcal{D}(\mathsf{Hybrid}_5)$.

Proof. Consider the programs $P_{KE'}$ (Hybrid_4) and $P_{KE''}$ (Hybrid_5). Note that the only place where the two programs may differ is on inputs where $t = t^*$. Then, for $t = t^*, h = h^*$, both programs output $\perp$. Moreover, because of functional equivalence of the punctured key $k^*\{h^*\}$ and k^* on all points where $h \neq h^*$, the programs are equivalent.

Suppose there exists a distinguisher $\mathcal{D}$ which distinguishes between these hybrids with non-negligible advantage. $\mathcal{D}$ can be used to break security of indistinguishability obfuscation via the following reduction. The reduction acts as challenger in the experiment of Hybrid_4. It obtains challenge set $\hat{S}$ from the distinguisher, and computes $h^* = H(\hat{S}|)$. Then, it constructs gives circuits $P_{KE'}, P_{KE''}$ as input to the Samp algorithm. Samp picks $C \xleftarrow{\$} \{P_{KE'}, P_{KE''}\}$ and sends $C' = iO(C)$ to the reduction. The reduction uses C' in place of $P_{KE'}$ in the experiment of Hybrid_4.

If $C = P_{KE'}$ then this is the experiment of Hybrid_4, and if $C = P_{KE''}$, then this is the experiment of Hybrid_5. Therefore, the reduction can mimic the output of the distinguisher between Hybrid_4 and Hybrid_5, thereby breaking security of the indistinguishability obfuscation with advantage $1/\mathsf{poly}(\kappa)$.

Lemma 8. *For all (non-uniform) PPT adversaries* $\mathcal{D}$, $\mathsf{Adv}_{\mathcal{D}}(\mathsf{Hybrid}_5) = \mathsf{negl}(\kappa)$.

Proof. Suppose there exists a distinguisher $\mathcal{D}$ which has non-negligible advantage in Hybrid_5. $\mathcal{D}$ can be used to break selective security of the puncturable PRF via the following reduction. The reduction acts as challenger in the experiment of Hybrid_5.

It obtains challenge set $\hat{S}$ from the distinguisher, and computes $h^* = H(\hat{S})$. Then, it gives h^* to the PRF challenger, and obtains punctured PRF key $k^*\{t^*\}$ and a challenge a, which is either chosen uniformly at random or is the output of the PRF at h^*. Then, the reduction continues the experiment of Hybrid_3 as challenger, except that he sets the shared key to a.

If $a = \mathsf{PRF}(k^*, h^*)$ then this corresponds to the correct shared key for group $\hat{S}$, whereas if a is chosen uniformly at random, this corresponds to a random key. Therefore, the reduction can mimic the output of the distinguisher, thereby breaking security of the puncturable PRF with advantage $1/\mathsf{poly}(\kappa)$.

References

1. Abusalah, H., Fuchsbauer, G., Pietrzak, K.: Constrained prfs for unbounded inputs. IACR Cryptology ePrint Archive 2014, p. 840 (2014). http://eprint.iacr.org/2014/840
2. Ananth, P., Boneh, D., Garg, S., Sahai, A., Zhandry, M.: Differing-inputs obfuscation and applications. IACR Cryptology ePrint Archive 2013, p. 689 (2013). http://eprint.iacr.org/2013/689
3. Barak, B., Goldreich, O., Impagliazzo, R., Rudich, S., Sahai, A., Vadhan, S.P., Yang, K.: On the (im)possibility of obfuscating programs. In: Kilian, J. (ed.) CRYPTO 2001. LNCS, vol. 2139, p. 1. Springer, Heidelberg (2001). http://dx.doi.org/10.1007/3-540-44647-8_1
4. Boneh, D., Silverberg, A.: Applications of multilinear forms to cryptography. IACR Cryptology ePrint Archive 2002, p. 80 (2002). http://eprint.iacr.org/2002/080
5. Boneh, D., Waters, B.: Constrained pseudorandom functions and their applications. IACR Cryptology ePrint Archive 2013, p. 352 (2013)
6. Boneh, D., Zhandry, M.: Multiparty key exchange, efficient traitor tracing, and more from indistinguishability obfuscation. In: Garay and Gennaro [21], pp. 480–499. http://dx.doi.org/10.1007/978-3-662-44371-2_27
7. Boyle, E., Chung, K.-M., Pass, R.: On extractability obfuscation. In: Lindell, Y. (ed.) TCC 2014. LNCS, vol. 8349, pp. 52–73. Springer, Heidelberg (2014). http://dx.doi.org/10.1007/978-3-642-54242-8_3
8. Boyle, E., Goldwasser, S., Ivan, I.: Functional signatures and pseudorandom functions. IACR Cryptology ePrint Archive 2013, p. 401 (2013)
9. Boyle, E., Pass, R.: Limits of extractability assumptions with distributional auxiliary input. IACR Cryptology ePrint Archive 2013, p. 703 (2013). http://eprint.iacr.org/2013/703
10. Brakerski, Z., Vaikuntanathan, V.: Efficient fully homomorphic encryption from (standard) LWE. SIAM J. Comput. 43(2), 831–871 (2014). http://dx.doi.org/10.1137/120868669
11. Canetti, R., Garay, J.A. (eds.): CRYPTO 2013, Part I. LNCS, vol. 8042. Springer, Heidelberg (2013). http://dx.doi.org/10.1007/978-3-642-40041-4
12. Canetti, R., Lin, H., Tessaro, S., Vaikuntanathan, V.: Obfuscation of probabilistic circuits and applications. In: Dodis and Nielsen [17], pp. 468–497. http://dx.doi.org/10.1007/978-3-662-46497-7_19
13. Cash, D., Kiltz, E., Shoup, V.: The twin diffie-hellman problem and applications. J. Cryptol. 22(4), 470–504 (2009). http://dx.doi.org/10.1007/s00145-009-9041-6

14. Coron, J., Lepoint, T., Tibouchi, M.: Practical multilinear maps over the integers. In: Canetti and Garay [11], pp. 476–493. http://dx.doi.org/10.1007/978-3-642-40041-4_26
15. Coron, J., Lepoint, T., Tibouchi, M.: New multilinear maps over the integers. IACR Cryptology ePrint Archive 2015, p. 162 (2015). http://eprint.iacr.org/2015/162
16. Diffie, W., Hellman, M.E.: New directions in cryptography. J. IEEE Trans. Inf. Theor. **22**(6), 644–654 (1976)
17. Dodis, Y., Nielsen, J.B. (eds.): TCC 2015, Part II. LNCS, vol. 9015. Springer, Heidelberg (2015). http://dx.doi.org/10.1007/978-3-662-46497-7
18. Dupont, R., Enge, A.: Provably secure non-interactive key distribution based on pairings. Discrete Appl. Math. **154**(2), 270–276 (2006). http://www.sciencedirect.com/science/article/pii/S0166218X05002337, Coding and Cryptography
19. Freire, E.S.V., Hofheinz, D., Kiltz, E., Paterson, K.G.: Non-interactive key exchange. IACR Cryptology ePrint Archive 2012, p. 732 (2012). http://eprint.iacr.org/2012/732
20. Freire, E.S.V., Hofheinz, D., Paterson, K.G., Striecks, C.: Programmable hash functions in the multilinear setting. In: Canetti and Garay [11], pp. 513–530. http://dx.doi.org/10.1007/978-3-642-40041-4_28
21. Garay, J.A., Gennaro, R. (eds.): CRYPTO 2014, Part I. LNCS, vol. 8616. Springer, Heidelberg (2014). http://dx.doi.org/10.1007/978-3-662-44371-2
22. Garg, S., Gentry, C., Halevi, S.: Candidate multilinear maps from ideal lattices. In: Johansson, T., Nguyen, P.Q. (eds.) EUROCRYPT 2013. LNCS, vol. 7881, pp. 1–17. Springer, Heidelberg (2013)
23. Garg, S., Gentry, C., Halevi, S., Raykova, M., Sahai, A., Waters, B.: Candidate indistinguishability obfuscation and functional encryption for all circuits. In: 54th Annual IEEE Symposium on Foundations of Computer Science, FOCS 2013, October 2013, Berkeley, CA, USA, pp. 40–49, 26–29. IEEE Computer Society (2013). http://dx.doi.org/10.1109/FOCS.2013.13
24. Garg, S., Gentry, C., Halevi, S., Raykova, M., Sahai, A., Waters, B.: Candidate indistinguishability obfuscation and functional encryption for all circuits. In: FOCS (2013)
25. Garg, S., Gentry, C., Halevi, S., Wichs, D.: On the implausibility of differing-inputs obfuscation and extractable witness encryption with auxiliary input. In: Garay and Gennaro [21], pp. 518–535. http://dx.doi.org/10.1007/978-3-662-44371-2_29
26. Gentry, C.: Fully homomorphic encryption using ideal lattices. In: Mitzenmacher, M. (ed.) Proceedings of the 41st Annual ACM Symposium on Theory of Computing, STOC 2009, Bethesda, MD, USA, May 31 - June 2, 2009, pp. 169–178. ACM (2009). http://doi.acm.org/10.1145/1536414.1536440
27. Gentry, C., Gorbunov, S., Halevi, S.: Graph-induced multilinear maps from lattices. In: Dodis and Nielsen [17], pp. 498–527. http://dx.doi.org/10.1007/978-3-662-46497-7_20
28. Goldreich, O., Goldwasser, S., Micali, S.: How to construct random functions (extended abstract). In: FOCS, pp. 464–479 (1984)
29. Hofheinz, D., Jager, T., Khurana, D., Sahai, A., Waters, B., Zhandry, M.: How to generate and use universal parameters. IACR Cryptology ePrint Archive 2014, p. 507 (2014). http://eprint.iacr.org/2014/507
30. Hubacek, P., Wichs, D.: On the communication complexity of secure function evaluation with long output. In: Roughgarden, T. (ed.) Proceedings of the 2015 Conference on Innovations in Theoretical Computer Science, ITCS 2015, Rehovot, Israel, January 11–13, 2015, pp. 163–172. ACM (2015). http://doi.acm.org/10.1145/2688073.2688105

31. Ishai, Y., Pandey, O., Sahai, A.: Public-coin differing-inputs obfuscation and its applications. In: Dodis and Nielsen [17], pp. 668–697. http://dx.doi.org/10.1007/978-3-662-46497-7_26
32. Joux, A.: Public-coin differing-inputs obfuscation and its applications. In: Bosma, W. (ed.) ANTS 2000. LNCS, vol. 1838. Springer, Heidelberg (2000). http://dx.doi.org/10.1007/10722028_23
33. Kiayias, A., Papadopoulos, S., Triandopoulos, N., Zacharias, T.: Delegatable pseudorandom functions and applications. IACR Cryptology ePrint Archive 2013, p. 379 (2013)
34. Koppula, V., Lewko, A.B., Waters, B.: Indistinguishability obfuscation for turing machines with unbounded memory. IACR Cryptology ePrint Archive 2014, p. 925 (2014). http://eprint.iacr.org/2014/925
35. Paterson, K.G., Srinivasan, S.: On the relations between non-interactive key distribution, identity-based encryption and trapdoor discrete log groups. Des. Codes Crypt. **52**(2), 219–241 (2009). http://dx.doi.org/10.1007/s10623-009-9278-y
36. Rao, V.: Adaptive multiparty non-interactive key exchange without setup in the standard model. IACR Cryptology ePrint Archive 2014, p. 910 (2014). http://eprint.iacr.org/2014/910
37. Sahai, A., Waters, B.: How to use indistinguishability obfuscation: deniable encryption, and more. In: Shmoys, D.B. (ed.) Symposium on Theory of Computing, STOC 2014, New York, NY, USA, May 31 - June 03, 2014, pp. 475–484. ACM (2014). http://doi.acm.org/10.1145/2591796.2591825
38. Sakai, R., Ohgishi, K., Kasahara, M.: Cryptosystems based on pairing. In: Symposium on Cryptography and Information Security SCIS (2000)
39. Yamakawa, T., Yamada, S., Hanaoka, G., Kunihiro, N.: Self-bilinear map on unknown order groups from indistinguishability obfuscation and its applications. Cryptology ePrint Archive, Report 2015/128 (2015). http://eprint.iacr.org/
40. Zhandry, M.: Adaptively secure broadcast encryption with small system parameters. IACR Cryptology ePrint Archive 2014, p.757 (2014). http://eprint.iacr.org/2014/757

PRFs and Hashes

Adaptively Secure Puncturable Pseudorandom Functions in the Standard Model

Susan Hohenberger[1]([✉]), Venkata Koppula[2], and Brent Waters[2]

[1] Johns Hopkins University, Baltimore, MD, USA
susan@cs.jhu.edu
[2] University of Texas at Austin, Austin, TX, USA
{kvenkata,bwaters}@cs.utexas.edu

Abstract. We study the adaptive security of constrained PRFs in the standard model. We initiate our exploration with puncturable PRFs. A puncturable PRF family is a special class of constrained PRFs, where the constrained key is associated with an element x' in the input domain. The key allows evaluation at all points $x \neq x'$.

We show how to build puncturable PRFs with adaptive security proofs in the standard model that involve only polynomial loss to the underlying assumptions. Prior work had either super-polynomial loss or applied the random oracle heuristic. Our construction uses indistinguishability obfuscation and DDH-hard algebraic groups of composite order.

More generally, one can consider a t-puncturable PRF: PRFs that can be punctured at any set of inputs S, provided the size of S is less than a fixed polynomial. We additionally show how to transform any (single) puncturable PRF family to a t-puncturable PRF family, using indistinguishability obfuscation.

1 Introduction

Pseudorandom functions (PRFs) are one of the fundamental building blocks in modern cryptography. A PRF system consists of a keyed function F and a set of keys $\mathcal{K}$ such that for a randomly chosen key $k \in \mathcal{K}$, the output of the function $F(k, x)$ for any input x in the input space "looks" random to a computationally bounded adversary, even when given polynomially many evaluations of $F(k, \cdot)$. Recently, the concept of *constrained pseudorandom functions*[1] was proposed in

S. Hohenberger—Supported by the National Science Foundation CNS-1228443 and CNS-1414023; the Defense Advanced Research Projects Agency (DARPA) and the Air Force Research Laboratory (AFRL) under contract FA8750-11-C-0080, the Office of Naval Research under contract N00014-14-1-0333, and a Microsoft Faculty Fellowship.

B. Waters—Supported by NSF CNS-1228599 and CNS-1414082, DARPA SafeWare, Google Faculty Research award, the Alfred P. Sloan Fellowship, Microsoft Faculty Fellowship, and Packard Foundation Fellowship.

[1] These were alternatively called *functional PRFs* [6] and *delegatable PRFs* [21].

© International Association for Cryptologic Research 2015
T. Iwata and J.H. Cheon (Eds.): ASIACRYPT 2015, Part I, LNCS 9452, pp. 79–102, 2015.
DOI: 10.1007/978-3-662-48797-6_4

the concurrent works of Boneh and Waters [4], Boyle, Goldwasser and Ivan [6] and Kiayias, Papadopoulos, Triandopoulos and Zacharias [21]. A constrained PRF system is associated with a family of boolean functions $\mathcal{F} = \{f\}$. As in standard PRFs, there exists a set of *master keys* $\mathcal{K}$ that can be used to evaluate the PRF F. However, given a master key k, it is also possible to derive a *constrained* key k_f associated with a function $f \in \mathcal{F}$. This constrained key k_f can be used to evaluate the function $F(k, \cdot)$ at all inputs x such that $f(x) = 1$. Intuitively, we would want that even if an adversary has k_f, the PRF evaluation at an input x not accepted by f looks random. Security is captured by an *adaptive* game between a PRF challenger and an adversary. The adversary is allowed to make multiple constrained key or point evaluation queries before committing to a challenge x^* not equal to any of the evaluation queries or accepted by any of the functions for which he obtained a constrained key.[2] The challenger either sends the PRF evaluation at x^* or an output chosen uniformly at random from the PRF range space, and the adversary wins if he can distinguish between these two cases.

Since their inception, constrained PRFs have found multiple applications. For example, Boneh and Waters [4] gave applications of broadcast encryption with optimal ciphertext length, identity-based key exchange, and policy-based key distribution. Sahai and Waters [24] used constrained PRFs as a central ingredient in their punctured programming methodology for building cryptosystems using indistinguishable obfuscation. Boneh and Zhandry [5] likewise applied constrained PRFs for realizing multi-party key exchange and broadcast systems.

Adaptive Security in Constrained PRFs. In their initial work, Boneh and Waters [4] showed constructions of constrained PRFs for different function families, including one for the class of all polynomial circuits (based on multilinear maps). However, all their constructions offer *selective security* - a weaker notion where the adversary must commit to the challenge input x^* *before* making any evaluation/constrained key queries.[3] Using complexity leveraging, one can obtain adaptive security by guessing the challenge input x^* before any queries are made. However, this results in exponential security loss. The works of [6,21] similarly dealt with selective security.

Recently, Fuchsbauer, Konstantinov, Pietrzak and Rao [11] showed adaptive security for *prefix-fixing* constrained PRFs, but with quasi-polynomial security loss. Also recently, Hofheinz [16] presented a novel construction that achieves adaptive security for *bit-fixing* constrained PRFs, but in the *random oracle model*.

While selective security has been sufficient for some applications of constrained PRFs, including many recent proofs leveraging the punctured programming [24] methodology (e.g., [2,5,19,24]), there are applications that demand

---

[2] This definition can be extended to handle multiple challenge points. See Sect. 3 for details.

[3] The prefix construction of [6,21] were also selective.

adaptive security, where the security game allows the adversary to query the PRF on many inputs before deciding on the point to puncture. For instance, [5] give a construction for multiparty key exchange that is semi-statically secure, and this construction requires adaptively secure constrained PRFs for circuits. We anticipate that the further realization of adaptively secure PRFs will introduce further applications of them.

Our Objective and Results. Our goal is to study adaptive security of constrained PRFs in the standard model. We initiate this exploration with *puncturable PRFs*, first explicitly introduced in [24] as a specialization of constrained PRFs. A puncturable PRF family is a special class of constrained PRFs, where the constrained key is associated with an element x' in the input domain. The key allows evaluation at all points $x \neq x'$. As noted by [4,6,21], the GGM tree-based construction of PRFs from one-way functions (OWFs) [14] can be modified to construct a puncturable PRF.[4] A selective proof of security follows via a hybrid argument, where the reduction algorithm uses the pre-determined challenge query x^* to "plant" its OWF challenge. However, such a technique does not seem powerful enough to obtain adaptive security with only a polynomial-factor security loss. The difficulty in proving adaptive security arises due to the fact that the reduction algorithm must respond to the evaluation queries, and then output a punctured key that is consistent with the evaluations. This means that the reduction algorithm must be able to evaluate the PRF at a large set S (so that all evaluation queries lie in S with non-negligible probability). However, S cannot be very large, otherwise the challenge x^* will lie in S, in which case the reduction algorithm cannot use the adversary's output.

In this work, we show new techniques for constructing adaptively-secure puncturable PRFs in the standard model. A central contribution is to overcome the conflict above, by allowing the reduction algorithm to commit to the evaluation queries, and at the same time, ensuring that the PRF output at the challenge point is unencumbered by the commitment.

Our main idea is to execute a delayed commitment to part of the PRF by partitioning. Initially, in our construction all points are tied to a single (Naor-Reingold [23] style) PRF. To prove security we begin by using the admissible hash function of Boneh and Boyen [3]. We partition the inputs into two distinct sets. The *evaluable set* which contains about $(1 - 1/q)$ fraction of inputs, and a *challenge set* which contains about $1/q$ fraction of inputs, where q is the number of point evaluation queries made by the attacker. Via a set of hybrid steps using the computational assumptions of indistinguishability obfuscation and subgroup hiding we modify the construction such that we use one Naor-Reingold PRF function to evaluate points in the evaluable set and a completely independent Naor-Reingold PRF to evaluate points in the challenge set.

After this separation has been achieved, there is a clearer path for our proof of security. At this point the reduction algorithm will create one PRF itself and

---

[4] In fact, the GGM PRF construction can be used to construct prefix-fixing constrained PRFs.

use it to answer any attacker point query in the evaluable set. If it is asked for a point x in the challenge set, it will simply abort. (The admissible hash function ensures that we get through without abort with some non-negligible probability.) Eventually, the attacker will ask for a punctured key on x^*, which defines x^* as the challenge input. Up until this point the reduction algorithm has made no commitments on what the second challenge PRF is. It then constructs the punctured key using the a freshly chosen PRF for the challenge inputs. However, when constructing this second PRF it now knows what the challenge x^* actually is and can fall back on selective techniques for completing the proof.

At a lower level our core PRF will be the Naor-Reingold PRF [23], but based in composite-order groups. Let $\mathbb{G}$ be a group of order $N = pq$, where p and q are primes. The master key consists of a group element $v \in \mathbb{G}$ and $2n$ exponents $d_{i,b} \in \mathbb{Z}_N$ (for $i = 1$ to n and $b \in \{0,1\}$). The PRF F takes as input a key $k = (v, \{d_{i,b}\})$, an ℓ-bit input x, uses a public admissible hash function $h : \{0,1\}^\ell \to \{0,1\}^n$ to compute $h(x) = b_1 \ldots b_n$ and outputs $v^{\prod_{j=1}^n d_{j,b_j}}$. A punctured key corresponding to x' derived from master key k is the obfuscation of a program P which has k, x' hardwired and outputs $F(k,x)$ on input $x \neq x'$, else it outputs $\perp$.

We will use a parameterized problem (in composite groups) to perform some of the separation step. Our assumption is that given $g, g^a, \ldots, g^{a^{n-1}}$ for randomly chosen $g \in \mathbb{G}$ and $a \in \mathbb{Z}_N^*$ it is hard to distinguish g^{a^n} from a random group element. While it is somewhat undesirable to base security on a parameterized assumption, we are able to use the recent results of Chase and Meiklejohn [8] to reduce this to the subgroup decision problem in DDH hard composite order groups.

t-puncturable PRFs. We also show how to construct t-puncturable PRFs: PRFs that can be punctured at any set of inputs S, provided $|S| \leq t$ (where $t(\cdot)$ is a fixed polynomial). We show how to transform any (single) puncturable PRF family to a t-puncturable PRF family, using indistinguishability obfuscation. In the security game for t-puncturable PRFs, the adversary is allowed to query for multiple t-punctured keys, each corresponding to a set S of size at most t. Finally, the adversary sends a challenge input x^* that lies in all the sets queried, and receives either the PRF evaluation at x^* or a uniformly random element of the range space.

In the construction, the setup and evaluation algorithm for the t-puncturable PRF are the same as those for the puncturable PRF. In order to puncture a key k at set S, the puncturing algorithm outputs the obfuscation of a program P that takes as input x, checks that $x \notin S$, and outputs $F(k,x)$.

For the proof of security, we observe that when the first t-punctured key query S_1 is made by the adversary, the challenger can guess the challenge $\tilde{x} \in S_1$. If this guess is incorrect, then the challenger simply aborts (which results in a $1/t$ factor security loss). However, if the guess is correct, then the challenger can now use the punctured key $K_{\tilde{x}}$ for all future evaluation/t-punctured key queries. From the security of puncturable PRFs, it follows that even after receiving evaluation/t-punctured key queries, the challenger will not be able to distinguish between $F(k, \tilde{x})$ and a random element in the range space.

We detail this transformation and its proof in Sect. 5.1. We also believe that we can use a similar approach to directly modify our main construction to handle multiple punctured points, however, we choose to focus on the generic transformation.

Related Works. Two recent works have explored the problem of adaptive security of constrained PRFs. Fuchsbauer, Konstantinov, Pietrzak and Rao [11] study the adaptive security of the GGM construction for prefix-free constrained PRFs. They show an interesting reduction to OWFs that suffers only a quasi-polynomial factor $q^{O(\log n)}$ loss, where q is the number of queries made by the adversary, and n is the length of the input. This beats the straightforward conversion from selective to adaptive security, which results in $O(2^n)$ security loss.

Hofheinz [16] shows a construction for bit-fixing constrained PRFs that is adaptively secure, assuming indistinguishability obfuscation and multilinear maps in the *random oracle model*. It also makes novel use of the random oracle for dynamically defining the challenge space based on the output of h. It is currently unclear whether such ideas could be adapted to the standard model.

Fuchsbauer et al. also show a negative result for the Boneh-Waters [4] construction of bit-fixing constrained PRFs. They show that any *simple* reduction from a static assumption to the adaptive security of the Boneh-Waters [4] bit-fixing constrained PRF construction must have an exponential factor security loss. More abstractly, using their techniques, one can show that any bit-fixing scheme that has the following properties will face this obstacle: (a) *fingerprinting queries* - By querying for a set of constrained keys, the adversary can obtain a *fingerprint* of the master key. (b) *checkability* - It is possible to efficiently check that any future evaluation/constrained key queries are consistent with the fingerprint. While these properties capture certain constructions, small perturbations to them could potentially circumvent checkability.

Partitioning type proofs have been used in several applications including identity-based encryption [1,3,17,25], verifiable random functions [20], and proofs of certain signature schemes [9,18,19]. We believe ours is the first to use partitioning for a delayed commitment to parameters. We note that our delayed technique is someway reminiscent to that of Lewko and Waters [22].

Recently, there has been a push to prove security for indistinguishability obfuscation from basic multilinear map assumptions. The recent work of Gentry, Lewko, Sahai and Waters [13] is a step in this direction, but itself requires the use of complexity leveraging. In the future work, we might hope for such reductions with just polynomial loss — perhaps for special cases of functionality. And thus give an end-to-end polynomial loss proof of puncturable PRFs from multilinear maps assumptions.

Two works have explored the notion of constrained verifiable random functions (VRFs). Fuchsbauer [10] and Chandran, Raghuraman and Vinayagamurthy [7] show constructions of selectively secure constrained VRFs for the class of all polynomial sized circuits. The construction in [7] is also delegatable.

Future Directions. A natural question is to construct adaptively-secure constrained PRFs for larger classes of functions in the standard model. Given the existing results of [11,16], both directions seem possible. While the techniques of [16] are intricately tied to the random oracle model, it is plausible there could be constructions in the standard model that evade the negative result of [11]. On the other hand, maybe the negative result of [11] (which is specific to the [4] construction) can be extended to show a similar lower bound for all constructions of constrained PRFs with respect to function family $\mathcal{F}$.

2   Preliminaries

First, we recall the notion of *admissible hash functions* due to Boneh and Boyen [3]. Here we state a simplified definition from [19]. Informally, an admissible hash function family is a function h with a 'partition sampling algorithm' AdmSample. This algorithm takes as input a parameter Q and outputs a 'random' partition of the outputs domain, where one of the partitions has $1/Q$ fraction of the points. Also, this partitioning has special structure which we will use in our proof.

Definition 1. *Let l, n and θ be efficiently computable univariate polynomials, $h : \{0,1\}^{l(\lambda)} \to \{0,1\}^{n(\lambda)}$ an efficiently computable function and AdmSample a PPT algorithm that takes as input 1^λ and an integer Q, and outputs a string $u \in \{0, 1, \bot\}^{n(\lambda)}$. For any $u \in \{0, 1, \bot\}^{n(\lambda)}$, define $P_u : \{0,1\}^{l(\lambda)} \to \{0,1\}$ as follows: $P_u(x) = 0$ if for all $1 \le j \le n(\lambda), h(x)_j \ne u_j$, else $P_u(x) = 1$.*

We say that $(h, \mathsf{AdmSample})$ is θ-admissible if the following condition holds: For any efficiently computable polynomial Q, for all $x_1, \ldots, x_{Q(\lambda)}, x^* \in \{0,1\}^{l(\lambda)}$, where $x^* \notin \{x_i\}_i$,

$$Pr[(\forall i \le Q(\lambda), P_u(x_i) = 1) \wedge P_u(x^*) = 0] \ge \frac{1}{\theta(Q(\lambda))}$$

where the probability is taken over $u \leftarrow \mathsf{AdmSample}(1^\lambda, Q(\lambda))$.

Theorem 1 (Admissible Hash Function Family [3], simplified proof in [9]). *For any efficiently computable polynomial l, there exist efficiently computable polynomials n, θ such that there exist θ-admissible function families mapping l bits to n bits.*

Note that the above theorem is information theoretic, and is not based on any cryptographic assumptions.

Next, we recall the definition of indistinguishability obfuscation from [12,24]. Let PPT denote probabilistic polynomial time.

Definition 2. *(Indistinguishability Obfuscation) A uniform PPT machine $i\mathcal{O}$ is called an indistinguishability obfuscator for a circuit class $\{\mathcal{C}_\lambda\}$ if it satisfies the following conditions:*

- *(Preserving Functionality) For all security parameters $\lambda \in \mathbb{N}$, for all $C \in \mathcal{C}_\lambda$, for all inputs x, we have that $C'(x) = C(x)$ where $C' \leftarrow i\mathcal{O}(\lambda, C)$.*

– *(Indistinguishability of Obfuscation) For any (not necessarily uniform) PPT distinguisher $\mathcal{B} = (Samp, \mathcal{D})$, there exists a negligible function $negl(\cdot)$ such that the following holds: if for all security parameters $\lambda \in \mathbb{N}, \Pr[\forall x, C_0(x) = C_1(x) : (C_0; C_1; \sigma) \leftarrow Samp(1^\lambda)] > 1 - negl(\lambda)$, then*

$$| \Pr[\mathcal{D}(\sigma, i\mathcal{O}(\lambda, C_0)) = 1 : (C_0; C_1; \sigma) \leftarrow Samp(1^\lambda)] -$$
$$\Pr[\mathcal{D}(\sigma, i\mathcal{O}(\lambda, C_1)) = 1 : (C_0; C_1; \sigma) \leftarrow Samp(1^\lambda)]| \leq negl(\lambda).$$

In a recent work, [12] showed how indistinguishability obfuscators can be constructed for the circuit class *P/poly*. We remark that $(Samp, \mathcal{D})$ are two algorithms that pass state, which can be viewed equivalently as a single stateful algorithm $\mathcal{B}$. In our proofs we employ the latter approach, although here we state the definition as it appears in prior work.

2.1 Assumptions

Let $\mathcal{G}$ be a PPT group generator algorithm that takes as input the security parameter 1^λ and outputs $(N, p, q, \mathbb{G}, \mathbb{G}_p, \mathbb{G}_q, g_1, g_2)$ where $p, q \in \Theta(2^\lambda)$ are primes, $N = pq$, $\mathbb{G}$ is a group of order N, $\mathbb{G}_p$ and $\mathbb{G}_q$ are subgroups of $\mathbb{G}$ of order p and q respectively, and g_1 and g_2 are generators of $\mathbb{G}_p$ and $\mathbb{G}_q$ respectively.

Assumption 1 (Subgroup Hiding for Composite DDH-Hard Groups).
Let $(N, p, q, \mathbb{G}, \mathbb{G}_p, \mathbb{G}_q, g_1, g_2) \leftarrow \mathcal{G}(1^\lambda)$ and $b \leftarrow \{0, 1\}$. Let $T \leftarrow \mathbb{G}$ if $b = 0$, else $T \leftarrow \mathbb{G}_p$. The advantage of algorithm $\mathcal{A}$ in solving Assumption 1 is defined as

$$\mathsf{Adv}_{\mathcal{A}}^{\mathrm{SGH}} = \left| \Pr[b \leftarrow \mathcal{A}(N, \mathbb{G}, \mathbb{G}_p, \mathbb{G}_q, g_1, g_2, T)] - \frac{1}{2} \right|$$

We say that Assumption 1 holds if for all PPT $\mathcal{A}$, $\mathsf{Adv}_{\mathcal{A}}^{\mathrm{SGH}}$ is negligible in λ.

Note that the adversary $\mathcal{A}$ gets generators for both subgroups $\mathbb{G}_p$ and $\mathbb{G}_q$. This is in contrast to bilinear groups, where, if given generators for both subgroups, the adversary can use the pairing to distinguish a random group element from a random subgroup element.

Analogously, we assume that no PPT adversary can distinguish between a random element of $\mathbb{G}$ and a random element of $\mathbb{G}_q$ with non-negligible advantage. This is essentially Assumption 1, where prime q is chosen instead of p, and $\mathbb{G}_q$ is chosen instead of $\mathbb{G}_p$.

Assumption 2. *This assumption is parameterized with an integer $n \in \mathbb{Z}$. Let $(N, p, q, \mathbb{G}, \mathbb{G}_p, \mathbb{G}_q, g_1, g_2) \leftarrow \mathcal{G}(1^\lambda)$, $g \leftarrow \mathbb{G}$, $a \leftarrow \mathbb{Z}_N^*$ and $b \leftarrow \{0, 1\}$. Let $D = (N, \mathbb{G}, \mathbb{G}_p, \mathbb{G}_q, g_1, g_2, g, g^a, \ldots, g^{a^{n-1}})$. Let $T = g^{a^n}$ if $b = 0$, else $T \leftarrow \mathbb{G}$. The advantage of algorithm $\mathcal{A}$ in solving Assumption 2 is defined as*

$$\mathsf{Adv}_{\mathcal{A}} = \left| \Pr[b \leftarrow \mathcal{A}(D, T)] - \frac{1}{2} \right|$$

We say that Assumption 2 holds if for all PPT $\mathcal{A}$, $\mathsf{Adv}_{\mathcal{A}}$ is negligible in λ.

We will use Assumption 2 for clarity in certain parts of our proof, but we do not give it a name because it is implied by other named assumptions. First, Assumption 2 is implied by the n-Power Decisional Diffie-Hellman Assumption [15]. Second, it is also implied by the non-parameterized Assumption 1. The recent results of Chase and Meiklejohn [8] essentially show this latter implication, but that work focuses on the target groups of bilinear maps, whereas our algebraic focus does not involve bilinear maps.

3   Constrained Pseudorandom Functions

In this section, we define the syntax and security properties of a constrained pseudorandom function family. This definition is similar to the one in Boneh-Waters [4], except that the keys are constrained with respect to a circuit family instead of a set system.

Let $\mathcal{K}$ denote the key space, $\mathcal{X}$ the input domain and $\mathcal{Y}$ the range space. The PRF is a function $F : \mathcal{K} \times \mathcal{X} \to \mathcal{Y}$ that can be computed by a deterministic polynomial time algorithm. We will assume there is a Setup algorithm F.setup that takes the security parameter λ as input and outputs a random secret key $k \in \mathcal{K}$.

A PRF $F : \mathcal{K} \times \mathcal{X} \to \mathcal{Y}$ is said to be *constrained* with respect to a circuit family $\mathcal{C}$ if there is an additional key space $\mathcal{K}_c$, and three algorithms F.setup, F.constrain and F.eval as follows:

- F.setup(1^λ) is a PPT algorithm that takes the security parameter λ as input and outputs a description of the key space $\mathcal{K}$, the constrained key space $\mathcal{K}_c$ and the PRF F.
- F.constrain(k, C) is a PPT algorithm that takes as input a PRF key $k \in \mathcal{K}$ and a circuit $C \in \mathcal{C}$ and outputs a constrained key $k_C \in \mathcal{K}_c$.
- F.eval(k_C, x) is a deterministic polynomial time algorithm that takes as input a constrained key $k_C \in \mathcal{K}_c$ and $x \in \mathcal{X}$ and outputs an element $y \in \mathcal{Y}$. Let k_C be the output of F.constrain(k, C). For correctness, we require the following:

$$F\text{.eval}(k_C, x) = \begin{cases} F(k, x) & \text{if } C(x) = 1 \\ \bot & \text{otherwise} \end{cases}$$

Security of Constrained Pseudorandom Functions: Intuitively, we require that even after obtaining several constrained keys, no polynomial time adversary can distinguish a truly random string from the PRF evaluation at a point not accepted by the queried circuits. This intuition can be formalized by the following security game between a challenger and an adversary A.

Let $F : \mathcal{K} \times \mathcal{X} \to \mathcal{Y}$ be a constrained PRF with respect to a circuit family $\mathcal{C}$. The security game consists of three phases.

Setup Phase. The challenger chooses a random key $k \leftarrow \mathcal{K}$ and $b \leftarrow \{0,1\}$.

Query Phase. In this phase, A is allowed to ask for the following queries:

- **Evaluation Query** A sends $x \in \mathcal{X}$, and receives $F(k, x)$.
- **Key Query** A sends a circuit $C \in \mathcal{C}$, and receives $F.\text{constrain}(k, C)$.
- **Challenge Query** A sends $x \in \mathcal{X}$ as a challenge query. If $b = 0$, the challenger outputs $F(k, x)$. Else, the challenger outputs a random element $y \leftarrow \mathcal{Y}$.

Guess. A outputs a guess b' of b.

Let $E \subset \mathcal{X}$ be the set of evaluation queries, $L \subset \mathcal{C}$ be the set of constrained key queries and $Z \subset \mathcal{X}$ the set of challenge queries. A wins if $b = b'$ and $E \cap Z = \phi$ and for all $C \in L, z \in Z, C(z) = 0$. The advantage of A is defined to be $\text{Adv}_A^F(\lambda) = Pr[A \text{ wins}]$.

Definition 3. *The PRF F is a secure constrained PRF with respect to $\mathcal{C}$ if for all PPT adversaries A $\text{Adv}_A^F(\lambda)$ is negligible in λ.*

3.1 Puncturable Pseudorandom Functions

In this section, we define the syntax and security properties of a puncturable pseudorandom function family. Puncturable PRFs are a special class of constrained pseudorandom functions.

A PRF $F : \mathcal{K} \times \mathcal{X} \rightarrow \mathcal{Y}$ is a puncturable pseudorandom function if there is an additional key space $\mathcal{K}_p$ and three polynomial time algorithms $F.\text{setup}$, $F.\text{eval}$ and $F.\text{puncture}$ as follows:

- $F.\text{setup}(1^\lambda)$ is a randomized algorithm that takes the security parameter λ as input and outputs a description of the key space $\mathcal{K}$, the punctured key space $\mathcal{K}_p$ and the PRF F.
- $F.\text{puncture}(k, x)$ is a randomized algorithm that takes as input a PRF key $k \in \mathcal{K}$ and $x \in \mathcal{X}$, and outputs a key $k_x \in \mathcal{K}_p$.
- $F.\text{eval}(k_x, x')$ is a deterministic algorithm that takes as input a punctured key $k_x \in \mathcal{K}_p$ and $x' \in \mathcal{X}$. Let $k \in \mathcal{K}$, $x \in \mathcal{X}$ and $k_x \leftarrow F.\text{puncture}(k, x)$. For correctness, we need the following property:

$$F.\text{eval}(k_x, x') = \begin{cases} F(k, x') & \text{if } x \neq x' \\ \bot & \text{otherwise} \end{cases}$$

Security of Puncturable PRFs: The security game between the challenger and the adversary A consists of the following four phases.

Setup Phase. The challenger chooses uniformly at random a PRF key $k \leftarrow \mathcal{K}$ and a bit $b \leftarrow \{0, 1\}$.

Evaluation Query Phase. A queries for polynomially many evaluations. For each evaluation query x, the challenger sends $F(k, x)$ to A.

Challenge Phase. A chooses a challenge $x^* \in \mathcal{X}$. The challenger computes $k_{x^*} \leftarrow$ F.puncture(k, x^*). If $b = 0$, the challenger outputs k_{x^*} and $F(k, x^*)$. Else, the challenger outputs k_{x^*} and $y \leftarrow \mathcal{Y}$ chosen uniformly at random.

Guess. A outputs a guess b' of b. Let $E \subset \mathcal{X}$ be the set of evaluation queries. A wins if $b = b'$ and $x^* \notin E$. The advantage of A is defined to be $\mathsf{Adv}_A^F(\lambda) = Pr[A \text{ wins}]$.

Definition 4. *The PRF F is a secure puncturable PRF if for all probabilistic polynomial time adversaries A $\mathsf{Adv}_A^F(\lambda)$ is negligible in λ.*

t-**Puncturable Pseudorandom Functions.** The notion of puncturable PRFs can be naturally extended to that of t-puncturable PRFs, where it is possible to derive a key punctured at any set S of size at most t. A formal definition of t-puncturable PRFs can be found in Sect. 5.

4   Construction

We now describe our puncturable PRF family. It consists of the PRF $F : \mathcal{K} \times \mathcal{X} \to \mathcal{Y}$ and the three algorithms F.setup, F.puncture and F.eval. The input domain is $\mathcal{X} = \{0, 1\}^\ell$, where $\ell = \ell(\lambda)$. We define the key space $\mathcal{K}$ and range space $\mathcal{Y}$ as part of the setup algorithm described next.

F.*setup*(1^λ) F.setup, on input 1^λ, runs $\mathcal{G}$ to compute $(N, p, q, \mathbb{G}, \mathbb{G}_p, \mathbb{G}_q, g_1, g_2) \leftarrow \mathcal{G}(1^\lambda)$. Let n, θ be polynomials such that there exists a θ-admissible hash function h mapping $\ell(\lambda)$ bits to $n(\lambda)$ bits. For simplicity of notation, we will drop the dependence of ℓ and n on λ.

The key space is $\mathcal{K} = \mathbb{G} \times (\mathbb{Z}_N^2)^n$ and the range is $\mathcal{Y} = \mathbb{G}$. The setup algorithm chooses $v \in \mathbb{G}$, $d_{i,b} \in \mathbb{Z}_N$, for $i = 1$ to n and $b \in \{0, 1\}$, and sets $k = (v, ((d_{1,0}, d_{1,1}), \dots, (d_{n,0}, d_{n,1})))$.

The PRF F for key k on input x is then computed as follows. Let $k = (v, ((d_{1,0}, d_{1,1}), \dots, (d_{n,0}, d_{n,1}))) \in \mathbb{G} \times (\mathbb{Z}_N^2)^n$ and $h(x) = (b_1, \dots, b_n)$, where $b_i \in \{0, 1\}$. Then,

$$F(k, x) = v^{\prod_{j=1}^n d_{j,b_j}}.$$

F.*puncture*$(\mathbf{k}, \mathbf{x}')$ F.puncture computes an obfuscation of $\mathsf{PuncturedKey}_{k,x'}$ (defined in Fig. 1); that is, $K_{x'} \leftarrow i\mathcal{O}(\lambda, \mathsf{PuncturedKey}_{k,x'})$ where $\mathsf{PuncturedKey}_{k,x'}$ is padded to be of appropriate size.[5]

F.*eval*$(\mathbf{K}_{x'}, \mathbf{x})$ The punctured key $K_{x'}$ is a program that takes an ℓ-bit input. We define

$$F.\mathsf{eval}(K_{x'}, x) = K_{x'}(x).$$

---

[5] Looking ahead, in the proof of security, the program $\mathsf{PuncturedKey}_{k,x'}$ will be replaced by $\mathsf{PuncturedKey}'_{V,w,D,u,x'}$, $\mathsf{PuncturedKeyAlt}_{u,k,k',x'}$ and $\mathsf{PuncturedKeyAlt}'_{u,W,E,k,x'}$ in subsequent hybrids. Since this transformation relies on $i\mathcal{O}$ being secure, we need that all programs have same size. Hence, all programs are padded appropriately to ensure that they have the same size.

PuncturedKey$_{k,x'}$

Input: $x \in \{0,1\}^{\ell}$

Constants : The group $\mathbb{G}$, $k = (v, ((d_{1,0}, d_{1,1}) \ldots (d_{n,0}, d_{n,1}))) \in \mathbb{G} \times (\mathbb{Z}_N^2)^n$

$\qquad x' \in \{0,1\}^{\ell}$

Compute $h(x) = b_1 \ldots b_n \in \{0,1\}^n$.

if $x = x'$ then

$\qquad$ Output $\perp$.

else

$\qquad$ Output $v^{\prod_{j=1}^n d_{j,b_j}}$.

end if

Fig. 1. Program PuncturedKey

4.1 Proof of Security

We will now prove that our construction is a secure puncturable PRF as defined in Definition 4. Specifically, the claim we show is:

Theorem 2 (Main Theorem). *Assuming $i\mathcal{O}$ is a secure indistinguishability obfuscator and the Subgroup Hiding Assumption holds for groups output by $\mathcal{G}$, the PRF F defined above, together with algorithms F.setup, F.puncture and F.eval, is a secure punctured pseudorandom function as defined in Definition 4.*

Proof. In order to prove this, we define the following sequence of games. Assume the adversary $\mathcal{A}$ makes $Q = Q(\lambda)$ evaluation queries (where $Q(\cdot)$ is a polynomial) before sending the challenge input.

Sequence of Games. We underline the primary changes from one game to the next.

Game 0. This game is the original security game from Definition 4 between the challenger and $\mathcal{A}$ instantiated by the construction under analysis. Here the challenger first chooses a random PRF key, then $\mathcal{A}$ makes evaluation queries and finally sends the challenge input. The challenger responds by sending a key punctured at the challenge input, and either a PRF evaluation at the challenged point or a random value.

1. Let $(N, p, q, \mathbb{G}, \mathbb{G}_p, \mathbb{G}_q, g_1, g_2) \leftarrow \mathcal{G}(1^{\lambda})$. Choose $v \in \mathbb{G}, d_{i,b} \in \mathbb{Z}_N$, for $i = 1$ to n and $b \in \{0,1\}$, and set $k = (v, ((d_{1,0}, d_{1,1}), \ldots, (d_{n,0}, d_{n,1})))$.
2. On any evaluation query $x_i \in \{0,1\}^{\ell}$, compute $h(x_i) = b_1^i \ldots b_n^i$ and output $v^{\prod_{j=1}^n d_{j,b_j^i}}$.
3. $\mathcal{A}$ sends challenge input x^* such that $x^* \neq x_i \ \forall \ i \leq Q$. Compute $K_{x^*} \leftarrow i\mathcal{O}(\lambda, \text{PuncturedKey}_{k,x^*})$ and $h(x^*) = b_1^* \ldots b_n^*$. Let $y_0 = v^{\prod_{j=1}^n d_{j,b_j^*}}$ and $y_1 \leftarrow \mathbb{G}$.

4. Flip coin $\beta \leftarrow \{0,1\}$. Output (K_{x^*}, y_β).
5. $\mathcal{A}$ outputs β' and wins if $\beta = \beta'$.

Game 1. This game is the same as the previous one, except that we simulate a partitioning game while the adversary operates and if an undesirable partition arises, we abort the game and decide whether or not the adversary "wins" by a coin flip. This partitioning game works as follows: the challenger samples $u \in \{0,1,\perp\}^n$ using AdmSample and aborts if either there exists an evaluation query x such that $P_u(x) = 0$ or the challenge query x^* satisfies $P_u(x^*) = 1$.

1. Let $(N, p, q, \mathbb{G}, \mathbb{G}_p, \mathbb{G}_q, g_1, g_2) \leftarrow \mathcal{G}(1^\lambda)$. Choose $v \in \mathbb{G}, d_{i,b} \in \mathbb{Z}_N$, for $i = 1$ to n and $b \in \{0,1\}$, and set $k = (v, ((d_{1,0}, d_{1,1}), \ldots, (d_{n,0}, d_{n,1})))$.
   Choose $u \leftarrow$ AdmSample$(1^\lambda, Q)$ and let $S_u = \{x : P_u(x) = 1\}$ (recall $P_u(x) = 0$ if $h(x)_j \neq u_j \ \forall 1 \leq j \leq n$).
2. On any evaluation query $x_i \in \{0,1\}^\ell$, check if $P_u(x_i) = 1$.
   If not, flip a coin $\gamma_i \leftarrow \{0,1\}$ and abort. $\mathcal{A}$ wins if $\gamma_i = 1$.
   Else compute $h(x_i) = b_1^i \ldots b_n^i$ and output $v^{\prod_{j=1}^n d_{j,b_j^i}}$.
3. $\mathcal{A}$ sends challenge input x^* such that $x^* \neq x_i \ \forall i \leq Q$. Check if $P_u(x^*) = 0$.
   If not, flip a coin $\gamma^* \leftarrow \{0,1\}$ and abort. $\mathcal{A}$ wins if $\gamma^* = 1$.
   Else compute $K_{x^*} \leftarrow i\mathcal{O}(\lambda, \mathsf{PuncturedKey}_{k,x^*})$ and $h(x^*) = b_1^* \ldots b_n^*$. Let $y_0 = v^{\prod_{j=1}^n d_{j,b_j^*}}$ and $y_1 \leftarrow \mathbb{G}$.
4. Flip coin $\beta \leftarrow \{0,1\}$. Output (K_{x^*}, y_β).
5. $\mathcal{A}$ outputs β' and wins if $\beta = \beta'$.

Game 2. In this game, the challenger modifies the punctured key and outputs an obfuscation of PuncturedKeyAlt defined in Fig. 2. On inputs x such that $P_u(x) = 1$, the altered punctured key uses the same master key k as before. However, if $P_u(x) = 0$, the altered punctured key uses a different master key k' that is randomly chosen from the key space.

1. Let $(N, p, q, \mathbb{G}, \mathbb{G}_p, \mathbb{G}_q, g_1, g_2) \leftarrow \mathcal{G}(1^\lambda)$. Choose $v \in \mathbb{G}, d_{i,b} \in \mathbb{Z}_N$, for $i = 1$ to n and $b \in \{0,1\}$.
   Set $k = (v, ((d_{1,0}, d_{1,1}), \ldots, (d_{n,0}, d_{n,1})))$.
   Choose $u \leftarrow$ AdmSample$(1^\lambda, Q)$.
2. On any evaluation query $x_i \in \{0,1\}^\ell$, check if $P_u(x_i) = 1$.
   If not, flip a coin $\gamma_i \leftarrow \{0,1\}$ and abort. $\mathcal{A}$ wins if $\gamma_i = 1$.
   Else compute $h(x_i) = b_1^i \ldots b_n^i$ and output $v^{\prod_{j=1}^n d_{j,b_j^i}}$.
3. $\mathcal{A}$ sends challenge input x^* such that $x^* \neq x_i \ \forall i \leq Q$. Check if $P_u(x^*) = 0$.
   If not, flip a coin $\gamma^* \leftarrow \{0,1\}$ and abort. $\mathcal{A}$ wins if $\gamma^* = 1$.
   Else choose $w \in \mathbb{G}, e_{i,b} \in \mathbb{Z}_N$, for $i = 1$ to n and $b \in \{0,1\}$.
   Set $k' = (w, ((e_{1,0}, e_{1,1}), \ldots, (e_{n,0}, e_{n,1})))$.
   Compute $K_{x^*} \leftarrow i\mathcal{O}(\lambda, \mathsf{PuncturedKeyAlt}_{u,k,k',x^*})$ and $h(x^*) = b_1^* \ldots b_n^*$. Let $y_0 = w^{\prod_{j=1}^n e_{j,b_j^*}}$ and $y_1 \leftarrow \mathbb{G}$.
4. Flip coin $\beta \leftarrow \{0,1\}$. Output (K_{x^*}, y_β).
5. $\mathcal{A}$ outputs β' and wins if $\beta = \beta'$.

PuncturedKeyAlt$_{u,k,k',x'}$

Input: $x \in \{0,1\}^\ell$

Constants : The group $\mathbb{G}, k = (v, ((d_{1,0}, d_{1,1}) \ldots (d_{n,0}, d_{n,1}))) \in \mathbb{G} \times (\mathbb{Z}_N^2)^n$

$k' = (w, ((e_{1,0}, e_{1,1}) \ldots (e_{n,0}, e_{n,1}))) \in \mathbb{G} \times (\mathbb{Z}_N^2)^n$

$x' \in \{0,1\}^\ell, u \in \{0,1,\perp\}^n$

Compute $h(x) = b_1 \ldots b_n$.
if $x = x'$ then
    Output $\perp$.
else if $P_u(x) = 0$ then
    output $w^{\overline{\prod_{j=1}^n e_{j,b_j}}}$.
else
    Output $v^{\prod_{j=1}^n d_{j,b_j}}$.
end if

Fig. 2. Program PuncturedKeyAlt

Game 3. In this game, the challenger changes how the master key k' is chosen so that some elements contain an a-factor, for use on inputs x where $P_u(x) = 0$.

1. Let $(N, p, q, \mathbb{G}, \mathbb{G}_p, \mathbb{G}_q, g_1, g_2) \leftarrow \mathcal{G}(1^\lambda)$. Choose $v \in \mathbb{G}, d_{i,b} \in \mathbb{Z}_N$, for $i = 1$ to n and $b \in \{0,1\}$, and set $k = (v, ((d_{1,0}, d_{1,1}), \ldots, (d_{n,0}, d_{n,1})))$. Choose $u \leftarrow$ AdmSample$(1^\lambda, Q)$.

2. On any evaluation query $x_i \in \{0,1\}^\ell$, check if $P_u(x_i) = 1$. If not, flip a coin $\gamma_i \leftarrow \{0,1\}$ and abort. $\mathcal{A}$ wins if $\gamma_i = 1$.
   Else compute $h(x_i) = b_1^i \ldots b_n^i$ and output $v^{\prod_j d_{j,b_j^i}}$.

3. $\mathcal{A}$ sends challenge input x^* such that $x^* \neq x_i \; \forall \; i \leq Q$. Check if $P_u(x^*) = 0$. If not, flip a coin $\gamma^* \leftarrow \{0,1\}$ and abort. $\mathcal{A}$ wins if $\gamma^* = 1$.
   Else choose $w \leftarrow \mathbb{G}, a \leftarrow \mathbb{Z}_N^*$ and $e'_{i,b} \leftarrow \mathbb{Z}_N$.
   Let $e_{i,b} = e'_{i,b} \cdot a$ if $\overline{h(x^*)_i = b}$, else $e_{i,b} = e'_{i,b}$.
   Let $k' = (w, ((e_{1,0}, e_{1,1}), \ldots, (e_{n,0}, e_{n,1}))), K_{x^*} \leftarrow i\mathcal{O}(\text{PuncturedKeyAlt}_{u,k,k',x^*})$.
   Let $h(x^*) = b_1^* \ldots b_n^*$ and $y_0 = w^{\prod_j e_{j,b_j^*}}$ and $y_1 \leftarrow \mathbb{G}$.

4. Flip coin $\beta \leftarrow \{0,1\}$. Output (K_{x^*}, y_β).

5. $\mathcal{A}$ outputs β' and wins if $\beta = \beta'$.

Game 4. This game is the same as the previous one, except that the altered punctured program contains the constants $\{w^{a^i}\}_{i=0}^n$ hardwired. These terms are used to compute the output of the punctured program. The punctured key is an obfuscation of PuncturedKeyAlt$'$ defined in Fig. 3.

1. Let $(N, p, q, \mathbb{G}, \mathbb{G}_p, \mathbb{G}_q, g_1, g_2) \leftarrow \mathcal{G}(1^\lambda)$. Choose $v \in \mathbb{G}, d_{i,b} \in \mathbb{Z}_N$, for $i = 1$ to n and $b \in \{0,1\}$, and set $k = (v, ((d_{1,0}, d_{1,1}), \ldots, (d_{n,0}, d_{n,1})))$. Choose $u \leftarrow$ AdmSample$(1^\lambda, Q)$.

2. On any evaluation query $x_i \in \{0,1\}^\ell$, check if $P_u(x_i) = 1$.
   If not, flip a coin $\gamma_i \leftarrow \{0,1\}$ and abort. $\mathcal{A}$ wins if $\gamma_i = 1$.
   Else compute $h(x_i) = b_1^i \ldots b_n^i$ and output $v^{\Pi_j\, d_{j,b_j^i}}$.
3. $\mathcal{A}$ sends challenge input x^* such that $x^* \neq x_i \;\forall\; i \leq Q$. Check if $P_u(x^*) = 0$.
   If not, flip a coin $\gamma^* \leftarrow \{0,1\}$ and abort. $\mathcal{A}$ wins if $\gamma^* = 1$.
   Else choose $w \leftarrow \mathbb{G}$, $a \leftarrow \mathbb{Z}_N^*$ and $e'_{i,b} \leftarrow \mathbb{Z}_N$.
   Let $W = (w, w^a, \ldots, w^{a^{n-1}})$, $E = ((e'_{1,0}, e'_{1,1}), \ldots, (e'_{n,0}, e'_{n,1}))$.
   Let $K''_{x^*} \leftarrow i\mathcal{O}(\mathsf{PuncturedKeyAlt}'_{u,W,E,k,x^*})$, $h(x^*) = b_1^* \ldots b_n^*$, $y_0 = \left(w^{a^n}\right)^{\Pi_j\, e'_{j,b_j^*}}$
   and $y_1 \leftarrow \mathbb{G}$.
4. Flip coin $\beta \leftarrow \{0,1\}$. Output (K''_{x^*}, y_β).
5. $\mathcal{A}$ outputs β' and wins if $\beta = \beta'$.

PuncturedKeyAlt$'_{u,W,E,k,x'}$

Input: $x \in \{0,1\}^\ell$

Constants : The group $\mathbb{G}$, $k = (v, ((d_{1,0}, d_{1,1}) \ldots (d_{n,0}, d_{n,1}))) \in \mathbb{G} \times (\mathbb{Z}_N^2)^n$
$W = (w_0, \ldots, w_{n-1}) \in \mathbb{G}^n$, $E = ((e'_{1,0}, e'_{1,1}), \ldots, (e'_{n,0}, e'_{n,1})) \in (\mathbb{Z}_N^2)^n$
$x' \in \{0,1\}^\ell$, $u \in \{0,1,\perp\}^n$

Compute $h(x) = b_1 \ldots b_n$ and $h(x') = b'_1 \ldots b'_n$. Let $t_{x'}(x) = |\{i : b_i = b'_i\}|$.
if $x = x'$ **then**
    Output $\perp$.
else if $P_u(x) = 0$ **then**
    output $\left(w_{t_{x'}(x)}\right)^{\Pi_{j=1}^n\, e'_{j,b_j}}$.
else
    Output $v^{\Pi_{j=1}^n\, d_{j,b_j}}$.
end if

Fig. 3. Program PuncturedKeyAlt'

Game 5. In this game, we replace the term w^{a^n} with a random element from $\mathbb{G}$. Hence, both y_0 and y_1 are random elements of $\mathbb{G}$, thereby ensuring that any adversary has zero advantage in this game.

1. Let $(N, p, q, \mathbb{G}, \mathbb{G}_p, \mathbb{G}_q, g_1, g_2) \leftarrow \mathcal{G}(1^\lambda)$. Choose $v \in \mathbb{G}$, $d_{i,b} \in \mathbb{Z}_N$, for $i = 1$ to
   n and $b \in \{0,1\}$, and set $k = (v, ((d_{1,0}, d_{1,1}), \ldots, (d_{n,0}, d_{n,1})))$.
   Choose $u \leftarrow \mathsf{AdmSample}(1^\lambda, Q)$.
2. On any evaluation query $x_i \in \{0,1\}^\ell$, check if $P_u(x_i) = 1$.
   If not, flip a coin $\gamma_i \leftarrow \{0,1\}$ and abort. $\mathcal{A}$ wins if $\gamma_i = 1$.
   Else compute $h(x_i) = b_1^i \ldots b_n^i$ and output $v^{\Pi_{j=1}^n\, d_{j,b_j^i}}$.

3. $\mathcal{A}$ sends challenge input x^* such that $x^* \neq x_i \ \forall \ i \leq Q$. Check if $P_u(x^*) = 0$. If not, flip a coin $\gamma^* \leftarrow \{0,1\}$ and abort. $\mathcal{A}$ wins if $\gamma^* = 1$.

   Else choose $w \leftarrow \mathbb{G}$, $a \leftarrow \mathbb{Z}_N^*$, and $e'_{i,b} \leftarrow \mathbb{Z}_N$. Let $W = (w, w^a, \ldots, w^{a^{n-1}})$, $E = ((e'_{1,0}, e'_{1,1}), \ldots, (e'_{n,0}, e'_{n,1}))$ and $K_{x^*} \leftarrow i\mathcal{O}(\lambda, \mathsf{PuncturedKeyAlt}'_{u,W,E,k,x^*})$.

   Let $h(x^*) = b_1^* \ldots b_n^*$. Choose $T \leftarrow \mathbb{G}$ and let $y_0 = (T)^{\prod_{j=1}^n e'_{j,b_j^*}}$ and $y_1 \leftarrow \mathbb{G}$.
4. Flip coin $\beta \leftarrow \{0,1\}$. Output (K_{x^*}, y_β).
5. $\mathcal{A}$ outputs β' and wins if $\beta = \beta'$.

Adversary's Advantage in These Games. Let $\mathsf{Adv}_{\mathcal{A}}^i$ denote the advantage of adversary $\mathcal{A}$ in Game i. We will now show that if an adversary $\mathcal{A}$ has non-negligible advantage in Game i, then $\mathcal{A}$ has non-negligible advantage in Game $i + 1$. Finally, we show that $\mathcal{A}$ has advantage 0 in Game 5.

Claim 1. *For any adversary $\mathcal{A}$, $\mathsf{Adv}_{\mathcal{A}}^1 \geq \mathsf{Adv}_{\mathcal{A}}^0/\theta(Q)$.*

Proof. This claim follows from the θ-admissibility of the hash function h. Recall h is θ-admissible if for all $x_1, \ldots, x_q, x^*, Pr[\ \forall i, P_u(x_i) = 1 \wedge P_u(x^*) = 0] \geq 1/\theta(Q)$, where the probability is only over the choice of $u \leftarrow \mathsf{AdmSample}(1^\lambda, Q)$. Therefore, if $\mathcal{A}$ wins with advantage ϵ in Game 0, then $\mathcal{A}$ wins with advantage at least $\epsilon/\theta(Q)$ in Game 1.

Claim 2. *Assuming $i\mathcal{O}$ is a secure indistinguishability obfuscator and the Subgroup Hiding Assumption holds, for any PPT adversary $\mathcal{A}$,*

$$\mathsf{Adv}_{\mathcal{A}}^1 - \mathsf{Adv}_{\mathcal{A}}^2 \leq negl(\lambda).$$

Clearly, the two programs in Game 1 and Game 2 are functionally different (they differ on 'challenge partition' inputs x where $P_u(x) = 0$), and therefore the proof of this claim involves multiple intermediate experiments. In the first hybrid experiment, we transform the program such that the program computes the output in a different manner, although the output is the same as in the original program. Next, the constants hardwired in the modified program are modified such that the output changes on all 'challenge partition' inputs (this step uses Assumption 2). Essentially, both programs use a different base for the challenge partition inputs. Next, using Subgroup Hiding Assumption and Chinese Remainder Theorem, even the exponents can be changed for the challenge partition, thereby ensuring that the original program and final program use different PRF keys for the challenge partition. The formal proof can be found in full version of this paper.

Claim 3. *For any PPT adversary $\mathcal{A}$, $\mathsf{Adv}_{\mathcal{A}}^3 = \mathsf{Adv}_{\mathcal{A}}^2$.*

Proof. Game 2 and Game 3 are identical, except for the manner in which the constants $e_{i,b}$ are chosen. In Game 2, $e_{i,b} \leftarrow \mathbb{Z}_N$, while in Game 3, the challenger first chooses $e'_{i,b} \leftarrow \mathbb{Z}_N$, $a \leftarrow \mathbb{Z}_N^*$, and sets $e_{i,b} = e'_{i,b} \cdot a$ if $h(x)_i = b$, else sets $e_{i,b} = e'_{i,b}$. Since $a \in \mathbb{Z}_N^*$ (and therefore is invertible), $e'_{i,b} \cdot a$ is also a uniformly random element in $\mathbb{Z}_N$ if $e'_{i,b}$ is. Hence the two experiments are identical.

Claim 4. *If there exists a PPT adversary $\mathcal{A}$ such that $\mathsf{Adv}_{\mathcal{A}}^3 - \mathsf{Adv}_{\mathcal{A}}^4$ is non-negligible in λ, then there exists a PPT distinguisher $\mathcal{B}$ that breaks the security of $i\mathcal{O}$ with advantage non-negligible in λ.*

Proof. Suppose there exists a PPT adversary $\mathcal{A}$ such that $\mathsf{Adv}_{\mathcal{A}}^3 - \mathsf{Adv}_{\mathcal{A}}^4 = \epsilon$. We will construct a PPT algorithm $\mathcal{B}$ that breaks the security of $i\mathcal{O}$ with advantage ϵ by interacting with $\mathcal{A}$. $\mathcal{B}$ first sets up the parameters, including u and k, and answers the evaluation queries of $\mathcal{A}$ exactly as in steps 1 and 2 of Game 3, which are identical to steps 1 and 2 of Game 4. When $\mathcal{A}$ sends $\mathcal{B}$ a challenge input x^*, $\mathcal{B}$ checks that $P_u(x^*) = 0$ and if not aborts (identical in both games).

Next $\mathcal{B}$ chooses further values to construct the circuits: $w \leftarrow \mathbb{G}$, $a \leftarrow \mathbb{Z}_N^*$ and $e'_{i,b} \leftarrow \mathbb{Z}_N$. Let $e_{i,b} = e'_{i,b} \cdot a$ if $h(x^*)_i = b$, else $e_{i,b} = e'_{i,b}$. Let $k' = (w, ((e_{1,0}, e_{1,1}), \dots, (e_{n,0}, e_{n,1})))$, $W = (w, w^a, \dots, w^{a^{n-1}})$ and $E = ((e'_{1,0}, e'_{1,1}), \dots, (e'_{n,0}, e'_{n,1}))$.

$\mathcal{B}$ constructs $C_0 = \mathsf{PuncturedKeyAlt}_{u,k,k',x^*}$, $C_1 = \mathsf{PuncturedKeyAlt}'_{u,W,E,k,x^*}$, and sends C_0, C_1 to the $i\mathcal{O}$ challenger. $\mathcal{B}$ receives $K_{x^*} \leftarrow i\mathcal{O}(C_b)$ from the challenger. It computes $h(x^*) = b_1^* \dots b_n^*$, $y_0 = w^{\Pi_j e_{j,b_j^*}}$, $y \leftarrow \mathbb{G}$, $\beta \leftarrow \{0,1\}$, sends (K_{x^*}, y_β) to $\mathcal{A}$ and receives β' in response. If $\beta = \beta'$, $\mathcal{B}$ outputs 0, else it outputs 1.

We will now prove that the circuits C_0 and C_1 have identical functionality. Consider any ℓ bit string x, and let $h(x) = b_1 \dots b_n$. Recall $t_{x^*}(x) = |\{i : b_i = b_i^*\}|$.

For any $x \in \{0,1\}^\ell$ such that $x = x^*$, both circuits output $\bot$.

For any $x \in \{0,1\}^\ell$ such that $x \neq x^*$ and $P_u(x) = 1$, both circuits output $v^{\Pi_{j=1}^n d_{j,b_j}}$.

For any $x \in \{0,1\}^\ell$ such that $x \neq x^*$ and $P_u(x) = 0$, we have

$$C_0(x) = \mathsf{PuncturedKeyAlt}_{u,k,k',x^*}(x) = w^{\Pi_{j=1}^n e_{j,b_j}} = w^{a^{t_{x^*}(x)} \Pi_{j=1}^n e'_{j,b_j}} =$$

$$\left(w_{t_{x^*}(x)}\right)^{\Pi_{j=1}^n e'_{j,b_j}} = \mathsf{PuncturedKeyAlt}'_{u,W,E,k,x^*}(x) = C_1(x).$$

As C_0 and C_1 have identical functionality, $Pr[\mathcal{B}\text{ wins }] = Pr[\mathcal{A}\text{ wins in Game 3}] - Pr[\mathcal{A}\text{ wins in Game 4}]$. If $\mathsf{Adv}_{\mathcal{A}}^3 - \mathsf{Adv}_{\mathcal{A}}^4 = \epsilon$, then $\mathcal{B}$ wins the $i\mathcal{O}$ security game with advantage ϵ.

Claim 5. *If there exists a PPT adversary $\mathcal{A}$ such that $\mathsf{Adv}_{\mathcal{A}}^4 - \mathsf{Adv}_{\mathcal{A}}^5$ is non-negligible in λ, then there exists a PPT adversary $\mathcal{B}$ that breaks Assumption 2 with advantage non-negligible in λ.*

Proof. Suppose there exists an adversary $\mathcal{A}$ such that $\mathsf{Adv}_{\mathcal{A}}^4 - \mathsf{Adv}_{\mathcal{A}}^5 = \epsilon$, then we can build an adversary that breaks Assumption 2 with advantage ϵ. The games are identical except that Game 5 replaces the term w^{a^n} with a random element of $\mathbb{G}$. On input an Assumption 2 instance $(N, \mathbb{G}, \mathbb{G}_p, \mathbb{G}_q, g_1, g_2, w, w^a, \dots, w^{a^{n-1}})$ together with challenge value T (which is either w^{a^n} or a random element in $\mathbb{G}$), use these parameters as in Game 5 with $\mathcal{A}$. If $\mathcal{A}$ guesses it was in Game 4, guess that $T = w^{a^n}$, else guess that T was random.

Observation 1. *For any adversary $\mathcal{A}$, $\mathsf{Adv}_{\mathcal{A}}^5 = 0$.*

Proof. If the challenger aborts either during the evaluation or challenge phase, then $\mathcal{A}$ has 0 advantage, since $\mathcal{A}$ wins with probability $1/2$. If the challenger does not abort during both these phases, then $\mathcal{A}$ receives (K_{x^*}, y_β), and $\mathcal{A}$ must guess β. However, both y_0 and y_1 are uniformly random elements in $\mathbb{G}$, and therefore, $\mathsf{Adv}_{\mathcal{A}}^5 = 0$.

Conclusion of the Main Proof. Given Claims 1–5 and Observation 1, we can conclude that if $i\mathcal{O}$ is a secure indistinguishability obfuscator and Assumption 1 holds (in the full version of this paper, we show that Assumption 1 implies Assumption 2), then any PPT adversary $\mathcal{A}$ has negligible advantage in the puncturable PRF security game (i.e., Game 0).

5 t-Puncturable PRFs

Let $t(\cdot)$ be a polynomial. A PRF $F_t : \mathcal{K} \times \mathcal{X} \to \mathcal{Y}$ is a t-puncturable pseudorandom function if there is an additional key space $\mathcal{K}_p$ and three polynomial time algorithms $F_t.\mathsf{setup}$, $F_t.\mathsf{eval}$ and $F_t.\mathsf{puncture}$ defined as follows.

- $F_t.\mathsf{setup}(1^\lambda)$ is a randomized algorithm that takes the security parameter λ as input and outputs a description of the key space $\mathcal{K}$, the punctured key space $\mathcal{K}_p$ and the PRF F_t.
- $F_t.\mathsf{puncture}(k, S)$ is a randomized algorithm that takes as input a PRF key $k \in \mathcal{K}$ and $S \subset \mathcal{X}$, $|S| \le t(\lambda)$, and outputs a t-punctured key $K_S \in \mathcal{K}_p$.
- $F_t.\mathsf{eval}(k_S, x')$ is a deterministic algorithm that takes as input a t-punctured key $k_S \in \mathcal{K}_p$ and $x' \in \mathcal{X}$. Let $k \in \mathcal{K}$, $S \subset \mathcal{X}$ and $k_S \leftarrow F_t.\mathsf{puncture}(k, S)$. For correctness, we need the following property:

$$F_t.\mathsf{eval}(k_S, x') = \begin{cases} F_t(k, x') & \text{if } x' \notin S \\ \bot & \text{otherwise} \end{cases}$$

The security game between the challenger and adversary is similar to the security game for puncturable PRFs. However, in this case, the adversary is allowed to make multiple challenge queries (as in the security game for constrained PRFs). The game consists of the following three phases.

Setup Phase. The challenger chooses a random key $k \leftarrow \mathcal{K}$ and $b \leftarrow \{0,1\}$.

Query Phase. In this phase, A is allowed to ask for the following queries:

- **Evaluation Query** A sends $x \in \mathcal{X}$, and receives $F_t(k, x)$.
- **Key Query** A sends a set $S \subset \mathcal{X}$, and receives $F_t.\mathsf{puncture}(k, S)$.
- **Challenge Query** A sends $x \in \mathcal{X}$ as a challenge query. If $b = 0$, the challenger outputs $F_t(k, x)$. Else, the challenger outputs a random element $y \leftarrow \mathcal{Y}$.

Guess A outputs a guess b' of b.

Let $x_1, \ldots, x_{q_1} \in \mathcal{X}$ be the evaluation queries, $S_1, \ldots, S_{q_2} \subset \mathcal{X}$ be the t-punctured key queries and $x_1^*, \ldots, x_s^*$ be the challenge queries. A wins if $\forall i \leq q_1, j \leq s$, $x_i \neq x_j^*$, $\forall i \leq q_2, j \leq s$, $x_j^* \in S_i$ and $b' = b$. The advantage of A is defined to be $\mathsf{Adv}_A^{F_t}(\lambda) = Pr[A \text{ wins}]$.

Definition 5. *The PRF F_t is a secure t-puncturable PRF if for all PPT adversaries A $\mathsf{Adv}_A^{F_t}(\lambda)$ is negligible in λ.*

5.1   Construction

In this section, we present our construction of t-puncturable PRFs from puncturable PRFs and indistinguishability obfuscation. Let $F : \mathcal{K} \times \mathcal{X} \rightarrow \mathcal{Y}$ be a puncturable PRF, and F.setup, F.puncture, F.eval the corresponding setup, puncturing and evaluation algorithms. We now describe our t-puncturable PRF F_t, and the corresponding algorithms F_t.setup, F_t.puncture and F_t.eval.

$\mathbf{F_t}$.setup$(\mathbf{1}^\lambda)$ F_t.setup is the same as F.setup.

$\mathbf{F_t}$.puncture$(\mathbf{k}, \mathbf{S})$ F_t.puncture(k, S) computes an obfuscation of the program PuncturedKey$^t{}_{k,S}$ defined in Fig. 4; that is, $K_S \leftarrow i\mathcal{O}(\lambda, \mathsf{PuncturedKey}^t{}_{k,S})$. As before, the program PuncturedKey$^t{}_{k,S}$ is padded to be of appropriate size.

PuncturedKey$^t{}_{k,S}$

Input: $x \in \mathcal{X}$

Constants : The function description $F, k \in \mathcal{K}, S \subset \mathcal{X}$ such that $|S| \leq t$

    if $x \in S$ **then**
        Output $\perp$.
    else
        Output $F(k, x)$.
    end if

Fig. 4. Program PuncturedKey^t

$\mathbf{F_t}$.eval$(\mathbf{K_S}, \mathbf{x})$ The punctured key K_S is a program that takes an input in $\mathcal{X}$. We define
$$F_t.\mathsf{eval}(K_S, x) = K_S(x).$$

5.2   Proof of Security

We will now prove that the above construction is a secure t-puncturable PRF as defined in Definition 5.

Theorem 3. *Assuming $i\mathcal{O}$ is a secure indistinguishability obfuscator and F, together with F.setup, F.puncture and F.eval is a secure puncturable PRF, the PRF F_t defined above, together with F_t.setup, F_t.puncture and F_t.eval, is a secure t-puncturable PRF.*

For simplicity, we will assume that the adversary makes q_1 evaluation queries, q_2 punctured key queries and 1 challenge query. As shown by [4], this can easily be extended to the general case of multiple challenge queries via a hybrid argument. We will first define the intermediate hybrid experiments.

Game 0. This game is the original security game between the challenger and adversary $\mathcal{A}$, where the challenger first chooses a PRF key, then $\mathcal{A}$ makes evaluation/t-punctured key queries and finally sends the challenge input. The challenger responds with either the PRF evaluation at challenge input, or sends a random element of the range space.

1. Choose a key $k \leftarrow \mathcal{K}$.
2. $\mathcal{A}$ makes evaluation/t-punctured key queries.
   (a) If $\mathcal{A}$ sends an evaluation query x_i, then output $F(k, x_i)$.
   (b) If $\mathcal{A}$ sends a t-punctured key query for set S_j, output the key $K_{S_j} \leftarrow i\mathcal{O}(\mathsf{PuncturedKey}^t{}_{k,S_j})$.
3. $\mathcal{A}$ sends challenge query x^* such that $x^* \neq x_i \; \forall i \leq q_1$ and $x^* \in S_j \; \forall j \leq q_2$. Choose $\beta \leftarrow \{0,1\}$. If $\beta = 0$, output $y = F(k, x^*)$, else output $y \leftarrow \mathcal{Y}$.
4. $\mathcal{A}$ sends β' and wins if $\beta = \beta'$.

Game 1. This game is the same as the previous one, except that the challenger introduces an abort condition. When the first t-punctured key query S_1 is made, the challenger guesses the challenge query $\tilde{x} \leftarrow S_1$. The challenger aborts if any of the evaluation queries are $\tilde{x}$, if any of the future t-punctured key queries does not contain $\tilde{x}$ or if the challenge query $x^* \neq \tilde{x}$.

1. Choose a key $k \leftarrow \mathcal{K}$.
2. $\mathcal{A}$ makes evaluation/t-punctured key queries.
   Let S_1 be the first t-punctured key query. Choose $\tilde{x} \leftarrow S_1$ and output key $K_{S_1} \leftarrow i\mathcal{O}(\lambda, \mathsf{PuncturedKey}^t{}_{k,S_1})$. For all evaluation queries x_i before S_1, output $F(k, x_i)$.
   For all queries after S_1, do the following.
   (a) If $\mathcal{A}$ sends an evaluation query x_i and $x_i = \tilde{x}$, abort.
       Choose $\gamma_i^1 \leftarrow \{0,1\}$. $\mathcal{A}$ wins if $\gamma_i^1 = 1$.
       Else if $x_i \neq \tilde{x}$, output $F(k, x_i)$.
   (b) If $\mathcal{A}$ sends a t-punctured key query for set S_j and $\tilde{x} \notin S_j$, abort.
       Choose $\gamma_i^2 \leftarrow \{0,1\}$. $\mathcal{A}$ wins if $\gamma_i^2 = 1$.
       Else if $\tilde{x} \in S_j$, output $K_{S_j} \leftarrow i\mathcal{O}(\lambda, \mathsf{PuncturedKey}^t{}_{k,S_j})$.
3. $\mathcal{A}$ sends challenge query x^* such that $x^* \neq x_i \; \forall i \leq q_1$ and $x^* \in S_j \; \forall j \leq q_2$. If $\tilde{x} \neq x^*$, abort. Choose $\gamma^* \leftarrow \{0,1\}$. $\mathcal{A}$ wins if $\gamma^* = 1$.
   Else if $\tilde{x} = x^*$, choose $\beta \leftarrow \{0,1\}$. If $\beta = 0$, output $y = F(k, x^*)$, else output $y \leftarrow \mathcal{Y}$.
4. $\mathcal{A}$ sends β' and wins if $\beta = \beta'$.

Next, we define q_2 games, Game 1_l, $1 \leq l \leq q_2$. Let Game $1_0 =$ Game 1.

Game 1_l. In this game, the first l punctured key queries use $K_{\tilde{x}}$, while the remaining use k.

1. Choose a key $k \leftarrow \mathcal{K}$.
2. $\mathcal{A}$ makes evaluation/t-punctured key queries.
   Let S_1 be the first t-punctured key query. Choose $\tilde{x} \leftarrow S_1$.
   Compute $K_{\tilde{x}} \leftarrow F.\mathsf{puncture}(k, \tilde{x})$.
   Output $K_{S_1} \leftarrow i\mathcal{O}(\lambda, \mathsf{PuncturedKeyAlt}^t{}_{K_{\tilde{x}}, S_1})$ (where $\mathsf{PuncturedKeyAlt}^t$ is defined in Fig. 5).
   For all evaluation queries x_i before S_1, output $F(k, x_i)$.
   For all queries after S_1, do the following.
   (a) If $\mathcal{A}$ sends an evaluation query x_i and $x_i = \tilde{x}$, abort. Choose $\gamma_i^1 \leftarrow \{0, 1\}$.
       $\mathcal{A}$ wins if $\gamma_i^1 = 1$.
       Else if $x_i \neq \tilde{x}$, output $F.\mathsf{eval}(K_{\tilde{x}}, x_i) = F(k, x_i)$.
   (b) If $\mathcal{A}$ sends a t-punctured key query for set S_j and $\tilde{x} \notin S_j$, abort. Choose
       $\gamma_i^2 \leftarrow \{0, 1\}$. $\mathcal{A}$ wins if $\gamma_i^2 = 1$.
       Else if $\tilde{x} \in S_j$ and $j \leq l$, output $K_{S_j} \leftarrow i\mathcal{O}(\lambda, \mathsf{PuncturedKeyAlt}^t{}_{K_{\tilde{x}}, S_j})$.
       Else output $K_{S_j} \leftarrow i\mathcal{O}(\lambda, \mathsf{PuncturedKey}^t{}_{k, S_j})$.
3. $\mathcal{A}$ sends challenge query x^* such that $x^* \neq x_i \; \forall i \leq q_1$ and $x^* \in S_j \; \forall j \leq q_2$.
   If $\tilde{x} \neq x^*$, abort. Choose $\gamma^* \leftarrow \{0, 1\}$. $\mathcal{A}$ wins if $\gamma^* = 1$.
   Else if $\tilde{x} = x^*$, choose $\beta \leftarrow \{0, 1\}$. If $\beta = 0$, output $y = F(k, x^*)$, else output
   $y \leftarrow \mathcal{Y}$.
4. $\mathcal{A}$ sends β' and wins if $\beta = \beta'$.

$\mathsf{PuncturedKeyAlt}^t{}_{K_{\tilde{x}}, S}$

Input: $x \in \mathcal{X}$

Constants : The function description $F, K_{\tilde{x}}, S \subset \mathcal{X}$ such that $|S| \leq t$

if $x \in S$ **then**
    Output $\perp$.
else
    Output $F.\mathsf{eval}(K_{\tilde{x}}, x)$.
end if

Fig. 5. Program $\mathsf{PuncturedKeyAlt}^t$

Game 2. In this game, the challenger outputs a random element as the response to the challenge query.

1. Choose a key $k \leftarrow \mathcal{K}$.
2. $\mathcal{A}$ makes evaluation/t-punctured key queries.
   Let S_1 be the first t-punctured key query. Choose $\tilde{x} \leftarrow S_1$ and compute
   $K_{\tilde{x}} \leftarrow F.\mathsf{puncture}(k, \tilde{x})$.
   Output $K_{S_1} \leftarrow i\mathcal{O}(\lambda, \mathsf{PuncturedKeyAlt}^t{}_{K_{\tilde{x}}, S_1})$.
   For all evaluation queries x_i before S_1, output $F(k, x_i)$.
   For all queries after S_1, do the following.

   (a) If $\mathcal{A}$ sends an evaluation query x_i and $x_i = \tilde{x}$, abort. Choose $\gamma_i^1 \leftarrow \{0, 1\}$.
      $\mathcal{A}$ wins if $\gamma_i^1 = 1$.
      Else if $x_i \neq \tilde{x}$, output $F.\mathsf{eval}(K_{\tilde{x}}, x_i) = F(k, x_i)$.
   (b) If $\mathcal{A}$ sends a t-punctured key query for set S_j and $\tilde{x} \notin S_j$, abort. Choose
      $\gamma_i^2 \leftarrow \{0, 1\}$. $\mathcal{A}$ wins if $\gamma_i^2 = 1$.
      Else if $\tilde{x} \in S_j$, output $K_{S_j} \leftarrow i\mathcal{O}(\lambda, \mathsf{PuncturedKeyAlt}^t{}_{K_{\tilde{x}}, S_j})$.
3. $\mathcal{A}$ sends challenge query x^* such that $x^* \neq x_i \;\forall i \leq q_1$ and $x^* \in S_j \;\forall j \leq q_2$.
   If $\tilde{x} \neq x^*$, abort. Choose $\gamma^* \leftarrow \{0, 1\}$. $\mathcal{A}$ wins if $\gamma^* = 1$.
   Else if $\tilde{x} = x^*$, choose $\beta \leftarrow \{0, 1\}$ and <u>output $y \leftarrow \mathcal{Y}$</u>.
4. $\mathcal{A}$ sends β' and wins if $\beta = \beta'$.

Adversary's Advantage in These Games. Let $\mathsf{Adv}_{\mathcal{A}}^i$ denote the advantage of adversary $\mathcal{A}$ in Game i.

Observation 2. *For any adversary $\mathcal{A}$, $\mathsf{Adv}_{\mathcal{A}}^1 \geq \mathsf{Adv}_{\mathcal{A}}^0 / t$.*

Proof. Since one of the elements of S_1 will be the challenge input, and $|S_1| \leq t$, the challenger's guess is correct with probability $1/|S_1| \geq 1/t$. Hence, $\mathsf{Adv}_{\mathcal{A}}^1 \geq \mathsf{Adv}_{\mathcal{A}}^0 / t$.

We will now show that Game 1_l and Game 1_{l+1} are computationally indistinguishable, assuming $i\mathcal{O}$ is secure.

Claim 6. *If there exists a PPT adversary $\mathcal{A}$ such that $\mathsf{Adv}_{\mathcal{A}}^{1_l} - \mathsf{Adv}_{\mathcal{A}}^{1_{l+1}}$ is non-negligible in λ, then there exists a PPT distinguisher $\mathcal{B}$ that breaks the security of $i\mathcal{O}$ with advantage non-negligible in λ.*

Proof. Note that the only difference between Game 1_l and Game 1_{l+1} is in the response to the $(l+1)th$ t-punctured key query. In Game 1_l, $\mathsf{PuncturedKey}^t{}_{k, S_{l+1}}$ is used to compute $K_{S_{l+1}}$, while in Game 1_{l+1}, $\mathsf{PuncturedKeyAlt}^t{}_{K, S_{l+1}}$ is used. Suppose there exists a PPT adversary $\mathcal{A}$ such that $\mathsf{Adv}_{\mathcal{A}}^{1_l} - \mathsf{Adv}_{\mathcal{A}}^{1_{l+1}} = \epsilon$. We will construct a PPT algorithm $\mathcal{B}$ that interacts with $\mathcal{A}$ and breaks the security of $i\mathcal{O}$ with advantage ϵ.

$\mathcal{B}$ chooses $k \leftarrow \mathcal{K}$ and for all evaluation queries x_i before the first t-punctured key query, outputs $F(k, x_i)$. On receiving the first t-punctured key query S_1, $\mathcal{B}$ chooses $\tilde{x} \leftarrow S_1$ and computes $K_{\tilde{x}} \leftarrow F.\mathsf{puncture}(k, \tilde{x})$. The evaluation queries are computed as in Game 1_l and 1_{l+1}. The first l t-punctured key queries are constructed using k, while the last $q_2 - l - 1$ t-punctured keys are constructed using $K_{\tilde{x}}$ (as in Game 1_l and Game 1_{l+1}). For the $(l+1)th$ query, $\mathcal{B}$ does the following. $\mathcal{B}$ sets $C_0 = \mathsf{PuncturedKey}^t{}_{k, S_{l+1}}$ and $C_1 = \mathsf{PuncturedKeyAlt}^t{}_{K_{\tilde{x}}, S_{l+1}}$, and sends C_0, C_1 to the $i\mathcal{O}$ challenger, and receives $K_{S_{l+1}}$ in response, which it sends to $\mathcal{A}$.

Finally, after all queries, the challenger sends the challenge query x^*. $\mathcal{B}$ checks that $\tilde{x} = x^*$, sets $y_0 = F(k, x^*)$ and chooses $y_1 \leftarrow \mathcal{Y}, \beta \leftarrow \{0, 1\}$. It outputs y_β and receives β' in response. If $\beta = \beta'$, $\mathcal{B}$ outputs 0, else it outputs 1.

From the correctness property of puncturable PRFs, it follows that $F.\mathsf{eval}$ $(K_{\tilde{x}}, x) = F(k, x)$ for all $x \notin S_{l+1}$. Hence, the circuits C_0 and C_1 are functionally identical. This completes our proof.

Next, we show that Game 1_{q_2} and Game 2 are computationally indistinguishable.

Claim 7. *If there exists a PPT adversary $\mathcal{A}$ such that $\mathsf{Adv}_{\mathcal{A}}^{1_{q_2}} - \mathsf{Adv}_{\mathcal{A}}^{2}$ is non-negligible in λ, then there exists a PPT distinguisher $\mathcal{B}$ that breaks the security of puncturable PRF F with advantage non-negligible in λ.*

Proof. We will use $\mathcal{A}$ to construct a PPT algorithm $\mathcal{B}$ that breaks the security of puncturable PRF F with advantage $\mathsf{Adv}_{\mathcal{A}}^{1_{q_2}} - \mathsf{Adv}_{\mathcal{A}}^{2}$. Observe that in Game 1_{q_2}, the challenger requires the master key k only for the evaluation queries before the first t-punctured key query. After the first t-punctured key query S_1, the challenger chooses $\tilde{x} \leftarrow S_1$, computes a punctured key $K_{\tilde{x}}$, and uses this to compute all future evaluation queries and t-punctured keys.

$\mathcal{B}$ begins interacting with $\mathcal{A}$. For each evaluation query x_i before the first t-punctured key query, $\mathcal{B}$ sends x_i to the puncturable PRF challenger, and receives y_i, which it forwards to $\mathcal{A}$. On receiving the first t-punctured key query S_1, $\mathcal{B}$ chooses $\tilde{x} \leftarrow S_1$ and sends $\tilde{x}$ as challenge input to the puncturable PRF challenger. $\mathcal{B}$ receives $K_{\tilde{x}}$ and y. It uses $K_{\tilde{x}}$ for all remaining queries. On receiving challenge x^* from $\mathcal{A}$, $\mathcal{B}$ checks $x^* = \tilde{x}$ and sends y. $\mathcal{B}$ sends $\mathcal{A}$'s response to the PRF challenger.

Note that until the challenge query is made, both games are identical and $\mathcal{B}$ simulates them perfectly. If y is truly random, then $\mathcal{A}$ receives a response as per Game 2, else it receives a response as per Game 1_{q_2}.

Finally, we have the following simple observation.

Observation 3. *For any adversary, $\mathsf{Adv}_{\mathcal{A}}^{3} = 0$.*

From the above claims and observations, we can conclude that if $i\mathcal{O}$ is a secure indistinguishability obfuscator as per Definition 2, and F, together with $F.\mathsf{setup}$, $F.\mathsf{puncture}$, $F.\mathsf{eval}$ is a secure puncturable PRF as per Definition 4, then any PPT adversary $\mathcal{A}$ has negligible advantage in Game 0.

6   Conclusion

Puncturable and t-puncturable PRFs have numerous cryptographic applications. This work provides the first constructions and proofs of adaptive security in the standard model. This is an interesting step forward in its own right, and we believe the techniques used to achieve adaptiveness from indistinguishability obfuscation may be useful elsewhere. Moreover, this work resolves for at least the puncturable PRF space, the larger question of characterizing which classes of functions admit an adaptively-secure constrained PRF in the standard model. As noted earlier, the results of [11] and [16] provide intuition both for and against whether this is indeed possible for many other function families.

References

1. Agrawal, S., Boneh, D., Boyen, X.: Efficient lattice (H)IBE in the standard model. In: Gilbert, H. (ed.) EUROCRYPT 2010. LNCS, vol. 6110, pp. 553–572. Springer, Heidelberg (2010)
2. Bitansky, N., Canetti, R., Paneth, O., Rosen, A.: Indistinguishability obfuscation vs. auxiliary-input extractable functions: One must fall. Cryptology ePrint Archive, Report 2013/641 (2013). http://eprint.iacr.org/
3. Boneh, D., Boyen, X.: Secure identity based encryption without random oracles. In: Franklin, M. (ed.) CRYPTO 2004. LNCS, vol. 3152, pp. 443–459. Springer, Heidelberg (2004)
4. Boneh, D., Waters, B.: Constrained pseudorandom functions and their applications. In: Sako, K., Sarkar, P. (eds.) ASIACRYPT 2013, Part II. LNCS, vol. 8270, pp. 280–300. Springer, Heidelberg (2013)
5. Boneh, D., Zhandry, M.: Multiparty key exchange, efficient traitor tracing, and more from indistinguishability obfuscation. In: Garay, J.A., Gennaro, R. (eds.) CRYPTO 2014, Part I. LNCS, vol. 8616, pp. 480–499. Springer, Heidelberg (2014)
6. Boyle, E., Goldwasser, S., Ivan, I.: Functional signatures and pseudorandom functions. In: Krawczyk, H. (ed.) PKC 2014. LNCS, vol. 8383, pp. 501–519. Springer, Heidelberg (2014)
7. Chandran, N., Raghuraman, S., Vinayagamurthy, D.: Constrained pseudorandom functions: Verifiable and delegatable. Cryptology ePrint Archive, Report 2014/522 (2014). http://eprint.iacr.org/
8. Chase, M., Meiklejohn, S.: Déjà Q: using dual systems to revisit q-type assumptions. In: Nguyen, P.Q., Oswald, E. (eds.) EUROCRYPT 2014. LNCS, vol. 8441, pp. 622–639. Springer, Heidelberg (2014)
9. Freire, E.S.V., Hofheinz, D., Paterson, K.G., Striecks, C.: Programmable hash functions in the multilinear setting. In: Canetti, R., Garay, J.A. (eds.) CRYPTO 2013, Part I. LNCS, vol. 8042, pp. 513–530. Springer, Heidelberg (2013)
10. Fuchsbauer, G.: Constrained verifiable random functions. In: Abdalla, M., De Prisco, R. (eds.) SCN 2014. LNCS, vol. 8642, pp. 95–114. Springer, Heidelberg (2014)
11. Fuchsbauer, G., Konstantinov, M., Pietrzak, K., Rao, V.: Adaptive security of constrained PRFs. In: Sarkar, P., Iwata, T. (eds.) ASIACRYPT 2014, Part II. LNCS, vol. 8874, pp. 82–101. Springer, Heidelberg (2014)
12. Garg, S., Gentry, C., Halevi, S., Raykova, M., Sahai, A., Waters, B.: Candidate indistinguishability obfuscation and functional encryption for all circuits. In: FOCS (2013)
13. Gentry, C., Lewko, A., Sahai, A., Waters, B.: Indistinguishability obfuscation from the multilinear subgroup elimination assumption. Cryptology ePrint Archive, Report 2014/309 (2014). http://eprint.iacr.org/
14. Goldreich, O., Goldwasser, S., Micali, S.: How to construct random functions (extended abstract). In: FOCS, pp. 464–479 (1984)
15. Golle, P., Jarecki, S., Mironov, I.: Cryptographic primitives enforcing communication and storage complexity. In: Financial Cryptography, pp. 120–135 (2002)
16. Hofheinz, D.: Fully secure constrained pseudorandom functions using random oracles. IACR Cryptology ePrint Archive, Report 2014/372 (2014)
17. Hofheinz, D., Kiltz, E.: Programmable hash functions and their applications. J. Cryptology $25(3)$, 484–527 (2012)

18. Hohenberger, S., Sahai, A., Waters, B.: Full domain hash from (Leveled) multilinear maps and identity-based aggregate signatures. In: Canetti, R., Garay, J.A. (eds.) CRYPTO 2013, Part I. LNCS, vol. 8042, pp. 494–512. Springer, Heidelberg (2013)
19. Hohenberger, S., Sahai, A., Waters, B.: Replacing a random oracle: full domain hash from indistinguishability obfuscation. In: Nguyen, P.Q., Oswald, E. (eds.) EUROCRYPT 2014. LNCS, vol. 8441, pp. 201–220. Springer, Heidelberg (2014)
20. Hohenberger, S., Waters, B.: Constructing verifiable random functions with large input spaces. In: Gilbert, H. (ed.) EUROCRYPT 2010. LNCS, vol. 6110, pp. 656–672. Springer, Heidelberg (2010)
21. Kiayias, A., Papadopoulos, S., Triandopoulos, N., Zacharias, T.: Delegatable pseudorandom functions and applications. In: ACM Conference on Computer and Communications Security, pp. 669–684 (2013)
22. Lewko, A., Waters, B.: New proof methods for attribute-based encryption: achieving full security through selective techniques. In: Safavi-Naini, R., Canetti, R. (eds.) CRYPTO 2012. LNCS, vol. 7417, pp. 180–198. Springer, Heidelberg (2012)
23. Naor, M., Reingold, O.: Number-theoretic constructions of efficient pseudo-random functions. J. ACM 51(2), 231–262 (2004)
24. Sahai, A., Waters, B.: How to use indistinguishability obfuscation: deniable encryption, and more. In: STOC, pp. 475–484 (2014)
25. Waters, B.: Efficient identity-based encryption without random oracles. In: Cramer, R. (ed.) EUROCRYPT 2005. LNCS, vol. 3494, pp. 114–127. Springer, Heidelberg (2005)

Multilinear and Aggregate Pseudorandom Functions: New Constructions and Improved Security

Michel Abdalla[✉], Fabrice Benhamouda, and Alain Passelègue

ENS, CNRS, INRIA, and PSL, Paris, France
{michel.abdalla,fabrice.ben.hamouda,alain.passelegue}@ens.fr
http://www.di.ens.fr/users/mabdalla
http://www.di.ens.fr/users/fbenhamo
http://www.di.ens.fr/users/passeleg

Abstract. Since its introduction, pseudorandom functions (PRFs) have become one of the main building blocks of cryptographic protocols. In this work, we revisit two recent extensions of standard PRFs, namely multilinear and aggregate PRFs, and provide several new results for these primitives. In the case of aggregate PRFs, one of our main results is a proof of security for the Naor-Reingold PRF with respect to read-once boolean aggregate queries under the standard Decision Diffie-Hellman problem, which was an open problem. In the case of multilinear PRFs, one of our main contributions is the construction of new multilinear PRFs achieving indistinguishability from random symmetric and skew-symmetric multilinear functions, which was also left as an open problem. In order to achieve these results, our main technical tool is a simple and natural generalization of the recent linear independent polynomial framework for PRFs proposed by Abdalla, Benhamouda, and Passelègue in Crypto 2015, that can handle larger classes of PRF constructions. In addition to simplifying and unifying proofs for multilinear and aggregate PRFs, our new framework also yields new constructions which are secure under weaker assumptions, such as the decisional k-linear assumption.

Keywords: Pseudorandom functions · Multilinear PRFs · Aggregate PRFs

1 Introduction

Pseudorandom functions (PRFs) are one of the most fundamental primitives in cryptography. One of the features that makes PRFs so useful is the fact that they behave as truly random functions with respect to computationally bounded adversaries. Since being introduced by Goldreich, Goldwasser, and Micali [15], PRFs have been used in many cryptographic applications, varying from symmetric encryption and authentication schemes to key exchange. In particular, they are very useful for modeling the security of concrete block ciphers, such as AES [4].

© International Association for Cryptologic Research 2015
T. Iwata and J.H. Cheon (Eds.): ASIACRYPT 2015, Part I, LNCS 9452, pp. 103–120, 2015.
DOI: 10.1007/978-3-662-48797-6_5

Given the large applicability of pseudorandom functions, several extensions have been proposed in the literature over the years, with the goal of providing additional functionalities to these functions. One concrete example of such an extension are constrained PRFs [9,11,17], which provides the owner of the secret key with the capability of delegating the computation of the pseudorandom function for different subsets of the input domain, without compromising the pseudorandomness property for the other points of the input domain. In this paper, we focus on two recent extensions of pseudorandom functions, namely multilinear PRFs [13], and aggregate PRFs [12], and solve several open problems related to the construction of these primitives.

Aggregate Pseudorandom Functions. Aggregate pseudorandom functions were introduced by Cohen, Goldwasser, and Vaikuntanathan in [12]. The main interest of an aggregate PRF is to provide the user with the possibility of aggregating the values of the function over *super-polynomially* many PRF values with only a *polynomial-time* computation, without enabling a polynomial-time adversary to distinguish the function from a truly random function. For instance, one such example of an aggregate query could be to compute the product of all the output values of the PRF corresponding to a given exponentially-sized interval of the input domain.

In addition to proposing the notion of aggregate PRFs, Cohen, Goldwasser, and Vaikuntanathan [12] also proposed new constructions for several different classes of aggregate queries, such as decision trees, hypercubes, and read-once boolean formulas, achieving different levels of expressiveness. Unfortunately, for most of the constructions proposed in [12], the proofs of security suffer from an exponential (in the input length) overhead in their running time and have to rely on the sub-exponential hardness of the Decisional Diffie-Hellman (DDH) problem.

Indeed, to prove the security of their constructions, the authors use a generic result which is simply saying the following: given an adversary $\mathscr{A}$ against the AGG-PRF security of a PRF F, one can build an adversary $\mathscr{B}$ against the standard PRF security of F. $\mathscr{B}$ simply queries all the values required to compute the aggregate values (or the PRF values), and computes the aggregate values itself before sending them to $\mathscr{A}$.

Clearly, this reduction proves that any secure PRF is actually also a secure aggregate PRF. However, this reduction is *not efficient*, since to answer to just one aggregate query, the adversary $\mathscr{B}$ may have to query an exponential number of values to its oracle. Hence, as soon as we can aggregate in one query a superpolynomial number of PRF values, this generic reduction does not run in polynomial time.

Multilinear Pseudorandom Functions. In order to overcome the shortcomings of the work of Cohen, Goldwasser, and Vaikuntanathan [12], Cohen and Holmgren introduced the concept of multilinear pseudorandom functions in [13]. Informally speaking, a multilinear pseudorandom function is a variant of the standard notion of pseudorandom functions, which works with vector spaces and which guarantees indistinguishability from random multilinear functions with the same domain and range. As shown in [13], multilinear pseudorandom functions can be used to prove the AGG-PRF security of the Naor-Reingold (NR) PRF [18] with a polynomial time reduction for the case of hypercubes and decision trees aggregations. Unfortunately, their technique does not extend to the more general case of read-once formulas aggregation, which is the most expressive form of aggregation in [12].

Our Techniques. In this work, we provide an alternative way of overcoming the limitations of the work of Cohen, Goldwasser, and Vaikuntanathan [12], based on a natural extension of the recent algebraic framework for pseudorandom functions proposed by Abdalla, Benhamouda, and Passelègue in [1], known as the linear independent polynomial (LIP) framework.

In a nutshell, the LIP framework essentially says that for any linearly independent polynomials $P_1, \ldots, P_q \in \mathbb{Z}_p[T_1, \ldots, T_n]$, the group elements

$$[P_1(\vec{a}) \cdot b], \ldots, [P_q(\vec{a}) \cdot b],$$

with $\vec{a} \xleftarrow{\$} \mathbb{Z}_p^n$ and $b \xleftarrow{\$} \mathbb{Z}_p$, are computationally indistinguishable from independent random group elements in $\mathbb{G}$, under the DDH assumption (when polynomials are multilinear) or the d-DDHI assumption (where d is the maximum degree of $P_1, \ldots, P_q$ in any indeterminate T_i). As a toy example, the LIP framework directly proves the security of the NR PRF defined as:

$$\mathsf{NR}((b, \vec{a}), x) = \left[b \prod_{i=1}^{n} a_i^{x_i} \right],$$

where $(b, \vec{a} = (a_1, \ldots, a_n)) \in \mathcal{K} = \mathbb{Z}_p \times \mathbb{Z}_p^n$ and $x \in \mathcal{D} = \{0,1\}^n$. Indeed, all the polynomials $P_x = b \prod_{i=1}^{n} a_i^{x_i}$ are linearly independent.

Unfortunately, the LIP framework is not enough to prove the security of multilinear PRFs or aggregate PRFs, as the outputs of the function (and the corresponding polynomials) may not be independent. To overcome these limitations, we provide a natural extension of the LIP framework, which we call polynomial linear pseudorandomness security (PLP), that can handle such dependences. Despite being a simple extension, the new PLP framework yields significant improvements over previous works on multilinear and aggregate PRFs. In particular, the multilinear constructions in [13] can be seen as a special case of our new PLP framework.

Main Results. Using our new PLP framework for pseudorandom functions, we obtain the following results.

First, we prove the security of the aggregate PRF for read-once formulas proposed in [12], under the DDH assumption and with a polynomial-time reduction. This in turn implies the security of all the other aggregate PRFs in [12], as the latter are particular cases of the aggregate PRFs for read-once formulas. The proof is very simple and based on linear algebra. Up to now, the only known reduction incurred an exponential blow-up in the length n of the input.

Second, we show that our PLP framework enables to very easily prove the security of the multilinear pseudorandom function construction in [13]. More importantly, it enables us to directly show the security of the symmetric variant of this construction, under the d-DDHI assumption, which was left as an open problem in [13].

Third, we extend all the above constructions to weaker assumptions, as the k-Lin assumption, which can hold in symmetric k-linear groups, contrary to DDH or d-DDHI. Again, these extensions are straightforward to prove thanks to our PLP framework.

Additionally, we solve two other open problems respectively in [12, end of Sects. 1 and 2.2] and in [13]: We show that unless NP=BPP, there cannot exist aggregate PRFs for DNF formulas, although satisfiability of DNF formulas can be tested in polynomial time; and we propose the first skew-symmetric multilinear PRF.

Additional Contributions. As a side contribution, we prove the hardness of $\mathcal{E}_{k,d}$-MDDH (defined in [1] and recalled in Sect. 2) in the generic (symmetric) k-linear group model, which was left as an open problem in [1] for $k > 2$ and $d > 1$. This result directly implies that all the results stated in [1] under the $\mathcal{E}_{2,d}$-MDDH now holds also for $\mathcal{E}_{k,d}$-MDDH, for any $k \geq 2$, which is also an interesting side contribution. To prove this result, we essentially need to prove there are no non-trivial polynomial relations of degree k between the elements of the assumptions (these elements being themselves polynomials), as in [8,10,14]. The proof is by induction over k: for the base case $k = 1$, the proof is straightforward as all the elements we consider are linearly independent; for the inductive case $k = 2$, we basically set some indeterminates to some carefully chosen values (for the polynomials defining the elements we consider) to come down to previous cases.

Paper Organization. The rest of the paper is composed of the following sections. In Sect. 2 and the full version [3], we give necessary background and notations. We introduce our general PLP security notion and explain our main result, termed PLP theorem (Theorem 1), in Sect. 3. We then present our new constructions and improved security bounds for aggregate and multilinear pseudorandom functions in Sect. 4 as well as some side results. The proofs of these results are detailed in the full version [3]. Finally, in the full version [3], we prove the hardness of our main assumption (the $\mathcal{E}_{k,d}$-MDDH assumption) in the generic k-linear group model.

2   Definitions

Notations and Conventions. We denote by κ the security parameter. Let $F \colon \mathcal{K} \times \mathcal{D} \to \mathcal{R}$ be a function that takes a key $K \in \mathcal{K}$ and an input $x \in \mathcal{D}$

and returns an output $F(K, x) \in \mathcal{R}$. The set of all functions $F: \mathcal{K} \times \mathcal{D} \to \mathcal{R}$ is denoted by $\mathsf{Fun}(\mathcal{K}, \mathcal{D}, \mathcal{R})$. Likewise, $\mathsf{Fun}(\mathcal{D}, \mathcal{R})$ denotes the set of all functions mapping $\mathcal{D}$ to $\mathcal{R}$. Also, if $\mathcal{D}$ and $\mathcal{R}$ are vector spaces, we denote by $\mathsf{L}(\mathcal{D}, \mathcal{R})$ the vector space of linear functions from $\mathcal{D}$ to $\mathcal{R}$. In addition, if $\mathcal{D}_1, \ldots, \mathcal{D}_n$ are n vector spaces, then $\mathsf{L}(\mathcal{D}_1 \otimes \cdots \otimes \mathcal{D}_n, \mathcal{R})$ is the vector space of n-linear functions from $\mathcal{D}_1 \times \cdots \times \mathcal{D}_n$ to $\mathcal{R}$.

If S is a set, then $|S|$ denotes its size. We denote by $s \xleftarrow{\$} S$ the operation of picking at random s in S. If $\vec{x}$ is a vector then we denote by $|\vec{x}|$ its length, so $\vec{x} = (x_1, \ldots, x_{|\vec{x}|})$. For a binary string x, we denote its length by $|x|$ so $x \in \{0, 1\}^{|x|}$, x_i its i-th bit, so $x = x_1 \| \ldots \| x_{|x|}$. For a matrix $\boldsymbol{A}$ of size $k \times m$, we denote by $a_{i,j}$ the coefficient of $\boldsymbol{A}$ in the i-th row and the j-th column. We denote by $\mathbb{Z}_p[T_1, \ldots, T_n]$ the subspace of multivariate polynomials in indeterminates $T_1, \ldots, T_n$, and by $\mathbb{Z}_p[T_1, \ldots, T_n]_{\leq d}$ the subring of polynomials of degree at most d in each indeterminate. For a polynomial $P \in \mathbb{Z}_p[T_1, \ldots, T_n]$, we denote by $P(\vec{T})$ the polynomial $P(T_1, \ldots, T_n)$ and by $P(\vec{a})$ its evaluation by setting $\vec{T}$ to $\vec{a}$, meaning that we set $T_1 = a_1, \ldots, T_n = a_n$.

We often implicitly consider a multiplicative group $\mathbb{G} = \langle g \rangle$ with public generator g of order p and we denote by $[a]$ the element g^a, for any $a \in \mathbb{Z}_p$. Similarly, if $\boldsymbol{A}$ is a matrix in $\mathbb{Z}_p^{k \times m}$, $[\boldsymbol{A}]$ is a matrix $\boldsymbol{U} \in \mathbb{G}^{k \times m}$, such that $u_{i,j} = [a_{i,j}]$ for $i = 1, \ldots, k$ and $j = 1, \ldots, m$. All vector spaces are implicitly supposed to be $\mathbb{Z}_p$-vector spaces.

We denote by **TestLin** a procedure which takes as inputs a list $\mathcal{L}$ of polynomials $(R_1, \ldots, R_L)$ (such that $R_1, \ldots, R_L$ are linearly independent as polynomials) and a polynomial R and which outputs:

$$
\begin{cases}
\perp & \text{if } R \text{ is linearly independent of the set } \{R_1, \ldots, R_L\} \\
\vec{\lambda} = (\lambda_1, \ldots, \lambda_L) & \text{otherwise, so that } R = \lambda_1 R_1 + \ldots + \lambda_L R_L
\end{cases}
$$

$\vec{\lambda}$ is uniquely defined since we assume that polynomials from the input list are linearly independent. No such procedure is known for multivariate polynomials, if we require the procedure to be deterministic and polynomial-time. However, it is easy to construct such a randomized procedure which is correct with overwhelming probability. Such a statistical procedure is sufficient for our purpose and was given in [2]. We recall this procedure in Fig. 1. This procedure is correct with probability at least $\frac{p-1}{p}$ as soon as $nd \leq \sqrt{p}$, where d is the maximum degree in one indeterminate and n is the number of indeterminates.

Games [5]. Most of our definitions and proofs use the code-based game-playing framework, in which a game has an **Initialize** procedure, procedures to respond to adversary oracle queries, and a **Finalize** procedure. In the case where the **Finalize** procedure is not explicitly defined, it is implicitly defined as the procedure that simply outputs its input. To execute a game G with an adversary $\mathscr{A}$, we proceed as follows. First, **Initialize** is executed and its outputs become the input of $\mathscr{A}$. When $\mathscr{A}$ executes, its oracle queries are answered by the corresponding procedures of G. When $\mathscr{A}$ terminates, its outputs become the input of **Finalize**. The output of the latter, denoted $\mathrm{G}^{\mathscr{A}}$ is called the output of the

procTestLin$(\mathcal{L}, R)$
 // $\mathcal{L}[\ell] = R_\ell$ for $\ell = 1, \ldots, L$ and $L = |\mathcal{L}|$
 $R_{L+1} \leftarrow R$
 $N \leftarrow 2L + 4$
 For $k = 1, \ldots, N$
     $\vec{\gamma_k} \xleftarrow{\$} \mathbb{Z}_p^n$
 M matrix over $\mathbb{Z}_p$ of $L + 1$ rows and N columns
 For $\ell = 1, \ldots, L + 1$
     For $k = 1, \ldots, N$
         $m_{\ell,k} \leftarrow R_\ell(\vec{\gamma_k})$
 Apply Gaussian elimination on M
 If M is full-rank then
     Return $\perp$
 Else
     Let $\vec{\lambda'}$ be the row vector such that $\vec{\lambda'} \cdot M = \vec{0}$
     $\vec{\lambda} \leftarrow (\lambda'_1/\lambda'_{L+1}, \ldots, \lambda'_L/\lambda'_{L+1})$
     Return $\vec{\lambda}$

Fig. 1. TestLin procedure

game, and we let "$\mathrm{G}^{\mathscr{A}} \Rightarrow 1$" denote the event that this game output takes the value 1. The running time of an adversary by convention is the worst case time for the execution of the adversary with any of the games defining its security, so that the time of the called game procedures is included.

Pseudorandom Functions. A PRF is an efficiently computable ensemble of functions $F: \mathcal{K} \times \mathcal{D} \to \mathcal{R}$, implicitly indexed by the security parameter κ, such that, when $K \xleftarrow{\$} \mathcal{K}$, the function $x \in \mathcal{D} \mapsto F(K, x) \in \mathcal{R}$ is indistinguishable from a random function. Formally, we say that F is a pseudorandom function if the advantage of any adversary $\mathscr{A}$ in attacking the standard PRF security of F is negligible, where this advantage is defined via

$$\mathbf{Adv}_F^{\mathrm{prf}}(\mathscr{A}) = \Pr\left[\mathrm{PRFReal}_F^{\mathscr{A}} \Rightarrow 1\right] - \Pr\left[\mathrm{PRFRand}_F^{\mathscr{A}} \Rightarrow 1\right],$$

where games $\mathrm{PRFReal}_F$ and $\mathrm{PRFRand}_F$ are depicted in Fig. 2.

Aggregation Function. Let $f: \mathcal{K} \times \mathcal{D} \to \mathcal{R}$ be a function. We define an aggregation function by describing two objects:

− a collection $\mathscr{S}$ of subsets S of the domain $\mathcal{D}$;
− an aggregation function $\Gamma: \mathcal{R}^* \to \mathcal{V}$ that takes as input a tuple of values from the range $\mathcal{R}$ of F and aggregates them to produce a value in an output set $\mathcal{V}$.

In addition, we require the set ensemble $\mathscr{S}$ to be *efficiently recognizable*, meaning that for any $S \in \mathscr{S}$, there exists a polynomial time procedure to check if $x \in S$, for any $x \in \mathcal{D}$. Also, we require the aggregation function Γ to be polynomial time and the output of the function not to depend on the order of the elements provided as inputs. Finally, we require all sets S to have a representation of size polynomial in the security parameter κ.

Given an aggregation function $(\mathscr{S}, \Gamma)$, we define the aggregate function $\text{AGG} = \text{AGG}_{f,\mathscr{S},\Gamma}$ as the function that takes as input a set $S \in \mathscr{S}$ and outputs the aggregation of all values $f(x)$ for all $x \in S$. That is, $\text{AGG}(S)$ outputs $\Gamma(f(x_1), \ldots, f(x_{|S|}))$, where $S = \{x_1, \ldots, x_{|S|}\}$. We will require the computation of AGG to be polynomial time (even if the input set S is exponentially large) if the function f provided is the pseudorandom function $F(K, \cdot)$ we consider, where K is some key.

Aggregate Pseudorandom Functions. Let $F \colon \mathcal{K} \times \mathcal{D} \to \mathcal{R}$ be a pseudorandom function and let $(\mathscr{S}, \Gamma)$ be an associated aggregation function. We say that F is an $(\mathscr{S}, \Gamma)$-aggregate pseudorandom function $((\mathscr{S}, \Gamma)\text{-AGG-PRF})$ if the advantage of any adversary in attacking the AGG-PRF security of F is negligible, where this advantage is defined via

$$\mathbf{Adv}_{F,\mathscr{S},\Gamma}^{\text{agg-prf}}(\mathscr{A}) = \Pr\left[\text{AGGPRFReal}_F^{\mathscr{A}} \Rightarrow 1\right] - \Pr\left[\text{AGGPRFRand}_F^{\mathscr{A}} \Rightarrow 1\right],$$

where games AGGPRFReal_F and AGGPRFRand_F are depicted in Fig. 2. Game AGGPRFRand_F may not be polynomial-time, as $\text{AGG}_{f,\mathscr{S},\Gamma}$ may not require to compute an exponential number of values $f(x)$. However, for all the aggregate PRFs that we consider, this game is statistically indistinguishable from a polynomial-time game, using the **TestLin** procedure, similarly to what is done in our new PLP security notion (see Sect. 3 and Fig. 3).

Multilinear Pseudorandom Functions. Multilinear pseudorandom functions are a variant of the standard notion of pseudorandom functions, which works with vector spaces. More precisely, a multilinear pseudorandom function $F \colon \mathcal{K} \times \mathcal{D} \to \mathcal{R}$, is an efficiently computable function with key space $\mathcal{K}$, domain $\mathcal{D} = \mathcal{D}_1 \times \cdots \times \mathcal{D}_n$ (a cartesian product of n vector spaces $\mathcal{D}_1, \ldots, \mathcal{D}_n$, for some integer n), range $\mathcal{R}$ which is a vector space, and which is indistinguishable from a random n-linear function with same domain and range. We say that F is a multilinear pseudorandom function (MPRF) if the advantage of any adversary in attacking the MPRF security of F is negligible, where this advantage is defined via

$$\mathbf{Adv}_F^{\text{mprf}}(\mathscr{A}) = \Pr\left[\text{MPRFReal}_F^{\mathscr{A}} \Rightarrow 1\right] - \Pr\left[\text{MPRFRand}_F^{\mathscr{A}} \Rightarrow 1\right],$$

where games MPRFReal_F and MPRFRand_F are depicted in Fig. 2. As explained in [13], Game MPRFRand_F can be implemented in polynomial time using a deterministic algorithm checking linearity of simple tensors [6]. Also, similarly to Game AGGPRFRand_F, it is also possible to implement a polynomial-time game that is statistically indistinguishable from MPRFRand_F using **TestLin**.

PRFReal$_F$	PRFRand$_F$
proc Initialize	**proc Initialize**
$K \xleftarrow{\$} \mathcal{K}$	$f \xleftarrow{\$} \mathsf{Fun}(\mathcal{D},\mathcal{R})$
proc Fn(x)	**proc Fn**(x)
Return $F(K,x)$	Return $f(x)$

AGGPRFReal$_F$	AGGPRFRand$_F$
proc Initialize	**proc Initialize**
$K \xleftarrow{\$} \mathcal{K}$	$f \xleftarrow{\$} \mathsf{Fun}(\mathcal{D},\mathcal{R})$
proc Fn(x)	**proc Fn**(x)
Return $F(K,x)$	Return $f(x)$
proc AGG(S)	**proc AGG**(S)
Return $\mathsf{AGG}_{F(K,\cdot),\mathscr{S},\Gamma}(S)$	Return $\mathsf{AGG}_{f,\mathscr{S},\Gamma}(S)$

MPRFReal$_F$	MPRFRand$_F$
proc Initialize	**proc Initialize**
$K \xleftarrow{\$} \mathcal{K}$	$f \xleftarrow{\$} \mathsf{L}(\mathcal{D}_1 \otimes \cdots \otimes \mathcal{D}_n, \mathcal{R})$
proc Fn$(\vec{x})$	**proc Fn**$(\vec{x})$
Return $F(K,\vec{x})$	Return $f(\vec{x})$

Fig. 2. Security games for (classical, aggregate, multilinear — from top to bottom) pseudorandom functions

Assumptions. Our main theorem is proven under the same MDDH assumption [14] introduced in [1] and termed $\mathcal{E}_{k,d}$-MDDH assumption. This MDDH assumption is defined by the matrix distribution $\mathcal{E}_{k,d}$ which samples matrices Γ as follows

$$\Gamma = \begin{pmatrix} A^0 \cdot B \\ A^1 \cdot B \\ \vdots \\ A^d \cdot B \end{pmatrix} \in \mathbb{Z}_p^{k(d+1)\times k} \quad \text{with } A, B \xleftarrow{\$} \mathbb{Z}_p^{k\times k}. \tag{1}$$

The advantage of an adversary $\mathscr{D}$ against the $\mathcal{E}_{k,d}$-MDDH assumption is

$$\mathbf{Adv}_{\mathbb{G}}^{\mathcal{E}_{k,d}\text{-mddh}}(\mathscr{D}) = \Pr\left[\,\mathscr{D}(g,[\Gamma],[\Gamma \cdot W])\,\right] - \Pr\left[\,\mathscr{D}(g,[\Gamma],[U])\,\right],$$

where $\Gamma \xleftarrow{\$} \mathcal{E}_{k,d}$, $W \xleftarrow{\$} \mathbb{Z}_p^{k\times 1}$, $U \xleftarrow{\$} \mathbb{Z}_p^{k(d+1)\times 1}$. This assumption is random self-reducible, as any other MDDH assumption (we will make use of this property in the proof of our main theorem, and recall this property in the full version [3]).

In Table 1, we summarize security results for $\mathcal{E}_{k,d}$-MDDH. For $k = 1$ or $d = 1$, the $\mathcal{E}_{k,d}$-MDDH assumption is implied by standard assumptions (DDH, DDHI, or k-Lin, as recalled in the full version [3]). $\mathcal{E}_{1,1}$-MDDH is actually exactly DDH.

Table 1. Security of $\mathcal{E}_{k,d}$-MDDH

	$k = 1$	$k = 2$	$k \geq 3$
$d = 1$	$= \mathbf{Adv}_{\mathbb{G}}^{\text{ddh}}$	$\lesssim 2 \cdot \mathbf{Adv}_{\mathbb{G}}^{\mathcal{U}_2\text{-mddh}}$	$\lesssim k \cdot \mathbf{Adv}_{\mathbb{G}}^{\mathcal{U}_k\text{-mddh}}$
$d \geq 2$	$\lesssim d \cdot \mathbf{Adv}_{\mathbb{G}}^{d-\text{ddhi}\,a}$	Generic bilinear group[b]	Generic k-linear group[c]

$\mathbf{Adv}_{\mathbb{G}}^{\text{ddh}}$, $\mathbf{Adv}_{\mathbb{G}}^{d\text{-ddhi}}$ and $\mathbf{Adv}_{\mathbb{G}}^{\mathcal{U}_k\text{-mddh}}$ are advantages for DDH, DDHI, and
$\mathcal{U}_k$-MDDH. This later assumption is weaker than k-Lin;
[a]proven in [1];
[b]proven in the generic (symmetric) bilinear group model [7] in [1]
[c]proven in the generic (symmetric) k-linear group model [16,19] in the full
   version [3].

In [1], the question of the hardness of the $\mathcal{E}_{k,d}$-MDDH problem in the generic
k-linear group model was left as an open problem when $d > 1$ and $k > 2$. One
of our contributions is to give a proof of hardness of these assumptions, which
is detailed in the full version [3].

3   Polynomial Linear Pseudorandomness Security

As we already mentioned in the introduction, while the LIP theorem from [1] is
quite powerful to prove the security of numerous constructions of pseudorandom
functions (and related-key secure pseudorandom functions), it falls short when
we need to prove the security of multilinear pseudorandom functions or aggregate
pseudorandom functions. Indeed, the LIP theorem requires that there is no linear
dependence between the outputs of the function. Thus, for the latter primitives,
it is clear that one cannot use the LIP theorem, since the main point of these
primitives is precisely that outputs can be related.

In order to deal with these primitives, we introduce a new security notion,
termed polynomial linear pseudorandomness security (PLP), which encompasses
the LIP security notion, but allows to handle multilinear pseudorandom functions and aggregate pseudorandom functions.

3.1   Intuition

Intuitively, the polynomial linear pseudorandomness security notion says that
for any polynomials $P_1, \ldots, P_q \in \mathbb{Z}_p[T_1, \ldots, T_n]$, the group elements

$$[P_1(\vec{a}) \cdot b], \ldots, [P_q(\vec{a}) \cdot b],$$

with $\vec{a} \xleftarrow{\$} \mathbb{Z}_p^n$ and $b \xleftarrow{\$} \mathbb{Z}_p$, are computationally indistinguishable from the group
elements:

$$[U(P_1)], \ldots, [U(P_q)],$$

with $U \xleftarrow{\$} \mathsf{L}(\mathbb{Z}_p[T_1, \ldots, T_n]_{\leq d}, \mathbb{Z}_p)$ being a random linear function from the poly-
nomial vector space $\mathbb{Z}_p[T_1, \ldots, T_n]_{\leq d}$ (with d the maximum degree of $P_1, \ldots, P_q$

in any indeterminate T_i) to the base field $\mathbb{Z}_p$. Our main theorem (Theorem 1) shows that this security notion holds under the $\mathcal{E}_{1,d}$-MDDH assumption (and thus also under DDH for $d = 1$ and d-DDHI for $d \geq 2$).

When $P_1, \ldots, P_q$ are linearly independent, $[U(P_1)], \ldots, [U(P_q)]$ are independent random group elements in $\mathbb{G}$. In that sense, the polynomial linear pseudorandomness security notion is a generalization of the LIP security notion.

We remark that, in the generic group model, the polynomial linear pseudorandomness security notion holds trivially, by definition. The difficulty of the work is to prove it under classical assumptions such as the $\mathcal{E}_{1,d}$-MDDH assumption.

Polynomial-Time Games. When we want to formally define the polynomial linear pseudorandomness security notion, we quickly face a problem: how to compute $[U(P_i)]$ for a random linear map $U \xleftarrow{\$} \mathsf{L}(\mathbb{Z}_p[T_1, \ldots, T_n]_{\leq d}, \mathbb{Z}_p)$? Such a map can be represented by a (random) vector with $(d + 1)^n$ entries. But doing so would make the game in the security notion exponential time. The idea is to define or draw U lazily: each time we need to evaluate it on a polynomial P_i linearly independent of all the previous polynomials P_j (with $j < i$), we define $U(P_i) \xleftarrow{\$} \mathbb{Z}_p$; otherwise, we compute $U(P_i)$ as a linear combination of $U(P_j)$. More precisely, if $P_i = \sum_{j=1}^{i-1} \lambda_j \cdot P_j$, $U(P_i) = \sum_{j=1}^{i-1} \lambda_j \cdot U(P_j)$. As explained in Sect. 2, no deterministic polynomial-time algorithm for checking linear dependency between polynomials in $\mathbb{Z}_p[T_1, \ldots, T_n]$ is known. But we can use one which is correct which overwhelming probability. We recall that we denote by **TestLin** such an algorithm.

On the Representation of the Polynomials. A second challenge is to define how the polynomials are represented. We cannot say they have to be given in their expanded form, because it would restrict us to polynomials with a polynomial number of monomials and forbid polynomials such as $\prod_{i=1}^{n}(a_i + 1)$.

Instead, we only suppose that polynomials can be (partially) evaluated, in polynomial time (in n and d, the maximum degree in each indeterminate). This encompasses polynomials defined by an expression (with $+$ and $\cdot$ operations, indeterminates, and scalars) of polynomial size (in n and d). Details are given in the full version [3].

Extension to Weaker Assumptions. Before, showing the formal definition and theorem, let us show an extension of our polynomial linear pseudorandomness security notion to handle weaker assumptions, namely $\mathcal{E}_{k,d}$-MDDH, with $k \geq 2$. In that case, we need to evaluate polynomials on matrices: $[P_i(\boldsymbol{A}) \cdot \boldsymbol{B}]$, with $\boldsymbol{A} \xleftarrow{\$} \mathbb{Z}_p^{k \times k}$ and $\boldsymbol{B} \xleftarrow{\$} \mathbb{Z}_p^{k \times m}$ (with $m \geq 1$ being a positive integer). As multiplication of matrices is not commutative, we need to be very careful. We therefore consider that T_n appears before T_{n-1} (in products), T_{n-1} before T_{n-2}, ... (or any other fixed ordering).

More formally, we suppose that polynomials are represented by an expression (similar to the case $k = 1$), such that in any subexpression $Q \cdot R$, if Q contains T_i

proc **Initialize**	proc **Pl**(P)
$\vec{A} \xleftarrow{\$} (\mathbb{Z}_p^{k \times k})^n$	If $b = 0$ then
$B \xleftarrow{\$} \mathbb{Z}_p^{k \times m}$	$\qquad Y \leftarrow P(\vec{A}) \cdot B$
$\mathcal{L}_1 \leftarrow$ empty list	Else
$\mathcal{L}_2 \leftarrow$ empty list	$\qquad \vec{\lambda} \leftarrow \textbf{TestLin}(\mathcal{L}_1, P)$
$L \leftarrow 0$	$\qquad$ If $\vec{\lambda} = \perp$ then
$b \xleftarrow{\$} \{0,1\}$	$\qquad\qquad Y \xleftarrow{\$} \mathbb{Z}_p^{k \times m}$
	$\qquad\qquad L \leftarrow L + 1$
proc **Finalize**(b')	$\qquad\qquad \mathcal{L}_1[L] \leftarrow P$
Return $b' = b$	$\qquad\qquad \mathcal{L}_2[L] \leftarrow Y$
	$\qquad$ Else
	$\qquad\qquad Y \leftarrow \sum_{i=1}^{L} \lambda_i \cdot \mathcal{L}_2[i]$
	Return $[Y]$

Fig. 3. Game defining the (n,d,k,m)-PLP security for a group $\mathbb{G}$

(formally as an expression and not just when the expression is expanded), then R contains no monomial T_j with $j > i$. Details are given in the full version [3].

3.2 Formal Security Notion and Theorem

Let $\mathbb{G} = \langle g \rangle$ be a group of prime order p. We define the advantage of an adversary $\mathscr{A}$ against the (n,d,k,m)-PLP security of $\mathbb{G}$, denoted $\textbf{Adv}_{\mathbb{G}}^{(n,d,k,m)\text{-plp}}(\mathscr{A})$ as the probability of success in the game defined in Fig. 3, with $\mathscr{A}$ being restricted to make queries $P \in \mathbb{Z}_p[T_1, \ldots, T_n]_{\leq d}$. When not specified, $m = 1$. When $k = m = 1$, we get exactly the intuitive security notion defined previously, as in that case $\vec{A} = \vec{a} \in \mathbb{Z}_p^n$ and $B = b \in \mathbb{Z}_p$.

Theorem 1 (PLP). *Let $\mathbb{G} = \langle g \rangle$ be a group of prime order p. Let $\mathscr{A}$ be an adversary against the (n,d,k,m)-PLP security of $\mathbb{G}$ that makes q oracle queries $P_1, \ldots, P_q$. Then we can design an adversary $\mathscr{B}$ against the $\mathcal{E}_{k,d}$-MDDH problem in $\mathbb{G}$, such that $\textbf{Adv}_{\mathbb{G}}^{(n,d,k,m)\text{-plp}}(\mathscr{A}) \leq n \cdot \textbf{Adv}_{\mathbb{G}}^{\mathcal{E}_{k,d}\text{-mddh}}(\mathscr{B}) + O(ndqN/p)$, where N is an integer polynomial in the size of the representations of the polynomials and $N = 1$ when $k = 1$ (see the full version [3] for details). The running time of $\mathscr{B}$ is that of $\mathscr{A}$ plus the time to perform a polynomial number (in q, n, and d) of operations in $\mathbb{Z}_p$ and $\mathbb{G}$.*

The proof of Theorem 1 is detailed in the full version [3]. It is similar to the proof of the LIP theorem (in the matrix case) in [1]. More precisely, we show a series of indistinguishable games where the first game corresponds to the (n,d,k,m)-PLP security game when $b = 0$, and the last game corresponds to this security game when $b = 1$. Basically, all the games except for the last two games are the same as in the proof of the LIP theorem. The two last games differ, as follows: for the LIP theorem, all polynomials are supposed to be linearly independent, and so in the last two games, all the returned values are drawn uniformly and independently, while for the PLP theorem, the returned values still have linear dependencies.

4   Applications

In this section, we describe how PLP theorem (Theorem 1) can be used to prove the security of aggregate pseudorandom functions as well as multilinear pseudorandom functions. In particular, we obtain polynomial-time reduction for all previous constructions of aggregate-pseudorandom, even for aggregate where only exponential-time reduction were known (read-once formulas). We also obtain a very simple proof of the multilinear pseudorandom function designed in [13]. Finally, we briefly explain how these results can be extended to build constructions based on weaker assumptions in an almost straightforward manner, by simply changing the key space. The proofs of security remain almost the same and consist in reducing the security to the adequate PLP security game.

4.1   Aggregate Pseudorandom Functions

In this subsection, we show that for all constructions proposed in [12], one can prove the AGG-PRF security with a polynomial time reduction, while proofs proposed in this seminal paper suffered from an exponential (in the input size) overhead in the running time of the reduction. Moreover, our reductions are almost straightforward via the PLP theorem.

A first attempt to solve the issue of the exponential time of the original reductions was done in [13]. By introducing multilinear pseudorandom functions and giving a particular instantiation, Cohen and Holmgren showed that one can prove the AGG-PRF security of NR with a polynomial time reduction for hypercubes and decision trees aggregation. However, their technique does not extend to the more general case of read-once formulas aggregation. Also, as we will show it the next subsection, their construction can be seen as a particular case of our main theorem, and then can be proven secure very easily using our result.

Here, we provide a polynomial time reduction for the general case of read-once formulas. This implies in particular the previous results on hypercubes and decision trees which are particular cases of read-once formulas.

Intuitively, if we consider the PLP security for $k = 1$ and aggregation with the Naor-Reingold PRF, our PLP theorem (Theorem 1) implicitly says that as long as the aggregate values can be computed as a group element whose discrete logarithm is the evaluation of a multivariate polynomial on the key, then, if the corresponding polynomials have a small representation, the PLP theorem guarantees the security (with a polynomial time reduction), even if the number of points aggregated is superpolynomial. Please notice that if these polynomials do not have any small representation (e.g. the smallest representation is exponential in the input size), then there is no point of considering such aggregation, since the whole point of aggregate pseudorandom function lies in the possibility of aggregating superpolynomially many PRF values with a very efficient computation.

Read-Once Formulas. A read-once formula is a circuit on $x = (x_1, \ldots, x_n) \in \{0,1\}^n$ composed of only AND, OR and NOT gates with fan-out 1, so that each input literal is fed into at most one gate and each gate output is fed into at most one other gate. We denote by ROF_n the family of all read-once boolean formulas over $x_1, \ldots, x_n$ variables. In order to ease the reading, we restrict these circuits to be in a standard form, so that they are composed of fan-in 2 and fan-out 1 AND and OR gates, and NOT gates occurring only at the inputs. This common restriction can be done without loss of generality. Hence, one can see such a circuit as a binary tree where each leaf is labeled by a variable x_i or its negation $\bar{x}_i$ and where each internal node has a label C and has two children with labels C_L and C_R and represents either an AND or an OR gate (with fan-in 2). We identify a formula (and the set it represents) with the label of its root C_ϕ.

Aggregation for Read-Once Formulas. We recall the definition of read-once formula aggregation used in [12]. For the sake of simplicity, we only consider the case of the Naor-Reingold PRF, defined as $\mathrm{NR}(\vec{a}, x) = [a_0 \prod_{i=1}^{n} a_i^{x_i}]$, where $a_0, \ldots, a_n \xleftarrow{\$} \mathbb{Z}_p$ and $x \in \{0,1\}^n$. We define the aggregation function for read-once formulas of length n as follows.

The collection $\mathscr{S}_{\mathrm{rof}} \subseteq \{0,1\}^n$ corresponds to all the subsets of $S \subseteq \{0,1\}^n$ such that there exists a read-once formula $C_\phi \in \mathrm{ROF}_n$ such that $S = \{x \in \{0,1\}^n \mid C_\phi(x) = 1\}$.

The aggregation function Γ_{rof} is defined as the product (assuming the group is a multiplicative group) of the values on such a subset. Hence, we have:

$$
\mathrm{AGG}_{\mathrm{NR}, \mathscr{S}_{\mathrm{rof}}, \Gamma_{\mathrm{rof}}}(C_\phi) = \prod_{x \mid C_\phi(x)=1} \left[a_0 \prod_{i=1}^{n} a_i^{x_i} \right] = \left[a_0 \sum_{x \mid C_\phi(x)=1} \prod_{i=1}^{n} a_i^{x_i} \right]
$$
$$
= \left[a_0 \cdot A_{C_\phi, 1}(\vec{a}) \right],
$$

where $A_{C,b}$ is the polynomial $\sum_{x \in \{0,1\}^n \mid C(x)=b} \prod_{i=1}^{n} T_i^{x_i}$ for any $C \in \mathrm{ROF}_n$ and $b \in \{0,1\}$.

Efficient Evaluation of $A_{C,b}$. One can efficiently compute $A_{C,b}$ recursively as follows:

- If C is a literal for variable x_i, then $A_{C,1} = T_i$ and $A_{C,0} = 1$ if $C = x_i$; and $A_{C,1} = 1$ and $A_{C,0} = T_i$ if $C = \bar{x}_i$;
- If C is an AND gate with C_L and C_R its two children, then we have:
$A_{C,1} = A_{C_L,1} \cdot A_{C_R,1}$
$A_{C,0} = A_{C_L,0} \cdot A_{C_R,0} + A_{C_L,1} \cdot A_{C_R,0} + A_{C_L,0} \cdot A_{C_R,1};$

– If C is an OR gate with C_L and C_R its two children, then we have:
$$A_{C,1} = A_{C_L,1} \cdot A_{C_R,1} + A_{C_L,1} \cdot A_{C_R,0} + A_{C_L,0} \cdot A_{C_R,1}$$
$$A_{C,0} = A_{C_L,0} \cdot A_{C_R,0}.$$

Now we have introduced everything, we can prove that NR (or more general constructions) is an $(\mathscr{S}_{\mathsf{rof}}, \Gamma_{\mathsf{rof}})$-AGG-PRF under the standard DDH assumption, as stated in the lemma below.

Lemma 2. *Let $\mathbb{G} = \langle g \rangle$ be a group of prime order p and NR be the Naor-Reingold PRF defined as $\mathsf{NR}(\vec{a}, x) = [a_0 \prod_{i=1}^n a_i^{x_i}]$, where the key is $(a_0, \dots, a_n) \xleftarrow{\$} \mathbb{Z}_p^{n+1}$ and the input is $x \in \{0,1\}^n$. Then one can reduce the $(\mathscr{S}_{\mathsf{rof}}, \Gamma_{\mathsf{rof}})$-AGG-PRF security of NR to the hardness of the DDH problem in $\mathbb{G}$, with a loss of a factor n. Moreover, the time overhead is polynomial in n and in the number of queries made by the adversary.*

The proof is straightforward using the PLP theorem: all queries in the security game for the aggregate PRF can be seen as a queries of the form $\mathbf{Pl}(P)$ for some polynomial P with a small representation: $\mathbf{Fn}(x)$ returns $\mathbf{Pl}(T_0 \prod_{i=1}^n T_i^{x_i})$ and $\mathrm{AGG}(C_\phi)$ returns $\mathbf{Pl}(T_0 \cdot A_{C_\phi,1}(\vec{T}))$. Details can be found in the full version [3].

Extensions. One can easily extend this result for k-Lin-based PRFs similar to NR using our main theorem. Also, one can easily use our PLP theorem (Theorem 1) to prove the security for any aggregate (for instance with NR) as soon as the aggregate values can be represented as group elements whose discrete logarithms are the evaluation of a (multivariate) polynomial on the key (and that this polynomial is efficiently computable).

Impossibility Result for CNF (Conjunctive Normal Form) and DNF (Disjunctive Normal Form) Formulas. In [12], the authors show that, unless NP=BPP, there does not exist an $(\mathscr{S}, \Gamma)$-aggregate pseudorandom function[1], with $\mathcal{D} = \{0,1\}^n$, $\mathscr{S}$ containing the following sets:

$$S_\phi = \{x \in \{0,1\}^n \mid \phi(x) = 1\}$$

with ϕ a CNF formula with n-bit input, and Γ a "reasonable" aggregate function, e.g., Γ_{rof} (assuming $\mathcal{R}$ is a cyclic group $\mathbb{G}$ of prime order p). The proof consists in showing that if such aggregate pseudorandom function exists, then we can solve SAT in polynomial time. More precisely, given a SAT instance, i.e., a CNF formula ϕ, we can compute $\mathrm{AGG}(\phi)$. If ϕ is not satisfiable, $\mathrm{AGG}(\phi) = 1 \in \mathbb{G}$, while otherwise $\mathrm{AGG}(\phi) = \prod_{x \in \{0,1\}^n, \ \phi(x)=1} F(K,x)$. This latter value is not 1 with high probability, otherwise we would get a non-uniform distinguisher against aggregate pseudorandomness.

---

[1] We suppose that the aggregate pseudorandomness security property holds non-uniformly. When $\mathscr{S}$ is expressive enough, we can also do the proof when this security property holds uniformly, see [12, Sect. 2.2] for details.

The case of DNF formulas (or more generally of any class for which satisfiability is tractable) was left as an important open problem in [12]. Here, we show that unless NP=BPP, there also does not exist an $(\mathscr{S}, \Gamma)$-aggregate pseudorandom function as above, when $\mathscr{S}$ contains S_ϕ for any DNF (instead of CNF) formula ϕ with n-bit input. For that, we first remark that the formula $\top$, always true, is a DNF formula (it is the disjunction of all the possible literals), and that the negation $\bar\phi$ of a CNF formula ϕ is a DNF formula. Then, given a SAT instance, a CNF formula ϕ, we compute $\mathrm{AGG}(\bar\phi)$ and $\mathrm{AGG}(\top)$. If ϕ is not satisfiable, $\bar\phi$ is always true and $\mathrm{AGG}(\bar\phi) = \mathrm{AGG}(\top)$, while otherwise, $\mathrm{AGG}(\bar\phi) = \mathrm{AGG}(\top)/\prod_{x\in\{0,1\}^n,\,\phi(x)=1} F(K,x)$. This latter value is not $\mathrm{AGG}(\top)$ with high probability, otherwise we would get a non-uniform distinguisher against aggregate pseudorandomness.

4.2 Multilinear Pseudorandom Functions

Here, we explain how our main theorem can be used to prove directly the security of the multilinear pseudorandom function built in [13]. We first recall their construction before explaining how to prove its security.

Cohen-Holmgren Multilinear Pseudorandom Function (CH). Let $\mathbb{G} = \langle g \rangle$ be a group of prime order p. The key space of the multilinear pseudorandom function is $\mathbb{Z}_p^{l_1} \times \cdots \times \mathbb{Z}_p^{l_n}$. The input space is the same as the key space. Given a key $(\vec{a_1}, \ldots, \vec{a_n})$ taken uniformly at random in the key space, the evaluation of the multilinear pseudorandom function on the input $(\vec{x_1}, \ldots, \vec{x_n})$) outputs:

$$\mathsf{CH}((\vec{a_1},\ldots,\vec{a_n}),(\vec{x_1},\ldots,\vec{x_n})) = \left[\prod_{i=1}^n \langle \vec{a_i}, \vec{x_i}\rangle\right]$$

where $\langle \vec{a}, \vec{x}\rangle$ denotes the canonical inner product $\langle \vec{a}, \vec{x}\rangle = \sum_{i=1}^l a_i \cdot x_i$, with l being the length of vectors $\vec{a}$ and $\vec{x}$.

In [13], Cohen and Holmgren prove that this construction is a secure multilinear pseudorandom function under the standard DDH assumption. One of their main contributions is to achieve a polynomial time reduction. Their technique can be seen as a special case of ours. In particular, using our main theorem, one can easily obtain the following lemma.

Lemma 3. *Let $\mathbb{G} = \langle g \rangle$ be a group of prime order p and CH: $(\mathbb{Z}_p^{l_1} \times \cdots \times \mathbb{Z}_p^{l_n}) \times (\mathbb{Z}_p^{l_1} \times \cdots \times \mathbb{Z}_p^{l_n}) \to \mathbb{G}$ denote the above multilinear pseudorandom function. Then we can reduce the multilinear PRF security of CH to the hardness of the DDH problem in $\mathbb{G}$, with a loss of a factor $l = \sum_{i=1}^n l_i$. Moreover, the time overhead is polynomial in l and in the number of queries made by the adversary.*

A detailed proof can be found in the full version [3], but we give an intuition of the proof in what follows.

Proof. Let $\vec{T} = (T_{1,1}, \ldots, T_{1,l_1}, \ldots, T_{n,1}, \ldots, T_{n,l_n})$ be a vector of indeterminates, and let $\vec{T_i} = (T_{i,1}, \ldots, T_{i,l_i})$. The PLP theorem shows that

$\mathsf{CH}(\overrightarrow{a_1}, \ldots, \overrightarrow{a_n}, \overrightarrow{x_1}, \ldots, \overrightarrow{x_n})$ (using a random key $\overrightarrow{a}$) is computationally indistinguishable from

$$\left[U\left(\prod_{i=1}^{n} \langle \overrightarrow{T_i}, \overrightarrow{x_i} \rangle \right) \right] = [f(\overrightarrow{x_1}, \ldots, \overrightarrow{x_n})]$$

with $U \xleftarrow{\$} \mathsf{L}(\mathbb{Z}_p[\overrightarrow{T}]_{\leq 1}, \mathbb{Z}_p)$ and

$$f : \left(\begin{array}{c} \mathbb{Z}_p^{l_1} \times \cdots \times \mathbb{Z}_p^{l_n} \to \mathbb{Z}_p \\ (\overrightarrow{x_1}, \ldots, \overrightarrow{x_n}) \mapsto U(\prod_{i=1}^{n} \langle \overrightarrow{T_i}, \overrightarrow{x_i} \rangle) \end{array} \right).$$

To conclude, we just need to prove that f is a random n-linear function in $\mathsf{L}(\mathbb{Z}_p^{l_1} \otimes \cdots \otimes \mathbb{Z}_p^{l_n}, \mathbb{Z}_p)$.

For that purpose, let us introduce the following n-linear application:

$$\psi : \left(\begin{array}{c} \mathbb{Z}_p^{l_1} \times \cdots \times \mathbb{Z}_p^{l_n} \to \mathbb{Z}_p[\overrightarrow{T}]_{\leq 1} \\ (\overrightarrow{x_1}, \ldots, \overrightarrow{x_n}) \mapsto \prod_{i=1}^{n} \langle \overrightarrow{T_i}, \overrightarrow{x_i} \rangle \end{array} \right).$$

We remark that f is the composition of U and ψ: $f = U \circ \psi$.

Furthermore, if we write $\overrightarrow{e_{i,l}} = (0, \ldots, 0, 1, 0, \ldots, 0)$ the i-th vector of the canonical base of $\mathbb{Z}_p^l$, then:

$$\psi(\overrightarrow{e_{i_1,l_1}}, \ldots, \overrightarrow{e_{i_n,l_n}}) = T_{1,i_1} \cdots T_{n,i_n};$$

and as the monomials $T_{1,i_1} \cdots T_{n,i_n}$ are linearly independent, ψ is injective. Since $f = U \circ \psi$ and $U \xleftarrow{\$} \mathsf{L}(\mathbb{Z}_p[\overrightarrow{T}]_{\leq 1}, \mathbb{Z}_p)$, the function f is a uniform random linear function from $\mathsf{L}(\mathbb{Z}_p^{l_1} \otimes \cdots \otimes \mathbb{Z}_p^{l_n}, \mathbb{Z}_p)$. This is exactly what we wanted to show. $\square$

Symmetric Multilinear Pseudorandom Function. In [12], constructing symmetric multilinear pseudorandom functions was left as an open problem. The definition of this notion is the same as the notion of multilinear pseudorandom function, except that we only require the function to be indistinguishable from a random *symmetric* multilinear function. In that case, we suppose that $l_1 = \cdots = l_n = l$, i.e., all the vectors $\overrightarrow{x_1}, \ldots, \overrightarrow{x_n}$ have the same size l. The authors wrote in [12] that the natural modification of the CH construction to obtain a symmetric construction consisting in setting $\overrightarrow{a_1} = \overrightarrow{a_2} = \cdots = \overrightarrow{a_n}$ (simply denoted $\overrightarrow{a}$ in what follows) leads to a symmetric multilinear pseudorandom function whose security is less clear, but claimed that it holds under the $\mathcal{E}_{1,n}$-MDDH assumption (which is exactly the n-Strong DDH assumption), when $l = |\overrightarrow{a}| = 2$. We show that this construction is actually secure under the same assumption for any $l = |\overrightarrow{a}| \geq 2$ as stated in the following lemma, whose proof is detailed in the full version [3] and is almost the same as the proof of Lemma 3.

Lemma 4. *Let* $\mathbb{G} = \langle g \rangle$ *be a group of prime order p and* $\mathsf{CH}_{\mathsf{sym}} : \mathbb{Z}_p^l \times (\mathbb{Z}_p^l)^n \to \mathbb{G}$ *that takes as input a key* $\overrightarrow{a} \in \mathbb{Z}_p^l$ *and an input* $\overrightarrow{x} = (\overrightarrow{x_1}, \ldots, \overrightarrow{x_n}) \in (\mathbb{Z}_p^l)^n$ *and outputs* $[\prod_{i=1}^{n} \langle \overrightarrow{a}, \overrightarrow{x_i} \rangle]$. *Then we can reduce the symmetric multilinear PRF security of* $\mathsf{CH}_{\mathsf{sym}}$ *to the hardness of the n-DDHI problem in $\mathbb{G}$, with a loss of a factor l. Moreover, the time overhead is polynomial in l and in the number of queries made by the adversary.*

Skew-Symmetric Multilinear Pseudorandom Function. In [12], the author left as an open problem the construction of a skew-symmetric multilinear pseudorandom function. The definition of this notion is the same as the notion of multilinear pseudorandom function, except that we only require the function to be indistinguishable from a random *skew-symmetric* multilinear function. We assume that $l_1 = \cdots = l_n = l = n$, i.e., all the vectors $\overrightarrow{x_1}, \ldots, \overrightarrow{x_n}$ have the same size $l = n$. We need $l = n$ because there is no skew-symmetric n-multilinear map from $\left(\mathbb{Z}_p^l\right)^n$ to $\mathbb{Z}_p$, when $l < n$.

We know that any skew-symmetric n-multilinear map f is of the form:

$$f(\overrightarrow{x_1}, \ldots, \overrightarrow{x_n}) = c \cdot \det(\overrightarrow{x_1}, \ldots, \overrightarrow{x_n}),$$

with c being a scalar in $\mathbb{Z}_p$ and det being the determinant function. Therefore, the function

$$F(a, (\overrightarrow{x_1}, \ldots, \overrightarrow{x_n})) = [a \cdot \det(\overrightarrow{x_1}, \ldots, \overrightarrow{x_n})]$$

is a skew-symmetric multilinear PRF with key $a \in \mathbb{Z}_p$. The proof is trivial since, $(\overrightarrow{x_1}, \ldots, \overrightarrow{x_n}) \mapsto F(a, (\overrightarrow{x_1}, \ldots, \overrightarrow{x_n}))$ is actually a random skew-symmetric n-multilinear map when a is a random scalar in $\mathbb{Z}_p$. No assumption is required. Our analysis shows that skew-symmetric multilinear PRFs are of limited interest, but our construction still solves an interesting open problem in [12].

Extensions. As for aggregate pseudorandom functions, it is very easy to build multilinear pseudorandom functions under k-Lin and to prove their security applying our PLP theorem (Theorem 1), for instance using the same construction but changing the key components from elements in $\mathbb{Z}_p$ to elements in $\mathbb{Z}_p^{k \times k}$ while keeping the same inputs space, and by defining $\langle \overrightarrow{A}, \overrightarrow{x} \rangle = \sum_{i=1}^{l} x_i \cdot A_i$, with $\overrightarrow{A} = (A_1, \ldots, A_l) \in (\mathbb{Z}_p^{k \times k})^l$ and $x = (x_1, \ldots, x_l) \in \mathbb{Z}_p^l$. This leads to the following construction:

$$F : \left(\begin{array}{c} \mathbb{Z}_p^{l_1} \times \cdots \times \mathbb{Z}_p^{l_n} \to \mathbb{G}^{k \times m} \\ (\overrightarrow{x_1}, \ldots, \overrightarrow{x_n}) \mapsto \left[(\prod_{i=1}^{n} \langle \overrightarrow{A_i}, \overrightarrow{x_i} \rangle) \cdot B \right] \end{array} \right)$$

with $(\overrightarrow{A_1}, \ldots, \overrightarrow{A_n}) \in (\mathbb{Z}_p^{k \times k})^{l_1} \times \cdots \times (\mathbb{Z}_p^{k \times k})^{l_n}$ and $B \in \mathbb{Z}_p^{k \times m}$.

References

1. Abdalla, M., Benhamouda, F., Passelègue, A.: An algebraic framework for pseudo-random functions and applications to related-key security. In: Gennaro, R., Robshaw, M.J.B. (eds.) CRYPTO 2015, Part I. LNCS, vol. 9215, pp. 388–409. Springer, Heidelberg (2015)
2. Abdalla, M., Benhamouda, F., Passelègue, A., Paterson, K.G.: Related-key security for pseudorandom functions beyond the linear barrier. In: Garay, J.A., Gennaro, R. (eds.) CRYPTO 2014, Part I. LNCS, vol. 8616, pp. 77–94. Springer, Heidelberg (2014)

120   M. Abdalla et al.

3. Abdalla, M., Benhamouda, F., Passelgue, A.: Multilinear and aggregate pseudorandom functions: new constructions and improved security, full version of this paper available at Cryptology ePrint Archive. http://eprint.iacr.org
4. Advanced Encryption Standard (AES). National Institute of Standards and Technology (NIST), FIPS PUB 197, U.S. Department of Commerce, November 2001
5. Bellare, M., Rogaway, P.: The security of triple encryption and a framework for code-based game-playing proofs. In: Vaudenay, S. (ed.) EUROCRYPT 2006. LNCS, vol. 4004, pp. 409–426. Springer, Heidelberg (2006)
6. Bogdanov, A., Wee, H.M.: A stateful implementation of a random function supporting parity queries over hypercubes. In: Jansen, K., Khanna, S., Rolim, J.D.P., Ron, D. (eds.) RANDOM 2004 and APPROX 2004. LNCS, vol. 3122, pp. 298–309. Springer, Heidelberg (2004)
7. Boneh, D., Boyen, X.: Short signatures without random oracles. In: Cachin, C., Camenisch, J.L. (eds.) EUROCRYPT 2004. LNCS, vol. 3027, pp. 56–73. Springer, Heidelberg (2004)
8. Boneh, D., Boyen, X., Goh, E.-J.: Hierarchical identity based encryption with constant size ciphertext. In: Cramer, R. (ed.) EUROCRYPT 2005. LNCS, vol. 3494, pp. 440–456. Springer, Heidelberg (2005)
9. Boneh, D., Waters, B.: Constrained pseudorandom functions and their applications. In: Sako, K., Sarkar, P. (eds.) ASIACRYPT 2013, Part II. LNCS, vol. 8270, pp. 280–300. Springer, Heidelberg (2013)
10. Boyen, X.: The uber-assumption family. In: Galbraith, S.D., Paterson, K.G. (eds.) Pairing 2008. LNCS, vol. 5209, pp. 39–56. Springer, Heidelberg (2008)
11. Boyle, E., Goldwasser, S., Ivan, I.: Functional signatures and pseudorandom functions. In: Krawczyk, H. (ed.) PKC 2014. LNCS, vol. 8383, pp. 501–519. Springer, Heidelberg (2014)
12. Cohen, A., Goldwasser, S., Vaikuntanathan, V.: Aggregate pseudorandom functions and connections to learning. In: Dodis, Y., Nielsen, J.B. (eds.) TCC 2015, Part II. LNCS, vol. 9015, pp. 61–89. Springer, Heidelberg (2015)
13. Cohen, A., Holmgren, J.: Multilinear Pseudorandom Functions. In: Halldórsson, M.M., Iwama, K., Kobayashi, N., Speckmann, B. (eds.) ICALP 2015. LNCS, vol. 9134, pp. 331–342. Springer, Heidelberg (2015)
14. Escala, A., Herold, G., Kiltz, E., Ràfols, C., Villar, J.: An algebraic framework for Diffie-Hellman assumptions. In: Canetti, R., Garay, J.A. (eds.) CRYPTO 2013, Part II. LNCS, vol. 8043, pp. 129–147. Springer, Heidelberg (2013)
15. Goldreich, O., Goldwasser, S., Micali, S.: How to construct random functions. J. ACM **33**(4), 792–807 (1986)
16. Hofheinz, D., Kiltz, E.: Secure hybrid encryption from weakened key encapsulation. In: Menezes, A. (ed.) CRYPTO 2007. LNCS, vol. 4622, pp. 553–571. Springer, Heidelberg (2007)
17. Kiayias, A., Papadopoulos, S., Triandopoulos, N., Zacharias, T.: Delegatable pseudorandom functions and applications. In: Sadeghi, A.R., Gligor, V.D., Yung, M. (eds.) ACM CCS 2013, pp. 669–684, ACM Press, November 2013
18. Naor, M., Reingold, O.: Number-theoretic constructions of efficient pseudo-random functions. In: 38th FOCS, pp. 458–467, IEEE Computer Society Press, October 1997
19. Shacham, H.: A cramer-shoup encryption scheme from the linear assumption and from progressively weaker linear variants. Cryptology ePrint Archive, Report 2007/074 (2007). http://eprint.iacr.org/2007/074

New Realizations of Somewhere Statistically Binding Hashing and Positional Accumulators

Tatsuaki Okamoto[1]([✉]), Krzysztof Pietrzak[2], Brent Waters[3], and Daniel Wichs[4]

[1] NTT Laboratories, Tokyo, Japan
okamoto.tatsuaki@lab.ntt.co.jp
[2] IST Austria, Klosterneuburg, Austria
pietrzak@ist.ac.at
[3] UT Austin, Austin, USA
bwaters@cs.utexas.edu
[4] Northeastern University, Boston, USA
wichs@ccs.neu.edu

Abstract. A *somewhere statistically binding (SSB) hash*, introduced by Hubáček and Wichs (ITCS '15), can be used to hash a long string x to a short digest $y = H_{hk}(x)$ using a public hashing-key hk. Furthermore, there is a way to set up the hash key hk to make it statistically binding on some arbitrary hidden position i, meaning that: (1) the digest y completely determines the i'th bit (or symbol) of x so that all pre-images of y have the same value in the i'th position, (2) it is computationally infeasible to distinguish the position i on which hk is statistically binding from any other position i'. Lastly, the hash should have a *local opening* property analogous to Merkle-Tree hashing, meaning that given x and $y = H_{hk}(x)$ it should be possible to create a short proof π that certifies the value of the i'th bit (or symbol) of x without having to provide the entire input x. A similar primitive called a *positional accumulator*, introduced by Koppula, Lewko and Waters (STOC '15) further supports dynamic updates of the hashed value. These tools, which are interesting in their own right, also serve as one of the main technical components in several recent works building advanced applications from indistinguishability obfuscation (iO).

The prior constructions of SSB hashing and positional accumulators required fully homomorphic encryption (FHE) and iO respectively. In this work, we give new constructions of these tools based on well studied number-theoretic assumptions such as DDH, Phi-Hiding and DCR, as well as a general construction from lossy/injective functions.

K. Pietrzak—Research supported by ERC starting grant (259668-PSPC)
B. Waters—Supported by NSF CNS-1228599 and CNS-1414082. DARPA SafeWare, Google Faculty Research award, the Alfred P. Sloan Fellowship, Microsoft Faculty Fellowship, and Packard Foundation Fellowship.
D. Wichs—Research supported by NSF grants CNS-1347350, CNS-1314722, CNS-1413964.

© International Association for Cryptologic Research 2015
T. Iwata and J.H. Cheon (Eds.): ASIACRYPT 2015, Part I, LNCS 9452, pp. 121–145, 2015.
DOI: 10.1007/978-3-662-48797-6_6

1 Introduction

SSB Hashing. A *somewhere statistically binding (SSB) hash*, introduced by Hubáček and Wichs [HW15], can be used to create a *short* digest $y = H_{hk}(x)$ of some *long* input $x = (x[0], \ldots, x[L-1]) \in \Sigma^L$, where Σ is some alphabet. The *hashing key* hk $\leftarrow$ Gen(i) can be chosen by providing a special "binding index" i and this ensures that the hash $y = H_{hk}(x)$ is statistically binding for the i'th symbol, meaning that it completely determines the value $x[i]$. In other words, even though y has many preimages x' such that $H_{hk}(x') = y$, all of these preimages agree in the i'th symbol $x'[i] = x[i]$. The index i on which the hash is statistically binding should remain computationally hidden given the hashing key hk. This is formalized analogously to semantic security so that for any indices i, i' the hashing keys hk $\leftarrow$ Gen(i) and hk$'$ $\leftarrow$ Gen(i') should be computationally indistinguishable. Moreover, we will be interested in SSB hash functions with a "local opening" property that allows us to prove that j'th symbol of x takes on some particular value $x[j] = u$ by providing a *short* opening π. This is analogous to Merkle-Tree hashing, where it is possible to open the j'th symbol of x by providing a proof π that consists of the hash values associated with all the sibling nodes along the path from the root of the tree to the j'th leaf. In the case of SSB hashing, when $j = i$ is the "binding index", there should (statistically) exist only one possible value that we can open $x[j]$ to by providing a corresponding proof.

Positional Accumulators. A related primitive called a *positional accumulator*, was introduced at the same time as SSB hashing by Koppula, Lewko and Waters [KLW15]. Roughly speaking, it includes the functionality of SSB hashing along with the ability to perform "local updates" where one can very efficiently update the hash $y = H_{hk}(x)$ if a particular position $x[j]$ is updated. Again, this is analogous to Merkle-Tree hashing, where it is possible to update the j'th symbol of x by only updating the hash values along the path from the root of the tree to the j'th leaf.[1]

Applications of SSB Hashing and Positional Accumulators. The above tools, which are interesting in their own right, turn out to be extremely useful in several applications when combines with indistinguishability obfuscation (iO) [BGI+12, GGH+13]. An iO scheme can be used to obfuscate a program (given by a circuit) so that the obfuscations of any two functionally equivalent programs are indistinguishable. Although this notion of obfuscation might a-priori

---

[1] The formal definitions of SSB hashing and positional accumulators as given in [HW15, KLW15] are technically incomparable. On a high level, the latter notion requires additional functionality in the form of updates but only insists on a weaker notion of security which essentially corresponds to "target collision resistance" where the target hash value is computed honestly. In this work, we construct schemes that achieve the best of both worlds, having the additional functionality and the stronger security.

seem too week to be useful, recent work has shown it to be surprisingly powerful (see e.g., [SW14]). Very recently, several results showed how to use iO in conjunction with SSB hashing and positional accumulators to achieve various advanced applications. The work of [HW15] uses SSB hashing and iO to construct the first general Multi-Party Computation (MPC) protocols in the semi-honest model where the communication complexity essentially matches that of the best insecure protocol for the same task. The work of [KLW15] uses positional accumulators and iO to construct succinct garbling for Turing Machines, and recent work extends this approach to RAM programs [CH15, CCC+15]. Lastly, the work of [Zha14] uses SSB hashing and iO to construct the first adaptively secure broadcast encryption with short system parameters.

Example: The Power of iO + SSB. To see the usefulness of combining iO and SSB hashing (or positional accumulators), let's take a simple illustrative example, adapted from [HW15].[2] Imagine that Alice has a (small) secret circuit C, and both Alice and Bob know a public value $x \in \Sigma^L$. Alice wishes to communicate the values $\{C(x[i])\}_{i \in [L]}$ to Bob while hiding some information about C. In particular, Bob shouldn't learn whether Alice has the circuit C or some other C' that satisfies $C(x[i]) = C'(x[i])$ for each $i \in [L]$. Note that C and C' may not be functionally equivalent and they only agree on the inputs $\{x[i]\}_{i \in [L]}$ but might disagree on other inputs. A naive secure solution would be for Alice to simply send the outputs $\{C(x[i])\}_{i \in [L]}$ to Bob, but this incurs communication proportional to L. An insecure but communication-efficient solution would be for Alice to just send the small circuit C to Bob. Can we get a secure solution with comparable communication independent of L? Simply sending an obfuscated copy of C is not sufficient since the circuits C, C' are not functionally equivalent and therefore their obfuscations might be easily distinguishable. However it is possible to achieve this with iO and SSB hashing. Alice can send an obfuscation of a circuit that has the hash $y = H_{hk}(x)$ hard-coded and takes as input a tuple (j, u, π): it checks that $j \in [L]$ and that π is a valid opening to $x[j] = u$ and if so outputs $C(u)$. Bob can evaluate this circuit on the values $\{x[j]\}_{j \in [L]}$ by providing the appropriate openings. It is possible to show that the above hides whether Alice started with C or C'. The proof proceeds in a sequence of L hybrids where in the i'th hybrid we obfuscate a circuit C_i that runs C' instead of C when $j \leq i$ and otherwise runs C. To go from hybrid i to $i+1$ we first switch the SSB hash key hk to be binding in position $i+1$ and then we can switch from obfuscating C_i to C_{i+1} by arguing that these are functionally equivalent; they only differ in the code they execute for inputs of the form $(j = i+1, u, \pi)$ where π is a valid proof but in this case, by the statistical binding property, the only possible value u for which a valid proof π exists is the unique value $u = x[j]$ for which both circuits produce the same output $C(x[j]) = C'(x[j])$.

---

[2] The contents of this paragraph and the notion of iO are not essential to understand the results of the paper, but we provide it to give some intuition for how SSB hashing and positional accumulators are used in conjunction with iO in prior works to get the various applications described above.

Prior Constructions of SSB and Positional Accumulators. The work of [HW15] constructed a SSB hash by relying on fully homomorphic encryption (FHE). Roughly speaking the construction combines FHE with Merkle Hash Trees. To hash some value $x = (x[0], \ldots, x[L-1])$ the construction creates a full binary tree of height $\log L$ (for simplicity, assume L is a power of 2) and deterministically associates a ciphertext with each node of the tree. The L leaf nodes will be associated with some deterministically created encryptions of the values $x[0], \ldots, x[L-1]$, say by using all 0s for the random coins of the encryption procedure. The hash key hk consists of an encryption of a path from the root of the tree to the i'th leaf where i is the binding index; concretely it contains $\log L$ FHE ciphertexts $(ct_1, \ldots, ct_{\log L})$ which encrypt bits $\beta_1, \ldots, \beta_{\log L}$ corresponding to the binary representation of the binding index i so that $\beta_i = 0$ denotes "left" and $\beta_i = 1$ denotes "right". The ciphertext associated with each non-leaf node are computed homomorphically to ensure that the value $x[i]$ is contained in each ciphertext along the path from the root to the i'th leaf. Concretely, the ciphertext associated with some node at level j is determined by a homomorphic computation which takes the two child ciphertexts c_0 (left) and c_1 (right) encrypting some values m_0, m_1 and the ciphertext ct_i contained in hk which encrypts β_i and homomorphically produces a ciphertext encrypting m_{β_i}. (For technical reasons, the actual construction is a bit more complicated and needs to use a different FHE key at each level of the tree – see [HW15] for full details.) This ensures that the binding index i is hidden by the semantic security of FHE and the statistically binding property follows by the correctness of FHE.

The work of [KLW15] constructs positional accumulators by also relying on a variant of Merkle Trees. However, instead of FHE, it relies on standard public-key encryption and iO. (It is relatively easy to see that the scheme of [HW15] would also yield an alternate construction of a positional accumulator).

1.1   Our Results

In this work we give new constructions of SSB hashing and positional accumulators from a wide variety of well studied number theoretic assumptions such as DDH, DCR (decisional composite residuocity), ϕ-hiding, LWE and others.

Two-to-One SSB. We first abstract out the common Merkle-tree style approach that is common to both SSB hashes and positional accumulators, and identify a basic underlying primitive that we call a *two-to-one* SSB hash, which can be used to instantiate this approach. Intuitively a two-to-one SSB hash takes as input $x = (x[0], x[1]) \in \Sigma^2$ consisting of just two alphabet symbols and outputs a value $y = H_{\mathsf{hk}}(x)$ which is not much larger than a single alphabet symbol. The key hk can be set up to be statistically binding on either position 0 or 1.

Instantiations of Two-to-One SSB. We show how to instantiate a two-to-one SSB hash from the DDH assumption and the decisional composite residuocity (DCR) assumption. More generally, we show how to instantiate a (slight variant of) two-to-one SSB hash from any lossy/injective function. This is a family of functions

$f_{\mathsf{pk}}(x)$ where the public key pk can be picked in one of two indistinguishable modes: in injective mode, the function $f_{\mathsf{pk}}(x)$ is an injective function and in lossy mode $f_{\mathsf{pk}}(x)$ it is a many-to-one function. To construct a two-to-one SSB hash from injective/lossy function we pick two public keys $\mathsf{hk} = (\mathsf{pk}_0, \mathsf{pk}_1)$ and define $H_{\mathsf{hk}}(x[0], x[1]) = h(f_{\mathsf{pk}_0}(x[0]), f_{\mathsf{pk}_1}(x[1]))$ where h is a universal hash function. To make the hk binding on index 0 we choose pk_0 to be injective and pk_1 to be lossy and to make is binding on index 1 we do the reverse. With appropriate parameters, we can ensure that the statistically binding property holds with overwhelming probability over the choice of h.

From Two-to-One SSB to Full SSB and Positional Accumulators. We can instantiate a (full) SSB hash with arbitrary input size Σ^L by combining two-to-one SSB hashes in a Merkle Tree, with a different key at each level. To make the full SSB binding at some location i, we choose the hash keys at each level to be binding on either the left or right child in such a way that they are binding along the path from the root of the tree to the leaf at position i. This allows us to "locally open" the j'th position of the input in the usual way, by giving the hash values of all the siblings along the path from the root to the j'th leaf. If $j = i$ is the binding index, then there is a unique value $x[j] = u$ for which there is a valid opening. To get positional accumulators, we use the fact that we can also locally update the hashed value by modifying one location $x[j]$ and only updating the hashes along the path from the root to the j'th leaf.

A Flatter Approach. We also explore a different approach for achieving SSB hashing from the ϕ-hiding assumption, which does not go through a Merkle-Tree type construction. Roughly our approach uses a construction is structurally similar to standard constructions RSA accumulators [BdM93]. However, we construct a modus N to be such that for some given prime exponent e we have that e divides $\phi(N)$. This means that if $y \in \mathbb{Z}_N$ is not an e-th residue mod N, then there exists no value $\pi \in \mathbb{Z}_N$ where $\pi^e = y$. This will lead to our statistical binding property as we will leverage this fact to make the value e related to an index we wish to be binding on. Index hiding will follow from the ϕ-hiding assumption.

2   Preliminaries

SSB Hash (with Local Opening). Our definition follows that of [HW15], but whereas that work only defined SSB hash which included the local opening requirement by default, it will be convenient for us to also separately define a weaker variant which does not require the local opening property.

Definition 2.1 (SSB Hash). *A somewhere statistically binding (SSB) hash consists of PPT algorithms $\mathcal{H} = (\mathsf{Gen}, H)$ and a polynomial $\ell(\cdot, \cdot)$ denoting the output length.*

- $\mathsf{hk} \leftarrow \mathsf{Gen}(1^\lambda, 1^s, L, i)$*: Takes as input a security parameter λ a block-length s an input-length $L \leq 2^\lambda$ and an index $i \in \{0, \dots, L-1\}$ (in binary) and* outputs a public hashing key hk. We let $\Sigma = \{0,1\}^s$ denote the block alphabet. The output size is $\ell = \ell(\lambda, s)$ and is independent of the input-length L.

- $H_{hk} : \Sigma^L \to \{0,1\}^\ell$: A deterministic poly-time algorithm that takes as input $x = (x[0], \ldots, x[L-1]) \in \Sigma^L$ and outputs $H_{hk}(x) \in \{0,1\}^\ell$.

We require the following properties:

Index Hiding: We consider the following game between an attacker $\mathcal{A}$ and a challenger:

- The attacker $\mathcal{A}(1^\lambda)$ chooses parameters $1^s, L$ and two indices $i_0, i_1 \in \{0, \ldots, L-1\}$.
- The challenger chooses a bit $b \leftarrow \{0,1\}$ and sets $hk \leftarrow \mathsf{Gen}(1^\lambda, 1^s, L, i_b)$.
- The attacker $\mathcal{A}$ gets hk and outputs a bit b'.

We require that for any PPT attacker $\mathcal{A}$ we have $|\Pr[b = b'] - \frac{1}{2}| \leq negl(\lambda)$ in the above game.

Somewhere Statistically Binding: We say that hk is statistically binding for an index $i \in [L]$ if there do not exist any values $x, x' \in \Sigma^L$ with $x[i] \neq x'[i]$ such that $H_{hk}(x) = H_{hk}(x')$. We require that for any parameters s, L and any integer $i \in \{0, \ldots, L-1\}$ we have:

$$\Pr[hk \text{ is statistically binding for index } i \ : \ hk \leftarrow \mathsf{Gen}(1^\lambda, 1^s, L, i)] \geq 1 - negl(\lambda).$$

We say that the hash is perfectly binding if the above probability is 1.

Definition 2.2 (SSB Hash with Local Opening). An SSB Hash with local opening $\mathcal{H} = (\mathsf{Gen}, H, \mathsf{Open}, \mathsf{Verify})$ consists of an SSB hash (Gen, H) with output size $\ell(\cdot, \cdot)$ along with two additional algorithms $\mathsf{Open}, \mathsf{Verify}$ and an opening size $p(\cdot, \cdot)$. The additional algorithms have the following syntax:

- $\pi \leftarrow \mathsf{Open}(hk, x, j)$: Given the hash key hk, $x \in \Sigma^L$ and an index $j \in \{0, \ldots, L-1\}$, creates an opening $\pi \in \{0,1\}^p$. The opening size $p = p(\lambda, s)$ is a polynomial which is independent of the input-length L.
- $\mathsf{Verify}(hk, y, j, u, \pi)$: Given a hash key hk a hash output $y \in \{0,1\}^\ell$, an integer index $j \in \{0, \ldots, L-1\}$, a value $u \in \Sigma$ and an opening $\pi \in \{0,1\}^p$, outputs a decision $\in \{\mathsf{accept}, \mathsf{reject}\}$. This is intended to verify that a pre-image x of $y = H_{hk}(x)$ has $x[j] = u$.

We require the following two additional properties.

Correctness of Opening: For any parameters s, L and any indices $i, j \in \{0, \ldots, L-1\}$, any $hk \leftarrow \mathsf{Gen}(1^\lambda, 1^s, L, i)$, $x \in \Sigma^L$, $\pi \leftarrow \mathsf{Open}(hk, x, j)$: we have $\mathsf{Verify}(hk, H_{hk}(x), j, x[j], \pi) = \mathsf{accept}$

Somewhere Statistically Binding w.r.t. Opening:[3] We say that hk is statistically binding w.r.t opening (abbreviated SBO) for an index i if there do not exist any values $y, u \neq u', \pi, \pi'$ s.t.

$$\mathsf{Verify}(hk, y, i, u, \pi) = \mathsf{Verify}(hk, y, i, u', \pi') = \mathsf{accept}.$$

---

[3] Note that the "somewhere stat. binding w.r.t. opening" property implies the basic "somewhere stat. binding" property of SSB hash.

We require that for any parameters s, L *and any index* $i \in \{0, \ldots, L-1\}$

$$\Pr[\mathsf{hk} \text{ is SBO for index } i \; : \; \mathsf{hk} \leftarrow \mathsf{Gen}(1^\lambda, 1^s, L, i)] \geq 1 - negl(\lambda).$$

We say that the hash is perfectly binding *w.r.t. opening* *if the above proba-bility is 1.*

Fixed-Parameter Variants. The above definitions allow for a flexible input-length L and block-length s specified by the user as inputs to the Gen algorithm. This will be the default throughout the paper, but we also consider variants of the above definition with a *fixed-input-length* L and/or *fixed-block-length* s where these values cannot be specified by the user as inputs to the Gen algorithm but are instead set to some fixed value (a constant or polynomial in the security parameter λ) determined by the scheme. In the case of a fixed-input-length variant, the definitions are non-trivial if the output-length ℓ and opening-size p satisfy $\ell, p < L \cdot s$.

Discussion. There are several constructions of SSB hash that do not provide local opening. For example, any PIR scheme can be used to realize an SSB hash without local opening. The hash key hk consists of a PIR query for index i and the hash $H_{\mathsf{hk}}(x)$ simply computes the PIR response using database x. Unfortunately, we do not know how to generically add a local opening capability to such SSB hash constructions.

3   Two-to-One SSB Hash

As our main building block, we rely on a notion of a "two-to-one SSB hash". Informally, this is a fixed-input-length and flexible-block-size SSB hash (we do not require local opening) that maps two input blocks ($L = 2$) to an output which is roughly the size of one block (up to some small multiplicative and additive factors).

Definition 3.1 (Two-to-One SSB Hash). *A two-to-one SSB hash is an SSB hash with a fixed input-length* $L = 2$ *and flexible block-length* s. *The output-length is* $\ell(\lambda, s) = s \cdot (1 + 1/\Omega(\lambda)) + \mathrm{POLY}(\lambda)$.

We give three constructions of a Two-to-One SSB Hash systems. Our first construction is built from the DDH-hard groups with compact representation. This construction achieves perfect binding. Our next construction is built from the DCR assumption. Lastly, we generalize our approach by showing a (variant of) Two-to-One SSB hashing that can work from *any* lossy function. We note that lossy functions can be built from a variety of number theoretic primitives including DDH (without compact representation), Learning with Errors, and the ϕ-hiding assumption.

Remark: Impossibility without Overhead. We note that the need for some "slack" is inherent in the above definition and we cannot get a two-to-one SSB hash where the output is exactly $\ell(\lambda, s) = s$ matching the size of one of the inputs. This is because in that case, if we choose $\mathsf{hk} \leftarrow \mathsf{Gen}(1^\lambda, 1^s, i = 0)$ then for each $x_0 \in \{0,1\}^s$ there is a unique choice of $y \in \{0,1\}^s$ such that $H_{\mathsf{hk}}(x_0, x_1) = y$ no matter what x_1 is. In other words, the function $H_{\mathsf{hk}}(x_0, x_1)$ does not depend on the argument x_1. Symmetrically, if $\mathsf{hk} \leftarrow \mathsf{Gen}(1^\lambda, 1^s, i = 1)$ then the function $H_{\mathsf{hk}}(x_0, x_1)$ does not depend on the argument x_0. These two cases are easy to distinguish.

3.1   Two-to-One SSB Hash from DDH

DDH Hard Groups and Representation Overhead. Let $\mathcal{G}$ be a PPT group generator algorithm that takes as input the security parameter 1^λ and outputs a pair $\mathbb{G}, p$ where $\mathbb{G}$ is a group description of prime order p for $p \in \Theta(2^\lambda)$.

Assumption 1 (Decision Diffie-Hellman Assumption). *Let* $(\mathbb{G}, p) \leftarrow \mathcal{G}(1^\lambda)$ *and* $b \leftarrow \{0,1\}$. *Choose a random generator* $g \in G$ *and random* $x, y \in \mathbb{Z}_p$ *Let* $T \leftarrow \mathbb{G}$ *if* $b = 0$, *else* $T \leftarrow g^{xy}$. *The advantage of algorithm* $\mathcal{A}$ *in solving the decision Diffie-Hellman problem is defined as*

$$\mathsf{Adv}_{\mathcal{A}} = \left| Pr[b \leftarrow \mathcal{A}(\mathbb{G}, p, g, g^x, g^y, T)] - \frac{1}{2} \right|.$$

We say that the Decision-Diffie Hallman assumption holds if for all PPT $\mathcal{A}$, $\mathsf{Adv}_{\mathcal{A}}$ *is negligible in* λ.

Representation Overhead of Group Elements. In this work we will be concerned with how efficiently (prime order) group elements are represented. We are interested in the difference between the number of bits to represent a group element and $\lfloor \lg(p) \rfloor$. In our definition we consider the bit representation of a group to be intrinsic to a particular group description.

Definition 3.2 (Representational Overhead). *Consider a family of prime order groups output from some group generation algorithm* $\mathcal{G}(1^\lambda)$ *that outputs a group of prime order* p *for* $2^\lambda < p < 2^{\lambda+1}$. *Notice that for a generator* g *in such a group that* $g^i \neq g^j$ *for* $i, j \in [0, 2^\lambda]$ *and* $i \neq j$. *(I.e. no "wraparound" happens.)*

We define the representational overhead $\delta(\lambda)$ *to be the function which expresses maximum difference between the number of bits used to represent a group element of* $\mathbb{G}$ *and* λ, *where* $\mathbb{G}, p \leftarrow \mathcal{G}(1^\lambda)$.

For this work we are interested in families of groups who representational overhead $\delta(\lambda)$ is some constant c. Examples of these include groups generated from strong primes and certain elliptic curve groups.

Construction of Two-to-One SSB. We now describe our Two-To-One SSB Hash. We will use a group generation algorithm $\mathcal{G}$ that has constant representational overhead c as defined in Definition 3.2. Consider a matrix $\mathbf{M}$ over $\mathbb{Z}_p$ and group generator g of order p we will use the notation $g^{\mathbf{M}}$ as short hand for giving out g raised to each element of $\mathbf{M}$.

The construction sets up a hash function key hk for a function that takes two s bit inputs x_A and x_B. If the index bit $\beta = 0$ it will be statistically binding on x_A; otherwise it is statistically binding on x_B. At a high level the construction setup is intuitively similar to the Lossy trapdoor function algorithms of Peikert and Waters [PW08] where the setup creates two functions — one injective and the other lossy and assigns whether the lossy function corresponds to the A or B input according to the index bit β.

There are two important differences from the basic PW construction. First the PW construction encrypted the input bit by bit. This low rate of encoding was needed in order to recover the input from a trapdoor in [PW08], but a trapdoor is not required for our hash function. Here we cram in as many bits into a group element as possible. This is necessary to satisfy the SSB output size properties. We note [BHK11] pack bits in a similar manner. The second property we have is that the randomness used to generate both the injective and lossy function is correlated such that we can intuitively combine the outputs of each into one output where the output length is both small and maintains the committing property of the injective function. We note that our description describes the procedures directly and the connection to injective and lossy functions is given for intuition, but not made formal.

$\mathsf{Gen}_{\text{Two-to-One}}(1^\lambda, 1^s, \beta \in \{0,1\})$
The generation algorithm first sets $t = \max(\lambda, \lfloor \sqrt{s \cdot c} \rfloor)$. (The variable t will be the number of bit each group element can uniquely represent.) It then calls $\mathcal{G}(1^t) \rightarrow (\mathbb{G}, p)$ with $2^t < p < 2^{t+1}$ and chooses a random generator $g \in \mathbb{G}$.

Next, it lets $d = \lceil \frac{s}{t} \rceil$. It then chooses random $w_1, \ldots, w_d \in \mathbb{Z}_p$, two random column vectors $\boldsymbol{a} = (a_1, \ldots, a_d) \in \mathbb{Z}_p^d$ and $\boldsymbol{b} = (b_1, \ldots, b_d) \in \mathbb{Z}_p^d$. We let $\tilde{\mathbf{A}}$ be the $d \times d$ matrix over $\mathbb{Z}_p$ where the (i,j)-th entry is $a_i \cdot w_j$ and $\tilde{\mathbf{B}}$ be the $d \times d$ matrix over $\mathbb{Z}_p$ where the (i,j)-th entry is $b_i \cdot w_j$. Finally, let $\mathbf{A}$ be $\tilde{\mathbf{A}} + (1 - \beta) \cdot \mathbf{I}$ and $\mathbf{B}$ be $\tilde{\mathbf{B}} + \beta \cdot \mathbf{I}$ where $\mathbf{I}$ is the identity matrix. (I.e. we add in the identity matrix to $\tilde{\mathbf{A}}$ to get the $\mathbf{A}$ matrix if the selection bit $\beta = 0$; otherwise, if $\beta = 1$ add in the identity matrix to $\tilde{\mathbf{B}}$ to get $\mathbf{B}$.)

The hash key is $\mathsf{hk} = (g^{\boldsymbol{a}}, g^{\boldsymbol{b}}, g^{\mathbf{A}}, g^{\mathbf{B}})$.

$H_{\mathsf{hk}} : \{0,1\}^s \times \{0,1\}^s \rightarrow \mathbb{G}^{d+1}$
The hash function algorithm takes in two inputs $x_A \in \{0,1\}^s$ and $x_B \in \{0,1\}^s$. We can view the bitstrings x_A and x_B each as consisting of d blocks each of t bits (except the last block which may be less). The function first parses these each as row vectors $\boldsymbol{x}_A = (x_{A,1}, \ldots, x_{A,d})$ and $\boldsymbol{x}_B = (x_{B,1}, \ldots, x_{B,d})$. These have the property that for $j \in [d]$ we have $x_{A,j}$ is an integer $< 2^t \leq p$ representing the $j - th$ block of bits as an integer.

Next, it computes

$$V = g^{x_A a + x_B b}, \ Y = g^{x_A \mathbf{A} + x_B \mathbf{B}}.$$

We observe that V is one group element in $\mathbb{G}$ and Y is a vector of d group elements. Thus the output size of the hash is $(d + 1) \cdot (t + c)$ bits.

Analysis. We now analyze the size overhead, index hiding and binding properties of the hash function.

Overhead. The output of the hash function is $d + 1$ group elements each of which takes $t + c$ bits to represent for a total output size of $(d+1)(t+c)$ bits. In the case where $\lfloor\sqrt{s \cdot c}\rfloor \geq \lambda$, we can plug in our parameter choices for t, d and see that the outputsize $\ell(\lambda, s) = s + \mathcal{O}(\sqrt{s})$, thus matching the requirements of Definition 3.1. In the case where $\lfloor\sqrt{s \cdot c}\rfloor < \lambda$ we have that $\ell(\lambda, s) = s + \mathcal{O}(\lambda)$ thus also matching our definition.

Somewhere Statistically Binding. We show that the hash function above is selectively binding respective to the bit β. We demonstrate this for the $\beta = 0$ case. The $\beta = 1$ case follows analogously.

Suppose a hash key hk were setup according to the process $\mathsf{Gen}_{\mathrm{Two-to-One}}$ as above with the input $\beta = 0$. Now consider the evaluation $H_{\mathsf{hk}}(x_A, x_B) = (V, Y = (Y_1, \ldots, Y_d))$. We have that for all $j \in [1, d]$ that $Y_j / V^{w_j} = g^{x_{A,j}}$. Let's verify this claim. First from the hash definition we can work out that

$$V = g^{\Sigma_{i \in [d]} x_{A,i} a_i + x_{B,i} b_i}$$

and

$$Y_j = g^{x_{A,j} + \Sigma_{i \in [d]} x_{A,i}(a_i w_j) + x_{B,i}(b_i w_j)} = g^{x_{A,j}} g^{w_j(\Sigma_{i \in [d]} x_{A,i} a_i + x_{B,i} b_i)}.$$

The claim that $Y_j / V^{w_j} = g^{x_{A,j}}$ follows immediately from these equations.

Now suppose that we are given two inputs (x_A, x_B) and (x'_A, x'_B) such that $x_A \neq x'_A$ There must then exist some j such that $x_{A,j} \neq x'_{A,j}$. Let $H_{\mathsf{hk}}(x_A, x_B) = (V, Y = (Y_1, \ldots, Y_d))$ and $H_{\mathsf{hk}}(x'_A, x'_B) = (V', Y' = (Y'_1, \ldots, Y'_d))$. From the above claim it follows that $Y_j / V^{w_j} = g^{x_{A,j}}$ and $Y_j / V^{w_j} = g^{x'_{A,j}}$. Therefore $(V, Y_j) \neq (V', Y'_j)$ and the outputs of the hashes are distinct.

Index Hiding. We now prove index hiding. To do this we define $\mathsf{Game}_{\mathrm{normal}}$ to be the normal index hiding game on the two-to-one construction and $\mathsf{Game}_{\mathrm{random}}$ to be the index hiding game, but where the matrices $\tilde{\mathbf{A}}$ and $\tilde{\mathbf{B}}$ are chosen randomly when constructing the hash function hk.

We first argue that if the decision Diffie-Hellman assumption holds, then the advantage of any PPT attacker $\mathcal{A}$ in $\mathsf{Game}_{\mathrm{normal}}$ must be negligibly close to its advantage in $\mathsf{Game}_{\mathrm{random}}$. To show this we apply a particular case of the decision matrix linear assumption family introduced by Naor and Segev [NS12]. They show (as part of a more general theorem) that if the decision Diffie-Hellman assumption holds that a PPT attacker cannot distinguish if a $2d \times (d+1)$ matrix $\mathbf{M}$ over $\mathbb{Z}_p$ was sampled randomly from the set of rank 1 matrices or rank $d+1$ matrices given $g^{\mathbf{M}}$.

Suppose that the difference of advantage for some attacker in Game $_{\text{normal}}$ and Game $_{\text{random}}$ is some non-negligible function of λ. Then we construct an algorithm $\mathcal{B}$ on the above decision matrix linear assumption. $\mathcal{B}$ receives a challenge $g^{\mathbf{M}}$ and breaks this into $g^{\mathbf{M}_A}$ and $g^{\mathbf{M}_B}$ where $\mathbf{M}_A$ is the top half of the matrix $\mathbf{M}$ and $\mathbf{M}_B$ is the bottom half. It then takes $g^{\mathbf{a}}$ from the first column of $g^{\mathbf{M}_A}$ and $g^{\tilde{\mathbf{A}}}$ as the remaining d columns. Similarly, $\mathcal{B}$ takes $g^{\mathbf{b}}$ from the first column of $g^{\mathbf{M}_B}$ and $g^{\tilde{\mathbf{B}}}$ as the remaining d columns. It then samples a random index $\beta \in \{0,1\}$ and continues to use these values in executing $\text{Gen}_{\text{Two-to-One}}$, giving the hash key hk to the attack algorithm.

If $g^{\mathbf{M}}$ were sampled as a rank 1 matrix, then the view of the attacker is the same as executing Game $_{\text{normal}}$. Otherwise, if $g^{\mathbf{M}}$ were sampled as a rank $d+1$ matrix the attacker's view is statistically close to Game $_{\text{random}}$ (as choosing a random rank $d+1$ matrix is statistically close to choosing a random matrix). If the attacker $\mathcal{A}$ correctly guesses $\beta' = \beta$, then $\mathcal{B}$ guesses the matrix was rank 1, else it guesses it was rank $d+1$. If the difference in advantage of $\mathcal{A}$ in the two games is non-neglgibile, then $\mathcal{B}$ has a non-negligible advantage in the decision matrix game.

Finally, we see that in Game $_{\text{random}}$ any attacker's advantage must be 0 as the distributions of the outputs are independent of β.

3.2   Two-to-One SSB Hash from DCR

We can also construct a two-to-one hash with perfect binding from the decisional composite residuocity (DCR) assumption. We do so by relying on the Damgård-Jurik cryptosystem [DJ01] which is itself a generalization of the Pallier cryptosystem based on the DCR assumption [Pai99]. We rely on the fact that this cryptosystem is additively homomorphic and "length flexible", meaning that it has a small ciphertext expansion. When we plug this construction of a two-to-one SSB hash into our full construction of SSB hash with local opening, we essentially get the private-information retrieval (PIR) scheme of Lipmaa [Lip05]. Note that, in general, PIR implies SSB hash but only without local opening. However, the particular PIR construction of [Lip05] already has a tree-like structure which enables efficient local opening.

Damgård-Jurik. The Damgård-Jurik cryptosystem consists of algorithms (KeyGen, Enc, Dec). The key generation $(\text{pk}, \text{sk}) \leftarrow \text{KeyGen}(1^{\lambda})$ generates a public key $\text{pk} = n = pq$ which is a product of two primes p, q and $\text{sk} = (p, q)$. For any (polynomial) w the scheme can be instantiated to have plaintext space $\mathbb{Z}_{n^w}$ and ciphertext space $\mathbb{Z}_{n^{w+1}}^*$. The encryption/decryption procedures $c = \text{Enc}_{\text{pk}}(m; r)$ and $\text{Dec}_{sk}(c)$ satisfy perfect correctness so that for all $m \in \mathbb{Z}_{n^w}$ and all possible choices of the randomness r we have $\text{Dec}_{sk}(\text{Enc}_{\text{pk}}(m; r)) = m$. Moreover the scheme is additively homomorphic, meaning that there is an operation $\oplus$

such that $\mathsf{Enc}_{\mathsf{pk}}(m; r) \oplus \mathsf{Enc}_{\mathsf{pk}}(m'; r') = \mathsf{Enc}_{\mathsf{pk}}(m + m'; r'')$ for some r'' (the operation $+$ is in the ring $\mathbb{Z}_{n^w}$). Similarly, we can define homomorphic subtraction $\ominus$. Furthermore, by performing repeated addition we can also implement an operation $\otimes$ that allows for multiplication by a plaitnext element $\mathsf{Enc}_{\mathsf{pk}}(m; r) \otimes m' = \mathsf{Enc}_{\mathsf{pk}}(m \cdot m'; r')$ for some r' (the operation $\cdot$ is in the ring $\mathbb{Z}_{n^w}$). The semantic security of the cryptosystem holds under the DCR assumption.

Construction of Two-to-One SSB. We use the Damgård-Jurik cryptosystem to construct an SSB hash as follows.

$\mathsf{hk} \leftarrow \mathsf{Gen}(1^\lambda, 1^s, \beta \in \{0,1\})$ Choose $(\mathsf{pk}, \mathsf{sk}) \leftarrow \mathsf{KeyGen}(1^\lambda)$ to be a Damgård-Jurik public/secret key. We assume (without loss of generality) that the modulus n satisfies $n > 2^\lambda$. Set the parameter w which determines the plaintext space $\mathbb{Z}_{n^w}$ and the ciphertext space $\mathbb{Z}^*_{n^{w+1}}$ to be $w = \lceil s/\log n \rceil$ so that we can interpret $\{0,1\}^s$ as a subset of $\mathbb{Z}_{n^w}$. Choose $c \leftarrow \mathsf{Enc}_{\mathsf{pk}}(\beta)$ and output $\mathsf{hk} = (\mathsf{pk}, c)$.

$H_{\mathsf{hk}}(x_0, x_1)$: Parse $\mathsf{hk} = (\mathsf{pk}, c)$ and interpret the values $x_0, x_1 \in \{0,1\}^s$ as ring elements $x_0, x_1 \in \mathbb{Z}_{n^w}$. Define the value $\mathbf{1}_{ct} = \mathsf{Enc}_{\mathsf{pk}}(1; r_0)$ to be a fixed encryption of 1 using some fixed randomness r_0 (say, all 0s). Compute $c^* := (x_1 \otimes c) \oplus (x_0 \otimes (\mathbf{1}_{ct} \ominus c))$. By the homomorphic properties of encryption, c^* is an encryption of x_β.

Theorem 3.3. *The above construction is a two-to-one SSB hash with perfect binding under the DCR assumption.*

Proof. The index hiding property follows directly from the semantic security of the Damgård-Jurik cryptosystem, which in turn follows from the DCR assumption.

The perfect binding property follows from the perfect correctness of the cryptosystem. In particular, if $\mathsf{hk} \leftarrow \mathsf{Gen}(1^\lambda, 1^s, \beta)$ then $y = H_{\mathsf{hk}}(x_0, x_1)$ satisfies $y = \mathsf{Enc}_{\mathsf{pk}}(x_\beta; r)$ for some r which perfectly determines x_β.

Lastly, the output size of the hash function is

$$\ell(s, \lambda) = (w + 1)\lceil \log n \rceil = (\lceil s/\log n \rceil + 1)\lceil \log n \rceil$$
$$\leq (1 + 1/\log n)s + O(\log n) = (1 + 1/\Omega(\lambda))s + \mathsf{poly}(\lambda).$$

3.3   SSB with Local Opening from Two-to-One SSB

We now show how to construct a SSB hash with local opening from a two-to-one SSB hash via the "Merkle Tree" construction. Assume that $\mathcal{H} = (\mathsf{Gen}, H)$ is a two-to-one SSB hash family with output length give by $\ell(s, \lambda)$. We use this hash function in a Merkle-Tree to construct an SSB hash with local opening $\mathcal{H}^* = (\mathsf{Gen}^*, H^*, \mathsf{Open}, \mathsf{Verify})$ as follows.

– $\mathsf{hk} \leftarrow \mathsf{Gen}^*(1^\lambda, 1^s, L, i)$: Let $(b_q, \ldots, b_1)$ be the binary representation of i (with b_1 being the least significant bit) where $q = \lceil \log L \rceil$. For $j \in [q]$ define the block-lengths $s_1, \ldots, s_q$ where $s_1 = s$ and $s_{j+1} = \ell(s_j, \lambda)$. Choose $\mathsf{hk}_j \leftarrow \mathsf{Gen}(1^\lambda, 1^{s_j}, b_j)$ and output $\mathsf{hk} = (\mathsf{hk}_1, \ldots, \mathsf{hk}_q)$.

- $y = H^*_{\mathsf{hk}}(x)$: For $x = (x[0], \ldots, x[L-1]) \in \Sigma^L$, $\mathsf{hk} = (\mathsf{hk}_1, \ldots, \mathsf{hk}_q)$ proceed as follows. Define T to be a complete binary tree of height q where level 0 of the tree denotes the leaves and level q denotes the root. We will assign a label to each vertex in the tree. The L leaf vertices are assigned the labels $x[0], \ldots, x[L-1]$. The rest of the labels are assigned inductively where each non-leaf vertex v at level j of the tree with children that have labels x'_0, x'_1 gets assigned the label $H_{\mathsf{hk}_j}(x'_0, x'_1)$. The output of the hash is the label y assigned to the root of the tree.
- $\pi = \mathsf{Open}(\mathsf{hk}, x, j)$: Compute the labeled tree T as above. Output the labels of all the sibling nodes along the path from the root to the j'th leaf.
- $\mathsf{Verify}(\mathsf{hk}, y, j, u, \pi)$: Recompute all of the labels of the nodes in the tree T that lie on the path from the root to the j'th leaf by using the value u for that leaf and the values given by π as the labels of all the sibling nodes along the path. Check that the recomputed label on the root of the tree is indeed y.

Theorem 3.4. *If $\mathcal{H}$ is a two-to-one SSB hash then $\mathcal{H}^*$ is a SSB hash with local opening.*

Proof. Firstly, the *index hiding* property of $\mathcal{H}^*$ follows directly from that of $\mathcal{H}$ via q hybrid arguments. In particular, if $i_0 = (b^0_q, \ldots, b^0_1)$ and $i_1 = (b^1_q, \ldots, b^1_1)$ are the two indices chosen by the attacker during the security game for index hiding, then we can prove the indistinguishability of $\mathsf{hk}^0 \leftarrow \mathsf{Gen}^*(1^\lambda, L, s, i_0)$ and $\mathsf{hk}^1 \leftarrow \mathsf{Gen}^*(1^\lambda, L, s, i_1)$ via q hybrid games where we switch the component keys $\mathsf{hk} = (\mathsf{hk}_1, \ldots, \mathsf{hk}_q)$ from being chosen as $\mathsf{hk}_j \leftarrow \mathsf{Gen}(1^\lambda, 1^s, b^0_j)$ to $\mathsf{hk}_j \leftarrow \mathsf{Gen}(1^\lambda, 1^s, b^1_j)$.

Secondly, to show that $\mathcal{H}^*$ is *somewhere statistically binding w.r.t. opening*, assume that there exist some $y, u \neq u', \pi, \pi'$ s.t. $\mathsf{Verify}(\mathsf{hk}, y, i, u, \pi) = \mathsf{Verify}(\mathsf{hk}, y, i, u', \pi') = \mathsf{accept}$. Recall that the verification procedure assigns labels to all the nodes along the path from the root to the i'th leaf. During the two runs of the verification procedure with the above inputs, let $0 < j \leq q$ be the lowest level at which both runs assign the same label w to the node at level j (this must exist since the root at level q is assigned the same label y in both runs and the leafs at level 0 are assigned different values u, u' in the two runs). Let v, v' be the two different labels assigned to the node at level $j-1$ by the two runs. Then $w = H_{\mathsf{hk}_j}(x) = H_{\mathsf{hk}_j}(x')$ for some $x, x' \in \Sigma^2$ such that $x[b_j] = v \neq x'[b_j] = v'$. This means that hk_j is not statistically binding on the index b_j, but this can only happen with negligible probability by the somewhere statistically binding property of the 2-to-1 SSB hash $\mathcal{H}$. Therefore $\mathcal{H}^*$ is somewhere statistically binding w.r.t. opening.

Lastly, the output length of $\mathcal{H}^*$ is given by $\ell^*(s, \lambda) = s_{q+1}$ where $s_1 = s$ and for each other $j \in [q]$, $s_{j+1} = \ell(s_j, \lambda)$. The output length of a SSB hash guarantees that $\ell(s_j, \lambda) = s_j(1 + 1/\Omega(\lambda)) + a(\lambda)$ where $a(\cdot)$ is some fixed polynomial. This ensures that

$$\ell^*(s, \lambda) = s(1 + 1/\Omega(\lambda))^q + a(\lambda) \sum_{j=0}^{q-1} (1 + 1/\Omega(\lambda))^j = O(s) + a(\lambda)O(\lambda)$$

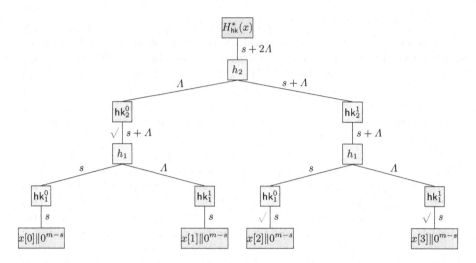

Fig. 1. Illustration of the SSB hash from a lossy function with key hk ← Gen*$(1^\lambda, 1^s, L = 2^q, i, \Lambda)$, i.e., $H_{hk}(x)$ perfectly binds $x[i = 2]$. For every level $j \in \{1, \ldots, q\}$ we sample a pairwise independent function $h_j : \{0,1\}^{2m'} \to \{0,1\}^m$, where $m = 2(s + q\Lambda) + \lambda$ for a statistical security parameter λ, and two functions $hk_j^0, hk_j^1 : \{0,1\}^m \to \{0,1\}^{m'}$ from an (m, Λ)-lossy family of functions, one lossy and one injective (we decide which one of the two is the injective one such that the path from the perfectly binded value – here $x[2]$ – to the root only contains injective functions). The injective and lossy functions are shown in green and red, respectively. The SBB hash is now a Merkle-hash with the hash function $H_j(a, b) = h_j(hk_j^0(a), hk_j^1(b))$ used in level j. An edge label t in the figure means that there are at most 2^t possible values at this point, e.g., there are 2^s values of the form $x[0]\|0^{m-s}$ and the output of a lossy function like hk_1^1 has at most 2^Λ values. To locally open a value, say $x[2]$, we reveal $x[2]$ all the siblings of the nodes on the path from $x[2]$ to the root, those are marked with $\sqrt{}$ in the figure.

is polynomial in s, λ. We rely on the fact that $q \leq \lambda$ to argue that $(1+1/\Omega(\lambda))^q \leq (1 + 1/\Omega(\lambda))^\lambda = O(1)$.

4   SSB Hash from Lossy Functions

In this section we describe a simple construction of an SSB Hash with local opening, the main tool we'll use are lossy functions, introduced by Peikert and Waters [PW08]. They actually introduced the stronger notion of lossy *trapdoor* functions, where a trapdoor allowed to invert functions with injective keys, we only need the lossiness property, but no trapdoors.

Definition 4.1. *An (m, Λ)-lossy function is given by a tuple of PPT algorithms*

- *For $m, \Lambda \in \mathbb{N}$ and* mode *$\in \{$injective $= 1$, lossy $= 0\}$,* $Gen_{LF}(m, \Lambda, mode)$ *outputs a key* hk.

- *Every such key* hk *defines a function* $\mathsf{hk}(.) : \{0,1\}^m \rightarrow \{0,1\}^{m'}$ *(for some $m' \geq m$).*

We have the following three properties:

injective: *If* hk $\leftarrow \mathsf{Gen}_{\mathsf{LF}}(m, \Lambda, \mathsf{injective})$, *then* $\mathsf{hk}(.)$ *is injective.*
lossy: *If* hk $\leftarrow \mathsf{Gen}_{\mathsf{LF}}(m, \Lambda, \mathsf{lossy})$, *then* $\mathsf{hk}(.)$*'s output domain has size $\leq 2^\Lambda$, i.e.*

$$|\{y \; : \; \exists x \in \{0,1\}^m, \mathsf{hk}(x) = y\}| \leq 2^\Lambda$$

indistinguishable: *Lossy and injective keys are computationally indistinguishable. More concretely, think of Λ as a security parameter and let $m = \mathsf{poly}(\Lambda)$, then the advantage of any PPT adversary in distinguishing* $\mathsf{Gen}_{\mathsf{LF}}(m, \Lambda, \mathsf{injective})$ *from* $\mathsf{Gen}_{\mathsf{LF}}(m, \Lambda, \mathsf{lossy})$ *is negligible in Λ.*

The Construction. Our construction $(\mathsf{Gen}^*, H^*, \mathsf{Open}, \mathsf{Verify})$ is illustrated in Fig. 1, we define it formally below.

- hk $\leftarrow \mathsf{Gen}^*(1^\lambda, 1^s, L = 2^q, i, \Lambda)$: Set $m = 2(s + q\Lambda) + \lambda$. For $i \in \{0, \dots, 2^q - 1\}$, let $(b_q, \dots, b_1)$ be the binary representation of i (with b_1 being the least significant bit).
  For every j: Choose $\mathsf{hk}_i^0 \leftarrow \mathsf{Gen}_{\mathsf{LF}}(m, \Lambda, 1 - b_j)$ and $\mathsf{hk}_i^1 \leftarrow \mathsf{Gen}_{\mathsf{LF}}(m, \Lambda, b_j)$. Sample a pairwise independent hash function $h_j : \{0,1\}^{2m'} \rightarrow \{0,1\}^m$ and let $\mathsf{hk}_j = (\mathsf{hk}_j^0, \mathsf{hk}_j^1, h_j)$. Each hk_j defines a mapping $H_j : \{0,1\}^{2m} \rightarrow \{0,1\}^m$ defined as
  $$H_j(a, b) = h_j(\mathsf{hk}_j^0(a), \mathsf{hk}_j^1(b))$$
  Output hk $= (\mathsf{hk}_1, \dots, \mathsf{hk}_q)$.
- $H_{\mathsf{hk}}^*(x)$: For $x = (x[0], \dots, x[2^q - 1]) \in \{0,1\}^{s \cdot 2^q}$, hk $= (\mathsf{hk}_1, \dots, \mathsf{hk}_q)$ proceed as follows. Define T to be a complete binary tree of height q where level 0 of the tree denotes the leaves and level q denotes the root. We will assign a label to each vertex in the tree. The 2^q leaf vertices are assigned the labels $x[0]\|0^{m-s}, \dots, x[2^q - 1]\|0^{m-s}$ (i.e., the input blocks padded to length m). The rest of the labels are assigned inductively where each non-leaf vertex v at level j of the tree with children that have labels x_0', x_1' gets assigned the label $y = H_j(x_0', x_1')$. The output $H_{\mathsf{hk}}^*(x)$ is the root of the tree.
- $\pi = \mathsf{Open}(\mathsf{hk}, x, j)$: Compute the labeled tree T as above. Output the labels of all the sibling nodes along the path from the root to the j'th leaf. Figure 1 the values to be opened to reveal $x[2]$ are marked with $\sqrt{}$.
- $\mathsf{Verify}(\mathsf{hk}, y, j, u, \pi)$: Recompute all of the labels of the nodes in the tree T that lie on the path from the root to the j'th leaf by using the value u for that leaf and the values given by π as the labels of all the sibling nodes along the path. Check that the recomputed label on the root of the tree is indeed y.

Theorem 4.2. *The construction of a SSB Hash (with local opening) described below, which maps $L = 2^q$ blocks of length s bits to a hash of size $m = 2(s +$*

$q\Lambda) + \lambda$ *bits where* λ *is a statistical security parameter and we assume* (m, Λ)-*lossy functions, is secure. More concretely, the somewhere statistically binding property holds with probability*

$$1 - q/2^\lambda$$

over the choice of the hash key, and the index hiding property can be reduced to the indistinguishability property of the lossy function losing a factor q.

Proof. The index hiding property follows immediately from the indsitinguishability of injective and lossy modes.

To show that the hash is somewhere statistically binding, consider a key hk $\leftarrow$ Gen*$(1^\lambda, 1^s, L = 2^q, i, \Lambda)$. We must prove that with overwhelming probability no hash $y \in \{0,1\}^{2(s+q\cdot\Lambda)+\lambda}$ exists where Verify(hk, y, i, u, π) = Verify(hk, y, i, u', π') = accept for some $u \neq u'$, that is, $x[i]$ can be opened to u and u'.

In a nutshell, the reason why with high probability (over the choice of hk) the hash H_{hk} is perfectly binding on its ith coordinate is that the value $x[i]$ at the leaf of the tree only passes through two kinds of functions on its way to the root: injective functions and pairwise independent hashes. Clearly, no information can be lost when passing through an injective function. And every time the value passes through some hash h_j, the other half of the input is the output of a lossy function, and thus can take at most 2^Λ possible values. Thus even as we arrive at the root, there are only $2^{s+q\cdot\Lambda}$ possible values. We now set the output length $m = 2(s + q \cdot \Lambda) + \lambda$ of the h_j's so that 2^m is a larger – by a factor 2^λ – than the square of the possible values. This then suffices to argue that every h_j will be injective on its possible inputs (recall that there are at most $2^{s+q\cdot\Lambda}$ of them) with probability $\geq 1 - 2^{-\lambda}$.

For the formal proof it's convenient to consider the case $i = 0$ (i.e., the leftmost value should be perfectly binding). Let $\pi = (w_1, \ldots, w_q)$ and $\pi' = (w'_1, \ldots, w'_q)$ be two openings for values $x[0] \neq x'[0]$, we'll prove that with probability $q/2^\lambda$ (over the choice of hk) the verification procedure will compute different hashes corresponding to any two such openings (i.e., for every opening $(\pi, x[0])$, there's at most one y which makes Verify(hk, $y, i = 0, x[0], \pi$) accept), and thus the hash is perfectly binding on index 0.

Let $v_0 = x[i]\|0^{m-s}$ and for $j = 1, \ldots, q$ define $v_j = h_j(\text{hk}_j^0(v_{j-1}), \text{hk}_j^1(w_j))$, the v'_j's are defined analogously for the other opening. Note that v_q is the final hash value, so we have to show that $v_q \neq v'_q$.

We will do so by induction, first, we claim that (for any hk) there are at most $2^{s+j\cdot\Lambda}$ possible values v_j can take. This is true for $j = 0$ as $v_0 = x[0]\|0^{m-s}$ can take exactly 2^s values by definition. Assume it holds for $j - 1$ and let $S_{j-1}, |S_{j-1}| \leq 2^{s+(j-1)\Lambda}$ denote the set of values v_{j-1} can take, then

$$|S_j| = |\{h_j(\text{hk}_j^0(v_{j-1}), \text{hk}_j^1(z)) \ : \ v_{j-1} \in S_{j-1}, z \in \{0,1\}^m)\}| \quad (1)$$

$$\leq |\{(v_{j-1}, \text{hk}_j^1(z)) \ : \ v_{j-1} \in S_{j-1}, z \in \{0,1\}^m))\}| \quad (2)$$

$$\leq |S_{j-1}| \cdot 2^\Lambda \quad (3)$$

$$\leq 2^{s+j\Lambda} \quad (4)$$

where the first step follows by definition of the set S_j, the second step follows as applying deterministic functions cannot increase the number of possible values, the third step follows as $\mathsf{hk}_j^1(.)$ is lossy and thus can take at most 2^Λ possible values. The last step follows by the induction hypothesis for $j-1$.

For the proof we will think of the hash key $\mathsf{hk} = (\mathsf{hk}_1, \ldots, \mathsf{hk}_q)$, where $\mathsf{hk}_j = (\mathsf{hk}_j^0, \mathsf{hk}_j^1, h_j)$, as being lazy sampled. Initially, we sample all the $\mathsf{hk}_j^0, \mathsf{hk}_j^1$ keys. Let $L_j \subset \{0,1\}^m$ denote the range of the (lossy) $\mathsf{hk}_j^1(.)$ functions, note that $|L_j| \leq 2^\Lambda$ for all j. The h_j's will be sampled one by one in each induction step below.

Assume so far he have sampled $h_1, \ldots, h_{j-1}$, and so far for any openings where $x[0] \neq x'[0]$ we had $v_j \neq v_j'$. For $j = 0$ this holds as $x[0] \neq x'[0]$ implies $v_0 = x[0]\|0^{m-s} \neq x'[0]\|0^{m-s}$.

The inputs to the function h_j (which is still to be sampled) are from $I_{j-1} = h_j^0(S_{j-1}) \times L_{j-1}$, which (as shown above) contains at most $|S_{j-1}| \cdot |L_{j-1}| \leq 2^{s+(j-1)\Lambda}2^\Lambda = 2^{s+j\cdot\Lambda}$ elements.

We now sample the pairwise independent hash h_j, as it has range 2^m the probability that any two elements $(v,l) \neq (v',l') \in I_{j-1}$ collide[4] is 2^{-m}, taking the union bound over all pairs of elements we get

$$2^{2(s+j\cdot\Lambda)}/2^m \leq 2^{-\lambda}$$

Taking the union bound, we get that the probability that the induction fails for any of the q steps is $q/2^\lambda$ as claimed.

5   SSB from ϕ-hiding

We now move on to building SSB from the ϕ-hiding assumption [CMS99]. This construction will be qualitatively different from the prior ones in that we will not employ a Merkle tree type structure for proving and verifying opens. In contrast a hash output will consist of two elements $\mathbb{Z}_{N_0}^*$ and $\mathbb{Z}_{N_1}^*$ for RSA primes N_0, N_1. An opening will consist of a single element of either $\mathbb{Z}_{N_0}^*$ or $\mathbb{Z}_{N_1}^*$.

Our construction is structurally similar to standard constructions RSA accumulators [BdM93]. Intuitively, the initial hash key will consist of two RSA moduli N_0, N_1 as well as two group elements h_0, h_1 and keys K_0, K_1 which hash to prime exponents. To compute the hash on input $x \in \{0,1\}^L$ let $S_0 = \{i : x[i] = 0\}$ be the set of all indices where the i-th bit is 0 and $S_1 = \{i : x[i] = 1\}$ be the set of indices where the i-th bit is 1. The function computes the output

$$y_0 = h_0^{\prod_{i \in S_0} F_{K_0}(i)} \mod N_0, \quad y_1 = h_1^{\prod_{i \in S_1} F_{K_1}(i)} \mod N_1.$$

To prove that the j-th bit was 0 the open algorithm will give the $F_{K_0}(j)$-th root of y_0. It computes this by letting $S_0 = \{i : x[i] = 0\}$ and setting

---

[4] Note that we prove something slightly stronger than required as we only need to consider pairs where $v \neq v'$.

$\pi = h_0^{\prod_{i \neq j \in S_0} F_{K_0}(i)}$ mod N_0. A proof can be checked by simply checking if $y_0 \overset{?}{=} \pi^{F_{K_0}(j)}$ mod N_0. (Proving an opening of 1 follows analogously).

The algorithms as described above very closely match a traditional RSA accumulator. The key distinction is that we can achieve statistical binding on index j by setting N_0 such that $K_0(j)$ divides $\phi(N_0)$ (and similarly for N_1). The idea is that in this setting if y_0 is not an $K_0(j)$-th residue then there will not exist a value π such that $y_0 \overset{?}{=} \pi^{F_{K_0}(j)}$ mod N_0. The index-hiding property will follow from the ϕ-hiding assumption.

5.1 RSA and ϕ-hiding Preliminaries

We begin by developing our notation and statement of the ϕ-hiding assumption both of which follow closely to Kiltz, O'Neill, and Smith [KOS10]. We let $\mathcal{P}_k$ denote the set of odd primes that are less than 2^k. In addition, we let $(N, p, q) \overset{\$}{\leftarrow} RSA_k$ be the process of choosing two primes p, q uniformly from $\mathcal{P}_k$ and letting $N = pq$. Further we let $(N, p, q) \overset{\$}{\leftarrow} RSA_k[p = 1 \mod e]$ be the be the process of choosing two primes p, q uniformly from $\mathcal{P}_k$ with the constraint that $p = 1$ mod e, then letting $N = pq$.

We can now state the ϕ-hiding assumption relative to some constant $0 < c < .5$. Consider the following distributions relative to a security parameter λ.

$$\mathcal{R} = \{(e, N) : e, e' \overset{\$}{\leftarrow} \mathcal{P}_{c\lambda}; (N, p, q) \overset{\$}{\leftarrow} RSA_\lambda[p = 1 \mod e']\}$$

$$\mathcal{L} = \{(e, N) : e \overset{\$}{\leftarrow} \mathcal{P}_{c\lambda}; (N, p, q) \overset{\$}{\leftarrow} RSA_\lambda[p = 1 \mod e]\}$$

Cachin, Micali and Stadler [CMS99] show that the two distributions can be efficiently sampled if the Extended Riemann Hypothesis holds. The ϕ-hiding assumption states that for all $c \in (0, .5)$ no PPT attacker can distinguish between the two distributions with better than negligible in λ probability.

5.2 Conforming Function

Before we give our construction we need one further abstraction. For any integer L we require the ability to sample a keyed hash function $F(K, \cdot)$ that hashes from an integer $i \in [0, L - 1]$ to a random prime in $\mathcal{P}_{c\lambda}$. Furthermore, the function should have the property that it is possible to sample the key K in such a way that for a single pair $i^* \in [0, L - 1]$ and $e^* \in \mathcal{P}_{c\lambda}$ $F(K, i^*) = e^*$. Moreover such programming should be undetectable if e^* is chosen at random from $\mathcal{P}_{c\lambda}$.

We give the definitions of such a function system here and show how to construct one in Appendix A. A conforming function system is parameterized by a constant $c \in (0, .5)$ and has three algorithms.

SAMPLE-NORMAL$(1^\lambda, L) \to K$
Takes in a security parameter λ and a length L (in binary) and outputs a function key K.

SAMPLE-PROGRAM$(1^\lambda, L, i^*, e^*) \to K$
Takes in a security parameter λ and a length L (in binary) as well as a program index $i^* \in [0, L-1]$ and $e^* \in \mathcal{P}_{c\lambda}$. It outputs a function key K.

$F_K : i \to \mathcal{P}_{c\lambda}$
If SAMPLE-NORMAL$(1^\lambda, L) \to K$, then F_K takes in an index $i \in [0, L-1]$ and outputs a prime from $\mathcal{P}_{c\lambda}$.

Properties. Such a system will have four properties:

Efficiency. The programs SAMPLE-NORMAL and SAMPLE-PROGRAM run in time polynomial in λ and L. Let SAMPLE-NORMAL$(1^\lambda, L) \to K$, then F_K runs in time polynomial in λ and $\lg(L)$.

Programming at i^*. For some λ, L, i^*, e^* let SAMPLE-PROGRAM$(1^\lambda, L, i^*, e^*) \to K$. Then $F_K(i^*) = e^*$ with all but negligible probability in λ.

Non colliding at i^*. For some λ, L, i^*, e^* let SAMPLE-PROGRAM$(1^\lambda, L, i^*, e^*) \to K$. Then for any $i \neq i^*$ the probability that $F_K(i^*) = F_K(i)$ is negligible in λ.

Indistinguishability of Setup. For any L, i^* consider the following two distributions:

$$\mathcal{R}_{L,i^*} = \{K : e^* \xleftarrow{\$} \mathcal{P}_{c\lambda}; \text{SAMPLE-NORMAL}(1^\lambda, L) \to K\}$$

$$\mathcal{L}_{L,i^*} = \{K : e^* \xleftarrow{\$} \mathcal{P}_{c\lambda}; \text{SAMPLE-PROGRAM}(1^\lambda, L, i^*, e^*) \to K\}$$

The indistinguishability of setup property states that all PPT adversaries have a most a negligible advantage in distinguishing between the two distributions for all L, i^*.

5.3 Our ϕ-hiding SSB Construction

We now present our ϕ-hiding based SSB construction. Our construction is for an alphabet of a single bit, thus s is implicitly 1 and omitted from our notation. In addition, the construction is parameterized relative to some constant $c \in (0, .5)$.

Gen$(1^\lambda, L, i^*)$
The generation algorithm first samples two random primes $e_0, e_1 \xleftarrow{\$} \mathcal{P}_{c\lambda}$. Next, it sets up two conforming functions as SAMPLE-PROGRAM $(1^\lambda, L, i^*, e_0) \to K_0$ and SAMPLE-PROGRAM $(1^\lambda, L, i^*, e_1) \to K_1$. Then it samples $(N_0, p_0, q_0) \xleftarrow{\$} \mathcal{RSA}_k[p_0 = 1 \mod e_0]$ and $(N_1, p_1, q_1) \xleftarrow{\$} \mathcal{RSA}_k[p_1 = 1 \mod e_1]$. Finally, it chooses $h_0 \in Z^*_{N_0}$ randomly with the constraint that $h_0^{(p_0-1)/e_0} \neq 1 \mod p_0$ and $h_1 \in Z^*_{N_1}$ randomly with the constraint that $h_1^{(p_1-1)/e_1} \neq 1 \mod p_1$.
It outputs the hash key as $\mathsf{hk} = \{L, (N_0, N_1), (K_0, K_1), (h_0, h_1)\}$.

$H_{\mathsf{hk}} : \{0,1\}^L \to Z^*_{N_0}, Z^*_{N_1}:$
On input $x \in \{0,1\}^L$ let $S_0 = \{i : x[i] = 0\}$ be the set of all indices where the

i-th bit is 0 and $S_1 = \{i : x[i] = 1\}$ be the set of indices where the i-th bit is 1. The function computes

$$y_0 = h_0^{\prod_{i \in S_0} F_{K_0}(i)} \mod N_0, \quad y_1 = h_1^{\prod_{i \in S_1} F_{K_1}(i)} \mod N_1.$$

The hash output is $y = (y_0, y_1)$.

We note that the computation in practice will be done by iteratively with repeated exponentiation as opposed to computing the large integer $\prod_{i \in S_0} F_{K_0}(i)$ up front.

Open(hk, x, j):
If $x_j = 0$ it first lets $S_0 = \{i : x[i] = 0\}$. Then it computes

$$\pi = h_0^{\prod_{i \neq j \in S_0} F_{K_0}(i)} \mod N_0.$$

Otherwise, if $x_j = 1$ it first lets $S_1 = \{i : x[i] = 1\}$. Then it computes

$$\pi = h_1^{\prod_{i \neq j \in S_1} F_{K_1}(i)} \mod N_1.$$

Verify($\mathsf{hk}, y = (y_0, y_1), j, b \in \{0, 1\}, \pi$):

The verify algorithm checks

$$y_b \stackrel{?}{=} \pi^{F_{K_b}(j)} \mod N_b.$$

Properties. We now show that the above construction meets the required properties for SSB with local opening. One minor difference from the original definition is that we weaken the statistically binding requirement. Previously, we wanted the binding property to hold for any hash digest y, even one which does not correspond to a correctly generated hash output. In the version we achieve here, we require that $y = H_{\mathsf{hk}}(x)$ for some x. We define the property formally below.

Weak Somewhere Statistically Binding w.r.t. Opening: We say that hk is *weak statistically binding w.r.t opening (abbreviates wSBO) for an index i* if there do not exist any values $x \in \Sigma^L, u' \neq x[i], \pi'$ s.t. Verify($\mathsf{hk}, H_{\mathsf{hk}}(x), i, u', \pi$) = accept. We require that for any parameters s, L and any index $i \in \{0, \ldots, L - 1\}$

$$\Pr[\mathsf{hk} \text{ is wSBO for index } i \ : \ \mathsf{hk} \leftarrow \mathsf{Gen}(1^\lambda, 1^s, L, i)] \geq 1 - \mathsf{negl}(\lambda).$$

We say that the hash is *perfectly binding w.r.t. opening* if the above probability is 1.

Correctness of Opening. Consider any hk generated from the setup algorithm and let π be the output from a call to Open(hk, x, j) for some x, j where that $x[j] = b \in \{0, 1\}$. Then $y_b = h_b^{\prod_{i \in S_b} F_{K_b}(i)}$ mod N_b and $\pi = h_b^{\prod_{i \neq j \in S_b} F_{K_b}(i)}$ mod N_b. It follows that $\pi^{F_{K_b}(j)} = y_b$ mod N_b.

Weak Somewhere Statistically Binding with Respect to Opening. Suppose that Gen($1^\lambda, L, i^*$) $\to$ hk. We argue that with all but negligible probability for all inputs $x \in \{0, 1\}^L$ that the function is statistically binding with respect to opening.

Consider a particular input x where $x[i^*] = 1 - b$ and $H_{hk}(x) = y = (y_0, y_1)$. We want to show that there does not exist a value π such that Verify(hk, $y = (y_0, y_1), i^*, b \in \{0, 1\}, \pi) = 1$. Let $e_b \in \mathcal{P}_{c\lambda}$ be the prime value chosen at hash function setup. By the setup process we have that $e_b | p_b - 1$ and that of $e_b = F_{K_b}(i^*)$. The latter follows from the Programming at i^* property of the conforming hash function. Therefore we have that $(\pi^{e_b})^{(p_b - 1)/e_b} = 1$ mod p_1 (i.e. π^{e_b} is an e_b-th residue mod p_b).

Recall that $y_b = h_b^{\prod_{i \in S_b} F_{K_b}(i)}$ mod N_b. Let $\alpha = \prod_{i \in S_b} F_{K_b}(i)$. By the non-colliding property of F coupled with the fact that $x[i^*] \neq b$ and thus $i^* \notin S_b$ with all but negligible probability for all $i \in S_b$ we have that $F_{K_b}(i)$ is a prime $\neq e_b$. Therefore α is relatively prime to e_b. Since h_b was chosen to not be a e_b-th residue mod p_b and α is relatively prime to e_b it follows that $y_b = h_b^\alpha$ is also not an e_b-th residue mod p_b. However, since π^{e_b} is an e_b-th residue mod p_b, it cannot be equal to y_b and the verification test will fail.

Index Hiding. We sketch a simple proof of index hiding via a sequence of games. We begin by defining the sequence.

- Game $_0$: The Index Hiding game on our construction.
- Game $_1$: Same as Game $_0$ except that an additional prime $e_0' \xleftarrow{\$} \mathcal{P}_{c\lambda}$ is sampled and $(N_0, p_0, q_0) \xleftarrow{\$} \mathcal{RSA}_\lambda[p = 1 \mod e_0']$. Note that we still sample $(1^\lambda, L, i_b, e_0) \to K_0$ where w.h.p $e_0 \neq e_0'$. additional prime $e_1' \xleftarrow{\$} \mathcal{P}_{c\lambda}$ is sampled and $(N_1, p_1, q_1) \xleftarrow{\$} \mathcal{RSA}_\lambda[p = 1 \mod e_1']$. Note that we still sample $(1^\lambda, L, i_b, e_1) \to K_1$ where w.h.p $e_1 \neq e_1'$.
- Game $_3$: Same as Game $_2$ except that K_0 is sampled as SAMPLE-NORMAL $(1^\lambda, L) \to K_0$.
- Game $_4$: Same as Game $_3$ except that K_1 is sampled as SAMPLE-NORMAL $(1^\lambda, L) \to K_1$.

It follows directly from the ϕ-hiding assumption that no PPT attacker can distinguish between Game $_1$ and Game $_1$ and that no attacker can distinguish between Game $_1$ and Game $_2$. At this point the primes e_0 and e_1 are used only in the programming of the hash function and are not reflected in the choice of the RSA moduli. For this reason we can now use the Indistinguishability of Setup property of the conforming has to show that no PPT attack can distinguish between Game $_2$ and Game $_3$ and Game $_3$ and Game $_4$. Finally, we observe that the index i_b is not used at Game $_4$ and thus the bit b is hidden from the attacker's view.

6   Positional Accumulators

In the full version of this paper, we also discuss how to extend some the above results to positional accumulators. In particular, we show how to construct positional accumulators from a (perfectly binding) two-to-one SSB hash. The construction can also be naturally extended to one based on lossy functions.

A   Constructing a Conforming Function

We now give our construction of a conforming hash function per the definition given in Sect. 5.2.

Recall, our goal is to construct a keyed hash function $F(K, \cdot)$ that hashes from an integer $i \in [0, L - 1]$ to a prime in $\mathcal{P}_{c\lambda}$. Furthermore, the function should have the property that it is possible to sample the key K in such a way that for a single pair $i^* \in [0, L - 1]$ and $e^* \in \mathcal{P}_{c\lambda}$ we have $F(K, i^*) = e^*$. (The constant $c \in (0, .5)$ is considered a parameter of the system.) Moreover, such programming should be undetectable if e^* is sampled at random from $\mathcal{P}_{c\lambda}$.

Our construction below is a simple implementation of this abstraction and all properties are statistically guaranteed (i.e. we do not require any computational assumptions).

SAMPLE-NORMAL$(1^\lambda, L) \to K$
We first let $B = 2^{\lfloor c\lambda \rfloor}$ and let $T = \lambda^2$. The algorithm chooses random $w_1, \ldots, w_T \in [0, B - 1]$. The key K is set as $K = (\lambda, w_1, \ldots, w_T)$.

SAMPLE-PROGRAM$(1^\lambda, L, i^*, e^*) \to K$
We first let $B = 2^{\lfloor c\lambda \rfloor}$ and let $T = \lambda^2$. Initialize a bit (local to this computation) PROGRAMMED to be 0. Then proceed in the following manner:

For $j = 1$ to T if PROGRAMMED $\overset{?}{=} 1$ choose v_j randomly in $[0, B - 1]$ and set $w_j = v_j - i^* \mod B$. This corresponds to the case where the value e^* was "already programmed". Else, if PROGRAMMED $\overset{?}{=} 0$, it first chooses v_j randomly in $[0, B - 1]$. If v_j is not prime it simply sets $w_j = v_j - i^* \mod B$. Otherwise, it sets $w_j = e^* - i^* \mod B$ and flips the bit PROGRAMMED to 1 so that e^* will not be programmed in again.

The key K is output as $K = (\lambda, w_1, \ldots, w_T)$.

$F_K : i \to \mathcal{P}_{c\lambda}$
The function proceeds as follows. Starting at $j = 1$ to T the function tests if $w_j + i$ is a prime (i.e. is in $\mathcal{P}_{c\lambda}$). If so it outputs $w_j + i$ and halts. Otherwise, it

increments j and tests again. If j goes past T and no primes have been found, the algorithm outputs a default prime $3 \in \mathcal{P}_{c\lambda}$.[5]

Properties. We now confirm that our function meets all the required properties.

Efficiency. The programs SAMPLE-NORMAL chooses T random values where T is polynomial in λ and SAMPLE-PROGRAM also chooses T random values as well as performing up to T primality tests. The keysizes of both are T integers in $[0, B]$. Thus the running times and keysizes are polynomial in λ and $\lg(L)$.

Programming at i^*. Consider a call to SAMPLE-PROGRAM$(1^\lambda, L, i^*, e^*) \to K$. The function $F_K(i^*)$ will resolve to the smallest j such that $w_j + i^*$ is a prime (if any of these are a prime). By the design of SAMPLE-PROGRAM this will be e^* since it puts in $w_j = e^* - i^* \mod B$ the first time a prime is sampled. In constructing the function if all v_j sampled were composite then $F_K(i^*) \neq e^*$, however, this will only occur with negligible probability since the probability of choosing T random integers in $2^{\lfloor c\lambda \rfloor}$ and none of them being prime is negligible.

Non colliding at i^*. For some λ, L, i^*, e^* let SAMPLE-PROGRAM$(1^\lambda, L, i^*, e^*) \to K$. Let's assume that $F_K(i^*) = e^*$. We first observe that the chances that there exist any pairs $(i_0, j_0) \neq (i_1, j_1)$ such that $w_{j_0} + i_0 = w_{j_1} + i_1$ is negligible. We consider the probability of this happening on an arbitrary pair and them apply the union bound.

Consider a pair $(i_0, j_0) \neq (i_1, j_1)$ If $j_0 = j_1$ this cannot happen since the two terms differ by $i_1 - i_0$. Otherwise, we notice that the probability of a particular pair colliding is at most $1/B$ (which is negligible) since v_{j_0} and v_{j_1} are chosen independently at random. Since there are at most a polynomial $\binom{T \cdot L}{2}$ such pairs the chances that any collide is negligible.

It follows that the chances of $F_K(i^*) = F_K(i)$ for $i = i^*$ is negligible since the above condition would be necessary for this to occur.

Indistinguishability of Setup. For any L, i^* consider the following two distributions:

$$\mathcal{R}_{L,i^*} = \{K : e^* \xleftarrow{\$} \mathcal{P}_{c\lambda}; \text{SAMPLE-NORMAL}(1^\lambda, L) \to K\}$$

$$\mathcal{L}_{L,i^*} = \{K : e^* \xleftarrow{\$} \mathcal{P}_{c\lambda}; \text{SAMPLE-PROGRAM}(1^\lambda, L, i^*, e^*) \to K\}$$

We argue that these two distributions are identical for all L, i^*. We show this by also considering an intermediate distribution $\mathcal{I}_{L,i^*}$. This distribution is generated by randomly sampling v_j in $[0, B-1]$ and setting $w_j = v_j - i^* \mod B$. This distribution is clearly equivalent to the SAMPLE-NORMAL distribution as for all j selecting w_j randomly and selecting v_j randomly and setting $w_j = v_j - i^* \mod B$ both result in w_j being chosen uniformly at random.

---

[5] Note there is nothing special about choosing 3. Any default prime would suffice.

We now argue that this intermediate distribution is equivalent to the $\mathcal{L}_{L,i^*}$ distribution which is equivalent to calling SAMPLE-PROGRAM with sampling e^* randomly from $\mathcal{P}_{c\lambda}$. We will step through an execution of SAMPLE-PROGRAM and argue that at each step j from $j = 1, \ldots, T$ v_j is chosen randomly from $[0, B-1]$ independently of all other $v_{j'}$ for $j' < j$. Consider an execution starting with $j = 1$ and PROGRAMMED $= 0$ and for our exposition let's consider that $e^* \in \mathcal{P}_{c\lambda}$ has not been sampled yet. While PROGRAMMED $\overset{?}{=} 0$ the algorithm samples v_j is sampled at random. If v_j is composite it is kept and put in the key, otherwise if it is prime in $\mathcal{P}_{c\lambda}$, v_j is replaced with e^* as another randomly sampled prime. Thus, for any composite value x the probability that $w_j + i^* = x$ is $1/B$ and for any prime value x the probability that $w_j + i^* = x$ is also $1/B$. The reason is that replacing any sampled prime with a different randomly sampled prime does not change the distribution.

After PROGRAMMED is set to 1 all further v_j values are chosen uniformly at random.

Remark A.1. We note that the Indistinguishability of Setup property holds perfectly while the programmability property holds statistically. One way to flip this is to always program $v_T = e^*$ at the end if if e^* has not been programmed in already.

References

[BdM93] Benaloh, J.C., de Mare, M.: One-way accumulators: a decentralized alternative to digital signatures. In: Helleseth, T. (ed.) EUROCRYPT 1993. LNCS, vol. 765, pp. 274–285. Springer, Heidelberg (1994)

[BGI+12] Barak, B., Goldreich, O., Impagliazzo, R., Rudich, S., Sahai, A., Vadhan, S.P., Yang, K.: On the (im)possibility of obfuscating programs. J. ACM **59**(2), 6 (2012)

[BHK11] Braverman, M., Hassidim, A., Kalai, Y.T.: Leaky pseudo-entropy functions. In: Proceedings of the Innovations in Computer Science - ICS 2010, Tsinghua University, Beijing, China, 7–9 January 2011, pp. 353–366 (2011)

[CCC+15] Chen, Y.C., Chow, S.S., Chung, K.M., Lai, R.W., Lin, W.K., Zhou, H.S.: Computation-trace indistinguishability obfuscation and its applications. Cryptology ePrint Archive, Report 2015/406 (2015). http://eprint.iacr.org/

[CH15] Canetti, R., Holmgren, J.: Fully succinct garbled RAM. Cryptology ePrint Archive, Report 2015/388 (2015). http://eprint.iacr.org/

[CMS99] Cachin, C., Micali, S., Stadler, M.A.: Computationally private information retrieval with polylogarithmic communication. In: Stern, J. (ed.) EUROCRYPT 1999. LNCS, vol. 1592, p. 402. Springer, Heidelberg (1999)

[DJ01] Damgård, I., Jurik, M.: A generalisation, a simplification and some applications of Paillier's probabilistic public-key system. In: Kim, K. (ed.) PKC 2001. LNCS, vol. 1992, pp. 119–136. Springer, Heidelberg (2001)

[GGH+13] Garg, S., Gentry, C., Halevi, S., Raykova, M., Sahai, A., Waters, B.: Candidate indistinguishability obfuscation and functional encryption for all circuits. In: FOCS (2013)

[HW15]   Hubacek, P., Wichs, D.: On the communication complexity of secure func-
         tion evaluation with long output. In: Proceedings of the 2015 Conference on
         Innovations in Theoretical Computer Science, ITCS 2015, Rehovot, Israel,
         11–13 January 2015, pp. 163–172 (2015)
[KLW15] Koppula, V., Lewko, A.B., Waters, B.: Indistinguishability obfuscation for
         turing machines with unbounded memory. In: STOC (2015)
[KOS10] Kiltz, E., O'Neill, A., Smith, A.: Instantiability of RSA-OAEP under
         chosen-plaintext attack. In: Rabin, T. (ed.) CRYPTO 2010. LNCS, vol.
         6223, pp. 295–313. Springer, Heidelberg (2010)
[Lip05] Lipmaa, H.: An oblivious transfer protocol with log-squared communica-
         tion. In: Zhou, J., López, J., Deng, R.H., Bao, F. (eds.) ISC 2005. LNCS,
         vol. 3650, pp. 314–328. Springer, Heidelberg (2005)
[NS12]   Naor, M., Segev, G.: Public-key cryptosystems resilient to key leakage.
         SIAM J. Comput. $41(4)$, 772–814 (2012)
[Pai99] Paillier, P.: Public-key cryptosystems based on composite degree residuos-
         ity classes. In: Stern, J. (ed.) EUROCRYPT 1999. LNCS, vol. 1592, pp.
         223–238. Springer, Heidelberg (1999)
[PW08]   Peikert, C., Waters, B.: Lossy trapdoor functions and their applications.
         In: Proceedings of the 40th Annual ACM Symposium on Theory of Com-
         puting, Victoria, British Columbia, Canada, 17–20 May 2008, pp. 187–196
         (2008)
[SW14]   Sahai, A., Waters, B.: How to use indistinguishability obfuscation: deniable
         encryption, and more. In: STOC, pp. 475–484 (2014)
[Zha14] Zhandry, M.: Adaptively secure broadcast encryption with small system
         parameters. In: IACR Cryptology ePrint Archive 2014, p. 757 (2014)

Discrete Logarithms and Number Theory

Computing Individual Discrete Logarithms Faster in GF(p^n) with the NFS-DL Algorithm

Aurore Guillevic[1,2]([✉])

[1] Inria Saclay, Palaiseau, France
[2] École Polytechnique/LIX, Palaiseau, France
guillevic@lix.polytechnique.fr

Abstract. The Number Field Sieve (NFS) algorithm is the best known method to compute discrete logarithms (DL) in finite fields $\mathbb{F}_{p^n}$, with p medium to large and $n \geq 1$ small. This algorithm comprises four steps: polynomial selection, relation collection, linear algebra and finally, individual logarithm computation. The first step outputs two polynomials defining two number fields, and a map from the polynomial ring over the integers modulo each of these polynomials to $\mathbb{F}_{p^n}$. After the relation collection and linear algebra phases, the (virtual) logarithm of a subset of elements in each number field is known. Given the target element in $\mathbb{F}_{p^n}$, the fourth step computes a preimage in one number field. If one can write the target preimage as a product of elements of known (virtual) logarithm, then one can deduce the discrete logarithm of the target.

As recently shown by the Logjam attack, this final step can be critical when it can be computed very quickly. But we realized that computing an individual DL is much slower in medium- and large-characteristic non-prime fields $\mathbb{F}_{p^n}$ with $n \geq 3$, compared to prime fields and quadratic fields $\mathbb{F}_{p^2}$. We optimize the first part of individual DL: the *booting step*, by reducing dramatically the size of the preimage norm. Its smoothness probability is higher, hence the running-time of the booting step is much improved. Our method is very efficient for small extension fields with $2 \leq n \leq 6$ and applies to any $n > 1$, in medium and large characteristic.

Keywords: Discrete logarithm · Finite field · Number field sieve · Individual logarithm

1 Introduction

1.1 Cryptographic Interest

Given a cyclic group $(G, \cdot)$ and a generator g of G, the discrete logarithm (DL) of $x \in G$ is the element $1 \leq a \leq \#G$ such that $x = g^a$. In well-chosen groups, the exponentiation $(g, a) \mapsto g^a$ is very fast but computing a from (g, x) is conjectured to be very difficult: this is the Discrete Logarithm Problem (DLP),

---

This research was partially funded by Agence Nationale de la Recherche grant ANR-12-BS02-0001.

T. Iwata and J.H. Cheon (Eds.): ASIACRYPT 2015, Part I, LNCS 9452, pp. 149–173, 2015.
DOI: 10.1007/978-3-662-48797-6_7

at the heart of many asymmetric cryptosystems. The first group proposed for DLP was the multiplicative group of a prime finite field. Nowadays, the group of points of elliptic curves defined over prime fields are replacing the prime fields for DLP-based cryptosystems. In pairing-based cryptography, the finite fields are still used, because they are a piece in the pairing mechanism. It is important in cryptography to know precisely the difficulty of DL computation in the considered groups, to estimate the security of the cryptosystems using them. Finite fields have a particularity: there exists a subexponential-time algorithm to compute DL in finite fields of medium to large characteristic: the Number Field Sieve (NFS). In small characteristic, this is even better: a quasi-polynomial-time algorithm was proposed very recently [7].

In May 2015, an international team of academic researchers revealed a surprisingly efficient attack against a Diffie-Hellman key exchange in TLS, the *Logjam* attack [2]. After a seven-day-precomputation stage (for relation collection and linear algebra of NFS-DL algorithm), it was made possible to compute any given individual DL in about one minute, for each of the two targeted 512-bit prime finite fields. This was fast enough for a man-in-the-middle attack. This experience shows how critical it can be to be able to compute individual logarithms very fast.

Another interesting application for fast individual DL is *batch-DLP*, and *delayed-target DLP*: in these contexts, an attacker aims to compute several DL in the same finite field. Since the costly phases of relation collection and linear algebra are only done one time for any fixed finite field, only the time for one individual DL is multiplied by the number of targets. This context usually arises in pairing-based cryptography and in particular in broadcast protocols and traitor tracing schemes, where a large number of DLP-based public/private key pairs are generated. The time to compute one individual DL is important in this context, even if parallelization is available.

1.2 The Number Field Sieve Algorithm for DL in Finite Fields

We recall that the NFS algorithm is made of four steps: polynomial selection, relation collection, linear algebra and finally, individual logarithm computation. *This last step is mandatory to break any given instance of a discrete logarithm problem.* The polynomial selection outputs two irreducible polynomials f and g defining two number fields K_f and K_g. One considers the rings $R_f = \mathbb{Z}[x]/(f(x))$ and $R_g = \mathbb{Z}[x]/(g(x))$. There exist two maps ρ_f, ρ_g to $\mathbb{F}_{p^n}$, as shown in the following diagram. Moreover, the monic polynomial defining the finite field is $\psi = \gcd(f, g) \mod p$, of degree n.

$$\mathbb{Z}[x]$$

$$R_f = \mathbb{Z}[x]/(f(x)) \qquad \mathbb{Z}[y]/(g(y)) = R_g$$

$$\rho_f : x \mapsto z \qquad\qquad \rho_g : y \mapsto z$$

$$\mathbb{F}_{p^n} = \mathbb{F}_p[z]/(\psi(z))$$

In the remaining of this paper, we will only use $\rho = \rho_f$, $K = K_f$ and R_f. After the relation collection and linear algebra phases, the (virtual) logarithm of a subset of elements in each ring R_f, R_g is known. The individual DL step computes a preimage in one of the rings R_f, R_g of the target element in $\mathbb{F}_{p^n}$. If one can write the target preimage as a product of elements of known (virtual) logarithm, then one can deduce the individual DL of the target. The key point of individual DL computation is finding a smooth decomposition in small enough factors of the target preimage.

1.3 Previous Work on Individual Discrete Logarithm

The asymptotic running time of NFS algorithm steps are estimated with the L-function:

$$L_Q[\alpha, c] = \exp\left(\big(c + o(1) \big) (\log Q)^\alpha (\log \log Q)^{1-\alpha} \right) \quad \text{with } \alpha \in [0,1] \text{ and } c > 0.$$

The α parameter measures the gap between polynomial time ($L_Q[\alpha = 0, c] = \log^c Q$) and exponential time ($L_Q[\alpha = 1, c] = Q^c$). When c is implicit, or obvious from the context, we simply write $L_Q[\alpha]$. When the complexity relates to an algorithm for a prime field $\mathbb{F}_p$, we write $L_p[\alpha, c]$.

Large Prime Fields. Many improvements for computing discrete logarithms first concerned prime fields. The first subexponential DL algorithm in prime fields was due to Adleman [1] and had a complexity of $L_p[1/2, 2]$. In 1986, Coppersmith, Odlyzko and Schroeppel [13] introduced a new algorithm (COS), of complexity $L_p[1/2, 1]$. They computed individual DL [13, Sect. 6] in $L_p[1/2, 1/2]$ in two steps (finding a boot of medium-sized primes, then finding relations of logarithms in the database for each medium prime). In these two algorithms, the factor basis was quite large (the smoothness bound was $L_p[1/2, 1/2]$ in both cases), providing a much faster individual DL compared to relation collection and linear algebra. This is where the common belief that individual logarithms are easy to find (and have a negligible cost compared with the prior relation collection and linear algebra phases) comes from.

In 1993, Gordon [15] proposed the first version of NFS–DL algorithm for prime fields $\mathbb{F}_p$ with asymptotic complexity $L_p[1/3, 9^{1/3} \simeq 2.08]$. However, with the $L_p[1/3]$ algorithm there are new difficulties, among them the individual DL phase. In this $L_p[1/3]$ algorithm, many fewer logarithms of small elements are known, because of a smaller smoothness bound (in $L_p[1/3]$ instead of $L_p[1/2]$). The relation collection is shortened, explaining the $L_p[1/3]$ running time. But in the individual DL phase, since some non-small elements in the decomposition of the target have an unknown logarithm, a dedicated sieving and linear algebra phase is done for each of them. Gordon estimated the running-time of individual DL computation to be $L_p[1/3, 9^{1/3} \simeq 2.08]$, i.e. *the same as the first two phases*. In 1998, Weber [24, Sect. 6] compared the NFS–DL algorithm to the COS algorithm for a 85 decimal digit prime and made the same observation about individual DL cost.

In 2003, ten years after Gordon's algorithm, Joux and Lercier [17] were the first to dissociate in NFS relation collection plus linear algebra on one side and individual DL on the other side. They used the *special-q* technique to find the logarithm of medium-sized elements in the target decomposition. In 2006, Commeine and Semaev [11] analyzed the Joux–Lercier method. They obtained an asymptotic complexity of $L_p[1/3, 3^{1/3} \simeq 1.44]$ for computing individual logarithms, independent of the relation collection and linear algebra phases. In 2013, Barbulescu [4, Sects. 4 and 7.3] gave a tight analysis of the individual DL computation for prime fields, decomposed in three steps: booting (also called smoothing), descent, and final combination of logarithms. The booting step has an asymptotic complexity of $L_p[1/3, 1.23]$ and the descent step of $L_p[1/3, 1.21]$. The final computation has a negligible cost.

Non-prime Fields of Medium to Large Characteristic. In 2006, Joux, Lercier, Smart and Vercauteren [19] computed a discrete logarithm in a cubic extension of a prime field. They used the special-q descent technique again. They proposed for large characteristic fields an equivalent of the *rational reconstruction* technique for prime fields and the *Waterloo* algorithm [8] for small characteristic fields, to improve the initializing step preceding the descent. For DLs in prime fields, the target is an integer modulo p. The rational reconstruction method outputs two integers of half size compared to p, such that their quotient is equal to the target element modulo p. Finding a smooth decomposition of the target modulo p becomes equivalent to finding a (simultaneous) smooth decomposition of two elements, each of half the size. We explain their method (that we call the JLSV fraction method in the following) for extension fields in Sect. 2.3.

Link with Polynomial Selection. The running-time for finding a smooth decomposition depends on the norm of the target preimage. The norm preimage depends on the polynomial defining the number field. In particular, the smaller the coefficients and degree of the polynomial, the smaller the preimage norm. Some polynomial selection methods output polynomials that produce much smaller norm. That may be one of the reasons why the record computation of Joux *et al.* [19] used another polynomial selection method, whose first polynomial has very small coefficients, and the second one has coefficients of size $O(p)$. Thanks to the very small coefficients of the first polynomial, their fraction technique was very useful. Their polynomial selection technique is now superseded by their JLSV$_1$ method [19, Sect. 2.3] for larger values of p. As noted in [19, Sect. 3.2], the fraction technique is useful in practice for small n. But for the JLSV$_1$ method and $n \geq 3$, this is already too slow (compared to not using it). In 2008, Zajac [25] implemented the NFS-DL algorithm for computing DLs in $\mathbb{F}_{p^6}$ with p of 40 bits (12 decimal digits (dd), i.e. $\mathbb{F}_{p^6}$ of 240 bits or 74 dd). He used the methods described in [19], with a first polynomial with very small coefficients and a second one with coefficients in $O(p)$. In this case, individual DL computation was possible (see the well-documented [25, Sect. 8.4.5]). In 2013, Hayasaka, Aoki, Kobayashi and Takagi [16] computed a DL in $\mathbb{F}_{p^{12}}$ with $p = 122663$ (p^n of 203 bits or 62 dd). We noted that all these records used the same polynomial selection method, so that one of the polynomials has very small coefficients (e.g. $f = x^3 + x^2 - 2x - 1$) whereas the second one has coefficients in $O(p)$.

In 2009, Joux, Lercier, Naccache and Thomé [18] proposed an attack of DLP in a protocol context. The relation collection is sped up with queries to an oracle. They wrote in [18, Sect. B] an extended analysis of individual DL computation. In their case, the individual logarithm phase of the NFS-DL algorithm has a running-time of $L_Q[1/3, c]$ where $c = 1.44$ in the large characteristic case, and $c = 1.62$ in the medium characteristic case. In 2014, Barbulescu and Pierrot [3] presented a multiple number field sieve variant (MNFS) for extension fields, based on Coppersmith's ideas [12]. The individual logarithm is studied in [3, Sect. A]. They also used a *descent* technique, for a global estimated running time in $L_Q[1/3, (9/2)^{1/3}]$, with a constant $c \approx 1.65$. Recently in 2014, Barbulescu, Gaudry, Guillevic and Morain [5,6] announced 160 and 180 decimal digit discrete logarithm records in quadratic fields. They also used a technique derived from the JLSV fraction method and a special-q descent technique, but did not give an asymptotic running-time. It appears that this technique becomes inefficient as soon as $n = 3$ or 4.

Overview of NFS-DL Asymptotic Complexities. The running-time of the relation collection step and the individual DL step rely on the smoothness probability of integers. An integer is said to be B-smooth if all its prime divisors are less than B. An ideal in a number field is said to be B-smooth if it factors into prime ideals whose norms are bounded by B. Usually, the relation collection and the linear algebra are balanced, so that they have both the same dominating asymptotic complexity. The NFS algorithm for DL in prime and large characteristic fields has a dominating complexity of $L_Q[1/3, (\frac{64}{9})^{1/3} \simeq 1.923]$. For the individual DL in a prime field $\mathbb{F}_p$, the norm of the target preimage in the number field is bounded by p. This bound gives the running time of this fourth step (much smaller than relation collection and linear algebra). Finding a smooth decomposition of the preimage and computing the individual logarithm (see [11]) has complexity $L_p[1/3, c]$ with $c = 1.44$, and $c = 1.23$ with the improvements of [4]. The booting step is dominating. In large characteristic fields, the individual DL has a complexity of $L_Q[1/3, 1.44]$, dominated by the booting step again ([18, Sect. B] for JLSV$_2$, Table 3 for gJL).

In generic medium characteristic fields, the complexity of the NFS algorithm is $L_Q[1/3, (\frac{128}{9})^{1/3} = 2.42]$ with the JLSV$_1$ method proposed in [19, Sect. 2.3], $L_Q[1/3, (\frac{32}{3})^{1/3} = 2.20]$ with the Conjugation method [6], and $L_Q[1/3, 2.156]$ with the MNFS version [23]. We focus on the individual DL step with the JLSV$_1$ and Conjugation methods. In these cases, the preimage norm bound is in fact much higher than in prime fields. Without any improvements, the dominating booting step has a complexity of $L_Q[1/3, c]$ with $c = 1.62$ [18, Sect. C] or $c = 1.65$ [3, Sect. A]. However, this requires to sieve over ideals of degree $1 < t < n$. For the Conjugation method, this is worse: the booting step has a running-time of $L_Q[1/3, 6^{1/3} \simeq 1.82]$ (see our computations in Table 3). Applying the JLSV fraction method lowers the norm bound to $O(Q)$ for the Conjugation method. The individual logarithm in this case has complexity $L_Q[1/3, 3^{1/3}]$ as for prime fields (without the improvements of [4, Sect. 4]). However, this method is not suited for number fields generated with the JLSV_1 method, for $n \geq 3$.

1.4 Our Contributions

In practice, we realized that the JLSV fraction method which seems interesting and sufficient because of the $O(Q)$ bound, is in fact not convenient for the gJL and Conjugation methods for n greater than 3. The preimage norm is much too large, so finding a smooth factorization is too slow by an order of magnitude. We propose a way to lift the target from the finite field to the number field, such that the norm is strictly smaller than $O(Q)$ for the gJL and Conjugation methods:

Theorem 1. *Let $n > 1$ and $s \in \mathbb{F}_{p^n}^*$ a random element (not in a proper subfield of $\mathbb{F}_{p^n}$). We want to compute its discrete logarithm modulo ℓ, where $\ell \mid \Phi_n(p)$, with Φ_n the n-th cyclotomic polynomial. Let K_f be the number field given by a polynomial selection method, whose defining polynomial has the smallest coefficient size, and $R_f = \mathbb{Z}[x]/(f(x))$.*

Then there exists a preimage $\boldsymbol{r}$ in R_f of some $r \in \mathbb{F}_{p^n}^*$, such that $\log \rho(\boldsymbol{r}) \equiv \log s \pmod{\ell}$ and such that the norm of $\boldsymbol{r}$ in K_f is bounded by $O(Q^e)$, where e is equal to

1. *$1 - \frac{1}{n}$ for the gJL and Conjugation methods;*
2. *$\frac{3}{2} - \frac{3}{2n}$ for the JLSV_1 method;*
3. *$1 - \frac{2}{n}$ for the Conjugation method, if K_f has a well-chosen quadratic subfield satisfying the conditions of Lemma 3;*
4. *$\frac{3}{2} - \frac{5}{2n}$ for the JLSV_1 method, if K_f has a well-chosen quadratic subfield satisfying the conditions of Lemma 3.*

Our method reaches the optimal bound of $Q^{\varphi(n)/n}$, with $\varphi(n)$ the Euler totient function, for $n = 2, 3, 4, 5$ combined with the gJL or the Conjugation method. We show that our method provides a dramatic improvement for individual logarithm computation for small n: the running-time of the booting step (finding boots) is $L_Q[1/3, c]$ with $c = 1.14$ for $n = 2, 4$, $c = 1.26$ for $n = 3, 6$ and $c = 1.34$ for $n = 5$. It generalizes to any n, so that the norm is always smaller than $O(Q)$ (the prime field case), hence the booting step running-time in $L_Q[1/3, c]$ always satisfies $c < 1.44$ for the two state-of-the-art variants of NFS for extension fields (we have $c = 1.44$ for prime fields). For the JLSV_1 method, this bound is satisfied for $n = 4$, where we have $c = 1.38$ (see Table 3).

1.5 Outline

We select three polynomial selection methods involved for NFS-DL in generic extension fields and recall their properties in Sect. 2.1. We recall a commonly used bound on the norm of an element in a number field (Sect. 2.2). We present in Sect. 2.3 a generalization of the JLSV fraction method of [19]. In Sect. 3.1 we give a proof of the booting step complexity stated in Lemma 1. We sketch in Sect. 3.2 the special-q descent technique and list the asymptotic complexities found in the literature according to the polynomial selection methods. We present in Sect. 4 our main idea to reduce the norm of the preimage in the number field, by reducing the preimage coefficient size with the LLL algorithm. We improve our technique in Sect. 5 by using a quadratic subfield when available, to finally complete the proof of Theorem 1. We provide practical examples in Sect. 6, for 180 dd finite fields in Sect. 6.1 and we give our running-time experiments for a 120 dd finite field $\mathbb{F}_{p^4}$ in Sect. 6.2.

2 Preliminaries

We recall the three polynomial selection methods we will study along this paper in Sect. 2.1. We give a common simple upper bound on the norm of an element in an number field in Sect. 2.2. We will need this formula to estimate a bound on the target preimage norm and the corresponding asymptotic running-time of the booting step of the individual logarithm computation.

We recall now an important property of the LLL algorithm [21] that we will widely use in this paper. Given a lattice $\mathcal{L}$ of $\mathbb{Z}^n$ defined by a basis given in an $n \times n$ matrix L, and parameters $\frac{1}{4} < \delta < 1$, $\frac{1}{2} < \eta < \sqrt{\delta}$, the LLL algorithm outputs a (η, δ)-reduced basis of the lattice. the coefficients of the first (shortest) vector are bounded by

$$(\delta - \eta^2)^{\frac{n-1}{4}} \det(L)^{1/n}.$$

With (η, δ) close to $(0.5, 0.999)$ (as in NTL or magma), the approximation factor $C = (\delta - \eta^2)^{\frac{n-1}{4}}$ is bounded by 1.075^{n-1} (see [10, Sect. 2.4.2])). Gama and Nguyen experiments [14] on numerous random lattices showed that on average, $C \approx 1.021^n$. In the remaining of this paper, we will simply denote by C this LLL approximation factor.

2.1 Polynomial Selection Methods

We will study the booting step of the NFS algorithm with these three polynomial selection methods:

1. the Joux–Lercier–Smart–Vercauteren (JLSV$_1$) method [19, Sect. 2.3];
2. the generalized Joux–Lercier (gJL) method [22, Sect. 2], [6, Sect. 3.2];
3. the Conjugation method [6, Sect. 3.3].

In a non-multiple NFS version, the JLSV$_2$ [19, Sect. 2.3] and gJL methods have the best asymptotic running-time in the large characteristic case, while the Conjugation method holds the best one in the medium characteristic case. However for a record computation in $\mathbb{F}_{p^2}$, the Conjugation method was used [6]. For medium characteristic fields of record size (between 150 and 200 dd), is seems also that the JLSV$_1$ method could be chosen [6, Sect. 4.5]. Since the use of each method is not fixed in practice, we study and compare the three above methods for the individual logarithm step of NFS. We recall now the construction and properties of these three methods.

Joux–Lercier–Smart–Vercauteren (JLSV$_1$) Method. This method was introduced in 2006. We describe it in Algorithm 1. The two polynomials f, g have degree n and coefficient size $O(p^{1/2})$. We set $\psi = \gcd(f, g) \bmod p$ monic of degree n. We will use ψ to represent the finite field extension $\mathbb{F}_{p^n} = \mathbb{F}_p[x]/(\psi(x))$.

Algorithm 1. Polynomial selection with the JLSV$_1$ method [19, Sect. 2.3]

Input: p prime and n integer
Output: f, g, ψ with $f, g \in \mathbb{Z}[x]$ irreducible and $\psi = \gcd(f \bmod p, g \bmod p)$ in
      $\mathbb{F}_p[x]$ irreducible of degree n
1 Select $f_1(x), f_0(x)$, two polynomials with small integer coefficients,
  $\deg f_1 < \deg f_0 = n$
2 **repeat**
3 | choose $y \approx \lceil \sqrt{p} \rceil$
4 **until** $f = f_0 + y f_1$ *is irreducible in* $\mathbb{F}_p[x]$
5 $(u, v) \leftarrow$ a rational reconstruction of y modulo p
6 $g \leftarrow v f_0 + u f_1$
7 **return** $(f, g, \psi = f \bmod p)$

Generalized Joux–Lercier (gJL) Method. This method was independently proposed in [22, Sect. 2] and [4, Sect. 8.3] (see also [6, Sect. 3.2]). This is a generalization of the Joux–Lercier method [17] for prime fields. We sketch this method in Algorithm 2. The coefficients of g have size $O(Q^{1/(d+1)})$ and those of f have size $O(\log p)$, with $\deg g = d \geq n$ and $\deg f = d + 1$.

Algorithm 2. Polynomial selection with the gJL method

Input: p prime, n integer and $d \geq n$ integer
Output: f, g, ψ with $f, g \in \mathbb{Z}[x]$ irreducible and $\psi = \gcd(f \bmod p, g \bmod p)$ in
      $\mathbb{F}_p[x]$ irreducible of degree n
1 Choose a polynomial $f(x)$ of degree $d + 1$ with small integer coefficients which
  has a monic irreducible factor $\psi(x) = \psi_0 + \psi_1 x + \cdots + x^n$ of degree n modulo p
2 Reduce the following matrix using LLL

$$M = \begin{bmatrix} p & & & & \\ & \ddots & & & \\ & & p & & \\ \psi_0 & \psi_1 & \cdots & 1 & \\ & \ddots & \ddots & & \ddots \\ & & \psi_0 & \psi_1 & \cdots & 1 \end{bmatrix} \begin{matrix} \left.\vphantom{\begin{matrix}a\\b\\c\end{matrix}}\right\} \deg \psi = n \\ \\ \left.\vphantom{\begin{matrix}a\\b\\c\end{matrix}}\right\} d+1-n \end{matrix} \text{, to get } \mathrm{LLL}(M) = \begin{bmatrix} g_0 & g_1 & \cdots & g_d \\ & & & \\ & & * & \\ & & & \end{bmatrix}.$$

return $(f, g = g_0 + g_1 x + \cdots + g_d x^d, \psi)$

Algorithm 3. Polynomial selection with the Conjugation method [6, Sect. 3.3]

Input: p prime and n integer
Output: f, g, ψ with $f, g \in \mathbb{Z}[x]$ irreducible and $\psi = \gcd(f \bmod p, g \bmod p)$ in $\mathbb{F}_p[x]$ irreducible of degree n

1 **repeat**
2   Select $g_1(x), g_0(x)$, two polynomials with small integer coefficients, $\deg g_1 < \deg g_0 = n$
3   Select $P_y(Y)$ a quadratic, monic, irreducible polynomial over $\mathbb{Z}$ with small coefficients
4 **until** $P_y(Y)$ has a root y in $\mathbb{F}_p$ and $\psi(x) = g_0(x) + y g_1(x)$ is irreducible in $\mathbb{F}_p[x]$
5 $f \leftarrow \mathrm{Res}_Y(P_y(Y), g_0(x) + Y g_1(x))$
6 $(u, v) \leftarrow$ a rational reconstruction of y
7 $g \leftarrow v g_0 + u g_1$
8 **return** (f, g, ψ)

Table 1. Properties: degree and coefficient size of the three polynomial selection methods for NFS-DL in $\mathbb{F}_{p^n}$. The coefficient sizes are in $O(X)$. To lighten the notations, we simply write the X term.

Method	$\deg f$	$\deg g$	$\|\|f\|\|_\infty$	$\|\|g\|\|_\infty$
JLSV$_1$	n	n	$Q^{1/2n}$	$Q^{1/2n}$
gJL	$d+1 > n$	$d \geq n$	$\log p$	$Q^{1/(d+1)}$
Conjugation	$2n$	n	$\log p$	$Q^{1/2n}$

Conjugation Method. This method was published in [6] and used for the discrete logarithm record in $\mathbb{F}_{p^2}$, with $f = x^4 + 1$. The coefficient size of f is in $O(\log p)$ and the coefficient size of g is in $O(p^{1/2})$. We describe it in Algorithm 3 (Table 1).

2.2 Norm Upper Bound in a Number Field

In Sect. 4 we will compute the norm of an element s in a number field K_f. We will need an upper bound of this norm. For all the polynomial selection methods chosen, f is monic, whereas g is not. We remove the leading coefficient of f from any formula involved with a monic f. So let f be a monic irreducible polynomial over $\mathbb{Q}$ and let $K_f = \mathbb{Q}[x]/(f(x))$ a number field. Write $s \in K_f$ as a polynomial in x, i.e. $s = \sum_{i=0}^{\deg f - 1} s_i x^i$. The norm is defined by a resultant computation:

$$\mathrm{Norm}_{K_f/\mathbb{Q}}(s) = \mathrm{Res}(f, s).$$

We use Kalkbrener's bound [20, Corollary 2] for an upper bound:

$$|\,\mathrm{Res}(f, s)| \leq \kappa(\deg f, \deg s) \cdot \|f\|_\infty^{\deg s} \|s\|_\infty^{\deg f},$$

where $\kappa(n, m) = \binom{n+m}{n}\binom{n+m-1}{n}$, and $\|f\|_\infty = \max_{0 \leq i \leq \deg f} |f_i|$ the absolute value of the greatest coefficient. An upper bound for $\kappa(n, m)$ is $(n+m)!$. We will use the following bound in Sect. 4:

$$| \operatorname{Norm}_{K_f/\mathbb{Q}}(s)| \leq (\deg f + \deg s)! ||f||_\infty^{\deg s} ||s||_\infty^{\deg f}. \tag{1}$$

2.3   Joux–Lercier–Smart–Vercauteren Fraction Method

Notation 1. *Row and column indices. In the following, we will define matrices of size $d \times d$, with $d \geq n$. For ease of notation, we will index the rows and columns from 0 to $d - 1$ instead of 1 to d, so that the $(i + 1)$-th row at index i, $L_i = [L_{ij}]_{0 \leq j \leq d-1}$, can be written in polynomial form $\sum_{j=0}^{d-1} L_{ij} x^j$, and the column index j coincides with the degree j of x^j.*

In 2006 was proposed in [19] a method to generalize to non-prime fields the rational reconstruction method used for prime fields. In the prime field setting, the target is an integer modulo p. The rational reconstruction method outputs two integers of half size compared to p and such that their quotient is equal to the target element modulo p. Finding a smooth decomposition of the target modulo p becomes equivalent to finding at the same time a smooth decomposition of two integers of half size each.

To generalize to extension fields, one writes the target preimage as a quotient of two number field elements, each with a smaller norm compared to the original preimage. We denote by s the target in the finite field $\mathbb{F}_{p^n}$ and by $\boldsymbol{s}$ a preimage (or lift) in K. Here is a first very simple preimage choice. Let $\mathbb{F}_{p^n} = \mathbb{F}_p[x]/(\psi(x))$ and $s = \sum_{i=0}^{\deg s} s_i x^i \in \mathbb{F}_{p^n}$, with $\deg s < n$. We lift the coefficients $s_i \in \mathbb{F}_p$ to $s_i \in \mathbb{Z}$ then we set a preimage of s in the number field K to be

$$\boldsymbol{s} = \sum_{i=0}^{\deg s} s_i X^i,$$

with X such that $K = \mathbb{Q}[X]/(f(X))$. (We can also write $\boldsymbol{s} = \sum_{i=0}^{\deg s} s_i \alpha^i$, with α a root of f in the number field: $K = \mathbb{Q}[\alpha]$.) We have $\rho(\boldsymbol{s}) = s$.

Now LLL is used to obtain a quotient whose numerator and denominator have smaller coefficients. We present here the lattice used with the JLSV$_1$ polynomial selection method. The number field K is of degree n. We define a lattice of dimension $2n$. For the corresponding matrix, each column of the left half corresponds to a power of X in the numerator; each column of the right half corresponds to a power of X in the denominator. The matrix is

$$L = \begin{bmatrix} p & & & & & & & 0 \\ & \ddots & & & & & & \vdots \\ & & p & & & & & n-1 \\ s_0 & \cdots & s_{n-1} & 1 & & & & n \\ \vdots & & \vdots & & \ddots & & & \vdots \\ sx^{n-1} \bmod \psi & & & & & 1 & & 2n-1 \end{bmatrix} \begin{matrix} \\ \\ \\ \\ \\ \\ 2n \times 2n \end{matrix}$$

The first n coefficients of the output vector, $u_0, u_1, \ldots, u_{n-1}$ give a numerator u and the last n coefficients give a denominator v, so that $s = a\frac{u(X)}{v(X)}$ with a a scalar in $\mathbb{Q}$. The coefficients u_i, v_i are bounded by $||u||_\infty, ||v||_\infty \leq Cp^{1/2}$ since the matrix determinant is $\det L = p^n$ and the matrix is of size $2n \times 2n$. However the product of the norms of each u, v in the number field K will be much larger than the norm of the single element s because of the large coefficients of f in the norm formula. We use formula (1) to estimate this bound:

$$\text{Norm}_{K/\mathbb{Q}}(u) \leq ||u||_\infty^{\deg f} ||f||_\infty^{\deg u} = O(p^{\frac{n}{2}} p^{\frac{n-1}{2}}) = O(p^{n-\frac{1}{2}}) = O(Q^{1-\frac{1}{2n}})$$

and the same for $\text{Norm}_{K/\mathbb{Q}}(v)$, hence the product of the two norms is bounded by $O(Q^{2-\frac{1}{n}})$. The norm of s is bounded by $\text{Norm}_{K/\mathbb{Q}}(s) \leq p^n p^{\frac{n-1}{2}} = Q^{\frac{3}{2}-\frac{1}{2n}}$ which is much smaller whenever $n \geq 3$. Finding a smooth decomposition of u and v at the same time will be much slower than finding one for s directly, for large p and $n \geq 3$. This is mainly because of the large coefficients of f (in $O(p^{1/2})$).

Application to gJL and Conjugation Method. The method of [19] to improve the smoothness of the target norm in the number field K_f has an advantage for the gJL and Conjugation methods. First we note that the number field degree is larger than n: this is $d + 1 \geq n + 1$ for the gJL method and $2n$ for the Conjugation method. For ease of notation, we denote by d_f the degree of f. We define a lattice of dimension $2d_f$. Hence there is more place to reduce the coefficient size of the target s.

We put p on the diagonal of the first $n - 1$ rows, then $x^i \psi(x)$ coefficients from row n to $d_f - 1$, where $0 \leq i < d_f - 1$ (ψ is of degree n and has $n + 1$ coefficients). The rows from index d_f to $2d_f$ are filled with $X^i s \mod f$ (these elements have d_f coefficients). We obtain a triangular matrix L.

$$L = \begin{bmatrix} p & & & & & & & \\ & \ddots & & & & & & \\ & & p & & & & & \\ \psi_0 & \cdots & \psi_{n-1} & 1 & & & & \\ & \ddots & & & \ddots & \ddots & & \\ & & \psi_0 & & \cdots & \psi_{n-1} & 1 & \\ s_0 & \cdots & s_{n-1} & & & & 1 & \\ \vdots & & & & & & & \ddots \\ X^{d_f - 1} s \mod f & & & & & & & 1 \end{bmatrix} \begin{matrix} 0 \\ \vdots \\ n-1 \\ n \\ \vdots \\ d_f - 1 \\ d_f \\ \vdots \\ 2d_f \end{matrix}$$

$$2d_f \times 2d_f$$

Since the determinant is $\det L = p^n$ and the matrix of dimension $2d_f \times 2d_f$, the coefficients obtained with LLL will be bounded by $Cp^{\frac{n}{2d_f}}$. The norm of the numerator or the denominator (with $s = u(X)/v(X) \in K_f$) is bounded by

$$\text{Norm}_{K_f/\mathbb{Q}}(u) \leq ||u||_\infty^{\deg f} ||f||_\infty^{\deg u} = O(p^{n/2}) = O(Q^{1/2}).$$

The product of the two norms will be bounded by $O(Q)$ hence we will have the same asymptotic running time as for prime fields, for finding a smooth decomposition of the target in a number field obtained with the gJL or Conjugation method. We will show in Sect. 4 that we can do even better.

3 Asymptotic Complexity of Individual DL Computation

3.1 Asymptotic Complexity of Initialization or Booting Step

In this section, we prove the following lemma on the booting step running-time to find a smooth decomposition of the norm preimage. This was already proven especially for an initial norm bound of $O(Q)$. We state it in the general case of a norm bound of Q^e. The smoothness bound $B = L_Q[2/3, \gamma]$ used here is not the same as for the relation collection step, where the smoothness bound was $B_0 = L_Q[1/3, \beta_0]$. Consequently, the special-q output in the booting step will be bounded by B.

Lemma 1 (Running-time of B-smooth decomposition). *Let $s \in \mathbb{F}_Q$ of order ℓ. Take at random $t \in [1, \ell - 1]$ and assume that the norm S_t of a preimage of $s^t \in \mathbb{F}_Q$, in the number field K_f, is bounded by $Q^e = L_Q[1, e]$. Write $B = L_Q[\alpha_B, \gamma]$ the smoothness bound for S_t. Then the lower bound of the expected running time for finding t s.t. the norm S_t of s^t is B-smooth is $L_Q[1/3, (3e)^{1/3}]$, obtained with $\alpha_B = 2/3$ and $\gamma = (e^2/3)^{1/3}$.*

First, we need a result on smoothness probability. We recall the definition of B-smoothness already stated in Sect. 1.4: an integer S is B-smooth if and only if all its prime divisors are less than or equal to B. We also recall the L-notation widely used for sub-exponential asymptotic complexities:

$$L_Q[\alpha, c] = \exp\left(\left(c + o(1)\right)(\log Q)^\alpha(\log\log Q)^{1-\alpha}\right) \quad \text{with } \alpha \in [0, 1] \text{ and } c > 0.$$

The Canfield–Erdős–Pomerance [9] theorem provides a useful result to measure smoothness probability:

Theorem 2 (B-smoothness probability). *Suppose $0 < \alpha_B < \alpha_S \leq 1$, $\sigma > 0$, and $\beta > 0$ are fixed. For a random integer S bounded by $L_Q[\alpha_S, \sigma]$ and a smoothness bound $B = L_Q[\alpha_B, \beta]$, the probability that S is B-smooth is*

$$\Pr(S \text{ is } B\text{-smooth}) = L_Q\left[\alpha_S - \alpha_B, -(\alpha_S - \alpha_B)\frac{\sigma}{\beta}\right] \tag{2}$$

for $Q \to \infty$.

We prove now the Lemma 1 that states the running-time of individual logarithm when the norm of the target in a number field is bounded by $O(Q^e)$.

Proof (of Lemma 1). From Theorem 2, the probability that S bounded by $Q^e = L_Q[1, e]$ is B-smooth with $B = L_Q[\alpha_B, \gamma]$ is $\Pr(S \text{ is } B\text{-smooth}) = L_Q[1 - \alpha_B, -(1 - \alpha_B)\frac{e}{\gamma}]$. We assume that a B-smoothness test with ECM takes time $L_B[1/2, 2^{1/2}] = L_Q[\frac{\alpha_B}{2}, (2\gamma\alpha_B)^{1/2}]$. The running-time for finding a B-smooth decomposition of S is the ratio of the time per test (ECM cost) to the B-smoothness probability of S:

$$L_Q\left[\frac{\alpha_B}{2}, (2\gamma\alpha_B)^{1/2}\right] L_Q\left[1 - \alpha_B, (1 - \alpha_B)\frac{e}{\gamma}\right].$$

We optimize first the α value, so that $\alpha \leq 1/3$ (that is, not exceeding the α of the two previous steps of the NFS algorithm): $\max(\alpha_B/2, 1 - \alpha_B) \leq \frac{1}{3}$. This gives the system $\begin{cases} \alpha_B \leq 2/3 \\ \alpha_B \geq 2/3 \end{cases}$ So we conclude that $\alpha_B = \frac{2}{3}$. The running-time for finding a B-smooth decomposition of S is therefore

$$L_Q\left[1/3, \left(\frac{4}{3}\gamma\right)^{1/2} + \frac{e}{3\gamma}\right].$$

The minimum[1] of the function $\gamma \mapsto (\frac{4}{3}\gamma)^{1/2} + \frac{e}{3\gamma}$ is $(3e)^{1/3}$, corresponding to $\gamma = (e^2/3)^{1/3}$, which yields our optimal running time, together with the special-q bound B:

$$L_Q\left[1/3, (3e)^{1/3}\right] \quad \text{with } q \leq B = L_Q\left[2/3, (e^2/3)^{1/3}\right].$$

$\square$

3.2 Running-Time of Special-q Descent

The second step of the individual logarithm computation is the *special-q descent*. This consists in computing the logarithms of the medium-sized elements in the factorization of the target in the number field. The first special-q is of order $L_Q[2/3, \gamma]$ (this is the boot obtained in the initialization step) and is the norm of a degree one prime ideal in the number field where the booting step was done (usually K_f). The idea is to *sieve* over linear combinations of degree one ideals, in K_f and K_g at the same time, whose norms for one side will be multiples of q by construction, in order to obtain a relation involving a degree one prime ideal of norm q and other degree one prime ideals of norm strictly smaller than q.

Here is the common way to obtain such a relation. Let $\mathfrak{q}$ be a degree one prime ideal of K_f, whose norm is q. We can write $\mathfrak{q} = \langle q, r_q \rangle$, with r_q a root of f modulo q (hence $|r_q| < q$). We need to compute two ideals $\mathfrak{q}_1, \mathfrak{q}_2 \in K_f$ whose respective norm is a multiple of q, and sieve over $a\mathfrak{q}_1 + b\mathfrak{q}_2$. The classical way to construct these two ideals is to reduce the two-dimensional lattice generated by q and $r_q - \alpha_f$, i.e. to compute LLL $\left(\begin{bmatrix} q & 0 \\ -r & 1 \end{bmatrix}\right) = \begin{bmatrix} u_1 & v_1 \\ u_2 & v_2 \end{bmatrix}$ to obtain two degree-one ideals $u_1 + v_1\alpha_f, u_2 + v_2\alpha_f$ with shorter coefficients. One sieves over

---

[1] One computes the derivative of the function $h_{a,b}(x) = a\sqrt{x} + \frac{b}{x}$: this is $h'_{a,b}(x) = \frac{a}{2\sqrt{x}} - \frac{b}{x^2}$ and find that the minimum of h for $x > 0$ is $h_{a,b}((\frac{2b}{a})^{2/3}) = 3(\frac{a^2 b}{4})^{1/3}$. With $a = 2/3^{1/2}$ and $b = e/3$, we obtain the minimum: $h((\frac{e}{3})^{1/3}) = (3e)^{1/3}$.

$\mathfrak{r}_f = (au_1 + bu_2) + (av_1 + bv_2)\alpha_f$ and $\mathfrak{r}_g = (au_1 + bu_2) + (av_1 + bv_2)\alpha_g$. The new ideals obtained in the relations will be treated as new special-qs until a relation of ideals of norm bounded by B_0 is found, where B_0 is the bound on the factor basis, so that the individual logarithms are finally known. The sieving is done in three stages, for the three ranges of parameters.

1. For $q = L_Q[2/3, \beta_1]$: large special-q;
2. For $q = L_Q[\lambda, \beta_2]$ with $1/3 < \lambda < 2/3$: medium special-q;
3. For $q = L_Q[1/3, \beta_3]$: small special-q.

The proof of the complexity is not trivial at all, and since this step is allegedly cheaper than the two main phases of sieving and linear algebra, whose complexity is $L_Q[1/3, (\frac{64}{9})^{1/3}]$, the proofs are not always expanded.

There is a detailed proof in [11, Sect. 4.3] and [4, Sect. 7.3] for prime fields $\mathbb{F}_p$. We found another detailed proof in [18, Sect. B] for large characteristic fields $\mathbb{F}_{p^n}$, however this was done for the polynomial selection of [19, Sect. 3.2] (which has the same main asymptotic complexity $L_Q[1/3, (\frac{64}{9})^{1/3}]$). In [22, Sect. 4, pp. 144–150] the NFS-DL algorithm is not proposed in the same order: the booting and descent steps (step (5) of the algorithm in [22, Sect. 2]) are done as a first sieving, then the relations are added to the matrix that is solved in the linear algebra phase. What corresponds to a booting step is proved to have a complexity bounded by $L_Q[1/3, 3^{1/3}]$ and there is a proof that the descent phase has a smaller complexity than the booting step. There is a proof for the JLSV$_1$ polynomial selection in [18, Sect. C] and [3, Sect. A] for a MNFS variant. We summarize in Table 2 the asymptotic complexity formulas for the booting step and the descent step that we found in the available papers.

Table 2. Complexity of the booting step and the descent step for computing one individual DL, in $\mathbb{F}_p$ and $\mathbb{F}_{p^n}$, in medium and large characteristic. The complexity is given by the formula $L_Q[1/3, c]$, only the constant c is given in the table for ease of notation. The descent of a medium special-q, bounded by $L_Q[\lambda, c]$ with $1/3 < \lambda < 2/3$, is proven to be negligible compared to the large and small special-q descents. In [18, Sects. B and C], the authors used a sieving technique over ideals of degree $t > 1$ for large and medium special-q descent.

Reference	Finite field	Polynomial selection	Target norm bound	Booting step	Descent step		
					Large	Med.	Small
[11, Sect. 4.3]	$\mathbb{F}_p$	JL03 [17]	p	1.44	<1.44		
[4, Table 7.1]	$\mathbb{F}_p$	JL03 [17]	p	1.23	1.21	neg	0.97
[22, Sect. 4]	$\mathbb{F}_{p^n}$, large p	gJL	Q	1.44	< 1.44		
[18, Sect. B]	$\mathbb{F}_{p^n}$, large p	JLSV$_2$	Q	1.44	–	neg	1.27
[18, Sect. C]	$\mathbb{F}_{p^n}$, med. p	JLSV$_1$ variant	$Q^{1+\alpha}$, $\alpha \simeq 0.4$	1.62	–	neg	0.93
[3, Sect. A]	$\mathbb{F}_{p^n}$, med. p	JLSV$_1$	$Q^{3/2}$	1.65	≤ 1.03		

Usually, the norm of the target is assumed to be bounded by Q (this is clearly the case for prime fields $\mathbb{F}_p$). The resulting initialization step (finding a boot for the descent) has complexity $L_Q[1/3, 3^{1/3} \approx 1.44]$. Since the large special-q descent complexity depends on the size of the largest special-q of the boot, lowering the norm, hence the booting step complexity *and* the largest special-q of the boot also decrease the large special-q descent step complexity. It would be a considerable project to rewrite new proofs for each polynomial selection method, according to the new booting step complexities. However, its seems to us that by construction, the large special-q descent step in these cases has a (from much to slightly) smaller complexity than the booting step. The medium special-q descent step has a negligible cost in the cases considered above. Finally, the small special-q descent step does not depend on the size of the boot but on the polynomial properties (degree, and coefficient size). We note that for the JLSV$_2$ polynomial selection, the constant of the complexity is 1.27. It would be interesting to know the constant for the gJL and Conjugation methods.

The third and final step of individual logarithm computation is very fast. It combines all of the logarithms computed before, to get the final discrete logarithm of the target.

4   Computing a Preimage in the Number Field

Our main idea is to compute a preimage in the number field with smaller degree (less than deg s) and/or of coefficients of reduced size, by using the subfield structure of $\mathbb{F}_{p^n}$. We at least have one non-trivial subfield: $\mathbb{F}_p$. In this section, we reduce the size of the coefficients of the preimage. This reduces its norm and give the first part of the proof of Theorem 1. In the following section, we will reduce the degree of the preimage when n is even, completing the proof.

Lemma 2. *Let $s \in \mathbb{F}_{p^n}^* = \sum_{i=0}^{\deg s} s_i x^i$, with $\deg s < n$. Let ℓ be a non-trivial divisor of $\Phi_n(p)$. Let $s' = u \cdot s$ with u in a proper subfield of $\mathbb{F}_{p^n}$. Then*

$$\log s' \equiv \log s \, mod \, \ell. \tag{3}$$

Proof. We start with $\log s' = \log s + \log u$ and since u is in a proper subfield, we have $u^{(p^n-1)/\Phi_n(p)} = 1$, then $u^{(p^n-1)/\ell} = 1$. Hence the logarithm of u modulo ℓ is zero, and $\log s' \equiv \log s \bmod \ell$.   □

Example 1 (Monic preimage). Let s' be equal to s divided by its leading term, $s' = \frac{1}{s_{\deg s}} s \in \mathbb{F}_{p^n}$. We have $\log s' \equiv \log s \bmod \ell$.

We assume in the following that the target s is monic since dividing by its leading term does not change its logarithm modulo ℓ.

4.1 Preimage Computation in the JLSV$_1$ Case

Let $s = \sum_{i=0}^{n-1} s_i x^i \in \mathbb{F}_{p^n}$ with $s_{n-1} = 1$. We define a lattice of dimension n by the $n \times n$ matrix

$$
L = \left[
\begin{array}{cccc}
p & & & \\
 & \ddots & & \\
 & & p & \\
s_0 & \cdots & s_{n-2} & 1
\end{array}
\right.
\left.
\begin{array}{c}
0 \\
\vdots \\
n-2 \\
n-1
\end{array}
\right]
\begin{array}{l}
\left.\rule{0pt}{18pt}\right\} n-1 \text{ rows} \\
\\
\} \text{row } n-1 \text{ with } \boldsymbol{s} \text{ coeffs}
\end{array}
$$
$$
{}_{n \times n}
$$

with p on the diagonal for the first $n-1$ rows (from 0 to $n-2$), and the coefficients of the monic element $\boldsymbol{s}$ on row $n-1$. Applying the LLL algorithm to M, we obtain a reduced element $\boldsymbol{r} = \sum_{i=0}^{n-1} r_i X^i \in K_f$ such that

$$
\boldsymbol{r} = \sum_{i=0}^{n-1} a_i L_i
$$

with L_i the vector defined by the i-th row of the matrix and a_i a scalar in $\mathbb{Z}$. We map this equality in $\mathbb{F}_{p^n}$ with ρ. All the terms cancel out modulo p except the line with $\boldsymbol{s}$:

$$
\rho(\boldsymbol{r}) \equiv \rho(a_{n-1}) \cdot \rho(\boldsymbol{s}) = u \cdot s \bmod (p, \psi)
$$

with $u = \rho(a_{n-1}) \in \mathbb{F}_p$. Hence, by Lemma 2,

$$
\log \rho(\boldsymbol{r}) \equiv \log s \bmod \ell. \tag{4}
$$

Moreover,

$$
||\boldsymbol{r}||_\infty \leq C p^{(n-1)/n}.
$$

It is straightforward, using Inequality (1), to deduce that

$$
\mathrm{Norm}_{K_f/\mathbb{Q}}(\boldsymbol{r}) = O\big(p^{\frac{3}{2}(n-1)}\big) = O\big(Q^{\frac{3}{2} - \frac{3}{2n}}\big).
$$

We note that this first simple improvement applied to the JLSV$_1$ construction is already better than doing nothing: in that case, $\mathrm{Norm}_{K_f/\mathbb{Q}}(s) = O(Q^{\frac{3}{2} - \frac{1}{2n}})$. The norm of $\boldsymbol{r}$ is smaller by a factor of size $Q^{\frac{1}{n}}$. For $n = 2$ we have $\mathrm{Norm}_{K_f/\mathbb{Q}}(\boldsymbol{r}) = O(Q^{\frac{3}{4}})$ but for $n = 3$, the bound is $\mathrm{Norm}_{K_f/\mathbb{Q}}(\boldsymbol{r}) = O(Q)$, and for $n = 4$, $O(Q^{11/8})$. This is already too large. We would like to obtain such a bound, strictly smaller than $O(Q)$, for any n.

4.2 Preimage Computation in the gJL and Conjugation Cases

Let $s = \sum_{i=0}^{n-1} s_i x^i \in \mathbb{F}_{p^n}$ with $s_{n-1} = 1$. In order to present a generic method for both the gJL and the Conjugation methods, we denote by d_f the degree of f. In the gJL case we have $d_f = d + 1 \geq n + 1$, while in the Conjugation case, $d_f = 2n$. We define the $d_f \times d_f$ matrix with p on the diagonal for the first $n - 1$

rows, and the coefficients of the monic element s on row $n-1$. The rows n to d_f are filled with the coefficients of the monic polynomial $x^j\psi$, with $0 \le j \le d_f - n$.

$$
L = \begin{bmatrix}
p & & & & & \\
& \ddots & & & & \\
& & p & & & \\
s_0 & \cdots & s_{n-2} & 1 & & \\
\psi_0 & \psi_1 & \cdots & \psi_{n-1} & 1 & \\
& \ddots & \ddots & & \ddots & \ddots \\
& & \psi_0 & \psi_1 & \cdots & \psi_{n-1}\ 1
\end{bmatrix}
\begin{array}{l}
0 \\
\vdots \\
n-2 \\
n-1 \\
n \\
\vdots \\
d_f - 1
\end{array}
\left.\begin{array}{l}
\\ \\ \\
\end{array}\right\} n-1 \text{ rows} \\
\left.\right\}\text{row } n-1 \text{ with } s \text{ coeffs} \\
\left.\begin{array}{l}\\ \\ \\ \end{array}\right\} d_f - n \text{ rows with } \psi \text{ coeffs}
$$

$$d_f \times d_f$$

Applying the LLL algorithm to L, we obtain a reduced element $r = \sum_{i=0}^{d_f-1} r_i X^i \in K_f$ such that $r = \sum_{i=0}^{d_f-1} a_i L_i$ where L_i is the i-th row vector of L and a_i is a scalar in $\mathbb{Z}$. We map this equality into $\mathbb{F}_{p^n}$ with ρ. All the terms cancel out modulo (p, ψ) except the one with s coefficients:

$$\rho(r) \equiv \rho(a_{n-1}) \cdot \rho(s) = u \cdot s \bmod (p, \psi)$$

with $u = \rho(a_{n-1}) \in \mathbb{F}_p$. Hence, by Lemma 2,

$$\log \rho(r) \equiv \log s \bmod \ell. \tag{5}$$

Moreover,

$$||r||_\infty \le C p^{(n-1)/d_f}.$$

It is straightforward, using Inequality (1), to deduce that

$$\mathrm{Norm}_{K_f/\mathbb{Q}}(r) = O(p^{n-1}) = O(Q^{1-1/n}).$$

Here we obtain a bound that is *always strictly smaller* than Q for any n. In the next section we show how to improve this bound to $O(Q^{1-2/n})$ when n is even and the number field defined by ψ has a well-suited quadratic subfield.

5   Preimages of Smaller Norm with Quadratic Subfields

Reducing the degree of s can reduce the norm size in the number field for the JLSV$_1$ polynomial construction. We present a way to compute $r \in \mathbb{F}_{p^n}$ of degree $n-2$ from $s \in \mathbb{F}_{p^n}$ of degree n in the given representation of $\mathbb{F}_{p^n}$, and r, s satisfying Lemma 2. We need n to be even and the finite field $\mathbb{F}_{p^n}$ to be expressed as a degree-$n/2$ extension of a quadratic extension defined by a polynomial of a certain form. We can define another lattice with r and get a preimage of degree $n-2$ instead of $n-1$ in the number field. This can be interesting with the JLSV$_1$ method. Combining this method with the previous one of Sect. 4 leads to our proof of Theorem 1.

5.1   Smaller Preimage Degree

In this section, we prove that when n is even and $\mathbb{F}_{p^n} = \mathbb{F}_p[X]/(\psi(X))$ has a quadratic base field $\mathbb{F}_{p^2}$ of a certain form, from a random element $s \in \mathbb{F}_{p^n}$ with $s_{n-1} \neq 0$, we can compute an element $r \in \mathbb{F}_{p^n}$ with $r_{n-1} = 0$, and $s = u \cdot r$ with $u \in \mathbb{F}_{p^2}$. Then, using Lemma 2, we will conclude that $\log r \equiv \log s \bmod \ell$.

Lemma 3. *Let $\psi(X)$ be a monic irreducible polynomial of $\mathbb{F}_p[X]$ of even degree n with a quadratic subfield defined by the polynomial $P_y = Y^2 + y_1 Y + y_0$. Moreover, assume that ψ splits over $\mathbb{F}_{p^2} = \mathbb{F}_p[Y]/(P_y(Y))$ as*

$$\psi(X) = (P_z(X) - Y)(P_z(X) - Y^p)$$
$$or \ \ \psi(X) = (P_z(X) - YX)(P_z(X) - Y^p X)$$

with P_z monic, of degree $n/2$ and coefficients in $\mathbb{F}_p$. Let $s \in \mathbb{F}_p[X]/(\psi(X))$ a random element, $s = \sum_{i=0}^{n-1} s_i X^i$.
Then there exists $r \in \mathbb{F}_{p^n}$ monic and of degree $n-2$ in X, and $u \in \mathbb{F}_{p^2}$, such that $s = u \cdot r$ in $\mathbb{F}_{p^n}$.

We first give an example for $s \in \mathbb{F}_{p^4}$ then present a constructive proof.

Example 2. Let $P_y = Y^2 + y_1 Y + y_0$ be a monic irreducible polynomial over $\mathbb{F}_p$ and set $\mathbb{F}_{p^2} = \mathbb{F}_p[Y]/(P_y(Y))$. Assume that $Z^2 - YZ + 1$ is irreducible over $\mathbb{F}_{p^2}$ and set $\mathbb{F}_{p^4} = \mathbb{F}_{p^2}[Z]/(Z^2 - YZ + 1)$. Let $\psi = X^4 + y_1 X^3 + (y_0 + 2)X^2 + y_1 X + 1$ be a monic reciprocal polynomial. By construction, ψ factors over $\mathbb{F}_{p^2}$ into $(X^2 - YX + 1)(X^2 - Y^p X + 1)$ and $\mathbb{F}_p[X]/(\psi(X))$ defines a quartic extension $\mathbb{F}_{p^4}$ of $\mathbb{F}_p$. We have these two representations for $\mathbb{F}_{p^4}$:

$$\mathbb{F}_{p^4} = \mathbb{F}_{p^2}[Z]/(Z^2 - YZ + 1) \quad \text{and} \quad \mathbb{F}_{p^4} = \mathbb{F}_p[X]/(X^4 + y_1 X^3 + (y_0+2)X^2 + y_1 X + 1)$$
$$\mathbb{F}_{p^2} = \mathbb{F}_p[Y]/(Y^2 + y_1 Y + y_0)$$
$$\mathbb{F}_p \qquad\qquad \mathbb{F}_p$$

Proof (of Lemma 3). Two possible extension field towers are:

$$\mathbb{F}_{p^n} = \mathbb{F}_{p^2}[Z]/(P_z(Z) - Y) \qquad \mathbb{F}_{p^n} = \mathbb{F}_{p^2}[Z]/(P_z(Z) - YZ)$$
$$\mathbb{F}_{p^2} = \mathbb{F}_p[Y]/(P_y(Y)) \quad \text{and} \quad \mathbb{F}_{p^2} = \mathbb{F}_p[Y]/(P_y(Y))$$
$$\mathbb{F}_p \qquad\qquad\qquad \mathbb{F}_p$$

We write s in the following representation to emphasize the subfield structure:

$$s = \sum_{i=0}^{n/2-1} (a_{i0} + a_{i1}Y)Z^i \text{ with } a_{ij} \in \mathbb{F}_p.$$

1. If $\psi = P_z(Z) - Y$ then we can divide s by $u_{LT} = a_{n/2,0} + a_{n/2,1}Y \in \mathbb{F}_{p^2}$ (the leading term in Z, i.e. the coefficient of $Z^{n/2}$) to make s monic in Z up to a subfield cofactor u_{LT}:

$$\frac{s}{u_{LT}} = \sum_{i=0}^{n/2-2} (b_{i0} + b_{i1}Y)Z^i \quad + Z^{n/2-1},$$

with the coefficients b_{ij} in the base field $\mathbb{F}_p$, and $b_{i0} + b_{i1}Y = (a_{i0} + a_{i1}Y)/u_{LT}$. Since $P_z(Z) = Y$ and $Z = X$ in $\mathbb{F}_{p^n}$ by construction, we replace Y by $P_z(Z)$ and Z by X to get an expression for s in X:

$$\frac{s}{u_{LT}} = \sum_{i=0}^{n/2-2} (b_{i0} + b_{i1}P_z(X))X^i + X^{n/2-1} = r(X).$$

The degree in X of r is $\deg r = \deg P_z(X)X^{n/2-2} = n - 2$ instead of $\deg s = n - 1$. We set $u = 1/u_{LT}$. By construction, $u \in \mathbb{F}_{p^2}$. We conclude that $s = ur \in \mathbb{F}_{p^n}$, with $\deg r = n - 2$ and $u \in \mathbb{F}_{p^2}$.

2. If $\psi = P_z(Z) - YZ$ then we can divide s by $u_{CT} = a_{00} + a_{01}Y \in \mathbb{F}_{p^2}$ (the constant term in Z) to make the constant coefficient of s to be 1:

$$\frac{s}{u_{CT}} = 1 + \sum_{i=1}^{n/2-1} (b_{i0} + b_{i1}Y)Z^i$$

with $b_{ij} \in \mathbb{F}_p$. Since $P_z(Z) = YZ$ and $Z = X$ in $\mathbb{F}_{p^n}$ by construction, we replace YZ by $P_z(Z)$ and Z by X to get

$$\frac{s}{u_{CT}} = 1 + \sum_{i=1}^{n/2-1} (b_{i0}X^i + b_{i1}P_z(X)X^{i-1}) = r(X).$$

The degree in X of r is $\deg r = \deg P_z(X)X^{n/2-1-1} = n - 2$ instead of $\deg s = n - 1$. We set $u = 1/u_{CT}$. By construction, $u \in \mathbb{F}_{p^2}$. We conclude that $s = ur \in \mathbb{F}_{p^n}$, with $\deg r = n - 2$ and $u \in \mathbb{F}_{p^2}$.   □

Now we apply the technique described in Sect. 4.1 to reduce the coefficient size of r in the JLSV$_1$ construction. We have $r_{n-1} = 0$ and we assume that $r_{n-2} = 1$. We define the lattice by the $(n-1) \times (n-1)$ matrix

$$L = \begin{bmatrix} p & & & & 0 \\ & \ddots & & & \vdots \\ & & p & & {\scriptstyle n-3} \\ r_0 & \cdots & r_{n-3} & 1 & {\scriptstyle n-2} \end{bmatrix} \begin{array}{l} \left.\rule{0pt}{3em}\right\} n-2 \text{ rows} \\ {\scriptstyle n-1\times n-1} \end{array} \}\text{row } n-2 \text{ with } \boldsymbol{r} \text{ coeffs}$$

After reducing the lattice with LLL, we obtain an element $\boldsymbol{r}'$ whose coefficients are bounded by $Cp^{\frac{n-2}{n-1}}$. The norm of $\boldsymbol{r}'$ in the number field K_f constructed with the JLSV$_1$ method is

$$\mathrm{Norm}_{K_f/\mathbb{Q}}(\boldsymbol{r}') = O(p^{\frac{3}{2}n-2-\frac{1}{n-1}}) = O(Q^{\frac{3}{2}-\frac{2}{n}-\frac{1}{n(n-1)}}).$$

This is better than the previous $O\left(Q^{\frac{3}{2}-\frac{3}{2n}}\right)$ case: the norm is smaller by a factor of size $O\left(Q^{\frac{1}{2}n+\frac{1}{n(n-1)}}\right)$. For $n = 4$, we obtain $\mathrm{Norm}_{K_f/\mathbb{Q}}(r') = O\left(Q^{\frac{11}{12}}\right)$, which is strictly less than $O(Q)$.

We can do even better by re-using the element r of degree $n - 2$ and the given one s of degree $n - 1$, and combining them.

Generalization to Subfields of Higher Degrees. It was pointed out to us by an anonymous reviewer that more generally, by standard linear algebra arguments, for $m \mid n$ and $s \in \mathbb{F}_{p^n}$, there exists a non-zero $u \in \mathbb{F}_{p^m}$ such that $s \cdot u$ is a polynomial of degree at most $n - m$.

5.2   Smaller Preimage Norm

First, suppose that the target element $s = \sum_{i=0}^{n-1} s_i x^i$ satisfies $s_{n-1} = 0$ and $s_{n-2} = 1$. We can define a lattice whose vectors, once mapped to $\mathbb{F}_{p^n}$, are either 0 (so vectors are sums of multiples of p and ψ) or are multiples of the initial target s, satisfying Lemma 2. The above r of degree $n - 2$ is a good candidate. The initial s also. If there is no initial s of degree $n - 1$, then simply take at random any u in a proper subfield of $\mathbb{F}_{p^n}$ which is not $\mathbb{F}_p$ itself and set $s = u \cdot r$. Then s will have $s_{n-1} \neq 0$. Then define the lattice

$$
L = \begin{bmatrix} p & & & & \\ & \ddots & & & \\ & & p & & \\ r_0 & \cdots & r_{n-3} & 1 & \\ s_0 & \cdots & s_{n-3} & s_{n-2} & 1 \end{bmatrix} \begin{matrix} 0 \\ \vdots \\ n-3 \\ n-2 \\ n-1 \end{matrix} \left. \begin{matrix} \\ \\ \end{matrix} \right\} n-2 \text{ rows}
$$

with $n-2$ marked as $\}$ row $n - 2$ with r coeffs, $n-1$ marked as $\}$ row $n - 1$ with s coeffs, dimension $n \times n$

and use it in place of the lattices of Sect. 4.1 or 4.2.

5.3   Summary of Results

We give in Table 3 the previous and new upper bounds for the norm of s in a number field K_f for three polynomial selection methods: the JLSV$_1$ method, the generalized Joux–Lercier method and the Conjugation method, and the complexity of the booting step to find a B-smooth decomposition of $\mathrm{Norm}_{K_f/\mathbb{Q}}(s)$. We give our practical results for small n, where there are the most dramatic improvements. We obtain the optimal norm size of $Q^{\varphi(n)/n}$ for $n = 2, 3, 5$ with the gJL method and also for $n = 4$ with the Conjugation method.

6   Practical Examples

We present an example for each of the three polynomial selection methods we decided to study. The Conjugation method provides the best timings for $\mathbb{F}_{p^2}$ at 180 dd [6]. We apply the gJL method to $\mathbb{F}_{p^3}$ according to [6, Fig. 3]. We decided to use the JLSV$_1$ method for $\mathbb{F}_{p^4}$ [6, Fig. 4].

Table 3. Norm bound of the preimage with our method, and booting step complexity.

$\mathbb{F}_{p^n}$	poly. selec.	norm bound			booting step $L_Q[\frac{1}{3},c]$		practical values of c				
		nothing	JLSV	this work	prev	this work	$n=2$	$n=3$	$n=4$	$n=5$	$n=6$
any $n>1$	gJL	$Q^{1+\frac{1}{n}}$	Q	$Q^{1-1/n}$	1.44	$(3(1-\frac{1}{n}))^{1/3}$	1.14	1.26	–	1.34	–
even $n\geq 4$				$Q^{1-2/n}$		$(3(1-\frac{2}{n}))^{1/3}$	–	–	1.14	–	1.26
any $n>1$	Conj	Q^2	Q	$Q^{1-1/n}$	1.44	$(3(1-\frac{1}{n}))^{1/3}$	1.14	1.26	–	1.34	–
even $n\geq 4$				$Q^{1-2/n}$		$(3(1-\frac{2}{n}))^{1/3}$	–	–	1.14	–	1.26
any $n>1$	JLSV$_1$	$Q^{\frac{3}{2}-\frac{1}{2n}}$	Q^2	$Q^{3/2-3/n}$	1.65	$(\frac{9}{2}(1-\frac{1}{n}))^{1/3}$	1.31	1.44	–	1.53	–
even $n\geq 4$				$Q^{3/2-5/n}$		$(\frac{3}{2}(3-\frac{5}{n}))^{1/3}$	–	–	1.38	–	1.48

6.1   Examples for Small n and p^n of 180 Decimal Digits (dd)

Example for $n=2$, Conjugation Method. We take the parameters of the record in [6]: p is a 90 decimal digit (300 bit) prime number, and f, ψ are computed with the Conjugation method. We choose a target s from the decimal digits of $\exp(1)$.

$p =$ 3141592653589793238462643383279502884197169399375105820974944592307816406286208998777709223

$f = x^4 + 1$

$\psi = x^2 +$ 1077815130958230186669898831022443948094122976438953490974106325080494553766987846916995931 $x + 1$

$s =$ 271828182845904523536028747135319858432320810108854154561922281807332337576949857498874314 x

$+$ 9588806625076732632114201657575319902277223541152654868480844097394920847119472461809069

We first compute $s' = \frac{1}{s_0}s$ then reduce

$$L = \begin{bmatrix} p & 0 & 0 & 0 \\ s'_0 & 1 & 0 & 0 \\ 1 & \psi_1 & 1 & 0 \\ 0 & 1 & \psi_1 & 1 \end{bmatrix}$$

then LLL(L) produces r of degree 3 and coefficient size $O(p^{1/4})$. Actually LLL outputs four short vectors, hence we get four small candidates for r, each of norm $\text{Norm}_{K_f/\mathbb{Q}}(r) = O(p) = O(Q^{1/2}) = O(Q^{\varphi(n)/n})$, i.e. 90 dd. To slightly improve the smoothness search time, we can compute linear combinations of these four reduced preimages.

360339728645720582847x^3 + 136790355536430097110787x^2 + 557746247085194895659481 x + 856176942703613067714

921946132448219081489381x^3 − 449817579633385492601321x^2 + 895775002549467382219831 x + 111788824169113006040911

282683909446241831417027x^3 + 569966674122622538525981x^2 − 178019404032168663329111 x + 544843224771048269684811

335216279294146314006081x^3 + 321258501223569290228721x^2 − 557063651808475912551311 x + 469265082905446625423271

The norm of the first element is

$\text{Norm}_{K_f/\mathbb{Q}}(r) =$ 2139882802952016861116904528030242843486696665709707576133759807076048534094867780016292111

of 90 decimal digits, as expected. For a close to optimal running-time of $L_Q[1/3, 1.14] \sim 2^{40}$ to find a boot, the special-q bound would be around 64 bits.

Example for $n = 3$, gJL Method. We take p of 60 dd (200 bits) so that $\mathbb{F}_{p^3}$ has size 180 dd (600 bits) as above. We took p a prime made of the 60 first decimal digits of π. We constructed f, ψ, g with the gJL method described in [6].

$$p = 314159265358979323846264338327950288419716939937510582723487$$

$$f = x^4 - x + 1$$

$$\psi = x^3 + {\scriptstyle 2271381442436423331299022877956647720436670532600892994785790} x^2$$

$$+ {\scriptstyle 1267980222014268054021867611104401101211578637915853289135650} x$$

$$+ {\scriptstyle 863983091574414435397918995177883881848539630718471155526380}$$

$$g = {\scriptstyle 2877670889871354566080333172463852249908214391} x^3 + {\scriptstyle 6099516524325575060821841620140470618863403881} x^2$$

$$- {\scriptstyle 1012353323483447331605328962316575643726729840} x + {\scriptstyle 2029073371791914965976041284208208450267120556}$$

$$s = {\scriptstyle 2718281828459045235360287471353198584323208101088541545619220} x^2$$

$$+ {\scriptstyle 2818073323375769498574988743140958880662507673263211420165750} x$$

$$+ {\scriptstyle 7531990227722354115265486848085895162649373929725913985987500}$$

We set $s' = \frac{1}{s_2} s$. The lattice to be reduced is

$$L = \begin{bmatrix} p & 0 & 0 & 0 \\ 0 & p & 0 & 0 \\ s'_0 & s'_1 & 1 & 0 \\ \psi_0 & \psi_1 & \psi_2 & 1 \end{bmatrix}$$

then $\mathrm{LLL}(L)$ computes four short vectors $\boldsymbol{r}$ of degree 3, of coefficient size $O(p^{1/2})$, and of norm size $\mathrm{Norm}_{K_f/\mathbb{Q}}(\boldsymbol{r}) = O(p^2) = O(Q^{2/3}) = O(Q^{\varphi(n)/n})$.

$$\scriptstyle 1597749306375059000093909307018 x^3 + {\scriptstyle 165819631832105094449987774814} x^2 + {\scriptstyle 1778281993224195536012663549040} x$$

$$- {\scriptstyle 1599127869369434884005903890195}$$

$$\scriptstyle 1365830293545209052324129410480 x^3 - {\scriptstyle 521269847225531188433352927453} x^2 + {\scriptstyle 322722415562853671586868492721} x$$

$$+ {\scriptstyle 255238068915917937217884608875}$$

$$\scriptstyle 1182890075989340687266630002660 x^3 + {\scriptstyle 499013489972894059858543976363} x^2 - {\scriptstyle 1050842208618441557970157136660} x$$

$$+ {\scriptstyle 535978811382585906107397024241}$$

$$\scriptstyle 4116038900545395001314743137730 x^3 - {\scriptstyle 240161030577722451131067159670} x^2 - {\scriptstyle 373289346204280810310169575030} x$$

$$- {\scriptstyle 389720783049275894296185820094}$$

The norm of the first element is

$$\mathrm{Norm}_{K_f/\mathbb{Q}}(\boldsymbol{r})$$
$$= {\scriptstyle 997840136509677868374734441582077227769466501519927620849763845265357390584602475858356409809239812991892769866071779}$$

of 117 decimal digits (with $\frac{2}{3} 180 = 120$ dd). For a close to optimal running-time of $L_Q[1/3, 1.26] \sim 2^{45}$ to find a boot, the special-q bound would be around 77 bits.

Example for $n = 4$, JLSV$_1$ Method.

p = 31415926535897932384626433832795028841998 0011

ℓ = 49348022005446793094172454999380755676651143247932834802731698819521755649884772819780061

$f = \psi = x^4 + x^3 + {}_{70898154036220641093162}x^2 + x + 1$

g = $_{10191609642706717156787 2}x^4 + {}_{10191609642706717156787 2}x^3 + {}_{2208063288740498985510 11}x^2$

  $+ {}_{10191609642706717156787 2}x + {}_{10191609642706717156787 2}$

s = $_{27182818284590452353602874713531985843232081 0}x^3 + {}_{10885415456192228180733233757694985749887431 4}x^2$

  $+ {}_{958880662507673263211420165757531990227722 35}x + {}_{411526548684808440973949208471275883919520 18}$

We set $s' = \frac{1}{s_3}s$. The subfield simplification for s gives

$r = x^2 + {}_{13496912239726310297974322691528235540016191 1}x + {}_{10464244064993775636854576533474104920712101 1}$.

We reduce the lattice defined by

$$L = \begin{bmatrix} p & 0 & 0 & 0 \\ 0 & p & 0 & 0 \\ r_0 & r_1 & 1 & 0 \\ s_0' & s_1' & s_2' & 1 \end{bmatrix}$$

then LLL(L) produces these four short vectors of degree 3, coefficient size $O(p^{1/2})$, and norm Norm$_{K_f/\mathbb{Q}}(r') = O(p^{\frac{7}{2}}) = O(Q^{7/8})$ (smaller than $O(Q)$).

$_{5842961997149263751946}x^3 + {}_{29073682733086101137 6}x^2 - {}_{5618779793817086743792}x + {}_{1092494800287557029045}$

$_{1640842643903161175359}x^3 + {}_{15552590269131889589575}x^2 - {}_{442548839416383827137 8}x - {}_{5734086421794811858814}$

$_{6450686906504525374853}x^3 + {}_{13768771242650957399419}x^2 + {}_{1061758394423409088057 9}x + {}_{1626161707916779758091 2}$

$_{1692913580413987886539 1}x^3 + {}_{6981855717048102583 44}x^2 + {}_{12799300411012246114079}x - {}_{2278728269871806528415 7}$

The norm of the first element is

Norm$_{K_f/\mathbb{Q}}(r')$ = $_{14521439292172711151668611104133579982787299949310242601944218977645007049527}\backslash$

$_{0123656021783074136945302749067576757516984664647990043605467452102146421782 85}$

of 155 decimal digits (with $\frac{7}{8}180 = 157.5$). For a close to optimal running-time of $L_Q[1/3, 1.34] \sim 2^{49}$ to find a boot, the special-q bound would be approximately of 92 bits. This is very large however.

6.2 Experiments: Finding Boots for $\mathbb{F}_{p^4}$ of 120 dd

We experimented our booting step method for $\mathbb{F}_{p^4}$ of 120 dd (400 bits). Without the quadratic subfield simplification, the randomized target norm is bounded by $Q^{9/8}$ of 135 dd (450 bits). The largest special-q in the boot has size $L_Q[2/3, 3/4]$ (25 dd, 82 bits) according to Lemma 1 with $e = 9/8$. The running-time to find one boot would be $L_Q[1/3, 1.5] \sim 2^{44}$.

We apply the quadratic subfield simplification. The norm of the randomized target is $Q^{7/8}$ of 105 dd ($\simeq 350$ bits). We apply Theorem 1 with $e = 7/8$. The size of the largest special-q in the boot will be approximately $L_Q[2/3, 0.634]$ which is

21 dd (69 bits). The running-time needed to find one boot with the special-q of no more than 21 dd is $L_Q[1/3, 1.38] \sim 2^{40}$ (to be compared with the dominating part of NFS-DL of $L_Q[1/3, 1.923] \sim 2^{57}$). We wrote a magma program to find boots, using GMP-ECM for q-smooth tests. We first set a special-q bound of 70 bits and obtained boots in about two CPU hours. We then reduced the special-q bound to a machine word size (64 bits) and also found boots in around two CPU hours. We used an Intel Xeon E5-2609 0 at 2.40 GHz with 8 cores.

7   Conclusion

We have presented a method to improve the booting step of individual logarithm computation, the final phase of the NFS algorithm. Our method is very efficient for small n, combined with the gJL or Conjugation methods; it is also usefull for the JLSV$_1$ method, but with a slower running-time. For the moment, the booting step remains the dominating part of the final individual discrete logarithm. If our method is improved, then special-q descent might become the new bottleneck in some cases. A lot of work remains to be done on final individual logarithm computations in order to be able to compute one individual logarithm as fast as was done in the Logjam [2] attack, especially for $n \geq 3$.

Acknowledgements. The author thanks the anonymous reviewers for their constructive comments and the generalization of Lemma 3. The author is grateful to Pierrick Gaudry, François Morain and Ben Smith.

References

1. Adleman, L.: A subexponential algorithm for the discrete logarithm problem with applications to cryptography. In: 20th FOCS, pp. 55–60. IEEE Computer Society Press, October 1979
2. Adrian, D., Bhargavan, K., Durumeric, Z., Gaudry, P., Green, M., Halderman, J.A., Heninger, N., Springall, D., Thomé, E., Valenta, L., VanderSloot, B., Wustrow, E., Zanella-Béguelin, S., Zimmermann, P.: Imperfect forward secrecy: how Diffie-Hellman fails in practice. In: CCS 2015, October 2015, to appear. https://weakdh. org/imperfect-forward-secrecy.pdf
3. Barbulescu, R., Pierrot, C.: The multiple number field sieve for medium- and high-characteristic finite fields. LMS J. Comput. Math. **17**, 230–246 (2014). http:// journals.cambridge.org/article_S1461157014000369
4. Barbulescu, R.: Algorithmes de logarithmes discrets dans les corps finis. Ph.D. thesis, Université de Lorraine (2013)
5. Barbulescu, R., Gaudry, P., Guillevic, A., Morain, F.: Discrete logarithms in $GF(p^2)$ - 180 digits (2014), announcement available at the NMBRTHRY archives. https://listserv.nodak.edu/cgi-bin/wa.exe?A2=NMBRTHRY;2ddabd4c.1406
6. Barbulescu, R., Gaudry, P., Guillevic, A., Morain, F.: Improving NFS for the discrete logarithm problem in non-prime finite fields. In: Oswald, E., Fischlin, M. (eds.) EUROCRYPT 2015. LNCS, vol. 9056, pp. 129–155. Springer, Heidelberg (2015)

7. Barbulescu, R., Gaudry, P., Joux, A., Thomé, E.: A heuristic quasi-polynomial algorithm for discrete logarithm in finite fields of small characteristic. In: Nguyen, P.Q., Oswald, E. (eds.) EUROCRYPT 2014. LNCS, vol. 8441, pp. 1–16. Springer, Heidelberg (2014)

8. Blake, I.F., Mullin, R.C., Vanstone, S.A.: Computing logarithms in $GF(2^n)$. In: Blakely, G.R., Chaum, D. (eds.) CRYPTO 1984. LNCS, vol. 196, pp. 73–82. Springer, Heidelberg (1985)

9. Canfield, E.R., Erdös, P., Pomerance, C.: On a problem of Oppenheim concerning "factorisatio numerorum". J. Number Theor. **17**(1), 1–28 (1983)

10. Chen, Y.: Réduction de réseau et sécurité concréte du chiffrement complétement homomorphe. Ph.D. thesis, Université Paris 7 Denis Diderot (2013)

11. Commeine, A., Semaev, I.A.: An algorithm to solve the discrete logarithm problem with the number field sieve. In: Yung, M., Dodis, Y., Kiayias, A., Malkin, T. (eds.) PKC 2006. LNCS, vol. 3958, pp. 174–190. Springer, Heidelberg (2006)

12. Coppersmith, D.: Modifications to the number field sieve. J. Cryptol. **6**(3), 169–180 (1993)

13. Coppersmith, D., Odlzyko, A.M., Schroeppel, R.: Discrete logarithms in GF(p). Algorithmica **1**(1–4), 1–15 (1986). http://dx.doi.org/10.1007/BF01840433

14. Gama, N., Nguyen, P.Q.: Predicting lattice reduction. In: Smart, N.P. (ed.) EURO-CRYPT 2008. LNCS, vol. 4965, pp. 31–51. Springer, Heidelberg (2008)

15. Gordon, D.M.: Discrete logarithms in GF(p) using the number field sieve. SIAM J. Discrete Math. **6**, 124–138 (1993)

16. Hayasaka, K., Aoki, K., Kobayashi, T., Takagi, T.: An experiment of number field sieve for discrete logarithm problem over GF(p^{12}). In: Fischlin, M., Katzenbeisser, S. (eds.) Buchmann Festschrift. LNCS, vol. 8260, pp. 108–120. Springer, Heidelberg (2013)

17. Joux, A., Lercier, R.: Improvements to the general number field for discrete logarithms in prime fields. Math. Comput. **72**(242), 953–967 (2003)

18. Joux, A., Lercier, R., Naccache, D., Thomé, E.: Oracle-assisted static Diffie-Hellman is easier than discrete logarithms. In: Parker, M.G. (ed.) Cryptography and Coding 2009. LNCS, vol. 5921, pp. 351–367. Springer, Heidelberg (2009)

19. Joux, A., Lercier, R., Smart, N.P., Vercauteren, F.: The number field sieve in the medium prime case. In: Dwork, C. (ed.) CRYPTO 2006. LNCS, vol. 4117, pp. 326–344. Springer, Heidelberg (2006)

20. Kalkbrener, M.: An upper bound on the number of monomials in determinants of sparse matrices with symbolic entries. Mathematica Pannonica **73**, 82 (1997)

21. Lenstra, A., Lenstra Jr., H.W., Lovász, L.: Factoring polynomials with rational coefficients. Mathematische Annalen **261**(4), 515–534 (1982). http://dx.doi.org/10.1007/BF01457454

22. Matyukhin, D.: Effective version of the number field sieve for discrete logarithms in the field GF(p^k) (in Russian). Trudy po Discretnoi Matematike **9**, 121–151 (2006)

23. Pierrot, C.: The multiple number field sieve with conjugation and generalized joux-lercier methods. In: Oswald, E., Fischlin, M. (eds.) EUROCRYPT 2015. LNCS, vol. 9056, pp. 156–170. Springer, Heidelberg (2015)

24. Weber, D.: Computing discrete logarithms with quadratic number rings. In: Nyberg, K. (ed.) EUROCRYPT 1998. LNCS, vol. 1403, pp. 171–183. Springer, Heidelberg (1998)

25. Zajac, P.: Discrete logarithm problem in degree six finite fields. Ph.D. thesis, Slovak University of Technology (2008)

Multiple Discrete Logarithm Problems with Auxiliary Inputs

Taechan Kim[(✉)]

NTT Secure Platform Laboratories, Tokyo, Japan
taechan.kim@lab.ntt.co.jp

Abstract. Let g be an element of prime order p in an abelian group and let $\alpha_1, \ldots, \alpha_L \in \mathbb{Z}_p$ for a positive integer L. First, we show that, if g, g^{α_i}, and $g^{\alpha_i^d}$ $(i = 1, \ldots, L)$ are given for $d \mid p - 1$, all the discrete logarithms α_i's can be computed probabilistically in $\widetilde{O}(\sqrt{L \cdot p/d} + \sqrt{L \cdot d})$ group exponentiations with $O(L)$ storage under the condition that $L \ll \min\{(p/d)^{1/4}, d^{1/4}\}$.

Let $f \in \mathbb{F}_p[x]$ be a polynomial of degree d and let ρ_f be the number of rational points over $\mathbb{F}_p$ on the curve determined by $f(x) - f(y) = 0$. Second, if $g, g^{\alpha_i}, g^{\alpha_i^2}, \ldots, g^{\alpha_i^d}$ are given for any $d \geq 1$, then we propose an algorithm that solves all α_i's in $\widetilde{O}(\max\{\sqrt{L \cdot p^2/\rho_f}, L \cdot d\})$ group exponentiations with $\widetilde{O}(\sqrt{L \cdot p^2/\rho_f})$ storage. In particular, we have explicit choices for a polynomial f when $d \mid p \pm 1$, that yield a running time of $\widetilde{O}(\sqrt{L \cdot p/d})$ whenever $L \leq \frac{p}{c \cdot d^3}$ for some constant c.

Keywords: Discrete logarithm problem · Multiple discrete logarithm · Birthday problem · Cryptanalysis

1 Introduction

Let G be a cyclic group of prime order p with a generator g. A discrete logarithm problem (DLP) aims to find the element α of $\mathbb{Z}_p$ when g and g^α are given. The DLP is a classical hard problem in computational number theory, and many encryption schemes, signatures, and key exchange protocols rely on the hardness of the DLP for their security.

In recent decades, many variants of the DLP have been introduced. These include the Weak Diffie–Hellman Problem [13], Strong Diffie–Hellman Problem [2], Bilinear Diffie–Hellman Inversion Problem [1], and Bilinear Diffie–Hellman Exponent Problem [3], and are intended to guarantee the security of many cryptosystems, such as traitor tracing [13], short signatures [2], ID-based encryption [1], and broadcast encryption [3]. These problems incorporate additional information to the original DLP problem. Although such additional information could weaken the problems, and their hardness is not well understood, these variants are widely used because they enable the construction of cryptosystems with various functionalities.

© International Association for Cryptologic Research 2015
T. Iwata and J.H. Cheon (Eds.): ASIACRYPT 2015, Part I, LNCS 9452, pp. 174–188, 2015.
DOI: 10.1007/978-3-662-48797-6_8

These variants can be considered as the problem of finding α when $g, g^{\alpha^{e_1}}, \ldots, g^{\alpha^{e_d}}$ are given for some $e_1, \ldots, e_d \in \mathbb{Z}$. This problem is called the discrete logarithm problem with auxiliary inputs (DLPwAI).

On the other hand, in the context of elliptic curve cryptography, because of large computational expense of generating a secure elliptic curve, a fixed curve is preferred to a random curve. One can choose a curve recommended by standards such as NIST. Then this causes an issue with the multiple DLP/DLPwAI and leads the following question. Can it be more efficient to solve them together than to solve each of instances individually when needed, if an adversary collects many instances of DLP/DLPwAI from one fixed curve?

In multiple discrete logarithm problem, an algorithm [11] computes L discrete logarithms in time $\widetilde{O}(\sqrt{L \cdot p})$ for $L \ll p^{1/4}$. Recently, it is proven that this algorithm is optimal in the sense that it requires at least $\Omega(\sqrt{L \cdot p})$ group operations to solve the multiple DLP in the generic group model [19].

On the other hand, an efficient algorithm for solving the DLPwAI is proposed by Cheon [5,6]. If g, g^{α}, and $g^{\alpha^d} \in G$ (resp. $g, g^{\alpha}, \ldots, g^{\alpha^{2d}} \in G$) are given, then one can solve the discrete logarithm $\alpha \in \mathbb{Z}_p$ in $O(\sqrt{p/d} + \sqrt{d})$ (resp. $O(\sqrt{p/d} + d)$) group operations in the case of $d \mid p - 1$ (resp. $d \mid p + 1$). Since solving the DLPwAI in the generic group model requires at least $\Omega(\sqrt{p/d})$ group operations [2], Cheon's algorithm achieves the lower bound complexity in the generic group model when $d \leq p^{1/2}$ (resp. $d \leq p^{1/3}$). Brown and Gallant [4] independently investigated an algorithm in the case of $d \mid p - 1$.

However, as far as we know, the DLPwAI algorithm in the multi-user setting has not been investigated yet. This paper proposes an algorithm to solve the multiple DLPwAI better than $O(L \cdot \sqrt{p/d})$ group operations in the case of $d \mid p \pm 1$, where L denotes the number of the target discrete logarithms.

Our Contributions. We propose two algorithms for the multiple DLPwAI. Our first algorithm is based on Cheon's $(p - 1)$-algorithm [5,6]. If g, g^{α_i}, and $g^{\alpha_i^d}$ $(i = 1, 2, \ldots, L)$ are given for $d \mid p - 1$, our algorithm solves L discrete logarithms probabilistically in $\widetilde{O}(\sqrt{L \cdot p/d} + \sqrt{L \cdot d})$ group operations with storages for $O(L)$ elements whenever $L \leq \min\{c_{p/d}(p/d)^{1/4}, c_d d^{1/4}\}$ (for some constants $0 < c_{p/d}, c_d < 1$). We also show a deterministic variant of this algorithm which applies for any $L > 0$ and has the running time of $\widetilde{O}(\sqrt{L \cdot p/d} + \sqrt{L \cdot d} + L)$, although it requires as large amount of the storage as the time complexity. However, an approach based on Cheon's $(p+1)$-algorithm does not apply to improve an algorithm in multi-user setting.

Our second algorithm is based on Kim and Cheon's algorithm [10]. The algorithm basically works for any $d > 0$. Let $f(x) \in \mathbb{F}_p[x]$ be a polynomial of degree d over $\mathbb{F}_p$ and define $\rho_f := |(x, y) \in \mathbb{F}_p \times \mathbb{F}_p : f(x) = f(y)|$. If $g, g^{\alpha_i}, g^{\alpha_i^2}, \ldots, g^{\alpha_i^d}$ $(i = 1, 2, \ldots, L)$ are given, the algorithm computes all α_i's in $\widetilde{O}(\max\{\sqrt{L \cdot p^2/\rho_f}, L \cdot d\})$ group operations with the storage for $\widetilde{O}(\sqrt{L \cdot p^2/\rho_f})$ elements.

In particular, if $L \cdot d \leq \sqrt{L \cdot p^2/\rho_f}$ (i.e. $L \leq \frac{p^2}{d^2 \cdot \rho_f}$), the time complexity is given by $\widetilde{O}(\sqrt{L \cdot p^2/\rho_f})$. Since $p \leq \rho_f \leq dp$, this value is always between $\widetilde{O}(\sqrt{L \cdot p/d})$ and $\widetilde{O}(\sqrt{L \cdot p})$. Explicitly, if $d \mid p-1$, one can choose the polynomial by $f(x) = x^d$ and in the case the complexity is given by the lower bound $\widetilde{O}(\sqrt{L \cdot p/d})$ whenever $L \leq p/d^3$. Similarly, in the case of $d \mid p+1$, if one takes the polynomial $f(x) = D_d(x,a)$, where $D_d(x,a)$ is the Dickson polynomial of degree d for some nonzero $a \in \mathbb{F}_p$, then it also has the running time of $\widetilde{O}(\sqrt{L \cdot p/d})$ for $L \lesssim p/(2d^3)$.

As far as the authors know, these two algorithms extend all existing DLPwAI-solving algorithms to the algorithms for multi-user setting.

Organization. This paper is organized as follows. In Sect. 2, we introduce several variants of DLP including a problem called discrete logarithm problem in the exponent (DLPX). We also show that several generic algorithms can be applied to solve the DLPX. In Sect. 3, we propose an algorithm solving the multiple DLPwAI based on Cheon's algorithm. In Sect. 4, we present another algorithm to solve the multiple DLPwAI using Kim and Cheon's algorithm. We conclude with some related open questions in Sect. 5.

2   Discrete Logarithm Problem and Related Problems

In this section, we introduce several problems related to the discrete logarithm problem. Throughout the paper, let $G = \langle g \rangle$ be a cyclic group of prime order p. Let $\mathbb{F}_q$ be a finite field with q elements for some prime power $q = p^r$. Let $\mathbb{Z}_N$ be the set of the residue classes of integers modulo an integer N.

- The *Discrete Logarithm Problem (DLP)* in G is: Given $g, g^\alpha \in G$, to solve $\alpha \in \mathbb{Z}_p$.
- The *Multiple Discrete Logarithm Problem (MDLP)* in G is: Given $g, g^{\alpha_1}, \ldots, g^{\alpha_L} \in G$, to solve all $\alpha_1, \ldots, \alpha_L \in \mathbb{Z}_p$.
- The $(e_1, \ldots, e_d)$ *-Discrete Logarithm Problem with Auxiliary Inputs (DLP-wAI)* in G is: Given $g, g^{\alpha^{e_1}}, g^{\alpha^{e_2}}, \ldots, g^{\alpha^{e_d}} \in G$, to solve $\alpha \in \mathbb{Z}_p$.
- The $(e_1, \ldots, e_d)$ *-Multiple Discrete Logarithm Problem with Auxiliary Inputs (MDLPwAI)* in G is: Given $g, g^{\alpha_i^{e_1}}, g^{\alpha_i^{e_2}}, \ldots, g^{\alpha_i^{e_d}} \in G$ for $i = 1, 2, \ldots, L$, to solve $\alpha_1, \ldots, \alpha_L \in \mathbb{Z}_p$.

In the case of $(e_1, e_2, \ldots, e_d) = (1, 2, \ldots, d)$, we simply denote $(1, 2, \ldots, d)$-(M)DLPwAI by d-(M)DLPwAI.

We also introduce the problem called $\mathbb{F}_p$ *-discrete logarithm problem in the exponent* $(\mathbb{F}_p\text{-}DLPX)$.

- The $\mathbb{F}_p$ *-Discrete Logarithm Problem in the Exponent* $(\mathbb{F}_p\text{ -}DLPX)$ in G is defined as follows: Let $\chi \in \mathbb{F}_p$ be an element of multiplicative order N, i.e. $N \mid p-1$. Given $g, g^{\chi^n} \in G$ and $\chi \in \mathbb{F}_p$, compute $n \in \mathbb{Z}_N$.
- The $\mathbb{F}_p$ *-Multiple Discrete Logarithm Problem in the Exponent* $(\mathbb{F}_p\text{-}MDLPX)$ in G is: Given $g, g^{\chi^{n_1}}, \ldots, g^{\chi^{n_L}} \in G$ and $\chi \in \mathbb{F}_p$, to solve $n_1, \ldots, n_L \in \mathbb{Z}_N$. In both cases, the $\mathbb{F}_p$-(M)DLPX is said to be defined over $\mathbb{Z}_N$.

Algorithm for DLPX. Observe that several DL-solving algorithms can be applied to solve the DLPX with the same complexity. For example, the baby-step-giant-step (BSGS) algorithm works as follows: Suppose that the DLPX is defined over $\mathbb{Z}_N$. Set an integer $K \approx \sqrt{N}$ and write $n = n_0 K + n_1$, where $0 \le n_0 \le N/K \approx \sqrt{N}$ and $0 \le n_1 < K$. For given $g, g^{\chi^n} \in G$ and $\chi \in \mathbb{F}_p$, compute and store the elements $g^{\chi^{i \cdot K}} = \left(g^{\chi^{(i-1) \cdot K}}\right)^{\chi^K}$ for all $i = 0, 1, \dots, N/K$. Then compute $\left(g^{\chi^n}\right)^{\chi^{-j}}$ for all $j = 0, 1, \dots, K-1$ and find a match between the stored elements. Then the discrete logarithm is given by $n = iK + j$ for the indices i and j corresponding to the match. It costs $O(\sqrt{N})$ group exponentiations by elements in $\mathbb{F}_p$ and $O(\sqrt{N})$ storage.

In a similar fashion, it is easy to check that the Pollard's lambda algorithm [15] also applies to solve the DLPX. It takes $O(\sqrt{N})$ group operations to solve the problem with small amount of storage. Also, check that the other algorithms such as Pohlig-Hellman algorithm [14] or the distinguished point method of Pollard's lambda algorithm [17] apply to solve the DLPX. The above observation was a main idea to solve the DLPwAI in [5,6].

3   Multiple DLPwAI: Cheon's Algorithm

In this section, we present an algorithm of solving the $(1, d)$-MDLPwAI based on Cheon's algorithm [5,6] when $d \mid p - 1$.

Workflow of This Section. Description of our algorithm is presented as follows. First, we recall how Cheon's algorithm solves the DLPwAI. In Sect. 3.1, we observed that the DLPwAI actually reduces to the DLPX (defined in Sect. 2) by Cheon's algorithm. It is, then, easy to check that to solve the MDLPwAI reduces to solve the MDLPX. So, we present an algorithm to solve the MDLPX in Sect. 3.2. Combined with the above results, we present an algorithm to solve the MDLPwAI in Sect. 3.3.

3.1   Reduction of DLPwAI to DLP in the Exponent Using Cheon's Algorithm

We briefly remind Cheon's algorithm in the case of $d \mid p - 1$. The algorithm solves $(1, d)$-DLPwAI. Let g, g^α, and g^{α^d} be given. Let ζ be a primitive element of $\mathbb{F}_p$ and $H = \langle \xi \rangle = \langle \zeta^d \rangle$ be a subgroup of $\mathbb{F}_p^*$ of order $\frac{p-1}{d}$. Since $\alpha^d \in H$, we have $\alpha^d = \xi^k$ for some $k \in \mathbb{Z}_{(p-1)/d}$. Our first task is to find such k. This is equivalent to solve the $\mathbb{F}_p$-DLPX defined over $\mathbb{Z}_{(p-1)/d}$, that is, to compute $k \in \mathbb{Z}_{(p-1)/d}$ for given $g, g^{\xi^k} \in G$ and $\xi \in \mathbb{F}_p$. Note that $g^{\xi^k} = g^{\alpha^d}$ is given from an instance of the DLPwAI and we know the value of ξ, since a primitive element in $\mathbb{F}_p$ can be efficiently found. As mentioned before, solving the DLPX over $\mathbb{Z}_{(p-1)/d}$ takes $O\left(\sqrt{p/d}\right)$ group exponentiations using BSGS algorithm or Pollard's lambda algorithm.

Continuously, if we write $\alpha \in \mathbb{F}_p$ as $\alpha = \zeta^\ell$, then since $\alpha^d = \zeta^{d\ell} = \zeta^{dk} = \xi^k$, it satisfies $\ell \equiv k \pmod{(p-1)/d}$, i.e. $\alpha\zeta^{-k} = (\zeta^{\frac{p-1}{d}})^m$ for some $m \in \mathbb{Z}_d$. Now we know the value of k, it remains to recover m. This is equivalent to solve $\mathbb{F}_p$-DLPX over $\mathbb{Z}_d$, that is, to solve $m \in \mathbb{Z}_d$ given the elements $g, g^{\mu^m} = (g^\alpha)^{\zeta^{-k}} \in G$ and $\mu \in \mathbb{F}_p$, where $\mu = \zeta^{\frac{p-1}{d}}$ is known. This step costs $O(\sqrt{d})$ group exponentiations. Overall, Cheon's $(p-1)$ algorithm reduces of solving two instances of DLP in the exponent with complexity $O(\sqrt{p/d} + \sqrt{d})$.

3.2   Algorithm for Multiple DLP in the Exponent

In this section, we describe an algorithm to solve L -*multiple DLP in the exponent*: Let L be a positive integer. Let χ be an element in $\mathbb{F}_p$ of multiplicative order N. The problem is to solve all $k_i \in \mathbb{Z}_N$ for given $g, y_1 := g^{\chi^{k_1}}, \ldots, y_L := g^{\chi^{k_L}}$ and χ.

We use Pollard's lambda-like algorithm. Define pseudo-random walk f from $y := g^{\chi^k}$ ($k \in \mathbb{Z}_N$) as follows. For an integer I, define a pseudo-random function $\iota : \{g^{\chi^n} : n \in \mathbb{Z}_N\} \to \{1, 2, \ldots, I\}$ and set $S := \{\chi^{s_1}, \ldots, \chi^{s_I}\}$ for some random integers s_i. For $y = g^{\chi^k}$, a pseudo-random walk f is defined by $f : y \mapsto y^{\chi^{s_\iota(y)}} = g^{\chi^{k+s_\iota(y)}}$.

Notice that Pollard's rho-like algorithm does not apply to solve the DLPX[1]. For instance, it seems hard to compute $g^{\chi^{2k}}$ from g^{χ^k} for unknown k if the Diffie-Hellman assumption holds in the group G. This is why we take Pollard's lambda-like approach.

The proposed algorithm is basically the same with the method by Kuhn and Struik [11]. It uses the distinguished point method of Pollard's rho (lambda) method [17]. Applying their method in the case of the DLPX, we describe the algorithm as follows.

Step 1. For $y_0 := g^{\chi^{k_0}}$ for $k_0 = N - 1$, compute the following chain until it reaches to a distinguished point d_0.

$$C_0 : y_0 \mapsto f(y_0) \mapsto f(f(y_0)) \mapsto \cdots \mapsto d_0.$$

Step 2. For $y_1 = g^{\chi^{k_1}}$, compute a chain until a distinguished point d_1 found.

$$C_1 : y_1 \mapsto f(y_1) \mapsto f(f(y_1)) \mapsto \cdots \mapsto d_1.$$

If we have a collision $d_1 = d_0$, then it reveals a discrete logarithm k_1. Otherwise, set $y_1' = y_1 \cdot g^{\chi^z}$ for known z and use it as a new starting point to compute a new chain to obtain a collision.

Step 3. Once we have found the discrete logarithm $k_1, \ldots, k_i$, then one iteratively computes the next discrete logarithm k_{i+1} as follows: Compute a chain as Step 2 with a starting point y_{i+1} until a distinguished point d_{i+1} is

---

[1] In the paper [16], they indeed consider Pollard's lambda algorithm rather than rho algorithm.

found. Then try to find a collision $d_{i+1} = d_j$ for some $1 \leq j \leq i$. It reveals the discrete logarithm of y_{i+1}. If it fails, compute a chain again with a new randomized starting point $y'_{i+1} = y_{i+1} \cdot g^{\chi^{z'}}$ for known z'.

By the analysis in [11], this algorithm has a running time of $\widetilde{O}(\sqrt{L \cdot N})$ operations for $L \leq c_N N^{1/4}$ (where $0 < c_N < 1$ is some constant depending on N) with storage for $O(L)$ elements of the distinguished points.

Remark 1. If we allow large amount of storage, then we have a deterministic algorithm solving the DLPX based on the BSGS method[2]. It works for any $L \geq 0$ as follows. First, choose an integer $K = \lceil \sqrt{N/L} \rceil$ and compute $g^{\chi^{K \cdot t}} = \left(g^{\chi^{K \cdot (t-1)}} \right)^{\chi^K}$ for all $t \leq \sqrt{L \cdot N}$ using $O(\sqrt{L \cdot N})$ group exponentiations and store all of the elements. Then, for each $i = 1, 2, \ldots, L$, compute $g^{\chi^{k_i - s}} = \left(g^{\chi^{k_i}} \right)^{\chi^{-s}}$ for all $s \leq \sqrt{N/L}$ and find a collision with the stored elements. It takes $O(L \cdot \sqrt{N/L})$ operations for all. If one has a collision, then we have $k_i = s + t \cdot K$ for the indices s and t corresponding to the collision.

Remark 2. There is a recent paper by [7] that claims that the MDLP can be solved in $\widetilde{O}(\sqrt{L \cdot N})$ for any L with small amount of storage. However, their analysis (Sect. 2, [7]) seems somewhat questionable.

In their analysis, they essentially assumed that a collision occurs independently from each different chains. The pseudo-random function, however, once it has been fixed, it becomes deterministic and not random. For example, assume that we have a collision between two chains, say C_1 and C_2. If a new chain C_3 also collides with C_1, then it deterministically collides with C_2, too. This contradicts with independency assumption. The event that the chain C_3 connects to the chain C_2 should be independent whether C_3 is connected to C_1 or not. This kind of heuristic might be of no problem when L is much smaller than compared to N. However, this is not the case for large L.

Several literatures focus on this *rigour* of pseudo-random function used in Pollard's algorithm. For further details on this, refer to [9].

3.3 Solving Multiple DLPwAI Using Cheon's Algorithm

Combined with the results from Sects. 3.1 and 3.2, we propose an algorithm solving the $(1, d)$-MDLPwAI in the case of $d \mid p - 1$. In Appendix A, we explain that Cheon's $(p + 1)$-algorithm does not help to solve the MDLPwAI in the case of $d \mid p + 1$.

Theorem 1 (Algorithm for $(1, d)$-MDLPwAI, $d \mid p - 1$). *Let the notations as above. Let $\alpha_1, \ldots, \alpha_L$ be randomly chosen elements from $\mathbb{Z}_p$. Assume that $d \mid p - 1$. For $L \leq \min\{c_{p/d}(p/d)^{1/4}, c_d d^{1/4}\}$ (where $0 < c_{p/d}, c_d < 1$ are some constants on p/d and d respectively), given the elements g, g^{α_i} and $g^{\alpha_i^d}$ for*

---

[2] The proof is contributed by Mehdi Tibouchi.

$i = 1, 2, \ldots, L$, we have an algorithm that computes α_i's in $\tilde{O}(\sqrt{L \cdot p/d} + \sqrt{L \cdot d})$ group exponentiations with storage for $O(L)$ elements in the set of the distinguished points.

Proof. Similarly as in Sect. 3.1, let $H = \langle \xi \rangle = \langle \zeta^d \rangle \subset G$ for a primitive element $\zeta \in \mathbb{F}_p$. Since $\alpha_i^d \in H$, we have $\alpha_i^d = \xi^{k_i}$ for some $k_1, \ldots, k_L$, where $k_i \in \mathbb{Z}_{(p-1)/d}$, and if we write $\alpha_i = \zeta^{\ell_i}$, then we have $\alpha_i \zeta^{-k_i} = \mu^{m_i}$ for $m_i \in \mathbb{Z}_d$. Thus the problem reduces of solving two multiple DLP in the exponent with instances $g^\xi, g^{\xi^{k_1}} = g^{\alpha_1^d}, \ldots, g^{\xi^{k_L}} = g^{\alpha_L^d}$ and $g^\mu, g^{\mu^{m_1}} = (g^{\alpha_1})^{\zeta^{-k_1}}, \ldots, g^{\mu^{m_L}} = (g^{\alpha_L})^{\zeta^{-k_L}}$, where ξ and μ are known. We compute α_i's as follows:

1. Given $g^\xi, g^{\alpha_1^d} = g^{\xi^{k_1}}, \ldots, g^{\alpha_L^d} = g^{\xi^{k_L}}$ for $k_i \in \mathbb{Z}_{(p-1)/d}$, compute k_i's using the algorithm in Sect. 3.2. It takes time $\tilde{O}(\sqrt{L \cdot p/d})$ with storage for $O(L)$ elements.
2. Given $g^{\alpha_1}, \ldots, g^{\alpha_L}$ and $k_1, \ldots, k_L$, compute $\zeta^{-k_1}, \ldots, \zeta^{-k_L}$ in $O(L)$ exponentiations in $\mathbb{F}_p$ and compute

$$g^{\mu^{m_1}} = (g^{\alpha_1})^{\zeta^{-k_1}}, \ldots, g^{\mu^{m_L}} = (g^{\alpha_L})^{\zeta^{-k_L}}$$

   in $O(L)$ exponentiations in G.
3. Compute $m_1, \ldots, m_L \in \mathbb{Z}_d$ from $g^{\mu^{m_1}}, \ldots, g^{\mu^{m_L}}$ using the algorithm in Sect. 3.2. It takes time $\tilde{O}(\sqrt{L \cdot d})$ with storage for $O(L)$ elements.

The overall complexity is given by $\tilde{O}(\sqrt{L \cdot p/d} + \sqrt{L \cdot d} + L)$. Since $L \leq \min\{p/d, d\}$ by the assumption, i.e. $L \leq \min\{\sqrt{L \cdot p/d}, \sqrt{L \cdot d}\}$, it is equivalent to $\tilde{O}(\sqrt{L \cdot p/d} + \sqrt{L \cdot d})$. ∎

Remark 3. Note that we can replace the algorithm to solve the MDLPX used in Step 1 and Step 3 with any algorithm solving the MDLPX. In that case, the complexity solving the MDLPwAI totally depends on that of the algorithm solving the MDLPX. For example, if we use the BSGS method described in Remark 1, then the proposed algorithm solves the MDLPwAI for any L in time complexity $O(\sqrt{L \cdot p/d} + \sqrt{L \cdot d} + L)$ with the same amount of storage.

4 Multiple DLPwAI: Kim and Cheon's Algorithm

In this section, we propose an approach to solve the d-MDLPwAI. The idea is basically based on Kim and Cheon's algorithm [10]. To analyze the complexity, we also need some discussion on non-uniform birthday problem.

4.1 Description of Algorithm

Let $G = \langle g \rangle$ be a group of prime order p. For $i = 1, 2, \ldots, L$, let $g, g^{\alpha_i}, \ldots, g^{\alpha_i^d}$ be given. We choose a polynomial $f(x) \in \mathbb{F}_p[x]$ of degree d and fix a positive integer ℓ which will be defined later. The proposed algorithm is described as follows:

Step 1. For each i, given $g, g^{\alpha_i}, \ldots, g^{\alpha_i^d}$ and $f(x)$, we compute and store a constant number of sets each of which is of form

$$S_i := \{g^{f(r_{i,1}\alpha_i)}, \ldots, g^{f(r_{i,\ell}\alpha_i)}\},$$

where $r_{i,j}$'s are randomly chosen from $\mathbb{F}_p$.

Step 2. We also compute and store a constant number of sets each of which consists of

$$S_0 := \{g^{f(s_1)}, \ldots, g^{f(s_\ell)}\},$$

where s_k's are known random values from $\mathbb{F}_p$.

Step 3. We construct a random graph with L vertices: we add an edge between vertices i and j, if S_i and S_j collide.

Step 4. If $f(r_{i,j}\alpha_i) = f(s_k)$ for some i, j and k, then α_i is one of d roots of the equation of degree d in variable α_i:

$$f(r_{i,j}\alpha_i) - f(s_k) = 0.$$

Step 5. If $f(r_{i,j}\alpha_i) = f(r_{i',j'}\alpha_{i'})$, for some i, j, i' and j', where α_i is known, then $\alpha_{i'}$ is one of d roots of the following equation of degree d in variable $\alpha_{i'}$:

$$\widetilde{f}(\alpha_{i'}) := f(r_{i,j}\alpha_i) - f(r_{i',j'}\alpha_{i'}) = 0.$$

We recover all α_i's when they are connected into a component with known discrete logs. In the next subsection, we analyze the complexity of the proposed algorithm more precisely.

4.2   Complexity Analysis

We analyze the complexity of the proposed algorithm.

Theorem 2 (Algorithm for d-MDLPwAI). *Let the notations as above. Let $f(x)$ be a polynomial of degree d over $\mathbb{F}_p$. Define $\rho_f := |\{(x,y) \in \mathbb{F}_p \times \mathbb{F}_p : f(x) = f(y)\}|$. Given $g, g^{\alpha_i}, \ldots, g^{\alpha_i^d}$ for $i = 1, 2, \ldots, L$, we have an algorithm that computes all α_i's in $\widetilde{O}(\max\{\sqrt{L \cdot p^2/\rho_f}, L \cdot d\})$ group exponentiations with storage for $\widetilde{O}(\sqrt{L \cdot p^2/\rho_f})$ elements in G.*

Proof. Consider the complexity of each step in the proposed algorithm. Throughout the paper, we denote $M(d)$ by the time complexity multiplying two polynomials of degree d over $\mathbb{F}_p$ (typically, we will take $M(d) = O(d \log d \log \log d)$ using the Schönhage-Strassen method).

In Step 2, we compute $f(s_1), \ldots, f(s_\ell)$ using fast multipoint evaluation method. It takes $O(\ell/d \cdot M(d) \log d) = O(\ell \log^2 d \log \log d)$ operations in $\mathbb{F}_p$ if $\ell \geq d$. Otherwise, the cost is bounded by $O(M(d) \log d) = O(d \log^2 d \log \log d)$ operations in $\mathbb{F}_p$. Then compute $g^{f(s_1)}, \ldots, g^{f(s_\ell)}$ in $O(\ell)$ exponentiations in G.

In Step 1, we use fast multipoint evaluation method in the exponent as described in [10, Theorem 2.1], which is the following: given $g^{F_0}, \ldots, g^{F_d}$, where

F_i is the coefficient of x^i of a polynomial $F(x) \in \mathbb{F}_p[x]$, and given random elements $r_1, \ldots, r_d \in \mathbb{F}_p$, it computes $g^{F(r_1)}, \ldots, g^{F(r_d)}$ in $O(M(d) \log d)$ operations in G.

In our case, for given $g, g^{\alpha_i}, \ldots, g^{\alpha_i^d}$ and $f(x) = a_0 + \cdots + a_d x^d$, we set $f_i(x) := f(\alpha_i x) = a_0 + (a_1 \alpha_i)x + \cdots + (a_d \alpha_i^d)x^d$ and compute $g^{a_0}, (g^{\alpha_i})^{a_1}, \ldots, (g^{\alpha_i^d})^{a_d}$ in $O(d)$ exponentiations in G for each i. It totally costs $O(L \cdot d)$ exponentiations for all $i = 1, \ldots, L$. Applying Theorem 2.1 in [10] to each polynomial $f_i(x)$, if $\ell \geq d$, we compute

$$S_i = \{g^{f_i(r_{i,1})}, \ldots, g^{f_i(r_{i,\ell})}\} = \{g^{f(r_{i,1}\alpha_i)}, \ldots, g^{f(r_{i,\ell}\alpha_i)}\}$$

in $O(\ell/d \cdot M(d) \log d)$ operations in G for each i. It costs $O(L \cdot \ell \log^2 d \log \log d)$ operations overall for all $i = 1, \ldots, L$. Otherwise, if $\ell \leq d$, then this step costs $O(L \cdot d \log^d \log \log d)$ operations.

In Step 4 and Step 5, the cost takes $O(M(d) \log d \log(dp))$ field operations on average [18] to compute roots of equation of degree d over $\mathbb{F}_p$. For each equation, we need to find α_i among at most d possible candidates. It takes $O(d)$ operations. These steps need to be done L times since we have L equations to be solved.

Consequently, to recover all α_i's, it takes overall $\widetilde{O}(\max\{L \cdot \ell, L \cdot d\})$ operations with $O(L \cdot \ell)$ storage. Now it remains to determine the value of ℓ. To this end, we need to clarify the probability of a collision between S_i and S_j (for $i \neq j$) in Step 3. It leads us to consider non-uniform birthday problem of two types. We will discuss on details for this in Appendix B.

We heuristically assume that the probability of a collision between S_i and S_j in Step 3 is equiprobable for any $i \neq j$ and we denote this probability by ω[3]. By Corollary 1 in Appendix B, the probability is given by $\omega = \Theta(\ell^2 \cdot \rho_f / p^2)$ for large p. Then the expected number of edges in the graph in Step 3 will be $\binom{L}{2} \cdot \omega \approx \frac{L^2 \omega}{2} \approx \frac{L^2 \ell^2}{2} \cdot \frac{\rho_f}{p^2}$. We require this value to be larger than $2L \ln L$ to connect all connected components in the graph (see [7]), i.e.

$$\ell \geq 2\sqrt{\frac{p^2}{\rho_f} \cdot \frac{\ln L}{L}}.$$

If we take $\ell = 2\sqrt{\frac{p^2}{\rho_f} \cdot \frac{\ln L}{L}}$, the overall time complexity becomes (without log terms) $\widetilde{O}(\max\{L \cdot \ell, L \cdot d\}) = \widetilde{O}\left(\max\left\{\sqrt{L \cdot p^2/\rho_f}, L \cdot d\right\}\right)$ with storage for $\widetilde{O}(L \cdot \ell) = \widetilde{O}(\sqrt{L \cdot p^2/\rho_f})$ elements in G. □

Remark 4. In general, the computation of ρ_f seems relatively not so obvious. However, for some functions f which are useful for our purpose, it can be efficiently computable. See Sect. 4.3.

---

[3] The assumption is reasonable, since every exponents of the elements in S_i's are randomly chosen from $\mathbb{F}_p$, i.e. the sets S_i's are independent from each other. Observe that this does not conflict with Remark 2.

If $L \leq \frac{p^2}{d^2 \cdot \rho_f}$, then the time complexity of the algorithm is given by $\tilde{O}(\sqrt{L \cdot p^2/\rho_f})$. Note that this value is always between $\tilde{O}(\sqrt{L \cdot \frac{p}{d}})$ and $\tilde{O}(\sqrt{L \cdot p})$. In the next subsection, we observe that one can find polynomials f with $\rho_f \approx C \cdot dp$ for some constant C in the case of $d \mid p \pm 1$. In such cases, the proposed algorithm has a running time of $\tilde{O}(\sqrt{L \cdot p/d})$ whenever $L \leq \frac{p}{C \cdot d^3}$.

It should be compared that application of Cheon's $(p+1)$-algorithm failed to achieve the lower bound complexity $\tilde{O}(\sqrt{L \cdot p/d})$ in the case of $d \mid p+1$ (see Appendix A).

4.3 Explicit Choices of Polynomials for Efficient Algorithms in the Case Of $d \mid P \pm 1$

For efficiency of the algorithm, we require a polynomial $f(x)$ with large ρ_f. In particular, ρ_f becomes larger as the map $x \mapsto f(x)$, restricted on $\mathbb{F}_p$ or a large subset of $\mathbb{F}_p$, has a smaller value set. See the examples below. For details on choices of these polynomials, refer to [10].

$d \mid p-1$**Case.** Let $f(x) = x^d$. Then the map by f is d-to-1 except at $x = 0$. Then we have $\rho_f = 1 + d(p-1) \approx dp$. In this case, the complexity of our algorithm becomes $\tilde{O}(\sqrt{L \cdot p/d})$ for $L \leq p/d^3$.

$d \mid p+1$**Case.** Let $f(x) = D_d(x, a)$ be the Dickson polynomial for a nonzero element $a \in \mathbb{F}_p$, where

$$D_d(x, a) = \sum_{k=0}^{\lfloor d/2 \rfloor} \frac{d}{d-k} \binom{d-k}{k} (-a)^k x^{d-2k}.$$

If $d \mid p+1$, then by [8,12], we have $\rho_f = \frac{(d+1)p}{2} + O(d^2) \approx \frac{dp}{2}$. In this case, our algorithm has the complexity of $\tilde{O}(\sqrt{L \cdot p/d})$ for $L \lesssim p/(2d^3)$.

5 Conclusion

In this paper, we proposed algorithms for the MDLPwAI based on two different approaches. These algorithms cover all extensions of existing DLPwAI-solving algorithms, since, up to our knowledge, there are only two (efficient) approaches solving the DLPwAI: Cheon's algorithm and Kim and Cheon's algorithm.

Our analysis shows that our algorithms have the best running time of either $\tilde{O}(\max\{\sqrt{L \cdot p/d}, \sqrt{L \cdot d}\})$ when $d \mid p-1$, or $\tilde{O}(\max\{\sqrt{L \cdot p/d}, L \cdot d\})$ when $d \mid p+1$. It shows that the choice of the prime p should be chosen carefully so that both of $p+1$ and $p-1$ have no small divisors. Readers might refer to [5,6] for careful choices of such prime p.

However, our second algorithm is based on some heuristics and requires relatively large amount of memory. Thus, it would be a challenging question either to reduce the storage requirement in the algorithm, or to make the algorithm more rigorous.

It would be also interesting to determine the lower bound complexity in the generic group model for solving the multiple DLPwAI. A very recent result [19] showed that at least $\Omega(\sqrt{L \cdot p})$ group operations are required to solve the L multiple DLP in the generic group model. Recall that the generic lower bound for the DLPwAI is $\Omega(\sqrt{p/d})$. Then it is natural to ask the following questions. What is the lower bound complexity in the generic group model to solve the multiple DLPwAI? Do we need at least $\Omega(\sqrt{L \cdot p/d})$ operations for solving the multiple DLPwAI?

Acknowledgement. The author would like to thank Pierre-Alain Fouque, Soojin Roh, Mehdi Tibouchi, and Aaram Yun for their valuable discussion. He also would like to extend his appreciation to anonymous reviewers who further improved this paper.

A   A Failed Approach for MDLPwAI When $d \mid P + 1$

$\mathbb{F}_{p^2}$**-Discrete Logarithm Problem in the Exponent.** To define $\mathbb{F}_{p^2}$-(M)DLPX, we introduce the following definition[4].

Definition 1. *Let $G = \langle g \rangle$ be a group of prime order p. Let $\mathbb{F}_{p^2} = \mathbb{F}_p[\theta] = \mathbb{F}_p[x]/(x^2 - \kappa)$ for some quadratic non-residue $\kappa \in \mathbb{F}_p$. For $\gamma = \gamma_0 + \gamma_1\theta \in \mathbb{F}_{p^2}$, we define $g^\gamma = (g^{\gamma_0}, g^{\gamma_1})$ with abuse of notations. For $\mathbf{g} := (g_0, g_1) \in G \times G$, we define*

$$\mathbf{g}^\gamma = \mathbf{g}^{\gamma_0 + \gamma_1\theta} = (g_0^{\gamma_0} g_1^{\kappa\gamma_1}, g_0^{\gamma_1} g_1^{\gamma_0}), \text{ where } \theta^2 = \kappa.$$

One can readily check that $(g^\gamma)^\delta = (g^{\gamma_0}, g^{\gamma_1})^\delta = (g^{\gamma_0\delta_0 + \kappa\gamma_1\delta_1}, g^{\gamma_0\delta_1 + \gamma_1\delta_0}) = g^{\gamma\delta}$, where $\delta = \delta_0 + \delta_1\theta$. Now we define $\mathbb{F}_{p^2}$-(M)DLPX.

- The $\mathbb{F}_{p^2}$ *-Discrete Logarithm Problem in the Exponent ($\mathbb{F}_{p^2}$-DLPX) in G* is defined as follows: Let $\chi \in \mathbb{F}_{p^2}$ be an element of multiplicative order N, i.e. $N \mid p^2 - 1$. Given $g \in G$ and $g^{\chi^n} \in G \times G$ and $\chi \in \mathbb{F}_{p^2}$, compute $n \in \mathbb{Z}_N$.
- The $\mathbb{F}_{p^2}$ *-Multiple Discrete Logarithm Problem in the Exponent ($\mathbb{F}_{p^2}$-MDLPX) in G* is: Given $g \in G$, $g^{\chi^{n_1}}, \ldots, g^{\chi^{n_L}} \in G \times G$ and $\chi \in \mathbb{F}_{p^2}$, to solve $n_1, \ldots, n_L \in \mathbb{Z}_N$. In both cases, the $\mathbb{F}_{p^2}$-(M)DLPX is said to be defined over $\mathbb{Z}_N$.

Observe that generic approaches to solve the (M)DLPX described in Sects. 2 and 3.2 also apply to solve the $\mathbb{F}_{p^2}$-(M)DLPX.

A Failed Approach when $d \mid p + 1$. We consider the MDLPwAI in the case of $d \mid p + 1$. Recall Cheon's $(p + 1)$ algorithm [5,6] which solves $2d$-DLPwAI. Let $g, g^{\alpha_i}, \ldots, g^{\alpha_i^{2d}}$, for $i = 1, 2, \ldots, L$, be given. We try to solve the problem as follows: For each $i = 1, 2, \ldots, L$, let $\beta_i := (1 + \alpha_i\theta)^{p-1} \in \mathbb{F}_{p^2} = \mathbb{F}_p[\theta]$ and let $\xi \in \mathbb{F}_{p^2}$ an element of multiplicative order $(p + 1)/d$. We compute $g_i := g^{(1-\kappa\alpha_i^2)^d}$ and $g_i^{\xi^{k_i}} = g_i^{\beta_i^d} := (g^{f_0(\alpha_i)}, g^{f_0(\alpha_i)})$ for the given elements

---

[4] This notion can be found in [5,6] when he solves DLPwAI using Pollard's lambda algorithm. We simply formalize them.

$g, g^{\alpha_i}, \ldots, g^{\alpha_i^{2d}}$, where $\beta_i^d = \frac{1}{(1-\kappa\alpha_i^2)^d}\{f_0(\alpha_i)+f_1(\alpha_i)\theta\}$. The remaining task is to solve $k_1, \ldots, k_L \in \mathbb{Z}_{(p+1)/d}$. This translates to solve L instances of the $\mathbb{F}_{p^2}$-DLPX, say $(g_1, g_1^{\xi^{k_1}}), (g_2, g_2^{\xi^{k_2}}), \ldots, (g_L, g_L^{\xi^{k_L}})$. Note that, however, these L instances cannot be solved efficiently in a batch computation based on our MDLPX algorithms, since all the bases of the instances are not the same.

B Non-uniform Birthday Problem: Girls and Boys

In this section, we consider the probability of a collision in Step 3, Sect. 4.1. More generally, we consider non-uniform birthday problem of two types. The main goal in this section is to prove the following theorem.

Theorem 3. *For a positive integer N and $i \in \{1,2,\ldots,N\}$, assume that the probability of a randomly chosen element from the set $\{1,2,\ldots,N\}$ to be i is ω_i. Let T_1 (respectively, T_2) be a set consisting of ℓ_1 (reps. ℓ_2) elements randomly chosen from $\{1,2,\ldots,N\}$. Then the probability ω that T_1 and T_2 have an element in common satisfies*

$$\ell_1\ell_2 \cdot \sum_{i=1}^N \omega_i^2 \geq \omega \geq \ell_1\ell_2 \cdot \sum_{i=1}^N \omega_i^2 - \left(\ell_1 \cdot \binom{\ell_2}{2} + \ell_2 \cdot \binom{\ell_1}{2}\right) \cdot \sum_{i=1}^N \omega_i^3$$
$$+ \binom{\ell_1}{2}\binom{\ell_2}{2}\left(\sum_{i=1}^N \omega_i^4 - \sum_{1\leq i<j\leq N} \omega_i^2\omega_j^2\right). \quad (1)$$

Proof. For each $i \in \{1,2,\ldots,N\}$, let $B_i^{(\ell_1,\ell_2)}$ be the event that two sets T_1 and T_2 have the element i in common. Then the probability ω that T_1 and T_2 have at least one element in common is given by

$$\omega = \Pr[B_1^{(\ell_1,\ell_2)} \cup \cdots \cup B_N^{(\ell_1,\ell_2)}].$$

From now on, we shall omit superscript in $B_i^{(\ell_1,\ell_2)}$ and simply denote it by B_i. To bound the value ω, we use Bonferroni inequality,

$$\sum_{i=1}^N \Pr[B_i] - \sum_{1\leq i<j\leq N} \Pr[B_i \cap B_j] \leq \omega \leq \sum_{i=1}^N \Pr[B_i].^5$$

---

[5] It is easy to check the lower bound inequality. Assume that $\Pr[B_1 \cup B_2] \geq \Pr[B_1] + \Pr[B_2] - \Pr[B_1 \cap B_2]$ (indeed the equality holds in this case). Then to see that

$$\Pr[(B_1 \cup B_2) \cup B_3] = \Pr[B_1 \cup B_2] + \Pr[B_3] - \Pr[(B_1 \cup B_2) \cap B_3]$$
$$\geq \Pr[B_1] + \Pr[B_2] + \Pr[B_3] - \Pr[B_1 \cap B_2] - \Pr[B_1 \cap B_3] - \Pr[B_2 \cap B_3],$$

it is enough to check that

$$\Pr[(B_1 \cup B_2) \cap B_3] = \Pr[(B_1 \cap B_3) \cup (B_2 \cap B_3)] \leq \Pr[B_1 \cap B_3] + \Pr[B_2 \cap B_3].$$

Now apply the induction on N.

We shall investigate bounds on $\Pr[B_i]$ and $\Pr[B_i \cap B_j]$ in the followings.

For each i, the set T_1 with ℓ_1 elements has the element i with probability $1 - (1 - w_i)^{\ell_1}$ and similarly for T_2. Thus both of T_1 and T_2 have the element i with probability $\Pr[B_i] = (1 - (1 - w_i)^{\ell_1}) \cdot (1 - (1 - w_i)^{\ell_2})$. Using the inequality $1 - nx \le (1 - x)^n \le 1 - nx + \binom{n}{2}x^2$ for $0 \le x \le 1$ and $n > 1$, we have

$$\left(\ell_1 w_i - \binom{\ell_1}{2}w_i^2 \right) \cdot \left(\ell_2 w_i - \binom{\ell_2}{2}w_i^2 \right) \le \Pr[B_i] \le \ell_1 \ell_2 \cdot w_i^2.$$

Furthermore, we have $\Pr[B_i \cap B_j] \le \binom{\ell_1}{2}\binom{\ell_2}{2}w_i^2 w_j^2$, since T_1 has the element i and j with probability at most $\binom{\ell_1}{2}w_i w_j$ and similarly for T_2.

Then the upper bound for w directly comes from the upper bound for $\Pr[B_i]$ and the lower bound comes from

$$w \ge \sum_{i=1}^{N} \Pr[B_i] - \sum_{1 \le i < j \le N} \Pr[B_i \cap B_j]$$

$$\ge \sum_{i=1}^{N} \left(\ell_1 w_i - \binom{\ell_1}{2}w_i^2 \right) \cdot \left(\ell_2 w_i - \binom{\ell_2}{2}w_i^2 \right) - \sum_{i<j} \binom{\ell_1}{2}\binom{\ell_2}{2}w_i^2 w_j^2.$$

This concludes the proof.   □

Corollary 1. *Let $W := \sum_{i=1}^{N} w_i^2$ in Theorem 3. If $\ell = \ell_1 = \ell_2 = O\left(\sqrt{\frac{1}{W}}\right)$ and $W \to 0$, we have*

$$\ell^2 W - \frac{(\ell^2 W)^2}{8} + O\left(\frac{1}{\sqrt{W}}\right) \le w \le \ell^2 W.$$

Proof. Evaluating $\ell = \ell_1 = \ell_2$ in the right most side of Eq. (1), we have

$$w \ge \ell^2 W - \ell^2(\ell - 1)\sum_{i=1}^{N} w_i^3 + \frac{\ell^2(\ell-1)^2}{4}\left(\frac{3}{2}\sum_i w_i^4 - \frac{1}{2}W^2 \right)$$

$$\ge \ell^2 W - \ell^3 \sum_{i=1}^{N} w_i^3 - \frac{1}{8}(\ell^2 W)^2.$$

In the first inequality, we used that $\sum_{i<j} w_i^2 w_j^2 = \frac{1}{2}\left[\left(\sum_i w_i^2\right)^2 - \sum_i w_i^4 \right]$. To see that $\ell^3 \sum_i w_i^3 \le O\left(\frac{1}{\sqrt{W}}\right)$, it is enough to check that $\sum_i w_i^3 = \sum_i w_i^2 w_i = \sum_i w_i^2 \cdot \sum_i w_i - \sum_{i \ne j} w_i^2 w_j \le \sum_i w_i^2 = W$ (recall that $\sum_i w_i = 1$).   □

Return to our interest. Intrinsically, in our application (Sect. 4), we consider the intersection between two sets $T_1 := \{t_1, \ldots, t_\ell\} = \{f(r_1), \ldots, f(r_\ell)\}$ and $T_2 := \{t'_1, \ldots, t'_\ell\} = \{f(r'_1), \ldots, f(r'_\ell)\}$ for a degree d polynomial $f(x) \in \mathbb{F}_p[x]$. This can be regarded as non-uniform birthday problem described in Theorem 3 similarly as in [10]: An element $t \in T_1$ (or $t' \in T_2$) is randomly chosen from $\mathbb{F}_p$ with the probability $\frac{|f^{-1}(t)|}{p}$. Let $R_i := |\{t \in \mathbb{F}_p : |f^{-1}(t)| = i\}|$ for a non-negative integer i. We have $R_i = 0$ for $i > d$ since $\deg(f) = d$. Then we might say that an element in T_1 (or T_2) is drown by following the probability distribution (with proper rearrange)

$$(\omega_1, \ldots, \omega_p) = (\underbrace{0, \ldots, 0}_{R_0}, \underbrace{\frac{1}{p}, \ldots, \frac{1}{p}}_{R_1}, \underbrace{\frac{2}{p}, \ldots, \frac{2}{p}}_{R_2}, \cdots, \underbrace{\frac{d}{p}, \ldots, \frac{d}{p}}_{R_d}).$$

Then $W = \sum_{i=1}^p \omega_i^2 = \frac{\sum_{i=1}^d i^2 R_i}{p^2} = \frac{\rho_f}{p^2}$, where $\rho_f := |\{(x,y) \in \mathbb{F}_p \times \mathbb{F}_p : f(x) = f(y)\}|$. In our case, we usually take $\ell = 2\sqrt{\frac{p^2}{\rho_f} \cdot \frac{\ln L}{L}} = O(\sqrt{1/W})$ (see the proof of Theorem 3), where L is the constant given by the number of the target discrete logarithms. Then, by Corollary 1, we roughly have $\ell^2 W - \frac{(\ell^2 W)^2}{8} \le \omega \le \ell^2 W$ for large enough p, i.e. $\omega = \Theta(\ell^2 W)$ (using $x - x^2/8 \ge (7/8)x$ for $0 \le x \le 1$). Consequently, this gives what we want for the analysis.

References

1. Boneh, D., Boyen, X.: Efficient selective-ID secure identity-based encryption without random oracles. In: Cachin, C., Camenisch, J.L. (eds.) EUROCRYPT 2004. LNCS, vol. 3027, pp. 223–238. Springer, Heidelberg (2004)
2. Boneh, D., Boyen, X.: Short signatures without random oracles. In: Cachin, C., Camenisch, J.L. (eds.) EUROCRYPT 2004. LNCS, vol. 3027, pp. 56–73. Springer, Heidelberg (2004)
3. Boneh, D., Gentry, C., Waters, B.: Collusion resistant broadcast encryption with short ciphertexts and private keys. In: Shoup, V. (ed.) CRYPTO 2005. LNCS, vol. 3621, pp. 258–275. Springer, Heidelberg (2005)
4. Brown, D.R.L., Gallant, R.P.: The static Diffie-Hellman problem. IACR Cryptology ePrint Archive (2004). http://eprint.iacr.org/2004/306
5. Cheon, J.H.: Security analysis of the strong Diffie-Hellman problem. In: Vaudenay, S. (ed.) EUROCRYPT 2006. LNCS, vol. 4004, pp. 1–11. Springer, Heidelberg (2006)
6. Cheon, J.H.: Discrete logarithm problems with auxiliary inputs. J. Cryptol. 23(3), 457–476 (2010)
7. Fouque, P.-A., Joux, A., Mavromati, C.: Multi-user collisions: applications to discrete logarithm, even-mansour and PRINCE. In: Sarkar, P., Iwata, T. (eds.) ASIACRYPT 2014. LNCS, vol. 8873, pp. 420–438. Springer, Heidelberg (2014)
8. Gomez-Calderon, J., Madden, D.J.: Polynomials with small value set over finite fields. J. Number Theory 28, 167–188 (1988)
9. Kijima, S., Montenegro, R.: Collision of random walks and a refined analysis of attacks on the discrete logarithm problem. In: Katz, J. (ed.) PKC 2015. LNCS, vol. 9020, pp. 127–149. Springer, Heidelberg (2015)
10. Kim, T., Cheon, J.H.: A new approach to discrete logarithm problem with auxiliary inputs. IACR Cryptology ePrint Archive (2012). http://eprint.iacr.org/2012/609

11. Kuhn, F., Struik, R.: Random walks revisited: extensions of Pollard's Rho algorithm for computing multiple discrete logarithms. In: Vaudenay, S., Youssef, A.M. (eds.) SAC 2001. LNCS, vol. 2259, pp. 212–229. Springer, Heidelberg (2001)
12. Mit'kin, D.A.: Polynomials with minimal set of values and the equation $f(x) = f(y)$ in a finite prime field. Matematicheskie Zametki **38**(1), 3–14 (1985)
13. Mitsunari, S., Sakai, R., Kasahara, M.: A new traitor tracing. IEICE Trans. Fundam. Electron. Commun. Comput. Sci. **85**(2), 481–484 (2002)
14. Pohlig, S.C., Hellman, M.E.: An improved algorithm for computing logarithms over $GF(p)$ and its cryptographic significance (corresp.). IEEE Trans. Inf. Theory **24**(1), 106–110 (1978)
15. Pollard, J.M.: Kangaroos, monopoly and discrete logarithms. J. Cryptol. **13**(4), 437–447 (2000)
16. Sakemi, Y., Izu, T., Takenaka, M., Yasuda, M.: Solving a DLP with auxiliary input with the ρ-algorithm. In: Jung, S., Yung, M. (eds.) WISA 2011. LNCS, vol. 7115, pp. 98–108. Springer, Heidelberg (2012)
17. van Oorschot, P.C., Wiener, M.J.: Parallel collision search with cryptanalytic applications. J. Cryptol. **12**(1), 1–28 (1999)
18. von zur Gathen, J., Gerhard, J.: Modern Computer Algebra. Cambridge University Press, Cambridge (2003)
19. Yun, A.: Generic hardness of the multiple discrete logarithm problem. In: Oswald, E., Fischlin, M. (eds.) EUROCRYPT 2015. LNCS, vol. 9057, pp. 817–836. Springer, Heidelberg (2015)

Solving Linear Equations Modulo Unknown Divisors: Revisited

Yao Lu[1,2], Rui Zhang[1][✉], Liqiang Peng[1], and Dongdai Lin[1]

[1] State Key Laboratory of Information Security (SKLOIS), Institute of Information Engineering, Chinese Academy of Sciences, Beijing 100093, China
lywhhit@gmail.com, {r-zhang,pengliqiang,ddlin}@iie.ac.cn
[2] The University of Tokyo, Tokyo, Japan

Abstract. We revisit the problem of finding small solutions to a collection of linear equations modulo an unknown divisor p for a known composite integer N. In CaLC 2001, Howgrave-Graham introduced an efficient algorithm for solving univariate linear equations; since then, two forms of multivariate generalizations have been considered in the context of cryptanalysis: modular multivariate linear equations by Herrmann and May (Asiacrypt'08) and simultaneous modular univariate linear equations by Cohn and Heninger (ANTS'12). Their algorithms have many important applications in cryptanalysis, such as factoring with known bits problem, fault attacks on RSA signatures, analysis of approximate GCD problem, etc.

In this paper, by introducing multiple parameters, we propose several generalizations of the above equations. The motivation behind these extensions is that some attacks on RSA variants can be reduced to solving these generalized equations, and previous algorithms do not apply. We present new approaches to solve them, and compared with previous methods, our new algorithms are more flexible and especially suitable for some cases. Applying our algorithms, we obtain the best analytical/experimental results for some attacks on RSA and its variants, specifically,

- We improve May's results (PKC'04) on small secret exponent attack on RSA variant with moduli $N = p^r q$ ($r \geq 2$).
- We experimentally improve Boneh et al.'s algorithm (Crypto'98) on factoring $N = p^r q$ ($r \geq 2$) with known bits problem.
- We significantly improve Jochemsz-May' attack (Asiacrypt'06) on Common Prime RSA.
- We extend Nitaj's result (Africacrypt'12) on weak encryption exponents of RSA and CRT-RSA.

Keywords: Lattice-based analysis · Linear modular equations · RSA

1 Introduction

Lattice-based cryptanalysis is a very useful tool in various cryptographic systems, e.g., historically, it was used to break the Merkle-Hellman knapsack cryptosystem [34]. The basic idea of the lattice-based approach is that if the system

© International Association for Cryptologic Research 2015
T. Iwata and J.H. Cheon (Eds.): ASIACRYPT 2015, Part I, LNCS 9452, pp. 189–213, 2015.
DOI: 10.1007/978-3-662-48797-6_9

parameters of the target problem can be transformed into a basis of a certain lattice, one can find some short vectors in the desired lattice using dedicated algorithms, like the *LLL*-algorithm [20]. One may then hope that the secret key can be recovered once the solutions from these short vectors are extracted. Although in most cases this assumption is not rigorous in theory, it usually works well in practice.

In the above approach, a key step is to construct the desired lattice. In 1997, Coppersmith [5] presented a subtle lattice construction method, and used it to find small roots of modular equations of special forms. Since then, this approach has been widely applied in the analysis of RSA. Among them, one of the most important applications is to solve approximate integer common divisor problem (ACDP), namely, given two integers that are near-multiples of a hidden integer, output that hidden integer. We note that ACDP was first introduced by Howgrave-Graham [15], which in turn has many important applications such as building fully homomorphic cryptosystems [37].

Let us briefly explain Howgrave-Graham's method. First, one reduces ACDP to solving a univariate modular polynomial:

$$f(x) = x + a \bmod p$$

where a is a given integer, and p $(p \geq N^\beta$ for some $0 < \beta \leq 1)$ is unknown that divides the known modulus N. Then he proposed a polynomial-time algorithm to find small roots of the univariate polynomial over integer. Note that this type of polynomial can also be applied in other RSA-related problems, such as factoring with known bits problem [21].

In 2003, May [21] generalized Howgrave-Graham's strategy by using a univariate linear polynomial to an arbitrary monic modular polynomial of degree δ, i.e. $f(x) = x^\delta + a_{\delta-1}x^{\delta-1} + \ldots + a_0 \bmod p$ where $\delta \geq 1$. As an important application, this algorithm can be used to solve the problem of factoring with known bits on Takagi's moduli $N = p^r q$ $(r > 1)$ [2].

In Asiacrypt'08, Herrmann and May [12] extended the univariate linear modular polynomial to polynomials with an arbitrary number of n variables. They presented a polynomial-time algorithm to find small roots of linear modular-polynomials

$$f(x_1, \ldots, x_n) = a_0 + a_1 x_1 + \cdots + a_n x_n \bmod p$$

where p is unknown and divides the known modulus N. Naturally, they applied their results to the problem of factoring with known bits for RSA modulus $N = pq$ where those unknown bits might spread across arbitrary number of blocks of p. Besides, Herrmann-May's algorithm also can be used to cryptanalyze Multiprime Φ-Hiding Assumption [11,19], and attack CRT-RSA signatures [6,7].

On the other hand, in 2012, Cohn and Heninger [4] generalized Howgrave-Graham's equations to the simultaneous modular univariate linear equations

$$\begin{cases} f_1(x_1) = a_1 + x_1 = 0 \bmod p \\ f_2(x_2) = a_2 + x_2 = 0 \bmod p \\ \quad\vdots \\ f_n(x_n) = a_n + x_n = 0 \bmod p \end{cases} \tag{1}$$

where $a_1, \ldots, a_n$ are given integers, and p $(p \geq N^\beta$ for some $0 < \beta < 1)$ is an unknown factor of known modulus N. These equations have many applications in public-key cryptanalysis. For example, in 2010, van Dijk et al. [37] introduced fully homomorphic encryption over the integers, which the security of their scheme is based on the hardness of solving Eq. (1). In 2011, Sarkar and Maitra [32] investigated implicit factorization problem [24] by solving Eq. (1). In 2012, Fouque et al. [10] proposed fault attacks on CRT-RSA signatures, which can also be reduced to solving Eq. (1).

1.1 Our Contributions

In this paper, we focus on the following three types of extensions of previous equations.

The first is an extension of Herrmann-May's equation, described in Sect. 3, we focus on the equations

$$f(x_1, x_2, \ldots, x_n) = a_0 + a_1 x_1 + \cdots + a_n x_n \bmod p^v \tag{2}$$

for some unknown divisor p^v $(v \geq 1)$ and known composite integer N $(N \equiv 0 \bmod p^u, u \geq 1)$. Here u, v are positive integers. Note that if $u = 1, v = 1$, that is exactly Herrmann-May's equation [12].

The second is a special case of Eq. (2): $a_0 = 0$, described in Sect. 4.

The last is a generalized version of Eq. (1), described in Sect. 5; we focus on the equations

$$\begin{cases} f_1(x_1) = a_1 + x_1 = 0 \bmod p^{r_1} \\ f_2(x_2) = a_2 + x_2 = 0 \bmod p^{r_2} \\ \qquad\qquad \vdots \\ f_n(x_n) = a_n + x_n = 0 \bmod p^{r_n} \end{cases} \tag{3}$$

where p $(p \geq N^\beta$ for some $0 < \beta < 1)$ is unknown that satisfies $N = 0 \bmod p^r$ and $a_1, \ldots, a_n, r, r_1, \ldots, r_n$ are given integers. Here $r, r_1, \ldots, r_n$ are positive integers. Note that if $r = r_1 = \cdots = r_n = 1$, that is exactly Eq. (1).

Notice that our generalized equations employ many parameters. The reason why we introduce these parameters is based on the fact that some attacks on RSA variants (such as Takagi's RSA variant [35]) can be reduced to solving this kind of equations. However, previous algorithms [4,12,23] do not seem to work in this situation. The difficulty lies in how to wisely embed this algebraic information in the lattice construction.

We solve the above equations by introducing new techniques. More precisely, we present a novel way to select appropriate polynomials in constructing desired lattice. Compared with previous algorithms, our algorithms are more flexible and especially suitable for some cases. Applying our algorithms, we obtain the best analytical/experimental results for some attacks on RSA and its variants. We elaborate them below. We further conjecture that our new algorithms may find new applications in various other contexts.

Small Secret Exponent Attack on Multi-power RSA. In multi-power RSA algorithm, suppose that the public key is (N, e), where $N = p^r q$ for some fixed $r \geq 2$ and p, q are of the same bit-size. The secret key d satisfies $ed \equiv 1 \mod \phi(N)$, where $\phi(N)$ is Euler's ϕ-function. In Crypto'99, Takagi [35] showed that when the secret exponent $d < N^{\frac{1}{2(r+1)}}$, one can factorize N. Later in PKC'04, May [22] improved Takagi's bound to $N^{\max\{\frac{r}{(r+1)^2}, \frac{(r-1)^2}{(r+1)^2}\}}$. In this paper, we further improve May's bound to $N^{\frac{r(r-1)}{(r+1)^2}}$, which is better than May's result when $r > 2$, and is also independent of the value of public exponent e. Similar as [22], our result also directly implies an improved partial key exposure attack for secret exponent d with known most significant bits (MSBs) or least significant bits (LSBs). Our improvements are based on our algorithm of solving the first type equations, with the observation that $\gcd(ed - 1, N) = p^{r-1}$ but $N \equiv 0 \mod p^r$.

Factoring Multi-power Moduli with Known Bits. In 1999, Boneh et al. [2] extended factoring with high bits problem to moduli of the form $N = p^r q (r \geq 2)$. They showed that this moduli can be factored in polynomial-time in the bit-length of N if $r = \Omega(\sqrt{\frac{\log N}{\log \log N}})$. Applying our algorithm of solving the first type equations, we can directly get another method to settle the problem of [2]. Though we can not get an asymptotic improvement, in practice, especially for large r, our new method performs better than [2].

Weak Encryption Exponents of RSA and CRT-RSA. In Africacrypt'12, Nitaj [26] presented some attacks on RSA and CRT-RSA (the public exponent e and the private CRT-exponents d_p and d_q satisfy $ed_p \equiv 1 \mod (p - 1)$ and $ed_q \equiv 1 \mod (q - 1)$). His attacks are based on Herrmann-May's technique [12] for finding small solutions of modular equations. In particular, he reduced his attacks to solving bivariate linear modular equations modulo unknown divisors: $ex + y \equiv 0 \mod p$ for some unknown p that divides the known modulus N. Noticing that his equations are homogeneous, we can improve his results with our algorithm of solving second type equations.

Small Secret Exponent Attack on Common Prime RSA. We give a simple but effective attack on an RSA variant called Common Prime RSA. This variant was originally introduced by Wiener [38] as a countermeasure for his continued fraction attack. He suggested to choose p and q such that $p - 1$ and $q - 1$ share a large common factor. In 2006, Hinek [13] revisited the security of Common Prime RSA, in the same year, Jochemsz and May [17] proposed a heuristic attack, and showed that parts of key space suggested by Hinek is insecure. In this paper, we further improve Jochemsz-May's bound by using our algorithm of solving third type equations.

Experimental Results. For all these attacks, we carry out experiments to verify the validity of our algorithms. These experimental results show that our attacks are effective.

2   Preliminary

In 1982, Lenstra, Lenstra and Lovász proposed the *LLL*-algorithm [20] that can find vectors in polynomial-time whose norm is small enough to satisfy the following condition.

Lemma 1 (LLL [20]). *Let $\mathcal{L}$ be a lattice of dimension w. Within polynomial-time, LLL-algorithm outputs a set of reduced basis vectors v_i, $1 \leqslant i \leqslant w$ that satisfies*

$$||v_1|| \leqslant ||v_2|| \leqslant \cdots \leqslant ||v_i|| \leqslant 2^{\frac{w(w-1)}{4(w+1-i)}} \det(\mathcal{L})^{\frac{1}{w+1-i}}$$

In practice, it is widely known that the *LLL*-algorithm tends to output the vectors whose norms are much smaller than theoretically predicted.

In 1997, Coppersmith [5] described a lattice-based technique to find small roots of modular and integer equations. Later, Howgrave-Graham [14] reformulated Coppersmith's ideas of finding modular roots. The main idea of Coppersmith's method is to reduce the problem of finding small roots of $f(x_1, \ldots, x_n)$ mod N to finding roots over the integers. Therefore, one can construct a collection of polynomials that share a common root modulo N^m for some well-chosen integer m. Then one can construct a lattice by defining a lattice basis via these polynomial's coefficient vectors. Using lattice basis reduction algorithms (like *LLL*-algorithm [20]), one can find a number of linear equations with sufficiently small norm. Howgrave-Graham [14] showed a sufficient condition to quantify the term sufficiently small. Next we review this useful lemma.

Let $g(x_1, \cdots, x_k) = \sum_{i_1, \cdots, i_k} a_{i_1, \cdots, i_k} x_1^{i_1} \cdots x_k^{i_k}$. We define the norm of g by the Euclidean norm of its coefficient vector: $||g||^2 = \sum_{i_1, \cdots, i_k} a_{i_1, \cdots, i_k}^2$.

Lemma 2 (Howgrave-Graham [14]). *Let $g(x_1, \cdots, x_k) \in \mathbb{Z}[x_1, \cdots, x_k]$ be an integer polynomial that consists of at most w monomials. Suppose that*

1. *$g(y_1, \cdots, y_k) = 0 \bmod p^m$ for $| y_1 | \leqslant X_1, \cdots, | y_k | \leqslant X_k$ and*
2. *$||g(x_1 X_1, \cdots, x_k X_k)|| < \frac{p^m}{\sqrt{w}}$*

Then $g(y_1, \cdots, y_k) = 0$ holds over integers.

Combining Lemmas 1 and 2, we can get following theorem.

Theorem 1 (Coppersmith [5], May [23]). *Let N be an integer of unknown factorization, which has a divisor $p \geq N^\beta$, $0 < \beta \leq 1$. Let $f(x)$ be a univariate monic polynomial of degree δ. Then we can find in time $\mathcal{O}(\epsilon^{-7}\delta^5 \log^9 N)$ all solutions x_0 for the equation*

$$f(x) = 0 \bmod p \quad with \quad |x_0| \leq N^{\frac{\beta^2}{\delta} - \epsilon}.$$

Additionally sometimes our attacks rely on a well-known assumption which was widely used in the literatures [1,9,12].

Assumption 1. *The lattice-based construction yields algebraically independent polynomials. The common roots of these polynomials can be efficiently computed using the* Gröbner *basis technique.*

Note that the time complexity of Gröbner basis computation is in general doubly exponential in the degree of the polynomials.

We would like to point out that our subsequent complexity considerations solely refer to our lattice basis reduction algorithm, that turns the polynomial $f(x_1,\ldots,x_n)$ mod N into the number of n polynomials over the integers. We assume that the running time of the Gröbner basis computation is negligible compared to the time complexity of the *LLL*-algorithm, since in general, our algorithm yields more than the number of n polynomials, so one can make use of these additional polynomials to speed up the Gröbner basis computation.

3   The First Type of Equations

In this section, we address how to solve $f_1(x) = a_0 + a_1 x$ mod p^v $(v \geq 1)$ for some unknown p where p^u divides a known modulus N (i.e. $N \equiv 0$ mod p^u, $u \geq 1$). In particular, Howgrave-Graham's result [15] can be viewed as a special case of our algorithm when $u = 1$, $v = 1$.

3.1   Our Main Result

Theorem 2. *For every $\epsilon > 0$, let N be a sufficiently large composite integer (of unknown factorization) with a divisor p^u $(p \geq N^\beta, u \geq 1)$. Let $f_1(x) \in \mathbb{Z}[x]$ be a univariate linear polynomial whose leading coefficient is coprime to N. Then one can find all the solutions y of the equation $f_1(x) = 0$ mod p^v with $v \geq 1$, $|y| \leq N^\gamma$ if $\gamma < uv\beta^2 - \epsilon$. The time complexity is $\mathcal{O}(\epsilon^{-7} v^2 \log^2 N)$.*

Proof. Consider the following univariate linear polynomial:

$$f_1(x) = a_0 + a_1 x \text{ mod } p^v$$

where N is known to be a multiple of p^u for known u and unknown p. Here we assume that $a_1 = 1$, since otherwise we can multiply f_1 by a_1^{-1} mod N. Let $f(x) = a_1^{-1} f_1(x)$ mod N.

We define a collection of polynomials as follows:

$$g_k(x) := f^k(x) N^{\max\{\lceil \frac{v(t-k)}{u} \rceil, 0\}}$$

for $k = 0, \ldots, m$ and integer parameters t and m with $t = \tau m$ $(0 \leq \tau < 1)$, which will be optimized later. Note that for all k, $g_k(y) \equiv 0$ mod p^{vt}.

Let $X := N^{uv\beta^2 - \epsilon}(= N^\gamma)$ be the upper bound on the desired root y. We will show that this bound can be achieved for any chosen value of ϵ by ensuring that $m \geq m^* := \lceil \frac{\beta(2u+v-uv\beta)}{\epsilon} \rceil - 1$

$$
\begin{array}{c}
N^4 \\
fN^4 \\
f^2N^3 \\
f^3N^2 \\
f^4N^2 \\
f^5N \\
f^6 \\
f^7 \\
f^8
\end{array}
\left(
\begin{array}{cccccccccc}
N^4 & & & & & & & & & \\
* & XN^4 & & & & & & & & \\
* & * & X^2N^3 & & & & & & & \\
* & * & * & X^3N^2 & & & \mathbf{0} & & & \\
* & * & * & * & X^4N^2 & & & & & \\
* & * & * & * & * & X^5N & & & & \\
* & * & * & * & * & * & X^6 & & & \\
* & * & * & * & * & * & * & X^7 & & \\
* & * & * & * & * & * & * & * & X^8 &
\end{array}
\right)
$$

Fig. 1. The matrix for the case $\beta = 0.25$, $u = 3$, $v = 2$, $t = 6$, $m = 8$

We build a lattice $\mathcal{L}$ of dimension $d = m + 1$ using the coefficient vectors of $g_k(xX)$ as basis vectors. We sort these polynomials according to the ascending order of g, i.e., $g_k < g_l$ if $k < l$. Figure 1 shows an example for the parameters $\beta = 0.25, u = 3, v = 2, t = 6, m = 8$.

From the triangular matrix of the lattice basis, we can compute the determinant as the product of the entries on the diagonal as $\det(\mathcal{L}) = X^s N^{s_N}$ where

$$
s = \sum_{k=0}^{m} k = \frac{m(m+1)}{2}
$$

$$
s_N = \sum_{k=0}^{t-1} \lceil \frac{v(t-k)}{u} \rceil = \sum_{k=0}^{t-1} \left(\frac{v(t-k)}{u} + c_k \right) = \frac{v\tau m(\tau m + 1)}{2u} + \sum_{k=0}^{t-1} c_k
$$

Here we rewrite $\lceil \frac{v(t-k)}{u} \rceil$ as $\left(\frac{v(t-k)}{u} + c_k \right)$ where $c_k \in [0, 1)$. To obtain a polynomial with short coefficients that contains all small roots over integer, we apply LLL-basis reduction algorithm to the lattice $\mathcal{L}$. Lemma 1 gives us an upper bound on the norm of the shortest vector in the LLL-reduced basis; if the bound is smaller than the bound given in Lemma 2, we can obtain the desired polynomial. We require the following condition:

$$
2^{\frac{d-1}{4}} \det(\mathcal{L})^{\frac{1}{d}} < \frac{N^{v\beta\tau m}}{\sqrt{d}}
$$

where $d = m + 1$. We plug in the values for $\det(\mathcal{L})$ and d, and obtain

$$
2^{\frac{m(m+1)}{4}} (m+1)^{\frac{m+1}{2}} X^{\frac{m(m+1)}{2}} < N^{v\beta\tau m(m+1) - \frac{v\tau m(\tau m+1)}{2u} - \sum_{k=0}^{t-1} c_k}
$$

To obtain the asymptotic bound, we let m grow to infinity. Note that for sufficiently large N the powers of 2 and $m + 1$ are negligible. Thus, we only consider the exponent of N. Then we have

$$
X < N^{2v\beta\tau - \frac{v\tau(\tau m+1)}{u(m+1)} - \frac{2\sum_{k=0}^{t-1} c_k}{m(m+1)}}
$$

Setting $\tau = u\beta$, and noting that $\sum_{k=0}^{t-1} c_k \leq t^1$, the exponent of N can be lower bounded by

$$uv\beta^2 - \frac{v\beta(1 - u\beta)}{m+1} - \frac{2u\beta}{m+1}$$

We appropriate the negative term $\frac{*}{m+1}$ by $\frac{*}{m}$ and obtain

$$uv\beta^2 - \frac{\beta(2u + v - uv\beta)}{m}$$

Enduring that $m \geq m^*$ will then gurantee that X satisfies the required bound for the chosen value of ϵ.

The running time of our method is dominated by LLL-algorithm, which is polynomial in the dimension of the lattice and in the maximal bit-size of the entries. We have a bound for the lattice d

$$d = m + 1 \geq \lceil \frac{\beta(2u + v - uv\beta)}{\epsilon} \rceil$$

Since $u\beta < 1$, then we obtain $d = \mathcal{O}(\epsilon^{-1})$. The maximal bit-size of the entries is bounded by

$$\max\{\frac{vt}{u} \log(N), duv\beta^2 \log(N)\} = \max\{v\beta d \log(N), duv\beta^2 \log(N)\}$$

Since $u\beta < 1$ and $d = \mathcal{O}(\epsilon^{-1})$, the bit-size of the entries can be upperbounded by

$$\max\{\mathcal{O}(v\beta\epsilon^{-1}) \log(N), \mathcal{O}(v\beta\epsilon^{-1}) \log(N)\} = \mathcal{O}(v\epsilon^{-1} \log(N))$$

Nguyen and Stehlé [25] proposed a modified version of the LLL-algorithm called L^2-algorithm. The L^2-algorithm achieves the same approximation quality for a shortest vectors as the LLL-algorithm, but has an improved worst case running time anlaysis. Its running time is $\mathcal{O}(d^5(d + \log b_d) \log b_d)$, where $\log b_d$ is the maximal bit-size of an entry in lattice. Thus, we can obtain the running time of our algorithm

$$\mathcal{O}\left(\left(\frac{1}{\epsilon}\right)^5 \left(\frac{1}{\epsilon} + \frac{v \log N}{\epsilon}\right) \frac{v \log N}{\epsilon}\right)$$

Therefore, the running time of our algorithm is $\mathcal{O}(\epsilon^{-7} v^2 \log^2 N)$. Eventually, the vector output by LLL-algorithm gives a univariate polynomial $g(x)$ such that $g(y) = 0$, and one can find the root of $g(x)$ over the integers. $\qquad\square$

Extension to Arbitrary Degree. We can generalize the result of Theorem 2 to univariate polynomials with arbitrary degree.

---

[1] This estimation is rough, we can do it more precisely for specific parameters u, v. For example, for $v = 1$, we can get $\sum_{k=0}^{t-1} c_k \leq \frac{t}{2} + 1$.

Theorem 3. *For every $\epsilon > 0$, let N be a sufficiently large composite integer (of unknown factorization) with a divisor p^u ($p \geq N^\beta$, $u \geq 1$). Let $f_1(x) \in \mathbb{Z}[x]$ be a univariate polynomial of degree δ whose leading coefficient is coprime to N. Then one can find all the solutions y of the equation $f_1(x) = 0$ (mod p^v) with $v \geq 1$, $|y| \leq N^\gamma$ if $\gamma < \frac{uv\beta^2}{\delta} - \epsilon$. The time complexity is $\mathcal{O}(\epsilon^{-7}\delta^5 v^2 \log^2 N)$.*

In the proof of Theorem 3, we use the following collection of polynomials:

$$g_k(x) := x^j f^k(x) N^{\max\{\lceil \frac{v(t-k)}{u} \rceil, 0\}}$$

for $k = 0, \ldots, m$, $j = 0, \ldots, \delta - 1$ and integer parameters t and m with $t = \tau m$ ($0 \leq \tau < 1$). The rest of the proof is the same as Theorem 2. We omit it here.

Specifically, the result in [23] can be viewed as a special case of our algorithm when $u = v$.

Extension to More Variables. We also generalize the result of Theorem 2 from univariate linear equations to an arbitrary number of n variables $x_1, \ldots, x_n$ ($n \geq 2$).

Proposition 1. *For every $\epsilon > 0$, let N be a sufficiently large composite integer (of unknown factorization) with a divisor p^u ($p \geq N^\beta$, $u \geq 1$). Furthermore, let $f_1(x_1, \ldots, x_n) \in \mathbb{Z}[x_1, \ldots, x_n]$ be a monic linear polynomial in $n(n \geq 2)$ variables. Under Assumption 1, we can find all the solutions $(y_1, \ldots, y_n)$ of the equation $f_1(x_1, \ldots, x_n) = 0$ (mod p^v) with $v \geq 1$, $|y_1| \leq N^{\gamma_1}, \ldots |y_n| \leq N^{\gamma_n}$ if*

$$\sum_{i=1}^{n} \gamma_i < \frac{v}{u}\left(1 - (1 - u\beta)^{\frac{n+1}{n}} - (n+1)(1 - u\beta)\left(1 - \sqrt[n]{1 - u\beta}\right)\right) - \epsilon$$

The running time of the algorithm is polynomial in ϵ^{-n} and $\epsilon^{-n} \log N$.

Proof. We define the following collection of polynomials which share a common root modulo p^{vt}:

$$g_{i_2, \ldots, i_n, k} = x_2^{i_2} \cdots x_n^{i_n} f_1^k N^{\max\{\lceil \frac{v(t-k)}{u} \rceil, 0\}}$$

for $k = 0, \ldots, m$ where $i_j \in \{0, \ldots, m\}$ such that $\sum_{j=2}^{n} i_j \leq m - k$, and the integer parameter $t = \tau m$ has to be optimized. The idea behind the above transformation is that we try to eliminate powers of N in the diagonal entries in order to keep the lattice determinant as small as possible.

Next we can construct the lattice $\mathcal{L}$ using the similar method of Herrmann-May [12], therefore, the lattice has triangular form, then the determinant $\det(\mathcal{L})$ is then simply the product of the entries on the diagonal:

$$\det(\mathcal{L}) = \prod_{i=1}^{n} X_i^{s_{x_i}} N^{s_N}$$

Let d denote the dimension of $\mathcal{L}$, $t = r \cdot h + c$ ($h, c \in \mathbb{Z}$ and $0 \leq c < r$). A straightforward but tedious computation yields that

$$s_{x_i} = \binom{m+n}{m-1} = \frac{1}{(n+1)!} m^{n+1} + o(m^{n+1})$$

$$s_N = \sum_{k=0}^{t-1} \sum_{0 \leq \sum_{j=2}^{n} i_j \leq m-k} \lceil \frac{v(t-k)}{u} \rceil$$

$$= \frac{v}{u} \frac{(n+1)\tau - 1 + (1-\tau)^{n+1}}{(n+1)!} m^{n+1} + o(m^{n+1})$$

$$d = \binom{m+n}{m} = \frac{1}{n!} m^n + o(m^n)$$

To obtain the number of n polynomial with short coefficients that contains all small roots over integer, we apply LLL-basis reduction algorithm to the lattice $\mathcal{L}$. Combining Lemma 1 with Lemma 2, we require the following condition:

$$2^{\frac{d(d-1)}{4(d+1-n)}} \det(\mathcal{L})^{\frac{1}{d-n+1}} < \frac{N^{v\beta\tau m}}{\sqrt{d}}$$

Let $X_i = N^{\gamma_i} (1 \leq i \leq n)$. Combining the values with the above condition, we obtain

$$\sum_{i=1}^{n} \gamma_i < \frac{v}{u} \left(1 - (1-\tau)^{n+1} \right) - \tau v(n+1)(\frac{1}{u} - \beta) - \epsilon$$

By setting $\tau = 1 - \sqrt[n]{1 - u\beta}$, the condition reduces to

$$\sum_{i=1}^{n} \gamma_i < \frac{v}{u} \left(1 - (1-u\beta)^{\frac{n+1}{n}} - (n+1)(1-u\beta) \left(1 - \sqrt[n]{1-u\beta} \right) \right) - \epsilon$$

The running time is dominated by the time to run LLL-lattice reduction on a basis matrix of dimension d and bit-size of the entries. Since $d = \mathcal{O}(\frac{m^n}{n!})$ and the parameter m depends on ϵ^{-1} only, therefore, our approach is polynomial in $\log N$ and ϵ^{-n}. Besides, our attack relies on Assumption 1.   □

3.2 Analysis of Multi-power RSA

We apply our algorithm to analyze an RSA variant, namely multi-power RSA, with moduli $N = p^r q$ ($r \geq 2$). Compared to the standard RSA, the multi-power RSA is more efficient in both key generation and decryption. Besides, moduli of this type have been applied in many cryptographic designs, e.g., the Okamoto-Uchiyama cryptosystem [27], or better known via EPOC and ESIGN [8], which uses the modulus $N = p^2 q$.

Using our algorithm of Theorem 2, we give two attacks on multi-power RSA: small secret exponent attack and factoring with known bits.

Small Secret Exponent Attack on Multi-power RSA. There are two variants of multi-power RSA. In the first variant $ed \equiv 1 \mod p^{r-1}(p-1)(q-1)$, while in the second variant $ed \equiv 1 \mod (p-1)(q-1)$. In [16], the authors proved that the second variant is vulnerable when $d < N^{\frac{2-\sqrt{2}}{r+1}}$.

In this section, we focus on the first variant. In Crypto'99, Takagi [35] proved that when the decryption exponent $d < N^{\frac{1}{2(r+1)}}$, one can factorize N in polynomial-time. Later, in PKC'04, May [22] improved Takagi's bound to $N^{\max\{\frac{r}{(r+1)^2}, \frac{(r-1)^2}{(r+1)^2}\}}$. Based on the technique of Theorem 2, we can further improve May's bound to $N^{\frac{r(r-1)}{(r+1)^2}}$.

Theorem 4. *Let $N = p^r q$, where $r \geq 2$ is a known integer and p, q are primes of the same bit-size. Let e be the public key exponent and d be the private key exponent, satisfying $ed \equiv 1 \mod \phi(N)$. For every $\epsilon > 0$, suppose that*

$$d < N^{\frac{r(r-1)}{(r+1)^2} - \epsilon}$$

then N can be factored in polynomial-time.

Proof. Since $\phi(N) = p^{r-1}(p-1)(q-1)$, we have the equation $ed - 1 = kp^{r-1}(p-1)(q-1)$ for some $k \in \mathbb{N}$. Then we want to find the root $y = d$ of the polynomial

$$f_1(x) = ex - 1 \mod p^{r-1}$$

with the known multiple (of unknown divisor p) N ($N \equiv 0 \mod p^r$). Let $d \approx N^\delta$. Applying Theorem 2, setting $\beta = \frac{1}{r+1}$, $u = r$, $v = r-1$, we obtain the final result $\delta < \frac{r(r-1)}{(r+1)^2} - \epsilon$ □

Recently, Sarkar [30,31] improved May's bound for modulus $N = p^r q$, however, unlike our method, his method can not applied for public key exponents e of arbitrary size. In addition, we get better experimental results for the case of $r > 2$ (see Sect. 3.2).

For small r, we provide the comparison of May's bound, Sarkar's bound, and our bound on δ in Table 1. Note that for $r = 2$, we obtain the same result as May's bound.

Table 1. Comparisons of May's bound, Sarkar's bound and ours on δ

r	2	3	4	5	6	7	8	9
May's bound	0.22	0.25	0.36	0.44	0.51	0.56	0.60	0.64
Sarkar's bound [30,31]	**0.39**	**0.46**	**0.50**	0.54	0.57	0.51	0.53	0.54
Our bound	0.22	0.37	0.48	**0.55**	**0.61**	**0.65**	**0.69**	**0.72**

Partial Key-Exposure Attacks on Multi-power RSA. Similar to the results of [22], the new attack of Theorem 4 immediately implies partial key exposure attacks for d with known MSBs/LSBs. Following we extend the approach of Theorem 4 to partial key exposure attacks.

Theorem 5 (MSBs). *Let $N = p^r q$, where $r \geq 2$ is a known integer and p, q are primes of the same bit-size. Let e be the public key exponent and d be the private key exponent, satisfying $ed = 1 \bmod \phi(N)$. For every $\epsilon > 0$, given $\tilde{d}$ such that $|d - \tilde{d}| < N^{\frac{r(r-1)}{(r+1)^2} - \epsilon}$, then N can be factored in polynomial-time.*

Proof. We have that

$$e(d - \tilde{d}) + e\tilde{d} - 1 \equiv 0 \bmod p^{r-1}$$

Then we want to find the root $y = d - \tilde{d}$ of the polynomial

$$f_1(x) = ex + e\tilde{d} - 1 \bmod p^{r-1}$$

with the known multiple (of unknown divisor p) N ($N \equiv 0 \bmod p^r$). Applying Theorem 2, setting $\beta = \frac{1}{r+1}$, $u = r$, $v = r - 1$, we obtain the final result. □

Theorem 6 (LSBs). *Let $N = p^r q$, where $r \geq 2$ is a known integer and p, q are primes of the same bit-size. Let e be the public key exponent and d be the private key exponent, satisfying $ed = 1 \bmod \phi(N)$. For every $\epsilon > 0$, given d_0, M with $d = d_0 \bmod M$ and $M > N^{\frac{3r+1}{(r+1)^2} + \epsilon}$, then N can be factored in polynomial-time.*

Proof. Rewrite $d = d_1 M + d_0$, then we have

$$ed_1 M + ed_0 - 1 \equiv 0 \bmod p^{r-1}$$

Then we want to find the root $y = d_1$ of the polynomial

$$f_1(x) = eMx + ed_0 - 1 \bmod p^{r-1}$$

with the known multiple (of unknown divisor p) N ($N \equiv 0 \bmod p^r$). Applying Theorem 2 and setting $\beta = \frac{1}{r+1}$, $u = r$, $v = r - 1$, we obtain the final result. □

We have implemented our algorithm in Magma 2.11 computer algebra system on our PC with Intel(R) Core(TM) Duo CPU (2.53GHz, 1.9GB RAM Windows 7). Table 2 shows the experimental results for multi-power RSA modulus N with 512-bit primes p, q. We compute the number of bits that one should theoretically be able to attack for d (column d-pred in Table 2). In all the listed experiments, we can recover the factorization of N. Note that our attack is also effective for large e.

In [31], for 1024-bit $N = p^3 q$, Sarkar considered $\delta = 0.27$ using a lattice with dimension 220, while we can achieve $\delta = 0.359$ using a lattice with dimension 41. Besides, Sarkar also stated that "for $r = 4, 5$, lattice dimension in our approach becomes very large to achieve better results. Hence in these cases we can not present experiment results to show the improvements over existing results." In Table 2, we can see that our experimental results are better than Sarkar's for $r > 2$.

Table 2. Experimental results of the attack from Theorem 4

N (bits)	r	e (bits)	d-pred (bits)	(m, t)	dim $(\mathcal{L})$	d-exp (bits)	δ	Time (sec)
1536	2	1536	341	$(30, 20)$	31	318	0.207	3155.687
2048	3	2048	768	$(20, 15)$	21	706	0.345	749.167
2048	3	4096	768	$(20, 15)$	21	706	0.345	745.170
2048	3	2048	768	$(40, 30)$	41	735	0.359	37800.462
2560	4	2560	1228	$(20, 16)$	21	1136	0.444	1245.754
2560	4	2560	1228	$(30, 24)$	31	1167	0.456	12266.749

Factoring Multi-power Moduli with Known Bits. In 1985, Rivest and Shamir [28] first introduced the factoring with high bits known problem, they presented an algorithm that factors $N = pq$ given $\frac{2}{3}$-fraction of the bits of p. Later, Coppersmith [5] gave a improved algorithm when half of the bits of p are known. In 1999, Boneh, Durfee and Howgrave-Graham [2] (referred as BDH method) extended Coppersmith's results to moduli $N = p^r q (r \geq 2)$. Basically, they considered the scenario that a few MSBs of the prime p are known to the attacker. Consider the univariate polynomial

$$f(x) = (\tilde{p} + x)^r \bmod p^r$$

For simplicity, we assume that p and q are of the same bit-size. Using the algorithm of Theorem 1, Boneh et al. showed that they can recover all roots x_0 with

$$|x_0| \leq N^{\frac{\beta^2}{\delta} - \epsilon} = N^{\frac{r}{(r+1)^2} - \epsilon}$$

in time $\mathcal{O}(\epsilon^{-7} \log^2 N)^2$. Thus we need a $\frac{1}{r+1}$-fraction of p in order to factor N in polynomial-time.

Applying our algorithm of Theorem 2, and setting $\beta = \frac{1}{r+1}, u = r, v = 1$, we can also find all roots x_0 with

$$|x_0| \leq N^{uv\beta^2 - \epsilon} = N^{\frac{r}{(r+1)^2} - \epsilon}$$

in time $\mathcal{O}(\epsilon^{-7} \log^2 N)$.

Note that we obtain the same asymptotic bound and running time complexity as BDH method. But, as opposed to BDH method, our algorithm is more flexible in choosing the lattice dimension. For example, in the case of $r = 10$, BDH method only works on the lattice dimension of $11 * m$ ($m \in \mathbb{Z}^+$) while our method can work on any lattice dimension m ($m \in \mathbb{Z}^+$). Figure 2 shows a comparsion of these two methods in terms of the size of $\tilde{p}$ ($\tilde{p} = N^\gamma$) that can be achieved. We can see that to achieve the same γ, we require smaller lattice dimensions than BDH method. Our algorithm is especially useful for large r. Actually our lattice is the same to the lattice of BDH method if the lattice dimensions are $11 * m$ ($m \in \mathbb{Z}^+$).

---

[2] Since this univariate equation is very special: $f(x) = (x + a)^r$, in fact we can remove the quantity r^5 from the time complexity of Theorem 1.

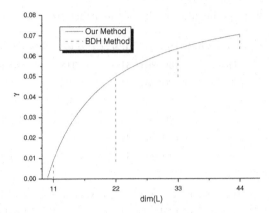

Fig. 2. Comparison of the achievable bound depending on the lattice dimension: the case of $r = 10$.

Table 3. Comparison of our experimental results with BDH method.

r	Theo.	Expt.	BDH method		Our method	
			Dim	Time (in seconds)	Dim	Time (in seconds)
5	84	164	30	112.914	26	29.281
5	84	134	48	2874.849	46	1343.683
10	46	186	44	670.695	34	259.298
10	46	166	44	1214.281	41	917.801

We also give some experimental results. Table 3 shows the experimental results for multi-power RSA modulus $N(N = p^r q)$ with 500-bit primes p, q. These experimental data confirmed our theoretical analysis. It is obvious that our method performs better than BDH method in practice.

4   The Second Type of Equations

In this section, we study the problem of finding small roots of homogeneous linear polynomials $f_2(x_1, x_2) = a_1 x_1 + a_2 x_2 \bmod p^v$ $(v \geq 1)$ for some unknown p where p^u divides a known modulus N (i.e. $N \equiv 0 \bmod p^u$, $u \geq 1$). Let (y_1, y_2) be a small solution of $f_2(x_1, x_2)$. We assume that we also know an upper bound $(X_1, X_2) \in \mathbb{Z}^2$ for the root such that $|y_1| \leq X_1, |y_2| \leq X_2$.

4.1   Our Main Result

Theorem 7. *For every $\epsilon > 0$, let N be a sufficiently large composite integer (of unknown factorization) with a divisor p^u $(p \geq N^\beta, u \geq 1)$. Let $f_2(x_1, x_2) \in \mathbb{Z}[x_1, x_2]$ be a homogeneous linear polynomial in two variables whose coefficients are coprime to N. Then one can find all the solutions (y_1, y_2) of the equation $f_2(x_1, x_2) = 0$ ($\bmod$ p^v) $(v \geq 1)$ with $\gcd(y_1, y_2) = 1$, $|y_1| \leq N^{\gamma_1}, |y_2| \leq N^{\gamma_2}$ if $\gamma_1 + \gamma_2 < uv\beta^2 - \epsilon$, and the time complexity of our algorithm is $\mathcal{O}(\epsilon^{-7}v^2 \log^2 N)$.*

Proof. Since the proof is similar to that of Theorem 2, we only give the sketch here. Consider the linear polynomial:

$$f_2(x_1, x_2) = a_1 x_1 + a_2 x_2 \bmod p^v$$

where N is known to be a multiple of p^u for known u and unknown p. Here we assume that $a_1 = 1$, since otherwise we can multiply f_2 by $a_1^{-1} \bmod N$. Let

$$f(x_1, x_2) = a_1^{-1} f_2(x_1, x_2) \bmod N$$

Fix $m := \lceil \frac{\beta(2u+v-uv\beta)}{\epsilon} \rceil$, and define a collection of polynomials as follows:

$$g_k(x_1, x_2) := x_2^{m-k} f^k(x_1, x_2) N^{\max\{\lceil \frac{v(t-k)}{u} \rceil, 0\}}$$

for $k = 0, \ldots, m$ and integer parameters t and m with $t = \tau m$ $(0 \leq \tau < 1)$, which will be optimized later. Note that for all k, $g_k(y_1, y_2) \equiv 0 \bmod p^{vt}$.

Let $X_1, X_2 (X_1 = N^{\gamma_1}, X_2 = N^{\gamma_2})$ be upper bounds on the desired root (y_1, y_2), and define $X_1 X_2 := N^{uv\beta^2 - \epsilon}$. We build a lattice $\mathcal{L}$ of dimension $d = m + 1$ using the coefficient vectors of $g_k(x_1 X_1, x_2 X_2)$ as basis vectors. We sort the polynomials according to the order as following: If $k < l$, then $g_k < g_l$.

From the triangular matrix of the lattice, we can easily compute the determinant as the product of the entries on the diagonal as $\det(\mathcal{L}) = X_1^{s_1} X_2^{s_2} N^{s_N}$ where

$$s_1 = s_2 = \sum_{k=0}^{m} k = \frac{m(m+1)}{2}$$

$$s_N = \sum_{k=0}^{t-1} \lceil \frac{v(t-k)}{u} \rceil = \sum_{k=0}^{t-1} \left(\frac{v(t-k)}{u} + c_k \right) = \frac{vt(t+1)}{2u} + \sum_{k=0}^{t-1} c_k$$

Here we rewrite $\lceil \frac{v(t-k)}{u} \rceil$ as $\left(\frac{v(t-k)}{u} + c_k \right)$ where $c_k \in [0, 1)$. Combining Lemmas 1 and 2, after some calculations, we can get the final result

$$\gamma_1 + \gamma_2 \leq uv\beta^2 - \frac{\beta(2u + v - uv\beta)}{m}$$

Similar to Theorem 2, the time complexity of our algorithm is $\mathcal{O}(\epsilon^{-7} v^2 \log^2 N)$.

The vector output by *LLL*-algorithm gives a polynomial $f'(x_1, x_2)$ such that $f'(y_1, y_2) = 0$. Let $z = x_1/x_2$, any rational root of the form y_1/y_2 can be found by extracting the rational roots of $f'(z) = 1/x_2^m f'(x_1, y_1)$ with classical methods. □

Comparisons with Previous Methods. For $u = 1, v = 1$, the upper bound $\delta_1 + \delta_2$ of Theorem 7 is β^2, that is exactly May's results [21] on univariate linear polynomial $f(x) = x + a$. Actually the problem of finding a small root of homogeneous polynomial $f(x_1, x_2)$ can be transformed to find small rational roots of univariate linear polynomial $F(z)$ i.e. $F(\frac{x_2}{x_1}) = f(x_1, x_2)/x_1$ (the discussions of the small rational roots can be found on pp. 413 of Joux's book [18]).

204 Y. Lu et al.

Our result improves Herrmann-May's bound $3\beta - 2 + 2(1 - \beta)^{\frac{3}{2}}$ up to β^2 if $a_0 = 0$. As a concrete example, for the case $\beta = 0.5$, our method improves the upper size of $X_1 X_2$ from $N^{0.207}$ to $N^{0.25}$.

Another important work to mention is that in [3], Castagnos, Joux, Laguillaumie and Nguyen also considered homogeneous polynomials. Their algorithm can be directly applied to our attack scenario. They consider the following bivariate homogeneous polynomial

$$f(x_1, x_2) = (a_1 x_1 + a_2 x_2)^{\frac{u}{v}} \bmod p$$

However, their algorithm can only deal with the cases $\frac{u}{v} \in \mathbb{Z}$, and our algorithm is more flexible: specially, for $\frac{u}{v}$-degree polynomial with $2^{\frac{u}{v}}$ monomials (the dimension of lattice is $\frac{u}{v} m$), whereas our algorithm is for linear polynomial with two monomials (the dimension of lattice is m). Besides, in [3], they formed a lattice using the coefficients of $g(x, y)$ instead of $g(xX, yY)$. This modification enjoys the benefits in terms of real efficiency, since their lattice has smaller determinant than in the classical bivariate approach. However, their algorithm fails when the solutions are significantly unbalanced ($X_1 \gg X_2$). We highlight the idea that the factor X, Y should not only be used to balance the size of different power of x, y but also to balance the variables x, y. That is why our algorithm is suitable for this unbalanced attack scenario.

Extension to More Variables. We generalize the result of Theorem 7 to an arbitrary number of n variables $x_1, \ldots, x_n$. The proof of the following result is similar to that for Proposition 1, so we state only the result itself.

Proposition 2. *For every $\epsilon > 0$, let N be a sufficiently large composite integer (of unknown factorization) with a divisor p^u ($p \geq N^\beta$, $u \geq 1$). Furthermore, let $f_2(x_1, \ldots, x_n) \in \mathbb{Z}[x_1, \ldots, x_n]$ be a homogeneous linear polynomial in $n(n \geq 3)$ variables. Under Assumption 1, we can find all the solutions $(y_1, \ldots, y_n)$ of the equation $f_2(x_1, \ldots, x_n) = 0 \mod p^v$ ($v \geq 1$) with $\gcd(y_1, \ldots, y_n) = 1$, $|y_1| \leq N^{\gamma_1}, \ldots |y_n| \leq N^{\gamma_n}$ if*

$$\sum_{i=1}^{n} \gamma_i < \frac{v}{u} \left(1 - (1 - u\beta)^{\frac{n}{n-1}} - n(1 - u\beta) \left(1 - \sqrt[n-1]{1 - u\beta} \right) \right) - \epsilon$$

The running time of the algorithm is polynomial in $\log N$ and $\epsilon^{-n} \log N$.

4.2 Applications

In Africacrypt'12, Nitaj [26] presented a new attack on RSA. His attack is based on Herrmann-May's method [12] for finding small roots of a bivariate linear equation. In particular, he showed that the public modulus N can be factored in polynomial-time for the RSA cryptosystem where the public exponent e satisfies an equation $ex + y \equiv 0 \pmod{p}$ with parameters x and y satisfying $ex + y \not\equiv 0 \pmod{N}$ $|x| < N^\gamma$ and $|y| < N^\delta$ with $\delta + \gamma \leq \frac{\sqrt{2}-1}{2}$.

Note that the equation of [26] is homogeneous, thus we can improve the upper bound of $\gamma + \delta$ using our result in Theorem 7. In [29], Sarkar proposed another method to extend Nitaj's weak encryption exponents. Here, the trick is to consider the fact that Nitaj's bound can be improved when the unknown variables in the modular equation are unbalanced (x and y are of different bit-size). In general, Sarkar's method is essentially Herrmann-May's method, whereas our algorithm is simpler (see Theorem 7). We present our result below.

Theorem 8. *Let $N = pq$ be an RSA modulus with $q < p < 2q$. Let e be a public exponent satisfying an equation $ex + y \equiv 0 \mod p$ with $|x| < N^\gamma$ and $|y| < N^\delta$. If $ex + y \not\equiv 0 \mod N$ and $\gamma + \delta \leq 0.25 - \epsilon$, N can be factored in polynomial-time.*

In [26], Nitaj also proposed a new attack on CRT-RSA. Let $N = pq$ be an RSA modulus with $q < p < 2q$. Nitaj showed that if $e < N^{\frac{\sqrt{2}}{2}}$ and $ed_p = 1 + k_p(p-1)$ for some d_p with $d_p < \frac{N^{\frac{\sqrt{2}}{4}}}{\sqrt{e}}$, N can be factored in polynomial-time. His method is also based on Herrmann-May's method. Similarly we can improve Nitaj's result in some cases using our idea as Theorem 7.

Theorem 9. *Let $N = pq$ be an RSA modulus with $q < p < 2q$. Let e be a public exponent satisfying $e < N^{0.75}$ and $ed_p = 1 + k_p(p-1)$ for some d_p with*

$$d_p < \frac{N^{\frac{0.75-\epsilon}{2}}}{\sqrt{e}}$$

Then, N can be factored in polynomial-time.

Proof. We rewrite the equation $ed_p = 1 + k_p(p-1)$ as

$$ed_p + k_p - 1 = k_p p$$

Then we focus on the equation modulo p

$$ex + y = 0 \mod p$$

with a root $(x_0, y_0) = (d_p, k_p - 1)$. Suppose that $e = N^\alpha$, $d_p = N^\delta$, then we get

$$k_p = \frac{ed_p - 1}{p - 1} < \frac{ed_p}{p - 1} < N^{\alpha + \delta - 0.5}$$

Applying Theorem 7 with the desired equation where $x_0 = d_p < N^\delta$ and $y_0 = k_p - 1 < N^{\alpha+\delta-0.5}$, setting $\beta = 0.5$, $u = 1$ and $v = 1$ we obtain

$$2\delta + \alpha < 0.75 - \epsilon$$

Note that $\gcd(x_0, y_0) = \gcd(d_p, k_p - 1) = 1$, $k_p < N^{\alpha+\delta-0.5} < N^{\alpha+2\delta-0.5} < N^{0.25} < p$, hence $ed_p + k_p - 1 \not\equiv 0 \mod N$. Then we can factorize N with $\gcd(N, ed_p + k_p - 1) = p$. □

Note that Theorem 9 requires the condition $e < N^{0.75}$ for $N = pq$, hence we cannot be using small CRT exponents both modulo p and modulo q. Our attack is valid for the case that the cryptographic algorithm has a small CRT-exponent modulo p, but a random CRT-exponent modulo q.

Table 4. Experimental results for weak encryption exponents

N (bit)	r	d_p-pred (bits)	(m,t)	dim $(\mathcal{L})$	d_p-exp (bits)	Time (sec)
1024	1	128	$(6,3)$	7	110	0.125
1024	1	128	$(10,5)$	11	115	1.576
1024	1	128	$(30,15)$	31	124	563.632

Experimental Results. Table 4 shows the experimental results for RSA modulus N with 512-bit primes p, q. In all of our experiments, we fix e's length as 512-bit, and so the scheme does not have a small CRT exponent modulo q. We also compute the number bits that one should theoretically be able to attack for d_p (column d_p-pred of Table 4).

That is actually the attack described in Theorem 9. In [26], the author showed that for a 1024-bit modulus N, the CRT-exponent d_p is typically of size at most 110. We obtain better results in our experiments as shown in Table 4.

5   The Third Type of Equations

In this section, we give our main algorithm to find small roots of extended simultaneous modular univariate linear equations. At first, we introduce this kind of equations.

Extended Simultaneous Modular Univariate Linear Equations. Given positive integers $r, r_1, \ldots, r_n$ and $N, a_1, \ldots, a_n$ and bounds $\gamma_1, \ldots, \gamma_n, \eta \in (0, 1)$. Suppose that $N = 0 \bmod p^r$ and $p \geq N^\eta$. We want to find all integers $(x_1^{(0)}, \ldots, x_n^{(0)})$ such that $|x_1^{(0)}| \leq N^{\gamma_1}, \ldots, |x_n^{(0)}| \leq N^{\gamma_n}$, and

$$
\begin{cases}
f_1(x_1^{(0)}) = a_1 + x_1^{(0)} = 0 \bmod p^{r_1} \\
f_2(x_2^{(0)}) = a_2 + x_2^{(0)} = 0 \bmod p^{r_2} \\
\quad\vdots \\
f_n(x_n^{(0)}) = a_n + x_n^{(0)} = 0 \bmod p^{r_n}
\end{cases}
$$

5.1   Our Main Result

Our main result is as follows:

Theorem 10. *Under Assumption 1, the above equations can be solved provided that*

$$
\sqrt[n]{\frac{\gamma_1 \cdots \gamma_n}{r r_1 \cdots r_n}} < \eta^{\frac{n+1}{n}} \quad and \quad \eta \gg \frac{1}{\sqrt{\log N}}
$$

The running time of the algorithm is polynomial in $\log N$ but exponential in n.

Proof. First, for every j ($j \in \{1, \ldots, n\}$), we check whether condition $\frac{\gamma_j}{r_j} \leq \eta$ is met. If there exists k such that $\frac{\gamma_k}{r_k} > \eta$, then we throw away this corresponding polynomial $f_k(x)$, since this polynomial could not offer any useful information. Here suppose that all the polynomials satisfy our criteria. Define a collection of polynomials as follows:

$$f_{[i_1,\ldots,i_n]}(x_1,\ldots,x_n) = (a_1 + x_1)^{i_1} \cdots (a_n + x_n)^{i_n} N^{\max\{\lceil \frac{t - \sum_{j=1}^{n} r_j i_j}{r} \rceil, 0\}}$$

Notice that for all indexes $i_1,\ldots,i_n$, $f_{[i_1,\ldots,i_n]}(x_1^{(0)},\ldots,x_n^{(0)}) = 0 \bmod p^t$. We select the collection of shift polynomials that satisfies

$$0 \le \sum_{j=1}^{n} \gamma_j i_j \le \eta t$$

The reason we select these shift polynomials is that we try to select as many helpful polynomials as possible by taking into account the sizes of the root bounds.

We define the polynomial order $\prec$ as $x_i^{i_1} x_2^{i_2} \cdots x_n^{i_n} \prec x_1^{i_1'} x_2^{i_2'} \cdots x_n^{i_n'}$ if

$$\sum_{j=1}^{n} i_j < \sum_{j=1}^{n} i_j' \quad \text{or} \quad \sum_{j=1}^{n} i_j = \sum_{j=1}^{n} i_j', \quad i_j = i_j'(j=1,\ldots,k), \quad i_{k+1} < i_{k+1}'$$

Ordered in this way, the basis matrices become triangular in general.

We compute the dimension of lattice $\mathcal{L}$ as w where

$$w = \dim(\mathcal{L}) = \sum_{0 \le \gamma_i i_1 + \cdots + \gamma_n i_n \le \beta t} 1 = \frac{(\eta t)^n}{n!} \frac{1}{\gamma_1 \cdots \gamma_n} + o(t^n)$$

and the determinate $\det(\mathcal{L}) = N^{s_N} X_1^{s_{X_1}} \cdots X_n^{s_{X_n}}$ where

$$s_N = \sum_{0 \le r_1 i_1 + \cdots + r_n i_n \le t} \lceil \frac{t - \sum_{j=1}^{n} r_j i_j}{r} \rceil = \frac{t^{n+1}}{(n+1)!} \frac{1}{r r_1 \cdots r_n} + o(t^{n+1})$$

$$s_{X_j} = \sum_{0 \le \gamma_1 i_1 + \cdots + \gamma_n i_n \le \eta t} i_j = \frac{t^{n+1}}{(n+1)!} \frac{1}{\gamma_1 \cdots \gamma_{j-1} \gamma_j^2 \gamma_{j+1} \cdots \gamma_n} + o(t^{n+1})$$

for each $s_{X_1}, s_{X_2}, \ldots, s_{X_n}$.

To obtain the number of n polynomials with short coefficients that contain all small roots over integer, we apply LLL basis reduction algorithm to the lattice $\mathcal{L}$. Lemma 1 gives us an upper bound on the norm of the shortest vector in the LLL-reduced basis; if the bound is smaller than the bound given in Lemma 2, we can obtain the desired polynomial. We require the following condition:

$$2^{\frac{w-1}{4}} \det(\mathcal{L})^{\frac{1}{w}} < \frac{N^{\eta t}}{\sqrt{w}} \tag{4}$$

Ignoring low order terms of m and the quantities that do not depend on N, we have the following result

$$s_N + \sum_{j=1}^{n} \gamma_j s_{X_j} < w\eta t$$

After some calculations, we can get the final result

$$\sqrt[n]{\frac{\gamma_1 \cdots \gamma_n}{rr_1 \cdots r_n}} < \eta^{\frac{n+1}{n}}$$

In particular, from the Eq. (4), in order to ignore the quantities that do not depend on N, we must have

$$2^{\frac{w}{4}} \ll N^{\eta t} \quad \text{and} \quad \det(\mathcal{L})^{\frac{1}{w}} < N^{\eta t}$$

and these inequations imply that

$$w \ll 4\eta t \log_2 N \quad \text{and} \quad \frac{s_N}{w} \log_2 N < \eta t \log_2 N$$

Finally we have

$$\frac{1}{4(n+1)rr_1 \cdots r_n} \ll \eta^2 \log_2 N$$

Furthermore, one can check that in order to let the value $2^{w/4}$ become negligible compared with $N^{\eta t}$, we must have

$$\eta^2 \log_2 N \gg 1$$

The running time is dominated by LLL-reduction, therefore, the total running time for this approach is polynomial in $\log N$ but exponential in n.   □

Like [4,36], we also consider the generalization to simultaneous linear equations of higher degree.

Extended Simultaneous Modular Univariate Equations. Suppose that $N = 0 \bmod p^r, p \geq N^\eta$, we consider the simultaneous modular univariate equations

$$\begin{cases} h_1(x_1) = x_1^{\delta_1} + a_{\delta_1} x_1^{\delta_1-1} + \cdots + a_1 = 0 \bmod p^{r_1} \\ h_2(x_2) = x_1^{\delta_2} + b_{\delta_2} x_1^{\delta_2-1} + \cdots + b_1 = 0 \bmod p^{r_2} \\ \qquad\qquad\qquad\vdots \\ h_n(x_n) = x_1^{\delta_n} + c_{\delta_n} x_1^{\delta_n-1} + \cdots + c_1 = 0 \bmod p^{r_n} \end{cases}$$

Here each equation $h_j(x_j)$ has one variable and the degree of $h_j(x_j)$ is δ_j. We give the following result.

Theorem 11. *Under Assumption 1, the above generalised problem can be solved provided that*

$$\sqrt[n]{\frac{\delta_1\gamma_1 \cdots \delta_n\gamma_n}{rr_1 \cdots r_n}} < \eta^{\frac{n+1}{n}} \quad \text{and} \quad \eta \gg \frac{1}{\sqrt{\log N}}$$

The running time of the algorithm is polynomial in $\log N$ but exponential in n.

The proof is very similar to [4,36], we omit it here.

5.2   Common Prime RSA

In [13], Hinek revisited a new variant of RSA, called Common Prime RSA, where the modulus $N = pq$ is chosen such that $p - 1$ and $q - 1$ have a large common factor. For convenience, we give a brief description on the property of Common Prime RSA. Without loss of generality, assume that $p = 2ga + 1$ and $q = 2gb + 1$, where $g \simeq N^\gamma$ and a, b are coprime integers, namely $\gcd(a, b) = 1$. The decryption exponent d and encryption exponent e satisfy that

$$ed \equiv 1 \bmod 2gab \tag{5}$$

where $e \simeq N^{1-\gamma}$ and $d \simeq N^\beta$.

For a better comparison with the previous attacks, we give a brief review on all known attacks.

Wiener's Attack [38]. Using a continued fraction attack, Wiener proved that given any valid Common Prime RSA public key (N, e) with private exponent $d < N^{\frac{1}{4}-\frac{\gamma}{2}}$, namely $\beta < \frac{1}{4} - \frac{\gamma}{2}$, one can factor N in polynomial-time.

Hinek's Attack [13]. Hinek revisited this problem and proposed two lattice-based attacks. Due to Hinek's work, when $\beta < \gamma^2$ or $\beta < \frac{2}{5}\gamma$, N can be factored in polynomial-time.

Jochemsz-May's Attack [17]. Jochemsz and May gave another look at the equation proposed by Hinek [13] and modified the unknown variables in the equation. The bound has been further improved as

$$\beta < \frac{1}{4}(4 + 4\gamma - \sqrt{13 + 20\gamma + 4\gamma^2}).$$

Sarkar-Maitra's Attack [33]. Sarkar and Maitra proposed two improved attacks, one attack worked when $\gamma \leq 0.051$, and another worked when $0.051 < \gamma \leq 0.2087$.

One can check that when $\gamma \geq 0.2087$, Jochemsz-May's attack [17] is superior to other attacks. We use the algorithm of Theorem 10 to make an improvement on previous attacks when $\gamma \geq 0.3872$. We give a comparison with Jochemsz-May's attack in Fig. 3.

Our results improve Jochemsz-May's attack dramatically when γ is large, for instance, when γ is close to 0.5, we improve the bound on β from 0.2752, which is the best result of previous attacks, to 0.5. Below is our main result.

Theorem 12. *Assume that there exists instance of Common Prime RSA $N = pq$ with the above-mentioned parameters. Under Assumption 1, N can be factored in polynomial-time provided*

$$\beta < 4\gamma^3 \quad and \quad \gamma > \frac{1}{4}$$

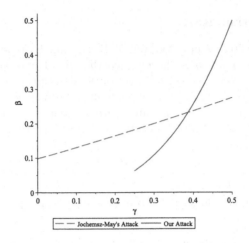

Fig. 3. Comparison of our theoretical bounds with Jochemsz-May's work.

Proof. According to the property of Common Prime RSA, we have $N = pq = (2ga + 1)(2gb + 1)$ which implies $N - 1 \equiv 0 \bmod g$. On the other hand, from Eq. (5) one can obtain

$$ed - 1 \equiv 0 \bmod g$$

Multiplying by the inverse of e modulo $N - 1$, we can obtain the following equation,

$$E - x \equiv 0 \bmod g$$

where E denotes the inverse of e modulo $N - 1$ and x denotes the unknown d. Moreover, since $(p - 1)(q - 1) = 4g^2 ab$, we have another equation,

$$N - y \equiv 0 \bmod g^2$$

where y denotes the unknown $p + q - 1$.

In summary, simultaneous modular univariate linear equations can be listed as

$$\begin{cases} E - x \equiv 0 \bmod g \\ N - y \equiv 0 \bmod g^2 \end{cases}$$

Note that $N - 1$ is a multiple of g and $(d, p + q - 1)$ is the desired solution of above equations, where $g \simeq N^\gamma$, $d \simeq N^\beta$ and $p + q - 1 \simeq N^{\frac{1}{2}}$. Obviously, this kind of modular equations is what we considered in Theorem 10. Setting

$$n = 2, r = 1, r_1 = 1, r_2 = 2, \gamma_1 = \beta, \gamma_2 = \frac{1}{2}, \eta = \gamma$$

We have

$$\gamma > \beta \quad \gamma > \frac{1}{4} \quad \beta < 4\gamma^3$$

Then we can obtain

$$\beta < 4\gamma^3 \quad \text{and} \quad \gamma > \frac{1}{4}$$

Under Assumption 1, one can solve the desired solution. This concludes the proof of Theorem 12. $\qquad\qquad\qquad\qquad\qquad\qquad\qquad\qquad\qquad\qquad\qquad\qquad$ □

Experimental Results. Some experimental data on the different size of g are listed in Table 5. Here we used 1000-bit N. Assumption 1 worked perfectly in all the cases. We always succeed to find out our desired roots.

Table 5. Comparison of our theoretical and experimental results with existing works.

γ	Theorem of [17]	Our result			
		Theo.	Expt.	Dim	Time (in seconds)
0.40	0.237	0.256	0.220	86	12321.521
0.42	0.245	0.294	0.260	113	53669.866
0.45	0.256	0.354	0.320	105	29128.554
0.48	0.268	0.415	0.390	98	15058.558

6   Conclusion

In this paper, we consider three type of generalized equations and propose some new techniques to find small root of these equations. Applying our algorithms, we obtain the best analytical/experimental results for some attacks on RSA and its variants. Besides, we believe that our new algorithms may find new applications in various other contexts.

Acknowledgments. We would like to thank the anonymous reviewers for helpful comments. This research was supported by CREST, JST. Part of this work was also supported by Strategic Priority Research Program of the Chinese Academy of Sciences (No. XDA06010703, No. XDA06010701 and No. XDA06010702), the National Key Basic Research Project of China (No. 2011CB302400 and No. 2013CB834203), and National Science Foundation of China (No. 61379139 and No. 61472417).

References

1. Boneh, D., Durfee, G.: Cryptanalysis of RSA with private key d less than $N^{0.292}$. IEEE Trans. Inf. Theor. **46**(4), 1339–1349 (2000)
2. Boneh, D., Durfee, G., Howgrave-Graham, N.: Factoring $N = p^r q$ for large r. In: Wiener, M. (ed.) CRYPTO 1999. LNCS, vol. 1666, p. 326. Springer, Heidelberg (1999)
3. Castagnos, G., Joux, A., Laguillaumie, F., Nguyen, P.Q.: Factoring pq^2 with quadratic forms: nice cryptanalyses. In: Matsui, M. (ed.) ASIACRYPT 2009. LNCS, vol. 5912, pp. 469–486. Springer, Heidelberg (2009)
4. Cohn, H., Heninger, N.: Approximate common divisors via lattices. ANTS-X (2012)

5. Coppersmith, D.: Small solutions to polynomial equations, and low exponent RSA vulnerabilities. J. Crypt. **10**(4), 233–260 (1997)
6. Coron, J.-S., Joux, A., Kizhvatov, I., Naccache, D., Paillier, P.: Fault attacks on RSA signatures with partially unknown messages. In: Clavier, C., Gaj, K. (eds.) CHES 2009. LNCS, vol. 5747, pp. 444–456. Springer, Heidelberg (2009)
7. Coron, J.-S., Naccache, D., Tibouchi, M.: Fault attacks against EMV signatures. In: Pieprzyk, J. (ed.) CT-RSA 2010. LNCS, vol. 5985, pp. 208–220. Springer, Heidelberg (2010)
8. The EPOC and the ESIGN Algorithms. IEEE P1363: Protocols from Other Families of Public-Key Algorithms (1998). http://grouper.ieee.org/groups/1363/StudyGroup/NewFam.html
9. Ernst, M., Jochemsz, E., May, A., de Weger, B.: Partial key exposure attacks on RSA up to full size exponents. In: Cramer, R. (ed.) EUROCRYPT 2005. LNCS, vol. 3494, pp. 371–386. Springer, Heidelberg (2005)
10. Fouque, P.A., Guillermin, N., Leresteux, D., Tibouchi, M., Zapalowicz, J.C.: Attacking RSA-CRT signatures with faults on montgomery multiplication. J. Cryptogr. Eng. **3**(1), 59–72 (2013). Springer
11. Herrmann, M.: Improved cryptanalysis of the multi-prime ϕ - hiding assumption. In: Nitaj, A., Pointcheval, D. (eds.) AFRICACRYPT 2011. LNCS, vol. 6737, pp. 92–99. Springer, Heidelberg (2011)
12. Herrmann, M., May, A.: Solving linear equations modulo divisors: on factoring given any bits. In: Pieprzyk, J. (ed.) ASIACRYPT 2008. LNCS, vol. 5350, pp. 406–424. Springer, Heidelberg (2008)
13. Hinek, M.J.: Another look at small RSA exponents. In: Pointcheval, D. (ed.) CT-RSA 2006. LNCS, vol. 3860, pp. 82–98. Springer, Heidelberg (2006)
14. Howgrave-Graham, N.: Finding small roots of univariate modular equations revisited. In: Darnell, M.J. (ed.) Cryptography and Coding 1997. LNCS, vol. 1355, pp. 131–142. Springer, Heidelberg (1997)
15. Howgrave-Graham, N.: Approximate integer common divisors. In: Silverman, J.H. (ed.) CaLC 2001. LNCS, vol. 2146, pp. 51–66. Springer, Heidelberg (2001)
16. Itoh, K., Kunihiro, N., Kurosawa, K.: Small secret key attack on a variant of RSA (due to Takagi). In: Malkin, T. (ed.) CT-RSA 2008. LNCS, vol. 4964, pp. 387–406. Springer, Heidelberg (2008)
17. Jochemsz, E., May, A.: A strategy for finding roots of multivariate polynomials with new applications in attacking RSA variants. In: Lai, X., Chen, K. (eds.) ASIACRYPT 2006. LNCS, vol. 4284, pp. 267–282. Springer, Heidelberg (2006)
18. Joux, A.: Algorithmic Cryptanalysis. Chapman & Hall/CRC, Boca Raton (2009)
19. Tosu, K., Kunihiro, N.: Optimal bounds for multi-prime Φ-hiding assumption. In: Susilo, W., Mu, Y., Seberry, J. (eds.) ACISP 2012. LNCS, vol. 7372, pp. 1–14. Springer, Heidelberg (2012)
20. Lenstra, A.K., Lenstra, H.W., Lovász, L.: Factoring polynomials with rational coefficients. Math. Ann. **261**(4), 515–534 (1982)
21. May, A.: New RSA vulnerabilities using lattice reduction methods. Ph.D. thesis (2003)
22. May, A.: Secret exponent attacks on RSA-type schemes with moduli $N = p^r q$. In: Bao, F., Deng, R., Zhou, J. (eds.) PKC 2004. LNCS, vol. 2947, pp. 218–230. Springer, Heidelberg (2004)
23. May, A.: Using LLL-reduction for solving RSA and factorization problems. In: Nguyen, P.Q., Vallée, B. (eds.) The LLL Algorithm, pp. 315–348. Springer, Heidelberg (2010)

24. May, A., Ritzenhofen, M.: Implicit factoring: on polynomial time factoring given only an implicit hint. In: Jarecki, S., Tsudik, G. (eds.) PKC 2009. LNCS, vol. 5443, pp. 1–14. Springer, Heidelberg (2009)
25. Nguên, P.Q., Stehlé, D.: Floating-point LLL revisited. In: Cramer, R. (ed.) EURO-CRYPT 2005. LNCS, vol. 3494, pp. 215–233. Springer, Heidelberg (2005)
26. Nitaj, A.: A new attack on RSA and CRT-RSA. In: Mitrokotsa, A., Vaudenay, S. (eds.) AFRICACRYPT 2012. LNCS, vol. 7374, pp. 221–233. Springer, Heidelberg (2012)
27. Okamoto, T., Uchiyama, S.: A new public-key cryptosystem as secure as factoring. In: Nyberg, K. (ed.) EUROCRYPT 1998. LNCS, vol. 1403, pp. 308–318. Springer, Heidelberg (1998)
28. Rivest, R.L., Shamir, A.: Efficient factoring based on partial information. In: Pichler, F. (ed.) EUROCRYPT 1985. LNCS, vol. 219, pp. 31–34. Springer, Heidelberg (1986)
29. Sarkar, S.: Reduction in lossiness of RSA trapdoor permutation. In: Bogdanov, A., Sanadhya, S. (eds.) SPACE 2012. LNCS, vol. 7644, pp. 144–152. Springer, Heidelberg (2012)
30. Sarkar, S.: Revisiting prime power RSA. Cryptology ePrint Archive, Report 2015/774 (2015). http://eprint.iacr.org/
31. Sarkar, S.: Small secret exponent attack on RSA variant with modulus $N = p^r q$. Des. Codes Cryptogr. **73**, 383–392 (2014)
32. Sarkar, S., Maitra, S.: Approximate integer common divisor problem relates to implicit factorization. IEEE Trans. Inf. Theor. **57**(6), 4002–4013 (2011)
33. Sarkar, S., Maitra, S.: Cryptanalytic results on Dual CRT and Common Prime RSA. Des. Codes Cryptgr. **66**(1–3), 157–174 (2013)
34. Shamir, A.: A polynomial time algorithm for breaking the basic Merkle-Hellman cryptosystem. In: FOCS 1982, pp. 145–152. IEEE (1982)
35. Takagi, T.: Fast RSA-type cryptosystem modulo $p^k q$. In: Krawczyk, H. (ed.) CRYPTO 1998. LNCS, vol. 1462, pp. 318–326. Springer, Heidelberg (1998)
36. Takayasu, A., Kunihiro, N.: Better lattice constructions for solving multivariate linear equations modulo unknown divisors. In: Boyd, C., Simpson, L. (eds.) ACISP. LNCS, vol. 7959, pp. 118–135. Springer, Heidelberg (2013)
37. van Dijk, M., Gentry, C., Halevi, S., Vaikuntanathan, V.: Fully homomorphic encryption over the integers. In: Gilbert, H. (ed.) EUROCRYPT 2010. LNCS, vol. 6110, pp. 24–43. Springer, Heidelberg (2010)
38. Wiener, M.J.: Cryptanalysis of short RSA secret exponents. IEEE Trans. Inf. Theor. **36**(3), 553–558 (1990)

FourQ: Four-Dimensional Decompositions on a Q-curve over the Mersenne Prime

Craig Costello[✉] and Patrick Longa

Microsoft Research, Redmond, USA
{craigco,plonga}@microsoft.com

Abstract. We introduce FourQ, a high-security, high-performance elliptic curve that targets the 128-bit security level. At the highest arithmetic level, cryptographic scalar multiplications on FourQ can use a four-dimensional Gallant-Lambert-Vanstone decomposition to minimize the total number of elliptic curve group operations. At the group arithmetic level, FourQ admits the use of extended twisted Edwards coordinates and can therefore exploit the fastest known elliptic curve addition formulas over large prime characteristic fields. Finally, at the finite field level, arithmetic is performed modulo the extremely fast Mersenne prime $p = 2^{127} - 1$. We show that this powerful combination facilitates scalar multiplications that are significantly faster than all prior works. On Intel's Haswell, Ivy Bridge and Sandy Bridge architectures, our software computes a variable-base scalar multiplication in 59,000, 71,000 cycles and 74,000 cycles, respectively; and, on the same platforms, our software computes a Diffie-Hellman shared secret in 92,000, 110,000 cycles and 116,000 cycles, respectively.

1 Introduction

This paper introduces a new, *complete* twisted Edwards [5] curve $\mathcal{E}(\mathbb{F}_{p^2})$: $-x^2 + y^2 = 1 + dx^2y^2$, where p is the Mersenne prime $p = 2^{127} - 1$, and d is a non-square in $\mathbb{F}_{p^2}$. This curve, dubbed "FourQ", arises as a special instance of recent constructions using Q-curves [27,46], and is thus equipped with an endomorphism ψ related to the p-power Frobenius map. In addition, it has *complex multiplication* (CM) by the order of discriminant $D = -40$, meaning it comes equipped with another efficient, low-degree endomorphism ϕ [47].

We built an elliptic curve cryptography (ECC) library that works inside the cryptographic subgroup $\mathcal{E}(\mathbb{F}_{p^2})[N]$, where N is a 246-bit prime. The endomorphisms ψ and ϕ do not give any practical speedup to Pollard's rho algorithm [42], which means the best known attack against the elliptic curve discrete logarithm problem (ECDLP) on $\mathcal{E}(\mathbb{F}_{p^2})[N]$ requires around $\sqrt{\pi N/4} \sim 2^{122.5}$ group operations on average. Thus, the cryptographic security of $\mathcal{E}$ (see Sect. 2.3 for more details) is closely comparable to other curves that target the 128-bit security level, e.g., [6,9,21,37].

Our choice of curve and the accompanying library offer a range of advantages over existing curves and implementations:

© International Association for Cryptologic Research 2015
T. Iwata and J.H. Cheon (Eds.): ASIACRYPT 2015, Part I, LNCS 9452, pp. 214–235, 2015.
DOI: 10.1007/978-3-662-48797-6_10

Speed: FourQ's library computes scalar multiplications significantly faster than all known software implementations of curve-based cryptographic primitives. It uses the endomorphisms ψ and ϕ to accelerate scalar multiplications via four-dimensional Gallant-Lambert-Vanstone (GLV)-style [22] decompositions. Four-dimensional decompositions have been used before [9,32,37], but not over *the* Mersenne prime[1]; this choice of field is significantly faster than any neighboring fields and several works have studied its arithmetic [13,21,36]. The combination of extremely fast modular reductions and four-dimensional scalar decompositions makes for highly efficient scalar multiplications on $\mathcal{E}$. Furthermore, we can exploit the fastest known addition formulas for elliptic curves over large characteristic fields [31], which are complete on $\mathcal{E}$ since the above d is non-square [31, Sect. 3]. In Sect. 2, we explain why four-dimensional decompositions and this special underlying field were not previously partnered at the 128-bit security level.

Simplicity and concrete correctness: Simplicity is a major priority in this work and in the development of our software; in some cases we sacrifice speed enhancements in order to design a more simple and compact algorithm (cf. Sect. 4.2).

On input of *any* point $P \in \mathcal{E}(\mathbb{F}_{p^2})[N]$, validated as in [14, Appendix A] if necessary, and *any* integer scalar $m \in [0, 2^{256})$, our software does the following (strictly in constant-time and without exception):

1. Computes $\phi(P)$, $\psi(P)$ and $\psi(\phi(P))$ using exactly[2] 68**M**, 27**S** and 49.5**A** – see Sect. 3.
2. Decomposes m (e.g., in less than 200 Sandy Bridge cycles) into a multiscalar $(a_1, a_2, a_3, a_4) \in \mathbb{Z}^4$ such that each a_i is positive and at most 64 bits – see Sect. 4.
3. Recodes the multiscalar (e.g., in less than 800 Sandy Bridge cycles) to ensure a simple and constant-time main loop – see Sect. 5.
4. Computes a lookup table of 8 elements using exactly 7 complete additions, before executing the main loop using exactly 64 complete twisted Edwards double-and-add operations, and finally outputting $[m]P = [a_1]P + [a_2]\phi(P) + [a_3]\psi(P) + [a_4]\psi\phi(P)$ – see Sect. 5.

This paper details each of the above steps explicitly, culminating in the full routine presented in Algorithm 2. Several prior works exploiting scalar decompositions have potential points of failure (cf. [30, Sect. 7], and Sect. 4.2), but crucially, and for the first time in the setting of four-dimensional decompositions, we accompany our routine with a robust proof of correctness – see Theorem 1.

---

[1] p stands alone as the only Mersenne prime suitable for high-security curves over quadratic extension fields. The next largest Mersenne prime is $2^{521} - 1$, which is suitable only for prime field curves targeting the 256-bit level.

[2] Here, and throughout, **I**, **M**, **S** and **A** are used to denote the respective costs of inversions, multiplications, squarings and additions in $\mathbb{F}_{p^2}$. We note that Frobenius operations amount to conjugations in $\mathbb{F}_p$, which are tallied as 0.5**A**.

Cryptographic versatility: FourℚQ is intended to be used in the same way, i.e., using the same model, same coordinates and same explicit formulas, irrespective of the cryptographic protocol or nature of the intended scalar multiplication. Unlike implementations using ladders [4,6,9,23], FourℚQ supports fast variable-base and fast fixed-base scalar multiplications, both of which use twisted Edwards coordinates; this serves as a basis for fast (ephemeral) Diffie-Hellman key exchange and fast Schnorr-like signatures. The presence of a single, complete addition law gives implementers the ability to easily wrap higher-level software and protocols around the FourℚQ's library exactly *as is*.

Public availability: Prior works exploiting four-dimensional decompositions have either made code available that did not attempt to run in constant-time [9], or not published code that did run in constant-time [18,37]. Our library, which is publicly available [15], is largely written in portable C and includes two modular implementations of the arithmetic over $\mathbb{F}_{p^2}$: a portable implementation written in C and a high-performance implementation for x64 platforms written in C and optional x64 assembly. The library also permits to select (at build time) whether the efficiently computable endomorphisms ψ and ϕ can be used or not for computing generic scalar multiplications. The code is accompanied by Magma scripts that can be used to verify the proofs of all claims and the claimed operation counts. Our aim is to make it easy for subsequent implementers to replicate the routine and, if desired, develop specialized code that is tailored to specific platforms for further performance gains or with different memory constraints.

When the NIST curves [40] were standardized in 1999, many of the landmark discoveries in ECC (e.g., [17,21,22,46]) were yet to be made. FourℚQ and its accompanying library represent the culmination of several of the best known ECC optimizations to date: it pulls together the extremely fast Mersenne prime, the fastest known large characteristic addition formulas [31], and the highest degree of scalar decompositions (there is currently no known way of achieving higher dimensional decompositions without exposing the ECDLP to attacks that are asymptotically much faster than Pollard rho). Subsequently, for generic scalar multiplications, FourℚQ performs around four to five times faster than the original NIST P-256 curve [26], between two and three times faster than curves that are currently under consideration as NIST alternatives, e.g., Curve25519 [4], and is also significantly faster than all of the other curves used to set previous speed records (see Sect. 6 for the comparisons). Interestingly, FourℚQ is still highly efficient if the endomorphisms ψ and ϕ are not used at all for computing generic scalar multiplications. In this case, FourℚQ performs about three times faster than the NIST P-256 curve and up to 1.5 times faster than Curve25519.

It is our belief that the demand for high-performance cryptography warrants the state-of-the-art in ECC to be part of the standardization discussion: this paper ultimately demonstrates the performance gains that are possible if such a curve was to be considered alongside the "conservative" choices.

The extended version. For space considerations, we have omitted the proofs of Propositions 1, 2, 4 and 5, Lemma 1 and Theorem 1, as well as several additional remarks. All of these, along with an appendix covering point validation, can be found in the extended version of this article [14].

2 The Curve: FourℚQ

This section describes the proposed curve, where we adopt Smith's notation [44,46] for the most part. We present the curve parameters in Sect. 2.1, shed some light on how the curve was found in Sect. 2.2, and discuss its cryptographic security in Sect. 2.3. Both Sects. 2.2 and 2.3 discuss that $\mathcal{E}$ is essentially one-of-a-kind, illustrating that there were no degrees of freedom in the choice of curve (see [14] for more details).

2.1 A Complete Twisted Edwards Curve

We will work over the quadratic extension field $\mathbb{F}_{p^2} := \mathbb{F}_p(i)$, where $p := 2^{127} - 1$ and $i^2 = -1$. We define $\mathcal{E}$ to be the twisted Edwards [5] curve

$$\mathcal{E}/\mathbb{F}_{p^2} : \ -x^2 + y^2 = 1 + dx^2y^2, \tag{1}$$

where $d := 125317048443780598345676279555970305165 \cdot i + 4205857648805777768770$.

The set of $\mathbb{F}_{p^2}$-rational points satisfying the affine model for $\mathcal{E}$ forms a group: the neutral element is $\mathcal{O}_{\mathcal{E}} = (0, 1)$ and the inverse of a point (x, y) is $(-x, y)$. The fastest set of explicit formulas for the addition law on $\mathcal{E}$ are due to Hisil, Wong, Carter and Dawson [31]: they use *extended twisted Edwards coordinates* to represent the affine point (x, y) on $\mathcal{E}$ by any projective tuple of the form $(X : Y : Z : T)$ for which $Z \neq 0$, $x = X/Z$, $y = Y/Z$ and $T = XY/Z$. Since d is not a square in $\mathbb{F}_{p^2}$, this set of formulas is also *complete* on $\mathcal{E}$ (see [5]), meaning that they will work without exception for all points in $\mathcal{E}(\mathbb{F}_{p^2})$.

The trace $t_{\mathcal{E}}$ of the p^2-power Frobenius endomorphism $\pi_{\mathcal{E}}$ of $\mathcal{E}$ is $t_{\mathcal{E}} = 136368062447564341573735631776713817674$, which reveals that

$$\#\mathcal{E}(\mathbb{F}_{p^2}) = p^2 + 1 - t_{\mathcal{E}} = 2^3 \cdot 7^2 \cdot N, \tag{2}$$

where N is a 246-bit prime. The cryptographic group we work with in this paper is $\mathcal{E}(\mathbb{F}_{p^2})[N]$.

2.2 Where did this Curve Come From?

The curve $\mathcal{E}$ above comes from the family of ℚ-curves of degree 2 – originally defined by Hasegawa [29] – that was recently used as one of the example families in Smith's general construction of ℚ-curve endomorphisms [44,46]. Certain examples of low-degree ℚ-curves (including this family) were independently obtained through a different construction by Guillevic and Ionica [27], who also studied 4-dimensional decompositions arising from such curves possessing CM.

In fact, $\mathcal{E}$ has a similar structure to the curve constructed in [27, Exercise 1], but is over the prime $p = 2^{127} - 1$.

For Δ a square-free integer, this family is defined over $\mathbb{Q}(\sqrt{\Delta})$ and is parameterized by $s \in \mathbb{Q}$ as

$$\tilde{\mathcal{E}}_{2,\Delta,s} \colon y^2 = x^3 - 6(5 - 3s\sqrt{\Delta})x + 8(7 - 9s\sqrt{\Delta}). \tag{3}$$

By definition [44, Definition 1], curves from this family are 2-isogenous (over $\mathbb{Q}(\Delta, \sqrt{-2})$) to their Galois conjugates ${}^\sigma\tilde{\mathcal{E}}_{2,\Delta,s}$. Smith reduces $\tilde{\mathcal{E}}_{2,\Delta,s}$ and ${}^\sigma\tilde{\mathcal{E}}_{2,\Delta,s}$ modulo primes p that are inert in $\mathbb{Q}(\sqrt{\Delta})$ to produce the curves $\mathcal{E}_{2,\Delta,s}$ and ${}^\sigma\mathcal{E}_{2,\Delta,s}$ defined over $\mathbb{F}_{p^2}$. He then composes the induced 2-isogeny from $\mathcal{E}_{2,\Delta,s}$ to ${}^\sigma\mathcal{E}_{2,\Delta,s}$ with the p-power Frobenius map from ${}^\sigma\mathcal{E}_{2,\Delta,s}$ back to $\mathcal{E}_{2,\Delta,s}$, which produces an efficiently computable degree $2p$ endomorphism ψ on $\mathcal{E}_{2,\Delta,s}$.

Recall that in this paper we fix $p = 2^{127} - 1$ for efficiency reasons. For this particular prime p and this family of $\mathbb{Q}$-curves, Smith's construction gives rise to precisely p non-isomorphic curves corresponding to each possible choice of $s \in \mathbb{F}_p$ [46, Proposition 1]. Varying s allows us to readily find curves belonging to this family with strong cryptographic group orders, each of which comes equipped with the endomorphism ψ that facilitates a two-dimensional scalar decomposition.

Seeking a four-dimensional (rather than two-dimensional) scalar decomposition on $\mathcal{E}_{2,\Delta,s}$ restricts us to a very small subset of possible s values. This is because we require the existence of another efficiently computable endomorphism on $\mathcal{E}_{2,\Delta,s}$, namely the low-degree GLV endomorphism ϕ on those instances of $\mathcal{E}_{2,\Delta,s}$ that possess CM over $\mathbb{Q}(\sqrt{\Delta})$. In [46, Sect. 9], Smith explains why there are only a handful of s values in any particular $\mathbb{Q}$-curve family that correspond to a curve with CM, before cataloging all such instances in the families of $\mathbb{Q}$-curves of degrees 2, 3, 5 and 7. In particular, up to isogeny and over any prime p, there are merely 13 values of s such that $\mathcal{E}_{2,\Delta,s}$ has CM over $\mathbb{Q}(\sqrt{\Delta})$. As is remarked in [46, Sect. 9], this scarcity of CM curves makes it highly unlikely that we will find a secure instance of a low-degree $\mathbb{Q}$-curve family with CM over any fixed prime p. This is the reason why other authors chasing high speeds at the 128-bit security level have previously sacrificed the fast Mersenne prime $p = 2^{127} - 1$ in favor of a four-dimensional decomposition [9,37]; one can always search through the small handfull of exceptional CM curves over many sub-optimal primes until a cryptographically secure instance is found. However, in the specific case of $p = 2^{127} - 1$, we actually get extremely lucky: our search through Smith's tables of exceptional $\mathbb{Q}$-curves with CM [46, Theorem 6] found one particular instance over $\mathbb{F}_{p^2}$ with a prime subgroup of 246-bits, namely $\mathcal{E}_{2,\Delta,s}$ with $s = \pm\frac{4}{9}$ and $\Delta = 5$. As is detailed in [46, Sect. 3], the specification of $\Delta = 5$ here does not dictate how we form the extension field $\mathbb{F}_{p^2}$ over $\mathbb{F}_p$; all quadratic extension fields of $\mathbb{F}_p$ are isomorphic, so we can take $s\sqrt{\Delta} = \pm\frac{4}{9}\sqrt{5}$ in (3) while still taking the reduction of $\tilde{\mathcal{E}}_{2,5,\pm\frac{4}{9}}$ modulo p to be $\mathcal{E}_{2,5,\pm\frac{4}{9}}/\mathbb{F}_{p^2}$ with $\mathbb{F}_{p^2} := \mathbb{F}_p(\sqrt{-1})$. To simplify notation, from hereon we fix $\tilde{\mathcal{E}}_W := \tilde{\mathcal{E}}_{2,5,\pm\frac{4}{9}}$ and define $\mathcal{E}_W$ as the reduction of $\tilde{\mathcal{E}}_W$ modulo p, given as

$$\mathcal{E}_W/\mathbb{F}_{p^2} : y^2 = x^3 - (30 - 8\sqrt{5})x + (56 - 32\sqrt{5}), \tag{4}$$

where the choice of the root $\sqrt{5}$ in $\mathbb{F}_{p^2}$ will be fixed in Sect. 3. We note that the short Weierstrass curve $\mathcal{E}_W$ is not isomorphic to our twisted Edwards curve $\mathcal{E}$, but rather to a twisted Edwards curve $\hat{\mathcal{E}}$ that is $\mathbb{F}_{p^2}$-isogenous to $\mathcal{E}$. The reason we work with $\mathcal{E}$ rather than $\hat{\mathcal{E}}$ is because the curve constant d on $\mathcal{E}$ is non-square in $\mathbb{F}_{p^2}$, which is not the case for the curve constant $\hat{d}$ on $\hat{\mathcal{E}}$; as we mentioned above, d being a non-square ensures that the fastest known addition formulas are also complete on $\mathcal{E}$. The isogenies between $\mathcal{E}$ and $\hat{\mathcal{E}}$ are made explicit as follows.

Proposition 1. *Let $\hat{\mathcal{E}}/K$ and $\mathcal{E}/K$ be the twisted Edwards curves defined by $\hat{\mathcal{E}}/K: -x^2 + y^2 = 1 + \hat{d}x^2 y^2$ and $\mathcal{E}/K: -x^2 + y^2 = 1 + dx^2 y^2$. If $d = -(1 + 1/\hat{d})$, then the map $\tau: \mathcal{E} \to \hat{\mathcal{E}}$, $(x,y) \mapsto \left(\frac{2xy}{(x^2 + y^2)\sqrt{\hat{d}}}, \frac{x^2 - y^2 + 2}{y^2 - x^2} \right)$ is a 4-isogeny, the dual of which is $\hat{\tau}: \hat{\mathcal{E}} \to \mathcal{E}$, $(x,y) \mapsto \left(\frac{2xy\sqrt{\hat{d}}}{x^2 - y^2 + 2}, \frac{y^2 - x^2}{y^2 + x^2} \right)$.*

We note at once that if $\hat{d}$ is a square in K, then τ and $\hat{\tau}$ are defined over K. Fortunately, while the twisted Edwards curve $\hat{\mathcal{E}}$ corresponding to $\mathcal{E}_W/\mathbb{F}_{p^2}$ has a square constant $\hat{d}$, our chosen isogenous curve $\mathcal{E}$ has the non-square constant $d = -(1 + 1/\hat{d})$. Our implementation will work solely in twisted Edwards coordinates on $\mathcal{E}$, but we will pass back and forth through $\mathcal{E}_W$ (via $\hat{\mathcal{E}}$) when deriving explicit formulas for the endomorphisms ϕ and ψ in Sect. 3. We note that Hamburg used 4-isogenies (also derived from [1]) to a similar effect in [28].

2.3   The Cryptographic Security of FourQ

Pollard's rho algorithm [42] is the best known way to solve the ECDLP in $\mathcal{E}(\mathbb{F}_{p^2})[N]$. An optimized version of this attack which uses the negation map [50] requires around $\sqrt{\pi N/4} \sim 2^{122.5}$ group operations on average. We note that, unlike some of the typical GLV [22] or GLS [21] endomorphisms that can be used to speed up Pollard's rho algorithm [16], both ψ and ϕ on $\mathcal{E}$ do not facilitate any known advantage; neither of these endomorphisms have a small orbit and they are both more expensive to compute than an amortized addition. Thus, the known complexity of the ECDLP on $\mathcal{E}$ is comparable to various other curves used in the speed-record literature; optimized implementations of Pollard rho against any of the fastest curves in [4,9,13,18,21,37,41] would require between $2^{124.8}$ and $2^{125.8}$ group operations on average. Ideally, we would prefer not to have the factor 7^2 dividing $\#\mathcal{E}(\mathbb{F}_{p^2})$, but the resulting ($\sim 2.8$ bit) security degradation is a small price to pay for having the fastest field at the 128-bit level in conjunction with a four-dimensional scalar decomposition. As we discuss further in [14], it was a long shot to try and find such a cryptographically secure Q-curve with CM over $\mathbb{F}_{p^2}$ in Smith's tables in the first place, let alone one that also had the necessary torsion to support a twisted Edwards model.

Since $\mathcal{E}(\mathbb{F}_{p^2})$ has rational 2-torsion, it is easy to write down the corresponding abelian surface over $\mathbb{F}_p$ whose Jacobian is isogenous to the Weil restriction of $\mathcal{E}$ – see [43, Lemma 2.1 and Lemma 3.1]. But since the best known algorithm to solve the discrete logarithm problem on such abelian surfaces is again Pollard's rho algorithm, the Weil descent philosophy (cf. [24]) does not pose a threat here. Furthermore, the embedding degree of $\mathcal{E}$ with respect to N is $(N-1)/2$, making it infeasible to reduce the ECDLP into a finite field [19,39].

We note that the largest prime factor dividing the group order of $\mathcal{E}$'s quadratic twist is 158 bits, but *twist-security* [4] is not an issue in this work: firstly, our software always validates input points (such validation is essentially free), and secondly, x-coordinate-only arithmetic (which is where twist-security makes sense) on $\mathcal{E}$ is not competitive with a four-dimensional decomposition that uses both coordinates.

In contrast to most currently standardized curves, the proposed curve is both defined over a quadratic extension field and has a small discriminant; one notable exception is secp256k1 in the SEC standard [11], which is used in the Bitcoin protocol and also has small discriminant. However, it is important to note that there is no better-than-generic attack known to date that can exploit either of these two properties on $\mathcal{E}$. In fact, with respect to ECDLP difficulty, Koblitz, Koblitz and Menezes [33, Sect. 11] point out that slower, large discriminant curves, like NIST P-256 and Curve25519, may turn out to be less conservative than specially chosen curves with small discriminant.

3   The Endomorphisms ψ and ϕ

In this section we derive explicit formulas for the two endomorphisms on $\mathcal{E}$. In what follows we use $c_{i,j,k,l}$ to denote the constant $i + j\sqrt{2} + k\sqrt{5} + l\sqrt{2}\sqrt{5}$ in $\mathbb{F}_{p^2}$, which is fixed by setting $\sqrt{2} := 2^{64}$ and $\sqrt{5} := 8739280708733697631800536882070724464 \cdot i$.

For both ψ and ϕ, we start by deriving the explicit formulas on the short Weierstrass model $\mathcal{E}_W$. As discussed in the previous section, we will pass back and forth between $\mathcal{E}$ and $\mathcal{E}_W$ via the twisted Edwards curve $\hat{\mathcal{E}}$ that is 4-isogenous to $\mathcal{E}$ over $\mathbb{F}_{p^2}$. The maps between $\mathcal{E}$ and $\hat{\mathcal{E}}$ are given in Proposition 1, and we take the maps $\delta \colon \mathcal{E}_W \to \hat{\mathcal{E}}$ and $\delta^{-1} \colon \hat{\mathcal{E}} \to \mathcal{E}_W$ from [46, Sect. 5] (tailored to our $\hat{\mathcal{E}}$) as $\delta \colon (x,y) \mapsto \left(\frac{\gamma(x-4)}{y}, \frac{x-4-c_{0,2,0,1}}{x-4+c_{0,2,0,1}} \right)$, and $\delta^{-1} \colon (x,y) \mapsto \left(\frac{c_{0,2,0,1}(y+1)}{1-y} + 4, \frac{c_{0,2,0,1}(y+1)\gamma}{x(1-y)} \right)$, where $\gamma^2 = c_{-12,-4,0,-2}$. The choice of the square root $\gamma \in \mathbb{F}_{p^2}$ becomes irrelevant in the compositions below.

3.1   Explicit Formulas for ψ

There is almost no work to be done in deriving ψ on $\mathcal{E}$, since this is Smith's $\mathbb{Q}$-curve endomorphism corresponding to the degree-2 family to which

$\mathcal{E}_W$ belongs. We start with $\psi_W\colon \mathcal{E}_W \to \mathcal{E}_W$, taken from [46, Sect. 5], as $\psi_W\colon (x,y) \mapsto \left(\left(-\frac{x}{2}-\frac{c_{9,0,4,0}}{x-4}\right)^p, \left(\frac{y}{i\sqrt{2}}\left(-\frac{1}{2}+\frac{c_{9,0,4,0}}{(x-4)^2}\right)\right)^p\right)$. With ψ_W as above, we define $\psi\colon \mathcal{E} \to \mathcal{E}$ as the composition $\psi = \hat{\tau}\delta\psi_W\delta^{-1}\tau$. In optimizing the explicit formulas for this composition, there is practically nothing to be gained by simplifying the full composition in the function field $\mathbb{F}_{p^2}(\mathcal{E})$. However, it is advantageous to optimize explicit formulas for the inner composition $(\delta\psi_W\delta^{-1})$ in the function field $\mathbb{F}_{p^2}(\hat{\mathcal{E}})$. In fact, for both ψ and ϕ, optimized explicit formulas for this inner composition are faster than the respective endomorphisms ψ_W and ϕ_W, and are therefore much faster than computing the respective compositions individually.

Simplifying the composition $\delta\psi_W\delta^{-1}$ in the function field $\mathbb{F}_{p^2}(\hat{\mathcal{E}})$ yields $(\delta\psi_W\delta^{-1})\colon \hat{\mathcal{E}} \to \hat{\mathcal{E}}$,

$$(x,y) \mapsto \left(\frac{2ix^p \cdot c_{-2,3,-1,0}}{y^p \cdot ((x^p)^2 \cdot c_{-140,99,0,0} + c_{-76,57,-36,24})}, \frac{c_{-9,-6,4,3}-(x^p)^2}{c_{-9,-6,4,3}+(x^p)^2}\right).$$

Note that each of the p-power Frobenius operations above amount to one $\mathbb{F}_p$ negation. As mentioned above, we compute the endomorphism $\psi = \hat{\tau}(\delta\psi_W\delta^{-1})\tau$ on $\mathcal{E}$ by computing τ and $\hat{\tau}$ separately; see Sect. 3.4 for the operation counts.

3.2 Deriving Explicit Formulas for ϕ

We now derive the second endomorphism ϕ that arises from $\mathcal{E}$ admitting CM by the order of discriminant $D = -40$. We start by pointing out that there is actually multiple routes that could be taken in defining and deriving ϕ (see the full version [14] for additional details). The possibility that we use in this paper produces an endomorphism of degree 5. This option was revealed to us in correspondence with Ben Smith, who pointed out that $\mathbb{Q}$-curves with CM can also be produced as the intersection of families of $\mathbb{Q}$-curves, and that our curve $\mathcal{E}$ is not only a degree-2 $\mathbb{Q}$-curve, but is also a degree-5 $\mathbb{Q}$-curve. Thus, the second endomorphism ϕ can be derived by first following the treatment in [46, Sect. 7] (see also [27, Sect. 3.3]) to derive ϕ_W as a 5-isogeny on $\mathcal{E}_W$, which we do below.

Working in $\mathbb{Q}(\sqrt{5})[x]$, the 5-division polynomial (cf. [20, Definition 9.8.4]) of $\tilde{\mathcal{E}}_W$ factors as $f(x)g(x)$, where $f(x) = x^2 + 4\sqrt{5}\cdot x + (18 - 4/5\sqrt{5})$ and $g(x)$ (which is of degree 10) are irreducible. The polynomial $f(x)$ defines the kernel of a 5-isogeny $\phi_W^\sigma\colon \tilde{\mathcal{E}}_W \to \tilde{\mathcal{E}}_W^\sigma$. We use this kernel to compute ϕ_W^σ via Vélu's formulae [49] (see also [34, Sect. 2.4]), reduce modulo p, and then compose with Frobenius $\pi_p\colon \mathcal{E}_W^\sigma \to \mathcal{E}_W$ to give $\phi_W\colon \mathcal{E}_W \to \mathcal{E}_W, (x,y) \mapsto (x_{\phi_W}, y_{\phi_W})$, where

$$x_{\phi_W} = \left(\frac{x^5 + 8\sqrt{5}x^4 + (40\sqrt{5}+260)x^3 + (720\sqrt{5}+640)x^2 + (656\sqrt{5}+4340)x + (1920\sqrt{5}+960)}{5\left((x^2+4\sqrt{5}x-1/5(4\sqrt{5}-90)\right)^2}\right)^p,$$

$$y_{\phi_W} = \left(\frac{-y\left(x^2+(4\sqrt{5}-8)x-12\sqrt{5}+26\right)\left(x^4+(8\sqrt{5}+8)x^3+28x^2-(48\sqrt{5}+112)x-32\sqrt{5}-124\right)}{\left(\sqrt{5}(x^2+4\sqrt{5}x-1/5(4\sqrt{5}-90))\right)^3}\right)^p,$$

As was the case with ψ in Sect. 3.1, it is advantageous to optimize formulas in $\mathbb{F}_{p^2}(\hat{\mathcal{E}})$ for the composition $(\delta\psi_W\delta^{-1})$, which gives $(\delta\psi_W\delta^{-1})\colon \hat{\mathcal{E}} \to \hat{\mathcal{E}}, (x,y) \mapsto$

(x_ϕ, y_ϕ), where

$$x_\phi = \left(\frac{c_{9,-6,4,-3} \cdot x \cdot (y^2 - c_{7,5,3,2} \cdot y + c_{21,15,10,7}) \cdot (y^2 + c_{7,5,3,2} \cdot y + c_{21,15,10,7})}{(y^2 + c_{3,2,1,1} \cdot y + c_{3,3,2,1}) \cdot (y^2 - c_{3,2,1,1} \cdot y + c_{3,3,2,1})} \right)^p,$$

$$y_\phi = \left(\frac{c_{15,10,6,4} \cdot (5y^4 + c_{120,90,60,40} \cdot y^2 + c_{175,120,74,54})}{5y \cdot (y^4 + c_{240,170,108,76} \cdot y^2 + c_{3055,2160,1366,966})} \right)^p.$$

Again, we use this to compute the full endomorphism $\psi = \hat{\tau}(\delta \psi_W \delta^{-1})\tau$ on $\mathcal{E}$ by computing τ and $\hat{\tau}$ separately; see Sect. 3.4 for the operation counts.

3.3 Eigenvalues

The eigenvalues of the two endomorphisms ψ and ϕ play a key role in developing scalar decompositions. In this subsection we write them in terms of the curve parameters. From [46, Theorem 2], and given that we used a 4-isogeny τ and its dual to pass back and forth to $\mathcal{E}_W$, the eigenvalues of ψ on $\mathcal{E}(\mathbb{F}_{p^2})[N]$ are $\lambda_\psi := 4 \cdot \frac{p+1}{r}$ (mod N) and $\lambda'_\psi := -\lambda_\psi$ (mod N), where r is an integer satisfying $2r^2 = 2p + t_\mathcal{E}$. To derive the eigenvalues for ϕ, we make use of the CM equation for $\mathcal{E}$, which (since $\mathcal{E}$ has CM by the order of discriminant $D = -40$) is $40V^2 = 4p^2 - t_\mathcal{E}^2$, for some integer V. We fix r and V to be the positive integers satisfying these equations, namely $V := 4929397548930634471175140323270296814$ and $r := 15437785290780909242$.

Proposition 2. *The eigenvalues of ϕ on $\mathcal{E}(\mathbb{F}_{p^2})[N]$ are*

$$\lambda_\phi := 4 \cdot \frac{(p-1)r^3}{(p+1)^2 V} \quad (\text{mod } N) \quad \text{and} \quad \lambda'_\phi := -\lambda_\phi \quad (\text{mod } N).$$

3.4 Section Summary

Table 1 summarizes the isogenies derived in this section, together with their exact operation counts. The reason that multiples of 0.5 appear in the additions column is that we count Frobenius operations (which amount to a negation in $\mathbb{F}_p$) as half an addition in $\mathbb{F}_{p^2}$. Four-dimensional scalar decompositions on $\mathcal{E}$ require the computation of $\phi(P)$, $\psi(P)$ and the composition $\psi(\phi(P))$; the ordering here is important since ψ is much faster than ϕ, meaning we actually compute ϕ once and ψ twice. We note that all sets of explicit formulas were derived assuming the inputs were projective points $(X : Y : Z)$ corresponding to a point $(X/Z, Y/Z)$ in the domain of the isogeny. Similarly, all explicit formulas output the point $(X' : Y' : Z')$ corresponding to $(X'/Z', Y'/Z')$ in the codomain, and in the special cases when the codomain is $\mathcal{E}$ (i.e., for $\hat{\tau}$, ϕ, ψ and $-\psi\phi$), we also output the coordinate T' (or a related variant) corresponding to $T' = X'Y'/Z'$, which facilitates faster subsequent group law formulas on $\mathcal{E}$ – see [14].

Table 1 reveals that, on input of a projective point in $\mathcal{E}(\mathbb{F}_{p^2})[N]$, the total cost of the three maps ϕ, ψ and $\psi\phi$ is $68\mathbf{M} + 27\mathbf{S} + 49.5\mathbf{A}$. Computing the maps using these explicit formulas requires the storage of 16 constants in $\mathbb{F}_{p^2}$, and at any stage of the endomorphism computations, requires the storage of at most 7 temporary variables.

Table 1. Summary of isogenies used in the derivation of the three endomorphisms ϕ, ψ and $\phi\psi$ on $\mathcal{E}$, together with the cost of their explicit formulas. Here **M**, **S** and **A** respectively denote the costs of one multiplication, one squaring and one addition in $\mathbb{F}_{p^2}$.

Isogeny	Domain & codomain	Degree	No. fixed constants	No. temp variables	Cost		
					M	**S**	**A**
τ	$\mathcal{E} \rightarrow \hat{\mathcal{E}}$	4	1	2	5	3	5
$\hat{\tau}$	$\hat{\mathcal{E}} \rightarrow \mathcal{E}$	4	1	2	5	3	4
$(\delta\phi_{\mathrm{W}}\delta^{-1})$	$\hat{\mathcal{E}} \rightarrow \hat{\mathcal{E}}$	$5p$	10	7	20	5	11.5
$(\delta\psi_{\mathrm{W}}\delta^{-1})$	$\hat{\mathcal{E}} \rightarrow \hat{\mathcal{E}}$	$2p$	4	2	9	2	5.5
ϕ		$80p$	11	7	30	11	20.5
ψ	$\mathcal{E} \rightarrow \mathcal{E}$	$32p$	5	2	19	8	14.5
$\psi\phi$		$2560p$	-	7	19	8	14.5
total cost $(\phi, \psi, \psi\phi)$		**16**		**7**	**68**	**27**	**49.5**

4   Optimal Scalar Decompositions

Let λ_ψ and λ_ϕ be as fixed in Sect. 3.3. In this section we show how to compute, for any integer scalar $m \in \mathbb{Z}$, a corresponding 4-dimensional multiscalar $(a_1, a_2, a_3, a_4) \in \mathbb{Z}^4$ such that $m \equiv a_1 + a_2\lambda_\phi + a_3\lambda_\psi + a_4\lambda_\phi\lambda_\psi \pmod{N}$, such that $0 \leq a_i < 2^{64} - 1$ for $i = 1, 2, 3, 4$, and such that a_1 is odd (which facilitates faster scalar recodings and multiplications – see Sect. 5). An excellent reference for general scalar decompositions in the context of elliptic curve cryptography is [45], where it is shown how to write down short lattice bases for scalar decompositions directly from the curve parameters. Here, we show how to further reduce such short bases into bases that are, in the context of multiscalar multiplications, *optimal*.

4.1   Babai Rounding and Optimal Bases

Following [45, Sect. 1], we define the *lattice of zero decompositions* as

$$\mathcal{L} := \langle (z_1, z_2, z_3, z_4) \in \mathbb{Z}^4 \mid z_1 + z_2\lambda_\phi + z_3\lambda_\psi + z_4\lambda_\phi\lambda_\psi \equiv 0 \pmod{N} \rangle,$$

so that the set of decompositions for $m \in \mathbb{Z}/N\mathbb{Z}$ is the lattice coset $(m, 0, 0, 0) + \mathcal{L}$. For a given basis $\mathbf{B} = (\mathbf{b}_1, \mathbf{b}_2, \mathbf{b}_3, \mathbf{b}_4)$ of $\mathcal{L}$, and on input of any $m \in \mathbb{Z}$, the *Babai rounding* technique [2] computes $(\alpha_1, \alpha_2, \alpha_3, \alpha_4) \in \mathbb{Q}^4$ as the unique solution to $(m, 0, 0, 0) = \sum_{i=1}^{4} \alpha_i \mathbf{b}_i$, and subsequently computes the multiscalar $(a_1, a_2, a_3, a_4) = (m, 0, 0, 0) - \sum_{i=1}^{4} \lfloor \alpha_i \rceil \cdot \mathbf{b}_i$. It follows that $(a_1, a_2, a_3, a_4) - (m, 0, 0, 0) \in \mathcal{L}$, so $m \equiv a_1 + a_2\lambda_\phi + a_3\lambda_\psi + a_4\lambda_\phi\lambda_\psi \pmod{N}$. Since $-1/2 \leq x - \lfloor x \rceil \leq 1/2$, this technique finds the unique element in $(m, 0, 0, 0) + \mathcal{L}$ that lies inside the parallelepiped[3] defined by $\mathcal{P}(\mathbf{B}) = \{\mathbf{B}\mathbf{x} \,|\, \mathbf{x} \in [-1/2, 1/2)^4\}$, i.e., Babai rounding maps $\mathbb{Z}$ onto $\mathcal{P}(\mathbf{B}) \cap \mathbb{Z}^4$. For a given m, the length of the corresponding multiscalar multiplication is then determined by the infinity norm, $||\cdot||_\infty$, of the corresponding element (a_1, a_2, a_3, a_4) in $\mathcal{P}(\mathbf{B}) \cap \mathbb{Z}^4$.

Since our scalar multiplications must run in time independent of m, the speed of the multiscalar exponentiations will depend on the worst case, i.e., on the maximal infinity norm taken across all elements in $\mathcal{P}(\mathbf{B}) \cap \mathbb{Z}^4$. Or, equivalently, the speed of routine will depend on the *width* of the smallest 4-cube whose convex body contains $\mathcal{P}(\mathbf{B}) \cap \mathbb{Z}^4$. This width depends only on the choice of $\mathbf{B}$, so this gives us a natural way of finding a basis that is optimal for our purposes. We make this concrete in the following definition, which is stated for an arbitrary lattice of dimension n. Definition 1 simplifies the situation by looking for the smallest n-cube containing $\mathcal{P}(\mathbf{B})$, rather than $\mathcal{P}(\mathbf{B}) \cap \mathbb{Z}^n$, but our candidate bases will always be orthogonal enough such that the conditions are equivalent in practice.

Definition 1 (Babai-optimal bases). We say that a basis $\mathbf{B}$ of a lattice $\mathcal{L} \in \mathbb{R}^n$ is *Babai-optimal* if the width of the smallest n-cube containing the parallelepiped $\mathcal{P}(\mathbf{B})$ is minimal across all bases for $\mathcal{L}$.

We note immediately that taking the n *successive minima* under $||\cdot||_\ell$, for any $\ell \in \{1, 2, \ldots, \infty\}$, will not be Babai-optimal in general. Indeed, for our specific lattice $\mathcal{L}$, neither the $||\cdot||_2$-reduced basis (output from LLL [35]) or the $||\cdot||_\infty$-reduced basis (in the sense of Lovász and Scarf [38]) are Babai-optimal.

For very low dimensions, such as those used in ECC scalar decompositions, we can find a Babai-optimal basis via straightforward *enumeration* as follows. Starting with any reasonably small basis $\mathbf{B}' = (\mathbf{b}_1', \ldots, \mathbf{b}_n')$, like the ones in [45], we compute the width, $w(\mathbf{B}')$, of the smallest n-cube whose convex body contains $\mathcal{P}(\mathbf{B}')$; by the definition of $\mathcal{P}$, this is $w(\mathbf{B}') = \max_{1 \le j \le n} \{\sum_{i=1}^n |\mathbf{b}_i'[j]|\}$. We then enumerate the set S of all vectors $\mathbf{v} \in \mathcal{L}$ such that $||\mathbf{v}||_\infty \le w(\mathbf{B}')$; any vector not in S cannot be in a basis whose width is smaller than $\mathbf{B}'$. We can then test all possible bases $\mathbf{B}$, that are formed as combinations of n linearly independent vectors in S, and choose one corresponding to the minimal value of $w(\mathbf{B})$.

Proposition 3. *A Babai optimal basis for our zero decomposition lattice $\mathcal{L}$ is given by* $\mathbf{B} := (\mathbf{b}_1, \mathbf{b}_2, \mathbf{b}_3, \mathbf{b}_4)$, *where*

$224 \cdot \mathbf{b}_1 := (16(-60\alpha + 13r - 10), \; 4(-10\alpha - 3r + 12), \; 4(-15\alpha + 5r - 13), \; -13\alpha - 6r + 3)$,

$8 \cdot \mathbf{b}_2 := (32(5\alpha - r), \; -8, \; 8, \; 2\alpha + r)$,

$224 \cdot \mathbf{b}_3 := (16(80\alpha - 15r + 18), \; 4(18\alpha - 3r - 16), \; 4(-15\alpha - 9r + 15), \; 15\alpha + 8r + 3\alpha)$,

$448 \cdot \mathbf{b}_4 := (16(-360\alpha + 77r + 42), \; 4(42\alpha + 17r + 72), \; 4(85\alpha - 21r - 77), \; (-77\alpha - 36r - 17))$,

for V and r as fixed in Sect. 3, and where $\alpha := V/r \in \mathbb{Z}$.

---

[3] This is a translate (by $-\frac{1}{2}(\sum_{i=1}^4 \mathbf{b}_i)$) of the *fundamental parallelepiped*, which is defined using $\mathbf{x} \in [0, 1)^4$.

Proof. Straightforward but lengthy calculations using the equations in Sect. 3.3 reveal that $\mathbf{b}_1$, $\mathbf{b}_2$, $\mathbf{b}_3$ and $\mathbf{b}_4$ are all in $\mathcal{L}$. Another direct calculation reveals that the determinant of $\langle \mathbf{b}_1, \mathbf{b}_2, \mathbf{b}_3, \mathbf{b}_4 \rangle$ is N, so $\mathbf{B}$ is a basis for $\mathcal{L}$. To show that $\mathbf{B}$ is Babai-optimal, we set $\mathbf{B}' = \mathbf{B}$ and compute $w(\mathbf{B}') = \max_{1 \leq j \leq 4} \left\{ \sum_{i=1}^{4} |\mathbf{b}_i'[j]| \right\}$, which (at $j = 1$) is $w(\mathbf{B}') = (245\alpha + 120r + 17)/448$. Enumeration under $||\cdot||_\infty$ yields exactly 128 vectors (up to sign) in $S = \{\mathbf{v} \in \mathcal{L} \mid ||\mathbf{v}||_\infty \leq w(\mathbf{B}')\}$; none of the rank 4 bases formed from S have a width smaller than $\mathbf{B}$. □

The size of the set S in the above proof depends on the quality of the initial basis $\mathbf{B}'$. For the proof, it suffices to start with the Babai-optimal basis $\mathbf{B}$ itself, but in practice we will usually start with a basis that is not optimal according to Definition 1. In our case we computed the basis in Proposition 3 by first writing down a short basis using Smith's methodology [45]. We input this into the LLL algorithm [35] to obtain an *LLL-reduced* basis $(\mathbf{b}_1, \mathbf{b}_2, \mathbf{b}_1 + \mathbf{b}_4, \mathbf{b}_3)$; these are also the four successive minima under $||\cdot||_2$. We then input this basis into the algorithm of Lovász and Scarf [38]; this forced the requisite changes to output a basis consisting of the four successive minima under $||\cdot||_\infty$, namely $(\mathbf{b}_1, \mathbf{b}_1 + \mathbf{b}_4, \mathbf{b}_2, \mathbf{b}_1 + \mathbf{b}_3)$. Using this as our input $\mathbf{B}'$ into the enumeration gave a set S of size 282, which we exhaustively searched to find $\mathbf{B}$.

We now describe a simple scalar decomposition that uses Babai rounding on the optimal basis above. Note that, since V and r are fixed, the four $\hat{\alpha}_i$ values below are fixed integer constants.

Proposition 4. *For a given integer m, and the basis $\mathbf{B} := (\mathbf{b}_1, \mathbf{b}_2, \mathbf{b}_3, \mathbf{b}_4)$ in Prop. 3, let $(\alpha_1, \alpha_2, \alpha_3, \alpha_4) \in \mathbb{Q}^4$ be the unique solution to $(m, 0, 0, 0) = \sum_{i=1}^{4} \alpha_i \mathbf{b}_i$, and let $(a_1, a_2, a_3, a_4) = (m, 0, 0, 0) - \sum_{i=1}^{4} \lfloor \alpha_i \rceil \cdot \mathbf{b}_i$. Then $m \equiv a_1 + a_2 \lambda_\phi + a_3 \lambda_\psi + a_4 \lambda_\psi \phi \pmod{N}$ and $|a_1|, |a_2|, |a_3|, |a_4| < 2^{62}$.*

4.2 Handling Round-Off Errors

The decomposition described in Proposition 4 requires the computation of four roundings $\lfloor \frac{\hat{\alpha}_i}{N} \cdot m \rceil$, where m is the input scalar and the four $\hat{\alpha}_i$ and N are fixed curve constants. Following [10, Sect. 4.2], one efficient way of performing these roundings is to choose a power of 2 greater than the denominator N, say μ, and precompute the fixed curve constants $\ell_i = \lfloor \frac{\hat{\alpha}_i}{N} \cdot \mu \rceil$, so that $\lfloor \frac{\hat{\alpha}_i}{N} \cdot m \rceil$ can be computed at runtime as $\lfloor \frac{\ell_i \cdot m}{\mu} \rfloor$, and the division by μ can be computed as a simple shift.

It is correctly noted in [10, Sect. 4.2] that computing the rounding in this way means the answer can be out by 1 in some cases, but it is further said that *"in practice this does not affect the size of the multiscalars"*. While this assertion may have been true in [10], in general this will not be the case, particularly when we wish to bound the size of the multiscalars as tightly as possible. We address this issue on $\mathcal{E}$ starting with Lemma 1.

Lemma 1. *Let $\hat{\alpha}$ be any integer, and let m, N and μ be positive integers with $m < \mu$. Then $\lfloor \frac{\hat{\alpha}m}{N} \rceil - \lfloor \lfloor \frac{\hat{\alpha}\mu}{N} \rceil \cdot \frac{m}{\mu} \rfloor$ is either 0 or 1.*

Lemma 1 says that, so long as we choose μ to be greater than the maximum size of our input scalars m, our fast method of approximating $\lfloor \frac{\hat{\alpha}_i}{N} \cdot m \rceil$ will either give the correct answer, or it will be $\lfloor \frac{\hat{\alpha}_i}{N} \cdot m \rceil - 1$. It is easy to see that larger choices of μ decrease the probability of a rounding error. For example, on 10 million random decompositions of integers between 0 and N with $\mu = 2^{246}$, roughly 2.2 million trials gave at least one error in the α_i; when $\mu = 2^{247}$, roughly 1.7 million trials gave at least one error; when $\mu = 2^{256}$, 4333 trials gave an error; and, taking $\mu = 2^{269}$ was the first power of two that gave no errors.

Prior works have seemingly addressed this problem by taking μ to be large enough so that the chance of roundoff errors are very (perhaps even exponentially) small. However, no matter how large μ is chosen, the existence of a permissible scalar whose decomposition gives a roundoff error is still a possibility[4], and this could violate constant-time promises.

In this work, and in light of Theorem 1, we instead choose to sacrifice some speed by guaranteeing that roundoff errors are always accounted for. Rather than assuming that $(a_1, a_2, a_3, a_4) = \sum_{i=1}^{4}(\alpha_i - \lfloor \alpha_i \rceil)\mathbf{b}_i$, we account for the approximation $\tilde{\alpha}_i$ to $\lfloor \alpha_i \rceil$ (described in Lemma 1) by allowing $(a_1, a_2, a_3, a_4) = \sum_{i=1}^{4}(\alpha_i - \tilde{\alpha}_i)\mathbf{b}_i = \sum_{i=1}^{4}(\alpha_i - (\lfloor \alpha_i \rceil - \epsilon_i))\mathbf{b}_i$, for all sixteen combinations arising from $\epsilon_i \in \{0, 1\}$, for $i = 1, 2, 3, 4$. This means that all integers less than μ will decompose to a multiscalar in $\mathbb{Z}^4$ whose coordinates lie inside the parallelepiped $\mathcal{P}_\epsilon(\mathbf{B}) := \{\mathbf{Bx} \mid \mathbf{x} \in [-1/2, 3/2)^4\}$. Theorem 1 permits scalars as any 256-bit strings, so we fix $\mu := 2^{256}$ from here on, which also means that division by μ will correspond to a shift of machine words. The edges of $\mathcal{P}_\epsilon(\mathbf{B})$ are twice as long as those of $\mathcal{P}(\mathbf{B})$, so the number of points in $\mathcal{P}_\epsilon(\mathbf{B}) \cap \mathbb{Z}^4$ is $\mathrm{vol}(\mathcal{P}_\epsilon) = 16N$. We note that, even though the number of permissible scalars far exceeds $16N$, the decomposition that maps integers in $[0, \mu)$ to multiscalars in $\mathcal{P}_\epsilon(\mathbf{B}) \cap \mathbb{Z}^4$ is certainly no longer *onto*; almost all of the μ scalars will map into $\mathcal{P}(\mathbf{B}) \cap \mathbb{Z}^4$, since the chance of roundoff errors that take us into $\mathcal{P}_\epsilon(\mathbf{B}) - \mathcal{P}(\mathbf{B})$ is small. Plainly, the width of smallest 4-cube containing $\mathcal{P}_\epsilon(\mathbf{B})$ is also twice that of the 4-cube containing $\mathcal{P}(\mathbf{B})$, so (in the sense of Definition 1) our basis is still Babai-optimal. Nevertheless, the bounds in Proposition 4 no longer apply, which is one of the issues addressed in the next subsection.

4.3   All-Positive Multiscalars

Many points in $\mathcal{P}_\epsilon(\mathbf{B}) \cap \mathbb{Z}^4$ have coordinates that are far greater than 2^{62} in absolute value, and in addition, the majority of them will have coordinates that are both positive and negative. Dealing with such signed multiscalars can require an additional iteration in the main loop of the scalar multiplication, so in this subsection we use an offset vector in $\mathcal{L}$ to find a translate of $\mathcal{P}_\epsilon(\mathbf{B})$ that contains points whose four coordinates are always positive. We note that this does not save the additional iteration mentioned above, but (at no cost) it does simplify

---

[4] This is not technically true: so long as the set of permissible scalars is finite, there will always be a μ large enough to round all scalar decompositions accurately, but finding or proving this is, to our knowledge, very difficult.

the scalar recoding, such that we do not have to deal with multiscalars that can have negative coordinates. Such offset vectors were used in two dimensions in [13, Sect. 4].

From the proof of Proposition 3, we have that the width of the smallest 4-cube containing $\mathcal{P}_\epsilon(\mathbf{B})$ is $2 \cdot (245\alpha + 120r + 17)/448$, which lies between 2^{63} and 2^{64}. Thus, the optimal situation is to translate of $\mathcal{P}_\epsilon(\mathbf{B})$ (using a vector in $\mathcal{L}$) that fits inside the convex body of the 4-cube $\mathcal{H} = \{2^{64} \cdot \mathbf{x} \,|\, \mathbf{x} \in [0,1]^4\}$. In fact, as we discuss in the next paragraph, we actually want to find two unique translates of $\mathcal{P}_\epsilon(\mathbf{B})$ inside $\mathcal{H}$.

The scalar recoding described in Sect. 5 requires that the first component of the multiscalar (a_1, a_2, a_3, a_4) is odd. In the case that a_1 is even, which happens around half of the time, previous works have employed this "odd-only" recoding by instead working with the multiscalar $(a_1 - 1, a_2, a_3, a_4)$, and adding the point P to the value output by the main loop (cf. [41, Algorithm 4] and [18, Algorithm 2]). Of course, in a constant-time routine, this scalar update and point addition must be performed regardless of the parity of a_1, and the correct scalars and results must be masked in and out of the main loop accordingly. In this work we simplify the situation by using offset vectors in $\mathcal{L}$ to achieve the same result; this has the added advantage of avoiding an extra point addition. We do this by finding two vectors $\mathbf{c}, \mathbf{c}' \in \mathcal{L}$ such that $\mathbf{c} + \mathcal{P}_\epsilon(\mathbf{B})$ and $\mathbf{c}' + \mathcal{P}_\epsilon(\mathbf{B})$ both lie inside $\mathcal{H}$, and such that precisely one of $(a_1, a_2, a_3, a_4) + \mathbf{c}$ and $(a_1, a_2, a_3, a_4) + \mathbf{c}'$ has a first component that is odd. This is made explicit in the full scalar decomposition described below.

Proposition 5 (Scalar Decompositions). *Let* $\mathbf{B} = (\mathbf{b}_1, \mathbf{b}_2, \mathbf{b}_3, \mathbf{b}_4)$ *be the basis in Proposition 3, let* $\mu = 2^{256}$, *and define the four curve constants* $\ell_i :=$ $\lfloor \hat{\alpha}_i \cdot \mu/N \rceil$ *for* $i = 1, 2, 3, 4$, *with the* $\hat{\alpha}_i$ *as given in Proposition 4. Let* $\mathbf{c} = 2\mathbf{b}_1 - \mathbf{b}_2 + 5\mathbf{b}_3 + 2\mathbf{b}_4$ *and* $\mathbf{c}' = 2\mathbf{b}_1 - \mathbf{b}_2 + 5\mathbf{b}_3 + \mathbf{b}_4$ *in* $\mathcal{L}$. *For any integer* $m \in [0, 2^{256})$, *let* $\tilde{\alpha}_i = \lfloor \ell_i m/\mu \rfloor$, *and let* $(a_1, a_2, a_3, a_4) = (m, 0, 0, 0) - \sum_{i=1}^4 \lfloor \tilde{\alpha}_i \rceil \cdot \mathbf{b}_i$. *Then, both of the multiscalars* $(a_1, a_2, a_3, a_4) + \mathbf{c}$ *and* $(a_1, a_2, a_3, a_4) + \mathbf{c}'$ *are valid decompositions of* m, *have all four coordinates positive and less than* 2^{64}, *and precisely one of them has a first coordinate that is odd.*

The scalar decomposition described in Proposition 5 outputs two multiscalars. Our decomposition routine uses a bitmask to select and output the one with an odd first coordinate in constant time.

5   The Scalar Multiplication

This section describes the full scalar multiplication of $P \in \mathcal{E}(\mathbb{F}_{p^2})$ by an integer $m \in [0, 2^{256})$, pulling together the endomorphisms and scalar decompositions derived in the previous two sections.

5.1   Recoding the Multiscalar

The "all-positive" multiscalar (a_1, a_2, a_3, a_4) that is obtained from the decomposition described in Proposition 5 could be fed *as is* into a simple 4-way multiexponentiation (e.g., the 4-dimensional version of [48]) to achieve an efficient scalar multiplication. However, more care needs to be taken to obtain an efficient routine that also runs in constant-time. For example, we need to guarantee that the main loop iterates in the same number of steps, which would not currently be the case since $\max_j(\log_2(|a_j|))$ can be several integers less than 64. As another example, a straightforward multiexponentiation could leak information in the case that the i-th bit of all four a_j values was 0, which would result in a "do-nothing" rather than a non-trivial addition.

To achieve an efficient constant-time routine, we adopt the general recoding Algorithm from [18, Algorithm 1], and tailor it to scalar multiplications on FourQ. This results in Algorithm 1 below, which is presented in two flavors: one that is geared towards the general reader and one that is geared towards implementers (we note that the lines do not coincide for the most part). On input of any multiscalar (a_1, a_2, a_3, a_4) produced by Proposition 5, Algorithm 1 outputs an equivalent multiscalar (b_1, b_2, b_3, b_4) with $b_j = \sum_{i=0}^{64} b_j[i] \cdot 2^i$ for $b_j[i] \in \{-1, 0, 1\}$ and $j = 1, 2, 3, 4$, such that we always have $b_1[64] = 1$ and such that $b_1[i]$ is non-zero for every $i = 0, \ldots, 63$. This fixes the length of the main loop and ensures that each addition step of the multiexponentiation requires an addition by something other than the neutral element.

Another benefit of Algorithm 1 is that $b_j[i] \in \{0, b_1[i]\}$ for $j = 2, 3, 4$; as was exploited in [18], this "sign-alignment" means that the lookup table used in our multiexponentiation only requires 8 elements, rather than the 16 that would be required in a naïve multiexponentiation that uses (a_1, a_2, a_3, a_4). More specifically, since $b_1[i]$ (which is to be multiplied by P) is always non-zero, every element of the lookup table T must contain P, so we have $T[u] := P + [u_0]\phi(P) + [u_1]\psi(P) + [u_2]\psi(\phi(P))$, where $u = (u_2, u_1, u_0)_2$ for $u = 0, \ldots, 7$. We point out that the recoding must itself be implemented in constant-time; the implementer-friendly version shows that Algorithm 1 indeed lends itself to such a constant-time implementation. We further note that the outputs of the two versions are formatted differently: the left side outputs the multiscalar (b_1, b_2, b_3, b_4), while the right side instead outputs the corresponding lookup table indices (the d_i) and the masks (the m_i) used to select the correct signs of the lookup elements. That is, $(m_{64}, \ldots, m_0)$ corresponds to the binary expansion of b_1 and $(d_{64}, \ldots, d_0)$ corresponds to the binary expansion of $b_2 + 2b_3 + 4b_4$.

5.2   The Full Routine

We now present Algorithm 2: the full scalar multiplication routine. This is accompanied by Theorem 1, the proof of which (see [14]) gives more details on the steps summarized in Algorithm 2; in particular, it specifies the representations of all points in order to state the total number of $\mathbb{F}_{p^2}$ operations. Algorithm 2 assumes that the input point P is in $\mathcal{E}(\mathbb{F}_{p^2})[N]$, i.e., has been validated according to [14, Appendix A].

Theorem 1. *For every point $P \in \mathcal{E}(\mathbb{F}_{p^2})[N]$ and every non-negative integer m less than 2^{256}, Algorithm 2 computes $[m]P$ correctly using a fixed sequence of field, integer and table-lookup operations.*

Algorithm 1. FourQ multiscalar recoding: reader-friendly (left) and implementer-friendly (right).

Input: four positive integers $a_j = (0, a_j[63], \ldots, a_j[0])_2 \in \{0,1\}^{65}$ less than 2^{64} for $1 \le j \le 4$ and with a_1 odd.

Output: four integers $b_j = \sum_{i=0}^{64} b_j[i] \cdot 2^i$, with $b_j[i] \in \{-1, 0, 1\}$.	**Output:** $(d_{64}, \ldots, d_0)$ with $0 \le d_i < 7$, and $(m_{64}, \ldots, m_0)$ with $m_i \in \{-1, 0\}$.
1: $b_1[64] = 1$	1: $m_{64} = -1$
2: **for** $i = 0$ **to** 64 **do**	2: **for** $i = 0$ **to** 63 **do**
3: **if** $i \ne 64$ **then**	3: $d_i = 0$
4: $b_1[i] = 2a_1[i+1] - 1$	4: $m_i = -a_1[i+1]$
5: **for** $j = 2$ **to** 4 **do**	5: **for** $j = 2$ **to** 4 **do**
6: $b_j[i] = b_1[i] \cdot a_j[0]$	6: $d_i = d_i + (a_j[0] \ll (j-2))$
7: $a_j = \lfloor a_j/2 \rfloor - \lfloor b_j[i]/2 \rfloor$	7: $c = (a_1[i+1] \,|\, a_j[0])^{\wedge} a_1[i+1]$
8: **return** $(b_j[64], \ldots, b_j[0])$ for $1 \le j \le 4$.	8: $a_j = (a_j \gg 1) + c$
	9: $d_{64} = a_2 + 2a_3 + 4a_4$
	10: **return** $(d_{64}, \ldots, d_0)$ and $(m_{64}, \ldots, m_0)$.

6 Performance Analysis and Results

This section shows that, at the 128-bit security level, FourQ is significantly faster than all other known curve-based primitives. We reiterate that our software runs in constant-time and is therefore fully protected against timing and cache attacks.

6.1 Operation Counts

We begin with a first-order comparison based on operation counts between FourQ and two other efficient curve-based primitives that are defined over large prime characteristic fields and that target the 128-bit security level: the twisted Edwards GLV+GLS curve defined over $\mathbb{F}_{p^2}$ with $p = 2^{127} - 5997$ proposed in [37], and the genus 2 Kummer surface defined over $\mathbb{F}_p$ with $p = 2^{127} - 1$ that was proposed in [25]; we dub these "GLV+GLS" and "Kummer" below. Both of these curves have recently set speed records on a variety of platforms (see [18] and [6]). Table 2 summarizes the operation counts for one variable-base scalar multiplication on FourQ, GLV+GLS and Kummer. In the right-most column we approximate the cost in terms of prime field operations (using the standard assumption that 1 base field squaring is approximately 0.8 base field multiplications), where we round each tally to the nearest integer. For the GLV+GLS and FourQ operation counts, we assume that one multiplication over $\mathbb{F}_{p^2}$ involves 3 multiplications and 5 additions/subtractions over $\mathbb{F}_p$ (when using Karatsuba) and one squaring over $\mathbb{F}_{p^2}$ involves 2 multiplications and 3 additions/subtractions over $\mathbb{F}_p$.

Table 2 shows that the GLV+GLS routine from [37] requires slightly fewer operations than FourQ. This can mainly be explained by the faster endomorphisms, but (as we will see in Table 3) this difference is more than made up

---

Algorithm 2. FourℚQ's scalar multiplication on $\mathcal{E}(\mathbb{F}_{p^2})[N]$.

Input: Point $P \in \mathcal{E}(\mathbb{F}_{p^2})[N]$ and integer scalar $m \in [0, 2^{256})$.
Output: $[m]P$.

Compute endomorphisms:
1: Compute $\phi(P)$, $\psi(P)$ and $\psi(\phi(P))$ using the explicit formulas summarized in Table 1.
Precompute lookup table:
2: Compute $T[u] = P + [u_0]\phi(P) + [u_1]\psi(P) + [u_2]\psi(\phi(P))$ for $u = (u_2, u_1, u_0)_2$ in $0 \leq u \leq 7$.
Scalar decomposition:
3: Decompose m into the multiscalar (a_1, a_2, a_3, a_4) as in Proposition 5.
Scalar recoding:
4: Recode (a_1, a_2, a_3, a_4) into $(d_{64}, \ldots, d_0)$ and $(m_{64}, \ldots, m_0)$ using Algorithm 1. Write $s_i = 1$ if $m_i = -1$ and $s_i = -1$ if $m_i = 0$.
Main loop:
5: $Q = s_{64} \cdot T[d_{64}]$
6: **for** $i = 63$ **to** 0 **do**
7:     $Q = [2]Q$
8:     $Q = Q + s_i \cdot T[d_i]$
9: **return** Q

Table 2. Operation counts for variable-base scalar multiplications on three different curves targeting the 128-bit security level. In the case of the Kummer surface, we additionally use a "word-mul" column to count the number of special multiplications of a general element in $\mathbb{F}_p$ by a small (i.e., one-word) constant – see [6].

primitive	prime char. p	op. count over $\mathbb{F}_{p^2}$				approx. op. count over $\mathbb{F}_p$			
		inv	mul	sqr	add	inv	mul	add	word-mul
FourℚQ	$2^{127} - 1$	1	842	283	950.5	1	3092	6960	-
GLV+GLS	$2^{127} - 5997$	1	833	191	769	1	2885	6278	-
Kummer	$2^{127} - 1$	-	-	-	-	1	4319	8032	2008

for by the faster modular arithmetic and superior simplicity of FourℚQ. Table 2 shows that FourℚQ requires far fewer operations (in the same ground field) than Kummer; it is therefore expected, in general, that implementations based on FourℚQ outperform Kummer implementations for computing variable-base scalar multiplications.

6.2   Experimental Results

To evaluate performance, we wrote a standalone library supporting Four ℚQ – see [15]. The library's design pursues modularity and code reuse, and leverages the simplicity of FourℚQ's arithmetic. It also facilitates the addition of specialized code for different platforms and applications: the core functionality of the library is fully written in portable C and works together with pluggable implementations of the arithmetic over $\mathbb{F}_{p^2}$ (and a few other complementary functions). The first release version of the library comes with two of those pluggable modules: a portable implementation written in C and a high-performance implementation for x64 platforms written in C and optional x64 assembly. The library computes all of the basic elliptic curve operations including variable-base and fixed-base scalar multiplications, making it suitable for a wide range of cryptographic protocols. In addition, the software permits the selection (at build time) of whether or not the endomorphisms ψ and ϕ are to be exploited in variable-based scalar multiplications.

In Table 3, we compare FourQ's performance with other state-of-the-art implementations documented in the literature. Our benchmarks cover a wide range of x64 processors, from high-end architectures (e.g., Intel's Haswell) to low-end architectures (e.g., Intel's Atom). To cast the performance numbers in the context of a real-world protocol, we choose to illustrate FourQ's performance in one round of an ephemeral Diffie-Hellman (DH) key exchange. This means that both parties can generate their public keys using a fixed-base scalar multiplication and generate the shared secret using a variable-base scalar multiplication. Exploiting such precomputations to generate *truly* ephemeral public keys agrees with the comments made by Bernstein and Lange in [8, Sect. 1], e.g., that *"forward secrecy is at its strongest when a key is discarded immediately after its use"*. Thus, Table 3 shows the execution time (in terms of clock cycles) for both variable-base and fixed-base scalar multiplications. We note that the laddered implementations in [4,6,9] only compute variable-base scalar multiplications, which is why we use the cost of two variable-base scalar multiplications to approximate the cost of ephemeral DH in those cases. For the FourQ and GLV+GLS implementations, precomputations for the fixed-base scalar multiplications occupied 7.5KB and 6KB of storage, respectively.

Table 3 shows that, in comparison with the "conservative" curves, FourQ is 2.1–2.7 times faster than the Curve25519 implementations in [3,12] and up to 5.4 times faster than the curve P-256 implementation in [26], when computing variable-base scalar multiplications. When considering the results for the DH key exchange, FourQ performs 1.8–3.5 times faster than Curve25519 and up to 4.2 times faster than curve P-256.

In terms of comparisons to the previously fastest implementations, variable-base scalar multiplications using our software are between 1.20 and 1.34 times faster than the Kummer [6,9] and the GLV+GLS [18] implementations on AMD's Kaveri and Intel's Atom Pineview, Sandy Bridge and Ivy Bridge. The Kummer implementation for Haswell in [6] is particularly fast because it takes advantage of the powerful AVX2 vector instructions. Nevertheless, our implementation (which does not currently exploit vector instructions to accelerate the field arithmetic) is still faster in the case of variable-base scalar multiplication. Moreover, in practice we expect a much larger advantage. For example, in the case of the DH key exchange, we leverage the efficiency of fixed-base scalar multiplications to achieve a factor 1.33x speedup over the Kummer implementation on Haswell. For the rest of platforms considered in Table 3, a DH shared secret

Table 3. Performance results (expressed in terms of thousands of clock cycles) of state-of-the-art implementations of various curves targeting the 128-bit security level on various x64 platforms. Benchmark tests were taken with Intel's TurboBoost and AMD's TurboCore disabled and the results were rounded to the nearest 1000 clock cycles. The benchmarks for the FourℚQ and GLV+GLS implementations were done on 1.66 GHz Intel Atom N570 Pineview, 3.4 GHz Intel Core i7-2600 Sandy Bridge, 3.4 GHz Intel Core i7-3770 Ivy Bridge, 3.4 GHz Intel Core i7-4770 Haswell and 3.1GHz AMD A8 PRO-7600B Kaveri. For the Kummer implementations [6,9] and Curve25519 implementation [3], Pineview, Sandy Bridge, Ivy Bridge and Haswell benchmarks were taken from eBACS [7] (machines h2atom, h6sandy, h9ivy and titan0), while AMD benchmarks were obtained by running eBACS' SUPERCOP toolkit on the corresponding targeted machine. The benchmarks for curve NIST P-256 were taken directly from [26] and the second set of Curve25519 benchmarks were taken directly from [12].

Processor	Operation	FourℚQ (this work)	GLV+GLS [18]	Kummer [9]	[6]	Curve25519 [3]	[12]	P-256 [26]
Atom Pineview	var-base	**442**	N/A	556	N/A	1,109	N/A	N/A
	fixed-base	**217**	N/A	-	N/A	-	N/A	N/A
	ephem. DH	**659**	N/A	1,112	N/A	2,218	N/A	N/A
Sandy Bridge	var-base	**74**	92	123	89	194	157	400
	fixed-base	**42**	51	-	-	-	54	90
	ephem. DH	**116**	143	246	178	388	211	490
Ivy Bridge	var-base	**71**	89	119	88	183	159	N/A
	fixed-base	**39**	49	-	-	-	52	N/A
	ephem. DH	**110**	138	238	176	366	211	N/A
Haswell	var-base	**59**	N/A	111	61	162	N/A	312
	fixed-base	**33**	N/A	-	-	-	N/A	67
	ephem. DH	**92**	N/A	222	122	324	N/A	379
AMD Kaveri	var-base	**122**	N/A	151	164	301	N/A	N/A
	fixed-base	**65**	N/A	-	-	-	N/A	N/A
	ephem. DH	**187**	N/A	302	328	602	N/A	N/A

using the FourℚQ software can be computed 1.5–1.8 times faster than a DH secret using the Kummer software in [6]. We note that the eBACS website [7] and [6] report different results for the same Kummer software on the same platform (i.e., Titan0): eBACS reports 60,556 Haswell cycles whereas [6] claims 54,389 Haswell cycles. This difference in performance raises questions regarding accuracy. The results that we obtained after running the eBACS' SUPERCOP toolkit on our own targeted Haswell machine seem to confirm that the results claimed in [6] for the Kummer were measured with TurboBoost enabled.

FourℚQ without endomorphisms. Our library can be built with a version of the variable-base scalar multiplication function that does not exploit the endomorphisms ψ and ϕ to accelerate computations (note that fixed-base scalar multiplications do not exploit these endomorphisms by default). In this case, FourℚQ computes one variable-base scalar multiplication in (respectively) 109, 131, 138 and 803 thousand cycles on the Haswell, Ivy Bridge, Sandy Bridge and Atom

Pineview processors used for our experiments. These results are up to 2.9 times faster than the corresponding results for NIST P-256 and up to 1.5 times faster than the corresponding results for Curve25519.

Acknowledgements. We thank Michael Naehrig for several discussions throughout this work, and Joppe Bos, Sorina Ionica and Greg Zaverucha for their comments on an earlier version of this paper. We are especially thankful to Ben Smith for pointing out the better option for ϕ in Sect. 3.2.

References

1. Ahmadi, O., Granger, R.: On isogeny classes of Edwards curves over finite fields. Cryptology ePrint Archive, Report 2011/135 (2011). http://eprint.iacr.org/
2. Babai, L.: On Lovász' lattice reduction and the nearest lattice point problem. Combinatorica **6**(1), 1–13 (1986)
3. Bernstein, D.J., Duif, N., Lange, T., Schwabe, P., Yang, B.-Y.: High-speed high-security signatures. In: Preneel, B., Takagi, T. (eds.) CHES 2011. LNCS, vol. 6917, pp. 124–142. Springer, Heidelberg (2011)
4. Bernstein, D.J.: Curve25519: new Diffie-Hellman speed records. In: Yung, M., Dodis, Y., Kiayias, A., Malkin, T. (eds.) PKC 2006. LNCS, vol. 3958, pp. 207–228. Springer, Heidelberg (2006)
5. Bernstein, D.J., Birkner, P., Joye, M., Lange, T., Peters, C.: Twisted Edwards curves. In: Vaudenay, S. (ed.) AFRICACRYPT 2008. LNCS, vol. 5023, pp. 389–405. Springer, Heidelberg (2008)
6. Bernstein, D.J., Chuengsatiansup, C., Lange, T., Schwabe, P.: Kummer strikes back: new DH speed records. In: Sarkar, P., Iwata, T. (eds.) ASIACRYPT 2014. LNCS, vol. 8873, pp. 317–337. Springer, Heidelberg (2014)
7. Bernstein, D.J., Lange, T.: eBACS: ECRYPT Benchmarking of Cryptographic Systems. http://bench.cr.yp.to/results-dh.html. Accessed on May 19 2015
8. Bernstein, D.J., Lange, T.: Hyper-and-elliptic-curve cryptography. LMS J. Comput. Math. **17**(A), 181–202 (2014)
9. Bos, J.W., Costello, C., Hisil, H., Lauter, K.: Fast cryptography in genus 2. In: Johansson, T., Nguyen, P.Q. (eds.) EUROCRYPT 2013. LNCS, vol. 7881, pp. 194–210. Springer, Heidelberg (2013)
10. Bos, J.W., Costello, C., Hisil, H., Lauter, K.: High-performance scalar multiplication using 8-dimensional GLV/GLS decomposition. In: Bertoni, G., Coron, J.-S. (eds.) CHES 2013. LNCS, vol. 8086, pp. 331–348. Springer, Heidelberg (2013)
11. Certicom Research. Standards for Efficient Cryptography 2: Recommended Elliptic Curve Domain Parameters, v2.0. Standard SEC2, Certicom (2010)
12. Chou, T.: Fastest Curve25519 implementation ever. In: Workshop on Elliptic Curve Cryptography Standards (2015). http://www.nist.gov/itl/csd/ct/ecc-workshop.cfm
13. Costello, C., Hisil, H., Smith, B.: Faster compact Diffie–Hellman: endomorphisms on the x-line. In: Nguyen, P.Q., Oswald, E. (eds.) EUROCRYPT 2014. LNCS, vol. 8441, pp. 183–200. Springer, Heidelberg (2014)
14. Costello, C., Longa, P.: FourQ: four-dimensional decompositions on a Q-curve over the Mersenne prime (extended version). Cryptology ePrint Archive, Report 2015/565 2015. http://eprint.iacr.org/

15. Costello, C., Longa, P.: FourQlib (2015). http://research.microsoft.com/fourqlib/
16. Duursma, I.M., Gaudry, P., Morain, F.: Speeding up the discrete log computation on curves with automorphisms. In: Lam, K.-Y., Okamoto, E., Xing, C. (eds.) ASIACRYPT 1999. LNCS, vol. 1716, pp. 103–121. Springer, Heidelberg (1999)
17. Edwards, H.: A normal form for elliptic curves. Bull. Am. Math. Soc. **44**(3), 393–422 (2007)
18. Faz-Hernández, A., Longa, P., Sánchez, A.H.: Efficient and secure algorithms for GLV-based scalar multiplication and their implementation on GLV-GLS curves (extended version). J. Cryptographic Eng. **5**(1), 31–52 (2015)
19. Frey, G., Müller, M., Rück, H.: The Tate pairing and the discrete logarithm applied to elliptic curve cryptosystems. IEEE Trans. Inf. Theor. **45**(5), 1717–1719 (1999)
20. Galbraith, S.D.: Mathematics of Public Key Cryptography. Cambridge University Press, Cambridge (2012)
21. Galbraith, S.D., Lin, X., Scott, M.: Endomorphisms for faster elliptic curve cryptography on a large class of curves. J. Cryptology **24**(3), 446–469 (2011)
22. Gallant, R.P., Lambert, R.J., Vanstone, S.A.: Faster point multiplication on elliptic curves with efficient endomorphisms. In: Kilian, J. (ed.) CRYPTO 2001. LNCS, vol. 2139, pp. 190–200. Springer, Heidelberg (2001)
23. Gaudry, P.: Fast genus 2 arithmetic based on Theta functions. J. Math. Cryptology **1**(3), 243–265 (2007)
24. Gaudry, P.: Index calculus for abelian varieties of small dimension and the elliptic curve discrete logarithm problem. J. Symbolic Comput. **44**(12), 1690–1702 (2009)
25. Gaudry, P., Schost, E.: Genus 2 point counting over prime fields. J. Symbolic Comput. **47**(4), 368–400 (2012)
26. Gueron, S., Krasnov, V.: Fast prime field elliptic curve cryptography with 256 bit primes. J. Cryptographic Eng. **5**(2), 141–151 (2015)
27. Guillevic, A., Ionica, S.: Four-dimensional GLV via the Weil restriction. In: Sako, K., Sarkar, P. (eds.) ASIACRYPT 2013, Part I. LNCS, vol. 8269, pp. 79–96. Springer, Heidelberg (2013)
28. Hamburg, M.: Twisting Edwards curves with isogenies. Cryptology ePrint Archive, Report 2014/027 (2014). http://eprint.iacr.org/
29. Hasegawa, Y.: Q-curves over quadratic fields. Manuscripta Math. **94**(1), 347–364 (1997)
30. Hisil, H., Costello, C.: Jacobian coordinates on genus 2 curves. In: Sarkar, P., Iwata, T. (eds.) ASIACRYPT 2014. LNCS, vol. 8873, pp. 338–357. Springer, Heidelberg (2014)
31. Hisil, H., Wong, K.K.-H., Carter, G., Dawson, E.: Twisted Edwards curves revisited. In: Pieprzyk, J. (ed.) ASIACRYPT 2008. LNCS, vol. 5350, pp. 326–343. Springer, Heidelberg (2008)
32. Hu, Z., Longa, P., Xu, M.: Implementing 4-dimensional GLV method on GLS elliptic curves with j-invariant 0. Des. Codes Cryptography **63**(3), 331–343 (2012)
33. Koblitz, A.H., Koblitz, N., Menezes, A.: Elliptic curve cryptography: the serpentine course of a paradigm shift. J. Number Theor. **131**(5), 781–814 (2011)
34. Kohel, D.: Endomorphism rings of elliptic curves over finite fields. Ph.D. thesis, University of California at Berkeley (1996)
35. Lenstra, A.K., Lenstra, H.W., Lovász, L.: Factoring polynomials with rational coefficients. Math. Ann. **261**(4), 515–534 (1982)
36. Longa, P., Gebotys, C.: Efficient techniques for high-speed elliptic curve cryptography. In: Mangard, S., Standaert, F.-X. (eds.) CHES 2010. LNCS, vol. 6225, pp. 80–94. Springer, Heidelberg (2010)

37. Longa, P., Sica, F.: Four-dimensional Gallant-Lambert-Vanstone scalar multiplication. J. Cryptology **27**(2), 248–283 (2014)
38. Lovász, L., Scarf, H.E.: The generalized basis reduction algorithm. Math. Oper. Res. **17**(3), 751–764 (1992)
39. Menezes, A., Vanstone, S.A., Okamoto, T.: Reducing elliptic curve logarithms to logarithms in a finite field. In: Koutsougeras, C., Vitter, J.S. (eds.) Proceedings of 23rd Annual ACM Symposium on Theory of Computing, pp. 80–89. ACM (1991)
40. National Institute of Standards and Technology (NIST). 186–2. Digital Signature Standard (DSS). Federal Information Processing Standards (FIPS) Publication (2000)
41. Oliveira, T., López, J., Aranha, D.F., Rodríguez-Henríquez, F.: Two is the fastest prime: lambda coordinates for binary elliptic curves. J. Cryptographic Eng. **4**(1), 3–17 (2014)
42. Pollard, J.M.: Monte Carlo methods for index computation (mod p). Math. Comput. **32**(143), 918–924 (1978)
43. Scholten, J.: Weil restriction of an elliptic curve over a quadratic extension (2004). http://citeseerx.ist.psu.edu/viewdoc/download?doi=10.1.1.118.7987&rep=rep1&type=pdf
44. Smith, B.: Families of fast elliptic curves from $\mathbb{Q}$-curves. In: Sako, K., Sarkar, P. (eds.) ASIACRYPT 2013, Part I. LNCS, vol. 8269, pp. 61–78. Springer, Heidelberg (2013)
45. Smith, B.: Easy scalar decompositions for efficient scalar multiplication on elliptic curves and genus 2 Jacobians. In: Contemporary Mathematics Series, vol. 637, p. 15. American Mathematical Society (2015)
46. Smith, B.: The $\mathbb{Q}$-curve construction for endomorphism-accelerated elliptic curves. J. Cryptology (2015, to appear)
47. Stark, H.M.: Class-numbers of complex quadratic fields. In: Kuijk, W. (ed.) Modular Functions of One Variable I, pp. 153–174. Springer, Heidelberg (1973)
48. Straus, E.G.: Addition chains of vectors. Am. Math. Mon. **70**(806–808), 16 (1964)
49. Vélu, J.: Isogénies entre courbes elliptiques. CR Acad. Sci. Paris Sér. AB **273**, A238–A241 (1971)
50. Wiener, M., Zuccherato, R.J.: Faster attacks on elliptic curve cryptosystems. In: Tavares, S., Meijer, H. (eds.) SAC 1998. LNCS, vol. 1556, pp. 190–200. Springer, Heidelberg (1999)

Signatures

Efficient Fully Structure-Preserving Signatures for Large Messages

Jens Groth$^{(\boxtimes)}$

University College London, London, UK
j.groth@ucl.ac.uk

Abstract. We construct both randomizable and strongly existentially unforgeable structure-preserving signatures for messages consisting of many group elements. To sign a message consisting of $N = mn$ group elements we have a verification key size of m group elements and signatures contain $n + 2$ elements. Verification of a signature requires evaluating $n + 1$ pairing product equations.

We also investigate the case of fully structure-preserving signatures where it is required that the secret signing key consists of group elements only. We show a variant of our signature scheme allowing the signer to pick part of the verification key at the time of signing is still secure. This gives us both randomizable and strongly existentially unforgeable fully structure-preserving signatures. In the fully structure preserving scheme the verification key is a single group element, signatures contain $m+n+1$ group elements and verification requires evaluating $n+1$ pairing product equations.

Keywords: Digital signatures · Pairing-based cryptography · Full structure-preservation

1 Introduction

Structure-preserving signatures are pairing-based signatures where verification keys, messages and signatures all consist solely of group elements and the verification algorithm relies on generic group operations such as multiplications and pairings to verify a signature. Structure-preserving signatures are interesting because they compose well with other structure-preserving primitives such as ElGamal encryption [ElG85] and Groth-Sahai proofs [GS12] for instance. By combining different structure-preserving components it is possible to build advanced cryptographic schemes in a modular manner. Applications of structure-preserving signatures include blind signatures [AFG+10,FV10], group signatures [AFG+10,FV10,LPY12], homomorphic signatures [LPJY13,

This research was supported by the Engineering and Physical Sciences Research Council grant EP/J009520/1 and the European Research Council under the European Union's Seventh Framework Programme (FP/2007-2013)/ERC Grant Agreement n. 307937.

© International Association for Cryptologic Research 2015
T. Iwata and J.H. Cheon (Eds.): ASIACRYPT 2015, Part I, LNCS 9452, pp. 239–259, 2015.
DOI: 10.1007/978-3-662-48797-6_11

ALP13], delegatable anonymous credentials [Fuc11], compact verifiable shuffles [CKLM12], network encoding [ALP12], oblivious transfer [GH08, CDEN12], tightly secure encryption [HJ12, ADK+13] and anonymous e-cash [ZLG12].

Since structure-preserving signatures are basic components when building cryptographic schemes it is crucial to make them as efficient as possible. All cryptographic protocols built on top of a structure-preserving signature scheme will be affected by its efficiency. There has therefore been a significant amount of research into finding barriers for how efficient structure-preserving signatures can be and constructing schemes achieving these bounds. Abe et al. [AGHO11] demonstrated a lower bound of 3 group elements for structure-preserving signatures (using Type III pairings, which is the most efficient type) and found matching constructions with 3 element signatures.

While the case of signing a single group element has been well studied, the question of signing larger messages has received less attention. Most structure-preserving schemes offering to sign many elements do so by increasing the size of the verification key linearly in the message to be signed. One could of course imagine chopping a large message into smaller pieces and signing each of them individually and then sign the resulting signatures to bind them together. However, this approach incurs a multiplicative overhead proportional to the size of the signatures we use, which due to the lower bound will be at least a factor 3. Also, such constructions would require the use of many pairing product equations in the verification of a signature.

Recently Abe et al. [AKOT15] introduced the notion of *fully* structure-preserving signatures. In a fully structure-preserving signature scheme also the secret key is required to consist of group elements only, which stands in contrast to most current structure-preserving signature schemes where the secret key consists of field elements. Fully structure-preservation is useful in several contexts, it is for instance often the case in a PKI that to get a public key certified one must demonstrate possession of a matching secret key. When the secret key consists of group elements it becomes possible to use Groth-Sahai proofs to give efficient proofs of knowledge of the secret key.

Abe et al. [AKOT15] also considered the question of signing messages that consist of many group elements. Surprisingly they showed that one can give fully structure-preserving signatures that only grow propotionately to the square root of the message size. The reason this is remarkable is that in structure-preserving signatures one cannot use collision-resistant hash-functions to reduce the message size since they are structure-destroying and furthermore it is known that size-reducing strictly structure-preserving commitments do not exist [AHO12]. They also showed a lower bound that says the combined length of the verification key and the signature size must be at least the square root of the message size, which holds regardless of whether the structure-preservation is full or not.

1.1   Our Contribution

As we said earlier it is crucial to optimize efficiency of structure-preserving signatures. In this paper we investigate the case of signing large messages and present very efficient structure-preserving signature schemes for signing many elements at once. Our signature schemes will be designed directly with large messages in mind and therefore be more efficient than constructions relying on the combination of multiple signature schemes.

We construct a structure-preserving signature scheme for messages consisting of $N = mn$ group elements. The verification key contains m elements and the signature size is $n + 2$ elements. This matches the best structure-preserving signature schemes for a single group element, in which case we would have a single group element verification key and a 3 element signature but unlike prior constructions our signature scheme scales very well for large messages. The verification process involves $n + 1$ pairing product equations, so also this matches state of the art for signing a single group element but scales well to handle larger messages.

Depending on the context, it may be desirable to use a strong signature scheme where it is not only infeasible to forge signatures on messages that have not been seen before but it is also infeasible to create a new different signatures on messages that have already been signed. In other circumstances, however, quite the opposite may be the case and it may be desirable to have signatures that can be randomized. In particular, when combining structure-preserving signatures with Groth-Sahai proofs, randomizability may be desirable since some of the signature elements can be revealed in the clear after being randomized.

Our signature scheme is very flexible in the sense that the same verification key can be used for both strong signatures and randomizable signatures at the same time. We define the notion of a combined signature scheme where the signer can choose for each message whether to make the signature strongly unforgeable or randomizable.

We also present a modified construction that is *fully* structure-preserving. In order to get full structure-preservation it is necessary for the signer to know discrete logarithms of group elements that are paired with the message since she does not know the discrete logarithms of the group elements in the message. Surprisingly this can be achieved in a simple way in our signature scheme by letting the signer pick most of the verification key herself. Due to this property we now get a fully structure-preserving signature scheme where the verification key is just a single group element and the signature consists of $m + n + 2$ group elements.

1.2 Related Work

The name "structure-preserving signature" was coined by Abe et al. [AFG+10] but there are earlier works giving structure-preserving signatures with the first being [Gro06].

Abe et al. [AGHO11] gave the first 3 element signature scheme for fully asymmetric pairings (Type III) and also proved that this is optimal. Abe et al. [AGOT14] give 2 element signatures based on partially asymmetric pairings (Type II) but Chatterjee and Menezes [CM15] showed that structure preserving signatures in the partially asymmetric setting are less efficient than

242   J. Groth

signatures based on fully asymmetric pairings. In this paper we therefore only consider the fully asymmetric setting, which gives the best efficiency and thus is the most relevant case to consider.

A line of research [HJ12, ACD+12, ADK+13, LPY15, BCPW15] has worked on basing structure-preserving signatures on standard assumptions such as the decision Diffie-Hellman or the decision linear assumptions. The fully structure-preserving signatures by Abe et al. [AKOT15] is based on the natural double pairing assumption, which is implied by the DDH assumption. However, Abe et al. [AGO11] has showed that 3 element signatures cannot be proven secure under a non-interactive assumption using black-box reductions, so strong assumptions are needed to get optimal efficiency. We will therefore base the security of our signatures on the generic group model [Nec94, Sho97] instead of aiming for security under a well-established assumption.

The signature scheme in Abe et al. [AGOT14] can be seen to be fully structure-preserving. It is a 3 group element signature scheme and is selectively randomiazable. Selective randomizability means that signatures are strong but the signer can choose to release a randomization token to make a signature randomizable. This notion is different from our notion of a combined signature scheme where the signer can choose to create randomizable or strong signatures. The advantage of selective randomizable signatures is that all signatures are verified with the same verification equation; the disadvantage is the need to issue randomization tokens when making a signature randomizable.

As discussed earlier the most directly related work is by Abe et al. [AKOT15] who introduced the notion of fully structure-preserving signatures and constructed a square root complexity scheme based on the double pairing assumption. We give a detailed performance comparison in Table 1. If we use $m \approx n \approx \sqrt{N}$ their verification key contains $11 + 6\sqrt{N}$ group elements, signatures contain $11 + 4\sqrt{N}$ group elements, and they require $5 + \sqrt{N}$ pairing product equations to verify a signature. In comparison, our fully structure-preserving signature scheme has a verification key with 1 group element, signatures consist of $2 + 2\sqrt{N}$ group elements, and we use $1 + \sqrt{N}$ pairing product equations to verify signatures.

Table 1. Comparison of structure-preserving signature schemes for messages consisting of $N = mn$ elements in $\mathbb{G}_2$. We display public parameter, verification key and signature sizes measured in group elements in $\mathbb{G}_1$ and $\mathbb{G}_2$ and number of pairing product equations required for verifying a signature. The public parameters also contain a description of the bilinear group. The public parameters can be reused for other cryptographic schemes so their cost can be amortized.

Scheme	Parameters	Verification key	Signature	PPE
[AKOT15]	$4\,\mathbb{G}_1, 4\,\mathbb{G}_2$	$1\,\mathbb{G}_1, 10 + 3m + 3n\,\mathbb{G}_2$	$7 + m + n\,\mathbb{G}_1, 4 + 2n\,\mathbb{G}_2$	$5 + n$
Our SPS	$1\,\mathbb{G}_1, n + 1\,\mathbb{G}_2$	$m\,\mathbb{G}_1$	$1\,\mathbb{G}_1, 1 + n\,\mathbb{G}_2$	$1 + n$
Our fully SPS	$1\,\mathbb{G}_1, n + m\,\mathbb{G}_2$	$1\,\mathbb{G}_1$	$m\,\mathbb{G}_1, 1 + n\,\mathbb{G}_2$	$1 + n$

2 Preliminaries

2.1 Bilinear Groups

Throughout the paper we let $\mathcal{G}$ be an asymmetric bilinear λ returns $(p, \mathbb{G}_1, \mathbb{G}_2, \mathbb{G}_T, e, G, H) \leftarrow \mathcal{G}(1^\lambda)$ with the following properties:

- $\mathbb{G}_1, \mathbb{G}_2, \mathbb{G}_T$ are groups of prime order p
- $e : \mathbb{G}_1 \times \mathbb{G}_2 \rightarrow \mathbb{G}_T$ is a bilinear map
- G generates $\mathbb{G}_1$, H generates $\mathbb{G}_2$ and $e(G, H)$ generates $\mathbb{G}_T$
- There are efficient algorithms for computing group operations, evaluating the bilinear map, comparing group elements and deciding membership of the groups

In a bilinear group we refer to deciding group membership, computing group operations in $\mathbb{G}_1, \mathbb{G}_2$ or $\mathbb{G}_T$, comparing group elements and evaluating the bilinear map as the generic group operations. In the signature schemes we construct we only use generic group operations.

Galbraith, Paterson and Smart [GPS08] distinguish between 3 types of bilinear group generators. In the Type I setting (also called the symmetric setting) $\mathbb{G}_1 = \mathbb{G}_2$, in the Type II setting there is an efficiently computable isomorphism $\psi : \mathbb{G}_2 \rightarrow \mathbb{G}_1$, and in the Type III setting no isomorphism that is efficiently computable in either direction between the source groups exists. Throughout the paper we will work in the Type III setting, which gives the most efficient operations and therefore is most important setting.

It will be useful to use the notation of Escala et al. [EHK+13] that keeps track of the discrete logarithm of group elements. They represent a group element X in $\mathbb{G}_1$ by $[x]_1$ when $X = G^x$ and a group element Y in $\mathbb{G}_2$ as $[y]_2$ when $Y = H^y$ and a group element $Z \in \mathbb{G}_T$ as $[z]_T$ when $Z = e(G, H)^z$. In this notation the source group generators G and H are $[1]_1$ and $[1]_2$.

The advantage of using this notation is that it highlights the underlying linear algebra performed on the exponents when we do group operations. Multiplying two group elements $X, Y \in \mathbb{G}_1$ to get XY for instance corresponds to $[x]_1 + [y]_1 = [x + y]_1$. Exponentiation of $X \in \mathbb{G}_1$ with $y \in \mathbb{Z}_p$ to get X^y can be written $y[x]_1 = [yx]_1$. Using the bilinear map on $X \in \mathbb{G}_1$ and $Y \in \mathbb{G}_2$ to get $e(X, Y)$ can be written as $[x]_1[y]_2 = [xy]_T$.

We can represent vectors of group elements $\boldsymbol{X} = (X_1, \ldots, X_n)$ in $\mathbb{G}_1$ as $[\boldsymbol{x}]_1$. The operations taking place in the groups have natural linear algebra equivalents, e.g., exponentiation of a vector of group elements to a matrix of exponents to get a new vector of group elements can be written $[\boldsymbol{x}]_1 A = [\boldsymbol{x}A]_1$. A pairing product $\prod_{i=1}^n e(X_i, Y_i)$ can be written $[\boldsymbol{x}]_1 \cdot [\boldsymbol{y}]_2 = [\boldsymbol{x} \cdot \boldsymbol{y}]_T$. Exponentiation of a number of group elements to the same exponent to get $(X_1^a, \ldots, X_n^a)$ can be written $[\boldsymbol{x}]_1 a = [\boldsymbol{x}a]_1$.

2.2 Signature Schemes

Our signature schemes work over an asymmetric bilinear group generated by $\mathcal{G}$. This group may be generated by the signer and included in the public verification key. In many cryptographic schemes it is convenient for the signer to work on top of a pre-existing bilinear group though. We will therefore in the description of our signatures explicitly distinguish between a setup algorithm **Setup** that produces public parameters pp and a key generation algorithm the signer uses to generate her own keys. The setup algorithm we use in our paper generates a bilinear group $(p, \mathbb{G}_1, \mathbb{G}_2, \mathbb{G}_T, e, [1]_1, [1]_2) \leftarrow \mathcal{G}(1^\lambda)$. It then extends the description of the bilinear group with additional randomly selected group elements. Our signature scheme does not rely on knowledge of the discrete logarithms of these random group elements, so the setup may be reused for many different signature schemes and other cryptographic schemes.

A signature scheme (with setup algorithm **Setup**) consists of efficient algorithms (**Setup, Gen, Sign, Vfy**).

Setup$(1^\lambda) \rightarrow pp$: The setup algorithm generates public parameters pp. They specify a message space $\mathcal{M}_{pp}$.

Gen$(pp) \rightarrow (vk, sk)$: The key generation algorithm takes public parameters pp as input and returns a public verification key vk and a secret signing key sk.

Sign$(pp, sk, m) \rightarrow \sigma$: The signing algorithm takes a signing key sk and a message $m \in \mathcal{M}_{pp}$ as input and returns a signature σ.

Vfy$(pp, vk, m, \sigma) \rightarrow 1/0$: The verification algorithm takes the verification key vk, a message m and a purported signature σ as input and returns either 1 (accept) or 0 (reject).

Definition 1 (Correctness). *The signature scheme* (**Setup, Gen, Sign, Vfy**) *is (perfectly) correct if for all security parameters* $k \in \mathbb{N}$

$$\Pr\left[\begin{array}{l} pp \leftarrow \textbf{Setup}(1^\lambda); (vk, sk) \leftarrow \textbf{Gen}(pp) \\ m \leftarrow \mathcal{M}_{pp}; \sigma \leftarrow \textbf{Sign}(pp, sk, m) \end{array} : \textbf{Vfy}(pp, vk, m, \sigma) = 1\right] = 1.$$

2.3 Structure-Preserving Signature Schemes

In this paper, we study structure-preserving signature schemes [AFG+10]. In a structure-preserving signature scheme the verification key, the messages and the signatures consist only of group elements from $\mathbb{G}_1$ and $\mathbb{G}_2$ and the verification algorithm evaluates the signature by deciding group membership of elements in the signature and by evaluating pairing product equations, which are equations of the form

$$\prod_i \prod_j e(X_i, X_j)^{a_{ij}} = 1,$$

where $X_1, X_2, \ldots \in \mathbb{G}_1$ are group elements appearing in pp, vk, m and σ and $a_{11}, a_{12}, \ldots \in \mathbb{Z}$ are constants.

Structure-preserving signatures are extremely versatile because they mix well with other pairing-based protocols. Groth-Sahai proofs [GS12] are for instance designed with pairing product equations in mind and can therefore easily be applied to structure-preserving signatures.

Definition 2 (Structure-preserving signatures). *A signature scheme is said to be structure preserving over bilinear group generator $\mathcal{G}$ if*

- *public parameters include a bilinear group $(p, \mathbb{G}_1, \mathbb{G}_2, \mathbb{G}_T, e, [1]_1, [1]_2) \leftarrow \mathcal{G}(1^\lambda)$,*
- *verification keys consist of group elements in $\mathbb{G}_1$ and $\mathbb{G}_2$,*
- *messages consist of group elements in $\mathbb{G}_1$ and $\mathbb{G}_2$,*
- *signatures consist of group elements in $\mathbb{G}_1$ and $\mathbb{G}_2$, and*
- *the verification algorithm only needs to decide membership in $\mathbb{G}_1$ and $\mathbb{G}_2$ and evaluate pairing product equations.*

Fully Structure Preserving Signatures. Abe et al. [AKOT15] argue that in several applications it is desirable that also the secret signing keys only contain source group elements. They define a structure-preserving signature scheme to be *fully* structure preserving if the signing key sk consists of group elements in $\mathbb{G}_1$ and $\mathbb{G}_2$ and the correctness of the secret signing key with respect to the public verification key can be verified using pairing product equations.

3 Randomizable and Strongly Unforgeable Signatures

A signature scheme is said to be existentially unforgeable if it is infeasible to forge a signature on a message that has not previously been signed. The standard definition of existential unforgeability allows the adversary to modify an existing signature on a message to a new signature on the same message. We say a signature scheme is randomizable if it is possible to randomize a signature on a message to get a new random signature on the same message. On the other hand, we say a signature scheme is *strongly* unforgeable when it is also infeasible to modify a signature, or more precisely it is infeasible to construct a valid message and signature pair that has not previously been seen.

Both strong signatures and randomizable signatures have many uses. We will therefore construct both strongly existentially unforgeable signatures and randomizable signatures. To capture the best of both worlds, we will define a combined signature scheme where the signer can decide whether a signature should be randomizable or strongly unforgeable. Randomizable signatures are constructed using signing algorithm $\mathbf{Sign}_0$ and verified by verification algorithm $\mathbf{Vfy}_0$. Strongly unforgeable signatures are constructed using signing algorithm $\mathbf{Sign}_1$ and verified by verification algorithm $\mathbf{Vfy}_1$.

A naïve combined signature scheme would have a verification key containing two verification keys, one for randomizable signatures and one for strong signatures. However, this solution has the disadvantage of increasing key size. Instead we will in this paper construct a combined signature scheme where the verification key is just a single group element that can be used to verify either type of signature. This dual use of the verification key means that we must carefully consider the security implications of combining two signature schemes though, so we will now define a combined signature scheme.

A combined signature scheme $(\mathbf{Setup}, \mathbf{Gen}, \mathbf{Sign}_0, \mathbf{Vfy}_0, \mathbf{Rand}, \mathbf{Sign}_1,$ $\mathbf{Vfy}_1)$ consists of 7 probabilistic polynomial time algorithms as described below.

$\mathbf{Setup}(1^\lambda, \text{size}) \to pp$: The setup algorithm takes the security parameter λ and description of the size of messages to be signed and generates public parameters. It defines a message space $\mathcal{M}_{pp}$ of messages that can be signed.

$\mathbf{Gen}(pp) \to (vk, sk)$: The key generation algorithm given public parameters generates a public verification key vk and a secret signing key sk.

$\mathbf{Sign}_0(pp, sk, m) \to \sigma$: The randomizable signature algorithm given the signing key and a message m returns a randomizable signature σ.

$\mathbf{Vfy}_0(pp, vk, m, \sigma) \to 1/0$: The randomizable signature verification algorithm given a message and a purported randomizable signature on it returns 1 if accepting the signature and 0 if rejecting the signature.

$\mathbf{Rand}(pp, vk, m, \sigma) \to \sigma'$: The randomization algorithm given a valid randomizable signature on a message returns a new randomized signature on the same message.

$\mathbf{Sign}_1(pp, sk, m) \to \sigma$: The strong signature algorithm given the signing key and a message m returns a strongly unforgeable signature σ.

$\mathbf{Vfy}_1(pp, vk, m, \sigma) \to 1/0$: The strong signature verification algorithm given a message and a purported strong signature on it returns 1 if accepting the signature and 0 if rejecting the signature.

We say a combined signature scheme has perfect correctness if the constituent randomizable and strongly unforgeable signature schemes $(\mathbf{Setup}, \mathbf{Gen}, \mathbf{Sign}_0, \mathbf{Vfy}_0)$ and $(\mathbf{Setup}, \mathbf{Gen}, \mathbf{Sign}_1, \mathbf{Vfy}_1)$ both are perfectly correct.

The combined signatures are perfectly randomizable if a randomized signature looks exactly like a fresh signature on the same message.

Definition 3 (Perfect randomizability). *The combined signature scheme is perfectly randomizable if for all $\lambda \in \mathbb{N}$ and all stateful adversaries $\mathcal{A}$*

$$\Pr\left[\begin{array}{l} pp \leftarrow \mathbf{Setup}(1^\lambda); (vk, sk) \leftarrow \mathbf{Gen}(pp) \\ m \leftarrow \mathcal{A}(pp, vk, sk); \sigma, \sigma_0 \leftarrow \mathbf{Sign}_0(pp, sk, m) : \mathcal{A}(\sigma, \sigma_b) = b \\ \sigma_1 \leftarrow \mathbf{Rand}(pp, vk, m, \sigma); b \leftarrow \{0, 1\} \end{array}\right] = \frac{1}{2},$$

where $\mathcal{A}$ outputs $m \in \mathcal{M}_{pp}$.

To capture the attacks that can occur against a combined signature scheme, we assume the adversary may arbitrarily query a signer for randomizable or strong signatures. We want the signature scheme to be combined existentially unforgeable in the sense that even seeing randomizable signatures does not help in breaking strong existential unforgeability and on the other hand seeing strong signatures does not help in producing randomizable signatures.

Definition 4 (Combined existential unforgeability under chosen message attack). *The combined signature scheme is combined existentially unforgeable under adaptive chosen message attack (C-EUF-CMA) if for all probabilistic polynomial time adversaries $\mathcal{A}$*

$$\Pr\left[\begin{array}{l} pp \leftarrow \mathbf{Setup}(1^\lambda); (vk, sk) \leftarrow \mathbf{Gen}(pp) \\ (m, \sigma) \leftarrow \mathcal{A}^{\mathbf{Sign}_0(pp, sk, \cdot), \mathbf{Sign}_1(pp, sk, \cdot)}(pp, vk) \end{array} : \begin{array}{l} \mathbf{Vfy}_0(pp, vk, m, \sigma) = 1 \wedge m \notin Q_0 \text{ or} \\ \mathbf{Vfy}_1(pp, vk, m, \sigma) = 1 \wedge (m, \sigma) \notin Q_1 \end{array}\right]$$

is negligible, where $\mathcal{A}$ outputs $m \in \mathcal{M}_{pp}$ and always queries on messages in $\mathcal{M}_{pp}$ and Q_0 is the set of messages that have been queried to $\mathbf{Sign}_0$ to get randomizable signatures and Q_1 is the set of message and signature pairs from queries to $\mathbf{Sign}_1$ to get strongly unforgeable signatures.

4   Structure-Preserving Combined Signature Scheme

Fig. 1 describes a structure-preserving combined signature scheme that can be used to sign messages consisting of $N = mn$ group elements in $\mathbb{G}_2$. It has a verification key size of m group elements, a signature size of $n+2$ group elements, and verification involves evaluating $n + 1$ pairing product equations.

In order to explain some of the design principles underlying the construction, let us first consider the special case where the message space is $\mathbb{G}_2$, i.e., we are signing a single group element and $N = m = n = 1$. The setup includes a random group element $[y]_2$, the verification key consists of a single group element $[v]_1$, and both randomizable and strongly unforgeable signatures are of the form $\sigma = ([r]_1, [s]_2, [t]_2)$.

For a randomizable signature there are two verification equations

$$[r]_1[s]_2 = [1]_1[y]_2 + [v]_1[1]_2 \qquad\qquad [r]_1[t]_2 = [1]_1[m]_2 + [v]_1[y]_2.$$

It is easy to see that we can randomize the factors in $[r]_1[s]_2$ and $[r]_1[t]_2$ into $(\frac{1}{\beta}[r]_1)(\beta[s]_2)$ and $(\frac{1}{\beta}[r]_1)(\beta[t]_2)$ without changing the products themselves, which gives us randomizability of the signatures.

Setup$(1^\lambda, m, n)$	**Vfy**$_b(pp, vk, [M]_2, \sigma)$
$gk = (p, \mathbb{G}_1, \mathbb{G}_2, \mathbb{G}_T, e, [1]_1, [1]_2) \leftarrow \mathcal{G}(1^\lambda)$	Parse $\sigma = ([r]_1, [s]_2, [\boldsymbol{t}]_2)$
$[\boldsymbol{y}]_2 \leftarrow \mathbb{G}_2^n$	Return 1 if and only if
Return $pp = (gk, m, n, [\boldsymbol{y}]_2)$	$\quad [M]_2 \in \mathbb{G}_2^{m \times n}$
	$\quad [r]_1 \in \mathbb{G}_1$
Gen(pp)	$\quad [s]_2 \in \mathbb{G}_2$
$\boldsymbol{u} \leftarrow \mathbb{Z}_p^{m-1}$, $v \leftarrow \mathbb{Z}_p$	$\quad [\boldsymbol{t}]_2 \in \mathbb{G}_2^n$
$vk = ([\boldsymbol{u}]_1, [v]_1)$	$\quad [r]_1[s]_2 = [1]_1[y_1]_2 + [v]_1[1]_2$
$sk = (\boldsymbol{u}, v)$	$\quad [r]_1[\boldsymbol{t}]_2 = [(\boldsymbol{u}, 1)]_1[M]_2 + [v]_1[\boldsymbol{y}]_2 + b[v]_1[s]_2 \boldsymbol{1}$
Return (vk, sk)	
	Rand(pp, vk, M, σ)
Sign$_b(pp, sk, [M]_2)$	Parse $\sigma = ([r]_1, [s]_2, [\boldsymbol{t}]_2)$
$z \leftarrow \mathbb{Z}_p^*$	$\beta \leftarrow \mathbb{Z}_p^*$
$r = \frac{1}{z}$	$[r']_1 = \frac{1}{\beta}[r]_1$
$[s]_2 = z([y_1]_2 + [v]_2)$	$[s']_2 = \beta[s]_2$
$[\boldsymbol{t}]_2 = z\left((\boldsymbol{u}, 1)[M]_2 + v[\boldsymbol{y}]_2 + bv[s]_2 \boldsymbol{1}\right)$	$[\boldsymbol{t}']_2 = \beta[\boldsymbol{t}]_2$
Return $\sigma = ([r]_1, [s]_2, [\boldsymbol{t}]_2)$	Return $\sigma' = ([r']_1, [s']_2, [\boldsymbol{t}']_2)$

Fig. 1. Structure-preserving combined signature scheme. The signature and verification algorithms for randomizable and strongly unforgeable signatures, respectively, are quite similar. We have there described them at the same time indicating the choice by $b = 0$ for randomizable signatures and $b = 1$ for strongly unforgeable signatures.

The first verification equation is designed to prevent the adversary from creating a forged signature from scratch after seeing the verification key only. An adversary *using only generic group operations* can do no better than computing $[r]_1 = \rho[1]_1 + \rho_v[v]_1$ and $[s]_2 = \sigma[1]_2 + \sigma_y[y]_2$ $\rho, \rho_v, \sigma, \sigma_y \in \mathbb{Z}_p$. Looking at the underlying discrete logarithms, the first verification equation then corresponds to the polynomial equation

$$(\rho + \rho_v v)(\sigma + \sigma_y y) = y + v$$

in the unknown discrete logarithms v and y. This equation is not solvable: Looking at the $\rho_v \sigma v = v$ terms we see $\sigma \neq 0$. Looking at the $\rho \sigma_y y = y$ terms we see $\rho \neq 0$. But this would leave us with a constant term $\rho \sigma \neq 0$.

Now, what if the adversary instead of creating a signature from scratch tries to modify an existing signature or combine many existing signatures? Well, due to the randomness in the choice of $z \leftarrow \mathbb{Z}_p^*$ in the signing protocol each signature query will yield a signature with a different random $[r_i]_1$. As it turns out this randomization used in each signature makes it hard for the adversary to combine multiple signatures, or even modify one signature, in a meaningful way with generic group operations. The intuition is that generic group operations allow the adversary to take linear combinations of elements it has seen, however, the verificaction equations are quadratic.

In order to prevent randomization and get strong existential unforgeability the combined signature scheme modifies the latter verification equation by adding a $[v]_1[s]_2$ term. This gives us the following verification equations for strongly unforgeable signatures

$$[r]_1[s]_2 = [1]_1[y]_2 + [v]_1[1]_2 \qquad [r]_1[t]_2 = [1]_1[m]_2 + [v]_1[y]_2 + [v]_1[s]_2.$$

Now the randomization technique fails because a randomization of $[s]_2$ means we must change $[t]_2$ in a way that counteracts this change in the second verification equation. However, $[t]_2$ is paired with $[r]_1$ that also changes when $[s]_2$ changes. The adversary is therefore faced with a non-linear modification of the signatures and gets stuck because generic group operations only enable it to do linear modifications of signature elements.

We can extend the one-element signature scheme to sign a vector $[m]_2$ with m group elements in $\mathbb{G}_2$ by extending the verification key by $m - 1$ random group elements $[u]_1 = [(u_1, \ldots, u_{m-1})]_1$. Now the verification equations become

$$[r]_1[s]_2 = [1]_1[y_1]_2 + [v]_1[1]_2 \qquad [r]_1[t]_2 = [(u, 1)]_1 \cdot [m]_2 + [v]_1[y]_2 + b[v]_1[s]_2,$$

where $b = 0$ for a randomizable signature and $b = 1$ for a strong signature. The idea is that the discrete logarithms of the elements in $[u]_1$ are unknown to the adversary making it hard to change either group element in a previously signed message to get a new message that will verify under the same signature.

Finally, to sign mn group elements in $\mathbb{G}_2$ instead of m group elements we keep the first verification equation, which does not involve the message, but add $n - 1$ extra verification equations similar to the second verification equation for a vector of group elements described above. This allows us to sign n vectors in parallel. In order to avoid linear combinations of message vectors and signature components being useful in other verification equations, we give each verification equation a separate $[v]_1[y_k]_2$ term, where $k = 1, \ldots, n$ is the number of the verification equation.

Theorem 1 *Fig. 1 gives a structure-preserving combined signature scheme that is C-EUF-CMA secure in the generic group model.*

Proof. Perfect correctness, perfect randomizability and structure-preservation follows by inspection. What remains now is to prove that the signature scheme is C-EUF-CMA secure in the generic group model. In the (Type III) generic bilinear group model the adversary may compute new group elements in either source group by taking arbitrary linear combinations of previously seen group elements in the same source group. We shall see that no such linear combination of group elements, viewed as formal Laurent polynomials in the variables picked by the key generator and the signing oracle, yields an existential forgery. It follows along the lines of the Uber assumption of Boneh, Boyen and Goh [BBG05] from the inability to produce forgeries when working with formal Laurent polynomials that the signature scheme is C-EUF-CMA secure in the generic bilinear group model.

Suppose the adversary makes q queries $[M_i]_2 \in \mathbb{G}_2^{m \times n}$ to get signatures

$$[r_i]_1 = [\frac{1}{z_i}]_1 \qquad [s_i]_2 = [z_i(y_1 + v)]_2 \qquad [t_i]_2 = [z_i((\boldsymbol{u}, 1)M_i + v\boldsymbol{y} + b_i z_i v(y_1 + v))]_2,$$

where $b_i = 0$ if query i is for a randomizable signature and $b_i = 1$ if query i is for a strong signature, and where M_i may depend on previously seen signature elements in $[s_j]_2, [t_j]_2$ for $j < i$.

Viewed as Laurent polynomials we have that a signature $([r]_1, [s]_2, [t]_2)$ generated by the adversary on $[M] \in \mathbb{G}_2^{m \times n}$ is of the form

$$r = \rho + v\rho_v + \boldsymbol{u}\boldsymbol{\rho}_u^\top + \sum_i \frac{1}{z_i}\rho_{r_i}$$

$$s = \sigma + \boldsymbol{\sigma}_y \boldsymbol{y}^\top + \sum_j \sigma_{s_j} z_j(y_1 + v) + \sum_j \sigma_{t_j} z_j((\boldsymbol{u}, 1)M_j + v\boldsymbol{y} + b_j z_j v(y_1 + v)\boldsymbol{1})$$

$$t = \boldsymbol{\tau} + \boldsymbol{y}T_y + \sum_j z_j(y_1 + v)\boldsymbol{\tau}_{s_j} + \sum_j z_j((\boldsymbol{u}, 1)M_j + v\boldsymbol{y} + b_j z_j v(y_1 + v)\boldsymbol{1})T_{t_j}$$

Similarly, all mn entries in M can be written on a form similar to s and all entries in queried matrices M_i can be written on a form similar to s where the sums are bounded by $j < i$.

For the first verification equation to be satisfied we must have $rs = y_1 + v$, i.e.,

$$\left(\begin{matrix} \rho + \boldsymbol{u}\boldsymbol{\rho}_u^\top \\ +v\rho_v + \sum_i \frac{1}{z_i}\rho_{r_i} \end{matrix} \right) \left(\begin{matrix} \sigma + \boldsymbol{\sigma}_y \boldsymbol{y}^\top + \sum_j \sigma_{s_j} z_j(y_1 + v) \\ +\sum_j \boldsymbol{\sigma}_{t_j} z_j\left((\boldsymbol{u}, 1)M_j + v\boldsymbol{y} + b_j v z_j(y_1 + v)\boldsymbol{1}\right) \end{matrix} \right)^\top = y_1 + v$$

We start by noting that $r \neq 0$ since otherwise rs cannot have the term y_1. Please observe that it is only in $\mathbb{G}_1$ that we have terms including indeterminates with negative power, i.e., $\frac{1}{z_i}$. In $\mathbb{G}_2$ all indeterminates have positive power, i.e., so s_j, t_j, M_j only contain proper multi-variate polynomials. Now suppose for a moment that $\rho_{r_i} = 0$ for all i. Then in order not to have a terms involving z_j's in rs we must have $\sum_j \sigma_{s_j} z_j (y_1 + v) + \sum_j \sigma_{t_j} z_j \left((\boldsymbol{u}, 1) M_j + v\boldsymbol{y} + b_j v z_j (y_1 + v) \mathbf{1} \right)^\top = 0$. The term y_1 now gives us $\rho \sigma_{y,1} = 1$ and the term v gives us $\rho_v \sigma = 1$. This means $\rho \neq 0$ and $\sigma \neq 0$ and therefore we reach a contradiction since the constant term should be $\rho\sigma = 0$. We conclude that there must exist some ℓ for which $\rho_{r_\ell} \neq 0$.

Now we have the term $\rho_{r_\ell} \sigma \frac{1}{z_\ell} = 0$, which shows us $\sigma = 0$. The terms $\rho_{r_\ell} \sigma_{y,k} \frac{y_k}{z_\ell} = 0$ for $k = 1, \ldots, n$ give us $\boldsymbol{\sigma}_y = \mathbf{0}$.

The polynomials corresponding to s_j and t_j contain the indeterminate z_j in all terms, so no linear combination of them can give us a term where the indeterminate component is $v y_k$ for some $k \in \{1, \ldots, n\}$. Since M_j is constructed as a linear combination of elements in the verification key and components in $\mathbb{G}_2$ from previously seen signatures, it too cannot contain a term where the indeterminate component is $v y_k$. The coefficient of $\frac{z_j}{z_\ell} v y_k$ is therefore $\rho_{r_\ell} \sigma_{t_j,k} = 0$ and therefore $\sigma_{t_j,k} = 0$ for every $j \neq \ell$ and $k \in \{1, \ldots, n\}$. This shows $\boldsymbol{\sigma}_{t_j} = \mathbf{0}$ for all $j \neq \ell$. Looking at the coefficients for $v y_k$ for $k = 1, \ldots, n$ we see that $\boldsymbol{\sigma}_{t_\ell} = \mathbf{0}$ too.

The terms $\rho_{r_\ell} \sigma_{s_j} \frac{z_j}{z_\ell} v$ give us $\sigma_{s_j} = 0$ for all $j \neq \ell$. In order to get a coefficient of 1 for the term y_1 we see that $\sigma_{s_\ell} = \frac{1}{\rho_{r_\ell}}$, which is non-zero. Our analysis has now shown that

$$ s = \frac{1}{\rho_{r_\ell}} z_\ell (y_1 + v). $$

Let us now analyze the structure of r. The term $\rho_v \sigma_\ell v^2 z_\ell = 0$ gives us $\rho_v = 0$. We know from our previous analysis that if there was a second $i \neq \ell$ for which $\rho_{r_i} \neq 0$ then also $\sigma_{\rho_\ell} = 0$, which it is not. Therefore for all $i \neq \ell$ we have $\rho_{r_i} = 0$. The term $\rho \sigma_{s_\ell} z_\ell y_1$ gives $\rho = 0$. The terms in $\sigma_{s_\ell} \boldsymbol{u} z_\ell v \boldsymbol{\rho}_u^\top$ give us $\boldsymbol{\rho}_u = \mathbf{0}$. Our analysis therefore shows

$$ r = \rho_{r_\ell} \frac{1}{z_\ell}. $$

We now turn to the second verification equation, which is $r t_1 = (\boldsymbol{u}, 1) \boldsymbol{m}^\top + v y_1 + b v s$, where $\boldsymbol{m}^\top$ is the first column vector of M. The message vector is of the form

$$ \boldsymbol{m} = \frac{\boldsymbol{\mu} + \boldsymbol{y} M_y + \sum_j \boldsymbol{\mu}_{s_j} z_j (y_1 + v)}{+ \sum_j z_j \left((\boldsymbol{u}, 1) M_j + v \boldsymbol{y} + b_j v z_j (y_1 + v) \mathbf{1} \right) M_{t_j}} $$

where $\boldsymbol{\mu}, M_y \boldsymbol{\mu}_{s_j}$ and M_{t_j} are suitably sized vectors and matrices with entries in $\mathbb{Z}_p$ chosen by the adversary. Similarly, we can write out $t_1 = \tau + \boldsymbol{\tau}_y \boldsymbol{y}^\top + \sum_j \tau_{s_j} z_j (y_1 + v) + \sum_j \boldsymbol{\tau}_{t_j} z_j \left((\boldsymbol{u}, 1) M_j + v \boldsymbol{y} + b_j v z_j (y_1 + v) \mathbf{1} \right)$ for elements and suitably sized vectors $\tau, \boldsymbol{\tau}_y, \boldsymbol{\tau}_{s_j}, \boldsymbol{\tau}_{t_j}$ with entries in $\mathbb{Z}_p$ chosen by the adversary.

Writing out the second verification equation we have

$$\rho_{r_\ell}\frac{1}{z_\ell}\left(\begin{array}{c}\tau + \boldsymbol{\tau}_y \boldsymbol{y}^\top + \sum_j \boldsymbol{\tau}_{s_j} z_j (y_1 + v)\\ + \sum_j \boldsymbol{\tau}_{t_j} z_j\left((\boldsymbol{u},1)M_j + v\boldsymbol{y} + b_j v z_j (y_1 + v)\mathbf{1}\right)\end{array}\right)$$

$$= vy_1 + bv\left(\frac{1}{\rho_{r_\ell}} z_\ell (y_1 + v)\right)$$

$$+ (\boldsymbol{u},1)\left(\begin{array}{c}\boldsymbol{\mu} + \boldsymbol{y} M_y + \sum_j \boldsymbol{\mu}_{s_j} z_j (y_1 + v)\\ + \sum_j z_j\left((\boldsymbol{u},1)M_j + v\boldsymbol{y} + b_j v z_j (y_1 + v)\mathbf{1}\right)M_{t_j}\end{array}\right)^\top.$$

Looking at the coefficients of terms involving $\frac{1}{z_\ell}$ and $\frac{y_k}{z_\ell}$ we get $\tau = 0$ and $\boldsymbol{\tau}_y = \mathbf{0}$. Looking at the terms in $\rho_{r_\ell}\boldsymbol{\tau}_{t_j}\frac{z_j}{z_\ell}v\boldsymbol{y}$ we get $\boldsymbol{\tau}_{t_j} = \mathbf{0}$ for all $j \neq \ell$. Similarly, the terms $\rho_{r_\ell}\boldsymbol{\tau}_{s_j}\frac{z_j}{z_\ell}v$ give us $\boldsymbol{\tau}_{s_j} = 0$ for all $j \neq \ell$. We are now left with

$$\rho_{r_\ell}\left(\boldsymbol{\tau}_{s_\ell}(y_1 + v) + \boldsymbol{\tau}_{t_\ell}\left((\boldsymbol{u},1)M_\ell + v\boldsymbol{y} + b_\ell v z_\ell (y_1 + v)\mathbf{1}\right)\right)$$

$$= vy_1 + bv\frac{1}{\rho_{r_\ell}}z_\ell (y_1 + v)$$

$$+ (\boldsymbol{u},1)\left(\begin{array}{c}\boldsymbol{\mu} + \boldsymbol{y} M_y + \sum_j \boldsymbol{\mu}_{s_j} z_j (y_1 + v)\\ + \sum_j z_j\left((\boldsymbol{u},1)M_j + v\boldsymbol{y} + b_j v z_j (y_1 + v)\mathbf{1}\right)M_{t_j}\end{array}\right)^\top.$$

Terms involving z_j and z_j^2 must cancel out, so we can assume $\boldsymbol{\mu}_{s_j} = \mathbf{0}$ and $M_{t_j} = 0$ for $j > \ell$. Since M_ℓ does not involve z_ℓ in any of its terms, we get from the terms in $(\boldsymbol{u},1)z_\ell v\boldsymbol{\mu}_{s_\ell}^\top$ that $\boldsymbol{\mu}_{s_\ell} = 0$. Since there can be no terms involving z_ℓ^2 we get $b_\ell \mathbf{1} M_{t_\ell}^\top = \mathbf{0}$. Looking at the coefficients for v we get $\boldsymbol{\tau}_{s_\ell} = 0$. This leaves us with

$$\rho_{r_\ell}\boldsymbol{\tau}_{t_\ell}\left((\boldsymbol{u},1)M_\ell + v\boldsymbol{y} + b_\ell v z_\ell (y_1 + v)\mathbf{1}\right)^\top$$

$$= vy_1 + bv\frac{1}{\rho_{r_\ell}}z_\ell (y_1 + v) + (\boldsymbol{u},1)z_\ell\left((\boldsymbol{u},1)M_\ell + v\boldsymbol{y}\right)M_{t_\ell})^\top$$

$$+ (\boldsymbol{u},1)\left(\begin{array}{c}\boldsymbol{\mu} + \boldsymbol{y} M_y + \sum_{j<\ell} \boldsymbol{\mu}_{s_j} z_j (y_1 + v)\\ + \sum_{j<\ell} z_j\left((\boldsymbol{u},1)M_j + v\boldsymbol{y} + b_j v z_j (y_1 + v)\mathbf{1}\right)M_{t_j}\end{array}\right)^\top.$$

Looking at the terms involving $z_\ell v^2$ we see $\rho_{r_\ell}\boldsymbol{\tau}_{t_\ell}b_\ell \mathbf{1}^\top = b\frac{1}{\rho_{r_\ell}}$. This cancels out the first two parts involving z_ℓ. The only remaining terms involving z_ℓ now give us $M_{t_\ell} = 0$. This gives us

$$\rho_{r_\ell}\boldsymbol{\tau}_{t_\ell}\left((\boldsymbol{u},1)M_\ell + v\boldsymbol{y}\right)^\top - y_1$$

$$= (\boldsymbol{u},1)\left(\begin{array}{c}\boldsymbol{\mu} + \boldsymbol{y} M_y + \sum_{j<\ell} \boldsymbol{\mu}_{s_j}^{(\ell)} z_j (y_1 + v)\\ + \sum_{j<\ell} z_j\left((\boldsymbol{u},1)M_j + v\boldsymbol{y} + b_j v z_j (y_1 + v)\mathbf{1}\right)M_{t_j}\end{array}\right)^\top.$$

Looking at the terms in $v\boldsymbol{y}$ we now get $\rho_{r_\ell}\boldsymbol{\tau}_{t_\ell} = (1,0,\ldots,0)$. Let the first column vector of M_ℓ be $\boldsymbol{m}_\ell^\top$ then we now have

$$(\boldsymbol{u},1)\boldsymbol{m}_\ell^\top = (\boldsymbol{u},1)\boldsymbol{m}^\top.$$

Writing

$$m' = \begin{aligned}&m_\ell - m = \mu' + yM'_y + \sum_{j<\ell}\mu'_{s_j}z_j(y_1 + v)\\&+\sum_{j<\ell}z_j\left((u,1)M_j + vy + b_jvz_j(y_1 + v)\mathbf{1}\right)M'_{t_j}\end{aligned}$$

we now have

$$(u,1)\left(\begin{aligned}&\mu' + yM'_y + \sum_{j<\ell}\mu'_{s_j}z_j(y_1 + v)\\&+\sum_{j<\ell}z_j\left((u,1)M_j + vy + b_jvz_j(y_1 + v)\mathbf{1}\right)M'_{t_j}\end{aligned}\right)^\top = 0.$$

The terms in $(u,1)\mu'^\top$ tell us $\mu' = 0$. Looking at terms involving u_iy_k or y_k gives us $M'_y = 0$. Terms with z_j^2 tell us $b_j\mathbf{1}M'_{t_j} = \mathbf{0}$ for all j. Terms in $(u,1)z_jv\mu'_{s_j}$ tell us $\mu'_{s_j} = 0$ for all j. Finally, terms in $(u,1)(vyM'_{t_j})$ give us $M'_{t_j} = 0$.

We have now deduced that $m' = 0$ and therefore $m_\ell = m$. This means the first column in M for which the adversary has produced a signature is a copy of the first column in the queried message M_ℓ. Using the same analysis on the last $n-1$ verification equations gives us that the other $n-1$ columns also match. This means a generic adversary can only produce valid signatures for previously queried messages, so we have EUF-CMA security.

Finally, let us consider the case where $b = 1$, i.e., we are doing a strong signature verification. We saw earlier that $\rho_{r_\ell}\tau_{t_\ell}b_\ell\mathbf{1}^\top = b_\ell = b\frac{1}{\rho_{r_\ell}}$ which can only be satisfied if $b_\ell = 1$ and $\rho_{r_\ell} = 1$. This means $s = s_\ell$ and $r = r_\ell$ and $M = M_\ell$ and therefore $t = t_\ell$. So the generic adversary can only satisfy the strong verification equation with $b = 1$ by copying both the message and signature from a previous query with $b_\ell = 1$.

On the other hand, if $b = 0$, i.e., we are verifying a randomizable signature, we see from $\rho_{r_\ell}\tau_{t_\ell}b_l\mathbf{1}^\top = b_\ell = b\frac{1}{\rho_{r_\ell}}$ that $b_\ell = 0$. So the adversary has randomized a signature intended for randomization.   □

5 Fully Structure-Preserving Combined Signature Scheme

The earlier structure-preserving signature scheme uses knowledge of the discrete logarithms of $[u]_1$ in a fundamental way since $[t]_2$ contains a $z(u,1)[M]_2$ component that could not be computed without these discrete logarithms. This situation is common for all structure-preserving signature schemes for messages that are vectors of group elements. The need to specify such discrete logarithms in the signing key therefore prevents them from being fully structure-preserving.

Abe et al. [AKOT15] get around this problem by only pairing message group elements with signature group elements where the signer knows the discrete logarithms. Inspired by their work, we will let the signer pick $[u]_1$ and include it in the signature.

To make this idea work we first make a minor modification to our signature scheme from before. We include a vector of $m-1$ group elements $[x]_2$ in the setup and we modify $[s]_2$ to have the form $[s]_2 = z([y_1]_2 + u \cdot [x]_2 + [v]_2)$. The first verification equation then becomes

$$[r]_1[s]_2 = [1]_1[y_1]_2 + [u]_1 \cdot [x]_2 + [v]_1[1]_2.$$

If this was the only modification we made it is not hard to see that the same security proof we gave earlier will work again, we are only modifying the verification equation by a random constant $[u \cdot x]_T$. The surprising thing though is that the signature scheme remains secure if we let the signer pick the $[u]_1$ part of the verification key herself and include it in the signature.

Letting the signer pick $[u]_1$ as part of the verification key means that she can know their discrete logarithms. Since she also picks $z \leftarrow \mathbb{Z}_p^*$ herself she can now use linear operations to compute the $z(u,1)[M]_2$ part of $[t]_2$. Furthermore, we have designed the scheme such that the rest can be computed with linear operations as well. To make randomizable signatures the signer just needs to know $[v]_2$ and $[vy]_2$. To make strong signatures she additionally needs to know $[vx]_2$ and $[v^2]_2$.

The resulting fully structure-preserving signature scheme is presented in Fig. 2 and can be used to sign messages consisting of $N = mn$ group elements in $\mathbb{G}_2$. It has a verification key size of 1 group elements, a signature size of $m + n + 1$ group elements, and verification involves evaluating $n + 1$ pairing product equations.

Theorem 2. *Fig. 2 gives a fully structure-preserving combined signature scheme that is C-EUF-CMA secure in the generic group model.*

Proof. Perfect correctness, perfect randomizability and structure-preservation follows by inspection. The secret key $sk = ([v]_2, [vx]_2, [vy]_2, [v^2]_2)$ consists of $m + n + 1$ group elements and we can verify that it matches the verification key

---

Setup$(1^\lambda, m, n)$
$gk = (p, \mathbb{G}_1, \mathbb{G}_2, \mathbb{G}_T, e, [1]_1, [1]_2) \leftarrow \mathcal{G}(1^\lambda)$
$[x]_2 \leftarrow \mathbb{G}_2^{m-1}$
$[y]_2 \leftarrow \mathbb{G}_2^n$
Return $pp = (gk, [x]_2, [y]_2)$

Gen(pp)
$v \leftarrow \mathbb{Z}_p$
$vk = [v]_1$
$sk = ([v]_2, [vx]_2, [vy]_2, [v^2]_2)$
Return (vk, sk)

Sign$_b(pp, sk, [M]_2)$
$u \leftarrow \mathbb{Z}_p^{m-1}$, $z \leftarrow \mathbb{Z}_p^*$, $r = \frac{1}{z}$
$[s]_2 = z([y_1]_2 + u \cdot [x]_2 + [v]_2)$
$[t]_2 = z \left(\begin{array}{c} (u,1)[M]_2 + [vy]_2 \\ +bz([vy_1]_2 + u \cdot [vx]_2 + [v^2]_2)1 \end{array} \right)$
Return $\sigma = ([u]_1, [r]_1, [s]_2, [t]_2)$

Vfy$_b(pp, vk, [M]_2, \sigma)$
Parse $\sigma = ([u]_1, [r]_1, [s]_2, [t]_2)$
Return 1 if and only if
$[M]_2 \in \mathbb{G}_2^{m \times n}$
$[r]_1 \in \mathbb{G}_1$, $[u]_1 \in \mathbb{G}_1^{m-1}$
$[s]_2 \in \mathbb{G}_2$, $[t]_2 \in \mathbb{G}_2^n$
$[r]_1[s]_2 = [1]_1[y_1]_2 + [u]_1 \cdot [x]_2 + [v]_1[1]_2$
$[r]_1[t]_2 = [(u,1)]_1[M]_2 + [v]_1[y]_2 + b[v]_1[s]_2 1$

Rand(pp, vk, M, σ)
Parse $\sigma = ([u]_1, [r]_1, [s]_2, [t]_2)$
$\alpha \leftarrow \mathbb{Z}_p^{m-1}$
$\beta \leftarrow \mathbb{Z}_p^*$
$[u']_1 = [u]_1 + \alpha[r]_1$
$[r']_1 = \frac{1}{\beta}[r]_1$
$[s']_2 = \beta([s]_2 + \alpha[x]_2)$
$[t']_2 = \beta([t]_2 + (\alpha, 0)[M]_2)$
Return $\sigma' = ([u']_1, [r']_1, [s']_2, [t']_2)$

Fig. 2. Fully structure-preserving combined signature scheme. Since they are quite similar we have described the randomizable signature and the strongly unforgeable signature algorithms jointly. Setting $b = 0$ gives the algorithms for randomizable signatures and setting $b = 1$ gives the algorithms for strongly unforgeable signatures.

$vk = [v]_1$ by checking the pairing product equations

$$[v]_1[1]_2 = [1]_1[v]_2 \quad [v]_1[\boldsymbol{x}]_2 = [1]_1[v\boldsymbol{x}]_2 \quad [v]_1[\boldsymbol{y}]_2 = [1]_1[v\boldsymbol{y}]_2 \quad [v]_1[v]_2 = [1]_1[v^2]_2,$$

so the signature scheme is fully structure preserving.

What remains now is to prove that the signature scheme is C-EUF-CMA secure in the generic group model. In the (Type III) generic bilinear group model the adversary may compute new group elements in either source group by taking arbitrary linear combinations of previously seen group elements in the same source group. We shall see that no such linear combination of group elements, viewed as formal Laurent polynomials in the variables picked by the key generator and the signing oracle, yields an existential forgery. It follows along the lines of the Uber assumption in [BBG05] this that the signature scheme is C-EUF-CMA secure in the generic bilinear group model.

Suppose the adversary makes q queries $[M_i]_2 \in \mathbb{G}_2^{m \times n}$ to get signatures

$$[\boldsymbol{u}_i]_1 \qquad [r_i]_1 = [\frac{1}{z_i}]_1 \qquad [s_i]_2 = [z_i(y_1 + \boldsymbol{u}_i \cdot \boldsymbol{x} + v)]_2$$
$$[\boldsymbol{t}_i]_2 = [z_i((\boldsymbol{u}_i, 1)M_i + v\boldsymbol{y} + b_i z_i v(y_1 + \boldsymbol{u}_i \cdot \boldsymbol{x} + v))]_2,$$

where $b_i = 0$ if query i is for a randomizable signature and $b_i = 1$ if query i is for a strong signature, and where M_i may depend on previously seen signature elements in $[s_j]_2, [\boldsymbol{t}_j]_2$ for $j < i$.

Viewed as Laurent polynomials we have that a signature $([\boldsymbol{u}]_1, [r]_1, [s]_2, [\boldsymbol{t}]_2)$ generated by the adversary on $[M] \in \mathbb{G}_2^{m \times n}$ is of the form

$$\boldsymbol{u} = \boldsymbol{\alpha} + v\boldsymbol{\alpha}_v + \sum_i \boldsymbol{u}_i A_i + \sum_i \frac{1}{z_i}\boldsymbol{\alpha}_{r_i}$$

$$r = \rho + v\rho_v + \sum_i \boldsymbol{u}_i \boldsymbol{\rho}_{\boldsymbol{u}_i}^\top + \sum_i \frac{1}{z_i}\rho_{r_i}$$

$$s = \sigma + \boldsymbol{\sigma}_x \boldsymbol{x}^\top + \boldsymbol{\sigma}_y \boldsymbol{y}^\top + \sum_j \sigma_{s_j} z_j(y_1 + \boldsymbol{u}_j \boldsymbol{x}^\top + v)$$
$$+ \sum_j \sigma_{t_j} z_j((\boldsymbol{u}_j, 1)M_j + v\boldsymbol{y} + b_j z_j v(y_1 + \boldsymbol{u}\boldsymbol{x}^\top + v)\boldsymbol{1})$$

$$t = \boldsymbol{\tau} + \boldsymbol{x}T_x + \boldsymbol{y}T_y + \sum_j z_j(y_1 + \boldsymbol{u}_j \boldsymbol{x}^\top + v)\boldsymbol{\tau}_{s_j}$$
$$+ \sum_j z_j((\boldsymbol{u}_j, 1)M_j + v\boldsymbol{y} + b_j z_j v(y_1 + \boldsymbol{u}\boldsymbol{x}^\top + v)\boldsymbol{1})T_{t_j}$$

Similarly, all mn entries in M can be written on a form similar to s and all entries in queried matrices M_i can be written on a form similar to s where the sums are bounded by $j < i$.

For the first verification equation to be satisfied we must have $rs = y_1 + \boldsymbol{ux}^\top + v$, i.e.,

$$\left(\begin{array}{c} \rho + \sum_i \boldsymbol{u}_i \boldsymbol{\rho}_{u_i}^\top \\ + v\rho_v + \sum_i \frac{1}{z_i} \rho_{r_i} \end{array} \right) \cdot \left(\begin{array}{c} \sigma + \boldsymbol{\sigma}_x \boldsymbol{x}^\top + \boldsymbol{\sigma}_y \boldsymbol{y}^\top + \sum_j \sigma_{s_j} z_j (y_1 + \boldsymbol{u}_j \boldsymbol{x}^\top + v) \\ + \sum_j \boldsymbol{\sigma}_{t_j} z_j \left((\boldsymbol{u}_j, 1) M_j + v\boldsymbol{y} + b_j v z_j (y_1 + \boldsymbol{u}_j \boldsymbol{x}^\top + v) \boldsymbol{1} \right)^\top \end{array} \right)$$

$$= y_1 + \left(\boldsymbol{\alpha} + v\boldsymbol{\alpha}_v + \sum_i \boldsymbol{u}_i A_i + \sum_i \frac{1}{z_i} \boldsymbol{\alpha}_{r_i} \right) \boldsymbol{x}^\top + v$$

We start by noting that $r \neq 0$ since otherwise rs cannot have the term y_1. Please observe that it is only in $\mathbb{G}_1$ that we have terms including indeterminates with negative power, i.e., $\frac{1}{z_i}$. In $\mathbb{G}_2$ all indeterminates have positive power, i.e., so s_j, t_j, M_j only contain proper multi-variate polynomials. Now suppose for a moment that $\rho_{r_i} = 0$ for all i. Then in order not to have a terms involving z_j's in rs we must have

$$\sum_j \sigma_{s_j} z_j (y_1 + \boldsymbol{u}_j \boldsymbol{x}^\top + v) + \sum_j \boldsymbol{\sigma}_{t_j} z_j \left((\boldsymbol{u}_j, 1) M_j + v\boldsymbol{y} + b_j v z_j (y_1 + \boldsymbol{u}_j \boldsymbol{x}^\top + v) \boldsymbol{1} \right)^\top = 0.$$

The term y_1 now gives us $\rho \sigma_{y,1} = 1$ and the term v gives us $\rho_v \sigma = 1$. This means $\rho \neq 0$ and $\sigma \neq 0$ and therefore we reach a contradiction since the constant term should be $\rho\sigma = 0$. We conclude that there must exist some ℓ for which $\rho_{r_\ell} \neq 0$.

Now we have the term $\rho_{r_\ell} \sigma \frac{1}{z_\ell} = 0$, which shows us $\sigma = 0$. The terms $\rho_{r_\ell} \sigma_{y,k} \frac{y_k}{z_\ell} = 0$ for $k = 1, \dots, n$ give us $\boldsymbol{\sigma}_y = \boldsymbol{0}$.

The polynomials corresponding to s_j and t_j contain the indeterminate z_j in all terms, so no linear combination of them can give us a term where the indeterminate component is vy_k for some $k \in \{1, \dots, n\}$. Since M_j is constructed as a linear combination of elements in the verification key and components in $\mathbb{G}_2$ from previously seen signatures, it too cannot contain a term where the indeterminate component is vy_k. The coefficient of $\frac{z_j}{z_\ell} vy_k$ is therefore $\rho_{r_\ell} \sigma_{t_j,k} = 0$ and therefore $\sigma_{t_j,k} = 0$ for every $j \neq \ell$ and $k \in \{1, \dots, n\}$. This shows $\boldsymbol{\sigma}_{t_j} = \boldsymbol{0}$ for all $j \neq \ell$. Looking at the coefficients for vy_k for $k = 1, \dots, n$ we see that $\boldsymbol{\sigma}_{t_\ell} = \boldsymbol{0}$ too.

The terms $\rho_{r_\ell} \sigma_{s_j} \frac{z_j}{z_\ell} v$ give us $\sigma_{s_j} = 0$ for all $j \neq \ell$. In order to get a coefficient of 1 for the term y_1 we see that $\sigma_{s_\ell} = \frac{1}{\rho_{r_\ell}}$, which is non-zero. Our analysis has now shown that

$$s = \boldsymbol{\sigma}_x \boldsymbol{x}^\top + \frac{1}{\rho_{r_\ell}} z_\ell (y_1 + \boldsymbol{u}_\ell \boldsymbol{x}^\top + v).$$

Let us now analyze the structure of r. The term $\rho_v \sigma_\ell v^2 z_\ell = 0$ gives us $\rho_v = 0$. We know from our previous analysis that if there was a second $i \neq \ell$ for which $\rho_{r_i} \neq 0$ then also $\sigma_{\rho_\ell} = 0$, which it is not. Therefore for all $i \neq \ell$ we have $\rho_{r_i} = 0$. The term $\rho \sigma_{s_\ell} z_\ell y_1$ gives $\rho = 0$. The terms in $\boldsymbol{\rho}_{u_i} \sigma_{s_\ell} \boldsymbol{u}_i z_\ell v$ give us $\boldsymbol{\rho}_{u_i} = \boldsymbol{0}$ for all i. Our analysis therefore shows

$$r = \rho_{r_\ell} \frac{1}{z_\ell}.$$

Finally, having simplifed r and s analysing the terms in $\boldsymbol{u}$ gives us

$$\boldsymbol{u} = \boldsymbol{u}_\ell + \rho_{r_\ell}\boldsymbol{\sigma}_x \frac{1}{z_\ell}.$$

We now turn to the second verification equation, which is $rt_1 = (\boldsymbol{u}, 1)\boldsymbol{m}^\top + vy_1 + bvs$, where $\boldsymbol{m}^\top$ is the first column vector of M. The message vector is of the form

$$\boldsymbol{m} = \begin{array}{l} \boldsymbol{\mu} + \boldsymbol{x}M_x + \boldsymbol{y}M_y + \sum_j \boldsymbol{\mu}_{s_j} z_j(y_1 + \boldsymbol{u}_j\boldsymbol{x}^\top + v) \\ + \sum_j z_j \left((\boldsymbol{u}_j, 1)M_j + v\boldsymbol{y} + b_j vz_j(y_1 + \boldsymbol{u}_j\boldsymbol{x}^\top + v)\boldsymbol{1}\right) M_{t_j} \end{array},$$

where $\boldsymbol{\mu}, M_x, M_y\boldsymbol{\mu}_{s_j}$ and M_{t_j} are suitably sized vectors and matrices with entries in $\mathbb{Z}_p$ chosen by the adversary. Similarly, we can write out $t_1 = \tau + \boldsymbol{\tau}_x\boldsymbol{x}^\top + \boldsymbol{\tau}_y\boldsymbol{y}^\top + \sum_j \tau_{s_j} z_j(y_1 + \boldsymbol{u}_j\boldsymbol{x}^\top + v) + \sum_j \boldsymbol{\tau}_{t_j} z_j \left((\boldsymbol{u}, 1)M_j + v\boldsymbol{y} + b_j vz_j(y_1 + \boldsymbol{u}_j\boldsymbol{x}^\top + v)\boldsymbol{1}\right)$ for elements and suitably sized vectors $\tau, \boldsymbol{\tau}_x, \boldsymbol{\tau}_y, \tau_{s_j}, \boldsymbol{\tau}_{t_j}$ with entries in $\mathbb{Z}_p$ chosen by the adversary.

Writing out the second verification equation we have

$$\rho_{r_\ell} \frac{1}{z_\ell} \left(\begin{array}{l} \tau + \boldsymbol{\tau}_x\boldsymbol{x}^\top + \boldsymbol{\tau}_y\boldsymbol{y}^\top + \sum_j \tau_{s_j} z_j(y_1 + \boldsymbol{u}_j\boldsymbol{x}^\top + v) \\ + \sum_j \boldsymbol{\tau}_{t_j} z_j \left((\boldsymbol{u}_j, 1)M_j + v\boldsymbol{y} + b_j vz_j(y_1 + \boldsymbol{u}_j\boldsymbol{x}^\top + v)\boldsymbol{1}\right)^\top \end{array} \right)$$

$$= vy_1 + bv \left(\boldsymbol{\sigma}_x\boldsymbol{x}^\top + \frac{1}{\rho_{r_\ell}} z_\ell(y_1 + \boldsymbol{u}_\ell\boldsymbol{x}^\top + v) \right)$$

$$+ \left(\boldsymbol{u}_\ell + \rho_{r_\ell}\boldsymbol{\sigma}_x\frac{1}{z_\ell}, 1 \right) \left(\begin{array}{l} \boldsymbol{\mu} + \boldsymbol{x}M_x + \boldsymbol{y}M_y + \sum_j \boldsymbol{\mu}_{s_j} z_j(y_1 + \boldsymbol{u}_j\boldsymbol{x}^\top + v) \\ + \sum_j z_j \left((\boldsymbol{u}_j, 1)M_j + v\boldsymbol{y} + b_j vz_j(y_1 + \boldsymbol{u}_j\boldsymbol{x}^\top + v)\boldsymbol{1}\right) M_{t_j} \end{array} \right)^\top.$$

Looking at the coefficients of terms involving $\frac{1}{z_\ell}$ we get the following equalities for all $j \neq \ell$: $\tau = \boldsymbol{\sigma}_x\boldsymbol{\mu}^\top$ $(\frac{1}{z_\ell})$, $\boldsymbol{\tau}_x = \boldsymbol{\sigma}_x M_x^\top$ $(\frac{x_k}{z_\ell})$, $\boldsymbol{\tau}_y = \boldsymbol{\sigma}_x M_y^\top$ $(\frac{y_k}{z_\ell})$, $\tau_{s_j} = \boldsymbol{\sigma}_x\boldsymbol{\mu}_{s_j}^\top$ $(\frac{vz_j}{z_\ell})$, $\boldsymbol{\tau}_{t_j} = \boldsymbol{\sigma}_x T_{t_j}^\top$ $(\frac{vy_k z_j}{z_\ell})$. Cancelling out these terms we are left with

$$\rho_{r_\ell} \left(\tau_{s_\ell}(y_1 + \boldsymbol{u}_\ell\boldsymbol{x}^\top + v) + \boldsymbol{\tau}_{t_\ell} \left((\boldsymbol{u}_\ell, 1)M_\ell + v\boldsymbol{y} + b_\ell vz_\ell(y_1 + \boldsymbol{u}_\ell\boldsymbol{x}^\top + v)\boldsymbol{1}\right)^\top \right)$$

$$= vy_1 + bv \left(\boldsymbol{\sigma}_x\boldsymbol{x}^\top + \frac{1}{\rho_{r_\ell}} z_\ell(y_1 + \boldsymbol{u}_\ell\boldsymbol{x}^\top + v) \right)$$

$$+ \rho_{r_\ell}\boldsymbol{\sigma}_x \left(\boldsymbol{\mu}_{s_\ell}(y_1 + \boldsymbol{u}_\ell\boldsymbol{x}^\top + v) + \left((\boldsymbol{u}_\ell, 1)M_\ell + v\boldsymbol{y} + b_\ell vz_\ell(y_1 + \boldsymbol{u}_\ell\boldsymbol{x}^\top + v)\boldsymbol{1}\right) M_{t_\ell} \right)^\top$$

$$+ (\boldsymbol{u}_\ell, 1) \left(\begin{array}{l} \boldsymbol{\mu} + \boldsymbol{x}M_x + \boldsymbol{y}M_y + \sum_j \boldsymbol{\mu}_{s_j} z_j(y_1 + \boldsymbol{u}_j\boldsymbol{x}^\top + v) \\ + \sum_j z_j \left((\boldsymbol{u}_j, 1)M_j + v\boldsymbol{y} + b_j vz_j(y_1 + \boldsymbol{u}_j\boldsymbol{x}^\top + v)\boldsymbol{1}\right) M_{t_j} \end{array} \right)^\top.$$

Terms involving z_j and z_j^2 must cancel out, so we can assume $\boldsymbol{\mu}_{s_j} = \boldsymbol{0}$ and $M_{t_j} = 0$ for $j > \ell$. Since M_ℓ does not involve z_ℓ in any of its terms, we get from the terms in $(\boldsymbol{u}_\ell, 1)z_\ell v\boldsymbol{\mu}_{s_\ell}^\top$ that $\boldsymbol{\mu}_{s_\ell} = \boldsymbol{0}$. Since there can be no terms involving z_ℓ^2 we get $b_\ell \boldsymbol{1} M_{t_\ell}^\top = \boldsymbol{0}$. Looking at the coefficients for v we get $\tau_{s_\ell} = \boldsymbol{\sigma}_x\boldsymbol{\mu}_{s_\ell}^\top$. This leaves us with

$$\rho_{r_\ell} \boldsymbol{\tau}_{t_\ell} \left((\boldsymbol{u}_\ell, 1) M_\ell + v\boldsymbol{y} + b_\ell v z_\ell (y_1 + \boldsymbol{u}_\ell \boldsymbol{x}^\top + v) \mathbf{1} \right)^\top$$

$$= vy_1 + bv \left(\boldsymbol{\sigma}_x \boldsymbol{x}^\top + \frac{1}{\rho_{r_\ell}} z_\ell (y_1 + \boldsymbol{u}_\ell \boldsymbol{x}^\top + v) \right)$$

$$+ \rho_{r_\ell} \boldsymbol{\sigma}_x \left(((\boldsymbol{u}_\ell, 1) M_\ell + v\boldsymbol{y}) M_{t_\ell} \right)^\top$$

$$+ (\boldsymbol{u}_\ell, 1) \left(\begin{array}{c} \boldsymbol{\mu} + \boldsymbol{x} M_x + \boldsymbol{y} M_y + \sum_{j<\ell} \boldsymbol{\mu}_{s_j} z_j (y_1 + \boldsymbol{u}_j \boldsymbol{x}^\top + v) \\ + \sum_{j<\ell} z_j \left((\boldsymbol{u}_j, 1) M_j + v\boldsymbol{y} + b_j v z_j (y_1 + \boldsymbol{u}_j \boldsymbol{x}^\top + v) \mathbf{1} \right) M_{t_j} \end{array} \right)^\top$$

$$+ (\boldsymbol{u}_\ell, 1) z_\ell \left((\boldsymbol{u}_\ell, 1) M_\ell + v\boldsymbol{y}) M_{t_\ell} \right)^\top .$$

Looking at the terms involving $z_\ell v^2$ we see $\rho_{r_\ell} \boldsymbol{\tau}_{t_\ell} b_\ell \mathbf{1}^\top = b\frac{1}{\rho_{r_\ell}}$. The only remaining terms involving z_ℓ now give us $M_{t_\ell} = 0$. This gives us

$$\rho_{r_\ell} \boldsymbol{\tau}_{t_\ell} \left((\boldsymbol{u}_\ell, 1) M_\ell + v\boldsymbol{y} \right)^\top$$

$$= vy_1 + bv\boldsymbol{\sigma}_x \boldsymbol{x}^\top$$

$$+ (\boldsymbol{u}_\ell, 1) \left(\begin{array}{c} \boldsymbol{\mu} + \boldsymbol{x} M_x + \boldsymbol{y} M_y + \sum_{j<\ell} \boldsymbol{\mu}_{s_j} z_j (y_1 + \boldsymbol{u}_j \boldsymbol{x}^\top + v) \\ + \sum_{j<\ell} z_j \left((\boldsymbol{u}_j, 1) M_j + v\boldsymbol{y} + b_j v z_j (y_1 + \boldsymbol{u}_j \boldsymbol{x}^\top + v) \mathbf{1} \right) M_{t_j} \end{array} \right)^\top$$

Looking at the terms in $v\boldsymbol{y}$ we now get $\rho_{r_\ell} \boldsymbol{\tau}_{t_\ell} = (1, 0, \ldots, 0)$. This means $(\boldsymbol{u}_\ell, 1) \boldsymbol{m}_\ell^\top = b\boldsymbol{\sigma}_x \boldsymbol{x}^\top + (\boldsymbol{u}_\ell, 1) \boldsymbol{m}^\top$, where $\boldsymbol{m}_\ell^\top$ is the first column of M_ℓ. Looking at the coefficients of vx_k we see that if $b\boldsymbol{\sigma}_x = \mathbf{0}$. Since $\boldsymbol{m}_\ell$ and $\boldsymbol{m}$ are independent of $\boldsymbol{u}_\ell$ this means $\boldsymbol{m} = \boldsymbol{m}_\ell$.

A similar argument can applied to the remaining $n-1$ verification equations showing us that in all columns M and M_ℓ match. This means $M = M_\ell$, so the signature scheme is existentially unforgeable both for randomizable signatures and strong signatures.

Finally, let us consider the case where $b = 1$, i.e., we are doing a strong signature verification. We have already seen that $b\boldsymbol{\sigma}_x = \mathbf{0}$ so when $b = 1$ this means $\boldsymbol{\sigma}_x = \mathbf{0}$. Since $\rho_{r_\ell} \boldsymbol{\tau}_{t_\ell} b_\ell \mathbf{1}^\top = b_\ell = b\frac{1}{\rho_{r_\ell}}$ we see that $b_\ell = 1$ and $\rho_{r_\ell} = 1$. This means $s = s_\ell$ and $r = r_\ell$ and $\boldsymbol{u} = \boldsymbol{u}_\ell$ and $M = M_\ell$ and therefore $\boldsymbol{t} = \boldsymbol{t}_\ell$. So the generic adversary can only satisfy the strong verification equation with $b = 1$ by copying both the message and signature from a previous query with $b_\ell = 1$.

On the other hand, if we have $b = 0$, i.e., we are verifying a randomizable signature, we see from $\rho_{r_\ell} \boldsymbol{\tau}_{t_\ell} b_l \mathbf{1}^\top = b_\ell = b\frac{1}{\rho_{r_\ell}}$ that $b_\ell = 0$. So the adversary has randomized a signature intended for randomization. □

Acknowledgment. We thank Masayuki Abe, Markulf Kohlweiss, Miyako Ohkubo and Mehdi Tibouchi for their comments and sharing an early version of [AKOT15] with us.

References

[ACD+12] Abe, M., Chase, M., David, B., Kohlweiss, M., Nishimaki, R., Ohkubo, M.: Constant-size structure-preserving signatures: generic constructions and simple assumptions. In: Wang, X., Sako, K. (eds.) ASIACRYPT 2012. LNCS, vol. 7658, pp. 4–24. Springer, Heidelberg (2012)

[ADK+13] Abe, M., David, B., Kohlweiss, M., Nishimaki, R., Ohkubo, M.: Tagged one-time signatures: tight security and optimal tag size. In: Kurosawa, K., Hanaoka, G. (eds.) PKC 2013. LNCS, vol. 7778, pp. 312–331. Springer, Heidelberg (2013)

[AFG+10] Abe, M., Fuchsbauer, G., Groth, J., Haralambiev, K., Ohkubo, M.: Structure-preserving signatures and commitments to group elements. In: Rabin, T. (ed.) CRYPTO 2010. LNCS, vol. 6223, pp. 209–236. Springer, Heidelberg (2010)

[AGHO11] Abe, M., Groth, J., Haralambiev, K., Ohkubo, M.: Optimal structure-preserving signatures in asymmetric bilinear groups. In: Rogaway, P. (ed.) CRYPTO 2011. LNCS, vol. 6841, pp. 649–666. Springer, Heidelberg (2011)

[AGO11] Abe, M., Groth, J., Ohkubo, M.: Separating short structure-preserving signatures from non-interactive assumptions. In: Lee, D.H., Wang, X. (eds.) ASIACRYPT 2011. LNCS, vol. 7073, pp. 628–646. Springer, Heidelberg (2011)

[AGOT14] Abe, M., Groth, J., Ohkubo, M., Tibouchi, M.: Unified, minimal and selectively randomizable structure-preserving signatures. In: Lindell, Y. (ed.) TCC 2014. LNCS, vol. 8349, pp. 688–712. Springer, Heidelberg (2014)

[AHO12] Abe, M., Haralambiev, K., Ohkubo, M.: Group to group commitments do not shrink. In: Pointcheval, D., Johansson, T. (eds.) EUROCRYPT 2012. LNCS, vol. 7237, pp. 301–317. Springer, Heidelberg (2012)

[AKOT15] Abe, M., Kohlweiss, M., Ohkubo, M., Tibouchi, M.: Fully structure-preserving signatures and shrinking commitments. In: Oswald, E., Fischlin, M. (eds.) EUROCRYPT 2015. LNCS, vol. 9057, pp. 35–65. Springer, Heidelberg (2015)

[ALP12] Attrapadung, N., Libert, B., Peters, T.: Computing on authenticated data: new privacy definitions and constructions. In: Wang, X., Sako, K. (eds.) ASIACRYPT 2012. LNCS, vol. 7658, pp. 367–385. Springer, Heidelberg (2012)

[ALP13] Attrapadung, N., Libert, B., Peters, T.: Efficient completely context-hiding quotable and linearly homomorphic signatures. In: Kurosawa, K., Hanaoka, G. (eds.) PKC 2013. LNCS, vol. 7778, pp. 386–404. Springer, Heidelberg (2013)

[BBG05] Boneh, D., Boyen, X., Goh, E.-J.: Hierarchical identity based encryption with constant size ciphertext. Cryptology ePrint Archive, Report 2005/015 (2005)

[BCPW15] Benhamouda, F., Couteau, G., Pointcheval, D., Wee, H.: Implicit zero-knowledge arguments and applications to the malicious setting. In: Gennaro, R., Robshaw, M. (eds.) CRYPTO 2015. LNCS, vol. 9216, pp. 107–129. Springer, Heidelberg (2015)

[CDEN12] Camenisch, J., Dubovitskaya, M., Enderlein, R.R., Neven, G.: Oblivious transfer with hidden access control from attribute-based encryption. In: Visconti, I., De Prisco, R. (eds.) SCN 2012. LNCS, vol. 7485, pp. 559–579. Springer, Heidelberg (2012)

[CKLM12] Chase, M., Kohlweiss, M., Lysyanskaya, A., Meiklejohn, S.: Malleable proof systems and applications. In: Pointcheval, D., Johansson, T. (eds.) EUROCRYPT 2012. LNCS, vol. 7237, pp. 281–300. Springer, Heidelberg (2012)

[CM15] Chatterjee, S., Menezes, A.: Type 2 structure-preserving signature schemes revisited. In: ASIACRYPT (2015)

[EHK+13] Escala, A., Herold, G., Kiltz, E., Ràfols, C., Villar, J.: An algebraic framework for Diffie-Hellman assumptions. In: Canetti, R., Garay, J.A. (eds.) CRYPTO 2013, Part II. LNCS, vol. 8043, pp. 129–147. Springer, Heidelberg (2013)

[ElG85] ElGamal, T.: A public key cryptosystem and a signature scheme based on discrete logarithms. IEEE Trans. Inf. Theor. **31**(4), 469–472 (1985)

[Fuc11] Fuchsbauer, G.: Commuting signatures and verifiable encryption. In: Paterson, K.G. (ed.) EUROCRYPT 2011. LNCS, vol. 6632, pp. 224–245. Springer, Heidelberg (2011)

[FV10] Fuchsbauer, G., Vergnaud, D.: Fair blind signatures without random oracles. In: Bernstein, D.J., Lange, T. (eds.) AFRICACRYPT 2010. LNCS, vol. 6055, pp. 16–33. Springer, Heidelberg (2010)

[GH08] Green, M., Hohenberger, S.: Universally composable adaptive oblivious transfer. In: Pieprzyk, J. (ed.) ASIACRYPT 2008. LNCS, vol. 5350, pp. 179–197. Springer, Heidelberg (2008)

[GPS08] Galbraith, S.D., Paterson, K.G., Smart, N.P.: Pairings for cryptographers. Discrete Appl. Math. **156**(16), 3113–3121 (2008)

[Gro06] Groth, J.: Simulation-sound NIZK proofs for a practical language and constant size group signatures. In: Lai, X., Chen, K. (eds.) ASIACRYPT 2006. LNCS, vol. 4284, pp. 444–459. Springer, Heidelberg (2006)

[GS12] Groth, J., Sahai, A.: Efficient noninteractive proof systems for bilinear groups. SIAM J. Comput. **41**(5), 1193–1232 (2012)

[HJ12] Hofheinz, D., Jager, T.: Tightly secure signatures and public-key encryption. In: Safavi-Naini, R., Canetti, R. (eds.) CRYPTO 2012. LNCS, vol. 7417, pp. 590–607. Springer, Heidelberg (2012)

[LPJY13] Libert, B., Peters, T., Joye, M., Yung, M.: Linearly homomorphic structure-preserving signatures and their applications. In: Canetti, R., Garay, J.A. (eds.) CRYPTO 2013, Part II. LNCS, vol. 8043, pp. 289–307. Springer, Heidelberg (2013)

[LPY12] Libert, B., Peters, T., Yung, M.: Group signatures with almost-for-free revocation. In: Safavi-Naini, R., Canetti, R. (eds.) CRYPTO 2012. LNCS, vol. 7417, pp. 571–589. Springer, Heidelberg (2012)

[LPY15] Libert, B., Peters, T., Yung, M.: Short group signatures via structure-preserving signatures: standard model security from simple assumptions. In: Gennaro, R., Robshaw, M. (eds.) CRYPTO 2015. LNCS, vol. 9216, pp. 296–316. Springer, Heidelberg (2015)

[Nec94] Nechaev, V.I.: Complexity of a determinate algorithm for the discrete logarithm. Mat. Zametki **55**(2), 91–101 (1994)

[Sho97] Shoup, V.: Lower bounds for discrete logarithms and related problems. In: Fumy, W. (ed.) EUROCRYPT 1997. LNCS, vol. 1233, pp. 256–266. Springer, Heidelberg (1997)

[ZLG12] Zhang, J., Li, Z., Guo, H.: Anonymous transferable conditional E-cash. In: Keromytis, A.D., Di Pietro, R. (eds.) SecureComm 2012. LNICST, vol. 106, pp. 45–60. Springer, Heidelberg (2013)

A Provably Secure Group Signature Scheme from Code-Based Assumptions

Martianus Frederic Ezerman, Hyung Tae Lee, San Ling,
Khoa Nguyen[✉], and Huaxiong Wang

Division of Mathematical Sciences, School of Physical and Mathematical Sciences,
Nanyang Technological University, Singapore, Singapore
{fredezerman,hyungtaelee,lingsan,khoantt,hxwang}@ntu.edu.sg

Abstract. We solve an open question in code-based cryptography by introducing the first provably secure group signature scheme from code-based assumptions. Specifically, the scheme satisfies the CPA-anonymity and traceability requirements in the random oracle model, assuming the hardness of the McEliece problem, the Learning Parity with Noise problem, and a variant of the Syndrome Decoding problem. Our construction produces smaller key and signature sizes than the existing post-quantum group signature schemes from lattices, as long as the cardinality of the underlying group does not exceed the population of the Netherlands ($\approx 2^{24}$ users). The feasibility of the scheme is supported by implementation results. Additionally, the techniques introduced in this work might be of independent interest: a new verifiable encryption protocol for the randomized McEliece encryption and a new approach to design formal security reductions from the Syndrome Decoding problem.

1 Introduction

1.1 Background and Motivation

Group signature [CvH91] is a fundamental cryptographic primitive with two intriguing features: On the one hand, it allows users of a group to anonymously sign documents on behalf of the whole group (*anonymity*); On the other hand, there is a tracing authority that can tie a given signature to the signer's identity should the need arise (*traceability*). These two properties make group signatures highly useful in various real-life scenarios such as controlled anonymous printing services, digital right management systems, e-bidding and e-voting schemes. Theoretically, designing secure and efficient group signature schemes is of deep interest since doing so typically requires a sophisticated combination of carefully chosen cryptographic ingredients. Numerous constructions of group signatures have been proposed, most of which are based on classical number-theoretic assumptions (e.g., [CS97, ACJT00, BBS04, BW06, LPY12]).

While number-theoretic-based group signatures could be very efficient (e.g., [ACJT00, BBS04]), such schemes would become insecure once the era of scalable quantum computing arrives [Sho97]. The search for post-quantum group

© International Association for Cryptologic Research 2015
T. Iwata and J.H. Cheon (Eds.): ASIACRYPT 2015, Part I, LNCS 9452, pp. 260–285, 2015.
DOI: 10.1007/978-3-662-48797-6_12

signatures, as a preparation for the future, has been quite active recently, with 6 published schemes [GKV10, CNR12, LLLS13, LLNW14, LNW15, NZZ15], all of which are based on computational assumptions from lattices. Despite their theoretical interest, those schemes involve significantly large key and signature sizes, and no implementation result has been given. Our evaluation shows that the lattice-based schemes listed above are indeed very far from being practical (see Sect. 1.2). This somewhat unsatisfactory situation highlights two interesting challenges: First, making post-quantum group signatures one step closer to practice; Second, bringing in more diversity with a scheme from another candidate for post-quantum cryptography (e.g., code-based, hash-based, multivariate-based). For instance, an easy-to-implement and competitively efficient code-based group signature scheme would be highly desirable.

A code-based group signature, in the strongest security model for static groups [BMW03], would typically require the following 3 cryptographic layers:

1. The first layer requires a secure (standard) signature scheme to sign messages[1]. We observe that the existing code-based signatures fall into two categories.

   The "hash-and-sign" category consists of the CFS signature [CFS01] and its modified versions [Dal08, Fin10, MVR12]. The known security proofs for schemes in this category, however, should be viewed with skepticism: the assumption used in [Dal08] was invalidated by distinguishing attacks [FGUO+13], while the new assumption proposed in [MVR12] lies on a rather fragile ground.

   The "Fiat-Shamir" category consists of schemes derived from Stern's identification protocol [Ste96] and its variant [Vér96, CVA10, MGS11] via the Fiat-Shamir transformation [FS86]. Although these schemes produce relatively large signatures (as the underlying protocol has to be repeated many times to make the soundness error negligibly small), their provable security (in the random oracle model) is well-understood.

2. The second layer demands a semantically secure encryption scheme to enable the tracing feature: the signer is constrained to encrypt its identifying information and to send the ciphertext as part of the group signature, so that the tracing authority can decrypt if and when necessary. This ingredient is also available in code-based cryptography, thanks to various CPA-secure and CCA-secure variants of the McEliece [McE78] and the Niederreiter [Nie86] cryptosystems (e.g., [NIKM08, DDMN12, Per12, MVVR12]).

3. The third layer, which is essentially bottleneck in realizing secure code-based group signatures, requires a zero-knowledge (ZK) protocol that connects the first two layers. Specifically, the protocol should demonstrate that a given signature is generated by a certain certified group user who honestly encrypts its identifying information. Constructing such a protocol is quite challenging. There have been ZK protocols involving the CFS and Stern's signatures,

---

[1] In most schemes in the [BMW03] model, a standard signature is also employed to issue users' secret keys. However, this is not necessarily the case: the scheme constructed in this paper is an illustrative example.

which yield identity-based identification schemes [CGG07, ACM11, YTM+14] and threshold ring signatures [MCG08, MCGL11]. There also have been ZK proofs of plaintext knowledge for the McEliece and the Niederreiter cryptosystems [HMT13]. Yet we are not aware of any efficient ZK protocol that *simultaneously* deals with both code-based signature and encryption schemes in the above sense.

Designing a provably secure group signature scheme, thus, is a long-standing open question in code-based cryptography (see, e.g., [CM10]).

1.2 Our Contributions

In this work, we construct a group signature scheme which is provably secure under code-based assumptions. Specifically, the scheme achieves the anonymity and traceability requirements ([BMW03, BBS04]) in the random oracle model, assuming the hardness of the McEliece problem, the Learning Parity with Noise problem, and a variant of the Syndrome Decoding problem.

Contributions to Code-Based Cryptography. By introducing the first provably secure code-based group signature scheme, we solve the open problem discussed earlier. Along the way, we introduce two new techniques for code-based cryptography, which might be of independent interest:

1. We design a ZK protocol for the randomized McEliece encryption scheme, that allows the prover to convince the verifier that a given ciphertext is well-formed, and that the hidden plaintext satisfies an additional condition. Such protocols, called *verifiable encryption protocols*, are useful not only in constructing group signatures, but also in much broader contexts [CS03]. It is worth noting that, prior to our work, verifiable encryption protocols for code-based cryptosystems only exist in the very basic form [HMT13] (where the plaintext is publicly given), which seem to have restricted applications.

2. In our security proof of the traceability property, to obtain a reduction from the hardness of the Syndrome Decoding (SD) problem, we come up with an approach that, as far as we know, has not been considered in the literature before. Recall that the (average-case) SD problem with parameters m, r, ω is as follows: given a *uniformly random* matrix $\widetilde{\mathbf{H}} \in \mathbb{F}_2^{r \times m}$ and a *uniformly random* syndrome $\tilde{\mathbf{y}} \in \mathbb{F}_2^r$, the problem asks to find a vector $\mathbf{s} \in \mathbb{F}_2^m$ that has Hamming weight ω (denoted by $\mathbf{s} \in \mathsf{B}(m, \omega)$) such that $\widetilde{\mathbf{H}} \cdot \mathbf{s}^\top = \tilde{\mathbf{y}}^\top$. In our scheme, the key generation algorithm produces public key containing matrix $\mathbf{H} \in \mathbb{F}_2^{r \times m}$ and syndromes $\mathbf{y}_j \in \mathbb{F}_2^r$, while users are given secret keys of the form $\mathbf{s}_j \in \mathsf{B}(m, \omega)$ such that $\mathbf{H} \cdot \mathbf{s}_j^\top = \mathbf{y}_j^\top$. In the security proof, since we would like to embed an SD challenge instance $(\widetilde{\mathbf{H}}, \tilde{\mathbf{y}})$ into the public key without being noticed with non-negligible probability by the adversary, we have to require that $\mathbf{H}$ and the $\mathbf{y}_j$'s produced by the key generation are indistinguishable from uniform.

One method to generate these keys is to employ the "hash-and-sign" technique from the CFS signature [CFS01]. Unfortunately, while the syndromes $\mathbf{y}_j$'s could be made uniformly random (as the outputs of the random oracle), the assumption that the CFS matrix $\mathbf{H}$ is *computationally* close to uniform (for practical parameters) is invalidated by distinguishing attacks [FGUO+13].

Another method, pioneered by Stern [Ste96], is to pick $\mathbf{H}$ and the $\mathbf{s}_j$'s uniformly at random. The corresponding syndromes $\mathbf{y}_j$'s could be made *computationally* close to uniform if the parameters are set such that ω is slightly smaller than the value ω_0 given by the Gilbert-Varshamov bound[2], i.e., ω_0 such that $\binom{m}{\omega_0} \approx 2^r$. However, for these parameters, it is not guaranteed with high probability that a uniformly random SD instance $(\widetilde{\mathbf{H}}, \widetilde{\mathbf{y}})$ has solutions, which would affect the success probability of the reduction algorithm.

In this work, we consider the case when ω is moderately larger than ω_0, so that two conditions hold: First, the uniform distribution over the set $\mathsf{B}(m, \omega)$ has sufficient min-entropy to apply the left-over hash lemma [GKPV10]; Second, the SD problem with parameters (m, r, ω) admits solutions with high probability, yet remains intractable[3] against the best known attacks [FS09, BJMM12]. This gives us a new method to generate uniformly random vectors $\mathbf{s}_j \in \mathsf{B}(m, \omega)$ and matrix $\mathbf{H} \in \mathbb{F}_2^{r \times m}$ so that the syndromes $\mathbf{y}_j$'s corresponding to the $\mathbf{s}_j$'s are *statistically* close to uniform. This approach, which somewhat resembles the technique used in [GPV08] for the Inhomogeneous Small Integer Solution problem, is helpful in our security proof (and generally, in designing formal security reductions from the SD problem).

Contributions to Post-Quantum Group Signatures. Our construction provides the first non-lattice-based alternative to provably secure post-quantum group signatures. The scheme features public key and signature sizes linear in the number of group users N, which is asymptotically not as efficient as the recently published lattice-based counterparts ([LLLS13, LLNW14, LNW15, NZZ15]). However, when instantiating with practical parameters, our scheme behaves much more efficiently than the scheme proposed in [NZZ15] (which is arguably the current most efficient lattice-based group signature in the asymptotic sense). Indeed, our estimation shows that our scheme gives public key and signature sizes that are 2300 times and 540 times smaller, respectively, for an average-size group of $N = 2^8$ users. As N grows, the advantage lessens, but our scheme remains more efficient even for a huge group of $N = 2^{24}$ users (which is comparable to the whole population of the Netherlands). The details of our estimation are given in Table 1.

Furthermore, we give implementation results - the first ones for post-quantum group signatures - to support the feasibility of our scheme (see Sect. 5). Our

---

[2] In this case, the function $f_{\mathbf{H}}(\mathbf{s}_j) = \mathbf{H} \cdot \mathbf{s}_j^\top$ acts as a pseudorandom generator [FS96].

[3] The variant of the SD problem considered in this work are not widely believed to be the hardest one [Ste96, Meu13], but suitable parameters can be chosen (e.g., see Sect. 5) such that the best known attacks run in exponential time.

Table 1. Efficiency comparison between our scheme and [NZZ15].

	N	Public key size		Signature size[b]	
Our scheme	2^8	5.13×10^6 bits	(642 KB)	8.57×10^6 bits	(1.07 MB)
	2^{16}	4.10×10^7 bits	(5.13 MB)	1.77×10^7 bits	(2.21 MB)
	2^{24}	9.23×10^9 bits	(1.16 GB)	2.36×10^9 bits	(294 MB)
[NZZ15][a]	$\leq 2^{24}$	1.18×10^{10} bits	(1.48 GB)	4.63×10^9 bits	(579 MB)

[a] The parameters of our scheme are set as in Sect. 5. For the [NZZ15] scheme, we choose the commonly used lattice dimension $n = 2^8$, and set parameters $m = 2^9 \times 150$ and $q = 2^{150}$ so that the requirements given in [NZZ15, Section 5.1] are satisfied. Both schemes achieve the CPA-anonymity notion [BBS04] and soundness error 2^{-80}.

[b] In our implementations presented in Sect. 5, the actual signature sizes could be reduced thanks to an additional technique.

results, while not yielding a truly practical scheme, would certainly help to bring post-quantum group signatures one step closer to practice.

1.3 Overview of Our Techniques

Let m, r, ω, n, k, t and ℓ be positive integers. We consider a group of size $N = 2^\ell$, where each user is indexed by an integer $j \in [0, N - 1]$. The secret signing key of user j is a vector $\mathbf{s}_j$ chosen uniformly at random from the set $\mathsf{B}(m, \omega)$. A uniformly random matrix $\mathbf{H} \in \mathbb{F}_2^{r \times m}$ and N syndromes $\mathbf{y}_0, \ldots, \mathbf{y}_{N-1} \in \mathbb{F}_2^r$, such that $\mathbf{H} \cdot \mathbf{s}_j^\top = \mathbf{y}_j^\top$, for all j, are made public. Let us now explain the development of the 3 ingredients used in our scheme.

The Signature Layer. User j can run Stern's ZK protocol [Ste96] to prove the possession of a vector $\mathbf{s} \in \mathsf{B}(m, \omega)$ such that $\mathbf{H} \cdot \mathbf{s}^\top = \mathbf{y}_j^\top$, where the constraint $\mathbf{s} \in \mathsf{B}(m, \omega)$ is proved in ZK by randomly permuting the entries of $\mathbf{s}$ and showing that the permuted vector belongs to $\mathsf{B}(m, \omega)$. The protocol is then transformed into a Fiat-Shamir signature [FS86]. However, such a signature is publicly verifiable only if the index j is given to the verifier.

The user can further hide its index j to achieve unconditional anonymity among all N users (which yields a *ring signature* [RST01] on the way, a la [BS13]), as follows. Let $\mathbf{A} = \left[\mathbf{y}_0^\top | \cdots | \mathbf{y}_j^\top | \cdots | \mathbf{y}_{N-1}^\top \right] \in \mathbb{F}_2^{r \times N}$ and let $\mathbf{x} = \delta_j^N$ - the N-dimensional unit vector with entry 1 at the j-th position. Observe that $\mathbf{A} \cdot \mathbf{x}^\top = \mathbf{y}_j^\top$, and thus, the equation $\mathbf{H} \cdot \mathbf{s}^\top = \mathbf{y}_j^\top$ can be written as

$$\mathbf{H} \cdot \mathbf{s}^\top \oplus \mathbf{A} \cdot \mathbf{x}^\top = \mathbf{0}, \tag{1}$$

where $\oplus$ denotes addition modulo 2. Stern's framework allows the user to prove in ZK the possession of $(\mathbf{s}, \mathbf{x})$ satisfying this equation, where the condition $\mathbf{x} = \delta_j^N$ can be justified using a random permutation.

The Encryption Layer. To enable the tracing capability of the scheme, we let user j encrypt the binary representation of j via the randomized McEliece encryption scheme [NIKM08]. Specifically, we represent j as vector $\mathsf{I2B}(j) = (j_0, \ldots, j_{\ell-1}) \in \{0,1\}^\ell$, where $\sum_{i=0}^{\ell-1} j_i 2^{\ell-1-i} = j$. Given a public encrypting key $\mathbf{G} \in \mathbb{F}_2^{k \times n}$, a ciphertext of $\mathsf{I2B}(j)$ is of the form:

$$\mathbf{c} = \big(\mathbf{u} \,\|\, \mathsf{I2B}(j) \big) \cdot \mathbf{G} \oplus \mathbf{e} \in \mathbb{F}_2^n, \tag{2}$$

where $(\mathbf{u}, \mathbf{e})$ is the encryption randomness, with $\mathbf{u} \in \mathbb{F}_2^{k-\ell}$, and $\mathbf{e} \in \mathsf{B}(n,t)$ (i.e., $\mathbf{e}$ is a vector in $\mathbb{F}_2^n$, that has weight t).

Connecting the Signature and Encryption Layers. User j must demonstrate that it does not cheat (e.g., by encrypting some string that does not point to j) without revealing j. Thus, we need a ZK protocol that allows the user to prove that the vector $\mathbf{x} = \delta_j^N$ used in (1) and the plaintext hidden in (2) both correspond to the same secret $j \in [0, N-1]$. The crucial challenge is to establish a connection (which is verifiable in ZK) between the "index representation" δ_j^N and the binary representation $\mathsf{I2B}(j)$. This challenge is well-handled by the following technique.

Instead of working with $\mathsf{I2B}(j) = (j_0, \ldots, j_{\ell-1})$, we consider an extension of $\mathsf{I2B}(j)$, defined as $\mathsf{Encode}(j) = (1 - j_0, j_0, \ldots, 1 - j_i, j_i, \ldots, 1 - j_{\ell-1}, j_{\ell-1}) \in \mathbb{F}_2^{2\ell}$. We then suitably insert ℓ zero-rows into matrix $\mathbf{G}$ to obtain matrix $\widehat{\mathbf{G}} \in \mathbb{F}_2^{(k+\ell) \times n}$ such that $\big(\mathbf{u} \,\|\, \mathsf{Encode}(j) \big) \cdot \widehat{\mathbf{G}} = \big(\mathbf{u} \,\|\, \mathsf{I2B}(j) \big) \cdot \mathbf{G}$. Let $\mathbf{f} = \mathsf{Encode}(j)$, then equation (2) can be rewritten as:

$$\mathbf{c} = \big(\mathbf{u} \,\|\, \mathbf{f} \big) \cdot \widehat{\mathbf{G}} \oplus \mathbf{e} \in \mathbb{F}_2^n. \tag{3}$$

Now, let $\mathsf{B2I} : \{0,1\}^\ell \to [0, N-1]$ be the inverse function of $\mathsf{I2B}(\cdot)$. For every $\mathbf{b} \in \{0,1\}^\ell$, we carefully design two classes of permutations $T_\mathbf{b} : \mathbb{F}_2^N \to \mathbb{F}_2^N$ and $T'_\mathbf{b} : \mathbb{F}_2^{2\ell} \to \mathbb{F}_2^{2\ell}$, such that for any $j \in [0, N-1]$, the following hold:

$$\mathbf{x} = \delta_j^N \iff T_\mathbf{b}(\mathbf{x}) = \delta_{\mathsf{B2I}(\mathsf{I2B}(j) \oplus \mathbf{b})}^N;$$
$$\mathbf{f} = \mathsf{Encode}(j) \iff T'_\mathbf{b}(\mathbf{f}) = \mathsf{Encode}(\mathsf{B2I}(\mathsf{I2B}(j) \oplus \mathbf{b})).$$

Given these equivalences, in the protocol, the user samples a uniformly random vector $\mathbf{b} \in \{0,1\}^\ell$, and sends $\mathbf{b}_1 = \mathsf{I2B}(j) \oplus \mathbf{b}$. The verifier, seeing that $T_\mathbf{b}(\mathbf{x}) = \delta_{\mathsf{B2I}(\mathbf{b}_1)}^N$ and $T'_\mathbf{b}(\mathbf{f}) = \mathsf{Encode}(\mathsf{B2I}(\mathbf{b}_1))$, should be convinced that $\mathbf{x}$ and $\mathbf{f}$ correspond to the same $j \in [0, N-1]$, yet the value of j is completely hidden from its view, because vector $\mathbf{b}$ essentially acts as a "one-time pad".

The technique extending $\mathsf{I2B}(j)$ into $\mathsf{Encode}(j)$ and then permuting $\mathsf{Encode}(j)$ in a "one-time pad" fashion is inspired by a method originally proposed by Langlois et al. [LLNW14] in a seemingly unrelated context, where the goal is to prove that the message being signed under the Bonsai tree signature [CHKP10] is of the form $\mathsf{I2B}(j)$, for some $j \in [0, N-1]$. Here, we adapt and develop their method to *simultaneously* prove two facts: the plaintext being encrypted under the randomized McEliece encryption is of the form $\mathsf{I2B}(j)$, and the unit vector $\mathbf{x} = \delta_j^N$ is used in the signature layer.

By embedding the above technique into Stern's framework, we obtain an interactive ZK argument system, in which, given the public input $(\mathbf{H}, \mathbf{A}, \mathbf{G})$, the user is able to prove the possession of a secret tuple $(j, \mathbf{s}, \mathbf{x}, \mathbf{u}, \mathbf{f}, \mathbf{e})$ satisfying (1) and (3). The protocol is repeated many times to achieve negligible soundness error, and then made non-interactive, resulting in a non-interactive ZK argument of knowledge Π. The final group signature is of the form $(\mathbf{c}, \Pi)$, where $\mathbf{c}$ is the ciphertext. In the random oracle model, the anonymity of the scheme relies on the zero-knowledge property of Π and the CPA-security of the randomized McEliece encryption scheme, while its traceability is based on the hardness of the variant of the SD problem discussed earlier.

1.4   Related Works and Open Questions

A group signature scheme based on the security of the ElGamal signature scheme and the hardness of decoding of linear codes was given in [MCK01]. In a concurrent and independent work, Alamélou et al. [ABCG15] also propose a code-based group signature scheme. These two works have yet to provide a *provably secure* group signature scheme based *solely* on code-based assumptions, which we achieve in the present paper.

Our work constitutes a foundational step in code-based group signatures. In the next steps, we will work towards improving the current construction in terms of efficiency (e.g., making the signature size less dependent on the number of group users), as well as functionality (e.g., achieving dynamic enrollment and efficient revocation of users). Another interesting open question is to construct a scheme achieving CCA-anonymity.

2   Preliminaries

NOTATIONS. We let λ denote the security parameter and $\mathsf{negl}(\lambda)$ denote a negligible function in λ. We denote by $a \xleftarrow{\$} A$ if a is chosen uniformly at random from the finite set A. The symmetric group of all permutations of k elements is denoted by S_k. We use bold capital letters, (e.g., $\mathbf{A}$), to denote matrices, and bold lowercase letters, (e.g., $\mathbf{x}$), to denote row vectors. We use $\mathbf{x}^\top$ to denote the transpose of $\mathbf{x}$ and $wt(\mathbf{x})$ to denote the (Hamming) weight of $\mathbf{x}$. We denote by $\mathsf{B}(m, \omega)$ the set of all vectors $\mathbf{x} \in \mathbb{F}_2^m$ such that $wt(\mathbf{x}) = \omega$. Throughout the paper, we define a function I2B which takes a non-negative integer a as an input, and outputs the binary representation $(a_0, \cdots, a_{\ell-1}) \in \{0,1\}^\ell$ of a such that $a = \sum_{i=0}^{\ell-1} a_i 2^{\ell-1-i}$, and a function B2I which takes as an input the binary representation $(a_0, \cdots, a_{\ell-1}) \in \{0,1\}^\ell$ of a, and outputs a. All logarithms are of base 2.

2.1   Background on Code-Based Cryptography

We first recall the Syndrome Decoding problem, which is well-known to be NP-complete [BMvT78], and is widely believed to be intractable in the average case for appropriate choice of parameters [Ste96, Meu13].

Definition 1 (The Syndrome Decoding problem). *The* $\mathsf{SD}(m, r, \omega)$ *problem is as follows: given a uniformly random matrix* $\mathbf{H} \in \mathbb{F}_2^{r \times m}$ *and a uniformly random syndrome* $\mathbf{y} \in \mathbb{F}_2^r$, *find a vector* $\mathbf{s} \in \mathsf{B}(m, \omega)$ *such that* $\mathbf{H} \cdot \mathbf{s}^\top = \mathbf{y}^\top$.

When $m = m(\lambda), r = r(\lambda), \omega = \omega(\lambda)$, *we say that the* $\mathsf{SD}(m, r, \omega)$ *problem is hard, if the success probability of any* PPT *algorithm in solving the problem is at most* $\mathsf{negl}(\lambda)$.

In our security reduction, the following variant of the left-over hash lemma for matrix multiplication over $\mathbb{F}_2$ is used.

Lemma 1 (Left-over hash lemma, adapted from [GKPV10]). *Let* D *be a distribution over* $\mathbb{F}_2^m$ *with min-entropy* e. *For* $\epsilon > 0$ *and* $r \leq e - 2\log(1/\epsilon) - \mathcal{O}(1)$, *the statistical distance between the distribution of* $(\mathbf{H}, \mathbf{H} \cdot \mathbf{s}^\top)$, *where* $\mathbf{H} \xleftarrow{\$} \mathbb{F}_2^{r \times m}$ *and* $\mathbf{s} \in \mathbb{F}_2^m$ *is drawn from distribution* D, *and the uniform distribution over* $\mathbb{F}_2^{r \times m} \times \mathbb{F}_2^r$ *is at most* ϵ.

In particular, if $\omega < m$ *is an integer such that* $r \leq \log \binom{m}{\omega} - 2\lambda - \mathcal{O}(1)$ *and* D *is the uniform distribution over* $\mathsf{B}(m, \omega)$ *(i.e.,* D *has min-entropy* $\log \binom{m}{\omega}$*), then the statistical distance between the distribution of* $(\mathbf{H}, \mathbf{H} \cdot \mathbf{s}^\top)$ *and the uniform distribution over* $\mathbb{F}_2^{r \times m} \times \mathbb{F}_2^r$ *is at most* $2^{-\lambda}$.

The Randomized McEliece Encryption Scheme. We employ a randomized variant of the McEliece [McE78] encryption scheme, suggested in [NIKM08], where a uniformly random vector is concatenated to the plaintext. The scheme is described as follows:

- $\mathsf{ME.Setup}(1^\lambda)$: Select parameters $n = n(\lambda), k = k(\lambda), t = t(\lambda)$ for a binary $[n, k, 2t + 1]$ Goppa code. Choose integers k_1, k_2 such that $k = k_1 + k_2$. Set the plaintext space as $\mathbb{F}_2^{k_2}$.
- $\mathsf{ME.KeyGen}(n, k, t)$: Perform the following steps:
  1. Produce a generator matrix $\mathbf{G}' \in \mathbb{F}_2^{k \times n}$ of a randomly selected $[n, k, 2t+1]$ Goppa code. Choose a random invertible matrix $\mathbf{S} \in \mathbb{F}_2^{k \times k}$ and a random permutation matrix $\mathbf{P} \in \mathbb{F}_2^{n \times n}$. Let $\mathbf{G} = \mathbf{S}\mathbf{G}'\mathbf{P} \in \mathbb{F}_2^{k \times n}$.
  2. Output encrypting key $\mathsf{pk_{ME}} = \mathbf{G}$ and decrypting key $\mathsf{sk_{ME}} = (\mathbf{S}, \mathbf{G}', \mathbf{P})$.
- $\mathsf{ME.Enc}(\mathsf{pk_{ME}}, \mathbf{m})$: To encrypt a message $\mathbf{m} \in \mathbb{F}_2^{k_2}$, sample $\mathbf{u} \xleftarrow{\$} \mathbb{F}_2^{k_1}$ and $\mathbf{e} \xleftarrow{\$} \mathsf{B}(n, t)$, then output the ciphertext $\mathbf{c} = (\mathbf{u} \| \mathbf{m}) \cdot \mathbf{G} \oplus \mathbf{e} \in \mathbb{F}_2^n$.
- $\mathsf{ME.Dec}(\mathsf{sk_{ME}}, \mathbf{c})$: Perform the following steps:
  1. Compute $\mathbf{c} \cdot \mathbf{P}^{-1} = ((\mathbf{u} \| \mathbf{m}) \cdot \mathbf{G} \oplus \mathbf{e}) \cdot \mathbf{P}^{-1}$ and then $\mathbf{m}' \cdot \mathbf{S} = Decode_{\mathbf{G}'}(\mathbf{c} \cdot \mathbf{P}^{-1})$ where $Decode$ is an error-correcting algorithm with respect to $\mathbf{G}'$. If $Decode$ fails, then return $\bot$.
  2. Compute $\mathbf{m}' = (\mathbf{m}'\mathbf{S}) \cdot \mathbf{S}^{-1}$, parse $\mathbf{m}' = (\mathbf{u} \| \mathbf{m})$, where $\mathbf{u} \in \mathbb{F}_2^{k_1}$ and $\mathbf{m} \in \mathbb{F}_2^{k_2}$, and return $\mathbf{m}$.

The scheme described above is CPA-secure in the standard model assuming the hardness of the $\mathsf{DMcE}(n, k, t)$ problem and the $\mathsf{DLPN}(k_1, n, \mathsf{B}(n, t))$ problem [NIKM08, Döt14]. We now recall these two problems.

Definition 2 (The Decisional McEliece problem). *The* $\mathsf{DMcE}(n,k,t)$ *problem is as follows: given a matrix* $\mathbf{G} \in \mathbb{F}_2^{k \times n}$, *distinguish whether* $\mathbf{G}$ *is a uniformly random matrix over* $\mathbb{F}_2^{k \times n}$ *or it is generated by algorithm* $\mathsf{ME.KeyGen}(n,k,t)$ *described above.*

When $n = n(\lambda), k = k(\lambda), t = t(\lambda)$, *we say that the* $\mathsf{DMcE}(n,k,t)$ *problem is hard, if the success probability of any* PPT *distinguisher is at most* $1/2 + \mathsf{negl}(\lambda)$.

Definition 3 (The Decisional Learning Parity with (fixed-weight) Noise problem). *The* $\mathsf{DLPN}(k,n,\mathsf{B}(n,t))$ *problem is as follows: given a pair* $(\mathbf{A},\mathbf{v}) \in \mathbb{F}_2^{k \times n} \times \mathbb{F}_2^n$, *distinguish whether* $(\mathbf{A},\mathbf{v})$ *is a uniformly random pair over* $\mathbb{F}_2^{k \times n} \times \mathbb{F}_2^n$ *or it is obtained by choosing* $\mathbf{A} \xleftarrow{\$} \mathbb{F}_2^{k \times n}$, $\mathbf{u} \xleftarrow{\$} \mathbb{F}_2^k$, $\mathbf{e} \xleftarrow{\$} \mathsf{B}(n,t)$ *and outputting* $(\mathbf{A}, \mathbf{u} \cdot \mathbf{A} \oplus \mathbf{e})$.

When $k = k(\lambda), n = n(\lambda), t = t(\lambda)$, *we say that the* $\mathsf{DLPN}(k,n,\mathsf{B}(n,t))$ *problem is hard, if the success probability of any* PPT *distinguisher is at most* $1/2 + \mathsf{negl}(\lambda)$.

2.2 Group Signatures

We follow the definition of group signatures provided in [BMW03] for the case of static groups.

Definition 4. A group signature $\mathcal{GS} = (\mathsf{KeyGen}, \mathsf{Sign}, \mathsf{Verify}, \mathsf{Open})$ is a tuple of four polynomial-time algorithms:

- KeyGen: This randomized algorithm takes as input $(1^\lambda, 1^N)$, where $N \in \mathbb{N}$ is the number of group users, and outputs $(\mathsf{gpk}, \mathsf{gmsk}, \mathsf{gsk})$, where gpk is the group public key, gmsk is the group manager's secret key, and $\mathsf{gsk} = \{\mathsf{gsk}[j]\}_{j \in [0,N-1]}$ with $\mathsf{gsk}[j]$ being the secret key for the group user of index j.
- Sign: This randomized algorithm takes as input a secret signing key $\mathsf{gsk}[j]$ for some $j \in [0, N-1]$ and a message M and returns a group signature Σ on M.
- Verify: This deterministic algorithm takes as input the group public key gpk, a message M, a signature Σ on M, and returns either 1 (Accept) or 0 (Reject).
- Open: This deterministic algorithm takes as input the group manager's secret key gmsk, a message M, a signature Σ on M, and returns an index $j \in [0, N-1]$ associated with a particular user, or $\bot$, indicating failure.

Correctness. The correctness of a group signature scheme requires that for all $\lambda, N \in \mathbb{N}$, all $(\mathsf{gpk}, \mathsf{gmsk}, \mathsf{gsk})$ produced by $\mathsf{KeyGen}(1^\lambda, 1^N)$, all $j \in [0, N-1]$, and all messages $M \in \{0,1\}^*$,

$$\mathsf{Verify}\big(\mathsf{gpk}, M, \mathsf{Sign}(\mathsf{gsk}[j], M)\big) = 1; \quad \mathsf{Open}\big(\mathsf{gmsk}, M, \mathsf{Sign}(\mathsf{gsk}[j], M)\big) = j.$$

Security Notions. A secure group signature scheme must satisfy two security notions:

- *Traceability* requires that all signatures, even those produced by a coalition of group users and the group manager, can be traced back to a member of the coalition.

– *Anonymity* requires that, signatures generated by two distinct group users are computationally indistinguishable to an adversary who knows all the user secret keys. In Bellare et al.'s model [BMW03], the anonymity adversary is granted access to an opening oracle (CCA-anonymity). Boneh et al. [BBS04] later proposed a relaxed notion, where the adversary cannot query the opening oracle (CPA-anonymity).

Formal definitions of CPA-anonymity and traceability are as follows.

Definition 5. *We say that a group signature* $\mathcal{GS} = $ (KeyGen, Sign, Verify, Open) *is* CPA-*anonymous if for all polynomial* $N(\cdot)$ *and any PPT adversaries* $\mathcal{A}$, *the advantage of* $\mathcal{A}$ *in the following experiment is negligible in* λ:

1. *Run* (gpk, gmsk, gsk) $\leftarrow$ KeyGen$(1^\lambda, 1^N)$ *and send* (gpk, gsk) *to* $\mathcal{A}$.
2. $\mathcal{A}$ *outputs two identities* $j_0, j_1 \in [0, N-1]$ *with a message* M. *Choose a random bit* b *and give* Sign(gsk$[j_b], M$) *to* $\mathcal{A}$. *Then,* $\mathcal{A}$ *outputs a bit* b'.

$\mathcal{A}$ *succeeds if* $b' = b$, *and the advantage of* $\mathcal{A}$ *is defined to* $\left| \Pr[\mathcal{A} \text{ succeeds}] - \dfrac{1}{2} \right|$.

Definition 6. *We say that a group signature* $\mathcal{GS} = $ (KeyGen, Sign, Verify, Open) *is traceable if for all polynomial* $N(\cdot)$ *and any PPT adversaries* $\mathcal{A}$, *the success probability of* $\mathcal{A}$ *in the following experiment is negligible in* λ:

1. *Run* (gpk, gmsk, gsk) $\leftarrow$ KeyGen$(1^\lambda, 1^N)$ *and send* (gpk, gmsk) *to* $\mathcal{A}$.
2. $\mathcal{A}$ *may query the following oracles adaptively and in any order:*
   – *A* $\mathcal{O}^{\text{Corrupt}}$ *oracle that on input* $j \in [0, N-1]$, *outputs* gsk$[j]$.
   – *A* $\mathcal{O}^{\text{Sign}}$ *oracle that on input* j, *a message* M, *returns* Sign(gsk$[j], M$).
   Let CU *be the set of identities queried to* $\mathcal{O}^{\text{Corrupt}}$.
3. *Finally,* $\mathcal{A}$ *outputs a message* M^* *and a signature* Σ^*.

$\mathcal{A}$ *succeeds if (1)* Verify(gpk, M^*, Σ^*) $= 1$ *and (2)* Sign(gsk$[j], M^*$) *was never queried for* $j \notin CU$, *yet (3)* Open(gmsk, M^*, Σ^*) $\notin CU$.

3   The Underlying Zero-Knowledge Argument System

Recall that a statistical zero-knowledge argument system is an interactive protocol where the soundness property holds for *computationally bounded* cheating provers, while the zero-knowledge property holds against *any* cheating verifier. In this section we present a statistical zero-knowledge argument system which will serve as a building block in our group signature scheme in Sect. 4.

Before describing the protocol, we first introduce several supporting notations and techniques. Let ℓ be a positive integer, and let $N = 2^\ell$.

1. For $\mathbf{x} = (x_0, x_1, \ldots, x_{N-1}) \in \mathbb{F}_2^N$ and for $j \in [0, N-1]$, we denote by $\mathbf{x} = \delta_j^N$ if $x_j = 1$ and $x_i = 0$ for all $i \neq j$.

2. We define an encoding function $\mathsf{Encode} : [0, N-1] \to \mathbb{F}_2^{2\ell}$, that encodes integer $j \in [0, N-1]$, whose binary representation is $\mathsf{I2B}(j) = (j_0, \ldots, j_{\ell-1})$, as vector:

$$\mathsf{Encode}(j) = (1 - j_0, j_0, \ldots, 1 - j_i, j_i, \ldots, 1 - j_{\ell-1}, j_{\ell-1}).$$

3. Given a vector $\mathbf{b} = (b_0, \ldots, b_{\ell-1}) \in \{0, 1\}^\ell$, we define the following 2 permutations:
   (a) $T_{\mathbf{b}} : \mathbb{F}_2^N \to \mathbb{F}_2^N$ that transforms $\mathbf{x} = (x_0, \ldots, x_{N-1})$ to $(x'_0, \ldots, x'_{N-1})$, where for each $i \in [0, N-1]$, we have $x_i = x'_{i^*}$, where $i^* = \mathsf{B2I}(\mathsf{I2B}(i) \oplus \mathbf{b})$.
   (b) $T'_{\mathbf{b}} : \mathbb{F}_2^{2\ell} \to \mathbb{F}_2^{2\ell}$ that transforms $\mathbf{f} = (f_0, f_1, \ldots, f_{2i}, f_{2i+1}, \ldots, f_{2(\ell-1)}, f_{2(\ell-1)+1})$ to $(f_{b_0}, f_{1-b_0}, \ldots, f_{2i+b_i}, f_{2i+(1-b_i)}, \ldots, f_{2(\ell-1)+b_{\ell-1}}, f_{2(\ell-1)+(1-b_{\ell-1})})$.

Observe that, for any $j \in [0, N-1]$ and any $\mathbf{b} \in \{0, 1\}^\ell$, we have:

$$\mathbf{x} = \delta_j^N \iff T_{\mathbf{b}}(\mathbf{x}) = \delta_{\mathsf{B2I}(\mathsf{I2B}(j) \oplus \mathbf{b})}^N; \tag{4}$$

$$\mathbf{f} = \mathsf{Encode}(j) \iff T'_{\mathbf{b}}(\mathbf{f}) = \mathsf{Encode}(\mathsf{B2I}(\mathsf{I2B}(j) \oplus \mathbf{b})). \tag{5}$$

Example: Let $N = 2^4$. Let $j = 6$, then $\mathsf{I2B}(j) = (0, 1, 1, 0)$ and $\mathsf{Encode}(j) = (1, 0, 0, 1, 0, 1, 1, 0)$. If $\mathbf{b} = (1, 0, 1, 0)$, then $\mathsf{B2I}(\mathsf{I2B}(j) \oplus \mathbf{b}) = \mathsf{B2I}(1, 1, 0, 0) = 12$, and we have:

$$T_{\mathbf{b}}(\delta_6^{16}) = \delta_{12}^{16} \text{ and } T'_{\mathbf{b}}(\mathsf{Encode}(6)) = (0, 1, 0, 1, 1, 0, 1, 0) = \mathsf{Encode}(12).$$

3.1 The Interactive Protocol

We now present our interactive zero-knowledge argument of knowledge (ZKAoK). Let $n, k, t, m, r, \omega, \ell$ be positive integers, and $N = 2^\ell$. The public input consists of matrices $\mathbf{G} \in \mathbb{F}_2^{k \times n}$, $\mathbf{H} \in \mathbb{F}_2^{r \times m}$; N syndromes $\mathbf{y}_0, \ldots, \mathbf{y}_{N-1} \in \mathbb{F}_2^r$; and a vector $\mathbf{c} \in \mathbb{F}_2^n$. The protocol allows prover $\mathcal{P}$ to *simultaneously* convince verifier $\mathcal{V}$ in zero-knowledge that $\mathcal{P}$ possesses a vector $\mathbf{s} \in \mathsf{B}(m, \omega)$ corresponding to certain syndrome $\mathbf{y}_j \in \{\mathbf{y}_0, \ldots, \mathbf{y}_{N-1}\}$ with hidden index j, *and* that $\mathbf{c}$ is a correct encryption of $\mathsf{I2B}(j)$ via the randomized McEliece encryption. Specifically, the secret witness of $\mathcal{P}$ is a tuple $(j, \mathbf{s}, \mathbf{u}, \mathbf{e}) \in [0, N-1] \times \mathbb{F}_2^m \times \mathbb{F}_2^{k-\ell} \times \mathbb{F}_2^n$ satisfying:

$$\begin{cases} \mathbf{H} \cdot \mathbf{s}^\top = \mathbf{y}_j^\top \ \wedge \ \mathbf{s} \in \mathsf{B}(m, \omega); \\ (\mathbf{u} \,\|\, \mathsf{I2B}(j)) \cdot \mathbf{G} \oplus \mathbf{e} = \mathbf{c} \ \wedge \ \mathbf{e} \in \mathsf{B}(n, t). \end{cases} \tag{6}$$

Let $\mathbf{A} = [\mathbf{y}_0^\top | \cdots | \mathbf{y}_j^\top | \cdots | \mathbf{y}_{N-1}^\top] \in \mathbb{F}_2^{r \times N}$ and $\mathbf{x} = \delta_j^N$. We have $\mathbf{A} \cdot \mathbf{x}^\top = \mathbf{y}_j^\top$, and thus, the equation $\mathbf{H} \cdot \mathbf{s}^\top = \mathbf{y}_j^\top$ can be written as $\mathbf{H} \cdot \mathbf{s}^\top \oplus \mathbf{A} \cdot \mathbf{x}^\top = \mathbf{0}$.

Let $\widehat{\mathbf{G}} \in \mathbb{F}_2^{(k+\ell) \times n}$ be the matrix obtained from $\mathbf{G} \in \mathbb{F}_2^{k \times n}$ by replacing its last ℓ rows $\mathbf{g}_{k-\ell+1}, \mathbf{g}_{k-\ell+2}, \ldots, \mathbf{g}_k$ by 2ℓ rows $\mathbf{0}^n, \mathbf{g}_{k-\ell+1}, \mathbf{0}^n, \mathbf{g}_{k-\ell+2}, \ldots, \mathbf{0}^n, \mathbf{g}_k$. We then observe that $(\mathbf{u} \,\|\, \mathsf{I2B}(j)) \cdot \mathbf{G} = (\mathbf{u} \,\|\, \mathsf{Encode}(j)) \cdot \widehat{\mathbf{G}}$.

Let $\mathbf{f} = \mathsf{Encode}(j)$, then (6) can be equivalently rewritten as:

$$\begin{cases} \mathbf{H} \cdot \mathbf{s}^\top \oplus \mathbf{A} \cdot \mathbf{x}^\top = \mathbf{0} \ \wedge \ \mathbf{x} = \delta_j^N \ \wedge \ \mathbf{s} \in \mathsf{B}(m,\omega); \\ (\mathbf{u} \| \mathbf{f}) \cdot \widehat{\mathbf{G}} \oplus \mathbf{e} = \mathbf{c} \ \wedge \ \mathbf{f} = \mathsf{Encode}(j) \ \wedge \ \mathbf{e} \in \mathsf{B}(n,t). \end{cases} \quad (7)$$

To obtain a ZKAoK for relation (7) in Stern's framework [Ste96], $\mathcal{P}$ proceeds as follows:

- To prove that $\mathbf{x} = \delta_j^N$ and $\mathbf{f} = \mathsf{Encode}(j)$ while keeping j secret, prover $\mathcal{P}$ samples a uniformly random vector $\mathbf{b} \in \{0,1\}^\ell$, sends $\mathbf{b}_1 = \mathsf{I2B}(j) \oplus \mathbf{b}$, and shows that:

$$T_\mathbf{b}(\mathbf{x}) = \delta_{\mathsf{B2I}(\mathbf{b}_1)}^N \ \wedge \ T_\mathbf{b}'(\mathbf{f}) = \mathsf{Encode}(\mathsf{B2I}(\mathbf{b}_1)).$$

By the equivalences observed in (4) and (5), the verifier will be convinced about the facts to prove. Furthermore, since $\mathbf{b}$ essentially acts as a "one-time pad", the secret j is perfectly hidden.
- To prove in zero-knowledge that $\mathbf{s} \in \mathsf{B}(m,\omega)$, $\mathcal{P}$ samples a uniformly random permutation $\pi \in \mathsf{S}_m$, and shows that $\pi(\mathbf{s}) \in \mathsf{B}(m,\omega)$. Similarly, to prove in zero-knowledge that $\mathbf{e} \in \mathsf{B}(n,t)$, a uniformly random permutation $\sigma \in \mathsf{S}_n$ is employed.
- Finally, to prove the linear equations in zero-knowledge, $\mathcal{P}$ samples uniformly random "masking" vectors $(\mathbf{r_s}, \mathbf{r_x}, \mathbf{r_u}, \mathbf{r_f}, \mathbf{r_e})$, and shows that:

$$\begin{cases} \mathbf{H} \cdot (\mathbf{s} \oplus \mathbf{r_s})^\top \oplus \mathbf{A} \cdot (\mathbf{x} \oplus \mathbf{r_x})^\top = \mathbf{H} \cdot \mathbf{r_s}^\top \oplus \mathbf{A} \cdot \mathbf{r_x}^\top; \\ (\mathbf{u} \oplus \mathbf{r_u} \| \mathbf{f} \oplus \mathbf{r_f}) \cdot \widehat{\mathbf{G}} \oplus (\mathbf{e} \oplus \mathbf{r_e}) \oplus \mathbf{c} = (\mathbf{r_u} \| \mathbf{r_f}) \cdot \widehat{\mathbf{G}} \oplus \mathbf{r_e}. \end{cases} \quad (8)$$

Now let $\mathsf{COM} : \{0,1\}^* \rightarrow \{0,1\}^\lambda$ be a collision-resistant hash function, to be modelled as a random oracle. Prover $\mathcal{P}$ and verifier $\mathcal{V}$ first perform the preparation steps described above, and then interact as described in Fig. 1.

3.2 Analysis of the Protocol

The properties of our protocol are summarized in the following theorem.

Theorem 1. *The interactive protocol described in Sect. 3.1 has perfect completeness, and has communication cost bounded by $\beta = (N + 3\log N) + m(\log m + 1) + n(\log n + 1) + k + 5\lambda$ bits. If COM is modelled as a random oracle, then the protocol is statistical zero-knowledge. If COM is a collision-resistant hash function, then the protocol is an argument of knowledge.*

Completeness. It can be seen that the given interactive protocol is perfectly complete, i.e., if $\mathcal{P}$ possesses a valid witness $(j, \mathbf{s}, \mathbf{u}, \mathbf{e})$ and follows the protocol, then $\mathcal{V}$ always outputs 1. Indeed, given $(j, \mathbf{s}, \mathbf{u}, \mathbf{e})$ satisfying (6), $\mathcal{P}$ can always obtain $(j, \mathbf{s}, \mathbf{x}, \mathbf{u}, \mathbf{f}, \mathbf{e})$ satisfying (7). Then, as discussed above, the following are true:

$$\begin{cases} \forall \pi \in \mathsf{S}_m : \pi(\mathbf{s}) \in \mathsf{B}(m,\omega); \ \forall \sigma \in \mathsf{S}_n : \sigma(\mathbf{e}) \in \mathsf{B}(n,t); \\ \forall \mathbf{b} \in \{0,1\}^\ell : T_\mathbf{b}(\mathbf{x}) = \delta_{\mathsf{B2I}(\mathsf{I2B}(j) \oplus \mathbf{b})}^N = \mathbf{w_x}; \ T_\mathbf{b}'(\mathbf{f}) = \mathsf{Encode}(\mathsf{B2I}(\mathsf{I2B}(j) \oplus \mathbf{b})) = \mathbf{w_f}. \end{cases}$$

1. **Commitment:** $\mathcal{P}$ samples the following uniformly random objects:
$$\begin{cases} \mathbf{b} \xleftarrow{\$} \{0,1\}^{\ell}; \ \pi \xleftarrow{\$} S_m; \ \sigma \xleftarrow{\$} S_n; \ \rho_1, \rho_2, \rho_3 \xleftarrow{\$} \{0,1\}^{\lambda}; \\ \mathbf{r_s} \xleftarrow{\$} \mathbb{F}_2^m; \ \mathbf{r_x} \xleftarrow{\$} \mathbb{F}_2^N; \ \mathbf{r_u} \xleftarrow{\$} \mathbb{F}_2^{k-\ell}; \ \mathbf{r_f} \xleftarrow{\$} \mathbb{F}_2^{2\ell}; \ \mathbf{r_e} \xleftarrow{\$} \mathbb{F}_2^n. \end{cases}$$

It then sends the commitment $\mathsf{CMT} := (c_1, c_2, c_3)$ to $\mathcal{V}$, where
$$\begin{cases} c_1 = \mathrm{COM}\big(\mathbf{b}, \pi, \sigma, \ \mathbf{H} \cdot \mathbf{r_s}^{\top} \oplus \mathbf{A} \cdot \mathbf{r_x}^{\top}, \ (\mathbf{r_u} \| \mathbf{r_f}) \cdot \widehat{\mathbf{G}} \oplus \mathbf{r_e}; \ \rho_1\big), \\ c_2 = \mathrm{COM}\big(\pi(\mathbf{r_s}), T_{\mathbf{b}}(\mathbf{r_x}), T_{\mathbf{b}}'(\mathbf{r_f}), \sigma(\mathbf{r_e}); \ \rho_2\big), \\ c_3 = \mathrm{COM}\big(\pi(\mathbf{s} \oplus \mathbf{r_s}), T_{\mathbf{b}}(\mathbf{x} \oplus \mathbf{r_x}), T_{\mathbf{b}}'(\mathbf{f} \oplus \mathbf{r_f}), \sigma(\mathbf{e} \oplus \mathbf{r_e}); \ \rho_3\big). \end{cases}$$

2. **Challenge:** Receiving CMT, $\mathcal{V}$ sends a challenge $\mathrm{Ch} \xleftarrow{\$} \{1,2,3\}$ to $\mathcal{P}$.
3. **Response:** $\mathcal{P}$ responds as follows:
   - If $\mathrm{Ch} = 1$: Reveal c_2 and c_3. Let $\mathbf{b}_1 = \mathsf{I2B}(j) \oplus \mathbf{b}$,
   $$\begin{cases} \mathbf{v_s} = \pi(\mathbf{r_s}), \\ \mathbf{w_s} = \pi(\mathbf{s}), \end{cases} \mathbf{v_x} = T_{\mathbf{b}}(\mathbf{r_x}), \ \mathbf{v_f} = T_{\mathbf{b}}'(\mathbf{r_f}), \text{ and } \begin{cases} \mathbf{v_e} = \sigma(\mathbf{r_e}), \\ \mathbf{w_e} = \sigma(\mathbf{e}). \end{cases}$$

   Send $\mathsf{RSP} := \big(\mathbf{b}_1, \mathbf{v_s}, \mathbf{w_s}, \mathbf{v_x}, \mathbf{v_f}, \mathbf{v_e}, \mathbf{w_e}; \rho_2, \rho_3\big)$ to $\mathcal{V}$.

   - If $\mathrm{Ch} = 2$: Reveal c_1 and c_3. Let
   $$\begin{cases} \mathbf{b}_2 = \mathbf{b}; \ \pi_2 = \pi; \ \sigma_2 = \sigma; \\ \mathbf{z_s} = \mathbf{s} \oplus \mathbf{r_s}; \ \mathbf{z_x} = \mathbf{x} \oplus \mathbf{r_x}; \ \mathbf{z_u} = \mathbf{u} \oplus \mathbf{r_u}; \ \mathbf{z_f} = \mathbf{f} \oplus \mathbf{r_f}; \ \mathbf{z_e} = \mathbf{e} \oplus \mathbf{r_e}. \end{cases}$$

   Send $\mathsf{RSP} := \big(\mathbf{b}_2, \pi_2, \sigma_2, \mathbf{z_s}, \mathbf{z_x}, \mathbf{z_u}, \mathbf{z_f}, \mathbf{z_e}; \rho_1, \rho_3\big)$ to $\mathcal{V}$.

   - If $\mathrm{Ch} = 3$: Reveal c_1 and c_2. Let
   $$\mathbf{b}_3 = \mathbf{b}; \ \pi_3 = \pi; \ \sigma_3 = \sigma; \ \mathbf{y_s} = \mathbf{r_s}; \ \mathbf{y_x} = \mathbf{r_x}; \ \mathbf{y_u} = \mathbf{r_u}; \ \mathbf{y_f} = \mathbf{r_f}; \ \mathbf{y_e} = \mathbf{r_e}.$$

   Send $\mathsf{RSP} := \big(\mathbf{b}_3, \pi_3, \sigma_3, \mathbf{y_s}, \mathbf{y_x}, \mathbf{y_u}, \mathbf{y_f}, \mathbf{y_e}; \rho_1, \rho_2\big)$ to $\mathcal{V}$.

Verification: Receiving RSP, $\mathcal{V}$ proceeds as follows:

- If $\mathrm{Ch} = 1$: Let $\mathbf{w_x} = \delta_{\mathsf{B2I}(\mathbf{b}_1)}^N \in \mathbb{F}_2^N$ and $\mathbf{w_f} = \mathsf{Encode}(\mathsf{B2I}(\mathbf{b}_1)) \in \mathbb{F}_2^{2\ell}$. Check that $\mathbf{w_s} \in \mathsf{B}(m, \omega)$, $\mathbf{w_e} \in \mathsf{B}(n, t)$, and that:
$$\begin{cases} c_2 = \mathrm{COM}\big(\mathbf{v_s}, \mathbf{v_x}, \mathbf{v_f}, \mathbf{v_e}; \rho_2\big), \\ c_3 = \mathrm{COM}\big(\mathbf{v_s} \oplus \mathbf{w_s}, \mathbf{v_x} \oplus \mathbf{w_x}, \mathbf{v_f} \oplus \mathbf{w_f}, \mathbf{v_e} \oplus \mathbf{w_e}; \rho_3\big). \end{cases}$$

- If $\mathrm{Ch} = 2$: Check that
$$\begin{cases} c_1 = \mathrm{COM}\big(\mathbf{b}_2, \pi_2, \sigma_2, \mathbf{H} \cdot \mathbf{z_s}^{\top} \oplus \mathbf{A} \cdot \mathbf{z_x}^{\top}, \ (\mathbf{z_u} \| \mathbf{z_f}) \cdot \widehat{\mathbf{G}} \oplus \mathbf{z_e} \oplus \mathbf{c}; \ \rho_1\big), \\ c_3 = \mathrm{COM}\big(\pi_2(\mathbf{z_s}), T_{\mathbf{b}_2}(\mathbf{z_x}), T_{\mathbf{b}_2}'(\mathbf{z_f}), \sigma_2(\mathbf{z_e}); \ \rho_3\big). \end{cases}$$

- If $\mathrm{Ch} = 3$: Check that
$$\begin{cases} c_1 = \mathrm{COM}\big(\mathbf{b}_3, \pi_3, \sigma_3, \mathbf{H} \cdot \mathbf{y_s}^{\top} \oplus \mathbf{A} \cdot \mathbf{y_x}^{\top}, \ (\mathbf{y_u} \| \mathbf{y_f}) \cdot \widehat{\mathbf{G}} \oplus \mathbf{y_e}; \ \rho_1\big), \\ c_2 = \mathrm{COM}\big(\pi_3(\mathbf{y_s}), T_{\mathbf{b}_3}(\mathbf{y_x}), T_{\mathbf{b}_3}'(\mathbf{y_f}), \sigma_3(\mathbf{y_e}); \ \rho_2\big). \end{cases}$$

In each case, $\mathcal{V}$ outputs 1 if and only if all the conditions hold. Otherwise, $\mathcal{V}$ outputs 0.

Fig. 1. The underlying zero-knowledge argument system of our group signature scheme.

As a result, $\mathcal{P}$ should always pass $\mathcal{V}$'s checks in the case Ch $= 1$. In the case Ch $= 2$, since the linear equations in (8) hold true, $\mathcal{P}$ should also pass the verification. Finally, in the case Ch $= 3$, it suffices to note that $\mathcal{V}$ simply checks for honest computations of c_1 and c_2.

Communication Cost. The commitment CMT has bit-size 3λ. If Ch $= 1$, then the response RSP has bit-size $3\ell + N + 2(m + n + \lambda)$. In each of the cases Ch $= 2$ and Ch $= 3$, RSP has bit-size $2\ell + N + m(\log m + 1) + n(\log n + 1) + k + 2\lambda$. Therefore, the total communication cost (in bits) of the protocol is less than the bound β specified in Theorem 1.

Zero-Knowledge Property. The following lemma says that our interactive protocol is statistically zero-knowledge if COM is modelled as a random oracle.

Lemma 2. *In the random oracle model, there exists an efficient simulator $\mathcal{S}$ interacting with a (possibly cheating) verifier $\widehat{\mathcal{V}}$, such that, given only the public input of the protocol, $\mathcal{S}$ outputs with probability negligibly close to $2/3$ a simulated transcript that is statistically close to the one produced by the honest prover in the real interaction.*

Argument of Knowledge Property. The next lemma states that our protocol satisfies the special soundness property of Σ-protocols, which implies that it is an argument of knowledge [Gro04].

Lemma 3. *Let COM be a collision-resistant hash function. Given the public input of the protocol, a commitment CMT and 3 valid responses $\mathsf{RSP}_1, \mathsf{RSP}_2, \mathsf{RSP}_3$ to all 3 possible values of the challenge Ch, one can efficiently construct a knowledge extractor $\mathcal{E}$ that outputs a tuple $(j', \mathbf{s}', \mathbf{u}', \mathbf{e}') \in [0, N - 1] \times \mathbb{F}_2^m \times \mathbb{F}_2^{k-\ell} \times \mathbb{F}_2^n$ such that:*

$$\begin{cases} \mathbf{H} \cdot \mathbf{s}'^{\mathsf{T}} = \mathbf{y}_{j'}^{\mathsf{T}} \;\wedge\; \mathbf{s}' \in \mathsf{B}(m, \omega); \\ \left(\mathbf{u}' \,\|\, \mathsf{I2B}(j') \right) \cdot \mathbf{G} \oplus \mathbf{e}' = \mathbf{c} \;\wedge\; \mathbf{e}' \in \mathsf{B}(n, t). \end{cases}$$

The proofs of Lemmas 2 and 3 employ the standard simulation and extraction techniques for Stern-type protocols (e.g., [Ste96, KTX08, LNSW13]). These proofs are omitted here due to space constraints. They can be found in the full version of this paper [ELL+15].

4   Our Code-Based Group Signature Scheme

4.1   Description of the Scheme

Our group signature scheme is described as follows:

KeyGen$(1^\lambda, 1^N)$: On input a security parameter λ and an expected number of group users $N = 2^\ell \in \mathsf{poly}(\lambda)$, for some positive integer ℓ, this algorithm first selects the following:

–Parameters $n = n(\lambda), k = k(\lambda), t = t(\lambda)$ for a binary $[n, k, 2t + 1]$ Goppa code.

–Parameters $m = m(\lambda), r = r(\lambda), \omega = \omega(\lambda)$ for the Syndrome Decoding problem, such that

$$r \leq \log \binom{m}{w} - 2\lambda - \mathcal{O}(1). \tag{9}$$

– Two collision-resistant hash functions, to be modelled as random oracles:
  1. $\mathsf{COM} : \{0,1\}^* \to \{0,1\}^\lambda$, to be used for generating zero-knowledge arguments.
  2. $\mathcal{H} : \{0,1\}^* \to \{1,2,3\}^\kappa$ (where $\kappa = \omega(\log \lambda)$), to be used in the Fiat-Shamir transformation.

The algorithm then performs the following steps:

1. Run $\mathsf{ME.KeyGen}(n, k, t)$ to obtain a key pair $\left(\mathsf{pk}_{\mathsf{ME}} = \mathbf{G} \in \mathbb{F}_2^{k \times n} \; ; \; \mathsf{sk}_{\mathsf{ME}}\right)$ for the randomized McEliece encryption scheme with respect to a binary $[n, k, 2t + 1]$ Goppa code. The plaintext space is $\mathbb{F}_2^\ell$.
2. Choose a matrix $\mathbf{H} \xleftarrow{\$} \mathbb{F}_2^{r \times m}$.
3. For each $j \in [0, N-1]$, pick $\mathbf{s}_j \xleftarrow{\$} \mathsf{B}(m, \omega)$, and let $\mathbf{y}_j \in \mathbb{F}_2^r$ be its syndrome, i.e., $\mathbf{y}_j^\top = \mathbf{H} \cdot \mathbf{s}_j^\top$.
   Remark 1. We note that, for parameters m, r, ω satisfying condition (9), the distribution of syndrome $\mathbf{y}_j$, for all $j \in [0, N-1]$, is statistically close to the uniform distribution over $\mathbb{F}_2^r$ (by Lemma 1).
4. Output

$$\left(\mathsf{gpk} = (\mathbf{G}, \mathbf{H}, \mathbf{y}_0, \ldots, \mathbf{y}_{N-1}), \; \mathsf{gmsk} = \mathsf{sk}_{\mathsf{ME}}, \; \mathsf{gsk} = (\mathbf{s}_0, \ldots, \mathbf{s}_{N-1})\right).(10)$$

$\mathsf{Sign}(\mathsf{gsk}[j], M)$: To sign a message $M \in \{0,1\}^*$ under gpk, the group user of index j, who possesses secret key $\mathbf{s} = \mathsf{gsk}[j]$, performs the following steps:

1. Encrypt the binary representation of j, i.e., vector $\mathsf{I2B}(j) \in \mathbb{F}_2^\ell$, under the randomized McEliece encrypting key $\mathbf{G}$. This is done by sampling $(\mathbf{u} \xleftarrow{\$} \mathbb{F}_2^{k-\ell}, \mathbf{e} \xleftarrow{\$} \mathsf{B}(n, t))$ and outputting the ciphertext:

$$\mathbf{c} = \left(\mathbf{u} \,\|\, \mathsf{I2B}(j)\right) \cdot \mathbf{G} \oplus \mathbf{e} \in \mathbb{F}_2^n.$$

2. Generate a $\mathsf{NIZKAoK}$ Π to simultaneously prove in zero-knowledge the possession of a vector $\mathbf{s} \in \mathsf{B}(m, \omega)$ corresponding to a certain syndrome $\mathbf{y}_j \in \{\mathbf{y}_0, \ldots, \mathbf{y}_{N-1}\}$ with hidden index j, and that $\mathbf{c}$ is a correct McEliece encryption of $\mathsf{I2B}(j)$. This is done by employing the interactive argument system in Sect. 3 with public input $(\mathbf{G}, \mathbf{H}, \mathbf{y}_0, \ldots, \mathbf{y}_{N-1}, \mathbf{c})$, and prover's witness $(j, \mathbf{s}, \mathbf{u}, \mathbf{e})$ that satisfies:

$$\begin{cases} \mathbf{H} \cdot \mathbf{s}^\top = \mathbf{y}_j^\top \; \wedge \; \mathbf{s} \in \mathsf{B}(m, \omega); \\ \left(\mathbf{u} \,\|\, \mathsf{I2B}(j)\right) \cdot \mathbf{G} \oplus \mathbf{e} = \mathbf{c} \; \wedge \; \mathbf{e} \in \mathsf{B}(n, t). \end{cases} \tag{11}$$

The protocol is repeated $\kappa = \omega(\log \lambda)$ times to achieve negligible soundness error, and then made non-interactive using the Fiat-Shamir heuristic. Namely, we have

$$\Pi = \left(\mathsf{CMT}^{(1)}, \ldots, \mathsf{CMT}^{(\kappa)}; \ (\mathsf{Ch}^{(1)}, \ldots, \mathsf{Ch}^{(\kappa)}); \ \mathsf{RSP}^{(1)}, \ldots, \mathsf{RSP}^{(\kappa)}\right), (12)$$

where $(\mathsf{Ch}^{(1)}, \ldots, \mathsf{Ch}^{(\kappa)}) = \mathcal{H}(M; \mathsf{CMT}^{(1)}, \ldots, \mathsf{CMT}^{(\kappa)}; \mathsf{gpk}, \mathbf{c}) \in \{1, 2, 3\}^\kappa$.

3. Output the group signature $\Sigma = (\mathbf{c}, \Pi)$.

Verify$(\mathsf{gpk}, M, \Sigma)$: Parse Σ as $(\mathbf{c}, \Pi)$ and parse Π as in (12). Then proceed as follows:

1. If $(\mathsf{Ch}^{(1)}, \ldots, \mathsf{Ch}^{(\kappa)}) \neq \mathcal{H}(M; \mathsf{CMT}^{(1)}, \ldots, \mathsf{CMT}^{(\kappa)}; \mathsf{gpk}, \mathbf{c})$, then return 0.
2. For $i = 1$ to κ, run the verification step of the interactive protocol in Sect. 3 with public input $(\mathbf{G}, \mathbf{H}, \mathbf{y}_0, \ldots, \mathbf{y}_{N-1}, \mathbf{c})$ to check the validity of $\mathsf{RSP}^{(i)}$ with respect to $\mathsf{CMT}^{(i)}$ and $\mathsf{Ch}^{(i)}$. If any of the verification conditions does not hold, then return 0.
3. Return 1.

Open$(\mathsf{gmsk}, M, \Sigma)$: Parse Σ as $(\mathbf{c}, \Pi)$ and run $\mathsf{ME.Dec}(\mathsf{gmsk}, \mathbf{c})$ to decrypt $\mathbf{c}$. If decryption fails, then return $\bot$. If decryption outputs $\mathbf{g} \in \mathbb{F}_2^\ell$, then return $j = \mathsf{B2I}(\mathbf{g}) \in [0, N-1]$.

The efficiency, correctness, and security aspects of the above group signature scheme are summarized in the following theorem.

Theorem 2. *The given group signature scheme is correct. The public key has size $nk + (m + N)r$ bits, and signatures have bit-size bounded by $\big((N + 3\log N) + m(\log m + 1) + n(\log n + 1) + k + 5\lambda\big)\kappa + n$. Furthermore, in the random oracle model:*

- *If the Decisional McEliece problem $\mathsf{DMcE}(n, k, t)$ and the Decisional Learning Parity with fixed-weight Noise problem $\mathsf{DLPN}(k - \ell, n, \mathsf{B}(n, t))$ are hard, then the scheme is CPA-anonymous.*
- *If the Syndrome Decoding problem $\mathsf{SD}(m, r, \omega)$ is hard, then the scheme is traceable.*

4.2 Efficiency and Correctness

Efficiency. It is clear from (10) that gpk has bit-size $nk + (m + N)r$. The length of the $\mathsf{NIZKAoK}$ Π is κ times the communication cost of the underlying interactive protocol. Thus, by Theorem 1, $\Sigma = (\mathbf{c}, \Pi)$ has bit-size bounded by $\big((N + 3\log N) + m(\log m + 1) + n(\log n + 1) + k + 5\lambda\big)\kappa + n$.

Correctness. To see that the given group signature scheme is correct, first observe that the honest user with index j, for any $j \in [0, N-1]$, can always obtain a tuple $(j, \mathbf{s}, \mathbf{u}, \mathbf{e})$ satisfying (11). Then, since the underlying interactive protocol is perfectly complete, Π is a valid $\mathsf{NIZKAoK}$ and algorithm $\mathsf{Verify}(\mathsf{gpk}, M, \Sigma)$ always outputs 1, for any message $M \in \{0, 1\}^*$.

Regarding the correctness of algorithm Open, it suffices to note that, if the ciphertext $\mathbf{c}$ is of the form $\mathbf{c} = \big(\mathbf{u}\,\|\,\mathsf{I2B}(j)\big) \cdot \mathbf{G} \oplus \mathbf{e}$, where $\mathbf{e} \in \mathsf{B}(n,t)$, then, by the correctness of the randomized McEliece encryption scheme, algorithm ME.Dec(gmsk, $\mathbf{c}$) will output $\mathsf{I2B}(j)$.

4.3   Anonymity

Let $\mathcal{A}$ be any PPT adversary attacking the CPA-anonymity of the scheme with advantage ϵ. We will prove that $\epsilon = \mathsf{negl}(\lambda)$ based on the ZK property of the underlying argument system, and the assumed hardness of the $\mathsf{DMcE}(n,k,t)$ and the $\mathsf{DLPN}(k-\ell, n, \mathsf{B}(n,t))$ problems. Specifically, we consider the following sequence of hybrid experiments $G_0^{(b)}, G_1^{(b)}, G_2^{(b)}, G_3^{(b)}$ and G_4.

Experiment $G_0^{(b)}$. This is the real CPA-anonymity game. The challenger runs KeyGen($1^\lambda, 1^N$) to obtain

$$\big(\mathsf{gpk} = (\mathbf{G}, \mathbf{H}, \mathbf{y}_0, \dots, \mathbf{y}_{N-1}), \; \mathsf{gmsk} = \mathsf{sk}_{\mathsf{ME}}, \; \mathsf{gsk} = (\mathsf{gsk}[0], \dots, \mathsf{gsk}[N-1])\,\big),$$

and then gives gpk and $\{\mathsf{gsk}[j]\}_{j \in [0, N-1]}$ to $\mathcal{A}$. In the challenge phase, $\mathcal{A}$ outputs a message M^* together with two indices $j_0, j_1 \in [0, N-1]$. The challenger sends back a challenge signature $\Sigma^* = (\mathbf{c}^*, \Pi^*) \leftarrow \mathsf{Sign}(\mathsf{gpk}, \mathsf{gsk}[j_b])$, where $\mathbf{c}^* = \big(\mathbf{u}\,\|\,\mathsf{I2B}(j_b)\big) \cdot \mathbf{G} \oplus \mathbf{e}$, with $\mathbf{u} \xleftarrow{\$} \mathbb{F}_2^{k-\ell}$ and $\mathbf{e} \xleftarrow{\$} \mathsf{B}(n,t)$. The adversary then outputs b with probability $1/2 + \epsilon$.

Experiment $G_1^{(b)}$. In this experiment, we introduce the following modification in the challenge phase: instead of faithfully generating the NIZKAoK Π^*, the challenger simulates it as follows:

1. Compute $\mathbf{c}^* \in \mathbb{F}_2^n$ as in experiment $G_0^{(b)}$.
2. Run the simulator of the underlying interactive protocol in Sect. 3 $t = \omega(\log \lambda)$ times on input $(\mathbf{G}, \mathbf{H}, \mathbf{y}_0, \dots, \mathbf{y}_{N-1}, \mathbf{c}^*)$, and then program the random oracle $\mathcal{H}$ accordingly.
3. Output the simulated NIZKAoK Π^*.

Since the underlying argument system is statistically zero-knowledge, Π^* is statistically close to the real NIZKAoK. As a result, the simulated signature $\Sigma^* = (\mathbf{c}^*, \Pi^*)$ is statistically close to the one in experiment $G_0^{(b)}$. It then follows that $G_0^{(b)}$ and $G_1^{(b)}$ are indistinguishable from $\mathcal{A}$'s view.

Experiment $G_2^{(b)}$. In this experiment, we make the following change with respect to $G_1^{(b)}$: the encrypting key $\mathbf{G}$ obtained from ME.KeyGen(n, k, t) is replaced by a uniformly random matrix $\mathbf{G} \xleftarrow{\$} \mathbb{F}_2^{k \times n}$. We will demonstrate in Lemma 4 that experiments $G_1^{(b)}$ and $G_2^{(b)}$ are computationally indistinguishable based on the assumed hardness of the $\mathsf{DMcE}(n,k,t)$ problem.

Lemma 4. *If $\mathcal{A}$ can distinguish experiments $G_1^{(b)}$ and $G_2^{(b)}$ with probability non-negligibly larger than $1/2$, then there exists an efficient distinguisher $\mathcal{D}_1$ solving the $\mathsf{DMcE}(n,k,t)$ problem with the same probability.*

Proof. An instance of the $\mathsf{DMcE}(n, k, t)$ problem is a matrix $\mathbf{G}^* \in \mathbb{F}_2^{k \times n}$ which can either be uniformly random, or be generated by $\mathsf{ME.KeyGen}(n, k, t)$. Distinguisher $\mathcal{D}_1$ receives a challenge instance $\mathbf{G}^*$ and uses $\mathcal{A}$ to distinguish between the two. It interacts with $\mathcal{A}$ as follows.

- Setup. Generate $(\mathbf{H}, \mathbf{y}_0, \ldots, \mathbf{y}_{N-1})$ and $(\mathsf{gsk}[0], \ldots, \mathsf{gsk}[N-1])$ as in the real scheme. Then, send the following to $\mathcal{A}$:

$$\left(\mathsf{gpk}^* = (\mathbf{G}^*, \mathbf{H}, \mathbf{y}_0, \ldots, \mathbf{y}_{N-1}), \ \mathsf{gsk} = (\mathsf{gsk}[0], \ldots, \mathsf{gsk}[N-1]) \right).$$

- Challenge. Receiving the challenge (M^*, j_0, j_1), $\mathcal{D}_1$ proceeds as follows:
  1. Pick $b \xleftarrow{\$} \{0, 1\}$, and compute $\mathbf{c}^* = \left(\mathbf{u} \, \| \, \mathsf{I2B}(j_b) \right) \cdot \mathbf{G}^* \oplus \mathbf{e}$, where $\mathbf{u} \xleftarrow{\$} \mathbb{F}_2^{k-\ell}$ and $\mathbf{e} \xleftarrow{\$} \mathsf{B}(n, t)$.
  2. Simulate the NIZKAoK Π^* on input $(\mathbf{G}^*, \mathbf{H}, \mathbf{y}_0, \ldots, \mathbf{y}_{N-1}, \mathbf{c}^*)$, and output $\Sigma^* = (\mathbf{c}^*, \Pi^*)$.

We observe that if $\mathbf{G}^*$ is generated by $\mathsf{ME.KeyGen}(n, k, t)$ then the view of $\mathcal{A}$ in the interaction with $\mathcal{D}_1$ is statistically close to its view in experiment $G_1^{(b)}$ with the challenger. On the other hand, if $\mathbf{G}^*$ is uniformly random, then $\mathcal{A}$'s view is statistically close to its view in experiment $G_2^{(b)}$. Therefore, if $\mathcal{A}$ can guess whether it is interacting with the challenger in $G_1^{(b)}$ or $G_2^{(b)}$ with probability non-negligibly larger than $1/2$, then $\mathcal{D}_1$ can use $\mathcal{A}$'s guess to solve the challenge instance $\mathbf{G}^*$ of the $\mathsf{DMcE}(n, k, t)$ problem, with the same probability. $\qquad \square$

Experiment $G_3^{(b)}$. Recall that in experiment $G_2^{(b)}$, we have

$$\mathbf{c}^* = \left(\mathbf{u} \, \| \, \mathsf{I2B}(j_b) \right) \cdot \mathbf{G} \oplus \mathbf{e} = (\mathbf{u} \cdot \mathbf{G}_1 \oplus \mathbf{e}) \oplus \mathsf{I2B}(j_b) \cdot \mathbf{G}_2,$$

where $\mathbf{G}_1 \in \mathbb{F}_2^{(k-\ell) \times n}$, $\mathbf{G}_2 \in \mathbb{F}_2^{\ell \times n}$ such that $\left[\frac{\mathbf{G}_1}{\mathbf{G}_2} \right] = \mathbf{G}$; and $\mathbf{u} \xleftarrow{\$} \mathbb{F}_2^{k-\ell}$, $\mathbf{e} \xleftarrow{\$} \mathsf{B}(n, t)$.

In experiment $G_3^{(b)}$, the generation of $\mathbf{c}^*$ is modified as follows: we instead let $\mathbf{c}^* = \mathbf{v} \oplus \mathsf{I2B}(j_b) \cdot \mathbf{G}_2$, where $\mathbf{v} \xleftarrow{\$} \mathbb{F}_2^n$. Experiments $G_2^{(b)}$ and $G_3^{(b)}$ are computationally indistinguishable based on the assumed hardness of the $\mathsf{DLPN}(k - \ell, n, \mathsf{B}(n, t))$ problem, as shown in Lemma 5.

Lemma 5. *If $\mathcal{A}$ can distinguish experiments $G_2^{(b)}$ and $G_3^{(b)}$ with probability non-negligibly larger than $1/2$, then there exists an efficient distinguisher $\mathcal{D}_2$ solving the $\mathsf{DLPN}(k - \ell, n, \mathsf{B}(n, t))$ problem with the same probability.*

Proof. An instance of the $\mathsf{DLPN}(k - \ell, n, \mathsf{B}(n, t))$ problem is a pair $(\mathbf{B}, \mathbf{v}) \in \mathbb{F}_2^{(k-\ell) \times n} \times \mathbb{F}_2^n$, where $\mathbf{B}$ is uniformly random, and $\mathbf{v}$ is either uniformly random or of the form $\mathbf{v} = \mathbf{u} \cdot \mathbf{B} \oplus \mathbf{e}$, for $(\mathbf{u} \xleftarrow{\$} \mathbb{F}_2^{k-\ell}; \mathbf{e} \xleftarrow{\$} \mathsf{B}(n, t))$. Distinguisher $\mathcal{D}_2$ receives a challenge instance $(\mathbf{B}, \mathbf{v})$ and uses $\mathcal{A}$ to distinguish between the two. It interacts with $\mathcal{A}$ as follows.

– Setup. Pick $\mathbf{G}_2 \xleftarrow{\$} \mathbb{F}_2^{\ell \times n}$ and let $\mathbf{G}^* = \left[\frac{\mathbf{B}}{\mathbf{G}_2}\right]$. Generate $(\mathbf{H}, \mathbf{y}_0, \ldots, \mathbf{y}_{N-1})$ and $(\mathsf{gsk}[0], \ldots, \mathsf{gsk}[N-1])$ as in the real scheme, and send the following to $\mathcal{A}$:

$$\big(\; \mathsf{gpk}^* = (\mathbf{G}^*, \mathbf{H}, \mathbf{y}_0, \ldots, \mathbf{y}_{N-1}), \;\; \mathsf{gsk} = (\mathsf{gsk}[0], \ldots, \mathsf{gsk}[N-1]) \;\big).$$

– Challenge. Receiving the challenge (M^*, j_0, j_1), $\mathcal{D}_2$ proceeds as follows:

1. Pick $b \xleftarrow{\$} \{0, 1\}$, and let $\mathbf{c}^* = \mathbf{v} \oplus \mathsf{I2B}(j_b) \cdot \mathbf{G}_2$, where $\mathbf{v}$ comes from the challenge DLPN instance.
2. Simulate the NIZKAoK Π^* on input $(\mathbf{G}^*, \mathbf{H}, \mathbf{y}_0, \ldots, \mathbf{y}_{N-1}, \mathbf{c}^*)$, and output $\Sigma^* = (\mathbf{c}^*, \Pi^*)$.

We observe that if $\mathcal{D}_2$'s input pair $(\mathbf{B}, \mathbf{v})$ is of the form $(\mathbf{B}, \mathbf{v} = \mathbf{u} \cdot \mathbf{B} \oplus \mathbf{e})$, where $\mathbf{u} \xleftarrow{\$} \mathbb{F}_2^{k-\ell}$ and $\mathbf{e} \xleftarrow{\$} \mathsf{B}(n, t)$, then the view of $\mathcal{A}$ in the interaction with $\mathcal{D}_2$ is statistically close to its view in experiment $G_2^{(b)}$ with the challenger. On the other hand, if the pair $(\mathbf{B}, \mathbf{v})$ is uniformly random, then $\mathcal{A}$'s view is statistically close to its view in experiment $G_3^{(b)}$. Therefore, if $\mathcal{A}$ can guess whether it is interacting with the challenger in $G_2^{(b)}$ or $G_3^{(b)}$ with probability non-negligibly larger than $1/2$, then $\mathcal{D}_2$ can use $\mathcal{A}$'s guess to solve the challenge instance of the DLPN$(k - \ell, \mathsf{B}(n, t))$ problem with the same probability.   □

Experiment G_4. In this experiment, we employ the following modification with respect to $G_3^{(b)}$: the ciphertext $\mathbf{c}^*$ is now set as $\mathbf{c}^* = \mathbf{r} \xleftarrow{\$} \mathbb{F}_2^n$. Clearly, the distributions of $\mathbf{c}^*$ in experiments $G_3^{(b)}$ and G_4 are identical. As a result, G_4 and $G_3^{(b)}$ are statistically indistinguishable. We note that G_4 no longer depends on the challenger's bit b, and thus, $\mathcal{A}$'s advantage in this experiment is 0.

The above discussion shows that experiments $G_0^{(b)}, G_1^{(b)}, G_2^{(b)}, G_3^{(b)}, G_4$ are indistinguishable, and that $\mathbf{Adv}_{\mathcal{A}}(G_4) = 0$. It then follows that the advantage of $\mathcal{A}$ in attacking the CPA-anonymity of the scheme, i.e., in experiment $G_0^{(b)}$, is negligible. This concludes the proof of the CPA-anonymity property.

4.4   Traceability

Let $\mathcal{A}$ be a PPT traceability adversary against our group signature scheme, that has success probability ϵ. We construct a PPT algorithm $\mathcal{F}$ that solves the SD(m, r, ω) problem with success probability polynomially related to ϵ.

Algorithm $\mathcal{F}$ receives a challenge SD(m, r, ω) instance, that is, a uniformly random matrix-syndrome pair $(\widetilde{\mathbf{H}}, \widetilde{\mathbf{y}}) \in \mathbb{F}_2^{r \times m} \times \mathbb{F}_2^r$. The goal of $\mathcal{F}$ is to find a vector $\mathbf{s} \in \mathsf{B}(m, \omega)$ such that $\widetilde{\mathbf{H}} \cdot \mathbf{s}^\top = \widetilde{\mathbf{y}}^\top$. It then proceeds as follows:

1. Pick a guess $j^* \xleftarrow{\$} [0, N-1]$ and set $\mathbf{y}_{j^*} = \widetilde{\mathbf{y}}$.
2. Set $\mathbf{H} = \widetilde{\mathbf{H}}$. For each $j \in [0, N-1]$ such that $j \neq j^*$, sample $\mathbf{s}_j \xleftarrow{\$} \mathsf{B}(m, \omega)$ and set $\mathbf{y}_j \in \mathbb{F}_2^r$ be its syndrome, i.e., $\mathbf{y}_j^\top = \mathbf{H} \cdot \mathbf{s}_j^\top$.
3. Run ME.KeyGen(n, k, t) to obtain a key pair $\big(\mathsf{pk}_{\mathsf{ME}} = \mathbf{G} \in \mathbb{F}_2^{k \times n} \;;\; \mathsf{sk}_{\mathsf{ME}}\big)$.
4. Send $\mathsf{gpk} = (\mathbf{G}, \mathbf{H}, \mathbf{y}_0, \ldots, \mathbf{y}_{N-1})$ and $\mathsf{gmsk} = \mathsf{sk}_{\mathsf{ME}}$ to $\mathcal{A}$.

We note that, since the parameters m, r, ω were chosen such that $r \leq \log \binom{m}{w} - 2\lambda - \mathcal{O}(1)$, by Lemma 1, the distribution of syndrome $\mathbf{y}_j$, for all $j \neq j^*$, is statistically close to the uniform distribution over $\mathbb{F}_2^r$. In addition, the syndrome $\mathbf{y}_{j^*} = \tilde{\mathbf{y}}$ is truly uniform over $\mathbb{F}_2^r$. It then follows that the distribution of $(\mathbf{y}_0, \ldots, \mathbf{y}_{N-1})$ is statistically close to that in the real scheme (see Remark 1). As a result, the distribution of $(\mathsf{gpk}, \mathsf{gmsk})$ is statistically close to the distribution expected by $\mathcal{A}$.

The forger $\mathcal{F}$ then initializes a set $CU = \emptyset$ and handles the queries from $\mathcal{A}$ as follows:

- Queries to the random oracle $\mathcal{H}$ are handled by consistently returning uniformly random values in $\{1, 2, 3\}^\kappa$. Suppose that $\mathcal{A}$ makes $Q_{\mathcal{H}}$ queries, then for each $\eta \leq Q_{\mathcal{H}}$, we let r_η denote the answer to the η-th query.
- $\mathcal{O}^{\mathsf{Corrupt}}(j)$, for any $j \in [0, N-1]$: If $j \neq j^*$, then $\mathcal{F}$ sets $CU := CU \cup \{j\}$ and gives $\mathbf{s}_j$ to $\mathcal{A}$; If $j = j^*$, then $\mathcal{F}$ aborts.
- $\mathcal{O}^{\mathsf{Sign}}(j, M)$, for any $j \in [0, N-1]$ and any message M:
  - If $j \neq j^*$, then $\mathcal{F}$ honestly computes a signature, since it has the secret key $\mathbf{s}_j$.
  - If $j = j^*$, then $\mathcal{F}$ returns a simulated signature Σ^* computed as in Sect. 4.3 (see Experiment $G_1^{(b)}$ in the proof of anonymity).

At some point, $\mathcal{A}$ outputs a forged group signature Σ^* on some message M^*, where

$$\Sigma^* = \left(\mathbf{c}^*, \left(\mathsf{CMT}^{(1)}, \ldots, \mathsf{CMT}^{(\kappa)}; \mathsf{Ch}^{(1)}, \ldots, \mathsf{Ch}^{(\kappa)}; \mathsf{RSP}^{(1)}, \ldots, \mathsf{RSP}^{(\kappa)}\right)\right).$$

By the requirements of the traceability experiment, one has $\mathsf{Verify}(\mathsf{gpk}, M^*, \Sigma^*) = 1$, and for all $j \in CU$, signatures of user j on M^* were never queried. Now $\mathcal{F}$ uses $\mathsf{sk}_{\mathsf{ME}}$ to open Σ^*, and aborts if the opening algorithm does not output j^*. It can be checked that $\mathcal{F}$ aborts with probability at most $(N-1)/N + (2/3)^\kappa$, because the choice of $j^* \in [0, N-1]$ is completely hidden from $\mathcal{A}$'s view, and $\mathcal{A}$ can violate the soundness of the argument system with probability at most $(2/3)^\kappa$. Thus, with probability at least $1/N - (2/3)^\kappa$, it holds that

$$\mathsf{Verify}(\mathsf{gpk}, M^*, \Sigma^*) = 1 \quad \wedge \quad \mathsf{Open}(\mathsf{sk}_{\mathsf{ME}}, M^*, \Sigma^*) = j^*. \tag{13}$$

Suppose that (13) holds. Algorithm $\mathcal{F}$ then exploits the forgery as follows. Denote by Δ the tuple $\left(M^*; \mathsf{CMT}^{(1)}, \ldots, \mathsf{CMT}^{(\kappa)}; \mathbf{G}, \mathbf{H}, \mathbf{y}_0, \ldots, \mathbf{y}_{N-1}, \mathbf{c}^*\right)$. Observe that if $\mathcal{A}$ has never queried the random oracle $\mathcal{H}$ on input Δ, then

$$\Pr\left[(\mathsf{Ch}^{(1)}, \ldots, \mathsf{Ch}^{(\kappa)}) = \mathcal{H}(\Delta)\right] \leq 3^{-\kappa}.$$

Therefore, with probability at least $\epsilon - 3^{-\kappa}$, there exists certain $\eta^* \leq Q_{\mathcal{H}}$ such that Δ was the input of the η^*-th query. Next, $\mathcal{F}$ picks η^* as the target forking point and replays $\mathcal{A}$ many times with the same random tape and input as in the original run. In each rerun, for the first $\eta^* - 1$ queries, $\mathcal{A}$ is given the same answers $r_1, \ldots, r_{\eta^*-1}$ as in the initial run, but from the η^*-th query onwards,

$\mathcal{F}$ replies with fresh random values $r'_{\eta*}, \dots, r'_{q_{\mathcal{H}}} \xleftarrow{\$} \{1, 2, 3\}^\kappa$. The Improved Forking Lemma of Pointcheval and Vaudenay [PV97, Lemma 7] implies that, with probability larger than $1/2$ and within less than $32 \cdot Q_{\mathcal{H}}/(\epsilon - 3^{-\kappa})$ executions of $\mathcal{A}$, algorithm $\mathcal{F}$ can obtain a 3-fork involving the tuple Δ. Now, let the answers of $\mathcal{F}$ with respect to the 3-fork branches be

$$r_{1,\eta*} = (\mathsf{Ch}_1^{(1)}, \dots, \mathsf{Ch}_1^{(\kappa)}); \quad r_{2,\eta*} = (\mathsf{Ch}_2^{(1)}, \dots, \mathsf{Ch}_2^{(\kappa)}); \quad r_{3,\kappa*} = (\mathsf{Ch}_3^{(1)}, \dots, \mathsf{Ch}_3^{(\kappa)}).$$

Then, by a simple calculation, one has:

$$\Pr\left[\exists i \in \{1, \dots, \kappa\} : \{\mathsf{Ch}_1^{(i)}, \mathsf{Ch}_2^{(i)}, \mathsf{Ch}_3^{(i)}\} = \{1, 2, 3\}\right] = 1 - (7/9)^\kappa.$$

Conditioned on the existence of such index i, one parses the 3 forgeries corresponding to the fork branches to obtain $(\mathsf{RSP}_1^{(i)}, \mathsf{RSP}_2^{(i)}, \mathsf{RSP}_3^{(i)})$. They turn out to be 3 *valid* responses with respect to 3 different challenges for the same commitment $\mathsf{CMT}^{(i)}$. Then, by using the knowledge extractor of the underlying interactive argument system (see Lemma 3), one can efficiently extract a tuple $(j', \mathbf{s}', \mathbf{u}', \mathbf{e}') \in [0, N-1] \times \mathbb{F}_2^m \times \mathbb{F}_2^{k-\ell} \times \mathbb{F}_2^n$ such that:

$$\begin{cases} \mathbf{H} \cdot \mathbf{s}'^\top = \mathbf{y}_{j'}^\top \ \wedge \ \mathbf{s}' \in \mathsf{B}(m, \omega); \\ (\mathbf{u}' \,\|\, \mathsf{I2B}(j')) \cdot \mathbf{G} \oplus \mathbf{e}' = \mathbf{c}^* \ \wedge \ \mathbf{e}' \in \mathsf{B}(n, t). \end{cases}$$

Since the given group signature scheme is correct, the equation $(\mathbf{u}' \,\|\, \mathsf{I2B}(j')) \cdot \mathbf{G} \oplus \mathbf{e}' = \mathbf{c}^*$ implies that $\mathsf{Open}(\mathsf{sk}_{\mathsf{ME}}, M^*, \Sigma^*) = j'$. On the other hand, we have $\mathsf{Open}(\mathsf{sk}_{\mathsf{ME}}, M^*, \Sigma^*) = j^*$, which leads to $j' = j^*$. Therefore, it holds that $\widetilde{\mathbf{H}} \cdot \mathbf{s}'^\top = \mathbf{H} \cdot \mathbf{s}'^\top = \mathbf{y}_{j^*}^\top = \widetilde{\mathbf{y}}^\top$, and that $\mathbf{s}' \in \mathsf{B}(m, \omega)$. In other words, $\mathbf{s}'$ is a valid solution to the challenge $\mathsf{SD}(m, r, \omega)$ instance $(\widetilde{\mathbf{H}}, \widetilde{\mathbf{y}})$.

Finally, the above analysis shows that, if $\mathcal{A}$ has success probability ϵ and running time T in attacking the traceability of our group signature scheme, then $\mathcal{F}$ has success probability at least $1/2(1/N - (2/3)^\kappa)(1 - (7/9)^\kappa)$ and running time at most $32 \cdot T \cdot Q_{\mathcal{H}}/(\epsilon - 3^{-\kappa}) + \mathsf{poly}(\lambda, N)$. This concludes the proof of the traceability property.

5   Implementation Results

This section presents our basic implementation results of the proposed code-based group signature to demonstrate its feasibility. The testing platform was a modern PC running at 3.5 GHz CPU with 16 GB RAM. We employed the NTL library [NTL] and the gf2x library [GF2] for efficient polynomial operations over a field of characteristic 2. To decode binary Goppa codes, the Paterson algorithm [Pat75] was used in our implementation of the McEliece encryption. We employed SHA-3 with various output sizes to realize several hash functions. To achieve 80-bit security, we chose the parameters as follows:

- The McEliece parameters were set to $(n, k, t) = (2^{11}, 1696, 32)$, as in [BS08].
- The parameters for Syndrome Decoding were set to $(m, r, \omega) = (2756, 550, 121)$ so that the distribution of $\mathbf{y}_0, \ldots, \mathbf{y}_{N-1}$ is 2^{-80}-close to the uniform distribution over $\mathbb{F}_2^r$ (by Lemma 1), and that the $\mathsf{SD}(m, r, \omega)$ problem is intractable with respect to the best known attacks. In particular, these parameters ensure that:
  1. The Information Set Decoding algorithm proposed in [BJMM12] has work factor more than 2^{80}. (See also [Sen14, Slide 3] for an evaluation formula.)
  2. The birthday attacks presented in [FS09] have work factors more than 2^{80}.
- The number of protocol repetitions κ was set to 140 to obtain soundness $1 - 2^{-80}$.

Table 2. Implementation results and sizes

N	PK size	Average signature size	Message	KeyGen	Sign	Verify	Open
2^4 (=16)	625 KB	111 KB	1 B	14.020	0.045	0.034	0.155
			1 GB		5.473	5.450	
2^8 (=256)	642 KB	114 KB	1 B	14.128	0.046	0.036	0.155
			1 GB		5.459	5.450	
2^{12} (=4,096)	906 KB	159 KB	1 B	14.255	0.059	0.044	0.155
			1 GB		5.474	5.462	
2^{16} (=65,536)	5.13 MB	876 KB	1 B	16.302	0.269	0.193	0.161
			1 GB		5.704	5.630	
2^{20} (=1,048,576)	72.8 MB	12.4 MB	1 B	52.084	3.734	2.605	0.155
			1 GB		9.196	8.055	
2^{24} (=16,777,216)	1.16 GB	196 MB	1 B	636.511	58.535	40.801	0.154
			1 GB		64.047	46.402	

Unit for time: second

Table 2 shows our implementation results, together with the public key and signature sizes with respect to various numbers of group users and different message sizes. To reduce the signature size, in the underlying zero-knowledge protocol, we sent a random seed instead of permutations when Ch = 2. Similarly, we sent a random seed instead of the whole response RSP when Ch = 3. Using this technique, the average signature sizes were reduced to about 159 KB for 4,096 users and 876 KB for 65,536 users, respectively. Our public key and signature sizes are linear in the number of group users N, but it does not come to the front while N is less than 2^{12} due to the size of parameters **G** and **H**.

Our implementation took about 0.27 and 0.20 seconds for 1 B message and about 5.70 and 5.60 seconds for 1 GB message, respectively, to sign a message and to verify a generated signature for a group of $65,536$ users. In our experiments, it takes about 5.40 seconds to hash 1 GB message and it leads to the differences of signing and verifying times between 1 B and 1 GB messages.

As far as we know, the implementation results presented here are the first ones for post-quantum group signatures. Our results, while not yielding a truly practical scheme, would certainly help to bring post-quantum group signatures one step closer to practice.

Acknowledgements. The authors would like to thank Jean-Pierre Tillich, Philippe Gaborit, Ayoub Otmani, Nicolas Sendrier, Nico Döttling, and anonymous reviewers of ASIACRYPT 2015 for helpful comments and discussions. The research was supported by Research Grant TL-9014101684-01 and the Singapore Ministry of Education under Research Grant MOE2013-T2-1-041.

References

[ABCG15] Alamélou, Q. Blazy, O., Cauchie, S., Gaborit, P.: A code-based group signature scheme. Presented at WCC, April 2015

[ACJT00] Ateniese, G., Camenisch, J.L., Joye, M., Tsudik, G.: A practical and provably secure coalition-resistant group signature scheme. In: Bellare, M. (ed.) CRYPTO 2000. LNCS, vol. 1880, pp. 255–270. Springer, Heidelberg (2000)

[ACM11] El Yousfi Alaoui, S.M., Cayrel, P.-L., Mohammed, M.: Improved identity-based identification and signature schemes using quasi-dyadic Goppa codes. In: Kim, T., Adeli, H., Robles, R.J., Balitanas, M. (eds.) ISA 2011. CCIS, vol. 200, pp. 146–155. Springer, Heidelberg (2011)

[BBS04] Boneh, D., Boyen, X., Shacham, H.: Short group signatures. In: Franklin, M. (ed.) CRYPTO 2004. LNCS, vol. 3152, pp. 41–55. Springer, Heidelberg (2004)

[BJMM12] Becker, A., Joux, A., May, A., Meurer, A.: Decoding random binary linear codes in $2^{n/20}$: how $1 + 1 = 0$ improves information set decoding. In: Pointcheval, D., Johansson, T. (eds.) EUROCRYPT 2012. LNCS, vol. 7237, pp. 520–536. Springer, Heidelberg (2012)

[BMvT78] Berlekamp, E., McEliece, R.J., van Tilborg, H.C.A.: On the inherent intractability of certain coding problems. IEEE Trans. Inf. Theor. **24**(3), 384–386 (1978)

[BMW03] Bellare, M., Micciancio, D., Warinschi, B.: Foundations of group signatures: formal definitions, simplified requirements, and a construction based on general assumptions. In: Biham, E. (ed.) EUROCRYPT 2003. LNCS, vol. 2656, pp. 614–629. Springer, Heidelberg (2003)

[BS08] Biswas, B., Sendrier, N.: McEliece cryptosystem implementation: theory and practice. In: Buchmann, J., Ding, J. (eds.) PQCrypto 2008. LNCS, vol. 5299, pp. 47–62. Springer, Heidelberg (2008)

[BS13] Bettaieb, S., Schrek, J.: Improved lattice-based threshold ring signature scheme. In: Gaborit, P. (ed.) PQCrypto 2013. LNCS, vol. 7932, pp. 34–51. Springer, Heidelberg (2013)

[BW06] Boyen, X., Waters, B.: Compact group signatures without random oracles. In: Vaudenay, S. (ed.) EUROCRYPT 2006. LNCS, vol. 4004, pp. 427–444. Springer, Heidelberg (2006)

[CFS01] Courtois, N.T., Finiasz, M., Sendrier, N.: How to achieve a McEliece-based digital signature scheme. In: Boyd, C. (ed.) ASIACRYPT 2001. LNCS, vol. 2248, pp. 157–174. Springer, Heidelberg (2001)

[CGG07] Cayrel, P. L., Gaborit, P., Girault, M.: Identity-based identification and signature schemes using correcting codes. In: WCC, pp. 69–78 (2007)

[CHKP10] Cash, D., Hofheinz, D., Kiltz, E., Peikert, C.: Bonsai trees, or how to delegate a lattice basis. In: Gilbert, H. (ed.) EUROCRYPT 2010. LNCS, vol. 6110, pp. 523–552. Springer, Heidelberg (2010)

[CM10] Cayrel, P.-L., Meziani, M.: Post-quantum cryptography: code-based signatures. In: Kim, T., Adeli, H. (eds.) AST/UCMA/ISA/ACN 2010. LNCS, vol. 6059, pp. 82–99. Springer, Heidelberg (2010)

[CNR12] Camenisch, J., Neven, G., Rückert, M.: Fully anonymous attribute tokens from lattices. In: Visconti, I., De Prisco, R. (eds.) SCN 2012. LNCS, vol. 7485, pp. 57–75. Springer, Heidelberg (2012)

[CS97] Camenisch, J., Stadler, M.A.: Efficient group signature schemes for large groups. In: Kaliski Jr, B.S. (ed.) CRYPTO 1997. LNCS, vol. 1294, pp. 410–424. Springer, Heidelberg (1997)

[CS03] Camenisch, J., Shoup, V.: Practical verifiable encryption and decryption of discrete logarithms. In: Boneh, D. (ed.) CRYPTO 2003. LNCS, vol. 2729, pp. 126–144. Springer, Heidelberg (2003)

[CVA10] Cayrel, P.-L., Véron, P., El Yousfi Alaoui, S.M.: A zero-knowledge identification scheme based on the q-ary syndrome decoding problem. In: Biryukov, A., Gong, G., Stinson, D.R. (eds.) SAC 2010. LNCS, vol. 6544, pp. 171–186. Springer, Heidelberg (2011)

[CvH91] Chaum, D., van Heyst, E.: Group signatures. In: Davies, D.W. (ed.) EUROCRYPT 1991. LNCS, vol. 547, pp. 257–265. Springer, Heidelberg (1991)

[Dal08] Dallot, L.: Towards a concrete security proof of Courtois, Finiasz and Sendrier signature scheme. In: Lucks, S., Sadeghi, A.-R., Wolf, C. (eds.) WEWoRC 2007. LNCS, vol. 4945, pp. 65–77. Springer, Heidelberg (2008)

[DDMN12] Döttling, N., Dowsley, R., Müller-Quade, J., Nascimento, A.C.A.: A CCA2 secure variant of the McEliece cryptosystem. IEEE Trans. Inf. Theor. **58**(10), 6672–6680 (2012)

[Döt14] Döttling, N.: Cryptography based on the hardness of decoding. Ph.D. thesis, Karlsruhe Institute of Technology (2014). https://crypto.iti.kit.edu/fileadmin/User/Doettling/thesis.pdf

[ELL+15] Ezerman, M.F., Lee, H.T., Ling, S., Nguyen, K., Wang, H.: A provably secure group signature scheme from code-based assumptions. In: IACR Cryptography ePrint Archive, Report 2015/479 (2015)

[FGUO+13] Faugere, J.-C., Gauthier-Umana, V., Otmani, A., Perret, L., Tillich, J.-P.: A distinguisher for high-rate McEliece cryptosystems. IEEE Trans. Inf. Theor. **59**(10), 6830–6844 (2013)

[Fin10] Finiasz, M.: Parallel-CFS. In: Biryukov, A., Gong, G., Stinson, D.R. (eds.) SAC 2010. LNCS, vol. 6544, pp. 159–170. Springer, Heidelberg (2011)

[FS86] Fiat, A., Shamir, A.: How to prove yourself: practical solutions to identification and signature problems. In: Odlyzko, A.M. (ed.) CRYPTO 1986. LNCS, vol. 263, pp. 186–194. Springer, Heidelberg (1987)

[FS96] Fischer, J.-B., Stern, J.: An efficient pseudo-random generator provably as secure as syndrome decoding. In: Maurer, U.M. (ed.) EUROCRYPT 1996. LNCS, vol. 1070, pp. 245–255. Springer, Heidelberg (1996)

[FS09]   Finiasz, M., Sendrier, N.: Security bounds for the design of code-based cryptosystems. In: Matsui, M. (ed.) ASIACRYPT 2009. LNCS, vol. 5912, pp. 88–105. Springer, Heidelberg (2009)

[GF2]   gf2x library, ver. 1.1. https://gforge.inria.fr/projects/gf2x/

[GKPV10]   Goldwasser, S., Kalai, Y., Peikert, C., Vaikuntanathan, V.: Robustness of the learning with errors assumption. In: ICS, pp. 230–240. Tsinghua University Press (2010)

[GKV10]   Gordon, S.D., Katz, J., Vaikuntanathan, V.: A group signature scheme from lattice assumptions. In: Abe, M. (ed.) ASIACRYPT 2010. LNCS, vol. 6477, pp. 395–412. Springer, Heidelberg (2010)

[GPV08]   Gentry, C., Peikert, C., Vaikuntanathan, V.: Trapdoors for hard lattices and new cryptographic constructions. In: STOC, pp. 197–206. ACM (2008)

[Gro04]   Groth, J.: Evaluating security of voting schemes in the universal composability framework. In: Jakobsson, M., Yung, M., Zhou, J. (eds.) ACNS 2004. LNCS, vol. 3089, pp. 46–60. Springer, Heidelberg (2004)

[HMT13]   Hu, R., Morozov, K., Takagi, T.: Proof of plaintext knowledge for code-based public-key encryption revisited. In: ASIA CCS, pp. 535–540. ACM (2013)

[KTX08]   Kawachi, A., Tanaka, K., Xagawa, K.: Concurrently secure identification schemes based on the worst-case hardness of lattice problems. In: Pieprzyk, J. (ed.) ASIACRYPT 2008. LNCS, vol. 5350, pp. 372–389. Springer, Heidelberg (2008)

[LLLS13]   Laguillaumie, F., Langlois, A., Libert, B., Stehlé, D.: Lattice-based group signatures with logarithmic signature size. In: Sako, K., Sarkar, P. (eds.) ASIACRYPT 2013, Part II. LNCS, vol. 8270, pp. 41–61. Springer, Heidelberg (2013)

[LLNW14]   Langlois, A., Ling, S., Nguyen, K., Wang, H.: Lattice-based group signature scheme with verifier-local revocation. In: Krawczyk, H. (ed.) PKC 2014. LNCS, vol. 8383, pp. 345–361. Springer, Heidelberg (2014)

[LNSW13]   Ling, S., Nguyen, K., Stehlé, D., Wang, H.: Improved zero-knowledge proofs of knowledge for the ISIS problem, and applications. In: Kurosawa, K., Hanaoka, G. (eds.) PKC 2013. LNCS, vol. 7778, pp. 107–124. Springer, Heidelberg (2013)

[LNW15]   Ling, S., Nguyen, K., Wang, H.: Group signatures from lattices: simpler, tighter, shorter, ring-based. In: Katz, J. (ed.) PKC 2015. LNCS, vol. 9020, pp. 427–449. Springer, Heidelberg (2015)

[LPY12]   Libert, B., Peters, T., Yung, M.: Scalable group signatures with revocation. In: Pointcheval, D., Johansson, T. (eds.) EUROCRYPT 2012. LNCS, vol. 7237, pp. 609–627. Springer, Heidelberg (2012)

[McE78]   McEliece, R.J.: A public-key cryptosystem based on algebraic coding theory. Deep Space Network Progress Report, vol. 44, pp. 114–116 (1978)

[MCG08]   Melchor, C.A., Cayrel, P.-L., Gaborit, P.: A new efficient threshold ring signature scheme based on coding theory. In: Buchmann, J., Ding, J. (eds.) PQCrypto 2008. LNCS, vol. 5299, pp. 1–16. Springer, Heidelberg (2008)

[MCGL11]   Melchor, C.A., Cayrel, P.-L., Gaborit, P., Laguillaumie, F.: A new efficient threshold ring signature scheme based on coding theory. IEEE Trans. Inf. Theor. 57(7), 4833–4842 (2011)

[MCK01]   Ma, J.F., Chiam, T.C., Kot, A.C: A new efficient group signature scheme based on linear codes. In: Networks, pp. 124–129. IEEE (2001)

[Meu13] Meurer, A.: A coding-theoretic approach to cryptanalysis. Ph.D. thesis, Ruhr University Bochum (2013). http://www.cits.rub.de/imperia/md/content/diss.pdf

[MGS11] Melchor, C.A., Gaborit, P., Schrek, J.: A new zero-knowledge code based identification scheme with reduced communication. CoRR, abs/1111.1644 (2011)

[MVR12] Mathew, K.P., Vasant, S., Rangan, C.P.: On provably secure code-based signature and signcryption scheme. In: IACR Cryptography ePrint Archive, Report 2012/585 (2012)

[MVVR12] Mathew, K.P., Vasant, S., Venkatesan, S., Pandu Rangan, C.: An efficient IND-CCA2 secure variant of the Niederreiter encryption scheme in the standard model. In: Susilo, W., Mu, Y., Seberry, J. (eds.) ACISP 2012. LNCS, vol. 7372, pp. 166–179. Springer, Heidelberg (2012)

[Nie86] Niederreiter, H.: Knapsack-type cryptosystems and algebraic coding theory. Probl. Control Inf. Theor. **15**(2), 159–166 (1986)

[NIKM08] Nojima, R., Imai, H., Kobara, K., Morozov, K.: Semantic security for the McEliece cryptosystem without random oracles. Des. Codes Cryptogr. **49**(1–3), 289–305 (2008)

[NTL] NTL: a library for doing number theory version 9.0.2. http://www.shoup.net/ntl/

[NZZ15] Nguyen, P.Q., Zhang, J., Zhang, Z.: Simpler efficient group signatures from lattices. In: Katz, J. (ed.) PKC 2015. LNCS, vol. 9020, pp. 401–426. Springer, Heidelberg (2015)

[Pat75] Patterson, N.J.: The algebraic decoding of Goppa codes. IEEE Trans. Inf. Theor. **21**(2), 203–207 (1975)

[Per12] Persichetti, E.: On a CCA2-secure variant of McEliece in the standard model. In: IACR Cryptography ePrint Archive, Report 2012/268 (2012)

[PV97] Pointcheval, D., Vaudenay, S.: On provable security for digital signature algorithms. Technical report LIENS-96-17, Laboratoire d'Informatique de Ecole Normale Superieure (1997)

[RST01] Rivest, R.L., Shamir, A., Tauman, Y.: How to leak a secret. In: Boyd, C. (ed.) ASIACRYPT 2001. LNCS, vol. 2248, pp. 552–565. Springer, Heidelberg (2001)

[Sen14] Sendrier, N.: QC-MDPC-McEliece: a public-key code-based encryption scheme based on quasi-cyclic moderate density parity check codes. In: Workshop "Post-Quantum Cryptography: Recent Results and Trends", Fukuoka, Japan, November 2014

[Sho97] Shor, P.: Polynomial-time algorithms for prime factorization and discrete logarithms on a quantum computer. SIAM J. Comput. **26**(5), 1484–1509 (1997)

[Ste96] Stern, J.: A new paradigm for public key identification. IEEE Trans. Inf. Theor. **42**(6), 1757–1768 (1996)

[Vér96] Véron, P.: Improved identification schemes based on error-correcting codes. Appl. Algebra Eng. Commun. Comput. **8**(1), 57–69 (1996)

[YTM+14] Yang, G., Tan, C.H., Mu, Y., Susilo, W., Wong, D.S.: Identity based identification from algebraic coding theory. Theor. Comput. Sci. **520**, 51–61 (2014)

Type 2 Structure-Preserving Signature Schemes Revisited

Sanjit Chatterjee[1]([✉]) and Alfred Menezes[2]

[1] Department of Computer Science and Automation,
Indian Institute of Science, Bengaluru, India
`sanjit@csa.iisc.ernet.in`
[2] Department of Combinatorics & Optimization, University of Waterloo,
Waterloo, Canada
`ajmeneze@uwaterloo.ca`

Abstract. At CRYPTO 2014, Abe et al. presented generic-signer structure-preserving signature schemes using Type 2 pairings. According to the authors, the proposed constructions are optimal with only two group elements in each signature and just one verification equation. The schemes beat the known lower bounds in the Type 3 setting and thereby establish that the Type 2 setting permits construction of cryptographic schemes with unique properties not achievable in Type 3.

In this paper we undertake a concrete analysis of the Abe et al. claims. By properly accounting for the actual structure of the underlying groups and subgroup membership testing of group elements in signatures, we show that the schemes are not as efficient as claimed. We present natural Type 3 analogues of the Type 2 schemes, and show that the Type 3 schemes are superior to their Type 2 counterparts in every aspect. We also formally establish that in the concrete mathematical structure of asymmetric pairing, *all* Type 2 structure-preserving signature schemes can be converted to the Type 3 setting without any penalty in security or efficiency, and show that the converse is false. Furthermore, we prove that the Type 2 setting does not allow one to circumvent the known lower bound result for the Type 3 setting. Our analysis puts the optimality claims for Type 2 structure-preserving signature in a concrete perspective and indicates an incompleteness in the definition of a generic bilinear group in the Type 2 setting.

1 Introduction

The terms 'Type 2' and 'Type 3' pairings were introduced by Galbraith, Paterson and Smart [16]. A bilinear map $e : \mathbb{G}_1 \times \mathbb{G}_2 \longrightarrow \mathbb{G}_T$ defined over prime-order groups is called Type 2 or Type 3 depending on whether or not an efficiently computable isomorphism from $\mathbb{G}_2$ to $\mathbb{G}_1$ is known. Their aptly titled paper "Pairings for cryptographers" begins with the observation that many research papers treat pairings as a "black box" and then develop schemes that "may not be realizable in practice, or may not be as efficient as the authors assume". A similar concern constitutes the central focus of the current work.

© International Association for Cryptologic Research 2015
T. Iwata and J.H. Cheon (Eds.): ASIACRYPT 2015, Part I, LNCS 9452, pp. 286–310, 2015.
DOI: 10.1007/978-3-662-48797-6_13

The term 'structure-preserving signature' (SPS) was coined in 2010 by Abe et al. [1] but such constructions existed even before (see, e.g., Groth [18]). These pairing-based signature schemes have the property that verification keys, messages, and signatures are all group elements. Moreover, signatures are verified by testing the equality of products of pairings of group elements; each such equality is called a product-of-pairings equation (PPE).

Unlike a standard digital signature, the *raison d'etre* for an SPS is not as a stand-alone scheme, but rather in the modular design of cryptographic protocols. They have been used in numerous cryptographic protocols (see [4] for a list). One of the primary reasons for the popularity of SPS schemes in protocol design is that they are fully compatible with the well-known Groth-Sahai (GS) constructions of pairing-based non-interactive witness-indistinguishable (NIWI) and non-interactive zero-knowledge (NIZK) proof systems [19].

In typical applications of structure-preserving signature schemes when used in conjunction with, say, GS proofs, a party has a signed message and wishes to convince a second party (the verifier) that it possesses the (valid) signed message without revealing the message or the signature.[1] Groth-Sahai NIWI and NIZK proofs allow a party (the prover) to convince a second party (the verifier) that it possesses a solution to a collection of PPEs. The complexity of verifying a GS proof is heavily dependent on the number of group elements in the signature and the number of PPEs in signature verification (see [11, Sect. 3.4]).

It is important to keep the above perspective in mind when investigating optimal constructions of structure-preserving signatures. In other words, having an optimal construction in terms of signature size (number of group elements) and verification complexity (number of PPEs and pairings) is useful for a protocol designer who cares for the *concrete* efficiency of a protocol designed on top of a structure-preserving signature. In contrast, if (at all) a structure-preserving signature finds application as a stand-alone primitive, then the high cost of pairing-based verifications can be easily mitigated by batching [10,15]. As can be expected, we have witnessed significant research to design structure-preserving signature schemes with the smallest possible number of group elements in a signature and with the smallest possible number of PPEs in signature verification (and recently, with the smallest possible number of pairings [9]).

Previous Work. At CRYPTO 2011, Abe et al. [2] presented a strongly secure SPS using Type 3 pairings. Verification has two PPEs, which was proven to be optimal in the sense that any Type 3 structure-preserving signature scheme with verification having a single PPE was shown to succumb to a random message attack. Moreover, signatures are comprised of three group elements, which was also shown to be optimal. In their lower bound results Abe et al. [2] used the notion of a 'generic signer'. A generic signer has access only to generic group

---

[1] We use the example of GS because many applications of structure-preserving signature schemes are in conjunction with such non-interactive proof systems. However, structure-preserving signatures are combined with other primitives too – see the work of Hanser and Slamanig [20].

operations and the same notion was used in later works including [4, 9] to prove lower bound results.

At TCC 2014, Abe et al. [3] extended the aforementioned optimality results to the Type 1 setting, thereby unifying the Type 1 and 3 settings. They also proposed a selectively randomizable SPS which is optimal in terms of signature size and verification complexity in both Type 1 and 3 settings.

At CRYPTO 2014, Abe et al. [4] continued their investigation of structure-preserving signature schemes in the Type 2 setting. They presented a strongly unforgeable structure-preserving signature scheme and a randomizable structure-preserving signature scheme using Type 2 pairings. Both schemes are claimed to have signatures that are comprised of only two group elements, have only one PPE in signature verification, and were proven secure in the generic group model for Type 2 pairings. The authors conclude that their schemes enjoy the smallest signature (in terms of number of group elements) and fastest signature verification. Furthermore, they claimed that their constructions in Type 2 are optimal in terms of signature size, number of verification equations and verification key (see Table 1 of [4]). In light of the aforementioned lower bounds on the number of group elements in signatures and the number of PPEs in signature verification for Type 3 structure-preserving signature schemes, they conclude that the Type 2 schemes have no analogues in the Type 3 setting. According to the authors [4]: "This is significant from a high level pairing-based cryptography perspective, as it provides a concrete example of a property that can be obtained in the Type 2 setting but not in the other settings." This is contrary to the arguments presented in [13] that any cryptographic protocol that employs Type 2 pairings has a natural counterpart in the Type 3 setting that does not suffer any loss in functionality, security or efficiency.

In a follow-up work, Barthe et al. [9] establish lower bounds on the number of pairings in the Type 2 setting. Using an automated tool they devise structure-preserving signatures that are 'strongly-optimal' – having one verification equation and minimum number of pairings in the Type 2 setting.

Concrete Differences Between Type 2 and Type 3 Pairings. Abe et al. [4] use the notion of 'generic algorithms' in their results that establish the claimed superiority of Type 2 setting for SPS. A bilinear group generator $\mathcal{G}$ is abstractly defined which takes input a security parameter and returns the descriptions of $\mathbb{G}_1, \mathbb{G}_2, \mathbb{G}_T$, a bilinear pairing $e : \mathbb{G}_1 \times \mathbb{G}_2 \longrightarrow \mathbb{G}_T$ along with an efficiently-computable isomorphism $\psi : \mathbb{G}_2 \longrightarrow \mathbb{G}_1$. In their abstraction all the relevant operations over $\mathbb{G}_1, \mathbb{G}_2, \mathbb{G}_T$ such as subgroup membership, computing group operations, and evaluating the maps ψ and e are treated as "black-box". Such an abstraction is useful *provided* it is able to capture all the essential properties of the concrete mathematical structure over which a Type 2 pairing is defined.

Type 2 and Type 3 pairings are concretely defined over certain elliptic curve groups [16]. As first pointed out in [16] and elaborated further in [13], each setting is constrained by the underlying mathematical structure. For example, no efficient method is known for hashing onto $\mathbb{G}_2$ in Type 2, whereas the isomorphism ψ, even though it exists in a mathematical sense, is not known to be efficiently computable in the Type 3 setting. Similarly, the structure of $\mathbb{G}_2$ in the Type 2 setting requires the evaluation of two pairings in subgroup membership tests for $\mathbb{G}_2$. All these are deemed to be *necessary* assumptions in the asymmetric pairing setting that a protocol designer needs to keep in mind if s/he is concerned with concrete instantiation of protocols in the real world.

Our Contributions. To critically evaluate the claimed advantages of Type 2 structure-preserving signature schemes, we deconstruct the Abe et al. proposals [4] in terms of the underlying concrete group structures. We show that the analysis of the Type 2 generic-signer structure-preserving signature schemes in [4] neglected to account for the concrete group structure and subgroup membership testing of group elements in a signature, leading to erroneous conclusions. Incorporating these subgroup membership tests into the signature verification increases the number of group elements in signatures and also increases the number of PPEs in signature verification. Next we examine whether the pairing-based subgroup membership tests can be discounted as verification equations when the signature scheme is composed with the Groth-Sahai proof system. Recall that such a modular composition is the primary motivation for structure-preserving signatures. Our analysis establishes that not all these pairing-based verifications can be dispensed with when the signature scheme is composed with such a proof system.

Furthermore, since GS proofs in the Type 2 setting are more costly than in the Type 3 setting, the Type 2 schemes are not as efficient as claimed in [4] in the stand-alone setting and significantly slower when composed with GS proofs. In support of this claim, two examples of Groth-Sahai NIWI proofs for verifying that the prover possesses a solution (X, Y) to the equation $e(A, X) \cdot e(B, Y) = t$ where e is a Type 2 or a Type 3 pairing are given in Appendix A. We present natural Type 3 analogues of the Type 2 schemes, and show that the Type 3 schemes are superior to their Type 2 counterparts in all aspects.

Continuing the process of deconstruction, we formally show that *all* Type 2 generic-signer structure-preserving signature schemes can be converted to Type 3 without any penalty in security and efficiency, but not all Type 3 schemes have a secure Type 2 counterpart. Further, we exhibit the impossibility of having a single pairing-based verification equation in the Type 2 setting even when messages are drawn from $\mathbb{G}_2$ and thereby put the lower bound results of [4] in the correct perspective. Our results demonstrate that any Type 2 structure-preserving signature scheme is merely an inefficient implementation of a corresponding Type 3 scheme. The claim of superiority of the Type 2 setting over Type 3 stems from an incomplete abstraction of the Type 2 setting in [4].

Organization. The remainder of the paper is organized as follows. In Sect. 2 we summarize the salient differences between Type 2 and Type 3 pairings derived from elliptic curves having even embedding degrees. In Sect. 3 we explain why, contrary to the claims, the strongly unforgeable structure-preserving signature scheme in [4] actually has signatures comprising of three group elements and has two PPEs in signature verification. We present a natural analogue of the scheme in the Type 3 setting, and show that it is more efficient than the Type 2 scheme. In Sect. 4, we present our Type 3 analogue of the Type 2 randomizable structure-preserving signature scheme in [4], and show that the Type 3 scheme is more efficient. In Sect. 5, we present our conversion framework for generic-signer structure-preserving signature schemes from the Type 2 setting to the Type 3 setting, the separation between Types 2 and 3, and the impossibility of having a single pairing-based verification equation in the Type 2 setting. We draw our conclusions in Sect. 6. Two instances of Groth-Sahai NIWI proofs in the Type 2 and Type 3 settings are given in Appendix A.

2   Asymmetric Bilinear Pairings

Let $\mathbb{F}_q$ be a finite field of characteristic $p \geq 5$, and let E be an ordinary elliptic curve defined over $\mathbb{F}_q$. Let n be a prime divisor of $\#E(\mathbb{F}_q)$ satisfying $\gcd(n, q) = 1$, and let k (the embedding degree) be the smallest positive integer such that $n \mid q^k - 1$. We will henceforth assume that k is even, since then some important speedups in pairing computations are applicable [7]. Some prominent families of elliptic curves with even embedding degree include the MNT [23], BN [8], KSS [22], and BLS [6] curves.

Since $k > 1$, we have $E[n] \subseteq E(\mathbb{F}_{q^k})$ where $E[n]$ denotes the n-torsion group of E. Let $G \in E(\mathbb{F}_q)[n]$ be an $\mathbb{F}_q$-rational point of order n, and define $\mathbb{G}_1 = \langle G \rangle$. Let $\mathbb{G}_T$ denote the order-n subgroup of the multiplicative subgroup of $\mathbb{F}_{q^k}$.

Type 3 Pairings. Following [16], we denote by D the CM discriminant of E and set $e = \gcd(k, 6)$ if $D = -3$, $e = \gcd(k, 4)$ if $D = -4$, $e = 2$ if $D < -4$, and $d = k/e$. For example, BN curves have $k = 12$, $e = 6$ and $d = 2$, whereas MNT curves have $k = 6$, $e = 2$ and $d = 3$. Now, E has a unique degree-e twist $\tilde{E}$ defined over $\mathbb{F}_{q^d}$ such that $n \mid \#\tilde{E}(\mathbb{F}_{q^d})$ [21]. Let $\tilde{I} \in \tilde{E}(\mathbb{F}_{q^d})$ be a point of order n, and let $\tilde{\mathbb{G}}_3 = \langle \tilde{I} \rangle$. Then there is a monomorphism $\phi : \tilde{\mathbb{G}}_3 \longrightarrow E(\mathbb{F}_{q^k})$ such that $I = \phi(\tilde{I}) \notin \mathbb{G}_1$. The group $\mathbb{G}_3 = \langle I \rangle$ is the Trace-0 subgroup of $E[n]$, so named because it consists of all points $P \in E[n]$ for which $\text{Tr}(P) = \sum_{i=0}^{k-1} \pi^i(P) = \infty$, where π denotes the q-th power Frobenius. The monomorphism ϕ can be defined so that $\phi : \tilde{\mathbb{G}}_3 \longrightarrow \mathbb{G}_3$ can be efficiently computed in both directions; therefore we can identify $\tilde{\mathbb{G}}_3$ and $\mathbb{G}_3$, and consequently the elements of $\mathbb{G}_3$ can be viewed as having coordinates in $\mathbb{F}_{q^d}$ (instead of in the larger field $\mathbb{F}_{q^k}$).

Non-degenerate bilinear pairings $e_3 : \mathbb{G}_1 \times \mathbb{G}_3 \longrightarrow \mathbb{G}_T$ are said to be of Type 3 because no efficiently-computable isomorphisms from $\mathbb{G}_1$ to $\mathbb{G}_3$ or from $\mathbb{G}_3$ to $\mathbb{G}_1$ are known [16]. There are several Type 3 pairings, of which the most efficient is Vercauteren's optimal pairing [24].

Type 2 Pairings. Let $H \in E[n]$ with $H \notin \mathbb{G}_1$ and $H \notin \mathbb{G}_3$. Then $\mathbb{G}_2 = \langle H \rangle$ is an order-n subgroup of $E(\mathbb{F}_{q^k})$ with $\mathbb{G}_2 \neq \mathbb{G}_1$ and $\mathbb{G}_2 \neq \mathbb{G}_3$. Non-degenerate bilinear pairings $e_2 : \mathbb{G}_1 \times \mathbb{G}_2 \longrightarrow \mathbb{G}_T$ are said to be of Type 2 because the map Tr is an efficiently-computable isomorphism from $\mathbb{G}_2$ to $\mathbb{G}_1$; note, however, that no efficiently-computable isomorphism from $\mathbb{G}_1$ to $\mathbb{G}_2$ is known. These pairings have the property that hashing onto $\mathbb{G}_2$ is infeasible (other than by multiplying H by a randomly selected integer).

The computation of e_2 is efficiently reduced to the task of computing Type 3 pairing e_3 [16]. Thus, the costs of computing e_2 and e_3 are approximately equal. To see this, define the maps $\psi : E[n] \longrightarrow \mathbb{G}_1$, $Q \mapsto \frac{1}{k}\mathrm{Tr}(Q)$ and $\rho : E[n] \longrightarrow \mathbb{G}_3$, $Q \mapsto Q - \psi(Q)$. Recall that e_2 and e_3 are restrictions of the (reduced) Tate pairing $\hat{e} : E[n] \times E[n] \longrightarrow \mathbb{G}_T$. Hence, for all $P \in \mathbb{G}_1$, $Q \in \mathbb{G}_2$, we have

$$e_2(P, Q) = \hat{e}(P, \psi(Q)) \cdot \hat{e}(P, \rho(Q)) = \hat{e}(P, \rho(Q)) = e_3(P, \rho(Q)). \tag{1}$$

Remark 1. Note that the Type 2 setting is equipped with not only the map $\psi : \mathbb{G}_2 \longrightarrow \mathbb{G}_1$ but also the map $\rho : \mathbb{G}_2 \longrightarrow \mathbb{G}_3$. The abstract definition of the Type 2 setting, e.g., in [4], does not capture the latter. However, as we show in the following sections, the map ρ plays a crucial role for a comparative study of the protocols in the Type 2 and Type 3 settings.

Comparing the Performance of Type 2 and Type 3 Pairings. Since points in $\mathbb{G}_2$ have coordinates in $\mathbb{F}_{q^k}$ whereas points in $\mathbb{G}_3$ have coordinates in the proper subfield $\mathbb{F}_{q^d}$, it would appear that the ratio of the bitlengths of points in $\mathbb{G}_2$ and $\mathbb{G}_3$ is k/d. Similarly, the ratio of the costs of addition in $\mathbb{G}_2$ and $\mathbb{G}_3$ can be expected to be k^2/d^2 bit operations (using naive methods for extension field arithmetic). These ratios are given in Table 3 of [16]. However, as observed in [12], points in $\mathbb{G}_2$ have a shorter representation which we describe next. We emphasize that this representation can be used for *all* order-n subgroups $\mathbb{G}_2$ of $E[n]$ different from $\mathbb{G}_1$ and $\mathbb{G}_3$.

Let H be an arbitrary point from $E[n]\backslash(\mathbb{G}_1 \cup \mathbb{G}_3)$, and set $\mathbb{G}_2 = \langle H \rangle$. Define $G = \frac{1}{k}\mathrm{Tr}(H)$ so that the map ψ restricted to $\mathbb{G}_2$ is an efficiently-computable isomorphism from $\mathbb{G}_2$ to $\mathbb{G}_1$ with $\psi(H) = G$. Finally, set $I = H - G$. Then $I \in \mathbb{G}_3$ and the map ρ restricted to $\mathbb{G}_2$ is an efficiently-computable isomorphism from $\mathbb{G}_2$ to $\mathbb{G}_3$ with $\rho(H) = I$.

Now, given a point $Q \in E[n]$, one can efficiently determine the unique points $Q_1 \in \mathbb{G}_1$ and $Q_2 \in \mathbb{G}_3$ such that $Q = Q_1 + Q_2$; namely, $Q_1 = \psi(Q)$ and $Q_2 = \rho(Q) = Q - Q_1$. Writing $D(Q) = (\psi(Q), \rho(Q))$ and letting $\mathbb{H}_2 \subseteq \mathbb{G}_1 \times \mathbb{G}_3$ denote the range of D applied to $\mathbb{G}_2$, we have an efficiently-computable isomorphism $D : \mathbb{G}_2 \longrightarrow \mathbb{H}_2$ whose inverse is also efficiently computable. Hence, without loss of generality, points $Q \in \mathbb{G}_2$ can be represented by a pair of points (Q_1, Q_2) with $Q_1 \in \mathbb{G}_1$ and $Q_2 \in \mathbb{G}_3$. Note that arithmetic in $\mathbb{G}_2$ with this representation is component-wise. Thus the ratio of the bitlengths of points in $\mathbb{G}_2$ and $\mathbb{G}_3$ is in fact $(d+1)/d$, whereas the ratio of the costs of addition in $\mathbb{G}_2$ and $\mathbb{G}_3$ is $(d^2+1)/d^2$. We also have the following simple condition for determining membership of a point $Q \in E[n]$ in $\mathbb{G}_2$.

Lemma 1. *Let $Q \in E[n]$, and let $Q_1 = \psi(Q)$ and $Q_2 = \rho(Q)$. Then $Q \in \mathbb{G}_2$ if and only if $\log_G Q_1 = \log_I Q_2$.*

Proof. Suppose that $Q \in \mathbb{G}_2$, so $Q = \ell H$ for some $\ell \in [0, n-1]$. Then $Q = \ell(G + I) = \ell G + \ell I$. Thus $Q_1 = \ell G$ and $Q_2 = \ell I$, whence $\log_G Q_1 = \log_I Q_2$. The converse is similar. $\square$

Table 2 of [12] lists the costs of performing basic operations in $\mathbb{G}_1$, $\mathbb{G}_2$ and $\mathbb{G}_3$ for a particular BN curve. The table confirms the expectation that basic operations in $\mathbb{G}_2$ are only marginally more expensive than the operations in $\mathbb{G}_3$. One notable exception is that testing membership in $\mathbb{G}_2$ is several times more expensive than testing membership in $\mathbb{G}_1$ and $\mathbb{G}_3$. To see this, let us consider the case of BN curves E defined over $\mathbb{F}_q$ where q and $n = \#E(\mathbb{F}_q)$ are prime; recall that these curves have embedding degree $k = 12$ and $d = 2$. Testing membership of a point Q in $\mathbb{G}_1$ is very efficient, and simply entails verifying that Q has coordinates in $\mathbb{F}_q$ and satisfies the equation that defines the curve, i.e., $Q \in E(\mathbb{F}_q)$. Testing membership of a point Q in $\mathbb{G}_3$ involves a fast check that $\phi^{-1}(Q)$ is in $\tilde{E}(\mathbb{F}_{q^2})$, followed by an exponentiation to verify that $nQ = \infty$. Testing membership in $\mathbb{G}_2$ is more costly since the known methods require two pairing computations. If the shorter representation (as elements of $\mathbb{G}_1 \times \mathbb{G}_3$) is used for $\mathbb{G}_2$ then, by Lemma 1, membership of (Q_1, Q_2) in $\mathbb{G}_2$ can be determined by first checking that $Q_1 \in \mathbb{G}_1$ and $Q_2 \in \mathbb{G}_3$, and then verifying that $e_3(Q_1, I) = e_3(G, Q_2)$ [14]. If the longer representation (as elements of $E(\mathbb{F}_{q^{12}})$) is used for $\mathbb{G}_2$, then membership of Q in $\mathbb{G}_2$ can be determined by first checking that $Q \in E(\mathbb{F}_{q^{12}})$ and $nQ = \infty$, and then verifying that $e_2(\psi(Q), H) = e_2(G, Q)$.

Remark 2. Unlike $\mathbb{G}_1$ and $\mathbb{G}_3$, the group $\mathbb{G}_2$ does not have any special structure, and all the $n - 1$ order-n subgroups of $E[n]$ other than $\mathbb{G}_1$ and $\mathbb{G}_3$ are candidates for $\mathbb{G}_2$. Subgroup membership testing in $\mathbb{G}_2$ is costly because given any arbitrary point Q, the task is to decide whether (i) $Q \in E[n]$ and then whether (ii) $\psi(Q)$ and $\rho(Q)$ have the same discrete log with respect to the generators G and I. Thus a pairing-based verification is assumed to be necessary for a subgroup membership test for $\mathbb{G}_2$ (unless one knows some other efficient method for testing equality of discrete logarithms in $\mathbb{G}_1$ and $\mathbb{G}_2$, e.g., by solving the discrete logarithm problem in $\mathbb{G}_1$ or $\mathbb{G}_2$). As the primary focus of our work is a concrete comparative study of structure-preserving signatures in Types 2 and 3, in the remainder of the paper we perform our analysis based on this reasonable assumption. However, for the sake of completeness, in Remark 5 we comment on why none of the superiority claims [4] of Type 2 structure-preserving signatures over Type 3 will hold even in the hypothetical scenario where an efficient subgroup membership testing in $\mathbb{G}_2$ that does not require pairing computation is discovered.

A Case for Concrete Treatment. Protocol designers usually assume the existence of a bilinear group generator which given a security parameter generates the relevant group descriptions and the bilinear map. This abstraction filters out the interconnection between Type 2 and 3 settings. For example, the existing generic definition of Type 2 pairings is oblivious to the fact that both Type 2 and 3 pairings can be defined over the same elliptic curve and are restrictions of the same function to different subgroups.

In contrast, comparative studies of Type 2 and Type 3 setting, as initiated in [16] or in follow-up works such as [13], are in the concrete security setting. In fact Galbraith et al. noted that the existence of a polynomial-time bilinear group generator assumed in the asymptotic treatment is not always automatic (see Sect. 2.1 of [16]), although it is not a problem in practice as one can efficiently generate a bilinear group description for any concrete security level of interest. For example, the BN family is optimized for the 128-bit security level and the notion of asymptotic security cannot be used in a meaningful way when the underlying pairing is derived from such family of curves.

In particular, when efficiency is being studied one cannot meaningfully distinguish between the Type 2 and Type 3 settings in the asymptotic sense. Clearly, it's the concrete efficiency (e.g., the number of group elements in a signature or the number of PPEs and pairings in verification) that Abe et al. [4] and Barthe et al. [9] are concerned with when they discuss the efficiency or optimality of their constructions of structure-preserving signature in the Type 2 setting.

Thus the focus here, as in [13,16], is on concrete security (along with functionality and efficiency) in the Type 2 and 3 settings. The Type 2 and 3 pairings (i.e., e_2 and e_3) are defined as restrictions of the (reduced) Tate pairing. In the performance comparison above we used the example of BN curves as they yield the most efficient pairings at the 128-bit security level. However, we note that our observations are without loss of generality and apply equally well to asymmetric pairings derived from other prominent families of elliptic curves such as MNT, KS and BLS. Readers are referred to Galbraith et al. [16] for a more general comparative treatment of the Type 2 and 3 settings including a discussion on the high cost of group membership testing for $\mathbb{G}_2$ in the Type 2 setting.[2]

In Sects. 3, 4 and 5, we use multiplicative notation for elements of $\mathbb{G}_1$, $\mathbb{G}_2$ and $\mathbb{G}_3$.

3   Strongly Unforgeable Structure-Preserving Signatures

We present the Type 2 strongly unforgeable SPS from [4] and our Type 3 analogue of it. The Type 3 scheme was obtained by following the general recipe given in [13] for converting a protocol from the Type 2 to the Type 3 setting.

3.1   Type 2 Strongly Unforgeable SPS [4]

1. *Setup.* Let $e_2 : \mathbb{G}_1 \times \mathbb{G}_2 \longrightarrow \mathbb{G}_T$ be a Type 2 pairing where $\mathbb{G}_1$, $\mathbb{G}_2$ and $\mathbb{G}_T$ have order n; G, H are fixed generators of $\mathbb{G}_1$, $\mathbb{G}_2$, respectively.

---

[2] Since this is how Type 2 and Type 3 pairings are currently defined, any concrete efficiency/security treatment must be based on that existing knowledge. No comparative study or claim of superiority of one setting over another will make sense based on hitherto undiscovered mathematical structure. If there is a completely new way to define Type 2 and Type 3 pairings in the future, then of course that will mandate a new concrete analysis of *all* asymmetric pairing-based protocols.

2. *Key generation.* The secret key is $v, w \in_R [1, n-1]$. The public key is (V, W) where $V = G^v$ and $W = G^w$.
3. *Signature generation.* To sign $M \in \mathbb{G}_2$, select $t \in_R [1, n-1]$ and compute $R = H^{t-w}$ and $S = M^{v/t} H^{1/t}$. The signature on M is (R, S).
4. *Signature verification.* To verify a signed message $(M, (R, S))$, check that (a) $M, R, S \in \mathbb{G}_2$; and (b) $e_2(W \psi(R), S) = e_2(V, M) \cdot e_2(G, H)$.

In [4, Theorem 2], the Type 2 scheme is proven strongly secure[3] against generic forgers. Signatures are comprised of two $\mathbb{G}_2$ elements. Signature verification requires three $\mathbb{G}_2$ membership tests and one PPE verification.

3.2 Type 3 Strongly Unforgeable SPS

1. *Setup.* Let $e_3 : \mathbb{G}_1 \times \mathbb{G}_3 \longrightarrow \mathbb{G}_T$ be a Type 3 pairing where $\mathbb{G}_1$, $\mathbb{G}_3$ and $\mathbb{G}_T$ have order n; G, I are fixed generators of $\mathbb{G}_1$, $\mathbb{G}_3$, respectively.
2. *Key generation.* The secret key is $v, w \in_R [1, n-1]$. The public key is (V, W) where $V = G^v$ and $W = G^w$.
3. *Signature generation.* To sign $M \in \mathbb{G}_3$, select $t \in_R [1, n-1]$ and compute $R_1 = G^{t-w}$, $R_2 = I^{t-w}$, and $S = M^{v/t} I^{1/t}$. The signature on M is (R_1, R_2, S).
4. *Signature verification.* To verify a signed message $(M, (R_1, R_2, S))$, check that
   (a) $R_1 \in \mathbb{G}_1$ and $M, R_2, S \in \mathbb{G}_3$;
   (b) $e_3(R_1, I) = e_3(G, R_2)$; and
   (c) $e_3(W R_1, S) = e_3(V, M) \cdot e_3(G, I)$.

It is easy to verify correctness of the Type 3 signature scheme. The security proof given in [4, Theorem 2] that the Type 2 scheme is strongly secure against generic forgers also applies (with minimal changes) to the Type 3 signature scheme. The reason that the proof carries over with minimal changes is that we follow the strategy of [13] in the conversion. The Type 3 scheme is obtained by first replacing all $\mathbb{G}_2$ elements by the corresponding $\mathbb{H}_2$ elements and then discarding the redundant $\mathbb{G}_1$ elements that are not used either in the construction or in security argument in the Type 2 setting.

Signatures for the Type 3 scheme are comprised of one $\mathbb{G}_1$ element and two $\mathbb{G}_3$ elements. Signature verification requires one $\mathbb{G}_1$ membership test, three $\mathbb{G}_3$ membership tests, and two PPE verifications.

We note that the verification step 4(b) of the Type 3 scheme cannot be omitted. Indeed, if this step is omitted then the scheme succumbs to the following key-only attack: $(1, (W^{-1}G, 1, I))$ is a valid forgery. Moreover, even if the message $M = 1$ is disallowed, the scheme succumbs to the following random message attack. The forger first obtains a signed message $(M, (R_1, R_2, S))$. It then computes $M' = MS^{-1}$ and $R_1' = R_1 V^{-1}$, thereby obtaining a valid forgery $(M', (R_1', R_2, S))$. We note that this attack is anticipated by the proof of Theorem 2 in [2] which establishes that any Type 3 structure-preserving signature scheme with a single verification equation is existentially forgeable under random message attack.

---

[3] A signature scheme is said to be *secure* if it is existentially unforgeable under chosen-message attack. If, in addition, it is infeasible to find a new signature for a message that has already been signed, then the signature scheme is said to be *strongly secure*.

3.3   Comparisons

Signature Size. Signatures in the Type 2 scheme are comprised of two $\mathbb{G}_2$ elements or, equivalently, two $\mathbb{G}_1$ and two $\mathbb{G}_3$ elements. Thus, signatures in the Type 3 scheme are smaller than signatures in the Type 2 scheme.

Signature Generation Cost. In signature generation, computing $R = H^{t-w}$ for the Type 2 scheme has exactly the same cost as computing $R_1 = G^{t-w}$ and $R_2 = I^{t-w}$ for the Type 3 scheme. However, the computation of $S = M^{v/t}H^{1/t}$ in the Type 2 scheme is significantly slower than in the Type 3 scheme since the computation takes place in $\mathbb{G}_2$ in the former and in $\mathbb{G}_3$ in the latter. Thus, signature generation is slower in the Type 2 scheme than in the Type 3 scheme.

Signature Verification Cost. Signature verification in the Type 2 scheme is significantly slower than in the Type 3 scheme. This is because, as explained in Sect. 2, the subgroup membership tests $M, R, S \in \mathbb{G}_2$ required in the Type 2 scheme each requires the verification of a PPE, whereas the subgroup membership tests $R_1 \in \mathbb{G}_1$ and $M, R_2, S \in \mathbb{G}_3$ in the Type 3 scheme are relatively inexpensive. Thus, signature verification in the Type 2 scheme requires *four* PPE verifications, whereas only *two* are needed in the Type 3 scheme. Note that the high cost of PPE verifications can be mitigated by batching [10, 15].

The costly subgroup membership tests in step 4(a) of the Type 2 scheme cannot be omitted for two reasons. First, if these tests are omitted then the security proof given in [4] is no longer applicable since the proof makes the assumption that $M, R, S \in \mathbb{G}_2$. Second, and more importantly, there are attacks on the scheme if the membership tests are omitted. For example, given a valid signed message $(M, (R, S))$, one can easily[4] select a second point $R' \in E[n]$ with $R' \neq R$ and $\psi(R') = \psi(R)$, thereby obtaining a second valid signed message $(M, (R', S))$. Similarly, given $(M, (R, S))$ one can obtain a second valid signed message $(M', (R, S))$ or $(M, (R, S'))$ if membership tests for M or S are omitted.

Cost of Signature Verification with Groth-Sahai Proofs. SPS schemes were not designed to be used as stand-alone primitives, but rather in conjunction with non-interactive proof systems like Groth-Sahai as explained in Sect. 1. Suppose that Groth-Sahai proof verification always requires subgroup membership tests for the group elements in commitment and proof as described in Appendix A. Now the pertinent question is whether in Type 2 it is possible to give a proof for a single PPE as opposed to two PPEs in Type 3. This may give some advantage to the Type 2 scheme because the cost of a Groth-Sahai proof depends heavily on the number of PPEs in signature verification.

Consider the Type 2 signature scheme of Abe et al. when used in conjunction with a Groth-Sahai proof. The prover provides a commitment of $(M, (R, S))$

---

[4] Given $R \in \mathbb{G}_2$, one computes $R_1 = \psi(R)$ and selects arbitrary $R_2' \in \mathbb{G}_3$ with $R_2' \neq R \cdot R_1^{-1}$. Then $R' = R_1 \cdot R_2'$ satisfies $\psi(R') = R_1$ and $R' \neq R$.

together with a proof that the committed values satisfy the following PPE:

$$e_2(W\psi(R), S) = e_2(V, M) \cdot e_2(G, H). \tag{2}$$

In this proof system, the group elements G, H, V and W are known to the verifier, whereas the variables are $M, R, S \in \mathbb{G}_2$. However, since Groth-Sahai proofs do not have a mechanism for incorporating the evaluation of $\psi(R)$, the variables in (2) are actually M, $\psi(R)$ and S. In other words, a Groth-Sahai proof for (2) only convinces a verifier that the prover knows $R_1 \in \mathbb{G}_1$ and $M, S \in \mathbb{G}_2$ that satisfy the following PPE:

$$e_2(WR_1, S) = e_2(V, M) \cdot e_2(G, H). \tag{3}$$

In particular, the proof does *not* establish that the prover knows $R \in \mathbb{G}_2$ such that $R_1 = \psi(R)$. As we have shown above, unless the prover establishes that s/he knows $R \in \mathbb{G}_2$ which has the same discrete logarithm to the base $H \in \mathbb{G}_2$ as R_1 to the base $G \in \mathbb{G}_1$, the signature scheme is insecure, i.e., not (strongly) unforgeable. Thus, as per the Groth-Sahai proof system, the prover needs to convince the verifier that it possesses a solution (M, R_1, R, S) to the following collection of PPEs:

$$e_2(WR_1, S) = e_2(V, M) \cdot e_2(G, H) \tag{4}$$

$$e_2(R_1, H) = e_2(G, R). \tag{5}$$

When composed with Groth-Sahai proof systems, the verification now has *two* PPEs (note that batching does not work in this scenario). This is in contrast to the claim made in [4] that the Type 2 signature scheme of Sect. 3.1 has only *one* PPE. Moreover, in addition to R, S, the prover has to commit to R_1 in the Groth-Sahai proof. So when composed with Groth-Sahai, signatures are comprised of *three* group elements, i.e., $R_1 \in \mathbb{G}_1$ must be included in the signature along with $R, S \in \mathbb{G}_2$.

Recall that the Type 3 signature scheme in Sect. 3.2 also has two PPEs in verification and signatures that are comprised of three group elements. Thus, it might appear at first glance that signature verification for the Type 2 and Type 3 schemes costs roughly the same when used in conjunction with Groth-Sahai proofs. However, the Groth-Sahai proofs for the Type 2 setting are based on hardness of the decisional linear (DLIN) problem in $\mathbb{G}_2$ [17], whereas Groth-Sahai proofs for the Type 3 setting can be based on hardness of the decisional Diffie-Hellman (DDH) problem in $\mathbb{G}_1$ and $\mathbb{G}_3$ [19]. Now, DLIN-based Groth-Sahai proofs are significantly more costly than DDH-based Groth-Sahai proofs in terms of commitment size, proof size, and the total number of pairing computations in proof verification. For example, one can see that the DLIN-based proof of knowledge of a solution (X, Y) to the equation $e_2(A, X) \cdot e_2(B, Y) = t$ in Appendix A.1 is significantly more costly than the DDH-based proof of knowledge of a solution (X, Y) to the equation $e_3(A, X) \cdot e_3(B, Y) = t$ in Appendix A.2; see also the performance estimates given in Sect. 3.4 of [11]. Thus, the Type 2 structure-preserving signature scheme will be significantly slower than its Type 3 counterpart when combined with Groth-Sahai proofs.

Conclusions. The Type 3 strongly unforgeable structure-preserving signature scheme is superior to its Type 2 counterpart with respect to signature size, signature generation cost, and signature verification cost when the schemes are used as stand-alone signature schemes and when used in conjunction with Groth-Sahai proofs. Moreover, the schemes have similar security proofs against generic forgers. Thus, the Type 2 scheme offers no advantages over the Type 3 scheme.

4   Randomizable Structure-Preserving Signatures

We present the Type 2 randomizable structure-preserving signature scheme from [4] and our Type 3 analogue of it. The Type 3 scheme was obtained by following the general recipe given in [13] for converting a protocol from the Type 2 setting to the Type 3 setting.

4.1   Type 2 Randomizable SPS [4]

1. *Setup.* Let $e_2 : \mathbb{G}_1 \times \mathbb{G}_2 \longrightarrow \mathbb{G}_T$ be a Type 2 pairing where $\mathbb{G}_1$, $\mathbb{G}_2$ and $\mathbb{G}_T$ have order n; G, H are fixed generators of $\mathbb{G}_1$, $\mathbb{G}_2$, respectively.
2. *Key generation.* The secret key is $v, w \in_R [1, n-1]$. The public key is (V, W) where $V = G^v$ and $W = G^w$.
3. *Signature generation.* To sign $M \in \mathbb{G}_2$, select $r \in_R [1, n-1]$ and compute $R = H^r$ and $S = M^v H^{r^2 + w}$. The signature on M is (R, S).
4. *Randomization.* To randomize $(M, (R, S))$, select $\alpha \in_R [1, n-1]$ and compute $R' = RH^\alpha$ and $S' = SR^{2\alpha} H^{\alpha^2}$. The randomized signature on M is (R', S').
5. *Signature verification.* To verify a signed message $(M, (R, S))$, check that (a) $M, R, S \in \mathbb{G}_2$; and (b) $e_2(G, S) = e_2(V, M) \cdot e_2(\psi(R), R) \cdot e_2(W, H)$.

In [4, Theorem 1], the Type 2 scheme is proven secure against generic forgers. Signatures are comprised of two $\mathbb{G}_2$ elements. Signature verification requires three $\mathbb{G}_2$ membership tests and one PPE verification.

4.2   Type 3 Randomizable SPS

1. *Setup.* Let $e_3 : \mathbb{G}_1 \times \mathbb{G}_3 \longrightarrow \mathbb{G}_T$ be a Type 3 pairing, where $\mathbb{G}_1$, $\mathbb{G}_3$ and $\mathbb{G}_T$ have order n; G, I are fixed generators of $\mathbb{G}_1$, $\mathbb{G}_3$, respectively.
2. *Key generation.* The secret key is $v, w \in_R [1, n-1]$. The public key is (V, W) where $V = G^v$ and $W = G^w$.
3. *Signature generation.* To sign $M \in \mathbb{G}_3$, select $r \in_R [1, n-1]$ and compute $R_1 = G^r$, $R_2 = I^r$ and $S = M^v I^{r^2 + w}$. The signature on M is (R_1, R_2, S).
4. *Randomization.* To randomize $(M, (R_1, R_2, S))$, select $\alpha \in_R [1, n-1]$ and compute $R_1' = R_1 G^\alpha$, $R_2' = R_2 I^\alpha$, and $S' = SR_2^{2\alpha} I^{\alpha^2}$. The randomized signature on M is (R_1', R_2', S').
5. *Signature verification.* To verify a signed message $(M, (R_1, R_2, S))$, check that
   (a) $R_1 \in \mathbb{G}_1$ and $M, R_2, S \in \mathbb{G}_3$;
   (b) $e_3(R_1, I) = e_3(G, R_2)$; and

(c) $e_3(G, S) = e_3(V, M) \cdot e_3(R_1, R_2) \cdot e_3(W, I)$.

It is easy to verify correctness of the Type 3 scheme. Following the strategy outlined in Sect. 3.2, the security proof given in [4, Theorem 1] that the Type 2 scheme is secure against generic forgers can be modified (with minimal changes) for the Type 3 signature scheme.

Signatures for the Type 3 scheme are comprised of one $\mathbb{G}_1$ element and two $\mathbb{G}_3$ elements. Signature verification requires one $\mathbb{G}_1$ membership test, three $\mathbb{G}_3$ membership tests, and two PPE verifications.

We note that the verification equation in step 5(b) of the Type 3 scheme cannot be omitted. Indeed, if this step is omitted then the scheme succumbs to the following random message attack. The forger first obtains a signed message $(M, (R_1, R_2, S))$. It then computes $M' = MR_2$ and $R_1' = R_1 V^{-1}$, thereby obtaining a valid forgery $(M', (R_1', R_2, S))$. Indeed, this attack is anticipated by the proof of Theorem 2 of [2].

4.3   Comparisons

The subgroup membership tests performed in step 5(a) of the Type 2 randomizable structure-preserving signature scheme cannot be omitted. If they are, then an attacker can proceed as follows. Having obtained a valid message-signature pair $(M, (R, S))$, she computes $M' = MR$ and $R' = RV^{-1}$. Note that $\rho(R') = \rho(R)$. Then $(M', (R', S))$ is a valid signed message since the term $e_2(V, M) \cdot e_2(\psi(R), R)$ in step 5(b) of signature verification remains unchanged:

$$
\begin{aligned}
e_2(V, M') \cdot e_2(\psi(R'), R') &= e_2(V, MR) \cdot e_2(\psi(R) \cdot \psi(V^{-1}), R') \\
&= e_2(V, M) \cdot e_2(V, R) \cdot e_2(\psi(R), R') \cdot e_2(\psi(V), R')^{-1} \\
&= e_2(V, M) \cdot e_3(V, \rho(R)) \cdot e_3(\psi(R), \rho(R)) \cdot e_3(V, \rho(R))^{-1} \\
&= e_2(V, M) \cdot e_2(\psi(R), R).
\end{aligned}
$$

The comparisons made between the Type 2 and Type 3 strongly unforgeable structure-preserving signature schemes in Sect. 3.3 are also valid for the Type 2 and Type 3 randomizable structure-preserving signature schemes in Sects. 4.1 and 4.2. Namely, the Type 3 scheme has smaller signatures, faster signature generation, faster signature verification in stand-alone applications (since it requires the verification of two PPEs instead of four PPEs for the Type 2 scheme), and faster signature verification when used with Groth-Sahai proofs (since both schemes have two PPEs and three group elements in signatures, but the Type 3 proofs are DDH-based instead of DLIN-based).

As mentioned in [4], randomizable structure-preserving signature schemes are useful in building anonymization protocols because the signature component that is uniformly distributed and independent of the message can be revealed without leaking any information about the message or the original signature from which the randomized signature was derived. In the Type 2 randomizable signature scheme of Sect. 4.1, the signature component R can be made public. In that case, only the single PPE in step 5(b) of signature verification needs to be transformed when used in conjunction with Groth-Sahai proofs (and the PPE is of the form described in Appendix A.1). Similarly, in the Type 3 randomizable signature scheme of Sect. 4.2, the signature components R_1 and R_2 can be made public. In that case, only the single PPE in step 5(c) of signature verification needs to be transformed when used in conjunction with Groth-Sahai proofs (and the PPE is of the form described in Appendix A.2).

In both situations, i.e., whether the message-independent signature components are made public or not, the Type 3 scheme is superior in all respects to its Type 2 counterpart.

4.4 Strongly-Optimal Signatures

In a recent paper, Barthe et al. [9] investigated the optimal number of pairings for structure-preserving signature. The question is indeed well motivated as the Groth-Sahai proof complexity also depends on the number of pairings in each PPE. Barthe et al. work in the Type 2 setting as that supposedly allows a single PPE based verification and explicitly disregard the PPEs in group membership testing for $\mathbb{G}_2$ elements in the verification. This is justified by stating that such tests "may require an amortizable (aka offline) pairing computation in practical instantiation". However, this is a *not* a valid assumption, particularly when the main goal of [9] is to find a lower bound on the "concrete number of pairings" and optimal construction meeting that bound. As we have already pointed out in the context of the Abe et al. constructions [4], one cannot in general ignore the pairing-based verification equations involved in $\mathbb{G}_2$ membership testing either in the stand-alone setting or in conjunction with Groth-Sahai proofs. It is also evident that these pairings cannot be treated as offline (and thereby, amortizable) since they involve message and/or signature elements.

Assuming that signature verification involves a single PPE, Barthe et al. [9] derive a lower bound of three pairings for CMA-secure construction and two pairings for RMA security in the generic Type 2 setting. They use an automated tool to obtain signature schemes matching these lower bounds which they term as "strongly optimal". However, when their abstract construction is translated to the concrete Type 2 setting, then we see that the CMA-secure scheme actually requires *six* more additional pairings, none of which can be made offline. Incidentally, following the general recipe of [13], they also propose a Type 3 counterpart that requires a total of five pairings of which only three are online.

More interesting is the case of their RMA-secure construction in the Type 2 setting which is claimed to have only two online pairings, whereas in concrete terms six additional online pairings will be required. Now consider the scenario when this signature scheme is composed with Groth-Sahai proofs. Given a signature $(R, S) \in \mathbb{G}_2^2$ for $M \in \mathbb{G}_2$, their verification equation[5] is of the form

$$e_2(\psi(S) \cdot W, H) = e(\psi(R) \cdot V, M).$$

---

[5] We correct a typo in [9] where R is used in the equation.

As the scheme is randomizable, the message-independent random group element R in the signature can be revealed but not the signature element $S \in \mathbb{G}_2$. As we already pointed out in the context of the Abe et al. strongly-unforgeable signature, Groth-Sahai proofs do not have any mechanism for incorporating the evaluation of $\psi(S)$. Hence, the signature now has an additional component $\psi(S) \in \mathbb{G}_1$ and verification involves one additional PPE:

$$e_2(\psi(S), H) = e_2(G, S).$$

Clearly, the signature contains three group elements and verification involves four online pairings that need to be counted when the scheme is composed with a Groth-Sahai proof.

5   A Closer Look at Type 2 Schemes

We first establish that *all* Type 2 generic-signer structure-preserving signature schemes can be transformed to the Type 3 setting without any penalty in security or efficiency.[6] Next, we demonstrate the impossibility of having signature verification with a single pairing-product equation in the Type 2 setting when messages are drawn from $\mathbb{G}_2$. Finally, we show a separation between the Type 2 and Type 3 settings by proposing a Type 3 signature scheme that has no secure Type 2 counterpart.

Based on the claimed optimality of their Type 2 schemes, Abe et al. [4] asserted that the Type 2 setting is different from Type 3 setting as it "permits the construction of cryptographic schemes with unique properties". This, according to [4], settles the open question in [13] of whether all Type 2 schemes can be converted to the Type 3 setting with no efficiency loss. In contrast, the results of this section formally establish that *all* Type 2 generic-signer structure-preserving signature schemes are merely Type 3 schemes in disguise and cannot beat the established lower bound results even when messages are drawn from $\mathbb{G}_2$.

5.1   Conversion from Type 2 to Type 3

Recall the definition of structure-preserving signatures (SPS) from [4, Definition 4]. Based on that definition, any generic-signer structure-preserving signature scheme with message space $\mathbb{G}_2$ can be described as follows. The conversion framework with message space $\mathbb{G}_1$ is analogous.

SPS-T2

1. *Setup.* Let $e_2 : \mathbb{G}_1 \times \mathbb{G}_2 \longrightarrow \mathbb{G}_T$ be a Type 2 pairing where $\mathbb{G}_1$, $\mathbb{G}_2$ and $\mathbb{G}_T$ have order n; G, H are fixed generators of $\mathbb{G}_1$, $\mathbb{G}_2$, respectively.

---

[6] Our transformation uses the concrete (and only known) mathematical structure over which Type 2 and Type 3 pairings are defined. This concreteness does not cause the transformation to lose its generality since *any* Type 2 structure-preserving signature scheme can be converted using our framework.

2. *Key generation.* The secret key contains elements $u_1, u_2, \ldots, v_1, v_2, \ldots \in_R [1, n-1]$. The public key contains elements $U_1, U_2, \ldots \in \mathbb{G}_1$, $V_1, V_2, \ldots \in \mathbb{G}_2$, where $U_i = G^{u_i}$ and $V_j = H^{v_j}$. Note that because the signer is generic, we can assume without loss of generality that the signer knows the discrete logarithm of the U_i and the V_j.

3. *Signature generation.* The message is $M \in \mathbb{G}_2$. However, unlike the public key, we cannot in general assume that the signer knows the discrete logarithm of $M = H^m$. The signing algorithm is restricted to generic group operations, so a generic signer can only construct signature elements of the form $S_i = \psi(M)^{\alpha_i} G^{\beta_i} \in \mathbb{G}_1$ and $T_j = M^{\gamma_j} H^{\delta_j}$ where $\alpha_i, \beta_i, \gamma_j, \delta_j \in [0, n-1]$ are independent of m. Finally, the algorithm outputs a signature containing elements $(S_1, S_2, \ldots) \in \mathbb{G}_1$ and $(T_1, T_2, \ldots) \in \mathbb{G}_2$.

4. *Signature verification.* Given message M and a corresponding signature of the form $(S_1, S_2, \ldots, T_1, T_2, \ldots)$, the verifier does the following:
   (a) check that $S_1, S_2, \ldots \in \mathbb{G}_1$;
   (b) check that $M \in \mathbb{G}_2$ and $T_1, T_2, \ldots \in \mathbb{G}_2$;
   (c) verify a collection of equations of the following form:

$$\prod_i \prod_j e_2(S_i, T_j)^{a_{qij}} \cdot \prod_i \prod_j e_2(S_i, V_j)^{b_{qij}} \cdot \prod_j e_2(\psi(M), T_j)^{c_{qj}}$$

$$\cdot \prod_j e_2(\psi(M), V_j)^{d_{qj}} \cdot \prod_i e_2(S_i, M)^{e_{qi}} \cdot \prod_i e_2(U_i, M)^{f_{qi}}$$

$$\cdot \prod_i \prod_j e_2(U_i, T_j)^{g_{qij}} \cdot e_2(\psi(M), M)^{h_q} = 1.$$

Note: We use the augmented set $S = \{S_1, S_2, \ldots\} \cup \{\psi(T_1), \psi(T_2), \ldots\}$ in the above verification equation. However, there is no need to consider the elements $\psi(V_j)$ separately because they can, without loss of generality, be included in the public key. The constant exponents $a_{qij}, b_{qij}, \ldots$ from $[0, n-1]$ used in the verification equations are specified as part of the signature verification algorithm.

We now propose the following transformation to convert SPS-T2 from the Type 2 to the Type 3 setting. The transformation uses the efficiently-computable isomorphism $D : \mathbb{G}_2 \longrightarrow \mathbb{H}_2$ given by $D(Q) = (\psi(Q), \rho(Q))$ where $\mathbb{H}_2 \subseteq \mathbb{G}_1 \times \mathbb{G}_3$ (see Sect. 2). Our strategy is very simple: apply D so that all $\mathbb{G}_2$ elements in SPS-T2 are replaced by their "shorter representation" as elements of $\mathbb{H}_2$. This strategy, together with the observation that the computation of a Type 2 pairing e_2 is efficiently reduced to the task of computing a Type 3 pairing e_3 (see Eq. (1)), immediately yields the following Type 3 structure-preserving signature scheme.

SPS-T3

1. *Setup.* Let $e_3 : \mathbb{G}_1 \times \mathbb{G}_3 \longrightarrow \mathbb{G}_T$ be a Type 3 pairing where $\mathbb{G}_1$, $\mathbb{G}_3$ and $\mathbb{G}_T$ have order n; G, I are fixed generators of $\mathbb{G}_1$, $\mathbb{G}_3$, respectively.

2. *Key generation.* For each element $V_j = H^{v_j}$ in SPS-T2, compute $V_{j_1} = G^{v_j}$ and $V_{j_2} = I^{v_j}$. The secret key contains elements $u_1, u_2, \ldots, v_1, v_2, \ldots \in_R [1, n-1]$. The public key contains elements $U_1, U_2, \ldots \in \mathbb{G}_1$ (as in SPS-T2) and $(V_{1_1}, V_{1_2}), (V_{2_1}, V_{2_2}), \ldots \in \mathbb{H}_2$.

3. *Signature generation.* The message $M = H^m$ in SPS-T2 can be written as $(M_1, M_2) = (G^m, I^m) \in \mathbb{H}_2$. Recall that using generic group operations, a generic signer in SPS-T2 can only construct $S_i = M_1^{\alpha_i} G^{\beta_i}$ and $T_j = M^{\gamma_j} H^{\delta_j}$ where $\alpha_i, \beta_i, \gamma_j, \delta_j$ are independent of m. Representing T_j as an element of $\mathbb{H}_2$ we have $T_j = (T_{j_1}, T_{j_2}) = (M_1^{\gamma_j} G^{\delta_j}, M_2^{\gamma_j} I^{\delta_j}) \in \mathbb{H}_2$. It is easy to see that a generic signer can compute the signature element $T_j \in \mathbb{G}_2$ if and only if she can compute $M_1^{\gamma_j} G^{\delta_j} \in \mathbb{G}_1$ and $M_2^{\gamma_j} I^{\delta_j} \in \mathbb{G}_3$. Using the above idea we can convert each signature element $T_j \in \mathbb{G}_2$ of SPS-T2 to $(T_{j_1}, T_{j_2}) \in \mathbb{H}_2$ and thereby obtain the corresponding signature elements in SPS-T3. Finally, the algorithm outputs a signature of the form $S_1, S_2, \ldots \in \mathbb{G}_1$ and $(T_{1_1}, T_{1_2}), (T_{2_1}, T_{2_2}), \ldots \in \mathbb{H}_2$.

4. *Signature verification.* Given a message (M_1, M_2) and corresponding signature $(S_1, S_2, \ldots, (T_{1_1}, T_{1_2}), (T_{2_1}, T_{2_2}), \ldots)$, the verifier does the following:
   (a) check that $S_1, S_2, \ldots \in \mathbb{G}_1$;
   (b) check that $(M_1, M_2), (T_{1_1}, T_{1_2}), (T_{2_1}, T_{2_2}), \ldots \in \mathbb{H}_2$;
   (c) verify a set of equations of the following form:

$$\prod_i \prod_j e_3(S_i, T_{j_2})^{a_{qij}} \cdot \prod_i \prod_j e_3(S_i, V_{j_2})^{b_{qij}} \cdot \prod_j e_3(M_1, T_{j_2})^{c_{qj}}$$

$$\cdot \prod_j e_3(M_1, V_{j_2})^{d_{qj}} \cdot \prod_i e_3(S_i, M_2)^{e_{qi}} \cdot \prod_i e_3(U_i, M_2)^{f_{qi}}$$

$$\cdot \prod_i \prod_j e_3(U_i, T_{j_2})^{g_{qij}} \cdot e_3(M_1, M_2)^{h_q} = 1.$$

Note: We use the augmented set $S = \{S_1, S_2, \ldots\} \cup \{T_{1_1}, T_{2_1}, \ldots\}$ in the above verification equation. As already observed in the context of SPS-T2, there is no need to consider the public key elements $V_{1_1}, V_{2_1}, \ldots$ separately and the constants in the exponent are specified in the verification algorithm.

Correctness of SPS-T3 follows directly from the correctness of SPS-T2. Moreover, SPS-T3 maintains all the claimed benefits of SPS-T2. We now show that SPS-T3 is as secure as its original Type 2 counterpart SPS-T2. For concreteness, the security argument is sketched for existential unforgeability under chosen message attack (EUF-CMA), but it is easy to see that the argument extends to other standard notions of security such as EUF-RMA and strong unforgeability under chosen/random message attack.

Claim 2. *SPS-T2 is* EUF-CMA-*secure if and only if SPS-T3 is* EUF-CMA-*secure.*

Proof. In the framework of the conversion described above, we have consistently replaced all $\mathbb{G}_2$ elements in SPS-T2 by the corresponding $\mathbb{H}_2$ elements to derive the corresponding algorithms of SPS-T3. Recall that $D : \mathbb{G}_2 \longrightarrow \mathbb{H}_2$ is an efficiently-computable isomorphism whose inverse is also efficiently computable. Hence, given an EUF-CMA adversary against SPS-T3, one can easily construct an EUF-CMA adversary against SPS-T2 and vice versa.   □

Remark 3. SPS-T3 does not have any efficiency gain (or loss) compared to SPS-T2. Further optimizations for SPS-T3 are usually possible by removing some redundant group elements after a careful scrutiny of the construction and its security argument as suggested in [13]. For example, the Type 3 schemes described in Sects. 3 and 4 are optimized versions of their Type 2 counterparts obtained by following the general recipe given above.

Remark 4. The subgroup membership tests described in step 4(b) of SPS-T2 and SPS-T3 involve pairing-based verification equations. We have observed in Sects. 3 and 4 that avoiding subgroup membership tests can lead to a random message attack in both the Type 2 and 3 settings. Apart from these pairing-based verifications of subgroup membership, signature verification will involve at least one more pairing product equation. See the proof of Theorem 3 for further details.

Remark 5. Consider the following hypothetical situation. Working within the mathematical structure of asymmetric pairings described in Sect. 2, someone in the future discovers an efficient method for membership testing in $\mathbb{G}_2$ that does not require a pairing computation. By Lemma 1, the pairing-based verifications in the Type 3 setting for testing whether $(Q_1, Q_2) \in \mathbb{H}_2$ (see step 4(b) in SPS-T3) will no longer be required. This simple observation together with Claim 2 immediately shows that if there exists, say, an EUF-CMA secure structure-preserving signature scheme in Type 2 with a single PPE-based verification, then there exists an EUF-CMA secure structure-preserving signature scheme in Type 3 with a single PPE-based verification. For example, if the membership testing in $\mathbb{G}_2$ in the verification step of the Type 2 randomizable SPS of [4] can be performed without pairing then the verification in step 5(b) of the Type 3 randomizable SPS of Sect. 4.2 can be replaced by a pairing-free check of $(R_1, R_2) \in \mathbb{H}_2$, leading to a single PPE-based verification in the Type 3 setting. Consequently, our hypothetical situation will refute the Abe et al. assertion [4] that, unlike the Type 2 setting, in the Type 3 setting no secure structure-preserving signature scheme can have a single PPE-based verification. Further, when read in conjunction with Claim 2 and Remark 3, it is easy to see that none of the superiority claims in [4] of a structure-preserving signature scheme in Type 2 over Type 3 will hold even in this hypothetical scenario.

5.2   Impossibility of Single PPE in Verification

In Theorem 2 of [2], Abe et al. showed that there is no Type 3 structure-preserving signature scheme with a single pairing-based verification equation that is existentially unforgeable under random message attack. The original argument was for messages in $\mathbb{G}_1$, but can be easily extended when messages are from $\mathbb{G}_3$. In Theorem 3 of [4], Abe et al. showed a similar impossibility result for Type 2 structure-preserving signature schemes with messages in $\mathbb{G}_1$.

Assuming that the hypothetical scenario discussed in Remark 5 does not occur[7], one can generalize the above results to show that the impossibility holds even when the messages are drawn from $\mathbb{H}_2$. As a corollary, one concludes that there is no Type 2 SPS scheme with a single pairing-based verification equation that is existentially unforgeable under random message attack.

Theorem 3. *No structure-preserving signature scheme with a single pairing-product equation based signature verification is secure in the sense of existential unforgeability under random message attack.*

Proof. The case of messages in $\mathbb{G}_1$ in the Type 3 setting (resp. the Type 2 setting) is proved in [2, Theorem 2] (resp. [4, Theorem 3]). The case of messages in $\mathbb{G}_3$ in the Type 3 setting is analogous to the proof of Theorem 2 in [2]. The case of the Type 1 setting was settled in [3, Theorem 4].

We now show the same impossibility for messages in $\mathbb{G}_2$. For ease of exposition, we will use the structure of SPS-T3, which we have already shown equivalent to SPS-T2, and the message space $\mathbb{H}_2$ (recall that $\mathbb{H}_2$ is isomorphic to $\mathbb{G}_2$, and that an element of $\mathbb{H}_2$ is comprised of a pair in $\mathbb{G}_1 \times \mathbb{G}_3$ the components of which have the same discrete logarithm with respect to the fixed generators G and I). Our argument closely follows the proof of Theorem 2 from [2] but needs to take care of additional complications due to the structure of $\mathbb{H}_2$.

Recall the signature verification for SPS-T3 where in step 4(c) we described the general form of a verification equation. Our claim is that having a *single* verification equation of the form 4(c) and omitting the subgroup membership test in step 4(b) lead to a random message attack. In other words, signature verification must involve more than one PPEs (some of which may be in the disguise of subgroup membership test for $\mathbb{H}_2$ i.e., $\mathbb{G}_2$). For simplicity, we assume that the signature contains two elements of $\mathbb{H}_2$. Note that Abe et al. claim that two group elements is the optimal signature size in Type 2 – see Table 1 of [4]. However, it is easy to see that our result holds for the more general case.

Consider a structure-preserving signature scheme for messages in $\mathbb{H}_2$ with verification key containing group elements $U_1, U_2, \ldots \in \mathbb{G}_1$, $V_1, V_2, \ldots \in \mathbb{G}_3$, and $Z \in \mathbb{G}_T$.[8] For simplicity, in the following we consider two U_i's and two V_i's in the verification key. A signature is of the form $(S_1, T_1), (S_2, T_2) \in \mathbb{H}_2$ and is verified

---

[7] Note that Theorem 2 of [2] has to be interpreted modulo the (implicit) assumption that the hypothetical scenario discussed in Remark 5 does not occur. Otherwise, the Type 3 randomizable structure-preserving signature scheme in Sect. 4.2 with message space $\mathbb{G}_3$ and its dual discussed later in Sect. 5.3 with message space $\mathbb{G}_1$ will contradict the impossibility result of [2].

[8] Here, as in [2], we have relaxed the original definition of structure-preserving signatures to allow the public verification key to contain an arbitrary element Z from $\mathbb{G}_T$ that appears in the verification equation. As already observed in [2], the relaxation strengthens the impossibility result.

by the following PPE:

$$e_3(S_1, T_1)^{a_{11}} \cdot e_3(S_1, T_2)^{a_{12}} \cdot e_3(S_2, T_1)^{a_{21}} \cdot e_3(S_2, T_2)^{a_{22}}$$
$$\cdot e_3(S_1, V_1)^{b_{11}} \cdot e_3(S_1, V_2)^{b_{12}} \cdot e_3(S_2, V_1)^{b_{21}} \cdot e_3(S_2, V_2)^{b_{22}}$$
$$\cdot e_3(M_1, T_1)^{c_{11}} \cdot e_3(M_1, T_2)^{c_{12}} \cdot e_3(M_1, V_1)^{d_{11}} \cdot e_3(M_1, V_2)^{d_{12}}$$
$$\cdot e_3(S_1, M_2)^{c_{21}} \cdot e_3(S_2, M_2)^{c_{22}} \cdot e_3(U_1, M_2)^{d_{21}} \cdot e_3(U_2, M_2)^{d_{22}}$$
$$\cdot e_3(U_1, T_1)^{e_{11}} \cdot e_3(U_1, T_2)^{e_{12}} \cdot e_3(U_2, T_1)^{e_{21}} \cdot e_3(U_2, T_2)^{e_{22}}$$
$$\cdot e_3(M_1, M_2)^f = Z.$$

Note that terms such as $e_3(U_i, V_j)$ can be incorporated in $Z \in \mathbb{G}_T$ without any loss of generality.

Given a signature $(S_1, T_1), (S_2, T_2) \in \mathbb{H}_2$ on a random message $(M_1, M_2) \in \mathbb{H}_2$, we isolate S_1, S_2 and M_2 in the verification equation to obtain:

$$A_1 = T_1^{a_{11}} T_2^{a_{12}} V_1^{b_{11}} V_2^{b_{12}} \qquad A_2 = T_1^{a_{21}} T_2^{a_{22}} V_1^{b_{21}} V_2^{b_{22}}$$
$$B_1 = M_1^f S_2^{c_{22}} U_1^{d_{21}} U_2^{d_{22}} \qquad B_2 = M_1^f S_1^{c_{21}} U_1^{d_{21}} U_2^{d_{22}}.$$

Suppose that $A_1 \neq M_2^{-c_{21}}$. We first rewrite the verification equation as

$$e_3(S_1, M_2)^{c_{21}} \cdot e_3(S_1, A_1) \cdot e_3(B_1, M_2) \cdot \hat{Z} = Z.$$

Note that $\hat{Z}$ does not contain the terms S_1 and M_2. If $c_{21} = 0$, then we set $S_1' = S_1 B_1^{-1}$ and $M_2' = M_2 A_1$. For the message (M_1, M_2') we have a forged signature $(S_1', T_1), (S_2, T_2)$.[9] If $c_{21} \neq 0$, then we set $S_1' = S_1^{-1} B_1^{-2/c_{21}}$ and $M_2' = M_2^{-1} A_1^{-2/c_{21}}$ and the corresponding forgery is $(S_1', T_1), (S_2, T_2)$ for message (M_1, M_2').

A similar attack works when $A_2 \neq M_2^{-c_{22}}$.

Suppose now that $A_1 M_2^{c_{21}} = 1$ and $A_2 M_2^{c_{22}} = 1$. So both S_1 and S_2 are cancelled from the verification equation and henceforth we will only consider the signature elements T_1, T_2. Now, the verification equation will be of the form

$$e_3(M_1, T_1)^{c_{11}} \cdot e_3(M_1, T_2)^{c_{12}} \cdot e_3(M_1, V_1)^{d_{11}} \cdot e_3(M_1, V_2)^{d_{12}}$$
$$\cdot e_3(U_1, M_2)^{d_{21}} \cdot e_3(U_2, M_2)^{d_{22}}$$
$$\cdot e_3(U_1, T_1)^{e_{11}} \cdot e_3(U_1, T_2)^{e_{12}} \cdot e_3(U_2, T_1)^{e_{21}} \cdot e_3(U_2, T_2)^{e_{22}}$$
$$\cdot e_3(M_1, M_2)^f = Z.$$

Proceeding as before, we isolate M_1 and M_2 to obtain

$$A_3 = T_1^{c_{11}} T_2^{c_{12}} V_1^{d_{11}} V_2^{d_{12}} \qquad B_3 = U_1^{d_{21}} U_2^{d_{22}}.$$

Suppose $A_3 \neq M_2^{-f}$. The verification equation can be written as

$$e_3(M_1, M_2)^f \cdot e_3(M_1, A_3) \cdot e_3(B_3, M_2) \cdot Z' = Z.$$

---

[9] The attack can be prevented by checking whether (M_1, M_2') and (S_1', T_1) are elements of $\mathbb{H}_2$ or not. However that requires *two* additional pairing-product equations in signature verification.

Note that Z' does not contain the elements M_1 and M_2. If $f = 0$, then setting $M_1' = M_1 B_3^{-1}$ and $M_2' = M_2 A_3$ yields the forgery (T_1, T_2) for (M_1', M_2'). If $f \neq 0$, then setting $M_1' = M_1^{-1} B_3^{-2/f}$ and $M_2' = M_2^{-1} A_3^{-2/f}$ yields the forgery (T_1, T_2) for (M_1', M_2').

Suppose now that $A_3 M_2^f = 1$; so the message element M_1 is also cancelled from the verification equation. Thus the signature verification is reduced to the form:

$$e_3(U_1, M_2)^{d_{21}} \cdot e_3(U_2, M_2)^{d_{22}} \cdot e_3(U_1, T_1)^{e_{11}} \cdot e_3(U_1, T_2)^{e_{12}}$$
$$\cdot e_3(U_2, T_1)^{e_{21}} \cdot e_3(U_2, T_2)^{e_{22}} = Z.$$

Producing a forgery is now trivial. The adversary obtains signatures (T_1, T_2) and (T_1', T_2') on random messages (M_1, M_2) and (M_1', M_2'). From these the adversary forms a signature $(T_1^2/T_1', T_2^2/T_2')$ on a new message $(M_1^2/M_1', M_2^2/M_2')$. □

5.3 Separation

We construct a Type 3 randomizable structure-preserving signature scheme that has no secure counterpart in the Type 2 setting. The Type 3 scheme is a "dual" of the scheme presented in Sect. 4.2 in the sense that the former has $V, W \in \mathbb{G}_1$ and $M, S \in \mathbb{G}_3$, whereas the latter has $V, W \in \mathbb{G}_3$ and $M, S \in \mathbb{G}_1$.

1. *Setup.* Let $e_3 : \mathbb{G}_1 \times \mathbb{G}_3 \longrightarrow \mathbb{G}_T$ be a Type 3 pairing, where $\mathbb{G}_1$, $\mathbb{G}_3$ and $\mathbb{G}_T$ have order n; G, I are fixed generators of $\mathbb{G}_1$, $\mathbb{G}_3$, respectively.
2. *Key generation.* The secret key is $v, w \in_R [1, n-1]$. The public key is (V, W) where $V = I^v$ and $W = I^w$.
3. *Signature generation.* To sign $M \in \mathbb{G}_1$, select $r \in_R [1, n-1]$ and compute $R_1 = G^r$, $R_2 = I^r$ and $S = M^v G^{r^2+w}$. The signature on M is (R_1, R_2, S).
4. *Randomization.* To randomize $(M, (R_1, R_2, S))$, select $\alpha \in_R [1, n-1]$ and compute $R_1' = R_1 G^\alpha$, $R_2' = R_2 I^\alpha$, and $S' = S R_1^{2\alpha} G^{\alpha^2}$. The randomized signature on M is (R_1', R_2', S').
5. *Signature verification.* To verify a signed message $(M, (R_1, R_2, S))$, check that
   (a) $M, R_1, S \in \mathbb{G}_1$ and $R_2 \in \mathbb{G}_3$;
   (b) $e_3(R_1, I) = e_3(G, R_2)$; and
   (c) $e_3(S, I) = e_3(M, V) \cdot e_3(R_1, R_2) \cdot e_3(G, W)$.

Because of the dual nature of the two schemes, the security proof against generic forgers for the Type 3 scheme indicated in Sect. 4.2 carries over to the Type 3 scheme described here when we swap the roles of the elements in $\mathbb{G}_1$ and $\mathbb{G}_3$.

However, the above Type 3 scheme does not have a secure and natural counterpart in the Type 2 setting. The natural Type 2 variant has public key $V = H^v$, $W = H^w$, signatures on a message $M \in \mathbb{G}_1$ comprising of $R = H^r$ and $S = M^v G^{r^2+w}$, and verification that checks $M, S \in \mathbb{G}_1$, $R \in \mathbb{G}_2$ and $e_2(S, H) = e_2(M, V) \cdot e_2(\psi(R), R) \cdot e_2(G, W)$. Now, given the public key

(V, W) an adversary can mount the following no-message attack. Select arbitrary $m, r \in [1, n-1]$ and compute a forged signature on $M = G^m$ as $R = H^r$ and $S = \psi(V)^m \psi(W) G^{r^2} = M^v G^{r^2+w}$. While the absence of an efficiently-computable isomorphism from $\mathbb{G}_3$ to $\mathbb{G}_1$ allows us to construct the secure Type 3 scheme described above, the availability of ψ in the Type 2 setting provides the adversary with the means to mount the no-message attack.

5.4 Type 2: A Designer's Artifact?

It is not the case that the Abe et al. [4] constructions and security arguments have any intrinsic weakness. However, their efficiency analysis as well as the optimality claims are incorrect. A similar observation holds for the optimality claims made in the follow-up work of Barthe et al. [9] and in various lower bound results of [4,9].[10] The central problem in the analysis of protocols in the generic Type 2 model and associated lower bound claims stems from an incomplete abstraction of the underlying mathematical structure.

In prime-order asymmetric pairing groups, a protocol designer has the choice of using elements from $\mathbb{G}_1$, $\mathbb{G}_3$ and $\mathbb{H}_2 \subseteq \mathbb{G}_1 \times \mathbb{G}_3$. However, the definition of a bilinear group generator in the generic Type 2 setting recognizes only $\mathbb{G}_1$, $\mathbb{G}_2$ and the isomorphism $\psi : \mathbb{G}_2 \longrightarrow \mathbb{G}_1$. See, for example, the definition of a bilinear group generator $\mathcal{G}$ in Sect. 2.1 of [4]. The definition does not take into account the fact that in concrete settings there may exist a group $\mathbb{G}_3$ and an efficiently-computable isomorphism $\rho : \mathbb{G}_2 \longrightarrow \mathbb{G}_3$. This incompleteness in the abstract definition has a significant bearing on the concrete analysis of pairing-based cryptographic protocols as we demonstrate in this paper.[11]

More generally, a protocol designer desiring to use the map ψ in a cryptographic protocol or the corresponding security argument unnecessarily restricts herself to $\mathbb{G}_1$ and $\mathbb{G}_2$ (i.e. $\mathbb{H}_2$). This design artifact introduces (costly) redundancy in the cryptographic scheme without any benefit in terms of functionality or security. This observation was first made in [13] based on a careful analysis of existing Type 2 schemes. However, [13] did not attempt a formal proof of the assertion that Type 2 pairings are "merely less efficient implementation of Type 3 pairings". Motivated by the erroneous claim of superiority of Type 2 over Type 3 in [4], in this paper we formally settle the relation between Type 2 and Type 3 settings in the context of generic-signer structure-preserving signatures.

---

[10] For example, Theorem 4 of [4] proves a lower bound of two group elements in the verification key under the assumption of a single verification equation. The theorem as stated is void because there is *no* secure structure-preserving signature with a single verification equation.

[11] Following the approach outlined here, we believe it is not difficult to devise a more comprehensive definition of generic bilinear group generator in the Type 2 setting. Such a definition should be able to better model the concrete properties of the Type 2 setting, such as infeasibility of hashing into $\mathbb{G}_2$ and the cost of subgroup membership testing in $\mathbb{G}_2$. However, we do not undertake such an exercise or, for that matter, a better model of Type 2 structure-preserving signature, since we don't see any concrete motivation for using the Type 2 setting in the first place.

6   Concluding Remarks

We presented natural Type 3 analogues of the Type 2 strongly unforgeable and randomizable structure-preserving signature schemes that were proposed in [4]. By properly accounting for subgroup membership testing of group elements in signatures, we have shown that the Type 3 schemes are superior to their Type 2 counterparts when the signature schemes are used in a stand-alone setting, and when used in conjunction with Groth-Sahai proofs. Finally, we show that all generic-signer Type 2 schemes are merely Type 3 schemes in disguise and cannot beat the existing lower bound results. On the other hand, not all Type 3 schemes have a secure Type 2 counterpart. We conclude that the question posed in [13] of the existence of a cryptographic protocol which necessarily has to be restricted to Type 2 for implementation or security reasons is still open.

Acknowledgements. We thank Jens Groth and Francisco Rodríguez-Henríquez for their comments on an earlier draft of the paper. We also thank the Asiacrypt reviewers for their helpful feedback.

A   Groth-Sahai Proofs

In this section, we use additive notation for elements of $\mathbb{G}_1$, $\mathbb{G}_2$ and $\mathbb{G}_3$.

A.1   DLIN-Based Proofs

Let $A, B \in \mathbb{G}_1$ and $t \in \mathbb{G}_T$. We present a Groth-Sahai non-interactive witness-indistinguishable proof of knowledge of $X, Y \in \mathbb{G}_2$ such that $e_2(A, X) \cdot e_2(B, Y) = t$. The NIWI proof is derived from the general description in Sect. 4.2 of [17]. It can also be used with Type 3 pairings. Security is based on the decisional linear (DLIN) assumption.

1. *Setup.* Let $e_2 : \mathbb{G}_1 \times \mathbb{G}_2 \longrightarrow \mathbb{G}_T$ be a Type 2 pairing.
2. *Common reference string.* Let H be a generator of $\mathbb{G}_2$. Let $a, b, i, j \in_R [1, n - 1]$, and define $U = aH$, $V = bH$, $I = iU$, $J = jV$, $K = (i+j)H$. The common reference string is (H, U, V, I, J, K).
3. *Commitment.* Select $s_{11}, s_{12}, s_{13}, s_{21}, s_{22}, s_{23} \in_R [1, n-1]$ and compute $d_{11} = s_{11}U + s_{13}I$, $d_{12} = s_{12}V + s_{13}J$, $d_{13} = X + s_{11}H + s_{12}H + s_{13}K$, $d_{21} = s_{21}U + s_{23}I$, $d_{22} = s_{22}V + s_{23}J$ and $d_{23} = Y + s_{21}H + s_{22}H + s_{23}K$. The commitment is $d = (d_{11}, d_{12}, d_{13}, d_{21}, d_{22}, d_{23})$.
4. *Proof.* Compute $\theta_1 = s_{11}A + s_{21}B$, $\theta_2 = s_{12}A + s_{22}B$ and $\theta_3 = s_{13}A + s_{23}B$. The proof is $\theta = (\theta_1, \theta_2, \theta_3)$.
5. *Verification.* Check that $\theta_1, \theta_2, \theta_3 \in \mathbb{G}_1$, $d_{11}, d_{12}, d_{13}, d_{21}, d_{22}, d_{23} \in \mathbb{G}_2$, and

$$e_2(A, d_{11}) \cdot e_2(B, d_{21}) = e_2(\theta_1, U) \cdot e_2(\theta_3, I)$$
$$e_2(A, d_{12}) \cdot e_2(B, d_{22}) = e_2(\theta_2, V) \cdot e_2(\theta_3, J)$$
$$e_2(A, d_{13}) \cdot e_2(B, d_{23}) = e_2(\theta_1, H) \cdot e_2(\theta_2, H) \cdot e_2(\theta_3, K) \cdot t.$$

A.2   DDH-Based Proofs

Let $A, B \in \mathbb{G}_1$ and $t \in \mathbb{G}_T$. We present a Groth-Sahai non-interactive witness-indistinguishable proof of knowledge of $X, Y \in \mathbb{G}_3$ such that $e_3(A, X) \cdot e_3(B, Y) = t$. The NIWI proof is derived from the general description in Sect. 4.1 of [17]. Security is based on the decisional Diffie-Hellman (DDH) assumption in $\mathbb{G}_3$. Since the decisional Diffie-Hellman problem is easy in $\mathbb{G}_2$, the NIWI proof has no counterpart with Type 2 pairings.

1. *Setup.* Let $e_3 : \mathbb{G}_1 \times \mathbb{G}_3 \longrightarrow \mathbb{G}_T$ be a Type 3 pairing.
2. *Common reference string.* Let I be a generator of $\mathbb{G}_3$. Let $a, b \in_R [1, n-1]$, and define $U = aI$, $V = bI$, $J = bU$. The common reference string is (I, U, V, J).
3. *Commitment.* Select $s_{11}, s_{12}, s_{21}, s_{22} \in_R [1, n-1]$ and compute $d_{11} = s_{11}I + s_{12}V$, $d_{12} = X + s_{11}U + s_{12}J$, $d_{21} = s_{21}I + s_{22}V$ and $d_{22} = Y + s_{21}U + s_{22}J$. The commitment is $d = (d_{11}, d_{12}, d_{21}, d_{22})$.
4. *Proof.* Compute $\theta_1 = s_{11}A + s_{21}B$ and $\theta_2 = s_{12}A + s_{22}B$. The proof is $\theta = (\theta_1, \theta_2)$.
5. *Verification.* Check that $\theta_1, \theta_2 \in \mathbb{G}_1$, $d_{11}, d_{12}, d_{21}, d_{22} \in \mathbb{G}_3$, and

$$e_3(A, d_{11}) \cdot e_3(B, d_{21}) = e_3(\theta_1, I) \cdot e_3(\theta_2, V)$$
$$e_3(A, d_{12}) \cdot e_3(B, d_{22}) = e_3(\theta_1, U) \cdot e_3(\theta_2, J) \cdot t.$$

References

1. Abe, M., Fuchsbauer, G., Groth, J., Haralambiev, K., Ohkubo, M.: Structure-preserving signatures and commitments to group elements. In: Rabin, T. (ed.) CRYPTO 2010. LNCS, vol. 6223, pp. 209–236. Springer, Heidelberg (2010)
2. Abe, M., Groth, J., Haralambiev, K., Ohkubo, M.: Optimal structure-preserving signatures in asymmetric bilinear groups. In: Rogaway, P. (ed.) CRYPTO 2011. LNCS, vol. 6841, pp. 649–666. Springer, Heidelberg (2011)
3. Abe, M., Groth, J., Ohkubo, M., Tibouchi, M.: Unified, minimal and selectively randomizable structure-preserving signatures. In: Lindell, Y. (ed.) TCC 2014. LNCS, vol. 8349, pp. 688–712. Springer, Heidelberg (2014)
4. Abe, M., Groth, J., Ohkubo, M., Tibouchi, M.: Structure-preserving signatures from type II pairings. In: Garay, J.A., Gennaro, R. (eds.) CRYPTO 2014, Part I. LNCS, vol. 8616, pp. 390–407. Springer, Heidelberg (2014)
5. Abe, M., Groth, J., Ohkubo, M., Tibouchi, M.: Structure-preserving signatures from type II pairings, full version of [4] (2014). http://eprint.iacr.org/2014/312
6. Barreto, P.S.L.M., Lynn, B., Scott, M.: Constructing elliptic curves with prescribed embedding degrees. In: Cimato, S., Galdi, C., Persiano, G. (eds.) SCN 2002. LNCS, vol. 2576, pp. 257–267. Springer, Heidelberg (2003)
7. Barreto, P., Lynn, B., Scott, M.: Efficient implementation of pairing-based cryptosystems. J. Cryptol. **17**, 321–334 (2004)
8. Barreto, P.S.L.M., Naehrig, M.: Pairing-friendly elliptic curves of prime order. In: Preneel, B., Tavares, S. (eds.) SAC 2005. LNCS, vol. 3897, pp. 319–331. Springer, Heidelberg (2006)

9. Barthe, G., Fagerholm, E., Fiore, D., Scedrov, A., Schmidt, B., Tibouchi, M.: Strongly-optimal structure preserving signatures from type II pairings: synthesis and lower bounds. In: Katz, J. (ed.) PKC 2015. LNCS, vol. 9020, pp. 355–376. Springer, Heidelberg (2015)

10. Bellare, M., Garay, J.A., Rabin, T.: Fast batch verification for modular exponentiation and digital signatures. In: Nyberg, K. (ed.) EUROCRYPT 1998. LNCS, vol. 1403, pp. 236–250. Springer, Heidelberg (1998)

11. Chase, M.: Efficient non-interactive zero-knowledge proofs for privacy applications. Ph.D. thesis, Brown University (2008)

12. Chatterjee, S., Hankerson, D., Knapp, E., Menezes, A.: Comparing two pairing-based aggregate signature schemes. Des. Codes Cryptogr. **55**, 141–167 (2010)

13. Chatterjee, S., Menezes, A.: On cryptographic protocols employing asymmetric pairings - the role of ψ revisited. Discrete Appl. Math. **159**, 1311–1322 (2011)

14. Chen, L., Cheng, Z., Smart, N.: Identity-based key agreement protocols from pairings. Inte. J. Inf. Secur. **6**, 213–241 (2007)

15. Ferrara, A.L., Green, M., Hohenberger, S., Pedersen, M.Ø.: Practical short signature batch verification. In: Fischlin, M. (ed.) CT-RSA 2009. LNCS, vol. 5473, pp. 309–324. Springer, Heidelberg (2009)

16. Galbraith, S., Paterson, K., Smart, N.: Pairings for cryptographers. Discrete Appl. Math. **156**, 3113–3121 (2008)

17. Ghadafi, E., Smart, N.P., Warinschi, B.: Groth–Sahai proofs revisited. In: Nguyen, P.Q., Pointcheval, D. (eds.) PKC 2010. LNCS, vol. 6056, pp. 177–192. Springer, Heidelberg (2010)

18. Groth, J.: Simulation-sound NIZK proofs for a practical language and constant size group signatures. In: Lai, X., Chen, K. (eds.) ASIACRYPT 2006. LNCS, vol. 4284, pp. 444–459. Springer, Heidelberg (2006)

19. Groth, J., Sahai, A.: Efficient noninteractive proof systems for bilinear groups. SIAM J. Comput. **41**, 1193–1232 (2012)

20. Hanser, C., Slamanig, D.: Structure-preserving signatures on equivalence classes and their application to anonymous credentials. In: Sarkar, P., Iwata, T. (eds.) ASIACRYPT 2014. LNCS, vol. 8873, pp. 491–511. Springer, Heidelberg (2014)

21. Hess, F., Smart, N., Vercauteren, F.: The eta pairing revisited. IEEE Trans. Inf. Theor. **52**, 4595–4602 (2006)

22. Kachisa, E.J., Schaefer, E.F., Scott, M.: Constructing Brezing-Weng pairing-friendly elliptic curves using elements in the cyclotomic field. In: Galbraith, S.D., Paterson, K.G. (eds.) Pairing 2008. LNCS, vol. 5209, pp. 126–135. Springer, Heidelberg (2008)

23. Miyaji, A., Nakabayashi, M., Tanako, S.: New explicit condition of elliptic curve trace for FR-reduction. IEICE Trans. Fundam. Electron. Commun. Comput. Sci. **E84–A**, 1234–1243 (2001)

24. Vercauteren, F.: Optimal pairings. IEEE Trans. Inf. Theor. **56**, 455–461 (2010)

Design Principles for HFEv- Based Multivariate Signature Schemes

Albrecht Petzoldt[1], Ming-Shing Chen[2,3], Bo-Yin Yang[2],
Chengdong Tao[4], and Jintai Ding[5,6](✉)

[1] Technische Universität Darmstadt, Darmstadt, Germany
[2] Academia Sinica, Taipei, Taiwan
[3] National Taiwan University, Taipei, Taiwan
[4] South China University of Technology, Guangzhou, China
[5] ChongQing University, Chongqing, China
[6] University of Cincinnati, Cincinnati, OH, USA
jintai.ding@gmail.com

Abstract. The Hidden Field Equations (HFE) Cryptosystem as proposed by Patarin is one of the best known and most studied multivariate schemes. While the security of the basic scheme appeared to be very weak, the HFEv- variant seems to be a good candidate for digital signature schemes on the basis of multivariate polynomials. However, the currently existing scheme of this type, the QUARTZ signature scheme, is hardly used in practice because of its poor efficiency. In this paper we analyze recent results from Ding and Yang about the degree of regularity of HFEv- systems and derive from them design principles for signature schemes of the HFEv- type. Based on these results we propose the new HFEv- based signature scheme Gui, which is more than 100 times faster than QUARTZ and therefore highly comparable with classical signature schemes such as RSA and ECDSA.

Keywords: Multivariate cryptography · Digital signatures · HFEv- · Design principles · Security · Performance

1 Introduction

Cryptographic techniques are an essential tool to guarantee the security of communication in modern society. Today, the security of nearly all of the cryptographic schemes used in practice is based on number theoretic problems such as factoring large integers and solving discrete logarithms. The best known schemes in this area are RSA [28], DSA [19] and ECC. However, schemes like these will become insecure as soon as large enough quantum computers arrive. The reason for this is Shor's algorithm [29], which solves number theoretic problems like integer factorization and discrete logarithms in polynomial time on a quantum computer. Therefore, one needs alternatives to those classical public key schemes, based on hard mathematical problems not affected by quantum computer attacks.

© International Association for Cryptologic Research 2015
T. Iwata and J.H. Cheon (Eds.): ASIACRYPT 2015, Part I, LNCS 9452, pp. 311–334, 2015.
DOI: 10.1007/978-3-662-48797-6_14

Besides lattice, code and hash based cryptosystems, multivariate cryptography is one of the main candidates for this [1]. Multivariate schemes are in general very fast and require only modest computational resources, which makes them attractive for the use on low cost devices like smart cards and RFID chips [5, 6]. Additionally, at least in the area of digital signatures, there exists a large number of practical multivariate schemes [10, 20].

In 2001, Patarin and Courtois proposed a multivariate signature scheme called QUARTZ [24], which is based on the concept of HFEv-. While QUARTZ produces very short signatures (128 bit), the signature generation process is very slow (at the time about 11 seconds per signature [6]). The main reason for this is the use of a high degree HFE polynomial (for QUARTZ this degree is given by $D = 129$), which makes the inversion of the central map very costly.

At the time of the design of the QUARTZ scheme, very little was known about the complexity of algebraic attacks against the HFE family of systems, in particular, the HFEv- schemes. Therefore, the authors of QUARTZ could not base their parameter choice on theoretical foundations. Recently, there has been a fundamental breakthrough in terms of understanding the behavior of algebraic attacks on the HFE family of systems [9, 11], which gives an upper bound on the degree of regularity of Gröbner basis attacks against those schemes.

In this paper, we review and analyze the results of Ding and Yang and derive from these results design criteria for HFEv- based signature schemes. In particular we show that we can, by increasing the numbers a of Minus equations and v of Vinegar variables, achieve adequate security even for low degree HFE polynomials and that the upper bound on the degree of regularity given by Ding and Yang is reasonably tight. Based on our analysis, we propose the new HFEv-based signature scheme Gui[1], which uses HFE polynomials of very low degree, namely $D \in \{5, 9, 17\}$. This enables us to speed up the signature generation process by a factor of more than 100 compared to QUARTZ, without weakening the security of the scheme. By doing so, we create a highly practical multivariate signature scheme, whose performance is comparable to that of classical signature schemes such as RSA and ECDSA.

The rest of this paper is organized as follows. In Sect. 2 we give an introduction into the area of multivariate cryptography and in particular Big-Field signature schemes. Section 3 introduces the HFEv-signature scheme and the changes made to this scheme by Patarin and Courtois when defining QUARTZ. Furthermore, in this section, we discuss the performance and the security of HFEv-based signature schemes. In Sect. 4 we analyze the results of Ding and Yang on the behaviour of direct attacks on HFEv- schemes by performing a large number of experiments and present the design criteria we derive from that. Based on these principles, we propose in Sect. 5 our new multivariate signature scheme Gui. Section 6 gives details on the implementation of the scheme and compares the efficiency of Gui with that of some standard signature schemes. Finally, Sect. 7 concludes the paper.

---

[1] We call our new scheme Gui, referring to earthenware pottery dating back to the 4000-year-old Longshan culture [31].

2   Multivariate Cryptography

The basic objects of multivariate cryptography are systems of multivariate quadratic polynomials (see Eq. (1)).

$$p^{(1)}(x_1,\ldots,x_n) = \sum_{i=1}^{n}\sum_{j=i}^{n} p_{ij}^{(1)} \cdot x_i x_j + \sum_{i=1}^{n} p_i^{(1)} \cdot x_i + p_0^{(1)}$$

$$p^{(2)}(x_1,\ldots,x_n) = \sum_{i=1}^{n}\sum_{j=i}^{n} p_{ij}^{(2)} \cdot x_i x_j + \sum_{i=1}^{n} p_i^{(2)} \cdot x_i + p_0^{(2)}$$

$$\vdots$$

$$p^{(m)}(x_1,\ldots,x_n) = \sum_{i=1}^{n}\sum_{j=i}^{n} p_{ij}^{(m)} \cdot x_i x_j + \sum_{i=1}^{n} p_i^{(m)} \cdot x_i + p_0^{(m)} \tag{1}$$

The security of multivariate schemes is based on the

MQ Problem: Given m multivariate quadratic polynomials $p^{(1)}(\mathbf{x}),\ldots,p^{(m)}(\mathbf{x})$ in n variables $x_1,\ldots,x_n$ as shown in Eq. (1), find a vector $\bar{\mathbf{x}} = (\bar{x}_1,\ldots,\bar{x}_n)$ such that $p^{(1)}(\bar{\mathbf{x}}) = \ldots = p^{(m)}(\bar{\mathbf{x}}) = 0$.

The MQ problem (for $m \approx n$) is proven to be NP-hard even for quadratic polynomials over the field GF(2) [15].

To build a public key cryptosystem based on the MQ problem, one starts with an easily invertible quadratic map $\mathcal{F} : \mathbb{F}^n \to \mathbb{F}^m$ (central map). To hide the structure of $\mathcal{F}$ in the public key, one composes it with two invertible affine (or linear) maps $\mathcal{S} : \mathbb{F}^m \to \mathbb{F}^m$ and $\mathcal{T} : \mathbb{F}^n \to \mathbb{F}^n$. The *public key* is therefore given by $\mathcal{P} = \mathcal{S} \circ \mathcal{F} \circ \mathcal{T}$. The *private key* consists of $\mathcal{S}$, $\mathcal{F}$ and $\mathcal{T}$ and therefore allows to invert the public key.

Note: Due to the above construction, the security of multivariate schemes is not only based on the MQ-Problem but also on the EIP-Problem ("Extended Isomorphism of Polynomials") of finding the composition of $\mathcal{P}$.

In this paper we concentrate on multivariate signature schemes of the Big-Field family. For this type of multivariate schemes, the map $\mathcal{F}$ is a specially chosen easily invertible map over a degree n extension field $\mathbb{E}$ of $\mathbb{F}$. One uses an isomorphism $\Phi : \mathbb{F}^n \to \mathbb{E}$ to transform $\mathcal{F}$ into a quadratic map

$$\bar{\mathcal{F}} = \Phi^{-1} \circ \mathcal{F} \circ \Phi \tag{2}$$

from $\mathbb{F}^n$ to itself. The public key of the scheme is therefore given by

$$\mathcal{P} = \mathcal{S} \circ \bar{\mathcal{F}} \circ \mathcal{T} = \mathcal{S} \circ \Phi^{-1} \circ \mathcal{F} \circ \Phi \circ \mathcal{T} : \mathbb{F}^n \to \mathbb{F}^n. \tag{3}$$

The standard signature generation and verification process of a multivariate BigField scheme works as shown in Fig. 1.

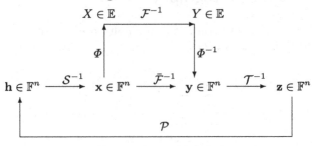

Fig. 1. General workflow of multivariate BigField signature schemes

Signature generation: To generate a signature for a message $\mathbf{h} \in \mathbb{F}^n$, one computes recursively $\mathbf{x} = \mathcal{S}^{-1}(\mathbf{h}) \in \mathbb{F}^n$, $X = \Phi(\mathbf{x}) \in \mathbb{E}$, $Y = \mathcal{F}^{-1}(X) \in \mathbb{E}$, $\mathbf{y} = \Phi^{-1}(Y) \in \mathbb{F}^n$ and $\mathbf{z} = \mathcal{T}^{-1}(\mathbf{y})$. The signature of the message $\mathbf{h}$ is $\mathbf{z} \in \mathbb{F}^n$.

Verification: To check the authenticity of a signature $\mathbf{z} \in \mathbb{F}^n$, one simply computes $\mathbf{h}' = \mathcal{P}(\mathbf{z}) \in \mathbb{F}^n$. If $\mathbf{h}' = \mathbf{h}$ holds, the signature is accepted, otherwise rejected.

A good overview on existing multivariate schemes can be found in [8].

Two widely used variations of multivariate BigField signature schemes are the Minus variation and the use of additional (Vinegar) variables.

Minus variation: The idea of this variation is to remove a small number of equations from the public key. The Minus-Variation was first used in schemes like SFLASH [25] to prevent Patarins Linearization Equations attack [26] against the Matsumoto-Imai cryptosystem [23].

Vinegar variation: In this variation one parametrizes the central map $\mathcal{F}$ by adding (a small set of) additional (Vinegar) variables. In the context of multivariate BigField signature schemes, the Vinegar variation can be used to increase the security of the scheme against direct and rank attacks.

3   The HFEv- Signature Scheme

In this section we introduce the HFEv- signature scheme, which is the basis of both QUARTZ and our new signature scheme Gui (see Sect. 5).

Let $\mathbb{F} = \mathbb{F}_q$ be a finite field with q elements and $\mathbb{E}$ be a degree n extension field of $\mathbb{F}$. Furthermore, we choose integers D, a and v. Let Φ be the canonical isomorphism between $\mathbb{F}^n$ and $\mathbb{E}$, i.e.

$$\Phi(x_1, \ldots, x_n) = \sum_{i=1}^{n} x_i \cdot X^{i-1}. \tag{4}$$

The central map $\mathcal{F}$ of the HFEv- scheme is a map from $\mathbb{E} \times \mathbb{F}^v$ to $\mathbb{E}$ of the form

$$
\mathcal{F}(X) = \sum_{\substack{0 \leq i \leq j}}^{q^i + q^j \leq D} \alpha_{ij} \cdot X^{q^i + q^j}
$$
$$
+ \sum_{i=0}^{q^i \leq D} \beta_i(v_1, \ldots, v_v) \cdot X^{q^i}
$$
$$
+ \gamma(v_1, \ldots, v_v), \tag{5}
$$

with $\alpha_{ij} \in \mathbb{E}$, $\beta_i : \mathbb{F}^v \to \mathbb{E}$ being linear and $\gamma : \mathbb{F}^v \to \mathbb{E}$ being a quadratic function.

Due to the special form of $\mathcal{F}$, the map $\bar{\mathcal{F}} = \Phi^{-1} \circ \mathcal{F} \circ \Phi$ is a quadratic polynomial map from $\mathbb{F}^{n+v}$ to $\mathbb{F}^n$. To hide the structure of $\bar{\mathcal{F}}$ in the public key, one combines it with two affine (or linear) maps $\mathcal{S} : \mathbb{F}^n \to \mathbb{F}^{n-a}$ and $\mathcal{T} : \mathbb{F}^{n+v} \to \mathbb{F}^{n+v}$ of maximal rank.

The *public key* of the scheme is the composed map $\mathcal{P} = \mathcal{S} \circ \bar{\mathcal{F}} \circ \mathcal{T} : \mathbb{F}^{n+v} \to \mathbb{F}^{n-a}$, the *private key* consists of $\mathcal{S}$, $\mathcal{F}$ and $\mathcal{T}$.

Signature generation: To generate a signature for a message $\mathbf{h} \in \mathbb{F}^{n-a}$, the signer performs the following three steps.

1. Compute a preimage $\mathbf{x} \in \mathbb{F}^n$ of $\mathbf{h}$ under the affine map $\mathcal{S}$.
2. Lift $\mathbf{x}$ to the extension field $\mathbb{E}$ (using the isomorphism Φ). Denote the result by X.
   Choose random values for the vinegar variables $v_1, \ldots, v_v \in \mathbb{F}$ and compute $\mathcal{F}_V = \mathcal{F}(v_1, \ldots, v_v)$.
   Solve the univariate polynomial equation $\mathcal{F}_V(Y) = X$ by Berlekamp's algorithm and compute $\mathbf{y}' = \Phi^{-1}(Y) \in \mathbb{F}^n$.
   Set $\mathbf{y} = (\mathbf{y}' || v_1 || \ldots || v_v)$.
3. Compute the signature $\mathbf{z} \in \mathbb{F}^{n+v}$ by $\mathbf{z} = \mathcal{T}^{-1}(\mathbf{y})$.

Signature verification: To check the authenticity of a signature $\mathbf{z} \in \mathbb{F}^{n+v}$, one simply computes $\mathbf{h}' = \mathcal{P}(\mathbf{z}) \in \mathbb{F}^{n-a}$. If $\mathbf{h}' = \mathbf{h}$ holds, the signature is accepted, otherwise rejected.

3.1 QUARTZ

In 2001, Patarin and Courtois proposed the multivariate signature scheme QUARTZ [24], which is based on the concept of HFEv-. Indeed, the public and private maps of QUARTZ are HFEv- maps with the parameters

$$
(\mathbb{F}, n, D, a, v) = (\mathrm{GF}(2), 103, 129, 3, 4).
$$

Due to this choice, the public key $\mathcal{P}$ of QUARTZ is a quadratic map from $\mathbb{F}^{107}$ to $\mathbb{F}^{100}$. The public key size of QUARTZ is 71 kB, the private key size 3 kB.

The input length of QUARTZ is only $n - a = 100$ bit. Therefore, it is possible for an attacker to use a birthday attack to find two different messages m_1 and

m_2 which map to the same input value $\mathbf{h} \in \mathbb{F}^{100}$ and therefore to the same signature.

To prevent this kind of attack, Patarin and Courtois developed a special procedure for the signature generation process of QUARTZ. Roughly spoken, one computes four HFEv- signatures (for the messages $\mathbf{h}$, $\mathcal{H}(\mathbf{h}||0x00)$, $\mathcal{H}(\mathbf{h}||0x01)$ and $\mathcal{H}(\mathbf{h}||0x02)$) and combines them to a single 128 bit signature of the message $\mathbf{h}$. Analogously, during the signature verification process, one has to use the public key $\mathcal{P}$ four times.

3.2   Performance

The most costly step during the signature generation process of HFEv- based signature schemes such as QUARTZ is the inversion of the univariate polynomial equation $\mathcal{F}_V$ over the extension field $\mathbb{E}$. This step is usually performed by Berlekamp's algorithm, whose complexity can be estimated by [27]

$$\mathcal{O}(D^3 + n \cdot D^2). \tag{6}$$

As can be seen from Eq. (6), the complexity of inverting $\mathcal{F}_V$ and therefore of the signature generation process of HFEv- based schemes is mainly determined by the degree D of the HFE polynomial. Due to the high degree of the HFE polynomial used in QUARTZ, the inversion of $\mathcal{F}_V$ is very costly. Furthermore, we have to perform this step four times during the signature generation of QUARTZ. Additionally, the design of QUARTZ requires the central equation $\mathcal{F}_V(Y) = X$ to have a unique root. Since, after choosing random values for Minus equations and Vinegar variables, $\mathcal{F}_V$ can be seen as a random function, this happens with probability about $\frac{1}{e}$. Altogether, we therefore have to run Berlekamp's algorithm about $4 \cdot e$ times during the signature generation process of QUARTZ. Thus, the QUARTZ signature scheme is rather slow and it takes about 11 seconds to generate a signature [6].

3.3   Security of HFEv- Based Schemes

The most important attacks against signature schemes of the HFEv- type are

- the MinRank attack and
- direct algebraic attacks.

The MinRank Attack on HFE. In this paragraph we describe the attack of Kipnis and Shamir [21] against the HFE cryptosystem. For the simplicity of our description we restrict ourselves to homogeneous maps $\mathcal{F}$ and $\mathcal{P}$.

The key idea of the attack is to lift the maps $\mathcal{S}$, $\mathcal{T}$ and $\mathcal{P}$ to functions $\mathcal{S}^\star$, $\mathcal{T}^\star$ and $\mathcal{P}^\star$ over the extension field $\mathbb{E}$. Since $\mathcal{S}$ and $\mathcal{T}$ are linear maps, $\mathcal{S}^\star$ and $\mathcal{T}^\star$ have the form

$$\mathcal{S}^\star(X) = \sum_{i=1}^{n-1} s_i \cdot X^{q^i} \text{ and } \mathcal{T}^\star(X) = \sum_{i=1}^{n-1} t_i \cdot X^{q^i}, \tag{7}$$

with coefficients s_i and $t_i \in \mathbb{E}$. The function $\mathcal{P}^\star$ can be expressed as

$$\mathcal{P}^\star(X) = \sum_{i=0}^{n-1}\sum_{j=0}^{n-1} p_{ij}^\star X^{q^i + q^j} = \underline{X} \cdot P^\star \cdot \underline{X}^T, \tag{8}$$

where $P^\star = [p_{ij}^\star]$ and $\underline{X} = (X^{q^0}, X^{q^1}, \ldots, X^{q^{n-1}})$. Due to the relation $\mathcal{P}^\star(X) = \mathcal{S}^\star \circ \mathcal{F} \circ \mathcal{T}^\star(X)$ we get $\mathcal{S}^{\star\,-1} \circ \mathcal{P}^\star(X) = \mathcal{F} \circ \mathcal{T}^\star(X)$ and

$$\tilde{P} = \sum_{k=0}^{n-1} s_k \cdot G^{\star k} = W \cdot F \cdot W^T \tag{9}$$

with $g_{ij}^{\star\,k} = (p_{i-k \bmod n, j-k \bmod n}^\star)^{q^k}$, $w_{ij} = s_{j-i \bmod n}^{q^i}$ and F being the $n \times n$ matrix representing the central map $\mathcal{F}$. Note that, due to the special structure of $\mathcal{F}$, the only non zero entries in the matrix F are located in the upper left $r \times r$ submatrix ($r = \lfloor \log_q D - 1 \rfloor + 1$).

Therefore, the rank of the matrix $W \cdot F \cdot W^T$ is less or equal to r, which means that we can determine the coefficients s_k of Eq. (9) by solving an instance of the MinRank problem.

In the setting of HFEv-, the rank of this matrix is, for odd characteristic, bounded from above by [11]

$$\text{Rank}(\widetilde{P}) \le r + a + v. \tag{10}$$

Under the assumption that the vinegar maps β_i look like random functions, we find that this bound is tight.

For fields of even characteristic we eventually have to decrease this rank by 1, since over those fields, the matrix $\widetilde{P}$ is always of even rank. The complexity of the MinRank attack against HFEv- based schemes is therefore given roughly by

$$\text{Complexity}_{\text{MinRank}} = \mathcal{O}(q^{n \cdot (r+v+a-1)} \cdot (n-a)^3). \tag{11}$$

In the paper [18] the authors showed that, due to the symmetry of the solutions of the equations for the MinRank problem in the Kipnis-Shamir attack and the fact that we work over a large extension field, the complexity of the Kipnis-Shamir attack is actually exponential in terms of the number of variables in the HFE system using known MinRank methods, and not polynomial as was originally stated. Though the theoretical argument underlying this observation does not apply directly to the generic MinRank problem, it demonstrates that the Kipnis-Shamir attack, for which one needs to solve a non-generic MinRank problem, has a much higher complexity than originally estimated. We therefore conclude, that the complexity of a MinRank attack against an HFEv- based signature scheme is, in practice, higher than the above estimation.

There is one other formulation of the MinRank problem. According to [13], solving a MinRank problem with $n \times n$ matrices to a rank of r' involves computing a Gröbner basis with degree of regularity $r'(n - r') + 1$, where the rank is given by $r' = r + v + a - 1$. When we raise the rank r' (by increasing $a + v$), this means that the attack complexity of the MinRank attack is much higher than that of a direct attack.

Direct Attacks. For the HFE family of schemes, the direct attack, namely the attack by directly solving the public equation $\mathcal{P}(\mathbf{x}) = \mathbf{h}$ by an algorithm like XL or a Gröbner basis method such as F_4 [12] is a major concern due to which happened to HFE challenge 1. At the time of the design of QUARTZ, very little was known theoretically about the complexity of algebraic attacks against the HFE family of systems, in particular, the HFEv- schemes. The authors of QUARTZ did not actually give an explanation for their selection of the parameters and therefore the parameter selection of their scheme was not supported by theoretical results. We need to point out that, as has been shown by experiments [22], the public systems of HFEv- based schemes can be solved easier than random systems.

Recently, there has been a fundamental breakthrough in terms of understanding how algebraic attacks on the HFE family of systems work [9,11]. In particular, we now have a solid insight what happens in the case of HFEv-. An upper bound for the degree of regularity of a Gröbner Basis attack against HFEv- systems is given by [11]

$$
d_{\mathrm{reg}} \leq \begin{cases} \frac{(q-1)\cdot(r-1+a+v)}{2} + 2 & q \text{ even and } r+a \text{ odd} \\ \frac{(q-1)\cdot(r+a+v)}{2} + 2 & \text{otherwise} \end{cases}, \tag{12}
$$

where r is given by $r = \lfloor \log_q(D-1) \rfloor + 1$.

Note: In [7] Courtois et al. estimated the complexity of a direct attack on QUARTZ by 2^{74} operations. However, they underestimated the degree of regularity of solving an HFEv- system drastically.

4 Design Principles for HFEv- Based Signature Schemes

The theoretical breakthrough mentioned in the previous subsection indicates that it might be possible to substantially improve the original design of QUARTZ without reducing the security of the scheme, if we adapt the number of Minus equations and Vinegar variables in an appropriate way. By reducing the degree of the central HFEv- polynomial we can speed up the operations of Berlekamp's algorithm and therefore the signature generation process of the HFEv- scheme. In this section, we analyze by experiments the behavior of direct attacks against HFEv- schemes and the tightness of the upper bound given by Eq. (12). From our results we derive design principles for the construction of HFEv- based signature schemes, which we later apply to our new signature scheme Gui presented in the next section.

In particular, we answer in this section the following questions.

1. Equation (12) shows a tradeoff between the degree D of the HFE polynomial and the sum $a+v$ of minus equations and vinegar variables. This would enable us to use low degree HFE polynomials in the construction of HFEv- based signature schemes and therefore to improve their performance drastically. Can we verify this by experiments?

2. Is the ratio between a and v important for the security of the scheme?
3. Is the upper bound on the degree of regularity given by equation (12) reasonably tight?
4. Does it help to guess some variables before applying a Gröbner basis algorithm to the system $\mathcal{P}$ (Hybrid Approach)?

To answer these questions, we performed a large number of experiments with the F_4 algorithm integrated in MAGMA. As we found, adding the field equations $\{x_i^2 - x_i\}$ to the system makes a huge difference regarding the degree of regularity and the running time of the attack.

4.1 Can We Use HFE Polynomials of Low Degree D?

To improve the efficiency of the signature generation process we are interested in decreasing the degree of the HFE polynomial in use as far as possible without weakening the security of the scheme. Doing so will reduce the complexity of Berlekamp's algorithm (see Eq. (6)) and therefore improve the performance of the scheme significantly. So, the first question we have to answer in this context is the following.

How should we choose the degree D of the HFE polynomial in order to obtain secure and efficient HFEv- based schemes?

- $D = 2, 3$: Such small values of D would lead to matrices F of rank 2. We therefore do not think that these schemes can be secure.
- $D = 5$: Although the plain HFE scheme with an HFE polynomial of degree 5 ($r = 3$) is highly insecure, we believe that the modified HFEv- scheme provides adequate security.
- $D = 9, 17$: Other promising values for the degree of the HFE polynomial in use are $D = 9$ and $D = 17$, which lead to values of r of 4 and 5 respectively.

In the first row of experiments we analyzed the behavior of direct attacks against HFEv- systems over GF(2) with different values of D. For this, we fixed the number of equations in the system. For different values of D, a and v we created HFEv- systems and fixed $a + v$ variables randomly to get determined systems. After adding the field equations $\{x_i^2 - x_i\}$ we solved the systems using MAGMA's implementation of the F_4 algorithm. For each parameter set we performed 10 experiments.

Table 1 shows the results of our experiments with determined HFEv- systems of 20 and 25 equations respectively. The degree of regularity of a random system of this size is 5 and 6 respectively. The table shows, for different values of D, the minimal values of a and v needed to reach this degree. Although, because of memory restrictions, we could not perform our experiments for larger values of n, we expect that similar results hold for arbitrary numbers of equations.

Table 1. Experiments with F_4 on determined HFEv- systems with 20 and 25 equations

D	r	20 equations				25 equations			
		Minimal a,v	d_{reg}	Time (s)	Memory (MB)	Minimal a, v	d_{reg}	Time (s)	Memory(MB)
129	8	$a = v = 0$	5	2.74	109.7	$a = v = 1$	6	276.2	7,621
65	7	$a = 0, v = 1$	5	2.73	110.2	$a = v = 2$	6	276.0	7,681
33	6	$a = v = 1$	5	2.75	109.7	$a = 2, v = 3$	6	273.4	7,762
17	5	$a = 1, v = 2$	5	2.72	109.7	$a = v = 3$	6	275.7	7,751
9	4	$a = v = 2$	5	2.73	109.9	$a = 3, v = 4$	6	276.4	7,693
5	3	$a = 2, v = 3$	5	2.73	109.6	$a = v = 4$	6	272.8	7,680
Random system			5	2.85	110.8		6	286.3	7,683

From the above experiments we obtain the following important observation

Let d be the degree of regularity of a direct attack against an HFEv-system with parameters D_1, n, a_1, v_1 and let $D_2 < D_1$.
By choosing large enough values for a_2 and v_2, we can obtain an HFEv-scheme with parameters D_2, n, a_2, v_2, such that the degree of regularity of a direct attack against this system is d, too.

From this observation we derive our first design principle for the construction of HFEv- based signature schemes.

Design Principle 1:
For the construction of HFEv- based signature schemes we use for efficiency reasons HFE polynomials of small degree D, namely $D \in \{5, 9, 17\}$. We then increase the numbers of Minus equations a and Vinegar variables v to obtain a secure scheme.

4.2 Is the Ratio Between a and v Important for the Security of the Scheme?

To answer this question, we performed experiments of the following type. For a fixed degree D of the HFE polynomial, a fixed number of equations and a fixed value s we created HFEv- systems with $a \in \{0, \ldots, s\}$ and $v = s - a$. After fixing $v + a$ variables to get a determined system and adding the field equations $\{x_i^2 - x_i\}$ we solved the systems by the F_4 algorithm integrated in MAGMA. For each parameter set we performed 10 experiments. The results are shown in Tables 2 and 3.

As the tables show, in particular for HFEv- schemes with low degree D, the number v of vinegar variables should not be too small. Especially, $v = 0$ (i.e. HFE-) seems to be a bad choice.

On the other hand, very high values of v do not increase the security of the scheme and increase the public key size of the scheme drastically. To achieve a good security and a moderate public key size, we therefore formulate our second design principle for HFEv- based signature schemes as follows.

Table 2. Experiments with F_4 on determined HFEv- systems with 20 equations

D = 5, a + v = 5					D = 9, a + v = 4					D = 17, a + v = 3				
a	v	d_{reg}	Time (s)	Memory (MB)	a	v	d_{reg}	Time (s)	Memory (MB)	a	v	d_{reg}	Time (s)	Memory (MB)
0	5	5	2.76	109.7	0	4	5	2.77	109.7	0	3	5	2.75	110.7
1	4	5	2.77	109.7	1	3	5	2.78	110.8	1	2	5	2.77	109.7
2	3	5	2.76	110.7	2	2	5	2.76	110.7	2	1	5	2.74	110.8
3	2	5	2.77	110.8	3	1	5	2.75	110.8	3	0	5	2.73	109.7
4	1	5	2.75	109.8	4	0	5	2.79	108.7	—				
5	0	4	**1.01**	**32.6**	—					—				

Table 3. Experiments with F_4 on determined HFEv- systems with 25 equations

D = 5, a + v = 8					D = 9, a + v = 7					D = 17, a + v = 6				
a	v	d_{reg}	Time (s)	Memory (MB)	a	v	d_{reg}	Time (s)	Memory (MB)	a	v	d_{reg}	Time (s)	Memory (MB)
0	8	6	246.6	7,582	0	7	6	248.9	7,582	0	6	6	247.0	7,581
1	7	6	246.2	7,579	1	6	6	247.4	7,582	1	5	6	247.6	7,581
2	6	6	246.6	7,580	2	5	6	248.0	7,580	2	4	6	247.6	7,581
3	5	6	248.1	7,581	3	4	6	246.4	7,593	3	3	6	248.3	7,579
4	4	6	247.1	7,581	4	3	6	248.3	7,578	4	2	6	246.5	7,580
5	3	6	248.3	7,582	5	2	6	248.5	7,579	5	1	6	248.8	7,580
6	2	6	248.3	7,554	6	1	6	247.3	7,581	6	0	6	247.9	7,581
7	1	5	**99.3**	**1,317**	7	0	5	**99.5**	**1,380**	—				
8	0	5	**88.3**	**1,509**	—					—				

Design Principle 2:
In the design of HFEv- based signature schemes we choose the number of Minus equations a and the number of Vinegar variables v to be as equal as possible, i.e. $v - a \leq 1$.

4.3 Is the Upper Bound on d_{reg} Given by Eq. (12) Reasonably Tight?

In this section we check by experiments if the upper bound on the degree of regularity given by Eq. (12) is tight. Due to memory restrictions, we can show the tightness of Eq. (12) only for some small values of D, a and v. However, for all values of D used in our scheme Gui ($D \in \{5, 9, 17\}$) we could find parameter sets for which the bound (12) is tight (see Table 4).

For most of the other parameter sets, we missed the upper bound on the degree of regularity given by Eq. (12) only by 1. We believe that, by increasing the number of equations in the systems, it would be possible to reach the upper bound for arbitrary values of (D, a, v). However, due to memory restrictions, we could not perform experiments with more than 38 equations.

Furthermore, as shown in Table 5, we could, for all of the proposed values of D, reach a degree of regularity of at least 7. These results are the basis of our parameter choice for Gui (see Sect. 5).

Table 4. Parameter sets, for which the upper bound (12) is tight

D	a	v	Upper bound for d_{reg} (12)	d_{reg} (experimental)	
5	0	0	3	3	for $n \geq 10$
	1	1	4	4	for $n \geq 23$
9	0	1	4	4	for $n \geq 23$
	1	1	4	4	for $n \geq 21$
17	0	0	4	4	for $n \geq 15$
	0	1	4	4	for $n \geq 12$

Table 5. Parameter sets which lead to $d_{reg} \geq 7$

D	a	v	d_{reg} (experimental)		Upper bound for d_{reg} (12)
5	6	6	7	for $n \geq 38$	9
9	5	5	7	for $n \geq 37$	8
17	4	4	7	for $n \geq 37$	8

4.4 Does it Help to Guess Some Variables Before Applying a Gröbner Basis Algorithm?

In the case of multivariate signature schemes such as HFEv- the public key $\mathcal{P}$ is an underdetermined system of quadratic equations. In our case this system consists of $n - a$ quadratic equations in $n + v$ variables. For the experiments presented in the previous subsections we fixed $a + v$ of the variables of the system to create a determined system before applying the F_4 algorithm.

However, for some multivariate systems, it is a good strategy to guess some additional variables before applying the Gröbner basis algorithm (Hybrid Approach [4]). The goal of this strategy is to create overdetermined systems which hopefully will be significantly easier to solve. When guessing k variables one has, to find a solution of the original system, to solve q^k instances of the simplified system, where q is the cardinality of the underlying field. To check whether this Hybrid approach helps to solve the public systems of the HFEv- scheme faster, we performed a number of experiments. For the three parameter sets $(D, a, v) \in \{(5, 6, 6), (9, 5, 5), (17, 4, 4)\}$ and varying numbers of n and k we created HFEv- systems and solved them with the F_4 algorithm integrated in MAGMA. Table 6 shows, for $k \in \{0, \ldots, 5\}$, the minimal value of n needed to reach a degree of regularity of 7.

As the table shows, we could, for each of the above parameter sets and each value $k \in \{0, \ldots, 5\}$, create a HFEv- system offering a good level of security, simply by increasing the number of equations in the system. In fact, the degree of regularity of a direct attack against such a system of $n-a$ quadratic equations in $n - a - k$ variables will be at least 7.

We therefore assume that, for large enough n, all the multivariate systems which have to be solved in the course of a direct/hybrid attack against our

Table 6. Experiments on HFEv- systems with the Hybrid Approach

# k of guessed variables	Minimal value of n to reach $d_{reg} \geq 7$		
	$D = 5$, $a = v = 6$	$D = 9$, $a = v = 5$	$D = 17$, $a = v = 4$
0	38	37	37
1	39	38	38
2	40	40	39
3	42	41	41
4	43	43	42
5	44	44	44

schemes, will have a degree of regularity of at least 7. This is the basis for our parameter selection presented in the next section.

5   The New Multivariate Signature Scheme Gui

Based on our experiments presented in the previous section we propose three different versions of our HFEv- based signature scheme Gui over the field GF(2):

- Gui-96 with $(n, D, a, v) = (96, 5, 6, 6)$ with 90 equations in 102 variables,
- Gui-95 with $(n, D, a, v) = (95, 9, 5, 5)$ with 90 equations in 100 variables and
- Gui-94 with $(n, D, a, v) = (94, 17, 4, 4)$ with 90 equations in 98 variables.

The complexity of direct attacks against these schemes can be estimated as follows.

According to our experiments (see Table 6), the degree of regularity of the F_4 algorithm (even with the Hybrid Approach) against these schemes will be at least 7.

For the complexity of a direct attack against one of our schemes (with guessing k variables) we have

$$\text{Complexity}_{F_4/F_5} \geq 3 \cdot \tau \cdot T^2, \tag{13}$$

where T is the number T of top-level monomials in the solving step of the F_4 algorithm and τ is the number of non zero elements in each equation. We get

$$\text{Compl}_{F_4/F_5} \geq 3 \cdot \tau \cdot T^2 = 2^k \cdot 3 \cdot \binom{n-a-k}{2} \cdot \binom{n-a-k}{d_{reg}}^2$$

$$= 3 \cdot \binom{n-a}{2} \cdot \binom{n-a}{d_{reg}}^2$$

$$\cdot \underbrace{2^k \cdot \frac{(n-a-k) \cdot (n-a-k-1)}{(n-a) \cdot (n-a-1)} \cdot \left(\frac{(n-a-k) \cdot \ldots \cdot (n-a-k-d_{reg}+1)}{(n-a) \cdot \ldots \cdot (n-a-d_{reg}+1)} \right)^2}_{\geq 1}$$

$$\geq 3 \cdot \binom{n-a}{2} \cdot \binom{n-a}{d_{reg}}^2 \geq 3 \cdot \binom{90}{7} \cdot \binom{90}{2} = 2^{80.7}. \tag{14}$$

Note that this number is very optimistic since we assume that the degree of regularity will not rise above 7.

Additionally, for better comparison to standard signature schemes, we propose a fourth version of Gui, Gui-127, with the parameters $(n, D, a, v) = (127, 9, 4, 6)$, providing a security level of 120 bits.

5.1 Signature Generation

The central component of the signature generation process of Gui is inverting the HFEv- core map.

To compute a pre-image of a $(n - a)$ bit digest $\mathbf{h}$, one first has to choose random values for the Minus equations and the Vinegar variables. In our concrete implementation, these values are the last $a + v$ bits of SHA-256($\mathbf{h}$). After that, one computes recursively $\mathbf{x} = \mathcal{S}^{-1}(\mathbf{h})$, $X = \Phi(x)$, $Y = \mathcal{F}_V^{-1}(X)$, $\mathbf{y} = (\Phi^{-1}(Y)||v_1|| \ldots ||v_v)$ and $\mathbf{z} = \mathcal{T}^{-1}(\mathbf{y})$ (see Fig. 2).

For the parameters of Gui, the length of the digest $\mathbf{h}$ is only $n - a = 90$ bits. To prevent birthday attacks, we therefore have to perform the above process several times (for different values of $\mathbf{h}$). We denote this repetition factor by k and set $k = 3$ for Gui-96 and Gui-95. For Gui-94 and Gui-127 the value k is chosen to be 4.

The signature generation process of Gui works as shown in Algorithm 1 and Fig. 3.

We initialize the $n - a$ vector S_0 to be $\mathbf{0}$ and compute the SHA-256 hash value $\mathbf{h}$ of the message. Let D_1 be the bitstring consisting of the first $(n - a)$ bits of $\mathbf{h}$. We compute the pre-image of D_1 under the HFEv- core (see above) and split the result into an $(n - a)$ bit string S_1 and an $a + v$ bit string X_1.

We set D_2 to be the string consisting of the first $(n - a)$ bits of SHA-256($\mathbf{h}$) and compute the HFEv- pre-image of $D_2 \oplus S_1$. Again, the result is split into the two parts S_2 ($n - a$ bits) and X_2 ($a + v$ bits). This process is repeated, until we have values S_i, X_i for $i = 1, \ldots, k$.

The final signature of the message is given by $\sigma = (S_k||X_k|| \ldots ||X_1)$. The resulting signature sizes for our schemes can be found in Table 7.

A detailed description, how the inversion of the central HFEv- map is performed in our implementation, can be found in Sect. 6.2. Due to some flaws in the SHA-1 algorithm, we replace the SHA-1 hash function used in the original QUARTZ design by SHA-256.

5.2 Signature Verification

To check the authenticity of a signature $\sigma \in \mathrm{GF}(2)^{(n-a)+k(a+v)}$ we parse σ into $S_k, X_k, \ldots, X_1$ and compute $D_1, \ldots, D_k$ as shown in Sect. 5.1. For $i = k - 1$ to 0 we compute recursively $S_i = \mathcal{P}(S_{i+1}||X_{i+1}) \oplus D_{i+1}$. The signature is accepted, if and only if $S_0 = \mathbf{0}$ holds.

By the above construction of the signature generation and verification process we prevent birthday attacks as follows. We consider an adversary A who wants to find two messages m_1 and m_2 which lead to the same signature σ.

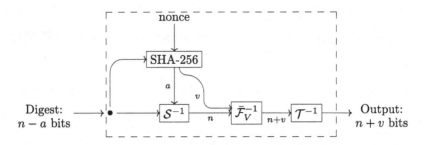

Fig. 2. Core operations of HFEv-

Algorithm 1. Signature Generation Process of Gui

Input: Gui private key $(\mathcal{S}, \mathcal{F}, \mathcal{T})$ message $\mathbf{d}$, repetition factor k
Output: signature $\sigma \in \mathrm{GF}(2)^{(n-a)+k(a+v)}$
1: $\mathbf{h} \leftarrow \text{SHA-256}(\mathbf{d})$
2: $S_0 \leftarrow \mathbf{0} \in \mathrm{GF}(2)^{n-a}$
3: **for** $i = 1$ to k **do**
4:      $D_i \leftarrow$ first $n - a$ bits of $\mathbf{h}$
5:      $(S_i, X_i) \leftarrow \text{HFEv-}^{-1}(D_i \oplus S_{i-1})$
6:      $\mathbf{h} \leftarrow \text{SHA-256}(\mathbf{h})$
7: **end for**
8: $\sigma \leftarrow (S_k || X_k || \ldots || X_1)$
9: **return** σ

Algorithm 2. Signature Verification Process of Gui

Input: Gui public key $\mathcal{P}$, message $\mathbf{d}$, repetition factor k, signature $\sigma \in$
        $\mathrm{GF}(2)^{(n-a)+k(a+v)}$
Output: TRUE or **FALSE**
1: $\mathbf{h} \leftarrow \text{SHA-256}(\mathbf{d})$
2: $(S_k, X_k, \ldots, X_1) \leftarrow \sigma$
3: **for** $i = 1$ to k **do**
4:      $D_i \leftarrow$ first $n - a$ bits of $\mathbf{h}$
5:      $\mathbf{h} \leftarrow \text{SHA-256}(\mathbf{h})$
6: **end for**
7: **for** $i = k - 1$ to 0 **do**
8:      $S_i \leftarrow \mathcal{P}(S_{i+1} || X_{i+1}) \oplus D_{i+1}$
9: **end for**
10: **if** $S_0 = \mathbf{0}$ **then**
11:      **return TRUE**
12: **else**
13:      **return FALSE**
14: **end if**

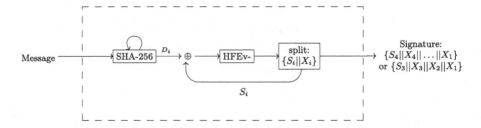

Fig. 3. Signature generation process of Gui

Table 7. Key and signature sizes of Gui-94, Gui-95, Gui-96, and Gui-127

Scheme	Core map HFEv- (n, D, a, v)	Public key size (byte)	Private key size (byte)	Repetition factor k	Signature size (bit)
Gui-96	(96, 5, 6, 6)	63036	3175	3	126
Gui-95	(95, 9, 5, 5)	60600	3053	3	120
Gui-94	(94, 17, 4, 4)	58212	2943	4	122
Gui-127	(127, 9, 4, 6)	142576	5350	4	163
QUARTZ	(103, 129, 3, 4)	75515	3774	4	128

For the plain HFEv- signature scheme it would be enough to find two messages m_1 and m_2 such that SHA-256$(m_1)_i =$ SHA-256$(m_2)_i$ for the first $n - a$ bits. If $(n - a) \leq 160$, the adversary can find m_1 and m_2 by a birthday attack.

In the context of our scheme Gui, the adversary now has to find messages m_1 and m_2 which lead to the same values of $D_1, \ldots, D_k$. For our values of the repetition factor k, this corresponds to finding a collision for a hash function of length 270, 360 and 492 bit (Gui-95/96, Gui-94 and Gui-127 respectively). This is, in general, assumed to be infeasible.

6   Implementation and Comparison

In this section we present the details of our implementation of the Gui signature scheme and compare the performance of our scheme with that of the original QUARTZ and other standard signature schemes.

6.1   Arithmetics Over Finite Fields

The first step in our implementation of the Gui signature scheme is to provide efficient arithmetics over the large binary fields in use. To speed up these computations, we use a set of new processor instructions for carry-less multiplication: PCLMULQDQ [30].

The instruction set PCLMULQDQ allows the efficient multiplication of two 64-bit polynomials over GF(2) resulting in an 128-bit polynomial. The PCLMULQDQ instructions are available on some new processors of Intel and AMD. Performance

Table 8. Performance of PCLMULQDQ on different platforms (source: [14,17])

Processor type		Latency cycles	Throughput cycles/multiplication
Intel	Sandy Bridge	14	8
	Ivy Bridge	14	8
	Hashwell	7	2
AMD	Bulldozer	12	7
	Piledriver	12	7
	Steamroller	11	7

data of PCLMULQDQ can be found in Table 8. In the case of Gui, the extension field $\mathbb{E}$ has less than 2^{128} elements. We represent an element of the field $\mathbb{E}$ as a polynomial over GF(2) which can be divided into two 64-bit polynomials.

A multiplication over the large field $\mathbb{E}$ is divided into two phases, namely a *multiplication* and a *reduction* phase.

In the *multiplication phase*, the multiplication of two 128-bit polynomials can be performed by 4 calls of PCLMULQDQ. With the help of the Karatsuba algorithm, we can avoid one call of PCLMULQDQ and therefore its long latency (see Table 8). To square an element of $\mathbb{E}$, we need only two calls of PCLMULQDQ since we are operating over a field of characteristic 2.

The *reduction phase* of the field multiplication heavily depends on the field representation. For the original QUARTZ scheme over the field GF(2^{103}) the authors used GF(2^{103}) := GF(2)[x]/($x^{103} + x^9 + 1$) [24]. For Gui, we choose the field representations

- GF(2^{94}) := GF(2)[x]/($x^{94} + x^{21} + 1$),
- GF(2^{95}) := GF(2)[x]/($x^{95} + x^{11} + 1$),
- GF(2^{96}) := GF(2)[x]/($x^{96} + x^{10} + x^9 + x^6 + 1$) and
- GF(2^{127}) := GF(2)[x]/($x^{127} + x + 1$) respectively.

The baseline for the reduction phase is two calls of PCLMULQDQ since, after the multiplication phase, the degree of the polynomial will be greater than 2×64. The irreducible polynomials above are chosen to contain only few terms of low degree. With few terms in the irreducible polynomials, we may replace the use of PCLMULQDQ by a few logic shifts and XOR instructions.

In the GF(2^{127}) case, for example, the reduction can be performed by only two 128-bit shifts for the x^{128} part and one conditional XOR for the x^{127} term, avoiding at least two calls of PCLMULQDQ while reducing the high 128 bit register.

Another technique is to represent elements as 128-bit polynomials while avoiding full reduction. This allows us to perform the reduction of degree 128–191 and 192–255 terms using only two calls of PCLMULQDQ without data dependency. In the GF(2^{96}) case, for example, we can perform the reduction phase by multiplying the degree 128–191 terms by $x^{128} = x^{42} + x^{41} + x^{38} + x^{32}$ and the degree 192–255 terms with $x^{192} = x^{20} + x^{18} + x^{12} + 1$. All the polynomials in use have degree ≤ 64, and we can perform the reduction by two calls of PCLMULQDQ.

The proposed implementation provides time-constant multiplication for preventing side channel leakage, since, regardless of the input, the same operations are performed. The same strategy is also applied to the calculation of multiplicative inverses. For example, for the sake of time-constant arithmetics, the inverse of an element $x \in \mathrm{GF}(2^{127})$ is calculated by raising x to $x^{2^{127}-2}$ instead of the faster extended Euclidean algorithm.

6.2 Inverting the HFEv- Core

In this section we describe how we can perform the inversion of the central HFEv- equation $\mathcal{F}_V(Y) = X$ efficiently. During the signature generation process of Gui we have to perform this step several times to avoid birthday attacks (see Sect. 5.1). Therefore it is extremely important to perform this step efficiently.

To invert the central HFEv- equation, we have to perform Berlekamp's algorithm to find the roots of the polynomial $\mathcal{F}_V(Y) - X$. Since the design of QUARTZ and Gui requires $\mathcal{F}_V(Y) - X$ to have a unique solution, we only have to perform the first step of Berlekamp's algorithm, i.e. the computation of

$$\gcd(\mathcal{F}_V(Y) - X, Y^{2^n} - Y). \tag{15}$$

We have

$$\gcd(\mathcal{F}_V(Y) - X, Y^{2^n} - Y)$$
$$= \gcd(\mathcal{F}_V(Y) - X, \prod_{i \in \mathbb{F}_{2^n}, i \neq 0} (Y - i)) = \prod_{i : \mathcal{F}_V(i) = X} (Y - i).$$

Therefore the main process in creating a signature consists in computing $\gcd(\mathcal{F}_V(Y) - X, Y^{2^n} - Y)$. The number of roots of $\mathcal{F}_V(Y) - X$ (as well as the only solution when that happens) can obviously be read off from the result.

Probability of a Unique Root. Every time we choose the values of Minus equations and Vinegar variables, we basically pick a random central equation $\mathcal{F}_V(Y) - X = 0$. The probability of this equation having a unique solution is about $1/e$. Therefore, in order to invert the HFEv- central equation, we have to perform the gcd computation about e times.

The repeated computation of the gcd (see Eq. 15) is probably the most detectable side channel leakage of our scheme. However, there are no known side channel attacks on big field schemes or HFEv- which use the information that one particular equation in the big field has no, respectively two or more solutions.

How Do We Optimize the Computation of the GCD? The main computation consumption in this step comes from the division of the extreme high power polynomial $Y^{2^n} - Y \mod \mathcal{F}_V(Y)$. A naive long division is unacceptable for this purpose due to its slow reduction phase. Instead of this, we choose to recursively raise the lower degree polynomial Y^{2^m} to the power of 2.

$$(Y^{2^m} \mod \mathcal{F}_V(\mathcal{Y}))^\in \mod \mathcal{F}_V(\mathcal{Y})$$
$$= (\sum_{i < 2^m} b_i Y^i)^2 \mod \mathcal{F}_V(\mathcal{Y}) = (\sum_{\rangle < \in^{\updownarrow}} \lfloor_{\rangle}^{\in} \mathcal{y}^{\in\rangle}) \mod \mathcal{F}_V(\mathcal{Y})$$

By multiplying Y to the naive relation $Y^D = \sum_{0 \le i \le j, 2^i + 2^j < D} a_{ij} Y^{2^i + 2^j}$, we can prepare a table for Y^{2i} mod $\mathcal{F}_V(\mathcal{Y})$ first. The rest of computation of the raising process is to square all the coefficients b_i in Y^{2^m} mod $\mathcal{F}_V(\mathcal{Y})$ and multiply them to the Y^{2i}s in the table.

Although the starting relation $\mathcal{F}_V(Y) = Y^D + \sum_{0 \le i \le j, 2^i + 2^j < D} a_{ij} Y^{2^i + 2^j}$ is a sparse polynomial, the polynomials become dense quickly in the course of the raising process. However, the number of terms in the polynomials is restricted by D because of mod $\mathcal{F}_V(\mathcal{Y})$. We expect the number of terms to be in average D during the computation.

We implemented Berlekamp's algorithm in such a way that it takes the same number of iterations in the main GCD loop and the same number of operations in the big field for each run at very low cost. Therefore it runs, independently from the input, at constant time.

The number of field multiplications needed to compute the Y^{2i} table is $O(2 \cdot D^2)$. To raise Y^{2^m} to Y^{2^n} we need $O((n-m) \cdot D)$ squarings and $O((n-m) \cdot D^2)$ multiplications. We can further reduce the number of computations needed for raising Y^{2^m} by using a higher degree Y^i table. For example, if we raise Y^{2^m} to $Y^{2^{4m}}$ in one step, we need only $O((n-m) \cdot D)$ squarings and $O(\frac{(n-m)}{2} \cdot D^2)$ multiplications. However, the computational effort for preparing the Y^i table increases. Table 9 shows the time needed to compute $\gcd(Y^{2^n} - Y, \mathcal{F}(Y))$ on three different CPUs.

6.3 Experiments and Comparison

Table 10 shows key and signature sizes as well as the running times of signature generation and verification of Gui and compares these data with those of some standard signature schemes. The data are benchmarked according to specifications given by the eBACS project [3].

We should note that the timings for Gui given by Table 10 are for C programs with a few intrinsic function calls of PCLMULQDQ. The PKCs benchmarked in the eBACs project also do not represent optimal implementations of RSA and ECC. We present these numbers in an effort to compare apples to apples by using only reference implementations.

Table 9. Key sizes of HFEv- schemes and running time of $\gcd(X^{2^n} - X, \mathcal{F}(X))$

Scheme	Security level (bit)	Public key size (kB)	Private key size (kB)	Time needed for inverting $\mathcal{F}$ (kilo-cycles)
HFEv- (96, 5, 6, 6)	80	61.6	3.1	72/76/55[a]
HFEv- (95, 9, 5, 5)	80	59.2	3.0	159/135/79
HFEv- (94, 17, 4, 4)	80	56.8	2.9	533/453/274
HFEv- (127, 9, 4, 6)	120	139.2	5.2	170/156/128
HFEv- (103, 129, 3, 4)	80	71.9	3.1	25,793/20,784/12,630

[a] AMD Opteron 6212, 2.5 GHz (Bulldozer)/Intel Xeon CPU E5-2620, 2.0 GHz (Sandy Bridge)/Intel Xeon E3-1245 v3, 3.4 GHz (Hashwell)

Table 10. Comparison between Gui and standard signature schemes

Scheme	Security level (bits)	Public key size (Bytes)	Private key size (Bytes)	Signature size (bits)	Signing time (k-cycles)[a]	Verification time (k-cycles)[a]
Gui-96 (96, 5, 6, 6)	80	63,036	3,175	126	603/569/238	97/70/62
Gui-95 (95, 9, 5, 5)	80	60,600	3,053	120	1,417/1,441/602	91/60/58
Gui-94 (94, 17, 4, 4)	80	58,212	2,943	124	5,800/5,480/2,495	118/74/71
Gui-127 (127, 9, 4, 6)	120	142,576	5,350	163	2,368/2,183/1,080	220/121/122
QUARTZ (103, 129, 3, 4)	80	73,626	3,174	128	302,882/315,716/128,736	145/84/86
RSA-1024	80	128	128	128	2,080/1,058/1,073	74/32/33
RSA-2048	112	256	256	256	8,834/5,347/4,625	138/76/61
ECDSA P160	80	40	60	320	1,283/558/588	1,448/635/652
ECDSA P192	96	48	72	384	1,513/773/697	1,715/867/779
ECDSA P256	128	64	96	512	830/388/342	2,111/920/816

[a] AMD Opteron 6212, 2.5 GHz (Bulldozer)/Intel Xeon CPU E5-2620, 2.0 GHz (Sandy Bridge)/Intel Xeon E3-1245 v3, 3.4 GHz (Hashwell)

6.4   Platforms Without `PCLMULQDQ`

We also optimized the arithmetics over large finite fields by SIMD table-lookup instructions for platforms without `PCLMULQDQ`. SIMD table-lookup instructions are common in contemporary CPUs, e.g., `PSHUFB` (Packed Shuffle Byte) on x86 and `VTBL` (Vector Table Lookup) on ARM platforms. For this we used an ARM Cortex-A9 processor with NEON instruction set, which is currently the most common version in smart phones. We use `PSHUFB` and `VTBL` as general table-lookup instructions with 4-bit index (although `VTBL` is capable of 5-bit indices), not necessarily restricted to the x86 platform. Since the length of indices in these instructions is only 4 or 5 bit, `PSHUFB` and `VTBL` were so far only applied to implementations over small fields, e.g., GF(16) and GF(256) [6]. For applying `PSHUFB` and `VTBL` to large finite fields, we have to represent the large field as an extension of the small field. In our case, we use for the implementation the following representation of $\mathrm{GF}(2^{96})$

- $\mathrm{GF}(16) := \mathrm{GF}(2)[y]/(y^4 + y + 1)$,
- $\mathrm{GF}(2^{96}) := \mathrm{GF}(16)[x]/(x^{24} + y^3 x^3 + x + y)$.

The multiplication in GF(16) is performed with `PSHUFB` and the multiplication in $\mathrm{GF}(2^{96})$ corresponds to a polynomial multiplication over GF(16). Furthermore, we use Karatsuba's technique for the computation of coefficients in different registers. To prevent the scheme from side channel leakage, we implement the multiplication in GF(16) with logarithm/exponential tables instead of multiplication tables, except for the multiplication with fixed values in the reduction phase of the polynomial multiplication. With logarithm tables, the multiplication in GF(16) is performed by an addition in the exponents of a multiplicative generator and therefore consists of two table lookups, addition and reduction. Although there is only one table lookup in a normal implementation of multiplication tables, an intentional cache miss would result in a time difference since the tables are loaded with the values of input operands.

Table 11. Average number of cycles for the arithmetics in $GF(2^{96})$ and $GF(2^{127})$ for various implementations.

	Implementation	Multiplication	Square	Inversion
$GF(2^{96})$	64-bit variables, school book	624/3392[a]	624/3384	68,752/357,728
	128-bit register, `PSHUFB/VTBL`	138/731	87/424	11,242/48,825
	128-bit register, `PCLMULQDQ`	12/-	8/-	2,489/-
$GF(2^{127})$	64-bit variables, school book	743/4,009	735/3,997	105,235/546,881
	128-bit register, `PSHUFB/VTBL`[b]	318/813	187/531	28,565/77,703
	128-bit register, `PCLMULQDQ`	15/-	9/-	3,257/-

[a] Intel Xeon E3-1245 v3, 3.4 GHz (Hashwell)/Xilinx Zynq 7020, 667 MHz (ARM Cortex-A9)
[b] $GF(2^{128})$

Performance data of these implementations as well as a reference implementation of school book multiplication for 64-bit variables, which is applicable to all platforms without these SIMD instructions, can be found in Table 11. Note that, in the sixth row of Table 11, we use the field $GF(2^{128}) := GF(16)[x]/(x^{32} + x^3 + x + y)$ instead of $GF(2^{127})$, since the shape of this field contains some restrictions in the extension from $GF(16)$.

Our ARM implementation takes on average 1,14 ms to invert the central map $\mathcal{F}_V(Y) = X$. Performance data for the implementation of Gui-96 on ARM platforms can be found in Table 12. As one can see, we are able to generate about 300 Gui signatures per second on a standard smart phone.

6.5 Grover's Algorithm and Potential Extension to Larger Fields

By Grover's algorithm [16] it might be possible to cut down the complexity of a brute-force search in an n-bit space to $O(2^{n/2})$. We believe that this is no major threat to HFEv- and in particular to Gui because of the large number of quantum bits (qubits) needed in this case: While we need only 1024 qubits to solve Discrete Logarithms on a 256-bit prime modulus elliptic curve and 6000 qubits to factorize 3000-bit RSA numbers using Shor's Algorithm, the number of qubits and quantum gates needed to attack Gui by Grover's algorithm is in the order of a million (n^3), since it implies the evaluation of n quadratic polynomials in n variables. Therefore, quantum algorithms can be used much more easy for the cryptanalysis of schemes such as RSA and ECC than for that of multivariate schemes such as Gui and we do not consider Grover's algorithm to be a major problem for our scheme. However, even if we have to take Grover's algorithm into account, there is an easy way to prevent this kind of attack, namely by choosing

Table 12. Performance data for Gui on ARM platforms (timings in 10^{-6} s)

Scheme	Key generation	Signature generation	Signature verification
Gui-96(96, 5, 6, 6)	99,555	3,291	102

the parameter n about twice as large while keeping all other parameters constant. In the implementation, this means an extra layer of the Karatsuba algorithm in the multiplication phase and therefore a factor of 3 slowdown. Furthermore, this increases the public key size by a factor of 8.

7   Conclusion and Future Work

In this paper, we analyzed the behavior of direct attacks against HFEv- based signature schemes. Experiments show that, even for low degree HFE polynomials in use, we can obtain adequate security levels by increasing the numbers a and v of Minus equations and Vinegar variables. Furthermore we find that the upper bound on the degree of regularity proposed by Ding and Yang in [11] is relatively tight. From our results we derive design principles for the construction of HFEv- based signature schemes, which lead to both secure and efficient schemes. We apply these principles to the construction of our new HFEv- based signature scheme Gui, which is more than 100 times faster than the original QUARTZ scheme. Furthermore we show that the performance of our scheme is highly comparable to that of standard signature schemes, including signatures on elliptic curves.

As future work we want to analyze the influence of the numbers a of Minus Equations and v of Vinegar variables on the security of HFEv- schemes further. Furthermore we plan to create for every common existing platform an optimal implementation of HFEv- (Gui) and compare it with some of the best optimized code for ECC and RSA, such as Ed25519 [2]. Another approach would be to verify such optimal Gui code for formal correctness. In short, we believe that there is still much work to be done on the HFEv- digital signature schemes.

Acknowledgements. We thank the anonymous reviewers of Asiacrypt for their comments which helped to improve the paper. Especially we want to thank the shepherd of our paper for his valuable advice. Due to this we included

- Further remarks on the complexity of the Kipnis-Shamir attack on HFE and its variants (Sect. 3.3).
- Additional experiments on the effect of the parameters a and v on the security of our scheme and the Hybrid approach (Sects. 4.2 and 4.4).
- Remarks on side channel leakage and countermeasures (Sects. 6.1 and 6.2).
- Implementation details of Gui on ARM platforms (Sect. 6.4).
- Remarks on how Grover's algorithm might affect our parameter choice (Sect. 6.5).

We would like to thank for partial support from the Charles Phelps Taft Research Center, the Center for Advanced Security Research Darmstadt (CASED), ECSPRIDE, Academia Sinica, the CAS/SAFEA International Partnership Program for Creative Research Teams, Taiwan's Ministry of Science and Technology, National Taiwan University and Intel Corporation under grands NIST 60NAN15D059, NSFC 61472054, MOST 103-2911-I-002-001, NTU-ICRP-104R7501 and NTU-ICRP-104R7501-1.

References

1. Bernstein, D.J., Buchmann, J., Dahmen, E. (eds.): Post Quantum Cryptography. Springer, Heidelberg (2009)
2. Bernstein, D.J., Duif, N., Lange, T., Schwabe, P., Yang, B.-Y.: High-speed high-security signatures. J. Cryptographic Eng. $2(2)$, 77–89 (2012)
3. Bernstein, D.J., Lange, T. (eds.): eBACS: ECRYPT Benchmarking of Cryptographic Systems. http://bench.cr.yp.to. Accessed 14 May 2014
4. Bettale, L., Faugère, J.C., Perret, L.: Hybrid approach for solving multivariate systems over finite fields. J. Math. Cryptol. 3, 177–197 (2009)
5. Bogdanov, A., Eisenbarth, T., Rupp, A., Wolf, C.: Time-area optimized public-key engines: $\mathcal{MQ}$-cryptosystems as replacement for elliptic curves? In: Oswald, E., Rohatgi, P. (eds.) CHES 2008. LNCS, vol. 5154, pp. 45–61. Springer, Heidelberg (2008)
6. Chen, A.I.-T., Chen, M.-S., Chen, T.-R., Cheng, C.-M., Ding, J., Kuo, E.L.-H., Lee, F.Y.-S., Yang, B.-Y.: SSE implementation of multivariate PKCs on modern x86 CPUs. In: Clavier, C., Gaj, K. (eds.) CHES 2009. LNCS, vol. 5747, pp. 33–48. Springer, Heidelberg (2009)
7. Courtois, N.T., Daum, M., Felke, P.: On the security of HFE, HFEv- and QUARTZ. In: Desmedt, Y.G. (ed.) PKC 2003. LNCS, vol. 2567, pp. 337–350. Springer, Heidelberg (2003)
8. Ding, J., Gower, J.E., Schmidt, D.S.: Multivariate Public Key Cryptosystems. Springer, New York (2006)
9. Ding, J., Kleinjung, T.: Degree of regularity for HFE-. IACR eprint 2011/570
10. Ding, J., Schmidt, D.: Rainbow, a new multivariable polynomial signature scheme. In: Ioannidis, J., Keromytis, A.D., Yung, M. (eds.) ACNS 2005. LNCS, vol. 3531, pp. 164–175. Springer, Heidelberg (2005)
11. Ding, J., Yang, B.-Y.: Degree of regularity for HFEv and HFEv-. In: Gaborit, P. (ed.) PQCrypto 2013. LNCS, vol. 7932, pp. 52–66. Springer, Heidelberg (2013)
12. Faugère, J.C.: A new efficient algorithm for computing Gröbner bases (F4). J. Pure Appl. Algebra **139**, 61–88 (1999)
13. Faugère, J.C., Safey el Din, M., Spaenlehauer, P.J.: On the complexity of the generalized MinRank problem. J. Symbolic Comput. **55**, 30–58 (2013)
14. Fog, A.: Instruction tables: Lists of instruction latencies, throughputs and micro-operation breakdowns for Intel, AMD and VIA CPUs, 7 December 2014. http://www.agner.org/optimize/
15. Garey, M.R., Johnson, D.S.: Computers and Intractability: A Guide to the Theory of NP-Completeness. W.H. Freeman and Company, New York (1979)
16. Grover, L.K.: A fast quantum mechanical algorithm for database search. In: Proceedings of STOC, pp. 212–219. ACM (1996)
17. Intel Corporation: Hashwell Cryptographic Performance. http://www.intel.com/content/dam/www/public/us/en/documents/white-papers/haswell-cryptographic-performance-paper.pdf
18. Jiang, X., Ding, J., Hu, L.: Kipnis-Shamir attack on HFE revisited. In: Pei, D., Yung, M., Lin, D., Wu, C. (eds.) Inscrypt 2007. LNCS, vol. 4990, pp. 399–411. Springer, Heidelberg (2008)
19. Kravitz, D.: Digital Signature Algorithm. US patent 5231668, July 1991
20. Kipnis, A., Patarin, J., Goubin, L.: Unbalanced oil and vinegar signature schemes. In: Stern, J. (ed.) EUROCRYPT 1999. LNCS, vol. 1592, pp. 206–222. Springer, Heidelberg (1999)

21. Kipnis, A., Shamir, A.: Cryptanalysis of the HFE public key cryptosystem by relinearization. In: Wiener, M. (ed.) CRYPTO 1999. LNCS, vol. 1666, pp. 19–30. Springer, Heidelberg (1999)
22. Mohamed, M.S.E., Ding, J., Buchmann, J.: Towards algebraic cryptanalysis of HFE challenge 2. In: Kim, T., Adeli, H., Robles, R.J., Balitanas, M. (eds.) ISA 2011. CCIS, vol. 200, pp. 123–131. Springer, Heidelberg (2011)
23. Matsumoto, T., Imai, H.: Public quadratic polynomial-tuples for efficient signature-verification and message-encryption. In: Günther, C.G. (ed.) EUROCRYPT 1988. LNCS, vol. 330, pp. 419–453. Springer, Heidelberg (1988)
24. Patarin, J., Courtois, N.T., Goubin, L.: QUARTZ, 128-bit long digital signatures. In: Naccache, D. (ed.) CT-RSA 2001. LNCS, vol. 2020, pp. 282–297. Springer, Heidelberg (2001)
25. Patarin, J., Courtois, N.T., Goubin, L.: FLASH, a fast multivariate signature algorithm. In: Naccache, D. (ed.) CT-RSA 2001. LNCS, vol. 2020, pp. 298–307. Springer, Heidelberg (2001)
26. Patarin, J.: Cryptanalysis of the Matsumoto and Imai public key scheme of Eurocrypt '88. In: Coppersmith, D. (ed.) CRYPTO 1995. LNCS, vol. 963, pp. 248–261. Springer, Heidelberg (1995)
27. Richards, C.: Algorithms for factoring square-free polynomials over finite fields. Master thesis, Simon Fraser University, Canada (2009)
28. Rivest, R.L., Shamir, A., Adleman, L.: A method for obtaining digital signatures and public-key cryptosystems. Commun. ACM $21(2)$, 120–126 (1978)
29. Shor, P.: Polynomial-time algorithms for prime factorization and discrete logarithms on a quantum computer. SIAM J. Comput. $26(5)$, 1484–1509 (1997)
30. Taverne, J., Faz-Hernández, A., Aranha, D.F., Rodríguez-Henríquez, F., Hankerson, D., López, J.: Software implementation of binary elliptic curves: impact of the carry-less multiplier on scalar multiplication. In: Preneel, B., Takagi, T. (eds.) CHES 2011. LNCS, vol. 6917, pp. 108–123. Springer, Heidelberg (2011)
31. http://en.wikipedia.org/wiki/File:CMOC_Treasures_of_Ancient_China_exhibit_white_pottery_gui_1.jpg

Multiparty Computation I

Oblivious Network RAM and Leveraging Parallelism to Achieve Obliviousness

Dana Dachman-Soled[1]([⊠]), Chang Liu[1], Charalampos Papamanthou[1], Elaine Shi[2], and Uzi Vishkin[1]

[1] University of Maryland, College Park, USA
danadach@ece.umd.edu, liuchang@cs.umd.edu,
cpap@umd.edu, vishkin@umiacs.umd.edu
[2] Cornell University, Ithaca, USA
runting@gmail.com

Abstract. Oblivious RAM (ORAM) is a cryptographic primitive that allows a trusted CPU to securely access untrusted memory, such that the access patterns reveal nothing about sensitive data. ORAM is known to have broad applications in secure processor design and secure multi-party computation for big data. Unfortunately, due to a logarithmic lower bound by Goldreich and Ostrovsky (Journal of the ACM, '96), ORAM is bound to incur a moderate cost in practice. In particular, with the latest developments in ORAM constructions, we are quickly approaching this limit, and the room for performance improvement is small.

In this paper, we consider new models of computation in which the cost of obliviousness can be fundamentally reduced in comparison with the standard ORAM model. We propose the Oblivious Network RAM model of computation, where a CPU communicates with multiple memory banks, such that the adversary observes only which bank the CPU is communicating with, but not the address offset within each memory bank. In other words, obliviousness within each bank comes for free—either because the architecture prevents a malicious party from observing the address accessed within a bank, or because another solution is used to obfuscate memory accesses within each bank—and hence we only need to obfuscate communication patterns between the CPU and the memory banks. We present new constructions for obliviously simulating general

D. Dachman-Soled—Work supported in part by NSF CAREER award #CNS-1453045 and by a Ralph E. Powe Junior Faculty Enhancement Award.

C. Liu—Work supported in part by NSF awards #CNS-1314857, #CNS-1453634, #CNS-1518765, #CNS-1514261, and Google Faculty Research Awards.

C. Papamanthou—Work supported in part by NSF award #CNS-1514261, by a Google Faculty Research Award and by the National Security Agency.

E. Shi—Work supported in part by NSF awards #CNS-1314857, #CNS-1453634, #CNS-1518765, #CNS-1514261, Google Faculty Research Awards, and a Sloan Fellowship. This work was done in part while a subset of the authors were visiting the Simons Institute for the Theory of Computing, supported by the Simons Foundation and by the DIMACS/Simons Collaboration in Cryptography through NSF award #CNS-1523467.

U. Vishkin—Work supported in part by NSF award #CNS-1161857.

T. Iwata and J.H. Cheon (Eds.): ASIACRYPT 2015, Part I, LNCS 9452, pp. 337–359, 2015.
DOI: 10.1007/978-3-662-48797-6_15

or parallel programs in the Network RAM model. We describe applications of our new model in secure processor design and in distributed storage applications with a network adversary.

1 Introduction

Oblivious RAM (ORAM), introduced by Goldreich and Ostrovsky [18,19], allows a *trusted* CPU (or a trusted computational node) to obliviously access *untrusted* memory (or storage) during computation, such that an adversary cannot gain any sensitive information by observing the data access patterns. Although the community initially viewed ORAM mainly from a theoretical perspective, there has recently been an upsurge in research on both new efficient algorithms (c.f. [8,13,22,36,39,43,46]) and practical systems [9,11,12,21,30,35,37,38,44,48] for ORAM. Still the most efficient ORAM implementations [10,37,39] require a relatively large bandwidth blowup, and part of this is inevitable in the standard ORAM model. Fundamentally, a well-known lower bound by Goldreich and Ostrovsky states that any ORAM scheme with constant CPU cache must incur at least $\Omega(\log N)$ blowup, where N is the number of memory words, in terms of bandwidth and runtime. To make ORAM techniques practical in real-life applications, we wish to further reduce its performance overhead. However, since latest ORAM schemes [39,43] have practical performance approaching the limit of the Goldreich-Ostrovsky lower bound, the room for improvement is small in the standard ORAM model. In this paper, we investigate the following question:

In what alternative, realistic models of computation can we significantly lower the cost of oblivious data accesses?

Motivated by practical applications, we propose the Network RAM (NRAM) model of computation and correspondingly, Oblivious Network RAM (O-NRAM). In this new model, one or more CPUs interact with M memory banks during execution. Therefore, each memory reference includes a *bank identifier*, and an *offset* within the specified memory bank. We assume that an *adversary cannot observe the address offset within a memory bank, but can observe which memory bank the CPU is communicating with*. In other words, obliviousness within each bank "comes for free". Under such a threat model, an Oblivious NRAM (O-NRAM) can be informally defined as an NRAM whose observable memory traces (consisting of the bank identifiers for each memory request) do not leak information about a program's private inputs (beyond the length of the execution). In other words, in an O-NRAM, the sequence of bank identifiers accessed during a program's execution must be provably obfuscated.

1.1 Practical Applications

Our NRAM models are motivated by two primary application domains:

Secure processor architecture. Today, secure processor architectures [1,12,30,35,40,41] are designed assuming that the memory system is *passive* and untrusted. In particular, an adversary can observe both memory contents and memory addresses during program execution. To secure against such an adversary, the trusted CPU must both encrypt data written to memory, and obfuscate memory access patterns.

Our new O-NRAM model provides a realistic alternative that has been mentioned in the architecture community [30,31] and was inspired by the Module Parallel Computer (MPC) model of Melhorn and Vishkin [32]. The idea is to introduce *trusted* decryption logic on the memory DIMMs (for decrypting memory addresses). This way, the CPU can encrypt the memory addresses before transmitting them over the insecure memory bus. In contrast with traditional passive memory, we refer to this new type of memory technology as *active memory*. In a simple model where a CPU communicates with a single active memory bank, obliviousness is automatically guaranteed, since the adversary can observe only *encrypted* memory contents and addresses. However, when there are multiple such active memory banks, we must obfuscate which memory bank the CPU is communicating with.

Distributed storage with a network adversary. Consider a scenario where a client (or a compute node) stores private, encrypted data on multiple distributed storage servers. We consider a setting where all endpoints (including the client and the storage servers) are *trusted*, but the network is an *untrusted* intermediary. In practice, trust in a storage server can be bootstrapped through means of trusted hardware such as the Trusted Platform Module (TPM) or as IBM 4758; and network communication between endpoints can be encrypted using standard SSL. Trusted storage servers have also been built in the systems community [3]. On the other hand, the untrusted network intermediary can take different forms in practice, e.g., an untrusted network router or WiFi access point, untrusted peers in a peer-to-peer network (e.g., Bitcoin, TOR), or packet sniffers in the same LAN. Achieving oblivious data access against such a network adversary is precisely captured by our O-NRAM model.

1.2   Background: The PRAM Model

Two of our main results deal with the parallel-RAM (PRAM) model, which is a synchronous generalization of the RAM model to the parallel processing setting. The PRAM model allows for an unbounded number of parallel processors with a shared memory. Each processor may access any shared memory cell and read/write conflicts are handled in various ways depending on the type of PRAM considered:

- **Exclusive Read Exclusive Write (EREW) PRAM:** A memory cell can be accessed by at most one processor in each time step.
- **Concurrent Read Exclusive Write (CREW) PRAM:** A memory cell can be read by multiple processors in a single time step, but can be written to by at most one processor in each time step.
- **Concurrent Read Concurrent Write (CRCW) PRAM:** A memory cell can be read and written to by multiple processors in a single time step. Reads

are assumed to complete prior to the writes of the same time step. Concurrent writes are resolved in one of the following ways: (1) *Common*—all concurrent writes must write the same value; (2) *Arbitrary*—an arbitrary write request is successful; (3) *Priority*—processor id determines which processor is successful.

1.3 Results and Contributions

We introduce the Oblivious Network RAM model, and conduct the first *systematic* study to understand the "cost of obliviousness" in this model. We consider running both *sequential* programs and *parallel* progams in this setting. We propose novel algorithms that exploit the "free obliviousness" within each bank, such that the obliviousness cost is significantly lower in comparison with the standard Oblivious (Parallel) RAMs. We give a summary of our results below.

First, observe that if there are only $O(1)$ number of memory banks, there is a trivial solution with $O(1)$ cost: just make one memory access (real or dummy) to each bank for each step of execution. On the other hand, if there are $\Omega(N)$ memory banks each of constant size (where N denotes the total number of memory words), then the problem approaches standard ORAM [18,19] or OPRAM [7]. The intermediate parameters are therefore the most interesting. For simplicity, in this section, we mainly state our results for the most interesting case when the number of banks $M = O(\sqrt{N})$, and each bank can store up to $O(\sqrt{N})$ words. In Sects. 3, 4 and 5, our results will be stated for more general parameter choices. We now state our results (see also Table 1 for an overview).

"Sequential-to-sequential" compiler. First, we show that any RAM program can be obliviously simulated on a Network RAM, consuming only $O(1)$ words of local CPU cache, with $\widehat{O}(\log N)$ blowup in both runtime and bandwidth, where–throughout the paper–when we say the complexity of our scheme is $\widehat{O}(f(N))$, we mean that for any choice of $h(N) = \omega(f(N))$, our scheme attains complexity $g(N) = O(h(N))$. Further, when the RAM program has $\Omega(\log^2 N)$ memory word size, it can be obliviously simulated on Network RAM with only $\widehat{O}(1)$ bandwidth blowup (assuming non-uniform memory word sizes as used by Stefanov et al. in [38]). In comparison, the best known (constant CPU cache) ORAM scheme has roughly $\widehat{O}(\log N)$ bandwidth blowup for $\Omega(\log^2 N)$ memory word size [43]. For smaller memory words, the best known ORAM scheme has $O(\log^2 / \log\log N)$ blowup in both runtime and bandwidth [25].

"Parallel-to-sequential" compiler. We demonstrate that parallelism can facilitate obliviousness, by showing that programs with a "sufficient degree of parallelism" – specifically, programs whose degree of parallelism $P = \omega(M \log N)$ – can be obliviously simulated in the Network RAM model with only $O(1)$ blowup in runtime and bandwidth. Here, we consider parallelism as a property of the program, but are not in fact executing the program on a parallel machine. The overhead stated above is for the sequential setting, i.e., considering that both NRAM and O-NRAM have single processor. Our compiler works when the underlying PRAM program is in the EREW, CREW, common CRCW or arbitrary CRCW model.

Table 1. A systematic study of "cost of obliviousness" in the Network ORAM model. W denotes the memory word size in # bits, N denotes the total number of memory words, and M denotes the number of memory banks. For simplicity, this table assumes that $M = O(\sqrt{N})$, and each bank has $O(\sqrt{N})$ words. Like implicit in existing ORAM works [19,25], small word size assumes at least $\log N$ bits per word—enough to store a virtual address of the word.

Setting	RAM to O-NRAM blowup	*c.f.* Best known ORAM blowup
Sequential-to-sequential compiler		
$W = $ small	$\widehat{O}(\log N)$	$O(\log^2 N / \log \log N)$ [25]
$W = \Omega(\log^2 N)$	Bandwidth: $\widehat{O}(1)$	Bandwidth: $\widehat{O}(\log N)$ [43]
	Runtime: $\widehat{O}(\log N)$	Runtime: $O(\log^2 N / \log \log N)$ [25]
$W = \Omega(N^\epsilon)$	$\widehat{O}(1)$	$\widehat{O}(\log N)$ [43]
Parallel-to-sequential compiler		
$\omega(M \log N)$-parallel	$O(1)$	Same as standard ORAM
Parallel-to-parallel compiler		
$M^{1+\delta}$-parallel	$O(\log^* N)$	Best known: poly $\log N$ [7]
for any const $\delta > 0$		Lower bound: $\Omega(\log N)$

Beyond the low overhead discussed above, our compiled sequential O-NRAM has the additional benefit that it allows for an extremely simple prefetching algorithm. In recent work, Yu et al. [49] proposed a dynamic prefetching algorithm for ORAM, which greatly improved the practical performance of ORAM. We note that our parallel-to-sequential compiler achieves prefetching essentially for free: Since the underlying PRAM program will make many parallel memory accesses to each bank, and since the compiler knows these memory addresses ahead of time, these memory accesses can automatically be prefetched. We note that a similar observation was made by Vishkin [42], who suggested leveraging parallelism for performance improvement by using (compile-time) prefetching in serial or parallel systems.

"Parallel-to-parallel" compiler. Finally, we consider oblivious simulation in the parallel setting. We show that for any parallel program executing in t parallel steps with $P = M^{1+\delta}$ processors, we can obliviously simulate the program on a Network PRAM with $P' = O(P / \log^* P)$ processors, running in $O(t \log^* P)$ time, thereby achieving $O(\log^* P)$ blowup in parallel time and bandwidth, and optimal work. In comparison, the best known OPRAM scheme has poly $\log N$ blowup in parallel time and bandwidth. The compiler works when the underlying program is in the EREW, CREW, common CRCW or arbitrary CRCW PRAM model. The resulting compiled program is in the arbitrary CRCW PRAM model.

1.4   Technical Highlights

Our most interesting technique is for the parallel-to-parallel compiler. We achieve this through an intermediate stepping stone where we first construct a parallel-to-sequential compiler (which may be of independent interest).

At a high level, the idea is to assign each virtual address to a pseudorandom memory bank (and this assignment stays the same during the entire execution). Suppose that a program is sufficiently parallel such that it always makes memory requests in $P = \omega(M \log N)$-sized batches. For now, assume that all memory requests within a batch operate on *distinct* virtual addresses – if not we can leverage a hash table to suppress duplicates, using an additional "scratch" bank as the CPU's working memory. Then, clearly each memory bank will in expectation serve P/M requests for each batch. With a simple Chernoff bound, we can conclude that each memory bank will serve $O(P/M)$ requests for each batch, except with *negligible* probability. In a sequential setting, we can easily achieve $O(1)$ bandwidth and runtime blowup: for each batch of memory requests, the CPU will sequentially access each bank $O(P/M)$ number of times, padding with dummy accesses if necessary (see Sect. 4).

However, additional difficulties arise when we try to execute the above algorithm in parallel. In each step, there is a batch of P memory requests, one coming from each processor. However, each processor cannot perform its own memory request, since the adversary can observe which processor is talking to which memory bank and can detect duplicates (note this problem did not exist in the sequential case since there was only one processor). Instead, we wish to

1. hash the memory requests into buckets according to their corresponding banks while suppressing duplicates; and
2. pad the number of accesses to each bank to a worst-case maximum – as mentioned earlier, if we suppressed duplicate addresses, each bank has $O(P/M)$ requests with probability $1 - \mathrm{negl}(N)$.

At this point, we can assign processors to the memory requests in a round-robin manner, such that which processor accesses which bank is "fixed". Now, to achieve the above two tasks in $O(\log^* P)$ parallel time, we need to employ non-trivial parallel algorithms for "colored compaction" [4] and "static hashing" [5,17], for the arbitrary CRCW PRAM model, while using a scratch bank as working memory (see Sect. 5).

1.5   Related Work

Oblivious RAM (ORAM) was first proposed in a seminal work by Goldreich and Ostrovsky [18,19] where they laid a vision of employing an ORAM-capable secure processor to protect software against piracy. In their work, Goldreich and Ostrovsky showed both a poly-logarithmic upper-bound (commonly referred to as the hierarchical ORAM framework) and a logarithmic lower-bound for ORAM—both under constant CPU cache. Goldreich and Ostrovsky's hierarchical construction was improved in several subsequent works [6,20,22,25,33,45–47].

Recently, Shi *et al.* proposed a new, tree-based paradigm for constructing ORAMs [36], thus leading to several new constructions that are simple and practically efficient [8,13,39,43]. Notably, Circuit ORAM [43] partially resolved the tightness of the Goldreich-Ostrovsky lower bound, by showing that certain stronger interpretations of their lower bound are indeed tight.

Theoretically, the best known ORAM scheme (with constant CPU cache) for small $O(\log N)$-sized memory words[1] is a construction by Kushilevitz *et al.* [25], achieving $O(\log^2 N/\log\log N)$ bandwidth and runtime blowup. Path ORAM (variant with $O(1)$ CPU cache [44]) and Circuit ORAM can achieve better bounds for bigger memory words. For example, Circuit ORAM achieves $O(\log N)\omega(1)$ bandwidth blowup for a word size of $\Omega(\log^2 N)$ bits; and for $O(\log N)\omega(1)$ runtime blowup for a memory word size of N^ϵ bits where $0 < \epsilon < 1$ is any constant within the specified range.

ORAMs with larger CPU cache sizes (caching up to N^α words for any constant $0 < \alpha < 1$) have been suggested for cloud storage outsourcing applications [20,38,47]. In this setting, Goodrich and Mitzenmacher [20] first showed how to achieve $O(\log N)$ bandwidth and runtime blowup.

Other than secure processors and cloud outsourcing, ORAM is also noted as a key primitive for scaling secure multi-party computation to big data [23,26,43,44]. In this context, Wang *et al.* [43,44] pointed out that the most relevant ORAM metric should be the circuit size rather than the traditionally considered bandwidth metrics. In the secure computation context, Lu and Ostrovsky [27] proposed a two-server ORAM scheme that achieves $O(\log N)$ runtime blowup. Similarly, ORAM can also be applied in other RAM-model cryptographic primitives such as (reusable) Garbled RAM [14–16,28,29].

Goodrich and Mitzenmacher [20] and Williams et al. [48] observed that computational tasks with inherent parallelism can be transformed into efficient, oblivious counterparts in the traditional ORAM setting—but our techniques apply to the NRAM model of computation. Finally, Oblivious RAM has been implemented in outsourced storage settings [37,38,45,47,48], on secure processors [9,11,12,30,31,35], and atop secure multiparty computation [23,43,44].

Comparison of our parallel-to-parallel compiler with the work of [7]. Recently, Boyle, Chung and Pass [7] proposed Oblivious Parallel RAM, and presented a construction for oblivious simulation of PRAMs in the PRAM model. Our result is incomparable to their result: Our security model is weaker than theirs since we assume obliviousness within each memory bank comes for free; on the other hand, we obtain far better asymptotical and concrete performance. We next elaborate further on the differences in the results and techniques of the two works. Reference [7] provide a compiler from the EREW, CREW and CRCW PRAM models to the EREW PRAM model. The security notion achieved by their compiler provides security against adversaries who see the entire access pattern, as in standard oblivious RAM. However, their compiled program incurs a poly log overhead in both the parallel time and total work. Our compiler is a compiler from the EREW, CREW, common CRCW and arbitrary CRCW

---

[1] Every memory word must be large enough to store the logical memory address.

PRAM models to the arbitrary CRCW PRAM model and the security notion we achieve is the weaker notion of oblivious network RAM, which protects against adversaries who see the bank being accessed, but not the offset within the bank. On the other hand, our compiled program incurs only a $\log^*$ time overhead and its work is asymptotically the *same* as the underlying PRAM. Both our work and the work of [7] leverage previous results and techniques from the parallel computing literature. However, our techniques are primarily from the CRCW PRAM literature, while [7] use primarily techniques from the low-depth circuit literature, such as highly efficient sorting networks.

2 Definitions

2.1 Background: Random Access Machines (RAM)

We consider RAM programs to be interactive stateful systems $\langle \Pi, \text{state}, D \rangle$, consisting of a memory array D of N memory words, a CPU state denoted state, and a next instruction function Π which given the current CPU state and a value rdata read from memory, outputs the next instruction I and an updated CPU state denoted state′:

$$(\text{state}', I) \leftarrow \Pi(\text{state}, \text{rdata})$$

Each instruction I is of the form $I = (\text{op}, \dots)$, where op is called the op-code whose value is read, write, or stop. The initial CPU state is set to $(\text{start}, *, \text{state}_{\text{init}})$. Upon input x, the RAM machine executes, computes output z and terminates. CPU state is reset to $(\text{start}, *, \text{state}_{\text{init}})$ when the computation on the current input terminates.

On input x, the execution of the RAM proceeds as follows. If state $= (\text{start}, *, \text{state}_{\text{init}})$, set state $:= (\text{start}, x, \text{state}_{\text{init}})$, and rdata $:= 0$. Now, repeat the doNext() till termination, where doNext() is defined as below:

---
doNext()

1. Compute $(I, \text{state}') = \Pi(\text{state}, \text{rdata})$. Set state $:=$ state′.
2. If $I = (\text{stop}, z)$ then terminate with output z.
3. If $I = (\text{write}, \text{vaddr}, \text{wdata})$ then set $D[\text{vaddr}] := \text{wdata}$.
4. If $I = (\text{read}, \text{vaddr}, \perp)$ then set rdata $:= D[\text{vaddr}]$.

2.2 Network RAM (NRAM)

Nework RAM. A Network RAM (NRAM) is the same as a regular RAM, except that memory is distributed across multiple banks, $\text{Bank}_1, \dots, \text{Bank}_M$. In an NRAM, every virtual address vaddr can be written in the format vaddr $:= (m, \textit{offset})$, where $m \in [M]$ is the bank identifier, and *offset* is the offset within the Bank_m.

Otherwise, the definition of NRAM is identical to the definition of RAM.

Probabilistic NRAM. Similar to the Probabilistic RAM notion formalized by Goldreich and Ostrovsky [18, 19], we additionally define a *Probabilistic NRAM*. A problablistic NRAM is an NRAM whose CPU state is initialized with randomness ρ (that is unobservable to the adversary). If an NRAM is deterministic, we can simply assume that the CPU's initial randomness is fixed to $\rho := 0$. Therefore, a deterministic NRAM can be considered as a special case of a Probabilistic NRAM.

Outcome of execution. Throughout the paper, we use the notation RAM(x) or NRAM(x) to denote the outcome of executing a RAM or NRAM on input x. Similarly, for a Probabilistic NRAM, we use the notation NRAM$_\rho(x)$ to denote the outcome of executing on input x, when the CPU's initial randomness is ρ.

2.3   Oblivious Network RAM (O-NRAM)

Observable traces. To define Oblivious Network RAM, we need to first specify which part of the memory trace an adversary is allowed to observe during a program's execution. As mentioned earlier in the introduction, each memory bank has trusted logic for encrypting and decrypting the memory offset. The offset within a bank is transferred in encrypted format on the memory bus. Hence, *for each memory access* op := *"read" or* op := *"write" to virtual address* vaddr := $(m, offset)$, *the adversary observes only the op-code* op *and the bank identifier m, but not the offset within the bank.*

Definition 1 (Observable traces). *For a probabilistic* NRAM, *we use the notation* Tr$_\rho$(NRAM, x) *to denote its observable traces upon input x, and initial CPU randomness ρ:*

$$\text{Tr}_\rho(\text{NRAM}, x) := \{(\text{op}_1, m_1), (\text{op}_2, m_2), \ldots, (\text{op}_T, m_T)\}$$

where T is the total execution time of the NRAM, *and* (op_i, m_i) *is the op-code and memory bank identifier during step $i \in [T]$ of the execution.*

We remark that one can consider a slight variant model where the opcodes $\{\text{op}_i\}_{i \in [T]}$ are also hidden from the adversary. Since to hide whether the operation is a read or write, one can simply perform one read and one write for each operation – the differences between these two models are insignificant for technical purposes. Therefore, in this paper, we consider the model whose observable traces are defined in Definition 1.

Oblivious Network RAM. Intuitively, an NRAM is said to be oblivious, if for any two inputs x_0 and x_1 resulting in the same execution time, their observable memory traces are computationally indistinguishable to an adversary.

For simplicity, we define obliviousness for NRAMs that run in deterministic T time regardless of the inputs and the CPU's initial randomness. One can also think of T as the worst-case runtime, and that the program is always padded to the worst-case execution time. Oblivious NRAM can also be similarly defined when its runtime is randomized – however we omit the definition in this paper.

Definition 2 (Oblivious Network RAM). *Consider an* NRAM *that runs in deterministic time* $T = \text{poly}(\lambda)$. *The* NRAM *is said to be computationally oblivious if no polynomial-time adversary* $\mathcal{A}$ *can win the following security game with more than* $\frac{1}{2} + \text{negl}(\lambda)$ *probability. Similarly, the* NRAM *is said to be statistically oblivious if no adversary, even computationally unbounded ones, can win the following game with more than* $\frac{1}{2} + \text{negl}(\lambda)$ *probability.*

- $\mathcal{A}$ *chooses two inputs* x_0 *and* x_1 *and submits them to a challenger.*
- *The challenger selects* $\rho \in \{0,1\}^\lambda$, *and a random bit* $b \in \{0,1\}$. *The challenger executes* NRAM *with initial randomness* ρ *and input* x_b *for exactly* T *steps, and gives the adversary* $\text{Tr}_\rho(\text{NRAM}, x_b)$.
- $\mathcal{A}$ *outputs a guess* b' *of* b, *and wins the game if* $b' = b$.

2.4 Notion of Simulation

Definition 3 (Simulation). *We say that a deterministic* RAM $:= \langle \Pi, \text{state}, D \rangle$ *can be correctly simulated by another probabilistic* NRAM $:= \langle \Pi', \text{state}', D' \rangle$ *if for any input* x *for any initial CPU randomness* ρ, $\text{RAM}(x) = \text{NRAM}_\rho(x)$. *Moreover, if* NRAM *is oblivious, we say that* NRAM *is an oblivious simulation of* RAM.

Below, we explain some subtleties regarding the model, and define the metrics for oblivious simulation.

Uniform vs. non-uniform memory word size. The O-NRAM simulation can either employ uniform memory word size or non-uniform memory word size. For example, the non-uniform word size model has been employed for recursion-based ORAMs in the literature [39,43]. In particular, Stefanov et al. describe a parametrization trick where they use a smaller word size for position map levels of the recursion [39].

Metrics for simulation overhead. In the ORAM literature, several performance metrics have been considered. To avoid confusion, we now explicitly define two metrics that we will adopt later. If an NRAM correctly simulates a RAM, we can quantify the overhead of the NRAM using the following metrics.

- **Runtime blowup.** If a RAM runs in time T, and its oblivious simulation runs in time T', then the runtime blowup is defined to be T'/T. This notion is adopted by Goldreich and Ostrovsky in their original ORAM paper [18,19].
- **Bandwidth blowup.** If a RAM transfers Y bits between the CPU and memory, and its oblivious simulation transfers Y' bits, then the bandwidth blowup is defined to be Y'/Y. Clearly, if the oblivious simulation is in a uniform word size model, then bandwidth blowup is equivalent to runtime blowup. However, bandwidth blowup may not be equal to runtime blowup in a non-uniform word size model.

In this paper, we consider oblivious simulation of RAMs in the NRAM model, and we focus on the case when the Oblivious NRAM has only $O(1)$ words of CPU cache.

3   Sequential Oblivious Simulation

We first consider oblivious (sequential) simulation of arbitrary RAMs in the NRAM model. The detailed proofs and algorithms for this section will appear in the full version. Most of the techniques used here (with the exception of how to obliviously store the position map in a separate bank) are inspired by the work on practical ORAM by Stefanov, Shi, and Song [38]. Here we describe how we have adjusted their techniques to fit the Network RAM model.

Let M denote the number of memory banks in our NRAM, where each bank has $O(N/M)$ capacity. For simplicity we first describe a simple Oblivious NRAM with $O(M)$ CPU private cache. In the beginning, every block $i \in [N]$ is assigned randomly to a bank $j \in [M]$. We also maintain locally (i) a position map that maps every block to each bank; (ii) a cache of M queues, which are initially empty. To read/write a block i:

- We retrieve its bank number x from the position map;
- We first look for block i in the local queue x. If it is not there, we send a dummy memory request to a random location. Otherwise we read and then remove block i from the memory bank x;
- We pick a fresh random memory bank x', and we push block i to the queue x' in the local cache.

To avoid the overflow of local queues, we use a background eviction technique from Stefanov, Shi, and Song [38], which ensures that the local queues do not grow too much, while still maintaining obliviousness. Although storing the position map takes $O(N \log M)$ bits of CPU cache, in the full version we describe a recursion technique [36,38] that can reduce this storage to $O(1)$. Finally, to further reduce the space from $O(M)$ to $O(1)$, we can store the CPU cache in a separate memory bank. However, this is challenging, as indicated below.

Main challenge. Placing the cache in a special memory bank to achieve constant client storage might violate obliviousness, since different operations to the cache might have different memory traces. The key challenge is to design a special data structure to store the cache inside the memory bank that ensures constant worst-case cost for each query—specifically, each queue in the eviction cache must support pop, push, ReadAndRm operations. Partly to design this special data structure, we modified the analysis of the deamortized Cuckoo hash table construction [2] to achieve negligible failure probability.

We defer details of our algorithms and techniques to the full version and next state our main theorem for our sequential-to-sequential compiler.

Theorem 1 (O-NRAM simulation of arbitrary RAM programs). *Any N-word RAM with a word size of $W = \Omega(\log^2 N)$ bits can be simulated by an Oblivious NRAM (with non-uniform word sizes) that consumes $O(W)$ bits of CPU cache, and with $O(M)$ memory banks each of $O(W \cdot (M + N/M + N^\delta))$ bits in size. Further, the oblivious NRAM simulation incurs $\widehat{\mathbf{O}}(1)$ bandwidth blowup and $\widehat{O}(\log N)$ run-time blowup.*

4   Sequential Oblivious Simulation of Parallel Programs

We are eventually interested in parallel oblivious simulation of parallel programs (Sect. 5). As a stepping stone, we first consider sequential oblivious simulation of parallel programs. However, we emphasize that the results in this section can be of independent interest. In particular, one way to interpret these results is that "parallelism facilitates obliviousness". Specifically, if a program exhibits a sufficient degree of parallelism, then this program can be made oblivious at only const overhead in the Network RAM model. The intuition for why this is so, is that instructions in each parallel time step can be executed in any order. Since subsequences of instructions can be executed in an arbitrary order during the simulation, many sequences of memory requests can be mapped to the same access pattern, and thus the request sequence is partially obfuscated.

4.1   Parallel RAM

To formally characterize what it means for a program to exhibit a sufficient degree of parallelism, we will formally define a P-parallel RAM. In this section, the reader should think of parallelism as a property of the program to be simulated – we actually characterize costs assuming both the non-oblivious and the oblivious programs are executed on a sequential machine (different from Sect. 5).

An P-parallel RAM machine is the same as a RAM machine, except the next instruction function outputs P instructions which can be executed in parallel.

Definition 4 (P-parallel RAM). *An P-Parallel RAM is a RAM which has a next instruction function $\Pi = \Pi_1, \ldots, \Pi_P$ such that on input (state $=$ state$_1\|\cdots\|$state$_P$, rdata $=$ rdata$_1\|\cdots\|$rdata$_P$), Π outputs P instructions $(I_1, \ldots, I_P)$ and P updated states state$'_1, \ldots,$ state$'_P$ such that for $p \in [P]$, $(I_p,$ state$'_p) = \Pi_p($state$_p,$ rdata$_p)$. The instructions $I_1, \ldots, I_P$ satisfy one of the following:*

- *All of $I_1, \ldots, I_P$ are set to (stop, z) (with the same z).*
- *All of $I_1, \ldots, I_P$ are either of the form. (read, vaddr, $\bot$) or (write, vaddr, wdata).*

Finally, the state state has size at most $O(P)$.

As a warmup exercise, we will first consider a special case where in each parallel step, the memory requests made by each processor in the underlying P-parallel RAM have distinct addresses—we refer to this model as a *restricted* PRAM. Later in Sect. 4.3, we will extend the result to the (arbitrary) CRCW PRAM case. Thus, our final compiler works when the underlying P-parallel RAM is in the EREW, CREW, common CRCW or arbitrary CRCW PRAM model.

Definition 5 (Restricted P-parallel RAM). *For a P-parallel RAM denoted* PRAM $:= \langle D,$ state$_1, \ldots,$ state$_P, \Pi_1, \ldots \Pi_P\rangle$, *if every batch of instructions $I_1, \ldots, I_P$ have unique* vaddr*'s, we say that* PRAM *is a* restricted *P-parallel RAM.*

4.2   Warmup: Restricted Parallel RAM to Oblivious NRAM

Our goal is to compile any P-parallel RAM (not necessarily restricted), into an efficient O-NRAM. As an intermediate step that facilitates presentation, we begin with a basic construction of O-NRAM from any *restricted*, parallel RAM. In the following section, we extend to a construction of O-NRAM from any parallel RAM (not necessarily restricted).

Let PRAM $:= \langle D, \mathsf{state}_1, \ldots, \mathsf{state}_P, \Pi_1, \ldots \Pi_P \rangle$ be a restricted P-Parallel RAM, for $P = \omega(M \log N)$. We now present an O-NRAM simulation of PRAM that requires $M + 1$ memory banks, each with $O(N/M + P)$ physical memory, where N is the database size.

Setup: Pseudorandomly assign memory words to banks. The setup phase takes the initial states of the PRAM, including the memory array D and the initial CPU state, and compiles them into the initial states of the Oblivious NRAM denoted ONRAM.

To do this, the setup algorithm chooses a secret key K, and sets ONRAM.state = PRAM.state$\|K$. Each memory bank of ONRAM will be initialized as a Cuckoo hash table. Each memory word in the PRAM's initial memory array D will be inserted into the bank numbered $(\mathsf{PRF}_K(\mathsf{vaddr}) \bmod M) + 1$, where vaddr is the virtual address of the word in PRAM. Note that the ONRAM's $(M + 1)$-th memory bank is reserved as a scratch bank whose usage will become clear later.

Simulating each step of the PRAM's execution. Each doNext() operation of the PRAM will be compiled into a sequence of instructions of the ONRAM. We now describe how this compilation works. Our presentation focuses on the case when the next instruction's op-codes are reads or writes. Wait or stop instructions are left unmodified during the compilation.

As shown in Fig. 1, for each doNext instruction, we first compute the batch of instructions $I_1, \ldots, I_P$, by evaluating the P parallel next-instruction circuits $\Pi_1, \ldots, \Pi_P$. This results in P parallel read or write memory operations. This batch of P memory operations (whose memory addresses are guaranteed to be distinct in the restricted parallel RAM model) will then be served using the subroutine Access.

We now elaborate on the Access subroutine. Each batch will have $P = \omega(M \log N)$ memory operations whose virtual addresses are distinct. Since each virtual address is randomly assigned to one of the M banks, in expectation, each bank will get $P/M = \omega(\log N)$ hits. Using a balls and bins analysis, we show that

---

doNext(): *//We only consider* read *and* write *instructions here but not* stop.

1: **For** $p := 1$ to P: $(\mathsf{op}_p, \mathsf{vaddr}_p, \mathsf{wdata}_p) := \Pi_p(\mathsf{state}_p, \mathsf{rdata}_p)$

2: $(\mathsf{rdata}_1, \mathsf{rdata}_2, \ldots, \mathsf{rdata}_p) := \mathsf{Access}\left(\left\{ \mathsf{op}_p, \mathsf{vaddr}_p, \mathsf{wdata}_p \right\}_{p \in [P]} \right)$

Fig. 1. Oblivious simulation of each step of the restricted parallel RAM

the number of hits for each bank is highly concentrated around the expectation. In fact, the probability of any constant factor, multiplicative deviation from the expectation is negligible in N. Therefore, we choose max $:= 2(P/M)$ for each bank, and make precisely max number of accesses to each memory bank. Specifically, the Access algorithm first scans through the batch of $P = \omega(M \log N)$ memory operations, and assigns them to M queues, where the m-th queue stores requests assigned to the m-th memory bank. Then, the Access algorithm sequentially serves the requests to memory banks $1, 2, \ldots, M$, padding the number of accesses to each bank to max. This way, the access patterns to the banks are guaranteed to be oblivious.

The description of Fig. 2 makes use of M queues with a total size of $P = \omega(M \log N)$ words. It is not hard to see that these queues can be stored in an additional scratch bank of size $O(P)$, incurring only constant number of accesses to the scratch bank per queue operation. Further, in Fig. 2, the time at which the queues are accessed, and the number of times they are accessed are not dependent on input data (notice that Line 7 can be done by linearly scanning through each queue, incurring a max cost each queue).

Cost analysis. Since max $= 2(P/M)$, in Fig. 2 (see Theorem 2), it is not hard to see each batch of $P = \omega(M \log N)$ memory operations will incur $\Theta(P)$ accesses to data banks in total, and $\Theta(P)$ accesses to the scratch bank. Therefore, the ONRAM incurs only a constant factor more total work and bandwidth than the underlying PRAM.

Access $\left(\{\mathsf{op}_p, \mathsf{vaddr}_p, \mathsf{wdata}_p\}_{p \in P}\right)$:

1: **for** $p = 1$ to P **do**
2:     $m \leftarrow (\mathsf{PRF}_K(\mathsf{vaddr}_p) \mod M) + 1$;
3:     queue$[m] :=$ queue$[m]$.push$(p, \mathsf{op}_p, \mathsf{vaddr}_p, \mathsf{wdata}_p)$;
       // queue *is stored in a separate scratch bank.*
4: **end for**
5: **for** $m = 1$ to M **do**
6:     **if** $|$queue$[m]| >$ max **then abort**
7:     Pad queue$[m]$ with dummy entries $(\bot, \bot, \bot, \bot)$ so that its size is max;
8:     **for** $i = 1$ to max **do**
9:         $(p, \mathsf{op}, \mathsf{vaddr}, \mathsf{wdata}) :=$ queue$[m]$.pop()
10:         rdata$_p :=$ ReadBank(m, vaddr)
         // *Each bank is a deamortized Cuckoo hash table.*
11:         **if** op $=$ write **then** wdata $:=$ rdata$_p$
12:         WriteBank$(m, \mathsf{vaddr}, \mathsf{wdata})$
13:     **end for**
14: **end for**
15: **return** (rdata$_1$, rdata$_2$, $\ldots$, rdata$_P$)

Fig. 2. Obliviously serving a batch of P memory requests with distinct virtual addresses.

Theorem 2. *Let* PRF *be a family of pseudorandom functions, and* PRAM *be a restricted P-Parallel RAM for* $P = \omega(M \log N)$. *Let* max $:= 2(P/M)$. *Then, the construction described above is an oblivious simulation of* PRAM *using M banks each of* $O(N/M + P)$ *words in size. Moreover, the oblivious simulation performs total work that is constant factor larger than that of the underlying* PRAM.

Proof. Assuming the execution never aborts (Line 6 in Fig. 2), then Theorem 2 follows immediately, since the access pattern is deterministic and independent of the inputs. Therefore, it suffices to show that the abort happens with negligible probability on Line 6. This is shown in the following lemma.

Lemma 1. *Let* max $:= 2(P/M)$. *For any* PRAM *and any input* x, *abort on Line 6 of Fig. 2 occurs only with negligible probability (over choice of the* PRF*).*

Proof. We first replace PRF with a truly random function f. Note that if we can prove the lemma for a truly random function, then the same should hold for PRF, since otherwise we obtain an adversary breaking pseudorandomness.

We argue that the probability that abort occurs on Line 6 of Fig. 2 in a particular step i of the execution is negligible. By taking a union bound over the (polynomial number of) steps of the execution, the lemma follows.

To upper bound the probability of abort in some step i, consider a thought experiment where we change the order of sampling the random variables: We run PRAM(x) to precompute all the PRAM's instructions up to and including the i-th step of the execution (independently of f), obtaining P distinct virtual addresses, and only then choose the outputs of the random function f on the fly. That is, when each virtual memory address vaddr_p in step i is serviced, we choose $m := f(\mathsf{vaddr}_p)$ uniformly and independently at random. Thus, in step i of the execution, there are P distinct virtual addresses (i.e., balls) to be thrown into M memory banks (i.e., bins). Due to standard Chernoff bounds, for $P = \omega(M \log N)$, we have $P/M = \omega(\log N)$ and so the probability that there exists a bin whose load exceeds $2(P/M)$ is $N^{-\omega(1)}$, which is negligible in N.

We note that in order for the above argument to hold, the input x cannot be chosen adaptively, and must be fixed before the PRAM emulation begins.

4.3   Parallel RAM to Oblivious NRAM

Use a hash table to suppress duplicates. In Sect. 4.2, we describe how to obliviously simulate a restricted parallel-RAM in the NRAM model. We now generalize this result to support any P-parallel RAM, not necessarily restricted ones. The difference is that for a generic P-parallel RAM, each batch of P memory operations generated by the next-instruction circuit need not have distinct virtual addresses. For simplicity, imagine that the entire batch of memory operations are reads. In the extreme case, if all $P = \omega(M \log N)$ operations correspond to the same virtual address residing in bank m, then the CPU should not read

---

Access $\left(\{\mathsf{op}_p, \mathsf{vaddr}_p, \mathsf{wdata}_p, p\}_{p \in P}\right)$:

/* HTable, queue, and result *data structures are stored in a scratch bank. For obliviousness,*
operations on these data structures must be padded to the worst-case cost as we elaborate in
the text./

1: **for** $p = 1$ to P: HTable[$\mathsf{op}_p, \mathsf{vaddr}_p$] := ($\mathsf{wdata}_p, p$) // hash table insertions
2: **for** $\{(\mathsf{op}, \mathsf{vaddr}), \mathsf{wdata}, p\} \in$ HTable **do** // iterate through hash table
3:      $m := (\mathsf{PRF}_K(\mathsf{vaddr}) \mod M) + 1$
4:      queue[m] := queue[m].push(op, vaddr, wdata);
5: **end for**
6: **for** $m = 1$ to M **do**
7:      **if** |queue[m]| > max **then abort**
8:      Pad queue[m] with dummy entries ($\bot, \bot, \bot$) so that its size is max;
9:      **for** $i = 1$ to max **do**
10:         (op, vaddr, wdata, p) := queue[m].pop()
11:         result[p] := ReadBank(m, vaddr)
12:         **if** op = write **then** wdata := rdata
13:         WriteBank(m, vaddr, wdata)
14:     **end for**
15: **end for**
16: **return** (result[1], ..., result[p]) // hash table lookups

---

Fig. 3. Obliviously serving a batch of P memory request, not necessarily with distinct virtual addresses.

bank m as many as P number of times. To address this issue, we rely on an additional Cuckoo hash table [34] denoted HTable to suppress the duplicate requests (see Fig. 3, and the doNext function is defined the same way as Sect. 4.2).

The HTable will be stored in the scratch bank. We can employ a standard Cuckoo hash table that need not be deamortized. As shown in Fig. 3, we need to support hash table insertions, lookups, and moreover, we need to be able to iterate through the hash table. We now make a few remarks important for ensuring obliviousness. Line 1 of Fig. 3 performs $P = \omega(M \log N)$ number of insertions into the Cuckoo hash table. Due to standard Cuckoo hash analysis, we know that these insertions will take $O(P)$ total time except with negligible probability. Therefore, to execute Line 1 obliviously, we simply need to pad with dummy insertions up to some $\mathsf{max}' = c \cdot P$, for an appropriate constant c.

Next, we describe how to execute the loop at Line 2 obliviously. The total size of the Cuckoo hash table is $O(P)$. To iterate over the hash table, we simply make a linear scan through the hash table. Some entries will correspond to dummy elements. When iterating over these dummy elements, we simply perform dummy operations for the **for** loop. Finally, observe that Line 16 performs lookups to the Cuckoo hash table, and each hash table lookup requires worst-case $O(1)$ accesses to the scratch bank.

Cost analysis. Since max $= 2(P/M)$ (see Theorem 2), it is not hard to see each batch of $P = \omega(M \log N)$ memory operations will incur $O(P)$ accesses to data banks in total, and $O(P)$ accesses to the scratch bank. Note that this takes into account the fact that Line 1 and the for-loop starting at Line 2 are padded with dummy accesses. Therefore, the ONRAM incurs only a constant factor more total work and bandwidth than the underlying PRAM.

Theorem 3. *Let* max $= 2(P/M)$. *Assume that* PRF *is a secure pseudorandom function, and* PRAM *is a P-Parallel RAM for* $P = \omega(M \log N)$. *Then, the above construction obliviously simulates* PRAM *in the NRAM model, incurring only a constant factor blowup in total work and bandwidth consumption.*

Proof. (sketch) Similar to the proof of Theorem 2, except that now we have the additional hash table. Note that obliviousness still holds, since, as discussed above, each batch of P memory requests requires $O(P)$ accesses to the scratch bank, and this can be padded with dummy accesses to ensure the number of scratch bank accesses remains the same in each execution.

5   Parallel Oblivious Simulation of Parallel Programs

In the previous section, we considered sequential oblivious simulation of programs that exhibit parallelism – there, we considered parallelism as being a property of the program which will actually be executed on a sequential machine. In this section we consider *parallel* and oblivious simulations of parallel programs. Here, the programs will actually be executed on a parallel machine, and we consider classical metrics such as parallel runtime and total work as in the parallel algorithms literature.

We introduce the *Network PRAM* model – informally, this is a Network RAM with parallel processing capability. Our goal in this section will be to compile a PRAM into an Oblivious Network PRAM (O-NPRAM), *a.k.a.*, the "parallel-to-parallel compiler".

Our O-NPRAM is the Network RAM analog of the Oblivious Parallel RAM (OPRAM) model by Boyle *et al.* [7]. Goldreich and Ostrovsky's logarithmic ORAM lower bound (in the sequential execution model) directly implies the following lower bound for standard OPRAM [7]: Let PRAM be an arbitrary PRAM with P processors running in parallel time t. Then, any P-parallel OPRAM simulating PRAM must incur $\Omega(t \log N)$ parallel time. Clearly, OPRAM would also work in our Network RAM model albeit not the most efficient, since it is not exploiting the fact that the addresses in each bank are inherently oblivious. In this section, we show how to perform oblivious parallel simulation of "sufficiently parallel" programs in the Network RAM model, incurring only $O(\log^* N)$ blowup in parallel runtime, and achieving optimal total work. Our techniques make use of fascinating results in the parallel algorithms literature [4,5,24].

5.1   Network PRAM (NPRAM) Definitions

Similar to our NRAM definition, an NPRAM is much the same as a standard PRAM, except that (1) memory is distributed across multiple banks, $\mathsf{Bank}_1, \ldots, \mathsf{Bank}_M$; and (2) every virtual address vaddr can be written in the format $\mathsf{vaddr} := (m, \textit{offset})$, where m is the bank identifier, and $\textit{offset}$ is the offset within the Bank_m. We use the notation P-parallel NPRAM to denote an NPRAM with P parallel processors, each with $O(1)$ words of cache. If processors are initialized with secret randomness unobservable to the adversary, we call this a Probabilistic NPRAM.

Observable traces. In the NPRAM model, we assume that an adversary can observe the following parts of the memory trace: (1) which processor is making the request; (2) whether this is a read or write request; and (3) which bank the request is going to. The adversary is unable to observe the offset within a memory bank.

Definition 6 (Observable traces for NPRAM). *For a probabilistic P-parallel* NPRAM*, we use* $\mathsf{Tr}_\rho(\mathsf{NPRAM}, x)$ *to denote its observable traces upon input* x*, and initial CPU randomness* ρ *(collective randomness over all processors):*

$$\mathsf{Tr}_\rho(\mathsf{NPRAM}, x) := \left[\left((\mathsf{op}_1^1, m_1^1), \ldots, (\mathsf{op}_1^P, m_1^P)\right), \ldots, \left((\mathsf{op}_T^1, m_T^1), \ldots, (\mathsf{op}_T^P, m_T^P)\right)\right]$$

where T *is the total parallel execution time of the* NPRAM*, and* $\{(\mathsf{op}_i^1, m_i^1), \ldots, (\mathsf{op}_i^P, m_i^P)\}$ *is of the op-codes and memory bank identifiers for each processor during parallel step* $i \in [T]$ *of the execution.*

Based on the above notion of observable memory trace, an Oblivious NPRAM can be defined in a similar manner as the notion of O-NRAM (Definition 2).

Metrics. We consider classical metrics adopted in the vast literature on parallel algorithms, namely, the parallel runtime and the total work. In particular, to characterize the oblivious simulation overhead, we will consider

- **Parallel runtime blowup.** The blowup of the parallel runtime comparing the O-NPRAM and the NPRAM.
- **Total work blowup.** The blowup of the total work comparing the O-NPRAM and the NPRAM. If the total work blowup is $O(1)$, we say that the O-NPRAM achieves *optimal* total work.

5.2   Construction of Oblivious Network PRAM

Preliminary: colored compaction. The colored compaction problem [4] is the following:

Given n objects of m different colors, initially placed in a single source array, move the objects to m different destination arrays, one for each color. In this paper, we assume that the *space for the m destination arrays are preallocated*. We use the notation d_i to denote the number of objects colored i for $i \in [m]$.

Lemma 2 (Log*-time parallel algorithm for colored compaction [4]).
There is a constant $\epsilon > 0$ such that for all given n, m, τ, $d_1, \ldots, d_m \in \mathbb{N}$, with $m = O(n^{1-\delta})$ for arbitrary fixed $\delta > 0$, and $\tau \geq \log^* n$, there exists a parallel algorithm (in the arbitrary CRCW PRAM model) for the colored compaction problem (assuming preallocated destination arrays) with n objects, m colors, and $d_1, \ldots, d_m$ number of objects for each color, executing in $O(\tau)$ time on $\lceil n/\tau \rceil$ processors, consuming $O(n + \sum_{i=1}^{m} d_i)$ space, and succeeding with probability at least $1 - 2^{-n^\epsilon}$.

Preliminary: parallel static hashing. We will also rely on a parallel, static hashing algorithm [5,24], by Bast and Hagerup. The static parallel hashing problem takes n elements (possibly with duplicates), and in parallel creates a hash table of size $O(n)$ of these elements, such that later each element can be visited in $O(1)$ time. In our setting, we rely on the parallel hashing to suppress duplicate memory requests. Bast and Hagerup show the following lemma:

Lemma 3 (Log*-time parallel static hashing [5,24]). *There is a constant $\epsilon > 0$ such that for all $\tau \geq \log^* n$, there is a parallel, static hashing algorithm (in the arbitrary CRCW PRAM model), such that hashing n elements (which need not be distinct) can be done in $O(\tau)$ parallel time, with $O(n/\tau)$ processors and $O(n)$ space, succeeding with $1 - 2^{-(\log n)^{\tau / \log^* n}} - 2^{-n^\epsilon}$ probability.*

Construction. We now present a construction that allows us to compile a P-parallel PRAM, where $P = M^{1+\delta}$ for any constant $\delta > 0$, into a $O(P/\log^* P)$-parallel Oblivious NPRAM. The resulting NPRAM has $O(\log^* P)$ blowup in parallel runtime, and is optimal in total amount of work.

In the original P-parallel PRAM, each of the P processors does constant amount of work in each step. In the oblivious simulation, this can trivially be simulated in $O(\log^* P)$ time with $O(P/\log^* P)$ processors. Therefore, clearly the key is how to obliviously fetch a batch of P memory accesses in parallel with $O(P/\log^* P)$ processors, and $O(\log^* P)$ time. We describe such an algorithm in Fig. 4. Using a scratch bank as working memory, we first call the parallel hashing algorithm to suppress duplicate memory requests. Next, we call the parallel colored compaction algorithm to assign memory request to their respective queues – depending on the destination memory bank. Finally, we make these memory accesses, including dummy ones, in parallel.

Theorem 4. *Let* PRF *be a secure pseudorandom function, let $M = N^\epsilon$ for any constant $\epsilon > 0$. Let* PRAM *be a P-parallel RAM for $P = M^{1+\delta}$, for constant $\delta > 0$. Then, there exists an Oblivious NPRAM simulation of* PRAM *with the following properties:*

- *The Oblivious NPRAM consumes M banks each of which $O(N/M + P)$ words in size.*
- *If the underlying* PRAM *executes in t parallel steps, then the Oblivious NPRAM executes in $O(t \log^* P)$ parallel steps utilizing $O(P/\log^* P)$ processors. We also say that the NPRAM has $O(\log^* P)$ blowup in parallel runtime.*

356   D. Dachman-Soled et al.

---

parAccess $\left(\{op_p, \text{vaddr}_p, \text{wdata}_p\}_{p \in P}\right)$:

/* All steps can be executed in $O(\log^* P)$ time with $P' = O(P/\log^* P)$ processors with all
but negligible probability. */

1: Using the scratch bank as memory, run the parallel hashing algorithm on the batch
   of $P = M^{1+\delta}$ memory requests to suppress duplicate addresses. Denote the resulting set as S, and pad S with dummy requests to the maximum length P.

2: In parallel, assign colors to each memory request in the array S. For each
   real memory access $\{op, \text{vaddr}, \text{wdata}\}$, its color is defined as $(\text{PRF}_K(\text{vaddr})$
   mod $M) + 1$. Each dummy memory access is assigned a random color. It is not
   hard to see that each color has no more than $\max := 2(P/M)$ requests except
   with negligible probability.

3: Using the scratch bank as memory, run the parallel colored compaction algorithm
   to assign the array S to M preallocated queues each of size $\max$ (residing in the
   scratch bank).

4: Now, each queue $i \in [M]$ contains $\max$ number of requests intended for bank i –
   some real, some dummy. Serve all memory requests in the M queues in parallel.
   Each processor $i \in [P']$ is assigned the k-th memory request iff $(k \mod P') = i$.
   Dummy requests incur accesses to the corresponding banks as well.
   For each request coming from processor p, the result of the fetch is stored in an
   array result$[p]$ in the scratch bank.

Fig. 4. Obliviously serving a batch of P memory requests using $P' :=
O(P/\log^* P)$ processors in $O(\log^* P)$ time. In Steps 1, 2, and 3, each processor
will make exactly one access to the scratch bank in each parallel execution step – *even
if the processor is idle in this step, it makes a dummy access to the scratch bank.* Steps
1 through 3are always padded to the worst-case parallel runtime.

– *The total work of the Oblivious NPRAM is asymptotically the same as the
  underlying* PRAM.

Proof. We note that our underlying PRAM can be in the EREW, CREW, common CRCW or arbitrary CRCW models. Our compiled oblivious NPRAM is in
the arbitrary CRCW model.

We now prove security and costs separately.

Security proof. Observe that Steps 1, 2, and 3 in Fig. 4 make accesses only to
the scratch bank. We make sure that each processor will make exactly one access
to the scratch bank in every parallel step – even if the processor is idle in this
step, it makes a dummy access. Further, Steps 1 through 3 are also padded to
the worst-case running time. Therefore, the observable memory traces of Steps
1 through 3 are perfectly simulatable without knowing secret inputs.

For Step 4 of the algorithm, since each of the M queues are of fixed length
max, and each element is assigned to each processor in a round-robin manner,
the bank number each processor will access is clearly independent of any secret
inputs, and can be perfectly simulated (recall that dummy request incur accesses
to the corresponding banks as well).

Costs. First, due to Lemma 1, each of the M queues will get at most $2(P/M)$ memory requests with probability $1 - \text{negl}(N)$. This part of the argument is the same as Sect. 4. Now, observe that the parallel runtime for Steps 2 and 4 are clearly $O(\log^* P)$ with $O(P/\log^* P)$ processors. Based on Lemmas 2 and 3, Steps 1 and 3 can be executed with a worst-case time of $O(\log^* P)$ on $O(P/\log^* P)$ processors as well. We note that the conditions $M = N^\epsilon$ and $P = M^{1+\delta}$ ensure $\text{negl}(N)$ failure probability.

References

1. Intel SGX for dummies (intel SGX design objectives). https://software.intel.com/en-us/blogs/2013/09/26/protecting-application-secrets-with-intel-sgx
2. Arbitman, Y., Naor, M., Segev, G.: De-amortized cuckoo hashing: provable worst-case performance and experimental results. In: Albers, S., Marchetti-Spaccamela, A., Matias, Y., Nikoletseas, S., Thomas, W. (eds.) ICALP 2009, Part I. LNCS, vol. 5555, pp. 107–118. Springer, Heidelberg (2009)
3. Bajaj, S., Sion, R.: Trusteddb: a trusted hardware-based database with privacy and data confidentiality. IEEE Trans. Knowl. Data Eng. **26**(3), 752–765 (2014)
4. Bast, H., Hagerup, T.: Fast parallel space allocation, estimation, and integer sorting. Inf. Comput. **123**(1), 72–110 (1995)
5. Bast, H., Hagerup, T.: Fast and reliable parallel hashing. In: SPAA, pp. 50–61 (1991)
6. Boneh, D., Mazieres, D., Popa, R.A.: Remote oblivious storage: making oblivious RAM practical (2011). http://dspace.mit.edu/bitstream/handle/1721.1/62006/MIT-CSAIL-TR-2011-018.pdf
7. Boyle, E., Chung, K.-M., Pass, R.: Oblivious parallel RAM. http://eprint.iacr.org/2014/594.pdf
8. Chung, K.-M., Liu, Z., Pass, R.: Statistically-secure oram with $\tilde{O}(\log^2 n)$ overhead. CoRR, abs/1307.3699 (2013)
9. Fletcher, C.W., van Dijk, M., Devadas, S.: A secure processor architecture for encrypted computation on untrusted programs. In: STC (2012)
10. Fletcher, C.W., Ren, L., Kwon, A., Van Dijk, M., Stefanov, E., Devadas, S.: Tiny ORAM: a low-latency, low-area hardware ORAM controller with integrity verification
11. Fletcher, C.W., Ren, L., Kwon, A., van Dijk, M., Stefanov, E., Devadas, S.: RAW path ORAM: a low-latency, low-area hardware ORAM controller with integrity verification. IACR Cryptology ePrint Archive 2014:431 (2014)
12. Fletcher, C.W., Ren, L., Yu, X., van Dijk, M., Khan, O., Devadas, S.: Suppressing the oblivious RAM timing channel while making information leakage and program efficiency trade-offs. In: HPCA, pp. 213–224 (2014)
13. Gentry, C., Goldman, K.A., Halevi, S., Julta, C., Raykova, M., Wichs, D.: Optimizing ORAM and using it efficiently for secure computation. In: De Cristofaro, E., Wright, M. (eds.) PETS 2013. LNCS, vol. 7981, pp. 1–18. Springer, Heidelberg (2013)
14. Gentry, C., Halevi, S., Lu, S., Ostrovsky, R., Raykova, M., Wichs, D.: Garbled RAM revisited. In: Nguyen, P.Q., Oswald, E. (eds.) EUROCRYPT 2014. LNCS, vol. 8441, pp. 405–422. Springer, Heidelberg (2014)
15. Gentry, C., Halevi, S., Raykova, M., Wichs, D.: Garbled RAM revisited, part i. Cryptology ePrint Archive, Report 2014/082 (2014). http://eprint.iacr.org/

16. Gentry, C., Halevi, S., Raykova, M., Wichs, D.: Outsourcing private RAM computation. IACR Cryptology ePrint Archive 2014:148 (2014)
17. Gil, J., Matias, Y., Vishkin, U.: Towards a theory of nearly constant time parallel algorithms. In: 32nd Annual Symposium on Foundations of Computer Science (FOCS), pp. 698–710 (1991)
18. Goldreich, O.: Towards a theory of software protection and simulation by oblivious RAMs. In: ACM Symposium on Theory of Computing (STOC) (1987)
19. Goldreich, O., Ostrovsky, R.: Software protection and simulation on oblivious RAMs. J. ACM **43**, 431–473 (1996)
20. Goodrich, M.T., Mitzenmacher, M.: Privacy-preserving access of outsourced data via oblivious RAM simulation. In: Aceto, L., Henzinger, M., Sgall, J. (eds.) ICALP 2011, Part II. LNCS, vol. 6756, pp. 576–587. Springer, Heidelberg (2011)
21. Goodrich, M.T., Mitzenmacher, M., Ohrimenko, O., Tamassia, R.: Practical oblivious storage. In: ACM Conference on Data and Application Security and Privacy (CODASPY) (2012)
22. Goodrich, M.T., Mitzenmacher, M., Ohrimenko, O., Tamassia, R.: Privacy-preserving group data access via stateless oblivious RAM simulation. In: SODA (2012)
23. Dov Gordon, S., Katz, J., Kolesnikov, V., Krell, F., Malkin, T., Raykova, M., Vahlis, Y.: Secure two-party computation in sublinear (amortized) time. In: ACM CCS (2012)
24. Hagerup, T.: The log-star revolution. In: Finkel, A., Jantzen, M. (eds.) STACS 1992. LNCS, vol. 577, pp. 259–278. Springer, Heidelberg (1992)
25. Kushilevitz, E., Lu, S., Ostrovsky, R.: On the (in)security of hash-based oblivious RAM and a new balancing scheme. In: SODA (2012)
26. Liu, C., Huang, Y., Shi, E., Katz, J., Hicks, M.: Automating efficient ram-model secure computation. In: IEEE S & P. IEEE Computer Society (2014)
27. Lu, S., Ostrovsky, R.: Distributed oblivious RAM for secure two-party computation. In: Sahai, A. (ed.) TCC 2013. LNCS, vol. 7785, pp. 377–396. Springer, Heidelberg (2013)
28. Lu, S., Ostrovsky, R.: How to garble RAM programs? In: Johansson, T., Nguyen, P.Q. (eds.) EUROCRYPT 2013. LNCS, vol. 7881, pp. 719–734. Springer, Heidelberg (2013)
29. Lu, S., Ostrovsky, R.: Garbled RAM revisited, part ii. Cryptology ePrint Archive, Report 2014/083 (2014). http://eprint.iacr.org/
30. Maas, M., Love, E., Stefanov, E., Tiwari, M., Shi, E., Asanovic, K., Kubiatowicz, J., Song, D.: Phantom: practical oblivious computation in a secure processor. In: CCS (2013)
31. Maas, M., Love, E., Stefanov, E., Tiwari, M., Shi, E., Asanovic, K., Kubiatowicz, J., Song, D.: A high-performance oblivious RAM controller on the convey HC-2ex heterogeneous computing platform. In: Workshop on the Intersections of Computer Architecture and Reconfigurable Logic (CARL) (2013)
32. Mehlhorn, K., Vishkin, U.: Randomized and deterministic simulations of prams by parallel machines with restricted granularity of parallel memories. Acta Inf. **21**, 339–374 (1984)
33. Ostrovsky, R., Shoup, V.: Private information storage (extended abstract). In: ACM Symposium on Theory of Computing (STOC) (1997)
34. Pagh, R., Rodler, F.F.: Cuckoo hashing. J. Algorithms **51**(2), 122–144 (2004)
35. Ren, L., Yu, X., Fletcher, C.W., van Dijk, M., Devadas, S.: Design space exploration and optimization of path oblivious RAM in secure processors. In: ISCA, pp. 571–582 (2013)

Oblivious Network RAM and Leveraging Parallelism   359

36. Shi, E., Chan, T.-H.H., Stefanov, E., Li, M.: Oblivious RAM with $O((\log N)^3)$ worst-case cost. In: Lee, D.H., Wang, X. (eds.) ASIACRYPT 2011. LNCS, vol. 7073, pp. 197–214. Springer, Heidelberg (2011)
37. Stefanov, E., Shi, E.: Oblivistore: high performance oblivious cloud storage. In: IEEE Symposium on Security and Privacy (S & P) (2013)
38. Stefanov, E., Shi, E., Song, D.: Towards practical oblivious RAM. In: NDSS (2012)
39. Stefanov, E., van Dijk, M., Shi, E., Hubert Chan, T.-H., Fletcher, C., Ren, L., Yu, X., Devadas, S.: Path ORAM: an extremely simple oblivious RAM protocol. In: ACM CCS (2013)
40. Edward Suh, G., Clarke, D., Gassend, B., van Dijk, M., Devadas, S.: Aegis: architecture for tamper-evident and tamper-resistant processing. In: International Conference on Supercomputing, ICS 2003, pp. 160–171 (2003)
41. Thekkath, D.L.C., Mitchell, M., Lincoln, P., Boneh, D., Mitchell, J., Horowitz, M.: Architectural support for copy and tamper resistant software. SIGOPS Oper. Syst. Rev. **34**(5), 168–177 (2000)
42. Vishkin, U.: Can parallel algorithms enhance seriel implementation? Commun. ACM **39**(9), 88–91 (1996)
43. Wang, X.S., Hubert Chan, T.-H., Shi, E.: Circuit ORAM: on tightness of the goldreich-ostrovksy lower bound. http://eprint.iacr.org/2014/672.pdf
44. Wang, X.S., Huang, Y., Chan, H.T-H., Shelat, A., Shi, E.: Scoram: oblivious ram for secure computation. http://eprint.iacr.org/2014/671.pdf
45. Williams, P., Sion, R.: Usable PIR. In: Network and Distributed System Security Symposium (NDSS) (2008)
46. Williams, P., Sion, R.: SR-ORAM: single round-trip oblivious RAM. In: ACM Conference on Computer and Communications Security (CCS) (2012)
47. Williams, P., Sion, R., Carbunar, B.: Building castles out of mud: practical access pattern privacy and correctness on untrusted storage. In: CCS (2008)
48. Williams, P., Sion, R., Tomescu, A.: Privatefs: a parallel oblivious file system. In: CCS (2012)
49. Yu, X., Haider, S.K., Ren, L., Fletcher, C.W., Kwon, A., van Dijk, M., Devadas, S.: Proram: dynamic prefetcher for oblivious RAM. In: Proceedings of the 42nd Annual International Symposium on Computer Architecture, Portland, OR, USA, 13–17 June 2015, pp. 616–628 (2015)

Three-Party ORAM for Secure Computation

Sky Faber[1], Stanislaw Jarecki[1(✉)], Sotirios Kentros[2], and Boyang Wei[1]

[1] University of California, Irvine, USA
{fabers,boyanw1}@uci.edu, stasio@ics.uci.edu
[2] Salem State University, Salem, USA
sotirios.kentros@salemstate.edu

Abstract. An Oblivious RAM (ORAM) protocol [13] allows a client to retrieve N-th element of a data array D stored by the server s.t. the server learns no information about N. A related notion is that of an *ORAM for Secure Computation* (SC-ORAM) [17], which is a protocol that securely implements a RAM functionality, i.e. given a secret-sharing of both D and N, it computes a secret-sharing of D[N]. SC-ORAM can be used as a subprotocol for implementing the RAM functionality for secure computation of RAM programs [7,14,17]. It can also implement a public database service which hides each client's access pattern even if a threshold of servers colludes with any number of clients.

Most previous works used two-party secure computation to implement each step of an ORAM client algorithm, but since secure computation of many functions becomes easier in the three-party honest-majority setting than in the two-party setting, it is natural to ask if the cost of an SC-ORAM scheme can be reduced if one was willing to use three servers instead of two and assumed an honest majority. We show a 3-party SC-ORAM scheme which is based on a variant of the *Binary Tree* Client-Server ORAM of Shi et al. [20]. However, whereas previous SC-ORAM implementations used general 2PC or MPC techniques like Yao's garbled circuits, e.g. [14,22], homomorphic encryption [11], or the SPDZ protocol for arithmetic circuits [15], our techniques are custom-made for the three-party setting, giving rise to a protocol which is secure against honest-but-curious faults using bandwidth and CPU costs which are comparable to those of the underlying Client-Server ORAM.

Keywords: Oblivious RAM · Secure computation · Private information retrieval

1 Introduction

Oblivious RAM for Secure Computation. An Oblivious RAM (ORAM) is a protocol between *client* and *server* which allows the client who can locally store only small amount of information to write and read from an outsourced memory in such a way that the server cannot tell which memory locations the client accesses, and the cost of each memory access is sublinear in the memory

© International Association for Cryptologic Research 2015
T. Iwata and J.H. Cheon (Eds.): ASIACRYPT 2015, Part I, LNCS 9452, pp. 360–385, 2015.
DOI: 10.1007/978-3-662-48797-6_16

size. Starting from the seminal work of Goldreich and Ostrovsky [13], there have been numerous improvements in ORAM techniques, achieving polylogarithmic client storage, computation, bandwidth, and server storage overheads, with the most recent proposal of Path ORAM by Stefanov et al. [21] practical enough to be implemented on secure processors [9,10,18].

The above classic formulation of the ORAM problem, which we will call a *client-server ORAM*, provides secure outsourced memory for a *single* client. The client-server ORAM notion can be generalized (and relaxed) by considering still a single client but $n > 1$ servers, and assuring client-access obliviousness only if at most $t < n$ of these servers collude. Such generalization was proposed and realized for $(t, n) = (1, 2)$ by Ostrovsky and Lu [17]. However, one can imagine a stronger notion, namely of a protocol which allows n servers to *emulate* an Oblivious RAM functionality so that a shared memory can be accessed by multiple clients, but their access patterns remain hidden even if up to t of these n servers collude, possibly with any coalition of the clients. Such multi-party ORAM emulator is equivalent to (multi-party) secure computation of the RAM functionality (called SC-ORAM for short), which given the secret-sharing of memory array D and a secret-sharing of a location i (and value v) outputs a secret-sharing of record $D[i]$ (and writes v at index i in the secret-shared D).

One source of interest in such ORAM emulation is that it can provide oblivious RAM access as subprotocol for secure computation of any RAM program [7,14,19]. Recall that secure computation allows a set of parties $S_1, ..., S_n$ to compute some (randomized) function f on their inputs, where each S_i inputs a private value x_i into the computation, in such a way that the protocol reveals nothing else but the final output value $y = f(x_1, ..., x_n)$ to the participants. A standard approach to secure computation is to represent function f as a Boolean circuit [3], an arithmetic circuit [1], or a decision diagram [16]. However, even for very simple functions, each of these representations can be impractically large. This is indeed necessarily so if some of the inputs x_i are very long, i.e. when some of the data involved in the computation of f is large. Consider any information retrieval task, where x_1 is a large database, x_2 is a search term, and f is a search algorithm. The circuit or decision tree representation of f is at least as long as x_1, and therefore secure computation of f using any of the above techniques must take time at least linear in the size of the database.

SC-ORAM Applications. On the other hand each information retrieval problem which has a practical solution does so because it has an efficient RAM program, and as Ostrovsky and Shoup were the first to point out [19], an ORAM emulator can be used to securely compute any RAM program, because each local computation step can be implemented using Yao's garbled circuit technique [24] while each memory access can be handled by the SC-ORAM subprotocol. Examples of such SC-ORAM usage were recently provided by Keller and Scholl [15], who used their SC-ORAM implementation to build MPC implementations of other datastructures, e.g. a priority queue, and then utilized them in MPC computation of various algorithms in the RAM computation model e.g. Dijkstra's shortest path algorithm. In general, SC-ORAM is well suited to secure computation of any information-retrieval algorithm because such algorithms rely very strongly on the

RAM model, e.g. by identifying database entries using hash tables of keywords. One application using SC-ORAM in this way could be provision of a shared database resource to *multiple clients* in a way that hides any client's access pattern even if all other clients collude. One can look at such usage of SC-ORAM as providing an alternative to Searchable Symmetric Encryption which requires interaction but hides all patterns of access to the encrypted data. In its most basic form of secure implementation of an array look-up, 3-server SC-ORAM also provides an interactive alternative to 3-server (symmetric) Private Information Retrieval [4,12], with support for both read and write, and with the added security property that the database itself is private to any (single) server.

Two-Party SC-ORAM Constructions. Apart of pointing out the usefulness of an SC-ORAM protocol, Ostrovsky and Shoup sketched a method for converting any client-server ORAM into a two-party SC-ORAM: One of the two parties can implement the server in the underlying client-server ORAM scheme, while all the work of the client can be jointly computed by the two parties using Yao's garbled circuit technique applied to the circuit representation of each step of the client's algorithm in any client-server ORAM scheme. This idea was also considered by Damgard et al. [7], and it was further developed by Gordon et al. [14] who showed an optimized two-party ORAM emulation protocol based on the *Binary Tree* client-server ORAM of Shi et al. [20], utilizing a novel subprotocol gadget for secure computation of a pseudorandom function (PRF) on secret-shared inputs. Gentry et al. [11] showed several space and computation saving modifications of the Binary Tree client-server ORAM of Shi et al., together with a very different two-party ORAM emulation protocol for it, which used a customized homomorphic encryption scheme instead of Yao's garbled circuits for the two-party computation of the ORAM client's algorithm. Other modifications of the Binary Tree ORAM were shown in [5], and in the *Path ORAM* proposal of Shi et al. [21], but even though they improve the client-server ORAM, it is not clear if these modifications translate into a faster ORAM emulator.

Indeed, Wang et al. [22] examined the circuit complexity of the client's algorithm in several proposed variants of the Binary Tree client-server ORAM, including [5,20], and they concluded that the client's algorithm in the original scheme of [20] has by far the smallest circuit. They also showed a set of modifications to the Binary Tree ORAM, building on the Path ORAM modifications, which result in roughly a factor of 10 reduction in the circuit representation of the client algorithm in a client-server ORAM, hence speeding up the two-party ORAM emulation protocol of [14] by the same factor. The SC-ORAM protocol for a database D containing 2^m records of size d each requires secure computation of a circuit of asymptotic size $O\left(m^3 + dm\right)$, resulting in $O\left(\kappa(m^3 + dm)\right)$ bound on protocol bandwidth where κ is a cryptographic security parameter. For m between 20 and 29 and $d = 4B$ this comes to between $4.6M$ and $13M$ gates, and its secure evaluation requires (on-line) as many hash or block cipher operations as the number of non-xor gates in the circuits. (In a work concurrent to ours Wang et al. [23] showed further reductions in the ORAM client circuit size, reporting $350K$ *and* gates for $m = 20$ and $d = 4B$.)

Multi-Party SC-ORAM and Our Contribution. Secure computation of many functionalities can be implemented with easier tools in the multi-party setting with honest majority than in the two-party setting. (In fact, assuming secure channels any function can be securely computed without further cryptographic assumptions [1,2].) Thus even as the search for a minimal circuit ORAM continues, one can ask if secure ORAM emulation can be made significantly easier by moving from the 2PC to the MPC setting. Keller and Scholl [15] showed one way to design such multi-party SC-ORAM protocols, using arithmetic circuit representation of an ORAM client and implementing it with the SPDZ MPC protocol of Damgard et al. [6,8]. Their implementation achieved significantly faster *on-line* times than the 2PC SC-ORAM implementation of [22]: [15] report 250 ms wall clock per access for $m = 20$ and $d = 5B$ for 2 machines with direct connection in the online stage, while Wang et al. [22] report 30 s of just CPU time for $m = 24$ and $d = 4B$. Moreover, the implementation of [15] is secure against malicious adversaries while that of [22] works only against honest-but-curious faults. On the other hand, this on-line speed-up comes at the cost of intensive precomputation required by the SPDZ MPC protocol, which [15] estimated at between 100 and 800 min for $m = 20$.

In this work we explore a different possibility for SC-ORAM design, specific to the setting of three parties with a single corrupted party, with security against honest-but-curious faults. The 3-party SC-ORAM protocol we propose uses a variant of the Binary-Tree ORAM as the underlying data-structure. The *access* part of the proposed SC-ORAM is based on the following observation: If P_1 and P_2 secret-share an array of (keyword,value) pairs (k, v) (this will be a path in the Binary-Tree ORAM) and a searched-for keyword k^* (this will be the searched-for address prefix), then a variant of the Conditional Disclosure of Secret protocol of [12] which we call *Secret-Shared Conditional OT* (SS-COT) allows P_3 to receive value v associated with keyword k^* at the cost roughly equal to the symmetric encryption and transmission of the array. Moreover, while SS-COT reveals the location of pair (k^*, v) in the array to P_3, this leakage can be easily masked if P_1, P_2 first shift the secret-shared array by a random offset. The *eviction* part of the SC-ORAM springs from an observation that instead of performing the eviction computation on all the data in the path retrieved at access, one can use garbled circuit to encode only the procedure determining eviction *movement logic*, i.e. determining which entries in each bucket should be shifted down the path. Then, if P_1, P_2 secret-share the retrieved path, and hence the bits which enter this computation, we can let P_3 evaluate this garbled circuit and learn the positions of the entries to be moved if (1) the eviction moves a constant number of entries in each bucket in a predictable way, e.g. one step down the path, and (2) P_1, P_2 randomly permute the entries in each bucket, so that P_3 always computes a fixed number of randomly distributed distinct indexes for each bucket. Computation of this movement logic uses only two input bits (appropriate direction bit and a full/empty flag) and 17 non-xor gates per bucket entry, so the garbled circuit is much smaller than if it coded the whole eviction procedure. Finally, the secret-shared data held by P_1, P_2 can be moved according to the movement matrix held by P_3 in another OT/CDS variant we call *Secret-Shared Shuffle OT* which uses only xor's and whose bandwidth is roughly four times the size of the secret-shared path.

Assuming constant record sizes the bandwidth of the resulting SC-ORAM protocol is O $\left(w(m^3 + \kappa m^2)\right)$ where w is the bucket size in the underlying Binary-Tree ORAM. Since the best exact bound on overflow probability we can give requires $w = \Omega(\lambda + m)$ where λ is a statistical security parameter, and since $m < \kappa$, this asymptotic bound is essentially the same as that of the 2-server SC-ORAM of [22]. However, the exact numbers for bandwidth and computation cost (measured in the number of block cipher or hash operations) are much lower, and this is because of two factors: First, even though we still use garbled circuits, the circuits involved have dramatically smaller complexity than in the 2PC implementations (see Table 1 below). Secondly, the cost of all operations *outside* the garbled circuits is a small factor away from the cost of transmission and decryption of server data in the underlying Binary-Tree ORAM algorithm. Concretely, the non-GC bandwidth is under $9P$ where P is the (total) size of tree *paths* retrieved by the Binary-Tree ORAM, and computation is bounded by symmetric encryption of roughly $20P$ bits, whereas for the underlying Binary-Tree ORAM both quantities are $2P$. Finally, stochastic evidence suggests that it suffices that $w = \Omega(\sqrt{\lambda + m})$, which for concrete parameters of $m = 36$ and $\lambda = 40$ reduces the required w, and hence all our protocol costs, by a further factor between 3 and 4.

We tested a preliminary Java implementation of our scheme, without making use of several possible optimizations, on three lowest-tier Amazon servers (t2.micro) connected through a LAN. Regarding the overall bandwidth, for a concrete value $m = 18$ (the largest size for which [22] give bandwidth data), $\lambda = 80$, and $d = 4B$, our scheme instantiated with $w = 128$, which satisfies exact security bounds for these m, λ values, uses $1.18MB$ bandwidth, a factor of 40 less than $45MB$ reported by [22]. Using the stochastic bound $w = \Omega(\sqrt{\lambda + m})$ which for the above case of m, λ is satisfied by $w = 32$ (see Sect. 5) would further

Table 1. Comparison of Circuit Size between this proposal (without optimizations) and the SC-ORAM scheme [22]. All numbers are reported as function of array size $|D| = 2^m$ for statistical security paramenter $\lambda = 80$. The first 3PORAM estimation uses bucket size $w = 128$ mandated by the strict bound implied by Lemma 2, while the second one uses bucket size $w = 32$ derived from the Markov Chain approximation. (We note Wang et al. [23] recently exhibited further reductions in ORAM circuit, e.g. reporting $350K$ *and* gates for $m = 20$. See Sect. 5 for further discussion.)

ORAM	Circuit Size (asymptotic bounds)	Circuit Size (gates)		Number of inputs	
		$m = 20$	$m = 29$	$m = 20$	$m = 29$
Path-SC ORAM	$\tilde{O}\left(m^3 + dm\right)\omega\left(1\right)$	37.2 M	111.7 M	0.2 M	0.3 M
SCORAM	N/A (heuristic)	4.6 M	13.0 M	0.3 M	0.9 M
3PORAM ($w = 128$)	$O\left(m^3\right)\omega\left(1\right)$	96.9K	213.9K	11.5K	25.3K
3PORAM ($w = 32$)	$O\left(m^3\right)\omega\left(1\right)$	28.5K	62.6K	3.4K	7.4K

decrease the bandwidth by a factor of about 2.5. Regarding computational costs, for $m = 36$ and $d = 4B$, using $w = 32$ justified by the stochastic evidence, our implementation takes 320 ms per access in the on-line phase and 1.3 s in the pre-computation phase. The SC-ORAM scheme of [22] reported only local execution times for m up to 26 while we tested our scheme on Amazon EC2 servers communicating over LAN for m up to 36, but for a conservative statistical security parameter $\lambda = 80$ the combined CPU cost of our implementation using $w = 128$ is a factor of about 50 less than that of [22], e.g. it is under 600 msec for $m = 24$ while that of [22] is 30 s.

In summary, we see our contributions as three-fold: First, we provide an immediate improvement to any application of SC-ORAM which can be done in the setting of three parties with an honest-majority. Secondly, the techniques we explore can be utilized in a different context, e.g. for a different "secure-computation friendly" eviction strategy for a Binary-Tree ORAM. Finally, the proposed protocol leaves several avenues for further improvements in 3-party SC-ORAM both on the level of system implementation and algorithm design.

Technical Overview. We base our implementation on the Shi *et al.* [20] hierarchical two party ORAM, and use a combination of three-party OT's and secure computation (using Yao garbled circuits [24]) in order to ensure privacy in the three party setting. Our protocol follows the same technical approach of two-party SC-ORAM schemes, i.e. of providing a secure computation protocol for access and eviction algorithms in a client-server ORAM. However, the existence of a third party allows us to greatly reduce the cost of this secure computation. Our main observation is that in Binary Tree access and eviction algorithms, like that of Gordon *et al.* [14], there is a separation in the role played by the input bits of the access or eviction circuit. Part of the bits are used to implement the logic of the circuit, but the majority are data that do not participate in the output of the logic and are, at best, just being moved between some locations based on the output of the logic. We exploit this separation in the three-party setting, by isolating the bits necessary for the logic, using Yao's garbled circuit to securely compute the logic only on those bits, and then use several variants of the (three-party) Oblivious Transfer (OT) protocol to move data to the locations pointed out by the output of the circuits. Since all these variants of OT can be implemented at a cost similar to just the *secure transmission* of the data the OT operates on, this leads to dramatic reductions in the cost of the resulting secure computation protocol. In addition, in the *access* protocol, as opposed to the *eviction*, we avoid using garbled circuits entirely, as the entire logic comes down to finding an index where two lists of n bitstrings contain a matching entry, which we implement using a three-party variant of *Conditional OT* which takes a single interaction round and costs roughly as much as encryption and transmission of these n bitstrings.

We make several modifications in the Binary-Tree ORAM of Shi *et al.* [20] to make it more efficient for the type of operations we are interested in. We use ideas from Gentry *et al.* [11] and Stefanov *et al.* [21]. In particular, we make the ORAM trees more shallow, as in Gentry *et al.* [11] by increasing how many entries in the ORAM will be mapped to each leaf in expectation and increasing the total capacity (in terms of entries) of the leaf nodes. To be more precise, for a tree that has a total capacity of 2^m entries and a capacity in each node of w, instead of having 2^m leafs in the ORAM tree, we have $\frac{2^m}{w}$ leafs instead. In order to ensure that overflow does not occur in the leafs of the tree we increase capacity of leaf nodes to $4w$. With this change we achieve linear overhead in terms of storage needed for the ORAM, meaning that now the total entries that can be stored in the ORAM are $O(2^m)$, in contrast with the $O(w \cdot 2^m)$ entries that the Shi et al. [20] ORAM had (note that for most settings $w = O(m)$). In addition, we observe that for internal nodes it is not necessary to increase their capacity, since the overflow of internal nodes is mandated by a difference probabilistic process that the one of leaf nodes. In contrast with the approach used in Gentry et al. [11], by only increasing the capacity of leaf nodes, we avoid doubling the bandwidth needed by the ORAM protocol (which is what happened in Gentry et al. [11], since they increase the capacity of all nodes, whether they are leafs or internal nodes).

We adopt the idea of eviction through a single path introduced by Gentry et al. [11]. The main problem we identified in the single path eviction, is that both Gentry et al. [11] and Stefanov et al. [21] evict all entries in all nodes of the path, as far down in the path as they can go. Although this is easy to do in a client-server ORAM where the client retrieves the whole path and performs all operations in the clear text data, in the setting of secure computation on the secret shared data, such eviction is very costly. For this reason, we modify the eviction to only evict at most two items from each node to the next node in the path, provided such items exist. This operation is limited enough to allow for simple garbled circuits. Moreover, it is an oblivious operation in the sense that always two entries are evicted to the next level (we evict empty entries if appropriate entries do not exist), which allows for its simple 3-party implementation. We choose not to increase the fun-out of nodes as Gentry et al. [11] do, since this would complicate both our circuits and the rest of the protocols. We also choose to avoid using the overflow cache used in Path ORAM of Stefanov et al. [21] in order to decrease the total space requirements for their ORAM, deciding instead to experiment at first with a design which maximally simplifies the eviction logic and the associated garbled circuits.

Lastly we briefly explain why it seems difficult to construct a (3-server) SC-ORAM scheme with competitive efficiency based on the *two-server* Client-Server ORAM of Lu and Ostrovsky [17]. Indeed, since in either 3-server or 2-server setting of SC-ORAM we rely on non-collusion between any two parties, we could use a two-server version of the underlying client-server ORAM. Because the two-server ORAM of [17] achieves $O(m)$ amortized overhead per query, the asymptotic running time of the SC-ORAM protocol based on this two-server ORAM could also be only linear in m, which would beat (at least asymptotically) anything based on the $O(m^3)$ single-server ORAM schemes of [11,20,21]. However, the Lu-Ostrovsky two-server ORAM has some features which adversely affect the practicality of the resulting SC-ORAM protocol. It is a hierarchical construction in the spirit of [13] with $O(m)$ levels where the i-th level contains $O(2^{i-1})$

encrypted memory entries. After every 2^t RAM accesses, the construction re-shuffles the first 2^t levels of the hierarchy, incurring $O(2^t)$ cost, which makes the running time of each access highly uneven. The scheme has other "MPC unfriendly" properties, e.g. the client's retrieval algorithm, which has to be emu-lated with secure computation, acts differently at a given level depending on whether the item has been found in a higher level. Also, the scheme seems to need oblivious computation of a PRF with secret-shared inputs (and possibly also out-puts), and the currently best protocol for such OPRF evaluation uses $O(t)$ expo-nentiations for a t-bit domain [14].

2   Baseline Client-Server ORAM Protocol

We describe the Client-Server ORAM which forms the basis of our three-party SC-ORAM protocol presented in Sect. 4. The scheme explained below is a variant of the *Binary Tree* client-server ORAM scheme of Shi et al. [20], with some optimizations adopted from a variant given by Gentry et al. [11]. The design principle behind our variant of the Binary Tree ORAM is two-fold: First, we want the client's algorithm to be "secure computation friendly". Secondly, we want to do so without increasing the parameters of the Binary Tree ORAM scheme (e.g. tree depth, tree size, tuple size, bucket size, etc.), as this would also negatively affect the efficiency of the resulting three-party ORAM emulator protocol.

ORAM Forest. Let D be an array of $|D| \leq 2^m$ records, where D[N] for every m-bit address N is a bitstring of fixed length d. For a given security parameter λ and cryptographic security parameter κ, the ORAM protocol needs only $O(\kappa)$ persistent storage and $O(m \cdot d)$ transient storage for client C, and $O(2^m \cdot d) = O(|D|)$ persistent storage of server S. Given m, d, λ, κ, an ORAM implementation is parametrized by two additional parameters $w, 2^\tau$ where $w = \max(\lambda, m)$ and τ is an integer divisor of m. Let $h = \log_{2^\tau} 2^m = m/\tau$. Server S stores an *ORAM Forest*, OTF $= (\mathrm{OT}_0, \mathrm{OT}_1, \ldots, \mathrm{OT}_h)$. Each *ORAM Tree* OT_i for $i > 0$ is a binary tree of height $\mathrm{d}_i = i\tau - \log w$ (if $i\tau \leq \log w$ then $\mathrm{d}_i = 0$). Let $\mathrm{N} = [\mathrm{N}^{(1)}|\ldots|\mathrm{N}^{(h)}]$ be the parsing of N into τ-bit segments and let $\mathrm{N}^i = [\mathrm{N}^{(1)}|\ldots|\mathrm{N}^{(i)}]$ be N's prefix of length τi. The last ORAM tree OT_h implements a look-up table F_h s.t. $\mathrm{F}_h(\mathrm{N}) = \mathrm{D}[\mathrm{N}]$, but the efficient retrieval of $\mathrm{F}_h(\mathrm{N})$ from OT_h is possible only given a *label* $\mathrm{L}^h \in \{0,1\}^{\mathrm{d}_h}$ which defines the leaf (or path, see below) in OT_h where value $\mathrm{F}_h(\mathrm{N})$ is stored. The way this label L^h can be found is that each ORAM tree OT_i for $i < h$ implements a look-up table F_i which maps N^i to label $\mathrm{L}^{i+1} \in \{0,1\}^{\mathrm{d}_{i+1}}$, and it is an invariant of OTF that for each i, $\mathrm{L}^{i+1} = \mathrm{F}_i(\mathrm{N}^i)$ defines a leaf (or path, see below) in OT_{i+1} which contains values of F_{i+1} on arguments $\mathrm{N}^{i+1} = \mathrm{N}^i|\mathrm{N}^{(i+1)}$ for all $\mathrm{N}^{(i+1)} \in \{0,1\}^\tau$. Therefore the ORAM *access* algorithm on input N proceeds recursively: The base tree OT_0 is a single vertex which contains values $\mathrm{L}^1 = \mathrm{F}_0(\mathrm{N}^1)$ for all $\mathrm{N}^1 \in \{0,1\}^\tau$, so the algorithm first retrieves $\mathrm{L}^1 = \mathrm{F}_0(\mathrm{N}^1)$ from OT_0, then using L^i it retrieves $\mathrm{L}^{i+1} = \mathrm{F}_i(\mathrm{N}^{i+1})$ from OT_i for $i = 1, \ldots, h-1$, and finally using L^h it retrieves $\mathrm{F}_h(\mathrm{N}) = \mathrm{D}[\mathrm{N}]$ from OT_h. For notational convenience we can think of an ORAM forest OTF as implementing function $F_{OTF} : \{0,1\}^m \rightarrow \{0,1\}^{d_1} \times \{0,1\}^{d_2} \times \ldots \times \{0,1\}^{d_h} \times \{0,1\}^d$, where $F_{OTF}(N) = (F_0(N^1), F_1(N^2), \ldots, F_{h-1}(N^h), F_h(N^h))$.

We now explain how F_i values are stored in binary tree OT_i. Let $\{0,1\}^{<m}$ denote the binary strings of length from 0 to $m-1$. The nodes of OT_i for $i > 0$ are formed as follows: Each *internal node*, indexed by $j \in \{0,1\}^{<d_i}$, stores a *bucket* B_j, while each *leaf node*, indexed by $j \in \{0,1\}^{d_i}$, is a set of *four* buckets $(B_{j00}, B_{j01}, B_{j10}, B_{j11})$. Tree OT_0 is an exception because it consists of a single root node B_{root}. (Constant 4 is chosen somewhat arbitrarily and can be adjusted, see Sect. 5.) Each bucket B_j is stored at node j in OT_i encrypted under the master key held by C. Each bucket is an array of w *tuples* of the form $T^i = (fb_i, N^i, L^i, A^i)$ where $fb_i \in \{0,1\}$, $N^i \in \{0,1\}^{i\tau}$, $L^i \in \{0,1\}^{d_i}$, and A^i for $i < h$ is an array containing 2^τ labels $L^{i+1} \in \{0,1\}^{d_{i+1}}$, while A^h is a record in D. The above invariant is maintained if for every N (or, if D is a sparse array, only for those N's for which D[N] is non-empty), there is a sequence of labels $(L^1, \ldots, L^h)$ (assume $L^0 = 0$ and $N^{(0)} = 0^\tau$) s.t. each OT_i contains a unique tuple of the form $T = (1, N^{(i)}, L^i, A^i)$ for some A^i and (1) this tuple is contained in some bucket along the path from the root to the leaf L^i in OT_i; and (2) if $i < h$ then $A^i [N^{(i+1)}] = L^{i+1}$, and $A^h = D[N]$. Observe that $(L^1, \ldots, L^h, D[N]) = F_{OTF}[N]$.

Access Procedure. To access location N in D, the client C performs the following loop sequentially for each $i = 0, \ldots, h$, given the recursively obtained label L^i: C sends L^i to the server S, who retrieves and sends to C the encrypted path $\overline{P_{L^i}}$ in tree OT_i from the root to the leaf L^i. C decrypts $\overline{P_{L^i}}$ using its master key into a corresponding plaintext path P_{L^i} which is a sequence of buckets $(B_1, \ldots, B_n)$ for $n = d_i + 4$ (recall that a leaf contains 4 buckets), finds the unique tuple T^i in this bucket sequence of the form $T^i = (1, N^{(i)}, L^i, A^i)$ and computes either label $L^{i+1} = A^i [N^{(i+1)}]$ if $i < h$, or, if $i = h$, outputs A^i as the record D[N].

Note that protocol reveals the vector of labels $(L^1, \ldots, L^h)$ to S. Therefore after each access C picks new random labels $((L^1)', \ldots, (L^h)')$, where $(L^i)'$ is a random bitstring of length d_i, and OTF needs to be updated so that $F_{OTF}(N) = ((L^1)', \ldots, (L^h)', D[N])$. To do this, C erases the tuple $T^i = (1, N^{(i)}, L^i, A^i)$ in the bucket in which it was found (by flipping the fb field to 0), replaces L^i with $(L^i)'$, sets $A^i [N^{(i+1)}]$ to $(L^{i+1})'$, and inserts this modified tuple $(T^i)'$ into the root bucket B_1. C then re-encrypts the buckets and sends the new encrypted path $(\overline{P_{L^i}})'$ to S to insert in place of $\overline{P_{L^i}}$ in OT_i.

Constrained Eviction Strategy. The above procedure works except for the fact that the root bucket fills up after w accesses. To ensure that this does not happen (with overwhelming probability), an eviction step is interjected into the access protocol before C re-encrypts $\overline{P_{L^i}}$ and sends it back to S. The aim of an eviction process is to move each tuple $T = (N^{(i)}, L, A)$ in an internal node of tree OT_i down towards its "destination leaf" L. Since in access C reads only the tuples in path P_{L^i}, this will only be done to the tuples found in the internal buckets of this path. Moreover, because we want our eviction strategy to be *secure-computation friendly*, i.e. to be as easy to compute securely in our three-party setting as possible, we restrict this eviction principle in two ways: we will attempt to move *at most two* tuples down in every path, and we will move them only one bucket down. Both of these two restrictions make no sense in the case of a client-server ORAM, where C sees all the buckets in P_{L^i} in the cleartext and can move all the tuples in this path as far down as they can go. However, in the context of the multi-party SC-ORAM protocol these constraints make the data movement pattern in the eviction process more predictable, and hence more easily implemented via a secure computation protocol which *does not* implement the whole eviction as a single (securely-computed) circuit.

Technically, this "constrained" eviction strategy works as follows: Consider a bucket B_j corresponding to an internal node in P_{L^i}, i.e. for $j \leq d_i$. We say that tuple $T = (fb, N^{(i)}, L, A)$ in B_j is *moveable* down the path towards leaf L^i if $fb = 1$ and the j-th bit in its label field L matches the j-th bit of leaf L^i. In every bucket B_j for $j \leq d_i$ we choose two random tuples which are moveable down towards leaf L^i. If there are no such tuples then we choose two random empty tuples instead (if they exist), and if there is only one then the second one is chosen as a random empty tuple (if it exists). In addition, we choose two random empty tuples (if they exist) among the $4w$ tuples contained in the four buckets B_j contained in the leaf node in P_{L^i}, i.e. for $j = d_i + 1, \ldots, d_i + 4$. Then, for each $i \leq d_i$, we take the two chosen tuples in bucket B_j and move them to the two spaces vacated in bucket B_{j+1} (except for $j = d_i$ where the two chosen spaces in the level below can be in any of the buckets $B_{d_i+1}, \ldots, B_{d_i+4}$).

Eviction fails in case of a bucket overflow i.e. if (1) some internal bucket in P_{L^i} does not contain two tuples which are either empty or moveable; or (2) the four buckets corresponding to the leaf node do not contain two empty tuples. As we argue in Sect. 5, both probabilities are negligible for $w = O(m)$, assuming the number *accesses* to D is polynomial in $|D|$.

Notation. In the 3-party SC-ORAM secure protocol for this Client-Server ORAM we will use notation $|P_L^i|$ to denote the length of any path in tree OT_i, and $|T^i|$ to denote the length of any tuple in such tree. Note that $|P_L^i| = (d_i + 4) \cdot w \cdot |T^i|$ and $|T^i| = 1 + i \cdot \tau + d_i + 2^\tau \cdot d_{i+1}$ if $i < h$ while $|T^h| = 1 + h \cdot \tau + d_h + d$ because $T^i = (fb, N^i, L^i, A^i)$, $|N^i| = i \cdot \tau$, $|L^i| = d_i$, and A^i for $i < h$ is an array holding 2^τ next-level leaf labels of length d_{i+1}, while $A^h = D[N]$ and $|D[N]| = d$.

3   Three-Party Protocol Building Blocks

Our SC-ORAM protocols Access, PostProcess, and Eviction of Sect. 4 rely on several variants of Oblivious Transfer (OT) or Conditional Disclosure of Secrets (CDS) protocols which we detail here. The efficiency of our SC-ORAM protocol relies on the fact that all these OT variants, including the OT variant employed in Yao's Garbled Circuit (GC) protocol, have significantly cheaper realizations in the 3-party setting. All presented protocols assume secure channels, although in many instances encryption overhead can be eliminated with simple protocol changes, e.g. using pairwise-shared keys in PRG's and PRF's.

Notation. Let κ denote the cryptographic security parameter, which will assume is both the key length and the block length of a symmetric cipher. Let G^ℓ be a PRG which outputs ℓ-bit strings given a seed of length κ. Let $\mathsf{F}_\mathsf{k}^\ell$ be a PRF which maps domain $\{0,1\}^\kappa$ onto $\{0,1\}^\ell$, for k randomly chosen in $\{0,1\}^\kappa$. We will write G and F_k when $\ell = \kappa$. In our implementation both F and G are implemented using counter-mode AES. If party A holds value a and party B holds value b s.t. $a \oplus b = v$ then we call pair (a, b) an "A/B secret-sharing" of v and denote it as $(\mathsf{s}_\mathsf{A}[v], \mathsf{s}_\mathsf{B}[v])$. Whenever we describe an intended output of some protocol as A/B secret-sharing of value v, we mean this to be a *random* xor-sharing of v i.e. pair $(r, r \oplus v)$ for r random in $\{0,1\}^{|v|}$. Let $s[j]$ denote the j-th bit of bitstring s, and let $[n]$ denote integer range $\{1, ..., n\}$.

3-Party Variants of Oblivious Transfer. We use several variants of the Oblivious Transfer problem in our three-party setting, namely Secret-Shared Conditional OT, SS-COT$^{[N]}$, Secret-Shared Index OT, SS-IOT$^{[2^\tau]}$, Shuffle OT, XOT $\binom{N}{k}$, Secret-Shared Shuffle OT, SS-XOT $\binom{N}{k}$, and Shift OT, Shift. We explain the functionality and our implementations of these OT variants below. The common feature of all our implementations is that they require one or two messages both in the pre-computation phase and in the online phase (except of Secret-Shared Shuffle OT which sends four messages), and the computational cost of each protocol for each party, both in pre-computation and on-line, is within a factor of 2 of the cost of secure transmission of the sender's inputs. We stress that all protocol we present form *secure computation* protocols of the corresponding functionalities assuming an *honest-but-curious adversary, secure channels*, and a *single corrupted player*. In each case the security proof is a straightforward simulation argument.

Algorithm 1. Secret-Shared Conditional OT Protocol SS-COT$^{[N]}$(S, R, H)

Input: S's input $(m_1, ..., m_N)$ and $(a_1, ..., a_N)$, H's input $(b_1, ..., b_N)$.
Output: R outputs pairs (t, m_t) s.t. $a_t = b_t$.
Parameters: ℓ and ℓ' s.t. $|m_t| = \ell$ and $|a_t| = |b_t| = \ell' \leq \kappa$ for all t.
Pre-computation phase: S, H share PRF F keys k, k' and κ-bit random nonces $r_1, ..., r_N$.

1: S sends $\{(e_t, v_t) = (\mathsf{G}^\ell(\mathsf{F}_\mathsf{k}^\kappa(x_t)) \oplus m_t, \mathsf{F}_{\mathsf{k}'}^\kappa(x_t))\}_{t=1}^N$ to R where $x_t = r_t \oplus [a_t | 0^{\kappa - \ell'}]$.
2: H sends $\{(p_t, w_t) = (\mathsf{F}_\mathsf{k}^\kappa(y_t), \mathsf{F}_{\mathsf{k}'}^\kappa(y_t))\}_{t=1}^N$ to R where $y_t = r_t \oplus [b_t | 0^{\kappa - \ell'}]$.
3: R outputs (t, m_t) where $m_t' = e_t \oplus \mathsf{G}^\ell(p_t)$ for each t s.t. $v_t = w_t$.

Secret-Shared Conditional OT, SS-COT$^{[N]}$(S, R, H), is a protocol where S inputs two lists, $(m_1, \ldots, m_N)$ and $(a_1, \ldots, a_N)$, H inputs a single list $(b_1, \ldots, b_N)$, and the protocol's goal is for R to output all pairs (t, m_t) s.t. $a_t = b_t$. This is a very close variant of the Conditional Disclosure of Secrets protocol of [12], and it can be implemented e.g. using modular arithmetic in a prime field. In Algorithm 1 we provide an alternative design which uses fewer (pseudo)random bits, and hence requires fewer PRG ops in pre-computation, but uses block ciphers in the on-line phase. (The algorithm proposed here was faster in our implementation even in

Algorithm 2. Shuffle OT Protocol $\mathsf{XOT}\begin{bmatrix}N\\k\end{bmatrix}(\mathrm{S},\mathrm{R},\mathrm{I})$

Input: S's input $(m_1,...,m_N)$ and I's input $(i_1,...,i_k)$ and $(\delta_1,...,\delta_k)$.
Output: R's output $(z_1,...,z_k)$ s.t. $z_\sigma = m_{i_\sigma} \oplus \delta_\sigma$ for all σ.
Parameters: Let $|m_t| = \ell$ for all t.
Pre-computation phase: I and S pick a random permutation π on $(1,...,N)$ and a sequence of ℓ-bit random pads $r_1,...,r_N$.

  1: S sends $(a_1,...,a_N) = (m_{\pi(1)} \oplus r_1,...,m_{\pi(N)} \oplus r_N)$ to R.
  2: I sends $(j_1,...,j_k) = (\pi^{-1}(i_1),...,\pi^{-1}(i_k))$ and $(p_1,...,p_k) = (r_{j_1} \oplus \delta_1,...,r_{j_k} \oplus \delta_k)$ to R.
  3: R outputs $(z_1,...,z_k) = (a_{j_1} \oplus p_1,...,a_{j_k} \oplus p_k)$.
    (Note that $z_\sigma = (m_{\pi(j_\sigma)} \oplus r_{j_\sigma}) \oplus (r_{j_\sigma} \oplus \delta_\sigma) = m_{i_\sigma} \oplus \delta_\sigma$ because $\pi(j_\sigma) = i_\sigma$.)

---

the on-line stage.) S and H share two PRF keys k,k', and for each t helper H sends to R a pair $(p_t,w_t) = (\mathsf{F}_\mathsf{k}(b_t),\mathsf{F}_{\mathsf{k}'}(b_t))$, while S sends (e_t,v_t) where e_t is an xor of message m_t and $\mathsf{G}(\mathsf{F}_\mathsf{k}(a_t))$ while $v_t = \mathsf{F}_{\mathsf{k}'}(a_t)$. For each t receiver R checks if $v_t = w_t$, and if so then it concludes that $a_t = b_t$ and outputs $m_t = e_t \oplus \mathsf{G}(p_t)$. To protect against collisions in (short) a_t,b_t values both within each protocol instance and across protocol instances each a_t and b_t is xor-ed by respectively S and H by a pre-shared one-time κ-bit random nonce r_t, with all nonces derived via a PRG on a seed shared by S and H.

Secret-Shared Index OT, $\mathsf{SS\text{-}IOT}^{[2^\tau]}(\mathrm{S},\mathrm{R},\mathrm{H})$, is a close variant of the Secret-Shared Conditional OT, where S holds a list of messages $(m_0,\ldots,m_{N-1})$ for $N = 2^\tau$ and an index share $j_\mathrm{S} \in \{0,1\}^\tau$ while H holds the other share $j_\mathrm{H} \in \{0,1\}^\tau$, and the aim of the protocol is for R to output (j,m_j) s.t. $j = j_\mathrm{S} \oplus j_\mathrm{H}$. Our protocol for $\mathsf{SS\text{-}IOT}^{[2^\tau]}$ executes similarly to $\mathsf{SS\text{-}COT}^{[N]}$ except H sends only two values, $(p,v) = (\mathsf{F}_\mathsf{k}(j_\mathrm{H}),\mathsf{F}_{\mathsf{k}'}(j_\mathrm{H}))$ and S's messages are computed as $e_t = \mathsf{G}(\mathsf{F}_\mathsf{k}(j_\mathrm{S} \oplus t))$ and $v_t = \mathsf{F}_{\mathsf{k}'}(j_\mathrm{S} \oplus t)$. Finally, to avoid correlations across protocol instances players H and S xor their PRF inputs with a single pre-shared random κ-bit nonce r.

Shuffle OT, $\mathsf{XOT}\begin{bmatrix}N\\k\end{bmatrix}(\mathrm{S},\mathrm{R},\mathrm{I})$, is a protocol between sender S, receiver R, and *indicator* I, where S inputs a sequence of messages $m_1,...,m_N$, I inputs a sequence of indexes $i_1,...,i_k$ and a sequence of masks $\delta_1,...,\delta_k$, and the protocol lets R output a sequence of messages $m_{i_1} \oplus \delta_1,...,m_{i_k} \oplus \delta_k$, without leaking anything else about S's and I's inputs. See Algorithm 2 for an implementation of this protocol.

Secret-Shared Shuffle OT, $\mathsf{SS\text{-}XOT}\begin{bmatrix}N\\k\end{bmatrix}(\mathrm{A},\mathrm{B},\mathrm{I})$, involves indicator I and two parties A and B. It is a close variant of the Shuffle OT above, where I holds indexes $i = (i_1,...,i_k)$, the pads $\delta_1,...,\delta_k$ are all set to zero, and both inputs $m_1,...,m_k$ and outputs $m_{i_1},...,m_{i_k}$ are secret-shared by A and B. We implement this protocol with two instances of $\mathsf{XOT}\begin{bmatrix}N\\k\end{bmatrix}$. The indicator I first chooses a sequence of random masks $\delta = (\delta_1,...,\delta_k)$, and inputs i,δ into both instances, where the first instance runs on A's input $(\mathsf{s}_\mathrm{A}[m_1],...,\mathsf{s}_\mathrm{A}[m_n])$, and lets B output $(\mathsf{s}_\mathrm{A}[m_{i_1}] \oplus \delta_1,...,\mathsf{s}_\mathrm{A}[m_{i_k}] \oplus \delta_k)$, while the second instance runs on B's inputs

$(s_B[m_1]$, and lets A output $(s_B[m_{i_1}] \oplus \delta_1, ..., s_B[m_{i_k}] \oplus \delta_k)$. It's easy to see that these outputs form a randomized A/B secret-sharing of $(m_{i_1}, ..., m_{i_k})$.

Shifting a Secret-Shared Sequence, Shift(A, B, H). As the access protocol traverses the forest of ORAM trees OTF $= (OT_0, OT_1, ..., OT_h)$, D and E recover the secret-sharing of path P_{L^i}, for $i = 1, ..., h$, and make several modifications to it. In particular, the buckets in the path are rotated by a random shift σ_i known to D and E. In the eviction protocol on this retrieved path we need a sub-protocol Shift to reverse this shift by transforming the secret-sharing of this path, which is a sequence of buckets, to a (fresh) secret-sharing of the same buckets but rotated back by σ_i positions. An inexpensive implementation of this task relies on the fact that in our three-party setting player D can act as a "helper" party and create, in pre-computation, *correlated* random inputs for E and C, which allows for an on-line protocol which consists of a few xor operations and a transmission of a single $|P_{L^i}|$-bit message from C to E. Since this protocol is a very close variant of protocol SS-IOT$^{[N]}$ given above we omit its description.

Yao's Garbled Circuit on Secret-Shared Inputs. The last component used in our ORAM construction is protocol GC[F](A, B, R), a Yao's garbled circuit solution for secure computation of an arbitrary function [24], executing on public inputs a circuit of function F, where the inputs X to this circuit are secret-shared between A and B, i.e. A inputs $s_A[X]$ and B inputs $s_B[X]$, and the protocol lets R compute $F(X)$. We stress that even though we do use Yao's garbled circuit evaluation as a subprotocol in our SC-ORAM scheme, we use it sparingly, and the computation involved is comparable, for realistic m values, to the necessary cost of decryption of paths P_{L^i} retrieved by the underlying Binary-Tree Client-Server ORAM scheme. The protocol is a simple modification of the delivery of the input-wire keys in Yao's protocol, adopted to the setting where the input X is secret-shared by parties A and B, while the third party R will compute the garbled circuit and get the $F(X)$. Let $n = |X|$ and let κ be the bitlength of the keys used in Yao's garbled circuit. In the off-line stage either A or B, say party A, prepares the garbled circuit for function F and sends it to R, and then for each input wire key pair (K_i^0, K_i^1) created by Yao's circuit garbling procedure, A picks random Δ_i in $\{0, 1\}^\kappa$, computes $(A_i^0, A_i^1) = (\Delta_i, K_i^0 \oplus K_i^1 \oplus \Delta_i)$ and $(B_i^0, B_i^1) = (K_i^0 \oplus \Delta_i, K_i^1 \oplus \Delta_i)$, and sends (B_i^0, B_i^1) to B. (To optimize pre-computation A can send to B a random seed from which $\{K_i^0, K_i^1, \Delta_i\}_{i=1}^n$ can be derived via a PRG.) In the on-line phase, for each $i = 1, ..., n$, party A on input bit $a = s_A[X_i]$ sends A_i^a to R, while party B on input bit $b = s_B[X_i]$ sends B_i^b. For each $i = 1, ..., n$, party R computes $K_i = A_i \oplus B_i$ for A_i, B_i received respectively from A and B, and then runs Yao's evaluation procedure inputting keys $K_1, ..., K_n$ into the garbled circuit received for F. Observe that $A_i \oplus B_i = K_i^v$ for $v = a \oplus b$, and hence if a, b is the XOR secret-sharing of the i-th input bit, i.e. if $a \oplus b = X_i$, then $K_i = K_i^0$ if $X_i = 0$ and $K_i = K_i^1$ if $X_i = 1$. The protocol is secure thanks to the random pad Δ_i, because for every X_i and every possible sharing (a, b) of X_i, values (A_i, B_i) sent to R are distributed as two random bitstrings s.t. $A_i \oplus B_i = K_i^v$ for $v = X_i$.

4   Three-Party SC-ORAM Protocol

We describe our three-party SC-ORAM protocol, which is a three-party secure computation of the Client-Server ORAM of Sect. 2. We refer to the three parties involved as C, D, and E. The basic idea for the protocol is to secret-share the datastructure OTF between two servers D and E, and have these two parties implement the Server's algorithm of the Client-Server ORAM scheme of Sect. 2, while the corresponding Client's algorithm will be implemented with a three-party secure computation involving parties C, D, E. In the description below we combine these two conceptually separate parts into a single protocol, but almost all of the protocol implements the three-party computation of the ORAM Client's algorithm, as the Server's side of this Client-Server ORAM consists only of retrieving (the shares of) path P_{L^i} from (the shares of) the i-th tree OT_i at the beginning of i-th iteration of the access procedure, and then writing (the shares of) a new path $P^\diamond_{L^i}$ in place of (the shares of) P_{L^i} at the end.

Given this secret-sharing scenario, the task of the three-party SC-ORAM protocol is to securely compute the following two functionalities:

1. The *access* functionality computes the next-tree label $L^{i+1} = F_i(N^{i+1})$ given the D/E secret-sharing of path P_{L^i}, for $L^i = F_{i-1}(N^i)$ and the D/E secret-sharing of address prefix N^{i+1};
2. The *eviction* functionality computes the D/E secret-sharing of path $P^\diamond_{L^i}$ output by the eviction algorithm applied to the D/E secret-shared path P_{L^i}, after the tuple containing the label identified by the access functionality is moved to the root node.

Both tasks can be computed using standard secure computation techniques but the protocol we show beats a generic one by a few orders of magnitude, and comes close to the computation cost of the underlying Client-Server ORAM itself. Note that the i-th iteration of the Client-Server ORAM needs a Server-to-Client transmission and decryption of path $P_L L^i$ and then encryption and Client-to-Server transmission of path $P^\diamond_{L^i}$. Therefore the base-line cost we want the SC-ORAM to come close to are $h+1$ rounds of Client-Server interaction with $2 \cdot |P^i_L|$ bandwidth and $(2/\kappa) \cdot |P^i_L|$ block cipher operations for $i = 0, ..., h$. The main idea which allows us to come close to these parameters is that if the inputs to either access or eviction functionalities, secret-shared by two parties, e.g. D and E, are shifted/permuted/rotated/masked in an appropriate way, then the correspondingly shifted/masked outputs of these functionalities can be revealed to the third party, e.g. C.

In the 3-party setting we separate the Client-Server access/eviction protocols into Access, PostProcess, and Eviction. Protocol Access contains all parts of the client-server access which have to be executed *sequentially*, i.e. the retrieval of sequence $F_{OTF}(N) = (L^1, L^2, ..., L^h, D[N])$ done by sequential identification (and removal from the OT_i trees) of the tuple sequence $(T^1, T^2, ..., T^h)$ where T^i is defined as path $P_L L^i$ of tree OT_i whose address field is equal to N's prefix N^i and whose A field contains label L^{i+1} at position $N^{(i+1)}$. Protocol PostProcess performs cleaning-up operations on each tuple T^i in this tuple sequence, by

Algorithm 3. Protocol Access[i] - Oblivious Retrieval of Next Label

Input: D, E's inputs: label L^i and secret-sharing of OT_i and $N^{i+1} = [N^i|N^{(i+1)}]$;
Output: (1) C outputs $L^{i+1} = A^i[N^{(i+1)}]$ where A^i is the A field of tuple T^i in P_{L^i} whose N field matches N^i; (2) C and E output a secret-sharing of T^i and $P_{L^i}^* = Rot^{[\sigma,\delta,\rho]}(P_{L^i}')$, where $P_{L^i}^*$ is P_{L^i} without tuple T^i; (3) D & E output σ, ρ;
Pre-computation phase: D & E's input: $(\sigma, \delta, \rho, p) \leftarrow [d_i+4] \times [w] \times \{0,1\}^\tau \times \{0,1\}^{|P_{L^i}|}$;
Parameters: $n = w(d_i+4)$.

1: D retrieves share $s_D[P_{L^i}]$ from $s_D[OT_i]$ and sets $s_D[Rot^{[\sigma,\delta,\rho]}(P_{L^i})]$ as the result of the three data-rotations using shifts (σ, δ, ρ) applied to $(s_D[P_{L^i}] \oplus p)$. E computes $s_E[Rot^{[\sigma,\delta,\rho]}(P_{L^i})]$ in the corresponding way.
2: D sends $s_D[Rot^{[\sigma,\delta,\rho]}(P_{L^i})]$ and $s_D[N^{i+1}] = (s_D[N^i]|s_D[N^{(i+1)}])$ to C.
3: D and E isolate in their shares of $Rot^{[\sigma,\delta,\rho]}(P_{L^i})$ a vector of shares of pairs (fb_j, N_j) for $j = 1, ..., n$ of fb and N fields of all tuples in this (rotated) path. E also isolates in $s_E[Rot^{[\sigma,\delta,\rho]}(P_{L^i})]$ shares $(s_E[Rot^{[\rho]}(A_1)], ..., s_E[Rot^{[\rho]}(A_n)])$ of the A field of all tuples. The parties then run SS-COT$^{[n]}$(E, C, D) on E's input $(m_1, ..., m_n)$ and $(a_1, ..., a_n)$ and D's input $(b_1, ..., b_n)$ where $m_t = s_E[Rot^{[\rho]}(A_t)] \oplus y$, $a_t = s_E[fb_t|N_t] \oplus [0|s_E[N^i]]$, and $b_t = s_D[fb_t|N_t] \oplus [1|s_D[N^i]]$. This subprotocol outputs $(j_1, \bar{e})$ for C s.t. $[fb_{j_1}|N_{j_1}] = [1|N^i]$ and $\bar{e} = y \oplus s_E[Rot^{[\rho]}(A_{j_1})]$. The client computes $z = \bar{e} \oplus \bar{d}$ where $\bar{d}$ is the A field in the j_1-th tuple in $s_D[Rot^{[\sigma,\delta,\rho]}(P_{L^i})]$. (Note that j_1-th tuple in $Rot^{[\sigma,\delta,\rho]}(P_{L^i})$ is equal to T^i, hence $A_{j_1} = A^i$ and $z \oplus y = Rot^{[\rho]}(A^i)$.)
4: Parties run SS-IOT$^{[2^\tau]}$(E, C, D) on E's input $(y_0, ..., y_{2^\tau-1})$ and $s_E[N^{(i+1)}]$ and D's input $s_D[N^{(i+1)}] \oplus \rho_i$, which outputs pair (j_2, y_{j_2}) for C.
5: Each party computes its output as follows:
    - C outputs $L^{i+1} = y_{j_2} \oplus z_{j_2}$ where z_{j_2} is j_2-th d_{i+1}-bit segment in z;
    - C and E form $(s_C[T^i], s_E[T^i])$ as $((1, s_D[N^i], 0^{d_i}, z), (0, s_E[N^i], L^i, y))$;
    - C and E form secret-sharing of $P_{L^i}^*$ by C setting its share to $s_D[Rot^{[\sigma,\delta,\rho]}(P_{L^i})]$ but with the j_1-th tuple modified by flipping bit fb and setting its other bits at random, and E setting its share to $s_E[Rot^{[\sigma,\delta,\rho]}(P_{L^i})]$;
    - D and E output (σ, ρ).

modifying its label field from L^i to $(L^i)'$ and modifying the label held at $N^{(i+1)}$-th position in the A^i array of this tuple from L^{i+1} to $(L^{i+1})'$. Importantly, the PostProcess and Eviction protocols can be done *in parallel* for all trees OT_i, which allows for a better CPU utilization in the protocol execution.

Access Protocol. Protocol Access runs on D/E secret-sharing of searched-for address N and the ORAM forest OTF, and it's goal is to compute a D/E secret-sharing of record D[N]. Protocol Access creates two additional outputs, for each $i = 0, ..., h$ (with some parts skipped in the edge cases of $i = 0$ and $i = h$): (1) C/E secret-sharing of the path P_{L^i} in OT_i, modified in the way we explain below, and with the tuple T^i defined above removed; and (2) whatever information needed for the PostProcess protocol to modify T^i into $(T^i)'$ which will be inserted into the root of OT_i in protocol Eviction.

Protocol Access proceeds by executing loop Access[i] *sequentially*, see Algorithm 3, for $i = 0, ..., h$. The inputs to Access[i] are: (1) D/E secret-sharing of OT_i; (2) D/E secret-sharing of address prefix $N^{i+1} = [N^i|N^{(i+1)}]$; (3) Leaf label L^i as the input of D and E (with N^0, $N^{(h+1)}$, and L^0 all empty strings). Its outputs are: (1) C's output the next leaf label $L^{i+1} = F_i(N^{i+1})$, for $i \neq h$, or the C/E secret-sharing of record $r = D[N]$, for $i = h$; (2) C/E secret-sharing of tuple T^i defined above; and (3) C/E secret-sharing of path $\mathsf{Rot}^{[\sigma_i, \delta_i, \rho_i]}(P'_{L^i})$ which results from rotating the data in P_{L^i} by three random shifts $(\sigma_i, \delta_i, \rho_i)$ known to E and D (and of removing T^i from P_{L^i}).

Data-Rotations and Conditional OT's. We first explain how E and D perform the three data-rotations on the secret-shared path P_{L^i} retrieved from the (shares of) the i-th level ORAM tree OT_i (and randomized by D and E xor-ing the shares of P_{L^i} retrieved from OT_i by a pre-agreed random pad). E and D pick three values during pre-processing, $\sigma_i, \delta_i, \rho_i$, at random in ranges resp. $\{1, ..., d_i + 4\}$, $\{1, ..., w\}$, and $\{0, 1\}^\tau$. The data-rotation defined by σ_i is performed on the bucket level, i.e. the $d_i + 4$ buckets in path P_{L^i} (recall that there are d_i internal nodes containing a bucket each and that the leaf node contains 4 buckets) are rotated clock-wise by σ_i positions. The data-rotation defined by δ_i is performed on the level of tuples within each bucket, i.e. in each of the $d_i + 4$ buckets in P_{L^i} the sequence of w tuples held in that bucket is rotated clock-wise by δ_i positions. Finally, the bit-vector ρ_i defines τ flips which will be applied to the array A in each of the $(d_i + 4) \cdot w$ tuples in the path. Namely, the A field in each tuple in the path is treated as a τ-dimensional cube whose content is flipped along the j-th dimension if the j-th bit in ρ_i is 1. Such τ flips define a permutation on elements of A where an element at position t moves to position $t \oplus \rho_i$, for each $t \in \{0, 1\}^\tau$. Note that E and D can perform all these data-rotations locally on their shares of the path P_{L^i}. We use $\mathsf{Rot}^{[\sigma_i, \delta_i, \rho_i]}(P_{L^i})$ to denote the resulting tree, and we use $\mathsf{Rot}^{[\rho_i]}(A)$ to denote the result of the permutation defined by $\rho_i \in \{0, 1\}^\tau$ on field A as explained above. After applying these data-rotations to P_{L^i} the parties run protocols $\mathsf{SS\text{-}COT}^{[n]}$ and $\mathsf{SS\text{-}IOT}^{[2^\tau]}$ described in Sect. 3, with E as the sender, D as the helper, and C as the receiver in both protocols. The goal of protocol $\mathsf{SS\text{-}COT}^{[n]}$, for $n = (d_i + 4) \cdot w$, is two-fold: (1) to let C compute the index $j_1 \in \{1, ..., n\}$ where path $\mathsf{Rot}^{[\sigma_i, \delta_i, \rho_i]}(P_{L^i})$ contains the unique tuple T^i defined above (i.e. the tuple that contains the searched-for address prefix N^i); and (2) to create a C/E secret-sharing of this tuple. The goal of $\mathsf{SS\text{-}IOT}^{[2^\tau]}$ is to let C compute the $N^{(i+1)}$-th entry in the A field of this secret-shared tuple T^i, because that field contains the next-tree label $L^{i+1} = F_i(N^{i+1})$.

Note that D and E hold the secret-sharing of N^i and for each $t = 1, ..., n$ they also hold the shares of the address N_t in the t-th tuple in $\mathsf{Rot}^{[\sigma_i, \delta_i, \rho_i]}(P_{L^i})$. If D and E form values a_t and b_t as an xor of these two sharings, i.e. $a_t = s_E[N^i \oplus s_E[N_t]]$ and $b_t = s_D[N^i \oplus s_D[N_t]]$ then $a_t = b_t$ if and only if $N_t = N^i$, i.e. if and only if t points to a unique tuple T^i in (rotated) path $\mathsf{Rot}^{[\sigma_i, \delta_i, \rho_i]}(P_{L^i})$ whose address field N equals the searched-for address N^i. Therefore if D and E run the Secret-Shared Conditional OT $\mathsf{SS\text{-}COT}^{[n]}$ on $(a_1, ..., a_n)$ and $(b_1, ..., b_n)$ defined above as their condition-share vectors, then C will compute the index j_1 to the searched-for tuple T^i contained in this path. Moreover, $\mathsf{SS\text{-}COT}^{[n]}$ will also compute the secret-sharing of T^i if E picks a random pad y of length $2^\tau \cdot d_{i+1}$, and defines the message vector it inputs to $\mathsf{SS\text{-}COT}^{[n]}$ as $(m_1, ..., m_n)$ where m_t is an xor of y

with E's share of the A field in the t-th tuple in $\mathsf{Rot}^{[\sigma_i,\delta_i,\rho_i]}(\mathsf{P}_{\mathsf{L}^i})$. Note that the A field in any entry in the rotated path corresponds to array $\mathsf{Rot}^{[\rho_i]}(\mathsf{A})$ where A was the field of the corresponding entry in the original path. Therefore C's output in this $\mathsf{SS\text{-}COT}^{[n]}$ instance will be j_2 together with $\bar{e} = y \oplus \mathsf{s_E}[\mathsf{Rot}^{[\rho_i]}(\mathsf{A}^i)]$ where the searched-for tuple T^i is defined as $(1, \mathsf{L}^i, \mathsf{N}^i, \mathsf{A}^i)$. Finally, D can send to C its share of the whole path $\mathsf{Rot}^{[\sigma_i,\delta_i,\rho_i]}(\mathsf{P}_{\mathsf{L}^i})$, so if C computes z as an xor of $\bar{e}$ with the A field in the j_1-th tuple in $\mathsf{s_D}[\mathsf{Rot}^{[\sigma_i,\delta_i,\rho_i]}(\mathsf{P}_{\mathsf{L}^i})]$ then (z,y) form a C/E secret-sharing of $\mathsf{Rot}^{[\rho_i]}(\mathsf{A}^i)$.

It remains for us to explain how $\mathsf{SS\text{-}IOT}^{[2^\tau]}$ computes an entry in this secret-shared field that corresponds to the next-level address chunk $\mathsf{N}^{(i+1)}$, because that's the entry which contains $\mathsf{L}^{i+1} = \mathsf{F}_i(\mathsf{N}^{i+1})$. Note that E and D hold the secret-sharing of $\mathsf{N}^{(i+1)}$ and that they also hold the bit-vector ρ_i s.t. the entry at t-th position in A^i is located at position $t \oplus \rho_i$ in $\mathsf{Rot}^{[\rho_i]}(\mathsf{A}^i)$. Since L^{i+1} sits at the t-th position in A^i for $t = \mathsf{N}^{(i+1)}$, we will find if we retrieve the j_2-th entry of $\mathsf{Rot}^{[\rho_i]}(\mathsf{A}^i)$ for $j_2 = \mathsf{N}^{(i+1)} \oplus \rho_i$. Note, however, that e.g. $\mathsf{s_D}[\mathsf{N}^{(i+1)}] \oplus \rho_i$ and $\mathsf{s_E}[\mathsf{N}^{(i+1)}]$ form a secret-sharing of j_2, and therefore the Secret-Shared Index OT protocol $\mathsf{SS\text{-}IOT}^{[2^\tau]}$ executed on sharing $(\mathsf{s_D}[\mathsf{N}^{(i+1)}] \oplus \rho_i, \mathsf{s_E}[\mathsf{N}^{(i+1)}])$ and E's data vector $y = (y_0, ..., y_{2^\tau-1})$, will let C output j_2 together with the j_2-th fragment y_{j_2} of y. Since (z,y) form the secret-sharing of $\mathsf{Rot}^{[\rho_i]}(\mathsf{A}^i)$, C can compute the j_2-th entry of $\mathsf{Rot}^{[\rho_i]}(\mathsf{A}^i)$, i.e. the next-level tree label L^{i+1}, by xor-ing y_{j_2} with j_2-th fragment of $z = (z_0, ..., z_{2^\tau-1})$.

Security Argument. This protocol is a secure computation of $\mathsf{Access}[i]$ functionality. Note that D and E do not receive any messages in this protocol, while C learns D's fresh random share $\mathsf{s_D}[\mathsf{Rot}^{[\sigma_i,\delta_i,\rho_i]}(\mathsf{P}_{\mathsf{L}^i})]$ of the rotated path, the index j_1 to the location of $\mathsf{T}^i = (1, \mathsf{N}^{i+1}, \mathsf{L}^i, \mathsf{Rot}^{[\rho_i]}(\mathsf{A}^i))$ in this rotated path, string $\bar{e} = y \oplus \mathsf{s_E}[\mathsf{Rot}^{[\rho_i]}(\mathsf{A}^i)]$, the index $j_2 = \mathsf{N}^{(i+1)} \oplus \rho_i$ where L^{i+1} is held in $\mathsf{Rot}^{[\rho_i]}(\mathsf{A}^i)$, and label $\mathsf{L}^{i+1} = \mathsf{F}_i(\mathsf{N}^{i+1})$. This view can be efficiently simulated given only L^{i+1} because (1) D's share of any path retrieved from OT_i is always a fresh random string because D and E randomize the sharing of $\mathsf{P}_{\mathsf{L}^i}$ after retrieving it from OT_i; (2)j_1 is a random integer in $\{1, ..., w \cdot (\mathsf{d}_i + 4)\}$ because the buckets are rotated by random $\sigma_i \in \{1, ..., \mathsf{d}_i + 4\}$ and the tuples within each bucket are rotated by random $\delta_i \in \{1, ..., w\}$; (3) $\bar{e}$ and j_2 are random bit-strings, because so are y and ρ_i; (4) C's view of $\mathsf{SS\text{-}COT}^{[n]}$ and $\mathsf{SS\text{-}IOT}^{[2^\tau]}$ can be simulated from their outputs.

Boundary Cases. Algorithm 3 shows protocol $\mathsf{Access}[i]$ for $0 < i < h$. For $i = 0$ tree OT_i contains a single node, shifts σ_0, δ_0 are not used, sub-protocol $\mathsf{SS\text{-}COT}^{[n]}$ is skipped, index j_1 is not used, and the outputs include only j_2 for C, ρ_0 for D and E, and the C/E secret-sharing of T^0 (with L^0 and N^0 set to empty strings). For $i = h$ the $\mathsf{SS\text{-}IOT}^{[2^\tau]}$ sub-protocol is skipped, shift ρ_h and index j_2 are not used, and (z,y) held by C and E form a secret-sharing of record $\mathsf{D}[\mathsf{N}]$.

Post-Process. The post-process protocol $\mathsf{PostProcess}$ transforms the C/E secret-shared tuples $\mathsf{T}^0, ..., \mathsf{T}^h$ output by Access to prepare the inputs for protocol $\mathsf{Eviction}$. It does so by executing a loop $\mathsf{PostProcessT}[i]$ in Algorithm 4 *in*

Algorithm 4. Protocol PostProcessT$[i]$ - Inserting New Labels into T^i

---

Input: C's input $\mathrm{s_C}[\mathrm{T}^i], \mathrm{L}^i, \mathrm{L}^{i+1}, j_2$; E's input $\mathrm{s_E}[\mathrm{T}^i]$;

Input known in pre-computation: C/D secret-sharing of labels $(\mathrm{L}^i)'$ and $(\mathrm{L}^{i+1})'$, where E forwards its shares to D;

Output: E/C secret-sharing of tuple $(\mathrm{T}^i)' = (1, \mathrm{N}^i, (\mathrm{L}^i)', \mathrm{A}')$ where $\mathrm{A}'[j_2] = (\mathrm{L}^{i+1})'$ and $\mathrm{A}'[t] = \mathrm{A}[t]$ for all $t \neq j_2$ where $\mathrm{T}^i = (1, \mathrm{N}^i, \mathrm{L}^i, \mathrm{A})$;

Pre-computation phase: D picks $r_1, ..., r_{2^\tau}$ in $\{0,1\}^{d_{i+1}}$ and α in $\{0,1\}^\tau$, and sends $\alpha, r_1, ..., r_{2^\tau}$ to C and $s_1, ..., s_{2^\tau}$ to E s.t. $s_\alpha = r_\alpha \oplus \mathrm{s_E}[(\mathrm{L}^{i+1})']$ and $s_t = r_t$ for $t \neq \alpha$.

1: C sends $\delta = \alpha - j_2 \pmod{2^\tau}$ to E.
2: C outputs $\mathrm{s_C}[(\mathrm{T}^i)'] = \mathrm{s_C}[\mathrm{T}^i] \oplus (0, 0^{i\tau}, \mathrm{L}^i \oplus \mathrm{s_C}[(\mathrm{L}^i)'], (c_1|...|c_{2^\tau}))$ where $c_t = r_{t+\delta \pmod{2^\tau}}$ for all $t \neq j_2$ and $c_t = r_{t+\delta \pmod{2^\tau}} \oplus \mathrm{L}^{i+1} \oplus \mathrm{s_C}[(\mathrm{L}^{i+1})']$ for $t = j_2$.
3: E outputs $\mathrm{s_E}[(\mathrm{T}^i)'] = \mathrm{s_E}[\mathrm{T}^i] \oplus (0, 0^{i\tau}, \mathrm{s_E}[(\mathrm{L}^i)'], (e_1|...|e_{2^\tau}))$ where $e_t = s_{t+\delta \pmod{2^\tau}}$.

---

parallel for $i = 0, ..., h-1$ (tuple T^h is not part of this step). The goal of post-processing is to replace the L^{i+1} value which sits at the j_2-th position in the A field of the *secret-shared* tuple T^i (where j_2 is an index C learns in Access$[i]$), with the *secret-shared* value $(\mathrm{L}^{i+1})'$. In other words, we need to inject a secret-shared value into a secret-shared array at a secret position known only to one party. However, we can utilize the fact that this secret-shared value to be injected can be chosen in pre-processing and that E's share of it can be revealed to D. Let $(c, e) = (\mathrm{s_C}[(\mathrm{L}^{i+1})'], \mathrm{s_E}[(\mathrm{L}^{i+1})'])$ and let E sends its share e to D in pre-processing. If D pre-computes two $|\mathrm{A}|$-long correlated random pads, one for C and one for E, with the known difference e between them at random location α known to C, then $e \oplus c$ can be injected at position j_2 into the C/E secret-sharing of T^i if (1) C sends $\delta = \alpha - j_2 \bmod 2^\tau$ to E, (2) both parties rotate the pads they receives from D counter-clockwise by δ positions, in this way placing the unique pad cells that differ by e at position j_2, (3) both parties xor their shares of T^i with these pads, with C injecting an xor with c at position j_2 into her share. (In addition C will also erase the previous leaf value at position j_2 in A field of T^i by adding L^{i+1} to that xor.)

Eviction Protocol. Protocol Eviction executes subprotocol Eviction$[i]$ in Algorithm 5 in *parallel* for each $i = 0, ..., h$. (For $i = 0$ protocol Eviction$[i]$ skips all the steps in Algorithm 5 except the last one.) Subprotocol Eviction$[i]$ performs an ORAM eviction procedure on path $\mathrm{P}_{\mathrm{L}^i}^*$, whose C/E secret-sharing is output by protocol Access. The protocol has two parts: First, using Yao's garbled circuit protocol GC (see Sect. 3) it allows D to identify two tuples in each internal bucket of $\mathrm{P}_{\mathrm{L}^i}^*$ which are either moveable one notch down this path or they are empty (see the eviction algorithm in Sect. 2). Another instance of GC will similarly find two empty tuples in the four buckets corresponding to the leaf in $\mathrm{P}_{\mathrm{L}^i}^*$. The reason these pairs of indices j_0, j_1 can be leaked to D is that (1) C and E randomly permute the tuples in each bucket in $\mathrm{P}_{\mathrm{L}^i}^*$ before using them in this protocol, and (2) index j_b computed for $b = 0, 1$ for each bucket in $\mathrm{P}_{\mathrm{L}^i}^*$ is defined as the first moveable tuple in that bucket after a random offset λ_b (counting the tuples cyclically), where shifts λ_0, λ_1 are chosen by E independently for each bucket at

Algorithm 5. Protocol Eviction[i] - Eviction in Path P_{L^i} of OT_i

Input: C/E secret-sharing of path $P^*_{L^i}$ and tuple $(T^i)'$; σ, ρ held by E, D;
Output: D/E secret-sharing of path $P^\diamond_{L^i}$ to be inserted into tree OT_i in place of P_{L^i}.
Notation: Let $W = \{1, ..., w\}$, $IB = \{0, ..., d_i - 1\}$, and $EB = \{d_i, d_i + 1, d_i + 2, d_i + 3\}$.
Pre-computation phase: C and E share random permutations $\pi_1, ..., \pi_{d_i}$ on set $[w]$, a random permutation π_{d_i+1} on set $[4 \cdot w]$, and a random pad ξ of length $|P_{L^i}|$;

1: Parties run protocol Shift(C, E, D) on inputs C/E-secret-sharing of $P^*_{L^i}$ and on D, E input a shift σ. The protocol outputs a C/E-secret-sharing of path identical to $P^*_{L^i}$ but with buckets shifted back by σ positions. In addition, for each $j \in IB$, C and E use π_j to permute (their shares of) the tuples in the j-th bucket in the resulting path, and they use π_{d_i+1} to permute (their shares of) the tuples in the four buckets corresponding to the leaf node. The resulting path, shared by C and E, is denoted $P^{**}_{L^i}$.

2: Let fb^j_ℓ and L^j_ℓ be the fb and L fields of the ℓ-th tuple in the j-th bucket in $P^{**}_{L^i}$. For each $j \in IB$, parties run protocol GC[F2FT](E, C, D), see Sec 3, on C's inputs $\{s_C[fb^j_\ell, L^j_\ell[j+1]]\}_{\ell \in W}$ and on E's inputs $\{s_E[fb^j_\ell], B_j\}_{\ell \in W}$ where $B_j = s_E[L^j_\ell[j+1]] \oplus 1 \oplus L^i[j+1]$. (Note that $s_C[L^j_\ell[j+1]] \oplus B_j = 1$ iff the secret-shared value L^j_ℓ and the public value L^i agree on $(j+1)$-st bit.) For each $j \in IB$, D defines $\alpha^1_j, \alpha^2_j \in [1, ..., w]$ as the indices of the two output wires of F2FT on which D received output bit 1 in the j-th instance of GC[F2FT].

3: The parties run protocol GC[F2ET](E, C, D) on E's inputs $\{s_E[fb^j_\ell]\}_{\ell \in W, j \in EB}$ and C's inputs $\{s_C[fb^j_\ell]\}_{\ell \in W, j \in EB}$. D defines $\alpha^1_{d_i}, \alpha^2_{d_i} \in [1, ..., 4 \cdot w]$ as the indices of the two output wires of F2ET on which D received output bit 1 in this instance of GC[F2ET].

4: D prepares a sequence of $k = w \cdot (d_i + 4)$ indices $I = (\beta_1, ..., \beta_k)$ s.t.

$$\beta_{w \cdot j + \ell} = \begin{cases} k+1, & \text{if } j = 0 \text{ and } \ell = \alpha^1_0 \\ k+2, & \text{if } j = 0 \text{ and } \ell = \alpha^2_0 \\ w \cdot (j-1) + \alpha^1_{j-1}, & \text{if } 1 \leq j \leq d_i - 1 \text{ and } \ell = \alpha^1_j \\ w \cdot (j-1) + \alpha^2_{j-1}, & \text{if } 1 \leq j \leq d_i - 1 \text{ and } \ell = \alpha^2_j \\ w \cdot (d_i - 1) + \alpha^1_{d_i - 1}, & \text{if } j \geq d_i \text{ and } w \cdot (j - d_i) + \ell = \alpha^1_{d_i} \\ w \cdot (d_i - 1) + \alpha^2_{d_i - 1}, & \text{if } j \geq d_i \text{ and } w \cdot (j - d_i) + \ell = \alpha^2_{d_i} \\ w \cdot j + \ell & \text{otherwise} \end{cases}$$

and then divides I into $d_i + 4$ chunks, each of which has w indices, and permutes each chunk with the corresponding ϱ_r.

5: C prepares a sequence of $k + 2$ shares $(s_C[a_1], ..., s_C[a_{k+2}])$ by setting $s_C[a_{w \cdot j + \ell}] = s_C[T^j_\ell]$ where T^j_ℓ is ℓ-th tuple in j-th bucket B^j in $P^{**}_{L^i}$, for $\ell \in W$ and $j \in IB \cup EB$, $s_C[a_{k+1}]$ as $s_C[(T^i)']$, and $s_C[a_{k+2}]$ as 0 concatenated with a random string of $i \cdot \tau + d_i + 2^\tau \cdot d_{i+1}$ bits. E prepares a sequence of $k + 2$ shares $(s_E[a_1], ..., s_E[a_{k+2}])$ in the corresponding way, using its shares of $P^{**}_{L^i}$ and $(T^i)'$.

6: The parties run protocol SS-XOT $\binom{k+2}{k}$ on C's input $(s_C[a_1], ..., s_C[a_{k+2}])$, E's input $(s_E[a_1], ..., s_E[a_{k+2}])$, and D's input I. C and E set their shares of path $P^\diamond_{L^i}$ to their output in this SS-XOT $\binom{k+2}{k}$ protocol xor'ed with string ξ.

7: C sends $s_C[P^\diamond_{L^i}]$ to D; D and E insert their shares of $P^\diamond_{L^i}$ into their shares of OT_i.

random in $\{1, ..., w\}$. The circuit computed for every internal bucket takes only $2w$ bits of input (one for bit fb and one for an agreement in the i-th bit of a leaf label in the tuple and the i-th bit of label L^i defining path $\mathsf{P}^*_{\mathsf{L}^i}$), and has only about $16w$ non-xor gates. Once D gets two indexes per each bucket in the path, it uses the Secret-Shared Shuffle OT protocol SS-XOT $\binom{k+2}{k}$ (see Section 3) to randomizes the secret-sharing of all tuples in $\mathsf{P}_{\mathsf{L}^i}$ while (1) moving the secret-shared tuple $(\mathsf{T}^i)'$ prepared by PostProcess into the root bucket, and (2) moving the two chosen tuples in each bucket to the space vacated by the two tuples chosen in the bucket below. Finally, C and E randomize their secret-sharing of the resulting path $\mathsf{P}^{**}_{\mathsf{L}^i}$ by xor-ing their shares with a pre-agreed random pad, C sends its share of $\mathsf{P}^{**}_{\mathsf{L}^i}$ to D, and D and E insert their respective shares of $\mathsf{P}^{**}_{\mathsf{L}^i}$ into their shares of OT_i, in place of the shares of the original path $\mathsf{P}_{\mathsf{L}^i}$ retrieved in the first step of Access$[i]$.

5   Protocol Analysis

Assuming constant record size the bandwidth of our protocol is $O\left(w(m^3 + \kappa m^2)\right)$, where w the bucket size of the nodes in our protocol, $|\mathsf{D}| = 2^m$, and κ is the cryptographic security parameter. The $O\left(wm^3\right)$ term comes from the fact that all our protocols except for the GC evaluation have bandwidth $O(|\mathsf{P}_{\mathsf{L}^i}|)$ where $\mathsf{P}_{\mathsf{L}^i}$ is a path accessed in OT_i. (The online part of our protocol requires 7 such transmissions per each OT_i). Each path $\mathsf{P}_{\mathsf{L}^i}$ in OT_i has length $O\left(w(\mathsf{d}_i)^2\right)$ where d_i is linear in i, and the summation is then done for i from 1 to $h = O(m)$. The $O\left(w\kappa m^2\right)$ term is the bandwidth for garbled circuits, since the inputs to the circuits for a path have $O(wm)$ bits and there are $O(m)$ paths retrieved during the traversal of the ORAM forest.

Each party's local cryptographic computation is $O\left(w\left(m^3/\kappa + m^2\right)\right)$ block cipher or hash operations. Note that the $O\left(wm^3/\kappa\right)$ factor comes already from secure transmission of data in the Client-Server ORAM, hence this cost seems cryptographically minimal. The GC computation contributes $O\left(wm^2\right)$ hash function operations, all performed by one party. Since $m < \kappa$, the $O\left(wm^2\right)$ term could dominate, and indeed we observe that the GC computation occupies a significant fraction of the overall CPU cost.

The performance of the scheme is linear in the bucket size parameter w, and the size of this parameter should be set so that the probability of overflow of any bucket throughout the execution of the scheme is bounded by $2^{-\lambda}$ for the desired statistical security parameter λ. The probability that an internal node overflows and the probability that a leaf node overflows are independent stochastic processes and for this reason we examine them separately. The analytical bounds we give for both cases are not optimal. For the leaf node overflow probability the bound we give in Lemma 1 could be made tight if the number of ORAM accesses N is equal to the number of memory locations 2^m, but for the general case of $\mathsf{N} > 2^m$ we use a simple union bound which adds a N factor. If a tighter analysis could be made, it could potentially reduce the required w by up to $\log(\mathsf{N})$ bits. The bound we give for the internal node overflow probability

in Lemma 2 is simplistic and clearly far the optimal. We amend this bound by
a discussion of a stochastic model which we used to approximate the eviction
process. If this approximation is close to the real stochastic process then the
scheme can be instantiated with much smaller bucket sizes than those implied
by Lemma 2.

Lemma 1. (Leaf Nodes) *If we have* N *accesses in an ORAM forest with the
total capacity for* 2^m *records and with leaf nodes which hold* $4w$ *entries, then the
probability that some leaf node overflows at some access is bounded by:*

$$\Pr[\textit{some leaf node in } \mathrm{OTF} \textit{ overflows}] \leq \mathsf{N} \cdot h^2 \cdot \frac{2^m}{w} \cdot 2^{-2w}$$

The proof of this lemma follows from a standard bins-and-balls argument.

To keep this probability below $2^{-\lambda}$ we need that $2w \geq \lambda + \log \mathsf{N} + m + 2\log m$.
It is easy to see that if you increase the number of buckets in a leaf node, the
constant of this linear relationship (which is roughly $\frac{1}{2}$ for 4 buckets per leaf)
decreases rapidly. For example if one uses 6 buckets per leaf, the constant of the
linear relationship between w and $m + \log \mathsf{N} + \lambda$ becomes $\frac{1}{6}$, allowing for much
smaller buckets. This means that by modifying the number of buckets per leaf,
we can ensure that it is the internal nodes that define the size of buckets. We
note that increasing the number of buckets per leaf increases the total space for
the ORAM forest OTF.

Lemma 2. (Internal Nodes) *If we have* N *accesses and subsequent evictions
in an ORAM forest with internal buckets of size* w, *then the probability that
some internal bucket overflows at some access is bounded by:*

$$\Pr[\textit{some internal bucket in } \mathrm{OTF} \textit{ overflows}] \leq \mathsf{N} \cdot h \cdot \mathrm{d}_h \cdot w \cdot 2^{-(w-1)}$$

We can prove Lemma 2 by assuming that there exists an internal node that
during all accesses and subsequent evictions is on the verge of overflowing (has
w or $w - 1$ entries in it). We also assume the worst case of each node always
receiving exactly two new entries, and we compute the probability that a node
is not able to evict two entries, thus causing an overflow.

To keep this probability below $2^{-\lambda}$, the lemma implies that $w - \log w \geq
\lambda + \log \mathsf{N} + 2\log m + 1$. For $w < 512$ this can be simplified as $w \geq \lambda + \log \mathsf{N} +
2\log m + 10$. For $\mathsf{N} \leq c \cdot 2^m$ this implies $w \geq \lambda + m + 2\log m + 10 + \log c$.

Stochastic Approximation. The above analysis is pessimistic, since it assumes
that there exists a critical bucket that is always full, having w or $w - 1$ entries
and bounds the probability of such a bucket having a "bad event". It does not
explore how difficult it is for a bucket to reach such a state, or how a congested
bucket is emptying over time. In order to better understand such behaviors we
observe that each internal node can be modeled as a Markov Chain, where the
state of the chain counts how many entries are currently in the node. The node
is initially empty. Whenever a node is selected in an eviction path it may receive
up to two entries depending on whether the parent node was able to evict one

or two entries. Moreover the node could evict up to two entries to its child that participates in the eviction path. The root always receives 1 entry and may evict up to two entries. Intuitively since the eviction path is picked at random and each entry is assigned to a random leaf node, each entry in a node in the eviction path can be evicted to the selected child node with probability $\frac{1}{2}$. So for this model we make the following relaxation: Instead of mapping an entry to a leaf node, when it is inserted for the first time in the root, we just let the leaf node be "defined" as the entry is pushed down the tree during eviction. In that sense we abstract entries and the only think we need to care for, is how many entries there exist in a given internal node at a given moment, which is expressed by the state of the Markov Chain.

This model needs one Markov Chain for each internal node. We make the following relaxation: We use one Markov Chain for each level of the tree. A Markov Chain starts empty. At each eviction step, a Markov Chain at level i may receive up to two entries depending on how many entries the Markov Chain in the previous level $i-1$ was able to evict. Moreover the Markov Chain at level i may evict up to two entries to level $i+1$. The Markov Chain for the root (level 0) always receives 1 entry. The state of a Markov Chain keeps tracks of how many entries are in it. At each eviction step an entry can be evicted with probability $\frac{1}{2}$ (the same as the probability we had for the previous model).

The final relaxation we do, is that we remove the direct relationship between a Markov Chain at level i evicting an entry and the Markov Chain at level $i+1$ receiving an entry. We first observe that on expectation at every level the Markov Chain receives at most 1 entry at each eviction step. Intuitively in order to prove this we observe that initially all nodes are empty. The root receives one entry in each eviction step, from there we can use a recursive argument that at any level i a node cannot be evicting more than 1 entry on expectation in each eviction step, which is what the node at level $i+1$ is receiving. Since the eviction probabilities only depend on the current state of a Markov Chain, the worst case for the Markov Chain, is when the variance of the input is maximized. This happens when with probability $\frac{1}{2}$ the node receives 0 entries and with probability $\frac{1}{2}$ the node receives 2 entries (also maximizes the expectation to 1).

We use this last model in order to bound the probability of overflow for internal nodes in our implementation and in order to set bucket sizes. In particular we generate a Markov Chain the has $w+2$ states, w for the bucket size, one empty state and one overflow state. The overflow state is a sink. We compute the probability of being in the overflow state after N accesses assuming the node was initially empty and perform a union bound on the number of nodes in all paths of the ORAM forest OTF. In Fig. 1 for different statistical security parameters λ equal with 20, 40 and 80, we show the minimum bucket sizes w for $\log N$ in the range $12, \ldots 36$ and $m = O(\log N)$. Generally, we observe that using the Markov Chain based approximation can lead to tighter bounds on the internal node sizes, from which we conjecture that the size of *internal nodes* can be reduced to $O\left(\sqrt{\lambda + \log N}\right)$ $(w > 2\sqrt{\lambda + \log N + 2\log m})$.

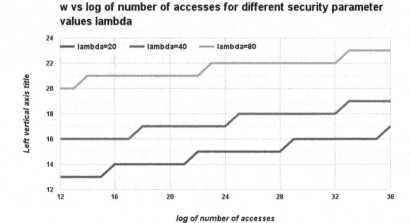

Fig. 1. w for different $\log N$

6   Implementation and Testing

We built and benchmarked a prototype JAVA implementation of the proposed 3-party SC-ORAM protocol. We tested this implementation on the entry-level Amazon EC2 t2.micro virtual servers, which have one hyperthread on a 2.5GHz CPU and 1GB RAM. Each of the three protocol participants C, D, E where co-located in the same availability zone and connected via a local area network. Here we will briefly show the most important findings, and we defer to the full version of the paper for more detailed performance data.

We measured the performance of the online and offline stages of our protocol separately, but our development effort was focused on optimizing the online stage so the offline timings provide merely a loose upper-bound on the precomputation overhead. We measured both wall clock and CPU times for each execution, where the wall clock time is defined as the maximum of the individual wall clocks, and the CPU time as the sum of the CPU times of the three parties. We tested our prototype for bit-length m of the RAM addresses ranging from 12 to 36, and for record size d ranging from 4 to 128 Bytes. Since the SC-ORAM protocol has two additional parameters, the bucket size w and the bitlength of RAM address segments τ, we tested the sensitivity of the performance to w using w equal to 16, 32, 64, or 128, and for each (m, w, d) tuple we searched for τ that minimize the wall clock (an optimal τ was always between 3 and 6 for the tested cases).

Figure 2 shows the wall clock time of the online stage as a function of the bitlength m of the RAM address space, for the two cases $(w, d) = (16, 4)$ and $(w, d) = (32, 4)$. We found that the CPU utilization in the online phase of our protocol is pretty stable, growing from about 25 % for smaller m's to 35 % for $m \geq 30$, hence the graph of the CPU costs as function of m has a very similar shape. Our testing showed that the influence of the record size d on the overall performance is very small for d less than $100B$, but higher payload sizes start

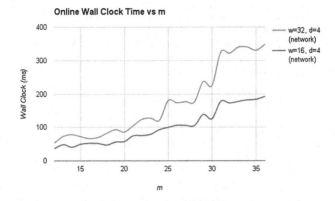

Fig. 2. Online Wall Clock vs RAM address size m

influencing the running time. Our testing confirms that the running time has clear linear relationship to the bucket size w: The wall clock for $w = 64$ grows by a factor close to 1.8 compared to $w = 32$, and for $w = 128$ by a factor close to 3.5 (for large m and small d). The offline wall clock time grows from 400 msec for $m = 12$ to 1300 msec for $m = 36$ for $w = 32$, but these numbers should be taken only as loose upper bounds on the precomputation overhead of our SC-ORAM. Finally, we profiled the code to measure the percentage of CPU time spent on different protocol components. We found that the fraction of the fraction of the total CPU costs of the online phase spent on Garbled Circuit evaluation decreases from $45\% - 50\%$ for $m = 12$ to 25% for $m = 36$. We also found that only about half of that cost is spent in SHA evaluation, i.e. that the Garbled Circuit evaluation protocol spends only about half its CPU time on decryption of the garbled gates. The fraction of the CPU cost spent on symmetric ciphers, which form the only cryptographic costs of all the non-GC part of our protocol, decreases from the already low figure of 10% for small m's to below 5% for $m = 36$. By contrast, the fraction of the CPU cost spent on handling message passing to and from TCP communication sockets grows from 12% for small m's to 30% for $m = 36$.

References

1. Ben-Or, M., Goldwasser, S., Wigderson, A.: Completeness theorems for non-cryptographic fault-tolerant distributed computation. In: Proceedings of the Twentieth Annual ACM Symposium on Theory of Computing, STOC 1988, pp. 1–10. ACM, New York (1988)
2. Chaum, D., Crépeau, C., Damgard, I.: Multiparty unconditionally secure protocols. In: Proceedings of the Twentieth Annual ACM Symposium on Theory of Computing, STOC 1988, pp. 11–19. ACM, New York (1988)
3. Choi, S.G., Hwang, K.-W., Katz, J., Malkin, T., Rubenstein, D.: Secure multiparty computation of boolean circuits with applications to privacy in on-line marketplaces. In: Dunkelman, O. (ed.) CT-RSA 2012. LNCS, vol. 7178, pp. 416–432. Springer, Heidelberg (2012)

4. Chor, B., Goldreich, O., Kushilevitz, E., Sudan, M.: Private information retrieval. In: Proceedings of 36th FOCS, pp. 41–50 (1995)
5. Chung, K.-M., Liu, Z., Pass, R.: Statistically-secure ORAM with $\tilde{O}(\log^2 n)$ overhead. In: Sarkar, P., Iwata, T. (eds.) ASIACRYPT 2014, Part II. LNCS, vol. 8874, pp. 62–81. Springer, Heidelberg (2014)
6. Damgård, I., Keller, M., Larraia, E., Pastro, V., Scholl, P., Smart, N.P.: Practical covertly secure MPC for dishonest majority – or: breaking the SPDZ limits. In: Crampton, J., Jajodia, S., Mayes, K. (eds.) ESORICS 2013. LNCS, vol. 8134, pp. 1–18. Springer, Heidelberg (2013)
7. Damgård, I., Meldgaard, S., Nielsen, J.B.: Perfectly secure oblivious RAM without random oracles. In: Ishai, Y. (ed.) TCC 2011. LNCS, vol. 6597, pp. 144–163. Springer, Heidelberg (2011)
8. Damgård, I., Pastro, V., Smart, N., Zakarias, S.: Multiparty computation from somewhat homomorphic encryption. In: Safavi-Naini, R., Canetti, R. (eds.) CRYPTO 2012. LNCS, vol. 7417, pp. 643–662. Springer, Heidelberg (2012)
9. Fletcher, C.: Ascend: an architecture for performing secure computation on encrypted data. In: MIT CSAIL CSG Technical Memo, p. 508 (2013)
10. Fletcher, C.W., van Dijk, M., Devadas, S.: A secure processor architecture for encrypted computation on untrusted programs. In: Proceedings of the Seventh ACM Workshop on Scalable Trusted Computing, STC 2012, pp. 3–8. ACM, New York (2012)
11. Gentry, C., Goldman, K.A., Halevi, S., Julta, C., Raykova, M., Wichs, D.: Optimizing ORAM and using it efficiently for secure computation. In: De Cristofaro, E., Wright, M. (eds.) PETS 2013. LNCS, vol. 7981, pp. 1–18. Springer, Heidelberg (2013)
12. Gertner, Y., Ishai, Y., Kushilevitz, E., Malkin, T.: Protecting data privacy in private information retrieval schemes. In: STOC (1998)
13. Goldreich, O., Ostrovsky, R.: Software protection and simulation on oblivious rams. J. ACM **43**(3), 431–473 (1996)
14. Gordon, S.D., Katz, J., Kolesnikov, V., Krell, F., Malkin, T., Raykova, M., Vahlis, Y.: Secure two-party computation in sublinear (amortized) time. In: Computer and Communications Security (CCS), CCS 2012, pp. 513–524 (2012)
15. Keller, M., Scholl, P.: Efficient, oblivious data structures for MPC. In: Sarkar, P., Iwata, T. (eds.) ASIACRYPT 2014, Part II. LNCS, vol. 8874, pp. 506–525. Springer, Heidelberg (2014)
16. Kruger, L., Jha, S., Goh, E.-J., Boneh, D.: Secure function evaluation with ordered binary decision diagrams. In: Conference on Computer and Communications Security, CCS 2006, pp. 410–420. ACM, New York (2006)
17. Lu, S., Ostrovsky, R.: Distributed oblivious RAM for secure two-party computation. In: Sahai, A. (ed.) TCC 2013. LNCS, vol. 7785, pp. 377–396. Springer, Heidelberg (2013)
18. Maas, M., Love, E., Stefanov, E., Tiwari, M., Shi, E., Asanovic, K., Kubiatowicz, J., Song, D.: Phantom: practical oblivious computation in a secure processor. In: Conference on Computer and Communications Security, CCS 2013, pp. 311–324. ACM, New York (2013)
19. Ostrovsky, R., Shoup, V.: Private information storage (extended abstract). In: Proceedings of the Twenty-Ninth Annual ACM Symposium on the Theory of Computing, El Paso, Texas, USA, 4–6 May 1997, pp. 294–303 (1997)
20. Shi, E., Chan, T.-H.H., Stefanov, E., Li, M.: Oblivious RAM with $O((\log N)^3)$ worst-case cost. In: Lee, D.H., Wang, X. (eds.) ASIACRYPT 2011. LNCS, vol. 7073, pp. 197–214. Springer, Heidelberg (2011)

21. Stefanov, E., Van Dijk, M., Shi, E., Fletcher, C., Ren, L., Yu, X., Devadas, S.: Path oram: an extremely simple oblivious ram protocol. In: Conference on Computer and Communications Security (CCS), CCS 2013, pp. 299–310 (2013)
22. Wang, X.S., Huang, Y., Chan, T.H., Shelat, A., Shi, E.: Scoram: oblivious ram for secure computation. In: Conference on Computer and Communications Security, CCS 2014, pp. 191–202. ACM (2014)
23. Wang, X.S., Hubert, T.-H., Shi, E.: Circuit oram: on tightness of the Goldreich-Ostrovsky lower bound. In: Eprint IACR Archive, 2015/672 (2014)
24. Yao, A.C.-C.: Protocols for secure computations (extended abstract). In: Proceedings of the 23rd Annual Symposium on Foundations of Computer Science, FOCS 1982, pp. 160–164 (1982)

On Cut-and-Choose Oblivious Transfer and Its Variants

Vladimir Kolesnikov[1]([✉]) and Ranjit Kumaresan[2]

[1] Bell Labs, Murray Hill, NJ, USA
kolesnikov@research.bell-labs.com
[2] MIT CSAIL, Cambridge, MA, USA
ranjit@csail.mit.edu

Abstract. Motivated by the recent progress in improving efficiency of secure computation, we study *cut-and-choose oblivious transfer*—a basic building block of state-of-the-art constant round two-party secure computation protocols that was introduced by Lindell and Pinkas (TCC 2011). In particular, we study the question of realizing cut-and-choose oblivious transfer and its variants in the OT-hybrid model. Towards this, we provide *new definitions* of cut-and-choose oblivious transfer (and its variants) that suffice for its application in cut-and-choose techniques for garbled circuit based two-party protocols. Furthermore, our definitions conceptually simplify previous definitions including those proposed by Lindell (Crypto 2013), Huang et al., (Crypto 2014), and Lindell and Riva (Crypto 2014). Our main result is an efficient realization (under our new definitions) of cut-and-choose OT and its variants with small concrete communication overhead in an OT-hybrid model. Among other things this implies that we can base cut-and-choose OT and its variants under a variety of assumptions, including those that are believed to be resilient to quantum attacks. By contrast, previous constructions of cut-and-choose OT and its variants relied on DDH and could not take advantage of OT extension. Also, our new definitions lead us to more efficient constructions for multistage cut-and-choose OT—a variant proposed by Huang et al. (Crypto 2014) that is useful in the multiple execution setting.

Keywords: Cut-and-choose oblivious transfer · OT extension · Concrete efficiency

1 Introduction

Secure two-party computation is rapidly moving from theory to practice. While the basic approach for semi-honest security, garbled circuits [33], is extensively studied and is largely settled, security against malicious players has recently

V. Kolesnikov—Supported by the Office of Naval Research under contract N00014-14-C-0113.
R. Kumaresan—Supported by Qatar Computing Research Institute and DARPA Grant number FA8750-11-2-0225. Work done in part while at the Technion.

© International Association for Cryptologic Research 2015
T. Iwata and J.H. Cheon (Eds.): ASIACRYPT 2015, Part I, LNCS 9452, pp. 386–412, 2015.
DOI: 10.1007/978-3-662-48797-6_17

seen significant improvements. The main technique for securing garbled circuit protocols against malicious adversaries is cut-and-choose, formalized and proven secure by Lindell and Pinkas [23]. A line of work [11,22–24,26,31] has focused on reducing the concrete overhead of the cut-and-choose approach: it is possible to guarantee probability of cheating $\leq 2^{-\sigma}$ using exactly σ garbled circuits.

The above works have been motivated by the impression that the major overhead of secure two-party computation arises from the generation, transmission, and evaluation of garbled circuits (especially for functions having large circuit size). Indeed, the work of Frederiksen and Nielsen [7] showed that the cost of the circuit communication and computation for oblivious two-party AES is approximately 80 % of the total cost; likewise, Kreuter et al. [19] showed that the circuit generation and evaluation for large circuits takes 99.999 % of the execution time.

Recent works of [10,21] consider the multiple-execution setting, where two parties compute the same function on possibly different inputs either in parallel or sequentially. These works show that to evaluate the same function t times, it is possible to reduce the number of garbled circuits to $O(\sigma/\log t)$. In concrete terms, this corresponds to a drastic reduction in the number of garbled circuits. For instance when $t = 3500$ and for $\sigma = 40$, the work of [10,21] shows a cut-and-choose technique that reduces the number of garbled circuits to less than 8 per execution. Thus it is reasonable to say that the overhead due to generation, transmission, and evaluation of garbled circuits has been significantly reduced.

However, state-of-the-art two-party secure computation protocols, both in the single-execution setting [22] and in the multiple-execution setting [10], suffer from major overheads due to use of public key operations for two reasons:

- Use of DDH-based zero-knowledge protocols to enforce circuit-generator's *input consistency*.
- Use of DDH-based cut-and-choose oblivious transfer protocols [5,10,21,22,24] to avoid "selective failure" attacks.

Of greater concern is the fact that these state-of-the-art protocols are unlikely to perform well in settings where the inputs of *even one of the parties* are large (because they use public key operations proportional to the total size of inputs of both parties). It is worthwhile to note that although techniques, most notably amortization via oblivious transfer (OT) extension [12,14,29], exist to reduce the number of public key operations required at least for one of the parties, the state-of-the-art two-party secure computation protocols simply are not able to take advantage of these amortization techniques.

If one restricts their attention to constant-round protocols with good concrete efficiency there are very few alternatives [23,26] that require reduced number of public key operations. For instance the protocols of [23,26] use public key operations only for the (seed) OTs (which can be amortized using OT extension). Furthermore, at least in the *single execution setting*, the techniques of [23,26] can be easily merged with state-of-the-art cut-and-choose techniques to reduce the number of public key operations. However, this results in a considerable overhead in the communication complexity (by factor σ) for proving input consistency of the circuit generator. More importantly the techniques of [23,26] do not adapt well to the state-of-the-art cut-and-choose techniques for the multiple executions setting, and require strong assumptions such as a *programmable* random oracle. Specifically, the "XOR-tree encoding schemes" technique employed in [23,26] to avoid the selective failure attack no longer appears to work with standard garbling techniques. On the other hand, a natural generalization of cut-and-choose OT, namely *multistage cut-and-choose OT* proposed in [10,21] *can* handle the selective failure attack in the multiple executions setting (cf. Sect. 1.2).

Unfortunately the only known constructions of cut-and-choose OT as well as its variants rely on DDH and consequently use public key operations proportional to the size of the cut-and-choose OT instance. This is further amplified by the fact that known cut-and-choose OT protocols require *regular exponentiations* which are more expensive relative to even fixed-base exponentiations. (Note that on the other hand the DDH-based zero-knowledge protocols to ensure input consistency used in [22,24] require only fixed-base exponentiations.)

Our Contributions. In this paper, we study cut-and-choose OT and its variants as independently interesting primitives. Motivated by the discussion above, our main goal will be to reduce the number of public key operations required to realize a cut-and-choose OT instance, while minimizing the concrete communication complexity. Towards this, we propose a new formulation of cut-and-choose OT and its variants that (1) is sufficient for its application to design secure two-party computation protocols, (2) allows a realization in an OT-hybrid model (as opposed to specific public key cryptosystems, and also provides alternative realizations which are resistant to quantum attacks), and (3) can be realized with low communication complexity in both concrete terms (roughly factor 4 overhead) as well as asymptotic terms. Furthermore, our formulation provides new insights into the design of multistage cut-and-choose OT protocols resulting in new constructions of the same that offer factor t (where t is the number of executions) improvement over prior work [10]. Note that the benefits of amortization in the multiple execution setting kick in for large t (e.g., 10X improvement when $t = 10^6$). Hence our protocols can offer significant gains in efficiency. Conceptually, our work can be considered as

- Pinning down the exact formulation of cut-and-choose OT and its variants that suffices for its applications.
- Basing cut-and-choose OT on a wide variety of assumptions (including LWE, RSA, DDH).
- Showing how to efficiently "extend" cut-and-choose OT (a la OT extension).
- An approach for porting "XOR-tree encoding schemes" to work in the multiple execution setting while preserving their efficiency.

Our new formulation of cut-and-choose OT has the following aspects:

- Treats cut-and-choose OT (and its variants) as a *reactive* functionality. This allows us to construct efficient protocols for *multistage* cut-and-choose OT.
- Requires ideal process simulation for corrupt receiver but *only privacy against corrupt sender*. This will allow us to realize cut-and-choose OT (and its variants) with low concrete communication complexity.

1.1   Cut-and-Choose Oblivious Transfer and Its Variants

We provide an overview of cut-and-choose OT and its variants. In the following, let λ (resp. σ) denote the computational (resp. statistical) security parameter.

Cut-and-Choose Oblivious Transfer. Cut-and-choose oblivious transfer (CCOT) [24], denoted $\mathcal{F}_{ccot}$ (see Fig. 1) is an extension of standard OT. The sender inputs n pairs of strings, and the receiver inputs n selection bits to select one string out of each pair of sender strings. However, the receiver also inputs a set J of size $n/2$ that consists of indices where it wants *both* the sender's inputs to be revealed. Note that for indices not contained in J, only those sender inputs that correspond to the receiver's selection bits are revealed.

Remark 1. Using a PRG it is possible to obtain OT on long strings given ideal access to OT on short strings of length λ [12]. This length extension technique is applicable to cut-and-choose and its variants. Furthermore, for applications to secure computation, sender input strings (i.e., garbled circuit keys) are of length λ. Therefore, we assume wlog that sender input strings are all of length λ.

Inputs:
- S inputs n pairs of strings $(x_{1,0}, x_{1,1}), \ldots, (x_{n,0}, x_{n,1}) \in \{0,1\}^{\lambda} \times \{0,1\}^{\lambda}$.
- R inputs a set of indices $J \subseteq [n]$ of size $n/2$, and selection bits $\{b_j\}_{j \notin J}$.

Outputs: If J is not of size $n/2$, then S and R receive $\perp$ as output.
- For every $j \in J$, party R receives $(x_{j,0}, x_{j,1})$.
- For every $j \notin J$, party R receives x_{j,b_j}.

Fig. 1. The cut-and-choose OT functionality $\mathcal{F}_{ccot}$ from [24].

Batch Single-Choice CCOT. In applications to secure computation, one needs *single-choice* CCOT, where the receiver is restricted to inputting the *same* selection bit in all the $n/2$ instances where it receives exactly one out of two sender strings. Furthermore, it is crucial that the subset J input by the receiver is the same across each instance of single-choice CCOT. This variant, called *batch single-choice* CCOT can be efficiently realized under DDH [24].

Modified Batch Single-Choice CCOT. Lindell [22] presented a variant of batch single-choice CCOT, denoted $\mathcal{F}_{ccot}^{\star}$, to address settings where the receiver's input set J may be of arbitrary size (i.e., not necessarily $n/2$). In addition to obtaining one of the sender's inputs, the receiver also obtains a "check value" for each index not in J. This variant can be realized under DDH [22].

Multistage CCOT. To handle the multiple (parallel) execution setting, a new variant of $\mathcal{F}_{ccot}^{\star}$ called batch single-choice *multi-stage* cut-and-choose oblivious transfer was proposed in [10]. For sake of simplicity, we refer to this primitive as *multistage cut-and-choose oblivious transfer* and denote it by $\mathcal{F}_{mcot}^{\star}$. At a high

---

Inputs:

- S inputs m sets of n pairs of strings $X^{(1)}, \ldots, X^{(m)}$ where $X^{(i)} = ((x_{1,0}^{(i)}, x_{1,1}^{(i)}),$ $\ldots, (x_{n,0}^{(i)}, x_{n,1}^{(i)}))$, and t "check values" vectors $\Phi^1 = (\phi_1^1, \ldots, \phi_n^1), \ldots, \Phi^t = (\phi_1^t,$ $\ldots, \phi_n^t)$, where each $x_{j,b}^{(i)} \in \{0,1\}^\lambda$ and each $\phi_j^k \in \{0,1\}^\sigma$.
- R inputs pairwise non-intersecting sets of indices $E_1, \ldots, E_t \subseteq [n]$, and selection bit vectors $\mathbf{b}_1 = (b_{1,1}, \ldots, b_{1,m}), \ldots, \mathbf{b}_t = (b_{t,1}, \ldots, b_{t,m})$.

Outputs: S receives no output. Define $J = [n] \backslash \cup_{k \in [t]} E_k$. R receives the following:

- For every $j \in J$, party R receives $\{ (x_{j,0}^{(i)}, x_{j,1}^{(i)}) \}_{i \in [m]}$.
- For every $k \in [t]$: For each (unique) $j \in E_k$, party R receives $\{x_{j,b_{k,i}}^{(i)}\}_{i \in [m]}$, and "check value" ϕ_j^k.

I.e., R obtains $\{(x_{j,0}^{(1)}, x_{j,1}^{(1)}), \ldots, (x_{j,0}^{(m)}, x_{j,1}^{(m)})\}_{j \in J}$ and $\{x_{j,b_{1,1}}^{(1)}, \ldots, x_{j,b_{1,m}}^{(m)}\}_{j \in E_1}, \ldots,$ $\{x_{j,b_{t,1}}^{(1)}, \ldots, x_{j,b_{t,m}}^{(m)}\}_{j \in E_t}$ and check values $\{\phi_j^1\}_{j \in E_1}, \ldots, \{\phi_j^t\}_{j \in E_t}$.

Fig. 2. Multistage cut-and-choose OT functionality $\mathcal{F}_{\text{mcot}}^\star$ [10,21].

level, this variant differs from $\mathcal{F}_{\text{ccot}}^\star$ in that the receiver can now input *multiple* sets $E_1, \ldots, E_t$ (where J is now implicitly defined as $[n] \setminus \cup_{k \in [t]} E_k$), and make independent selections for each $E_1, \ldots, E_t$. In fact the above definition reflects the cut-and-choose technique employed in [10,21] for the multiple execution setting. The technique proceeds by first choosing a subset of the n garbled circuits to be checked, and then partitioning the remaining garbled circuits into t evaluation "buckets". An information-theoretic reduction of $\mathcal{F}_{\text{mcot}}^\star$ to t instances of $\mathcal{F}_{\text{ccot}}^\star$ with total communication cost $O(nt^2\lambda)$ was shown in [10].

For lack of space, we present only the multistage cut-and-choose OT functionality in Fig. 2. Note that $\mathcal{F}_{\text{mcot}}^\star$ generalizes *modified batch single-choice CCOT* of [22] (simply by setting $t = 1$) as well as *batch single-choice CCOT* of [24] (by setting $t = 1$ and forcing $|J| = n/2$ and setting all ϕ_j^k values to 0^σ).

1.2 Selective Failure Attacks

In garbled circuit protocols, OT is used to enable the circuit generator (referred to as the sender) S to transfer input keys for the garbled circuit corresponding to the circuit evaluator (referred to as the receiver) R's inputs. However, when S is malicious, this can lead to a "selective failure" attack. To explain this problem in more detail, consider the following naïve scheme. For simplicity assume that R has only one input bit b. Let the keys corresponding to R's input be $(x_{j,0}, x_{j,1})$ in the j-th garbled circuit. In the following, let com be a commitment scheme.

- S sends $(\text{com}(x_{1,0}), \text{com}(x_{1,1})), \ldots, (\text{com}(x_{n,0}), \text{com}(x_{n,1}))$ to R.
- S and R participate in a *single* instance of $\mathcal{F}_{\text{OT}}$ where S's input $((d_{1,0}, \ldots, d_{n,0}), (d_{1,1}, \ldots, d_{n,1}))$ where $d_{j,c}$ is the decommitment corresponding to $\text{com}(x_{j,c})$, and R's input is b. R obtains $(d_{1,b}, \ldots, d_{n,b})$ from $\mathcal{F}_{\text{OT}}$.
- Then R sends check indices $J \subseteq [n]$ to S.
- S sends $\{d_{j,0}, d_{j,1}\}_{j \in J}$ to R.

The selective failure attack operates in the following way: S supplies $(d_{1,0}, \ldots, d_{n,0}), (d'_{1,1}, \ldots, d'_{n,1})$ where $d_{i,0}$ is a valid decommitment for $\mathsf{com}(x_{i,0})$ while $d'_{i,1}$ is not a valid decommitment for $\mathsf{com}(x_{i,1})$. Then when R sends check indices, S responds with $\{d_{j,0}, d_{j,1}\}_{j \in J}$ where $d_{j,0}$ and $d_{j,1}$ are valid decommitments for $\mathsf{com}(x_{j,0})$ and $\mathsf{com}(x_{j,1})$ respectively. Suppose R's input equals 0. In this case, R does not detect any inconsistency, and continues the protocol, and obtains output. Suppose R's input equals 1. Now R will not obtain $x_{j,1}$ for all $j \in [n]$ since it receives invalid decommitments. If R aborts then S knows that R's input bit equals 1. In any case, R cannot obtain the final output. I.e., the ideal process and the real process can be distinguished when R's input equals 1, and the protocol is insecure since S can force an abort depending on R's input.

Approaches Based on "XOR-Tree Encoding Schemes". The first solution to the selective failure attack was proposed in [6,15,23] where the idea was to randomly encode R's input and then augment the circuit with a supplemental subcircuit (e.g., "XOR-tree") that performs the decoding to compute R's actual input. Note that the "selective failure"-type attack can still be applied by S but the use of encoding ensures that the event that R aborts due to the attack is almost statistically independent of its actual input. The basic XOR-tree encoding scheme incurs a multiplicative overhead of σ in the number of OTs and increases the circuit size by σ XOR gates. The "random combinations" XOR-tree encoding [23,25,30] incurs a total overhead of $m' = \max(4m, 8\sigma)$ in the number of OTs where m is the length of R's inputs, and an additional $0.3\,mm'$ XOR gates. (Note that use of the free-XOR technique [18] can lead to nullifying the cost of the additional XOR gates.) [13] uses σ-wise independent generators to provide a rate-1 encoding of inputs which can be decoded using an $\mathbf{NC}^0$ circuit.

Approaches Based on CCOT. CCOT forces S to "commit" to all keys corresponding to R's input and reveals a subset of these keys corresponding to R's input but without the knowledge of which subset of keys were revealed. This allows us to intertwine the OT and the circuit checks and avoids the need to augment the original circuit with a supplemental decoding subcircuit. I.e., selective failure attacks are "caught" along with check for incorrectly constructed circuits, and this results in a simpler security analysis.

Approaches for the Multiple Execution Setting. While either approach seems sufficient to solve the selective failure attack, the CCOT based approach offers a qualitative advantage in the multiple parallel execution setting. First let us provide an overview of the cut-and-choose technique in the multiple execution setting [10,21]. S sends n garbled circuits, and R picks a check set $J \subseteq [n]$. The garbled circuits corresponding to check sets will eventually be opened by S. The garbled circuits which are not check circuits are *randomly* partitioned into t evaluation "buckets" denoted by $E_1, \ldots, E_t$. We now explain the difficulty in adapting XOR-tree encoding schemes to this cut-and-choose technique.

Observe that when using standard garbling schemes [23,33] in a 2-party garbled circuits protocol, the OT step needs to be carried out before the garbled circuits are sent. This is necessary for the simulator to generate correctly faked garbled circuits (using R's inputs extracted from the OT) in the simulation for corrupt R. For simplicity assume that R has exactly one input bit (which may vary across different executions). Now when using XOR-tree encoding schemes we need to enforce that in each execution, R inputs the same choice in all the OTs. Batching the OTs together for each execution can be implemented if S knows which circuits are going to be evaluation circuits for each execution, but R cannot reveal which circuits are evaluation circuits because this allows a corrupt S to transmit well-formed check circuits and ill-formed evaluation circuits. Thus it is unclear how to apply the XOR-tree encoding schemes and ensure that corrupt R chooses the same inputs for the evaluation circuits within an execution.

A generalization of CCOT called *multistage CCOT* (Fig. 2) is well-suited to the multiple parallel execution setting. Indeed, multistage CCOT $\mathcal{F}^{\star}_{\mathrm{mcot}}$ takes as inputs (1) from S: all input keys corresponding to R's inputs in each of the n garbled circuits, and (2) from R: the sets $E_1, \ldots, E_t$ along with independent choice bits for each of the t executions. Thus $\mathcal{F}^{\star}_{\mathrm{mcot}}$ avoids the selective failure attack in the same way as CCOT does it in the single execution setting. Further, it ensures that R is forced to choose the same inputs within each execution.

Remark 2. Surprisingly, CCOT has a significant advantage over XOR-tree encoding schemes only in the parallel execution setting. In the sequential execution setting, it is unclear how to use CCOT since R's inputs for each of its executions are not available at the beginning of the protocol. It appears necessary to do the OT for each execution after all the garbled circuits are sent. Then one may use *adaptively secure garbling schemes* [2,3] (e.g., in the *programmable* random oracle model) to enable the simulator to generate correctly faked garbled circuits in the simulation for corrupt R. Assuming that the garbling is adaptively secure, XOR-tree encoding schemes suffice to circumvent the selective failure attack in the multiple sequential setting. This also applies to the multiple parallel setting.

1.3   Overview of Definitions and Constructions

As mentioned in the Introduction, all known constructions of CCOT rely on DDH and thus make heavy use of public key operations. A natural approach to remedy the above situation is try and construct CCOT in a OT-hybrid model and then use OT extension techniques [12,29].

Basing CCOT on OT. A first idea is to use general OT-based 2PC (e.g., [14]) to realize CCOT but it is not clear if this would result in a CCOT protocol with good concrete efficiency. Note that the circuit implementing CCOT has very small depth, and that S's inputs are of length $O(n\lambda)$ while R's is of length $O(n)$ (where the big-Oh hides small constants). Protocols of [23,26] do not perform well since there's a multiplicative overhead of (at least) $\lambda\sigma$ over the instance size (i.e., $O(n\lambda)$) simply because of garbling (factor λ) and cut-and-choose (factor σ). Protocols of [10,22,24] already rely on CCOT and the instance size of CCOT required inside these 2PC protocols are larger than the CCOT instance we wish to realize. Since the circuit has very small constant depth it is possible to employ non-constant round solutions [29] but this still incurs a factor λ overhead due to use of authenticated OTs. Employing information-theoretic garbled circuit variants [15,17] in the protocols of [23,26] still incur a factor σ overhead due to cut-and-choose. In summary, none of the above are satisfactory for implementing CCOT as they incur at least concrete factor $\min(\lambda, \sigma)$ multiplicative overhead.

To explain the intuition behind our definitions and constructions, we start with the seemingly close relationship between CCOT and 2-out-of-3 OT. At first glance, it seems that it must be easy to construct CCOT from 2-out-of-3 OT. For example, for each index, we can let S input the pair of real input keys along with a "dummy check value" as its 3 inputs to 2-out-of-3 OT, and then let R pick two out of the three values (i.e., both keys if it's a check circuit, or the dummy check value along with the key that corresponds to R's real input). There are multiple issues with making this idea work in the presence of malicious adversaries. Perhaps the most important issue is that this idea still wouldn't help us achieve our goal of showing a reduction from CCOT to 1-out-of-2 OT. More precisely, we do not know how to construct efficient protocols for 2-out-of-3 OT from 1-out-of-2 OT. Consider the following toy example for the same.

INPUTS: S holds (x_0, x_1, x_2) and R holds $b_1 \in \{0, 2\}, b_2 \in \{1, 2\}$.

TOY PROTOCOL:

- S sends (x_0, x_2) to $\mathcal{F}_{\mathrm{OT}}$ and R sends b_1 to $\mathcal{F}_{\mathrm{OT}}$.
- S sends (x_1, x_2) to $\mathcal{F}_{\mathrm{OT}}$ and R sends b_2 to $\mathcal{F}_{\mathrm{OT}}$.

OUTPUTS: S outputs nothing. R outputs x_{b_1}, x_{b_2}.
The problem with the protocol above is that simulation extraction will fail with probability $1/2$ since a malicious S may input different values for x_2 in each of the two queries to $\mathcal{F}_{\mathrm{OT}}$. Note that even enforcing S to send $h = \tilde{H}(x_2)$ to R where $\tilde{H}$ is a collision-resistant hash function (or an extractable commitment) does not help the simulator. On the other hand this hash value does enable R to detect an inconsistency if (1) S supplied two different values for x_2 in each of the two queries to $\mathcal{F}_{\mathrm{OT}}$ and (2) R picked the x_2 value which is not consistent with h. However, if R aborts on detection of inconsistency this leaks information.

Our main observation is that the attacks on the toy protocol are very similar to the selective failure attacks discussed in Sect. 1.2. Motivated by this one may attempt to use "XOR-tree encoding schemes" to avoid the selective failure attacks, and attempt to construct CCOT directly from 1-out-of-2 OT. However, note that the encoding schemes alone do not suffice to prevent selective failure attacks; they need to be used along with a supplemental decoding circuit. Here our main observation is that known encoding schemes (possibly with the exception of [32]) used to prevent selective failure attacks [13,23] can be decoded using (a circuit that performs) only XOR operations. Thus, one may use the free-XOR technique [18] to get rid of the need for a supplemental decoding circuit, and instead perform XOR operations directly on strings. Indeed the above idea can be successfully applied to prevent selective failure attacks that could be mounted on the toy protocol, and can also be extended to yield a protocol for

CCOT. Although the resulting CCOT protocol is simulatable against a malicious receiver, unfortunately we do not know how to simulate a corrupt sender (specifically, extract sender's input).

Relaxing CCOT. Our main observation is that for application to 2PC, full simulation against a corrupt sender is not required. It is only privacy that is required. This is because S's inputs to the 2PC are extracted typically via ZK (or the mechanism used for input consistency checks), and the inputs to the CCOT are just random garbled keys which are unrelated to its real input. Note that in 2PC protocols that use CCOT [10,22,24] the following three steps happen after the CCOT protocol is completed: (1) S sends all the garbled circuits, and (2) then R reveals the identity of the evaluation circuits, and (3) then S reveals its keys corresponding to its input for the evaluation circuits. Consider the second step mentioned above, namely that R reveals the identity of the evaluation circuits. This is a relatively subtle step since a malicious R may claim (a) that a check circuit is an evaluation circuit, or (b) that an evaluation circuit is a check circuit. Both these conditions need to be handled carefully since in case (a) corrupt R, upon receiving S's input keys in step (3) will be able to evaluate the garbled circuits on several inputs of its choice. Case (b) is problematic while simulating a corrupt R as the simulator does not know which circuits to generate correctly and which ones to fake. Therefore, 2PC protocols that use CCOT require R to "prove" the identity of the check/evaluation circuits. In [10,22], this is done via "check values" and "checkset values". We use similar ideas in our protocols: if $j \in [n]$ is such that $j \notin J$, then R receives some dummy check value ϕ_j, and if $j \in J$ then R receives "checkset values" $x_{j,0}, x_{j,1}$ which correspond to S's inputs. Thus, R can prove the identity of check/evaluation circuits simply by sending the "check values" $\{\phi_j\}_{j\notin J}$ and "checkset values" $\{x_{j,0}, x_{j,1}\}_{j\in J}$. Observe that this step does not reveal any information about R's input bits $\{b_j\}_{j\notin J}$ to S. To do this, we would need to include a "reveal" step.

Motivated by the discussions above, we formulate a new definition for CCOT and its variants. Our definitions pose CCOT and its variants as *reactive functionalities*, and in particular include a "reveal phase" where R's evaluation set $[n] \setminus J$ is simply revealed to S by the functionality. More precisely, in the reveal phase we allow R to decide whether it wants to abort or reveal J. Note that for the case of $\mathcal{F}^{\star}_{\mathrm{mcot}}$, the evaluation sets $E_1, \ldots, E_t$ is revealed to S by the functionality. This in particular allows us to eliminate the "check values" in the definitions of $\mathcal{F}^{\star}_{\mathrm{ccot}}$ [22] and $\mathcal{F}^{\star}_{\mathrm{mcot}}$ [10], and allows us to present protocols for (the reactive variant of) $\mathcal{F}^{\star}_{\mathrm{mcot}}$ that is more efficient than prior constructions [10]. We formulate CCOT as a reactive functionality because step (1) where S sends all the garbled circuits happens immediately after the CCOT step and before step (2) where R reveals the identity of the evaluation circuits. It is easy to see that this relaxed formulation suffices for applications to secure computation.

Discussion. Such relaxed definitions, in particular requiring only privacy against corrupt sender, is not at all uncommon for OT and its variants (cf. [1,28]) or PIR (cf. [4,20]). Similarly, [8] propose "keyword OT" protocols in a client-server setting, and require one to simulate the server's (which acts as the sender) view alone, without considering its joint distribution with the honest client's output. For another example, consider [11] who use a CDH-based OT protocol that achieves privacy (but is not known to be simulatable) against a malicious sender, and yet this suffices for their purposes to construct efficient 2PC protocols.

2   Definitions

We formulate CCOT and its variants as *reactive* functionalities and provide relaxed definitions formally. Recall that the main differences from prior formulations is that we require (1) only privacy against corrupt sender, and (2) R to provide the check set J and evaluation sets $E_1, \ldots, E_t$ to S at the end of the protocol. We emphasize that privacy against corrupt sender must hold even after $J, E_1, \ldots, E_t$ is revealed. Due to space constraints we describe our new formulation only for the case of multistage CCOT denoted $\mathcal{F}^+_{\mathrm{mcot}}$ in Fig. 3. (The extensions to all other variants is straightforward.)

Input phase:
- S inputs m sets of n pairs of strings $X^{(1)}, \ldots, X^{(m)}$ where $X^{(i)} = ((x^{(i)}_{1,0}, x^{(i)}_{1,1}), \ldots, (x^{(i)}_{n,0}, x^{(i)}_{n,1}))$.
- R inputs pairwise non-intersecting sets of indices $E_1, \ldots, E_t \subseteq [n]$, and selection bits $\{b_{k,i}\}_{k \in [t], i \in [m]}$.

Output phase: S receives no output. Define $J = [n] \setminus \cup_{k \in [t]} E_k$.
- For every $j \in J$, party R receives $\{ (x^{(i)}_{j,0}, x^{(i)}_{j,1}) \}_{i \in [m]}$.
- For every $k \in [t]$: For each (unique) $j \in E_k$, party R receives $\{x^{(i)}_{j,b_{k,i}}\}_{i \in [m]}$.

Reveal Phase. Upon receiving "reveal" from R, sender S receives $E_1, \ldots, E_t$.

Fig. 3. The reactive multistage CCOT functionality $\mathcal{F}^+_{\mathrm{mcot}}$.

We will be using the following definitions (loosely based on analogous definitions for keyword OT [8]) for CCOT as well as its variants. Therefore for convenience we will define these as security notions for an arbitrary functionality F, and then in our theorem statements we will refer to F as being CCOT or one of its variants.

Definition 1 (Correctness). *If both parties are honest, then, after running the protocol on inputs (X, Y), the receiver outputs Z such that $Z = F(X, Y)$.*

Definition 2 (Receiver's privacy: indistinguishability). *Let σ be a statistical security parameter. Then, for any PPT S' executing the sender's part and for any inputs X, Y, Y', the statistical distance between the views that S' sees on input X, in the case that the receiver inputs Y and the case that it inputs Y' is bound by $2^{-\sigma + O(1)}$.*

Definition 3 (Sender's privacy: comparison with the ideal model). *For every* PPT *machine R' substituting the receiver in the real protocol, there exists a* PPT *machine R'' that plays the receiver's role in the ideal implementation, such that on any inputs (X, Y), the view of R' is computationally indistinguishable from the output of R''. (In the semi-honest model $R' = R$.)*

Definition 4. *A protocol π securely realizes functionality F with sender-simulatability and receiver-privacy if it satisfies Definitions 1, 2, and 3.*

XOR-Tree Encoding Schemes. Selective failure attacks essentially correspond to letting a corrupt sender learn a disjunctive predicate of the receiver's input. We define an XOR-tree encoding scheme consisting of a tuple $(\mathsf{En}, \mathsf{De}, \mathsf{En}', \mathsf{De}')$ of randomized algorithms (implicitly parameterized with statistical security parameter σ, and possibly public randomness ω_0) as satisfying:

1. Algorithm En takes input $\{(x_0^i, x_1^i)\}_{i \in [m]}$ and produces pairs of random λ-bit strings $\{u_0^\ell, u_1^\ell\}_{\ell \in [m']}$ s.t. for each $\ell, \ell' \in [m']$, it holds that $u_0^{\ell'} \oplus u_1^{\ell'} = u_0^\ell \oplus u_1^\ell$.
2. Algorithm En' takes input $\mathbf{b} = (b_1, \ldots, b_m) \in \{0,1\}^m$ and outputs $\{b_\ell'\}_{\ell \in [m']}$.
3. For every $\mathbf{b} = (b_1, \ldots, b_m) \in \{0,1\}^m$ and every $\{(x_0^i, x_1^i)\}_{i \in [m]}$ it holds that

$$\Pr\left[\begin{array}{l} \{b_\ell'\}_{\ell \in [m']} \leftarrow \mathsf{En}'(\mathbf{b}); \\ \{(u_0^\ell, u_1^\ell)\}_{\ell \in [m']} \leftarrow \mathsf{En}(\{(x_0^i, x_1^i)\}_{i \in [m]}) \end{array} : \mathsf{De}(\{u_{b_\ell'}^\ell\}_{\ell \in [m']}) = \{x_{b_i}^i\}_{i \in [m]} \right] = 1.$$

We sometimes abuse notation and allow De to take sets of pairs of strings as input in which case we require that for every $\{(x_0^i, x_1^i)\}_{i \in [m]}$ it holds that

$$\Pr\left[\begin{array}{l} \{(u_0^\ell, u_1^\ell)\}_{\ell \in [m']} \leftarrow \mathsf{En}(\{(x_0^i, x_1^i)\}_{i \in [m]}) : \\ \mathsf{De}(\{(u_0^\ell, u_1^\ell)\}_{\ell \in [m']}) = \{(x_0^i, x_1^i)\}_{i \in [m]} \end{array}\right] = 1.$$

4. For every $\mathbf{b}$, it holds that $\Pr[\mathsf{De}'(\mathsf{En}'(\mathbf{b})) = \mathbf{b}] = 1$.
5. Algorithms $\mathsf{De}, \mathsf{De}'$ can be implemented by using (a tree of) XOR gates only.
6. For every disjunctive predicate $P(\cdot)$, the following holds: (1) If P involves at most $\sigma - 1$ literals, then $\Pr[P(\mathsf{En}'(\mathbf{b})) = 1]$ is completely independent of $\mathbf{b}$. (2) Otherwise, $\Pr[P(\mathsf{En}'(\mathbf{b})) = 1] \geq 1 - 2^{-\sigma + 1}$.
7. For every $\{(x_0^i, x_1^i)\}_{i \in [m]}$ and for every (possibly unbounded) adversary $\mathcal{A}'$ and for every $\{b_\ell'\}_{\ell \in [m']} \in \{0,1\}^{m'}$, there exists a PPT algorithm $\mathcal{S}'$ such that the following holds:

$$\Pr[\{(u_0^\ell, u_1^\ell)\}_{\ell \in [m']} \leftarrow \mathsf{En}(\{(x_0^i, x_1^i)\}_{i \in [m]}) : \mathcal{A}'(\{b_\ell'\}_{\ell \in [m']}, \{u_{b_\ell'}^\ell\}_{\ell \in [m']}) = 1] =$$

$$\Pr\left[\begin{array}{l} (b_1, \ldots, b_m) \leftarrow \mathsf{De}'(\{b_\ell'\}_{\ell \in [m']}); \\ \{\tilde{u}^\ell\}_{\ell \in [m']} \leftarrow \mathcal{S}'(\{b_\ell'\}_{\ell \in [m']}, \{x_{b_i}^i\}_{i \in [m]}) \end{array} : \mathcal{A}'(\{b_\ell'\}_{\ell \in [m']}, \{\tilde{u}^\ell\}_{\ell \in [m']}) = 1 \right].$$

(This in particular, implies that $\mathcal{A}$ obtains no information about $\{x_{1-b_i}^i\}_{i \in [m]}$.)

Algorithms $(\mathsf{En}, \mathsf{De}, \mathsf{En}', \mathsf{De}')$ for the basic XOR-tree encoding scheme [23] are simple and implicit in our basic CCOT construction (cf. Fig. 4). For the random combinations XOR-tree encoding [23] algorithm En' is simply a random linear mapping (i.e., public randomness ω_0 defines this random linear mapping, see e.g., [23,30] for more details). Finally, for the σ-wise independent generators XOR-tree encoding the algorithm En' depends on the generator (i.e., public randomness ω_0 defines this generator) which can be implemented only using XOR gates [27]. Note that in all of the above, En' essentially creates a $(\sigma - 1)$-independent encoding of its input, and thus Property 6 holds (see also Lemma 1). In all our constructions, En simply maps its inputs to a pairs of random strings such that the XOR of the two strings within a pair is always some fixed Δ. Algorithms $\mathsf{De}, \mathsf{De}'$ are deterministic and function to simply reverse the respective encoding algorithms $\mathsf{En}, \mathsf{En}'$. Note that $\mathsf{De}, \mathsf{De}'$ (acting respectively on outputs of $\mathsf{En}, \mathsf{En}'$) are naturally defined by the supplemental decoding circuit that decodes the XOR-tree encoding, and thus can be implemented using XOR gates only. We point out that algorithm De' is used only in the simulation to extract R's input from its XOR-tree encoded form. Finally, Property 7 is justified by the fact that XOR-tree encoding schemes that are useful in standard two-party secure computation protocols, the receiver R obtains only one of two keys corresponding to the encoding (via OTs), and these keys reveal the output keys of the supplemental decoding circuit (that correspond exactly to the output of the decoding) and nothing else.

3 Constructions

CCOT FROM OT. See the protocol in Fig. 4 for the CCOT protocol that uses the basic XOR-tree encoding scheme of [23] in order to implement CCOT when $n = 1$. The case when $n \geq 1$ is handled by parallel repetition. While we prove that the resulting CCOT protocol is simulatable against a malicious receiver, unfortunately we do not know how to extract corrupt sender's input. To see this, note that a corrupt sender may supply values for some $\ell, \ell' \in [\sigma]$ values $u_0^\ell, u_1^\ell, u_0^{\ell'}, u_1^{\ell'}$ such that $u_0^\ell \oplus u_1^\ell \neq u_0^{\ell'} \oplus u_1^{\ell'}$. Needless to say, such a deviation is caught by R when $J \neq \emptyset$. However, this deviation goes undetected when R's input $J = \emptyset$. Note that the simulator for a corrupt sender needs to extract S's input without knowing R's input; however when S provides inconsistent inputs to $\mathcal{F}_{OT}$, it is unclear how to extract S's inputs. We prove that the protocol in Fig. 4 securely realizes $\mathcal{F}_{ccot}$ with sender-simulatability and receiver-privacy. We start by observing that correctness follows from inspection of the protocol.

Simulating Corrupt Receiver. Assume that H is modeled as a (non-programmable) random oracle. Acting as $\mathcal{F}_{OT}$ the simulator does the following:

- Chooses random $\Delta', \{u_0^\ell, u_2^\ell\}_{\ell \in [\sigma]}$ and sets for all $\ell \in [\sigma]$, value $u_1^\ell = u_0^\ell \oplus \Delta'$.
- For each $\ell \in [\sigma]$, acting as $\mathcal{F}_{OT}$ obtain values $\{c_0^\ell, c_1^\ell\}$ and return answers from $\{u_0^\ell, u_1^\ell, u_2^\ell\}$ exactly as in the protocol.
- If there exists $\ell \in [\sigma]$ such that $c_0^\ell = c_1^\ell = 0$, then set $J = \{1\}$ and send J to the trusted party and receive back (x_0, x_1). Now set $\Delta_0' = x_0 \oplus \bigoplus_\ell u_0^\ell$ and $\Delta_1' = x_1 \oplus \Delta' \oplus \bigoplus_\ell u_0^\ell$. Pick random Δ_ϕ' and random $h' \leftarrow \{0,1\}^\lambda$. Finally, send $\Delta_0', \Delta_1', \Delta_\phi', h'$ to R.

Inputs: S holds (x_0, x_1) and R holds $J \subseteq \{1\}$ and b_1.

Protocol:

1. S chooses check value $\phi \in \{0,1\}^\sigma$ at random. Then S chooses $\Delta, \Delta'_0, \Delta'_\phi, \{(u_0^\ell, u_2^\ell)\}_{\ell \in [\sigma]}$ at random such that $\bigoplus_\ell u_0^\ell = x_0 \oplus \Delta'_0$ and $\bigoplus_\ell u_2^\ell = \phi \oplus \Delta'_\phi$. Then S sets for all $\ell \in [\sigma]$, $u_1^\ell = u_0^\ell \oplus \Delta$.

2. R sets $\{(c_0^\ell, c_1^\ell)\}_{\ell \in [\sigma]}$ as follows:
   - If $J \neq \emptyset$, then set $c_0^\ell = c_1^\ell = 0$ for all $\ell \in [\sigma]$.
   - Else choose $\{b'_\ell\}_{\ell \in [\sigma]}$ at random such that $\bigoplus_\ell b'_\ell = b_1$, and for each $\ell \in [\sigma]$ set $c_{b'_\ell}^\ell = 0$ and $c_{1-b'_\ell}^\ell = 1$.

3. Then for $\ell \in [\sigma]$, S and R do the following:
   - S sends (u_0^ℓ, u_2^ℓ) to $\mathcal{F}_{OT}$, and R sends c_0^ℓ to $\mathcal{F}_{OT}$.
   - S sends (u_1^ℓ, u_2^ℓ) to $\mathcal{F}_{OT}$, and R sends c_1^ℓ to $\mathcal{F}_{OT}$.

   Note R receives $\{(u_0^\ell, u_1^\ell)\}_{\ell \in [\sigma]}$ if $J \neq \emptyset$, and otherwise receives $\{(u_{b'_\ell}^\ell, u_2^\ell)\}_{\ell \in [\sigma]}$.

4. S sends $\Delta'_0, \Delta'_1 = x_0 \oplus x_1 \oplus \Delta \oplus \Delta'_0, \Delta'_\phi$, and $h = H(\phi)$ to R.

5. If $J \neq \emptyset$, then R reconstructs $\tilde{x}_0 = \Delta'_0 \oplus \bigoplus_\ell u_0^\ell$ and $\tilde{x}_1 = \Delta'_1 \oplus \bigoplus_\ell u_1^\ell$. Else R reconstructs $\tilde{x}_{b_1} = \Delta'_{b_1} \oplus \bigoplus_\ell u_{b'_\ell}^\ell$ and $\tilde{\phi} = \Delta'_\phi \oplus \bigoplus_\ell u_2^\ell$.

6. R initializes $\mathbf{J} = J$, $\Psi = \emptyset$, and $\Phi = \bot$, and then does the following:
   - If $J \neq \emptyset$ and if for all $\ell, \ell' \in [\sigma]$ it holds that $u_0^\ell \oplus u_1^\ell = u_0^{\ell'} \oplus u_1^{\ell'}$ then R sets $\Psi = (\tilde{x}_0, \tilde{x}_1)$.
   - If $J = \emptyset$ and $h = H(\tilde{\phi})$: R sets $\Phi = \tilde{\phi}$.
   - If $(|J| = 1$ and $\Psi = \emptyset)$ or $(|J| = 0$ and $\Phi = \emptyset)$, then set $\mathbf{J} = \Psi = \Phi = \emptyset$.

Reveal: R sends $(\mathbf{J}, \Psi, \Phi)$ to S, else sends $\bot$. S aborts if these values are inconsistent with its inputs and check values.

Fig. 4. CCOT via the basic XOR-tree encoding scheme.

- Else if for all $\ell \in [\sigma]$, it holds that $c_0^\ell \neq c_1^\ell$, then for each $\ell \in [\sigma]$ compute b'_ℓ such that $c_{b'_\ell}^\ell = 0$. Extract $b' = \bigoplus_\ell b'_\ell$. Set $J = \emptyset$, send $(J, b = b')$ to the trusted party and receive back x_b. If $b = 0$, set $\Delta'_0 = x_0 \oplus \bigoplus_\ell u_0^\ell$. Else if $b = 1$, set $\Delta'_1 = x_1 \oplus \Delta' \oplus \bigoplus_\ell u_0^\ell$. Pick random $\Delta'_{1-b}, \Delta'_\phi \leftarrow \{0,1\}^\lambda$, and set $h' = H(\Delta_\phi \oplus \bigoplus_\ell u_2^\ell)$. Finally, send $\Delta'_0, \Delta'_1, \Delta'_\phi, h'$ to R.
- Else set $J = \emptyset$ and choose random $b' \leftarrow \{0,1\}$ and send $(J, b = b')$ to the trusted party. Receive back x_b. If $b = 0$, set $\Delta'_0 = x_0 \oplus \bigoplus_\ell u_0^\ell$. Else if $b = 1$, set $\Delta'_1 = x_1 \oplus \Delta' \oplus \bigoplus_\ell u_0^\ell$. Pick random $\Delta'_{1-b}, \Delta'_\phi \leftarrow \{0,1\}^\lambda$, and set $h' = H(\Delta_\phi \oplus \bigoplus_\ell u_2^\ell)$. Finally, send $\Delta'_0, \Delta'_1, \Delta'_\phi, h'$ to R.
- In the reveal phase, if R sends (J', Ψ', Φ') such that $J' \neq J$ or the values Ψ', Φ' are not consistent with the values above, then abort the reveal phase. Else, send "reveal" to the trusted party.

It is easy to see that the above simulation is indistinguishable from the real execution. Indeed if there exists any ℓ such that $c_0^\ell = c_1^\ell = 0$, then in this case corrupt R learns Δ but does not obtain u_2^ℓ. Therefore, in this case it misses at least one additive share of ϕ and since $h = H(\phi)$ does not reveal information (unless H is queried on ϕ), ϕ is statistically hidden from corrupt R. Thus, this case corresponds to $J \neq \emptyset$ since R could potentially know both x_0 and x_1 (since it knows Δ and potentially at least one of u_0^ℓ, u_1^ℓ for each $\ell \in [\sigma]$) but not ϕ. On the other hand, if for all $\ell \in [\sigma]$ it holds that $c_0^\ell \neq c_1^\ell$ then it is easy to see that the extracted input b' equals R's input b_1 and that the rest of the simulation is indistinguishable from the real execution. Finally the remaining case (i.e., there exists $\ell \in [\sigma]$ such that $c_0^\ell = c_1^\ell = 1$ and there does not exist $\ell' \in [\sigma]$ such that $c_0^{\ell'} = c_1^{\ell'} = 0$) is when R obtains only ϕ and neither x_0 nor x_1. This case is rather straightforward to handle; the simulator supplies $J = \emptyset$ (since R knows ϕ) and a random choice bit b'. This works because there exists some $\ell \in [\sigma]$ such that R neither obtains u_0^ℓ nor u_1^ℓ. As a result both x_0 and x_1 are information-theoretically hidden from it.

Privacy Against Corrupt Sender. Note that except in the reveal phase, information flows only from S to R. If S is honest, then reveals made by R do not leak any information. (Recall J is revealed to S in the real as well as the ideal execution.) We have to show that even a corrupt S does not learn any information about b_1. Clearly when $J \neq \emptyset$, R's actions are independent of its input b_1 and thus does not leak any information. On the other hand when $J = \emptyset$, observe that R does not reveal $\tilde{x}_{b_1}$, and thus S only learns whether $\Psi = \Phi = \emptyset$ or not. This translates to learning information about R's input b_1 only if for some (possibly many) $\ell \in [\sigma]$, S provided (u_0^ℓ, u_2^ℓ) in one instance of $\mathcal{F}_{OT}$ and $(u_1^\ell, \hat{u}_2^\ell)$ in the other instance with $u_2^\ell \neq \hat{u}_2^\ell$. This is because such a strategy would allow S to learn whether R input $c_0^\ell = 1$ (in which case R does not abort) or $c_1^\ell = 1$ (in which case R does abort), and consequently leak information about b_ℓ' (i.e., depending on which of c_0^ℓ, c_1^ℓ was 0 when $J = \emptyset$). More generally, such a strategy allows S to learn any disjunctive predicate of R's selections $\{c_0^\ell, c_1^\ell\}_\ell$. To prove that such a strategy does not help S we use the following easy lemma.

Lemma 1 ([13]). *Let* $\mathsf{En}' : \{0,1\}^m \to \{0,1\}^{m'}$ *be such that for any* $\mathbf{b} \in \{0,1\}^m$, *it holds that* $\mathsf{En}'(\mathbf{b})$ *is a* κ-wise independent encoding of $\mathbf{b}$. *Then for every disjunctive predicate* $P(\cdot)$ *the following holds: (1) If* P *involves at most* κ *literals, then* $\Pr[P(\mathsf{En}'(\mathbf{b})) = 1]$ *is completely independent of* $\mathbf{b}$. *(2) Otherwise,* $\Pr[P(\mathsf{En}'(\mathbf{b})) = 1] \geq 1 - 2^{-\kappa}$.

To apply the lemma in our context, note that En' here corresponds to the "XOR-tree encoding", i.e.,. encoding of b_1 into $\{b_\ell'\}_\ell$. Clearly, En' is a $\kappa = (\sigma-1)$-wise independent encoding of b_1. Thus we have that if S supplied inconsistent values (i.e., $u_2^\ell, \hat{u}_2^\ell$) in at most $(\sigma - 1)$ instances, then the S does not learn any information about b_1 in the reveal phase. Further, even if S supplied inconsistent values in all instances, then with all but negligible probability (exponentially negligible in σ) R will abort in the reveal phase (irrespective of R's true input b_1). This concludes the proof of privacy against corrupt sender.

SINGLE-CHOICE CCOT. Next, we consider the case of single-choice CCOT, where S holds $(x_{1,0}, x_{1,1}), \ldots, (x_{n,0}, x_{n,1})$ and R holds $J \in [n]$ and a single choice bit b_1. At the end of the protocol, R receives $\{(x_{j,0}, x_{j,1})\}_{j \in J}$ and $\{x_{j,b_1}\}_{j \notin J}$. That is, this is exactly the same as CCOT except we enforce that R inputs the same choice across all n pairs of strings held by S. Our protocol in Fig. 5 enforces this using a symmetric-key encryption scheme denoted $(\mathsf{Enc}, \mathsf{Dec})$.

Inputs: S holds $(x_{1,0}, x_{1,1}), \ldots, (x_{n,0}, x_{n,1})$ and R holds $J \subseteq [n]$ and b_1.

Protocol:

1. S does the following:
   - Choose $\{(\Delta'_{j,0}, \Delta'_{j,\phi})\}_{j \in [n]}$ uniformly at random from $\{0,1\}^\lambda$.
   - Choose $\{u^\ell_{j,0}\}_{j \in [n], \ell \in [\sigma]}$ at random such that for all $j \in [n]$ it holds that $\bigoplus_\ell u^\ell_{j,0} = x_{j,0} \oplus \Delta'_{j,0}$.
   - Choose $\phi_1, \ldots, \phi_n, \{u^\ell_{j,2}\}_{j \in [n], \ell \in [\sigma]}$ at random such that for all $j \in [n]$ it holds that $\bigoplus_\ell u^\ell_{j,2} = \phi_j \oplus \Delta'_{j,\phi}$.
   - Choose $\{(K_{\ell,0}, K_{\ell,1})\}_{\ell \in [\sigma]}$ at random where each $K_{\ell,0}, K_{\ell,1} \leftarrow \{0,1\}^\lambda$.
   - Choose $\Delta_1, \ldots, \Delta_n \leftarrow \{0,1\}^\lambda$ at random and set $u^\ell_{j,1} = u^\ell_{j,0} \oplus \Delta_j$.

2. R does the following:
   - Choose $\{b'_\ell\}_{\ell \in [\sigma]}$ such that $\bigoplus_\ell b'_\ell = b_1$.
   - For each $j \in [n]$, R sets $\{(c^\ell_{j,0}, c^\ell_{j,1})\}_{\ell \in [\sigma]}$ as follows:
     - If $j \in J$, then set $c^\ell_{j,0} = c^\ell_{j,1} = 0$ for all $\ell \in [\sigma]$.
     - Else for each $\ell \in [\sigma]$ set $c^\ell_{j,b'_\ell} = 0$ and $c^\ell_{j,1-b'_\ell} = 1$.

3. For each $\ell \in [\sigma]$ do: S sends $(K_{\ell,0}, K_{\ell,1})$ to $\mathcal{F}_{\text{OT}}$ and R sends b'_ℓ to $\mathcal{F}_{\text{OT}}$. R receives $\{K_{\ell,b'_\ell}\}_{\ell \in [\sigma]}$ from $\mathcal{F}_{\text{OT}}$.

4. For each $j \in [n]$ and each $\ell \in [\sigma]$:
   - S sends $(u^\ell_{j,0}, e^\ell_{j,1} = \text{Enc}(K_{\ell,1}, u^\ell_{j,2}))$ to $\mathcal{F}_{\text{OT}}$, and R sends $c^\ell_{j,0}$ to $\mathcal{F}_{\text{OT}}$.
   - S sends $(u^\ell_{j,1}, e^\ell_{j,0} = \text{Enc}(K_{\ell,0}, u^\ell_{j,2}))$ to $\mathcal{F}_{\text{OT}}$, and R sends $c^\ell_{j,1}$ to $\mathcal{F}_{\text{OT}}$.

   That is, R receives $\{(u^\ell_{j,0}, u^\ell_{j,1})\}_{\ell \in [\sigma]}$ if $j \in J$, and otherwise receives $\{(u^\ell_{j,b'_\ell}, e^\ell_{j,b'_\ell})\}_{\ell \in [\sigma]}$.

5. For each $j \in [n]$: S sends $\Delta'_{j,0}, \Delta'_{j,1} = x_{j,0} \oplus x_{j,1} \oplus \Delta_j \oplus \Delta'_{j,0}, \Delta'_{j,\phi}, h_j = H(\phi_j)$ to R.

6. For each $j \in [n]$, R reconstructs the following:
   - If $j \in J$, then compute $\tilde{x}_{j,0} = \Delta'_{j,0} \oplus \bigoplus_\ell u^\ell_{j,0}$ and $\tilde{x}_{j,1} = \Delta'_{j,1} \oplus \bigoplus_\ell u^\ell_{j,1}$.
   - Else, compute $\tilde{x}_{j,b_1} = \Delta'_{j,b_1} \oplus \bigoplus_\ell u^\ell_{j,b'_\ell}$ and $\tilde{\phi}_j = \Delta'_{j,\phi} \oplus \bigoplus_\ell \text{Dec}(K_{\ell,b'_\ell}, e_{j,b'_\ell})$.

7. R sets $\mathbf{J} = J$, $\Psi = \emptyset$, and $\Phi = \emptyset$, and does the following:
   - If $\forall j \in J$ and if $\forall \ell, \ell' \in [\sigma]$ it holds that $u^\ell_{j,0} \oplus u^\ell_{j,1} = u^{\ell'}_{j,0} \oplus u^{\ell'}_{j,1}$ then R sets $\Psi = \{(\tilde{x}_{j,0}, \tilde{x}_{j,1})\}_{j \in J}$.
   - If for every $j \notin J$ it holds that $h_j = H(\tilde{\phi}_j)$ then R sets $\Phi = \{\tilde{\phi}_j\}_{j \notin J}$.
   - If $(|J| > 0$ and $\Psi = \emptyset)$ or $(|J| < n$ and $\Phi = \emptyset)$, then set $\mathbf{J} = \Psi = \Phi = \emptyset$.

Reveal phase: R sends $(\mathbf{J}, \Psi, \Phi)$ to S. S aborts if these values are inconsistent with its inputs and check values.

Fig. 5. Realizing single-choice CCOT in the $\mathcal{F}_{\text{OT}}$-hybrid model.

We prove that the protocol in Fig. 5 securely realizes single-choice CCOT with sender-simulatability and receiver-privacy. We start by observing that correctness follows from inspection of the protocol.

Simulating Corrupt Receiver. The simulation is quite similar to the simulation of CCOT construction presented in Fig. 4. Obviously the main difference now is that R may attempt to use different b'_ℓ values for $j, j' \in [n]$ (where b'_ℓ is defined as the value R inputs to $\mathcal{F}_{OT}$ in Step 3). However the key observation is that it receives only one key K_{ℓ,b'_ℓ} in $\{K_{\ell,0}, K_{\ell,1}\}$. Therefore, even if it attempts to deviate and and obtain e^ℓ_{j',b''_ℓ} for $b''_\ell \neq b'_\ell$, it still cannot decrypt since it does not possess the secret key K_{ℓ,b''_ℓ}. Semantic security of the encryption allows us to argue that if such a deviation happens then the value of $\phi_{j'}$ is hidden from S. Therefore in this case, the simulator can simply add j' to J, and the simulation can be completed. It is instructive to note that when such a deviation happens, R will be neither be able to provide $\{x_{j',0}, x_{j',1}\}$ nor the value $\phi_{j'}$, and thus will get rejected by S during the reveal phase.

We proceed to the formal simulation. Assume that H is modeled as a (non-programmable) random oracle. Acting as $\mathcal{F}_{OT}$ the simulator does the following:

- Chooses random $\{\Delta'_j\}_{j\in[n]}, \{K_{\ell,0}, K_{\ell,1}\}_{\ell\in[\sigma]}, \{u^\ell_{j,0}, u^\ell_{j,2}\}_{j\in[n],\ell\in[\sigma]}$ and sets for all $j \in [n], \ell \in [\sigma]$, value $u^\ell_{j,1} = u^\ell_{j,0} \oplus \Delta'_j$.
- For each $\ell \in [\sigma]$, acting as $\mathcal{F}_{OT}$ obtain values b_ℓ, return key K_{ℓ,b'_ℓ}, and set $e^\ell_{j,b'_\ell} = \mathsf{Enc}(K_{\ell,b'_\ell}, u^\ell_{j,2})$, $e^\ell_{j,1-b'_\ell} = \mathsf{Enc}(K_{\ell,1-b'_\ell}, \mathbf{0})$. Compute $b' = \bigoplus_\ell b'_\ell$.
- For each $j \in [n], \ell \in [\sigma]$, acting as $\mathcal{F}_{OT}$ obtain values $\{c^\ell_{j,0}, c^\ell_{j,1}\}$ and return answers using values $u^\ell_{j,0}, u^\ell_{j,1}, e^\ell_{j,b'_\ell}, e^\ell_{j,1-b'_\ell}$ (computed as above) exactly as in the protocol.
- Initialize $J = \emptyset$. For each $j \in [n]$: If there exists $\ell \in [\sigma]$ such that $c^\ell_{j,0} = c^\ell_{j,1} = 0$, then add j to J.
- Initialize $\mathsf{flag} = 0$. For each $j \notin J$: If there exists $\ell \in [\sigma]$ such that either $c^\ell_{j,b'_\ell} = 0$ or $c^\ell_{j,1-b'_\ell} = 1$ do not hold, then add j to J and set $\mathsf{flag} = 1$.
- Send (J, b') to the trusted-party and receive $\{x_{j,0}, x_{j,1}\}_{j\in J}$ and $\{x_{j,b'}\}_{j\notin J}$.
- For each $j \in J$, do: (1) set $\Delta'_{j,0} = x_{j,0}\oplus\bigoplus_\ell u^\ell_{j,0}$ and $\Delta'_{j,1} = x_{j,1}\oplus\Delta'\oplus\bigoplus_\ell u^\ell_{j,0}$, and (2) pick random $\Delta'_{j,\phi}$ and random $h'_j \leftarrow \{0,1\}^\lambda$.
- For each $j \notin J$, do: (1) if $b' = 0$, set $\Delta'_{j,0} = x_{j,0}\oplus\bigoplus_\ell u^\ell_0$, (2) else if $b' = 1$, set $\Delta'_{j,1} = x_{j,1}\oplus\Delta'_j\oplus\bigoplus_\ell u^\ell_0$, and (3) pick random $\Delta'_{j,1-b'}, \Delta'_{j,\phi} \leftarrow \{0,1\}^\lambda$, and set $h'_j = H(\Delta'_{j,\phi}\oplus\bigoplus_\ell u^\ell_{j,2})$.
- Send $\{\Delta'_{j,0}, \Delta'_{j,1}, \Delta'_{j,\phi}, h'_j\}_{j\in[n]}$ to R.
- In the reveal phase, if $\mathsf{flag} = 1$ or if R sends (J', Ψ', Φ') such that $J' \neq J$ or the values Ψ', Φ' are not consistent with the values above, then abort the reveal phase. Else, send "reveal" to the trusted party.

We show that the simulation is indistinguishable from the real execution. First, note that if for some $j \in [n], \ell \in [\sigma]$, it holds that $c^\ell_{j,0} = c^\ell_{j,1} = 0$, then R receives both $u^\ell_{j,0}$ and $u^\ell_{j,1}$, and therefore knows Δ_j. In this case, it is safe to presume that R will end up knowing both $x_{j,0}$ as well as $x_{j,1}$ (since for any ℓ', if it receives even one of $u^{\ell'}_{j,0}, u^{\ell'}_{j,1}$ it will know the other as well since it knows Δ_j). Therefore, $\mathcal{S}$ includes j in J and obtains both $x_{j,0}, x_{j,1}$ from the trusted party. Now $\mathcal{S}$ can carry out the simulation whether or not R obtained both $x_{j,0}$ and $x_{j,1}$.

Next, suppose that for some $j \in [n]$ such that for no $\ell \in [\sigma]$, it holds that $c^\ell_{j,0} = c^\ell_{j,1} = 0$, and yet there exists some $\ell \in [\sigma]$ such that either $c^\ell_{j,b'_\ell} = 0$ or $c^\ell_{j,1-b'_\ell} = 1$ does not hold. In this case, it is easy to see that R will not be able to produce both $x_{j,0}, x_{j,1}$ (since it is missing one of $u^\ell_{j,0}, u^\ell_{j,1}$) in the real execution. Further, it can be shown that except with negligible probability R cannot produce ϕ_j either. This is because (1) R does not obtain e^ℓ_{j,b'_ℓ}, and (2) R has no information about the plaintext encrypted as $e^\ell_{j,1-b'_\ell}$, and (3) $h_j = H(\phi_j)$ does not reveal any information about ϕ_j except with statistically negligible probability (i.e., unless H is queried on ϕ_j). Point (2) above trivially holds in the simulation because $e^\ell_{j,1-b'_\ell}$ encrypts $\mathbf{0}$ instead of $u^\ell_{j,2}$. On the other hand, in the real execution, observe that R does not possess the key $K_{\ell,1-b'_\ell}$. It follows from a straightforward reduction to the semantic security of the encryption scheme that the real execution is indistinguishable from the simulation. In particular, in this case R will not be able to produce (J', Ψ', Φ') that will be accepted by S in the real execution, and is equivalent to S sending abort in the ideal execution.

Finally, suppose that for every $j \in [n]$, either (1) for all $\ell \in [\sigma]$ it holds that $c^\ell_{j,0} = c^\ell_{j,1} = 1$, or (2) for all $\ell \in [\sigma]$ it holds that $c^\ell_{j,b'_\ell} = 0$ and $c^\ell_{j,1-b'_\ell} = 1$. This indeed corresponds to honest behavior on the part of R. Specifically, in case (1), we have $j \in J$, and in case (2), we have $j \notin J$. This is exactly how the simulator constructs J. It remains to be shown that in this case, any reveal (J', Ψ', Φ') such that $J' \neq J$ or Ψ', Φ' is not consistent with the simulation will be rejected by S in the real execution. This follows from: (a) Any $j \in J$ cannot be claimed by R to not be in the checkset. This is because in this case, R does not have any information about ϕ_j (other than $H(\phi_j)$ which leaks no information unless H is queried on ϕ_j). (b) Any $j \notin J$ cannot be claimed by R to be in the checkset. This is because in this case, R obtains exactly one of $\{u^\ell_{j,0}, u^\ell_{j,1}\}$ for every $\ell \in [\sigma]$ and thus is able to reconstruct at most one of $\{x_{j,0}, x_{j,1}\}$. This concludes the proof of security against corrupt receiver.

Privacy Against Corrupt Sender. The proof of privacy against corrupt sender is very similar to the corresponding proof for (the basic) CCOT. Specifically, note that except in the reveal phase, information flows only from S to R. Next note that if S is honest, then the reveals made by R in the reveal phase do not leak any information about R's input b_1. (Recall J is revealed to S in the real as well as the ideal execution.) It remains to be shown that even a corrupt S does not learn any information about b_1. Clearly for $j \in J$, R's actions are independent of its input b_1 and thus does not leak any information. On the other hand for $j \notin J$, observe that R does not reveal $\tilde{x}_{j,b_1}$ (i.e., in the reveal phase), and thus the only information learnt by S is whether $\Psi = \Phi = \emptyset$ or not. This translates to learning information about R's input b_1 only if for some (possibly many) $j \in [n], \ell \in [\sigma]$, S provided $(u^\ell_{j,0}, u^\ell_{j,2})$ in one instance of $\mathcal{F}_{OT}$ and $(u^\ell_{j,1}, \hat{u}^\ell_{j,2})$ in the other instance with $u^\ell_{j,2} \neq \hat{u}^\ell_{j,2}$. This is because such a strategy would allow S to learn whether R input $c^\ell_{j,0} = 1$ (in which case R does not abort) or $c^\ell_{j,1} = 1$ (in which case R does abort), and consequently leak information about b'_ℓ (i.e., depending on which of $c^\ell_{j,0}, c^\ell_{j,1}$ was 0 when $j \notin J$). More generally, such a strategy allows S to learn any disjunctive predicate of R's selections $\{c^\ell_{j,0}, c^\ell_{j,1}\}_\ell$.

To prove that such a strategy does not help S we once again make use of Lemma 1. As before, to apply the lemma in our context, note that En' here corresponds to the "XOR-tree encoding", i.e.,. encoding of b_1 into $\{b'_\ell\}_\ell$. Clearly, En' is a $\kappa = (\sigma - 1)$-wise independent encoding of b_1. Thus we have that if S supplied inconsistent values (i.e., $u_2^\ell, \hat{u}_2^\ell$) in at most $(\sigma-1)$ instances, then S does not learn any information about b_1 in the reveal phase. Further, even if S supplied inconsistent values in all instances, then with all but negligible probability (exponentially negligible in σ) R will abort in the reveal phase (irrespective of R's true input b_1). This concludes the proof of privacy against corrupt sender.

BATCH SINGLE-CHOICE CCOT. This functionality, which has actually been used directly in 2PC constructions of [24] is our next stepping stone. (The description can be obtained by modifying $\mathcal{F}^\star_{\mathrm{mcot}}$ Fig. 2 by setting $t = 1$, setting $|J| = n/2$ and setting all ϕ_j^k values to 0^σ.) The construction of this primitive follows easily merely by repeating the single-choice CCOT protocol batch-wise in parallel. That is, in the m-th (parallel) execution, S and R participate in a single-choice CCOT where S holds $(x_{1,0}^{(i)}, x_{1,1}^{(i)}), \ldots, (x_{n,0}^{(i)}, x_{n,1}^{(i)})$ while R holds $J \subseteq [n]$ and b_i. Obviously the main difficulty is in enforcing that R supplies the same check set J in each execution. However, this is easily enforceable in the following way. Recall that in the reveal phase of each execution of single-choice CCOT (which are now executed in parallel), R will have to reveal (E_i, Ψ_i, Φ_i). In addition to checking whether these values are consistent with its inputs and check value, S additionally checks if $E_i = E_{i'}$ for every $i, i' \in [m]$.

Using more efficient "XOR-tree" encoding schemes. Observe that the construction for batch single-choice CCOT described above incurs a multiplicative overhead of (exactly) σ simply because the underlying single-choice CCOT protocol makes use of the basic XOR-tree encoding scheme. Fortunately, the batch setting makes it possible to apply more sophisticated encodings whose overhead is much lower. More concretely, using encoding schemes based on random combinations approach [23], the overhead can be as low as an additional $\leq 6 \cdot \max(4m, 8\sigma)$ while using encoding schemes based on σ-wise independent generators [13] one can obtain rate-$1/6$ communication complexity (and likely to be practical when $m \gg \sigma$). We show constructions of batch single-choice CCOT using abstract encoding schemes.

We describe a protocol for $\mathcal{F}^{\mathrm{bat,sin}}_{\mathrm{ccot}}$ in the $\mathcal{F}_{\mathrm{OT}}$-hybrid model that makes use of an arbitrary XOR-tree encoding scheme in Fig. 6. The protocol itself is a straightforward extension combining ideas from protocols in Figs. 4 and 5 while abstracting away the underlying encoding scheme. We now prove that the protocol $\pi^{\mathrm{bat,sin}}_{\mathrm{ccot}}$ described in Fig. 6 realizes batch single-choice CCOT with sender-simulatability and receiver-privacy. We start by observing that correctness follows from correctness properties of the XOR-tree encoding schemes (specifically, Property 3).

Simulating Corrupt Receiver. The simulation is quite similar to the simulation of single-choice CCOT construction presented in Fig. 5. Obviously the main difference now is that we need to deal with encodings over R's entire input. We proceed to the formal simulation. Assume that H is modeled as a (non-programmable)

Inputs: S holds $X^{(1)}, \ldots, X^{(m)}$ where $X^{(i)} = (x_{1,0}^{(i)}, x_{1,1}^{(i)}), \ldots, (x_{n,0}^{(i)}, x_{n,1}^{(i)})$; R holds $J \subseteq [n]$ and $\{b_i\}_{i \in [m]}$.

Protocol:

1. R picks randomness ω_0 for the encoding scheme and sends to S.

2. S does the following for each $j \in [n]$:
   - Choose $\Delta'_{j,\phi}, \{(\Delta_{j,0}^{(i)}, \Delta_{j,1}^{(i)})\}_{i \in [m]}$ uniformly at random.
   - Choose randomness ω_j and compute $\{(u_{j,0}^\ell, u_{j,1}^\ell)\}_{\ell \in [m']}$ $\leftarrow$ $\mathsf{En}_{\omega_0}(\{(x_{j,0}^{(i)} \oplus \Delta_{j,0}^{(i)}, x_{j,1}^{(i)} \oplus \Delta_{j,1}^{(i)})\}_{i \in [m]}; \omega_j)$.
   - Choose $\phi_1, \ldots, \phi_n, \{u_{j,2}^\ell\}_{j \in [n], \ell \in [m']}$ at random such that for all $j \in [n]$ it holds that $\bigoplus_\ell u_{j,2}^\ell = \phi_j \oplus \Delta'_{j,\phi}$.
   - Choose $\{(K_{\ell,0}, K_{\ell,1})\}_{\ell \in [m']}$ at random where each $K_{\ell,0}, K_{\ell,1} \leftarrow \{0,1\}^\lambda$.

3. R does the following:
   - Choose random ω', compute $\{b_\ell'\}_{\ell \in [m']} \leftarrow \mathsf{En}'_{\omega_0}((b_1, \ldots, b_m); \omega')$.
   - For each $j \in [n]$, R sets $\{(c_{j,0}^\ell, c_{j,1}^\ell)\}_{\ell \in [m']}$ as follows:
     - If $j \in J$, then set $c_{j,0}^\ell = c_{j,1}^\ell = 0$ for all $\ell \in [m']$.
     - Else for each $\ell \in [m']$ set $c_{j,b_\ell'}^\ell = 0$ and $c_{j,1-b_\ell'}^\ell = 1$.

4. For each $\ell \in [m']$: S sends $(K_{\ell,0}, K_{\ell,1})$ to $\mathcal{F}_{\mathrm{OT}}$ and R sends b_ℓ' to $\mathcal{F}_{\mathrm{OT}}$. R receives $\{K_{\ell,b_\ell'}\}_{\ell \in [m']}$ from $\mathcal{F}_{\mathrm{OT}}$.

5. For each $j \in [n]$ and each $\ell \in [m']$:
   - S sends $(u_{j,0}^\ell, e_{j,1}^\ell = \mathsf{Enc}(K_{\ell,1}, u_{j,2}^\ell))$ to $\mathcal{F}_{\mathrm{OT}}$, and R sends $c_{j,0}^\ell$ to $\mathcal{F}_{\mathrm{OT}}$.
   - S sends $(u_{j,1}^\ell, e_{j,0}^\ell = \mathsf{Enc}(K_{\ell,0}, u_{j,2}^\ell))$ to $\mathcal{F}_{\mathrm{OT}}$, and R sends $c_{j,1}^\ell$ to $\mathcal{F}_{\mathrm{OT}}$.

   That is, R receives $\{(u_{j,0}^\ell, u_{j,1}^\ell)\}_{\ell \in [m']}$ if $j \in J$, and otherwise receives $\{(u_{j,b_\ell'}^\ell, e_{j,b_\ell'}^\ell)\}_{\ell \in [m']}$.

6. For each $j \in [n]$: S sends $\{(\Delta_{j,0}^{(i)}, \Delta_{j,1}^{(i)})\}_{i \in [m]}, \Delta'_{j,\phi}, h_j = H(\phi_j)$ to R.

7. For each $j \in [n]$, R reconstructs the following:
   - If $j \in J$, then compute $\{(\tilde{x}_{j,0}^{(i)} \oplus \Delta_{j,0}^{(i)}, \tilde{x}_{j,1}^{(i)} \oplus \Delta_{j,1}^{(i)})\}_{i \in [m]}$ $\leftarrow$ $\mathsf{De}_{\omega_0}(\{(u_{j,0}^\ell, u_{j,1}^\ell)\}_{\ell \in [m']})$.
   - Else, compute $\{\tilde{x}_{j,b_i}^{(i)} \oplus \Delta_{j,b_i}^{(i)}\}_{i \in [m]} = \mathsf{De}_{\omega_0}(\{u_{j,b_\ell'}^\ell\}_{\ell \in [m']})$ and $\tilde{\phi}_j = \Delta'_{j,\phi} \oplus \bigoplus_\ell \mathsf{Dec}(K_{\ell,b_\ell'}, e_{j,b_\ell'})$.

8. R sets $\mathbf{J} = J, \Psi = \emptyset$, and $\Phi = \emptyset$, and does the following:
   - If $\forall j \in J$ and if $\forall \ell, \ell' \in [m']$ it holds that $u_{j,0}^\ell \oplus u_{j,1}^\ell = u_{j,0}^{\ell'} \oplus u_{j,1}^{\ell'}$ then R sets $\Psi = \{(\tilde{x}_{j,0}^{(i)}, \tilde{x}_{j,1}^{(i)})\}_{j \in J, i \in [m]}$.
   - If for every $j \notin J$ it holds that $h_j = H(\tilde{\phi}_j)$ then R sets $\Phi = \{\tilde{\phi}_j\}_{j \notin J}$.
   - If $(|J| > 0$ and $\Psi = \emptyset)$ or $(|J| < n$ and $\Phi = \emptyset)$, then set $\mathbf{J} = \Psi = \Phi = \emptyset$.

Reveal phase: R sends $(\mathbf{J}, \Psi, \Phi)$ to S. S aborts if these values are inconsistent with its inputs and check values.

Fig. 6. Protocol $\pi_{\mathrm{ccot}}^{\mathrm{bat,sin}}$ realizing batch single-choice CCOT.

random oracle. S first receives public randomness ω_0 for the XOR-tree encoding scheme $(\mathsf{En}, \mathsf{De}, \mathsf{En}', \mathsf{De}')$. Acting as $\mathcal{F}_{\mathrm{OT}}$ the simulator does the following:

- Samples for each $j \in [n]$, uniform $\{(\hat{x}_{j,0}^{(i)}, \hat{x}_{j,1}^{(i)})\}_{i \in [m]}$, uniformly random ω_j and computes $\{(u_{j,0}^\ell, u_{j,1}^\ell)\}_{\ell \in [m']} \leftarrow \mathsf{En}_{\omega_0}(\{(\hat{x}_{j,0}^{(i)}, \hat{x}_{j,1}^{(i)})\}_{i \in [m]}; \omega_j)$.

- Chooses random $\{(K_{\ell,0}, K_{\ell,1})\}_{\ell \in [m']}, \{u_{j,2}^{\ell}\}_{j \in [n], \ell \in [m']}$.
- For each $\ell \in [m']$, acting as $\mathcal{F}_{\text{OT}}$ obtain values b_{ℓ}', return key $K_{\ell,b_{\ell}'}$, and set
  $e_{j,b_{\ell}'}^{\ell} = \text{Enc}(K_{\ell,b_{\ell}'}, u_{j,2}^{\ell})$, $e_{j,1-b_{\ell}'}^{\ell} = \text{Enc}(K_{\ell,1-b_{\ell}'}, \mathbf{0})$.
  Compute $(b_1, \ldots, b_m) = \text{De}'(\{b_{\ell}'\}_{\ell \in [m']})$.
- For each $j \in [n], \ell \in [m']$, acting as $\mathcal{F}_{\text{OT}}$ obtain values $\{c_{j,0}^{\ell}, c_{j,1}^{\ell}\}$ and return
  answers using values $u_{j,0}^{\ell}, u_{j,1}^{\ell}, e_{j,b_{\ell}'}^{\ell}, e_{j,1-b_{\ell}'}^{\ell}$ (computed as above) exactly as
  in the protocol.
- Initialize $J = \emptyset$. For each $j \in [n]$: If there exists $\ell \in [m']$ such that $c_{j,0}^{\ell} = c_{j,1}^{\ell} = 0$, then add j to J.
- Initialize $\text{flag} = 0$. For each $j \notin J$: If there exists $\ell \in [m']$ such that either
  $c_{j,b_{\ell}'}^{\ell} = 0$ or $c_{j,1-b_{\ell}'}^{\ell} = 1$ do not hold, then add j to J and set $\text{flag} = 1$.
- Send $(J, \{b_i\}_{i \in [m]})$ to the trusted party and receive back $\{(x_{j,0}^{(i)}, x_{j,1}^{(i)})\}_{i \in [m], j \in J}$
  and $\{x_{j,b_i}^{(i)}\}_{i \in [m], j \notin J}$.
- For each $j \in J$, do: (1) for each $i \in [m]$, set $\hat{\Delta}_{j,0}^{(i)} = \hat{x}_{j,0}^{(i)} \oplus x_{j,0}^{(i)}$ and $\hat{\Delta}_{j,1}^{(i)} = \hat{x}_{j,1}^{(i)} \oplus x_{j,1}^{(i)}$, and (2) pick random $\Delta_{j,\phi}'$ and random $h_j' \leftarrow \{0,1\}^{\lambda}$.
- For each $j \notin J$, do: (1) for each $i \in [m]$: set $\hat{\Delta}_{j,b_i}^{(i)} = \hat{x}_{j,b_i}^{(i)} \oplus x_{j,b_i}^{(i)}$ and pick random
  $\hat{\Delta}_{j,1-b_i}^{(i)}$, and (2) pick random $\Delta_{j,\phi}' \leftarrow \{0,1\}^{\lambda}$, and set $h_j' = H(\Delta_{j,\phi}' \oplus \bigoplus_{\ell} u_{j,2}^{\ell})$.
- For each $j \in [n]$: send $\{\hat{\Delta}_{j,0}^{(i)}, \hat{\Delta}_{j,1}^{(i)}\}_{i \in [m]}, \Delta_{j,\phi}', h_j'$ to R.
- In the reveal phase, if $\text{flag} = 1$ or if R sends (J', Ψ', Φ') such that $J' \neq J$ or the
  values Ψ', Φ' are not consistent with the values above, then abort the reveal
  phase. Else, send "reveal" to the trusted party.

We show that the simulation is indistinguishable from the real execution. First,
note that if for some $j \in [n], \ell \in [m']$, it holds that $c_{j,0}^{\ell} = c_{j,1}^{\ell} = 0$, then R
receives both $u_{j,0}^{\ell}$ and $u_{j,1}^{\ell}$, but does not obtain $u_{j,2}^{\ell}$. In this case, it is safe to
presume that R will end up knowing both $x_{j,0}^{(i)}$ as well as $x_{j,1}^{(i)}$ but R definitely
misses an additive share of (and consequently has no information about) ϕ_j.
Therefore, $\mathcal{S}$ includes j in J and obtains both $x_{j,0}^{(i)}, x_{j,1}^{(i)}$ from the trusted party.
This allows $\mathcal{S}$ to carry out a correct simulation irrespective of whether or not R
obtained both $x_{j,0}^{(i)}$ and $x_{j,1}^{(i)}$.

Next, suppose that for some $j \in [n]$ such that for no $\ell \in [m']$ it holds that
$c_{j,0}^{\ell} = c_{j,1}^{\ell} = 0$, and yet there exists some $\ell \in [m']$ such that either $c_{j,b_{\ell}'}^{\ell} = 0$ or
$c_{j,1-b_{\ell}'}^{\ell} = 1$ does not hold. In this case, we claim that R will not be able to produce both $x_{j,0}^{(i)}, x_{j,1}^{(i)}$ in the real execution. This follows from the properties of the
XOR-tree encoding schemes and the fact that R misses one of $u_{j,0}^{\ell}, u_{j,1}^{\ell}$. Further,
it can be shown that except with negligible probability R cannot produce ϕ_j
either. This is because (1) R does not obtain $e_{j,b_{\ell}'}^{\ell}$, and (2) R has no information
about the plaintext encrypted as $e_{j,1-b_{\ell}'}^{\ell}$, and (3) $h_j = H(\phi_j)$ does not reveal
any information about ϕ_j with statistically negligible probability (i.e., unless H
is queried on ϕ_j). Point (2) mentioned above trivially holds in the simulation
because $e_{j,1-b_{\ell}'}^{\ell}$ encrypts $\mathbf{0}$ instead of $u_{j,2}^{\ell}$. On the other hand, in the real execution, observe that R does not possess the key $K_{\ell,1-b_{\ell}'}$. It then follows from

a straightforward reduction to the semantic security of the encryption scheme that the real execution is indistinguishable from the simulation. In particular, in this case R will not be able to produce (J', Ψ', Φ') that will be accepted by S in the real execution, and is equivalent to S sending abort in the ideal execution.

Finally, suppose that for every $j \in [n]$, either (1) for all $\ell \in [m']$ it holds that $c_{j,0}^\ell = c_{j,1}^\ell = 0$, or (2) for all $\ell \in [m']$ it holds that $c_{j,b_\ell'}^\ell = 0$ and $c_{j,1-b_\ell'}^\ell = 1$. This indeed corresponds to honest behavior on the part of R. Specifically, in case (1), we have $j \in J$, and in case (2), we have $j \notin J$. This is exactly how the simulator constructs J. It remains to be shown that in this case, any reveal (J', Ψ', Φ') such that $J' \neq J$ or Ψ', Φ' is not consistent with the simulation will be rejected by S in the real execution. This follows from: (a) Any $j \in J$ cannot be claimed by R to not be in the checkset. This is because in this case, R does not have any information about ϕ_j (other than $H(\phi_j)$ which leaks no information unless H is queried on ϕ_j). (b) Any $j \notin J$ cannot be claimed by R to be in the checkset. This is because in this case, R obtains exactly one of $\{u_{j,0}^\ell, u_{j,1}^\ell\}$ for every $\ell \in [m']$ and thus by Property 7 of XOR-tree encoding schemes, is able to reconstruct at most one of $\{x_{j,0}^{(i)}, x_{j,1}^{(i)}\}$.

Privacy Against Corrupt Sender. The proof of privacy against corrupt sender is very similar to the corresponding proof for (the basic) CCOT. Specifically, note that except in the reveal phase, information flows only from S to R. Next note that if S is honest, then the reveals made by R in the reveal phase do not leak any information about R's input $b_1, \ldots, b_m$. (Recall J is revealed to S in the real as well as the ideal execution.) It remains to be shown that even a corrupt S does not learn any information about $b_1, \ldots, b_m$. Clearly for $j \in J$, R's actions are independent of its inputs $b_1, \ldots, b_m$ and thus does not leak any information. On the other hand for $j \notin J$, observe that R does not reveal any information about $\{\tilde{x}_{j,b_i}^{(i)}\}_{i \in [m]}$ (i.e., in the reveal phase), and thus the only information learnt by S is whether $\Psi = \Phi = \emptyset$ or not. This translates to learning information about R's inputs $b_1, \ldots, b_m$ only if for some (possibly many) $j \in [n], \ell \in [\sigma]$, S provided $(u_{j,0}^\ell, u_{j,2}^\ell)$ in one instance of $\mathcal{F}_{OT}$ and $(u_{j,1}^\ell, \hat{u}_{j,2}^\ell)$ in the other instance with $u_{j,2}^\ell \neq \hat{u}_{j,2}^\ell$. This is because such a strategy would allow S to learn whether R input $c_{j,0}^\ell = 1$ (in which case R does not abort) or $c_{j,1}^\ell = 1$ (in which case R does abort), and consequently leak information about b_ℓ' (i.e., depending on which of $c_{j,0}^\ell, c_{j,1}^\ell$ was 0 when $j \notin J$). More generally, such a strategy allows S to learn any disjunctive predicate of R's selections $\{c_{j,0}^\ell, c_{j,1}^\ell\}_\ell$.

To prove that such a strategy does not help S we make use of Property 6 of XOR-tree encoding schemes. Thus we have that if S supplied inconsistent values (i.e., $u_2^\ell, \hat{u}_2^\ell$) in at most $(\sigma - 1)$ instances, then S does not learn any information about $b_1, \ldots, b_m$ in the reveal phase. Further, even if S supplied inconsistent values in all instances, then with all but negligible probability (exponentially negligible in σ) R will abort in the reveal phase (irrespective of R's true input $b_1, \ldots, b_m$). This concludes the proof of privacy against corrupt sender.

It is easy to see that our construction of batch single-choice CCOT described above is also a realization of modified batch single-choice CCOT.

MULTISTAGE CCOT. Note that now R has several evaluation sets $E_1, \ldots, E_t$ (corresponding to t executions). To realize $\mathcal{F}_{\text{mcot}}^+$, we will rely on the protocol $\pi_{\text{ccot}}^{\text{bat,sin}}$ designed for realizing $\mathcal{F}_{\text{ccot}}^{\text{bat,sin}}$ presented previously. Indeed as in the protocol designed in [10] we will run $\pi_{\text{ccot}}^{\text{bat,sin}}$ t times to obtain a protocol for $\mathcal{F}_{\text{mcot}}^+$. Our protocol for $\mathcal{F}_{\text{mcot}}^+$ is described in Fig. 7. Unlike protocols for other variants of CCOT, here we improve over prior work by using the reactive functionality relaxation (as opposed to receiver-privacy relaxation) to obtain a simpler protocol secure against corrupt receiver. Prior work [10] required an overhead of t^2 while our protocol requires only a factor t overhead. We prove that the protocol in Fig. 7 securely realizes multistage CCOT with sender-simulatability and receiver-privacy. We start by observing that correctness follows from correctness of each instance of $\pi_{\text{ccot}}^{\text{bat,sin}}$.

Simulating Corrupt Receiver. Using the simulator of $\pi_{\text{ccot}}^{\text{bat,sin}}$, the simulator $\mathcal{S}$ first extracts for all $k \in [t]$, the check sets $[n] \setminus E_k'$ and the selection bits $b_{k,1}, \ldots, b_{k,m}$. Note that a malicious R may supply sets $E_1', \ldots, E_t'$ that may overlap. The simulation extraction for $\mathcal{F}_{\text{mcot}}^{\star}$ first initializes each of $E_1, \ldots, E_t$ to $\emptyset$, flag to 0 (flag $= 1$ indicates whether $\mathcal{S}$ will choose to abort in the reveal phase), and proceeds as follows

- For every $j \in [n]$ such that there exists unique $\alpha \in [t]$ such that $j \in E_\alpha'$, then add j to E_α.
- For every $j \in [n]$ such that there exists $\alpha, \beta \in [t]$ such that $j \in E_\alpha' \cap E_\beta'$, then add j to J and set flag $= 1$.

It is easy to see that $E_1, \ldots, E_t$ are disjoint sets. The simulator then sends $E_1, \ldots, E_t$ (as obtained above), and the values $\{b_{k,i}\}_{k \in [t], i \in [m]}$ (as obtained from the t invocations of the simulator of $\pi_{\text{ccot}}^{\text{bat,sin}}$) to the ideal functionality $\mathcal{F}_{\text{mcot}}^+$. Then upon receiving R's output from $\mathcal{F}_{\text{mcot}}^+$, $\mathcal{S}$ additively secret shares each values in R's output to obtain t additive shares of each value, and then feeds the k-th share of each value to the k-th copy of the invoked simulator for $\pi_{\text{ccot}}^{\text{bat,sin}}$. Then the simulator uses the k-th copy of the invoked simulator for $\pi_{\text{ccot}}^{\text{bat,sin}}$ to complete the simulation of each of the t parallel instances of $\pi_{\text{ccot}}^{\text{bat,sin}}$. Then in the reveal phase, the simulator sends abort to $\mathcal{F}_{\text{mcot}}^+$ if flag $= 1$. On the other hand if flag $= 0$, then the simulator receives $(E_1'', \ldots, E_t'', \Psi, \Phi)$ from R. If $E_1'', \ldots, E_t''$ are pairwise nonintersecting, and further for every $k \in [t]$ it holds that $E_k = E_k''$, then $\mathcal{S}$ sends reveal to $\mathcal{F}_{\text{mcot}}^+$, else sends abort. This completes the description of the simulation. To see why the above simulation works, first note that each of the t copies of the invoked simulator for $\pi_{\text{ccot}}^{\text{bat,sin}}$ (each of which independently guarantee correct simulation of a single instance of $\pi_{\text{ccot}}^{\text{bat,sin}}$) are run on random $(t-1)$-wise independent values. Since $\mathcal{S}$ generates these $(t-1)$-wise independent values correctly using the output received from $\mathcal{F}_{\text{mcot}}^+$, it follows that the t copies of the invoked simulator for $\pi_{\text{ccot}}^{\text{bat,sin}}$ taken together also guarantee correct simulation of the protocol realizing $\mathcal{F}_{\text{mcot}}^+$. In particular, at the end of the output phase, the view of the adversary in real protocol is indistinguishable from that in the simulation. It then remains to be shown that (except with statistically negligible probability) a corrupt R will not be able to reveal $(E_1'', \ldots, E_t'', \Psi, \Phi)$ that is accepted by the sender in the real protocol and yet $(E_1'', \ldots, E_t'') \neq (E_1, \ldots, E_t)$, where $E_1, \ldots, E_t$ are the sets constructed by S as described above. This follows from observing that for every $j \in [n]$:

- If $j \in E_\alpha'$ for some unique $\alpha \in [t]$, then R does not have any information about ϕ_j^β for any $\beta \neq \alpha$. Thus, it can successfully reveal $(E_1'', \ldots, E_t'')$ with $j \in E_\beta''$ for $\beta \neq \alpha$ only with probability negligible in σ. More precisely in this case R will not be able to provide Φ_β consistent with $(E_1'', \ldots, E_t'')$. That is if $j \in E_\alpha'$ for some unique $\alpha \in [t]$, then for every reveal $(E_1'', \ldots, E_t'')$ that is accepted by the sender it must hold that $j \in E_\alpha''$. Stated differently, if for every $j \in [n]$, there exists unique $\alpha \in [t]$ such that $j \in E_\alpha'$, then R can successfully reveal $(E_1'', \ldots, E_t'')$ only for $(E_1'', \ldots, E_t'') = (E_1', \ldots, E_t')$. Recall that in this case, the simulator S set flag $= 0$ and thus will reveal $(E_1, \ldots, E_t) = (E_1', \ldots, E_t')$ in the reveal phase. Therefore, in this case it holds that the real protocol is indistinguishable from the ideal simulation.

- If $j \in E_\alpha' \cap E_\beta'$ for $\alpha \neq \beta$, then one of $x_{j,0}^{(i,\beta)}, x_{j,1}^{(i,\beta)}$ (alternatively one of $x_{j,0}^{(i,\alpha)}, x_{j,1}^{(i,\alpha)}$) is information-theoretically hidden from R. Thus, it can successfully reveal $(E_1'', \ldots, E_t'')$ with $j \in E_\beta''$ (resp. $j \in E_\alpha''$) only if it guesses the missing value, i.e., with probability negligible in λ. More precisely in this case R will not be able to provide Ψ_β (resp. Ψ_α) consistent with $(E_1'', \ldots, E_t'')$. In other words, if $j \in E_\alpha' \cap E_\beta'$ for $\alpha \neq \beta$, then for every reveal $(E_1'', \ldots, E_t'')$ that is accepted by the sender it must hold that $j \notin \cup_k E_k''$. Indeed, it can be observed that any reveal by R will be rejected by S. In particular, R cannot reveal $j \notin \cup_k E_k''$ either, since in this case it will be required to produce both $x_{j,0}^{(i,k)}, x_{j,1}^{(i,k)}$ for every $k \in [t]$. As pointed out earlier, R cannot do this except with negligible probability for $k \in \{\alpha, \beta\}$. Recall that in this case, the simulator S set flag $= 1$ and thus will abort in the reveal phase. Therefore, in this case it holds that the real protocol is indistinguishable from the ideal simulation.

Privacy Against Corrupt Sender. First observe that in the output phase of each instance of $\pi_{\text{ccot}}^{\text{bat,sin}}$ information flows only from the sender to the receiver during the output phase. Thus privacy at the end of the output phase trivially holds, and in particular S has no information about the sets $E_1, \ldots, E_t$. It remains to be shown that the information revealed by R to S in the reveal phase does not leak any information about R's input bits $\mathbf{b}_1, \ldots, \mathbf{b}_t$. For simplicity, first consider the case when S is honest. In this case, observe that in a given instance of $\pi_{\text{ccot}}^{\text{bat,sin}}$, say the k-th instance, R's reveal message depends on input $\mathbf{b}_k$ and is independent of $\{\mathbf{b}_\alpha\}_{\alpha \neq k}$. Privacy then follows from the privacy guaranteed by (each instance of) $\pi_{\text{ccot}}^{\text{bat,sin}}$. On the other hand, when S is corrupt, R's reveal message in the k-th instance of $\pi_{\text{ccot}}^{\text{bat,sin}}$ depends on its input $\mathbf{b}_k$ and whether S's cheating attempt (if any) was detected in any instance. Privacy follows from the fact that each instance of $\pi_{\text{ccot}}^{\text{bat,sin}}$ preserves privacy of R's inputs.

Inputs: S holds $X^{(1)}, \ldots, X^{(m)}$ where $X^{(i)} = (x_{1,0}^{(i)}, x_{1,1}^{(i)}), \ldots, (x_{n,0}^{(i)}, x_{n,1}^{(i)})$. R holds pairwise non-intersecting sets of indices $E_1, \ldots, E_t \subseteq [n]$ and selection bits $\mathbf{b}_1 = (b_{1,1}, \ldots, b_{1,m}), \ldots, \mathbf{b}_t = (b_{t,1}, \ldots, b_{t,m})$.

Protocol:

1. S performs a t-out-of-t additive sharing of each $x_{j,b}^{(i)}$ value. Denote the shares of $x_{j,b}^{(i)}$ by $\{x_{j,b}^{(i,k)}\}_{k \in [t]}$.

2. S and R participate in t parallel instances of protocol $\pi_{ccot}^{bat,sin}$ in the following way: In the k-th instance:
   - S inputs $X^{(1,k)}, \ldots, X^{(m,k)}$ where $X^{(i,k)} = (x_{1,0}^{(i,k)}, x_{1,1}^{(i,k)}), \ldots, (x_{n,0}^{(i,k)}, x_{n,1}^{(i,k)})$ and internally uses "check values" $\phi_1^k, \ldots, \phi_n^k$.
   - R inputs $[n] \setminus E_k$ as the check set, along with selection bits $b_{k,1}, \ldots, b_{k,m}$.
   - At the end of the output phase, R receives the following:
     - For each $j \in [n] \setminus E_k$ the values $\{(\tilde{x}_{j,0}^{(i,k)}, \tilde{x}_{j,1}^{(i,k)})\}_{i \in [m]}$ and "checkset value" Ψ_k.
     - For each $j \in E_k$ the values $\{\tilde{x}_{j,b_{k,i}}^{(i,k)}\}_{i \in [m]}$ and "check value" Φ_k.

3. R sets $\mathbf{J} = (E_1, \ldots, E_t)$, $\Psi = \emptyset$, $\Phi = \emptyset$, and does the following:
   - If there exists $k \in [t]$ such that $E_k \neq \emptyset$ but $\Psi_k = \Phi_k = \emptyset$ holds, then set $\mathbf{J} = \Psi = \Phi = \emptyset$.
   - Else, set $\Psi = \{\Psi_k\}_{k \in [t]}$ and $\Phi = \{\Phi_k\}_{k \in [t]}$.

4. $\forall k \in [t]$, $\forall j \in E_k$, $\forall i \in [m]$, R reconstructs $x_{j,b_{k,i}}^{(i)} = \bigoplus_{\ell \in [t]} x_{j,b_{k,i}}^{(i,\ell)}$.

5. $\forall j \in [n] \setminus \cup_k E_k$, $\forall i \in [m]$, $\forall b \in \{0,1\}$, R reconstructs $x_{j,b}^{(i)} = \bigoplus_{\ell \in [t]} x_{j,b}^{(i,\ell)}$.

Reveal phase: R sends $(\mathbf{J}, \Psi, \Phi)$ to S. S aborts if these values are not consistent with its input/check values.

Fig. 7. Realizing $\mathcal{F}_{mcot}^{+}$ in the $\mathcal{F}_{OT}$-hybrid model.

Additional Optimizations. Instead of sending the values $\Psi = \{\tilde{x}_{j,0}, \tilde{x}_{j,1}\}_{j \in J}$ and $\Phi = \{\tilde{\phi}_j\}_{j \notin J}$, R could send $(J, H'(\Psi), H''(\Phi))$ to S, where H', H'' are modeled as collision-resistant hash functions (alternatively, random oracles). Note that these optimizations are applicable in a straightforward way in other constructions we present. We omit detailing them to keep the exposition more clear.

In applications to secure computation, full receiver simulation in CCOT is also not required. We require only privacy, i.e., we do not need to consider the joint distribution of receiver's view and sender's inputs. This is because sender's inputs are just random keys for the garbled circuits, and in the simulation of the 2PC protocol, it is the simulator that will generate these keys. On the other hand, extracting receiver inputs is very crucial in order to enable the simulator to generate correctly faked garbled circuits. However our definitions will require full receiver simulation (including extraction). Fortunately, achieving full receiver simulation comes only with a small multiplicative overhead.

Summary of Efficiency. All our protocols are presented in the $\mathcal{F}_{OT}$-hybrid model and thus can take advantage of OT extension techniques. Further, using standard leveraging techniques (such as ones used in [9]), OT extension of [14], the XOR-tree encoding scheme of [13], and the constructions in

Figs. 4, 5, 6, and 7, one can obtain a rate-1/6 construction for $\mathcal{F}_{\text{ccot}}^{\text{bat,sin}}$ (in the non-programmable RO model) with sender-simulatability and receiver-privacy as in Definition 4. In concrete terms, it is easy to verify that the additional overhead of realizing $\mathcal{F}_{\text{ccot}}^{\text{bat,sin}}$ is $\leq 6 \cdot \max(4m, 8\sigma)$. The efficiency of our CCOT protocol in the single execution setting is comparable to that of XOR-tree encodings of [23], but is clearly better than DDH-based CCOT [22,24] since we take advantage of OT extension (under the assumption that correlation-robust hash functions exist [12,14,29]). Finally, we can realize $\mathcal{F}_{\text{mcot}}^{+}$ (in the non-programmable random oracle model) with sender-simulatability and receiver-privacy as in Definition 4 while bearing an overhead at most t over the cost of realizing $\mathcal{F}_{\text{ccot}}^{\text{bat,sin}}$ where t denotes the number of executions.

References

1. Aiello, W., Ishai, Y., Reingold, O.: Priced oblivious transfer: how to sell digital goods. In: Pfitzmann, B. (ed.) EUROCRYPT 2001. LNCS, vol. 2045, pp. 119–135. Springer, Heidelberg (2001)

2. Applebaum, B., Ishai, Y., Kushilevitz, E., Waters, B.: Encoding functions with constant online rate or how to compress garbled circuits keys. In: Canetti, R., Garay, J.A. (eds.) CRYPTO 2013, Part II. LNCS, vol. 8043, pp. 166–184. Springer, Heidelberg (2013)

3. Bellare, M., Hoang, V.T., Rogaway, P.: Adaptively secure garbling with applications to one-time programs and secure outsourcing. In: Wang, X., Sako, K. (eds.) ASIACRYPT 2012. LNCS, vol. 7658, pp. 134–153. Springer, Heidelberg (2012)

4. Cachin, C., Micali, S., Stadler, M.A.: Computationally private information retrieval with polylogarithmic communication. In: Stern, J. (ed.) EUROCRYPT 1999. LNCS, vol. 1592, pp. 402–414. Springer, Heidelberg (1999)

5. David, B.M., Nishimaki, R., Ranellucci, S., Tapp, A.: Generalizing efficient multiparty computation. In: Lehmann, A., Wolf, S. (eds.) Information Theoretic Security. LNCS, vol. 9063, pp. 15–32. Springer, Heidelberg (2015)

6. Feige, U., Kilian, J., Naor, M.: A minimal model for secure computation (extended abstract). In: STOC, pp. 554–563 (1994)

7. Frederiksen, T.K., Jakobsen, T.P., Nielsen, J.B.: Faster maliciously secure two-party computation using the GPU. In: Abdalla, M., De Prisco, R. (eds.) SCN 2014. LNCS, vol. 8642, pp. 358–379. Springer, Heidelberg (2014)

8. Freedman, M.J., Ishai, Y., Pinkas, B., Reingold, O.: Keyword search and oblivious pseudorandom functions. In: Kilian, J. (ed.) TCC 2005. LNCS, vol. 3378, pp. 303–324. Springer, Heidelberg (2005)

9. Garay, J.A., Ishai, Y., Kumaresan, R., Wee, H.: On the complexity of UC commitments. In: Nguyen, P.Q., Oswald, E. (eds.) EUROCRYPT 2014. LNCS, vol. 8441, pp. 677–694. Springer, Heidelberg (2014)

10. Huang, Y., Katz, J., Kolesnikov, V., Kumaresan, R., Malozemoff, A.J.: Amortizing garbled circuits. In: Garay, J.A., Gennaro, R. (eds.) CRYPTO 2014, Part II. LNCS, vol. 8617, pp. 458–475. Springer, Heidelberg (2014)

11. Huang, Y., Katz, J., Evans, D.: Efficient secure two-party computation using symmetric cut-and-choose. In: Canetti, R., Garay, J.A. (eds.) CRYPTO 2013, Part II. LNCS, vol. 8043, pp. 18–35. Springer, Heidelberg (2013)

12. Ishai, Y., Kilian, J., Nissim, K., Petrank, E.: Extending oblivious transfers efficiently. In: Boneh, D. (ed.) CRYPTO 2003. LNCS, vol. 2729, pp. 145–161. Springer, Heidelberg (2003)
13. Ishai, Y., Kushilevitz, E., Ostrovsky, R., Prabhakaran, M., Sahai, A.: Efficient non-interactive secure computation. In: Paterson, K.G. (ed.) EUROCRYPT 2011. LNCS, vol. 6632, pp. 406–425. Springer, Heidelberg (2011)
14. Ishai, Y., Prabhakaran, M., Sahai, A.: Founding cryptography on oblivious transfer – efficiently. In: Wagner, D. (ed.) CRYPTO 2008. LNCS, vol. 5157, pp. 572–591. Springer, Heidelberg (2008)
15. Kilian, J.: Founding cryptography on OT. In: STOC, pp. 20–31 (1988)
16. Kolesnikov, V., Kumaresan, R.: Improved OT extension for transferring short secrets. In: Canetti, R., Garay, J.A. (eds.) CRYPTO 2013, Part II. LNCS, vol. 8043, pp. 54–70. Springer, Heidelberg (2013)
17. Kolesnikov, V.: Gate evaluation secret sharing and secure one-round two-party computation. In: Roy, B. (ed.) ASIACRYPT 2005. LNCS, vol. 3788, pp. 136–155. Springer, Heidelberg (2005)
18. Kolesnikov, V., Schneider, T.: Improved garbled circuit: free XOR gates and applications. In: Aceto, L., Damgård, I., Goldberg, L.A., Halldórsson, M.M., Ingólfsdóttir, A., Walukiewicz, I. (eds.) ICALP 2008, Part II. LNCS, vol. 5126, pp. 486–498. Springer, Heidelberg (2008)
19. Kreuter, B., Shelat, A., Shen, C.: Billion-gate secure computation with malicious adversaries. In: USENIX (2012)
20. Kushilevitz, E., Ostrovsky, R.: Replication is NOT needed: SINGLE database, computationally-private information retrieval. In: FOCS, pp. 364–373 (1997)
21. Lindell, Y., Riva, B.: Cut-and-Choose Yao-Based secure computation in the online/offline and batch settings. In: Garay, J.A., Gennaro, R. (eds.) CRYPTO 2014, Part II. LNCS, vol. 8617, pp. 476–494. Springer, Heidelberg (2014)
22. Lindell, Y.: Fast cut-and-choose based protocols for malicious and covert adversaries. In: Canetti, R., Garay, J.A. (eds.) CRYPTO 2013, Part II. LNCS, vol. 8043, pp. 1–17. Springer, Heidelberg (2013)
23. Lindell, Y., Pinkas, B.: An efficient protocol for secure two-party computation in the presence of malicious adversaries. In: Naor, M. (ed.) EUROCRYPT 2007. LNCS, vol. 4515, pp. 52–78. Springer, Heidelberg (2007)
24. Lindell, Y., Pinkas, B.: Secure two-party computation via cut-and-choose oblivious transfer. In: Ishai, Y. (ed.) TCC 2011. LNCS, vol. 6597, pp. 329–346. Springer, Heidelberg (2011)
25. Lindell, Y., Pinkas, B., Smart, N.P.: Implementing two-party computation efficiently with security against malicious adversaries. In: Ostrovsky, R., De Prisco, R., Visconti, I. (eds.) SCN 2008. LNCS, vol. 5229, pp. 2–20. Springer, Heidelberg (2008)
26. Mohassel, P., Riva, B.: Garbled circuits checking garbled circuits: more efficient and secure two-party computation. In: Canetti, R., Garay, J.A. (eds.) CRYPTO 2013, Part II. LNCS, vol. 8043, pp. 36–53. Springer, Heidelberg (2013)
27. Mossel, E., Shpilka, A., Trevisan, L.: On e-biased generators in NC0. In: FOCS, pp. 136–145 (2003)
28. Naor, M., Pinkas, B.: Efficient oblivious transfer protocols. In: SODA, pp. 448–457 (2001)
29. Nielsen, J.B., Nordholt, P.S., Orlandi, C., Burra, S.S.: A new approach to practical active-secure two-party computation. In: Safavi-Naini, R., Canetti, R. (eds.) CRYPTO 2012. LNCS, vol. 7417, pp. 681–700. Springer, Heidelberg (2012)

30. Pinkas, B., Schneider, T., Smart, N.P., Williams, S.C.: Secure two-party computation is practical. In: Matsui, M. (ed.) ASIACRYPT 2009. LNCS, vol. 5912, pp. 250–267. Springer, Heidelberg (2009)
31. Shelat, A., Shen, C.: Two-output secure computation with malicious adversaries. In: Paterson, K.G. (ed.) EUROCRYPT 2011. LNCS, vol. 6632, pp. 386–405. Springer, Heidelberg (2011)
32. Shelat, A., Shen, C.: Fast two-party secure computation with minimal assumptions. In: CCS, pp. 523–534 (2013)
33. Yao, A.: How to generate and exchange secrets. In: FOCS, pp. 162–167 (1986)

Public Key Encryption

An Asymptotically Optimal Method for Converting Bit Encryption to Multi-Bit Encryption

Takahiro Matsuda$^{(\boxtimes)}$ and Goichiro Hanaoka

National Institute of Advanced Industrial Science and Technology (AIST),
Tokyo, Japan
{t-matsuda,hanaoka-goichiro}@aist.go.jp

Abstract. Myers and Shelat (FOCS 2009) showed how to convert a chosen ciphertext secure (CCA secure) PKE scheme that can encrypt only 1-bit plaintexts into a CCA secure scheme that can encrypt arbitrarily long plaintexts (via the notion of key encapsulation mechanism (KEM) and hybrid encryption), and subsequent works improved efficiency and simplicity. In terms of efficiency, the best known construction of a CCA secure KEM from a CCA secure 1-bit PKE scheme, has the public key size $\Omega(k) \cdot |pk|$ and the ciphertext size $\Omega(k^2) \cdot |c|$, where k is a security parameter, and $|pk|$ and $|c|$ denote the public key size and the ciphertext size of the underlying 1-bit scheme, respectively.

In this paper, we show a new CCA secure KEM based on a CCA secure 1-bit PKE scheme which achieves the public key size $2 \cdot |pk|$ and the ciphertext size $(2k + o(k)) \cdot |c|$. These sizes are asymptotically optimal in the sense that they are the same as those of the simplest "bitwise-encrypt" construction (seen as a KEM by encrypting a k-bit random session-key) that works for the chosen plaintext attack and non-adaptive chosen ciphertext attack settings. We achieve our main result by developing several new techniques and results on the "double-layered" construction (which builds a KEM from an inner PKE/KEM and an outer PKE scheme) by Myers and Shelat and on the notion of detectable PKE/KEM by Hohenberger, Lewko, and Waters (EUROCRYPT 2012).

1 Introduction

1.1 Background and Motivation

In this paper, we revisit the problem of how to construct a chosen ciphertext secure (CCA2, or just CCA) public key encryption (PKE) scheme that can encrypt plaintexts of arbitrary length from a CCA secure PKE scheme whose plaintext space is only 1-bit. (Hereafter, we call a PKE scheme whose plaintext space is $\{0,1\}^n$ an *n-bit PKE scheme*.) It is well-known that if we only consider chosen plaintext attack (CPA) and non-adaptive chosen ciphertext attack (CCA1) settings, then the simple(st) *"bitwise-encrypt"* construction suffices, in which a plaintext is encrypted bit-by-bit (under the same public key) by a 1-bit

© International Association for Cryptologic Research 2015
T. Iwata and J.H. Cheon (Eds.): ASIACRYPT 2015, Part I, LNCS 9452, pp. 415–442, 2015.
DOI: 10.1007/978-3-662-48797-6_18

PKE scheme, and the concatenation of all ciphertexts is regarded as a ciphertext of the construction. However, for the CCA setting, until recently, the simple question of how (and even whether) one can realize such a "1-bit-to-multi-bit" conversion had been left open.

This open problem was resolved affirmatively by Myers and Shelat [20]. They actually constructed a CCA secure key encapsulation mechanism (KEM) which encrypts a random session-key, and can be used together with a CCA secure symmetric key encryption (SKE) scheme to achieve a full-fledged CCA secure PKE scheme via hybrid encryption [8]. One of the important steps of the approach by Myers and Shelat is to consider the "*double-layered*" construction of a KEM from an "inner" PKE scheme and an "outer" PKE scheme, where the inner ciphertext encrypts a plaintext (or a session-key if one wants to construct a KEM) and a randomness used for outer encryption, and the outer ciphertext encrypts the inner ciphertext using the randomness encrypted in the inner ciphertext. To decrypt a ciphertext, one first decrypts the outer ciphertext, and then the resulting inner ciphertext, to recover a plaintext and a randomness (for outer encryption), and the plaintext is output if the re-encryption of the inner ciphertext using the recovered randomness results in the outer ciphertext. Myers and Shelat showed that if the outer scheme that is built from a 1-bit scheme satisfies the security notion called "unquoted CCA" (UCCA) security (which is a weaker security notion than CCA security that can be considered only for a PKE scheme constructed based on 1-bit PKE scheme), and the inner scheme satisfies "1-wise non-malleability against UCCA" (which has a similar flavor to 1-bounded CCA security [7]), the resulting construction achieves CCA security.

The efficiency and simplicity of the construction by Myers and Shelat were improved by Hohenberger, Lewko, and Waters [16]. Specifically, they introduced the notion of a *detectable PKE* scheme, which is a PKE scheme that has an efficiently computable predicate F as part of the syntax, and whose security notions are defined with respect to this F. In particular, they introduced the notions of *detectable CCA* (DCCA) security (which is a relaxed variant of CCA security) and *unpredictability*, and considered a construction which has a mixed flavor of the double-layered construction of Myers and Shelat, and the double (parallel) encryption of Naor and Yung [21] (this construction has two PKE schemes for the outer encryption). They showed that if the "inner" PKE scheme satisfies DCCA security and unpredictability, and the "outer" PKE schemes are CPA secure and 1-bounded CCA secure [7], respectively, then the resulting PKE scheme is CCA secure. They also showed that the "bitwise-encrypt" construction based on a CCA secure 1-bit PKE scheme yields a DCCA secure and unpredictable detectable PKE scheme for long plaintexts, and thus achieves a 1-bit-to-multi-bit conversion for CCA security. (In their construction, in fact a 1-bit scheme satisfying only DCCA security and unpredictability suffices as the building block.) The efficiency of the construction in [16] was further improved by Matsuda and Hanaoka [19] using the ideas and techniques of hybrid encryption.

Despite the elegant ideas employed in [16,19,20], however, even in the best construction of [19] (in terms of efficiency), the public key size is $\Omega(k) \cdot |pk|$ and the ciphertext size (when seen as a KEM) is $\Omega(k^2) \cdot |c|$, where k is a security parameter, and $|pk|$ and $|c|$ denote the public key size and the ciphertext size of a CCA secure 1-bit scheme, respectively. On the other hand, for constructing a CPA (resp. CCA1) secure KEM from a CPA (resp. CCA1) secure 1-bit scheme, one can use the above mentioned bitwise-encrypt construction in which one encrypts a k-bit random string and regards this as a session-key of a KEM. Note that the public key size of this KEM is just $|pk|$ and the ciphertext size is $k \cdot |c|$. Compared to this simplest and most straightforward method, in the CCA setting, the known constructions have the public key size and the ciphertext size that are at least $\Omega(k)$ times larger.

Motivated by the above, in this paper we study the following question: *How efficient can a 1-bit-to-multi-bit conversion for CCA security be?*

1.2 Our Contributions

As our main result, we show a new 1-bit-to-multi-bit construction for the CCA setting, i.e., a construction of a CCA secure KEM based on a CCA secure 1-bit PKE scheme, with much better asymptotic efficiency than the existing constructions. Specifically, our construction achieves the public key size $2 \cdot |pk|$, and the ciphertext size $(2k + o(k)) \cdot |c| = O(k) \cdot |c|$, which are asymptotically optimal in the sense that these sizes are (except for a constant factor) the same as for the simple bitwise-encrypt construction for CPA and CCA1 security.

We achieve our main result by developing several new techniques and results on the double-layered construction of Myers and Shelat [20] and on the notion of detectable PKE/KEM by Hohenberger, Lewko, and Waters [16]. Our technical contributions in this paper lie in (1) coming up with appropriate security notions for detectable PKE/KEM so that we can conduct CCA security proofs for the double-layered construction using the language of detectable PKE/KEM (without addressing the details of how each of the inner and outer schemes is constructed) which we believe helps us understanding our proposed construction (and more generally the double-layered approach itself) in a clearer manner, and (2) showing how one can realize the inner and outer schemes (satisfying the requirements of our security proofs) from a CCA secure 1-bit PKE scheme, so that the resulting CCA secure KEM achieves asymptotically optimal efficiency with respect to the bitwise-encrypt construction.

Below we explain more technical details of our results.

New Security Notions for Detectable PKE/KEM. In Sect. 3, we introduce new security notions for detectable PKE and detectable KEMs. Recall that DCCA security of [16] is defined like ordinary CCA security, except that in the security experiment, the decryption oracle is restricted according to the predicate F (which is a part of the syntax of detectable PKE/KEM): an adversary is not allowed to query a ciphertext c such that $F(c^*, c) = 1$ where c^* is the challenge ciphertext. The first notion we introduce is a weak form of *non-malleability*

[3,12,22] under DCCA that we simply name wNM-DCCA *security*, which is defined like DCCA security except that we allow an adversary to make one "unrestricted" decryption query (which is not affected by the restriction of F). We also introduce an even weaker variant, which is a "replayable"-CCA-analogue [4] of wNM-DCCA security, which we call wRNM-DCCA *security*, that is defined like wNM-DCCA security except that the final unrestricted decryption query (and only this query) is answered like a decryption query in the replayable CCA security.

We also introduce a new security notion for detectable PKE/KEM that we call *randomness-inextractability*. Recall that a DCCA secure detectable PKE scheme is meaningful only if it also satisfies another security notion that prevents the predicate F from outputting 1 for every input (which makes DCCA security equivalent to CPA security). *Unpredictability* [16] is one example of a security notion that prevents DCCA security from being trivial, which ensures that a ciphertext c satisfying $F(c^*, c) = 1$ is hard to find without seeing c^*. Randomness-inextractability is another such security notion for detectable PKE: Informally, it requires that if an adversary is given a ciphertext c^* (that encrypts a plaintext m of the adversary's choice), it cannot come up with a pair of a (possibly different) plaintext m' and randomness r' such that $F(c^*, c') = 1$, where c' is the encryption of m' generated using the randomness r'. We also show that randomness-inextractability and unpredictability do not imply each other, even if we combine one notion with wNM-DCCA security. See Sect. 3 for the details.

New CCA Security Proofs for the Double-Layered Construction Based on Detectable PKE/KEM. In Sect. 4, we show our main technical results: two new CCA security proofs for the double-layered construction of Myers and Shelat [20]. Our first security proof shows that if the inner KEM is a detectable KEM satisfying DCCA security and unpredictability, and the outer PKE scheme is a detectable PKE scheme satisfying wRNM-DCCA security and randomness-inextractability, then the KEM obtained from the double-layered construction is CCA secure. Our main result with asymptotically optimal efficiency is obtained from this security proof.

Our second security proof shows that if the inner KEM is wNM-DCCA secure and unpredictable, and the outer PKE scheme is DCCA secure and randomness-inextractable, then the KEM obtained from the double-layered construction is CCA secure. Interestingly, this security proof can be seen as a generalization of Myers-Shelat's original security proof of their construction [20].

Both of the security proofs have similar flavors to the security proofs of [16,19]. Namely, DCCA security of the inner KEM guarantees that a session-key (hidden in the challenge ciphertext) is random as long as an adversary does not submit a "dangerous" decryption query (which are defined with respect to the predicate F from the inner detectable KEM), and we then upperbound the probability that the adversary comes up with such "dangerous" decryption queries to be negligible by the combination of the security properties of the outer PKE scheme and the inner KEM. However, unlike the previous works [16,19] that use a "detectable" primitive only for the inner scheme, we employ a detectable primitive also for the outer scheme. Consequently, we have to deal with two types of "dangerous" decryption queries in the security proofs: an "inner-dangerous" query and an "outer-dangerous" query, which, as the names indicate, are related to the inner KEM and the outer PKE scheme, respectively. Our two security proofs differ in the treatment of the inner- and outer-dangerous queries, which lead to the difference between which of the inner KEM or the outer PKE scheme needs to be "non-malleable" under DCCA. In both of the proofs, randomness-inextractability of the outer PKE scheme is used to show that the adversary's outer-dangerous queries do not help.

We also show an evidence that indicates that our reliance on "non-malleability" under DCCA for either the inner KEM or the outer PKE scheme would be unavoidable, by showing a counterexample for the double-layered construction that does not achieve CCA security if the inner and outer schemes only satisfy DCCA security, unpredictability, and randomness-inextractability. For the details, see Sect. 4.

A Detectable PKE Scheme Satisfying wRNM-DCCA *Security and Randomness-Inextractability from CCA Secure 1-bit PKE.* In Sect. 5, we show a construction of a detectable PKE scheme satisfying wRNM-DCCA security and randomness-inextractability, using a CCA secure 1-bit PKE scheme and a *non-malleable code* [13] for "bitwise-tampering and bit-level permutations" [1,2]. The idea of this construction is based on the recent result by Agrawal et al. [2] who showed how to transform a 1-bit commitment scheme secure against chosen commitment attacks (CCA) into a non-malleable string commitment scheme: We first encode a plaintext by a non-malleable code, and then do "bitwise-encryption" of the encoded value by a CCA secure 1-bit PKE scheme. (Due to its structure, we call this construction the "*Encode-then-Bitwise-Encrypt*" (EtBE) construction.) Our contribution regarding this construction is to clarify that the approach of [2] also works well for detectable PKE as we require.

Agrawal et al. [1] recently constructed a non-malleable code for the above mentioned class of functions with "optimal rate", meaning that the ratio between the length n of a codeword and the length k of a message can be made arbitrarily close to 1 (i.e. $n = k + o(k)$). We employ this non-malleable code to achieve the asymptotic efficiency of our proposed KEM.

The Proposed 1-Bit-to-Multi-Bit Conversion, and More. Our main result, i.e. a CCA secure KEM from a CCA secure 1-bit PKE scheme that achieves optimal asymptotic efficiency in terms of the public key and ciphertext sizes, is obtained by using the above mentioned detectable PKE scheme (together with some hybrid encryption techniques) as the outer PKE scheme, and using the bitwise-encrypt construction of a detectable KEM as the inner KEM, in the double-layered construction, via our first security proof. In Sect. 6, we show the full description of our construction. As noted above, our construction uses only two key pairs of the underlying 1-bit PKE scheme.

Interestingly, there we also show that if a 2-bit PKE scheme can be used instead of a 1-bit PKE scheme, then one can construct a CCA secure KEM (with almost the same construction as our main construction) that uses only one key pair.

On the Necessity of Two Key Pairs. As mentioned above, our proposed KEM from a 1-bit PKE scheme uses two key pairs of the underlying CCA secure 1-bit PKE scheme. Given this, it is natural to ask if the number 2 of key pairs of the underlying 1-bit scheme is optimal for 1-bit-to-multi-bit constructions for CCA security. Although we could not answer this question affirmatively or negatively, we show that the one-key variant of our proposed construction is vulnerable to a CCA attack. (This result is shown in the full version.) This negative result shows a necessity of different techniques and ideas than ours towards answering the question. It also contrasts strikingly with our 2-bit-to-multi-bit construction for CCA security that uses only one key pair of the underlying 2-bit scheme.

We leave it as an open problem to clarify whether one can achieve a 1-bit-to-multi-bit conversion using only one key pair of the underlying 1-bit scheme, or it is generally impossible.

1.3 Related Work

The double-layered construction [16,20], and extension of the plaintext space of encryption schemes based on it, have been used in several works: Lin and Tessaro [18] showed how to turn a 1-bit PKE scheme whose correctness is not perfect and which only satisfies weak CCA security (weak in the sense that an adversary may have bounded but non-negligible CCA advantage), into a PKE scheme (with a large plaintext space) satisfying ordinary CCA security, via the construction of [16]. Dachman-Soled et al. [9] studied the notion of "enhanced" CCA security for PKE schemes with randomness recovery property, where the decryption oracle in the security experiment returns not only the decryption result of a queried ciphertext but also a randomness that is consistent with the ciphertext, and (among other things) showed that the construction of [16] can be used to achieve a 1-bit-to-multi-bit conversion for enhanced CCA security. Most recently, Kitagawa et al. [17] showed that a simpler variant of the double-layered construction which does not have validity check by re-encryption in the decryption algorithm, can be used to extend the plaintext space of PKE satisfying key-dependent message (KDM) security against CCA with respect to projection functions (projection-KDM-CCA security).

Very recently, Coretti et al. [6] showed a 1-bit-to-multi-bit conversion for a PKE scheme. However, the security notion considered in their construction is so-called "self-destruct" CCA security, which is defined like ordinary CCA security except that in the security experiment, once an adversary submits an invalid ciphertext (which does not decrypt to a valid plaintext) as a decryption query, the decryption oracle "self-destructs", i.e. it will not answer to subsequent decryption queries. This security notion is strictly weaker than ordinary CCA security. Furthermore, in another recent work, Coretti et al. [5] considered non-malleability under self-destruct CCA, which is also strictly weaker than ordinary CCA security, and showed a 1-bit-to-multi-bit conversion for a PKE scheme satisfying this security notion. The 1-bit-to-multi-bit constructions of [5,6] share the same idea with Agrawal et al.'s conversion (and hence with our "outer" PKE scheme): first encode a plaintext by a suitable non-malleable code, and then do bitwise encryption. The main differences between these works [5,6] and our "double-layered" construction are: (1) Ours achieves ordinary (full) CCA security, while they achieve weaker security notions. (2) Our construction uses only two key pairs of the underlying 1-bit scheme, while the constructions in [5,6] use $O(k)$ key pairs, of the building block 1-bit scheme. (3) The requirements of the used non-malleable codes are all different: [5,6] need stronger form of non-malleability called "continuous" non-malleability [15] (and its extension), while we only need the original definition of non-malleability in [13] that captures "one-time" tampering.; The tampering functions with respect to which non-malleability is considered in [5,6] are based on bit-wise tampering (extended to take into account continuous non-malleability), while ours requires additionally non-malleability against bit-level permutation (as in [1,2]).

Paper Organization. The rest of this paper is organized as follows: Sect. 2 reviews the basic notation and definitions of cryptographic primitives. In Sect. 3, we define new security notions for detectable PKE, and also show several facts on them. In Sect. 4, we show our main technical result: new security proofs for the "double-layered" construction. We also explain some evidence that justifies our reliance on non-malleability under DCCA. In Sect. 5, we show how to build a detectable PKE scheme satisfying our new security notions based on a CCA secure 1-bit PKE scheme and a non-malleable code. In Sect. 6, we provide the full description of our proposed 1-bit-to-multi-bit construction. There we also explain our 2-bit-to-multi-bit construction with a single key pair. We give a comparison among 1-bit-to-multi-bit constructions in Sect. 7.

Due to space limitation, the proofs of the theorems and lemmas in this paper are omitted and will be given in the full version, and we only give proof sketches or intuitive explanations.

2   Preliminaries

In this section, we review the basic notation and the definitions for cryptographic primitives.

Basic Notation. $\mathbb{N}$ denotes the set of all natural numbers. For $n \in \mathbb{N}$, we define $[n] := \{1, \ldots, n\}$. "$x \leftarrow y$" denotes that x is chosen uniformly at random from y if y is a finite set, x is output from y if y is a function or an algorithm, or y is assigned to x otherwise. If x and y are strings, then "$|x|$" denotes the bit-length of x, "$x\|y$" denotes the concatenation x and y, and "$(x \overset{?}{=} y)$" is defined to be 1 if $x = y$ and 0 otherwise. "(P)PTA" stands for a *(probabilistic) polynomial time algorithm*. For a finite set S, "$|S|$" denotes its size. If $\mathcal{A}$ is a probabilistic algorithm then "$y \leftarrow \mathcal{A}(x; r)$" denotes that $\mathcal{A}$ computes y as output by taking x as input and using r as randomness. If furthermore $\mathcal{O}$ is an algorithm, then "$\mathcal{A}^{\mathcal{O}}$" denotes that $\mathcal{A}$ has oracle access to $\mathcal{O}$. A function $\epsilon(\cdot) : \mathbb{N} \to [0,1]$ is said to be *negligible* if for all positive polynomials $p(k)$ and all sufficiently large $k \in \mathbb{N}$, we have $\epsilon(k) < 1/p(k)$. Throughout this paper, we use the character "k" to denote a security parameter.

2.1   (Detectable) Public Key Encryption

A public key encryption (PKE) scheme Π consists of the three PPTAs (PKG, Enc, Dec) with the following interface:

Key Generation:	Encryption:	Decryption:
$(pk, sk) \leftarrow \mathsf{PKG}(1^k)$	$c \leftarrow \mathsf{Enc}(pk, m)$	$m \text{ (or } \perp) \leftarrow \mathsf{Dec}(sk, c)$

where Dec is a deterministic algorithm, (pk, sk) is a public/secret key pair, and c is a ciphertext of a plaintext m under pk. We say that a PKE scheme satisfies *correctness* if for all $k \in \mathbb{N}$, all keys (pk, sk) output from $\mathsf{PKG}(1^k)$, and all plaintexts m, it holds that $\mathsf{Dec}(sk, \mathsf{Enc}(pk, m)) = m$.

Detectable PKE. In this paper, we use the notion of *detectable PKE* as defined in [16]. It is a PKE scheme that has a predicate F that tests whether two ciphertexts c and c' are "related" in the sense that to decrypt c, the information of the decryption result of c' is useful (and hence, revealing the decryption result of c' is "dangerous"). This predicate F is used to define multiple security notions of the primitive, and hence we explicitly define it as a part of the syntax of the primitive (this approach is also taken in [16,19]).

Formally, a tuple of PPTAs $\Pi = (\mathsf{PKG}, \mathsf{Enc}, \mathsf{Dec}, \mathsf{F})$ is said to be a *detectable PKE* scheme if $(\mathsf{PKG}, \mathsf{Enc}, \mathsf{Dec})$ constitutes PKE, and F is a predicate that takes a public key pk and two ciphertexts c, c' as input, and outputs either 0 or 1.

$\mathsf{Expt}_{\Pi,\mathcal{A}}^{\mathsf{ATK}}(k)$:
 $(pk, sk) \leftarrow \mathsf{PKG}(1^k)$
 $(m_0, m_1, \mathsf{st}) \leftarrow \mathcal{A}_1^{\mathcal{O}(\cdot)}(pk)$
 $b \leftarrow \{0, 1\}$
 $c^* \leftarrow \mathsf{Enc}(pk, m_b)$
 $b' \leftarrow \mathcal{A}_2^{\mathcal{O}(\cdot)}(\mathsf{st}, c^*)$
 Return $(b' \stackrel{?}{=} b)$

$\mathsf{Expt}_{\Pi,\mathcal{A}}^{\mathsf{UNP}}(k)$:
 $(pk, sk) \leftarrow \mathsf{PKG}(1^k)$
 $(m, c) \leftarrow \mathcal{A}^{\mathcal{O}(\cdot)}(pk)$
 $c^* \leftarrow \mathsf{Enc}(pk, m)$
 Return $\mathsf{F}(pk, c^*, c)$

$\mathsf{Expt}_{\Gamma,\mathcal{A}}^{\mathsf{ATK}}(k)$:
 $(pk, sk) \leftarrow \mathsf{KKG}(1^k)$
 $(c^*, K_1^*) \leftarrow \mathsf{Encap}(pk)$
 $K_0^* \leftarrow \mathcal{K}$
 $b \leftarrow \{0, 1\}$
 $b' \leftarrow \mathcal{A}^{\mathcal{O}(\cdot)}(pk, c^*, K_b^*)$
 Return $(b' \stackrel{?}{=} b)$

$\mathsf{Expt}_{\Gamma,\mathcal{A}}^{\mathsf{UNP}}(k)$:
 $(pk, sk) \leftarrow \mathsf{KKG}(1^k)$
 $c \leftarrow \mathcal{A}^{\mathcal{O}(\cdot)}(pk)$
 $(c^*, K^*) \leftarrow \mathsf{Encap}(pk)$
 Return $\mathsf{F}(pk, c^*, c)$

$\mathsf{Expt}_{C,\mathcal{A}}^{\mathcal{F}\text{-}\mathsf{NM}}(k)$:
 $(f, m_0, m_1, \mathsf{st}) \leftarrow \mathcal{A}_1(1^k)$
 $b \leftarrow \{0, 1\}$
 $s^* \leftarrow \mathsf{E}(1^k, m_b)$
 $s' \leftarrow f(s^*)$
 $m' \leftarrow \mathsf{D}(1^k, s')$
 If $m' \in \{m_0, m_1\}$ then
                          $m' \leftarrow \mathsf{same}$
 $b' \leftarrow \mathcal{A}_2(\mathsf{st}, m')$
 Return $(b' \stackrel{?}{=} b)$

Fig. 1. The experiments for defining the security of detectable PKE (left-top/bottom), of detectable KEM (center-top/bottom), and of an $\mathcal{F}$-non-malleable code (right). In the $\mathsf{ATK}(\in \{\mathsf{CCA}, \mathsf{DCCA}\})$ and UNP experiments for PKE (resp. KEM), $\mathcal{O}(\cdot)$ is the decryption oracle $\mathsf{Dec}(sk, \cdot)$ (resp. decapsulation oracle $\mathsf{Decap}(sk, \cdot)$). In the CCA (resp. DCCA) experiment for PKE, $\mathcal{A}_2$ is not allowed to query c^* (resp. ciphertexts c such that $\mathsf{F}(pk, c^*, c) = 1$). Similar restrictions apply to $\mathcal{A}$ in the $\mathsf{CCA}/\mathsf{DCCA}$ experiment for KEMs.

We require that for all $k \in \mathbb{N}$, all public keys pk output by $\mathsf{PKG}(1^k)$, and all ciphertexts c output by $\mathsf{Enc}(pk, \cdot)$, we have $\mathsf{F}(pk, c, c) = 1$.[1]

Security Notions. Here we recall *chosen ciphertext security (CCA security)* for PKE, and *detectable CCA (DCCA) security* and *unpredictability* for detectable PKE [16].

Let $\mathsf{ATK} \in \{\mathsf{CCA}, \mathsf{DCCA}\}$. For a (detectable) PKE scheme Π and an adversary $\mathcal{A} = (\mathcal{A}_1, \mathcal{A}_2)$, consider the ATK experiment $\mathsf{Expt}_{\Pi,\mathcal{A}}^{\mathsf{ATK}}(k)$ described in Fig. 1 (left-top). In the experiment, it is required that $|m_0| = |m_1|$, and $\mathcal{A}_2$ is not allowed to submit the "prohibited" queries to the decryption oracle: If $\mathsf{ATK} = \mathsf{CCA}$, then the prohibited query is c^*, and if $\mathsf{ATK} = \mathsf{DCCA}$, then the prohibited queries are c satisfying $\mathsf{F}(pk, c^*, c) = 1$. We say that a (detectable) PKE scheme Π is ATK secure if for all PPTAs $\mathcal{A}$, $\mathsf{Adv}_{\Pi,\mathcal{A}}^{\mathsf{ATK}}(k) := 2 \cdot |\Pr[\mathsf{Expt}_{\Pi,\mathcal{A}}^{\mathsf{ATK}}(k) = 1] - 1/2|$ is negligible.

For a detectable PKE scheme Π (with predicate F) and an adversary $\mathcal{A}$, consider the unpredictability experiment $\mathsf{Expt}_{\Pi,\mathcal{A}}^{\mathsf{UNP}}(k)$ described in Fig. 1 (left-bottom). We say that a detectable PKE scheme Π is *unpredictable* if for all PPTAs $\mathcal{A}$, $\mathsf{Adv}_{\Pi,\mathcal{A}}^{\mathsf{UNP}}(k) := \Pr[\mathsf{Expt}_{\Pi,\mathcal{A}}^{\mathsf{UNP}}(k) = 1]$ is negligible.

2.2 (Detectable) Key Encapsulation Mechanism

A key encapsulation mechanism (KEM) Γ consists of the three PPTAs (KKG, $\mathsf{Encap}, \mathsf{Decap}$) with the following interface:

Key Generation:	Encapsulation:	Decapsulation:
$(pk, sk) \leftarrow \mathsf{KKG}(1^k)$	$(c, K) \leftarrow \mathsf{Encap}(pk)$	K (or $\perp$) $\leftarrow \mathsf{Decap}(sk, c)$

where Decap is a deterministic algorithm, (pk, sk) is a public/secret key pair that defines a session-key space $\mathcal{K}$, and c is a ciphertext of a session-key $K \in \mathcal{K}$ under pk. We say that a KEM satisfies *correctness* if for all $k \in \mathbb{N}$, all keys (pk, sk) output from $\mathsf{KKG}(1^k)$ and all ciphertext/session-key pairs (c, K) output from $\mathsf{Encap}(pk)$, it holds that $\mathsf{Decap}(sk, c) = K$.

We also define a KEM-analogue of detectable PKE, which we call *detectable KEM*, as a KEM that has an efficiently computable predicate F whose interface is exactly the same as that of detectable PKE.

Security Notions. Here we review the definition of *CCA security* for a KEM, and the definitions of *DCCA security* and *unpredictability* for a detectable KEM.

Let $\mathsf{ATK} \in \{\mathsf{CCA}, \mathsf{DCCA}\}$. For a (detectable) KEM Γ and an adversary $\mathcal{A}$, consider the ATK experiment $\mathsf{Expt}_{\Gamma,\mathcal{A}}^{\mathsf{ATK}}(k)$ described in Fig. 1 (center-top). In the experiment, $\mathcal{A}$ is not allowed to submit the "prohibited" queries that are defined in the same way as those for the PKE case. We say that a (detectable) KEM Γ is ATK secure if for all PPTAs $\mathcal{A}$, $\mathsf{Adv}_{\Gamma,\mathcal{A}}^{\mathsf{ATK}}(k) := 2 \cdot |\Pr[\mathsf{Expt}_{\Gamma,\mathcal{A}}^{\mathsf{ATK}}(k) = 1] - 1/2|$ is negligible.

---

[1] This requirement is not explicitly defined in [16], but is actually necessary for DCCA security to be meaningful. Without this requirement, DCCA security is unachievable, as an adversary can submit the challenge ciphertext to the decryption oracle.

For a detectable KEM Γ (with predicate F) and an adversary $\mathcal{A}$, consider the unpredictability experiment $\mathsf{Expt}^{\mathsf{UNP}}_{\Gamma,\mathcal{A}}(k)$ described in Fig. 1 (center-bottom). We say that a detectable KEM Γ is *unpredictable* if for all PPTAs $\mathcal{A}$, $\mathsf{Adv}^{\mathsf{UNP}}_{\Gamma,\mathcal{A}}(k) :=$ $\Pr[\mathsf{Expt}^{\mathsf{UNP}}_{\Gamma,\mathcal{A}}(k) = 1]$ is negligible.

2.3 Non-malleable Codes

Here, we recall the definition of *non-malleable codes* [13].

A code $\mathcal{C}$ with message length $\kappa = \kappa(k)$ and codeword length $n = n(k)$ (called also an (n, κ)-code) consists of the two PPTAs (E, D): E is the encoding algorithm that takes 1^k and a message $m \in \{0,1\}^\kappa$ as input, and outputs a codeword $c \in \{0,1\}^n$.; D takes 1^k and c as input, and outputs $m \in \{0,1\}^\kappa$ or the special symbol $\perp$ indicating that c is invalid. We require for all $k \in \mathbb{N}$ and all messages $m \in \{0,1\}^\kappa$, it holds that $\mathsf{D}(1^k, \mathsf{E}(1^k, m)) = m$.

Non-malleability. Non-malleability for codes, formalized by Dziembowski et al. [13], is defined with respect to a class of tampering functions $\mathcal{F}$. Intuitively, non-malleability guarantees that if an encoding c of a message m is modified into $c' = f(c)$ by a function $f \in \mathcal{F}$, then the decoded value m' of c' is either the original message m itself, or a completely unrelated message (or $\perp$). Here we recall the indistinguishability-based definition which is most convenient for us to work with, which is called the "alternative-non-malleability" in [14, Definition A.1]. It was shown in [14] that this definition is equivalent to the original simulation-based definition for codes whose message length κ is superlogarithmic in k.

Let $n, \kappa : \mathbb{N} \to \mathbb{N}$ be positive polynomials of k such that $n(k) \geq \kappa(k)$. For an (n, κ)-code $\mathcal{C} = (\mathsf{E}, \mathsf{D})$, a class of functions $\mathcal{F} = \{\mathcal{F}_k : \{0,1\}^k \to \{0,1\}^k\}_{k\in\mathbb{N}}$, and an adversary $\mathcal{A} = (\mathcal{A}_1, \mathcal{A}_2)$, consider the $\mathcal{F}$-NM experiment $\mathsf{Expt}^{\mathcal{F}\text{-NM}}_{\mathcal{C},\mathcal{A}}(k)$ described in Fig. 1 (right). In the experiment, "same" is the special symbol indicating that the decoded message m' was either m_0 or m_1, and it is required that $f \in \mathcal{F}_n$ and $|m_0| = |m_1| = \kappa(k)$. We say that $\mathcal{C}$ is *non-malleable with respect to the function class $\mathcal{F}$* ($\mathcal{F}$-non-malleable, for short) if for all PPTAs[2] $\mathcal{A}$, $\mathsf{Adv}^{\mathcal{F}\text{-NM}}_{\mathcal{C},\mathcal{A}}(k) :=$ $2 \cdot |\Pr[\mathsf{Expt}^{\mathcal{F}\text{-NM}}_{\mathcal{C},\mathcal{A}}(k) = 1] - 1/2|$ is negligible. We also say that $\mathcal{C}$ is an $\mathcal{F}$-non-malleable code.

Classes of Tampering Functions. In this paper, we consider the following classes of functions.

Composition of "Bitwise Tampering" and "Bit-Level Permutation" $\mathcal{P}$:
Let $\mathsf{set}, \mathsf{reset}, \mathsf{forward}, \mathsf{toggle} : \{0,1\} \to \{0,1\}$ be the functions over a bit, defined by $\mathsf{set}(x) := 1$, $\mathsf{reset}(x) := 0$, $\mathsf{forward}(x) := x$, and $\mathsf{toggle}(x) := 1 - x$. We define $\mathcal{F}_{\mathsf{BIT}} := \{\mathsf{set}, \mathsf{reset}, \mathsf{forward}, \mathsf{toggle}\}$.
Let $\mathcal{P} = \{\mathcal{P}_n\}_{n\in\mathbb{N}}$ be the class of functions which first perform "bitwise tampering" to an input, followed by a "bit-level permutation." Namely, $\mathcal{P}_n$ is the set of all functions $f : \{0,1\}^n \to \{0,1\}^n$ that can be described by

---

[2] The original definition [13] considered security against computationally unbounded adversaries. In this paper, however, we only need security against PPTAs.

using n bitwise-tampering functions $f_1, \ldots, f_n \in \mathcal{F}_{\mathtt{BIT}}$ and a permutation $\pi : [n] \to [n]$, as follows:

$$x = (x_1 \| \ldots \| x_n) \overset{f}{\mapsto} \left(f_{\pi^{-1}(1)}(x_{\pi^{-1}(1)}) \| \cdots \| f_{\pi^{-1}(n)}(x_{\pi^{-1}(n)}) \right).$$

"Bit-Fixing" or "Quoting an Input without Duplicated Positions"
$\mathcal{Q}$: Let $\mathtt{one} : \{0,1\}^n \to \{0,1\}$ and $\mathtt{zero} : \{0,1\}^n \to \{0,1\}$ be the constant functions that output 1 and 0 for any n-bit inputs, respectively. Furthermore, for $j \in [n]$, let $\mathtt{quote}^j : \{0,1\}^n \to \{0,1\}$ be the "quoting" function that always outputs the j-th bit of its input.

Let $\mathcal{Q} = \{\mathcal{Q}_n\}_{n \in \mathbb{N}}$ be the class of functions each of whose output bits is either a "fixed value" or "quoting the input without duplicated positions." More formally, $\mathcal{Q}_n$ is the set of all functions $f : \{0,1\}^n \to \{0,1\}^n$ that can be decomposed to n functions $f_1, \ldots, f_n : \{0,1\}^n \to \{0,1\}$ so that $f(x) = (f_1(x) \| \ldots \| f_n(x))$ for all $x \in \{0,1\}^n$, and furthermore it holds that for every $i \in [n]$:

$$f_i \in \{\mathtt{one}, \mathtt{zero}\} \cup \left(\{\mathtt{quote}^j\}_{j \in [n]} \backslash \{f_j\}_{j \in [i-1]} \right).$$

Note that the above guarantees that there exist no indices $i, i', j \in [n]$ such that $f_i = f_{i'} = \mathtt{quote}^j$ and $i \neq i'$. We call this condition the *no duplicated quoting condition*.

Agrawal et al. [1] showed the following elegant result, which is crucial for the efficiency of our proposed KEM:

Lemma 1. [1] *There exists an explicit (n,k)-code such that (1) it is $\mathcal{P}$-non-malleable, and (2) its "rate", defined by k/n, asymptotically approaches to 1 as k increases (and hence $n = k + o(k)$).*

Furthermore, the following is implicitly used by Agrawal et al. [2], and also is useful for our purpose. (Although it is almost straightforward from the definitions of $\mathcal{P}$ and $\mathcal{Q}$, we will show its formal proof in the full version.)

Lemma 2. *For all $n \in \mathbb{N}$, $\mathcal{Q}_n \subseteq \mathcal{P}_n$. (This holds even if $\mathcal{F}_{\mathtt{BIT}}$ does not contain* $\mathtt{toggle}$.*) Hence, any $\mathcal{P}$-non-malleable code is also $\mathcal{Q}$-non-malleable.*

2.4 Other Standard Primitives

In this paper we also use a pseudorandom generator (PRG) G, and a $\mathtt{CCA}$ secure deterministic symmetric key encryption (SKE) $E = (\mathsf{SEnc}, \mathsf{SDec})$: For notation, encryption of a plaintext m using a key $K \in \{0,1\}^k$ is denoted by "$c \leftarrow \mathsf{SEnc}(K, m)$" where c is a ciphertext, and decryption of c using K is denoted by "$m \leftarrow \mathsf{SDec}(K, c)$" where m could be the invalid symbol $\perp$. Since their security definitions are standard, we omit them in the proceedings version.

3   New Security Notions for Detectable PKE and KEM

In this section, we introduce new security notions for detectable PKE: wNM-DCCA *security* and wRNM-DCCA *security* in Sect. 3.1, and *randomness-inextractability* in Sect. 3.2. We also show some useful facts regarding the new security notions in Sect. 3.3.

We also define wNM-DCCA security and randomness-inextractability for detectable KEMs. Since their definitions are straightforward KEM-analogues of those for detectable PKE in this section, we omit them here and formally provide them in the full version.

3.1   "Weak" Non-malleability Under DCCA and Its "Replayable" Variant

Here, we define a "weak" form of non-malleability against DCCA for detectable PKE, which we call wNM-DCCA *security*, that captures the intuition that a DCCA adversary who works in the DCCA experiment cannot come up with a ciphertext that is "meaningfully related" to the challenge ciphertext. Recall that the original definitions of non-malleability for PKE [3,12,22] ensure that an adversary cannot come up with even a vector of ciphertexts that are "meaningfully related" to the challenge ciphertext, while our notion here only requires that it cannot come up with only a single related ciphertext. Technically, following the formalizations in [3,20,22], we formalize wNM-DCCA security by modifying the original DCCA experiment (in which originally the usage of the decryption oracle is restricted according the predicate F of detectable PKE), so that at the end of the experiment an adversary is allowed to make a single "unrestricted" decryption query, regardless of F. Thus, it is like "1-bounded" CCA security [7], albeit an adversary has additionally access to DCCA decryption oracle. Myers and Shelat [20] defined a security notion for PKE-to-PKE constructions called "q-wise-non-malleability under UCCA." Our definition of wNM-DCCA security is a detectable-PKE-analogue of their 1-wise-non-malleability.

We also define a weaker variant of wNM-DCCA security, in the security experiment of which the final "unrestricted" decryption query is answered like a decryption query in the "replayable" CCA experiment [4], namely, if the decryption result is one of the challenge plaintexts that an adversary uses, then the adversary is only informed so and is not given the actual decryption result. Due to the lack of a better name, we call it wRNM-DCCA security (where R stands for "**R**eplayable").

Fomally, for a detectable PKE scheme $\Pi = (\mathsf{PKG}, \mathsf{Enc}, \mathsf{Dec}, \mathsf{F})$ and an adversary $\mathcal{A} = (\mathcal{A}_1, \mathcal{A}_2, \mathcal{A}_3)$, we define the wNM-DCCA experiment $\mathsf{Expt}_{\Pi,\mathcal{A}}^{\mathsf{wNM\text{-}DCCA}}(k)$ and the wRNM-DCCA experiment $\mathsf{Expt}_{\Pi,\mathcal{A}}^{\mathsf{wRNM\text{-}DCCA}}(k)$ described in Fig. 2 (left and center, respectively). In both of the experiments, it is required that $|m_0| = |m_1|$, and as in the DCCA experiment, $\mathcal{A}_2$ is not allowed to submit a decryption query c satisfying $\mathsf{F}(pk, c^*, c) = 1$ to the decryption oracle. The adversary's final "unrestricted" decryption query is captured by the ciphertext c' that is finally output by $\mathcal{A}_2$, and naturally it is required that $c' \neq c^*$. *However, we allow c' to be such* *that* $F(pk, c^*, c') = 1$. In the wRNM–DCCA experiment, "same" is the special symbol (which is distinguished from $\perp$) that indicates that $Dec(sk, c') \in \{m_0, m_1\}$.

Definition 1. *We say that a detectable PKE scheme Π is* wNM–DCCA *secure if for all PPTAs $\mathcal{A}$,* $Adv_{\Pi,\mathcal{A}}^{\text{wNM–DCCA}}(k) := 2 \cdot |\Pr[\text{Expt}_{\Pi,\mathcal{A}}^{\text{wNM–DCCA}}(k) = 1] - 1/2|$ *is negligible. We define* wRNM–DCCA *security analogously.*

3.2 Randomness-Inextractability

Here we introduce another security notion for detectable PKE that we call *randomness-inextractability.* Roughly, this security notion ensures that given the challenge ciphertext c^* (which is an encryption of a plaintext of an adversary's choice), an adversary cannot come up with a pair (m', r') of a plaintext and a randomness such that $F(pk, c^*, \text{Enc}(pk, m'; r')) = 1$. If the predicate $F(pk, c^*, c')$ tests the equality $(c^* \stackrel{?}{=} c')$, then this notion exactly demands that the randomness used in c^* cannot be recovered, and hence we use the name "randomness-inextractability" (although we allow more general predicates for F).

Formally, for a detectable PKE scheme $\Pi = (\text{PKG}, \text{Enc}, \text{Dec}, F)$ and an adversary $\mathcal{A} = (\mathcal{A}_1, \mathcal{A}_2)$, consider the R-Inext experiment described in Fig. 2 (right).

Definition 2. *We say that a detectable PKE scheme Π satisfies* randomness-inextractability *if for all PPTAs $\mathcal{A}$,* $Adv_{\Pi,\mathcal{A}}^{\text{R-Inext}}(k) := \Pr[\text{Expt}_{\Pi,\mathcal{A}}^{\text{R-Inext}}(k) = 1]$ *is negligible.*

Remark. We could have defined the randomness-inextractability experiment so that we let an adversary choose its challenge message m after given a public key pk. This makes the security stronger. However, we do not need this stronger variant for our results.

3.3 Useful Facts

Stretching a Session-Key. As in the case of ordinary KEMs, for a detectable KEM, session-keys can be stretched by using a PRG. More formally, let $\Gamma = (\text{KKG}, \text{Encap}, \text{Decap}, F)$ be a detectable KEM whose session-key space is $\{0, 1\}^k$. Let $G : \{0, 1\}^k \rightarrow \{0, 1\}^\ell$ be a PRG with $\ell = \ell(k) > k$, where for convenience we define $G(\perp) := \perp$. Then, consider the detectable KEM $\Gamma' = (\text{KKG}, \text{Encap}', \text{Decap}', F)$ whose session-key space is $\{0, 1\}^\ell$, which is naturally constructed by combining Γ and G: $\text{Encap}'(pk)$ runs $(c, K) \leftarrow \text{Encap}(pk)$ and outputs a ciphertext/session key pair $(c, G(K))$.; We define $\text{Decap}'(sk, c) := G(\text{Decap}(sk, c))$. The following is straightforward, and thus its proof is omitted.

Lemma 3. *If the detectable KEM Γ satisfies randomness-inextractability (resp. unpredictability), then so does the detectable KEM Γ'. Furthermore, if Γ is* DCCA *(resp.* wNM–DCCA*) secure and G is a PRG, then Γ' is* DCCA *(resp.* wNM–DCCA*) secure.*

$\mathsf{Expt}_{\Pi,\mathcal{A}}^{\text{wNM-DCCA}}(k):$
$\quad (pk, sk) \leftarrow \mathsf{PKG}(1^k)$
$\quad (m_0, m_1, \mathsf{st}) \leftarrow \mathcal{A}_1^{\mathcal{O}(\cdot)}(pk)$
$\quad b \leftarrow \{0, 1\}$
$\quad c^* \leftarrow \mathsf{Enc}(pk, m_b)$
$\quad (c', \mathsf{st}') \leftarrow \mathcal{A}_2^{\mathcal{O}(\cdot)}(\mathsf{st}, c^*)$
$\quad m' \leftarrow \mathsf{Dec}(sk, c')$
$\quad b' \leftarrow \mathcal{A}_3(\mathsf{st}', m')$
$\quad \text{Return } (b' \overset{?}{=} b).$

$\mathsf{Expt}_{\Pi,\mathcal{A}}^{\text{wRNM-DCCA}}(k):$
$\quad (pk, sk) \leftarrow \mathsf{PKG}(1^k)$
$\quad (m_0, m_1, \mathsf{st}) \leftarrow \mathcal{A}_1^{\mathcal{O}(\cdot)}(pk)$
$\quad b \leftarrow \{0, 1\}$
$\quad c^* \leftarrow \mathsf{Enc}(pk, m_b)$
$\quad (c', \mathsf{st}') \leftarrow \mathcal{A}_2^{\mathcal{O}(\cdot)}(\mathsf{st}, c^*)$
$\quad m' \leftarrow \mathsf{Dec}(sk, c')$
$\quad \text{If } m' \in \{m_0, m_1\} \text{ then}$
$\qquad\qquad\qquad m' \leftarrow \mathsf{same}$
$\quad b' \leftarrow \mathcal{A}_3(\mathsf{st}', m')$
$\quad \text{Return } (b' \overset{?}{=} b).$

$\mathsf{Expt}_{\Pi,\mathcal{A}}^{\text{R-Inext}}(k):$
$\quad (m, \mathsf{st}) \leftarrow \mathcal{A}_1(1^k)$
$\quad (pk, sk) \leftarrow \mathsf{PKG}(1^k)$
$\quad c^* \leftarrow \mathsf{Enc}(pk, m)$
$\quad (m', r') \leftarrow \mathcal{A}_2^{\mathcal{O}(\cdot)}(\mathsf{st}, pk, c^*)$
$\quad c' \leftarrow \mathsf{Enc}(pk, m'; r')$
$\quad \text{Return } \mathsf{F}(pk, c^*, c').$

Fig. 2. Security experiments for wNM-DCCA security (left), wRNM-DCCA security (center), and randomness-inextractability (right). In the experiments, $\mathcal{O}(\cdot)$ is the decryption oracle $\mathsf{Dec}(sk, \cdot)$, and in the wNM/wRNM-CCA experiments, the decryption oracle for $\mathcal{A}_2$ has the same restriction as in the DCCA experiment.

Hybrid Encryption. For a detectable PKE scheme, a straightforward application of hybrid encryption preserves w(R)NM-DCCA security and randomness-inextractability, when combined with a CCA secure SKE scheme. Since a CCA secure SKE scheme with "zero" ciphertext overhead can be realized from a strong pseudorandom permutation [23] (which is in turn realized based on any one-way function), the ciphertext overhead of a detectable PKE scheme with w(R)NM-DCCA security and randomness-inextractability, can be as small as the ciphertext size of the scheme for encrypting a random session-key (usually a k-bit string).

Formally, let $\Pi = (\mathsf{PKG}, \mathsf{Enc}, \mathsf{Dec}, \mathsf{F})$ be a detectable PKE scheme where the randomness space of Enc is $\{0, 1\}^\ell$, and let $E = (\mathsf{SEnc}, \mathsf{SDec})$ be a deterministic SKE scheme (i.e. its encryption algorithm SEnc is deterministic). Then, we naturally construct the detectable PKE scheme $\Pi_{\mathsf{HYB}} = (\mathsf{PKG}_{\mathsf{HYB}}, \mathsf{Enc}_{\mathsf{HYB}}, \mathsf{Dec}_{\mathsf{HYB}}, \mathsf{F}_{\mathsf{HYB}})$ via hybrid encryption, as in Fig. 3. (We describe the randomness of $\mathsf{Enc}_{\mathsf{HYB}}$ explicitly so that it is convenient to consider its randomness-inextractability.) The randomness space of $\mathsf{Enc}_{\mathsf{HYB}}$ is $\{0, 1\}^{\ell+k}$.

$\mathsf{PKG}_{\mathsf{HYB}}(1^k):$	$\mathsf{Enc}_{\mathsf{HYB}}(pk, m; R):$	$\mathsf{Dec}_{\mathsf{HYB}}(sk, C):$	$\mathsf{F}_{\mathsf{HYB}}(pk, C^*, C'):$
$(pk, sk) \leftarrow \mathsf{PKG}(1^k)$	Parse R as (r, K)	$(c, \widehat{c}) \leftarrow C$	$(c^*, \widehat{c^*}) \leftarrow C^*$
Return (pk, sk).	$\quad \in \{0,1\}^\ell \times \{0,1\}^k$.	$K \leftarrow \mathsf{Dec}(sk, c)$	$(c', \widehat{c'}) \leftarrow C'$
	$c \leftarrow \mathsf{Enc}(pk, K; r)$	If $K = \bot$ then	$b \leftarrow \mathsf{F}(pk, c^*, c')$
	$\widehat{c} \leftarrow \mathsf{SEnc}(K, m)$	$\quad$ return $\bot$.	Return b.
	$C \leftarrow (c, \widehat{c})$	$m \leftarrow \mathsf{SDec}(K, \widehat{c})$	
	Return C.	Return m.	

Fig. 3. Hybrid encryption Π_{HYB} for detectable PKE.

Regarding the security of the hybrid encryption construction, the following lemma is straightforward to see.

Lemma 4. *If the detectable PKE scheme Π is* wNM–DCCA *secure (resp.* wRNM–DCCA *secure) and the SKE scheme E is* CCA *secure, then the detectable PKE scheme Π_{HYB} in Fig. 3 is* wNM–DCCA *secure (resp.* wRNM–DCCA *secure). Furthermore, if Π satisfies randomness-inextractability (resp. unpredictability), then so does Π_{HYB}.*

From wRNM–DCCA *Security to* wNM–DCCA *Security.* Canetti, Krawczyk, and Nielsen [4] showed how to convert a "replayable" CCA secure PKE scheme into an ordinary CCA secure KEM, using a message authentication code (MAC), with almost no overhead. This method can be used for converting a wRNM–DCCA secure detectable PKE scheme into a wNM–DCCA secure detectable KEM. We review this transformation in the full version.

On the Non-triviality of Randomness-Inextractability. One might wonder whether there is an implication from randomness-inextractability to unpredictability and/or vice versa (especially in case if a detectable PKE scheme already satisfies wNM–DCCA security). We show that this is not the case, for both directions. Specifically, (via artificial counterexamples) we can show the following lemma that shows the non-triviality of these notions, which we formally show in the full version.

Lemma 5. *A detectable PKE scheme satisfying* wNM–DCCA *security and unpredictability simultaneously does not necessarily satisfy randomness-inextractability. Furthermore, a detectable PKE scheme satisfying* wNM–DCCA *security and randomness-inextractability simultaneously does not necessarily satisfy unpredictability.*

4   Chosen Ciphertext Security of the Double-Layered Construction

In this section, we show our main result: two new CCA security proofs for the "double-layered" construction Γ_{DL} (of a KEM) constructed from the "inner" detectable KEM Γ_{in} and the "outer" detectable PKE scheme Π_{out}. We also show a partial evidence that we need to rely on "non-malleability" that we defined in the previous section.

The Double-Layered Construction. Let $\Pi_{\text{out}} = (\text{PKG}_{\text{out}}, \text{Enc}_{\text{out}}, \text{Dec}_{\text{out}}, \text{F}_{\text{out}})$ be a detectable PKE scheme. We assume the plaintext space of Π_{out} to be $\{0,1\}^n$ (where $n = n(k)$ is determined below), and the randomness space of Enc_{out} to be $\{0,1\}^\ell$ for some positive polynomial $\ell = \ell(k)$. Let $\Gamma_{\text{in}} = (\text{KKG}_{\text{in}}, \text{Encap}_{\text{in}}, \text{Decap}_{\text{in}}, \text{F}_{\text{in}})$ be a detectable KEM such that the ciphertext length is n bit, and the session-key space is $\{0,1\}^{\ell+k}$. Then we construct the "double-layered" KEM $\Gamma_{\text{DL}} = (\text{KKG}_{\text{DL}}, \text{Encap}_{\text{DL}}, \text{Decap}_{\text{DL}})$ as in Fig. 4. For convenience, we occasionally call Γ_{in} the *inner* KEM and Π_{out} the *outer* PKE scheme.

Our First Security Proof. The CCA security of Γ_{DL} can be shown as follows.

$\mathsf{KKG_{DL}}(1^k)$:	$\mathsf{Decap_{DL}}(SK, c)$:
$\quad (pk_{\mathrm{in}}, sk_{\mathrm{in}}) \leftarrow \mathsf{KKG_{in}}(1^k)$	$\quad (sk_{\mathrm{in}}, sk_{\mathrm{out}}, PK) \leftarrow SK$
$\quad (pk_{\mathrm{out}}, sk_{\mathrm{out}}) \leftarrow \mathsf{PKG_{out}}(1^k)$	$\quad (pk_{\mathrm{in}}, pk_{\mathrm{out}}) \leftarrow PK$
$\quad PK \leftarrow (pk_{\mathrm{in}}, pk_{\mathrm{out}})$	$\quad c_{\mathrm{in}} \leftarrow \mathsf{Dec_{out}}(sk_{\mathrm{out}}, c)$
$\quad SK \leftarrow (sk_{\mathrm{in}}, sk_{\mathrm{out}}, PK)$	$\quad$ If $c_{\mathrm{in}} = \bot$ then return $\bot$.
$\quad$ Return (PK, SK).	$\quad \alpha \leftarrow \mathsf{Decap_{in}}(sk_{\mathrm{in}}, c_{\mathrm{in}})$
$\mathsf{Encap_{DL}}(PK)$:	$\quad$ If $\alpha = \bot$ then return $\bot$.
$\quad (pk_{\mathrm{in}}, pk_{\mathrm{out}}) \leftarrow PK$	$\quad$ Parse α as $(r, K) \in \{0,1\}^{\ell} \times \{0,1\}^k$.
$\quad (c_{\mathrm{in}}, \alpha) \leftarrow \mathsf{Encap_{in}}(pk_{\mathrm{in}})$	$\quad$ If $\mathsf{Enc_{out}}(pk_{\mathrm{out}}, c_{\mathrm{in}}; r) = c$
$\quad$ Parse α as $(r, K) \in \{0,1\}^{\ell} \times \{0,1\}^k$.	$\quad\quad$ then return K else return $\bot$.
$\quad c \leftarrow \mathsf{Enc_{out}}(pk_{\mathrm{out}}, c_{\mathrm{in}}; r)$	
$\quad$ Return (c, K).	

Fig. 4. The double-layered KEM construction Γ_{DL} from a detectable PKE scheme Π_{out} and a detectable KEM Γ_{in}.

Theorem 1. *Assume that the "outer" PKE scheme Π_{out} is a detectable PKE scheme satisfying* wRNM–DCCA *security and randomness-inextractability, and the "inner" KEM Γ_{in} is a detectable KEM satisfying* DCCA *security and unpredictability. Then, the KEM Γ_{DL} in Fig. 4 is* CCA *secure.*

The structure of the proof is similar to the security proofs for the constructions by Hohenberger et al. [16] and by Matsuda and Hanaoka [19]. However, the details differ due to the difference in the construction and the used assumptions.

We explain the ideas for the proof of Theorem 1. (Here, the values with asterisk ($*$) represent those related to the challenge ciphertext c^*.) As the first step, note that since a session-key K of Γ_{DL} is part of a session-key $\alpha = (r\|K)$ of the DCCA secure inner KEM Γ_{in}, unless a CCA adversary $\mathcal{A}$ submits a decapsulation query c that simultaneously satisfies (1) $\mathsf{Dec_{out}}(sk_{\mathrm{out}}, c) = c_{\mathrm{in}} \neq \bot$ and (2) $\mathsf{F_{in}}(pk_{\mathrm{in}}, c_{\mathrm{in}}^*, c_{\mathrm{in}}) = 1$, $\mathcal{A}$ has no chance in distinguishing the real session-key K_1^* from a random K_0^*. Following [16,19], we call this type of decapsulation query a *dangerous* query. If the probability that $\mathcal{A}$ comes up with a dangerous query is negligible, then we can finish the proof. Furthermore, observe that since Γ_{in} satisfies unpredictability, if we can ensure that the information of the inner ciphertext c_{in}^* is hidden from $\mathcal{A}$'s view, then the probability that $\mathcal{A}$ comes up with a dangerous query is negligible.

To show that the probability that $\mathcal{A}$ comes up with a dangerous query in the original security game is negligibly close to that in the security game in which $\mathcal{A}$'s view does not contain c_{in}^* at all (and hence we can invoke the unpredictability of Γ_{in}), we rely on the security properties of the outer PKE scheme Π_{out} to gradually change the security game for $\mathcal{A}$ so that in the final game, c^* as well as other values in $\mathcal{A}$'s view contain no information on c_{in}^*. Note that in the actual encapsulation algorithm $\mathsf{Encap_{DL}}$, the randomness r used for outer encryption is also a part of the session-key α of the inner KEM. Thus, once we invoke the DCCA security of the inner KEM Γ_{in} (which we have already done as the first step), not only the real session-key K_1^* but also the randomness r^* used to generate the challenge ciphertext c^* are made uniformly random values, which enables us to rely on the security properties of Π_{out} from that point on.

Now, intuitively, the DCCA security (which is implied by wRNM-DCCA security) of Π_{out} guarantees that c_{in}^* is hidden from $\mathcal{A}$'s view as long as $\mathcal{A}$ only submits a decapsulation query c such that $\mathsf{F}_{\text{out}}(pk_{\text{out}}, c^*, c) = 0$. However, $\mathcal{A}$ is free to choose its own decapsulation query, and may submit c such that $\mathsf{F}_{\text{out}}(pk_{\text{out}}, c^*, c) = 1$. As mentioned in Sect. 1.2, this is another type of "dangerous" query, in the sense that the condition $\mathsf{F}_{\text{out}}(pk_{\text{out}}, c^*, c) = 1$ prevents us from relying on the DCCA security of the outer PKE scheme Π_{out}. To distinguish this from the above mentioned type of dangerous queries with respect to the inner KEM, let us use the names "*inner-dangerous* queries" and "*outer-dangerous* queries" which are associated with the inner KEM and the outer PKE scheme, respectively.

In the full proof, we will show that the randomness-inextractability of the outer PKE scheme allows us to reject decapsulation queries c satisfying $\mathsf{F}_{\text{out}}(pk_{\text{out}}, c^*, c) = 1$, without being noticed by $\mathcal{A}$. Intuitively, this is possible because in order for $\mathcal{A}$ to notice the difference between a security game in which a decryption query c with $\mathsf{F}_{\text{out}}(pk_{\text{out}}, c^*, c) = 1$ is not rejected and a security game in which such c is rejected, $\mathcal{A}$ has to come up with a "valid" query c satisfying $\mathsf{F}_{\text{out}}(pk_{\text{out}}, c^*, c) = 1$ and $\mathsf{Decap}_{\text{DL}}(SK, c) \neq \bot$. However, the latter condition implies $\mathsf{Dec}_{\text{out}}(sk_{\text{out}}, c) = c_{\text{in}} \neq \bot$, $\mathsf{Decap}_{\text{in}}(sk_{\text{in}}, c_{\text{in}}) = (r\|K) \neq \bot$, and $\mathsf{Enc}_{\text{out}}(pk_{\text{out}}, c_{\text{in}}; r) = c$, among which the combination of $\mathsf{F}_{\text{out}}(pk_{\text{out}}, c^*, c) = 1$ and $\mathsf{Enc}_{\text{out}}(pk_{\text{out}}, c_{\text{in}}; r) = c$ is exactly the condition of violating randomness-inextractability, and thus such a valid query c must be hard to find.

If we can safely reject an outer-dangerous query, one might wonder why we need non-malleability for the outer PKE scheme, and why ordinary DCCA security is not sufficient. The reason is that although DCCA security of Π_{out} intuitively ensures that $\mathcal{A}$ cannot "see" the inner challenge ciphertext c_{in}^*, it does not prevent $\mathcal{A}$ from coming up with an inner-dangerous decapsulation query c such that $\mathsf{F}_{\text{out}}(pk_{\text{out}}, c^*, c) = 1$. From the viewpoint of the security proof, we may be able to come up with a DCCA adversary (a reduction algorithm) for Π_{out} that perfectly simulates the security game (in which queries c with $\mathsf{F}_{\text{out}}(pk_{\text{out}}, c^*, c) = 1$ are rejected) for $\mathcal{A}$. However, such DCCA adversary cannot check if $\mathcal{A}$'s query satisfying $\mathsf{F}_{\text{out}}(pk_{\text{out}}, c^*, c) = 1$ is an inner-dangerous query due to the restriction on the decryption oracle.

This is the place where the non-malleability of the outer PKE scheme comes into play. Note that an inner ciphertext is a "plaintext" of the outer PKE scheme, and the notion of "inner-dangerous queries" is a "meaningful relation" between c_{in}^* and another inner ciphertext. Therefore, the wRNM-DCCA security of Π_{out} ensures that $\mathcal{A}$ cannot come up with even a single inner-dangerous query c, as long as $\mathcal{A}$ can only observe the decapsulation results of queries c' satisfying $\mathsf{F}_{\text{out}}(pk_{\text{out}}, c^*, c') = 0$. From the viewpoint of the security proof, if a reduction algorithm is a wRNM-DCCA adversary for Π_{out}, it can check if $\mathcal{A}$'s query c is inner-dangerous by its final "unrestricted" decryption query, even if $\mathsf{F}_{\text{out}}(pk_{\text{out}}, c^*, c) = 1$ holds. This enables us to finally show that the probability that $\mathcal{A}$ comes up with an inner-dangerous query in the original security game, is negligibly close to the probability that $\mathcal{A}$ does so in the game in which $\mathcal{A}$'s view does not contain the information on c_{in}^*.

Hence, combining all the security properties of the building blocks leads to CCA security. However, the explanation so far hides some technical subtleties that arise due to the "replayable-CCA"-like nature of wRNM-DCCA security, and the treatment of the cases where $\mathcal{A}$'s decapsulation query c satisfies $\mathsf{Dec}_{\mathsf{out}}(sk_{\mathsf{out}}, c) = c_{\mathsf{in}}^*$, etc. For the details, see the proof in the full version.

Our Second Security Proof. We show an alternative security proof for the double-layered construction based on slightly different assumptions on the building blocks.

Theorem 2. *Assume that the "outer" PKE scheme Π_{out} is a detectable PKE scheme satisfying* DCCA *security and randomness-inextractability, and the "inner" KEM Γ_{in} is a detectable KEM satisfying* wNM-DCCA *security and unpredictability. Then, the KEM Γ_{DL} in Fig. 4 is* CCA *secure.*

Recall that Myers and Shelat's original double-layered construction uses an "unquoted" CCA (UCCA) secure construction of a PKE scheme for the outer PKE scheme and a construction of a KEM which is "1-wise-non-malleable under UCCA" for the inner KEM, where UCCA security and its non-malleable variant are security notions considered for PKE-to-PKE constructions (i.e. constructions that use another PKE scheme as a building block). Recall also that DCCA security is an abstraction of UCCA security [16], from a security notion for a PKE-to-PKE construction to that of a wider notion of detectable PKE. Analogously, our definition of wNM-DCCA security can be seen as an abstraction of Myers and Shelat's "1-wise non-malleability under UCCA". Furthermore, we can easily see that the actual instantiations of the inner KEM and the outer PKE scheme used in the original Myers-Shelat construction [20], when respectively seen as a detectable KEM and a detectable PKE scheme, satisfy unpredictability and randomness-inextractability. Therefore, Theorem 2 can be seen as a generalization of Myers and Shelat's result.

The structure of the proof of Theorem 2 is similar to our first proof. However, there are several subtle but crucial differences. In particular, the definitions of "inner/outer-dangerous queries" are different from those used in the proof of Theorem 1, and correspondingly we consider a different ordering of the sequence of games for this proof. Furthermore, the role of the "non-malleability" in this proof and that of the proof of Theorem 1 are different. Informally speaking, in this proof, the wNM-DCCA security of the inner detectable KEM Γ_{in} is used to ensure that the probability that a CCA adversary comes up with an outer-dangerous query is not noticeably different between the games in which we invoke (the indistinguishability property of) the DCCA security of the inner KEM.

Can We Avoid w(R)NM-DCCA *Security?* Both of our security proofs for the CCA security of the double-layered construction require either the inner detectable KEM or the outer detectable PKE scheme to be "non-malleable" under DCCA.

Looking ahead, in the next section, we will see that the simplest "bitwise-encrypt" construction based on CCA secure 1-bit PKE satisfies DCCA security, unpredictability, and randomness-inextractability. Thus, a natural question would be whether we can prove the CCA security of the double-layered construction without using the non-malleability notions for both of the building blocks (and instead only requiring DCCA security). If such a security proof were possible, then one can use the bitwise-encrypt-based construction both for the inner KEM and the outer PKE scheme, and the resulting CCA secure KEM would be fairly simple.

Unfortunately, however, we show that such a security proof for the double-layered construction is impossible, as there is a counterexample.

Theorem 3. *Assume there exists a detectable PKE scheme which is DCCA secure and unpredictable. Then, there exist a detectable KEM Γ_{in} and a detectable PKE scheme Π_{out} such that the following simultaneously hold: (1) Γ_{in} is DCCA secure and unpredictable. (2) Π_{out} is DCCA secure and randomness-inextractable. (3) The double-layered KEM Γ_{DL} constructed using Γ_{in} as the inner KEM and Π_{out} as the outer PKE scheme, is not CCA secure (in fact, not secure in the sense of one-wayness under 1-bounded CCA).*

Our counterexample is based on an observation that the combination of DCCA security, unpredictability, and randomness-inextractability, does not rule out a double-layered KEM with the following property: A ciphertext C is of the form $C = (c_1, c_2)$ and the corresponding session-key K is of the form $K = (K_1, K_2)$, and furthermore it is "blockwise" consistent, meaning that each pair (c_i, K_i) is individually consistent as a ciphertext/session-key pair of the double-layered construction. Thus, the decapsulation result of the "swapped" ciphertext $\widehat{C} = (c_2, c_1)$ is the "swapped" session-key $\widehat{K} = (K_2, K_1)$. Such a KEM is clearly malleable, and its one-wayness is broken by just a single decapsulation query.

5   Concrete Instantiations of Building Blocks

In this section, we show how to construct a detectable PKE scheme, which we call "encode-then-bitwise-encrypt" (EtBE) construction, that uses a CCA secure 1-bit PKE scheme and a $\mathcal{Q}$-non-malleable code as building blocks and simultaneously satisfies wRNM-DCCA security and randomness-inextractability. Since it is much easier to understand it if we first review the simple "bitwise-encrypt" construction, we first review it in Sect. 5.1 together with its security properties, and then we show the EtBE construction in Sect. 5.2.

5.1   Bitwise-Encrypt Construction

Here, we show that the "bitwise-encrypt" construction of a detectable PKE scheme based on a 1-bit PKE scheme, in which each bit of a plaintext is encrypted in a bit-by-bit fashion by the underlying 1-bit scheme, can be shown to satisfy

$\mathsf{Enc}_{\mathrm{BE}}^n(pk, m; r):$
  Parse r as $(r_1, \ldots, r_n) \in (\{0,1\}^\ell)^n.$
  View m as $(m_1 \| \ldots \| m_n) \in \{0,1\}^n.$
  $\forall i \in [n]: c_i \leftarrow \mathsf{Enc}_1(pk, m_i; r_i)$
  Return $C \leftarrow (c_1, \ldots, c_n).$

$\mathsf{Dec}_{\mathrm{BE}}^n(sk, C):$
  $(c_1, \ldots, c_n) \leftarrow C$
  $\forall i \in [n]: m_i \leftarrow \mathsf{Dec}_1(sk, c_i)$
  If $\exists i \in [n]: m_i = \bot$ then return $\bot.$
  Return $m \leftarrow (m_1 \| \ldots \| m_n).$

$\mathsf{F}_{\mathrm{BE}}^n(pk, C^*, C'):$
  $(c_1^*, \ldots, c_n^*) \leftarrow C^*$
  $(c_i', \ldots, c_n') \leftarrow C'$
  If $\exists i, j \in [n]: c_i^* = c_j'$
    then return 1 else return 0.

$\mathsf{Enc}_{\mathrm{EtBE}}(pk, m; R):$
  Parse R as $(r, \widehat{r}) \in \{0,1\}^{\ell \cdot n} \times \{0,1\}^{\widehat{\ell}}.$
  $s = (s_1 \| \ldots \| s_n) \leftarrow \mathsf{E}(1^k, m; \widehat{r})$
  $C = (c_1, \ldots, c_n) \leftarrow \mathsf{Enc}_{\mathrm{BE}}^n(pk, s; r)$
  If $\mathrm{DUPCHK}(C) = 1$ then return $\bot.^\dagger$
  Return $C.$

$\mathsf{Dec}_{\mathrm{EtBE}}(sk, C):$
  If $\mathrm{DUPCHK}(C) = 1$ then return $\bot.$
  $s \leftarrow \mathsf{Dec}_{\mathrm{BE}}^n(sk, C)$
  If $s = \bot$ then return $\bot.$
  Return $m \leftarrow \mathsf{D}(1^k, s).$

$\mathsf{F}_{\mathrm{EtBE}}(pk, C^*, C'):$
  If (a) $\wedge$ (b) then return 1 else return 0:
    (a) $\mathrm{DUPCHK}(C^*) = \mathrm{DUPCHK}(C') = 0$
    (b) $\mathsf{F}_{\mathrm{BE}}^n(pk, C^*, C') = 1$

Fig. 5. The "bitwise-encrypt" (n-bit) construction Π_{BE}^n (left), and the "encode-then-bitwise-encrypt" (EtBE) construction Π_{EtBE} (right), both based on a 1-bit PKE scheme Π_1. The key generation algorithms for Π_{BE}^n and Π_{EtBE} are the key generation algorithm PKG_1 of the underlying scheme $\Pi_1.^\dagger$ Regarding the case in which $\mathsf{Enc}_{\mathrm{EtBE}}$ returns $\bot$, see the explanation in the text.

randomness-inextractability, DCCA security, and unpredictability, if the underlying 1-bit PKE scheme is CCA secure.

Let $\Pi_1 = (\mathsf{PKG}_1, \mathsf{Enc}_1, \mathsf{Dec}_1)$ be a 1-bit PKE scheme, and the randomness space of whose encryption algorithm Enc_1 is $\{0,1\}^\ell$ (where $\ell = \ell(k)$ is some positive polynomial). Then, for a polynomial $n = n(k) > 0$, consider the "bitwise-encrypt" construction $\Pi_{\mathrm{BE}}^n = (\mathsf{PKG}_{\mathrm{BE}}^n := \mathsf{PKG}_1, \mathsf{Enc}_{\mathrm{BE}}^n, \mathsf{Dec}_{\mathrm{BE}}^n, \mathsf{F}_{\mathrm{BE}}^n)$ of an n-bit detectable PKE scheme described in Fig. 5 (left). The key generation algorithm $\mathsf{PKG}_{\mathrm{BE}}^n$ is actually PKG_1 itself, and we do not show it in the figure. The randomness space of $\mathsf{Enc}_{\mathrm{BE}}^n$ is $\{0,1\}^{\ell \cdot n}$. In the figure, we make the randomness used by $\mathsf{Enc}_{\mathrm{BE}}^n$ explicit so that it is convenient to consider randomness-inextractability.

The following result was shown by Hohenberger et al. [16]:

Lemma 6. [16] *Let $n = n(k) > 0$ be a polynomial. If the 1-bit PKE scheme Π_1 is CCA secure, then the detectable PKE scheme Π_{BE}^n scheme satisfies DCCA security and unpredictability.*

We show a similar statement regarding randomness-inextractability.

Lemma 7. *Let $n = n(k) > 0$ be a polynomial. If the PKE scheme Π_1 is CCA secure, then the detectable PKE scheme Π_{BE}^n satisfies randomness-inextractability.*

Here we explain an intuition why Lemma 7 is true, which is quite straightforward: Suppose an adversary $\mathcal{A}$, given a public key pk and the challenge ciphertext $C^* = (c_1^*, \ldots, c_n^*)$ and access to the decryption oracle, succeeds in outputting a plaintext $m' = (m_1' \| \ldots \| m_n')$ and a randomness $r' = (r_1', \ldots, r_n')$ such that

$F_{BE}^n(pk, C^*, C') = 1$ with $C' = (c_1', \ldots, c_n') = \text{Enc}_{BE}^n(pk, m'; r')$. Then, by definition, there must be a position $i \in [n]$ such that $c_i^* = c_j'$ holds for some $j \in [n]$, where $c_a' = \text{Enc}_1(pk, m_a'; r_a')$ for each $a \in [n]$. Note that such $\mathcal{A}$ is in fact "extracting" the randomness used for generating c_i^*. Note also that extracting a randomness used for generating a ciphertext is a harder task than breaking indistinguishability. Thus, it is easy to construct another CCA adversary (a reduction algorithm) $\mathcal{B}$ for Π_1 that initially guesses the position i such that $c_i^* = c_j'$ holds with some j, embeds $\mathcal{B}$'s challenge ciphertext into the i-th position of the challenge ciphertext for $\mathcal{A}$, and has the CCA advantage at least $1/n$ times that of $\mathcal{A}$'s advantage in breaking randomness-inextractability.

5.2 Encode-then-Bitwise-Encrypt Construction

Here, we show the construction of detectable PKE that we call "Encode-then-Bitwise-Encrypt" (EtBE) construction, which simultaneously achieves wRNM-DCCA security and randomness-inextractability, based on the security properties of the bitwise-encrypt construction (which are in turn based on the underlying CCA secure 1-bit scheme) and a $\mathcal{Q}$-non-malleable code. Our construction is actually a direct "PKE"-analogue of the transformation of a CCA secure 1-bit commitment scheme into a non-malleable string commitment scheme by Agrawal et al. [2]. We adapt their construction into the (detectable) PKE setting.

Let $\mathcal{C} = (\text{E}, \text{D})$ be a code with message length k and codeword length $n = n(k) \geq k$. Let $\Pi_1 = (\text{PKG}_1, \text{Enc}_1, \text{Dec}_1)$ be a 1-bit PKE scheme. Let $\Pi_{BE}^n = (\text{PKG}_{BE}^n = \text{PKG}_1, \text{Enc}_{BE}^n, \text{Dec}_{BE}^n, F_{BE}^n)$ be the bitwise-encrypt construction based on Π_1. For convenience, we introduce the procedure "DUPCHK($\cdot$)" which takes a ciphertext $C = (c_1, \ldots, c_n)$ of Π_{BE}^n as input, and returns 1 if there exist distinct $i, j \in [n]$ such that $c_i = c_j$, and returns 0 otherwise. (That is, DUPCHK(C) checks a duplication in the component ciphertexts $(c_i)_{i \in [n]}$.)

Using $\mathcal{C}$, Π_{BE}^n (and Π_1), and DUPCHK, the EtBE construction $\Pi_{EtBE} = (\text{PKG}_{EtBE} := \text{PKG}_1, \text{Enc}_{EtBE}, \text{Dec}_{EtBE}, F_{EtBE})$ is constructed as in Fig. 5 (right). Like Π_{BE}^n, the key generation algorithm PKG_{EtBE} is PKG_1 itself, and we do not show it in the figure. The plaintext space of Π_{EtBE} is $\{0,1\}^k$.

On the Correctness of Π_{EtBE}. Note that the encryption algorithm Enc_{EtBE} returns $\perp$ if it happens to be the case that DUPCHK(C) = 1. This check is to ensure that a valid ciphertext does not have "duplicated" components, which is required due to our use of a $\mathcal{Q}$-non-malleable code whose non-malleability can only take care of a "non-duplicated" quoting. Since the probability (over the randomness of Enc_{EtBE}) that Enc_{EtBE} outputs $\perp$ is not zero, our construction Π_{EtBE} does not satisfy correctness in a strict sense. (The exactly same problem arises in the construction of string commitments in [2].) However, it is easy to show that if Π_1 satisfies CCA security (or even CPA security), the probability of Enc_{EtBE} outputting $\perp$ is negligible, and thus it does not do any harm in practice. (In practice, for example, in case $\perp$ is output, one can re-execute Enc_{EtBE} with a fresh randomness. The expected execution time of Enc_{EtBE} is negligibly close to 1.) Furthermore, if one needs standard correctness, then instead of letting Enc_{EtBE} output $\perp$ in case $\mathrm{DUPCHK}(C) = 1$, one can let it output a plaintext m (being encrypted) as an "irregular ciphertext", so that if the decryption algorithm $\mathrm{Dec}_{\mathrm{EtBE}}$ takes an irregular ciphertext C as input, it outputs C as a "decryption result" of C. (In order to actually implement this, in case $\mathrm{DUPCHK}(C) = 1$ occurs, $m \in \{0,1\}^k$ needs to be padded to the length $n \cdot |c|$ of an ordinary ciphertext, and we furthermore need to put a prefix for every ciphertext that tells the decryption algorithm whether the received ciphertext should be treated as a normal ciphertext or an irregular one.) Such a modification also does no harm to the security properties of Π_{EtBE} (it only contributes to increasing an adversary's advantage negligibly), thanks to the CCA security of the building block Π_1. For simplicity, in this paper we focus on the current construction of Π_{EtBE}.

Security of Π_{EtBE}. The security properties of the EtBE construction is guaranteed by the following lemmas.

Lemma 8. *Assume that Π_1 is CCA secure and C is a Q-non-malleable code. Then, the detectable PKE scheme Π_{EtBE} in Fig. 5 (right) is wRNM–DCCA secure.*

Lemma 9. *If Π_1 is CCA secure, then the detectable PKE scheme Π_{EtBE} scheme in Fig. 5 (right) satisfies unpredictability and randomness-inextractabilty.*

The proof of Lemma 9 is straightforward given the unpredictability (Lemma 6) and randomness-inextractability (Lemma 7) of the bitwise-encrypt construction Π_{BE}^n, and thus omitted.

The proof of Lemma 8 follows essentially the same story line as the security proof of the non-malleable string commitment by Agrawal et al. [2]. A high-level idea is as follows: In the wRNM–DCCA experiment, an adversary $\mathcal{A} = (\mathcal{A}_1, \mathcal{A}_2, \mathcal{A}_3)$ is allowed to submit a single "unrestricted" decryption query $C' = (c_1', \ldots c_n')$, which is captured by the ciphertext finally output by $\mathcal{A}_2$. In order for this query to be valid, however, C' has to satisfy $\mathrm{DUPCHK}(C') = 0$, which guarantees that C' does not have duplicated components. Thus, since each component is a ciphertext of the CCA secure scheme Π_1, the best $\mathcal{A}$ can do to generate C' that is "related" to the challenge ciphertext $C^* = (c_1^*, \ldots, c_n^*)$ is to "quote" some of c_i^*'s into C' in such a way that no c_i^* appears more than once. However, such "quoting without duplicated positions" is exactly the function class Q with respect to which the code C is non-malleable. Specifically, the Q-non-malleability of C guarantees that even if an adversary observes the decryption result of such C' that quotes some of components of C^* without duplicated positions, $\mathcal{A}$ gains essentially no information of the original content m_b of the encoding s^* encrypted in C^*, and hence no information of the challenge bit b. Actually, it might be the case that $\mathcal{A}$ succeeds in generating C' so that $\mathrm{Dec}_{\mathrm{BE}}^n(sk, C')$ is s^* itself (and hence its decoded value is exactly the challenge plaintext m_b). According to the rule of the wRNM–DCCA experiment, however, in such a case $\mathcal{A}$ is not given the actual decryption result $\mathrm{Dec}_{\mathrm{EtBE}}(sk, C')$ directly but is given the symbol same which only informs that the decryption result is either m_0 or m_1. Furthermore, all other queries without quoting do not leak the information of the challenge bit b because of the DCCA security of the bitwise-encrypt construction Π_{BE}^n (Lemma 6).

These ideas lead to wRNM-DCCA security of Π_{EtBE}. For the details, see the proof in the full version.

6   Full Description of Our 1-bit-to-Multi-bit Conversion

Given the results in the previous sections, we are now ready to describe our proposed 1-bit-to-multi-bit conversion, i.e. a CCA secure KEM from a CCA secure 1-bit PKE scheme. Let $\Pi_1 = (\mathsf{PKG}_1, \mathsf{Enc}_1, \mathsf{Dec}_1)$ be a 1-bit PKE scheme whose public key size is "$|pk|$", the ciphertext size is "$|c|$", and the randomness space of whose encryption algorithm Enc_1 is $\{0,1\}^\ell$. Let $\mathcal{C} = (\mathsf{E}, \mathsf{D})$ be a $\mathcal{Q}$-non-malleable (n,k)-code with $n = n(k) \geq k$, and the randomness space of whose encoding algorithm E is $\{0,1\}^{\widehat{\ell}}$. Let $\ell' = n \cdot \ell + \widehat{\ell} + 2k$, and $\mathsf{G}: \{0,1\}^k \to \{0,1\}^{\ell'}$ be a PRG. Finally, let $E = (\mathsf{SEnc}, \mathsf{SDec})$ be a deterministic SKE scheme whose plaintext space is $\{0,1\}^{k \cdot |c|}$, and it has zero ciphertext overhead (i.e. its ciphertext size is the same as that of a plaintext).

From these building blocks, consider the following detectable KEM Γ_{in} and detectable PKE scheme Π_{out}:

Γ_{in}: Consider the bitwise-encrypt construction Π_{BE}^k (Fig. 5) based on the PKE scheme Π_1, and regard it as a detectable KEM by encrypting a random k-bit string as a session-key. For this detectable KEM, use the PRG G with the method explained in the first paragraph of Sect. 3.3 to stretch its session-key into ℓ' bits. Γ_{in} is the resultant KEM.
   The public key size of Γ_{in} is $|pk|$, its ciphertext size is $k \cdot |c|$, and its session-key space is $\{0,1\}^{\ell'}$. Due to Lemmas 3 and 6, Γ_{in} satisfies DCCA security and unpredictability based on the CCA security of Π_1 and the security of G.

Π_{out}: Consider the EtBE construction Π_{EtBE} based on the code $\mathcal{C}$ and the bitwise-encrypt construction Π_{BE}^n (which is in turn based on Π_1) (Fig. 5). Combine this detectable PKE scheme with the SKE scheme E by the method explained in the second paragraph of Sect. 3.3 (see Fig. 3). Π_{out} is the resultant PKE scheme.
   The public key size of Π_{out} is $|pk|$, its ciphertext overhead (the difference between the total ciphertext size minus the plaintext size) is $n \cdot |c|$, its plaintext space is $\{0,1\}^{k \cdot |c|}$, and the randomness space of its encryption algorithm is $\{0,1\}^{\ell'-k}$. Due to Lemmas 4, 6, 7, 8, and 9, Π_{out} satisfies wRNM-DCCA security and randomness-inextractability, based on the CCA security of Π_1, $\mathcal{Q}$-non-malleability of $\mathcal{C}$, and the CCA security of E.

Our proposed KEM $\widetilde{\Gamma} = (\widetilde{\mathsf{KKG}}, \widetilde{\mathsf{Encap}}, \widetilde{\mathsf{Decap}})$ is then obtained from the double-layered construction Γ_{DL} in which the inner KEM is Γ_{in} and the outer PKE scheme is Π_{out} explained above. More concretely, the description of $\widetilde{\Gamma}$ is as in Fig. 6.

   The public key size of $\widetilde{\Gamma}$ is $2 \cdot |pk|$, and its ciphertext size is $(n+k) \cdot |c|$ (where Γ_{in} contributes $k \cdot |c|$ and Π_{out} contributes $n \cdot |c|$). Using the $\mathcal{P}$-non-malleable code with "optimal rate" (Lemma 1) by Agrawal et al. [1] which also satisfies

$\overline{\text{KKG}}(1^k)$:
 $(pk_{\text{in}}, sk_{\text{in}}) \leftarrow \text{PKG}_1(1^k)$
 $(pk_{\text{out}}, sk_{\text{out}}) \leftarrow \text{PKG}_1(1^k)$
 $PK \leftarrow (pk_{\text{in}}, pk_{\text{out}})$
 $SK \leftarrow (sk_{\text{in}}, sk_{\text{out}}, PK)$
 Return (PK, SK).

$\overline{\text{Encap}}(PK)$:
 $(pk_{\text{in}}, pk_{\text{out}}) \leftarrow PK$
 $K_{\text{in}} = (K_{\text{in}}^{(1)} \| \dots \| K_{\text{in}}^{(k)}) \leftarrow \{0,1\}^k$
 $\forall i \in [k] : c_{\text{in}}^{(i)} \leftarrow \text{Enc}_1(pk_{\text{in}}, K_{\text{in}}^{(i)})$
 $\alpha \leftarrow \text{G}(K_{\text{in}})$
 Parse α as $(r_1, \dots, r_n, \widehat{r}, K_{\text{out}}, K)$
  $\in (\{0,1\}^\ell)^n \times \{0,1\}^{\widehat{\ell}} \times (\{0,1\}^k)^2$.
 $s = (s_1 \| \dots \| s_n) \leftarrow \text{E}(1^k, K_{\text{out}}; \widehat{r})$
 $\forall i \in [n] : c_i \leftarrow \text{Enc}_1(pk_{\text{out}}, s_i; r_i)$
 If $\text{DUPCHK}((c_i)_{i \in [n]}) = 1$ then return $\perp$.
 $\widehat{c} \leftarrow \text{SEnc}(K_{\text{out}}, (c_{\text{in}}^{(1)} \| \dots \| c_{\text{in}}^{(k)}))$
 $C \leftarrow (c_1, \dots, c_n, \widehat{c})$
 Return (C, K).

$\overline{\text{Decap}}(SK, C)$:
 $(sk_{\text{in}}, sk_{\text{out}}, PK) \leftarrow SK$
 $(pk_{\text{in}}, pk_{\text{out}}) \leftarrow PK$; $(c_1, \dots, c_n, \widehat{c}) \leftarrow C$
 If $\text{DUPCHK}((c_i)_{i \in [n]}) = 1$ then return $\perp$.
 $\forall i \in [n] : s_i \leftarrow \text{Dec}_1(sk_{\text{out}}, c_i)$
 If $\exists i \in [n] : s_i = \perp$ then return $\perp$.
 $K_{\text{out}} \leftarrow \text{D}(1^k, s = (s_1 \| \dots \| s_n))$
 If $K_{\text{out}} = \perp$ then return $\perp$.
 $(c_{\text{in}}^{(1)} \| \dots \| c_{\text{in}}^{(k)}) \leftarrow \text{SDec}(K_{\text{out}}, \widehat{c})$
 If SDec has returned $\perp$ then return $\perp$.
 $\forall i \in [k] : K_{\text{in}}^{(i)} \leftarrow \text{Dec}_1(sk_{\text{in}}, c_{\text{in}}^{(i)})$
 If $\exists i \in [k] : K_{\text{in}}^{(i)} = \perp$ then return $\perp$.
 $\alpha \leftarrow \text{G}(K_{\text{in}} = (K_{\text{in}}^{(1)} \| \dots \| K_{\text{in}}^{(k)}))$
 Parse α as $(r_1, \dots, r_n, \widehat{r}, K'_{\text{out}}, K)$
  $\in (\{0,1\}^\ell)^n \times \{0,1\}^{\widehat{\ell}} \times (\{0,1\}^k)^2$.
 If (a) $\wedge$ (b) $\wedge$ (c) then return K
                                      else return $\perp$:
 (a) $\forall i \in [n] : \text{Enc}_1(pk_{\text{out}}, s_i; r_i) = c_i$
 (b) $\text{E}(1^k, K'_{\text{out}}; \widehat{r}) = s$
 (c) $\text{SEnc}(K'_{\text{out}}, (c_{\text{in}}^{(1)} \| \dots \| c_{\text{in}}^{(k)})) = \widehat{c}$

Fig. 6. The proposed "1-bit-to-multi-bit" construction (KEM) $\widetilde{\Gamma}$.

$\mathcal{Q}$-non-malleability by Lemma 2, we have $n = k + o(k)$. Thus, the ciphertext size of $\widetilde{\Gamma}$ can be made asymptotically $(2k + o(k)) \cdot |c|$.

The following statement is obtained as a corollary of the combination of Theorem 1 and Lemmas 1, 2, 3, 4, 6, 7, 8, and 9.

Theorem 4. *Assume that the PKE scheme Π_1 is CCA secure, $\mathcal{C}$ is a $\mathcal{Q}$-non-malleable code, G is a PRG, and the SKE scheme E is CCA secure. Then, the KEM $\widetilde{\Gamma}$ in Fig. 6 is CCA secure.*

2-bit-to-multi-bit Construction with a Single Key Pair. Note that our proposed 1-bit-to-multi-bit conversion $\widetilde{\Gamma}$ uses two key pairs. It turns out that if we can use a 2-bit PKE scheme as a building block instead of a 1-bit scheme, then we can construct a CCA secure KEM that uses only one key pair of the underlying 2-bit scheme, with a very similar way to $\widetilde{\Gamma}$. The idea of this 2-bit-to-multi-bit conversion is to use the additional 1-bit of the plaintext space as the "indicator bit" that indicates whether each component ciphertext is generated for the inner layer or the outer layer. That is, each inner ciphertext $c_{\text{in}}^{(i)}$ is an encryption of $(1 \| K_{\text{in}}^{(i)})$, and each outer ciphertext c_i is an encryption of $(0 \| s_i)$, and in the decapsulation algorithm, we check whether the component ciphertexts $\{c_i\}_{i \in [n]}$ and $\{c_{\text{in}}^{(i)}\}_{i \in [k]}$ have appropriate indicator bits ("1" for the inner layer and "0" for the outer layer). This additional indicator bit and its check prevent a quoting of an inner ciphertext into the outer layer and vice versa, and thus make the encryption/decryption operations for the inner layer and those of the outer layer virtually independent, as if each layer has an individual key pair. This enables us to conduct the security proof in essentially the same way as that of $\widetilde{\Gamma}$. Due to the lack of space, we detail it in the full version.

On the Necessity of Two Key Pairs. As mentioned in Introduction, our positive results on the 1-/2-bit-to-multi-bit constructions for CCA security raise an interesting question in terms of the number of public keys: Is it necessary to use two key pairs in 1-bit-to-multi-bit constructions for CCA security? Motivated by this question, in the full version we consider the one-key variant of our proposed KEM $\widetilde{\Gamma}$, and show that it is vulnerable to a CCA attack. Hence, using two key pairs of the underlying 1-bit scheme is essential for our proposed construction $\widetilde{\Gamma}$. Clarifying the optimality of the number of key pairs in 1-bit-to-multi-bit constructions would be an interesting open problem.

7 Comparison

Table 1 compares the public key size and ciphertext size of the existing "1-bit-to-multi-bit" constructions that achieve CCA security (or related security). Specifically, in the table, "MS" represents the construction by Myers and Shelat [20].; "HLW" represents the construction by Hohenberger et al. [16] which uses a CPA secure PKE scheme, a 1-bounded CCA secure [7] PKE scheme, and a detectable PKE scheme satisfying DCCA security and unpredictability. We assume that for the 1-bounded CCA secure scheme, the construction by Dodis and Fiore [11, Appendix C] is used, which constructs such a scheme from a CPA secure scheme and a one-time signature scheme, and we also assume that its detectable scheme and the CPA secure scheme are realized by the bitwise-encrypt construction Π_{BE}^k. (If we need to encrypt a value longer than k-bit, then we assume that hybrid encryption is used everywhere possible by encrypting a k-bit random session-key and using it as a key for SKE (where the length of SKE ciphertexts are assumed to be the same as a plaintext [23]), which we do the same for the constructions explained below.); "MH" represents the construction by Matsuda and Hanaoka [19], which can be seen as an efficient version of HLW [16] due to hybrid encryption techniques, and we assume that the building blocks similar to HLW are used.; "CMTV" represents the construction by Coretti et al. [6], the size parameters of which are taken from the introduction of [6].; "CDTV" represents the construction by Coretti et al. [5], where the size parameters are estimated according to the explanations in [5, Sections 4.2 & 4.3].; "Ours" is the KEM $\widetilde{\Gamma}$ shown in Fig. 6 in Sect. 6.

As is clear from Table 1, if one starts from a CCA secure 1-bit PKE scheme (and assuming that building blocks implied by one-way functions are available for free), then "Ours" achieves asymptotically the best efficiency. Notably, the public size and the ciphertext size of "Ours" are asymptotically "optimal" in the sense that they are asymptotically the same as the bitwise-encrypt construction

Table 1. Comparison among the 1-bit-to-multi-bit constructions for CCA (and related) security.

Scheme	PK size	Ciphertext size	Sec. of Π_1	Add. Bld. Blk
MS [20]	$(20k^2 + 1)|pk|$	$(10k^3|c| + |vk| + |\sigma|)|c|$	CCA	Sig., PRG
HLW [16]	$(2k + 2)|pk|$	$(k^2 + 3k)|c| + |vk| + |\sigma| + 6k$	DCCA &UNP	Sig., PRG, SKE
MH [19]	$(2k + 2)|pk|$	$(k^2 + 2k)|c| + |vk| + |\sigma|$	DCCA &UNP	Sig., PRG, SKE
CMTV[†] [6]	$\approx k|pk|$	$\approx 5k|c|$	SDA	—
CDTV[†] [5]	$O(k)|pk|$	$O(k)|c|$	NM-SDA	—
Ours (Sect. 6)	$2|pk|$	$(2k + o(k))|c|$	CCA	PRG, SKE

In the columns "PK Size" and "Ciphertext Size", $|pk|$ and $|c|$ denote the public key size and the ciphertext size of the underlying 1-bit PKE scheme Π_1, respectively, and $|vk|$ and $|\sigma|$ denote the size of a verification key and that of a signature of the one-time signature scheme used as a building block, respectively. The column "Sec. on Π_1" shows the assumption on the security of the underlying 1-bit PKE scheme required to show the CCA (or the related) security of the entire construction. Here, "SDA" and "NM-SDA" denote "(indistinguishability against) self-destruct CCA" [6] and "non-malleability against SDA" [5], respectively. The column "Add. Bld. Blk." shows the additional building blocks (used in each construction) that can be realized only from the existence of a one-way function. Here, "Sig" stands for a one-time signature scheme. (†) As explained in Introduction, CMTV [6] and CTDV [5] only achieve SDA security and NM-SDA security, respectively, which are both implied by ordinary CCA security but are strictly weaker than it

Π_{BE}^k that works as a 1-bit-to-multi-bit conversion for the CPA and non-adaptive CCA (CCA1) settings. Note also that all the previous constructions that achieve ordinary CCA security have the public key size $\Omega(k) \cdot |pk|$, and the ciphertext size $\Omega(k^2) \cdot |c|$.

We note that, as mentioned in Sect. 1.3, CMTV [6] and CDTV [5] achieve only indistinguishability under self-destruct CCA (SDA) and non-malleability under self-destruct CCA (NM-SDA), respectively, which are both implied by ordinary CCA security but are strictly weaker than it. Nonetheless, "Ours" actually achieves better asymptotic efficiency than them.

However, for fairness we note that our construction requires CCA security for the underlying 1-bit PKE scheme Π_1, while HLW [16] and MH [19] only require DCCA security and unpredictability, and the constructions CMTV [6] and CDTV [5] only require SDA and NM-SDA security for Π_1, respectively, and thus there is a tradeoff among the assumptions on the building block Π_1.

Acknowledgement. The authors would like to thank the members of the study group "Shin-Akarui-Angou-Benkyou-Kai," and the anonymous reviewers of ASIACRYPT 2015 for their helpful comments and suggestions.

References

1. Agrawal, S., Gupta, D., Maji, H.K., Pandey, O., Prabhakaran, M.: A rate-optimizing compiler for non-malleable codes against bit-wise tampering and permutations. In: Dodis, Y., Nielsen, J.B. (eds.) TCC 2015, Part I. LNCS, vol. 9014, pp. 375–397. Springer, Heidelberg (2015)

2. Agrawal, S., Gupta, D., Maji, H.K., Pandey, O., Prabhakaran, M.: Explicit non-malleable codes against bit-wise tampering and permutations. In: Gennaro, R., Robshaw, M. (eds.) CRYPTO 2015. LNCS, vol. 9216, pp. 538–557. Springer, Heidelberg (2015)

3. Bellare, M., Sahai, A.: Non-malleable encryption: equivalence between two notions, and an indistinguishability-based characterization. In: Wiener, M. (ed.) CRYPTO 1999. LNCS, vol. 1666, pp. 519–536. Springer, Heidelberg (1999)

4. Canetti, R., Krawczyk, H., Nielsen, J.B.: Relaxing chosen-ciphertext security. In: Boneh, D. (ed.) CRYPTO 2003. LNCS, vol. 2729, pp. 565–582. Springer, Heidelberg (2003)

5. Coretti, S., Dodis, Y., Tackmann, B., Venturi, D.: Non-malleable encryption: simpler, shorter, stronger (2015). http://eprint.iacr.org/2015/772

6. Coretti, S., Maurer, U., Tackmann, B., Venturi, D.: From single-bit to multi-bit public-key encryption via non-malleable codes. In: Dodis, Y., Nielsen, J.B. (eds.) TCC 2015, Part I. LNCS, vol. 9014, pp. 532–560. Springer, Heidelberg (2015)

7. Cramer, R., Hanaoka, G., Hofheinz, D., Imai, H., Kiltz, E., Pass, R., Shelat, A., Vaikuntanathan, V.: Bounded CCA2-secure encryption. In: Kurosawa, K. (ed.) ASIACRYPT 2007. LNCS, vol. 4833, pp. 502–518. Springer, Heidelberg (2007)

8. Cramer, R., Shoup, V.: Design and analysis of practical public-key encryption schemes secure against adaptive chosen ciphertext attack. SIAM J. Comput. $33(1)$, 167–226 (2003)

9. Dachman-Soled, D., Fuchsbauer, G., Mohassel, P., O'Neill, A.: Enhanced chosen-ciphertext security and applications. In: Krawczyk, H. (ed.) PKC 2014. LNCS, vol. 8383, pp. 329–344. Springer, Heidelberg (2014)

10. Dodis, Y., Fiore, D.: Interactive encryption and message authentication. In: Abdalla, M., De Prisco, R. (eds.) SCN 2014. LNCS, vol. 8642, pp. 494–513. Springer, Heidelberg (2014)

11. Dodis, Y., Fiore, D.: Interactive encryption and message authentication (2013). Full version of [10]. http://eprint.iacr.org/2013/817

12. Dolev, D., Dwork, C., Naor, M.:Non-malleable cryptography. In: STOC 1991, pp. 542–552. ACM (1991)

13. Dziembowski, S., Pietrzak, K., Wichs, D.: Non-malleable codes. In: ICS 2010, pp. 434–452 (2010)

14. Dziembowski, S., Pietrzak, K., Wichs, D.: Non-malleable codes. Full version of [13]. http://eprint.iacr.org/2009/608

15. Faust, S., Mukherjee, P., Nielsen, J.B., Venturi, D.: Continuous non-malleable codes. In: Lindell, Y. (ed.) TCC 2014. LNCS, vol. 8349, pp. 465–488. Springer, Heidelberg (2014)

16. Hohenberger, S., Lewko, A., Waters, B.: Detecting dangerous queries: a new approach for chosen ciphertext security. In: Pointcheval, D., Johansson, T. (eds.) EUROCRYPT 2012. LNCS, vol. 7237, pp. 663–681. Springer, Heidelberg (2012)

17. Kitagawa, F., Matsuda, T., Hanaoka, G., Tanaka, K.: Completeness of single-bit projection-kdm security for public key encryption. In: Nyberg, K. (ed.) CT-RSA 2015. LNCS, vol. 9048, pp. 201–219. Springer, Heidelberg (2015)

18. Lin, H., Tessaro, S.: Amplification of chosen-ciphertext security. In: Johansson, T., Nguyen, P.Q. (eds.) EUROCRYPT 2013. LNCS, vol. 7881, pp. 503–519. Springer, Heidelberg (2013)

19. Matsuda, T., Hanaoka, G.: Achieving chosen ciphertext security from detectable public key encryption efficiently via hybrid encryption. In: Sakiyama, K., Terada, M. (eds.) IWSEC 2013. LNCS, vol. 8231, pp. 226–243. Springer, Heidelberg (2013)

20. Myers, S., Shelat, A.: Bit encryption is complete. In: FOCS 2009, pp. 607–616 (2009)
21. Naor, M., Yung, M.: Public-key cryptosystems provably secure against chosen ciphertext attacks. In: STOC 1990, pp. 427–437. ACM (1990)
22. Pass, R., Shelat, A., Vaikuntanathan, V.: Relations among notions of non-malleability for encryption. In: Kurosawa, K. (ed.) ASIACRYPT 2007. LNCS, vol. 4833, pp. 519–535. Springer, Heidelberg (2007)
23. Phan, D.H., Pointcheval, D.: About the security of ciphers (semantic security and pseudo-random permutations). In: Handschuh, H., Hasan, M.A. (eds.) SAC 2004. LNCS, vol. 3357, pp. 182–197. Springer, Heidelberg (2004)

Selective Opening Security for Receivers

Carmit Hazay[1]($\boxtimes$), Arpita Patra[2], and Bogdan Warinschi[3]

[1] Faculty of Engineering, Bar-Ilan University, Ramat Gan, Israel
carmit.hazay@biu.ac.il
[2] Department of Computer Science & Automation,
Indian Institute of Science, Bengaluru, India
arpita@csa.iisc.ernet.in
[3] Department of Computer Science, University of Bristol, Bristol, UK
csxbw@bristol.ac.uk

Abstract. In a selective opening (SO) attack an adversary breaks into a subset of honestly created ciphertexts and tries to learn information on the plaintexts of some untouched (but potentially related) ciphertexts. Contrary to intuition, standard security notions do not always imply security against this type of adversary, making SO security an important standalone goal. In this paper we study *receiver security*, where the attacker is allowed to obtain the decryption keys corresponding to some of the ciphertexts.

First we study the relation between two existing security definitions, one based on simulation and the other based on indistinguishability, and show that the former is strictly stronger. We continue with feasibility results for both notions which we show can be achieved from (variants of) non-committing encryption schemes. In particular, we show that indistinguishability-based SO security can be achieved from a tweaked variant of non-committing encryption which, in turn, can be instantiated from a variety of basic, well-established, assumptions. We conclude our study by showing that SO security is however strictly weaker than all variants of non-committing encryption that we consider, leaving potentially more efficient constructions as an interesting open problem.

Keywords: Selective opening attacks · Encryption schemes · Non-committing encryption

1 Introduction

Security notions for encryption come in many forms that reflect different attacker goals (e.g. one-wayness, indistinguishability for plaintexts or non-malleability of ciphertexts), variations in possible attack scenarios (e.g. chosen plaintext or ciphertext attacks) and definitional paradigms (e.g. through games or simulation). A class of attacks motivated by practical considerations are those where the adversary may perform *selective openings* (SO). Here, an adversary is allowed to break into a subset of honestly created ciphertexts leaving untouched other (potentially related) ciphertexts.

© International Association for Cryptologic Research 2015
T. Iwata and J.H. Cheon (Eds.): ASIACRYPT 2015, Part I, LNCS 9452, pp. 443–469, 2015.
DOI: 10.1007/978-3-662-48797-6_19

This attack scenario was first identified in the context of adaptively secure multi-party computation (MPC) where communication is over encrypted channels visible to the adversary. The standard trust model for MPC considers an adversary who, based on the information that he sees, can decide to corrupt parties and learn their internal state. In turn, this may allow the attacker to determine the parties' long term secret keys and/or the randomness used to create the ciphertexts. The broader context of Internet communication also naturally gives rise to SO attacks. Attackers that access and store large amount of encrypted Internet traffic are a reality, and getting access to the internal states of honest parties can be done by leveraging design or implementation weaknesses of deployed systems. For example the Heartbleed attack allowed a remote party to extract (among other things) the encryption keys used to protect OpenSSL connections.

Security against SO attacks comes in several distinct flavors. Depending on the attack scenario, we distinguish two settings that fall under the general idea of SO attacks. In *sender security*, we have n senders and one receiver. The receiver holds a secret key relative to a public key known to all the senders. The senders encrypt messages for the receiver and the adversary is allowed to corrupt some of the senders (and learn the messages and randomness underlying some of the ciphertexts). The concern is that the messages sent by uncorrupted senders stay secret. The second scenario deals with *receiver security*. Here we consider one sender and n receivers who hold independently generated public and secret keys. The attacker is allowed to corrupt some of the receivers (and learn the secret keys that decrypt some of the observed ciphertexts). Security in this setting is concerned with the messages received by uncorrupted receivers. For each of these settings, security can be defined using either the standard indistinguishability paradigm or simulation-based definitions. Importantly, both scenarios capture realistic attacks in secure computation where usually every party acts as either a sender or a receiver at some point of time during a protocol execution.

Since most of the existent encryption schemes have been analyzed w.r.t. traditional notions of security (e.g. indistinguishability under chosen plaintext or chosen ciphertext attacks (**ind-cpa, ind-cca**)), a central question in this area is to understand how security against SO attacks relates to the established definitions. Despite compelling intuition that the only information that an adversary obtains is what it can glean from the opened plaintexts, progress towards confirming or disproving this conjecture has been rather slow. Perhaps the most interesting and surprising results are due to Bellare et al. [1,2] who showed that selective sender security as captured via *simulation* based definitions is strictly stronger than indistinguishability under chosen plaintext attacks [15] (denoted by **ind-cpa** security). The gap between standard notions of security and SO security is uncomfortable: while SO attacks may naturally occur we do not have a clear understanding of the level of security that existing constructions offer nor do we have many ideas on how to achieve security against such attacks.

In this paper we study receiver security. This setting is less studied than sender security yet it corresponds to more plausible attacks (e.g. the Heartbleed attack). In a nutshell, we clarify the relation between various security notions for receiver security and propose novel constructions. Before we describe our contributions in detail we overview existing work in the area and take this opportunity to introduce more carefully the different security notions of SO security.

1.1 Related Work

Selective opening attacks were first introduced in [12] in the context of commitment schemes. In the context of encryption schemes, the first rigorous definitions were proposed by Bellare, Hofheinz and Yilek [2]. They studied SO security for public key encryption (PKE), for both the receiver and the sender settings and for each setting proposed two types of definitions, indistinguishability-based and simulation-based ones.

Very roughly, the indistinguishability-based definition (denoted by **ind-so**) requires that an adversary that sees a vector of ciphertexts cannot distinguish the true plaintexts of the unopened ciphertexts from independently sampled plaintexts. This is required even with access to the randomness used for generating the opened ciphertexts (in the sender corruption setting), or with access to the secret keys that decrypt the opened ciphertexts (in the receiver corruption setting). This definition requires messages to come from a distribution that is *efficiently resamplable*. A stronger security variant that does not restrict the message distribution called *full* **ind-so** has been introduced later by Böhl, Hofheinz and Kraschewski [5]. The simulation based notion (denoted by **sim-so**) is reminiscent of the definitional approach of Dwork et al. [12] and requires computational indistinguishability between an idealized execution and the real one.

The first feasibility results for security against SO attacks are for the sender setting and leverage an interesting relation with lossy encryption: a lossy PKE implies **ind-so** for sender security [2]. Furthermore, if the PKE scheme has an efficient opening algorithm of ciphertexts, then the scheme also satisfies **sim-so** security. The work of Hemenway et al. [18] shows that lossy (and therefore **ind-so**) PKE can be constructed based on several generic cryptographic primitives.

For primitives that benefit from multiple security notions, a central question is to understand how these notions relate to each other. This type of results are important as they clarify the limitations of some of the notions and enable trade-offs between security and efficiency (to gain efficiency, a scheme with weaker guarantees may be employed, if the setting allows it). The relation between traditional security notions of encryption and security against SO attacks was a long-standing open problem that was solved by Bellare et al. [1]. Their result is that standard **ind-cpa** security *does not* imply **sim-so** (neither in the sender nor in the receiver setting). There is no fully satisfactory result concerning the relation between **ind-cpa** and **ind-so**. Here, the best result is that these two notions imply each other in the generic group model [19] and that for the chosen-chiphertext attacks variant (CCA) the two notions are distinct.

Relations between the different notions for selective opening have mainly been studied in the sender setting. Böhl et al. establish that full **ind-so** and **sim-so** are incomparable. Recently, [23] introduced an even stronger variant of the full **ind-so** definition, and showed that many **ind-cpa**, **ind-so** and **sim-so** secure encryption schemes are insecure according to their new notion. They further showed that **sim-so** definition does not imply lossy encryption even without efficient openability. Finally, SO security has been considered for CCA attacks [13,20] and in the identity-based encryption scenario [3].

1.2   Our Contribution

With only two exceptions [1,2] prior work on SO security has addressed mainly the sender setting. We concentrate on the receiver setting. Though theoretically the feasibility for SO security for the receiver is implied by the existence of non-committing encryption schemes [6,8,9,22], the state of the art constructions still leave many interesting open problems in terms of relations between notions and feasibility results. This is the focus of this work.

For relation between notions, similarly to prior separating results in the SO setting [5,19,23], we demonstrate the existence of a separating scheme that is based on generic assumptions and can be instantiated under various concrete assumptions. For constructions, we find it useful to leverage the close relation between (variants of) non-committing encryption and security under SO attacks. For example, we show that **ind-so** security follows from a tweaked variant of non-committing encryption which, in turn, we show how to instantiate from a variety of standard assumptions. Interestingly, we also show a separation between SO security and non-committing encryption (which leaves open the question of potentially more efficient constructions that meet the former notion but not the latter). Below, we elaborate on our results in details.

Notation-wise, we denote the indistinguishability and simulation-based definitions in the receiver setting by **rind-so** and **rsim-so**, respectively. For the corresponding notions in the sender setting we write **sind-so** and **ssim-so**, respectively. That is, we prepend "s" or "r" to indicate if the definition is for sender security or receiver security.

The relation between **rind-so** *and* **rsim-so**. First, we study the relation between the indistinguishability and simulation-based security notions in the receiver setting. We establish that the **rind-so** notion is strictly weaker (and therefore easier to realize) than the notion of **rsim-so**, by presenting a concrete pubic key scheme that meets the former but not the latter level of security. Loosely speaking, a ciphertext includes a commitment to the plaintext together with encryptions of the opening information of this commitment (namely, the plaintext and the corresponding randomness). We then prove that when switching to an alternative fake mode the hiding properties of our building blocks (commitment and encryption schemes) imply that the ciphertext does not contain any information about the plaintext. Nevertheless, simulation always fails since it would require breaking the binding property of the commitment. Applying the observation that **rsim-so** implies **rind-so** security,[1] we obtain the result that **rind-so** is strictly weaker.

In more details, our separating scheme is built from a commitment scheme and a primitive called non-committing encryption for the receiver (NCER) [7] that operates in two indistinguishable ciphertexts modes: valid and fake, where a fake ciphertext can be decrypted into any plaintext using a proper secret key. This property is referred to as *secret key equivocation* and is implied by the fact that fake ciphertexts are lossy which, in turn, implies **rind-so** security. Specifically, the security of our scheme implies that:

Theorem 1.1 (Informal). *There exists a PKE that is* **rind-so** *secure but is not* **rsim-so** *secure.*

Somewhat related to our work, [1] proved that the standard **ind-cpa** security does not imply **rsim-so** security via the notion of *decryption verifiability* – the idea that it is hard to decrypt a ciphertext into two distinct messages (even using two different secret keys). Specifically, [1] showed that any **ind-cpa** secure PKE that is decryption verifiable cannot be **rsim-so** secure. Compared with their result, our result implies that **rsim-so** security is strictly stronger than **rind-so** security (which may turn out to be stronger than **ind-cpa** security).

The feasibility of **rind-so** *and* **rsim-so**. We recall that in the sender setting, the notions **sind-so** and **ssim-so** are achievable from lossy encryption and lossy encryption with efficient openability.[2] We identify a security notion (and a variant) which plays for receiver security the role that lossy encryption plays in sender security. Specifically, we prove that NCER implies **rsim-so** and that a variant of NCER, which we refer as *tweaked NCER* (formally defined in Tweaked NCER subsection of Sect. 3), implies **rind-so**. Loosely speaking, the security of tweaked NCER is formalized as follows. Similarly to NCER, tweaked NCER has the ability to create fake ciphertexts that are computationally indistinguishable from real ciphertexts. Nevertheless, while in NCER a fake ciphertext can be efficiently decrypted to any plaintext (by producing a matching secret key), in tweaked NCER a fake ciphertext can only be efficiently decrypted to a concrete predetermined plaintext. Informally, our results are captured by the following theorem:

Theorem 1.2 (Informal). *Assume the existence of tweaked NCER and NCER, then there exist PKE schemes that are* **rind-so** *and* **rsim-so** *secure, respectively.*

---

[1] This can be derived from the fact that the adversary's view is identical for any two simulated executions with different sets of unopened messages, as the simulator never gets to see these messages.

[2] Recall that a lossy encryption scheme is a public key encryption with the additional ability to generate fake indistinguishable public keys so that a fake ciphertext (that is generated using a fake public key) is lossy and is a non-committing ciphertext with respect to the plaintext. A lossy encryption implies the existence of an opening algorithm (possibly inefficient) that can compute a randomness for a given fake ciphertext and a message.

Interestingly, we show that the converse implications do not hold. That is, a **rsim-so** secure PKE is not necessarily a tweaked NCER or a NCER. This further implies that a **rind-so** secure PKE is not necessarily a tweaked NCER or NCER. This result is reminiscent of the previous result that **sim-so** and **rind-so** secure PKE do not imply lossy encryption even without efficient openability [23].

Our separating scheme is based on an arbitrary key-simulatable PKE scheme. Intuitively, in such schemes, it is possible to produce a public key without sampling the corresponding secret key. The set of obliviously sampled public keys may be larger than the the set of public keys sampled together with their associated secret key, yet it is possible to explain a public key sampled along with a secret key as one sampled without. In these schemes we also require that the two type of keys are also computationally indistinguishable. Our proof holds for the case that the set of obliviously sampled keys is indeed larger, so that not every obliviously sampled public key can be explained to possess a secret key. In summary, we prove that:

Theorem 1.3 (Informal). *Assume the existence of key-simulatable PKE, then there exists a PKE scheme that is* **rsim-so** *secure but is neither tweaked NCER nor NCER.*

Our constructions show that **rsim-so** (and **rind-so**) security can be achieved under the same assumptions as key-simulatable PKE – there are results that show that the latter can be constructed from a variety of hardness assumptions such as Decisional Diffie-Hellman (DDH) and Decisional Composite Residuosity (DCR). They also show that we can construct schemes from any hardness assumption that implies simulatable PKE [9] (where both public keys and ciphertexts can be obliviously sampled).

Realizing tweaked NCER. Finally, we demonstrate the broad applicability of this primitive and show how to construct it from various important primitives: key-simultable PKE, two-round honest-receiver statistically-hiding $\binom{2}{1}$ oblivious transfer (OT) and hash proof systems (HPS). We stress that it is not known how to build NCER under these assumptions (or any other generic assumption), which implies that tweaked NCER is much easier to realize. In addition, we prove that the two existing NCER schemes [7] with security under the DDH and DCR hardness assumptions imply the tweaked NCER notion, where surprisingly, the former construction that is a secure NCER for only polynomial-size message spaces, is a tweaked NCER for exponential-size message spaces (this further hints that tweaked NCER may be constructed more efficiently than NCER). These results imply that tweaked NCER (and thus **rind-so**) can be realized based on DDH, DCR, RSA, factoring and learning with errors (LWE) hardness assumptions.

Our results are summarized in Fig. 1.

The relation between **sind-so** *and* **ssim-so**. As a side result, we study the relation between the indistinguishability and simulation based security definitions in the sender setting. We show that **sind-so** is strictly weaker than the notion of **ssim-so**

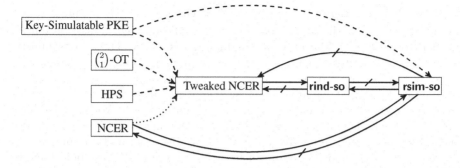

Fig. 1. The arrows can be read as follows: *solid arrows* denote implication, *crossed arrows* denote counterexamples, *dashed arrows* denote assumption-wise implication and *dotted arrows* denote implication with respect to concrete instances (where the implication may not hold in general). The implication of receiver indistinguishability security by simulation security is a known result.

by presenting a concrete public key scheme that meets the former but not the latter level of security. Our separating scheme is built using the two primitives lossy public key encryption and commitment scheme. We exploit the hiding properties of these building blocks to prove that our scheme implies **sind-so** security. On the other hand, simulation always fails since it implies breaking the binding property of the commitment scheme. Informally, we prove the following theorem:

Theorem 1.4 (Informal). *There exists a PKE that is* **sind-so** *secure but is not* **ssim-so** *secure.*

We stress that this was already demonstrated indirectly in [4] (by combining two separation results). Here we design a concrete counter example to demonstrate the same in a simpler manner. A similar result has been shown for *full* **ind-so** and **sim-so** in [5], demonstrating that these definitions do *not* imply each other in the sender setting.

To sum up, we study the different levels for receiver security in the presence of SO attacks. We clarify the relation between these notions and provide constructions that meet them using the close conceptual relation between SO security and non-committing encryption. From a broader perspective, our results position more precisely SO security for the receiver in the spectrum of security notions for encryption.

2   Preliminaries

Basic notations. For $x, y \in \mathbb{N}$ with $x < y$, let $[x] := \{1, \ldots, x\}$ and $[x, y] := \{x, x + 1, \ldots, y\}$. We denote the computational security parameter by k and statistical security parameter by s. A function $\mu(\cdot)$ is *negligible* in security parameter κ if for every polynomial $p(\cdot)$ there exists a value N such that for all

$\kappa > N$ it holds that $\mu(k) < \frac{1}{p(\kappa)}$, where κ is either k or s. For a finite set S, we denote by $s \leftarrow S$ the process of sampling s uniformly. For a distribution X, we denote by $x \leftarrow X$ the process of sampling x from X. For a deterministic algorithm A, we write $a \leftarrow A(x)$ the process of running A on input x and assigning y the result. For a randomized algorithm A, we write $a \leftarrow A(x; r)$ the process of running A on input x and randomness r and assigning a the result. At times we skip r in the parenthesis to avoid mentioning it explicitly. We write PPT for probabilistic polynomial-time. For a PKE (or commitment) scheme C, we use the notation $\mathcal{M}_C$ and respectively $\mathcal{R}_C$ to denote the input and the randomness space of the encryption (or commitment) algorithm of C. We use bold fonts to denote vectors. If $\mathbf{m}$ is an n dimensional vector, we write $\mathbf{m}_i$ for the i-th entry in $\mathbf{m}$; if $\mathcal{I} \subseteq [n]$ is a set of indices we write $\mathbf{m}_\mathcal{I}$ for the vector of dimension $|\mathcal{I}|$ obtained by projecting $\mathbf{m}$ on the coordinates in $\mathcal{I}$.

2.1 Public Key Encryption

A public key encryption (PKE) scheme PKE with message space $\mathcal{M}$ consists of three PPT algorithms (Gen, Enc, Dec). The key generation algorithm $\mathsf{Gen}(1^k)$ outputs a public key pk and a secret key sk. The encryption algorithm $\mathsf{Enc}_{pk}(m; r)$ takes pk and a message $m \in \mathcal{M}$ and randomness $r \in \mathcal{R}$, and outputs a ciphertext c. The decryption algorithm $\mathsf{Dec}_{sk}(c)$ takes sk and a ciphertext c and outputs a message m. For correctness, we require that $m = \mathsf{Dec}_{sk}(c)$ for all $m \in \mathcal{M}$ and all $(pk, sk) \leftarrow \mathsf{Gen}(1^k)$ and all $c \leftarrow \mathsf{Enc}_{pk}(m)$. The standard notion of security for PKE is indistinguishability under chosen plaintext attacks, denoted by **ind-cpa** [15] (and the corresponding experiment is denoted as $\mathbf{Exp}^{\mathbf{ind}}\textbf{-cpa}_{\mathsf{PKE}}$). *As a general remark, we note that whenever we refer to a secret key, we refer to the randomness used to generate it by the key generation algorithm.*

2.2 Selective Opening Security

Depending on the attack scenario, we distinguish two settings that fall under the general idea of SO attacks. In *sender security*, we have n senders and one receiver. The receiver holds a secret key relative to a public key known to all senders. The senders send messages to the receiver and the adversary is allowed to corrupt some of the senders (and learn the messages and randomness underlying some of the ciphertexts). The concern is that the messages sent by uncorrupted users stay secret. The second scenario deals with *receiver security*. Here we consider one sender and n receivers who hold independently generated public and secret keys. The attacker is allowed to learn the secret keys of some of the receivers. Security is concerned with the messages received by uncorrupted receivers.

For each of these settings we consider two types of definitions from the literature [2]: (1) an indistinguishability based definition and (2) a simulation based definition. Indistinguishability-based definitions require that an adversary that sees a vector of ciphertexts cannot distinguish the true plaintexts of the ciphertexts from independently sampled plaintexts, even in the presence of the randomness used for generating the opened ciphertexts (in the sender corruption setting), or the secret keys that decrypt the opened ciphertexts (in the receiver corruption setting). The indistinguishability based definitions use the notion of *efficiently resamplable* message distributions which we recall next following [5].

Definition 2.1 (Efficiently Resamplable Distribution). *Let $n = n(k) > 0$ and let Dist be a joint distribution over $\left(\{0,1\}^k\right)^n$. We say that Dist is efficiently resamplable if there is a PPT algorithm $\mathsf{Resamp}_{\mathsf{Dist}}$ such that for any $\mathcal{I} \subseteq [n]$ and any partial vector $\mathbf{m}'_{\mathcal{I}} \in \left(\{0,1\}^k\right)^{|\mathcal{I}|}$, $\mathsf{Resamp}_{\mathsf{Dist}}(\mathbf{m}'_{\mathcal{I}})$ returns a vector $\mathbf{m}$ sampled from $\mathsf{Dist}|\mathbf{m}'_{\mathcal{I}}$, i.e. $\mathbf{m}'$ is sampled from Dist conditioned on $\mathbf{m}_{\mathcal{I}} = \mathbf{m}'_{\mathcal{I}}$.*

Below, we recall indistinguishability and simulation based definitions for security in the presence of selective opening attacks[3]. We present the definitions for sender and receiver security. To avoid heavy notation we follow the following conventions when naming the security notions: we use "ind" or "sim" to indicate if the definition is indistinguishability-based or simulation-based, and prepend "s" or "r" to indicate if the definition is for sender security or receiver security; we keep "so" in the name of the notion to indicate that we deal with selective opening attacks. We also note that we consider chosen plaintext attacks only, but avoid showing this explicitly in the names of the security notions.

Experiment 1 $\mathbf{Exp}^{\mathsf{sind}}_{\mathsf{PKE}}\text{-}\mathsf{so}(\mathsf{A}, k)$

$b \leftarrow \{0, 1\}$
$(pk, sk) \leftarrow \mathsf{Gen}(1^k)$
$(\mathsf{Dist}, \mathsf{Resamp}_{\mathsf{Dist}}, \mathsf{state}_1) \leftarrow \mathsf{A}(pk)$
$\mathbf{m} := (m_i)_{i \in [n]} \leftarrow \mathsf{Dist}$
$\mathbf{r} := (r_i)_{i \in [n]} \leftarrow \mathcal{R}^n_{\mathsf{PKE}}$
$\mathbf{e} := (e_i)_{i \in [n]} \leftarrow (\mathsf{Enc}_{pk}(m_i; r_i))_{i \in [n]}$
$(\mathcal{I}, \mathsf{state}_2) \leftarrow \mathsf{A}(\mathbf{e}, \mathsf{state}_1)$
$\mathbf{m}' \leftarrow \mathsf{Resamp}(\mathbf{m}_{\mathcal{I}})$
$\mathbf{m}^* = \mathbf{m}$ *if* $b = 0$, *else* $\mathbf{m}^* = \mathbf{m}'$
$b' \leftarrow \mathsf{A}(\mathbf{r}_{\mathcal{I}}, \mathbf{m}^*, \mathsf{state}_2)$,
Return 1 if $b = b'$, and 0 otherwise.

Experiment 2 $\mathbf{Exp}^{\mathsf{rind}}_{\mathsf{PKE}}\text{-}\mathsf{so}(\mathsf{A}, k)$

$b \leftarrow \{0, 1\}$
$(\mathbf{pk}, \mathbf{sk}) := (pk_i, sk_i) \leftarrow (\mathsf{Gen}(1^k))_{i \in [n]}$
$(\mathsf{Dist}, \mathsf{Resamp}_{\mathsf{Dist}}, \mathsf{state}_1) \leftarrow \mathsf{A}(\mathbf{pk})$
$\mathbf{m} := (m_i)_{i \in [n]} \leftarrow \mathsf{Dist}$
$\mathbf{r} := (r_i)_{i \in [n]} \leftarrow \mathcal{R}^n_{\mathsf{PKE}}$
$\mathbf{e} := (e_i)_{i \in [n]} \leftarrow (\mathsf{Enc}_{pk_i}(m_i; r_i))_{i \in [n]}$
$(\mathcal{I}, \mathsf{state}_2) \leftarrow \mathsf{A}(\mathbf{e}, \mathsf{state}_1)$
$\mathbf{m}' \leftarrow \mathsf{Resamp}(\mathbf{m}_{\mathcal{I}})$
$\mathbf{m}^* = \mathbf{m}$ *if* $b = 0$, *else* $\mathbf{m}^* = \mathbf{m}'$
$b' \leftarrow \mathsf{A}(\mathbf{sk}_{\mathcal{I}}, \mathbf{m}^*, \mathsf{state}_2)$
Return 1 if $b = b'$, and 0 otherwise.

Definition 2.2 (Indistinguishability Based SO Security). *For a PKE scheme $\mathsf{PKE} = (\mathsf{Gen}, \mathsf{Enc}, \mathsf{Dec})$, a polynomially bounded function $n = n(k) > 0$ and a stateful PPT adversary A, consider the following two experiments; the left experiment corresponds to sender corruptions, whereas, the right experiment corresponds to receiver corruptions.*

In the above experiments we only assume adversaries that are well-behaved in that they always output efficiently resamplable distributions together with resampling algorithms.

---

[3] We remark that a stronger security notion that does not does require efficient resamplability is possible, but no constructions that satisfy this stronger notion are known.

We say that PKE is **sind-so** secure if for a well-behaved PPT A there exists a negligible function $\mu = \mu(k)$ such that

$$\mathbf{Adv^{sind}\text{-}so}_{PKE}(A, k) := 2 \left| \Pr[\mathbf{Exp^{sind}_{PKE}\text{-}so}(A, k) = 1] - \frac{1}{2} \right| \le \mu.$$

We say that PKE is **rind-so** secure if for a well-behaved PPT A there exists a negligible function $\mu = \mu(k)$ such that

$$\mathbf{Adv^{rind}\text{-}so}_{PKE}(A, k) := 2 \left| \Pr[\mathbf{Exp^{rind}_{PKE}\text{-}so}(A, k) = 1] - \frac{1}{2} \right| \le \mu.$$

$\Pr\left[\mathbf{Exp^{sind}_{PKE}\text{-}so}(A, k) = 1 \right]$ and $\Pr\left[\mathbf{Exp^{rind}_{PKE}\text{-}so}(A, k) = 1 \right]$ denote the **winning probability** of A in the respective experiments.

Simulation based security is defined, as usual, by comparing an idealized execution with the real one. Again, we consider both sender and receiver security.

Experiment 3 $\mathbf{Exp^{SSIM\text{-}SO}_{PKE}\text{-}real}(A, k)$

$(pk, sk) \leftarrow \mathsf{Gen}(1^k)$
$(\mathsf{Dist}, state_1) \leftarrow A(pk)$
$\mathbf{m} := (m_i)_{i \in [n]} \leftarrow \mathsf{Dist}$
$\mathbf{r} := (r_i)_{i \in [n]} \leftarrow \mathcal{R}^n_{PKE}$
$\mathbf{e} := (e_i)_{i \in [n]} \leftarrow (\mathsf{Enc}_{pk}(m_i; r_i))_{i \in [n]}$
$(\mathcal{I}, state_2) \leftarrow A(\mathbf{e}, state_1)$
output $\leftarrow A(\mathbf{r}_\mathcal{I}, \mathbf{m}_\mathcal{I}, state_2)$
Return $(\mathbf{m}, \mathsf{Dist}, \mathcal{I}, output)$.

Experiment 4 $\mathbf{Exp^{SSIM\text{-}SO}_{PKE}\text{-}ideal}(S, k)$

$(\mathsf{Dist}, state_1) \leftarrow S(\cdot)$
$\mathbf{m} := (m_i)_{i \in [n]} \leftarrow \mathsf{Dist}$
$(\mathcal{I}, state_2) \leftarrow S(state_1)$
output $\leftarrow S(\mathbf{m}_\mathcal{I}, state_2)$
Return $(\mathbf{m}, \mathsf{Dist}, \mathcal{I}, output)$.

Experiment 5 $\mathbf{Exp^{rsim\text{-}so}_{PKE}\text{-}real}(A, k)$

$(\mathbf{pk}, \mathbf{sk}) := (pk_i, sk_i) \leftarrow (\mathsf{Gen}(1^k))_{i \in [n]}$
$(\mathsf{Dist}, state_1) \leftarrow A(\mathbf{pk})$
$\mathbf{m} := (m_i)_{i \in [n]} \leftarrow \mathsf{Dist}$
$\mathbf{r} := (r_i)_{i \in [n]} \leftarrow \mathcal{R}^n_{PKE}$
$\mathbf{e} := (e_i)_{i \in [n]} \leftarrow (\mathsf{Enc}_{pk_i}(m_i; r_i))_{i \in [n]}$
$(\mathcal{I}, state_2) \leftarrow A(\mathbf{e}, state_1)$
output $\leftarrow A(\mathbf{sk}_\mathcal{I}, \mathbf{m}_\mathcal{I}, state_2)$
Return $(\mathbf{m}, \mathsf{Dist}, \mathcal{I}, output)$.

Experiment 6 $\mathbf{Exp^{rsim\text{-}so}_{PKE}\text{-}ideal}(S, k)$

$(\mathsf{Dist}, state_1) \leftarrow S(\cdot)$
$\mathbf{m} := (m_i)_{i \in [n]} \leftarrow \mathsf{Dist}$
$(\mathcal{I}, state_2) \leftarrow S(state_1)$
output $\leftarrow S(\mathbf{m}_\mathcal{I}, state_2)$
Return $(\mathbf{m}, \mathsf{Dist}, \mathcal{I}, output)$.

Definition 2.3 (Simulation Based SO Security). *For a PKE scheme* PKE = (Gen, Enc, Dec), *a polynomially bounded function* $n = n(k) > 0$, *a PPT adversary* A *and a PPT algorithm* S, *we define the following pairs of experiments.*

We say that PKE *is* **ssim-so** *secure iff for every PPT* A *there is a PPT algorithm* S, *a PPT distinguisher* D *with binary output and a negligible function* $\mu = \mu(k)$ *such that*

$\mathbf{Adv^{ssim}\text{-}so}_{PKE}(D, k) :=$

$$\left| \Pr[1 \leftarrow D(\mathbf{Exp^{SSIM\text{-}SO}_{PKE}\text{-}real}(A, k))] - \Pr[1 \leftarrow D(\mathbf{Exp^{SSIM\text{-}SO}_{PKE}\text{-}ideal}(S, k))] \right| \le \mu.$$

We say that PKE *is* **rsim-so** *secure iff for every PPT* A *there is a PPT algorithm* S, *a PPT distinguisher* D *with binary output and a negligible function* $\mu = \mu(k)$ *such that*

$\mathbf{Adv^{rsim}\text{-}so}_{PKE}(D, k) :=$

$$\left| \Pr[1 \leftarrow D(\mathbf{Exp^{rsim\text{-}so}_{PKE}\text{-}real}(A, k))] - \Pr[1 \leftarrow D(\mathbf{Exp^{rsim\text{-}so}_{PKE}\text{-}ideal}(S, k))] \right| \le \mu.$$

Our definitions consider non-adaptive attacks, where the adversary corrupts the parties in one go. Our results remain unaffected even in the face of an adaptive adversary [5].

3   Building Blocks

Our constructions employ a number of fundamental cryptographic building blocks as well as a new primitive which we call tweaked NCER.

Commitment Schemes. We define a non-interactive statistically hiding commitment scheme (NISHCOM).

Definition 3.1 (NISHCOM). *A non-interactive commitment scheme* nisCom *consists of two algorithms* (nisCommit, nisOpen) *defined as follows. Given a security parameter* k, *message* $m \in \mathcal{M}_{\text{nisCom}}$ *and random coins* $r \in \mathcal{R}_{\text{nisCom}}$, *PPT algorithm* nisCommit *outputs commitment* c. *Given* k, *commitment* c *and message* m, *(possibly inefficient) algorithm* nisOpen *outputs* r. *We require the following properties:*

- Correctness. *We require that* $c = \text{nisCommit}(m; r)$ *for all* $m \in \mathcal{M}_{\text{nisCom}}$ *and* $r \leftarrow \text{nisOpen}(c, m)$.
- Security. *A NISHCOM* nisCom *is* **stat-hide** *secure if commitments of two distinct messages are statistically indistinguishable. Specifically, for any unbounded powerful adversary* A, *there exists a negligible function* $\mu = \mu(s)$ *such that*
  $\mathbf{Adv}_{\text{nisCom}}^{\text{stat}}\text{-hide}(A, k) := |\Pr[1 \leftarrow A(c_0)] - \Pr[1 \leftarrow A(c_1)]| \leq \mu$
  for $c_i \leftarrow \text{nisCommit}(m_i)$, $i \in \{0, 1\}$ *and* $m_0, m_1 \in \mathcal{M}_{\text{nisCom}}$.
  A NISHCOM nisCom *is* **comp-bind** *secure if no commitment can be opened to two different messages in polynomial time. Specifically, the advantage* $\mathbf{Adv}_{\text{nisCom}}^{\text{comp}}\text{-bind}(A, k)$ *of* A *is defined by* $\Pr[(m_0, r_0, m_1, r_1) \leftarrow A(k)$: $\text{nisCommit}(prm, m_0; r_0) = \text{nisCommit}(prm, m_1; r_1)]$ *(with the probability over the choice of the coins of* A*) is smaller than some negligible function* $\mu = \mu(k)$. *A NISHCOM* nisCom *is called* **secure** *it is* {**stat-hide, comp-bind**} *secure.*

Non-committing Encryption for Receiver (NCER). A non-committing encryption for receiver [7,21] is a PKE scheme with the property that there is a way to generate fake ciphertexts which can then be decrypted (with the help of a trapdoor) to any plaintext. Intuitively, fake ciphertexts are generated in a lossy way so that the plaintext is no longer well defined given the ciphertext and the public key. This leaves enough entropy for the secret key to be sampled in a way that determines the desired plaintext. We continue with a formal definition of NCER and its security notion referred as **ind-ncer** security.

Definition 3.2 (NCER). *An NCER* nPKE *consists of five PPT algorithms* (nGen, nEnc, nEnc*, nDec, nOpen) *defined as follows. Algorithms* (nGen, nEnc, nDec) *form a PKE. Given the public key* pk, *the fake encryption algorithm* nEnc* *outputs a ciphertext* e^* *and a trapdoor* t. *Given the secret key* sk, *the public key* pk, *fake ciphertext* e^*, *trapdoor* t *and plaintext* m, *algorithm* nOpen *outputs* sk^*.

- Correctness. *We require that* $m = \text{nDec}_{sk}(c)$ *for all* $m \in \mathcal{M}$, *all* $(pk, sk) \leftarrow \text{nGen}(1^k)$ *and all* $c \leftarrow \text{nEnc}_{pk}(m)$.

- Security. *An NCER scheme* nPKE *is* **ind-ncer** *secure if the real and fake ciphertexts are indistinguishable. Specifically, for a PPT adversary* A, *consider the experiment* $\mathbf{Exp}^{ind}_{nPKE}$**-ncer** *defined as follows.*

> **Experiment 7** $\mathbf{Exp}^{ind}_{nPKE}$**-ncer**$(A, k)$
> $b \leftarrow \{0, 1\}$
> $(pk, sk_0) \leftarrow \mathsf{nGen}(1^k)$
> $m \leftarrow A(pk)$
> $e_0 \leftarrow \mathsf{nEnc}_{pk}(m)$
> $(e_1, t) \leftarrow \mathsf{nEnc}^*_{pk}(1^k)$, $sk_1 \leftarrow \mathsf{nOpen}(sk_0, pk, e_1, t, m)$
> $b' \leftarrow A(sk_b, e_b)$
> *Return 1 if* $b = b'$, *and 0 otherwise.*

We say that nPKE *is* **ind-ncer**-*secure if for a PPT adversary* A, *there exists a negligible function* $\mu = \mu(k)$ *such that*

$$\mathbf{Adv}^{ind\text{-}ncer}_{nPKE}(A, k) := 2 \left| \Pr[\mathbf{Exp}^{ind}\text{-}ncer_{nPKE}(A, k) = 1] - \frac{1}{2} \right| \le \mu.$$

An NCER nPKE *is* **secure** *if it is* **ind-ncer** *secure.*

Tweaked NCER. We introduce a variant of NCER which modifies the definition of NCER in the following two ways. First, the opening algorithm nOpen may be *inefficient*. In addition, the fake encryption algorithm is required to output a fake ciphertext e^* given the secret key sk and a plaintext m, so that decryption is "correct" with respect to e^* and m. We call the resulting notion, which we formalize below, *tweaked NCER*.

Definition 3.3 (Tweaked NCER). *A tweaked NCER scheme* tPKE *is a PKE that consists of five algorithms* $(\mathsf{tGen}, \mathsf{tEnc}, \mathsf{tEnc}^*, \mathsf{tDec}, \mathsf{tOpen})$ *defined as follows. Algorithms* $(\mathsf{tGen}, \mathsf{tEnc}, \mathsf{tDec})$ *form a PKE. Given the secret key* sk *and the public key* pk, *and a plaintext* m, *the PPT fake encryption algorithm* tEnc^* *outputs a ciphertext* e^*. *Given the secret key* sk *and the public key* pk, *fake ciphertext* e^* *such that* $e^* \leftarrow \mathsf{tEnc}^*_{pk}(sk, m')$ *for some* $m' \in \mathcal{M}_{tPKE}$ *and a plaintext* m, *the inefficient algorithm* tOpen *outputs* sk^* *such that* $m = \mathsf{tDec}_{sk^*}(e^*)$.

- Correctness. *We require that* $m = \mathsf{tDec}_{sk}(c)$ *for all* $m \in \mathcal{M}$, *all* $(pk, sk) \leftarrow \mathsf{tGen}(1^k)$ *and all* $c \leftarrow \mathsf{tEnc}_{pk}(m)$.
- Security. *A tweaked NCER scheme* tPKE *is* **ind-tcipher** *secure if real and fake ciphertexts are indistinguishable. Specifically, for a PPT adversary* A, *consider the experiment* $\mathbf{Exp}^{ind}_{tPKE}$**-tcipher** *defined as follows.*

> **Experiment 8** $\mathbf{Exp}^{ind}_{tPKE}$**-tcipher**$(A, k)$
> $b \leftarrow \{0, 1\}$
> $(pk, sk) \leftarrow \mathsf{tGen}(1^k)$
> $m \leftarrow A(pk)$
> $e_0 \leftarrow \mathsf{tEnc}_{pk}(m)$
> $e_1 \leftarrow \mathsf{tEnc}^*_{pk}(sk, m)$
> $b' \leftarrow A(sk, e_b)$
> *Return 1 if* $b = b'$, *and 0 otherwise.*

> **Experiment 9** $\mathbf{Exp}^{ind}_{tPKE}$**-tncer**$(A, k)$
> $b \leftarrow \{0, 1\}$
> $(pk, sk_0) \leftarrow \mathsf{tGen}(1^k)$
> $m \leftarrow A(pk)$
> $e_0 \leftarrow \mathsf{tEnc}^*_{pk}(sk_0, m)$
> $e_1 \leftarrow \mathsf{tEnc}^*_{pk}(sk_0, m')$ *for* $m' \in \mathcal{M}_{tPKE}$
> $sk_1 \leftarrow \mathsf{tOpen}(e_1, m)$
> $b' \leftarrow A(sk_b, e_b)$
> *Return 1 if* $b = b'$, *and 0 otherwise.*

We say that tPKE *is* **ind-tcipher** *secure if for a PPT adversary* A, *there exists a negligible function* $\mu = \mu(k)$ *such that*

$$\mathbf{Adv}_{\mathsf{tPKE}}^{\mathsf{ind\text{-}tcipher}}(\mathsf{A}, k) := 2 \big| \Pr[\mathbf{Exp}^{\mathsf{ind}}\text{-}\mathbf{tcipher}_{\mathsf{tPKE}}(\mathsf{A}, k) = 1] - \frac{1}{2} \big| \le \mu.$$

We say that tPKE *is* **ind-tncer** *secure if for an unbounded adversary* A, *there exists a negligible function* $\mu = \mu(s)$ *such that*

$$\mathbf{Adv}_{\mathsf{tPKE}}^{\mathsf{ind\text{-}tncer}}(\mathsf{A}, k) := 2 \big| \Pr[\mathbf{Exp}^{\mathsf{ind}}\text{-}\mathbf{tncer}_{\mathsf{tPKE}}(\mathsf{A}, k) = 1] - \frac{1}{2} \big| \le \mu.$$

A tweaked NCER tPKE *is* **secure** *if it is* {**ind-tcipher**, **ind-tncer**} *secure.*

Key-Simulatable PKE. A key-simulatable public key encryption scheme is a PKE in which the public keys can be generated in two modes. In the first mode a public key is picked together with a secret key, whereas the second mode implies an oblivious public key generation without the secret key. Let $\mathcal{V}$ denote the set of public keys generated in the first mode and $\mathcal{K}$ denote the set of public keys generated in the second mode. Then it is possible that $\mathcal{K}$ contains $\mathcal{V}$ (i.e., $\mathcal{V} \subseteq \mathcal{K}$). Moreover, in case $\mathcal{V} \subset \mathcal{K}$ the set of public keys from $\mathcal{K} \backslash \mathcal{V}$ is not associated with any secret key. We respectively denote the keys in $\mathcal{V}$ and $\mathcal{K} \backslash \mathcal{V}$ as *valid* and *invalid* public keys. In addition to the key generation algorithms, key-simulatable PKE also consists of an efficient key faking algorithm that explains a public key from $\mathcal{V}$, that was generated in the first mode, as an obliviously generated public key from $\mathcal{K}$ that was generated without the corresponding secret key. The security requirement asserts that it is hard to distinguish a random element from $\mathcal{K}$ from a random element from $\mathcal{V}$. The formal definition follows. We note that the notion of key-simulatable PKE is very similar to the simulatable PKE [9] notion with the differences that the latter notion assumes that $\mathcal{K} = \mathcal{V}$ and further supports oblivious ciphertext generation and ciphertext faking.

Definition 3.4 (Key-simulatable PKE). *A key-simulatable public key encryption* sPKE *consists of five PPT algorithms* $(\mathsf{sGen}, \mathsf{sEnc}, \mathsf{sDec}, \widetilde{\mathsf{sGen}}, \widetilde{\mathsf{sGen}}^{-1})$ *defined as follows. Algorithms* $(\mathsf{sGen}, \mathsf{sEnc}, \mathsf{sDec})$ *form a PKE. Given the security parameter* k, *the oblivious public key generator* $\widetilde{\mathsf{sGen}}$ *returns a public key* pk' *from* $\mathcal{K}$ *and the random coins* r' *used to sample* pk'. *Given a public key* $pk \in \mathcal{V}$, *the key faking algorithm returns some random coins* r.

- Correctness. *We require that* $m = \mathsf{sDec}_{sk}(c)$ *for all* $m \in \mathcal{M}$, *all* $(pk, sk) \leftarrow \mathsf{sGen}(1^k)$ *and all* $c \leftarrow \mathsf{sEnc}_{pk}(m)$.
- Security. *A key-simulatable scheme* sPKE *is* **ind-cpa** *secure if* $(\mathsf{sGen}, \mathsf{sEnc}, \mathsf{sDec})$ *is* **ind-cpa** *secure. It is called* **ksim** *secure if it is hard to distinguish an obliviously generated key from a legitimately generated key. Specifically, for a PPT adversary* A, *there exists a negligible function* $\mu = \mu(k)$ *such that* $\mathbf{Adv}_{\mathsf{sPKE}}^{\mathsf{ksim}}(\mathsf{A}, k) := \big| \Pr[1 \leftarrow \mathsf{A}(r, pk)] - \Pr[1 \leftarrow \mathsf{A}(r', pk')] \big| \le \mu$ *where* $(pk, sk) \leftarrow \mathsf{sGen}(1^k)$, $r \leftarrow \widetilde{\mathsf{sGen}}^{-1}(pk)$ *and* $(pk', r') \leftarrow \widetilde{\mathsf{sGen}}(1^k)$.
A key-simulatable scheme sPKE *is* **secure** *if it is* {**ind-cpa**, **ksim**} *secure.*

An *extended* key-simulatable PKE is a secure key-simulatable where in addition $\mathcal{V} \subset \mathcal{K}$ and it holds that $\Pr\left[pk \in \mathcal{K} \backslash \mathcal{V} \mid (pk, r) \leftarrow \widetilde{\mathsf{sGen}}(1^k)\right]$ is non-negligible.

4   Selective Opening Security for the Receiver

In this section we provide negative and positive results regarding security for the receiver in the presence of selective opening attacks. First, we show that **rind-so** is strictly weaker than **rsim-so** security by constructing a scheme that meets the former but not the latter level of security. We then relate the different forms of security under SO attacks with non-committing encryption (for the receiver). Specifically, we show that secure NCER implies **rsim-so** and that secure tweaked NCER implies **rind-so**. Interestingly, we show that the converse implications do not hold. In terms of constructions, we show that tweaked NCER can be constructed from various primitives such as key-simulatable PKE, two-round honest-receiver statistically-hiding $\binom{2}{1}$-OT protocol, secure HPS and NCER. The DDH based secure NCER scheme of [7] that works for polynomial message space turns out to be secure tweaked NCER for exponential message space.

4.1   **rind-so** Secure PKE $\not\Rightarrow$ **rind-so** Secure PKE

Our construction is built from an **ind-ncer** secure scheme nPKE and a {**stat-hide**, **comp-bind**} secure NISHCOM nisCom that satisfy a compatibility condition. Specifically, we require that the message and randomness spaces of nisCom, denoted by $\mathcal{M}_{\mathsf{nisCom}}$ and $\mathcal{R}_{\mathsf{nisCom}}$, are compatible with the message space $\mathcal{M}_{\mathsf{nPKE}}$ of nPKE.

Definition 4.1. *An* **ind-ncer** *secure NCER* nPKE *and a* {**stat-hide**, **comp-bind**} *secure NISHCOM* nisCom *are said to be* compatible *if* $\mathcal{M}_{\mathsf{nPKE}} = \mathcal{M}_{\mathsf{nisCom}} = \mathcal{R}_{\mathsf{nisCom}}$.

Theorem 4.2. *Assume there exist an* **ind-ncer** *secure NCER and a* {**stat-hide**, **comp-bind**} *secure NISHCOM that are compatible. Then, there exists a PKE that is* **rind-so** *secure but is not* **rsim-so** *secure.*

Proof: We describe our separating encryption scheme first. Consider a scheme nPKE = (nGen, nEnc, nEnc*, nDec, nOpen) that is secure NCER (cf. Definition 3.2) and an NISHCOM nisCom = (nisCommit, nisOpen) (cf. Definition 3.1) that are compatible. We define the encryption scheme PKE = (Gen, Enc, Dec) as follows.

$\mathsf{Gen}(1^k)$	$\mathsf{Enc}_{pk}(m)$	$\mathsf{Dec}_{sk}(e)$
$(pk_0, sk_0) \leftarrow \mathsf{nGen}(1^k)$	$c \leftarrow \mathsf{nisCommit}(m, r)$	$e := (e_0, e_1, c)$
$(pk_1, sk_1) \leftarrow \mathsf{nGen}(1^k)$	$e_0 \leftarrow \mathsf{nEnc}_{pk_0}(m)$	$m = \mathsf{nDec}_{sk_0}(e_0)$
$pk = (pk_0, pk_1)$	$e_1 \leftarrow \mathsf{nEnc}_{pk_1}(r)$	$r = \mathsf{nDec}_{sk_1}(e_1)$
$sk = (sk_0, sk_1)$	Return $e = (e_0, e_1, c)$	if $c = \mathsf{nisCommit}(m, r)$
Return (pk, sk)		Return m
		else Return $\perp$

The proof follows from Lemmas 4.3 and 4.7 below which formalize that PKE is **rind-so** secure but not **rsim-so** secure. ∎

Lemma 4.3. *Assume that* nPKE *is* **ind-ncer** *secure and* nisCom *is* {**stat-hide**, **comp-bind**} *secure, then* PKE *is* **rind-so** *secure.*

Proof: More precisely we show that for any PPT adversary A attacking PKE there exist a PPT adversary B and an unbounded powerful adversary C such that

$$\mathbf{Adv}_{\mathsf{PKE}}^{\mathsf{rind}}\text{-}\mathbf{so}(\mathsf{A}, k) \leq n \left(4 \cdot \mathbf{Adv}_{\mathsf{nPKE}}^{\mathsf{ind}}\text{-}\mathbf{ncer}(\mathsf{B}, k) + \mathbf{Adv}_{\mathsf{nisCom}}^{\mathsf{stat}}\text{-}\mathbf{hide}(\mathsf{C}, k) \right).$$

We prove this lemma using the following sequence of experiments.

- $\mathbf{Exp}_0 = \mathbf{Exp}_{\mathsf{PKE}}^{\mathsf{rind}}\text{-}\mathbf{so}$.
- $\mathbf{Exp}_1$ is identical to $\mathbf{Exp}_0$ except that the first component of each ciphertext in the vector $\mathbf{e}$ is computed using nEnc^* of nPKE. That is, for all $i \in [n]$ ciphertext e_i is defined by (e_{i0}^*, e_{i1}, c_i) such that $(e_{i0}^*, t_{i0}) \leftarrow \mathsf{nEnc}_{pk_{i0}}^*(1^k)$. Furthermore, if $i \in \mathcal{I}$ (i.e., A asks to open the ith ciphertext), then $\mathbf{Exp}_1$ computes $sk_{i0}^* \leftarrow \mathsf{nOpen}(sk_{i0}, e_{i0}^*, t_{i0}, m_i)$ and hands (sk_{i0}^*, sk_{i1}) to A.
- $\mathbf{Exp}_2$ is identical to $\mathbf{Exp}_1$ except that the second component of each ciphertext in the vector $\mathbf{e}$ is computed using nEnc^* of nPKE, That is, for all $i \in [n]$ ciphertext e_i is defined by $(e_{i0}^*, e_{i1}^*, c_i)$ such that $(e_{i1}^*, t_{i1}) \leftarrow \mathsf{nEnc}_{pk_{i1}}^*(1^k)$. Furthermore, if $i \in \mathcal{I}$ (i.e., A asks to open the ith ciphertext), then $\mathbf{Exp}_2$ computes $sk_{i1}^* \leftarrow \mathsf{nOpen}(sk_{i1}, e_{i1}^*, t_{i1}, r_i)$ and hands (sk_{i0}^*, sk_{i1}^*) to A, where r_i is the randomness used to compute c_i.
- $\mathbf{Exp}_3$ is identical to $\mathbf{Exp}_2$ except that the third component of each ciphertext in the vector $\mathbf{e}$ is a commitment of a dummy message. That is, for all $i \in [n]$ ciphertext e_i is defined by $(e_{i0}^*, e_{i1}^*, c_i^*)$ such that $c_i^* \leftarrow \mathsf{nisCommit}(m_i^*; r_i^*)$, where m_i^* is a dummy message from $\mathcal{M}_{\mathsf{nisCom}}$ and $r_i^* \leftarrow \mathcal{R}_{\mathsf{nisCom}}$. Furthermore, if $i \in \mathcal{I}$ then $\mathbf{Exp}_3$ first computes $r_i \leftarrow \mathsf{nisOpen}(c_i^*, m_i)$. Then it computes $sk_{i1}^* \leftarrow \mathsf{nOpen}(sk_{i1}, e_{i1}^*, t_{i1}, r_i)$ and hands (sk_{i0}^*, sk_{i1}^*) to A, where r_i is the randomness returned by nisOpen.

We note that although the third experiment is not efficient (the experiment needs to equivocate the commitment without a trapdoor), it does not introduce a problem in our proof: an adversary that distinguishes between $\mathbf{Exp}_2$ and $\mathbf{Exp}_3$ gives rise to an unbounded adversary that breaks the statistical hiding property of the commitment scheme used by our construction.

Let ϵ_j be the advantage of A in $\mathbf{Exp}_j$, i.e. $\epsilon_j := 2 \left| \Pr[\mathbf{Exp}_j(\mathsf{A}, k) = 1] - \frac{1}{2} \right|$. We first note that $\epsilon_3 = 0$ since in experiment $\mathbf{Exp}_3$ the adversary receives a vector of ciphertexts that are statistically independent of the encrypted plaintexts, implying that the adversary (even with unbounded computing power) outputs the correct bit b with probability $1/2$. Next we show that $|\epsilon_0 - \epsilon_1| \leq 2n\Delta_{\mathsf{ind}}\text{-}\mathbf{ncer}$ and $|\epsilon_1 - \epsilon_2| \leq 2n\Delta_{\mathsf{ind}}\text{-}\mathbf{ncer}$, where $\Delta_{\mathsf{ind}}\text{-}\mathbf{ncer} = \mathbf{Adv}_{\mathsf{nPKE}}^{\mathsf{ind}}\text{-}\mathbf{ncer}(\mathsf{B}, k)$ for a PPT adversary B. Finally, we argue that $|\epsilon_2 - \epsilon_3| \leq n\Delta_{\mathsf{stat}}\text{-}\mathbf{hide}$ where $\Delta_{\mathsf{stat}}\text{-}\mathbf{hide} =$

$\mathbf{Adv}^{\mathsf{stat}}_{\mathsf{nisCom}}\text{-}\mathbf{hide}(\mathsf{C}, k)$ for an unbounded powerful adversary C. All together this implies that $|\epsilon_0 - \epsilon_3| \le 4n\Delta_{\mathsf{ind}}\text{-}\mathbf{ncer} + n\Delta_{\mathsf{stat}}\text{-}\mathbf{hide}$ and that $\epsilon_0 \le 4n\Delta_{\mathsf{ind}}\text{-}\mathbf{ncer} + n\Delta_{\mathsf{stat}}\text{-}\mathbf{hide}$, which proves the lemma.

Claim 4.4. $|\epsilon_0 - \epsilon_1| \le 2n\Delta_{\mathsf{ind}}\text{-}\mathbf{ncer}$, *where* $\Delta_{\mathsf{ind}}\text{-}\mathbf{ncer} = \mathbf{Adv}^{\mathsf{ind}}_{\mathsf{nPKE}}\text{-}\mathbf{ncer}(\mathsf{B}, k)$.

Proof: We prove the claim by introducing n intermediate hybrids experiments between $\mathbf{Exp}_0$ and $\mathbf{Exp}_1$; the difference between two consequent hybrids is bounded by a reduction to **ind-ncer** security of nPKE. More specifically, we introduce $n - 1$ intermediate hybrid experiments so that $E_0 = \mathbf{Exp}_0$, $E_n = \mathbf{Exp}_1$ and the ith hybrid experiment E_i is defined recursively. That is,

- $E_0 = \mathbf{Exp}_0$.
- For $i = [n]$, E_i is identical to E_{i-1} except that the ith ciphertext e_i is computed by (e^*_{i0}, e_{i1}, c_i) where $(e^*_{i0}, t_{i0}) \leftarrow \mathsf{nEnc}^*_{pk_{i0}}(1^k)$. Furthermore, if $i \in \mathcal{I}$ (i.e., if A asks to open the ith ciphertext), then E_i computes $sk^*_{i0} \leftarrow \mathsf{nOpen}(sk_{i0}, e^*_{i0}, t_{i0}, m_i)$ and hands (sk^*_{i0}, sk_{i1}) to A.

Clearly $E_n = \mathbf{Exp}_1$ where the first component of all ciphertext is computed using nEnc^*. Let γ_i define the advantage of A in E_i, i.e. $\gamma_i := 2\left|\Pr[E_i(\mathsf{A}, k) = 1] - \frac{1}{2}\right|$. Next we show that $|\gamma_{i-1} - \gamma_i| \le 2\Delta_{\mathsf{ind}}\text{-}\mathbf{ncer}$ for all $i \in [n]$. This implies that $|\gamma_0 - \gamma_n| \le 2n\Delta_{\mathsf{ind}}\text{-}\mathbf{ncer}$. Now, since $\gamma_0 = \epsilon_0$ and $\gamma_n = \epsilon_1$ we get $|\epsilon_0 - \epsilon_1| \le 2n\Delta_{\mathsf{ind}}\text{-}\mathbf{ncer}$, thus proving the claim.

We fix $i \in [n]$ and prove that $|\gamma_{i-1} - \gamma_i| \le 2\Delta_{\mathsf{ind}}\text{-}\mathbf{ncer}$. Specifically, we show that any adversary B that wishes to distinguish a real ciphertext from a fake one relative to nPKE can utilize the power of adversary A. Upon receiving pk from experiment $\mathbf{Exp}^{\mathsf{ind}}_{\mathsf{nPKE}}\text{-}\mathbf{ncer}$ and i, B interacts with A as follows.

1. B samples first a bit $b \leftarrow \{0, 1\}$ and sets $pk_{i0} = pk$. It then uses nGen to generate the rest of the public keys to obtain $\mathbf{pk}$ (and all but the $(i0)$th secret key).[4] Finally, it hands $\mathbf{pk}$ to A that returns Dist and $\mathsf{Resamp}_{\mathsf{Dist}}$.
2. B samples $\mathbf{m} \leftarrow \mathsf{Dist}(1^k)$ and outputs m_i to $\mathbf{Exp}^{\mathsf{ind}}_{\mathsf{nPKE}}\text{-}\mathbf{ncer}$ that returns (sk, e). B then sets $sk_{i0} = sk$. (Note that this completes vector $\mathbf{sk}$ since B generated the rest of the secret keys in the previous step).
   - For $j \in [i - 1]$, B computes the first component of ciphertext e_j by $(e_{j0}, t_{j0}) \leftarrow \mathsf{nEnc}^*_{pk_{j0}}(1^k)$. B completes e_j honestly (i.e., exactly as specified in Enc).
   - For $j = i$, B sets the first component of e_j to be e. B completes e_j honestly.
   - For $j \in [i + 1, n]$, B computes ciphertext e_j honestly.
   Let $\mathbf{e} = (e_j)_{j\in[n]}$. B hands $\mathbf{e}$ to A that returns $\mathcal{I}$.
3. B resamples $\mathbf{m}' \leftarrow \mathsf{Resamp}_{\mathsf{Dist}}(\mathbf{m}_{\mathcal{I}})$. Subsequently it hands $\mathbf{m}^*$ to A as well as secret keys for all the indices that are specified in $\mathcal{I}$, where $\mathbf{m}^* = \mathbf{m}$ if $b = 0$, $\mathbf{m}^* = \mathbf{m}'$ otherwise. That is,
   - If $j \in \mathcal{I}$ lies in $[i - 1]$, then B computes $sk^*_{j0} \leftarrow \mathsf{nOpen}(sk_{j0}, e_{j0}, t_{j0}, m_j)$ and hands (sk^*_{j0}, sk_{j1}).

---

[4] Recall that each public key within $\mathbf{pk}$ includes two public keys relative to nPKE.

- If $j \in \mathcal{I}$ equals i, then B hands (sk_{j0}, sk_{j1}) where sk_{j0} is same as sk that B had received from $\mathbf{Exp}^{\mathsf{ind}}_{\mathsf{nPKE}}\text{-}\mathbf{ncer}$.
- If $j \in \mathcal{I}$ lies in $[i+1, n]$, then B returns (sk_{j0}, sk_{j1}).

4. B outputs 1 in experiment $\mathbf{Exp}^{\mathsf{ind}}_{\mathsf{nPKE}}\text{-}\mathbf{ncer}$ if A wins.

Next, note that B perfectly simulates E_{i-1} if it received a real ciphertext e within (sk, e). Otherwise, B perfectly simulates E_i. This ensures that the probability that B outputs 1 in $\mathbf{Exp}^{\mathsf{ind}}_{\mathsf{nPKE}}\text{-}\mathbf{ncer}$ given a real ciphertext is at least as good as the probability that A wins in E_{i-1}. On the other hand, the probability that B outputs 1 in $\mathbf{Exp}^{\mathsf{ind}}_{\mathsf{nPKE}}\text{-}\mathbf{ncer}$ given a fake ciphertext is at least as good as the probability that A wins in E_i. Since the advantage of A in E_i is γ_i, its winning probability (cf. Definition 2.2) $\Pr[E_i(\mathsf{A}, k) = 1]$ in the experiment is $\frac{\gamma_i}{2} + \frac{1}{2}$. Similarly, the winning probability of A in experiment E_{i-1} is $\frac{\gamma_{i-1}}{2} + \frac{1}{2}$. Denoting the bit picked in $\mathbf{Exp}^{\mathsf{ind}}\text{-}\mathbf{ncer}_{\mathsf{nPKE}}$ by c we get,

$$\underbrace{\Pr\left[1 \leftarrow \mathsf{B}(sk, e) \mid (pk, sk) \leftarrow \mathsf{nGen}(1^k) \wedge e \leftarrow \mathsf{nEnc}_{pk}(m_i)\right]}_{=\Pr[1 \leftarrow \mathsf{B} \mid c = 0]} \geq \frac{\gamma_{i-1}}{2} + \frac{1}{2} \quad \text{and}$$

$$\underbrace{\Pr\left[1 \leftarrow \mathsf{B}(sk, e) \mid (pk, sk) \leftarrow \mathsf{nGen} \wedge (e, t_e) \leftarrow \mathsf{nEnc}^*_{pk} \wedge sk \leftarrow \mathsf{nOpen}(sk, e, t_e, m_i)\right]}_{=\Pr[1 \leftarrow \mathsf{B} \mid c = 1]}$$

$$\geq \frac{\gamma_i}{2} + \frac{1}{2}.$$

This implies that

$$\Delta_{\mathsf{ind}}\text{-}\mathbf{ncer} = \mathbf{Adv}^{\mathsf{ind}\text{-}\mathsf{ncer}}_{\mathsf{nPKE}}(\mathsf{B}, k) = 2\left|\Pr[\mathbf{Exp}^{\mathsf{ind}}\text{-}\mathbf{ncer}_{\mathsf{nPKE}}(\mathsf{B}, k) = 1] - \frac{1}{2}\right|$$

$$= 2\left|\Pr[0 \leftarrow \mathsf{B} \mid c = 0]\underbrace{\Pr(c = 0)}_{=1/2} + \Pr[1 \leftarrow \mathsf{B} \mid c = 1]\underbrace{\Pr(c = 1)}_{=1/2} - \frac{1}{2}\right|$$

$$= \left|\Pr[0 \leftarrow \mathsf{B} \mid c = 0] + \Pr[1 \leftarrow \mathsf{B} \mid c = 1] - 1\right|$$

$$= \left|\Pr[1 \leftarrow \mathsf{B} \mid c = 0] - \Pr[1 \leftarrow \mathsf{B} \mid c = 1]\right| \geq \frac{|\gamma_{i-1} - \gamma_i|}{2}.$$

$\square$

The following claim follows by a similar hybrid argument as described above.

Claim 4.5. $|\epsilon_1 - \epsilon_2| \leq 2n\Delta_{\mathsf{ind}}\text{-}\mathbf{ncer}$, where $\Delta_{\mathsf{ind}}\text{-}\mathbf{ncer} = \mathbf{Adv}^{\mathsf{ind}}_{\mathsf{nPKE}}\text{-}\mathbf{ncer}(\mathsf{B}, k)$.

Finally, we prove the following claim.

Claim 4.6. $|\epsilon_2 - \epsilon_3| \leq n\Delta_{\mathsf{stat}}\text{-}\mathbf{hide}$, where $\Delta_{\mathsf{stat}}\text{-}\mathbf{hide} = \mathbf{Adv}^{\mathsf{stat}}_{\mathsf{nisCom}}\text{-}\mathbf{hide}(\mathsf{C}, k)$.

Proof: We prove the claim by introducing n intermediate hybrids experiments between $\mathbf{Exp}_2$ and $\mathbf{Exp}_3$; we show that each pair of consecutive experiments is statistically indistinguishable based on **stat-hide** security of the NISHCOM. These hybrid experiments are defined as follows:

- $H_0 = \mathbf{Exp}_2$.
- For $i = [n]$, H_i is identical to H_{i-1} except that the ith ciphertext e_i in $\mathbf{e}$ is computed as $(e_{i0}^*, e_{i1}^*, c_i^*)$ where $c_i^* \leftarrow \mathsf{nisCommit}(m_i^*; r_i^*)$, where m_i^* is a dummy message from $\mathcal{M}_{\mathsf{nisCom}}$ and $r^* \leftarrow \mathcal{R}_{\mathsf{nisCom}}$. Furthermore, if $i \in \mathcal{I}$, then H_i computes $r_i \leftarrow \mathsf{nisOpen}(c_i^*, m_i)$ and hands (sk_{i0}^*, sk_{i1}^*) to A.

We remark again that the hybrid experiments defined above are not efficient, but this is not an issue as we rely on the statistical security of the underlying NISHCOM.

Clearly, $H_n = \mathbf{Exp}_3$ where the third component of each ciphertext within $\mathbf{e}$ is computed using dummy messages. Let ν_i be the advantage of A in H_i, i.e., $\nu_i := 2 \left| \Pr[H_i(\mathsf{A}, k) = 1] - \frac{1}{2} \right|$. Next, we show that $|\nu_{i-1} - \nu_i| \leq \Delta_{\mathsf{stat}}\text{-}\mathbf{hide}$ for all $i \in [n]$, where $\Delta_{\mathsf{stat}}\text{-}\mathbf{hide} = \mathbf{Adv}_{\mathsf{nisCom}}^{\mathsf{stat}}\text{-}\mathbf{hide}(\mathsf{C}, k)$. All together, this implies that $|\nu_0 - \nu_n| \leq n\Delta_{\mathsf{stat}}\text{-}\mathbf{hide}$. Since $\nu_0 = \epsilon_2$ and $\nu_n = \epsilon_3$ we get that $|\epsilon_2 - \epsilon_3| \leq n\Delta_{\mathsf{stat}}\text{-}\mathbf{hide}$ which proves the claim.

Fix $i \in [n]$. The only difference between experiments H_{i-1} and H_i is relative to the third component of ciphertext e_i. Namely, in H_{i-1}, the third component in e_i is a commitment to m_i where m_i is the ith element in $\mathbf{m}$. On the other hand, in H_i it is a commitment to a dummy message from $\mathcal{M}_{\mathsf{nisCom}}$. As the underlying NISHCOM satisfies statistical hiding property, even an unbounded adversary C cannot distinct H_{i-1} and H_i with probability better than $\Delta_{\mathsf{stat}}\text{-}\mathbf{hide}$, so $|\nu_{i-1} - \nu_i| \leq \Delta_{\mathsf{stat}}\text{-}\mathbf{hide}$ as desired.   □

We conclude with the proof of the following lemma.

Lemma 4.7. PKE *is not* **rsim-so** *secure.*

Proof: We then rely on a result of [1] which establishes that no decryption verifiable **ind-cpa** secure is **rsim-so**. Informally, decryption verifiability implies the existence of an algorithm W (that either outputs accept or reject), such that it is hard to find pk, sk_0, sk_1, distinct m_0, m_1 and a ciphertext e where both $W(pk, sk_0, e, m_0)$ and $W(pk, sk_1, e, m_1)$ accept. Note that it is hard to find two valid secret keys and plaintexts as required since decryption follows successfully only if the commitment that is part of the ciphertext is also correctly opened. In particular, an adversary that produces a ciphertext that can be successfully decrypted into two distinct plaintexts (under two different keys) must break the **comp-bind** security of the underlying commitment scheme.[5] This implies that PKE is not **rsim-so** secure.   □

Compatible Secure NCER and Secure NISHCOM. We instantiate the commitment scheme with the Paillier based scheme of Damgård and Nielsen [10, 11], which is comprised of the following algorithms that use public parameters (N, g) where N is a k-bit RSA composite and $g = x^N \bmod N^2$ for an uniformly random $x \leftarrow \mathbb{Z}_N^*$.

---

[5] Recall that the decryption algorithm verifies first whether the commitment within the ciphertext is consistent with the decrypted ciphertexts (that encrypt the committed message and its corresponding randomness for commitment).

- nisCommit, given N, g and message $m \in \mathbb{Z}_N$, pick $r \leftarrow \mathbb{Z}_N^*$ and compute $g^m \cdot r^N \bmod N^2$.
- nisOpen, given commitment c and message m, compute randomness r such that $c = g^m \cdot r^N \bmod N^2$. Namely, find first $\tilde{r}$ such that $c = \tilde{r}^N \bmod N^2$. This implies that $\tilde{r}^N = (x^N)^m \cdot r^N \bmod N^2$ for some $r \in \mathbb{Z}_N^*$, since we can fix $r = \tilde{r}/x^m$.

This scheme is computationally binding, as a commitment is simply a random Paillier encryption of zero. Furthermore, opening to two different values implies finding the Nth root of g (which breaks the underlying assumption of Paillier, i.e., DCR). Finally, the NCER can be instantiated with the scheme from [7] that is also based on the DCR assumption. The message space of these two primitives is $\mathbb{Z}_N$. In addition, the randomness of the commitment scheme is $\mathbb{Z}_N^*$ and thus can be made consistent with the plaintext spaces, as it is infeasible to find an element in $\mathbb{Z}_N/\mathbb{Z}_N^*$.

4.2 Secure Tweaked NCER $\Longrightarrow$ rind-so Secure PKE

In this section we prove that every secure tweaked NCER is a **rind-so** secure PKE. Intuitively, this holds since real ciphertexts are indistinguishable from fake ones, and fake ciphertexts do not commit to any fixed plaintext. This implies that the probability of distinguishing an encryption of one message from another is exactly half, even for an unbounded adversary.

Theorem 4.8. *Assume there exists an* {**ind-tcipher**, **ind-tncer**} *secure tweaked NCER, then there exists a PKE that is* **rind-so** *secure.*

Proof: More precisely, let tPKE $=$ (tGen, tEnc, tEnc*, tDec, tOpen) denote a secure tweaked NCER. Then we prove that tPKE is **rind-so** secure, by proving that for any PPT adversary A attacking tPKE in the **rind-so** experiment there exist a PPT adversary B and an unbounded powerful adversary C such that

$$\mathbf{Adv}_{\mathsf{tPKE}}^{\mathsf{rind}}\text{-}\mathbf{so}(\mathsf{A}, k) \leq 2n \left(\mathbf{Adv}_{\mathsf{tPKE}}^{\mathsf{ind}}\text{-}\mathbf{tcipher}(\mathsf{B}, k) + \mathbf{Adv}_{\mathsf{tPKE}}^{\mathsf{ind}}\text{-}\mathbf{tncer}(\mathsf{C}, k) \right).$$

We modify experiment **rind-so** step by step, defining a sequence of $2n + 1$ experiments and bound the advantage of A in the last experiment. The proof is then concluded by proving that any two intermediate consecutive experiments are indistinguishable due to either **ind-tcipher** security or **ind-tncer** security of tPKE. Specifically, we define a sequence of hybrid experiments $\{\mathbf{Exp}_i\}_{i=0}^{2n}$ as follows.

- $\mathbf{Exp}_0 = \mathbf{Exp}_{\mathsf{tPKE}}^{\mathsf{rind}}\text{-}\mathbf{so}$.
- For all $i \in [n]$, $\mathbf{Exp}_i$ is identical to $\mathbf{Exp}_{i-1}$ except that the ith ciphertext in vector $\mathbf{e}$ is computed by $e_i^* \leftarrow \mathsf{tEnc}_{pk_i}^*(sk_i, m_i)$, so that if $i \in \mathcal{I}$ then $\mathbf{Exp}_i$ outputs the secret key sk_i computed by tGen and hands sk_i to adversary A (here we rely on the additional property of tEnc*).

– For all $i \in [n]$, $\mathbf{Exp}_{n+i}$ is identical to $\mathbf{Exp}_{n+i-1}$ except that the ith ciphertext in vector $\mathbf{e}$ is computed by sampling a random message $m_i^* \in \mathcal{M}_{\mathsf{tPKE}}$ first and then computing $e_i^* \leftarrow \mathsf{tEnc}_{pk_i}^*(sk_i, m_i^*)$. Next, if $i \in \mathcal{I}$ then $\mathbf{Exp}_{n+i}$ computes a secret key $sk_i^* \leftarrow \mathsf{tOpen}(e_i^*, m_i)$ and hands sk_i^* to A.

Let ϵ_i denote the advantage of A in experiment $\mathbf{Exp}_i$ i.e., $\epsilon_i := |\Pr[\mathbf{Exp}_i(\mathsf{A}, k) = 1] - \frac{1}{2}|$. We first note that $\epsilon_{2n} = 0$ since in experiment $\mathbf{Exp}_{2n}$ the adversary receives a vector of ciphertexts that are statistically independent of the encrypted plaintexts, implying that the adversary outputs the correct bit b with probability $1/2$. We next show that $|\epsilon_{i-1} - \epsilon_i| \leq 2\Delta_{\mathsf{ind}}\text{-}\mathsf{tcipher}$ for any $i \in [n]$, where $\Delta_{\mathsf{ind}}\text{-}\mathsf{tcipher} = \mathbf{Adv}_{\mathsf{tPKE}}^{\mathsf{ind}}\text{-}\mathsf{tcipher}(\mathsf{B}, k)$ for a PPT adversary B. Finally, we prove that $|\epsilon_{n+i-1} - \epsilon_{n+i}| \leq 2\Delta_{\mathsf{ind}}\text{-}\mathsf{tncer}$ for any $i \in [n]$, where $\Delta_{\mathsf{ind}}\text{-}\mathsf{tncer} = \mathbf{Adv}_{\mathsf{tPKE}}^{\mathsf{ind}}\text{-}\mathsf{tncer}(\mathsf{C}, k)$ for an unbounded powerful adversary C. Together this implies that $|\epsilon_0 - \epsilon_{2n}| \leq 2n(\Delta_{\mathsf{ind}}\text{-}\mathsf{tcipher} + \Delta_{\mathsf{ind}}\text{-}\mathsf{tncer})$. So we conclude that $\epsilon_0 \leq n(\Delta_{\mathsf{ind}}\text{-}\mathsf{tcipher} + \Delta_{\mathsf{ind}}\text{-}\mathsf{tncer}) + \epsilon_{2n} = 2n(\Delta_{\mathsf{ind}}\text{-}\mathsf{tcipher} + \Delta_{\mathsf{ind}}\text{-}\mathsf{tncer})$ which concludes the proof of the theorem for all $i \in [n]$. □

Claim 4.9. $|\epsilon_{i-1} - \epsilon_i| \leq 2n\Delta_{\mathsf{ind}}\text{-}\mathsf{tcipher}$, where $\Delta_{\mathsf{ind}}\text{-}\mathsf{tcipher} = \mathbf{Adv}_{\mathsf{tPKE}}^{\mathsf{ind}}\text{-}\mathsf{tcipher}(\mathsf{B}, k)$.

Proof: In the following, we prove that one can design an adversary B that distinguishes a real ciphertext from a fake one in $\mathbf{Exp}_{\mathsf{tPKE}}^{\mathsf{ind}}\text{-}\mathsf{tcipher}$, using adversary A. B interacts with A as follows:

1. Upon receiving pk from $\mathbf{Exp}_{\mathsf{tPKE}}^{\mathsf{ind}}\text{-}\mathsf{tcipher}$ and an integer i, B sets $pk_i = pk$. It picks a bit b randomly. It then generates the rest of the public and secret key pairs using tGen for all $j \in [n] \backslash i$, obtaining $\mathbf{pk}$. It hands $\mathbf{pk}$ to A who returns Dist and $\mathsf{Resamp}_{\mathsf{Dist}}$.
2. B samples $\mathbf{m} \leftarrow \mathsf{Dist}(1^k)$ and hands m_i to $\mathbf{Exp}_{\mathsf{tPKE}}^{\mathsf{ind}}\text{-}\mathsf{tcipher}$ which returns (sk, e). B fixes $e_i = e$ and completes $\mathbf{sk}$ by setting $sk_i = sk$. Next, for $j \in [i-1]$ it computes $e_j \leftarrow \mathsf{tEnc}_{pk_j}^*(sk_j, m_j)$, whereas for $j \in [i+1, n]$ it samples randomness $r_j \leftarrow \mathcal{R}_{\mathsf{tPKE}}$ and computes $e_j \leftarrow \mathsf{tEnc}_{pk_j}(m_j; r_j)$. Let $\mathbf{e} = (e_i)_{i \in [n]}$. B hands $\mathbf{e}$ to A who returns $\mathcal{I}$.
3. B samples $\mathbf{m}' \leftarrow \mathsf{Resamp}(\mathbf{m}_{\mathcal{I}})$ and hands A $\mathbf{m}^*$ and the following secret keys for all the indices that are specified in $\mathcal{I}$. Here $\mathbf{m}^*$ is $\mathbf{m}$ if $b = 0$ and $\mathbf{m}'$ otherwise. That is,
   – If $j \in \mathcal{I}$ lies in $[i-1]$ or in $[i+1, n]$, then B returns sk_j.
   – If $j \in \mathcal{I}$ equals i, then B returns sk.
4. B outputs 1 in $\mathbf{Exp}_{\mathsf{tPKE}}^{\mathsf{ind}}\text{-}\mathsf{tcipher}$ if A wins.

Next, note that B perfectly simulates $\mathbf{Exp}_{i-1}$ if it receives a real ciphertext e within (sk, e). On the other hand, B perfectly simulates $\mathbf{Exp}_i$ if e is a fake ciphertext. This ensures that the probability that B outputs 1 given a real ciphertext is at least as good as the probability that A wins in $\mathbf{Exp}_{i-1}$. On the other hand, the probability that B outputs 1 given a fake ciphertext is at least as good as the probability that A wins in $\mathbf{Exp}_i$. Since the advantage of A in $\mathbf{Exp}_i$ is ϵ_i, its winning probability (cf. Definition 2.2) $\Pr[\mathbf{Exp}_i(\mathsf{A}, k) = 1]$ in the experiment is

$\frac{\epsilon_i}{2} + \frac{1}{2}$. Similarly, the winning probability of A in experiment $\mathbf{Exp}_{i-1}$ is $\frac{\epsilon_{i-1}}{2} + \frac{1}{2}$. Denoting the bit picked in $\mathbf{Exp^{ind}\text{-}tcipher}_{tPKE}$ by c,

$$\underbrace{\Pr\left[1 \leftarrow B(pk, sk, e, m_i) \mid (pk, sk) \leftarrow \mathsf{tGen}(1^k) \wedge e \leftarrow \mathsf{tEnc}_{pk}(m_i)\right]}_{=\Pr[1\leftarrow B \mid c=0]} \geq \frac{\epsilon_{i-1}}{2} + \frac{1}{2} \text{ and}$$

$$\underbrace{\Pr\left[1 \leftarrow B(pk, sk, e^*, m_i) \mid (pk, sk) \leftarrow \mathsf{tGen}(1^k) \wedge e^* \leftarrow \mathsf{tEnc}_{pk}^*(sk, m_i)\right]}_{=\Pr[1\leftarrow B \mid c=1]} \geq \frac{\epsilon_i}{2} + \frac{1}{2}.$$

This implies that

$$
\begin{aligned}
\Delta_{\mathsf{ind}}\text{-}\mathbf{tcipher} \quad &= \mathsf{Adv}_{tPKE}^{\mathsf{ind\text{-}tcipher}}(B, k) = 2\left|\Pr[\mathbf{Exp^{ind}\text{-}tcipher}_{tPKE}(B, k) = 1] - \frac{1}{2}\right| \\
&= 2\left|\Pr[0 \leftarrow B \mid c = 0]\underbrace{\Pr(c = 0)}_{=1/2} + \Pr[1 \leftarrow B \mid c = 1]\underbrace{\Pr(c = 1)}_{=1/2} - \frac{1}{2}\right| \\
&= |\Pr[0 \leftarrow B \mid c = 0] + \Pr[1 \leftarrow B \mid c = 1] - 1| \\
&= |\Pr[1 \leftarrow B \mid c = 0] - \Pr[1 \leftarrow B \mid c = 1]| \geq \frac{|\epsilon_{i-1} - \epsilon_i|}{2}
\end{aligned}
$$

$\square$

Claim 4.10. $|\epsilon_{n+i-1} - \epsilon_{n+i}| \leq 2n\Delta_{\mathsf{ind}}\text{-}\mathbf{tcipher}$ *for all* $i \in [n]$, *where* $\Delta_{\mathsf{ind}}\text{-}\mathbf{tncer} = \mathsf{Adv}_{tPKE}^{\mathsf{ind}}\text{-}\mathbf{tncer}(C, k)$.

Proof: We prove that one can design an unbounded adversary C that distinguishes the two views generated in experiment **ind-tncer**, using adversary A. C interacts with A:

1. Upon receiving pk from $\mathbf{Exp^{ind}\text{-}tncer}_{tPKE}$ and an integer i, C sets $pk_i = pk$ and picks a bit b. It then generates the rest of the public and secret key pairs using tGen for all $j \in [n]\setminus\{i\}$, obtaining $\mathbf{pk}$. It hands $\mathbf{pk}$ to A who returns Dist and $\mathsf{Resamp}_{\mathsf{Dist}}$.
2. C samples $\mathbf{m} \leftarrow \mathsf{Dist}(1^k)$ and hands m_i to $\mathbf{Exp^{ind}\text{-}tncer}_{tPKE}$ which returns (sk, e). C fixes $e_i = e$ and completes $\mathbf{sk}$ by setting $sk_i = sk$. Next, for $j \in [i-1]$ it samples $\mathbf{m}_j^* \leftarrow \mathcal{M}_{tPKE}$ and computes $e_j \leftarrow \mathsf{tEnc}_{pk_j}^*(sk_j, m_j^*)$, whereas for $j \in [i+1, n]$ it computes $e_j \leftarrow \mathsf{tEnc}_{pk_j}^*(sk_j, m_j)$. Let $\mathbf{e} = (e_j)_{j \in [n]}$. C hands $\mathbf{e}$ to A receiving $\mathcal{I}$.
3. C samples $\mathbf{m}' \leftarrow \mathsf{Resamp}(\mathbf{m}_{\mathcal{I}})$ and hands $\mathbf{m}^*$ to A and the following secret keys for all the indices that are specified in $\mathcal{I}$. Here $\mathbf{m}^*$ is $\mathbf{m}$ if $b = 0$ and $\mathbf{m}'$ otherwise. That is,
   - If $j \in \mathcal{I}$ lies in $[i - 1]$, then C returns sk_j such that $sk_j = \mathsf{tOpen}(e_j, m_j)$.
   - If $j \in \mathcal{I}$ equals i, then C returns sk.
   - If $j \in \mathcal{I}$ lies in $[i + 1, n]$, then C returns sk_j.

4. C outputs 1 in $\mathbf{Exp^{ind}}$-$\mathbf{tncer}_{tPKE}$ if A wins.

Next, note that B perfectly simulates $\mathbf{Exp}_{n+i-1}$ if it receives a real ciphertext e within (sk, e). On the other hand, B perfectly simulates $\mathbf{Exp}_{n+i}$ if e is a fake ciphertext and sk is a secret key returned by tOpen. This ensures that the probability that B outputs 1 given a real ciphertext is at least as good as the probability that A wins in $\mathbf{Exp}_{n+i-1}$. On the other hand, the probability that B outputs 1 given a fake ciphertext is at least as good as the probability that A wins in $\mathbf{Exp}_{n+i}$. Since the advantage of A in $\mathbf{Exp}_i$ is ϵ_{n+i}, its winning probability (c.f Definition 2.2) $\Pr[\mathbf{Exp}_i(A, k) = 1]$ in the experiment is $\frac{\epsilon_{n+i}}{2} + \frac{1}{2}$. Similarly, the winning probability of A in experiment $\mathbf{Exp}_{n+i-1}$ is $\frac{\epsilon_{n+i-1}}{2} + \frac{1}{2}$. Denoting the bit picked in $\mathbf{Exp^{ind}}$-$\mathbf{tncer}_{tPKE}$ by c we get,

$$\underbrace{\Pr\left[1 \leftarrow C(sk, e) \mid (pk, sk) \leftarrow \mathsf{tGen}(1^k) \wedge e \leftarrow \mathsf{tEnc}^*_{pk}(sk, m_i)\right]}_{=\Pr[1 \leftarrow C \mid c=0]} \geq \frac{\epsilon_{n+i-1}}{2} + \frac{1}{2} \text{ and}$$

$$\underbrace{\Pr\left[1 \leftarrow C(sk^*, e^*) \mid (pk, sk) \leftarrow \mathsf{tGen} \wedge e^* \leftarrow \mathsf{tEnc}^*_{pk}(sk, m^*) \wedge sk^* \leftarrow \mathsf{tOpen}(e^*, sk, m_i)\right]}_{=\Pr[1 \leftarrow C \mid c=1]}$$
$$\geq \frac{\epsilon_{n+i}}{2} + \frac{1}{2}.$$

Following a similar argument as in the previous claim, we conclude that $2\Delta_{\mathsf{ind}}$-$\mathsf{tncer} \geq |\epsilon_{n+i-1} - \epsilon_{n+i}|$.   ∎

4.3   Secure NCER $\implies$ rsim-so Secure PKE

In this section we claim that secure NCER implies selective opening security in the presence of receiver corruption. Our theorem is stated for the stronger simulation based security definition but holds for the indistinguishability definition as well. The proof is given in the full version [17].

Theorem 4.11. *Assume there exists an* **ind-ncer** *secure PKE, then there exists a PKE that is* **rsim-so** *secure.*

4.4   rsim-so Secure PKE $\nRightarrow$ Secure NCER and Tweaked NCER

In this section we prove that **rsim-so** does not imply both tweaked NCER and NCER by providing a concrete counter example based on an extended key-simulatable PKE (cf. see Key-Simulatable PKE subsubsection of Sect. 3). The key point in our proof is that in some cases simulatable public keys cannot be explained as valid public keys. Formally,

Theorem 4.12. *Assume there exists an* {**ind-cpa**, **ksim**} *secure extended key-simulatable PKE, then there exists a PKE that is* **rsim-so** *secure but is not a* {**ind-tcipher**, **ind-tncer**} *secure tweaked NCER nor a* **ind-ncer** *secure NCER.*

Proof: We describe our separating encryption scheme first; the complete proof is given in the full version [17]. Given an extended key-simulatable PKE sPKE = $(\mathsf{sGen}, \mathsf{sEnc}, \mathsf{sDec}, \widetilde{\mathsf{sGen}}, \widetilde{\mathsf{sGen}}^{-1})$ for a plaintext space $\mathcal{M}_{\mathsf{sPKE}}$, we construct a new scheme PKE = (Gen, Enc, Dec) with a binary plaintext space that is **rsim-so** secure, and thus also **rind-so** secure, yet it does not imply tweaked NCER. For simplicity, we assume that $\mathcal{M}_{\mathsf{sPKE}}$ is the binary space $\{0,1\}$. The DDH based instantiation of sPKE with $\mathcal{V} \subset \mathcal{K}$ from see Realizing Key-Simulatable and Extended Key-Simulatable PKE subsection of Sect. 4.4 is defined with respect to this space. ∎

$$
\begin{array}{lll}
\underline{\mathsf{Gen}(1^k)} & \underline{\mathsf{Enc}_{pk}(b)} & \\
\alpha \leftarrow \{0,1\} & e_0 \leftarrow \mathsf{Enc}_{pk_0}(b) & \underline{\mathsf{Dec}_{sk}(e)} \\
(pk_\alpha, sk_\alpha) \leftarrow \mathsf{sGen}(1^k) & e_1 \leftarrow \mathsf{Enc}_{pk_1}(b) & sk = (\alpha, sk_\alpha, r_{1-\alpha}) \\
(pk_{1-\alpha}, r_{1-\alpha}) \leftarrow \widetilde{\mathsf{sGen}}(1^k) & \text{Return } e = (e_0, e_1) & e := (e_0, e_1) \\
pk = (pk_0, pk_1) & & b = \mathsf{Dec}_{sk_\alpha}(e_\alpha) \\
sk = (\alpha, sk_\alpha, r_{1-\alpha}) & & \text{Return } b \\
\text{Return } (pk, sk) & &
\end{array}
$$

Realizing Key-Simulatable and Extended Key-Simulatable PKE. An example of a {**ind-cpa, ksim**} secure key-simulatable PKE is the ElGamal PKE [14] where we set $\mathcal{K}$ to be equal to the set of *valid* public keys, i.e. $\mathcal{K} = \mathcal{V}$. In addition, note that any simulatable PKE as defined in [9] is also {**ind-cpa, ksim**} secure key-simulatable PKE.

Below we provide an example of extended key-simulatable PKE with security under the DDH assumption. For simplicity we consider a binary plaintext space. Let $(g_0, g_1, p) \leftarrow \mathcal{G}(1^k)$ be an algorithm that given a security parameter k returns a group description $\mathbb{G} = \mathbb{G}_{g_0,g_1,p}$ specified by its generators g_0, g_1 and its order p. Furthermore, we set $\mathcal{K} = \mathbb{G}^2$ and $\mathcal{V} = \{(g_0^x, g_1^x) \in \mathbb{G}^2 \mid x \in \mathbb{Z}_p\}$. Then define the following extended key-simulatable PKE,

- sGen, given the security parameter k, set $(g_0, g_1, p) \leftarrow \mathcal{G}(1^k)$. Choose uniformly random $x \leftarrow \mathbb{Z}_p$ and compute $h_i = g_i^x$ for all $i \in \{0,1\}$. Output the secret key $sk = x$ and the public key $pk = (h_0, h_1)$.
- sEnc, given the public key pk and plaintext $m \in \{0,1\}$, choose a uniformly random $s, t \leftarrow \mathbb{Z}_p$. Output the ciphertext $(g_0^s g_1^t, g_0^m \cdot (h_0^s h_1^t))$.
- sDec, given the secret key x and ciphertext (g_c, h_c), output $h_c \cdot (g_c^x)^{-1}$.
- $\widetilde{\mathsf{sGen}}$, given 1^k, output two random elements from $\mathbb{G}$ and their bit sequence as the randomness.
- $\widetilde{\mathsf{sGen}}^{-1}$, given a legitimate public key h_0, h_1, simply returns the bit strings of h_0, h_1 as the randomness used to sample them from $\mathbb{G}^2$ by $\widetilde{\mathsf{sGen}}$.

We remark that a public key chosen randomly from $\mathbb{G}^2$ does not necessarily correspond to a secret key. Furthermore, $\Pr\left[pk \in \mathcal{K}\backslash\mathcal{V} \mid pk \leftarrow \widetilde{\mathsf{sGen}}(1^k)\right]$ is non-negligible. This is a key property in our proof from Sect. 4.4.

4.5   Realizing Tweaked NCER

Based on key-simulatable PKE. We prove that secure tweaked NCER can be
built based on any secure key-simulatable PKE with $\mathcal{K} = \mathcal{V}$ (cf. Definition see
Key-Simulatable PKE subsubsection of Sect. 3). Specifically, our construction is
based on the separating scheme presented in Sect. 4.4. In addition, we define the
fake encryption algorithm so that it outputs two ciphertexts that encrypt two
distinct plaintexts rather than the same plaintext twice (implying that ciphertext
indistinguishability follows from the **ind-cpa** security of the underlying encryption scheme). More formally, the fake encryption algorithm can be defined as fol-
lows. Given $sk = (\alpha, sk_\alpha, r_{1-\alpha})$ and message b, a fake encryption of b is computed
by $e^* = (\mathsf{sEnc}_{pk_0}(b), \mathsf{sEnc}_{pk_1}(1-b))$ if $\alpha = 0$ and $e^* = (\mathsf{sEnc}_{pk_0}(1-b), \mathsf{sEnc}_{pk_1}(b))$
otherwise. It is easy to verify that given sk, the decryption of e^* returns b and
that e^* is computationally indistinguishable from a valid encryption even given
the secret key. Next, we discuss the details of the non-efficient opening algorithm
which is required to generate a secret key for a corresponding public key given a
fake ciphertext and a message b'. In more details, assuming $sk = (\alpha, sk_\alpha, r_{1-\alpha})$
and $pk = (pk_0, pk_1)$,

$$\mathsf{tOpen}(sk, pk, (e_0^*, e_1^*), b') = \begin{cases} (\alpha, sk_\alpha, r_{1-\alpha}) & \text{if } e_\alpha^* = \mathsf{sEnc}_{pk_\alpha}(b') \\ (1-\alpha, sk_{1-\alpha}, r_\alpha) & \text{else, where } r_\alpha \leftarrow \widetilde{\mathsf{sGen}}^{-1}(pk_\alpha) \\ & \text{and } sk_{1-\alpha} \text{ is a valid secret key} \\ & \text{of } pk_{1-\alpha}. \end{cases}$$

Note that since it holds that $\mathcal{V} = \mathcal{K}$ for the underlying sPKE scheme, there exists
a secret key that corresponds to $pk_{1-\alpha}$ and it can be computed (possibly in an
inefficient way). Encryption schemes for larger plaintext spaces can be obtained
by repeating this basic scheme sufficiently many times.[6] Finally, we note that the
scheme is {**ind-tcipher, ind-tncer**} secure. Recalling that any simulatable PKE
with $\mathcal{K} = \mathcal{V}$ is a key-simulatable PKE [8,9], we conclude that secure tweaked
NCER for a binary plaintext space can be built relying on DDH, RSA, factoring
and LWE assumptions.

An additional realization based on statistically-hiding $\binom{2}{1}$-OT in presented in
the full version [17]. These two implementations support binary plaintext space.
Below presented new constructions that support exponential plaintext spaces.

Based on NCER. We show that the DCR based secure NCER of [7] is also
a secure tweaked NCER. Let $(p', q') \leftarrow \mathcal{G}(1^n)$ be an algorithm that given a
security parameter k returns two random n bit primes p' and q' such that $p =
2p' + 1$ and $q = 2q' + 1$ are also primes. Let $N = pq$ and $N' = p'q'$. Define
$(\mathsf{tGen}, \mathsf{tEnc}, \mathsf{tEnc}^*, \mathsf{tDec}, \mathsf{tOpen})$ by,

– tGen, given the security parameter k, run $(p', q') \leftarrow \mathcal{G}(1^n)$ and set $p = 2p' + 1$,
   $q = 2q' + 1$, $N = pq$ and $N' = p'q'$. Choose random $x_0, x_1 \leftarrow \mathbb{Z}_{N^2/4}$ and a

---

[6] We note that this construction was discussed in [16] in the context of weak hash
proof systems and leakage resilient PKE.

random $g' \in \mathbb{Z}_{N^2}^*$ and compute $g_0 = g'^{2N}$, $h_0 = g_0^{x_0}$ and $h_1 = g_0^{x_1}$. Output public key $pk = (N, g_0, h_0, h_1)$ and secret key $sk = (x_0, x_1)$.

– tEnc, given the public key pk and a plaintext $m \in \mathbb{Z}_N$, choose a uniformly random $t \leftarrow \mathbb{Z}_{N/4}$ and output ciphertext

$$c \leftarrow \mathsf{tEnc}_{pk}(m; t) = \left(g_0^t \bmod N^2, (1+N)^m h_0^t \bmod N^2, h_1^t \bmod N^2\right).$$

– tDec, given the secret key (x_0, x_1) and a ciphertext (c_0, c_1, c_2), check whether $c_0^{2x_1} = (c_2)^2$; if not output $\perp$. Then set $\hat{m} = (c_1/c_0^{x_0})^{N+1}$. If $\hat{m} = 1 + mN$ for some $m \in \mathbb{Z}_N$, then output m; else output $\perp$.

– tEnc*, given the public key pk, secret key sk and a message m, choose uniformly random $t \leftarrow \mathbb{Z}_{\phi(N)/4}$, compute the fake ciphertext (where all the group elements are computed $\bmod N^2$) $c^* \leftarrow (c_0^*, c_1^*, c_2^*) = ((1+N) \cdot g_0^t, (1+N)^m \cdot (c_0^*)^{x_0}, (c_0^*)^{x_1})$.

– tOpen, given N', (x_0, x_1), a triple (c_0, c_1, c_2) such that $(c_0, c_1, c_2) \leftarrow \mathsf{tEnc}_{pk}^*(sk, m)$ and a plaintext $m^* \in \mathbb{Z}_N$, output $sk^* = (x_0^*, x_1)$, where $x_0^* \leftarrow \mathbb{Z}_{NN'}$ is the unique solution to the equations $x_0^* = x \bmod N'$ and $x_0^* = x_0 + m - m^* \bmod N$. These equations have a unique solution due to the fact that $gcd(N, N') = 1$ and the solution can be obtained employing Chinese Remainder Theorem. It can be verified that the secret key sk^* matches the public key pk and also decrypts the 'simulated' ciphertext to the required message m^*. The first and third components of pk remain the same since x_1 has not been changed. Now $g^{x_0^*} = g^{x_0^* \bmod N'} = g^{x_0 \bmod N'} = g^{x_0} = h_0$. Using the fact that the order of $(1+N)$ in $\mathbb{Z}_{N^2}^*$ is N, we have

$$\left(\frac{c_1}{c_0^{x_0^*}}\right)^{N+1} = \left(\frac{(1+N)^{x_0+m} g_0^{tx_0}}{(1+N)^{x_0^*} g_0^{tx_0^*}}\right)^{N+1}$$

$$= \left((1+N)^{x_0+m-x_0^* \bmod N}\right)^{N+1} = ((1+N)^m)^{N+1} = (1+mN).$$

It is easy to verify that real and fake ciphertexts are computationally indistinguishable under the DCR assumption since the only difference is with respect to the first element (which is an $2N$th power in a real ciphertext and not an $2N$th power in a simulated ciphertext). The other two elements are powers of the first element. Furthermore $sk = (x_0, x_1)$ and $sk^* = (x_0^*, x_1)$ are statistically close since $x_0 \leftarrow \mathbb{Z}_{N^2/4}$ and $x_0^* \leftarrow \mathbb{Z}_{NN'}$ and the uniform distribution over $\mathbb{Z}_{NN'}$ and $\mathbb{Z}_{N^2/4}$ is statistically close.

5   Selective Opening Security for the Sender

In this section we prove **sind-so** is strictly weaker than **ssim-so** security by constructing a scheme that meets the former but not the latter level of security. Our starting point is a lossy encryption scheme loPKE = (loGen, loGen*, loEnc, loDec). We then modify loPKE by adding a (statistically hiding) commitment to each ciphertext such that the new scheme, denoted by PKE, becomes committing.

Next, we prove that PKE is **sind-so** secure by showing that the scheme remains lossy and is therefore **sind-so** secure according to [2]. Finally, using the result from [1] we claim that PKE is not **ssim-so** secure. The following theorem is proven in the full version [17].

Theorem 5.1. *Assume there exists a* {**ind-lossy**, **ind-lossycipher**} *secure lossy PKE and a* {**stat-hide**, **comp-bind**} *secure NISHCOM that are compatible. Then, there exists a PKE that is* **sind-so** *secure but is not* **ssim-so** *secure.*

Acknowledgements. Carmit Hazay acknowledges support from the Israel Ministry of Science and Technology (grant No. 3-10883). Arpita Patra acknowledges support from project entitled 'ISEA - Part II' funded by Department of Electronics and Information Technology of Govt. of India. Part of this work was carried out while Bogdan Warinschi was visiting Microsoft Research, Cambridge, UK and IMDEA, Madrid, Spain. He has been supported in part by ERC Advanced Grant ERC-2010-AdG-267188-CRIPTO, by EPSRC via grant EP/H043454/1, and has received funding from the European Union Seventh Framework Programme (FP7/2007-2013) under grant agreement 609611 (PRACTICE).

References

1. Bellare, M., Dowsley, R., Waters, B., Yilek, S.: Standard security does not imply security against selective-opening. In: Pointcheval, D., Johansson, T. (eds.) EURO-CRYPT 2012. LNCS, vol. 7237, pp. 645–662. Springer, Heidelberg (2012)
2. Bellare, M., Hofheinz, D., Yilek, S.: Possibility and impossibility results for encryption and commitment secure under selective opening. In: Joux, A. (ed.) EURO-CRYPT 2009. LNCS, vol. 5479, pp. 1–35. Springer, Heidelberg (2009)
3. Bellare, M., Waters, B., Yilek, S.: Identity-based encryption secure against selective opening attack. In: Ishai, Y. (ed.) TCC 2011. LNCS, vol. 6597, pp. 235–252. Springer, Heidelberg (2011)
4. Bellare, M., Yilek, S.: Encryption schemes secure under selective opening attack. In: IACR Cryptology ePrint Archive 2009, p. 101 (2009)
5. Pinkas, B., Schneider, T., Smart, N.P., Williams, S.C.: Secure two-party computation is practical. In: Matsui, M. (ed.) ASIACRYPT 2009. LNCS, vol. 5912, pp. 250–267. Springer, Heidelberg (2009)
6. Canetti, R., Friege, U., Goldreich, O., Naor, M.: Adaptively secure multi-party computation. In: STOC, pp. 639–648 (1996)
7. Canetti, R., Halevi, S., Katz, J.: Adaptively-secure, non-interactive public-key encryption. In: Kilian, J. (ed.) TCC 2005. LNCS, vol. 3378, pp. 150–168. Springer, Heidelberg (2005)
8. Choi, S.G., Dachman-Soled, D., Malkin, T., Wee, H.: Improved non-committing encryption with applications to adaptively secure protocols. In: Matsui, M. (ed.) ASIACRYPT 2009. LNCS, vol. 5912, pp. 287–302. Springer, Heidelberg (2009)
9. Damgård, I.B., Nielsen, J.B.: Improved non-committing encryption schemes based on a general complexity assumption. In: Bellare, M. (ed.) CRYPTO 2000. LNCS, vol. 1880, pp. 432–450. Springer, Heidelberg (2000)
10. Damgård, I.B., Nielsen, J.B.: Perfect hiding and perfect binding universally composable commitment schemes with constant expansion factor. In: Yung, M. (ed.) CRYPTO 2002. LNCS, vol. 2442, pp. 581–596. Springer, Heidelberg (2002)

11. Damgård, I.B., Nielsen, J.B.: Universally composable efficient multiparty computation from threshold homomorphic encryption. In: Boneh, D. (ed.) CRYPTO 2003. LNCS, vol. 2729, pp. 247–264. Springer, Heidelberg (2003)
12. Dwork, C., Naor, M., Reingold, O., Stockmeyer, L.J.: Magic functions. J. ACM 50(6), 852–921 (2003)
13. Fehr, S., Hofheinz, D., Kiltz, E., Wee, H.: Encryption schemes secure against chosen-ciphertext selective opening attacks. In: Gilbert, H. (ed.) EUROCRYPT 2010. LNCS, vol. 6110, pp. 381–402. Springer, Heidelberg (2010)
14. El Gamal, T.: A public key cryptosystem and a signature scheme based on discrete logarithms. IEEE Trans. Inf. Theor. 31(4), 469–472 (1985)
15. Goldwasser, S., Micali, S.: Probabilistic encryption. J. Comput. Syst. Sci. 28(2), 270–299 (1984)
16. Hazay, C., López-Alt, A., Wee, H., Wichs, D.: Leakage-resilient cryptography from minimal assumptions. In: Johansson, T., Nguyen, P.Q. (eds.) EUROCRYPT 2013. LNCS, vol. 7881, pp. 160–176. Springer, Heidelberg (2013)
17. Hazay, C., Patra, A., Warinschi, B. Selective opening security for receivers. In: IACR Cryptology ePrint Archive 2015, p. 860 (2015)
18. Hemenway, B., Libert, B., Ostrovsky, R., Vergnaud, D.: Lossy encryption: constructions from general assumptions and efficient selective opening chosen ciphertext security. In: Lee, D.H., Wang, X. (eds.) ASIACRYPT 2011. LNCS, vol. 7073, pp. 70–88. Springer, Heidelberg (2011)
19. Hofheinz, D., Rupp, A.: Standard versus selective opening security: separation and equivalence results. In: Lindell, Y. (ed.) TCC 2014. LNCS, vol. 8349, pp. 591–615. Springer, Heidelberg (2014)
20. Huang, Z., Liu, S., Qin, B.: Sender-equivocable encryption schemes secure against chosen-ciphertext attacks revisited. In: Kurosawa, K., Hanaoka, G. (eds.) PKC 2013. LNCS, vol. 7778, pp. 369–385. Springer, Heidelberg (2013)
21. Jarecki, S., Lysyanskaya, A.: Adaptively secure threshold cryptography: introducing concurrency, removing erasures (extended abstract). In: Preneel, B. (ed.) EUROCRYPT 2000. LNCS, vol. 1807, p. 221. Springer, Heidelberg (2000)
22. Nielsen, J.B.: Separating random oracle proofs from complexity theoretic proofs: the non-committing encryption case. In: Yung, M. (ed.) CRYPTO 2002. LNCS, vol. 2442, pp. 111–126. Springer, Heidelberg (2002)
23. Ostrovsky, R., Rao, V., Visconti, I.: On selective-opening attacks against encryption schemes. In: Abdalla, M., De Prisco, R. (eds.) SCN 2014. LNCS, vol. 8642, pp. 578–597. Springer, Heidelberg (2014)

Function-Hiding Inner Product Encryption

Allison Bishop[1]($\boxtimes$), Abhishek Jain[2], and Lucas Kowalczyk[1]

[1] Columbia University, New York, USA
{allison,luke}@cs.columbia.edu
[2] Johns Hopkins University, Baltimore, USA
abhishek@cs.jhu.edu

Abstract. We extend the reach of functional encryption schemes that are provably secure under simple assumptions against unbounded collusion to include function-hiding inner product schemes. Our scheme is a private key functional encryption scheme, where ciphertexts correspond to vectors $\vec{x}$, secret keys correspond to vectors $\vec{y}$, and a decryptor learns $\langle \vec{x}, \vec{y} \rangle$. Our scheme employs asymmetric bilinear maps and relies only on the SXDH assumption to satisfy a natural indistinguishability-based security notion where arbitrarily many key and ciphertext vectors can be simultaneously changed as long as the key-ciphertext dot product relationships are all preserved.

1 Introduction

Functional encryption (FE) [8,23,25] is an exciting paradigm for non-interactively computing on encrypted data. In a functional encryption scheme for a family $\mathcal{F}$, it is possible to derive "special-purpose" decryption keys K_f for any function $f \in \mathcal{F}$ from a master secret key. Given such a decryption key K_f and an encryption of some input x, a user should be able to learn $f(x)$ and nothing else about x.

A driving force behind FE has been to understand what class of functions $\mathcal{F}$ can be supported and what notions of security can be achieved. In terms of functionality, research in FE started with the early works on attribute-based encryption [16,25], progressively evolving to support more expressive classes of functions, leading to the state of art works that are now able to support computation of general polynomial-size circuits [11,14,15,24]. In terms of security, most of the prior work in this area focuses on the privacy of (encrypted) *messages* (see, e.g., [8,10,23] for various security definitions considered in the literature for message privacy).

In many application scenarios, however, it is important to also consider privacy of the *function* being computed. Consider the following motivating example: suppose a hospital subscribes to a cloud service provider to store medical records of its patients. To protect the privacy of the data, these records are stored in an

---

A. Bishop—Supported in part by NSF CNS 1413971 and NSF CCF 1423306.
A. Jain—Supported in part by a DARPA/ARL Safeware Grant W911NF-15-C-0213 and NSF CNS-1414023.
L. Kowalczyk—Supported by an NSF Graduate Research Fellowship DGE-11-44155.

T. Iwata and J.H. Cheon (Eds.): ASIACRYPT 2015, Part I, LNCS 9452, pp. 470–491, 2015.
DOI: 10.1007/978-3-662-48797-6_20

encrypted form. At a later point in time, the hospital can request the cloud to perform some analysis on the encrypted records by releasing a decryption key K_f for a function f of its choice. If the FE scheme in use does not guarantee any hiding of the function (which is the case for many existing FE schemes), then the key K_f might reveal f completely to the cloud, which is undesirable when f itself contains sensitive information.

This has motivated the study of function privacy in FE, starting with the work of Shen et al. [26], and more recently by [2,6,7,9]. Intuitively speaking, function privacy requires that given a decryption key K_f for a function f, one should not be able to learn any unnecessary information about f. Using the analogy to secure computation, function private FE can be seen as the non-interactive analogue of private function evaluation (which guarantees the privacy of both the input x and the function f being computed on x) just like standard FE can be seen as the non-interactive analogue of secure function evaluation (which only guarantees privacy of the input x). One may also observe that the notion of function privacy is similar in spirit to program obfuscation [5,11]. Indeed, in the public-key setting, function private FE, in fact, implies program obfuscation.[1] In the secret-key setting, however, no such implication is known.

In this work, we continue the study of function privacy in FE. In particular, we focus on the inner product functionality $\mathcal{IP}$: a function $\mathrm{IP}_{\vec{y}} \in \mathcal{IP}$ in this function family is parametrized by a vector $\vec{y}$ in the finite field $\mathbb{Z}_p$. On an input $\vec{x} \in \mathbb{Z}_p$, $\mathrm{IP}_{\vec{y}}(\vec{x}) = \langle \vec{x}, \vec{y} \rangle$, where $\langle \vec{x}, \vec{y} \rangle$ denotes the inner product $\sum_{i=1}^{n} x_i y_i \in \mathbb{Z}_p$. Inner product is a particularly useful function for statistical analysis. In particular, in the context of FE, (as shown by [17]) it enables computation of conjunctions, disjunctions, CNF/DNF formulas, polynomial evaluation and exact thresholds.

Prior work on FE for inner product can be cast into the following two categories:

- *Generic constructions:* By now, we have a large sequence of works [4,10–12,14, 15,24,27] on FE that support computation of general circuits. Very recently, Brakerski and Segev [9] give a general transformation from any FE scheme for general circuits into one that achieves function privacy. Then, by combining [9] with the aforementioned works, one can obtain a function-private FE scheme for inner product as a special case.
  We note, however, that these generic FE constructions use heavy-duty tools for secure computation (such as fully-homomorphic encryption [13] and program obfuscation [5,11]) and are therefore extremely inefficient. Furthermore, in order to achieve collusion-resistance – one of the central goals in FE since its inception – the above solution would rely on indistinguishability obfuscation [5,11], which is a strong assumption.

---

[1] Here, the security definition for function privacy determines the security notion of program obfuscation that we obtain. See, e.g., [6] for further discussion on this connection.

– *Direct constructions:* To the best of our knowledge, the only "direct" construction of FE for inner product that avoids the aforementioned expensive tools is due to the recent work of Abdalla et al. [1]. Their work, however, does not consider function privacy.

We clarify that our formulation of inner product FE is different from that considered in the works of [3, 7, 17, 18, 21, 22, 26]. Very briefly, these works study inner product in the context of predicate encryption, where a message m is encrypted along with a tag $\vec{x}$ and decryption with a key K_y yields m iff $\langle \vec{x}, \vec{y} \rangle = 0$. In contrast, as discussed above, we are interested in learning the actual inner product value (in $\mathbb{Z}_p$).

In summary, the state of the art leaves open the problem of constructing a collusion-resistant, function-private FE scheme for inner product from standard assumptions. We stress that unless we put some restrictions on the distribution of the messages (as in the work of [6, 7]), this question only makes sense in the secret-key setting.

Our Results. In this work, we resolve this open problem. Specifically, we construct a function-private secret-key FE scheme for the inner product functionality that supports any arbitrary polynomial number of key queries and message queries. Our construction makes use of asymmetric bilinear maps and is significantly more efficient than the generic solutions discussed earlier. The security notion we prove for our construction is a natural indistinguishability-based notion, and we establish it under the Symmetric External Diffie-Hellman Assumption (SXDH). To obtain correctness for our scheme, we assume that inner products will be contained in a polynomially-sized range. This assumption is quite reasonable for statistical applications, where the average or count of some bounded quantity over a polynomially-sized database will naturally be in a polynomial range.

Our Techniques. We begin with the basic idea for inner product encryption developed in [17], which is the observation that one can place two vectors in the exponents on opposite sides of a bilinear group and compute the dot product via the pairing. This already provides some protection for the vectors as discrete log is thought to be hard in these groups, but without further randomization, this is vulnerable to many attacks, such as guessing the vector or learning whether two coordinates of a vector are the same. For this reason, the construction in [17] multiplies each of the exponent vectors by a random scalar value and uses additional subgroups in a composite order group to supply more randomization. The use of composite order groups in this way is by no means inherent, as subsequent works [18, 21, 22] (for example) demonstrate how to supply a sufficient amount of randomization in prime order bilinear groups using dual pairing vector spaces. However, in all of these schemes, the random scalars prevent a decryptor from learning the actual value of the inner product. Of course this is intentional and required, as these are predicate encryption schemes where the prescribed functionality only depends on whether the inner product is zero or nonzero.

To adapt these methods to allow a decryptor to learn the inner product, we must augment the construction with additional group elements that produce the same product of scalars in the exponent. Then the decryptor can produce the value by finding a ratio between the exponents of two group elements. This will be efficiently computable when the value of the inner product is in a known polynomially-sized range. Crucially, we must prove that these additional group elements do not reveal any unintended information.

We note that the construction in [17] is not known to be function-hiding. And since we are further allowing the inner product itself to be learned, function-hiding for our scheme means something different than function-hiding for schemes such as [17]. In particular, function hiding for a public key scheme in our setting would be impossible: one could simply create ciphertexts for a basis of vectors and test decryption of one's key against all of them to fully reconstruct the vector embedded in the key. It is thus fundamental that the public key scheme in [1] is not function-hiding. Indeed, their secret keys include vectors given in the clear and have no hiding properties.

To prove function-hiding for our construction, we thus leverage our private key setting to obtain a perfect symmetry between secret keys and ciphertexts, both in our construction and in our security reduction. Since no public parameters for encryption need to be published, the same techniques that we use to hide the underlying vectors in the ciphertexts can be flipped to argue that function-hiding holds for the secret keys.

The core of our security argument is an information-theoretic step (in the setting of dual pairing vector spaces as introduced by Okamoto and Takashima [19,20]). Essentially, our master secret key consists of two dual orthonormal bases that will be employed in the exponents to encode the vectors for ciphertexts and secret keys respectively. Secret keys and ciphertexts thus correspond to linear combinations of these basis vectors in the exponent. Since the bases themselves are never made public, if all of the secret keys (for example) are orthogonal to a particular vector, then there is a hidden "dimension" in the bases that can be used to argue that the ciphertext vector can be switched to another vector that has the same inner products with the provided keys. In fact, if we did not want any function privacy and instead only wanted to hide whether a single ciphertext corresponded to a vector $\vec{x}_0$ or $\vec{x}_1$ while giving out secret keys for vectors $\vec{y}$ orthogonal to $\vec{x}_0 - \vec{x}_1$, then we would do this information-theoretically. When we instead have many ciphertexts and we also demand function privacy for the keys, we use a hybrid argument, employing various applications of the SXDH assumption to move things around in the exponent bases and isolate a single ciphertext or key in a particular portion of the bases to apply our information-theoretic argument. The symmetry between keys and ciphertexts in our construction allows us to perform the same hybrid argument to obtain function privacy as in the case of multiple-ciphertext security.

2   Preliminaries

2.1   Functional Encryption Specifications and Security Definitions

In the rest of this paper, we will consider a specialization of the general definition of functional encryption to the particular functionality of computing dot products of n-length vectors over a finite field $\mathbb{Z}_p$. A private key functional encryption scheme for this class of functions will have the following PPT algorithms:

$Setup(1^\lambda, n) \rightarrow \mathrm{PP}, \mathrm{MSK}$ The setup algorithm will take in the security parameter λ and the vector length parameter n (a positive integer that is polynomial in λ). It will produce a master secret key MSK and public parameters PP. (Note that this is not a public key scheme, so the PP are not sufficient to encrypt - they are just parameters that do not need to be kept secret.)

$Encrypt(\mathrm{MSK}, \mathrm{PP}, \vec{x}) \rightarrow \mathrm{CT}$ The encryption algorithm will take in the master secret key MSK, the public parameters PP, and a vector $\vec{x} \in \mathbb{Z}_p^n$. It produces a ciphertext CT.

$KeyGen(\mathrm{MSK}, \mathrm{PP}, \vec{y}) \rightarrow \mathrm{SK}$ The key generation algorithm will take in the master secret key MSK, the public parameters PP, and a vector $\vec{y} \in \mathbb{Z}_p^n$. It produces a secret key SK.

$Decrypt(\mathrm{PP}, \mathrm{CT}, \mathrm{SK}) \rightarrow m \in \mathbb{Z}_p \, or \perp$ The decryption algorithm will take in the public parameters PP, a ciphertext CT, and a secret key SK. It will output either a value $m \in \mathbb{Z}_p$ or $\perp$.

For correctness, we will require the following. We suppose that $\mathrm{PP}, \mathrm{MSK}$ are the result of calling Setup$(1^\lambda, n)$, and CT, SK are then the result of calling Encrypt(MSK, PP, $\vec{x}$) and KeyGen(MSK, PP, $\vec{y}$) respectively. We then require that the output of Decrypt(PP, CT, SK) must be either $m = \langle \vec{x}, \vec{y} \rangle$ or $\perp$. We will only require that it is $\langle \vec{x}, \vec{y} \rangle$ and not $\perp$ when $\langle \vec{x}, \vec{y} \rangle$ is from a fixed polynomial range of values inside $\mathbb{Z}_p$, as this will allow a decryption algorithm to compute it as a discrete log in a group where discrete log is generally hard.

Security. We will consider an indistinguishability-based security notion defined by a game between a challenger and an attacker. At the beginning of the game, the challenger calls Setup$(1^\lambda, n, B)$ to produce MSK, PP. It gives PP to the attacker. The challenger also selects a random bit b.

Throughout the game, the attacker can (adaptively) make two types of a queries. To make a *key query*, it submits two vectors $\vec{y}^0, \vec{y}^1 \in \mathbb{Z}_p^n$ to the challenger, who then runs KeyGen(MSK, PP, $\vec{y}^b$) and returns the resulting SK to the attacker. To make a *ciphertext query*, the attacker submits two vectors $\vec{x}^0, \vec{x}^1 \in \mathbb{Z}_p^n$ to the challenger, who then runs Encrypt(MSK, PP, $\vec{x}^b$) and returns the resulting ciphertext to the attacker. The attacker can make any polynomial number of key and ciphertext queries throughout the game. At the end of the game, the attacker must submit a guess b' for the bit b. We require that for all key queries $\vec{y}^0, \vec{y}^1$ and all ciphertext queries $\vec{x}^0, \vec{x}^1$, it must hold that

$$\langle \vec{y}^0, \vec{x}^0 \rangle = \langle \vec{y}^0, \vec{x}^1 \rangle = \langle \vec{y}^1, \vec{x}^0 \rangle = \langle \vec{y}^1, \vec{x}^1 \rangle$$

The attacker's advantage is defined to be the probability that $b' = b$ minus $\frac{1}{2}$.

Definition 1. *We say a private key functional encryption scheme for dot products over $\mathbb{Z}_p^n$ satisfies function-hiding indistinguishability-based security if any PPT attacker's advantage in the above game is negligible as a function of the security parameter λ.*

Remark 1. We note that the attacker can trivially win the security game if we allowed a key query $\vec{y}^0, \vec{y}^1$ and ciphertext query $\vec{x}^0, \vec{x}^1$ such that $\langle \vec{y}^0, \vec{x}^0 \rangle \neq \langle \vec{y}^1, \vec{x}^1 \rangle$. Our stronger requirement that $\langle \vec{y}^0, \vec{x}^1 \rangle$ and $\langle \vec{y}^1, \vec{x}^0 \rangle$ is used for our hybrid security proof, but it might be possible to remove it by developing different proof techniques.

2.2 Asymmetric Bilinear Groups

We will construct our scheme in aymmetric bilinear groups. We let $\mathcal{G}$ denote a group generator - an algorithm which takes a security parameter λ as input and outputs a description of prime order groups G_1, G_2, G_T with a bilinear map $e : G_1 \times G_2 \to G_T$. We define $\mathcal{G}$'s output as (p, G_1, G_2, G_T, e), where p is a prime, G_1, G_2 and G_T are cyclic groups of order p, and $e : G_1 \times G_2 \to G_T$ is a map with the following properties:

1. (Bilinear) $\forall g_1 \in G_1, g_2 \in G_2, a, b \in \mathbb{Z}_p, e(g_1^a, g_2^b) = e(g_1, g_2)^{ab}$
2. (Non-degenerate) $\exists g_1 \in G_1, g_2 \in G_2$ such that $e(g_1, g_2)$ has order p in G_T.

We refer to G_1 and G_2 as the source groups and G_T as the target group. We assume that the group operations in G_1, G_2, and G_T and the map e are computable in polynomial time with respect to λ, and the group descriptions of G_1, G_2, and G_T include a generator of each group.

The SXDH Assumption. The security of our construction relies on the hardness of the SXDH assumption. Given prime order groups $(p, G_1, G_2, G_T, e) \leftarrow \mathcal{G}(\lambda)$, we define the SXDH problem as distinguishing between the following two distributions:

$$D_1 = (g_1, g_1^a, g_1^b, g_1^{ab}, g_2)$$

and

$$D_2 = (g_1, g_1^a, g_1^b, g_1^{ab+r}, g_2)$$

where g_1, g_2 are generators of G_1, G_2, and $a, b, r \leftarrow \mathbb{Z}_p$.

The SXDH Assumption states that no polynomial-time algorithm can achieve non-negligible advantage in deciding between D_1 and D_2. It also states that the same is true for the analogous distributions formed from switching the roles of G_1, G_2 (that is, $D_1 = (g_2, g_2^a, g_2^b, g_2^{ab}, g_1)$ and $D_2 = (g_2, g_2^a, g_2^b, g_2^{ab+r}, g_1)$)

2.3   Dual Pairing Vector Spaces

In addition to referring to individual elements of G_1 and G_2, we will also consider "vectors" of group elements. For $\vec{v} = (v_1, ..., v_m) \in \mathbb{Z}_p^m$ and $g_1 \in G_1$, we write $g_1^{\vec{v}}$ to denote the m-tuple of elements of G_1:

$$g_1^{\vec{v}} := (g_1^{v_1}, ..., g_1^{v_m})$$

We can also perform scalar multiplication and exponentiation in the exponent. For any $a \in \mathbb{Z}_p$ and $\vec{v}, \vec{w} \in \mathbb{Z}_p^m$, we have:

$$g_1^{a\vec{v}} := (g_1^{av_1}, ..., g_1^{av_m})$$
$$g_1^{\vec{v}+\vec{w}} = (g_1^{v_1+w_1}, ..., g_1^{v_m+w_m})$$

We abuse notation slightly and also let e denote the product of the component wise pairings:

$$e(g_1^{\vec{v}}, g_2^{\vec{w}}) := \prod_{i=1}^{m} e(g_1^{v_i}, g_2^{w_i}) = e(g_1, g_2)^{\langle \vec{v}, \vec{w} \rangle}$$

Here, the dot product is taken modulo p.

Dual Pairing Vector Spaces. We will employ the concept of dual pairing vector spaces from [19,20]. We will choose two random sets of vectors: $\mathbb{B} := \{\vec{b}_1, \ldots, \vec{b}_m\}$ and $\mathbb{B}^* = \{\vec{b}_1^*, \ldots, \vec{b}_m^*\}$ of $\mathbb{Z}_p^m$ subject to the constraint that they are "dual orthonormal" in the following sense:

$$\langle \vec{b}_i, \vec{b}_i^* \rangle = 1 \pmod{p} \text{ for all } i$$
$$\langle \vec{b}_i, \vec{b}_j^* \rangle = 0 \pmod{p} \text{ for all } j \neq i.$$

We note that choosing sets $(\mathbb{B}, \mathbb{B}^*)$ at random from sets satisfying these dual orthonormality constraints can be realized by choosing a set of m vectors $\mathbb{B}$ uniformly at random from $\mathbb{Z}_p^m$ (these vectors will be linearly independent with high probability), then determining each vector of $\mathbb{B}^*$ from its orthonormality constraints. We will denote choosing random dual orthonormal sets this way as: $(\mathbb{B}, \mathbb{B}^*) \leftarrow Dual(\mathbb{Z}_p^m)$.

3   Construction

We now present our construction in asymmetric bilinear groups. We will choose dual orthonormal bases $\mathbb{B}$ and $\mathbb{B}^*$ that will be used in the exponent to encode ciphertext and key vectors respectively. Vectors will be encoded twice to create space for a hybrid security proof and will be additionally masked by random scalars (these basic features are also present in [17]). We will use additional dual bases $\mathbb{D}, \mathbb{D}^*$ to separately encode these same scalars in the exponent so that their effect can be removed from the final decryption result. We view it as a core feature of our construction that the structure of keys and ciphertexts in our scheme is perfectly symmetric, just on different sides of dual orthonormal bases. This enables us to prove function hiding for the keys with exactly the same techniques we use to prove indistinguishability security for the ciphertexts.

$Setup(1^\lambda, n), \to \text{MSK}, \text{PP}$ The setup algorithm takes in the security parameter λ and a positive integer n specifying the desired length of vectors for the keys and ciphertexts. It chooses an asymmetric bilinear group consisting of G_1, G_2, G_T, all with prime order p. It fixes generators g_1, g_2 of G_1 and G_2 respectively. It then samples dual orthonormal bases $\mathbb{B}, \mathbb{B}^* \leftarrow Dual(\mathbb{Z}_p^{2n})$ and dual orthonormal bases $\mathbb{D}, \mathbb{D}^* \leftarrow Dual(\mathbb{Z}_p^2)$. It defines the master secret key as $\text{MSK} := \mathbb{B}, \mathbb{B}^*, \mathbb{D}, \mathbb{D}^*$. The groups G_1, G_2, G_T, the generators g_1, g_2, and p are set to be public parameters.

$Encrypt(\text{MSK}, \text{PP}, \vec{x}) \to \text{CT}$ The encryption algorithm takes in the master secret key $\mathbb{B}, \mathbb{B}^*, \mathbb{D}, \mathbb{D}^*$, the public parameters, and a vector $\vec{x} \in \mathbb{Z}_p^n$. It chooses two independent and uniformly random elements $\alpha, \tilde{\alpha} \in \mathbb{Z}_p$. It then computes:

$$C_1 := g_1^{\alpha(x_1\vec{b}_1^* + \cdots + x_n\vec{b}_n^*) + \tilde{\alpha}(x_1\vec{b}_{n+1}^* + \cdots + x_n\vec{b}_{2n}^*)}$$

$$C_2 := g_1^{\alpha\vec{d}_1^* + \tilde{\alpha}\vec{d}_2^*}.$$

The ciphertext $\text{CT} = \{C_1, C_2\}$.

$KeyGen(\text{MSK}, \text{PP}, \vec{y}) \to \text{SK}$ The secret key generation algorithm takes in the master secret key $\mathbb{B}, \mathbb{B}^*, \mathbb{D}, \mathbb{D}^*$, the public parameters, and a vector $\vec{y} \in \mathbb{Z}_p^n$. It chooses two independent and uniformly random elements $\beta, \tilde{\beta} \in \mathbb{Z}_p$. It then computes:

$$K_1 := g_2^{\beta(y_1\vec{b}_1 + \cdots + y_n\vec{b}_n) + \tilde{\beta}(y_1\vec{b}_{n+1} + \cdots + y_n\vec{b}_{2n})}$$

$$K_2 := g_2^{\beta\vec{d}_1 + \tilde{\beta}\vec{d}_2}.$$

The secret key $\text{SK} = \{K_1, K_2\}$.

$Decrypt(\text{PP}, \text{CT}, \text{SK}) \to m \in \mathbb{Z}_p or \perp$ The decryption algorithm takes in the public parameters, the ciphertext C_1, C_2, and the secret key K_1, K_2. It computes:

$$D_1 := e(C_1, K_1)$$

$$D_2 := e(C_2, K_2).$$

It then computes an m such that $D_2^m = D_1$ as elements of G_T. It outputs m. We note that we can guarantee that the decryption algorithm runs in polynomial time when we restrict to checking a fixed, polynomially size range of possible values for m and output $\perp$ when none of them satisfy the criterion $D_2^m = D_1$.

Correctness. We observe that for a ciphertext formed by calling Encrypt(MSK, PP, $\vec{x}$) and a key formed by calling $KeyGen$(MSK, PP, $\vec{y}$), we have

$$D_1 = e(C_1, K_1) = e(g_1, g_2)^{\alpha\beta\langle\vec{x},\vec{y}\rangle + \tilde{\alpha}\tilde{\beta}\langle\vec{x},\vec{y}\rangle} = e(g_1, g_2)^{(\alpha\beta + \tilde{\alpha}\tilde{\beta})\langle\vec{x},\vec{y}\rangle}$$

$$\text{and } D_2 = e(C_2, K_2) = e(g_1, g_2)^{\alpha\beta + \tilde{\alpha}\tilde{\beta}}.$$

This follows immediately from the definitions of C_1, C_2, K_1, K_2 and the fact that $\mathbb{B}, \mathbb{B}^*$ and $\mathbb{D}, \mathbb{D}^*$ are dual orthonormal bases pairs. Thus, if $\langle\vec{x},\vec{y}\rangle$ is in the polynomial range of possible values for m that the decryption algorithm checks, it will output $m := \langle\vec{x},\vec{y}\rangle$ as desired.

4   Security Proof

Our security proof is structured as a hybrid argument over a series of games which differ in how the ciphertext and keys are constructed. Intuitively, if there were only one ciphertext, we could embed the difference of the two possible ciphertext vectors, namely $\vec{x}^0 - \vec{x}^1$, into the definition of the bases $\mathbb{B}, \mathbb{B}^*$ to argue that this difference is hidden when only key vectors orthogonal to $\vec{x}^0 - \vec{x}^1$ are provided. In other words, there is ambiguity in the choice of $\mathbb{B}, \mathbb{B}^*$ left conditioned on the provided keys, and this can be exploited to switch $\vec{x}^0$ for $\vec{x}^1$. But there is a limited amount of such ambiguity, so to re-purpose it for many ciphertexts, we employ a hybrid argument that isolates each ciphertext in turn in a portion of the basis. Since keys and ciphertexts are constructed and treated symmetrically in our scheme, we can apply the same hybrid argument over keys to prove function hiding, just reversing the roles of $\mathbb{B}$ and $\mathbb{B}^*$.

Notice that a normal ciphertext for a vector $\vec{x}$ contains two parallel copies of $\vec{x}$ in the exponent of C_1: one attached to $\vec{b}_i^*$'s and one attached to $\vec{b}_{n+i}^*$'s for $i = 1, ..., n$. We will refer to this as a type-$(\vec{x}, \vec{x})$ ciphertext. We will use this notation to define a type-$(\vec{0}, \vec{x})$ ciphertext - one which is normally formed but has no $\vec{b}_i^*$ components for $i = 1, ..., n$ and no $\vec{d}_1^*$ component in C_2. We will also use the same terminology to refer to keys (i.e.: type-$(\vec{y}, \vec{y})$ / type-$(\vec{0}, \vec{y})$ / type-$(\vec{y}, \vec{0})$ keys).

Letting Q_1 denote the total number of ciphertext queries the attacker makes, we define 7 games for each $j = 0, ..., Q_1$:

$Game^1_{j,Z}$ In $Game^1_{j,Z}$ all ciphertexts before the jth ciphertext are of type-$(\vec{0}, \vec{x}_i^1)$, the jth ciphertext is of type-$(\vec{0}, \vec{x}_i^0)$, all ciphertexts after the jth ciphertext are also type-$(\vec{0}, \vec{x}_i^0)$ ciphertexts, and all keys are of type-$(\vec{y}_i^0, \vec{y}_i^0)$.

$Game^2_{j,Z}$ $Game^2_{j,Z}$ is the same as $Game^1_{j,Z}$ except that the jth ciphertext is now of type-$(\vec{x}_j^0, \vec{x}_j^0)$.

$Game^3_{j,Z}$ $Game^3_{j,Z}$ is the same as $Game^2_{j,Z}$ except that the jth ciphertext is now of type-$(\vec{x}_j^0, \vec{x}_j^1)$.

$Game_{j,Z}^4$ $Game_{j,Z}^4$ is the same as $Game_{j,Z}^3$ except that all ciphertexts before the jth ciphertext are now of type-$(\vec{x}_i^1, \vec{x}_i^1)$ and all ciphertexts after the jth ciphertext are now type-$(\vec{x}_i^0, \vec{x}_i^0)$ ciphertexts.

$Game_{j,Z}^5$ $Game_{j,Z}^5$ is the same as $Game_{j,Z}^4$ except that all ciphertexts before the jth ciphertext are now of type-$(\vec{x}_i^1, \vec{0})$ and all ciphertexts after the jth ciphertext are now type-$(\vec{x}_i^0, \vec{0})$ ciphertexts.

$Game_{j,Z}^6$ $Game_{j,Z}^6$ is the same as $Game_{j,Z}^5$ except that the jth ciphertext is now of type-$(\vec{x}_j^1, \vec{x}_j^1)$.

$Game_{j,Z}^7$ $Game_{j,Z}^7$ is the same as $Game_{j,Z}^6$ except that all ciphertexts before the jth ciphertext are now of type-$(\vec{x}_i^1, \vec{x}_i^1)$ and all ciphertexts after the jth ciphertext are now type-$(\vec{x}_i^0, \vec{x}_i^0)$ ciphertexts.
Letting Q_2 denote the total number of key requests the attacker makes, we define 7 additional games for each $j = 0, ..., Q_2$:

$Game_{O,j}^1$ In $Game_{O,j}^1$ all keys before the jth key are of type-$(\vec{0}, \vec{y}_i^1)$, the jth key is of type-$(\vec{0}, \vec{y}_i^0)$, all keys after the jth key are also type-$(\vec{0}, \vec{y}_i^0)$ keys, and all ciphertexts are of type-$(\vec{x}_i^1, \vec{x}_i^1)$.

$Game_{O,j}^2$ $Game_{O,j}^2$ is the same as $Game_{O,j}^1$ except that the jth key is now of type-$(\vec{y}_j^0, \vec{y}_j^0)$.

$Game_{O,j}^3$ $Game_{O,j}^3$ is the same as $Game_{O,j}^2$ except that the jth key is now of type-$(\vec{y}_j^0, \vec{y}_j^1)$.

$Game_{O,j}^4$ $Game_{O,j}^4$ is the same as $Game_{O,j}^3$ except that all keys before the jth key are now of type-$(\vec{y}_i^1, \vec{y}_i^1)$ and all keys after the jth key are now type-$(\vec{y}_i^0, \vec{y}_i^0)$ keys.

$Game_{O,j}^5$ $Game_{O,j}^5$ is the same as $Game_{O,j}^4$ except that all keys before the jth key are now of type-$(\vec{y}_i^1, \vec{0})$ and all keys after the jth key are now type-$(\vec{y}_i^0, \vec{0})$ keys.

$Game_{O,j}^6$ $Game_{O,j}^6$ is the same as $Game_{O,j}^5$ except that the jth key is now of type-$(\vec{y}_j^1, \vec{y}_j^1)$.

$Game_{O,j}^7$ $Game_{O,j}^7$ is the same as $Game_{O,j}^6$ except that all keys before the jth key are now of type-$(\vec{y}_i^1, \vec{y}_i^1)$ and all keys after the jth key are now type-$(\vec{y}_i^0, \vec{y}_i^0)$ keys.
Note that $Game_{0,Z}^7$ is the real security game played with $b = 0$ and $Game_{O,Q_2}^7$ is the real security game played with $b = 1$. Note also that $Game_{Q_1,Z}^7$ and $Game_{O,0}^7$ are identical.

We will use a hybrid argument to transition between the two to show that no polynomial attacker can achieve non-negligible advantage in the security game (distinguishing between $b = 0$ and $b = 1$.). Our hybrid works in two parts, first transitioning all ciphertexts from type-$(\vec{x}_i^0, \vec{x}_i^0)$ to type-$(\vec{x}_i^1, \vec{x}_i^1)$ (using the Game$_{j,Z}^i$'s), then transitioning all keys from type-$(\vec{y}_i^0, \vec{y}_i^0)$ to type-$(\vec{y}_i^1, \vec{y}_i^1)$ (using the Game$_{O,j}^i$'s).

First we will transition from Game$_{0,Z}^7$ (the real security game played with $b = 0$) to Game$_{1,Z}^1$. We then transition from Game$_{1,Z}^1$ to Game$_{1,Z}^2$, to Game$_{1,Z}^3$, ..., to Game$_{1,Z}^7$, to Game$_{2,Z}^1$ etc. until reaching Game$_{Q_1,Z}^7$, where all ciphertexts are of type $(\vec{x}_i^1, \vec{x}_i^1)$ (but all keys are still of type $(\vec{y}_i^0, \vec{y}_i^0)$). Recall that Game$_{Q_1,Z}^7$ is identical to Game$_{O,0}^7$. We will then transition from Game$_{Q_1,Z}^7 = $ Game$_{O,0}^7$ to Game$_{O,1}^1$, to Game$_{O,1}^2$, ..., to Game$_{O,1}^7$, to Game$_{O,2}^1$, etc. until reaching Game$_{O,Q_2}^7$, where all keys are of type $(\vec{y}_i^1, \vec{y}_i^1)$ (and all ciphertexts are of type $(\vec{x}_i^1, \vec{x}_i^1)$). This is identical to the real security game played with $b = 1$.

We begin the first transition in a hybrid over the Q_1 ciphertexts. Recall that the real security game played with $b = 0$ is identical to Game$_{0,Z}^7$, so in particular, the following lemma allows us to make the first transition from Game$_{0,Z}^7$ to Game$_{1,Z}^1$.

Lemma 1. *No polynomial-time attacker can achieve a non-negligible difference in advantage between* Game$_{(j-1),Z}^7$ *and* Game$_{j,Z}^1$ *for* $j = 1, ..., Q_1$ *under the SXDH assumption.*

Proof. Given an attacker that achieves non-negligible difference in advantage between Game$_{(j-1),Z}^7$ and Game$_{j,Z}^1$ for some $j \in [1, Q_1]$, we could achieve non-negligible advantage in deciding the SXDH problem as follows:

Given SXDH instance $g_1, g_1^a, g_1^b, T = g_1^{ab+r}, g_2$ where either $r \leftarrow \mathbb{Z}_p$ or $r = 0$, use g_1, g_2 as the generators of the same name used to form ciphertexts and keys respectively. Generate bases $(\mathbb{F}, \mathbb{F}^*) \leftarrow Dual(\mathbb{Z}_p^{2n})$, $(\mathbb{H}, \mathbb{H}^*) \leftarrow Dual(\mathbb{Z}_p^2)$ and implicitly define new bases $(\mathbb{B}, \mathbb{B}^*)$, $(\mathbb{D}, \mathbb{D}^*)$ as the following:

$$\vec{b}_i = \vec{f}_i - a\vec{f}_{n+i} \text{ for } i = 1, ..., n$$

$$\vec{b}_{n+i} = \vec{f}_{n+i} \text{ for } i = 1, ..., n$$

$$\vec{b}_i^* = \vec{f}_i^* \text{ for } i = 1, ..., n$$

$$\vec{b}_{n+i}^* = \vec{f}_{n+i}^* + a\vec{f}_i^* \text{ for } i = 1, ..., n$$

$$\vec{d}_1 = \vec{h}_1 - a\vec{h}_2$$

$$\vec{d}_2 = \vec{h}_2$$

$$\vec{d}_1^* = \vec{h}_1^*$$

$$\vec{d}_2^* = \vec{h}_2^* + a\vec{h}_1^*$$

Note that these bases are distributed exactly the same as those output by $Dual(\mathbb{Z}_p^{2n})$ and $Dual(\mathbb{Z}_p^2)$ respectively (they are created by applying an invertible linear transformation to the output of $Dual(\mathbb{Z}_p^{2n})$ and $Dual(\mathbb{Z}_p^2)$).

To construct any key (for, say, vector $\vec{y}^0$), generate random $\beta, \tilde{\beta}' \leftarrow \mathbb{Z}_p$, implicitly define $\tilde{\beta} = \beta a + \tilde{\beta}'$, and compute:

$$K_1 = g_2^{\beta(y_1^0 \vec{f}_1 + \cdots + y_n^0 \vec{f}_n) + \tilde{\beta}'(y_1^0 \vec{f}_{n+1} + \cdots + y_n^0 \vec{f}_{2n})}$$

$$= g_2^{\beta(y_1^0 \vec{f}_1 + \cdots + y_n^0 \vec{f}_n) + (\tilde{\beta} - \beta a)(y_1^0 \vec{f}_{n+1} + \cdots + y_n^0 \vec{f}_{2n})}$$

$$= g_2^{\beta(y_1^0(\vec{f}_1 - a\vec{f}_{n+1}) + \cdots + y_n^0(\vec{f}_n - a\vec{f}_{2n})) + \tilde{\beta}(y_1^0 \vec{f}_{n+1} + \cdots + y_n^0 \vec{f}_{2n})}$$

$$= g_2^{\beta(y_1^0 \vec{b}_1 + \cdots + y_n^0 \vec{b}_n) + \tilde{\beta}(y_1^0 \vec{b}_{n+1} + \cdots + y_n^0 \vec{b}_{2n})}$$

$$K_2 = g_2^{\beta \vec{h}_1 + \tilde{\beta}' \vec{h}_2}$$

$$= g_2^{\beta \vec{h}_1 + (\tilde{\beta} - \beta a)\vec{h}_2}$$

$$= g_2^{\beta(\vec{h}_1 - a\vec{h}_2) + \tilde{\beta}\vec{h}_2}$$

$$= g_2^{\beta \vec{d}_1 + \tilde{\beta}\vec{d}_2}$$

a properly distributed type-$(\vec{y}^0, \vec{y}^0)$ key (as expected in both $\text{Game}_{(j-1),Z}^7$ and $\text{Game}_{j,Z}^1$).

For the jth ciphertext and all ciphertexts after, draw $\alpha_i', \tilde{\alpha}_i' \leftarrow \mathbb{Z}_p$ and compute:

$$C_{1,i} = (g_1^b)^{\alpha_i'(x_{1,i}^0 \vec{f}_{n+1}^* + \cdots + x_{n,i}^0 \vec{f}_{2n}^*)} (T)^{\alpha_i'(x_{1,i}^0 \vec{f}_1^* + \cdots + x_{n,i}^0 \vec{f}_n^*)} g_1^{\tilde{\alpha}_i'(x_{1,i}^0 \vec{f}_{n+1}^* + \cdots + x_{n,i}^0 \vec{f}_{2n}^*)}$$

$$\cdot (g_1^a)^{\tilde{\alpha}_i'(x_{1,i}^0 \vec{f}_1^* + \cdots + x_{n,i}^0 \vec{f}_n^*)}$$

$$= g_1^{\alpha_i' r(x_{1,i}^0 \vec{f}_1^* + \cdots + x_{n,i}^0 \vec{f}_n^*) + (\tilde{\alpha}_i' + \alpha_i' b)(x_{1,i}^0 (\vec{f}_{n+1}^* + a\vec{f}_1^*) + \cdots + x_{n,i}^0 (\vec{f}_{2n}^* + a\vec{f}_n^*))}$$

$$= g_1^{\alpha_i' r(x_{1,i}^0 \vec{b}_1^* + \cdots + x_{n,i}^0 \vec{b}_n^*) + (\tilde{\alpha}_i' + \alpha_i' b)(x_{1,i}^0 \vec{b}_{n+1}^* + \cdots + x_{n,i}^0 \vec{b}_{2n}^*)}$$

$$C_{2,i} = (g_1^b)^{\alpha_i' \vec{h}_2^*} (T)^{\alpha_i' \vec{h}_1^*} g_1^{\tilde{\alpha}_i' \vec{h}_2^*} (g_1^a)^{\tilde{\alpha}_i' \vec{h}_1^*}$$

$$= g_1^{\alpha_i' r \vec{h}_1^* + (\tilde{\alpha}_i' + \alpha_i' b)(\vec{h}_2^* + a\vec{h}_1^*)}$$

$$= g_1^{\alpha_i' r \vec{d}_1^* + (\tilde{\alpha}_i' + \alpha_i' b)\vec{d}_2^*}$$

For ciphertexts before the jth ciphertext, draw $\alpha_i', \tilde{\alpha}_i' \leftarrow \mathbb{Z}_p$ and compute:

$$C_{1,i} = (g_1^b)^{\alpha_i'(x_{1,i}^1 \vec{f}_{n+1}^* + \cdots + x_{n,i}^1 \vec{f}_{2n}^*)} (T)^{\alpha_i'(x_{1,i}^1 \vec{f}_1^* + \cdots + x_{n,i}^1 \vec{f}_n^*)} g_1^{\tilde{\alpha}_i'(x_{1,i}^1 \vec{f}_{n+1}^* + \cdots + x_{n,i}^1 \vec{f}_{2n}^*)}$$

$$\cdot (g_1^a)^{\tilde{\alpha}_i'(x_{1,i}^1 \vec{f}_1^* + \cdots + x_{n,i}^1 \vec{f}_n^*)}$$

$$= g_1^{\alpha_i' r(x_{1,i}^1 \vec{f}_1^* + \cdots + x_{n,i}^1 \vec{f}_n^*) + (\tilde{\alpha}_i' + \alpha_i' b)(x_{1,i}^1 (\vec{f}_{n+1}^* + a\vec{f}_1^*) + \cdots + x_{n,i}^1 (\vec{f}_{2n}^* + a\vec{f}_n^*))}$$

$$= g_1^{\alpha_i' r(x_{1,i}^1 \vec{b}_1^* + \cdots + x_{n,i}^1 \vec{b}_n^*) + (\tilde{\alpha}_i' + \alpha_i' b)(x_{1,i}^1 \vec{b}_{n+1}^* + \cdots + x_{n,i}^1 \vec{b}_{2n}^*)}$$

$$C_{2,i} = (g_1^b)^{\alpha_i' \vec{h}_2^*} (T)^{\alpha_i' \vec{h}_1^*} g_1^{\tilde{\alpha}_i' \vec{h}_2^*} (g_1^a)^{\tilde{\alpha}_i' \vec{h}_1^*}$$

$$= g_1^{\alpha_i' r \vec{h}_1^* + (\tilde{\alpha}_i' + \alpha_i' b)(\vec{h}_2^* + a\vec{h}_1^*)}$$

$$= g_1^{\alpha_i' r \vec{d}_1^* + (\tilde{\alpha}_i' + \alpha_i' b)\vec{d}_2^*}$$

482   A. Bishop et al.

(The only difference from the prior ciphertext construction is that $\vec{x}^1$ is used instead of $\vec{x}^0$).

When $r \leftarrow \mathbb{Z}_p$, all ciphertexts before the jth ciphertext are properly distributed type-$(\vec{x}_i^1, \vec{x}_i^1)$ ciphertexts and the remaining ciphertexts are properly distributed type-$(\vec{x}_i^0, \vec{x}_i^0)$ ciphertexts where $\alpha_i = \alpha_i' r$ and $\tilde{\alpha}_i = \tilde{\alpha}_i' + \alpha_i' b$. This is as would be expected in $\text{Game}_{(j-1),Z}^7$.

When $r = 0$, all ciphertexts before the jth ciphertext are properly distributed type-$(\vec{0}, \vec{x}_i^1)$ ciphertexts and the remaining ciphertexts are properly distributed type-$(\vec{0}, \vec{x}_i^0)$ ciphertexts where $\tilde{\alpha}_i = \tilde{\alpha}_i' + \alpha_i' b$. This is as would be expected in $\text{Game}_{j,Z}^1$.

Our simulation is therefore identical to either $\text{Game}_{(j-1),Z}^7$ or $\text{Game}_{j,Z}^1$ depending on the SXDH challenge $T = g_1^{ab+r}$ having $r \leftarrow \mathbb{Z}_p$ or $r = 0$ respectively. Therefore, by playing the security game in this manner with the supposed attacker and using the attacker's output as an answer to the SXDH challenge, we will enjoy the same non-negligible advantage as the supposed attacker in deciding the SXDH problem.

By the SXDH assumption, this is not possible, so no such adversary can exist.

Lemma 2. *No polynomial-time attacker can achieve a non-negligible difference in advantage between $\text{Game}_{j,Z}^1$ and $\text{Game}_{j,Z}^2$ for $j = 1, ..., Q_1$ under the SXDH assumption.*

Proof. Given an attacker that achieves non-negligible difference in advantage between $\text{Game}_{j,Z}^1$ and $\text{Game}_{j,Z}^2$ for some $j \in [1, Q_1]$, we could achieve non-negligible advantage in deciding the SXDH problem as follows:

Given SXDH instance $g_1, g_1^a, g_1^b, T = g_1^{ab+r}, g_2$ where either $r \leftarrow \mathbb{Z}_p$ or $r = 0$, use g_1, g_2 as the generators of the same name used to form ciphertexts and keys respectively.. Generate bases $(\mathbb{F}, \mathbb{F}^*) \leftarrow Dual(\mathbb{Z}_p^{2n})$, $(\mathbb{H}, \mathbb{H}^*) \leftarrow Dual(\mathbb{Z}_p^2)$ and implicitly define new bases $(\mathbb{B}, \mathbb{B}^*)$, $(\mathbb{D}, \mathbb{D}^*)$ as the following:

$$\vec{b}_i = \vec{f}_i - a\vec{f}_{n+i} \text{ for } i = 1, ..., n$$
$$\vec{b}_{n+i} = \vec{f}_{n+i} \text{ for } i = 1, ..., n$$
$$\vec{b}_i^* = \vec{f}_i^* \text{ for } i = 1, ..., n$$
$$\vec{b}_{n+i}^* = \vec{f}_{n+i}^* + a\vec{f}_i^* \text{ for } i = 1, ..., n$$

$$\vec{d}_1 = \vec{h}_1 - a\vec{h}_2$$
$$\vec{d}_2 = \vec{h}_2$$
$$\vec{d}_1^* = \vec{h}_1^*$$
$$\vec{d}_2^* = \vec{h}_2^* + a\vec{h}_1^*$$

Note that these bases are distributed exactly the same as those output by $Dual(\mathbb{Z}_p^{2n})$ and $Dual(\mathbb{Z}_p^2)$ respectively (they are created by applying an invertible linear transformation to the output of $Dual(\mathbb{Z}_p^{2n})$ and $Dual(\mathbb{Z}_p^2)$).

To construct any key (for, say, vector $\vec{y}^0$), generate random $\beta, \tilde{\beta}' \leftarrow \mathbb{Z}_p$, implicitly define $\tilde{\beta} = \beta a + \tilde{\beta}'$, and compute:

$$
\begin{aligned}
K_1 &= g_2^{\beta(y_1^0 \vec{f}_1 + \cdots + y_n^0 \vec{f}_n) + \tilde{\beta}'(y_1^0 \vec{f}_{n+1} + \cdots + y_n^0 \vec{f}_{2n})} \\
&= g_2^{\beta(y_1^0 \vec{f}_1 + \cdots + y_n^0 \vec{f}_n) + (\tilde{\beta} - \beta a)(y_1^0 \vec{f}_{n+1} + \cdots + y_n^0 \vec{f}_{2n})} \\
&= g_2^{\beta(y_1^0(\vec{f}_1 - a\vec{f}_{n+1}) + \cdots + y_n^0(\vec{f}_n - a\vec{f}_{2n})) + \tilde{\beta}(y_1^0 \vec{f}_{n+1} + \cdots + y_n^0 \vec{f}_{2n})} \\
&= g_2^{\beta(y_1^0 \vec{b}_1 + \cdots + y_n^0 \vec{b}_n) + \tilde{\beta}(y_1^0 \vec{b}_{n+1} + \cdots + y_n^0 \vec{b}_{2n})} \\
K_2 &= g_2^{\beta \vec{h}_1 + \tilde{\beta}' \vec{h}_2} \\
&= g_2^{\beta \vec{h}_1 + (\tilde{\beta} - \beta a) \vec{h}_2} \\
&= g_2^{\beta(\vec{h}_1 - a\vec{h}_2) + \tilde{\beta} \vec{h}_2} \\
&= g_2^{\beta \vec{d}_1 + \tilde{\beta} \vec{d}_2}
\end{aligned}
$$

a properly distributed type-$(\vec{y}^0, \vec{y}^0)$ key (as expected in both $\text{Game}_{j,Z}^1$ and $\text{Game}_{j,Z}^2$).

For ciphertexts before the jth ciphertext draw random $\tilde{\alpha}_i \leftarrow \mathbb{Z}_p$ and compute:

$$
\begin{aligned}
C_{1,i} &= g_1^{\tilde{\alpha}_i(x_{1,i}^1 \vec{f}_{n+1}^* + \cdots + x_{n,i}^1 \vec{f}_{2n}^*)}(g_1^a)^{\tilde{\alpha}_i(x_{1,i}^1 \vec{f}_1^* + \cdots + x_{n,i}^1 \vec{f}_n^*)} \\
&= g_1^{\tilde{\alpha}_i(x_{1,i}^1(\vec{f}_{n+1}^* + a\vec{f}_1^*) + \cdots + x_{n,i}^1(\vec{f}_{2n}^* + a\vec{f}_n^*))} \\
&= g_1^{\tilde{\alpha}_i(x_{1,i}^1 \vec{b}_{n+1}^* + \cdots + x_{n,i}^1 \vec{b}_{2n}^*)} \\
C_{2,i} &= g_1^{\tilde{\alpha}_i \vec{h}_2^*}(g_1^a)^{\tilde{\alpha}_i \vec{h}_1^*} \\
&= g_1^{\tilde{\alpha}_i(\vec{h}_2^* + a\vec{h}_1^*)} \\
&= g_1^{\tilde{\alpha}_i \vec{d}_2^*}
\end{aligned}
$$

a properly distributed type-$(\vec{0}, \vec{x}_i^1)$ ciphertext (as expected in both $\text{Game}_{j,Z}^1$ and $\text{Game}_{j,Z}^2$).

For ciphertexts after the jth ciphertext draw random $\tilde{\alpha}_i \leftarrow \mathbb{Z}_p$ and compute:

$$
\begin{aligned}
C_{1,i} &= g_1^{\tilde{\alpha}_i(x_{1,i}^0 \vec{f}_{n+1}^* + \cdots + x_{n,i}^0 \vec{f}_{2n}^*)}(g_1^a)^{\tilde{\alpha}_i(x_{1,i}^0 \vec{f}_1^* + \cdots + x_{n,i}^0 \vec{f}_n^*)} \\
&= g_1^{\tilde{\alpha}_i(x_{1,i}^0(\vec{f}_{n+1}^* + a\vec{f}_1^*) + \cdots + x_{n,i}^0(\vec{f}_{2n}^* + a\vec{f}_n^*))} \\
&= g_1^{\tilde{\alpha}_i(x_{1,i}^0 \vec{b}_{n+1}^* + \cdots + x_{n,i}^0 \vec{b}_{2n}^*)} \\
C_{2,i} &= g_1^{\tilde{\alpha}_i \vec{h}_2^*}(g_1^a)^{\tilde{\alpha}_i \vec{h}_1^*} \\
&= g_1^{\tilde{\alpha}_i(\vec{h}_2^* + a\vec{h}_1^*)} \\
&= g_1^{\tilde{\alpha}_i \vec{d}_2^*}
\end{aligned}
$$

a properly distributed type-$(\vec{0}, \vec{x}_i^0)$ ciphertext (as expected in both $\text{Game}_{j,Z}^1$ and $\text{Game}_{j,Z}^2$). Note that this construction is the same as that of the first $j-1$ ciphertexts except for the $\vec{x}_i^0$ components used instead of $\vec{x}_i^1$.

For the jth ciphertext, compute:

$$C_{1,j} = (g_1^b)^{x_{1,j}^0 \vec{f}_{n+1}^* + \cdots + x_{n,j}^0 \vec{f}_{2n}^*} (T)^{x_{1,j}^0 \vec{f}_1^* + \cdots + x_{n,j}^0 \vec{f}_n^*}$$

$$= g_1^{r(x_{1,j}^0 \vec{f}_1^* + \cdots + x_{n,j}^0 \vec{f}_n^*) + b(x_{1,j}^0 (\vec{f}_{n+1}^* + a\vec{f}_1^*) + \cdots + x_{n,j}^0 (\vec{f}_{2n}^* + a\vec{f}_n^*))}$$

$$= g_1^{r(x_{1,j}^0 \vec{b}_1^* + \cdots + x_{n,j}^0 \vec{b}_n^*) + b(x_{1,j}^0 \vec{b}_{n+1}^* + \cdots + x_{n,j}^0 \vec{b}_{2n}^*)}$$

$$C_{2,j} = (g_1^b)^{\vec{h}_2^*} (T)^{\vec{h}_1^*}$$

$$= g_1^{r\vec{h}_1^* + b(\vec{h}_2^* + a\vec{h}_1^*)}$$

$$= g_1^{r\vec{d}_1^* + b\vec{d}_2^*}$$

When $r = 0$, this is a properly distributed type-$(\vec{0}, \vec{x}_j^0)$ ciphertext where $\tilde{\alpha}_j = b$ (as would be expected in $\text{Game}_{j,Z}^1$).

When $r \leftarrow \mathbb{Z}_p$, this is a properly distributed type-$(\vec{x}_j^0, \vec{x}_j^0)$ ciphertext where $\alpha_j = r$ and $\tilde{\alpha}_j = b$ (as would be expected in $\text{Game}_{j,Z}^2$).

Our simulation is therefore identical to either $\text{Game}_{j,Z}^1$ or $\text{Game}_{j,Z}^2$ depending on the SXDH challenge $T = g_1^{ab+r}$ having $r = 0$ or $r \leftarrow \mathbb{Z}_p$ respectively. Therefore, by playing the security game in this manner with the supposed attacker and using the attacker's output as an answer to the SXDH challenge, we will enjoy the same non-negligible advantage as the supposed attacker in deciding the SXDH problem.

By the SXDH assumption, this is not possible, so no such adversary can exist.

Lemma 3. *No attacker can achieve a non-negligible difference in advantage between $\text{Game}_{j,Z}^2$ and $\text{Game}_{j,Z}^3$ for $j = 1, ..., Q_1$.*

Proof. This lemma is unconditionally true because both games are information-theoretically identical.

In the simulation of the security $\text{Game}_{j,Z}^2$, one draws bases $(\mathbb{B}, \mathbb{B}^*) \leftarrow Dual(\mathbb{Z}_p^{2n})$. However, imagine knowing the jth ciphertext vectors $\vec{x}_j^0, \vec{x}_j^1$ ahead of time and drawing the bases in the following way: first, draw $\mathbb{B}, \mathbb{B}^* \leftarrow Dual(\mathbb{Z}_p^{2n})$ then apply following invertible linear transformation to get $(\mathbb{F}, \mathbb{F}^*)$, which is used (along with a normally drawn $(\mathbb{D}, \mathbb{D}^*) \leftarrow Dual(\mathbb{Z}_p^2)$) as the basis:

$$\vec{f}_i = \vec{b}_i \text{ for } i = 1, ..., n$$

$$\vec{f}_{n+i} = \vec{b}_{n+i} + \frac{\tilde{\alpha}_j(x_{i,j}^0 - x_{i,j}^1)}{\alpha_j x_{1,j}^0} \vec{b}_1 \text{ for } i = 1, ..., n$$

$$\vec{f}_1^* = \vec{b}_1^* - \sum_{i=1}^{n} \frac{\tilde{\alpha}_j(x_{i,j}^0 - x_{i,j}^1)}{\alpha_j x_{1,j}^0} \vec{b}_{n+i}^*$$

$$\vec{f}_i^* = \vec{b}_i^* \text{ for } i = 2, ..., 2n$$

where $\alpha_j, \tilde{\alpha}_j$ are randomly drawn values used in the creation of the jth cipher-text. Recall that the distribution of the $(\mathbb{F}, \mathbb{F}^*)$ produced this way is identical to that produced by $Dual(\mathbb{Z}_p^{2n}))$.

Consider simulating $\text{Game}_{j,Z}^2$ with this basis. Any key (for, say, vector $\vec{y}^0$) looks like:

$$K_1 = g_2^{\beta(y_1^0 \vec{f}_1 + \cdots + y_n^0 \vec{f}_n) + \tilde{\beta}(y_1^0 \vec{f}_{n+1} + \cdots + y_n^0 \vec{f}_{2n})}$$

$$K_2 = g_2^{\beta \vec{d}_1 + \tilde{\beta} \vec{d}_2}.$$

where:

$$K_1 = g_2^{\beta(y_1^0 \vec{f}_1 + \cdots + y_n^0 \vec{f}_n) + \tilde{\beta}(y_1^0 \vec{f}_{n+1} + \cdots + y_n^0 \vec{f}_{2n})}$$

$$= g_2^{\beta(y_1^0 \vec{b}_1 + \cdots + y_n^0 \vec{b}_n) + \tilde{\beta}(y_1^0(\vec{b}_{n+1} + \frac{\tilde{\alpha}_j(x_{1,j}^0 - x_{1,j}^1)}{\alpha_j x_{1,j}^0} \vec{b}_1) + \cdots + y_n^0(\vec{b}_{2n} + \frac{\tilde{\alpha}_j(x_{n,j}^0 - x_{n,j}^1)}{\alpha_j x_{1,j}^0} \vec{b}_1))}$$

$$= g_2^{\beta(y_1^0 \vec{b}_1 + \cdots + y_n^0 \vec{b}_n) + \tilde{\beta}(y_1^0 \vec{b}_{n+1} + \cdots + y_n^0 \vec{b}_{2n}) + \frac{\tilde{\beta}\tilde{\alpha}_j \langle \vec{y}^0, \vec{x}_j^0 - \vec{x}_j^1 \rangle}{\alpha_j x_{1,j}^0} \vec{b}_1}$$

$$= g_2^{\beta(y_1^0 \vec{b}_1 + \cdots + y_n^0 \vec{b}_n) + \tilde{\beta}(y_1^0 \vec{b}_{n+1} + \cdots + y_n^0 \vec{b}_{2n})}$$

The last step above comes from the fact that $\langle \vec{y}^0, \vec{x}_j^0 - \vec{x}_j^1 \rangle = 0$.
($\vec{x}_j^0, \vec{x}_j^1$ are vectors requested in the game where we require $\langle \vec{y}^0, \vec{x}_j^0 \rangle = \langle \vec{y}^0, \vec{x}_j^1 \rangle \implies \langle \vec{y}^0, \vec{x}_j^0 - \vec{x}_j^1 \rangle = 0$). This is a type-$(\vec{y}^0, \vec{y}^0)$ key in the $(\mathbb{B}, \mathbb{B}^*)$ basis (as expected in both $\text{Game}_{j,Z}^2$ and $\text{Game}_{j,Z}^3$).

All ciphertexts created before the jth ciphertext look like properly distributed type-$(\vec{0}, \vec{x}_i^1)$ ciphertexts in the $(\mathbb{B}, \mathbb{B}^*)$ basis:

$$C_{1,i} = g_1^{\tilde{\alpha}_i(x_{1,i}^1 \vec{f}_{n+1}^* + \cdots + x_{n,i}^1 \vec{f}_{2n}^*)}$$

$$= g_1^{\tilde{\alpha}_i(x_{1,i}^1 \vec{b}_{n+1}^* + \cdots + x_{n,i}^1 \vec{b}_{2n}^*)}$$

$$C_{2,i} = g_1^{\tilde{\alpha}_i \vec{d}_2^*}$$

Similarly, all ciphertexts created after the jth ciphertext look like properly distributed type-$(\vec{0}, \vec{x}_i^0)$ ciphertexts in the $(\mathbb{B}, \mathbb{B}^*)$ basis:

$$C_{1,i} = g_1^{\tilde{\alpha}_i(x_{1,i}^0 \vec{f}_{n+1}^* + \cdots + x_{n,i}^0 \vec{f}_{2n}^*)}$$

$$= g_1^{\tilde{\alpha}_i(x_{1,i}^0 \vec{b}_{n+1}^* + \cdots + x_{n,i}^0 \vec{b}_{2n}^*)}$$

$$C_{2,i} = g_1^{\tilde{\alpha}_i \vec{d}_2^*}$$

However, the jth ciphertext constructed (as a type-$(\vec{x}_j^0, \vec{x}_j^0)$ ciphertext) looks like a type-$(\vec{x}_j^0, \vec{x}_j^1)$ ciphertext in the $(\mathbb{B}, \mathbb{B}^*)$ basis:

$$C_{1,j} = g_1^{\alpha_j(x_{1,j}^0 \vec{f}_1^* + \cdots + x_{n,j}^0 \vec{f}_n^*) + \tilde{\alpha}_j(x_{1,j}^0 \vec{f}_{n+1}^* + \cdots + x_{n,j}^0 \vec{f}_{2n}^*)}$$

$$\alpha_j (x_{1,j}^0 (\vec{b}_1^* - \sum_{i=1}^{n} \frac{\tilde{\alpha}_j (x_{i,j}^0 - x_{i,j}^1)}{\alpha_j x_{1,j}^0} \vec{b}_{n+i}^*) + \cdots + x_{n,j}^0 \vec{b}_n^*) + \tilde{\alpha}_j (x_{1,j}^0 \vec{b}_{n+1}^* + \cdots + x_{n,j}^0 \vec{b}_{2n}^*)$$
$$= g_1$$

$$= g_1^{\alpha_j (x_{1,j}^0 \vec{b}_1^* + \cdots + x_{n,j}^0 \vec{b}_n^*) + \tilde{\alpha}_j (x_{1,j}^1 \vec{b}_{n+1}^* + \cdots + x_{n,j}^1 \vec{b}_{2n}^*)}$$

$$C_{2,j} = g_1^{\alpha_j \vec{d}_1^* + \tilde{\alpha}_j \vec{d}_2^*}$$

This construction of the jth ciphertext is the only difference between $\text{Game}_{j,Z}^2$ and $\text{Game}_{j,Z}^3$. So, drawing $(\mathbb{B}, \mathbb{B}^*)$ directly from $Dual(\mathbb{Z}_p^{2n})$ and using it to play $\text{Game}_{j,Z}^2$ results in $\text{Game}_{j,Z}^2$ in the $(\mathbb{B}, \mathbb{B}^*)$ basis. However, transforming this basis to $(\mathbb{F}, \mathbb{F}^*)$ and using it instead results in playing $\text{Game}_{j,Z}^3$ with the $(\mathbb{B}, \mathbb{B}^*)$ basis. However, since $(\mathbb{B}, \mathbb{B}^*)$ and $(\mathbb{F}, \mathbb{F}^*)$ have the same distribution (the one produced by $Dual(\mathbb{Z}_p^{2n})$), this means that the two Games are actually information-theoretically identical. Therefore, no attacker can achieve non-negligible difference advantage in distinguishing between $\text{Game}_{j,Z}^2$ and $\text{Game}_{j,Z}^3$.

Lemma 4. *No polynomial-time attacker can achieve a non-negligible difference in advantage between $\text{Game}_{j,Z}^3$ and $\text{Game}_{j,Z}^4$ for $j = 1, ..., Q_1$ under the SXDH assumption.*

Proof. Given an attacker that achieves non-negligible difference in advantage between $\text{Game}_{j,Z}^3$ and $\text{Game}_{j,Z}^4$ for some $j \in [1, Q_1]$, we could achieve non-negligible advantage in deciding the SXDH problem as follows:

Given SXDH instance $g_1, g_1^a, g_1^b, T = g_1^{ab+r}, g_2$ where either $r \leftarrow \mathbb{Z}_p$ or $r = 0$, use g_1, g_2 as the generators of the same name used to form ciphertexts and keys respectively. Generate bases $(\mathbb{F}, \mathbb{F}^*) \leftarrow Dual(\mathbb{Z}_p^{2n})$, $(\mathbb{H}, \mathbb{H}^*) \leftarrow Dual(\mathbb{Z}_p^2)$ and implicitly define new bases $(\mathbb{B}, \mathbb{B}^*)$, $(\mathbb{D}, \mathbb{D}^*)$ as the following:

$$\vec{b}_i = \vec{f}_i - a\vec{f}_{n+i} \text{ for } i = 1, ..., n$$
$$\vec{b}_{n+i} = \vec{f}_{n+i} \text{ for } i = 1, ..., n$$
$$\vec{b}_i^* = \vec{f}_i^* \text{ for } i = 1, ..., n$$
$$\vec{b}_{n+i}^* = \vec{f}_{n+i}^* + a\vec{f}_i^* \text{ for } i = 1, ..., n$$

$$\vec{d}_1 = \vec{h}_1 - a\vec{h}_2$$
$$\vec{d}_2 = \vec{h}_2$$
$$\vec{d}_1^* = \vec{h}_1^*$$
$$\vec{d}_2^* = \vec{h}_2^* + a\vec{h}_1^*$$

Note that these bases are distributed exactly the same as those output by $Dual(\mathbb{Z}_p^{2n})$ and $Dual(\mathbb{Z}_p^2)$ respectively (they are created by applying an invertible linear transformation to the output of $Dual(\mathbb{Z}_p^{2n})$ and $Dual(\mathbb{Z}_p^2)$).

To construct any key (for, say, vector $\vec{y}^0$), generate random $\beta, \tilde{\beta}' \leftarrow \mathbb{Z}_p$, implicitly define $\tilde{\beta} = \beta a + \tilde{\beta}'$, and compute:

$$
\begin{aligned}
K_1 &= g_2^{\beta(y_1^0 \vec{f}_1 + \cdots + y_n^0 \vec{f}_n) + \tilde{\beta}'(y_1^0 \vec{f}_{n+1} + \cdots + y_n^0 \vec{f}_{2n})} \\
&= g_2^{\beta(y_1^0 \vec{f}_1 + \cdots + y_n^0 \vec{f}_n) + (\tilde{\beta} - \beta a)(y_1^0 \vec{f}_{n+1} + \cdots + y_n^0 \vec{f}_{2n})} \\
&= g_2^{\beta(y_1^0(\vec{f}_1 - a\vec{f}_{n+1}) + \cdots + y_n^0(\vec{f}_n - a\vec{f}_{2n})) + \tilde{\beta}(y_1^0 \vec{f}_{n+1} + \cdots + y_n^0 \vec{f}_{2n})} \\
&= g_2^{\beta(y_1^0 \vec{b}_1 + \cdots + y_n^0 \vec{b}_n) + \tilde{\beta}(y_1^0 \vec{b}_{n+1} + \cdots + y_n^0 \vec{b}_{2n})} \\
K_2 &= g_2^{\beta \vec{h}_1 + \tilde{\beta}' \vec{h}_2} \\
&= g_2^{\beta \vec{h}_1 + (\tilde{\beta} - \beta a)\vec{h}_2} \\
&= g_2^{\beta(\vec{h}_1 - a\vec{h}_2) + \tilde{\beta}\vec{h}_2} \\
&= g_2^{\beta \vec{d}_1 + \tilde{\beta}\vec{d}_2}
\end{aligned}
$$

a properly distributed type-$(\vec{y}^0, \vec{y}^0)$ key (as expected in both $\text{Game}_{j,Z}^3$ and $\text{Game}_{j,Z}^4$).

For the jth ciphertext, draw $\alpha_j, \tilde{\alpha}_j \leftarrow \mathbb{Z}_p$ and compute:

$$
\begin{aligned}
C_{1,j} &= g_1^{\alpha_j(x_{1,j}^0 \vec{f}_1^* + \cdots + x_{n,j}^0 \vec{f}_n^*) + \tilde{\alpha}_j(x_{1,j}^1 \vec{f}_{n+1}^* + \cdots + x_{n,j}^1 \vec{f}_{2n}^*)} (g_1^a)^{\tilde{\alpha}_j(x_{1,j}^1 \vec{f}_1^* + \cdots + x_{n,j}^1 \vec{f}_n^*)} \\
&= g_1^{\alpha_j(x_{1,j}^0 \vec{f}_1^* + \cdots + x_{n,j}^0 \vec{f}_n^*) + \tilde{\alpha}_j(x_{1,j}^1(\vec{f}_{n+1}^* + a\vec{f}_1^*) + \cdots + x_{n,j}^1(\vec{f}_{2n}^* + a\vec{f}_n^*))} \\
&= g_1^{\alpha_j(x_{1,j}^0 \vec{b}_1^* + \cdots + x_{n,j}^0 \vec{b}_n^*) + \tilde{\alpha}_j(x_{1,j}^1 \vec{b}_{n+1}^* + \cdots + x_{n,j}^1 \vec{b}_{2n}^*)} \\
C_{2,j} &= g_1^{\alpha_j \vec{h}_1^*} g_1^{\tilde{\alpha}_j \vec{h}_2^*} (g_1^a)^{\tilde{\alpha}_j \vec{h}_1^*} \\
&= g_1^{\alpha_j \vec{h}_1^* + \tilde{\alpha}_j(\vec{h}_2^* + a\vec{h}_1^*)} \\
&= g_1^{\alpha_j \vec{d}_1^* + \tilde{\alpha}_j \vec{d}_2^*}
\end{aligned}
$$

a properly distributed type-$(\vec{x}_j^0, \vec{x}_j^1)$ ciphertext (as expected in both $\text{Game}_{j,Z}^3$ and $\text{Game}_{j,Z}^4$).

For all ciphertexts after the jth ciphertext, draw $\alpha_i', \tilde{\alpha}_i' \leftarrow \mathbb{Z}_p$ and compute:

$$
\begin{aligned}
C_{1,i} &= (g_1^b)^{\alpha_i'(x_{1,i}^0 \vec{f}_{n+1}^* + \cdots + x_{n,i}^0 \vec{f}_{2n}^*)} (T)^{\alpha_i'(x_{1,i}^0 \vec{f}_1^* + \cdots + x_{n,i}^0 \vec{f}_n^*)} g_1^{\tilde{\alpha}_i'(x_{1,i}^0 \vec{f}_{n+1}^* + \cdots + x_{n,i}^0 \vec{f}_{2n}^*)} \\
&\quad \cdot (g_1^a)^{\tilde{\alpha}_i'(x_{1,i}^0 \vec{f}_1^* + \cdots + x_{n,i}^0 \vec{f}_n^*)} \\
&= g_1^{\alpha_i' r(x_{1,i}^0 \vec{f}_1^* + \cdots + x_{n,i}^0 \vec{f}_n^*) + (\tilde{\alpha}_i' + \alpha_i' b)(x_{1,i}^0(\vec{f}_{n+1}^* + a\vec{f}_1^*) + \cdots + x_{n,i}^0(\vec{f}_{2n}^* + a\vec{f}_n^*))} \\
&= g_1^{\alpha_i' r(x_{1,i}^0 \vec{b}_1^* + \cdots + x_{n,i}^0 \vec{b}_n^*) + (\tilde{\alpha}_i' + \alpha_i' b)(x_{1,i}^0 \vec{b}_{n+1}^* + \cdots + x_{n,i}^0 \vec{b}_{2n}^*)} \\
C_{2,i} &= (g_1^b)^{\alpha_i' \vec{h}_2^*} (T)^{\alpha_i' \vec{h}_1^*} g_1^{\tilde{\alpha}_i' \vec{h}_2^*} (g_1^a)^{\tilde{\alpha}_i' \vec{h}_1^*} \\
&= g_1^{\alpha_i' r \vec{h}_1^* + (\tilde{\alpha}_i' + \alpha_i' b)(\vec{h}_2^* + a\vec{h}_1^*)} \\
&= g_1^{\alpha_i' r \vec{d}_1^* + (\tilde{\alpha}_i' + \alpha_i' b)\vec{d}_2^*}
\end{aligned}
$$

For ciphertexts before the jth ciphertext, draw $\alpha'_i, \tilde{\alpha}'_i \leftarrow \mathbb{Z}_p$ and compute:

$$C_{1,i} = (g_1^b)^{\alpha'_i(x^1_{1,i}\vec{f}^*_{n+1}+\cdots+x^1_{n,i}\vec{f}^*_{2n})}(T)^{\alpha'_i(x^1_{1,i}\vec{f}^*_{1}+\cdots+x^1_{n,i}\vec{f}^*_{n})}g_1^{\tilde{\alpha}'_i(x^1_{1,i}\vec{f}^*_{n+1}+\cdots+x^1_{n,i}\vec{f}^*_{2n})}$$

$$\cdot (g_1^a)^{\tilde{\alpha}'_i(x^1_{1,i}\vec{f}^*_{1}+\cdots+x^1_{n,i}\vec{f}^*_{n})}$$

$$= g_1^{\alpha'_i r(x^1_{1,i}\vec{f}^*_{1}+\cdots+x^1_{n,i}\vec{f}^*_{n})+(\tilde{\alpha}'_i+\alpha'_i b)(x^1_{1,i}(\vec{f}^*_{n+1}+a\vec{f}^*_{1})+\cdots+x^1_{n,i}(\vec{f}^*_{2n}+a\vec{f}^*_{n}))}$$

$$= g_1^{\alpha'_i r(x^1_{1,i}\vec{b}^*_{1}+\cdots+x^1_{n,i}\vec{b}^*_{n})+(\tilde{\alpha}'_i+\alpha'_i b)(x^1_{1,i}\vec{b}^*_{n+1}+\cdots+x^1_{n,i}\vec{b}^*_{2n})}$$

$$C_{2,i} = (g_1^b)^{\alpha'_i \vec{h}^*_{2}}(T)^{\alpha'_i \vec{h}^*_{1}}g_1^{\tilde{\alpha}'_i \vec{h}^*_{2}}(g_1^a)^{\tilde{\alpha}'_i \vec{h}^*_{1}}$$

$$= g_1^{\alpha'_i r\vec{h}^*_{1}+(\tilde{\alpha}'_i+\alpha'_i b)(\vec{h}^*_{2}+a\vec{h}^*_{1})}$$

$$= g_1^{\alpha'_i r\vec{d}^*_{1}+(\tilde{\alpha}'_i+\alpha'_i b)\vec{d}^*_{2}}$$

(The only difference from the prior ciphertext construction is that $\vec{x}^1$ is used instead of $\vec{x}^0$).

When $r = 0$, all ciphertexts before the jth ciphertext are properly distributed type-$(\vec{0}, \vec{x}^1_i)$ ciphertexts and all ciphertexts after the jth are properly distributed type-$(\vec{0}, \vec{x}^0_i)$ ciphertexts where $\tilde{\alpha}_i = \tilde{\alpha}'_i + \alpha'_i b$. This is as would be expected in $\text{Game}^3_{j,Z}$.

When $r \leftarrow \mathbb{Z}_p$, all ciphertexts before the jth are properly distributed type-$(\vec{x}^1_i, \vec{x}^1_i)$ ciphertexts and all ciphertexts after the jth are properly distributed type-$(\vec{x}^0_i, \vec{x}^0_i)$ ciphertexts where $\alpha_i = \alpha'_i r$ and $\tilde{\alpha}_i = \tilde{\alpha}'_i + \alpha'_i b$. This is as would be expected in $\text{Game}^4_{j,Z}$.

Our simulation is therefore identical to either $\text{Game}^3_{j,Z}$ or $\text{Game}^4_{j,Z}$ depending on the SXDH challenge $T = g_1^{ab+r}$ having $r = 0$ or $r \leftarrow \mathbb{Z}_p$ respectively. Therefore, by playing the security game in this manner with the supposed attacker and using the attacker's output as an answer to the SXDH challenge, we will enjoy the same non-negligible advantage as the supposed attacker in deciding the SXDH problem.

By the SXDH assumption, this is not possible, so no such adversary can exist.

Lemma 5. *No polynomial-time attacker can achieve a non-negligible difference in advantage between $\text{Game}^4_{j,Z}$ and $\text{Game}^5_{j,Z}$ for $j = 1, ..., Q_1$ under the SXDH assumption.*

Proof. The proof of this lemma is symmetric to that of the previous Lemma 4, just flipping the role of the two parallel bases.

Lemma 4 showed how to create keys of type-$(\vec{y}^0, \vec{y}^0)$ and a type-$(\vec{x}^0_j, \vec{x}^1_j)$ ciphertext while having the ciphertexts before and after the jth be type-$(\vec{0}, \vec{x}^1_i)$ and type-$(\vec{0}, \vec{x}^0_i)$ or type-$(\vec{x}^1_i, \vec{x}^1_i)$ and type-$(\vec{x}^0_i, \vec{x}^0_i)$ respectively based on the challenge elements of the SXDH problem. By symmetrically applying the same embedding to the opposite halves of the parallel bases, we can achieve the same result, but with the ciphertexts before and after the jth being type-$(\vec{x}^1_i, \vec{x}^1_i)$ and type-$(\vec{x}_i^0, \vec{x}_i^0)$ or type-$(\vec{x}_i^1, \vec{0})$ and type-$(\vec{x}_i^0, \vec{0})$ respectively based on the challenge elements of the SXDH problem, which is what we need for this transition.

Lemma 6. *No attacker can achieve a non-negligible difference in advantage between $\text{Game}_{j,Z}^5$ and $\text{Game}_{j,Z}^6$ for $j = 1, ..., Q_1$.*

Proof. The proof of this lemma is information-theoretic and similarly symmetric to that of Lemma 3, just flipping the role of the two parallel bases (just like the previous lemma did with Lemma 4).

Lemma 7. *No polynomial-time attacker can achieve a non-negligible difference in advantage between $\text{Game}_{j,Z}^6$ and $\text{Game}_{j,Z}^7$ for $j = 1, ..., Q_1$ under the SXDH assumption.*

Proof. The proof of this lemma is nearly identical to that of Lemma 5 (which transitioned ciphertexts before the jth between type-$(\vec{x}_i^1, \vec{0})$ and type-$(\vec{x}_i^1, \vec{x}_i^1)$ and ciphertexts after the jth between type-$(\vec{x}_i^0, \vec{0})$ and type-$(\vec{x}_i^0, \vec{x}_i^0)$) but instead constructing the jth ciphertext as type-$(\vec{x}_i^1, \vec{x}_i^1)$ instead of type-$(\vec{x}_i^0, \vec{x}_i^1)$.

The previous lemmas let us transition to $\text{Game}_{Q_1,Z}^7$, where all ciphertexts are of type-$(\vec{x}_i^1, \vec{x}_i^1)$ and all keys are type-$(\vec{y}_i^0, \vec{y}_i^0)$ keys. Notice that $\text{Game}_{Q_1,Z}^7$ is identical to $\text{Game}_{O,0}^7$. It is easy to see that we can now use a symmetric set of hybrids to transition all keys to type-$(\vec{y}_i^1, \vec{y}_i^1)$ in a similar manner: starting by transitioning from $\text{Game}_{Q_1,Z}^7 = \text{Game}_{O,0}^7$ to $\text{Game}_{O,1}^7$, then proceeding through the $\text{Game}_{O,j}^i$ using the same methods as in the $\text{Game}_{j,Z}^i$, just switching the roles of the basis vectors (switching all $\vec{b}^*$ with $\vec{b}$, $\vec{d}^*$ with $\vec{d}$, α with β, $\tilde{\alpha}$ with $\tilde{\beta}$, and $\vec{x}_i$ with $\vec{y}_i$, while always producing type $(\vec{x}_i^1, \vec{x}_i^1)$ ciphertexts (instead of always producing type $(\vec{y}_i^0, \vec{y}_i^0)$ keys).

These symmetric arguments let us transition to Game_{O,Q_2}^7, where all ciphertexts are of type-$(\vec{x}^1, \vec{x}^1)$ and all keys are type-$(\vec{y}_i^1, \vec{y}_i^1)$ keys. This is identical to the original security game when $b = 1$. So, by a hybrid argument, we have shown that no polynomial-time attacker gan achieve non-negligible difference in advantage in the security game when $b = 0$ ($\text{Game}_{0,Z}^7$) vs when $b = 1$ (Game_{O,Q_2}^7) under the SXDH assumption, so our construction is therefore secure.

References

1. Abdalla, M., Bourse, F., De Caro, A., Pointcheval, D.: Simple functional encryption schemes for inner products. In: Katz, J. (ed.) PKC 2015. LNCS, vol. 9020, pp. 733–751. Springer, Heidelberg (2015)
2. Agrawal, S., Agrawal, S., Badrinarayanan, S., Kumarasubramanian, A., Prabhakaran, M., Sahai, A.: On the practical security of inner product functional encryption. In: Katz, J. (ed.) PKC 2015. LNCS, vol. 9020, pp. 777–798. Springer, Heidelberg (2015)
3. Agrawal, S., Freeman, D.M., Vaikuntanathan, V.: Functional encryption for inner product predicates from learning with errors. In: Lee, D.H., Wang, X. (eds.) ASIACRYPT 2011. LNCS, vol. 7073, pp. 21–40. Springer, Heidelberg (2011)

4. Ananth, P., Brakerski, Z., Segev, G., Vaikuntanathan, V.: From selective to adaptive security in functional encryption. In: Gennaro, R., Robshaw, M. (eds.) CRYPTO 2015. LNCS, vol. 9216, pp. 657–677. Springer, Heidelberg (2015)

5. Barak, B., Goldreich, O., Impagliazzo, R., Rudich, S., Sahai, A., Vadhan, S.P., Yang, K.: On the (im)possibility of obfuscating programs. In: Kilian, J. (ed.) CRYPTO 2001. LNCS, vol. 2139, pp. 1–18. Springer, Heidelberg (2001)

6. Boneh, D., Raghunathan, A., Segev, G.: Function-private identity-based encryption: hiding the function in functional encryption. In: Canetti, R., Garay, J.A. (eds.) CRYPTO 2013, Part II. LNCS, vol. 8043, pp. 461–478. Springer, Heidelberg (2013)

7. Boneh, D., Raghunathan, A., Segev, G.: Function-private subspace-membership encryption and its applications. In: Sako, K., Sarkar, P. (eds.) ASIACRYPT 2013, Part I. LNCS, vol. 8269, pp. 255–275. Springer, Heidelberg (2013)

8. Boneh, D., Sahai, A., Waters, B.: Functional encryption: definitions and challenges. In: Ishai, Y. (ed.) TCC 2011. LNCS, vol. 6597, pp. 253–273. Springer, Heidelberg (2011)

9. Brakerski, Z., Segev, G.: Function-private functional encryption in the private-key setting. In: Dodis, Y., Nielsen, J.B. (eds.) TCC 2015, Part II. LNCS, vol. 9015, pp. 306–324. Springer, Heidelberg (2015)

10. De Caro, A., Iovino, V., Jain, A., O'Neill, A., Paneth, O., Persiano, G.: On the achievability of simulation-based security for functional encryption. In: Canetti, R., Garay, J.A. (eds.) CRYPTO 2013, Part II. LNCS, vol. 8043, pp. 519–535. Springer, Heidelberg (2013)

11. Garg, S., Gentry, C., Halevi, S., Raykova, M., Sahai, A., Waters, B.: Candidate indistinguishability obfuscation and functional encryption for all circuits. In: FOCS, pp. 40–49 (2013)

12. Garg, S., Gentry, C., Halevi, S., Zhandry, M.: Fully secure functional encryption without obfuscation. Cryptology ePrint Archive, report 2014/666 (2014). http://eprint.iacr.org/

13. Gentry, C.: Fully homomorphic encryption using ideal lattices. In: STOC, p. 169–178 (2009)

14. Goldwasser, S., Kalai, Y.T., Popa, R.A., Vaikuntanathan, V., Zeldovich, N.: How to run turing machines on encrypted data. In: Canetti, R., Garay, J.A. (eds.) CRYPTO 2013, Part II. LNCS, vol. 8043, pp. 536–553. Springer, Heidelberg (2013)

15. Gorbunov, S., Vaikuntanathan, V., Wee, H.: Functional encryption with bounded collusions via multi-party computation. In: Safavi-Naini, R., Canetti, R. (eds.) CRYPTO 2012. LNCS, vol. 7417, pp. 162–179. Springer, Heidelberg (2012)

16. Goyal, V., Pandey, O., Sahai, A., Waters, B.: Attribute based encryption for fine-grained access control of encrypted data. In: ACM conference on Computer and Communications Security, pp. 89–98 (2006)

17. Katz, J., Sahai, A., Waters, B.: Predicate encryption supporting disjunctions, polynomial equations, and inner products. In: Smart, N.P. (ed.) EUROCRYPT 2008. LNCS, vol. 4965, pp. 146–162. Springer, Heidelberg (2008)

18. Lewko, A., Okamoto, T., Sahai, A., Takashima, K., Waters, B.: Fully secure functional encryption: attribute-based encryption and (hierarchical) inner product encryption. In: Gilbert, H. (ed.) EUROCRYPT 2010. LNCS, vol. 6110, pp. 62–91. Springer, Heidelberg (2010)

19. Okamoto, T., Takashima, K.: Homomorphic encryption and signatures from vector decomposition. In: Galbraith, S.D., Paterson, K.G. (eds.) Pairing 2008. LNCS, vol. 5209, pp. 57–74. Springer, Heidelberg (2008)

20. Okamoto, T., Takashima, K.: Hierarchical predicate encryption for inner-products. In: Matsui, M. (ed.) ASIACRYPT 2009. LNCS, vol. 5912, pp. 214–231. Springer, Heidelberg (2009)
21. Okamoto, T., Takashima, K.: Fully secure functional encryption with general relations from the decisional linear assumption. In: Rabin, T. (ed.) CRYPTO 2010. LNCS, vol. 6223, pp. 191–208. Springer, Heidelberg (2010)
22. Okamoto, T., Takashima, K.: Fully secure unbounded inner-product and attribute-based encryption. In: Wang, X., Sako, K. (eds.) ASIACRYPT 2012. LNCS, vol. 7658, pp. 349–366. Springer, Heidelberg (2012)
23. O'Neill, A.: Definitional issues in functional encryption. IACR Cryptology ePrint Archive (2010)
24. Sahai, A., Seyalioglu, H.: Worry-free encryption: functional encryption with public keys. In CCS, pp. 463–472 (2010)
25. Sahai, A., Waters, B.: Fuzzy identity-based encryption. In: Cramer, R. (ed.) EUROCRYPT 2005. LNCS, vol. 3494, pp. 457–473. Springer, Heidelberg (2005)
26. Shen, E., Shi, E., Waters, B.: Predicate privacy in encryption systems. In: Reingold, O. (ed.) TCC 2009. LNCS, vol. 5444, pp. 457–473. Springer, Heidelberg (2009)
27. Waters, B.: A punctured programming approach to adaptively secure functional encryption. In: Gennaro, R., Robshaw, M. (eds.) CRYPTO 2015. LNCS, vol. 9216, pp. 678–697. Springer, Heidelberg (2015)

ABE and IBE

Idealizing Identity-Based Encryption

Dennis Hofheinz[1], Christian Matt[2 (✉)], and Ueli Maurer[2]

[1] Karlsruhe Institute of Technology (KIT), Karlsruhe, Germany
dennis.hofheinz@kit.edu
[2] Department of Computer Science, ETH Zurich, Zurich, Switzerland
{mattc,maurer}@inf.ethz.ch

Abstract. We formalize the standard application of identity-based encryption (IBE), namely non-interactive secure communication, as realizing an ideal system which we call delivery controlled channel (DCC). This system allows users to be registered (by a central authority) for an identity and to send messages securely to other users only known by their identity.

Quite surprisingly, we show that existing security definitions for IBE are not sufficient to realize DCC. In fact, it is impossible to do so in the standard model. We show, however, how to adjust any IBE scheme that satisfies the standard security definition IND-ID-CPA to achieve this goal in the random oracle model.

We also show that the impossibility result can be avoided in the standard model by considering a weaker ideal system that requires all users to be registered in an initial phase before any messages are sent. To achieve this, a weaker security notion, which we introduce and call IND-ID1-CPA, is actually sufficient. This justifies our new security definition and might open the door for more efficient schemes. We further investigate which ideal systems can be realized with schemes satisfying the standard notion and variants of selective security.

As a contribution of independent interest, we show how to model features of an ideal system that are potentially available to dishonest parties but not guaranteed, and which such features arise when using IBE.

Keywords: Identity-based encryption · Definitions · Impossibility results · Composability

1 Introduction

1.1 Motivation

Identity-based encryption (IBE) is a generalization of public-key encryption where messages can be encrypted using a master public key and the *identity* of a user, which can be an arbitrary bit string, such as the user's e-mail address. Ciphertexts can be decrypted with a user secret key for the corresponding identity, where user secret keys are derived from a master secret key, which is generated together with the master public key.

ⓒ International Association for Cryptologic Research 2015
T. Iwata and J.H. Cheon (Eds.): ASIACRYPT 2015, Part I, LNCS 9452, pp. 495–520, 2015.
DOI: 10.1007/978-3-662-48797-6_21

The apparent standard application of IBE is non-interactive secure communication. More specifically, we assume a setting with many parties, and the goal is to enable each party to send any other party (known only by his/her identity) messages in a secure way. This secure communication should be non-interactive (or "one-shot") in the sense that the sending party should not be required to, e.g., look up a public key of the receiving party, or to communicate in any other way (beyond of course sending one message to the receiver). In fact, our requirements and expectations can be described as follows. We define a "resource" (or "ideal functionality" [1,5,10,12,13,15,19]) that provides the following basic services (via appropriate calls to the resource):

Registration. Each party is able to *register* his/her identity *id*. (Intuitively, an identity could be an email address or telephone number, that—presumably uniquely—identifies the registering party.)

Communication. Each party is able to *send* a message *m* to another party with identity *id*.

While an IBE scheme can be used in an obvious way to syntactically realize this functionality, the application is only secure if the IBE scheme satisfies a suitable security definition. Investigating the suitability of different security definitions for this task is the purpose of this paper.

The Semantics of Security Definitions. We point out that security definitions for cryptographic primitives can serve two entirely different purposes, which are often not clearly distinguished. The first is to serve as a (technical) reference point, on one hand for devising schemes provably satisfying the definition based on a weak assumption, and on the other hand for building more sophisticated primitives from any scheme satisfying the definition. For instance, the one-way function definition serves this purpose excellently.

In this work, we are interested in a second purpose of security definitions, namely assuring the security of a certain type of application when a scheme satisfying the (technical) security definition is used. While definitions are usually devised with much intuition for what is needed in a certain application, a conventional technical security definition for a cryptographic primitive generally cannot directly imply the security of an associated application. Guaranteeing the security of an application can be seen as giving an application-semantics to a security definition.

1.2   Identity-Based Encryption and Its Security

The concept of identity-based encryption has been conceived as early as 1984 [20]. A first candidate of an IBE scheme was presented in 1991 in [14], although without a detailed security model. In the 2000s, however, both a detailed security model [3] and a number of concrete IBE schemes (with security proofs under various assumptions) emerged, e.g., [3,7,9,21].

Both standard IBE security notions (IND-ID-CPA and IND-ID-CCA) are formalized as a security game. In this game, a hypothetical adversary $\mathcal{A}$ chooses an identity id^*, and messages m_0^* and m_1^*, and tries to distinguish an encryption of m_0^* from an encryption of m_1^* (both prepared for receiver identity id^*). Besides, $\mathcal{A}$ may (adaptively) ask for arbitrary user secret keys for identities $id \neq id^*$. (In case of IND-ID-CCA security, $\mathcal{A}$ additionally gets access to a decryption oracle for arbitrary identities.) If no efficient $\mathcal{A}$ can successfully distinguish these ciphertexts, we consider the system secure.

At this point, we note that these game-based notions of security do allow for a form of adaptivity (in the sense that $\mathcal{A}$ may adaptively ask for user secret keys), but do not directly consider a concrete communication scenario.

1.3 Contributions

In this work, we investigate the goal of non-interactive communication, and in particular the use of IBE schemes to achieve that goal. Perhaps surprisingly, it turns out that the standard notions of IBE security do *not* imply non-interactive communication in the standard model. However, we prove that standard IBE security notions do imply non-interactive communication in the random oracle model and also weaker forms of non-interactive communication in the standard model. (Loosely speaking, standard IBE security notions achieve non-interactive communication in a setting in which registrations always occur *before* any attempt is made to send messages to the respective receiving party.) Furthermore, we introduce a new security notion that is weaker than the standard notion, but still implies a very natural weaker notion of non-interactive communication in the standard model.

To formalize our results, we use the constructive cryptography (CC) framework due to Maurer and Renner [12,13]. We stress, however, that our results do not depend on that particular formal model. Specifically, the reason that standard IBE security does not imply non-interactive communication is not tied to the specifics of CC. (We give a more detailed explanation of this reason below, and we will hint at the differences to a potential formulation in Canetti's universal composability framework [5] where appropriate.)

A More Technical View. A little more technically, we model non-interactive communication as a "delivery controlled channels" resource DCC.[1] This resource has a number of interfaces, called A, $B_1, \ldots, B_n$, and C, to the involved users. Intuitively, interface C is used to register parties, A is used to send messages[2], and the interfaces B_i are used to receive messages by different parties.

More specifically, our resource admits the following types of queries:

---

[1] The name "delivery controlled channels" indicates that a user can specify (or, control) to which recipient the message should be delivered.

[2] In this work, we focus on passive attacks (i.e., on eavesdropping adversaries). In particular, we will not consider adversarially sent messages. Thus, for simplicity, we will assume that all incoming requests to *send* a message arrive at a single interface A.

- *Registration* queries (made at interface C) register an interface B_i for receiving messages sent to an identity id. (Depending on the envisioned physical registration process, the *fact* that B_i was registered under identity id may become public. We model this by leaking the pair (id, i) at all interfaces B_j.)
- *Send* queries (at interface A) send a message m to a given identity id. (The message will then be delivered to all interfaces which have been registered for this identity. Besides, any interface B_i which is *later* registered for that identity id will also receive m upon registration.)
- When thinking of an IBE scheme as realizing DCC, we cannot prevent dishonest parties from sharing their keys in the real world. As a result, also the messages sent to that party are shared with every party that got the key. Our ideal system DCC has to make this explicit, so we admit *share* queries (at any interface B_i) that cause all messages sent to this interface to be *potentially*[3] published at all other interfaces B_j that have also made a *share* query.

Furthermore, all parties (i.e., all interfaces B_i) at the beginning (potentially) receive an honestly generated random string (that corresponds to the randomness in the public master key of an IBE scheme that can potentially be extracted). We deem an IBE scheme secure if it implements this resource (when used in the straightforward way) in the sense of constructive cryptography. (In particular, this means that the view of any given party using the real IBE scheme can be simulated efficiently with access to the ideal non-interactive communication resource only.) We note that we do not model secret keys or ciphertexts in our ideal resource.

We remark that a possible ideal functionality in the UC setting would not use interfaces, but instead restrict the registration, send, and share queries to different parties. That is, only a designated "master party" could *register* other parties for receiving messages under certain identities. Every party P could *send* messages, and also issue a *share* query (with the same consequences as in our CC-based formulation).

Why Current Game-Based Definitions Do Not Realize DCC. Our first observation is that existing game-based definitions of IBE security (such as IND-ID-CPA or IND-ID-CCA) do not appear to realize the above resource. To explain the reason, suppose that one party P performs its own registration (under an arbitrary identity and at an arbitrary interface B_i) *after* messages are sent to P. (Naturally, P will not be able to receive these messages before obtaining his/her own user secret key during registration.) Now we claim that P's view in that scenario cannot be simulated efficiently. Concretely, observe that P's view with a real IBE scheme essentially consists of two elements: first, a ciphertext c of a yet-unknown message m sent by another party; and second, a user secret key usk

---

[3] Sharing is not guaranteed because our real system does not include channels between the B_i (since they are not needed). When composed with other systems, it might however be the case that such channels become available, so sharing cannot be excluded in a composable framework.

that allows to decrypt c to m. In order to simulate P's view, a simulator must thus first produce a ciphertext c at a point at which P is not registered as a receiving party. Since at that point, m is not yet known to P, c must in fact be simulated without knowledge of m. Later on, however, the simulator must also produce a user secret key usk that opens c as an encryption of m.

Put differently, the simulation thus faces a commitment problem: first, it has to commit to a ciphertext c, and later explain this ciphertext as an encryption of an arbitrary message m. For technically very similar reasons, public-key encryption cannot be simulated in the face of *adaptive* corruptions [17]. (However, we stress that in our case, no adaptive corruptions occur; see also the remark below.) As a consequence, we can show that non-interactive communication (as formalized by our resource DCC) cannot be achieved in the standard model. (We also note that this argument applies verbatim to the potential UC-based formulation sketched above.)

Weaker Notions of Non-interactive Communication. Our negative result for the above resource DCC raises the question what we can do to achieve *some* form of non-interactive communication and also what existing, game-based IBE security notions actually achieve.

Recall that the commitment problem that arises with DCC occurs only when identities are registered *after* messages have been sent to this identity. A natural way to avoid this scenario is to assume first a registration phase (in which no message transmissions are allowed), and second a transmission phase (in which no registrations are allowed). This separation into two phases can be modeled as a resource st2DCC that only allows message transmissions (and from then on ignores registration attempts) after a specific input at the "registration" interface C.[4] We can show that st2DCC *can* be achieved by IND-ID-CPA secure IBE schemes. In that sense, the commitment problem of DCC is the *only* reason why we cannot achieve that resource. Interestingly, achieving st2DCC actually corresponds to a game-based notion of IBE security that we introduce and call IND-ID1-CPA security and that is weaker than IND-ID-CPA security.

We also show that IND-ID-CPA security exactly corresponds to a resource stDCC which only allows registrations of identities to which no message has been sent so far. (In that sense, stDCC implements a "local" version of the two-phase separation of st2DCC. Again, we stress that it is the responsibility of the implementation to enforce such a local separation.)

Finally, we provide relaxed resources preDCC and pre2DCC that are "selective" versions of stDCC and st2DCC, respectively. (Here, "selective" means that the set of identities id that can be registered has to be specified initially, over interface A.) We proceed to show that resource preDCC is achieved precisely

---

[4] While this separation is easily modeled as a resource, we stress that it is the responsibility of the (designer of the) implementation to physically enforce this separation. For instance, in face of a passive adversary, such a separation into phases could be enforced simply by telling honest parties not to send any messages until the second phase.

by selective IND-ID-CPA secure IBE schemes. Similarly, the resource pre2DCC is equivalent to a selective version of the game-based notion associated with the resource st2DCC. The relations among security definitions and the achieved constructions are summarized in Fig. 1.

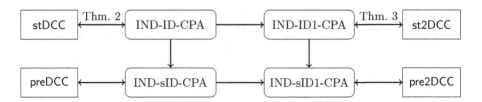

Fig. 1. Implications among security definitions and the constructed resources. Security definitions are drawn in boxes with rounded corners and resources are shown in rectangular boxes. The figure says for example that by Theorem 2, an IBE scheme can be used to construct the resource stDCC if and only if it is IND-ID-CPA secure, while IND-ID-CPA security implies IND-sID-CPA security and IND-ID1-CPA security. The equivalences of the selective security variants and the corresponding constructions are shown in the full version.

Relevance of the Impossibility Result. While it perhaps appears natural to process all registrations before messages for the corresponding identities are sent, this restriction substantially weakens the usefulness of IBE. For example, if IBE is used in a large context to encrypt emails where the encryption service is independent of the email providers, it seems desirable to be able to send encrypted emails to anyone with a valid email address, without knowing whether they have already registered for the encryption service. In fact, if one has to "ask" whether a user has already received his key before being able to send him a message, one gives up non-interactivity and does not gain much compared to standard public-key encryption.

Moreover, an interesting application, which was suggested in [3], is impossible: Assume the key authority every day publishes a key for the identity that corresponds to the current date. One should now be able to send a message "to the future" by encrypting it for the identity corresponding to, e.g., the following day. We are here precisely in the situation where a ciphertext is received before the corresponding key, so standard IBE does not guarantee the security of this application[5] (our construction in the random oracle model, however, does provide this guarantee).

---

[5] One can give a less technical argument why standard definitions are insufficient for this application than the inability to simulate: It is not excluded by IND-ID-CPA or IND-ID-CCA that first providing a ciphertext and later the user secret key for the corresponding identity yields a binding commitment (maybe only for some specific subset of the message space). In this case, a dishonest recipient Bob of a ciphertext for the following day can use this ciphertext to commit himself (to some third party) to the encrypted value, and open the commitment on the next day. Note that Bob committed himself to a value *he did not know*, misleading the third party

On Dishonest Senders. The results in this paper only consider passive attacks, i.e., we assume only honest parties send messages. This makes our impossibility result only stronger, and all positive results can in principle be lifted to a setting with potentially dishonest senders by replacing the CPA-definitions with their (R)CCA-counterparts. However, this leads to some subtleties in the modeling. For example, one needs to simulate a dishonest sender sending some nonsensical bit string (which does not constitute a valid ciphertext) to a dishonest receiver. Furthermore, the two phases in the results with a separate registration and transmission phase become intermixed, because only honest parties are prevented from sending during the registration phase. To avoid such technicalities and simplify the presentation, we formulate all results only for honest senders.

1.4   Related Work

On the Difference to the IBE Ideal Functionality of Nishimaki Et al. We note that an ideal functionality for IBE has already been presented by Nishimaki et al. [18] in the UC framework. However, unlike our resources (when interpreted as UC functionalities as sketched above), their functionality was constructed directly along the IBE *algorithms*, and not to model the *goal* of non-interactive communication. Besides, their functionality does not guarantee secrecy for ciphertexts generated before the respective receiver has been initialized. (This relaxed guarantee corresponds to our relaxed resource stDCC that disallows registrations after communication attempts.)

As a consequence, [18] could indeed show that the standard game-based definition of security for IBE schemes is equivalent to realizing their ideal functionality. Specifically, their IBE abstraction thus compares differently from ours to game-based IBE security notions.

Relation to Functional Encryption. Identity-based encryption is known to be a special case of functional encryption [4], which has already been modeled in the constructive cryptography framework [11]. However, the results from that paper cannot directly be applied to the context of non-interactive communication as studied in our paper. One reason is that a different goal was modeled in [11] (namely adding access control to a public repository), where only three parties are considered. More importantly, we analyze security definitions which are specific to IBE, while [11] only considers (simulation based) security definitions for general functional encryption, which are more involved. We note, however, that the same commitment problem arises in the context of functional encryption [4].

Relation to Adaptive Corruptions in the Public-Key Setting. As noted, *technically*, the commitment problem we encounter is very similar to the commitment problem faced in adaptively secure public-key encryption [17]. There,

---

into believing he knew it, which is not possible when an ideal "sending-to-the-future" functionality is used.

a simulation would have to first produce a ciphertext (without knowing the supposed plaintext). Later, upon an adaptive corruption of the respective receiver, the simulation would have to provide a secret key that opens that ciphertext suitably.

However, in our case, the actual *setting* in which the problem occurs is not directly related to corruptions. Namely, in our setting, a similar commitment problem occurs because messages may be sent to an identity prior to an "activation" of the corresponding communication channel. (In fact, since the mapping of receiving parties to identities may not be clear beforehand, prior to such an activation it is not even clear where to route the corresponding sent messages.) Hence, we can argue that the commitment problem we face is inherent to the IBE setting, independently of adaptive corruptions (all results in this paper are actually formulated for static corruptions).

2   Preliminaries

Constructive Cryptography. The results in this paper are formulated using a simulation-based notion of security. There are many protocol frameworks based on such a simulation-based security notion (e.g., [1,5,10,12,13,15,19]). However, in this work, we use the constructive cryptography (CC) framework [12,13].

Briefly, CC makes statements about *constructions* of *resources* from other resources. A resource is a system with interfaces via which the resource interacts with its environment and which can be thought of as being assigned to parties. *Converters* are systems that can be attached to an interface of a resource to change the inputs and outputs at that interface, which yields another resource. The protocols of honest parties and simulators correspond to converters. Dishonest behavior at an interface is captured by *not* applying the protocol (instead of modeling an explicit adversary). An ideal resource is *constructed* from a real resource by a protocol, if the real resource with the protocol converters attached at the honest interfaces is indistinguishable from the ideal resource with the simulators attached at the dishonest interfaces.

We introduce the relevant concepts in more detail, following [13], in the following subsections. For readers more familiar with the Universal Composability (UC) framework [5], we also include explanations of how the presented concepts relate to similar concepts in UC.

Efficiency and Security Parameters. Negligibility and efficiency is defined with respect to a security parameter and the complexity of all algorithms and systems in this paper is polynomial in this security parameter. Thus, distinguishing advantages and advantages in winning a game are functions of this parameter. To simplify notation, we will omit security parameters and not provide them as additional inputs.

Notation for Algorithms and Systems. The algorithms and systems in this paper are described by pseudocode using the following conventions: For variables

x and y, $x \leftarrow y$ denotes the assignment after which x has the value of y. For a finite set $\mathcal{S}$, $x \leftarrow \mathcal{S}$ denotes the assignment of a uniformly random element in $\mathcal{S}$ to x. If A is an algorithm, $x \leftarrow \text{A}(\ldots)$ denotes executing $\text{A}(\ldots)$ and assigning the returned value to x. For a probabilistic algorithm A and a (sufficiently long) bit string r, $\text{A}(r;\ldots)$ denotes the execution of A with randomness r. We denote the length of a bit string s by $|s|$ and for s_1, s_2, $|(s_1, s_2)|$ denotes the bit length of (some fixed) unique encoding of (s_1, s_2).

2.1   Resources, Converters, and Distinguishers

We consider different types of *systems*, which are objects with *interfaces* via which they interact with their environment. Interfaces are denoted by uppercase letters. One can compose two systems by connecting one interface of each system. The composed object is again a system.

Two types of systems we consider here are *resources* and *converters*. Resources are denoted by bold uppercase letters or sans serif fonts and have a finite set $\mathcal{I}$ of interfaces. Resources with interface set $\mathcal{I}$ are called $\mathcal{I}$-*resources*. Converters have one *inside* and one *outside interface* and are denoted by lowercase Greek letters or sans serif fonts. The inside interface of a converter α can be connected to interface $I \in \mathcal{I}$ of a resource $\mathbf{R}$. The outside interface of α then serves as the new interface I of the composed resource, which is denoted by $\alpha^I \mathbf{R}$. We also write $\alpha_I \mathbf{R}$ instead of $\alpha_I^I \mathbf{R}$ for a converter α_I. For a vector of converters $\alpha = (\alpha_{I_1}, \ldots, \alpha_{I_n})$ with $I_1, \ldots, I_n \in \mathcal{I}$ and a set $\mathcal{P} \subseteq \{I_1, \ldots, I_n\}$ of interfaces, $\alpha_{\mathcal{P}} \mathbf{R}$ denotes the $\mathcal{I}$-resource that results from connecting α_I to interface I of $\mathbf{R}$ for every $I \in \mathcal{P}$. Moreover, $\alpha_{\overline{\mathcal{P}}} \mathbf{R}$ denotes the $\mathcal{I}$-resource one gets when α_I is connected to interface I of $\mathbf{R}$ for every $I \in \{I_1, \ldots, I_n\} \setminus \mathcal{P}$. For $\mathcal{I}$-resources $\mathbf{R}_1, \ldots, \mathbf{R}_m$, the *parallel composition* $[\mathbf{R}_1, \ldots, \mathbf{R}_m]$ is defined as the $\mathcal{I}$-resource where each interface $I \in \mathcal{I}$ allows to access the corresponding interfaces of all sub-systems $\mathbf{R}_i$ as sub-interfaces. Similarly, for converters $\alpha_1, \ldots, \alpha_m$, we define the *parallel composition* $[\alpha_1, \ldots, \alpha_m]$ via $[\alpha_1, \ldots, \alpha_m]^I [\mathbf{R}_1, \ldots, \mathbf{R}_m] := [\alpha_1^I \mathbf{R}_1, \ldots, \alpha_m^I \mathbf{R}_m]$.

A *distinguisher* $\mathbf{D}$ for resources with n interfaces is a system with $n+1$ interfaces, where n of them connect to the interfaces of a resource and a bit is output at the remaining one. We write $\Pr[\mathbf{DR} = 1]$ to denote the probability that $\mathbf{D}$ outputs the bit 1 when connected to resource $\mathbf{R}$. The goal of a distinguisher is to distinguish two resources by outputting a different bit when connected to a different resource. Its success is measured by the distinguishing advantage.

Definition 1. *The* distinguishing advantage *of a distinguisher* $\mathbf{D}$ *for resources* $\mathbf{R}$ *and* $\mathbf{S}$ *is defined as*

$$\Delta^{\mathbf{D}}(\mathbf{R}, \mathbf{S}) := |\Pr[\mathbf{DR} = 1] - \Pr[\mathbf{DS} = 1]|.$$

If $\Delta^{\mathbf{D}}(\mathbf{R}, \mathbf{S}) = 0$ *for all distinguishers* $\mathbf{D}$, *we say* $\mathbf{R}$ *and* $\mathbf{S}$ *are* equivalent, *denoted as* $\mathbf{R} \equiv \mathbf{S}$. *If the distinguishing advantage is negligible for all efficient distinguishers, we say* $\mathbf{R}$ *and* $\mathbf{S}$ *are* computationally indistinguishable, *denoted as* $\mathbf{R} \approx \mathbf{S}$.

We introduce two special converters **1** and $\perp$. The converter **1** forwards all inputs at one of its interfaces to the other one. We thus have for all $\mathcal{I}$-resources **R** and all $I \in \mathcal{I}$

$$\mathbf{1}^I \mathbf{R} \equiv \mathbf{R}.$$

One can equivalently understand connecting **1** to interface I of a resource as not connecting any converter to that interface. Moreover, the converter $\perp$ blocks all inputs at the connected interface. That is, interface I of $\perp^I \mathbf{R}$ does not accept any inputs and there are no outputs at this interface.

Relation to UC Concepts. In UC, systems as above can correspond to protocols, ideal functionalities, or simulators that interact with the protocol environment. More specifically, resources correspond to ideal functionalities, while converters can correspond to real or hybrid protocols, or to simulators. Namely, a UC protocol can be viewed as a way to convert calls to that protocol to calls to an underlying communication infrastructure (or hybrid functionality). Conversely, a UC simulator can be viewed as a way to convert the network interface of one protocol into that of another one. (In CC, there is no a-priori distinction between I/O and network interfaces; hence, both UC protocols and UC simulators correspond to converters.) Distinguishers as above correspond to the UC protocol environments.

2.2   Filtered Resources

In some situations, specific interactions with a resource might not be guaranteed but only potentially available. To model such situations, we extend the concept of a resource. Let **R** be an $\mathcal{I}$-resource and let $\phi = (\phi_I)_{I \in \mathcal{I}}$ be a vector of converters. We define the *filtered resource* $\mathbf{R}_\phi$ as a resource with the same set of interfaces $\mathcal{I}$. For a party connected to interface I of $\mathbf{R}_\phi$, interactions through the converter ϕ_I are guaranteed to be available, while interactions with **R** directly are only potentially available to dishonest parties. The converter ϕ_I can be seen as a filter shielding specific functionality of interface I. Dishonest parties can potentially remove the filter to get access to all features of the resource **R**. Formally, $\mathbf{R}_\phi$ is defined as the set of all resources that allows all interactions allowed $\phi_{\mathcal{I}}\mathbf{R}$ but not more than allowed by **R**; see [13] for more details.

2.3   Communication Resources

An important example of resources are communication channels, which allow the sender A to send messages from the message space $\mathcal{M} := \{0,1\}^*$ to the receiver B. We define two such channels, which differ in the capabilities of the adversary E. If a channel is used in a context with several potentially dishonest parties, all of them have access to interface E.

Definition 2. *An authenticated channel from A to B, denoted as $\mathsf{AUT}^{A,B}$, and a secure channel from A to B, denoted as $\mathsf{SEC}^{A,B}$, are resources with three* *interfaces A, B, and E. On input a message $m \in \mathcal{M}$ at interface A, they both output the same message m at interface B. Additionally, $\mathsf{AUT}^{A,B}$ outputs m at interface E and $\mathsf{SEC}^{A,B}$ outputs the length $|m|$ of the message at interface E. Other inputs are ignored. Both channels allow arbitrarily many messages to be sent.*

Remark 1. Alternatively, one could define authenticated and secure channels such that E also has the ability to delete messages. The results in this paper can be adapted to such a setting, but our assumption that sent messages are always delivered allows to simplify the presentation.

For authenticated channels, we do not want to guarantee that an adversary learns the message, it is rather not excluded. Similarly, secure channels should not guarantee that the length of the message leaks. To model this, we introduce filters that block all outputs at interface E. We then have that a secure channel is also authenticated, i.e., the set of (filtered) secure channels is a subset of the set of (filtered) authenticated channels.

Definition 3. *Let $\phi^{\mathsf{AUT}} = \phi^{\mathsf{SEC}} := (\mathbf{1}, \mathbf{1}, \perp)$. We will consider the filtered resources $\mathsf{AUT}^{A,B}_{\phi^{\mathsf{AUT}}}$ and $\mathsf{SEC}^{A,B}_{\phi^{\mathsf{SEC}}}$.*

Note that

$$\phi^{\mathsf{AUT}}_{\{A,B,E\}}\mathsf{AUT}^{A,B} = \mathbf{1}^A\mathbf{1}^B\perp^E\mathsf{AUT}^{A,B} \equiv \mathbf{1}^A\mathbf{1}^B\perp^E\mathsf{SEC}^{A,B} = \phi^{\mathsf{SEC}}_{\{A,B,E\}}\mathsf{SEC}^{A,B}$$

accepts messages at interface A and outputs them at interface B where interface E is inactive.

We finally introduce a more advanced communication resource that has many interfaces and allows a sender to send messages to all other interfaces. It is authenticated in the sense that the messages cannot be modified and everyone receives the same message.

Definition 4. *The* broadcast *resource $\mathsf{BCAST}^{A,\mathcal{B}}$ for a set $\mathcal{B}$ has interface set $\{A\} \cup \mathcal{B}$. On input a message $m \in \mathcal{M}$ at interface A, the same message is output at all interfaces $B \in \mathcal{B}$. Other inputs are ignored.*

Relation to UC Concepts. The presented resources directly correspond to UC ideal functionalities for authenticated, secure, or broadcast channels. The different interfaces of the presented resources correspond to what different parties in UC could send or receive. (Here we note a common design difference in UC and CC: in UC, typically one would assume parties as fixed entities, and model communication and interfaces around them. In CC, one would typically start with the interfaces that reflect the semantic types of in- and outputs of a resource, and only later think of connecting entities like parties.)

2.4   Construction of Resources

A *protocol* is a vector of converters with the purpose of constructing a so-called ideal resource from an available real resource. Depending on which parties are considered potentially dishonest, we get a different notion of construction.

As an example from [8], consider the setting for public-key encryption with honest A and B where we want to construct a secure channel $\mathsf{SEC}^{A,B}_{\phi^{\mathsf{SEC}}}$ from authenticated channels $\mathsf{AUT}^{B,A}_{\phi^{\mathsf{AUT}}}$ and $\mathsf{AUT}^{A,B}_{\phi^{\mathsf{AUT}}}$ in presence of a dishonest eavesdropper E. Here, the real resource is $\left[\mathsf{AUT}^{B,A}_{\phi^{\mathsf{AUT}}}, \mathsf{AUT}^{A,B}_{\phi^{\mathsf{AUT}}}\right]$ and the ideal resource is $\mathsf{SEC}^{A,B}_{\phi^{\mathsf{SEC}}}$. In this setting, a protocol $\pi = (\pi_A, \pi_B, \pi_E)$ constructs $\mathbf{S}$ from $\mathbf{R}$ with potentially dishonest E if there exists a converter σ_E (called *simulator*) such that

$$\pi_A \pi_B \pi_E \left[\phi^{\mathsf{AUT}}_E \mathsf{AUT}^{B,A}, \phi^{\mathsf{AUT}}_E \mathsf{AUT}^{A,B}\right] \approx \phi^{\mathsf{SEC}}_E \mathsf{SEC}^{A,B}$$

$$\text{and} \quad \pi_A \pi_B \left[\mathsf{AUT}^{B,A}, \mathsf{AUT}^{A,B}\right] \approx \sigma_E \mathsf{SEC}^{A,B},$$

where σ_E provides a sub-interface to the distinguisher for each channel that constitutes the real resource. The first condition ensures that the protocol implements the required functionality and the second condition ensures that whatever Eve can do when connected to the real resource without necessarily following the protocol, she could do as well when connected to the ideal resource by using the simulator σ_E. Since Eve is here only a hypothetical entity, we typically have $\pi_E = \bot$.

In this paper, we consider the more general setting that includes several potentially dishonest parties that (in contrast to Eve in the above example) also get certain guarantees if they are honest while unable to do more than specified by the ideal resource even if they are dishonest. We define a secure construction as follows.

Definition 5. *Let $\mathbf{R}_\phi$ and $\mathbf{S}_\psi$ be filtered $\mathcal{I}$-resources and let $\pi = (\pi_I)_{I \in \mathcal{I}}$ be a protocol. Further let $\mathcal{U} \subseteq \mathcal{I}$ be the set of interfaces with potentially dishonest behavior. We say π constructs $\mathbf{S}_\psi$ from $\mathbf{R}_\phi$ with potentially dishonest $\mathcal{U}$, denoted by*

$$\mathbf{R}_\phi \overset{\pi}{\underset{\mathcal{U}}{\longmapsto}} \mathbf{S}_\psi,$$

if there exist converters $\sigma = (\sigma_U)_{U \in \mathcal{U}}$ such that

$$\forall \mathcal{P} \subseteq \mathcal{U} : \pi_{\overline{\mathcal{P}}} \phi_{\overline{\mathcal{P}}} \mathbf{R} \approx \sigma_{\mathcal{P}} \psi_{\overline{\mathcal{P}}} \mathbf{S}.$$

The converters σ_U are called simulators.

For $\mathcal{U} = \mathcal{I}$, this definition corresponds to the abstraction notion from [13], which considers all parties as potentially dishonest. The construction notion is composable in the following sense:

$$\mathbf{R}_\phi \overset{\pi}{\underset{\mathcal{U}}{\longmapsto}} \mathbf{S}_\psi \wedge \mathbf{S}_\psi \overset{\pi'}{\underset{\mathcal{U}}{\longmapsto}} \mathbf{T}_\tau \implies \mathbf{R}_\phi \overset{\pi'\pi}{\underset{\mathcal{U}}{\longmapsto}} \mathbf{T}_\tau,$$

where $\pi'\pi$ is the protocol that corresponds to first applying π and then π' to the resource.

To apply the above definition to an unfiltered resource $\mathbf{R}$, one can formally introduce trivial filters $\phi_I := \mathbf{1}$ for $I \in \mathcal{I}$ and consider the filtered resource $\mathbf{R}_\phi$ which is identical to $\mathbf{R}$. In such cases, we will omit the filters. We refer the reader to [13] for more details.

Relation to UC Concepts. The "constructs" notion presented above directly corresponds to the UC notion of secure realization. (The combination of π and $\mathbf{R}$ corresponds to the real protocol in UC, while $\mathbf{S}$ matches the UC ideal protocol.) The "constructs" notion does not consider an explicit adversary on the real protocol. (Instead, in UC terms, a dummy adversary is considered without loss of generality.) There is a difference, however, in the modeling of corruptions. Generally, in UC, adaptive corruptions are considered. In the CC modeling above, only static corruptions of parties are considered. Moreover, instead of modeling corruptions through special "corrupt" messages sent from the adversary or environment, in CC corruptions are modeled simply be letting the distinguisher connect to the interfaces of corrupted parties.

Finally, a subtle difference between CC and UC security is that CC security requires "local" simulators for each interface, whereas in UC, one simulator is required that handles all parties (resp. interfaces) at once. While this makes CC security a stricter notion than UC security, this difference will not be relevant to our results. (In particular, our negative result has nothing to do with the fact that CC security requires local simulation.)

3   Delivery Controlled Channels

A broadcast channel allows a sender A to send messages to recipients $B_1, \ldots, B_n$. One can understand the application of an IBE scheme to add some form of delivery control to such a channel. More specifically, the enhanced channel allows A to send a message for some identity id in an identity space $\mathcal{ID}$ such that only the B_i that are registered for this identity receive the message, even if several other B_i are dishonest. We assume this registration is managed by a central authority C. We formalize this by a *delivery controlled channel* DCC. This resource also allows the registration of identities after messages have been sent for this identity. In this case, the corresponding user after registration learns all such messages.

Because the public key and each ciphertext contain randomness, during initialization and for each sent message, all parties (potentially) receive common randomness. Moreover, when someone gets registered for an identity, this identity together with a corresponding user secret key is sent to this party over a secure channel. By definition, a secure channel can leak the length of the transmitted messages. Since the length of user secret keys can depend on the identity for which the key has been generated and also on the used randomness, dishonest users potentially learn which identity has just been registered for whom and potentially even which randomness was used to generate the corresponding secret key. Furthermore, dishonest recipients can share their secret keys with others in the real world, which has the effect in the ideal world that the other recipients also learn the messages sent for an identity that has been registered for the user who shared his keys. We model this by a special symbol $\mathtt{share}$ that B_i can input. A message sent for identity id is then received by B_i if id has been registered for B_i or if there is a B_j such that B_i and B_j have input $\mathtt{share}$ and id has been registered for B_j.

Definition 6. *Let $n, \rho \in \mathbb{N}$, $\mathcal{M} := \{0,1\}^*$, and let $\mathcal{ID}$ be a nonempty set. The resource $\mathsf{DCC}^{n,\mathcal{ID},\rho}$ has the interfaces A, C, and B_i for $i \in \{1,\ldots,n\}$. The resource internally manages the set $S \subseteq \{B_1,\ldots,B_n\}$ of interface names that want to share their identities and for each $i \in \{1,\ldots,n\}$, the set $I_i \subseteq \mathcal{ID}$ of identities registered for interface B_i. Initially, both sets are empty. The resource works as follows:*

Initialization

$j \leftarrow 1$
$r \leftarrow \{0,1\}^\rho$
for all $i \in \{1,\ldots,n\}$ **do**
    output r at interface B_i

Interface A

Require: $(id_j, m_j) \in \mathcal{ID} \times \mathcal{M}$
$r_j \leftarrow \{0,1\}^\rho$
for all $i \in \{1,\ldots,n\}$ **do**
    if $id_j \in I_i$ **or** $\left(B_i \in S \text{ and } id_j \in \bigcup_{k \in S} I_k\right)$ **then**
        output (id_j, m_j, r_j) at interface B_i
    else
        output $(id_j, |m_j|, r_j)$ at interface B_i
$j \leftarrow j + 1$

Interface B_i

Require: $\mathtt{share}$
$S \leftarrow S \cup \{B_i\}$

Interface C

Require: $(id, i) \in \mathcal{ID} \times \{1,\ldots,n\}$
$I_i \leftarrow I_i \cup \{id\}$
$r \leftarrow \{0,1\}^\rho$
for all $k \in \{1,\ldots,n\}$ **do**
    output (id, i, r) at interface B_k
    if $k = i$ **or** $\{B_i, B_k\} \subseteq S$ **then**
        for all $l \in \{1,\ldots,j-1\}$ such that $id_l = id$ **do**
            output m_l at interface B_k

All inputs not matching the given format are ignored.

The randomness that the B_i get corresponds to randomness one can potentially extract from the public key, the ciphertexts, and the length of the user secret keys of an IBE scheme. Honest users are not guaranteed to receive this randomness, we rather cannot exclude that dishonest parties do so. Similarly, we cannot exclude that dishonest parties share their identities, that they learn the identity for which a message is designated and the length of the message without being registered for that identity, and that they learn who gets registered for which identity. To model that these interactions are not guaranteed, we introduce filters to block inputs and outputs at interfaces B_i for honest parties: For $i \in \{1, \ldots, n\}$, let $\phi_{B_i}^{\mathsf{DCC}}$ be the converter that on input $(id, m, r) \in \mathcal{ID} \times \mathcal{M} \times \{0,1\}^\rho$ at its inside interface, outputs (id, m) at its outside interface, on input $m \in \mathcal{M}$ at its inside interface, outputs m at its outside interface, and on input $(id, k, r) \in \mathcal{ID} \times \{1, \ldots, n\} \times \{0,1\}^\rho$ with $k = i$ at its inside interface, outputs id at its outside interface. All other inputs at any of its interfaces are ignored and thereby blocked. Further let $\phi_A^{\mathsf{DCC}} = \phi_C^{\mathsf{DCC}} := \mathbf{1}$ be the converter that forwards all inputs at one of its interfaces to the other one and let $\phi^{\mathsf{DCC}} := (\phi_A^{\mathsf{DCC}}, \phi_C^{\mathsf{DCC}}, \phi_{B_1}^{\mathsf{DCC}}, \ldots, \phi_{B_n}^{\mathsf{DCC}})$. We will consider the filtered resource $\mathsf{DCC}_{\phi^{\mathsf{DCC}}}^{n, \mathcal{ID}, \rho}$.

Remark 2. The resource defined above assumes that a central authority C registers all identities and allows one party to have more than one identity and one identity to be registered for several users. That resource can now be used in larger context where this registration process is regulated. For example, one can have a protocol programmed on top of DCC that requires B_i to send his identity together with a copy of his passport to C. Moreover, C could ensure that each identity is registered for at most one user. In such an application, the resource DCC could directly be used without considering how it was constructed. Due to composition of the constructive cryptography framework, we can thus focus on the construction of DCC and decouple confidentiality from the actual registration process.

Static Identity Management. We now define a more restricted resource that only allows the registration of an identity as long as no message has been sent for this identity.

Definition 7. *Let $n, \rho \in \mathbb{N}$, $\mathcal{M} := \{0,1\}^*$, and let $\mathcal{ID}$ be a nonempty set. The resource $\mathsf{stDCC}^{n, \mathcal{ID}, \rho}$ is identical to $\mathsf{DCC}^{n, \mathcal{ID}, \rho}$ except that inputs $(id, i) \in \mathcal{ID} \times \{1, \ldots, n\}$ at interface C are ignored if $id \in \bigcup_{k=1}^{j-1} \{id_k\}$. We will use the same filters as above and consider the resource $\mathsf{stDCC}_{\phi^{\mathsf{DCC}}}^{n, \mathcal{ID}, \rho}$.*

The above resource prevents identities for which messages have been sent to be registered, but other identities can still be registered. The following resource restricts the registration process further and operates in two phases: Initially, only registrations are allowed and no messages can be sent. At any point, C can end the registration phase and enable A to send messages.

Definition 8. *Let* $n, \rho \in \mathbb{N}$, $\mathcal{M} := \{0,1\}^*$, *and let* $\mathcal{ID}$ *be a nonempty set. The resource* st2DCC$^{n,\mathcal{ID},\rho}$ *behaves as* DCC$^{n,\mathcal{ID},\rho}$ *except that it initially ignores all inputs at interface A. On input the special symbol* **end registration** *at interface C, the resource outputs* **registration ended** *at interfaces* $B_1, \ldots, B_n$,[6] *and from then on ignores all inputs at interface C and allows inputs at interface A. We will consider the filtered resource* st2DCC$^{n,\mathcal{ID},\rho}_{\phi^{\text{DCC}}}$.

Note that when using stDCC, A can prevent the registration of an identity by sending a message for this identity. On the other hand, st2DCC gives C full control over the registration process while being less dynamic. Depending on the application, one of these resources might be preferable.

Predetermined Identities. We finally introduce two resources that additionally require all identities that are used be determined at the beginning. This allows us to capture the guarantees provided by selectively secure IBE schemes (see Definition 11).

Definition 9. *Let* $n, \rho \in \mathbb{N}$, $\mathcal{M} := \{0,1\}^*$, *and let* $\mathcal{ID}$ *be a nonempty set. The resources* preDCC$^{n,\mathcal{ID},\rho}$ *and* pre2DCC$^{n,\mathcal{ID},\rho}$ *have the interfaces A, C, and* B_i *for* $i \in \{1,\ldots,n\}$. *Before the resources output anything or accept any input, they wait for the input of a finite set* $\mathcal{S} \subseteq \mathcal{ID}$ *(encoded as a list of its elements) at interface A. On this input, they output* **ok** *at interfaces* $B_1, \ldots, B_n$. *Afterwards,* preDCC$^{n,\mathcal{ID},\rho}$ *behaves identically to* stDCC$^{n,\mathcal{ID},\rho}$ *and* pre2DCC$^{n,\mathcal{ID},\rho}$ *behaves identically to* st2DCC$^{n,\mathcal{ID},\rho}$ *with the exception that they only accept inputs* $(id_j, m_j) \in \mathcal{S} \times \mathcal{M}$ *at interface A (there is no restriction on inputs at interface C). We will again consider the filtered resources* preDCC$^{n,\mathcal{ID},\rho}_{\phi^{\text{DCC}}}$ *and* pre2DCC$^{n,\mathcal{ID},\rho}_{\phi^{\text{DCC}}}$.[7]

4   IBE Schemes and Protocols

4.1   IBE Schemes and Their Security

Identity-Based Encryption. An *identity-based encryption (IBE) scheme* $\mathcal{E}$ with message space $\mathcal{M}$ and identity space $\mathcal{ID}$ consists of four PPT algorithms. Key generation Gen() outputs a master public key mpk and a master secret key msk. Extraction Ext(msk, id) (for a master secret key msk and an identity $id \in \mathcal{ID}$) outputs a user secret key usk_{id}. Encryption Enc(mpk, id, m) (for

---

[6] Note that ϕ^{DCC} blocks this output for honest users, i.e., it is not necessarily guaranteed that everyone learns that the registration has ended. It is not excluded by our protocol since C there informs A that messages may now be sent, and this communication could be observed by dishonest users. If it is desirable in an application that everyone learns that the registration has ended, one can still use st2DCC$^{n,\mathcal{ID},\rho}$ by letting C explicitly send that information to all B_i via an additional channel. This would happen outside of the resource st2DCC$^{n,\mathcal{ID},\rho}$ as a separate construction.

[7] Again, the filter ϕ^{DCC} blocks the outputs **ok** and **registration ended** at interfaces B_i.

Experiment $\mathsf{Exp}^{\mathsf{ind\text{-}id\text{-}cpa}}_{\mathcal{E},\mathcal{A}}$:
  $(mpk, msk) \leftarrow \mathsf{Gen}()$
  $(st, id, m_0, m_1) \leftarrow \mathcal{A}^{\mathsf{Ext}(msk,\cdot)}(mpk)$
  $b \leftarrow \{0, 1\}$
  $c^* \leftarrow \mathsf{Enc}(mpk, id, m_b)$
  $b' \leftarrow \mathcal{A}^{\mathsf{Ext}(msk,\cdot)}(st, c^*)$
  Return 1 if $b' = b$, else return 0

Experiment $\mathsf{Exp}^{\mathsf{ind\text{-}sid\text{-}cpa}}_{\mathcal{E},\mathcal{A}}$:
  $(st, id) \leftarrow \mathcal{A}()$
  $(mpk, msk) \leftarrow \mathsf{Gen}()$
  $(st', m_0, m_1) \leftarrow \mathcal{A}^{\mathsf{Ext}(msk,\cdot)}(st, mpk)$
  $b \leftarrow \{0, 1\}$
  $c^* \leftarrow \mathsf{Enc}(mpk, id, m_b)$
  $b' \leftarrow \mathcal{A}^{\mathsf{Ext}(msk,\cdot)}(st', c^*)$
  Return 1 if $b' = b$, else return 0

Fig. 2. The IND-(s)ID-CPA experiment with scheme $\mathcal{E}$ and adversary $\mathcal{A}$.

a master public key mpk, an identity $id \in \mathcal{ID}$, and a message $m \in \mathcal{M}$) outputs a ciphertext c. Decryption $\mathsf{Dec}(usk_{id}, id, c)$ (for a user secret key usk_{id}, an identity $id \in \mathcal{ID}$, and a ciphertext c) outputs a message $m \in \mathcal{M} \cup \{\bot\}$. For correctness, we require that for all $(mpk, msk) \leftarrow \mathsf{Gen}()$, all $id \in \mathcal{ID}$, all $m \in \mathcal{M}$, all $c \leftarrow \mathsf{Enc}(mpk, id, m)$, and all $usk_{id} \leftarrow \mathsf{Ext}(msk, id)$, we always have $\mathsf{Dec}(usk_{id}, id, c) = m$.

Standard Security Definitions for IBE Schemes. We first provide the standard security definition for IBE schemes against passive attacks:

Definition 10 (IND-ID-CPA security). *Consider the experiment* $\mathsf{Exp}^{\mathsf{ind\text{-}id\text{-}cpa}}_{\mathcal{E},\mathcal{A}}$ *in Fig. 2 for an IBE scheme* $\mathcal{E} = (\mathsf{Gen}, \mathsf{Ext}, \mathsf{Enc}, \mathsf{Dec})$ *and an algorithm* $\mathcal{A}$. *In this experiment,* $\mathcal{A}$ *is not allowed to output an identity* id *that it has queried to its* Ext *oracle, or to later query* id *to* Ext. *Furthermore,* $\mathcal{A}$ *must output* m_0, m_1 *of equal length. Let*

$$\mathsf{Adv}^{\mathsf{ind\text{-}id\text{-}cpa}}_{\mathcal{E},\mathcal{A}} := \Pr\left[\mathsf{Exp}^{\mathsf{ind\text{-}id\text{-}cpa}}_{\mathcal{E},\mathcal{A}} = 1\right] - 1/2.$$

We say that $\mathcal{E}$ *has indistinguishable ciphertexts under chosen-plaintext attacks (is IND-ID-CPA secure) if* $\mathsf{Adv}^{\mathsf{ind\text{-}id\text{-}cpa}}_{\mathcal{E},\mathcal{A}}$ *is negligible for all PPT* $\mathcal{A}$.

We further consider a weaker security notion introduced in [6] where the adversary has to specify the identity he wants to attack at the beginning of the experiment.

Definition 11 (IND-sID-CPA security). *Consider experiment* $\mathsf{Exp}^{\mathsf{ind\text{-}sid\text{-}cpa}}_{\mathcal{E},\mathcal{A}}$ *in Fig. 2 for an IBE scheme* $\mathcal{E} = (\mathsf{Gen}, \mathsf{Ext}, \mathsf{Enc}, \mathsf{Dec})$ *and an algorithm* $\mathcal{A}$. *In this experiment,* $\mathcal{A}$ *is not allowed to query* id *to* Ext *and has to output* m_0, m_1 *of equal length. Let*

$$\mathsf{Adv}^{\mathsf{ind\text{-}sid\text{-}cpa}}_{\mathcal{E},\mathcal{A}} := \Pr\left[\mathsf{Exp}^{\mathsf{ind\text{-}sid\text{-}cpa}}_{\mathcal{E},\mathcal{A}} = 1\right] - 1/2.$$

We say that $\mathcal{E}$ *has indistinguishable ciphertexts under selective identity, chosen-plaintext attacks (is IND-sID-CPA secure) if* $\mathsf{Adv}^{\mathsf{ind\text{-}sid\text{-}cpa}}_{\mathcal{E},\mathcal{A}}$ *is negligible for all PPT* $\mathcal{A}$.

Experiment $\mathsf{Exp}_{\mathcal{E},\mathcal{A}}^{\text{ind-id1-cpa}}$:

$(mpk, msk) \leftarrow \mathbf{Gen}()$

$st \leftarrow \mathcal{A}^{\mathbf{Ext}(msk,\cdot)}()$

$(st', id, m_0, m_1) \leftarrow \mathcal{A}(st, mpk)$

$b \leftarrow \{0, 1\}$

$c^* \leftarrow \mathbf{Enc}(mpk, id, m_b)$

$b' \leftarrow \mathcal{A}(st', c^*)$

Return 1 if $b' = b$, else return 0

Experiment $\mathsf{Exp}_{\mathcal{E},\mathcal{A}}^{\text{ind-sid1-cpa}}$:

$(st, id) \leftarrow \mathcal{A}()$

$(mpk, msk) \leftarrow \mathbf{Gen}()$

$st' \leftarrow \mathcal{A}^{\mathbf{Ext}(msk,\cdot)}(st)$

$(st'', m_0, m_1) \leftarrow \mathcal{A}(st', mpk)$

$b \leftarrow \{0, 1\}$

$c^* \leftarrow \mathbf{Enc}(mpk, id, m_b)$

$b' \leftarrow \mathcal{A}(st'', c^*)$

Return 1 if $b' = b$, else return 0

Fig. 3. The IND-(s)ID1-CPA experiment with scheme $\mathcal{E}$ and adversary $\mathcal{A}$.

Non-adaptive Security. We introduce two novel security notions for IBE schemes that loosely correspond to variants of the standard definitions under "lunchtime attacks" [16]. While CCA1 in contrast to CCA allows the adversary only to ask decryption queries in an initial phase, our definitions restrict the adversary to ask **Ext** queries only in an initial phase.

Definition 12 (IND-(s)ID1-CPA security). *Consider the two experiments* $\mathsf{Exp}_{\mathcal{E},\mathcal{A}}^{\text{ind-id1-cpa}}$ *and* $\mathsf{Exp}_{\mathcal{E},\mathcal{A}}^{\text{ind-sid1-cpa}}$ *for an IBE scheme* $\mathcal{E} = (\mathbf{Gen}, \mathbf{Ext}, \mathbf{Enc}, \mathbf{Dec})$ *and an algorithm* $\mathcal{A}$ *in Fig. 3. In these experiments,* $\mathcal{A}$ *is only considered valid if all queries to its* **Ext** *oracle are different from id and if* $|m_0| = |m_1|$. *Let*

$$\mathsf{Adv}_{\mathcal{E},\mathcal{A}}^{\text{ind-id1-cpa}} := \Pr\left[\mathsf{Exp}_{\mathcal{E},\mathcal{A}}^{\text{ind-id1-cpa}} = 1\right] - 1/2 \quad and$$

$$\mathsf{Adv}_{\mathcal{E},\mathcal{A}}^{\text{ind-sid1-cpa}} := \Pr\left[\mathsf{Exp}_{\mathcal{E},\mathcal{A}}^{\text{ind-sid1-cpa}} = 1\right] - 1/2.$$

We say that $\mathcal{E}$ *has indistinguishable ciphertexts under non-adaptive chosen-plaintext attacks (is IND-ID1-CPA secure) if* $\mathsf{Adv}_{\mathcal{E},\mathcal{A}}^{\text{ind-id1-cpa}}$ *is negligible for all valid PPT* $\mathcal{A}$ *and* $\mathcal{E}$ *has indistinguishable ciphertexts under selective identity, non-adaptive chosen-plaintext attacks (is IND-sID1-CPA secure) if* $\mathsf{Adv}_{\mathcal{E},\mathcal{A}}^{\text{ind-sid1-cpa}}$ *is negligible for all valid PPT* $\mathcal{A}$.

4.2 Using IBE Schemes in Constructions

In this section, we define the real resources we assume to be available and describe the protocol converters that are designed to construct the resources defined in Sect. 3 using an IBE scheme. Whether these constructions are achieved according to Definition 5 depends on the security properties of the IBE scheme, which we analyze in Sect. 5.

Delivery Controlled Channels. To construct a delivery controlled channel from a broadcast channel[8], we use an IBE scheme in a straightforward way: The

---

[8] Note that we consider the sender to be honest in this paper. Hence, assuming a broadcast channel to be available is not a strong assumption.

party at interface C generates all keys, sends the public key authentically to A and the user secret keys securely to the corresponding B_i. To send a message, A broadcasts an encryption thereof and the B_i with matching identity decrypt it. Hence, we need in addition to the broadcast channel an authenticated channel from C to A to transmit the public key and secure channels from C to each B_i. We abbreviate the network consisting of these channels as

$$\mathsf{NW} := \left[\mathsf{BCAST}^{A,\{B_1,\dots,B_n\}}, \mathsf{AUT}^{C,A}, \mathsf{SEC}^{C,B_1}, \dots, \mathsf{SEC}^{C,B_n} \right].$$

The real resource in our construction corresponds to the filtered resource $\mathsf{NW}_{\phi^{\mathsf{NW}}}$ where $\phi^{\mathsf{NW}} := (\phi_A^{\mathsf{NW}}, \phi_C^{\mathsf{NW}}, \phi_{B_1}^{\mathsf{NW}}, \dots, \phi_{B_n}^{\mathsf{NW}})$ with $\phi_I^{\mathsf{NW}} := [\mathbf{1}, \phi_I^{\mathsf{AUT}}, \phi_I^{\mathsf{SEC}}, \dots, \phi_I^{\mathsf{SEC}}]$ for $I \in \{A, C, B_1, \dots, B_n\}$.[9]

For an IBE scheme $\mathcal{E}$, we define protocol converters enc, dec, and reg as follows and let $\mathsf{IBE} := (\mathsf{enc}, \mathsf{reg}, \mathsf{dec}, \dots, \mathsf{dec})$: The converter enc first expects to receive a master public key mpk at its inside interface and stores it internally. On input a message and identity $(id, m) \in \mathcal{ID} \times \mathcal{M}$ at its outside interface, it computes $c \leftarrow \mathtt{Enc}(mpk, id, m)$ and outputs (id, c) at its inside sub-interface to $\mathsf{BCAST}^{A,\{B_1,\dots,B_n\}}$. The converter dec on input an identity and a corresponding user secret key (id, usk_{id}) at its inside interface, stores this tuple internally and outputs id at its outside interface. For all pairs (id_j, c_j) with $id_j = id$ stored internally, dec computes $m_j \leftarrow \mathtt{Dec}(usk_{id}, id, c_j)$ and outputs m_j at its outside interface. On input an identity and a ciphertext (id, c) at its inside interface, it stores (id, c) internally and if it has stored a user secret key for the identity id, computes $m \leftarrow \mathtt{Dec}(usk_{id}, id, c)$ and outputs (id, m) at its outside interface. The converter reg initially computes $(mpk, msk) \leftarrow \mathtt{Gen}()$, stores msk internally, and outputs mpk at its inside sub-interface to $\mathsf{AUT}_{\phi^{\mathsf{AUT}}}^{C,A}$. On input (id, i) at its outside interface, it computes $usk_{id} \leftarrow \mathtt{Ext}(msk, id)$ and outputs (id, usk_{id}) at its inside sub-interface to $\mathsf{SEC}_{\phi^{\mathsf{SEC}}}^{C,B_i}$.

Static Identity Management. To construct stDCC, the protocol at interface C has to reject registration requests for identities for which messages have already been sent. To be able to do so, it needs to know for which identities this is the case. We thus assume there is an additional authenticated channel from A to C that is used to inform C about usage of identities. The real resource is then $\mathsf{NW}_{\phi^{\mathsf{NW}+}}^+$ for

$$\mathsf{NW}^+ := \left[\mathsf{BCAST}^{A,\{B_1,\dots,B_n\}}, \mathsf{AUT}^{A,C}, \mathsf{AUT}^{C,A}, \mathsf{SEC}^{C,B_1}, \dots, \mathsf{SEC}^{C,B_n} \right]$$

---

[9] In this context, the channel SEC^{C,B_i} is a resource with $n+2$ interfaces where interface C corresponds to interface A of the resource in Definition 2, interface B_i corresponds to interface B, and interfaces B_j for $j \neq i$ correspond to copies of interface E. Similarly, ϕ_C^{SEC} corresponds to ϕ_A^{SEC} in Definition 3, $\phi_{B_i}^{\mathsf{SEC}}$ corresponds to ϕ_B^{SEC}, and $\phi_{B_j}^{\mathsf{SEC}}$ to ϕ_E^{SEC} for $j \neq i$. For simplicity, we do not introduce a different notation for the different filters.

and $\phi^{\mathsf{NW}^+} := (\phi_A^{\mathsf{NW}^+}, \phi_C^{\mathsf{NW}^+}, \phi_{B_1}^{\mathsf{NW}^+}, \dots, \phi_{B_n}^{\mathsf{NW}^+})$ where for $I \in \{A, C, B_1, \dots, B_n\}$, $\phi_I^{\mathsf{NW}} := [1, \phi_I^{\mathsf{AUT}}, \phi_I^{\mathsf{AUT}}, \phi_I^{\mathsf{SEC}}, \dots, \phi_I^{\mathsf{SEC}}]$.

We define the protocol $\mathsf{IBE}^{\mathsf{s}} := (\mathsf{enc}^{\mathsf{s}}, \mathsf{reg}^{\mathsf{s}}, \mathsf{dec}^{\mathsf{s}}, \dots, \mathsf{dec}^{\mathsf{s}})$ by describing the differences from IBE as follows: On input $(id, m) \in \mathcal{ID} \times \mathcal{M}$ at its outside interface, $\mathsf{enc}^{\mathsf{s}}$ additionally outputs id at its inside interface to $\mathsf{AUT}_{\phi^{\mathsf{AUT}}}^{A,C}$. The converter $\mathsf{reg}^{\mathsf{s}}$ on input id at its inside interface, stores this identity internally. It subsequently ignores inputs (id, i) at its outside interface if it has stored id.

Note that it is crucial for this construction that $\mathsf{AUT}^{A,C}$ cannot be interrupted or delayed. Otherwise an attacker could prevent C from learning that some identity has already been used to send messages and this identity could still be registered. In practice, one could realize such channel by letting C acknowledge the receipt while A sends the message only after receiving this acknowledgment. This would, however, contradict the goal of non-interactivity.

If such reliable channel is not available, we can still construct st2DCC from NW using the protocol $\mathsf{IBE}^{\mathsf{2s}} := (\mathsf{enc}^{\mathsf{2s}}, \mathsf{reg}^{\mathsf{2s}}, \mathsf{dec}^{\mathsf{2s}}, \dots, \mathsf{dec}^{\mathsf{2s}})$ defined as follows: It works as IBE, except that $\mathsf{reg}^{\mathsf{2s}}$ initially does not send mpk to A. On input $\mathtt{end\ registration}$ at its outside interface, $\mathsf{reg}^{\mathsf{2s}}$ sends mpk to A and ignores further inputs. The converter $\mathsf{enc}^{\mathsf{2s}}$ ignores all inputs until it receives mpk at its inside interface and from then on handles all inputs as enc.

Remark 3. Note that sending mpk is here used to signal A that it can now start sending messages. Since we assume that the sender is always honest, we do not need to require, e.g., that mpk cannot be computed from user secret keys; as long as mpk has not been sent, A will not send any messages.

Predetermined Identities. To construct $\mathsf{preDCC}_{\phi^{\mathsf{DCC}}}$ from $\mathsf{NW}_{\phi^{\mathsf{NW}^+}}^+$, we define the protocol $\mathsf{IBE}^{\mathsf{p}} = (\mathsf{enc}^{\mathsf{p}}, \mathsf{reg}^{\mathsf{p}}, \mathsf{dec}^{\mathsf{p}}, \dots, \mathsf{dec}^{\mathsf{p}})$ that uses a selectively secure IBE scheme. The protocol is almost identical to $\mathsf{IBE}^{\mathsf{s}}$ with the difference that $\mathsf{enc}^{\mathsf{p}}$ initially expects a finite set $\mathcal{S} \subseteq \mathcal{ID}$ (encoded as a list of its elements) as input at its outside interface. On this input, it stores $\mathcal{S}$ internally, sends ok to C via $\mathsf{AUT}_{\phi^{\mathsf{AUT}}}^{A,C}$, and subsequently ignores all inputs (id, m) for $id \notin \mathcal{S}$. The converter $\mathsf{reg}^{\mathsf{p}}$ initially waits and ignores all inputs at its outside interface until it receives the input ok at its inside interface. It then sends mpk to A and from then on behaves identically to $\mathsf{reg}^{\mathsf{2s}}$.

Similarly, we define a protocol $\mathsf{IBE}^{\mathsf{2p}} = (\mathsf{enc}^{\mathsf{2p}}, \mathsf{reg}^{\mathsf{2p}}, \mathsf{dec}^{\mathsf{2p}}, \dots, \mathsf{dec}^{\mathsf{2p}})$ to construct $\mathsf{pre2DCC}_{\phi^{\mathsf{DCC}}}$ from $\mathsf{NW}_{\phi^{\mathsf{NW}^+}}^+$. It works as IBE except that $\mathsf{enc}^{\mathsf{2p}}$ initially expects a finite set $\mathcal{S} \subseteq \mathcal{ID}$ (encoded as a list of its elements) as input at its outside interface. On this input, it stores $\mathcal{S}$ internally, sends ok to C via $\mathsf{AUT}_{\phi^{\mathsf{AUT}}}^{A,C}$, and ignores all further inputs until it receives mpk over $\mathsf{AUT}_{\phi^{\mathsf{AUT}}}^{C,A}$. From then on, it handles all inputs as enc, but ignores inputs (id, m) for $id \notin \mathcal{S}$. The converter $\mathsf{reg}^{\mathsf{2p}}$ initially waits and ignores all inputs at its outside interface until it receives the input ok at its inside interface. It then accepts registration requests at its outside interface as reg. On input $\mathtt{end\ registration}$ at its outside interface, $\mathsf{reg}^{\mathsf{2p}}$ sends mpk to A and ignores further inputs.

Remark 4. While both IBE^p and IBE^2p need $\mathsf{AUT}^{A,C}_{\phi\mathsf{AUT}}$, IBE^2p uses this channel only once in the beginning to let A send ok to C. The availability of such channel only at the beginning might be easier to guarantee in practice.

5 Constructing Delivery Controlled Channels

5.1 Impossibility of Construction

We now show that there is no IBE scheme that can be used to construct $\mathsf{DCC}_{\phi\mathsf{DCC}}$ from $\mathsf{NW}_{\phi\mathsf{NW}}$.

Theorem 1. *Let* $n > 0$, $\mathcal{ID}$ *a nonempty set, and let* $\rho \in \mathbb{N}$. *Then there is no IBE scheme such that we have for the corresponding protocol* IBE

$$\mathsf{NW}_{\phi\mathsf{NW}} \xrightarrow[\{B_1,\ldots,B_n\}]{\mathsf{IBE}} \mathsf{DCC}^{n,\mathcal{ID},\rho}_{\phi\mathsf{DCC}}.$$

Proof. This proof closely resembles Nielsen's impossibility proof of non-committing public-key encryption [17]. Assume $\mathsf{IBE} = (\mathsf{enc}, \mathsf{reg}, \mathsf{dec}, \ldots, \mathsf{dec})$ achieves the construction and let $\mathcal{P} := \{B_1\}$. Then there exists a converter σ_{B_1} such that $\mathsf{IBE}_{\overline{\mathcal{P}}}\phi^{\mathsf{NW}}_{\overline{\mathcal{P}}}\mathsf{NW} \approx \sigma_{\mathcal{P}}\phi^{\mathsf{DCC}}_{\overline{\mathcal{P}}}\mathsf{DCC}^{n,\mathcal{ID},\rho}$. Let $id \in \mathcal{ID}$, let ν be an upper bound on the length of the output of $\mathsf{Ext}(\cdot, id)$, and consider the following distinguisher: The distinguisher $\mathbf{D}$ chooses $m \in \{0,1\}^{\nu+1}$ uniformly at random and inputs (id, m) at interface A. Let (id, c) be the resulting output at interface B_1 (if there is no such output, $\mathbf{D}$ returns 0). Then, $\mathbf{D}$ inputs $(id, 1)$ at interface C. Let (id, usk) be the resulting output at interface B_1 and return 0 if there is no such output or if $|usk| > \nu$. Finally, $\mathbf{D}$ inputs first (id, c) and then (id, usk) at the inside interface of dec and returns 1 if dec outputs id and m at its outside interface, and 0 otherwise.

Correctness of the IBE scheme implies that $\mathbf{D}$ always outputs 1 if connected to the real resource. In the ideal world, c is generated independently of m only given $|m|$ because σ_{B_1} does not learn m until $(id, 1)$ is input at interface C. Moreover, there are at most 2^ν possible values for usk such that $|usk| \leq \nu$. Hence, there are at most 2^ν values of m such that there exists a usk that decrypts c to m with probability more than $\frac{1}{2}$. Since m was chosen uniformly from $\{0,1\}^{\nu+1}$, $\mathbf{D}$ outputs 1 with probability at most $\frac{1}{2} + \frac{1}{2} \cdot \frac{1}{2} = \frac{3}{4}$ when connected to the ideal resource. Thus, the distinguishing advantage is at least $\frac{1}{4}$, which is a contradiction. $\square$

5.2 Equivalence of IND-ID-CPA Security and Construction of Statically Delivery Controlled Channels

While no IBE scheme constructs $\mathsf{DCC}_{\phi\mathsf{DCC}}$ from $\mathsf{NW}_{\phi\mathsf{NW}}$, we show that IND-ID-CPA security is sufficient to construct $\mathsf{stDCC}_{\phi\mathsf{DCC}}$ from $\mathsf{NW}^+_{\phi\mathsf{NW}+}$. See the full version for a proof.

Lemma 1. *Let ρ be an upper bound on the randomness used in one invocation of Gen, Ext, and Enc. Then, there exist efficient converters $\sigma_{B_1}, \ldots, \sigma_{B_n}$ such that for all $\mathcal{P} \subseteq \{B_1, \ldots, B_n\}$ and for all efficient distinguishers $\mathbf{D}$ that input at most q messages at interface A, there exists an efficient algorithm $\mathcal{A}$ such that*

$$\Delta^{\mathbf{D}} \left(\mathsf{IBE}^{\mathsf{s}}_{\overline{\mathcal{P}}} \phi^{\mathsf{NW}^+}_{\overline{\mathcal{P}}} \mathsf{NW}^+, \sigma_{\mathcal{P}} \phi^{\mathsf{DCC}}_{\overline{\mathcal{P}}} \mathsf{stDCC}^{n,\mathcal{ID},\rho} \right) = 2q \cdot \left| \mathsf{Adv}^{\mathsf{ind\text{-}id\text{-}cpa}}_{\mathcal{E},\mathcal{A}} \right|.$$

We now prove conversely that IND-ID-CPA security is also necessary for the construction:

Lemma 2. *Let $\rho \in \mathbb{N}$ and $\mathcal{P} \subseteq \{B_1, \ldots, B_n\}, \mathcal{P} \neq \emptyset$. Then, for all valid IND-ID-CPA adversaries $\mathcal{A}$ and for all efficient converters σ_{B_i} for $B_i \in \mathcal{P}$, there exists an efficient distinguisher $\mathbf{D}$ such that*

$$\left| \mathsf{Adv}^{\mathsf{ind\text{-}id\text{-}cpa}}_{\mathcal{E},\mathcal{A}} \right| = \Delta^{\mathbf{D}} \left(\mathsf{IBE}^{\mathsf{s}}_{\overline{\mathcal{P}}} \phi^{\mathsf{NW}^+}_{\overline{\mathcal{P}}} \mathsf{NW}^+, \sigma_{\mathcal{P}} \phi^{\mathsf{DCC}}_{\overline{\mathcal{P}}} \mathsf{stDCC}^{n,\mathcal{ID},\rho} \right).$$

Proof. Let $\mathcal{A}$ be a valid IND-ID-CPA adversary and let σ_{B_i} be efficient converters for $B_i \in \mathcal{P}$. Further let $B_i \in \mathcal{P}$. We now define two distinguishers, $\mathbf{D}_0$ and $\mathbf{D}_1$. Let mpk be the initial output at interface B_i of the resource connected to the distinguisher (if nothing is output, let mpk be some default value[10]). Both distinguishers then invoke $\mathcal{A}(mpk)$. The oracle query id' of $\mathcal{A}$ is answered as follows by both distinguishers: They input (id', i) at interface C and let the answer to the query be $usk_{id'}$ where $(id', usk_{id'})$ is the resulting output of the resource at interface B_i (and let $usk_{id'}$ be some default value if there is no such output). If $\mathcal{A}$ returns $(, id, m_0, m_1)$, $\mathbf{D}_0$ and $\mathbf{D}_1$ input (id, m_0) and (id, m_1) at interface A, respectively. Now let (id, c^*) be the resulting output at the sub-interface of B_i corresponding to $\mathsf{BCAST}^{A,\{B_1,\ldots,B_n\}}$ (and let c^* be some default value if there is no such output). Both distinguishers then invoke $\mathcal{A}$ on input $(, c^*)$. Oracle queries are answered as above. Note that id will not be queried since $\mathcal{A}$ is a valid IND-ID-CPA adversary and therefore inputs at interface C will be handled as before. Finally, $\mathbf{D}_0$ and $\mathbf{D}_1$ output the bit returned by $\mathcal{A}$.

Note that for all $\beta \in \{0,1\}$

$$\Pr \left[\mathbf{D}_\beta \left(\mathsf{IBE}^{\mathsf{s}}_{\overline{\mathcal{P}}} \phi^{\mathsf{NW}^+}_{\overline{\mathcal{P}}} \mathsf{NW}^+ \right) = 1 \right] = \Pr \left[\mathsf{Exp}^{\mathsf{ind\text{-}id\text{-}cpa}}_{\mathcal{E},\mathcal{A}} = \beta \mid b = \beta \right]$$

because the outputs of the real system are precisely generated as the corresponding values in the IND-ID-CPA experiment. Further note that we have

$$\Pr \left[\mathbf{D}_0 \left(\sigma_{\mathcal{P}} \phi^{\mathsf{DCC}}_{\overline{\mathcal{P}}} \mathsf{stDCC}^{n,\mathcal{ID},\rho} \right) = 1 \right] = \Pr \left[\mathbf{D}_1 \left(\sigma_{\mathcal{P}} \phi^{\mathsf{DCC}}_{\overline{\mathcal{P}}} \mathsf{stDCC}^{n,\mathcal{ID},\rho} \right) = 1 \right]$$

since $\mathbf{D}_0$ and $\mathbf{D}_1$ only differ in the message they input and σ_{B_i} only learns the length of that message, which is the same for the two messages (since $\mathcal{A}$ is a valid IND-ID-CPA adversary), so its output does not depend on the choice of

---

[10] Note that this is only possible in the ideal system if σ_{B_i} is flawed. Hence, one could distinguish better in this case, but we do not need that for the proof.

the message. Now let $\mathbf{D}$ be the distinguisher that chooses $\beta \in \{0,1\}$ uniformly at random, runs $\mathbf{D}_\beta$, and outputs the XOR of $\mathbf{D}_\beta$'s output and β. We conclude

$$\left| \mathsf{Adv}_{\mathcal{E},\mathcal{A}}^{\text{ind-id-cpa}} \right| = \left| \Pr\left[\mathsf{Exp}_{\mathcal{E},\mathcal{A}}^{\text{ind-id-cpa}} = 1 \right] - \frac{1}{2} \right|$$

$$= \frac{1}{2} \left| \Pr\left[\mathsf{Exp}_{\mathcal{E},\mathcal{A}}^{\text{ind-id-cpa}} = 1 \,\middle|\, b = 0 \right] + \Pr\left[\mathsf{Exp}_{\mathcal{E},\mathcal{A}}^{\text{ind-id-cpa}} = 1 \,\middle|\, b = 1 \right] - 1 \right|$$

$$= \frac{1}{2} \left| \Pr\left[\mathsf{Exp}_{\mathcal{E},\mathcal{A}}^{\text{ind-id-cpa}} = 0 \,\middle|\, b = 0 \right] - \Pr\left[\mathsf{Exp}_{\mathcal{E},\mathcal{A}}^{\text{ind-id-cpa}} = 1 \,\middle|\, b = 1 \right] \right|$$

$$= \frac{1}{2} \left| \Pr\left[\mathbf{D}_0 \left(\mathsf{IBE}_{\overline{\mathcal{P}}}^{\mathsf{s}} \phi_{\overline{\mathcal{P}}}^{\mathsf{NW}^+} \mathsf{NW}^+ \right) = 1 \right] - \Pr\left[\mathbf{D}_1 \left(\mathsf{IBE}_{\overline{\mathcal{P}}}^{\mathsf{s}} \phi_{\overline{\mathcal{P}}}^{\mathsf{NW}^+} \mathsf{NW}^+ \right) = 1 \right] \right|$$

$$= \frac{1}{2} \left| \Pr\left[\mathbf{D}_0 \left(\mathsf{IBE}_{\overline{\mathcal{P}}}^{\mathsf{s}} \phi_{\overline{\mathcal{P}}}^{\mathsf{NW}^+} \mathsf{NW}^+ \right) = 1 \right] + \Pr\left[\mathbf{D}_1 \left(\mathsf{IBE}_{\overline{\mathcal{P}}}^{\mathsf{s}} \phi_{\overline{\mathcal{P}}}^{\mathsf{NW}^+} \mathsf{NW}^+ \right) = 0 \right] \right.$$

$$\left. - \Pr\left[\mathbf{D}_0 \left(\sigma_{\mathcal{P}} \phi_{\overline{\mathcal{P}}}^{\mathsf{DCC}} \mathsf{stDCC}^{n,\mathcal{ID},\rho} \right) = 1 \right] - \Pr\left[\mathbf{D}_1 \left(\sigma_{\mathcal{P}} \phi_{\overline{\mathcal{P}}}^{\mathsf{DCC}} \mathsf{stDCC}^{n,\mathcal{ID},\rho} \right) = 0 \right] \right|$$

$$= \Delta^{\mathbf{D}} \left(\mathsf{IBE}_{\overline{\mathcal{P}}}^{\mathsf{s}} \phi_{\overline{\mathcal{P}}}^{\mathsf{NW}^+} \mathsf{NW}^+, \sigma_{\mathcal{P}} \phi_{\overline{\mathcal{P}}}^{\mathsf{DCC}} \mathsf{stDCC}^{n,\mathcal{ID},\rho} \right). \qquad \square$$

Lemma 1 and 2 together imply the following theorem:

Theorem 2. *Let ρ be an upper bound on the randomness used in one invocation of Gen, Ext, and Enc. We then have*

$$\mathsf{NW}^+_{\phi^{\mathsf{NW}^+}} \overset{\mathsf{IBE}^{\mathsf{s}}}{\underset{\{B_1,\ldots,B_n\}}{\Longmapsto}} \mathsf{stDCC}^{n,\mathcal{ID},\rho}_{\phi^{\mathsf{DCC}}}$$

$$\Longleftrightarrow \quad \textit{the underlying IBE scheme is IND-ID-CPA-secure.}$$

The following theorem can be proven very similarly by observing that the reductions used to prove Theorem 2 translate queries to the Ext oracle by the adversary to inputs at interface C by the distinguisher and vice versa and that $\mathsf{NW}_{\phi^{\mathsf{NW}}}$ and $\mathsf{st2DCC}^{n,\mathcal{ID},\rho}_{\phi^{\mathsf{DCC}}}$ restrict such inputs exactly as $\mathcal{A}$ is restricted in $\mathsf{Exp}_{\mathcal{E},\mathcal{A}}^{\text{ind-id1-cpa}}$.

Theorem 3. *Let ρ be an upper bound on the randomness used in one invocation of Gen, Ext, and Enc. We then have*

$$\mathsf{NW}_{\phi^{\mathsf{NW}}} \overset{\mathsf{IBE}^{\mathsf{2s}}}{\underset{\{B_1,\ldots,B_n\}}{\Longmapsto}} \mathsf{st2DCC}^{n,\mathcal{ID},\rho}_{\phi^{\mathsf{DCC}}}$$

$$\Longleftrightarrow \quad \textit{the underlying IBE scheme is IND-ID1-CPA-secure.}$$

Selective Security. We similarly show the equivalence of IND-sID-CPA security and the construction of statically delivery controlled channels with predetermined identities in the full version.

6   Construction with Random Oracles

6.1   Random Oracles

We show how any IND-ID-CPA secure IBE scheme $\mathcal{E} = (\mathsf{Gen}, \mathsf{Ext}, \mathsf{Enc}, \mathsf{Dec})$ can be used to construct DCC from the resource $\mathsf{NW}^{\mathsf{RO}}$, which corresponds to our network together with a random oracle. A random oracle is a uniform random function $\{0,1\}^* \to \{0,1\}^k$ for some k to which all parties have access. The heuristic to model a hash function as a random oracle was proposed by Bellare and Rogaway [2]. Theorem 1 implies that no hash function can be used to instantiate the random oracle in this construction. However, if a random oracle is actually available, e.g., via a trusted party or secure hardware, the overall construction is sound. For our purpose, it is sufficient to consider random oracles with binary co-domain.

Definition 13. *The resource* RO *has interfaces A, C, and $B_1, \ldots, B_n$. On input $x \in \{0,1\}^*$ at interface $I \in \{A, C, B_1, \ldots, B_n\}$, if x has not been input before (at any interface), RO chooses $y \in \{0,1\}$ uniformly at random and outputs y at interface I; if x has been input before and the resulting output was y, RO outputs y at interface I.*

Programmability. For our construction, we will assume that a random oracle is available as part of the real resource. Our protocol then constructs an ideal resource that does not give the honest parties access to the random oracle. Thus, the simulators in the ideal world can answer queries to the random oracle arbitrarily as long as they are consistent with previous answers and are indistinguishable from uniform bits. This gives the simulators additional power which allows us to overcome the impossibility result from Theorem 1. Since the simulators can in some sense "reprogram" the random oracle, we are in a scenario that is often referred to as *programmable random oracle model.*

6.2   Construction of Delivery Controlled Channels

Our protocol $\mathsf{IBE}^{\mathsf{ro}}$ uses the same idea as Nielsen's scheme [17] and essentially corresponds to the transformation from [4, Section 5.3] (see also [11]) applied to an IBE scheme. At a high level, it works as follows: To send a message m for identity id, choose a bit string r (of sufficient length, say λ) uniformly at random, input $(r, 1), \ldots, (r, |m|)$ to the random oracle to obtain a uniform value r' with $|r'| = |m|$. Finally encrypt r with the IBE scheme for identity id and send the resulting ciphertext together with $m \oplus r'$. The security proof exploits that the one-time pad is non-committing and the random oracle is programmable.

A detailed description of the protocol and the involved resources as well as a proof sketch of the following theorem can be found in the full version.

Theorem 4. *Let ρ be an upper bound on the randomness used in one invocation of* Gen*,* Ext *and* Enc*. If $\mathcal{E}$ is IND-ID-CPA secure, we have*

$$\mathsf{NW}^{\mathsf{RO}}_{\phi^{\mathsf{NWRO}}} \xrightarrow[\{B_1, \ldots, B_n\}]{\mathsf{IBE}^{\mathsf{ro}}} \left[\mathsf{DCC}^{n, \mathcal{ID}, \rho + \lambda}_{\phi^{\mathsf{DCC}}}, \mathsf{RO}_{\phi^{\mathsf{RO}}} \right].$$

Acknowledgments. Ueli Maurer was supported by the Swiss National Science Foundation (SNF), project no. 200020-132794. Dennis Hofheinz was supported by DFG grants HO 4534/2-2 and HO 4534/4-1.

References

1. Beaver, D.: Foundations of secure interactive computing. In: Feigenbaum, J. (ed.) CRYPTO 1991. LNCS, vol. 576, pp. 377–391. Springer, Heidelberg (1992)
2. Bellare, M., Rogaway, P.: Random oracles are practical: a paradigm for designing efficient protocols. In: Proceedings of the 1st ACM Conference on Computer and Communications Security, CCS 1993, pp. 62–73. ACM, New York (1993)
3. Boneh, D., Franklin, M.: Identity-based encryption from the weil pairing. In: Kilian, J. (ed.) CRYPTO 2001. LNCS, vol. 2139, pp. 213–229. Springer, Heidelberg (2001)
4. Boneh, D., Sahai, A., Waters, B.: Functional encryption: definitions and challenges. In: Ishai, Y. (ed.) TCC 2011. LNCS, vol. 6597, pp. 253–273. Springer, Heidelberg (2011)
5. Canetti, R.: Universally composable security: a new paradigm for cryptographic protocols. In: Proceedings of FOCS 2001, pp. 136–145. IEEE Computer Society (2001)
6. Canetti, R., Halevi, S., Katz, J.: A forward-secure public-key encryption scheme. In: Biham, E. (ed.) EUROCRYPT 2003. LNCS, vol. 2656, pp. 255–271. Springer, Heidelberg (2003)
7. Cocks, C.: An Identity based encryption scheme based on quadratic residues. In: Honary, B. (ed.) Cryptography and Coding 2001. LNCS, vol. 2260, pp. 360–363. Springer, Heidelberg (2001)
8. Coretti, S., Maurer, U., Tackmann, B.: Constructing confidential channels from authenticated channelspublic-key encryption revisited. In: Sako, K., Sarkar, P. (eds.) Advances in Cryptology - ASIACRYPT 2013. Lecture Notes in Computer Science, vol. 8269, pp. 134–153. Springer, Heidelberg (2013)
9. Gentry, C., Peikert, C., Vaikuntanathan, V.: Trapdoors for hard lattices and new cryptographic constructions. In: Proceedings of STOC 2008, pp. 197–206. ACM (2008)
10. Goldreich, O., Micali, S., Wigderson, A.: How to play any mental game or a completeness theorem for protocols with honest majority. In: Proceedings of STOC 1987, pp. 218–229. ACM (1987)
11. Matt, C., Maurer, U.: A definitional framework for functional encryption. Cryptology ePrint Archive, Report 2013/559 (2013)
12. Maurer, U.: Constructive cryptography – a new paradigm for security definitions and proofs. In: Mödersheim, S., Palamidessi, C. (eds.) TOSCA 2011. LNCS, vol. 6993, pp. 33–56. Springer, Heidelberg (2012)
13. Maurer, U., Renner, R.: Abstract cryptography. In: Chazelle, B. (ed.) The Second Symposium on Innovations in Computer Science, ICS 2011, pp. 1–21. Tsinghua University Press January 2011
14. Maurer, U.M., Yacobi, Y.: Non-interactive public-key cryptography. In: Davies, D.W. (ed.) EUROCRYPT 1991. LNCS, vol. 547, pp. 498–507. Springer, Heidelberg (1991)
15. Micali, S., Rogaway, P.: Secure computation. In: Feigenbaum, J. (ed.) CRYPTO 1991. LNCS, vol. 576, pp. 392–404. Springer, Heidelberg (1992)
16. Naor, M., Yung, M.: Public-key cryptosystems provably secure against chosen ciphertext attacks. In: STOC, pp. 427–437. ACM (1990)

17. Nielsen, J.B.: Separating random oracle proofs from complexity theoretic proofs: the non-committing encryption case. In: Yung, M. (ed.) CRYPTO 2002. LNCS, vol. 2442, p. 111. Springer, Heidelberg (2002)
18. Nishimaki, R., Manabe, Y., Okamoto, T.: Universally composable identity-based encryption. In: Nguyên, P.Q. (ed.) VIETCRYPT 2006. LNCS, vol. 4341, pp. 337–353. Springer, Heidelberg (2006)
19. Pfitzmann, B., Waidner, M.: A model for asynchronous reactive systems and its application to secure message transmission. In: Proceedings of IEEE Symposium on Security and Privacy 2001, pp. 184–200. IEEE Computer Society (2001)
20. Shamir, A.: Identity-based cryptosystems and signature schemes. In: Blakely, G.R., Chaum, D. (eds.) CRYPTO 1984. LNCS, vol. 196, pp. 47–53. Springer, Heidelberg (1985)
21. Waters, B.: Efficient identity-based encryption without random oracles. In: Cramer, R. (ed.) EUROCRYPT 2005. LNCS, vol. 3494, pp. 114–127. Springer, Heidelberg (2005)

A Framework for Identity-Based Encryption with Almost Tight Security

Nuttapong Attrapadung, Goichiro Hanaoka, and Shota Yamada[(⊠)]

National Institute of Advanced Industrial Science and Technology (AIST),
Tokyo, Japan
{n.attrapadung,hanaoka-goichiro,yamada-shota}@aist.go.jp

Abstract. We show a framework for constructing identity-based encryption (IBE) schemes that are (almost) tightly secure in the multi-challenge and multi-instance setting. In particular, we formalize a new notion called *broadcast encoding*, analogously to encoding notions by Attrapadung (Eurocrypt 2014) and Wee (TCC 2014). We then show that it can be converted into such an IBE. By instantiating the framework using several encoding schemes (new or known ones), we obtain the following:
- We obtain (almost) tightly secure IBE in the multi-challenge, multi-instance setting, both in composite and prime-order groups. The latter resolves the open problem posed by Hofheinz et al. (PKC 2015).
- We obtain the first (almost) tightly secure IBE with sub-linear size public parameters (master public keys). In particular, we can set the size of the public parameters to constant at the cost of longer ciphertexts and private keys. This gives a partial solution to the open problem posed by Chen and Wee (Crypto 2013).

By applying (a variant of) the Canetti-Halevi-Katz transformation to our schemes, we obtain several CCA-secure PKE schemes with tight security in the multi-challenge, multi-instance setting. One of our schemes achieves very small ciphertext overhead, consisting of less than 12 group elements. This significantly improves the state-of-the-art construction by Libert et al. (in ePrint Archive) which requires 47 group elements. Furthermore, by modifying one of our IBE schemes obtained above, we can make it anonymous. This gives the first anonymous IBE whose security is almost tightly shown in the multi-challenge setting.

Keywords: Tight security reduction · Identity-based encryption · Multi-challenge security · Chosen ciphertext security

1 Introduction

1.1 Backgrounds

In the context of provable security, we reduce the security of a given scheme to the hardness of a computational problem, in order to gain confidence in the security of the scheme. Namely, we assume an adversary $\mathcal{A}$ who breaks the scheme and

© International Association for Cryptologic Research 2015
T. Iwata and J.H. Cheon (Eds.): ASIACRYPT 2015, Part I, LNCS 9452, pp. 521–549, 2015.
DOI: 10.1007/978-3-662-48797-6_22

then show another adversary $\mathcal{B}$ who solves the (assumed) hard problem using $\mathcal{A}$. Such a reduction should be as *tight* as possible, in the sense that $\mathcal{B}$'s success probability is as large as $\mathcal{A}$. In this paper, we mostly focus on the tight security reduction in identity-based encryption (IBE) [47].

IBE is an advanced form of public key encryption in which one can encrypt a message for a user identity, rather than a public key. The first fully secure (or often called, adaptively secure) construction in the standard model was given in [11]. Later, further developments were made [8,29,48,49]. All the above mentioned papers only deal with the single-challenge, single-instance case. Since it is known that the security in the (much more realistic) multi-challenge and multi-instance setting can be reduced to the security in the single-challenge and single-instance setting [7], these schemes are secure in the former setting in asymptotic sense. However, this reduction incurs $O(\mu Q_c)$ security loss, where Q_c is the number of challenge queries made by the adversary and μ is the number of instances. Since all the above schemes already loose at least $O(Q_k)$ security in the reductions, where Q_k is the number of key extraction queries made by $\mathcal{A}$, theses schemes loose at least $O(\mu Q_c Q_k)$ security in total.

Recently and somewhat surprisingly, Chen and Wee [17,19] showed the first IBE scheme (CW scheme) whose reduction cost is independent of Q_k, resolving an important open question posed in [48]. Subsequently, Blazy et al. [9] were able to obtain anonymous IBE and hierarchical IBE with the same security guarantee. The drawback of these schemes is its large public parameters (master public keys): It is proportional to the security parameter and thus rather large. Note that they only consider the single-challenge and single-instance setting. Very recently, further important development was made by Hofheinz, Koch, and Striecks [31] who extended the proof technique of Chen and Wee in a novel way and proposed the first IBE scheme (HKS scheme) whose reduction cost is independent from all of μ, Q_c, and Q_k. However, they only give a construction in composite-order groups and explicitly mention that the construction in prime-order groups remains open. We focus on the following two important open problems in this paper:

- *Can we construct a fully, (almost) tightly secure IBE scheme in the multi-challenge and multi-instance setting from a static assumption in the prime-order groups?*
- *Can we construct a fully, (almost) tightly secure IBE scheme from a static assumption with constant-size public parameters even in the single-challenge and single-instance setting?*

1.2 Our Results

New Tightly-Secure IBE Schemes. In this paper, to tackle the above problems, we revisit the proof technique in [17,31] and propose a framework for constructing almost tightly secure IBE. The almost tight security means that the reduction cost is independent from μ, Q_c, and Q_k, and is a small polynomial in the security parameter. In particular, we formalize the notion of broadcast encoding analogously to Attrapadung [4] and Wee [50]. Then we show that it can be converted into fully, (almost) tightly secure IBE scheme, in the multi-challenge and multi-instance setting. We propose such conversions both in prime-order and composite-order groups. Furthermore, we propose two broadcast encoding schemes satisfying our requirement. By instantiating our generic conversion with these schemes, we obtain several new IBE schemes. In particular,

- We obtain the first IBE scheme in *prime-order groups* with almost tight security in the multi-challenge and multi-instance setting. The security of our scheme can be shown under the decisional linear (DLIN) assumption. This resolves the first question above.
- We obtain the first IBE scheme with almost tight security in the multi-challenge and multi-instance setting and with *sub-linear public parameter-size* (but at the cost of larger private key and ciphertext size). An IBE scheme with almost tight security and sub-linear public parameter size is not known, even in the single-challenge setting. This partially answers the second question above.

Application to Chosen-Ciphertext Secure Public Key Encryption. By applying a variant of Canetti-Halevi-Katz transformation to the new IBE schemes, we obtain several new chosen-ciphertext (CCA) secure public key encryption (PKE) schemes. The conversion is tightness-preserving, namely, if the original IBE is tightly secure in the multi-challenge and multi-instance setting, the resulting PKE scheme is also tightly secure in the same setting. One of our schemes achieves very compact ciphertext size. The ciphertext overhead of the scheme only consists of 10 group elements and 2 elements in $\mathbb{Z}_p$. This is much shorter than the state-of-the-art construction of PKE scheme with the same security guarantee [34]: their scheme requires 47 group elements.

Extension to Anonymous IBE. Furthermore, by modifying one of the new IBE schemes obtained above, we obtain the first anonymous IBE scheme with (almost) tight security reduction in the multi-challenge settings for the first time. The security proof is done by carefully combining information-theoretic argument due to Chen et al. [16] and a computational argument.

See Table 1 for overview of our schemes.

1.3   Our Techniques

Difficulties. To solve the first question above, a natural starting point would be trying to apply the frameworks for composite-order-to-prime-order-conversion dedicated to identity/attribute-based encryption [2,3,16,18,35] to the HKS scheme [31]. However, security proofs for CW and HKS schemes significantly deviate from the most standard form of dual system encryption methodology [4,37,39,50], only for which the above mentioned frameworks can be applied.

Table 1. Comparison of almost tight IBE from static assumptions

Schemes	$\|pp\| + \|mpk\|$	$\|CT\|$	$\|sk_{ID}\|$	Anon?	Multi-challenge?	Underlying group	Security assumption
CW13 [17]	$O(\kappa)$	$O(1)$	$O(1)$	No	No	Composite	SGD, CW
HKS15 [31]	$O(\kappa)$	$O(1)$	$O(1)$	No	Yes	Composite	SGD, HKS
Ours: Φ_{cc}^{comp}	$O(\kappa)$	$O(1)$	$O(1)$	No	Yes	Composite	SGD, Problem 5
Ours: Φ_{slp}^{comp}	$O(\kappa^{1-c})$	$O(\kappa^c)$	$O(\kappa^c)$	No	Yes	Composite	SGD, DLIN
CW13 [17]†	$O(\kappa)$	$O(1)$	$O(1)$	No	No	Prime	DLIN
BKP14 [9]*†	$O(\kappa)$	$O(1)$	$O(1)$	Yes	No	Prime	DLIN
Ours: Φ_{cc}^{prime}	$O(\kappa)$	$O(1)$	$O(1)$	No	Yes	Prime	DLIN
Ours: Φ_{slp}^{prime}	$O(\kappa^{1-c})$	$O(\kappa^c)$	$O(\kappa^c)$	No	Yes	Prime	DLIN
Ours: Φ_{anon}	$O(\kappa)$	$O(1)$	$O(1)$	Yes	Yes	Prime	DLIN

We compare IBE schemes focusing tight security reduction from static assumptions in the standard model. $\|pp\| + \|mpk\|$, $\|CT\|$, and $\|sk_{ID}\|$ show the size of the master public keys and public parameters, ciphertexts, and private keys, respectively. To measure the efficiency, we count the number of group elements. In the table, κ denotes the security parameter. "Anon" shows whether the scheme is anonymous. "Multi-Challenge?" asks whether (almost) tight security reduction in the multi-challenge setting is shown. "SGD" stands for sub-group decision assumptions. "CW" and "HKS" denote specific assumptions used in the corresponding papers. For Φ_{slp}^{comp} and Φ_{slp}^{prime}, we can assign any $0 \le c \le 1$.
* This is the only scheme that can be generalized to HIBE.
† These schemes can be generalized to be secure under the k-linear assumption (k-LIN) [28,46] for any $k \in \mathbb{N}$. In such a case, $\|pp\| + \|mpk\|$, $\|CT\|$, and $\|sk_{ID}\|$ are changed to be $O(k^2\kappa)$, $O(k)$, and $O(k)$, respectively. Note that the DLIN assumption corresponds to the 2-LIN assumption.

Another approach is to try to convert specific assumptions they use into prime-order. In fact, Chen and Wee [17] were able to accomplish such a conversion for their scheme. However, their technique is non-generic and therefore it is highly unclear whether the same argument is possible for the assumptions that HKS use.

Next, we explain the difficulty of the second question. The reason why all IBE schemes featuring (almost) tight security reduction in previous works [9,17,31] require large public parameters is that they use (randomized version of) Naor-Reingold PRF [40] in their construction. Note that the Naor-Reingold PRF requires seed length which is linear in the input size, which in turn implies rather long public parameters in the IBE schemes. A natural approach to improve the efficiency would be, as noted by Chen and Wee [17,19], to reduce the seed length of the Naor-Reingold PRF. However, this is a long-standing open problem and turns out to be quite difficult.

Our Strategy. In this paper, we introduce new proof techniques for IBE schemes (with almost tight security) that rely *only on the subgroup decision assumptions*[1] This allows us to use frameworks for composite-order-to-prime-order conversions in the literature [2,3,16,22,23,26,35,42] (to name only a few) which converts subgroup decision assumption into a static assumption in prime-order groups,

---

[1] In fact, we also require the decisional bilinear Diffie-Hellman (DBDH) assumption on the composite-order groups (Problem 5) in addition to the subgroup decision assumptions. However, the assumption does not use the power of composite-order groups. In other words, it does *not* imply the factoring assumption. Therefore, it is ready to be converted into prime-order.

such as the DLIN assumption. Therefore, using these techniques, we are able to convert a variant of HKS scheme into prime-order. This answers the first question above. Note that in the security proof of HKS (and CW), they rely on some specific assumptions in composite-order groups in addition to subgroup decision assumptions. Because of these, it is unclear how to convert HKS scheme into prime-order.

As for the second question, we view Chen and Wee's scheme as being constructed from, somewhat surprisingly, *broadcast encryption* mechanism, instead of (Naor-Reingold) PRF, and hence can avoid the above difficulty regarding PRF. More precisely, we show that the task of constructing almost tightly secure IBE scheme is essentially reduced to a construction of broadcast encryption, and based on this idea, we are able to obtain the first IBE scheme with sub-linear size public parameters and almost tight security. In the following, we explain our technique.

Detailed Overview of Our Technique. Let us start from the following variant of the Chen and Wee's IBE scheme. Let the identity space of the scheme be $\{0,1\}^\ell$. For $i \in \{1,2,3\}$, let g_i be the generator of a subgroup of order p_i of $\mathbb{G}$, which is bilinear groups of composite order $N = p_1 p_2 p_3$. Let also h be a generator of $\mathbb{G}$. The master public key, a ciphertext, and a private key for an identity ID are in the following form:

$$\mathsf{mpk} = \left(g_1, g_1^{w_{1,0}}, g_1^{w_{1,1}}, \ldots, g_1^{w_{\ell,0}}, g_1^{w_{\ell,1}}, e(g_1,h)^\alpha\right),$$

$$\mathsf{CT_{ID}} = \left(g_1^s, \; g_1^{s\sum_{i\in[1,\ell]} w_{i,\mathsf{ID}_i}}, \; e(g_1,h)^{s\alpha} \cdot \mathsf{M}\right), \quad \mathsf{sk_{ID}} = \left(h^\alpha \cdot g_1^{r\sum_{i\in[1,\ell]} w_{i,\mathsf{ID}_i}}, \; g_1^{-r}\right)$$

where ID_i is the i-th bit of ID and M is the message.[2] Now we are going to show the security. We only consider the single-challenge and single-instance case here for simplicity. In the security proof, at first, the challenge ciphertext is changed to the following form using a subgroup decision assumption:

$$\left(g_1^s \cdot g_2^{\hat{s}}, \; g_1^{s\sum_{i\in[1,\ell]} w_{i,\mathsf{ID}_i}} \cdot g_2^{\hat{s}\sum_{i\in[1,\ell]} w_{i,\mathsf{ID}_i}}, \; e(g_1^s \cdot g_2^{\hat{s}}, h^\alpha) \cdot \mathsf{M}\right).$$

Then, we consider ℓ hybrid games. In Game_i, all private keys are in the following form:

$$\left(h^\alpha \cdot \boxed{g_2^{\widehat{R}_i(\mathsf{ID}|_i)}} \cdot g_1^{r\sum_{i\in[1,\ell]} w_{i,\mathsf{ID}_i}}, \; g_1^{-r}\right)$$

where $\mathsf{ID}|_i$ is the length i prefix of the identity ID and $\widehat{R}_i : \{0,1\}^i \to N$ is a random function. Intuitively, through these hybrid games, the randomizing part of the key (highlighted in the box) are gradually randomized and made dependent on more and more bits of each identity. Finally, in Game_ℓ, we can argue that

---

[2] In the actual scheme, $\mathsf{sk_{ID}}$ is randomized by elements of $\mathbb{G}_{p_3}$, but we do not care about this point in this overview.

any adversary cannot obtain the information on the message M, because these randomizing parts prevent it.

A crucial part of the security proof is to establish the indistinguishability between Game_{i^*-1} and Game_{i^*} for all $i^* \in [1, \ell]$. For the target identity ID^* (recall that we are considering the single-challenge and single-instance case for now), we assume that $b^* := \mathsf{ID}_{i^*}^*$ is known to the reduction algorithm in advance, since it can be guessed with probability $1/2$. At the core of the proof for this is an indistinguishability of the following distributions:

$$\text{Given} \quad \left(g_1^s \cdot g_2^{\hat{s}}, \; g_1^{s \sum_{i \in [1,\ell]} w_{i,\mathsf{ID}_i^*}} \cdot g_2^{\hat{s} \sum_{i \in [1,\ell]} w_{i,\mathsf{ID}_i^*}} \right),$$

$$\left(g_1^{r \sum_{i \in [1,\ell]} w_{i,\mathsf{ID}_i}}, g_1^{-r} \right) \stackrel{c}{\approx} \left(\boxed{g_2^{\hat{\alpha}}} \cdot g_1^{r \sum_{i \in [1,\ell]} w_{i,\mathsf{ID}_i}}, g_1^{-r} \right) \tag{1}$$

for all ID such that $\mathsf{ID}_{i^*} \neq b^*$, where $\hat{\alpha} \stackrel{\$}{\leftarrow} \mathbb{Z}_N$. Indistinguishability of Game_{i^*-1} and Game_{i^*} is reduced to Eq. (1). The reduction algorithm can create the challenge ciphertext using the first term in Eq. (1). It can also set private key as

$$\begin{cases} h^\alpha \cdot g_2^{\widehat{\mathsf{R}}_{i^*-1}(\mathsf{ID}|_{i^*-1})} \cdot g_1^{r \sum_{i \in S} w_{i,\mathsf{ID}_i}}, g_1^{-r} & \text{if } \mathsf{ID}_{i^*} = b^* \\ h^\alpha \cdot g_2^{\widehat{\mathsf{R}}_{i^*-1}(\mathsf{ID}|_{i^*-1})} \cdot \boxed{g_2^{\hat{\alpha}}} \cdot g_1^{r \sum_{i \in S} w_{i,\mathsf{ID}_i}}, g_1^{-r} & \text{if } \mathsf{ID}_{i^*} \neq b^* \end{cases}$$

where $\hat{\alpha} = 0$ or $\hat{\alpha} \stackrel{\$}{\leftarrow} \mathbb{Z}_N$. It is clear that the game corresponds to Game_{i^*-1} if $\hat{\alpha} = 0$. On the other hand, if $\hat{\alpha} \stackrel{\$}{\leftarrow} \mathbb{Z}_N$, it corresponds to Game_{i^*} with

$$\widehat{\mathsf{R}}_{i^*}(\mathsf{ID}|_{i^*}) = \begin{cases} \widehat{\mathsf{R}}_{i^*-1}(\mathsf{ID}|_{i^*-1}) & \text{if } \mathsf{ID}_{i^*} = b^* \\ \widehat{\mathsf{R}}_{i^*-1}(\mathsf{ID}|_{i^*-1}) + \hat{\alpha} & \text{if } \mathsf{ID}_{i^*} \neq b^* \end{cases}.$$

If $\hat{\alpha}$ is freshly chosen for every distinct $\mathsf{ID}|_{i^*}$, the simulation is perfect. Therefore, our task of the security proof is reduced to establish Eq. (1). To understand better, we decompose the private key in Eq. (1) and restate it again in a slightly stronger form:

$$\text{Given} \quad \left(g_1^s \cdot g_2^{\hat{s}}, \; g_1^{s \sum_{i \in [1,\ell]} w_{i,\mathsf{ID}_i^*}} \cdot g_2^{\hat{s} \sum_{i \in [1,\ell]} w_{i,\mathsf{ID}_i^*}} \right),$$

$$\left(g_1^{r w_{i^*,1-b^*}}, g_1^{-r}, \{g_1^{r w_{j,b}}\}_{(j,b)\neq(i^*,1-b^*)} \right)$$
$$\stackrel{c}{\approx} \left(\boxed{g_2^{\hat{\alpha}}} \cdot g_1^{r w_{i^*,1-b^*}}, g_1^{-r}, \{g_1^{r w_{j,b}}\}_{(j,b)\neq(i^*,1-b^*)} \right).$$

Let us consider a bijection map $f : \{(i,b)\}_{i \in [1,\ell], b \in \{0,1\}} \rightarrow [1, 2\ell]$ and replace (i,b) with $f((i,b))$. We can further restate the requirement as:

$$\text{Given} \quad \left(g_1^s \cdot g_2^{\hat{s}}, \; g_1^{s \sum_{j \in S^*} w_j} \cdot g_2^{\hat{s} \sum_{j \in S^*} w_j} \right),$$

$$\left(g_1^{r w_{\tau^*}}, g_1^{-r}, \{g_1^{r w_j}\}_{j \neq \tau^*} \right) \stackrel{c}{\approx} \left(\boxed{g_2^{\hat{\alpha}}} \cdot g_1^{r w_{\tau^*}}, g_1^{-r}, \{g_1^{r w_j}\}_{j \neq \tau^*} \right) \tag{2}$$

where $S^* = \{f(i, \mathsf{ID}_i^*)\}_{i \in [\ell]}$, $\tau^* = f((i^*, 1 - b^*))$, and thus $\tau^* \notin S^*$. We call the terms in the second line above as the challenge terms. (It should not be confused with challenge ciphertext.) At this point, we can now see a similarity to broadcast encryption. We consider the following broadcast encryption which captures the essence of the above requirement. Let the set of user index be $[1, 2\ell]$.

$$\mathsf{mpk} = (g_1, g_1^{w_1}, \ldots, g_1^{w_{2\ell}}, e(g_1, h)^{\alpha}),$$

$$\mathsf{CT}_S = (g_1^s, g_1^{s \sum_{j \in S} w_j}, e(g_1, h)^{s\alpha} \cdot \mathsf{M}), \quad \mathsf{sk}_\tau = (h^\alpha g_1^{rw_\tau}, g_1^{-r}, \{g_1^{rw_j}\}_{j \in [2\ell] \setminus \{\tau\}})$$

where CT_S is a ciphertext for a set $S \subseteq [2\ell]$ and sk_τ is a private key for a user index $\tau \in [2\ell]$. This is in fact a variant of the broadcast encryption by Gentry and Waters [25]! Indeed, Eq. (2) can be interpreted as a security condition for this broadcast encryption scheme (in the sense of encoding analogous to [4,50]). It says that given semi-functional ciphertext for a set $S^\star$, a normal private key for $\tau^\star \notin S^\star$ is indistinguishable from a semi-functional private key for $\tau^\star$. At this point, we are able to understand the core technique in Chen and Wee in terms of broadcast encryption scheme.

However, we have not finished yet. In order to make the proof go through, we argue that an adversary cannot distinguish challenge terms in Eq. (2), even if these are given to the adversary *unbounded many times* with freshly chosen randomness $\hat{\alpha}, r$. Such an indistinguishability can be shown by a standard technique [4,36,50] if the challenge term is given to the adversary *only once*. This can be accomplished by the combination of subgroup decision assumption and the parameter-hiding argument. In parameter-hiding argument, a value which is information-theoretically hidden is used to make normal private key semi-functional [4,36,37,50]. At the first glance, this argument does not seem to be extended to the case where many challenge terms are given to the adversary, since entropy of hidden parameters (in this case, $w_1, \ldots, w_{2\ell} \mod p_2$) is limited. However, we have to simulate unbounded number of challenge terms. Chen and Wee [17] resolve this problem by using computational argument instead of information-theoretic argument as above. Namely, they assume a variant of the DDH assumption on $\mathbb{G}_{p_2}$[3] and embed the problem instance into the above challenge terms. Indistinguishability of multiple challenge terms are tightly reduced to the assumption, using the random self-reducibility of the assumption. On the other hand, our technique for boosting to multi-challenge is much simpler. Our key observation is that the challenge term in Eq. (2) can be easily randomized by picking $a \xleftarrow{\$} \mathbb{Z}_N$ and computing

$$\left((g_2^{\hat{\alpha}} \cdot g_1^{rw_{\tau^\star}})^a, (g_1^{-r})^a, \{(g_1^{rw_j})^a\}_{j \neq \tau^\star} \right) = \left(g_2^{\hat{\alpha}'} \cdot g_1^{r'w_{\tau^\star}}, g_1^{-r'}, \{g_1^{r'w_j}\}_{j \neq \tau^\star} \right) \quad (3)$$

where $r' = ar$ and $\hat{\alpha}' = a\hat{\alpha}$. It is easy to see that $r' \mod p_1$ is uniformly random and independent from anything. We can also see that $\hat{\alpha}' \mod p_2 = 0$ if $\hat{\alpha} = 0$ and $\hat{\alpha}' \mod p_2$ is uniformly random if $\hat{\alpha} \neq 0 \mod p_2$. By this argument, we can see that indistinguishability of the single-challenge-term case implies that for the

---

[3] Of course, in symmetric bilinear groups, the DDH assumption does not hold. They considered a DDH assumption on $\mathbb{G}_{p_2}$ where each term is perturbed by a random element in $\mathbb{G}_{p_3}$, which prevents trivial attack against the assumption.

multi-challenge-term case. Based on all the above discussion, we are able to show the security for the above scheme *only using the subgroup decision assumption.*

Overview of Our Framework. We refine the idea above and combine it with the technique by HKS to propose our framework for constructing IBE schemes that are (almost) tightly secure in the multi-challenge and multi-instance setting, in both composite and prime-order groups. We first define a broadcast encoding, which is an abstraction of broadcast encryption. The syntax of it is a special case of "pair encoding" in [4] (also similar to "predicate encoding" in [50]). Then, we define perfect master-key hiding (PMH) security and computational-master-key hiding (CMH) security for it. These security notions are also similar to those of [4,50]. The former is statistical requirement for the encoding, and the latter is computational requirement. We can easily show that the former implies the latter. Then, we also introduce intermediate notion multi-master-key hiding (MMH) security for the encoding. This is more complex notion compared to the PMH and CMH-security, but implied by these, thanks to our boosting technique above. Then, we show that broadcast encoding satisfying the MMH security requirement can be converted into IBE scheme. All these reductions are (almost) tightness-preserving, namely, if the original broadcast encoding is tightly PMH/CMH secure, the resulting IBE scheme is also tightly secure in the multi-challenge and multi-instance setting. Finally, we provide broadcast encoding schemes that satisfy our requirement. One is implicit in Gentry-Waters broadcast encryption scheme [25] and the other is completely new. By instantiating our general framework with the latter construction, we obtain IBE scheme with almost tight security and with sub-linear master public key size.

1.4   Related Works

Related Works on IBE. The first realizations of IBE in the random oracle model were given in [13,20,45]. Later, realization in the standard model [10,14] were given. In the random oracle model, it is possible to obtain efficient and tightly secure IBE scheme [5]. Gentry [24] proposed a tightly secure anonymous IBE scheme under a non-static, parametrized assumption. Chen and Wee proposed the first almost tightly secure IBE scheme under static and simple assumptions [17,19]. Attrapadung [4] proposed an IBE scheme whose security loss only depends on the number of key queries before the challenge phase. Jutla and Roy [32] constructed very efficient IBE scheme from the SXDH assumption, based on a technique related to NIZK. Blazy, Kiltz, and Pan [9] further generalized the idea and show that a message authentication code with a certain specific algebraic structure implies (H)IBE. They further obtained almost tightly secure anonymous IBE and (non-anonymous) HIBE via the framework. Note that all above mentioned schemes only focus on the single-challenge setting.

Related Works on the Multi-Challenge CCA-Secure PKE. Bellare, Boldyreva, and Micali [7] gave a tight reduction for the Cramer-Shoup encryption [21] in the multi-instance (multi-user) and the single-challenge setting. They posed an important open question of whether it is possible to construct tightly CCA-secure PKE scheme in the multi-instance and the multi-challenge setting. The first PKE scheme satisfying the requirement was proposed by Hofheinz and Jager [30]. Their scheme requires hundreds of group elements in the ciphertexts. Subsequently, Abe et al. [1] reduced the size by improving the efficiency of the underlying one-time signature. Libert et al. [33] greatly reduced the ciphertext and made it constant-size for the first time. The ciphertext overhead of their scheme consist of 68 group elements. Very recently, Libert et al. [34] further reduced it to 47 group elements. Concurrently and independently to us, Hofheinz [27] proposes the first PKE scheme with the same security guarantee and fully compact parameters, which means all parameters are constant-size. While the ciphertext-size (which consists of 60 group elements) is longer than construction in [34], it achieves much shorter public parameters. We note that while the technique is very powerful, it is unclear how to extend it to the IBE setting.

Due to space limitations, many definitions and proofs are omitted from this version. These can be found in the full version of the paper [6].

2   Preliminaries

Notation. Vectors will be treated as either row or column vector matrices. When unspecified, we shall let it be a row vector. We denote by $\mathbf{e}_i$ the i-th unit (row) vector: its i-th component is one, all others are zero. $\mathbf{0}$ denotes the zero vector or zero matrix. For an integer $n \in \mathbb{N}$ and a field $\mathbb{F}$, $\mathbb{GL}_n(\mathbb{F})$ denotes the set of all invertible matrix in $\mathbb{F}^{n \times n}$. For a multiplicative group $\mathbb{G}$, we denote by $\mathbb{G}^*$ a set of all *generators* in $\mathbb{G}$. We also denote by $[a, b]$ a set $\{a, \ldots, b\}$ for any integer a and b and $[n] = [1, n]$ for any $n \in \mathbb{N}$. We denote by $u \xleftarrow{\$} U$ the fact that u is picked uniformly at random from a finite set U.

2.1   Identity-Based Encryption

In this section, we define the syntax and security of IBE (in the multi-challenge, multi-instance setting).

Syntax. An IBE scheme with identity space $\mathcal{ID}$ and message space $\mathcal{M}$ consists of the following algorithms:

Par(1^κ) → (pp, sp): The parameter sampling algorithm takes as input a security
    parameter 1^κ and outputs a public parameter pp and a secret parameter sp.
Gen(pp, sp) → (mpk, msk): The key generation algorithm takes pp and sp as
    input and outputs a master public key mpk and master secret key msk.
Ext(msk, mpk, ID) → sk$_\mathsf{ID}$: The user private key extraction algorithm takes as
    input the master secret key msk, the master public key mpk, and an identity
    ID $\in \mathcal{ID}$. It outputs a private key sk$_\mathsf{ID}$.

$\mathsf{Enc}(\mathsf{mpk}, \mathsf{ID}, \mathsf{M}) \to \mathsf{CT}$: The encryption algorithm takes as input a master public key mpk, an identity ID, and a message $\mathsf{M} \in \mathcal{M}$. It will output a ciphertext CT.

$\mathsf{Dec}(\mathsf{sk}_{\mathsf{ID}}, \mathsf{CT}) \to \mathsf{M}$: The decryption algorithm takes as input a private key $\mathsf{sk}_{\mathsf{ID}}$ and a ciphertext CT. It outputs a message M or $\bot$ which indicates that the ciphertext is not in a valid form.

We refer (standard) notion of correctness of IBE to [6].

In our constructions, we will set identity space $\mathcal{ID} = \{0,1\}^{\ell}$ for some $\ell \in \mathbb{N}$. Note that the restriction on the identity space can be easily removed by applying a collision resistant hash function $\mathsf{CRH} : \{0,1\}^* \to \{0,1\}^{\ell}$ to an identity. Typically, we would set $\ell = \Theta(\kappa)$ to avoid the birthday attack.

Security Model. We now define (μ, Q_c, Q_k)-security for an IBE $\Phi = (\mathsf{Par}, \mathsf{Gen}, \mathsf{Ext}, \mathsf{Enc}, \mathsf{Dec})$. This security notion is defined by the following game between a challenger and an attacker $\mathcal{A}$.

Setup. The challenger runs $(\mathsf{pp}, \mathsf{sp}) \xleftarrow{\$} \mathsf{Par}(1^{\kappa})$ and $(\mathsf{mpk}^{(j)}, \mathsf{msk}^{(j)}) \xleftarrow{\$} \mathsf{Gen}(\mathsf{pp}, \mathsf{sp})$ for $j \in [\mu]$. The challenger also picks random coin coin $\xleftarrow{\$} \{0,1\}$ whose value is fixed throughout the game. Then, $(\mathsf{pp}, \{\mathsf{mpk}^{(j)}\}_{j \in [\mu]})$ is given to $\mathcal{A}$.

In the following, $\mathcal{A}$ adaptively makes the following two types of queries in an arbitrary order.

–**Key Extraction Query.** The adversary $\mathcal{A}$ submits $(\texttt{Extraction}, j \in [\mu],$ $\mathsf{ID} \in \mathcal{ID})$ to the challenger. Then, the challenge runs $\mathsf{sk}_{\mathsf{ID}}^{(j)} \xleftarrow{\$} \mathsf{Ext}(\mathsf{msk}^{(j)},$ $\mathsf{mpk}^{(j)}, \mathsf{ID})$ and returns $\mathsf{sk}_{\mathsf{ID}}^{(j)}$ to $\mathcal{A}$.

–**Challenge Query.** The adversary $\mathcal{A}$ submits $(\texttt{Challenge}, j \in [\mu], \mathsf{ID} \in$ $\mathcal{ID}, \mathsf{M}_0, \mathsf{M}_1 \in \mathcal{M})$ to the challenger. Then, the challenger runs $\mathsf{CT} \xleftarrow{\$}$ $\mathsf{Enc}(\mathsf{mpk}^{(j)}, \mathsf{ID}, \mathsf{M}_{\mathsf{coin}})$ and returns CT to $\mathcal{A}$.

Guess. At last, $\mathcal{A}$ outputs a guess coin' for coin. The advantage of an attacker $\mathcal{A}$ in the game is defined as $\mathsf{Adv}_{\mathcal{A},\Phi,(\mu,Q_c,Q_k)}^{\mathsf{IBE}}(\kappa) = |\Pr[\mathsf{coin}' = \mathsf{coin}] - \frac{1}{2}|$.

We say that the adversary $\mathcal{A}$ is valid if and only if $\mathcal{A}$ never queries $(\texttt{Extraction}, j, \mathsf{ID})$ such that it has already queried $(\texttt{Challenge}, j, \mathsf{ID}, \mathsf{M}_0, \mathsf{M}_1)$ for the same (j, ID) (and vice versa); $\mathcal{A}$ has made at most Q_c challenge queries; and $\mathcal{A}$ has made at most Q_k key extraction queries.

Definition 1. *We say that IBE Φ is secure if $\mathsf{Adv}_{\mathcal{A},\Phi,(\mu,Q_c,Q_k)}^{\mathsf{IBE}}(\kappa)$ is negligible for any polynomially bounded μ, Q_c, Q_k, and any valid PPT adversary $\mathcal{A}$.*

Anonymity. We also consider anonymity for the IBE scheme. To define (μ, Q_c, Q_k)-anonymity for an IBE scheme, we change the form of challenge queries in the above game as follows.

–Challenge Query. The adversary $\mathcal{A}$ submits (Challenge, $j \in [\mu]$, ID_0, $\mathsf{ID}_1 \in \mathcal{ID}$, M_0, $\mathsf{M}_1 \in \mathcal{M}$) to the challenger. Then, the challenger runs CT $\xleftarrow{\$}$ Enc(mpk$^{(j)}$, $\mathsf{ID}_{\mathsf{coin}}$, $\mathsf{M}_{\mathsf{coin}}$) and returns CT to $\mathcal{A}$.

We say that the adversary $\mathcal{A}$ is valid if $\mathcal{A}$ never queries (Extraction, j, ID) such that it has already queried (Challenge, j, ID_0, ID_1, M_0, M_1) for the same j and $\mathsf{ID} \in \{\mathsf{ID}_0, \mathsf{ID}_1\}$ (and vice versa); $\mathcal{A}$ has made at most Q_c challenge queries; and $\mathcal{A}$ has made at most Q_k key extraction queries. We define the advantage of $\mathcal{A}$ in this modified game as $\mathsf{Adv}^{\mathsf{AIBE}}_{\mathcal{A}, \Phi, (\mu, Q_c, Q_k)}(\kappa) := |\Pr[\mathsf{coin}' = \mathsf{coin}] - \frac{1}{2}|$.

Definition 2. *We say that IBE Φ is anonymous if* $\mathsf{Adv}^{\mathsf{AIBE}}_{\mathcal{A}, \Phi, (\mu, Q_c, Q_k)}(\kappa)$ *is negligible for any polynomially bounded* μ, Q_c, Q_k, *and any valid PPT adversary* $\mathcal{A}$.

2.2 Composite-Order Bilinear Groups

We will use bilinear group $(\mathbb{G}, \mathbb{G}_T)$ of composite order $N = p_1 p_2 p_3 p_4$, where p_1, p_2, p_3, p_4 are four distinct prime numbers, with efficiently computable and non-degenerate bilinear map $e(\cdot) : \mathbb{G} \times \mathbb{G} \to \mathbb{G}_T$. For each $d | N$, $\mathbb{G}$ has unique subgroup of order d denoted by $\mathbb{G}_d$. We let g_i be a generator of $\mathbb{G}_{p_i}$. For our purpose, we define a (composite order) bilinear group generator $\mathcal{G}_{\mathsf{comp}}$ that takes as input a security parameter 1^κ and outputs $(N, \mathbb{G}, \mathbb{G}_T, g_1, g_2, g_3, g_4, e(\cdot))$. Any $h \in \mathbb{G}$ can be expressed as $h = g_1^{a_1} g_2^{a_2} g_3^{a_3} g_4^{a_4}$, where a_i is uniquely determined modulo p_i. We call $g_i^{a_i}$ the $\mathbb{G}_{p_i}$ component of h. We have that $e(g^a, h^b) = e(g, h)^{ab}$ for any $g, h \in \mathbb{G}$, $a, b \in \mathbb{Z}$ and $e(g, g) = 1_{\mathbb{G}_T}$ for $g \in \mathbb{G}_{p_i}$ and $h \in \mathbb{G}_{p_j}$ with $i \neq j$.

Let $(N, \mathbb{G}, \mathbb{G}_T, g_1, g_2, g_3, g_4, e(\cdot)) \xleftarrow{\$} \mathcal{G}_{\mathsf{comp}}(1^\kappa)$ and $g \xleftarrow{\$} \mathbb{G}^*$. We define advantage function $\mathsf{Adv}^{\mathsf{Pxx}}_{\mathcal{A}}(\kappa)$ for Problem xx for any adversary $\mathcal{A}$ as

$$\mathsf{Adv}^{\mathsf{Pxx}}_{\mathcal{A}}(\kappa) = |\Pr[\mathcal{A}(g_1, g_4, g, D, T_0) \to 1] - \Pr[\mathcal{A}(g_1, g_4, g, D, T_1) \to 1]|.$$

In each problem, D, T_0, and T_1 are defined as follows. In the following, for $i, j \in [1, 4]$, g_{ij} is chosen as $g_{ij} \xleftarrow{\$} \mathbb{G}^*_{p_i p_j}$.

Problem 1. $D = \emptyset$, $T_0 \xleftarrow{\$} \mathbb{G}^*_{p_1}$, and $T_1 \xleftarrow{\$} \mathbb{G}^*_{p_1 p_2}$.

Problem 2. $D = (g_{12}, g_3, g_{24})$, $T_0 \xleftarrow{\$} \mathbb{G}^*_{p_1 p_4}$, and $T_1 \xleftarrow{\$} \mathbb{G}^*_{p_1 p_2 p_4}$.

Problem 3. $D = (g_{13}, g_2, g_{34})$, $T_0 \xleftarrow{\$} \mathbb{G}^*_{p_1 p_4}$, and $T_1 \xleftarrow{\$} \mathbb{G}^*_{p_1 p_3 p_4}$.

Problem 4. $D = (g_{12}, g_{23})$, $T_0 \xleftarrow{\$} \mathbb{G}^*_{p_1 p_2}$, and $T_1 \xleftarrow{\$} \mathbb{G}^*_{p_1 p_3}$.

Problem 5. $D = (g_2, g_3, g_2^x, g_2^y, g_2^z)$, $T_0 = e(g_2, g_2)^{xyz}$, and $T_1 = e(g_2, g_2)^{xyz+\gamma}$, where $x, y, z \xleftarrow{\$} \mathbb{Z}_N$ and $\gamma \xleftarrow{\$} \mathbb{Z}_N^*$.

Problems 1, 2, 3, and 4 are called sub-group decision problems. Problem 5 is called the decisional bilinear Diffie-Hellman problem.

Matrix-in-the-Exponent. Given any vector $\mathbf{w} = (w_1, \ldots, w_n) \in \mathbb{Z}_N^n$ and a group element g, we write $g^{\mathbf{w}} \in \mathbb{G}^n$ to denote $(g^{w_1}, \ldots, g^{w_n}) \in \mathbb{G}^n$: we define $g^{\mathbf{A}}$ for a matrix $\mathbf{A}$ in a similar way. $g^{\mathbf{A}} \cdot g^{\mathbf{B}}$ denotes componentwise product: $g^{\mathbf{A}} \cdot g^{\mathbf{B}} = g^{\mathbf{A}+\mathbf{B}}$. Note that given $g^{\mathbf{A}}$ and a matrix $\mathbf{B}$ of "exponents", one can efficiently compute $g^{\mathbf{BA}}$ and $g^{\mathbf{AB}} = (g^{\mathbf{A}})^{\mathbf{B}}$. Furthermore, if there is an efficiently computable map $e : \mathbb{G} \times \mathbb{G} \to \mathbb{G}_T$, then given $g^{\mathbf{A}}$ and $g^{\mathbf{B}}$, one can efficiently compute $e(g,g)^{\mathbf{A}^\top \mathbf{B}}$ via $(e(g,g)^{\mathbf{A}^\top \mathbf{B}})_{i,j} = \prod_k e(g^{A_{k,i}}, g^{B_{k,j}})$ where $A_{i,j}$ and $B_{i,j}$ denote the (i,j)-th coefficient of $\mathbf{A}$ and $\mathbf{B}$ respectively. We will use $e(g^{\mathbf{A}}, g^{\mathbf{B}}) = e(g,g)^{\mathbf{A}^\top \mathbf{B}}$ to denote this operation.

3 Broadcast Encoding: Definitions and Reductions

In this section, we define the syntax and the security notions for broadcast encoding. The syntax of our definition corresponds to a special case of "pair encoding" defined in [4] and is also similar to "predicate encoding" in [50]. As for the security requirement for the encoding, ours are slightly different from both. We define several flavours of the security requirement: perfect master-key hiding security (PMH), computational-master-key hiding (CMH) security, and the multi-master-key hiding (MMH) security. The last one is useful, since we can obtain IBE scheme from broadcast encoding scheme satisfying the security notion, as we will explain in Sect. 4. However, MMH security is defined by relatively complex game and may not be easy to show. Later in this section, we will see that MMH security can be tightly reduced to much simpler CMH and PMH security.

3.1 Broadcast Encoding: Syntax

The broadcast encoding Π consists of the following four deterministic algorithms.

Param$(n, N) \to d_1$: It takes as input an integer n and N and outputs $d_1 \in \mathbb{N}$ which specifies the number of common variables in CEnc and KEnc. For the default notation, $\mathbf{w} = (w_1, \ldots, w_{d_1})$ denotes the list of common variables.

KEnc$(\tau, N) \to (\mathbf{k}, d_2')$: It takes as input $\tau \in [n]$, $N \in \mathbb{N}$, and outputs a vector of polynomials $\mathbf{k} = (k_1, \ldots, k_{d_2})$ with coefficients in $\mathbb{Z}_N$, and $d_2' \in \mathbb{N}$ that specifies the number of its own variables. We assume that d_2 and d_2' only depend on n and do not depend on τ without loss of generality. We require that each polynomials $\mathbf{k}$ is a *linear combination of monomials* α, r_j, $w_k r_j$ where $\alpha, r_1, \ldots, r_{d_2'}, w_1, \ldots, w_{d_1}$ are variables. More precisely, it outputs $\{b_\iota\}_{\iota \in [d_2]}$, $\{b_{\iota,j}\}_{(\iota,j) \in [d_2] \times [d_2']}$, and $\{b_{\iota,j,k}\}_{(\iota,j,k) \in [d_2] \times [d_2'] \times [d_1]}$ in $\mathbb{Z}_N$ such that

$$k_\iota\left(\alpha, r_1, \ldots, r_{d_2'}, w_1, \ldots, w_{d_1}\right)$$
$$= b_\iota \alpha + \left(\sum_{j \in [d_2']} b_{\iota,j} r_j\right) + \left(\sum_{(j,k) \in [d_2'] \times [d_1]} b_{\iota,j,k} w_k r_j\right) \quad (4)$$

for $\iota \in [d_2]$.

$\mathsf{CEnc}(S, N) \to (\mathbf{c}, d_3')$: It takes as input $S \subseteq [n]$, $N \in \mathbb{N}$, and outputs a vector of polynomials $\mathbf{c} = (c_1, \ldots, c_{d_3})$ with coefficients in $\mathbb{Z}_N$, and $d_3' \in \mathbb{N}$ that specifies the number of its own variables. We require that polynomials $\mathbf{c}$ in variables $s_0, s_1, \ldots, s_{d_3'}, w_1, \ldots, w_{d_1}$ have the following form:
There exist (efficiently computable) set of coefficients $\{a_{\iota,j}\}_{(\iota,j) \in [d_3] \times [0,d_3']}$ and $\{a_{\iota,j,k}\}_{(\iota,j,k) \in [d_3] \times [0,d_3'] \times [d_1]}$ in $\mathbb{Z}_N$ such that

$$c_\iota \left(s_0, s_1, \ldots, s_{d_3'}, w_1, \ldots, w_{d_1} \right)$$
$$= \left(\sum_{j \in [0,d_3']} a_{\iota,j} s_j \right) + \left(\sum_{(j,k) \in [0,d_3'] \times [d_1]} a_{\iota,j,k} w_k s_j \right) \qquad (5)$$

for $\iota \in [d_3]$. We also require that $c_1 = s_0$.
$\mathsf{Pair}(\tau, S, N) \to \mathbf{E}$: It takes as input $\tau \in [n]$, $S \subseteq [n]$, and $N \in \mathbb{N}$ and outputs a matrix $\mathbf{E} = (E_{i,j})_{i \in [d_2], j \in [d_3]} \in \mathbb{Z}_N^{d_2 \times d_3}$.

Correctness. The correctness requirement is as follows.

- We require that for any n, N, $d_1 \leftarrow \mathsf{Param}(n, N)$, $\mathbf{k} \leftarrow \mathsf{KEnc}(\tau, N)$, $\mathbf{c} \leftarrow \mathsf{CEnc}(S, N)$, and $\mathbf{E} \leftarrow \mathsf{Pair}(\tau, S, N)$, we have that

$$\mathbf{k}\mathbf{E}\mathbf{c}^\top = \alpha s_0 \qquad \text{whenever} \qquad \tau \in S.$$

The equation holds symbolically, or equivalently, as polynomials in variables $\alpha, r_1, \ldots, r_{d_2'}, s_0, s_1, \cdots, s_{d_3'}, w_1, \ldots, w_{d_1}$.
- For p that divides N, if we let $\mathsf{KEnc}(\tau, N) \to (\mathbf{k}, d_2')$ and $\mathsf{KEnc}(\tau, p) \to (\mathbf{k}', d_2'')$, then it holds that $d_2' = d_2''$ and $\mathbf{k} \bmod p = \mathbf{k}'$. The requirement for CEnc is similar.

Note that since $\mathbf{k}\mathbf{E}\mathbf{c}^\top = \sum_{(i,j) \in [d_2] \times [d_3]} E_{i,j} k_i c_j$, the first requirement amounts to check if there is a linear combination of $k_i c_j$ terms summed up to αs_0. In the descriptions of proposed broadcast encoding schemes, which will appear later in this paper, we will not explicitly write down $\mathbf{E}$. Instead, we will check this condition.

3.2 Broadcast Encoding: Security

Here, we define two flavours of security notions for broadcast encoding: perfect security and computational security. As we will see, the former implies the latter. In what follows, we denote $\mathbf{w} = (w_1, \ldots, w_{d_1})$, $\mathbf{r} = (r_1, \ldots, r_{d_2'})$, and $\mathbf{s} = (s_0, s_1, \ldots, s_{d_3'})$.

(Perfect Security). The pair encoding scheme $\Pi = (\mathsf{Param}, \mathsf{KEnc}, \mathsf{CEnc}, \mathsf{Pair})$ is Q-perfectly master-key hiding (Q-PMH) if the following holds. For any $n \in \mathbb{N}$, prime $p \in \mathbb{N}$, $\tau \in [n]$, and $S_1, \ldots, S_Q \subset [n]$ such that $\tau \notin S_j$ for all $j \in [Q]$, let $\mathsf{Param}(n,p) \to d_1$, $(\mathbf{k}_\tau, d_2') \leftarrow \mathsf{KEnc}(\tau, p)$, and $(\mathbf{c}_{S_j}, d_{3,j}') \leftarrow \mathsf{CEnc}(S_j, p)$ for $j \in [Q]$, then the following two distributions are identical:

$$\{\ \{\mathbf{c}_{S_j}(\mathbf{s}_j, \mathbf{w})\}_{j \in [Q]},\ \mathbf{k}_\tau(0, \mathbf{r}, \mathbf{w})\ \}\ \text{ and }\ \{\ \{\mathbf{c}_{S_j}(\mathbf{s}_j, \mathbf{w})\}_{j \in [Q]},\ \mathbf{k}_\tau(\alpha, \mathbf{r}, \mathbf{w})\ \}$$

where $\mathbf{w} \xleftarrow{\$} \mathbb{Z}_p^{d_1}$, $\alpha \xleftarrow{\$} \mathbb{Z}_p$, $\mathbf{r} \xleftarrow{\$} (\mathbb{Z}_p^*)^{d_2'}$, $\mathbf{s}_j \xleftarrow{\$} \mathbb{Z}_p^{d_3'+1}$ for $j \in [Q]$.

(Computational Security on $\mathbb{G}_{p_2}$). We define Q-computational-master-key hiding (Q-CMH[4]) security on $\mathbb{G}_{p_2}$ for a broadcast encoding $\Pi = (\mathsf{Param}, \mathsf{KEnc}, \mathsf{CEnc}, \mathsf{Pair})$ by the following game. At the beginning of the game, an (stateful) adversary $\mathcal{A}$ is given $(1^\kappa, n)$ and chooses $\tau^* \in [n]$. Then, parameters are chosen as $(N, \mathbb{G}, \mathbb{G}_T, g_1, g_2, g_3, g_4, e(\cdot)) \xleftarrow{\$} \mathcal{G}_{\mathsf{comp}}(1^\kappa)$, $\mathsf{Param}(n, N) \to d_1$, and $\hat{\mathbf{w}} \xleftarrow{\$} \mathbb{Z}_N^{d_1}$. The advantage of $\mathcal{A}$ is defined as

$$\mathsf{Adv}_{\mathcal{A},\Pi,Q,\mathbb{G}_{p_2}}^{\mathsf{CMH}}(\kappa) = |\Pr[\mathcal{A}(1^\kappa, n) \to \tau^*,\ \mathcal{A}(g_1, g_2, g_3, g_4)^{\mathcal{O}_{\tau^*,\hat{\mathbf{w}}}^{\mathsf{CMH,C}}(\cdot), \mathcal{O}_{\tau^*,\hat{\mathbf{w}},0}^{\mathsf{CMH,K}}(\cdot)} \to 1] -$$

$$\Pr[\mathcal{A}(1^\kappa, n) \to \tau^*,\ \mathcal{A}(g_1, g_2, g_3, g_4)^{\mathcal{O}_{\tau^*,\hat{\mathbf{w}}}^{\mathsf{CMH,C}}(\cdot), \mathcal{O}_{\tau^*,\hat{\mathbf{w}},1}^{\mathsf{CMH,K}}(\cdot)} \to 1]|.$$

In the above, $\mathcal{O}_{\tau^*,\hat{\mathbf{w}},b}^{\mathsf{CMH,K}}(\cdot)$ for $b \in \{0,1\}$ are called only once while $\mathcal{O}_{\tau^*,\hat{\mathbf{w}}}^{\mathsf{CMH,C}}(\cdot)$ can be called at most Q times. These oracles can be called in any order.

- $\mathcal{O}_{\tau^*,\hat{\mathbf{w}}}^{\mathsf{CMH,C}}(\cdot)$ takes $S \subset [n]$ such that $\tau^* \notin S$ as input. It then runs $\mathsf{CEnc}(S, N) \to (\mathbf{c}, d_3')$, picks $\hat{\mathbf{s}} = (\hat{s}_0, \hat{s}_1, \ldots, \hat{s}_{d_3'}) \xleftarrow{\$} \mathbb{Z}_N^{d_3'+1}$, and returns $g_2^{\mathbf{c}(\hat{\mathbf{s}}, \hat{\mathbf{w}})}$. We note that $\hat{\mathbf{s}}$ is *freshly chosen* every time the oracle is called.
- $\mathcal{O}_{\tau^*,\hat{\mathbf{w}},b}^{\mathsf{CMH,K}}(\cdot)$ ignores its input. When it is called, it first runs $\mathsf{KEnc}(\tau^*, N) \to (\mathbf{k}, d_2')$ and picks $\hat{\mathbf{r}} = (\hat{r}_1, \ldots, \hat{r}_{d_2'}) \xleftarrow{\$} \mathbb{Z}_N^{d_2'}$ and $\hat{\alpha} \xleftarrow{\$} \mathbb{Z}_N$. Then it returns

$$g_2^{\mathbf{k}(b\cdot\hat{\alpha}, \hat{\mathbf{r}}, \hat{\mathbf{w}})} = \begin{cases} g_2^{\mathbf{k}(0, \hat{\mathbf{r}}, \hat{\mathbf{w}})} & \text{if } b = 0 \\ g_2^{\mathbf{k}(\hat{\alpha}, \hat{\mathbf{r}}, \hat{\mathbf{w}})} & \text{if } b = 1. \end{cases}$$

We say that the broadcast encoding is Q-CMH secure on $\mathbb{G}_{p_2}$ if $\mathsf{Adv}_{\mathcal{A},\Pi,Q,\mathbb{G}_{p_2}}^{\mathsf{CMH}}(\kappa)$ is negligible for all PPT adversary $\mathcal{A}$.

(Computational Security on $\mathbb{G}_{p_3}$). We define $\mathsf{Adv}_{\mathcal{A},\Pi,Q,\mathbb{G}_{p_3}}^{\mathsf{CMH}}(\kappa)$ and Q-CMH security on $\mathbb{G}_{p_3}$ via similar game, by swapping g_2 and g_3 in the above.

COMPARISON WITH DEFINITION IN [4]. By setting $Q = 1$, the Q-PMH and the Q-CMH security defined as above almost correspond to the perfect security and the co-selective security defined in [4] respectively. We need to deal with the case of $Q \gg 1$ in order to handle the multi-challenge setting. Another difference is

---

[4] Here, we use CMH to stand for "computational-master-key hiding" (for broadcast encoding), while in [4], CMH refers to "co-selective master-key hiding" (for pair encoding). We hope that this should not be confusing, since our notion of 1-CMH security is in fact almost the same as the notion of co-selective master-key hiding security (for broadcast predicate) anyway.

that we use groups with the order being a product of four primes, while they deal with a product of three primes.

We have the following lemma which indicates that Q-PMH security unconditionally implies Q-CMH security on both of $\mathbb{G}_{p_2}$ and $\mathbb{G}_{p_3}$.

Lemma 1. *Assume that a broadcast encoding* Π *satisfies* Q-PMH *security for some* $Q \in \mathbb{N}$. *Then it follows that* $\mathsf{Adv}^{\mathsf{CMH}}_{\mathcal{A},\Pi,Q,\mathbb{G}_{p_i}}(\kappa) \leq d_2'/p_i$ *for* $i \in \{2,3\}$.

3.3 Multi-master-key Hiding Security in Composite Order Groups

Here, we define multi-master-key hiding security for a broadcast encoding, which is more complex security notion compared to the CMH security. A broadcast encoding scheme that satisfies the security notion can be converted into an IBE scheme as we will see in Sect. 4.

Multi-master-key Hiding Security (on $\mathbb{G}_{p_2}$). We define (Q_c, Q_k)-multi-master-key hiding $((Q_c, Q_k)$-MMH) security on $\mathbb{G}_{p_2}$ for a broadcast encoding $\Pi = (\mathsf{Param}, \mathsf{KEnc}, \mathsf{CEnc}, \mathsf{Pair})$. The security is defined by the following game. At the beginning of the game, $\mathcal{A}$ is given $(1^\kappa, n)$ and chooses $\tau^\star \in [n]$. Then, parameters are chosen as $(N, \mathbb{G}, \mathbb{G}_T, g_1, g_2, g_3, g_4, e(\cdot)) \overset{\$}{\leftarrow} \mathcal{G}_{\mathsf{comp}}(1^\kappa)$, $g_{24} \overset{\$}{\leftarrow} \mathbb{G}^*_{p_2 p_4}$, $d_1 \leftarrow \mathsf{Param}(n, N)$, and $\mathbf{w} \overset{\$}{\leftarrow} \mathbb{Z}_N^{d_1}$. The advantage of $\mathcal{A}$ is defined as

$$\mathsf{Adv}^{\mathsf{MMH}}_{\mathcal{A},\Pi,(Q_c,Q_k),\mathbb{G}_{p_2}}(\kappa) =$$

$$| \Pr[\mathcal{A}(1^\kappa, n) \to \tau^\star, \ \mathcal{A}(g_1, g_1^{\mathbf{w}}, g_3^{\mathbf{w}}, g_{24}, g_3, g_4)^{\mathcal{O}^{\mathsf{MMH},\mathsf{C}}_{\tau^\star,\mathbf{w}}(\cdot), \mathcal{O}^{\mathsf{MMH},\mathsf{K}}_{\tau^\star,\mathbf{w},0}(\cdot)} \to 1] -$$

$$\Pr[\mathcal{A}(1^\kappa, n) \to \tau^\star, \ \mathcal{A}(g_1, g_1^{\mathbf{w}}, g_3^{\mathbf{w}}, g_{24}, g_3, g_4)^{\mathcal{O}^{\mathsf{MMH},\mathsf{C}}_{\tau^\star,\mathbf{w}}(\cdot), \mathcal{O}^{\mathsf{MMH},\mathsf{K}}_{\tau^\star,\mathbf{w},1}(\cdot)} \to 1] |.$$

In the above, $\mathcal{O}^{\mathsf{MMH},\mathsf{C}}_{\tau^\star,\mathbf{w}}(\cdot)$ and $\mathcal{O}^{\mathsf{MMH},\mathsf{K}}_{\tau^\star,\mathbf{w},b}(\cdot)$ for $b \in \{0,1\}$ can be called at most Q_c times and Q_k times, respectively. They can be called in any order.

- $\mathcal{O}^{\mathsf{MMH},\mathsf{C}}_{\tau^\star,\mathbf{w}}(\cdot)$ takes $S \subset [n]$ such that $\tau^\star \notin S$ as input. It then runs $\mathsf{CEnc}(S, N) \to (\mathbf{c}, d_3')$, picks $\mathbf{s} \overset{\$}{\leftarrow} \mathbb{Z}_N^{d_3'+1}$ and $\hat{\mathbf{s}} \overset{\$}{\leftarrow} \mathbb{Z}_N^{d_3'+1}$ and returns $g_1^{\mathbf{c}(\mathbf{s},\mathbf{w})} \cdot g_2^{\mathbf{c}(\hat{\mathbf{s}},\mathbf{w})}$.
- $\mathcal{O}^{\mathsf{MMH},\mathsf{K}}_{\tau^\star,\mathbf{w},b}(\cdot)$ ignores its input. When it is called, it first runs $\mathsf{KEnc}(\tau^\star, N) \to (\mathbf{k}, d_2')$, picks $\hat{\alpha} \overset{\$}{\leftarrow} \mathbb{Z}_N$, $\mathbf{r} \overset{\$}{\leftarrow} \mathbb{Z}_N^{d_2'}$, $\boldsymbol{\delta} \overset{\$}{\leftarrow} \mathbb{Z}_N^{d_2}$. Then it returns

$$g_1^{\mathbf{k}(0,\mathbf{r},\mathbf{w})} \cdot g_2^{\mathbf{k}(b \cdot \hat{\alpha},0,0)} \cdot g_4^{\boldsymbol{\delta}} = \begin{cases} g_1^{\mathbf{k}(0,\mathbf{r},\mathbf{w})} \cdot g_4^{\boldsymbol{\delta}} & \text{if } b = 0 \\ g_1^{\mathbf{k}(0,\mathbf{r},\mathbf{w})} \cdot g_2^{\mathbf{k}(\hat{\alpha},0,0)} \cdot g_4^{\boldsymbol{\delta}} & \text{if } b = 1. \end{cases}$$

In the above, $\mathbf{r}$, $\hat{\alpha}$, and $\boldsymbol{\delta}$ as well as $\mathbf{s}$ and $\hat{\mathbf{s}}$ are all *freshly chosen* every time the corresponding oracle is called. We say that the broadcast encoding is (Q_c, Q_k)-MMH secure on $\mathbb{G}_{p_2}$ if $\mathsf{Adv}^{\mathsf{MMH}}_{\mathcal{A},\Pi,(Q_c,Q_k),\mathbb{G}_{p_2}}(\kappa)$ is negligible for all PPT adversary $\mathcal{A}$.

Multi-master-key Hiding Security (on $\mathbb{G}_{p_3}$). We define (Q_c, Q_k)-MMH security on $\mathbb{G}_{p_3}$ and $\mathsf{Adv}^{\mathsf{MMH}}_{\mathcal{A},\Pi,(Q_c,Q_k),\mathbb{G}_{p_3}}(\kappa)$ similarly to the above. The difference is the following.

- The input to $\mathcal{A}$ is replaced with $(g_1, g_1^{\mathbf{w}}, g_2^{\mathbf{w}}, g_{34}, g_2, g_4)$.
- $g_1^{\mathbf{c}(\mathbf{s},\mathbf{w})} \cdot g_2^{\mathbf{c}(\hat{\mathbf{s}},\mathbf{w})}$ in the above is replaced with $g_1^{\mathbf{c}(\mathbf{s},\mathbf{w})} \cdot g_3^{\mathbf{c}(\hat{\mathbf{s}},\mathbf{w})}$.
- $g_1^{\mathbf{k}(0,r,\mathbf{w})} \cdot g_2^{\mathbf{k}(b \cdot \hat{\alpha},0,0)} \cdot g_4^{\delta}$ is replaced with $g_1^{\mathbf{k}(0,r,\mathbf{w})} \cdot g_3^{\mathbf{k}(b \cdot \hat{\alpha},0,0)} \cdot g_4^{\delta}$.

3.4 Reduction from MMH Security to CMH Security

We can prove the following theorem that indicates that the (Q_c, Q_k)-MMH security for a broadcast encoding on $\mathbb{G}_{p_2}$ (resp. $\mathbb{G}_{p_3}$) can be tightly reduced to its Q_c-CMH security on $\mathbb{G}_{p_2}$ (resp. $\mathbb{G}_{p_3}$) and the hardness of the Problem 2 (resp. 3).

Theorem 1. *For any $i \in \{2, 3\}$, broadcast encoding Π, and adversary $\mathcal{A}$, there exist adversaries $\mathcal{B}_1$ and $\mathcal{B}_2$ such that*

$$\mathsf{Adv}_{\mathcal{A},\Pi,(Q_c,Q_k),\mathbb{G}_{p_i}}^{\mathsf{MMH}}(\kappa) \leq \mathsf{Adv}_{\mathcal{B}_1,\Pi,Q_c,\mathbb{G}_{p_i}}^{\mathsf{CMH}}(\kappa) + 2\mathsf{Adv}_{\mathcal{B}_2}^{\mathsf{P}_{\mathsf{xx}}} + \frac{1}{p_i}$$

and $\max\{\mathsf{Time}(\mathcal{B}_1), \mathsf{Time}(\mathcal{B}_2)\} \approx \mathsf{Time}(\mathcal{A}) + (Q_k + Q_c) \cdot \mathsf{poly}(\kappa, n)$ *where* $\mathsf{poly}(\kappa, n)$ *is independent of* $\mathsf{Time}(\mathcal{A})$. *In the above,* $\mathsf{P}_{\mathsf{xx}} = \mathsf{P}_2$ *if* $i = 2$ *and* $\mathsf{P}_{\mathsf{xx}} = \mathsf{P}_3$ *if* $i = 3$.

4 Almost Tight IBE from Broadcast Encoding in Composite-Order Groups

In this section, we show a generic conversion from a broadcast encoding scheme to an IBE scheme. An important property of the resulting IBE scheme is that (μ, Q_c, Q_k)-security of the scheme can be almost tightly reduced to the Q_c-CMH security of the underlying broadcast encoding scheme (and Problems 1, 2, 3, 4, and 5). In particular, the reduction only incurs small polynomial security loss, which is independent of μ and Q_k. Therefore, if the underlying broadcast encoding scheme is tightly Q_c-CMH secure, which is the case for all of our constructions, the resulting IBE scheme obtained by the conversion is almost tightly secure. Note that in the following construction, we have $\mathsf{sp} = \bot$. This mean that the key generation algorithm Par *does not output any secret parameter*. This property will be needed to convert our IBE scheme into CCA secure PKE scheme in Sect. 8.

Construction. Here, we construct an IBE scheme Φ^{comp} from a broadcast encoding $\Pi = (\mathsf{Param}, \mathsf{KEnc}, \mathsf{CEnc}, \mathsf{Pair})$. Let the identity space of the scheme be $\mathcal{ID} = \{0, 1\}^{\ell}$ and the message space be $\mathcal{M} = \{0, 1\}^m$. We also let $\mathcal{H}$ be a family of pairwise independent hash functions $\mathsf{H} : \mathbb{G}_T \rightarrow \mathcal{M}$. We assume that $\sqrt{\frac{2^m}{p_2}} = 2^{-\Omega(\kappa)}$ so that the left-over hash lemma can be applied in the security proof.

$\mathsf{Par}(1^{\kappa})$: It first runs $(N, \mathbb{G}, \mathbb{G}_T, g_1, g_2, g_3, g_4, e(\cdot)) \xleftarrow{\$} \mathcal{G}_{\mathsf{comp}}(1^{\kappa})$ and $\mathsf{Param}(2\ell, N) \rightarrow d_1$. Then it picks $\mathbf{w} \xleftarrow{\$} \mathbb{Z}_N^{d_1}$, $a \xleftarrow{\$} \mathbb{Z}_N^*$, $\mathsf{H} \xleftarrow{\$} \mathcal{H}$ and sets $h := (g_1 g_2 g_3 g_4)^a$. Finally, it outputs $\mathsf{pp} = (g_1, g_1^{\mathbf{w}}, g_4, h, \mathsf{H})$ and $\mathsf{sp} = \bot$.

$\mathsf{Gen}(\mathsf{pp}, \mathsf{sp})$: It picks $\alpha \xleftarrow{\$} \mathbb{Z}_N$ and outputs $\mathsf{mpk} = (\mathsf{pp}, e(g_1, h)^\alpha)$ and $\mathsf{msk} = \alpha$.

$\mathsf{Ext}(\mathsf{msk}, \mathsf{mpk}, \mathsf{ID})$: It first sets $S = \{2i - \mathsf{ID}_i | i \in [\ell]\}$ where $\mathsf{ID}_i \in \{0, 1\}$ is the i-th bit of $\mathsf{ID} \in \{0, 1\}^\ell$. Then it runs $\mathsf{KEnc}(j, N) \to (\mathbf{k}_j, d_2')$ and picks $\mathbf{r}_j \xleftarrow{\$} \mathbb{Z}_N^{d_2'}$ and $\boldsymbol{\delta}_j \xleftarrow{\$} \mathbb{Z}_N^{d_2}$ for all $j \in S$. It also picks random $\{\alpha_j \in \mathbb{Z}_N\}_{j \in S}$ subject to constraint that $\alpha = \sum_{j \in S} \alpha_j$. Then, it computes $g_1^{\mathbf{k}_j(0, \mathbf{r}_j, \mathbf{w})}$, $\mathsf{Pair}(j, S, N) \to \mathbf{E}_j$, and

$$\mathsf{sk}_j = h^{\mathbf{k}_j(\alpha_j, 0, 0)} \cdot g_1^{\mathbf{k}_j(0, \mathbf{r}_j, \mathbf{w})} \cdot g_4^{\boldsymbol{\delta}_j}$$

for all $j \in S$. Note that $g_1^{\mathbf{k}_j(0, \mathbf{r}_j, \mathbf{w})}$ can be computed from $g_1^{\mathbf{w}}$ and $\mathbf{r}_j = (r_{j,1}, \dots, r_{j,d_2'})$ efficiently because $\mathbf{k}_j(0, \mathbf{r}_j, \mathbf{w})$ contains only linear combinations of monomials $r_{j,i}$, $r_{j,i} w_{j'}$. Finally, it outputs private key $\mathsf{sk}_{\mathsf{ID}} = \prod_{j \in S}(\mathsf{sk}_j)^{\mathbf{E}_j}$.

$\mathsf{Enc}(\mathsf{mpk}, \mathsf{ID}, \mathsf{M})$: It first sets $S = \{2i - \mathsf{ID}_i | i \in [\ell]\}$. Then it runs $\mathsf{CEnc}(S, N) \to (\mathbf{c}, d_3')$, picks $\mathbf{s} = (s_0, s_1, \dots, s_{d_3'}) \xleftarrow{\$} \mathbb{Z}_N^{d_3'+1}$, and computes $g_1^{\mathbf{c}(\mathbf{s}, \mathbf{w})}$. Note that $g_1^{\mathbf{c}(\mathbf{s}, \mathbf{w})}$ can be computed from $g_1^{\mathbf{w}}$ and $\mathbf{s}$ efficiently because $\mathbf{c}(\mathbf{s}, \mathbf{w})$ contains only linear combinations of monomials s_i, $s_i w_j$. Finally, it outputs

$$\mathsf{CT} = \left(C_1 = g_1^{\mathbf{c}(\mathbf{s}, \mathbf{w})}, \quad C_2 = \mathsf{H}\big(e(g_1, h)^{s_0 \alpha}\big) \oplus \mathsf{M} \right).$$

Here, $\oplus$ denotes bitwise exclusive OR of two bit strings.

$\mathsf{Dec}(\mathsf{sk}_{\mathsf{ID}}, \mathsf{CT})$: It parses $\mathsf{CT} \to (C_1, C_2)$ and computes $e(\mathsf{sk}_{\mathsf{ID}}^\top, C_1^\top) = e(g_1, h)^{s_0 \alpha}$. Then, it recovers the message by $\mathsf{M} = C_2 \oplus \mathsf{H}(e(g_1, h)^{s_0 \alpha})$.

CORRECTNESS. We show the correctness of the scheme. It suffices to show the following.

$$
\begin{aligned}
e(\mathsf{sk}_{\mathsf{ID}}^\top, C_1^\top) &= e\big((\prod_{j \in S}(\mathsf{sk}_j)^{\mathbf{E}_j})^\top, g_1^{\mathbf{c}(\mathbf{s}, \mathbf{w})^\top}\big) = \prod_{j \in S} e(g_1, g_1)^{\mathbf{k}_j(a\alpha_j, \mathbf{r}_j, \mathbf{w}) \mathbf{E}_j \mathbf{c}(\mathbf{s}, \mathbf{w})^\top} \\
&= \prod_{j \in S} e(g_1, g_1)^{s_0 a \alpha_j} = \prod_{j \in S} e(g_1, h)^{s_0 \alpha_j} = e(g_1, h)^{s_0 \alpha}.
\end{aligned}
$$

The third equation above follows from the correctness of the broadcast encoding.

Security. The following theorem indicates that the security of the IBE is (almost) tightly reduced to the MMH security of the underlying broadcast encoding on $\mathbb{G}_{p_2}$ and $\mathbb{G}_{p_3}$ and Problems 1, 4, and 5. Combining the theorem with Theorem 1, the security of the scheme can be almost tightly reduced to the Q_c-CMH security of the underlying encoding (and Problems 1, 2, 3, 4, and 5). The reduction only incurs $O(\ell)$ security loss.

Theorem 2. *For any adversary $\mathcal{A}$, there exist adversaries $\mathcal{B}_i$ for $i \in [1, 5]$ such that*

$$\mathsf{Adv}^{\mathsf{IBE}}_{\mathcal{A}, \Phi^{\mathrm{comp}}, (\mu, Q_c, Q_k)}(\kappa) \leq \mathsf{Adv}^{\mathsf{P}_1}_{\mathcal{B}_1}(\kappa) + \mathsf{Adv}^{\mathsf{P}_5}_{\mathcal{B}_2}(\kappa) + Q_c \cdot 2^{-\Omega(\kappa)}$$

$$+ \ell \left(2\mathsf{Adv}^{\mathsf{P}_4}_{\mathcal{B}_3}(\kappa) + \mathsf{Adv}^{\mathsf{MMH}}_{\mathcal{B}_4, \Pi, (Q_c, Q_k), \mathbb{G}_{p_2}}(\kappa) + \mathsf{Adv}^{\mathsf{MMH}}_{\mathcal{B}_5, \Pi, (Q_c, Q_k), \mathbb{G}_{p_3}}(\kappa) \right)$$

and $\max\{\mathsf{Time}(\mathcal{B}_i)|i \in [1,5]\} \approx \mathsf{Time}(\mathcal{A}) + (\mu + Q_c + Q_k) \cdot \mathsf{poly}(\kappa, \ell)$ *where* $\mathsf{poly}(\kappa, \ell)$ *is independent of* $\mathsf{Time}(\mathcal{A})$.

5 Framework for Constructions in Prime-Order Groups

In Sects. 3 and 4, we show our framework to construct almost tightly secure IBE in composite-order groups. Since we carefully constructed the framework so that we only use the subgroup decision assumptions and the DBDH assumption in the security proof, we can apply recent composite-order-to-prime-order conversion techniques in the literature [2,3,16,18] to the framework. We choose to use [3], but other choices might be possible. In this section, we show our framework for constructing almost tightly secure IBE in prime-order groups. Our framework is almost parallel to that in composite-order groups. Namely, we define CMH security and MMH security in prime-order groups. Then, we show reduction between them. Finally, we show a generic construction of IBE scheme from broadcast encoding and show that the scheme is (almost) tightly secure if the underlying encoding is tightly CMH secure.

In the following, we will use asymmetric bilinear group $(\mathbb{G}_1, \mathbb{G}_2, \mathbb{G}_T)$ of prime order p with efficiently computable and non-degenerate bilinear map $e(\cdot) : \mathbb{G}_1 \times \mathbb{G}_2 \to \mathbb{G}_T$. For our purpose, we define a prime-order bilinear group generator $\mathcal{G}_{\mathsf{prime}}$ that takes as input a security parameter 1^κ and outputs $(p, \mathbb{G}_1, \mathbb{G}_2, \mathbb{G}_T, g, h, e(\cdot))$ where g and h are random generator of $\mathbb{G}_1$ and $\mathbb{G}_2$, respectively. Let $\pi_1 : \mathbb{Z}_p^{4\times 4} \to \mathbb{Z}_p^{4\times 2}$, $\pi_2 : \mathbb{Z}_p^{4\times 4} \to \mathbb{Z}_p^{4\times 1}$, and $\pi_3 : \mathbb{Z}_p^{4\times 4} \to \mathbb{Z}_p^{4\times 1}$ be the projection maps that map a 4×4 matrix to the leftmost 2 columns, the third column, and the fourth column, respectively.

Intuition. In prime-order groups, we work with 4×4 matrix. The first two dimensions serve as "normal space" (corresponding to $\mathbb{G}_{p_1}$), while the third and the fourth dimension serve as *double* "semi-functional spaces" (corresponding to $\mathbb{G}_{p_2}$ and $\mathbb{G}_{p_3}$). There is no corresponding dimension to $\mathbb{G}_{p_4}$. While the use of 4×4 matrices is similar to Chen and Wee [17,19][5], conceptually, our techniques are quite different from theirs. They use the first two dimensions as a normal space and the last two dimensions as *single* semi-functional space. In contrast, we introduce additional semi-functional space to be able to prove the multi-challenge security rather than single-challenge security. Furthermore, due to our new proof technique, these semi-functional spaces are smaller compared to those of [17,19].

5.1 Preparation

Here, we introduce definitions and notations needed to describe our result. Let p be a prime number and $\mathbf{k}$ and $\mathbf{c}$ be vectors output by $\mathsf{KEnc}()$ and $\mathsf{CEnc}()$ on

---

[5] They showed a construction that is secure under the k-LIN assumption for any k, using $2k \times 2k$ matrices. When $k = 2$, the scheme is secure under the DLIN assumption.

some input respectively. Here, we assign each variable w_i in the vector a matrix $\mathbf{W}_i \in \mathbb{Z}_p^{4\times4}$ for $i \in [d_1]$ (rather than assigning a scalar value), variable α a column vector $\boldsymbol{\alpha} \in \mathbb{Z}_p^{4\times1}$, variable r_i a vector $\mathbf{x}_i \in \mathbb{Z}_p^{4\times1}$ for $i \in [d_2']$, and variable s_i a vector $\mathbf{y}_i \in \mathbb{Z}_p^{4\times1}$ for $i \in [0, d_3']$. The evaluation of polynomials $\mathbf{k_Z}$ and $\mathbf{c_B}$, which are indexed by an invertible matrix $\mathbf{B} \in \mathbb{Z}_p^{4\times4}$ and $\mathbf{Z} \in \mathbb{Z}_p^{4\times4}$, are defined as follows. In the following, we denote

$$\mathbb{W} = (\mathbf{W}_1,\ldots,\mathbf{W}_{d_1}) \in (\mathbb{Z}_p^{4\times4})^{d_1}, \quad \mathbf{X} = (\mathbf{x}_1,\ldots,\mathbf{x}_{d_2'}) \in \mathbb{Z}_p^{4\times d_2'}$$

$$\mathbf{Y} = (\mathbf{y}_0,\mathbf{y}_1,\ldots,\mathbf{y}_{d_3'}) \in \mathbb{Z}_p^{4\times(d_3'+1)}, \qquad \mathbf{Z} = (\mathbf{B}^{-1})^\top \cdot \mathbf{D}.$$

where $\mathbf{D} \in \mathbb{Z}_p^{4\times4}$ is a full-rank diagonal matrix with the entries $(3,3)$ and $(4,4)$ being 1.

Let $\mathbf{k} = (k_1,\ldots,k_{d_2})$ be a vector of polynomials in variables $\alpha,r_1,\ldots,r_{d_2'},w_1,\ldots,w_{d_1}$ with coefficients in $\mathbb{Z}_p$ defined as Eq. (4). We define $\mathbf{k_Z}(\alpha,\mathbf{X},\mathbb{W}) \in \mathbb{Z}_p^{4\times d_2}$ as $\mathbf{k_Z}(\alpha,\mathbf{X},\mathbb{W}) = \{k_{\mathbf{Z},\iota}(\alpha,\mathbf{X},\mathbb{W})\}_{\iota\in[d_2]} =$

$$\left\{ b_\iota\boldsymbol{\alpha} + \Big(\sum_{j\in[d_2']} b_{\iota,j}\mathbf{Z}\mathbf{x}_j\Big) + \Big(\sum_{(j,k)\in[d_2']\times[d_1]} b_{\iota,j,k}\mathbf{W}_k^\top\mathbf{Z}\mathbf{x}_j\Big) \in \mathbb{Z}_p^{4\times1} \right\}_{\iota\in[d_2]}.$$

Let $\mathbf{c} = (c_1,\ldots,c_{d_3})$ be a vector of polynomials in variables $s_0,s_1,\ldots,s_{d_3'}, w_1,\ldots,w_{d_1}$ with coefficients in $\mathbb{Z}_p$ defined as Eq. (5). We define $\mathbf{c_B}(\mathbf{Y},\mathbb{W}) \in \mathbb{Z}_p^{4\times d_3}$ as $\mathbf{c_B}(\mathbf{Y},\mathbb{W}) = \{c_{\mathbf{B},\iota}(\mathbf{Y},\mathbb{W})\}_{\iota\in[d_3]} =$

$$\left\{ \Big(\sum_{j\in[0,d_3']} a_{\iota,j}\mathbf{B}\mathbf{y}_j\Big) + \Big(\sum_{(j,k)\in[0,d_3']\times[d_1]} a_{\iota,j,k}\mathbf{W}_k\mathbf{B}\mathbf{y}_j\Big) \in \mathbb{Z}_p^{4\times1} \right\}_{\iota\in[d_3]}.$$

Restriction on the Encoding. In our framework for prime-order constructions, we define and require *regularity* of encoding similarly to [3], which is needed to prove the security of our IBE obtained from the broadcast encoding. We omit the definition and defer to the full version for the details [6].

Correctness of Encoding. Let $\tau \in [n]$ and $S \subseteq [n]$ be an index and a set such that $\tau \in S$. Let also $\mathsf{KEnc}(\tau,p) \rightarrow (\mathbf{k},d_2')$, $\mathsf{CEnc}(S,p) \rightarrow (\mathbf{c},d_3')$, and $\mathsf{Pair}(\tau,S,p) \rightarrow \mathbf{E} = (E_{\eta,\iota})_{(\eta,\iota)\in[d_2]\times[d_3]} \in \mathbb{Z}_p^{d_2\times d_3}$. Then, by the correctness of the broadcast encoding, we have $\sum_{(\eta,\iota)\in[d_2]\times[d_3]} E_{\eta,\iota}k_\eta c_\iota = \alpha s_0$ (the equation holds symbolically). From this, we have the following. (Note that the claim is shown similarly to Claim 15 in [3].)

Lemma 2. *We have* $\sum_{(\eta,\iota)\in[d_2]\times[d_3]} E_{\eta,\iota} \cdot k_{\mathbf{Z},\eta}(\alpha,\mathbf{X},\mathbb{W})^\top c_{\mathbf{B},\iota}(\mathbf{Y},\mathbb{W}) = \boldsymbol{\alpha}^\top\mathbf{B}\mathbf{y}_0.$

CMH and MMH Security. In the full version [6], we define the Q-CMH security for broadcast encoding on prime-order groups, analogously to the corresponding notion on composite-order groups. We also define the (Q_c, Q_k)-MMH security for broadcast encoding on prime-order groups. The former is (unconditionally) implied by the Q-PMH security. Furthermore, we can show that the latter is tightly reduced to the former, similarly to the case in composite-order groups.

5.2 Almost Tightly Secure IBE from Broadcast Encoding in Prime Order Groups

Here, we construct an IBE scheme Φ^{prime} from broadcast encoding scheme $\Pi = (\mathsf{Param}, \mathsf{KEnc}, \mathsf{CEnc}, \mathsf{Pair})$. Let the identity space of Φ^{prime} be $\mathcal{ID} = \{0,1\}^\ell$ and the message space $\mathcal{M}$ be $\mathcal{M} = \mathbb{G}_T$. We will not use pairwise independent hash function differently from our construction in composite-order groups. We note that similarly to our construction in composite-order groups, we have $\mathsf{sp} = \bot$ in the following.

$\mathsf{Par}(1^\kappa, \ell)$: It first runs $(p, \mathbb{G}_1, \mathbb{G}_2, \mathbb{G}_T, g, h, e(\cdot)) \xleftarrow{\$} \mathcal{G}_{\mathsf{prime}}(1^\kappa)$ and $\mathsf{Param}(2\ell, p) \to d_1$. Then it picks $\mathbf{B} \xleftarrow{\$} \mathbb{GL}_4(\mathbb{Z}_p)$, $\mathbb{W} = (\mathbf{W}_1, \ldots, \mathbf{W}_{d_1}) \xleftarrow{\$} (\mathbb{Z}_p^{4 \times 4})^{d_1}$ and a random full-rank diagonal matrix $\mathbf{D} \in \mathbb{Z}_p^{4 \times 4}$ with the entries $(3,3)$ and $(4,4)$ being 1. Finally, it sets $\mathbf{Z} = \mathbf{B}^{-\top}\mathbf{D}$ and outputs

$$\mathsf{pp} = \begin{pmatrix} g, \ g^{\pi_1(\mathbf{B})}, \ g^{\pi_1(\mathbf{W}_1\mathbf{B})}, \ \ldots, \ g^{\pi_1(\mathbf{W}_{d_1}\mathbf{B})} \\ h, \ h^{\pi_1(\mathbf{Z})}, \ h^{\pi_1(\mathbf{W}_1^\top\mathbf{Z})}, \ \ldots, \ h^{\pi_1(\mathbf{W}_{d_1}^\top\mathbf{Z})} \end{pmatrix} \quad \text{and} \quad \mathsf{sp} = \bot.$$

In the following, we will omit subscript $\mathbf{B}$ and $\mathbf{Z}$ from $\mathbf{c}_\mathbf{B}(\mathbf{S}, \mathbb{W})$ and $\mathbf{k}_\mathbf{Z}(\alpha, \mathbf{R}, \mathbb{W})$ and just denote $\mathbf{c}(\mathbf{S}, \mathbb{W})$ and $\mathbf{k}(\alpha, \mathbf{R}, \mathbb{W})$ for ease of notation. $\mathbf{B}$ and $\mathbf{Z}$ are fixed in the following and clear from the context.

$\mathsf{Gen}(\mathsf{pp})$: It picks $\alpha \xleftarrow{\$} \mathbb{Z}_p^{4 \times 1}$ and outputs $\mathsf{mpk} = (\mathsf{pp}, e(g, h)^{\alpha^\top \pi_1(\mathbf{B})})$ and $\mathsf{msk} = \alpha$.

$\mathsf{Ext}(\mathsf{msk}, \mathsf{mpk}, \mathsf{ID})$: It first sets $S = \{2i - \mathsf{ID}_i | i \in [\ell]\}$ where $\mathsf{ID}_i \in \{0,1\}$ is the i-th bit of $\mathsf{ID} \in \{0,1\}^\ell$. Then it runs $\mathsf{KEnc}(j, p) \to (\mathbf{k}_j, d_2')$, picks $\mathbf{r}_{j,1}, \ldots, \mathbf{r}_{j,d_2'} \xleftarrow{\$} \mathbb{Z}_p^{2 \times 1}$, and sets $\mathbf{R}_j = \left(\begin{pmatrix} \mathbf{r}_{j,1} \\ 0 \\ 0 \end{pmatrix}, \cdots, \begin{pmatrix} \mathbf{r}_{j,d_2'} \\ 0 \\ 0 \end{pmatrix} \right) \in \mathbb{Z}_p^{4 \times d_2'}$ for all $j \in S$. It also picks random $\{\alpha_j \in \mathbb{Z}_p^{4 \times 1}\}_{j \in S}$ subject to constraint that $\alpha = \sum_{j \in S} \alpha_j$. Then, it computes $\mathsf{Pair}(j, S, p) \to \mathbf{E}_j = (E_{j,\eta,\iota})_{(\eta,\iota) \in [d_2] \times [d_3]}$ and

$$\mathsf{sk}_j = h^{\mathbf{k}_j(\alpha_j, \mathbf{R}_j, \mathbb{W})} = \{\mathsf{sk}_{j,\eta} = h^{k_{j,\eta}(\alpha_j, \mathbf{R}_j, \mathbb{W})}\}_{\eta \in [d_2]}$$

for all $j \in S$. Note that $h^{\mathbf{k}_j(\alpha_j, \mathbf{R}_j, \mathbb{W})}$ can be computed from α_j, $h^{\pi_1(\mathbf{Z})}$, and $\{g^{\pi_1(\mathbf{W}_i^\top \mathbf{Z})}\}_{i \in [d_1]}$ efficiently because $\mathbf{k}_j(\alpha_j, \mathbf{R}_j, \mathbb{W}) = \{k_{j,\iota}(\alpha_j, \mathbf{R}_j, \mathbb{W})\}_{\iota \in [d_2]}$ contains only linear combination of α_j, $\mathbf{Z}\begin{pmatrix} \mathbf{r}_i \\ 0 \\ 0 \end{pmatrix} = \pi_1(\mathbf{Z})\mathbf{r}_i$, and $\mathbf{W}_i^\top \mathbf{Z}\begin{pmatrix} \mathbf{r}_{j'} \\ 0 \\ 0 \end{pmatrix} = \pi_1(\mathbf{W}_i^\top \mathbf{Z})\mathbf{r}_{j'}$. Finally, it outputs private key $\mathsf{sk}_{\mathsf{ID}} = \left\{ \prod_{j \in S, \eta \in [d_2]} \mathsf{sk}_{j,\eta}^{E_{j,\eta,\iota}} \right\}_{\iota \in [d_3]}$.

$\mathsf{Enc}(\mathsf{mpk}, \mathsf{ID}, \mathsf{M})$: It first sets $S = \{2i - \mathsf{ID}_i | i \in [\ell]\}$. Then it runs $\mathsf{CEnc}(S, p) \rightarrow (\mathbf{c}, d_3')$, picks $\mathbf{s}_0, \mathbf{s}_1, \ldots, \mathbf{s}_{d_3'} \xleftarrow{\$} \mathbb{Z}_p^{2 \times 1}$, and sets $\mathbf{S} = \left(\begin{pmatrix} \mathbf{s}_0 \\ 0 \\ 0 \end{pmatrix}, \begin{pmatrix} \mathbf{s}_1 \\ 0 \\ 0 \end{pmatrix}, \cdots, \begin{pmatrix} \mathbf{s}_{d_3'} \\ 0 \\ 0 \end{pmatrix} \right) \in \mathbb{Z}_p^{4 \times (d_3'+1)}$. Then it returns

$$\mathsf{CT} = \left(C_1 = g^{\mathbf{c}(\mathbf{S}, \mathbb{W})}, \quad C_2 = e(g, h)^{\boldsymbol{\alpha}^\top \pi_1(\mathbf{B})\mathbf{s}_0} \cdot \mathsf{M} \right).$$

Note that $g^{\mathbf{c}(\mathbf{S}, \mathbb{W})}$ can be computed from $g^{\pi_1(\mathbf{B})}$ and $\{g^{\pi_1(\mathbf{W}_i \mathbf{B})}\}_{i \in [d_1]}$ efficiently because $\mathbf{c}(\mathbf{S}, \mathbb{W})$ contains only linear combinations of $\mathbf{B} \begin{pmatrix} \mathbf{s}_i \\ 0 \\ 0 \end{pmatrix} = \pi_1(\mathbf{B})\mathbf{s}_i$ and $\mathbf{W}_i \mathbf{B} \begin{pmatrix} \mathbf{s}_j \\ 0 \\ 0 \end{pmatrix} = \pi_1(\mathbf{W}_i \mathbf{B})\mathbf{s}_j$. C_2 can be computed from $e(g, h)^{\boldsymbol{\alpha}^\top \pi_1(\mathbf{B})}$.

$\mathsf{Dec}(\mathsf{sk}_{\mathsf{ID}}, \mathsf{CT})$: Let CT be $\mathsf{CT} = (C_1, C_2)$. From $C_1 = g^{\mathbf{c}(\mathbf{S}, \mathbb{W})} = \{g^{c_\iota(\mathbf{S}, \mathbb{W})}\}_{\iota \in [d_3]}$, it computes

$$\prod_{\iota \in [d_3]} e \left(g^{c_\iota(\mathbf{S}, \mathbb{W})}, \prod_{j \in S, \eta \in [d_2]} \mathsf{sk}_{j,\eta}^{E_{\eta, \iota}} \right) = e(g, h)^{\boldsymbol{\alpha}^\top \pi_1(\mathbf{B})\mathbf{s}_0} \tag{6}$$

and recovers the message by $C_2 / e(g, h)^{\boldsymbol{\alpha}^\top \pi_1(\mathbf{B})\mathbf{s}_0} = \mathsf{M}$.

CORRECTNESS. To see correctness of the scheme, it suffices to show Eq. (6).

$$\prod_{\iota \in [d_3]} e \left(g^{c_\iota(\mathbf{S}, \mathbb{W})}, \prod_{j \in S, \eta \in [d_2]} \mathsf{sk}_{j,\eta}^{E_{\eta, \iota}} \right)$$
$$= \prod_{j \in S} e(g, h)^{\sum_{(\iota, \eta) \in [d_3, d_2]} E_{\eta, \iota} k_{j,\eta} (\boldsymbol{\alpha}_j, \mathbf{R}_j, \mathbb{W})^\top c_\iota(\mathbf{S}, \mathbb{W})}$$
$$= \prod_{j \in S} e(g, h)^{\boldsymbol{\alpha}_j^\top \mathbf{B} \begin{pmatrix} \mathbf{s}_0 \\ 0 \\ 0 \end{pmatrix}} = e(g, h)^{\boldsymbol{\alpha}^\top \pi_1(\mathbf{B})\mathbf{s}_0}$$

The second equation above follows from the correctness of the underlying broadcast encoding.

Security. Assume that the broadcast encoding satisfies regularity requirement. Then, we can show that the security of the above IBE is reduced to the hardness of the (standard) decisional linear assumption and the (Q_c, Q_k)-MMH security of the underlying broadcast encoding on prime-order groups. The reduction only incurs $O(\ell)$ security loss. Since the Q_c-CMH security tightly implies (Q_c, Q_k)-MMH security, the above IBE scheme is (almost) tightly secure if the underlying broadcast encoding is tightly Q_c-CMH. The details will appear in the full version [6].

6   Construction of Broadcast Encoding Schemes

In this section, we show two broadcast encoding schemes Π_{cc} and Π_{slp}. For these schemes, we can tightly prove the Q_c-CMH security for any Q_c. Therefore, by applying the conversion in Sects. 4 and 5, we obtain IBE schemes with almost tight security in the multi-challenge and multi-instance setting both in prime and composite-order groups. An IBE obtained from Π_{cc} achieves constant-size ciphertexts, but at the cost of requiring public parameters with the number of group elements being linear in the security parameter. Our second broadcast encoding scheme Π_{slp} partially compensate for this. By appropriately setting parameters, we can realize trade-off between size of ciphertexts and public parameters. For example, from the encoding, we obtain the first almost tightly secure IBE with all communication cost (the size of pp and CT) being $O(\sqrt{\kappa})$. Such a scheme is not known even in the single-challenge setting [9,17]. While the structure of Π_{cc} is implicit in [25], Π_{slp} is new. The construction of Π_{slp} is inspired by recent works on unbounded attribute-based encryption schemes [38,43,44]. However, the security proof for the encoding is completely different.

6.1   Broadcast Encoding with Constant-Size Ciphertexts

At first, we show the following broadcast encoding scheme that we call Π_{cc}. The scheme has the same structure as the broadcast encryption scheme proposed by Gentry and Waters [25]. For Π_{cc}, we can prove Q-PMH security for any Q. By Lemma 1, we have that Q-CMH security of Π_{cc} on $\mathbb{G}_{p_2}$ and $\mathbb{G}_{p_3}$ can be tightly proven unconditionally. Similar implication holds in prime-order groups.

$\mathsf{Param}(n, N) \to d_1$: It outputs $d_1 = n$.
$\mathsf{KEnc}(\tau, N) \to (\mathbf{k}, d_2')$: It outputs $\mathbf{k} = (\alpha + rw_\tau, rw_1, \ldots, rw_{\tau-1}, r, rw_{\tau+1}, \ldots,$
$rw_n)$ and $d_2' = 1$ where $\mathbf{r} = r$.
$\mathsf{CEnc}(S, N) \to (\mathbf{c}, d_3')$: Let $S \subseteq [n]$. It outputs $\mathbf{c} = (s, \sum_{j \in S} sw_j)$ and $d_3' = 0$
where $\mathbf{s} = s$.

CORRECTNESS. Let $\tau \in S$. Then, we have

$$s \cdot \left((\alpha + rw_\tau) + \Big(\sum_{j \in S \setminus \{\tau\}} rw_j \Big) \right) - \Big(\sum_{j \in S} sw_j \Big) \cdot r = s\alpha.$$

Lemma 3. Π_{cc} *defined above is Q-PMH secure for any $Q \in \mathbb{N}$.*

Proof. Let $\tau \notin \cup_{j \in [Q]} S_j$. It is clear that information on w_τ is not leaked given $\{\mathbf{c}_{S_j}(\mathbf{s}_j, \mathbf{w})\}_{j \in [Q]}$. Thus, α is information-theoretically hidden from $\mathbf{k}_\tau(\alpha, \mathbf{r}, \mathbf{w})$, because α is masked by rw_τ which is uniformly random over $\mathbb{Z}_p$. Thus, the lemma follows.

6.2   Encoding with Sub-linear Parameters

We propose the following broadcast encoding scheme that we call Π_{slp}. We can realize trade-off between sizes of parameters by setting n_1. For the encoding scheme, we are not able to show the Q-PMH security. Instead, we show the Q-CMH security.

$\mathsf{Param}(n, N) \rightarrow d_1$: It outputs $d_1 = 2n_1 + 3$. We let $n_2 = \lceil n/n_1 \rceil$. For ease of the notation, we will denote $\mathbf{w} = (u_1, \ldots, u_{n_1}, v, u'_1, \ldots, u'_{n_1}, v', w)$ in the following.

$\mathsf{KEnc}(\tau, N) \rightarrow (\mathbf{k}, d'_2)$: It computes unique $\tau_1 \in [n_1]$ and $\tau_2 \in [n_2]$ such that $\tau = \tau_1 + (\tau_2 - 1) \cdot n_1$. Then it sets $d'_2 = 1$ and $\mathbf{r} = r$ and outputs

$$\mathbf{k} = \big(\alpha + rw, r, r(v + \tau_2 u_{\tau_1}), \{ru_i\}_{i \in [n_1] \backslash \{\tau_1\}}, r(v' + \tau_2 u'_{\tau_1}), \{ru'_i\}_{i \in [n_1] \backslash \{\tau_1\}} \big).$$

$\mathsf{CEnc}(S, N) \rightarrow (\mathbf{c}, d'_3)$: It first defines $\tilde{S}_j$ and S_j for $j \in [n_2]$ as

$$\tilde{S}_j = S \cap [(j-1)n_1 + 1, jn_1], \quad S_j = \{j' - (j-1)n_1 \mid j' \in \tilde{S}_j\},$$

sets $\mathbf{s} = (s_0, t_1, \ldots, t_{n_2}, t'_1, \ldots, t'_{n_2})$ and $d'_3 = 2n_2 + 1$, and outputs

$$\mathbf{c} = \Big(s_0, \ \{ s_0 w + t_i \big(v + i \sum_{j \in S_i} u_j \big) + t'_i \big(v' + i \sum_{j \in S_i} u'_j \big), \quad t_i, \quad t'_i \ \}_{i \in [n_2]} \Big).$$

CORRECTNESS. Let $\tau \in S$ and τ_1, τ_2 be defined as above. Then, we have $\tau_1 \in S_{\tau_2}$ and

$$s_0 \cdot (\alpha + rw) - \Big(s_0 w + t_{\tau_2} \big(v + \tau_2 \sum_{j \in S_{\tau_2}} u_j \big) + t'_{\tau_2} \big(v' + \tau_2 \sum_{j \in S_{\tau_2}} u'_j \big) \Big) \cdot r$$

$$+ t_{\tau_2} \Big(r(v + \tau_2 u_{\tau_1}) + \tau_2 \cdot \big(\sum_{j \in S_{\tau_2} \backslash \{\tau_1\}} ru_j \big) \Big) + t'_{\tau_2} \Big(r(v' + \tau_2 u'_{\tau_1}) + \tau_2 \cdot \big(\sum_{j \in S_{\tau_2} \backslash \{\tau_1\}} ru'_j \big) \Big)$$

$$= s_0 \alpha.$$

We can tightly prove the Q-CMH security of Π_{slp} on composite-order (resp. prime-order) groups assuming the DLIN assumption on the composite-order (resp. prime-order) group. The details can be found in the full version [6].

6.3   Implications

For Π_{xx}, we call an IBE scheme obtained by applying the conversion in Sect. 4 to Π_{xx} $\Phi_{\mathsf{xx}}^{\mathsf{comp}}$. Similarly, we call a scheme obtained by the conversion in Sect. 5.2 $\Phi_{\mathsf{xx}}^{\mathsf{prime}}$. $\Phi_{\mathsf{cc}}^{\mathsf{prime}}$ and $\Phi_{\mathsf{slp}}^{\mathsf{prime}}$ are the first IBE schemes that are (almost) tightly secure in the multi-challenge and multi-instance setting, from a static assumption in prime-order groups (the DLIN assumption). $\Phi_{\mathsf{cc}}^{\mathsf{comp}}$ and $\Phi_{\mathsf{cc}}^{\mathsf{prime}}$ achieve constant-size ciphertext, meaning the number of group elements in ciphertexts is constant. The drawback of the schemes is their long public parameters. In $\Phi_{\mathsf{slp}}^{\mathsf{comp}}$ and $\Phi_{\mathsf{slp}}^{\mathsf{prime}}$, we can trade-off the size of ciphertexts and public parameters. For example, by setting $n_1 = \sqrt{n}$, we obtain the first almost tightly secure IBE scheme such that all communication cost (the size of the public parameters, the master public keys, and the ciphertexts) is sub-linear in the security parameter. Such a scheme is not known in the literature, even in the single-challenge and single-instance setting. Also see Table 1 in Sect. 1 for the overview of the obtained schemes.

7 Anonymous IBE with Tight Security Reduction

All our IBE schemes obtained so far is not anonymous. In these schemes, one can efficiently check that a ciphertext is in a specific form using pairing computation, which leads to an attack against anonymity. In this section, we show that $\Phi_{\mathsf{cc}}^{\mathsf{prime}}$ can be modified to be anonymous, by removing all group elements in $\mathbb{G}_2$ from the public parameter pp and put these in sp instead. We call the resulting scheme Φ_{anon}. This is the first IBE scheme whose anonymity is (almost) tightly proven in the multi-challenge setting. While our technique for making the scheme anonymous is similar to that in [16], the security proof for our scheme requires some new ideas. This is because [16] only deals with the *single-challenge* setting whereas we prove tight security in the *multi-challenge* setting. In the security proof, we introduce new combination of information-theoretic argument (as in [16]) and computational argument.

Construction. Let the identity space of the scheme be $\{0,1\}^\ell$ and the message space be $\mathbb{G}_T$. We note that we have $\mathsf{sp} \neq \bot$ in the following, differently from other constructions in this paper.

$\mathsf{Par}(1^\kappa, \ell)$: It first runs $(p, \mathbb{G}_1, \mathbb{G}_2, \mathbb{G}_T, g, h, e(\cdot)) \xleftarrow{\$} \mathcal{G}_{\mathsf{prime}}(1^\kappa)$. Then it picks
$\mathbf{B} \xleftarrow{\$} \mathrm{GL}_4(\mathbb{Z}_p)$, $\mathbf{W}_1, \ldots, \mathbf{W}_{2\ell} \xleftarrow{\$} \mathbb{Z}_p^{4 \times 4}$ and a random full-rank diagonal
matrix $\mathbf{D} \in \mathbb{Z}_p^{4 \times 4}$ with the entries $(3,3)$ and $(4,4)$ being 1. Finally, it
sets $\mathbf{Z} = \mathbf{B}^{-\top}\mathbf{D}$ and returns $\mathsf{pp} = (g, g^{\pi_1(\mathbf{B})}, g^{\pi_1(\mathbf{W}_1\mathbf{B})}, \ldots, g^{\pi_1(\mathbf{W}_{2\ell}\mathbf{B})})$ and
$\mathsf{sp} = (h, h^{\pi_1(\mathbf{Z})}, h^{\pi_1(\mathbf{W}_1^\top \mathbf{Z})}, \ldots, g^{\pi_1(\mathbf{W}_{2\ell}^\top \mathbf{Z})})$.
$\mathsf{Gen}(\mathsf{pp}, \mathsf{sp})$: It picks $\boldsymbol{\alpha} \xleftarrow{\$} \mathbb{Z}_p^{4 \times 1}$ and outputs $\mathsf{mpk} = (\mathsf{pp}, e(g, h)^{\boldsymbol{\alpha}^\top \pi_1(\mathbf{B})})$ and
$\mathsf{msk} = (\boldsymbol{\alpha}, \mathsf{sp})$.
$\mathsf{Ext}(\mathsf{msk}, \mathsf{mpk}, \mathsf{ID})$: It first sets $S = \{2i - \mathsf{ID}_i | i \in [\ell]\}$ where $\mathsf{ID}_i \in \{0,1\}$ is
the i-th bit of $\mathsf{ID} \in \{0,1\}^\ell$. Then it picks random $\mathbf{r} \xleftarrow{\$} \mathbb{Z}_p^{2 \times 1}$ and returns
$\mathsf{sk}_{\mathsf{ID}} = (K_1 = h^{\boldsymbol{\alpha} + \sum_{i \in S} \pi_1(\mathbf{W}_i^\top \mathbf{Z})\mathbf{r}}, K_2 = h^{-\pi_1(\mathbf{Z})\mathbf{r}})$.
$\mathsf{Enc}(\mathsf{mpk}, \mathsf{ID}, \mathsf{M})$: It first sets $S = \{2i - \mathsf{ID}_i | i \in [\ell]\}$. Then it picks random
$\mathbf{s} \xleftarrow{\$} \mathbb{Z}_p^{2 \times 1}$ and returns $\mathsf{CT} = (C_1 = g^{\pi_1(\mathbf{B})\mathbf{s}}, C_2 = g^{\sum_{i \in S} \pi_1(\mathbf{W}_i \mathbf{B})\mathbf{s}}, C_3 = e(g, h)^{\boldsymbol{\alpha}^\top \pi_1(\mathbf{B})\mathbf{s}} \cdot \mathsf{M})$.
$\mathsf{Dec}(\mathsf{sk}_{\mathsf{ID}}, \mathsf{CT})$: It parses the ciphertext CT as $\mathsf{CT} \to (C_1, C_2, C_3)$, and com-
putes $e(C_1, K_1)e(C_2, K_2) = e(g, h)^{\boldsymbol{\alpha}^\top \pi_1(\mathbf{B})\mathbf{s}}$. Then, it recovers the message
by $C_3 / e(g, h)^{\boldsymbol{\alpha}^\top \pi_1(\mathbf{B})\mathbf{s}} = \mathsf{M}$.

Remark. We have to ensure that the key extraction algorithm Ext always use the same randomness $\mathbf{r}$ for the same identity, in order to (tightly) prove the security of the scheme. This can be easily accomplished, for example, using PRF [24]. For the sake of simplicity, we do not incorporate this change into the description of our scheme.

Security. We can prove $(1, Q_c, Q_k)$-anonymity of Φ_{anon} under the DLIN assumption (single instance case). The reduction cost is $O(\ell)$, which is independent from Q_c and Q_k. While we think that it is not difficult to extend the result to the multi-instance setting, we do not treat it in this paper.

8   Application to CCA Secure Public Key Encryption

Here, we discuss that our IBE schemes with almost tight security reduction in the multi-instance and multi-challenge setting yield almost tightly CCA secure PKE in the same setting via simple modification of Canetti-Halevi-Katz (CHK) transformation [15]. The difference from the ordinary CHK transformation is that we use (tightly secure) Q-fold one-time signature introduced and constructed in [30]. Another difference is that we need a restriction on the original IBE scheme. That is, we require that the key generation algorithm Gen of the IBE scheme does not output any secret parameter. Namely, $\mathsf{sp} = \bot$. Roughly speaking, this is needed since the syntax of the PKE does not allow key generation algorithm to take any secret parameter. Note that this condition is satisfied by all of our constructions except for that in Sect. 7.

By applying the above conversion to $\Phi_{\mathsf{slp}}^{\mathsf{prime}}$ and $\Phi_{\mathsf{cc}}^{\mathsf{prime}}$, we obtain new PKE schemes that we call $\Psi_{\mathsf{slp}}^{\mathsf{prime}}$ and $\Psi_{\mathsf{cc}}^{\mathsf{prime}}$. The former allows flexible trade-off between the size of public parameters and ciphertexts. The latter achieves very short ciphertext-size: The ciphertext overhead of our scheme only consists of 10 group elements and 2 elements in $\mathbb{Z}_p$. This significantly improves previous results [1,27,30,33,34] on PKE scheme with the same security guarantee in terms of the ciphertext-size. Note that state-of-the-art construction by [27,34] require 47 and 59 group elements of ciphertext overhead, respectively. Namely, ciphertext overhead of our scheme is (at least) 74 % shorter, compared to theirs. On the other hand, the size of public parameter of the scheme in [27] is much shorter than ours (and those of [33,34]). The former only requires 17 group elements, but the latter requires many more.

The reason why we can achieve very short ciphertext size is that our strategy to obtain PKE scheme is quite different from other works. Roughly speaking, all of the previous constructions [1,27,30,33,34] follow the template established by Hofheinz and Jager [30]. They first construct (almost) tightly-secure signature. Then, they use the signature to construct (almost) tightly-secure unbounded simulation sound (quasi-adaptive) NIZK. Finally, they follow the Naor-Yung paradigm [41] and convert the CPA-secure PKE with tight security reduction [12] into CCA-secure one using the NIZK. On the other hand, our construction is much more direct and simpler. Our conversion only requires very small amount of overhead in public parameters and ciphertexts.

Acknowledgement. We thank the members of Shin-Akarui-Ango-Benkyo-Kai for valuable comments. We also thank anonymous reviewers for their constructive comments.

References

1. Abe, M., David, B., Kohlweiss, M., Nishimaki, R., Ohkubo, M.: Tagged one-time signatures: tight security and optimal tag size. In: Kurosawa, K., Hanaoka, G. (eds.) PKC 2013. LNCS, vol. 7778, pp. 312–331. Springer, Heidelberg (2013)
2. Agrawal, S., Chase, M.: A study of Pair Encodings: Predicate Encryption in prime order groups. IACR Cryptology ePrint Archive, Report 2015/390
3. Attrapadung, N.: Dual System Encryption Framework in Prime-Order Groups. IACR Cryptology ePrint Archive, Report 2015/390
4. Attrapadung, N.: Dual system encryption via doubly selective security: framework, fully secure functional encryption for regular languages, and more. In: Nguyen, P.Q., Oswald, E. (eds.) EUROCRYPT 2014. LNCS, vol. 8441, pp. 557–577. Springer, Heidelberg (2014)
5. Attrapadung, N., Furukawa, J., Gomi, T., Hanaoka, G., Imai, H., Zhang, R.: Efficient identity-based encryption with tight security reduction. In: Pointcheval, D., Mu, Y., Chen, K. (eds.) CANS 2006. LNCS, vol. 4301, pp. 19–36. Springer, Heidelberg (2006)
6. Attrapadung, N., Hanaoka, G., Yamada, S.: A framework for identity-based encryption with almost tight security. IACR Cryptology ePrint Archive 2015:566 (2015)
7. Bellare, M., Boldyreva, A., Micali, S.: Public-key encryption in a multi-user setting: security proofs and improvements. In: Preneel, B. (ed.) EUROCRYPT 2000. LNCS, vol. 1807, pp. 259–274. Springer, Heidelberg (2000)
8. Bellare, M., Ristenpart, T.: Simulation without the artificial abort: simplified proof and improved concrete security for waters' IBE scheme. In: Joux, A. (ed.) EUROCRYPT 2009. LNCS, vol. 5479, pp. 407–424. Springer, Heidelberg (2009)
9. Blazy, O., Kiltz, E., Pan, J.: (Hierarchical) Identity-based encryption from affine message authentication. In: Garay, J.A., Gennaro, R. (eds.) CRYPTO 2014, Part I. LNCS, vol. 8616, pp. 408–425. Springer, Heidelberg (2014)
10. Boneh, D., Boyen, X.: Efficient selective-ID secure identity-based encryption without random oracles. In: Cachin, C., Camenisch, J.L. (eds.) EUROCRYPT 2004. LNCS, vol. 3027, pp. 223–238. Springer, Heidelberg (2004)
11. Boneh, D., Boyen, X.: Secure identity based encryption without random oracles. In: Franklin, M. (ed.) CRYPTO 2004. LNCS, vol. 3152, pp. 443–459. Springer, Heidelberg (2004)
12. Boneh, D., Boyen, X., Shacham, H.: Short group signatures. In: Franklin, M. (ed.) CRYPTO 2004. LNCS, vol. 3152, pp. 41–55. Springer, Heidelberg (2004)
13. Boneh, D., Franklin, M.: Identity-based encryption from the weil pairing. In: Kilian, J. (ed.) CRYPTO 2001. LNCS, vol. 2139, pp. 213–229. Springer, Heidelberg (2001)
14. Canetti, R., Halevi, S., Katz, J.: A forward-secure public-key encryption scheme. In: Biham, E. (ed.) EUROCRYPT 2003. LNCS, vol. 2656, pp. 255–271. Springer, Heidelberg (2003)

15. Canetti, R., Halevi, S., Katz, J.: Chosen-ciphertext security from identity-based encryption. In: Cachin, C., Camenisch, J.L. (eds.) EUROCRYPT 2004. LNCS, vol. 3027, pp. 207–222. Springer, Heidelberg (2004)
16. Chen, J., Gay, R., Wee, H.: Improved dual system ABE in prime-order groups via predicate encodings. In: Oswald, E., Fischlin, M. (eds.) EUROCRYPT 2015. LNCS, vol. 9057, pp. 595–624. Springer, Heidelberg (2015)
17. Chen, J., Wee, H.: Fully, (almost) tightly secure IBE from standard assumptions. IACR Cryptology ePrint Archive, Report 2013/803
18. Chen, J., Wee, H.: Dual system groups and its applications - compact HIBE and more. IACR Cryptology ePrint Archive, Report 2014/265
19. Chen, J., Wee, H.: Fully, (Almost) Tightly Secure IBE and Dual System Groups. CRYPTO,pp. 435–460 (2013). A merge of two papers [19, 20]
20. Cocks, C.: An identity based encryption scheme based on quadratic residues. In: Honary, B. (ed.) Cryptography and Coding 2001. LNCS, vol. 2260, pp. 360–363. Springer, Heidelberg (2001)
21. Cramer, R., Shoup, V.: A practical public key cryptosystem provably secure against adaptive chosen ciphertext attack. In: Krawczyk, H. (ed.) CRYPTO 1998. LNCS, vol. 1462, pp. 13–25. Springer, Heidelberg (1998)
22. Escala, A., Herold, G., Kiltz, E., Ràfols, C., Villar, J.: An algebraic framework for diffie-hellman assumptions. In: Canetti, R., Garay, J.A. (eds.) CRYPTO 2013, Part II. LNCS, vol. 8043, pp. 129–147. Springer, Heidelberg (2013)
23. Freeman, D.M.: Converting pairing-based cryptosystems from composite-order groups to prime-order groups. In: Gilbert, H. (ed.) EUROCRYPT 2010. LNCS, vol. 6110, pp. 44–61. Springer, Heidelberg (2010)
24. Gentry, C.: Practical identity-based encryption without random oracles. In: Vaudenay, S. (ed.) EUROCRYPT 2006. LNCS, vol. 4004, pp. 445–464. Springer, Heidelberg (2006)
25. Gentry, C., Waters, B.: Adaptive security in broadcast encryption systems (with short ciphertexts). In: Joux, A. (ed.) EUROCRYPT 2009. LNCS, vol. 5479, pp. 171–188. Springer, Heidelberg (2009)
26. Herold, G., Hesse, J., Hofheinz, D., Ràfols, C., Rupp, A.: Polynomial spaces: a new framework for composite-to-prime-order transformations. In: Garay, J.A., Gennaro, R. (eds.) CRYPTO 2014, Part I. LNCS, vol. 8616, pp. 261–279. Springer, Heidelberg (2014)
27. Hofheinz, D.: Algebraic partitioning: fully compact and (almost) tightly secure cryptography. IACR Cryptology ePrint Archive, Report 2015/499
28. Hofheinz, D., Kiltz, E.: Secure hybrid encryption from weakened key encapsulation. In: Menezes, A. (ed.) CRYPTO 2007. LNCS, vol. 4622, pp. 553–571. Springer, Heidelberg (2007)
29. Hofheinz, D., Kiltz, E.: Programmable hash functions and their applications. In: Wagner, D. (ed.) CRYPTO 2008. LNCS, vol. 5157, pp. 21–38. Springer, Heidelberg (2008)
30. Hofheinz, D., Jager, T.: Tightly secure signatures and public-key encryption. In: Safavi-Naini, R., Canetti, R. (eds.) CRYPTO 2012. LNCS, vol. 7417, pp. 590–607. Springer, Heidelberg (2012)
31. Hofheinz, D., Koch, J., Striecks, C.: Identity-based encryption with (almost) tight security in the multi-instance, multi-ciphertext setting. In: Katz, J. (ed.) PKC 2015. LNCS, vol. 9020, pp. 799–822. Springer, Heidelberg (2015)

32. Jutla, C.S., Roy, A.: Shorter quasi-adaptive NIZK proofs for linear subspaces. In: Sako, K., Sarkar, P. (eds.) ASIACRYPT 2013, Part I. LNCS, vol. 8269, pp. 1–20. Springer, Heidelberg (2013)

33. Libert, B., Joye, M., Yung, M., Peters, T.: Concise multi-challenge CCA-secure encryption and signatures with almost tight security. In: Sarkar, P., Iwata, T. (eds.) ASIACRYPT 2014, Part II. LNCS, vol. 8874, pp. 1–21. Springer, Heidelberg (2014)

34. Libert, B., Joye, M., Yung, M., Peters, T.: Compactly Hiding Linear Spans: Tightly Secure Constant-Size Simulation-Sound QA-NIZK Proofs and Applications. IACR Cryptology ePrint Archive, Report 2015/242

35. Lewko, A.: Tools for simulating features of composite order bilinear groups in the prime order setting. In: Pointcheval, D., Johansson, T. (eds.) EUROCRYPT 2012. LNCS, vol. 7237, pp. 318–335. Springer, Heidelberg (2012)

36. Lewko, A., Okamoto, T., Sahai, A., Takashima, K., Waters, B.: Fully secure functional encryption: attribute-based encryption and (hierarchical) inner product encryption. In: Gilbert, H. (ed.) EUROCRYPT 2010. LNCS, vol. 6110, pp. 62–91. Springer, Heidelberg (2010)

37. Lewko, A., Waters, B.: New techniques for dual system encryption and fully secure hibe with short ciphertexts. In: Micciancio, D. (ed.) TCC 2010. LNCS, vol. 5978, pp. 455–479. Springer, Heidelberg (2010)

38. Lewko, A., Waters, B.: Unbounded HIBE and attribute-based encryption. In: Paterson, K.G. (ed.) EUROCRYPT 2011. LNCS, vol. 6632, pp. 547–567. Springer, Heidelberg (2011)

39. Lewko, A., Waters, B.: New proof methods for attribute-based encryption: achieving full security through selective techniques. In: Safavi-Naini, R., Canetti, R. (eds.) CRYPTO 2012. LNCS, vol. 7417, pp. 180–198. Springer, Heidelberg (2012)

40. Naor, M., Reingold, O.: Number-theoretic constructions of efficient pseudo-random functions. J. ACM $51(2)$, 231–262 (2004)

41. Naor, M., Yung, M.: Public-key cryptosystems provably secure against chosen ciphertext attacks. In: STOC, pp. 427–437 (1990)

42. Okamoto, T., Takashima, K.: Fully secure functional encryption with general relations from the decisional linear assumption. In: Rabin, T. (ed.) CRYPTO 2010. LNCS, vol. 6223, pp. 191–208. Springer, Heidelberg (2010)

43. Okamoto, T., Takashima, K.: Fully secure unbounded inner-product and attribute-based encryption. In: Wang, X., Sako, K. (eds.) ASIACRYPT 2012. LNCS, vol. 7658, pp. 349–366. Springer, Heidelberg (2012)

44. Rouselakis, Y., Waters, B.: Practical constructions and new proof methods for large universe attribute-based encryption. In: ACM-CCS, pp. 463–474 (2013)

45. Sakai, R., Ohgishi, K., Kasahara, M.: Cryptosystems based on pairing over elliptic curve. In: The 2001 Symposium on Cryptography and Information Security (2001). (in Japanese)

46. Shacham, H.: A Cramer-Shoup encryption scheme from the linear assumption and from progressively weaker linear variants, IACR Cryptology ePrint Archive, Report 2007/074

47. Shamir, A.: Identity-based cryptosystems and signature schemes. In: Blakely, G.R., Chaum, D. (eds.) CRYPTO 1984. LNCS, vol. 196, pp. 47–53. Springer, Heidelberg (1985)

48. Waters, B.: Efficient identity-based encryption without random oracles. In: Cramer, R. (ed.) EUROCRYPT 2005. LNCS, vol. 3494, pp. 114–127. Springer, Heidelberg (2005)
49. Waters, B.: Dual system encryption: realizing fully secure IBE and HIBE under simple assumptions. In: Halevi, S. (ed.) CRYPTO 2009. LNCS, vol. 5677, pp. 619–636. Springer, Heidelberg (2009)
50. Wee, H.: Dual system encryption via predicate encodings. In: Lindell, Y. (ed.) TCC 2014. LNCS, vol. 8349, pp. 616–637. Springer, Heidelberg (2014)

Riding on Asymmetry: Efficient ABE for Branching Programs

Sergey Gorbunov[1]([✉]) and Dhinakaran Vinayagamurthy[2]

[1] Aikicrypt, Boston, USA
sergey@aikicrypt.com
[2] University of Waterloo, Waterloo, Canada
dvinayag@uwaterloo.ca

Abstract. In an Attribute-Based Encryption (ABE) scheme the ciphertext encrypting a message μ, is associated with a public attribute vector $\mathbf{x}$ and a secret key sk_P is associated with a predicate P. The decryption returns μ if and only if $P(\mathbf{x}) = 1$. ABE provides efficient and simple mechanism for data sharing supporting fine-grained access control. Moreover, it is used as a critical component in constructions of succinct functional encryption, reusable garbled circuits, token-based obfuscation and more.

In this work, we describe a new efficient ABE scheme for a family of branching programs with short secret keys and from a mild assumption. In particular, in our construction the size of the secret key for a branching program P is $|P| + \text{poly}(\lambda)$, where λ is the security parameter. Our construction is secure assuming the standard Learning With Errors (LWE) problem with approximation factors $n^{\omega(1)}$. Previous constructions relied on $n^{O(\log n)}$ approximation factors of LWE (resulting in less efficient parameters instantiation) or had large secret keys of size $|P| \times \text{poly}(\lambda)$. We rely on techniques developed by Boneh et al. (EUROCRYPT'14) and Brakerski et al. (ITCS'14) in the context of ABE for circuits and fully-homomorphic encryption.

1 Introduction

Attribute-Based Encryption (ABE) was introduced by Sahai and Waters [40] in order to realize the vision of fine-grained access control to encrypted data. Using ABE, a user can encrypt a message μ with respect to a public attribute-vector $\mathbf{x}$ to obtain a ciphertext $\mathsf{ct}_{\mathbf{x}}$. Anyone holding a secret key sk_P, associated with an access policy P, can decrypt the message μ if $P(\mathbf{x}) = 1$. Moreover, the security notion guarantees that no collusion of adversaries holding secret keys $\mathsf{sk}_{P_1}, \ldots, \mathsf{sk}_{P_t}$ can learn anything about the message μ if none of the individual keys allow to decrypt it. Until recently, candidate constructions of ABE were limited to restricted classes of access policies that test for equality (IBE), boolean formulas and inner-products: $[1, 2, 8, 12, 14, 15, 30\text{–}32, 41]$.

Work done while at MIT, supported by Microsoft PhD fellowship.
Work done while at University of Toronto.

T. Iwata and J.H. Cheon (Eds.): ASIACRYPT 2015, Part I, LNCS 9452, pp. 550–574, 2015.
DOI: 10.1007/978-3-662-48797-6_23

In recent breakthroughs Gorbunov, Vaikuntanathan and Wee [26] and Garg, Gentry, Halevi, Sahai and Waters [20] constructed ABE schemes for arbitrary boolean predicates. The GVW construction is based on the standard Learning With Errors (LWE) problem with sub-exponential approximation factors, whereas GGHSW relies on hardness of a (currently) stronger assumptions over existing multilinear map candidates [16,18,21]. But in both these ABE schemes, the size of the secret keys had a multiplicative dependence on the size of the predicate: $|P| \cdot \text{poly}(\lambda, d)$ (where d is the depth of the circuit representation of the predicate). In a subsequent work, Boneh et al. [10] showed how to construct ABE for arithmetic predicates with short secret keys: $|P| + \text{poly}(\lambda, d)$, also assuming hardness of LWE with sub-exponential approximation factors. However, in [26], the authors also showed an additional construction for a family of branching programs under a milder and quantitatively better assumption: hardness of LWE with polynomial approximation factors. Basing the security on LWE with polynomial approximation factors, as opposed to sub-exponential, results in two main advantages. First, the security of the resulting construction relies on the hardness of a much milder LWE assumption. But moreover, the resulting instantiation has better parameter – small modulo q – leading directly to practical efficiency improvements.

In this work, we focus on constructing an ABE scheme under milder security assumptions and better performance guarantees. We concentrate on ABE for a family of branching programs which is sufficient for most existing applications such as medical and multimedia data sharing [5,33,36].

First, we summarize the two most efficient results from learning with errors problem translated to the setting of branching programs (via standard Barrington's theorem [7]). Let L be the length of a branching program P and let λ denote the security parameter. Then,

- [26]: There exists an ABE scheme for length L branching programs with *large secret keys* based on the security of *LWE with polynomial approximation factors*. In particular, in the instantiation $|\text{sk}_P| = |L| \times \text{poly}(\lambda)$ and $q = \text{poly}(L, \lambda)$.
- [10]: There exists an ABE scheme for length L branching programs with *small secret keys* based on the security of *LWE with quasi-polynomial approximation factors*. In particular, $|\text{sk}_P| = |L| + \text{poly}(\lambda, \log L)$, $q = \text{poly}(\lambda)^{\log L}$.

To advance the state of the art for both theoretical and practical reasons, the natural question that arises is whether we can obtain the best of both worlds and:

Construct an ABE for branching programs with small secret keys based on the security of LWE with polynomial approximation factors?

1.1   Our Results

We present a new efficient construction of ABE for branching programs from a mild LWE assumption. Our result can be summarized in the following theorem.

Theorem 1 (Informal). *There exists a selectively-secure Attribute-Based Encryption for a family of length-L branching programs with* small secret keys *based on the security of* LWE *with polynomial approximation factors. More formally, the size of the secret key* sk_P *is* $L + \mathrm{poly}(\lambda, \log L)$ *and modulo* $q = \mathrm{poly}(L, \lambda)$, *where* λ *is the security parameter.*

Furthermore, we can extend our construction to support arbitrary length branching programs by setting q to some small super-polynomial.

As an additional contribution, our techniques lead to a new efficient constructing of homomorphic signatures for branching programs. In particular, Gorbunov et al. [28] showed how to construct homomorphic signatures for circuits based on the simulation techniques of Boneh et al. [10] in the context of ABE. Their resulting construction is secure based on the short integer solution (SIS) problem with *sub-exponential approximation factors* (or quasi-polynomial in the setting of branching programs). Analogously, our simulation algorithm presented in Sect. 3.4 can be used directly to construct homomorphic signatures for branching programs based on SIS with *polynomial approximation factors*.

Theorem 2 (Informal). *There exists a homomorphic signatures scheme for the family of length-L branching programs based on the security of* SIS *with polynomial approximation factors.*

High Level Overview. The starting point of our ABE construction is the ABE scheme for circuits with short secret keys by Boneh et al. [10]. At the heart of their construction is a fully key-homomorphic encoding scheme.

It encodes $a \in \{0, 1\}$ with respect to a public key $\mathbf{A} \xleftarrow{\$} \mathbb{Z}_q^{n \times m}$ in a "noisy" sample:

$$\psi_{\mathbf{A}, a} = (\mathbf{A} + a \cdot \mathbf{G})^\mathsf{T} \mathbf{s} + \mathbf{e}$$

where $\mathbf{s} \xleftarrow{\$} \mathbb{Z}_q^n$ and $\mathbf{G} \in \mathbb{Z}_q^{n \times m}$ are fixed across all the encodings and $\mathbf{e} \xleftarrow{\$} \chi^m$ (for some noise distribution χ) is chosen independently every time. The authors show that one can turn such a key-homomorphic encoding scheme, where homomorphism is satisfied over the encoded values and over the public keys simultaneously, into an attribute based encryption scheme for circuits.

Our first key observation is the asymmetric noise growth in their homomorphic multiplication over the encodings. Consider ψ_1, ψ_2 to be the encodings of a_1, a_2 under public keys $\mathbf{A}_1, \mathbf{A}_2$. To achieve multiplicative homomorphism, their first step is to achieve homomorphism over a_1 and a_2 by computing

$$a_1 \cdot \psi_2 = (a_1 \cdot \mathbf{A}_2 + (a_1 a_2) \cdot \mathbf{G})^\mathsf{T} \mathbf{s} + a_1 \mathbf{e}_2 \qquad (1)$$

Now, since homomorphism over the public key matrices must also be satisfied in the resulting encoding *independently* of a_1, a_2 we must replace $a_1 \cdot \mathbf{A}_2$ in Eq. 1 with operations over $\mathbf{A}_1, \mathbf{A}_2$ only. To do this, we can use the first encoding $\psi_1 = (\mathbf{A}_1 + a_1 \cdot \mathbf{G})^\top + \mathbf{e}_1$ and replace $a_1 \cdot \mathbf{G}$ with $a_1 \cdot \mathbf{A}_2$ as follows. First, compute $\widetilde{\mathbf{A}}_2 \in \{0, 1\}^{m \times m}$ such that $\mathbf{G} \cdot \widetilde{\mathbf{A}}_2 = \mathbf{A}_2$. (Finding such $\widetilde{\mathbf{A}}_2$ is possible since the "trapdoor" of $\mathbf{G}$ is known publicly). Then compute

$$
\begin{aligned}
(\widetilde{\mathbf{A}}_2)^\top \cdot \psi_1 &= \widetilde{\mathbf{A}}_2^\top \cdot ((\mathbf{A}_1 + a_1 \cdot \mathbf{G})^\top \mathbf{s} + \mathbf{e}_1) \\
&= \left(\mathbf{A}_1 \widetilde{\mathbf{A}}_2 + a_1 \cdot \mathbf{G} \widetilde{\mathbf{A}}_2 \right)^\top \mathbf{s} + \widetilde{\mathbf{A}}_2 \mathbf{e}_1 \\
&= \left(\mathbf{A}_1 \widetilde{\mathbf{A}}_2 + a_1 \cdot \mathbf{A}_2 \right)^\top \mathbf{s} + \mathbf{e}_1'
\end{aligned}
\tag{2}
$$

Subtracting Eq. 2 from 1, we get $\left(-\mathbf{A}_1 \widetilde{\mathbf{A}}_2 + (a_1 a_2) \cdot \mathbf{G} \right)^\top \mathbf{s} + \mathbf{e}'$ which is an encoding of $a_1 a_2$ under the public key $\mathbf{A}^\times := -\mathbf{A}_1 \widetilde{\mathbf{A}}_2$. Thus,

$$
\psi_{\mathbf{A}^\times, a^\times} := a_1 \cdot \psi_2 - \widetilde{\mathbf{A}}_2^\top \cdot \psi_1
$$

where $a^\times := a_1 a_2$. Here, $\mathbf{e}'$ remains small enough because $\widetilde{\mathbf{A}}_2$ has small (binary) entries. We observe that the new noise $\mathbf{e}' = a_1 \mathbf{e}_2 - \widetilde{\mathbf{A}}_2 \mathbf{e}_1$ grows asymmetrically. That is, the poly(n) multiplicative increase always occurs with respect to the first noise $\mathbf{e}_1$. Naïvely evaluating k levels of multiplicative homomorphism results in a noise of magnitude poly(n)^k. Can we manage the noise growth by some careful design of the order of homomorphic operations?

To achieve this, comes our second idea: design evaluation algorithms for a "sequential" representation of a matrix branching program to carefully manage the noise growth following the Brakerski-Vaikuntanathan paradigm in the context of fully-homomorphic encryption [13].

First, to generate a ciphertext with respect to an attribute vector $\mathbf{x} = (x_1, \ldots, x_\ell)$ we publish encodings of its individual bits:

$$
\psi_i \approx (\mathbf{A}_i + x_i \cdot \mathbf{G})^\top \mathbf{s}
$$

We also publish encoding of an initial start state 0^1:

$$
\psi_0^v \approx (\mathbf{A}_0^v + v_0 \cdot \mathbf{G})^\top \mathbf{s}
$$

The message μ is encrypted under encoding $\mathbf{u}^\top \mathbf{s} + e$ (where $\mathbf{u}$ is treated as the public key) and during decryption the user should obtain a value $\approx \mathbf{u}^\top \mathbf{s}$ from $\{\psi_i\}_{i \in [\ell]}, \psi_0^v$ iff $P(\mathbf{x}) = 1$.

Now, suppose the user wants to evaluate a branching program P on the attribute vector $\mathbf{x}$. Informally, the evaluation of a branching program proceeds in steps updating a special state vector. The next state is determined by the current state and one of the input bits (pertaining to this step). Viewing the

---

[1] Technically, we need to publish encodings of 5 states, but we simplify the notation in the introduction for conceptual clarify.

sequential representation of the branching program allows us to update the state using only a *single* multiplication and a few additions. Suppose v_t represents the state of the program P at step t and the user holds its corresponding encoding ψ_t^v (under some public key). To obtain ψ_{t+1}^v the user needs to use ψ_i (for some i determined by the program). Leveraging on the asymmetry, the state can be updated by multiplying ψ_i with the matrix $\widetilde{\mathbf{A}}_t^v$ corresponding to the encoding ψ_t^v (and then following a few simple addition steps). Since ψ_i always contains a "fresh" noise (which is never increased as we progress evaluating the program), the noise in ψ_{t+1}^v increases from the noise in ψ_t^v only by a *constant additive* factor! As a result, after k steps in the evaluation procedure the noise will be bounded by $k \cdot \text{poly}(n)$. Eventually, if $P(\mathbf{x}) = 1$, the user will learn $\approx \mathbf{u}^\mathsf{T}\mathbf{s}$ and be able to recover μ (we refer the reader to the main construction for details).

The main challenge in "riding on asymmetry" for attribute-based encryption is the requirement for satisfying parallel homomorphic properties: we must design separate homomorphic algorithms for operating over the public key matrices and over the encodings that allow for correct decryption. First, we define and design an algorithm for public key homomorphic operations that works specially for branching programs. Second, we design a homomorphic algorithm that works over the encodings that *preserves the homomorphism over public key matrices and the bits*[2] and *carefully manages the noise growth* as illustrated above. To prove the security, we need to argue that no collusion of users is able to learn anything about the message given many secret keys for programs that do not allow for decryption individually. We design a separate public-key simulation algorithm to accomplish this.

1.2   Applications

We summarize some of the known applications of attribute-based encryption. Parno, Raykova and Vaikuntanathan [37] showed how to use ABE to design (publicly) verifiable two-message delegation delegation scheme with a pre-processing phase. Goldwasser, Kalai, Popa, Vaikuntanathan and Zeldovich [24] showed how to use ABE as a critical building block to construct succinct one-query functional encryption, reusable garbled circuits, token-based obfuscation and homomorphic encryption for Turing machines. Our efficiency improvements for branching programs can be carried into all these applications.

1.3   Other Related Work

A number of works optimized attribute-based encryption for boolean formulas: Attrapadung et al. [6] and Emura et al. [17] designed ABE schemes with constant size ciphertext from bilinear assumptions. For arbitrary circuits, Boneh et al. [10] also showed an ABE with constant size ciphertext from multilinear assumptions. ABE can also be viewed as a special case of functional encryptions [9]. Gorbunov et al. [25] showed functional encryption for arbitrary functions

---

[2] These bits represent the bits of the attribute vector in the ABE scheme.

in a bounded collusion model from standard public-key encryption scheme. Garg et al. [19] presented a functional encryption for unbounded collusions for arbitrary functions under a weaker security model from multilinear assumptions. More recently, Gorbunov et al. exploited a the asymmetry in the noise growth in [10] in a different context of design of a predicate encryption scheme based on standard LWE [27].

1.4 Organization

In Sect. 2 we present the lattice preliminaries, definitions for ABE and branching programs. In Sect. 3 we present our main evaluation algorithms and build our ABE scheme in Sect. 4. We present a concrete instantiation of the parameters in Sect. 5. Finally, we outline the extensions in Sect. 6.

2 Preliminaries

Notation. Let PPT denote probabilistic polynomial-time. For any integer $q \geq 2$, we let $\mathbb{Z}_q$ denote the ring of integers modulo q and we represent $\mathbb{Z}_q$ as integers in $(-q/2, q/2]$. We let $\mathbb{Z}_q^{n \times m}$ denote the set of $n \times m$ matrices with entries in $\mathbb{Z}_q$. We use bold capital letters (e.g. $\mathbf{A}$) to denote matrices, bold lowercase letters (e.g. $\mathbf{x}$) to denote vectors. The notation $\mathbf{A}^\mathsf{T}$ denotes the transpose of the matrix $\mathbf{A}$. If $\mathbf{A}_1$ is an $n \times m$ matrix and $\mathbf{A}_2$ is an $n \times m'$ matrix, then $[\mathbf{A}_1 \| \mathbf{A}_2]$ denotes the $n \times (m + m')$ matrix formed by concatenating $\mathbf{A}_1$ and $\mathbf{A}_2$. A similar notation applies to vectors. When doing matrix-vector multiplication we always view vectors as column vectors. Also, $[n]$ denotes the set of numbers $1, \ldots, n$.

2.1 Lattice Preliminaries

Learning with Errors (LWE) Assumption The LWE problem was introduced by Regev [39], who showed that solving it *on the average* is as hard as (quantumly) solving several standard lattice problems *in the worst case*.

Definition 1 (LWE). *For an integer* $q = q(n) \geq 2$ *and an error distribution* $\chi = \chi(n)$ *over* $\mathbb{Z}_q$, *the learning with errors problem* $\mathsf{dLWE}_{n,m,q,\chi}$ *is to distinguish between the following pairs of distributions:*

$$\{\mathbf{A}, \mathbf{A}^\mathsf{T}\mathbf{s} + \mathbf{x}\} \quad and \quad \{\mathbf{A}, \mathbf{u}\}$$

where $\mathbf{A} \xleftarrow{\$} \mathbb{Z}_q^{n \times m}, \mathbf{s} \xleftarrow{\$} \mathbb{Z}_q^n, \mathbf{x} \xleftarrow{\$} \chi^m, \mathbf{u} \xleftarrow{\$} \mathbb{Z}_q^m.$

Connection to Lattices. Let $B = B(n) \in \mathbb{N}$. A family of distributions $\chi = \{\chi_n\}_{n \in \mathbb{N}}$ is called B-bounded if

$$\Pr[\chi \in \{-B, \ldots, B-1, B\}] = 1.$$

There are known quantum [39] and classical [38] reductions between $\mathsf{dLWE}_{n,m,q,\chi}$ and approximating short vector problems in lattices in the worst case, where

χ is a B-bounded (truncated) discretized Gaussian for some appropriate B. The state-of-the-art algorithms for these lattice problems run in time nearly exponential in the dimension n [4,35]; more generally, we can get a 2^k-approximation in time $2^{\tilde{O}(n/k)}$. Throughout this paper, the parameter $m = \text{poly}(n)$, in which case we will shorten the notation slightly to $\text{LWE}_{n,q,\chi}$.

Trapdoors for Lattices and LWE

Gaussian Distributions. Let $D_{\mathbb{Z}^m, \sigma}$ be the truncated discrete Gaussian distribution over $\mathbb{Z}^m$ with parameter σ, that is, we replace the output by $\mathbf{0}$ whenever the $|| \cdot ||_\infty$ norm exceeds $\sqrt{m} \cdot \sigma$. Note that $D_{\mathbb{Z}^m, \sigma}$ is $\sqrt{m} \cdot \sigma$-bounded.

Lemma 1 (Lattice Trapdoors [3,22,34]). *There is an efficient randomized algorithm* $\text{TrapSamp}(1^n, 1^m, q)$ *that, given any integers* $n \geq 1$, $q \geq 2$, *and sufficiently large* $m = \Omega(n \log q)$, *outputs a parity check matrix* $\mathbf{A} \in \mathbb{Z}_q^{n \times m}$ *and a 'trapdoor' matrix* $\mathbf{T_A} \in \mathbb{Z}^{m \times m}$ *such that the distribution of* $\mathbf{A}$ *is* negl(n)-*close to uniform.*

Moreover, there is an efficient algorithm SampleD *that with overwhelming probability over all random choices, does the following: For any* $\mathbf{u} \in \mathbb{Z}_q^n$, *and large enough* $s = \Omega(\sqrt{n \log q})$, *the randomized algorithm* $\text{SampleD}(\mathbf{A}, \mathbf{T_A}, \mathbf{u}, s)$ *outputs a vector* $\mathbf{r} \in \mathbb{Z}^m$ *with norm* $||\mathbf{r}||_\infty \leq ||\mathbf{r}||_2 \leq s\sqrt{n}$ *(with probability 1). Furthermore, the following distributions of the tuple* $(\mathbf{A}, \mathbf{T_A}, \mathbf{U}, \mathbf{R})$ *are within* negl(n) *statistical distance of each other for any polynomial* $k \in \mathbb{N}$:

- $(\mathbf{A}, \mathbf{T_A}) \leftarrow \text{TrapSamp}(1^n, 1^m, q)$; $\mathbf{U} \leftarrow \mathbb{Z}_q^{n \times k}$; $\mathbf{R} \leftarrow \text{SampleD}(\mathbf{A}, \mathbf{T_A}, \mathbf{U}, s)$.
- $(\mathbf{A}, \mathbf{T_A}) \leftarrow \text{TrapSamp}(1^n, 1^m, q)$; $\mathbf{R} \leftarrow (D_{\mathbb{Z}^m, s})^k$; $\mathbf{U} := \mathbf{AR} \pmod{q}$.

Sampling Algorithms We will use the following algorithms to sample short vectors from specific lattices. Looking ahead, the algorithm SampleLeft [1,14] will be used to sample keys in the real system, while the algorithm SampleRight [1] will be used to sample keys in the simulation.

Algorithm $\text{SampleLeft}(\mathbf{A}, \mathbf{B}, \mathbf{T_A}, \mathbf{u}, \alpha)$:

> *Inputs:* a full rank matrix $\mathbf{A}$ in $\mathbb{Z}_q^{n \times m}$, a "short" basis $\mathbf{T}_A$ of $\Lambda_q^\perp(\mathbf{A})$, a matrix $\mathbf{B}$ in $\mathbb{Z}_q^{n \times m_1}$, a vector $\mathbf{u} \in \mathbb{Z}_q^n$, and a Gaussian parameter α.
> *Output:* Let $\mathbf{F} := (\mathbf{A} \parallel \mathbf{B})$. The algorithm outputs a vector $\mathbf{e} \in \mathbb{Z}^{m+m_1}$ in the coset $\Lambda_{\mathbf{F}+\mathbf{u}}$.

Theorem 3 ([1, Theorem 17], [14, Lemma 3.2]). *Let* $q > 2$, $m > n$ *and* $\alpha > ||\mathbf{T_A}||_{\text{GS}} \cdot \omega(\sqrt{\log(m + m_1)})$. *Then* $\text{SampleLeft}(\mathbf{A}, \mathbf{B}, \mathbf{T_A}, \mathbf{u}, \alpha)$ *taking inputs as in (3) outputs a vector* $\mathbf{e} \in \mathbb{Z}^{m+m_1}$ *distributed statistically close to* $D_{\Lambda_{\mathbf{F}+\mathbf{u}}, \alpha}$, *where* $\mathbf{F} := (\mathbf{A} \parallel \mathbf{B})$.

where $||\mathbf{T}||_{\text{GS}}$ refers to the norm of Gram-Schmidt orthogonalisation of $\mathbf{T}$. We refer the readers to [1] for more details.

Algorithm $\text{SampleRight}(\mathbf{A}, \mathbf{G}, \mathbf{R}, \mathbf{T_G}, \mathbf{u}, \alpha)$:

Inputs: matrices $\mathbf{A}$ in $\mathbb{Z}_q^{n \times k}$ and $\mathbf{R}$ in $\mathbb{Z}^{k \times m}$, a full rank matrix $\mathbf{G}$ in $\mathbb{Z}_q^{n \times m}$, a "short" basis $\mathbf{T_G}$ of $\Lambda_q^\perp(\mathbf{G})$, a vector $\mathbf{u} \in \mathbb{Z}_q^n$, and a Gaussian parameter α.

Output: Let $\mathbf{F} := (\mathbf{A} \parallel \mathbf{AR} + \mathbf{G})$. The algorithm outputs a vector $\mathbf{e} \in \mathbb{Z}^{m+k}$ in the coset $\Lambda_{\mathbf{F}+\mathbf{u}}$.

Often the matrix $\mathbf{R}$ given to the algorithm as input will be a random matrix in $\{1, -1\}^{m \times m}$. Let S^m be the m-sphere $\{\mathbf{x} \in \mathbb{R}^{m+1} : \|\mathbf{x}\| = 1\}$. We define $s_R := \|\mathbf{R}\| := \sup_{\mathbf{x} \in S^{m-1}} \|\mathbf{R} \cdot \mathbf{x}\|$.

Theorem 4 ([1, Theorem 19]). *Let* $q > 2, m > n$ *and* $\alpha > \|\mathbf{T_G}\|_{\mathsf{GS}} \cdot s_R \cdot \omega(\sqrt{\log m})$. *Then* SampleRight$(\mathbf{A}, \mathbf{G}, \mathbf{R}, \mathbf{T_G}, \mathbf{u}, \alpha)$ *taking inputs as in (3) outputs a vector* $\mathbf{e} \in \mathbb{Z}^{m+k}$ *distributed statistically close to* $D_{\Lambda_{\mathbf{F}+\mathbf{u}},\alpha}$, *where* $\mathbf{F} := (\mathbf{A} \parallel \mathbf{AR} + \mathbf{G})$.

Primitive Matrix We use the primitive matrix $\mathbf{G} \in \mathbb{Z}_q^{n \times m}$ defined in [34]. This matrix has a trapdoor $\mathbf{T_G}$ such that $\|\mathbf{T_G}\|_\infty = 2$.

We also define an algorithm invG $: \mathbb{Z}_q^{n \times m} \to \mathbb{Z}_q^{m \times m}$ which deterministically derives a pre-image $\widetilde{\mathbf{A}}$ satisfying $\mathbf{G} \cdot \widetilde{\mathbf{A}} = \mathbf{A}$. From [34], there exists a way to get $\widetilde{\mathbf{A}}$ such that $\widetilde{\mathbf{A}} \in \{0, 1\}^{m \times m}$.

2.2 Attribute-Based Encryption

An attribute-based encryption scheme $\mathcal{ABE}$ [30] for a class of circuits $\mathcal{C}$ with ℓ bit inputs and message space $\mathcal{M}$ consists of a tuple of p.p.t. algorithms (Params, Setup, Enc, KeyGen, Dec):

Params$(1^\lambda) \to$ pp: The parameter generation algorithm takes the security parameter 1^λ and outputs a public parameter pp which is implicitly given to all the other algorithms of the scheme.

Setup$(1^\ell) \to$ (mpk, msk): The setup algorithm gets as input the length ℓ of the input index, and outputs the master public key mpk, and the master key msk.

Enc(mpk, $x, \mu) \to$ ct$_\mathbf{x}$: The encryption algorithm gets as input mpk, an index $x \in \{0, 1\}^\ell$ and a message $\mu \in \mathcal{M}$. It outputs a ciphertext ct$_\mathbf{x}$.

KeyGen(msk, $C) \to$ sk$_C$: The key generation algorithm gets as input msk and a predicate specified by $C \in \mathcal{C}$. It outputs a secret key sk$_C$.

Dec(ct$_\mathbf{x}$, sk$_C) \to \mu$: The decryption algorithm gets as input ct$_\mathbf{x}$ and sk$_C$, and outputs either $\perp$ or a message $\mu \in \mathcal{M}$.

Definition 2 (Correctness). *We require that for all* $(\mathbf{x}, C)$ *such that* $C(\mathbf{x}) = 1$ *and for all* $\mu \in \mathcal{M}$, *we have* $\Pr[\mathsf{ct_x} \leftarrow \mathsf{Enc}(\mathsf{mpk}, \mathbf{x}, \mu); \mathsf{Dec}(\mathsf{ct_x}, \mathsf{sk}_C) = \mu)] = 1$ *where the probability is taken over* pp $\leftarrow$ Params(1^λ), (mpk, msk) $\leftarrow$ Setup(1^ℓ) *and the coins of all the algorithms in the expression above.*

Definition 3 (Security). *For a stateful adversary* $\mathcal{A}$, *we define the advantage function* $\mathsf{Adv}_{\mathcal{A}}^{\mathrm{ABE}}(\lambda)$ *to be*

$$
\Pr \left[b = b' : \begin{array}{l} \mathbf{x}, d_{\max} \leftarrow \mathcal{A}(1^\lambda, 1^\ell); \\ \mathsf{pp} \leftarrow \mathsf{Params}(1^\lambda, 1^{d_{\max}}); \\ (\mathsf{mpk}, \mathsf{msk}) \leftarrow \mathsf{Setup}(1^\lambda, 1^\ell, \mathbf{x}^*); \\ (\mu_0, \mu_1) \leftarrow \mathcal{A}^{\mathsf{Keygen(msk,\cdot)}}(\mathsf{mpk}), \\ |\mu_0| = |\mu_1|; \\ b \xleftarrow{\$} \{0,1\}; \\ \mathsf{ct}_\mathbf{x} \leftarrow \mathsf{Enc}(\mathsf{mpk}, \mathbf{x}, \mu_b); \\ b' \leftarrow \mathcal{A}^{\mathsf{Keygen(msk,\cdot)}}(\mathsf{ct}_\mathbf{x}) \end{array} \right] - \frac{1}{2}
$$

with the restriction that all queries y *that* $\mathcal{A}$ *makes to* $\mathsf{Keygen(msk,\cdot)}$ *satisfies* $C(\mathbf{x}^*) = 0$ *(that is,* sk_C *does not decrypt* $\mathsf{ct}_\mathbf{x}$*). An attribute-based encryption scheme is* selectively secure *if for all PPT adversaries* $\mathcal{A}$, *the advantage* $\mathsf{Adv}_{\mathcal{A}}^{\mathrm{ABE}}(\lambda)$ *is a negligible function in* λ.

2.3   Branching Programs

We define branching programs similar to [13]. A width-w branching program BP of length L with input space $\{0,1\}^\ell$ and s states (represented by $[s]$) is a sequence of L tuples of the form $(\mathsf{var}(t), \sigma_{t,0}, \sigma_{t,1})$ where

- $\sigma_{t,0}$ and $\sigma_{t,1}$ are injective functions from $[s]$ to itself.
- $\mathsf{var} : [L] \rightarrow [\ell]$ is a function that associates the t-th tuple $\sigma_{t,0}, \sigma_{t,1}$ with the input bit $x_{\mathsf{var}(t)}$.

The branching program BP on input $\mathbf{x} = (x_1, \ldots, x_\ell)$ computes its output as follows. At step t, we denote the state of the computation by $\eta_t \in [s]$. The initial state is $\eta_0 = 1$. In general, η_t can be computed recursively as

$$
\eta_t = \sigma_{t, x_{\mathsf{var}(t)}}(\eta_{t-1})
$$

Finally, after L steps, the output of the computation $\mathsf{BP}(\mathbf{x}) = 1$ if $\eta_L = 1$ and 0 otherwise.

As done in [13], we represent states with bits rather than numbers to bound the noise growth. In particular, we represent the state $\eta_t \in [s]$ by a unit vector $\mathbf{v}_t \in \{0,1\}^s$. The idea is that $\mathbf{v}_t[i] = 1$ if and only if $\sigma_{t, x_{\mathsf{var}(t)}}(\eta_{t-1}) = i$. Note that we can also write the above expression as $\mathbf{v}_t[i] = 1$ if and only if either:

- $\mathbf{v}_{t-1}\left[\sigma_{t,0}^{-1}(i)\right] = 1$ and $x_{\mathsf{var}(t)} = 0$
- $\mathbf{v}_{t-1}\left[\sigma_{t,1}^{-1}(i)\right] = 1$ and $x_{\mathsf{var}(t)} = 1$

This latter form will be useful for us since it can be captured by the following formula. For $t \in [L]$ and $i \in [s]$,

$$
\begin{aligned}
\mathbf{v}_t[i] &:= \mathbf{v}_{t-1}\left[\sigma_{t,0}^{-1}(i)\right] \cdot (1 - x_{\mathsf{var}(t)}) + \mathbf{v}_{t-1}\left[\sigma_{t,1}^{-1}(i)\right] \cdot x_{\mathsf{var}(t)} \\
&= \mathbf{v}_{t-1}\left[\gamma_{t,i,0}\right] \cdot (1 - x_{\mathsf{var}(t)}) + \mathbf{v}_{t-1}\left[\gamma_{t,i,1}\right] \cdot x_{\mathsf{var}(t)}
\end{aligned}
$$

where $\gamma_{t,i,0} := \sigma_{t,0}^{-1}(i)$ and $\gamma_{t,i,1} = \sigma_{t,1}^{-1}(i)$ can be publicly computed from the description of the branching program. Hence, $\{\mathsf{var}(t), \{\gamma_{t,i,0}, \gamma_{t,i,1}\}_{i\in[s]}\}_{t\in[L]}$ is also valid representation of a branching program BP.

For clarity of presentation, we will deal with width-5 permutation branching programs, which is shown to be equivalent to the circuit class $\mathcal{NC}^1$ [7]. Hence, we have $s = w = 5$ and the functions σ_0, σ_1 are permutations on [5].

3   Our Evaluation Algorithms

In this section we describe the key evaluation and encoding (ciphertext) evaluation algorithms that will be used in our ABE construction. The algorithms are carefully designed to manage the noise growth in the LWE encodings *and* to preserve parallel homomorphism over the public keys and the encoded values.

3.1   Basic Homomorphic Operations

We first describe basic homomorphic addition and multiplication algorithms over the public keys and encodings (ciphertexts) based on the techniques developed by Boneh et al. [10].

Definition 4 (LWE Encoding). *For any matrix* $\mathbf{A} \xleftarrow{\$} \mathbb{Z}_q^{n\times m}$, *we define an LWE encoding of a bit* $a \in \{0,1\}$ *with respect to a (public) key* $\mathbf{A}$ *and randomness* $\mathbf{s} \xleftarrow{\$} \mathbb{Z}_q^n$ *as*

$$\psi_{\mathbf{A},\mathbf{s},a} = (\mathbf{A} + a \cdot \mathbf{G})^\mathsf{T}\mathbf{s} + \mathbf{e} \in \mathbb{Z}_q^m$$

for error vector $\mathbf{e} \xleftarrow{\$} \chi^m$ *and an (extended) primitive matrix* $\mathbf{G} \in \mathbb{Z}_q^{n\times m}$.

In our construction, however, all encodings will be under the same LWE secret $\mathbf{s}$, hence for simplicity we will simply refer to such an encoding as $\psi_{\mathbf{A},a}$.

Definition 5 (Noise Function). *For every* $\mathbf{A} \in \mathbb{Z}_q^{n\times m}, \mathbf{s} \in \mathbb{Z}_q^n$ *and encoding* $\psi_{\mathbf{A},a} \in \mathbb{Z}_q^m$ *of a bit* $a \in \{0,1\}$ *we define a noise function as*

$$\mathsf{Noise}_\mathbf{s}(\psi_{\mathbf{A},a}) := ||\psi_{\mathbf{A},a} - (\mathbf{A} + a \cdot \mathbf{G})^\mathsf{T}\mathbf{s} \mod q||_\infty$$

Looking ahead, in Lemma 8 we show that if the noise obtained after applying homomorphic evaluation is $\leq q/4$, then our ABE scheme will decrypt the message correctly. Now we define the basic additive and multiplicative operations on the encodings of this form, as per [10]. In their context, they refer to a matrix $\mathbf{A}$ as the "public key" and $\psi_{\mathbf{A},a}$ as a ciphertext.

Homomorphic Addition   This algorithm takes as input two encodings $\psi_{\mathbf{A},a}, \psi_{\mathbf{A}',a'}$ and outputs the sum of them. Let $\mathbf{A}^+ = \mathbf{A} + \mathbf{A}'$ and $a^+ = a + a'$.

$$\mathsf{Add}_\mathsf{en}(\psi_{\mathbf{A},a}, \psi_{\mathbf{A}',a'}): \text{ Output } \psi_{\mathbf{A}^+,a^+} := \psi_{\mathbf{A},a} + \psi_{\mathbf{A}',a'} \mod q$$

Lemma 2 (Noise Growth in $\mathsf{Add_{en}}$). *For any two valid encodings* $\psi_{\mathbf{A},a}, \psi_{\mathbf{A}',a'} \in \mathbb{Z}_q^m$, *let* $\mathbf{A}^+ = \mathbf{A} + \mathbf{A}'$ *and* $a^+ = a + a'$ *and* $\psi_{\mathbf{A}^+,a^+} = \mathsf{Add_{en}}(\psi_{\mathbf{A},a}, \psi_{\mathbf{A}',a'})$, *then we have*

$$\mathsf{Noise}_{\mathbf{A}^+,a^+}(\psi_{\mathbf{A}^+,a^+}) \leq \mathsf{Noise}_{\mathbf{A},a}(\psi_{\mathbf{A},a}) + \mathsf{Noise}_{\mathbf{A}',a'}(\psi_{\mathbf{A}',a'})$$

Proof Given two encodings we have,

$$\begin{aligned}
\psi_{\mathbf{A}^+,a^+} &= \psi_{\mathbf{A},a} + \psi_{\mathbf{A}',a'} \\
&= ((\mathbf{A} + a \cdot \mathbf{G})^\mathsf{T}\mathbf{s} + \mathbf{e}) + ((\mathbf{A}' + a' \cdot \mathbf{G})^\mathsf{T}\mathbf{s} + \mathbf{e}') \\
&= ((\mathbf{A} + \mathbf{A}') + (a + a') \cdot \mathbf{G})^\mathsf{T}\mathbf{s} + (\mathbf{e} + \mathbf{e}') \\
&= (\mathbf{A}^+ + a^+ \cdot \mathbf{G})^\mathsf{T}\mathbf{s} + (\mathbf{e} + \mathbf{e}')
\end{aligned}$$

Thus, from the definition of the noise function, it follows that

$$\mathsf{Noise}_{\mathbf{A}^+,a^+}(\psi_{\mathbf{A},a} + \psi_{\mathbf{A}',a'}) \leq \mathsf{Noise}_{\mathbf{A},a}(\psi_{\mathbf{A},a}) + \mathsf{Noise}_{\mathbf{A}',a'}(\psi_{\mathbf{A}',a'})$$

Homomorphic Multiplication This algorithm takes in two encodings $\psi_{\mathbf{A},a} = (\mathbf{A} + a \cdot \mathbf{G})^\mathsf{T}\mathbf{s} + \mathbf{e}_1$ and $\psi_{\mathbf{A}',a'} = (\mathbf{A}' + a' \cdot \mathbf{G})^\mathsf{T}\mathbf{s} + \mathbf{e}_2$ and outputs an encoding $\psi_{\mathbf{A}^\times,a^\times}$ where $\mathbf{A}^\times = -\mathbf{A}\mathbf{A}'$ and $a^\times = aa'$ as follows:

$$\mathsf{Multiply_{en}}(\psi_{\mathbf{A},a}, \psi_{\mathbf{A}',a'}) : \text{ Output } \psi_{\mathbf{A}^\times,a^\times} := -\widetilde{\mathbf{A}'}^\mathsf{T} \cdot \psi + a \cdot \psi'.$$

Note that this process requires the knowledge of the attribute a in clear.

Lemma 3 (Noise Growth in $\mathsf{Multiply_{en}}$). *For any two valid encodings* $\psi_{\mathbf{A},a}, \psi_{\mathbf{A}',a'} \in \mathbb{Z}_q^m$, *let* $\mathbf{A}^\times = -\mathbf{A}\mathbf{A}'$ *and* $a^\times = aa'$ *and* $\psi_{\mathbf{A}^\times,a^\times} = \mathsf{Multiply_{en}}(\psi_{\mathbf{A},a}, \psi_{\mathbf{A}',a'})$ *then we have*

$$\mathsf{Noise}_{\mathbf{A}^\times,a^\times}(\psi_{\mathbf{A}^\times,a^\times}) \leq m \cdot \mathsf{Noise}_{\mathbf{A},a}(\psi_{\mathbf{A},a}) + a \cdot \mathsf{Noise}_{\mathbf{A}',a'}(\psi_{\mathbf{A}',a'})$$

Proof Given two valid encodings, we have

$$\begin{aligned}
\psi_{\mathbf{A}^\times,a^\times} &= -\widetilde{\mathbf{A}'}^\mathsf{T} \cdot \psi + a \cdot \psi' \\
&= -\widetilde{\mathbf{A}'}^\mathsf{T}((\mathbf{A} + a \cdot \mathbf{G})^\mathsf{T}\mathbf{s} + \mathbf{e}) + a \cdot ((\mathbf{A}' + a' \cdot \mathbf{G})^\mathsf{T}\mathbf{s} + \mathbf{e}') \\
&= \left((-\mathbf{A}\widetilde{\mathbf{A}'} - a \cdot \mathbf{A}')^\mathsf{T}\mathbf{s} - \widetilde{\mathbf{A}'}^\mathsf{T}\mathbf{e}\right) + \left((a \cdot \mathbf{A}' + aa' \cdot \mathbf{G})^\mathsf{T}\mathbf{s} + a \cdot \mathbf{e}'\right) \\
&= ((\underbrace{-\mathbf{A}\widetilde{\mathbf{A}}_2}_{\mathbf{A}^\times}) + \underbrace{aa'}_{a^\times} \cdot \mathbf{G})^\mathsf{T}\mathbf{s} + (\underbrace{-\widetilde{\mathbf{A}'}^\mathsf{T}\mathbf{e} + a \cdot \mathbf{e}'}_{\mathbf{e}^\times})
\end{aligned}$$

Thus, from the definition of the noise function, we must bound the noise $\mathbf{e}^\times$. Hence,

$$\|\mathbf{e}^\times\|_\infty \leq \left\|\widetilde{\mathbf{A}'}^\mathsf{T}\mathbf{e}\right\|_\infty + a \cdot \|\mathbf{e}'\|_\infty \leq m \cdot \|\mathbf{e}\|_\infty + a \cdot \|\mathbf{e}'\|_\infty$$

where the last inequality holds since $\widetilde{\mathbf{A}'} \in \{0,1\}^{m \times m}$.

Note: This type of homomorphism is different from a standard fully homomorphic encryption (FHE) mainly for the following two reasons.

- To perform multiplicative homomorphism, here we need one of the input values *in clear* but the FHE homomorphic operations are performed without the knowledge of the input values.
- The other big difference is that, here we require the output public key matrices $\mathbf{A}^+, \mathbf{A}^\times$ to be independent of the input values a_1, a_2. More generally, when given an arbitrary circuit with AND and OR gates along with the matrices corresponding to its input wires, one should be able to determine the matrix corresponding to the output wire without the knowledge of the values of the input wires. But, this property is not present in any of the existing FHE schemes.

3.2 Our Public Key Evaluation Algorithm

We define a (public) key evaluation algorithm $\mathsf{Eval}_{\mathsf{pk}}$. The algorithm takes as input a description of the branching program BP, a collection of public keys $\{\mathbf{A}_i\}_{i \in [\ell]}$ (one for each attribute bit $\mathbf{x}_i$), a collection of public keys $\mathbf{V}_{0,i}$ for initial state vector and an auxiliary matrix $\mathbf{A}^c$. The algorithm outputs an "evaluated" public key corresponding to the branching program:

$$\mathsf{Eval}_{\mathsf{pk}}(\mathsf{BP}, \{\mathbf{A}_i\}_{i \in [\ell]}, \{\mathbf{V}_{0,i}\}_{i \in [5]}, \mathbf{A}^c) \to \mathbf{V}_{\mathsf{BP}}$$

The auxiliary matrix $\mathbf{A}^c$ can be thought of as the public key we use to encode a constant 1. We also define $\mathbf{A}'_i := \mathbf{A}^c - \mathbf{A}_i$, as a public key that will encode $1 - \mathbf{x}_i$. The output $\mathbf{V}_{\mathsf{BP}} \in \mathbb{Z}_q^{n \times m}$ is the homomorphically defined public key $\mathbf{V}_{L,1}$ at position 1 of the state vector at the Lth step of the branching program evaluation.

The algorithm proceeds as follows. Recall the description of the branching program BP represented by tuples $\big(\mathsf{var}(t), \{\gamma_{t,i,0}, \gamma_{t,i,1}\}_{i \in [5]}\big)$ for $t \in [L]$. The initial state vector is always taken to be $\mathbf{v}_0 := [1, 0, 0, 0, 0]$. And for $t \in [L]$,

$$\mathbf{v}_t[i] = \mathbf{v}_{t-1}[\gamma_{t,i,0}] \cdot (1 - x_{\mathsf{var}(t)}) + \mathbf{v}_{t-1}[\gamma_{t,i,1}] \cdot x_{\mathsf{var}(t)}$$

Our algorithm calculates $\mathbf{V}_{\mathsf{BP}}$ inductively as follows. Assume at time $t-1 \in [L]$, the state public keys $\{\mathbf{V}_{t-1,i}\}_{i \in [5]}$ are already assigned. We assign state public keys $\{\mathbf{V}_{t,i}\}_{i \in [5]}$ at time t as follows.

1. Let $\gamma_0 := \gamma_{t,i,0}$ and $\gamma_1 := \gamma_{t,i,1}$.
2. Let $\mathbf{V}_{t,i} = -\mathbf{A}'_{\mathsf{var}(t)} \widetilde{\mathbf{V}}_{t-1,\gamma_0} - \mathbf{A}_{\mathsf{var}(t)} \widetilde{\mathbf{V}}_{t-1,\gamma_1}$.

It is important to note that the public key defined at each step of the state vector is *independent of any input attribute vector*. Now, let $\mathbf{V}_{L,1}$ be the public key assigned at position 1 at step L of the branching program. We simply output $\mathbf{V}_{\mathsf{BP}} := \mathbf{V}_{L,1}$.

3.3   Our Encoding Evaluation Algorithm

We also define an encoding evaluation algorithm $\mathsf{Eval_{en}}$ which we will use in the decryption algorithm of our ABE scheme. The algorithm takes as input the description of a branching program BP, an attribute vector $\mathbf{x}$, a set of encodings for the attribute (with corresponding public keys) $\{\mathbf{A}_i, \psi_i := \psi_{\mathbf{A}_i, x_i}\}_{i \in [\ell]}$, encodings of the initial state vector $\{\mathbf{V}_{0,i}, \psi_{0,i} := \psi_{\mathbf{V}_{0,i}, \mathbf{v}_0[i]}\}_{i \in [5]}$ and an encoding of a constant "1" $\psi^c := \psi_{\mathbf{A}^c, 1}$. (From now on, we will use the simplified notations $\psi_i, \psi_{0,i}, \psi^c$ for the encodings). $\mathsf{Eval_{en}}$ outputs an encoding of the result $y := \mathsf{BP}(\mathbf{x})$ with respect to a homomorphically derived public key $\mathbf{V}_{\mathsf{BP}} := \mathbf{V}_{L,1}$.

$$\mathsf{Eval_{en}}\big(\mathsf{BP}, \mathbf{x}, \{\mathbf{A}_i, \psi_i\}_{i \in [\ell]}, \{\mathbf{V}_{0,i}, \psi_{0,i}\}_{i \in [5]}, \mathbf{A}^c, \psi^c\big) \to \psi_{\mathsf{BP}}$$

Recall that for $t \in [L]$, we have for all $i \in [5]$:

$$\mathbf{v}_t[i] = \mathbf{v}_{t-1}\left[\gamma_{t,i,0}\right] \cdot (1 - x_{\mathsf{var}(t)}) + \mathbf{v}_{t-1}\left[\gamma_{t,i,1}\right] \cdot x_{\mathsf{var}(t)}$$

The evaluation algorithm proceeds inductively to update the encoding of the state vector for each step of the branching program. The key idea to obtain the desired noise growth is that we only multiply the *fresh encodings* of the attribute bits with the binary decomposition of the public keys. The result is then be added to update the encoding of the state vector. Hence, at each step of the computation the noise in the encodings of the state will only grow by some fixed additive factor.

The algorithm proceeds as follows. We define $\psi'_i := \psi_{\mathbf{A}'_i, (1-x_i)} = (\mathbf{A}'_i + (1 - x_i) \cdot \mathbf{G})^\top \mathbf{s} + \mathbf{e}'_i$ to denote the encoding of $1 - x_i$ with respect to $\mathbf{A}'_i = \mathbf{A}^c - \mathbf{A}_i$. Note that it can be computed using $\mathsf{Add_{en}}(\psi_{\mathbf{A}^c, 1}, -\psi_{\mathbf{A}_i, x_i})$. Assume at time $t - 1 \in [L]$ we hold encodings of the state vector $\{\psi_{\mathbf{V}_{t-1,i}, \mathbf{v}_t[i]}\}_{i \in [5]}$. Now, we compute the encodings of the new state values:

$$\psi_{t,i} = \mathsf{Add_{en}}\left(\mathsf{Multiply_{en}}(\psi'_{\mathsf{var}(t)}, \psi_{t-1,\gamma_0}), \mathsf{Multiply_{en}}(\psi_{\mathsf{var}(t)}, \psi_{t-1,\gamma_1})\right)$$

where $\gamma_0 := \gamma_{t,i,0}$ and $\gamma_1 := \gamma_{t,i,1}$. As we show below (in Lemma 4), this new encoding has the form $\left(\mathbf{V}_{t,i} + \mathbf{v}_t[i] \cdot \mathbf{G}\right)^\top \mathbf{s} + \mathbf{e}_{t,i}$ (for a small enough noise term $\mathbf{e}_{t,i}$).

Finally, let $\psi_{L,1}$ be the encoding obtained at the Lth step corresponding to state value at position "1" by this process. As we show in Lemma 5, noise term $\mathbf{e}_{\mathsf{BP}}$ has "low" infinity norm enabling correct decryption (Lemma 8). The algorithm outputs $\psi_{\mathsf{BP}} := \psi_{L,1}$.

Correctness and Analysis

Lemma 4. *For any valid set of encodings* $\psi_{\mathsf{var}(t)}, \psi'_{\mathsf{var}(t)}$ *for the bits* $x_{\mathsf{var}(t)}, (1 - x_{\mathsf{var}(t)})$ *and* $\{\psi_{t-1,i}\}_{i \in [5]}$ *for the state vector* $\mathbf{v}_{t-1}$ *at step* $t - 1$, *the output of the function*

$$\mathsf{Add_{en}}\left(\mathsf{Multiply_{en}}(\psi'_{\mathsf{var}(t)}, \psi_{t-1,\gamma_0}), \mathsf{Multiply_{en}}(\psi_{\mathsf{var}(t)}, \psi_{t-1,\gamma_1})\right) \to \psi_{t,i}$$

where $\psi_{t,i} = \left(\mathbf{V}_{t,i} + \mathbf{v}_t[i] \cdot \mathbf{G}\right)^\top \mathbf{s} + \mathbf{e}_{t,i}$, *for some noise term* $\mathbf{e}_{t,i}$.

Proof. Given valid encodings $\psi_{\mathsf{var}(t)}, \psi'_{\mathsf{var}(t)}$ and $\{\psi_{t-1,i}\}_{i \in [5]}$, we have:

$$
\begin{aligned}
\psi_{t,i} &= \mathsf{Add}_{\mathsf{en}}\Big(\mathsf{Multiply}_{\mathsf{en}}(\psi'_{\mathsf{var}(t)}, \psi_{t-1,\gamma_0}), \mathsf{Multiply}_{\mathsf{en}}(\psi_{\mathsf{var}(t)}, \psi_{t-1,\gamma_1})\Big) \\
&= \mathsf{Add}_{\mathsf{en}}\Big(\big[(-\mathbf{A}'_{\mathsf{var}(t)}\widetilde{\mathbf{V}}_{t-1,\gamma_0} + (\mathbf{v}_t[\gamma_0] \cdot (1 - x_{\mathsf{var}(t)})) \cdot \mathbf{G})^{\mathsf{T}}\mathbf{s} + \mathbf{e}_1)\big], \\
&\qquad\qquad \big[(-\mathbf{A}_{\mathsf{var}(t)}\widetilde{\mathbf{V}}_{t-1,\gamma_1} + (\mathbf{v}_t[\gamma_1] \cdot x_{\mathsf{var}(t)}) \cdot \mathbf{G})^{\mathsf{T}}\mathbf{s} + \mathbf{e}_2)\big]\Big) \\
&= \Big[\underbrace{\big(-\mathbf{A}'_{\mathsf{var}(t)}\widetilde{\mathbf{V}}_{t-1,\gamma_0} - \mathbf{A}_{\mathsf{var}(t)}\widetilde{\mathbf{V}}_{t-1,\gamma_1}\big)}_{\mathbf{v}_{t,i}} + \underbrace{(\mathbf{v}_t[\gamma_0] \cdot (1 - x_{\mathsf{var}(t)}) + \mathbf{v}_t[\gamma_1] \cdot x_{\mathsf{var}(t)})}_{\mathbf{v}_t[i]} \cdot \mathbf{G}\Big]^{\mathsf{T}}\mathbf{s} + \mathbf{e}_{t,i}
\end{aligned}
$$

where the first step follows from the correctness of $\mathsf{Multiply}_{\mathsf{en}}$ algorithm and last step from that of $\mathsf{Add}_{\mathsf{en}}$ with $\mathbf{e}_{t,i} = \mathbf{e}_1 + \mathbf{e}_2$ where $\mathbf{e}_1 = -\big(\widetilde{\mathbf{V}}_{t-1,\gamma_0}\big)^{\mathsf{T}} \mathbf{e}'_{\mathsf{var}(t)} - (1 - x_{\mathsf{var}(t)}) \cdot \mathbf{e}_{t-1,\gamma_0}$ and $\mathbf{e}_2 = -\big(\widetilde{\mathbf{V}}_{t-1,\gamma_1}\big)^{\mathsf{T}} \mathbf{e}_{\mathsf{var}(t)} - x_{\mathsf{var}(t)} \cdot \mathbf{e}_{t-1,\gamma_1}$.

Lemma 5. *Let* $\mathsf{Eval}_{\mathsf{en}}\big(\mathsf{BP}, \mathbf{x}, \{\mathbf{A}_i, \psi_i\}_{i \in [\ell]}, \{\mathbf{V}_{0,i}, \psi_{0,i}\}_{i \in [5]}, \mathbf{A}^c, \psi^c\big) \to \psi_{\mathsf{BP}}$ *such that* *all* *the* *noise* *terms,* $\big\{\mathsf{Noise}_{\mathbf{A}_i, x_i}(\psi_i)\big\}_{i \in [\ell]}, \mathsf{Noise}_{\mathbf{A}^c, 1}(\psi^c), \big\{\mathsf{Noise}_{\mathbf{V}_{0,i}, \mathbf{v}_0[i]}(\psi_{0,i})\big\}_{i \in [5]}$ *are bounded by* B, *then*

$$\mathsf{Noise}_{\mathbf{V}_{\mathsf{BP}}, y}(\psi_{\mathsf{BP}}) \leq 3m \cdot L \cdot B + B$$

Proof. We will prove this lemma by induction. That is, we will prove that at any step t,

$$\mathsf{Noise}_{\mathbf{V}_{t,i}, \mathbf{v}_t[i]}(\psi_{t,i}) \leq 3m \cdot t \cdot B + B$$

for $i \in [5]$. For the base case, $t = 0$, we operate on *fresh* encodings for the initial state vector $\mathbf{v}_0$. Hence, we have that, $\mathsf{Noise}_{\mathbf{V}_{0,i}, \mathbf{v}_0[i]}(\psi_{0,i}) \leq B$, for all $i \in [5]$. Let $\{\psi_{t-1,i}\}_{i \in [5]}$ be the encodings of the state vector $\mathbf{v}_{t-1}$ at step $t-1$ such that

$$\mathsf{Noise}_{\mathbf{V}_{t-1,i}, \mathbf{v}_{t-1}[i]}(\psi_{t-1,i}) \leq 3m \cdot (t-1) \cdot B + B$$

for $i \in [5]$. We know that $\psi_{t,i} = \mathsf{Add}_{\mathsf{en}}\Big(\mathsf{Multiply}_{\mathsf{en}}(\psi'_{\mathsf{var}(t)}, \psi_{t-1,\gamma_0}), \mathsf{Multiply}_{\mathsf{en}}(\psi_{\mathsf{var}(t)}, \psi_{t-1,\gamma_1})\Big)$. Hence, from Lemmas 2 and 3, we get:

$$
\begin{aligned}
\mathsf{Noise}_{\mathbf{V}_{t,i}, \mathbf{v}_t[i]}(\psi_{t,i}) &\leq \Big(m \cdot \mathsf{Noise}_{\mathbf{A}'_{\mathsf{var}(t)}, (1 - x_{\mathsf{var}(t)})}(\psi'_{\mathsf{var}(t)}) + (1 - x_{\mathsf{var}(t)}) \cdot \mathsf{Noise}_{\mathbf{V}_{t-1,\gamma_0}, \mathbf{v}_{t-1}[\gamma_0]}\Big) \\
&\quad + \Big(m \cdot \mathsf{Noise}_{\mathbf{A}_{\mathsf{var}(t)}, x_{\mathsf{var}(t)}}(\psi_{\mathsf{var}(t)}) + x_{\mathsf{var}(t)} \cdot \mathsf{Noise}_{\mathbf{V}_{t-1,\gamma_1}, \mathbf{v}_{t-1}[\gamma_1]}\Big) \\
&= (m \cdot 2B + (1 - x_{\mathsf{var}(t)}) \cdot (3m(t-1)B + B)) \\
&\quad + (m \cdot B + x_{\mathsf{var}(t)} \cdot (3m(t-1)B + B)) \\
&= 3m \cdot t \cdot B + B
\end{aligned}
$$

where

$$\mathsf{Noise}_{\mathbf{A}'_{\mathsf{var}(t)}, (1 - x_{\mathsf{var}(t)})}(\psi'_{\mathsf{var}(t)}) \leq \mathsf{Noise}_{\mathbf{A}^c, 1}(\psi^c) + \mathsf{Noise}_{-\mathbf{A}_{\mathsf{var}(t)}, -x_{\mathsf{var}(t)}}(-\psi_{\mathsf{var}(t)}) \leq B + B = 2B$$

by Lemma 2. With ψ_{BP} being an encoding at step L, we have $\mathsf{Noise}_{\mathbf{V}_{\mathsf{BP}}, y}(\psi_{\mathsf{BP}}) \leq 3m \cdot L \cdot B + B$. Thus, $\mathsf{Noise}_{\mathbf{V}_{\mathsf{BP}}, y}(\psi_{\mathsf{BP}}) = O(m \cdot L \cdot B)$.

3.4 Our Simulated Public Key Evaluation Algorithm

During simulation, we will use a different procedure for assigning public keys to each wire of the input and the state vector. In particular, $\mathbf{A}_i = \mathbf{A} \cdot \mathbf{R}_i - x_i \cdot \mathbf{G}$ for some *shared* public key $\mathbf{A}$ and some low norm matrix $\mathbf{R}_i$. Similarly, the state public keys $\mathbf{V}_{t,i} = \mathbf{A} \cdot \mathbf{R}_{t,i} - \mathbf{v}_t[i] \cdot \mathbf{G}$. The algorithm thus takes as input the description of the branching program BP, the attribute vector $\mathbf{x}$, two collection of low norm matrices $\{\mathbf{R}_i\}, \{\mathbf{R}_{0,i}\}$ corresponding to the input public keys and initial state vector, a low norm matrix $\mathbf{R}^c$ for the public key of constant 1 and a shared matrix $\mathbf{A}$. It outputs a homomorphically derived low norm matrix $\mathbf{R}_{\mathsf{BP}}$.

$$\mathsf{Eval}_{\mathsf{SIM}}(\mathsf{BP}, \mathbf{x}, \{\mathbf{R}_i\}_{i \in [\ell]}, \{\mathbf{R}_{0,i}\}_{i \in [5]}, \mathbf{R}^c, \mathbf{A}) \to \mathbf{R}_{\mathsf{BP}}$$

The algorithm will ensure that the output $\mathbf{R}_{\mathsf{BP}}$ satisfies $\mathbf{A} \cdot \mathbf{R}_{\mathsf{BP}} - \mathsf{BP}(\mathbf{x}) \cdot \mathbf{G} = \mathbf{V}_{\mathsf{BP}}$, where $\mathbf{V}_{\mathsf{BP}}$ is the homomorphically derived public key.

The algorithm proceeds inductively as follows. Assume at time $t - 1 \in [L]$, the we hold a collection of low norm matrices $\mathbf{R}_{t-1,i}$ and public keys $\mathbf{V}_{t-1,i} = \mathbf{A} \cdot \mathbf{R}_{t-1,i} - \mathbf{v}_t[i] \cdot \mathbf{G}$ for $i \in [5]$ corresponding to the state vector. Let $\mathbf{R}'_i = \mathbf{R}^c - \mathbf{R}_i$ for all $i \in [\ell]$. We show how to derive the low norm matrices $\mathbf{R}_{t,i}$ for all $i \in [5]$:

1. Let $\gamma_0 := \gamma_{t,i,0}$ and $\gamma_1 := \gamma_{t,i,1}$.
2. Compute

$$\mathbf{R}_{t,i} = \left(-\mathbf{R}'_{\mathsf{var}(t)} \tilde{\mathbf{V}}_{t-1,\gamma_0} + (1 - x_{\mathsf{var}(t)}) \cdot \mathbf{R}_{t-1,\gamma_0}\right) + \left(-\mathbf{R}_{\mathsf{var}(t)} \tilde{\mathbf{V}}_{t-1,\gamma_1} + x_{\mathsf{var}(t)} \cdot \mathbf{R}_{t-1,\gamma_1}\right)$$

Finally, let $\mathbf{R}_{L,1}$ be the matrix obtained at the Lth step corresponding to state value "1" by the above algorithm. Output $\mathbf{R}_{\mathsf{BP}} := \mathbf{R}_{L,1}$. Below, we show that the norm of $\mathbf{R}_{\mathsf{BP}}$ remains small and that homomorphically computed public key $\mathbf{V}_{\mathsf{BP}}$ using $\mathsf{Eval}_{\mathsf{pk}}$ satisfies that $\mathbf{V}_{\mathsf{BP}} = \mathbf{A} \cdot \mathbf{R}_{\mathsf{BP}} - \mathsf{BP}(\mathbf{x}) \cdot \mathbf{G}$.

Lemma 6 (Correctness of $\mathsf{Eval}_{\mathsf{SIM}}$). *For any set of valid inputs to $\mathsf{Eval}_{\mathsf{SIM}}$, we have*

$$\mathsf{Eval}_{\mathsf{SIM}}(\mathsf{BP}, \mathbf{x}, \{\mathbf{R}_i\}_{i \in [\ell]}, \{\mathbf{R}_{0,i}\}_{i \in [5]}, \mathbf{R}^c, \mathbf{A}) \to \mathbf{R}_{\mathsf{BP}}$$

where $\mathbf{V}_{\mathsf{BP}} = \mathbf{A}\mathbf{R}_{\mathsf{BP}} - \mathsf{BP}(\mathbf{x}) \cdot \mathbf{G}$.

Proof. We will prove this lemma by induction. That is, we will prove that at any step t,

$$\mathbf{V}_{t,i} = \mathbf{A}\mathbf{R}_{t,i} - \mathbf{v}_t[i] \cdot \mathbf{G}$$

for any $i \in [5]$. For the base case $t = 0$, since the inputs are valid, we have that $\mathbf{V}_{0,i} = \mathbf{A}\mathbf{R}_{0,i} - \mathbf{v}_0[i] \cdot \mathbf{G}$, for all $i \in [5]$. Let $\mathbf{V}_{t-1,i} = \mathbf{A}\mathbf{R}_{t-1,i} - \mathbf{v}_{t-1}[i] \cdot \mathbf{G}$ for $i \in [5]$. Hence, we get:

$$
\begin{aligned}
\mathbf{A}\mathbf{R}_{t,i} &= \left(-\mathbf{A}\mathbf{R}'_{\mathsf{var}(t)} \tilde{\mathbf{V}}_{t-1,\gamma_0} + (1 - x_{\mathsf{var}(t)}) \cdot \mathbf{A}\mathbf{R}_{t-1,\gamma_0}\right) + \left(-\mathbf{A}\mathbf{R}_{\mathsf{var}(t)} \tilde{\mathbf{V}}_{t-1,\gamma_1} + x_{\mathsf{var}(t)} \cdot \mathbf{A}\mathbf{R}_{t-1,\gamma_1}\right) \\
&= \left(-(\mathbf{A}'_{\mathsf{var}(t)} + (1 - x_{\mathsf{var}(t)}) \cdot \mathbf{G}) \tilde{\mathbf{V}}_{t-1,\gamma_0} + (1 - x_{\mathsf{var}(t)}) \cdot (\mathbf{V}_{t-1,\gamma_0} + \mathbf{v}_{t-1}[\gamma_0] \cdot \mathbf{G})\right) \\
&\quad + \left(-(\mathbf{A}_{\mathsf{var}(t)} + x_{\mathsf{var}(t)} \cdot \mathbf{G}) \tilde{\mathbf{V}}_{t-1,\gamma_1} + x_{\mathsf{var}(t)} \cdot (\mathbf{V}_{t-1,\gamma_1} + \mathbf{v}_{t-1}[\gamma_1] \cdot \mathbf{G})\right) \\
&= \left(-\mathbf{A}'_{\mathsf{var}(t)} \tilde{\mathbf{V}}_{t-1,\gamma_0} - (1 - x_{\mathsf{var}(t)}) \cdot \mathbf{V}_{t-1,\gamma_0} + (1 - x_{\mathsf{var}(t)}) \cdot \mathbf{V}_{t-1,\gamma_0} + ((1 - x_{\mathsf{var}(t)})\mathbf{v}_{t-1}[\gamma_0]) \cdot \mathbf{G}\right) \\
&\quad + \left(-\mathbf{A}_{\mathsf{var}(t)} \tilde{\mathbf{V}}_{t-1,\gamma_1} - x_{\mathsf{var}(t)} \cdot \mathbf{V}_{t-1,\gamma_1} + x_{\mathsf{var}(t)} \cdot \mathbf{V}_{t-1,\gamma_1} + (x_{\mathsf{var}(t)}\mathbf{v}_{t-1}[\gamma_1]) \cdot \mathbf{G}\right) \\
&= \underbrace{(-\mathbf{A}'_{\mathsf{var}(t)} \tilde{\mathbf{V}}_{t-1,\gamma_0} - \mathbf{A}_{\mathsf{var}(t)} \tilde{\mathbf{V}}_{t-1,\gamma_1})}_{\mathbf{V}_{t,i}} + \underbrace{((1 - x_{\mathsf{var}(t)})\mathbf{v}_{t-1}[\gamma_0] + (x_{\mathsf{var}(t)}\mathbf{v}_{t-1}[\gamma_1])) \cdot \mathbf{G}}_{\mathbf{v}_t[i]}
\end{aligned}
$$

Hence, we have $\mathbf{V}_{t,i} = \mathbf{AR}_{t,i} - \mathbf{v}_t[i] \cdot \mathbf{G}$. Thus, at the Lth step, we have by induction that

$$\mathbf{V}_{\mathsf{BP}} = \mathbf{V}_{L,1} = \mathbf{AR}_{L,1-\mathbf{v}_t[i] \cdot \mathbf{G}} = \mathbf{AR}_{\mathsf{BP}} - \mathbf{v}_t[i] \cdot \mathbf{G}$$

Lemma 7. *Let* $\mathsf{Eval}_{\mathsf{SIM}}\left(\mathsf{BP}, \mathbf{x}, \{\mathbf{R}_i\}_{i \in [\ell]}, \{\mathbf{R}_{0,i}\}_{i \in [5]}, \mathbf{R}^c, \mathbf{A}\right) \to \mathbf{R}_{\mathsf{BP}}$ *such that all the "$\mathbf{R}$" matrices are sampled from* $\{-1, 1\}^{m \times m}$, *then*

$$\|\mathbf{R}_{\mathsf{BP}}\|_\infty \leq 3m \cdot L + 1$$

Proof. This proof is very similar to that of Lemma 5. We will prove this lemma also by induction. That is, we will prove that at any step t,

$$\|\mathbf{R}_{t,i}\|_\infty \leq 3m \cdot t + 1$$

for $i \in [5]$. For the base case, $t = 0$, the input $\mathbf{R}_{0,i}$s are such that, $\|\mathbf{R}_{t,0}\|_\infty = 1$, for all $i \in [5]$. Let $\|\mathbf{R}_{t-1,i}\|_\infty \leq 3m \cdot (t-1) + 1$ for $i \in [5]$. We know that

$$\mathbf{R}_{t,i} = \left(-\mathbf{R}'_{\mathsf{var}(t)} \widetilde{\mathbf{V}}_{t-1,\gamma_0} + (1 - x_{\mathsf{var}(t)}) \cdot \mathbf{R}_{t-1,\gamma_0}\right) + \left(-\mathbf{R}_{\mathsf{var}(t)} \widetilde{\mathbf{V}}_{t-1,\gamma_1} + x_{\mathsf{var}(t)} \cdot \mathbf{R}_{t-1,\gamma_1}\right)$$

Hence, we have:

$$\|\mathbf{R}_{t,i}\|_\infty \leq \left(m \cdot \left\|\widetilde{\mathbf{V}}_{t-1,\gamma_0}\right\|_\infty \cdot \left\|\mathbf{R}'_{\mathsf{var}(t)}\right\|_\infty + (1 - x_{\mathsf{var}(t)}) \cdot \|\mathbf{R}_{t-1,\gamma_0}\|_\infty\right)$$
$$+ \left(m \cdot \left\|\widetilde{\mathbf{V}}_{t-1,\gamma_0}\right\|_\infty \cdot \left\|\mathbf{R}_{\mathsf{var}(t)}\right\|_\infty + x_{\mathsf{var}(t)} \cdot \|\mathbf{R}_{t-1,\gamma_1}\|_\infty\right)$$
$$= (m \cdot 1 \cdot 2 + (1 - x_{\mathsf{var}(t)}) \cdot 3m \cdot (t-1)) + (m \cdot 1 \cdot 1 + x_{\mathsf{var}(t)} \cdot 3m \cdot (t-1))$$
$$= 3m \cdot t + 1$$

where $\|\mathbf{R}'_i\|_\infty \leq \|\mathbf{R}^c + \mathbf{R}_i\|_\infty \leq \|\mathbf{R}^c\|_\infty + \|\mathbf{R}_i\|_\infty \leq 1 + 1 = 2$. With $\mathbf{R}_{\mathsf{BP}}$ being at step L, we have $\|\mathbf{R}_{\mathsf{BP}}\|_\infty \leq 3m \cdot L + 1$. Thus, $\|\mathbf{R}_{\mathsf{BP}}\|_\infty = O(m \cdot L)$.

4   Our Attribute-Based Encryption

In this section we describe our attribute-based encryption scheme for branching programs. We present the scheme for a bounded length branching programs, but note that we can trivially support unbounded length by setting modulo q to a small superpolynomial. For a family of branching programs of length bounded by L and input space $\{0,1\}^\ell$, we define the $\mathcal{ABE}$ algorithms (Params, Setup, KeyGen, Enc, Dec) as follows.

- Params($1^\lambda, 1^L$): For a security parameter λ and length bound L, let the LWE dimension be $n = n(\lambda)$ and let the LWE modulus be $q = q(n, L)$. Let χ be an error distribution over $\mathbb{Z}$ and let $B = B(n)$ be an error bound. We additionally choose two Gaussian parameters: a "small" Gaussian parameter $s = s(n)$ and a "large" Gaussian parameter $\alpha = \alpha(n)$. Both these parameters are polynomially bounded (in λ, L). The public parameters $\mathsf{pp} = (\lambda, L, n, q, m, \chi, B, s, \alpha)$ are implicitly given as input to all the algorithms below.

- Setup(1^ℓ): The setup algorithm takes as input the length of the attribute vector ℓ.
    1. Sample a matrix with a trapdoor: $(\mathbf{A}, \mathbf{T_A}) \leftarrow \mathsf{TrapSamp}(1^n, 1^m, q)$.
    2. Let $\mathbf{G} \in \mathbb{Z}_q^{n \times m}$ be the primitive matrix with the public trapdoor basis $\mathbf{T_G}$.
    3. Choose $\ell + 6$ matrices $\{\mathbf{A}_i\}_{i \in [\ell]}, \{\mathbf{V}_{0,1}\}_{i \in [5]}, \mathbf{A}^c$ at random from $\mathbb{Z}_q^{n \times m}$. First, ℓ matrices form the LWE "public keys" for the bits of attribute vector, next 5 form the "public keys" for the initial configuration of the state vector, and the last matrix as a "public key" for a constant 1.
    4. Choose a vector $\mathbf{u} \in \mathbb{Z}_q^n$ at random.
    5. Output the master public key

$$\mathsf{mpk} := \big(\mathbf{A}, \mathbf{A}^c, \{\mathbf{A}_i\}_{i \in [\ell]}, \{\mathbf{V}_{0,i}\}_{i \in [5]}, \mathbf{G}, \mathbf{u}\big)$$

    and the master secret key $\mathsf{msk} := (\mathbf{T_A}, \mathsf{mpk})$.
- Enc($\mathsf{mpk}, \mathbf{x}, \mu$): The encryption algorithm takes as input the master public key mpk, the attribute vector $\mathbf{x} \in \{0,1\}^\ell$ and a message μ.
    1. Choose an LWE secret vector $\mathbf{s} \in \mathbb{Z}_q^n$ at random.
    2. Choose noise vector $\mathbf{e} \xleftarrow{\$} \chi^m$ and compute $\psi_0 = \mathbf{A}^\mathsf{T}\mathbf{s} + \mathbf{e}$.
    3. Choose a random matrix $\mathbf{R}^c \leftarrow \{-1, 1\}^{m \times m}$ and let $\mathbf{e}^c = (\mathbf{R}^c)^\mathsf{T}\mathbf{e}$. Now, compute an encoding of a constant 1:

$$\psi^c = (\mathbf{A}^c + \mathbf{G})^\mathsf{T}\mathbf{s} + \mathbf{e}^c$$

    4. Encode each bit $i \in [\ell]$ of the attribute vector:
        (a) Choose random matrices $\mathbf{R}_i \leftarrow \{-1, 1\}^{m \times m}$ and let $\mathbf{e}_i = \mathbf{R}_i^\mathsf{T}\mathbf{e}$.
        (b) Compute $\psi_i = (\mathbf{A}_i + x_i \cdot \mathbf{G})^\mathsf{T}\mathbf{s} + \mathbf{e}_i$.
    5. Encode the initial state configuration vector $\mathbf{v}_0 = [1, 0, 0, 0, 0]$: for all $i \in [5]$,
        (a) Choose a random matrix $\mathbf{R}_{0,i} \leftarrow \{-1, 1\}^{m \times m}$ and let $\mathbf{e}_{0,i} = \mathbf{R}_{0,i}^\mathsf{T}\mathbf{e}$.
        (b) Compute $\psi_{0,i} = (\mathbf{V}_{0,i} + \mathbf{v}_0[i] \cdot \mathbf{G})^\mathsf{T}\mathbf{s} + \mathbf{e}_{0,i}$.
    6. Encrypt the message μ as $\tau = \mathbf{u}^\mathsf{T}\mathbf{s} + e + \lfloor q/2 \rfloor \mu$, where $e \leftarrow \chi$.
    7. Output the ciphertext

$$\mathsf{ct_x} = \big(\mathbf{x}, \psi_0, \psi^c, \{\psi_i\}_{i \in [\ell]}, \{\psi_{0,i}\}_{i \in [5]}, \tau\big)$$

- KeyGen($\mathsf{msk}, \mathsf{BP}$): The key-generation algorithm takes as input the master secret key msk and a description of a branching program:

$$\mathsf{BP} := \big(\mathbf{v}_0, \{\mathsf{var}(t), \{\gamma_{t,i,0}, \gamma_{t,i,1}\}_{i \in [5]}\}_{t \in [L]}\big)$$

The secret key $\mathsf{sk_{BP}}$ is computed as follows.
    1. Homomorphically compute a "public key" matrix associated with the branching program:

$$\mathbf{V}_{\mathsf{BP}} \leftarrow \mathsf{Eval}_{\mathsf{pk}}(\mathsf{BP}, \{\mathbf{A}_i\}_{i \in [\ell]}, \{\mathbf{V}_{0,i}\}_{i \in [5]}, \mathbf{A}^c)$$

2. Let $\mathbf{F} = [\mathbf{A}||(\mathbf{V}_{\mathsf{BP}} + \mathbf{G})] \in \mathbb{Z}_q^{n \times 2m}$. Compute $\mathbf{r}_{\mathrm{out}} \leftarrow \mathsf{SampleLeft}(\mathbf{A},$ $(\mathbf{V}_{\mathsf{BP}} + \mathbf{G}), \mathbf{T_A}, \mathbf{u}, \alpha)$ such that $\mathbf{F} \cdot \mathbf{r}_{\mathrm{out}} = \mathbf{u}$.
3. Output the secret key for the branching program as

$$\mathsf{sk}_{\mathsf{BP}} := (\mathsf{BP}, \mathbf{r}_{\mathrm{out}})$$

– $\mathsf{Dec}(\mathsf{sk}_{\mathsf{BP}}, \mathsf{ct_x})$: The decryption algorithm takes as input the secret key for a branching program $\mathsf{sk}_{\mathsf{BP}}$ and a ciphertext $\mathsf{ct_x}$. If $\mathsf{BP}(\mathbf{x}) = 0$, output $\bot$. Otherwise,

1. Homomorphically compute the encoding of the result $\mathsf{BP}(\mathbf{x})$ associated with the public key of the branching program:

$$\psi_{\mathsf{BP}} \leftarrow \mathsf{Eval}_{\mathsf{en}}(\mathsf{BP}, \mathbf{x}, \{\mathbf{A}_i, \psi_i\}_{i \in [\ell]}, \{\mathbf{V}_{0,i}, \psi_{0,i}\}_{i \in [5]}, (\mathbf{A}^{\mathsf{c}}, \psi^{\mathsf{c}}))$$

2. Finally, compute $\phi = \mathbf{r}_{\mathrm{out}}^{\mathsf{T}} \cdot [\psi||\psi_{\mathsf{BP}}]$. As we show in Lemma 8, $\phi = \mathbf{u}^{\mathsf{T}}\mathbf{s} + \lfloor q/2 \rfloor \mu + e_\phi \pmod q$, for a short e_ϕ.
3. Output $\mu = 0$ if $|\tau - \phi| < q/4$ and $\mu = 1$ otherwise.

4.1 Correctness

Lemma 8. *Let $\mathcal{BP}$ be a family of width-5 permutation branching programs with their length bounded by L and let $\mathcal{ABE} = (\mathsf{Params}, \mathsf{Setup}, \mathsf{KeyGen}, \mathsf{Enc}, \mathsf{Dec})$ be our attribute-based encryption scheme. For a LWE dimension $n = n(\lambda)$, the parameters for $\mathcal{ABE}$ are instantiated as follows (according to Sect. 5):*

$$\chi = D_{\mathbb{Z}, \sqrt{n}} \qquad\qquad B = O(n) \qquad\qquad m = O(n \log q)$$
$$q = \tilde{O}(n^7 \cdot L^2) \qquad\qquad \alpha = \tilde{O}(n \log q)^2 \cdot L$$

then the scheme $\mathcal{ABE}$ is correct, according to the definition in Sect. 2.2.

Proof. We have to show that the decryption algorithm outputs the correct message μ, given a valid set of a secret key and a ciphertext.

From Lemma 4, we have that $\psi_{\mathsf{BP}} = (\mathbf{V}_{\mathsf{BP}} + \mathbf{G})^{\mathsf{T}}\mathbf{s} + \mathbf{e}_{\mathsf{BP}}$ since $\mathsf{BP}(\mathbf{x}) = 1$. Also, from Lemma 5, we know that $\|\mathbf{e}_{\mathsf{BP}}\|_\infty = O(m \cdot L \cdot (m \cdot B)) = O(m^2 \cdot L \cdot B)$ since our input encodings have noise terms bounded by $m \cdot B$. Thus, the noise term in ϕ is bounded by:

$$\|e_\phi\|_\infty = m \cdot \left(\mathsf{Noise}_{\mathbf{A},0}(\psi) + \mathsf{Noise}_{\mathbf{V}_{\mathsf{BP}},1}(\psi_{\mathsf{BP}})\right) \cdot \|\mathbf{r}_{\mathrm{out}}\|_\infty$$
$$= m \cdot (B + O(m^2 \cdot L \cdot B)) \cdot \tilde{O}(n \log q)^2 \cdot L\sqrt{m}$$
$$= O((n \log q)^6 \cdot L^2 \cdot B)$$

where $m = O(n \log q)$ and $\|\mathbf{r}_{\mathrm{out}}\|_\infty \leq \alpha\sqrt{m} = \tilde{O}(n \log q)^2 \cdot L\sqrt{m}$ according to Sect. 5. Hence, we have

$$|\tau - \phi| \leq \|e\|_\infty + \|e_\phi\|_\infty = O((n \log q)^6 \cdot L^2 \cdot B) \leq q/4$$

Clearly, the last inequality is satisfied when $q = \tilde{O}(n^7 \cdot L^2)$. Hence, the decryption proceeds correctly outputting the correct μ.

4.2   Security Proof

Theorem 5. *For any ℓ and any length bound L, $\mathcal{ABE}$ scheme defined above satisfies selective security game 3 for any family of branching programs BP of length L with ℓ-bit inputs, assuming hardness of $\mathsf{dLWE}_{n,q,\chi}$ for sufficiently large $n = \mathrm{poly}(\lambda), q = \tilde{O}(n^7 \cdot L^2)$ and $\mathrm{poly}(n)$ bounded error distribution χ. Moreover, the size of the secret keys grows polynomially with L (and independent of the width of BP).*

Proof. We define a series of hybrid games, where the first and the last games correspond to the real experiments encrypting messages μ_0, μ_1 respectively. We show that these games are indistinguishable except with negligible probability. Recall that in a selective security game, the challenge attribute vector $\mathbf{x}^*$ is declared before the Setup algorithm and all the secret key queries that adversary makes must satisfy $\mathsf{BP}(\mathbf{x}^*) = 0$. First, we define auxiliary simulated $\mathcal{ABE}^*$ algorithms.

- Setup$^*(1^\lambda, 1^\ell, 1^L, \mathbf{x}^*)$: The simulated setup algorithm takes as input the security parameter λ, the challenge attribute vector $\mathbf{x}^*$, its length ℓ and the maximum length of the branching program L.
  1. Choose a random matrix $\mathbf{A} \leftarrow \mathbb{Z}_q^{n \times m}$ and a vector $\mathbf{u}$ at random.
  2. Let $\mathbf{G} \in \mathbb{Z}_q^{n \times m}$ be the primitive matrix with the public trapdoor basis $\mathbf{T_G}$.
  3. Choose $\ell + 6$ random matrices $\{\mathbf{R}_i\}_{i \in [\ell]}, \{\mathbf{R}_{0,i}\}_{i \in [5]}, \mathbf{R}^c$ from $\{-1, 1\}^{m \times m}$ and set
     (a) $\mathbf{A}_i = \mathbf{A} \cdot \mathbf{R}_i - x^* \mathbf{G}$ for $i \in [\ell]$,
     (b) $\mathbf{V}_{0,i} = \mathbf{A} \cdot \mathbf{R}_{0,i} - \mathbf{v}_0[i] \cdot \mathbf{G}$ for $i \in [5]$ where $\mathbf{v}_0 = [1, 0, 0, 0, 0]$,
     (c) $\mathbf{A}^c = \mathbf{A} \cdot \mathbf{R}^c - \mathbf{G}$.
  4. Output the master public key

$$\mathsf{mpk} := \left(\mathbf{A}, \mathbf{A}^c, \{\mathbf{A}_i\}_{i \in [\ell]}, \{\mathbf{V}_{0,i}\}_{i \in [5]}, \mathbf{G}, \mathbf{u}\right)$$

  and the secret key

$$\mathsf{msk} := \left(\mathbf{x}^*, \mathbf{A}, \mathbf{R}^c, \{\mathbf{R}_i\}_{i \in [\ell]}, \{\mathbf{R}_{0,i}\}_{i \in [5]}\right)$$

- Enc$^*(\mathsf{mpk}, \mathbf{x}^*, \mu)$: The simulated encryption algorithm takes as input $\mathsf{mpk}, \mathbf{x}^*$ and the message μ. It computes the ciphertext using the knowledge of short matrices $\{\mathbf{R}_i\}, \{\mathbf{R}_{0,i}\}, \mathbf{R}^c$ as follows.
  1. Choose a vector $\mathbf{s} \in \mathbb{Z}_q^n$ at random.
  2. Choose noise vector $\mathbf{e} \xleftarrow{\$} \chi^m$ and compute $\psi_0 = \mathbf{A}^\mathsf{T} \mathbf{s} + \mathbf{e}$.
  3. Compute an encoding of an identity as $\psi^c = (\mathbf{A}^c)^\mathsf{T} \mathbf{s} + (\mathbf{R}^c)^\mathsf{T} \mathbf{e}$.
  4. For all bits of the attribute vector $i \in [\ell]$ compute

$$\psi_i = (\mathbf{A}_i + x_i \cdot \mathbf{G})^\mathsf{T} \mathbf{s} + \mathbf{R}_i^\mathsf{T} \mathbf{e}$$

5. For all $i \in [5]$, encode the bits of the initial state configuration vector $\mathbf{v}_0 = [1, 0, 0, 0, 0]$

$$\psi_{0,i} = (\mathbf{V}_{0,i} + \mathbf{v}_0[i] \cdot \mathbf{G})^\mathsf{T}\mathbf{s} + \mathbf{R}_{0,i}^\mathsf{T}\mathbf{e}$$

6. Encrypt the message μ as $\tau = \mathbf{u}^\mathsf{T}\mathbf{s} + e + \lfloor q/2 \rceil \mu$, where $e \leftarrow \chi$.
7. Output the ciphertext

$$\mathsf{ct} = (\mathbf{x}, \psi_0, \{\psi_i\}_{i\in[\ell]}, \psi^c, \{\psi_{0,i}\}_{i\in[5]}, \tau)$$

– KeyGen*(msk, BP): The simulated key-generation algorithm takes as input the master secret key msk and the description of the branching program BP. It computes the secret key sk_BP as follows.

1. Obtain a short homomorphically derived matrix associated with the output public key of the branching program:

$$\mathbf{R}_\mathsf{BP} \leftarrow \mathsf{Eval}_\mathsf{SIM}(\mathsf{BP}, \mathbf{x}^*, \{\mathbf{R}_i\}_{i\in[\ell]}, \{\mathbf{R}_{0,i}\}_{i\in[5]}, \mathbf{R}^c, \mathbf{A})$$

2. By the correctness of $\mathsf{Eval}_\mathsf{SIM}$, we have $\mathbf{V}_\mathsf{BP} = \mathbf{A}\mathbf{R}_\mathsf{BP} - \mathsf{BP}(\mathbf{x}^*) \cdot \mathbf{G}$. Let $\mathbf{F} = [\mathbf{A}||(\mathbf{V}_\mathsf{BP} + \mathbf{G})] \in \mathbb{Z}_q^{n \times 2m}$. Compute $\mathbf{r}_\mathsf{out} \leftarrow \mathsf{SampleRight}(\mathbf{A}, \mathbf{G}, \mathbf{R}_\mathsf{BP}, \mathbf{T}_\mathbf{G}, \mathbf{u}, \alpha)$ such that $\mathbf{F} \cdot \mathbf{r}_\mathsf{out} = \mathbf{u}$ (this step relies on the fact that $\mathsf{BP}(\mathbf{x}^*) = 0$).
3. Output the secret key for the branching program

$$\mathsf{sk}_\mathsf{BP} := (\mathsf{BP}, \mathbf{r}_\mathsf{out})$$

Game Sequence. We now define a series of games and then prove that all games **Game i** and **Game i+1** are either statistically or computationally indistinguishable.

– **Game 0:** The challenger runs the real ABE algorithms and encrypts message μ_0 for the challenge index $\mathbf{x}^*$.
– **Game 1:** The challenger runs the simulated ABE algorithms Setup*, KeyGen*, Enc* and encrypts message μ_0 for the challenge index $\mathbf{x}^*$.
– **Game 2:** The challenger runs the simulated ABE algorithms Setup*, KeyGen*, but chooses a uniformly random element of the ciphertext space for the challenge index $\mathbf{x}^*$.
– **Game 3:** The challenger runs the simulated ABE algorithms Setup*, KeyGen*, Enc* and encrypts message μ_1 for the challenge index $\mathbf{x}^*$.
– **Game 4:** The challenger runs the real ABE algorithms and encrypts message μ_1 for the challenge index $\mathbf{x}^*$.

Lemma 9. *The view of an adversary in* **Game 0** *is statistically indistinguishable from* **Game 1**. *Similarly, the view of an adversary in* **Game 4** *is statistically indistinguishable from* **Game 3**.

Proof. We prove for the case of **Game 0** and **Game 1**, as the other case is identical. First, note the differences between the games:

– In **Game 0**, matrix $\mathbf{A}$ is sampled using TrapSamp algorithm and matrices $\mathbf{A}_i, \mathbf{A}^c, \mathbf{V}_{0,j} \in \mathbb{Z}_q^{n \times m}$ are randomly chosen for $i \in [\ell], j \in [5]$. In **Game 1**, matrix $\mathbf{A} \in \mathbb{Z}_p^{n \times m}$ is chosen uniformly at random and matrices $\mathbf{A}_i = \mathbf{AR}_i - x_i^* \cdot \mathbf{G}$, $\mathbf{A}^c = \mathbf{AR}^c - \mathbf{G}$, $\mathbf{V}_{0,j} = \mathbf{AR}_{0,j} - \mathbf{v}_0[j] \cdot \mathbf{G}$ for randomly chosen $\mathbf{R}_i, \mathbf{R}^c, \mathbf{R}_{0,j} \in \{-1,1\}^{m \times m}$.
– In **Game 0**, each ciphertext component is computed as:

$$\psi_i = (\mathbf{A}_i + x_i^* \cdot \mathbf{G})^\intercal \mathbf{s} + \mathbf{e}_i = (\mathbf{A}_i + x_i^* \cdot \mathbf{G})^\intercal \mathbf{s} + \mathbf{R}_i^\intercal \mathbf{e}$$
$$\psi^c = (\mathbf{A}^c + \mathbf{G})^\intercal \mathbf{s} + \mathbf{e}^1 = (\mathbf{A}^c + \mathbf{G})^\intercal \mathbf{s} + (\mathbf{R}^c)^\intercal \mathbf{e}$$
$$\psi_{0,j} = (\mathbf{V}_{0,j} + \mathbf{v}_0[j] \cdot \mathbf{G})^\intercal \mathbf{s} + \mathbf{e}_i = (\mathbf{V}_{0,j} + \mathbf{v}_0[j] \cdot \mathbf{G})^\intercal \mathbf{s} + \mathbf{R}_{0,j}^\intercal \mathbf{e}$$

On the other hand, in **Game 1** each ciphertext component is computed as:

$$\psi_i = (\mathbf{A}_i + x_i^* \cdot \mathbf{G})^\intercal \mathbf{s} + \mathbf{R}_i^\intercal \mathbf{e} = (\mathbf{AR}_i)^\intercal \mathbf{s} + \mathbf{R}_i^\intercal \mathbf{e} = \mathbf{R}_i^\intercal (\mathbf{A}^\intercal \mathbf{s} + \mathbf{e})$$

Similarly, $\psi^c = (\mathbf{R}^c)^\intercal (\mathbf{A}^\intercal \mathbf{s} + \mathbf{e})$ and $\psi_{0,j} = \mathbf{R}_{0,j}^\intercal (\mathbf{As} + \mathbf{e})$.
– Finally, in **Game 0** the vector $\mathbf{r}_{\text{out}}$ is sampled using SampleLeft, whereas in **Game 1** it is sampled using SampleRight algorithm.

For sufficiently large α (See Sect. 5), the distributions produced in two games are statistically indistinguishable. This follows readily from [2, Lemma 4.3], Theorems 3 and 4. Please refer to the full version [29] for a detailed proof.

Lemma 10. *If the decisional* LWE *assumption holds, then the view of an adversary in* **Game 1** *is computationally indistinguishable from* **Game 2**. *Similarly, if the decisional* LWE *assumption holds, then the view of an adversary in* **Game 3** *is computationally indistinguishable from* **Game 2**.

Proof. Assume there exist an adversary Adv that distinguishes between **Game 1** and **Game 2**. We show how to break LWE problem given a challenge $\{(\mathbf{a}_i, y_i)\}_{i \in [m+1]}$ where each y_i is either a random sample in $\mathbb{Z}_q$ or $\mathbf{a}_i^\intercal \cdot \mathbf{s} + e_i$ (for a fixed, random $\mathbf{s} \in \mathbb{Z}_q^n$ and a noise term sampled from the error distribution $e_i \leftarrow \chi$). Let $\mathbf{A} = [\mathbf{a}_1, \mathbf{a}_2, \ldots, \mathbf{a}_m] \in \mathbb{Z}_q^{n \times m}$ and $\mathbf{u} = \mathbf{a}_{m+1}$. Let $\psi_0^* = [y_1, y_2, \ldots, y_m]$ and $\tau = y_{m+1} + \mu \lfloor q/2 \rfloor$.

Now, run the simulated Setup* algorithm where $\mathbf{A}, \mathbf{u}$ are as defined above. Run the simulated KeyGen* algorithm. Finally, to simulate the challenge ciphertext set ψ_0^*, τ as defined above and compute

$$\psi_i = \mathbf{R}_i^\intercal \cdot \psi_0^* = \mathbf{R}_i^\intercal (\mathbf{A}^\intercal \mathbf{s} + \mathbf{e})$$

for $i \in [\ell]$. Similarly, $\psi^c = (\mathbf{R}^c)^\intercal (\mathbf{A}^\intercal \mathbf{s} + \mathbf{e})$ and $\psi_{0,j} = \mathbf{R}_{0,j}^\intercal (\mathbf{A}^\intercal \mathbf{s} + \mathbf{e})$, for $j \in [5]$. Note that if y_i's are LWE samples, then this corresponds exactly to the **Game 1**. Otherwise, the ciphertext corresponds to an independent random sample as in **Game 2** by the left-over hash lemma. Thus, an adversary which distinguishes between **Game 1** and **Game 2** can also be used to break the decisional LWE assumption with almost the same advantage. The computational indistinguishability of **Game 3** and **Game 2** follows from the same argument.

Thus, **Game 0** and **Game 4** are computationally indistinguishable by the standard hybrid argument and hence no adversary can distinguish between encryptions of μ_0 and μ_1 with non-negligible advantage establishing the selective security of our ABE scheme.

5 Parameter Selection

This section provides a concise description on the selection of parameters for our scheme, so that both correctness (see Lemma 8) and security (see Theorem 5) of our scheme are satisfied.

For a family of width-5 permutation branching programs $\mathcal{BP}$ of bounded length L, with the LWE dimension n, the parameters can be chosen as follows: (we start with an arbitrary q and we will instantiate it later)

- The parameter $m = O(n \log q)$. The error distribution $\chi = D_{\mathbb{Z}, \sqrt{n}}$ with parameter $\sigma = \sqrt{n}$. And, the error bound $B = O(\sigma \sqrt{n}) = O(n)$.
- The "large" Gaussian parameter $\alpha = \alpha(n, L)$ is chosen such that the output of the SampleLeft and the SampleRight algorithms are statistically indistinguishable from each other, when provided with the same set of inputs $\mathbf{F}$ and $\mathbf{u}$. The SampleRight algorithm (Algorithm 3) requires

$$\alpha > \|\mathbf{T_G}\|_{\mathsf{GS}} \cdot \|\mathbf{R}_{\mathsf{BP}}\| \cdot \omega(\sqrt{\log m}) \tag{3}$$

From Lemma 7, we have that $\|\mathbf{R}_{\mathsf{BP}}\|_\infty = O(m \cdot L)$. Then, we get:

$$\|\mathbf{R}_{\mathsf{BP}}\| := \sup_{\mathbf{x} \in S^{m-1}} \|\mathbf{R}_{\mathsf{BP}} \cdot \mathbf{x}\| \leq m \cdot \|\mathbf{R}_{\mathsf{BP}}\|_\infty \leq O(m^2 \cdot L)$$

Finally, from Eq. 3, the value of α required for the SampleRight algorithm is

$$\alpha \geq O(m^2 \cdot L) \cdot \omega(\sqrt{\log m}) \tag{4}$$

The value of the parameter α required for the SampleLeft algorithm (Algorithm 3) is

$$\alpha \geq \|\mathbf{T_A}\|_{\mathsf{GS}} \cdot \omega(\sqrt{\log 2m}) \geq O(\sqrt{n \log q}) \cdot \omega(\sqrt{\log 2m}) \tag{5}$$

Thus, to satisfy both Eqs. 4 and 5, we set the parameter

$$\alpha \geq O(m^2 \cdot L) \cdot \omega(\sqrt{\log m}) = \tilde{O}(n \log q)^2 \cdot L$$

When our scheme is instantiated with these parameters, the correctness (see Lemma 8) of the scheme is satisfied when $O((n \log q)^6 \cdot L^2 \cdot B) < q/4$. Clearly, this condition is satisfied when $q = \tilde{O}(n^7 L^2)$.

6 Extensions

We note a few possible extensions on our basic construction that lead to further efficiency improvements. First, we can support arbitrary width branching programs by appropriately increasing the dimension of the state vector in the encryption. Second, we can switch to an arithmetic setting, similarly as it was done in [10].

References

1. Agrawal, S., Boneh, D., Boyen, X.: Efficient lattice (H)IBE in the standard model. In: Gilbert [23], pp. 553–572
2. Agrawal, S., Freeman, D.M., Vaikuntanathan, V.: Functional encryption for inner product predicates from learning with errors. In: Lee, D.H., Wang, X. (eds.) ASIACRYPT 2011. LNCS, vol. 7073, pp. 21–40. Springer, Heidelberg (2011)
3. Ajtai, M.: Generating hard instances of the short basis problem. In: Wiedermann, J., Van Emde Boas, P., Nielsen, M. (eds.) ICALP 1999. LNCS, vol. 1644, pp. 1–9. Springer, Heidelberg (1999)
4. Ajtai, M., Kumar, R., Sivakumar, D.: A sieve algorithm for the shortest lattice vector problem. In: Vitter, J.S., Spirakis, P.G., Yannakakis, M. (eds.) STOC, pp. 601–610, ACM (2001)
5. Akinyele, J.A., Pagano, M.W., Green, M.D., Lehmann, C.U., Peterson, Z.N.J., Rubin, A.D.: Securing electronic medical records using attribute-based encryption on mobile devices. In: Jiang, X., Bhattacharya, A., Dasgupta, P., Enck, W. (eds.) SPSM, pp. 75–86, ACM (2011)
6. Attrapadung, N., Libert, B., de Panafieu, E.: Expressive key-policy attribute-based encryption with constant-size ciphertexts. In: Catalano, D., Fazio, N., Gennaro, R., Nicolosi, A. (eds.) PKC 2011. LNCS, vol. 6571, pp. 90–108. Springer, Heidelberg (2011)
7. Barrington, D.A.M.: Bounded-width polynomial-size branching programs recognize exactly those languages in nc^1. In: Hartmanis, J. (ed.) STOC, pp. 1–5, ACM (1986)
8. Boneh, D., Franklin, M.K.: Identity-based encryption from the Weil pairing. SIAM J. Comput. $32(3)$, 586–615 (2003)
9. Boneh, D., Sahai, A., Waters, B.: Functional encryption: a new vision for public-key cryptography. Commun. ACM $55(11)$, 56–64 (2012)
10. Boneh, D., Gentry, C., Gorbunov, S., Halevi, S., Nikolaenko, V., Segev, G., Vaikuntanathan, V., Vinayagamurthy, D.: Fully key-homomorphic encryption, arithmetic circuit ABE and compact garbled circuits. In: Nguyen, P.Q., Oswald, E. (eds.) EUROCRYPT 2014. LNCS, vol. 8441, pp. 533–556. Springer, Heidelberg (2014)
11. Boneh, D., Roughgarden, T., Feigenbaum, J. (eds.): Symposium on Theory of Computing Conference, STOC 2013, Palo Alto, CA, USA, ACM, 1–4 Jun 2013
12. Boyen, X.: Attribute-based functional encryption on lattices. In: Sahai, A. (ed.) TCC 2013. LNCS, vol. 7785, pp. 122–142. Springer, Heidelberg (2013)
13. Brakerski, Z., Vaikuntanathan, V.: Lattice-based FHE as secure as PKE. In: Naor, M. (ed.) ITCS, pp. 1–12, ACM (2014)
14. Cash, D., Hofheinz, D., Kiltz, E., Peikert, C.: Bonsai trees, or how to delegate a lattice basis. J. Cryptol. $25(4)$, 601–639 (2012)
15. Cocks, C.: An identity based encryption scheme based on quadratic residues. In: Honary, B. (ed.) Cryptography and Coding 2001. LNCS, vol. 2260, pp. 360–363. Springer, Heidelberg (2001)
16. Coron, J.-S., Lepoint, T., Tibouchi, M.: Practical multilinear maps over the integers. In: Canetti, R., Garay, J.A. (eds.) CRYPTO 2013, Part I. LNCS, vol. 8042, pp. 476–493. Springer, Heidelberg (2013)
17. Emura, K., Miyaji, A., Nomura, A., Omote, K., Soshi, M.: A ciphertext-policy attribute-based encryption scheme with constant ciphertext length. In: Bao, F., Li, H., Wang, G. (eds.) ISPEC 2009. LNCS, vol. 5451, pp. 13–23. Springer, Heidelberg (2009)

18. Garg, S., Gentry, C., Halevi, S.: Candidate multilinear maps from ideal lattices. In: Johansson, T., Nguyen, P.Q. (eds.) EUROCRYPT 2013. LNCS, vol. 7881, pp. 1–17. Springer, Heidelberg (2013)

19. Garg, S., Gentry, C., Halevi, S., Raykova, M., Sahai, A., Waters, B.: Candidate indistinguishability obfuscation and functional encryption for all circuits. In: FOCS, pp. 40–49, IEEE Computer Society (2013)

20. Garg, S., Gentry, C., Halevi, S., Sahai, A., Waters, B.: Attribute-based encryption for circuits from multilinear maps. In: Canetti, R., Garay, J.A. (eds.) CRYPTO 2013, Part II. LNCS, vol. 8043, pp. 479–499. Springer, Heidelberg (2013)

21. Gentry, C., Gorbunov, S., Halevi, S.: Graph-induced multilinear maps from lattices. In: Dodis, Y., Nielsen, J.B. (eds.) TCC 2015, Part II. LNCS, vol. 9015, pp. 498–527. Springer, Heidelberg (2015)

22. Gentry, C., Peikert, C., Vaikuntanathan, V.: Trapdoors for hard lattices and new cryptographic constructions. In: Dwork, C. (ed.) STOC, pp. 197–206, ACM (2008)

23. Gilbert, H. (ed.): EUROCRYPT 2010. LNCS, vol. 6110. Springer, Heidelberg (2010)

24. Goldwasser, S., Kalai, Y.T., Popa, R.A., Vaikuntanathan, V., Zeldovich, N.: Reusable garbled circuits and succinct functional encryption. In: Boneh et al. [11], pp. 555–564

25. Gorbunov, S., Vaikuntanathan, V., Wee, H.: Functional encryption with bounded collusions via multi-party computation. In: Safavi-Naini, R., Canetti, R. (eds.) CRYPTO 2012. LNCS, vol. 7417, pp. 162–179. Springer, Heidelberg (2012)

26. Gorbunov, S., Vaikuntanathan, V., Wee, H.: Attribute-based encryption for circuits. In: Boneh et al. [11], pp. 545–554

27. Gorbunov, S., Vaikuntanathan, V., Wee, H.: Predicate encryption for circuits from LWE. In: Gennaro, R., Robshaw, M. (eds.) CRYPTO 2015. LNCS, vol. 9216, pp. 503–523. Springer, Heidelberg (2015)

28. Gorbunov, S., Vaikuntanathan, V., Wichs, D.: Leveled fully homomorphic signatures from standard lattices. In: Servedio, R.A., Rubinfeld, R. (eds.) STOC, pp. 469–477, ACM (2015)

29. Gorbunov, S., Vinayagamurthy, D.: Riding on asymmetry: efficient abe for branching programs. Cryptology ePrint Archive, Report 2014/819 (2014)

30. Goyal, V., Pandey, O., Sahai, A., Waters, B.: Attribute-based encryption for fine-grained access control of encrypted data. In: Juels, A., Wright, R.N., di Vimercati, S.D.C. (eds.) CCS, pp. 89–98, ACM (2006)

31. Lewko, A.B., Okamoto, T., Sahai, A., Takashima, K., Waters, B.: Fully secure functional encryption: attribute-based encryption and (hierarchical) inner product encryption. In: Gilbert [23], pp. 62–91

32. Lewko, A., Waters, B.: New techniques for dual system encryption and fully secure HIBE with short ciphertexts. In: Micciancio, D. (ed.) TCC 2010. LNCS, vol. 5978, pp. 455–479. Springer, Heidelberg (2010)

33. Li, M., Yu, S., Zheng, Y., Ren, K., Lou, W.: Scalable and secure sharing of personal health records in cloud computing using attribute-based encryption. IEEE Trans. Parallel Distrib. Syst. 24(1), 131–143 (2013)

34. Micciancio, D., Peikert, C.: Trapdoors for lattices: simpler, tighter, faster, smaller. In: Pointcheval, D., Johansson, T. (eds.) EUROCRYPT 2012. LNCS, vol. 7237, pp. 700–718. Springer, Heidelberg (2012)

35. Micciancio, D., Voulgaris, P.: A deterministic single exponential time algorithm for most lattice problems based on voronoi cell computations. In: Schulman, L.J. (ed.) STOC, pp. 351–358, ACM (2010)

36. Papanis, J., Papapanagiotou, S., Mousas, A., Lioudakis, G., Kaklamani, D., Venieris, I.: On the use of attribute-based encryption for multimedia content protection over information-centric networks. Trans. Emerg. Telecommun. Technol. **25**(4), 422–435 (2014)
37. Parno, B., Raykova, M., Vaikuntanathan, V.: How to delegate and verify in public: verifiable computation from attribute-based encryption. In: Cramer, R. (ed.) TCC 2012. LNCS, vol. 7194, pp. 422–439. Springer, Heidelberg (2012)
38. Peikert, C.: Public-key cryptosystems from the worst-case shortest vector problem: extended abstract. In: Mitzenmacher, M. (ed.) STOC, pp. 333–342, ACM (2009)
39. Regev, O.: On lattices, learning with errors, random linear codes, and cryptography. J. ACM **56**(6), Article No. 34 (2009)
40. Sahai, A., Waters, B.: Fuzzy identity-based encryption. In: Cramer, R. (ed.) EUROCRYPT 2005. LNCS, vol. 3494, pp. 457–473. Springer, Heidelberg (2005)
41. Waters, B.: Dual system encryption: realizing fully secure IBE and HIBE under simple assumptions. In: Halevi, S. (ed.) CRYPTO 2009. LNCS, vol. 5677, pp. 619–636. Springer, Heidelberg (2009)

Conversions Among Several Classes of Predicate Encryption and Applications to ABE with Various Compactness Tradeoffs

Nuttapong Attrapadung, Goichiro Hanaoka, and Shota Yamada$^{(\boxtimes)}$

National Institute of Advanced Industrial Science and Technology (AIST),
Tokyo, Japan
{n.attrapadung,hanaoka-goichiro,yamada-shota}@aist.go.jp

Abstract. Predicate encryption is an advanced form of public-key encryption that yields high flexibility in terms of access control. In the literature, many predicate encryption schemes have been proposed such as fuzzy-IBE, KP-ABE, CP-ABE, (doubly) spatial encryption (DSE), and ABE for arithmetic span programs. In this paper, we study relations among them and show that some of them are in fact equivalent by giving conversions among them. More specifically, our main contributions are as follows:

- We show that monotonic, small universe KP-ABE (CP-ABE) with bounds on the size of attribute sets and span programs (or linear secret sharing matrix) can be converted into DSE. Furthermore, we show that DSE implies non-monotonic CP-ABE (and KP-ABE) with the same bounds on parameters. This implies that monotonic/non-monotonic KP/CP-ABE (with the bounds) and DSE are all equivalent in the sense that one implies another.
- We also show that if we start from KP-ABE without bounds on the size of span programs (but bounds on the size of attribute sets), we can obtain ABE for arithmetic span programs. The other direction is also shown: ABE for arithmetic span programs can be converted into KP-ABE. These results imply, somewhat surprisingly, KP-ABE without bounds on span program sizes is in fact equivalent to ABE for arithmetic span programs, which was thought to be more expressive or at least incomparable.

By applying these conversions to existing schemes, we obtain many non-trivial consequences. We obtain the first non-monotonic, large universe CP-ABE (that supports span programs) with constant-size ciphertexts, the first KP-ABE with constant-size private keys, the first (adaptively-secure, multi-use) ABE for arithmetic span programs with constant-size ciphertexts, and more. We also obtain the first attribute-based signature scheme that supports non-monotone span programs and achieves constant-size signatures via our techniques.

Keywords: Attribute-based encryption · Doubly spatial encryption · Generic conversion · Constant-size ciphertexts · Constant-size keys · Arithmetic span programs

© International Association for Cryptologic Research 2015
T. Iwata and J.H. Cheon (Eds.): ASIACRYPT 2015, Part I, LNCS 9452, pp. 575–601, 2015.
DOI: 10.1007/978-3-662-48797-6_24

1 Introduction

Predicate encryption (PE) is an advanced form of public-key encryption that allows much flexibility. Instead of encrypting data to a target recipient, a sender will specify in a more general way about who should be able to view the message. In predicate encryption for a predicate R, a sender can associate a ciphertext with a ciphertext attribute X while a private key is associated with a key attribute Y. Such a ciphertext can then be decrypted by such a key if the predicate evaluation $R(X, Y)$ holds true.

There exist many classes of PE, each is defined by specifying a corresponding class of predicates. One notable class is attribute-based encryption (ABE) [24,38] for span programs (or equivalently, linear secret sharing schemes), of which predicate is defined over key attributes being a span program and ciphertext attributes being a set of attributes, and its evaluation holds true if the span program accepts the set. This is called key-policy ABE (KP-ABE). There is also ciphertext-policy ABE (CP-ABE), where the roles of key and ciphertext attributes are exchanged. Another important class is doubly spatial encryption (DSE) [25], of which predicate is defined over both key and ciphertext attributes being affine subspaces, and its evaluation holds true if both subspaces intersect. Very recently, a new important class of PE, that is called attribute encryption for arithmetic span programs is defined in [28]. They showed such a PE scheme is useful by demonstrating that the scheme can be efficiently converted into ABE for arithmetic branching programs for both zero-type and non-zero type predicates. If the scheme satisfies a certain requirement for efficiency (namely, encryption cost is at most linear in ciphertext predicate size), it is also possible to obtain a publicly verifiable delegation scheme for arithmetic branching programs, by exploiting a conversion shown in [36]. Furthermore, they gave a concrete construction of such scheme.

Compared to specific constructions of predicate encryption [19,21,30–32,40] (to name just a few) that focus on achieving more expressive predicates and/or stronger security guarantee, relations among predicate encryption schemes are much less investigated. The purpose of this paper is to improve our understanding of relations among them.

1.1 Our Results

Relations among PE. Towards the goal above, we study relations among PE and show that some of them are in fact equivalent by giving generic conversion among them. We first investigate the relation among ABE with some bounds on parameters (the size of attribute sets and the size of span programs) and DSE. We have the following results:

- First, we show a conversion from KP-ABE (or CP-ABE) with the bounds on parameters into DSE (without key delegation, in Sect. 3). Such an implication is not straightforward in the first place. Intuitively, one reason stems from the different nature between both predicates: while DSE can be considered as an algebraic object that involves affine spaces, ABE can be seen as a somewhat more combinatorial object that involves sets (of attributes). Our approach involves some new technique for "programming" a set associated to a ciphertext and a span program associated to a private key in the KP-ABE scheme so that they can emulate the relation for doubly spatial encryption.

- We then extend the result of [25], which showed that DSE implies CP/KP-ABE with large universes. We provide a new conversion from DSE (without delegation) to non-monotonic CP/KP-ABE with large universes (in Sect. 4). We note that the resulting schemes obtained by the above conversions have some bounds on parameters. In the conversion, we extensively use a special form of polynomial introduced in [29] and carefully design a matrix so that DSE can capture a relation for ABE.

Somewhat surprisingly, by combining the above results, we obtain generic conversions that can boost the functionality of (bounded) ABE: from monotonic to non-monotonic, and from small-universe to large-universe; moreover, we also obtain conversions which transform ABE to its dual (key-policy to ciphertext-policy, and vice versa). This implies that they are essentially equivalent in some sense. See Fig. 1 for the details.

So far, we have considered ABE schemes with bounds on parameters, especially on the size of span programs. We then proceed to investigate relation among ABE schemes without bounds on the size of span programs (but with a bound on the size of attribute sets) and ABE for arithmetic span programs recently introduced and studied by Ishai and Wee [28]. We call the latter key-policy ABE for arithmetic span programs (KASP), since in the latter, a ciphertext is associated with a vector while a private key is associated with an arithmetic span program which specifies a policy. By exchanging key and ciphertext attribute, we can also define ciphertext-policy version of ABE for arithmetic span program (CASP). We have the following results:

- We show that monotonic KP-ABE with small universe (without bound on the size of span programs) can be converted into KASP (in Sect. 5). The idea for the conversion is similar to that in Sect. 3.
- In the full version of the paper [4], we also investigate the converse direction. In fact, we show somewhat stronger result. That is, KASP can be converted into non-monotonic KP-ABE with large universe, which trivially implies monotonic KP-ABE with small universe. The idea for the conversion is similar to that in Sect. 4.

Given the above results, we have all of the following are equivalent: monotonic KP-ABE with small universe, non-monotonic KP-ABE with large universe, and KASP. Similar implications hold for the case of CP-ABE and CASP. However, we do not have a conversion from KP-ABE to CP-ABE in this case. Again, see Fig. 1 for the details.

Direct Applications: New Instantiations. By applying our conversions to existing schemes, we obtain many new instantiations. Most of them have new properties that were not achieved before. These include

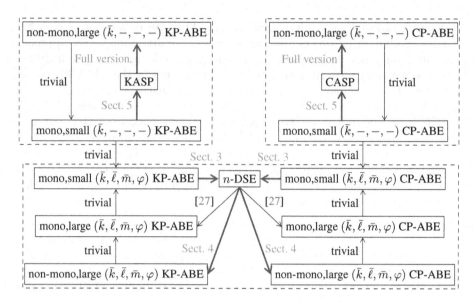

Fig. 1. Relations among predicate encryption primitives. In this figure, arrows indicate conversions that transform the primitive of the starting point to that of the end point. The red arrows indicate our results in this paper. For ABE, 'mono' and 'non-mono' indicate whether it is monotonic or non-monotonic, while 'small' and 'large' indicate whether the attribute universes are large (i.e., exponentially large) or small (i.e., polynomially bounded). $(\bar{k}, \bar{\ell}, \bar{m}, \varphi)$ specify bounds on size of sets of attributes and span programs. See Sect. 2.1 for details. As a result, primitives inside each dashed box are all equivalent in the sense there is a conversion between each pair (Color figure online).

- the first DSE with constant-size public key,
- the first DSE with constant-size ciphertexts,
- the first DSE with constant-size private keys,
- the first non-monotonic, large-universe CP-ABE with constant-size ciphertexts,
- the first non-monotonic, large-universe KP-ABE with constant-size keys,
- the first KASP, CASP with constant-size public key,
- the first KASP, CASP with adaptive security and unbounded multi-use,
- the first KASP with constant-size ciphertexts,
- the first CASP with constant-size keys,

which together offer various compactness tradeoffs. Previously, all DSE schemes require linear (or more) sizes in all parameters [14,17,25]. Previous CP-ABE with constant-size ciphertexts [12,13,18,20] can only deal with threshold or even more limited expressiveness. As for KP-ABE, to the best of our knowledge, there were no constructions with constant-size keys.[1] Previous KASP and

---

[1] KP-ABE with (asymptotically) short keys was also proposed in [9]. Compared to ours, their key size is not constant but they focus on more expressive ABE, namely ABE for circuits.

CASP [16,28] require linear sizes in all parameters. Moreover, the adaptively secure schemes [16] support only attribute one-use. See Sect. 6 and tables therein for our instantiations and comparisons.

Application to Attribute-Based Signatures. Our technique is also useful in the settings of attribute-based signatures (ABS) [33,34]. We first define a notion that we call predicate signature (PS) which is a signature analogue of PE. Then, we construct a specific PS scheme with constant-size signatures such that a signature is associated with a set of attributes while a private key is associated with a policy (or monotone span programs). This is in some sense a dual notion of ordinary ABS in which a signature is associated with a policy and a private key with a set. By using the technique developed in the above, we can convert the PS scheme into an ABS scheme. As a result, we obtain the first ABS scheme with constant-size signatures. Previous ABS schemes with constant-size signatures [12,26] only support threshold or more limited policies.

Finally, we remark that although our conversions are feasible, they often introduce polynomial-size overheads to some parameters. Thus, in most cases, above schemes obtained by the conversions should be seen as feasibility results in the sense that they might not be totally efficient. As a future direction, it would be interesting to construct more efficient schemes directly.

1.2 Related Works

There are several previous works investigating relations among PE primitives. In [23], a black box separation between threshold predicate encryption (fuzzy IBE) and IBE was shown. They also rule out certain natural constructions of PE for $\mathbf{NC}^1$ from PE for $\mathbf{AC}^0$. In [15], it was shown that hierarchical inner product encryption is equivalent to spatial encryption, which is a special case of doubly spatial encryption.

[22] showed a generic conversion from KP-ABE supporting threshold formulae to CP-ABE supporting threshold formulae. Their result and ours are incomparable. Our KP-ABE to CP-ABE conversion requires the original KP-ABE to support monotone span programs, which is a stronger requirement than [22]. On the other hand, the resulting scheme obtained by our conversion supports non-monotone span programs, which is a wider class than threshold formulae[2]. Thus, by applying our conversion, we can obtain new schemes (such as CP-ABE supporting non-monotone span programs with constant-size ciphertext) that is not possible to obtain by the conversion by [22].

In recent works [2,6], it is shown that PE satisfying certain specific template can be converted into PE for its dual predicate. In particular, it yields KP-ABE-to-CP-ABE conversion. Again, their result and ours are incomparable. On the one hand, schemes obtained from their conversion are typically more efficient than ours. On the other hand, their conversion only works for schemes with the

---

[2] While it is known that monotone span programs contain threshold formulae [24], the converse is not known to be true.

template while our conversion is completely generic. Furthermore, since they essentially exchange key and ciphertext components in the conversion, the size of keys and ciphertexts are also exchanged. For example, if we start from KP-ABE with constant-size ciphertexts, they obtain CP-ABE with *constant-size private keys* while we obtain CP-ABE with *constant-size ciphertexts*.

We also remark that in the settings where PE for general circuit is available, we can easily convert any KP-ABE into CP-ABE by using universal circuits as discussed in [19, 21]. However, in the settings where only PE for span programs is available, this technique is not known to be applicable. We note that all existing PE schemes for general circuits [9, 19, 21] are quite inefficient and based on strong assumptions (e.g., existence of secure multi-linear map or hardness of certain lattice problems for an exponential approximation factor). In [7], in the context of quantum computation, Belovs studies a span program that decides whether two spaces intersect or not. The problem and its solution considered there is very similar to that in Sect. 3 of our paper. However, he does not consider application to cryptography and the result is not applicable to our setting immediately since the syntax of span programs is slightly different.

Concurrent and Independent Work. Concurrently and independently to our work, Aggrawal and Chase [1] show specific construction of CP-ABE scheme with constant-size ciphertexts. Compared to our CP-ABE scheme with constant-size ciphertexts, which is obtained by our conversion, their scheme only supports monotone access structure over large universe, whereas our scheme supports non-monotonic access structure over large universe. Furthermore, we can obtain adaptively secure scheme whereas their scheme is only selectively secure. On the other hand, their scheme has shorter keys.

2   Preliminaries

Notation. Throughout the paper, p denotes a prime number. We will treat a vector as a column vector, unless stated otherwise. For a vector $\mathbf{a} \in \mathbb{Z}_p^n$, $\mathbf{a}[i] \in \mathbb{Z}_p$ represents i-th element of the vector. Namely, $\mathbf{a} = (\mathbf{a}[1], \ldots, \mathbf{a}[n])^\top$. For $\mathbf{a}, \mathbf{b} \in \mathbb{Z}_p^n$, we denote their inner product as $\langle \mathbf{a}, \mathbf{b} \rangle = \mathbf{a}^\top \mathbf{b} = \sum_{i=1}^n \mathbf{a}[i] \cdot \mathbf{b}[i]$. We denote by e_i the i-th unit vector: its i-th component is one, all others are zero. $\mathbf{I}_n$ and $\mathbf{0}_{n \times m}$ represent an identity matrix in $\mathbb{Z}_p^{n \times n}$ and zero matrix in $\mathbb{Z}_p^{n \times m}$ respectively. We also define $\mathbf{1}_n = (1, 1, \ldots, 1)^\top \in \mathbb{Z}_p^n$ and $\mathbf{0}_n = \mathbf{0}_{n \times 1}$. We often omit the subscript if it is clear from the context. We denote by $[a, b]$ a set $\{a, a+1, \ldots, b\}$ for $a, b \in \mathbb{Z}$ such that $a \le b$ and $[b]$ denotes $[1, b]$. For a matrix $\mathbf{X} \in \mathbb{Z}_p^{n \times d}$, $\mathrm{span}(\mathbf{X})$ denotes a linear space $\{\mathbf{X} \cdot \mathbf{u} | \mathbf{u} \in \mathbb{Z}_p^d\}$ spanned by columns of $\mathbf{X}$. For matrices $\mathbf{A} \in \mathbb{Z}_p^{n_1 \times m}$ and $\mathbf{B} \in \mathbb{Z}_p^{n_2 \times m}$, $[\mathbf{A}; \mathbf{B}] \in \mathbb{Z}_p^{(n_1+n_2) \times m}$ denotes $[\mathbf{A}^\top, \mathbf{B}^\top]^\top$ i.e., the vertical concatenation of them.

2.1   Definition of Predicate Encryption

Here, we define the syntax of predicate encryption. We emphasize that we do not consider attribute hiding in this paper[3].

Syntax. Let $R = \{R_N : A_N \times B_N \to \{0, 1\} \mid N \in \mathbb{N}^c\}$ be a relation family where A_N and B_N denote "ciphertext attribute" and "key attribute" spaces and c is some fixed constant. The index $N = (n_1, n_2, \ldots, n_c)$ of R_N denotes the numbers of bounds for corresponding parameters. A predicate encryption (PE) scheme for R is defined by the following algorithms:

Setup$(\lambda, N) \to$ (mpk, msk): The setup algorithm takes as input a security parameter λ and an index N of the relation R_N and outputs a master public key
  mpk and a master secret key msk.
Encrypt(mpk, M, X) $\to C$: The encryption algorithm takes as input a master
  public key mpk, the message M, and a ciphertext attribute $X \in A_N$. It will
  output a ciphertext C.
KeyGen(msk, mpk, Y) $\to$ sk$_Y$: The key generation algorithm takes as input the
  master secret key msk, the master public key mpk, and a key attribute $Y \in$
  B_N. It outputs a private key sk$_Y$.
Decrypt(mpk, C, X, sk$_Y$, Y) $\to$ M or $\bot$: We assume that the decryption algorithm is deterministic. The decryption algorithm takes as input the master
  public key mpk, a ciphertext C, ciphertext attribute $X \in A_N$, a private
  key sk$_Y$, and private key attribute Y. It outputs the message M or $\bot$ which
  represents that the ciphertext is not in a valid form.

We refer (standard) definitions of correctness and security of PE to [2,4].

2.2   (Arithmetic) Span Program, ABE, and Doubly Spatial Encryption

Definition of Span Program. Let $\mathcal{U} = \{u_1, \ldots, u_t\}$ be a set of variables. For each u_i, denote $\neg u_i$ as a new variable. Intuitively, u_i and $\neg u_i$ correspond to positive and negative attributes, respectively. Also let $\mathcal{U}' = \{\neg u_1, \ldots, \neg u_t\}$. A span program over $\mathbb{Z}_p$ is specified by a pair $(\mathbf{L}, \rho)$ of a matrix and a labelling function where

$$\mathbf{L} \in \mathbb{Z}_p^{\ell \times m} \qquad\qquad \rho : [\ell] \to \mathcal{U} \cup \mathcal{U}'$$

for some integer ℓ, m. Intuitively, the map ρ labels row i with attribute $\rho(i)$.

A span program accepts or rejects an input by the following criterion. For an input $\delta \in \{0, 1\}^t$, we define the sub-matrix $\mathbf{L}_\delta$ of $\mathbf{L}$ to consist of the rows whose labels are set to 1 by the input δ. That is, it consists of either rows labelled by some u_i such that $\delta_i = 1$ or rows labelled by some $\neg u_i$ such that $\delta_i = 0$. We say that

$$(\mathbf{L}, \rho) \text{ accepts } \delta \text{ iff } (1, 0, \ldots, 0) \text{ is in the row span of } \mathbf{L}_\delta.$$

---

[3] This is called "public-index" predicate encryption, categorized in [11].

We can write this also as $e_1 \in \text{span}(\mathbf{L}_\delta^\top)$. A span program is called *monotone* if the labels of the rows consist of only the positive literals, in $\mathcal{U}$.

Key-Policy and Ciphertext-Policy Attribute-Based Encryption. Let $\mathcal{U}$ be the universe of attributes. We define a relation R^{KP} on any span programs $(\mathbf{L}, \rho)$ over $\mathbb{Z}_p$ and any sets of attributes $S \subseteq \mathcal{U}$ as follows. For $S \subseteq \mathcal{U}$, we define $\delta \in \{0,1\}^t$ as an indicator vector corresponding to S. Namely, $\delta_i = 1$ if $u_i \in S$ and $\delta_i = 0$ if $u_i \notin S$. We define

$$R^{\mathsf{KP}}(S, (\mathbf{L}, \rho)) = 1 \text{ iff } (\mathbf{L}, \rho) \text{ accepts } \delta.$$

Similarly, R^{CP} is defined as $R^{\mathsf{CP}}((\mathbf{L}, \rho), S) = 1$ iff $(\mathbf{L}, \rho)$ accepts δ.

A KP-ABE scheme may require some bounds on parameters: we denote

$$\bar{k} = \text{ the maximum size of } k \text{ (the size of attribute set } S),$$
$$\bar{\ell} = \text{ the maximum size of } \ell \text{ (the number of rows of } \mathbf{L}),$$
$$\bar{m} = \text{ the maximum size of } m \text{ (the number of columns of } \mathbf{L}),$$
$$\varphi = \text{ the maximum size of allowed repetition in } \{\rho(1), \ldots, \rho(\ell)\}.$$

These bounds define the index $N = (\bar{k}, \bar{\ell}, \bar{m}, \varphi)$ for the predicate family. When there is no restriction on corresponding parameter, we represent it by "−" such as $(\bar{k}, -, -, -)$. We define A_N and B_N as the set of all attribute sets and the set of all span programs whose sizes are restricted by N, respectively. KP-ABE is a predicate encryption for $R_N^{\mathsf{KP}} : A_N \times B_N \to \{0,1\}$, where R_N^{KP} is restricted on N in a natural manner. CP-ABE is defined dually with A_N and B_N swapped.

Let $t := |\mathcal{U}|$. We say the scheme supports small universe if t is polynomially bounded and large universe if t is exponentially large. The scheme is monotonic if span programs are restricted to be monotone, and non-monotonic otherwise.

Attribute-Based Encryption for Arithmetic Span Programs [28]. In this predicate, the index N for the family is specified by an integer n. We call it the dimension of the scheme. We define $A_N = \mathbb{Z}_p^n$. An arithmetic span program of dimension n is specified by a tuple $(\mathbf{Y}, \mathbf{Z}, \rho)$ of two matrices $\mathbf{Y}, \mathbf{Z} \in \mathbb{Z}_p^{m \times \ell}$ and a map $\rho : [\ell] \to [n]$, for some integers ℓ, m. There is no restriction on ℓ and m. If ρ is restricted to injective, we say that the scheme supports only *attribute one-use*. Otherwise, if there is no restriction on ρ, we say that it is *unbounded multi-use*. We let B_N be the set of all arithmetic span programs of dimension n. We then define

$$R_N^{\mathsf{KASP}}(\mathbf{x}, (\mathbf{Y}, \mathbf{Z}, \rho)) = 1 \text{ iff } e_1 \in \text{span}\{\mathbf{x}[\rho(j)] \cdot \mathbf{y}_j + \mathbf{z}_j\}_{j \in [\ell]},$$

where here $e_1 = (1, 0, \ldots, 0)^\top \in \mathbb{Z}_p^m$ and $\mathbf{x}[\rho(j)]$ is the $\rho(j)$-th term of $\mathbf{x}$, while $\mathbf{y}_j$ and $\mathbf{z}_j$ are the j-th column of $\mathbf{Y}$ and $\mathbf{Z}$ respectively. We call predicate encryption for R^{KASP} key-policy attribute-based encryption for arithmetic span program (KASP). Ciphertext-policy ASP (CASP) can be defined dually with A_N and B_N swapped.

Doubly Spatial Encryption. In this predicate, the index N for the family is specified by an integer n (the dimension of the scheme). We define the domains as $A_N = B_N = \mathbb{Z}_p^n \times (\cup_{0 \leq d \leq n} \mathbb{Z}_p^{n \times d})$. We define

$$R_N^{\mathsf{DSE}}\big((x_0, \mathbf{X}), (y_0, \mathbf{Y})\big) = 1 \text{ iff } \big(x_0 + \mathrm{span}(\mathbf{X})\big) \cap \big(y_0 + \mathrm{span}(\mathbf{Y})\big) \neq \emptyset.$$

Doubly spatial encryption is PE for relation R_N^{DSE} equipped with additional key delegation algorithm. The key delegation algorithm takes a private key for some affine space as an input and outputs a private key for another affine space, which is a subset of the first one. We require that the distribution of a key obtained by the delegation is the same as that of a key directly obtained by the key generation algorithm. We refer to [4,17] for the formal definition.

2.3   Embedding Lemma for PE

The following useful lemma from [10] describes a sufficient criterion for implication from PE for a given predicate to PE for another predicate. The lemma is applicable to any relation family.

We consider two relation families:

$$R_N^{\mathsf{F}} : A_N \times B_N \to \{0,1\}, \qquad R_{N'}^{\mathsf{F}'} : A'_{N'} \times B'_{N'} \to \{0,1\},$$

which is parametrized by $N \in \mathbb{N}^c$ and $N' \in \mathbb{N}^{c'}$ respectively. Suppose that there exists three efficient mappings

$$f_{\mathsf{p}} : \mathbb{Z}^{c'} \to \mathbb{Z}^c \qquad f_{\mathsf{e}} : A'_{N'} \to A_{f_{\mathsf{p}}(N')} \qquad f_{\mathsf{k}} : B'_{N'} \to B_{f_{\mathsf{p}}(N')}$$

which maps parameters, ciphertext attributes, and key attributes, respectively, such that for all $X' \in A'_{N'}, Y' \in B'_{N'}$,

$$R_{N'}^{\mathsf{F}'}(X', Y') = 1 \Leftrightarrow R_{f_{\mathsf{p}}(N')}^{\mathsf{F}}(f_{\mathsf{e}}(X'), f_{\mathsf{k}}(Y')) = 1. \tag{1}$$

We can then construct a PE scheme $\Pi' = \{\mathsf{Setup}', \mathsf{Encrypt}', \mathsf{KeyGen}', \mathsf{Decrypt}'\}$ for predicate $R_{N'}^{\mathsf{F}'}$ from a PE scheme $\Pi = \{\mathsf{Setup}, \mathsf{Encrypt}, \mathsf{KeyGen}, \mathsf{Decrypt}\}$ for predicate R_N^{F} as follows. Let $\mathsf{Setup}'(\lambda, N') = \mathsf{Setup}(\lambda, f_{\mathsf{p}}(N'))$ and

$$\mathsf{Encrypt}'(\mathsf{mpk}, M, X') = \mathsf{Encrypt}(\mathsf{mpk}, M, f_{\mathsf{e}}(X')),$$
$$\mathsf{KeyGen}'(\mathsf{msk}, \mathsf{mpk}, Y') = \mathsf{KeyGen}(\mathsf{msk}, \mathsf{mpk}, f_{\mathsf{k}}(Y')),$$

and $\mathsf{Decrypt}'(\mathsf{mpk}, C, X', \mathsf{sk}_{Y'}, Y') = \mathsf{Decrypt}(\mathsf{mpk}, C, f_{\mathsf{e}}(X'), \mathsf{sk}_{Y'}, f_{\mathsf{k}}(Y'))$.

Lemma 1 (Embedding lemma [10]). *If Π is correct and secure, then so is Π'. This holds for selective security and adaptive security.*

Intuitively, the forward and backward direction of Relation (1) ensure that the correctness and the security are preserving, respectively.

3   Conversion from ABE to DSE

In this section, we show how to construct DSE for dimension n from monotonic KP-ABE (with bounds on the size of attribute sets and span programs). We note that by simply swapping key and ciphertext attributes, we can also obtain CP-ABE-to-DSE conversion. We first describe the conversion, then explain the intuition behind the conversion later below.

3.1   The Conversion

Mapping Parameters. We map $f_p^{\mathsf{DSE}\to\mathsf{KP}} : n \mapsto (\bar{k}, \bar{\ell}, \bar{m}, \psi)$ where

$$\bar{k} = n(n+1)\kappa + 1, \qquad\qquad \bar{\ell} = 2(n\kappa + 1)(n+1),$$
$$\bar{m} = (n\kappa + 1)(n+1) + 1, \qquad \psi = 2(n+1),$$

where we define $\kappa := \lceil \log_2 p \rceil$. Moreover, we set the universe $\mathcal{U}$ as follows.

$$\mathcal{U} = \Big\{ \mathsf{Att}[i][j][k][b] \;\Big|\; (i,j,k,b) \in [0,n] \times [1,n] \times [1,\kappa] \times \{0,1\} \Big\} \cup \{\mathsf{D}\},$$

where D is a dummy attribute which will be assigned for all ciphertext. Hence, the universe size is $|\mathcal{U}| = 2n(n+1)\kappa + 1$. Intuitively, $\mathsf{Att}[i][j][k][b]$ represents an indicator for the condition "the k-th least significant bit of the binary representation of the j-th element of the vector x_i is $b \in \{0,1\}$".

Mapping Ciphertext Attributes. For $\mathrm{x}_0 \in \mathbb{Z}_p^n$ and $\mathbf{X} = [\mathrm{x}_1, \ldots, \mathrm{x}_{d_1}] \in \mathbb{Z}_p^{n \times d_1}$ such that $d_1 \leq n$, we map $f_e^{\mathsf{DSE}\to\mathsf{KP}} : (\mathrm{x}_0, \mathbf{X}) \mapsto S$ where

$$S = \Big\{ \mathsf{Att}[i][j][k][b] \;\Big|\; (i,j,k) \in [0,d_1] \times [1,n] \times [1,\kappa],\ b = \mathrm{x}_i[j][k] \Big\} \cup \{\mathsf{D}\}.$$

Here, we define $\mathrm{x}_i[j][k] \in \{0,1\}$ so that they satisfy

$$\mathrm{x}_i[j] = \sum_{k=1}^{\kappa} 2^{k-1} \cdot \mathrm{x}_i[j][k].$$

Namely, $\mathrm{x}_i[j][k]$ is the k-th least significant bit of the binary representation of $\mathrm{x}_i[j]$.

Mapping Key Attributes. For $\mathrm{y}_0 \in \mathbb{Z}_p^n$ and $\mathbf{Y} = [\mathrm{y}_1, \ldots, \mathrm{y}_{d_2}] \in \mathbb{Z}_p^{n \times d_2}$ such that $d_2 \leq n$, we map $f_k^{\mathsf{DSE}\to\mathsf{KP}} : (\mathrm{y}_0, \mathbf{Y}) \mapsto (\mathbf{L}, \rho)$ as follows. Let the numbers of rows and columns of $\mathbf{L}$ be

$$\ell = (2n\kappa + 1)(n+1) + d_2 + 1, \qquad m = (n\kappa + 1)(n+1) + 1,$$

respectively. We then define

$$\mathbf{L} = \begin{pmatrix} \mathrm{e}_1\ \mathrm{e}_1 + \mathrm{e}_{d_2+2}\ \mathrm{y}_0^\top \\ \qquad\quad \mathbf{Y}^\top \\ \qquad\quad \mathbf{E}\ \ \mathbf{J} \\ \qquad\quad \mathbf{E}\quad \mathbf{J} \\ \qquad\quad \vdots \qquad \ddots \\ \qquad\quad \mathbf{E} \qquad\qquad \mathbf{J} \end{pmatrix} \in \mathbb{Z}_p^{\ell \times m}, \tag{2}$$

of which each sub-matrix $\mathbf{E}$ and $\mathbf{J}$ both appears $n + 1$ times, where we define

$$\mathbf{E} = \begin{pmatrix} g & & & \\ & g & & \\ & & \ddots & \\ & & & g \\ 0 & 0 & \dots & 0 \end{pmatrix} \in \mathbb{Z}_p^{(2n\kappa+1)\times n}, \quad \mathbf{J} = \begin{pmatrix} -1 & & & & & \\ -1 & & & & & \\ & -1 & & & & \\ & -1 & & & & \\ & & \ddots & & & \\ & & & -1 & \\ & & & -1 \\ 1 & 1 & \dots & 1 \end{pmatrix} \in \mathbb{Z}_p^{(2n\kappa+1)\times n\kappa} \quad (3)$$

where $g = (0, 1, 0, 2, \dots, 0, 2^i, \dots, 0, 2^{\kappa-1})^\top \in \mathbb{Z}_p^{2\kappa}$.

Next, we define the map $\rho : [1, \ell] \to \mathcal{U}$ as follows.

- If $i \le d_2 + 1$, we set $\rho(i) := \mathsf{D}$.
- Else, we have $i \in [d_2 + 2, \ell]$. We then write

$$i = (d_2 + 1) + (2n\kappa + 1)i' + i''$$

with a unique $i' \in [0, n+1]$ and a unique $i'' \in [0, 2n\kappa]$.
  - $i'' = 0$, we again set $\rho(i) = \mathsf{D}$.
  - Else, we have $i'' \in [1, 2n\kappa]$. We then write

$$i'' = 2\kappa j' + 2k' + b' + 1$$

with unique $j' \in [0, n-1]$, $k' \in [0, \kappa - 1]$, and $b' \in \{0, 1\}$. We finally set

$$\rho(i) = \mathsf{Att}[i'][j'+1][k'+1][b'].$$

Intuition. We explain the intuition behind the conversion. S can be seen as a binary representation of the information of $(\mathsf{x}_0, \mathbf{X})$. In the span program $(\mathbf{L}, \rho)$, $\mathbf{E}$ is used to reproduce the information of $(\mathsf{x}_0, \mathbf{X})$ in the matrix while $\mathbf{J}$ is used to constrain the form of linear combination among rows to a certain form.[4] In some sense, the roll of the lower part of the matrix $\mathbf{L}$ (the last $(2n\kappa + 1)(n+1)$ rows) is similar to universal circuit while the upper part of the matrix contains the information of $(\mathsf{y}_0, \mathbf{Y})$.

3.2 Correctness of the Conversion

We show the following theorem. The implication from KP-ABE to DSE would then follow from the embedding lemma (Lemma 1).

---

[4] A somewhat similar technique to ours that restricts the form of linear combination of vectors was used in [8] in a different context (for constructing a monotone span program that tests co-primality of two numbers).

Theorem 1. *For $n \in \mathbb{N}$, for any $x_0 \in \mathbb{Z}_p^n$, $\mathbf{X} \in \mathbb{Z}_p^{n \times d_1}$, $y_0 \in \mathbb{Z}_p^n$ and $\mathbf{Y} \in \mathbb{Z}_p^{n \times d_2}$, it holds that*

$$R_N^{\mathsf{KP}}(S, (\mathbf{L}, \rho)) = 1 \Leftrightarrow R_n^{\mathsf{DSE}}\big((x_0, \mathbf{X}), (y_0, \mathbf{Y})\big) = 1$$

with $N = f_{\mathsf{p}}^{\mathsf{DSE} \to \mathsf{KP}}(n)$, $S = f_{\mathsf{e}}^{\mathsf{DSE} \to \mathsf{KP}}(x_0, \mathbf{X})$, and $(\mathbf{L}, \rho) = f_{\mathsf{k}}^{\mathsf{DSE} \to \mathsf{KP}}(y_0, \mathbf{Y})$.

Proof. Define $I \subset [\ell]$ as $I := \{i | \rho(i) \in S\}$ and define $\mathbf{L}_I$ as the sub-matrix of $\mathbf{L}$ formed by all the rows of which index is in I. From the definition of $f_{\mathsf{e}}^{\mathsf{DSE} \to \mathsf{KP}}$, we have that $\mathbf{L}_I$ is in the form of

$$\mathbf{L}_I = \begin{pmatrix} e_1 & e_1 + e_{d_2+2} & y_0^\top \\ & & \mathbf{Y}^\top \\ & & \mathbf{E}_0 & \mathbf{J}' \\ & & \mathbf{E}_1 & & \mathbf{J}' \\ & & \vdots & & & \ddots \\ & & \mathbf{E}_{d_1} & & & & \mathbf{J}' \\ & & & & & & & 1_{n\kappa}^\top \\ & & & & & & & & \ddots \\ & & & & & & & & & 1_{n\kappa}^\top \end{pmatrix} \in \mathbb{Z}_p^{\ell_I \times m_I}$$

where $\ell_I := (n\kappa + 1)(d_1 + 1) + n - d_1 + d_2 + 1$ and $m_I := (n\kappa + 1)(n+1) + 1$ and

$$\mathbf{E}_i = \begin{pmatrix} g_{i,1} & & & \\ & g_{i,2} & & \\ & & \ddots & \\ & & & g_{i,n} \\ 0 & 0 & \dots & 0 \end{pmatrix} \in \mathbb{Z}_p^{(n\kappa+1) \times n}, \quad \mathbf{J}' = \begin{pmatrix} -1 & & & \\ & -1 & & \\ & & \ddots & \\ & & & -1 \\ 1 & 1 & \dots & 1 \end{pmatrix} \in \mathbb{Z}_p^{(n\kappa+1) \times n\kappa}.$$

for $i \in [0, d_1]$, where

$$g_{i,j} = \Big(x_i[j][1], \ 2x_i[j][2], \ \dots, 2^{\kappa-1}x_i[j][\kappa]\Big)^\top \in \mathbb{Z}_p^\kappa.$$

We remark that it holds that $\langle 1_\kappa, g_{i,j} \rangle = x_i[j]$ by the definition of $x_i[j][k]$ and thus $\mathbf{E}_i^\top \cdot 1_{n\kappa+1} = x_i$ holds. We also remark that if $v^\top \mathbf{J}' = \mathbf{0}$ holds for some $v \in \mathbb{Z}_p^{n\kappa+1}$, then there exists $v \in \mathbb{Z}_p$ such that $v = v 1_{n\kappa+1}$. These properties will be used later.

To prove the theorem statement is now equivalent to prove that

$$e_1 \in \mathrm{span}(\mathbf{L}_I^\top) \quad \Leftrightarrow \quad \big(x_0 + \mathrm{span}(\mathbf{X})\big) \cap \big(y_0 + \mathrm{span}(\mathbf{Y})\big) \neq \emptyset.$$

Forward Direction ($\Rightarrow$). Suppose $e_1 \in \mathrm{span}(\mathbf{L}_I^\top)$. Then, there exists $\mathbf{u} \in \mathbb{Z}_p^{\ell_I}$ such that $\mathbf{u}^\top \mathbf{L}_I = e_1^\top$. We write $\mathbf{u}$ as

$$\mathbf{u}^\top = \big(\underbrace{v}_{1}, \underbrace{v^\top}_{d_2}, \underbrace{u_0^\top}_{n\kappa+1}, \underbrace{u_1^\top}_{n\kappa+1}, \dots, \underbrace{u_{d_1}^\top}_{n\kappa+1}, \underbrace{u_{d_1+1}}_{1}, \dots, \underbrace{u_n}_{1} \big).$$

We then write

$$\mathbf{u}^\top \mathbf{L}_I = \left(v,\ \left(v + \langle \mathbf{u}_0, \mathbf{e}_1 \rangle \right),\ \left(v\mathbf{y}_0^\top + \mathbf{v}^\top \mathbf{Y}^\top + \sum_{i=0}^{d_1} \mathbf{u}_i^\top \mathbf{E}_i \right),\ \left(\mathbf{u}_0^\top \cdot \mathbf{J}' \right),\ \ldots, \right.$$

$$\left. \left(\mathbf{u}_{d_1}^\top \cdot \mathbf{J}' \right),\ \left(u_{d_1+1} \mathbf{1}_{n\kappa+1}^\top \right),\ \ldots,\ \left(u_n \mathbf{1}_{n\kappa+1}^\top \right) \right)$$

Since $\mathbf{u}^\top \mathbf{L}_I = \mathbf{e}_1^\top$, we have $u_{d_1+1} = \cdots = u_n = 0$, by comparing each element of the vector. Furthermore, since $\mathbf{u}_i^\top \cdot \mathbf{J}' = \mathbf{0}$ for $i \in [0, d_1]$, there exist $\{u_i \in \mathbb{Z}_p\}_{i \in [0,d_1]}$ such that $\mathbf{u}_i = u_i \mathbf{1}_{n\kappa+1}$. By comparing the first and the second element of the vector, we obtain $v = 1$ and $v + \langle \mathbf{u}_0, \mathbf{e}_1 \rangle = 1 + u_0 \langle \mathbf{1}_{n\kappa+1}^\top, \mathbf{e}_1 \rangle = 1 + u_0 = 0$. Hence, $u_0 = -1$. Finally, we have that $\sum_{i=0}^{d_1} \mathbf{u}_i^\top \mathbf{E}_i + v\mathbf{y}_0^\top + \mathbf{v}^\top \mathbf{Y}^\top = \mathbf{0}$ and thus

$$-\sum_{i=0}^{d_1} \mathbf{E}_i^\top \mathbf{u}_i = \mathbf{y}_0 + \mathbf{Y} \cdot \mathbf{v}.$$

The left hand side of the equation is

$$-\sum_{i=0}^{d_1} \mathbf{E}_i^\top \mathbf{u}_i = -u_0 \mathbf{E}_0^\top \cdot \mathbf{1}_{n\kappa+1} - \sum_{i=1}^{d_1} u_i \mathbf{E}_i^\top \cdot \mathbf{1}_{n\kappa+1}$$

$$= \mathbf{x}_0 - \sum_{i=1}^{d_1} u_i \cdot \mathbf{x}_i \in \left(\mathbf{x}_0 + \mathrm{span}(\mathbf{X}) \right).$$

while the right hand side is $\mathbf{y}_0 + \mathbf{Y} \cdot \mathbf{v} \in (\mathbf{y}_0 + \mathrm{span}(\mathbf{Y}))$. This implies that $\left(\mathbf{x}_0 + \mathrm{span}(\mathbf{X}) \right) \cap \left(\mathbf{y}_0 + \mathrm{span}(\mathbf{Y}) \right) \neq \emptyset$.

Converse Direction ($\Leftarrow$). Suppose $\left(\mathbf{x}_0 + \mathrm{span}(\mathbf{X}) \right) \cap \left(\mathbf{y}_0 + \mathrm{span}(\mathbf{Y}) \right) \neq \emptyset$. Hence, there exist sets $\{u_i \in \mathbb{Z}_p\}_{i \in [1,d_1]}$ and $\{v_i \in \mathbb{Z}_p\}_{i \in [1,d_2]}$ such that $\mathbf{x}_0 + \sum_{i=1}^{d_1} u_i \mathbf{x}_i = \mathbf{y}_0 + \sum_{i=1}^{d_2} v_i \mathbf{y}_i$. We set a vector $\mathbf{u}$ as

$$\mathbf{u}^\top = \big(1, \underbrace{v_1, \ldots, v_{d_2}}_{d_2}, \underbrace{-\mathbf{1}_{n\kappa+1}^\top, -u_1 \mathbf{1}_{n\kappa+1}^\top, \ldots, -u_{d_1} \mathbf{1}_{n\kappa+1}^\top}_{(n\kappa+1)(d_1+1)}, \underbrace{0, \ldots, 0}_{n-d_1} \big) \big).$$

Therefore, we have

$$\mathbf{u}^\top \mathbf{L}_I = \left(1,\ 1-1,\ \left(\mathbf{y}_0^\top + \sum_{i=1}^{d_2} v_i \mathbf{y}_i^\top - \mathbf{1}_{n\kappa+1}^\top (\mathbf{E}_0 + \sum_{i=1}^{d_1} u_i \mathbf{E}_i) \right), \right.$$

$$\left. \left(-\mathbf{1}_{n\kappa+1}^\top \mathbf{J}' \right),\ \left(-u_1 \mathbf{1}_{n\kappa+1}^\top \mathbf{J}' \right),\ \ldots,\ \left(-u_n \mathbf{1}_{n\kappa+1}^\top \mathbf{J}' \right), 0 \ldots, 0 \right)$$

$$= \left(1, 0, \left(\mathbf{y}_0^\top + \sum_{i=1}^{d_2} v_i \mathbf{y}_i^\top \right) - \left(\mathbf{x}_0^\top + \sum_{i=1}^{d_1} u_i \mathbf{x}_i^\top \right), 0 \ldots, 0 \right) = \mathbf{e}_1^\top$$

as desired. This concludes the proof of the theorem.

4   From DSE to Non-Monotonic ABE

In [25], it is shown that DSE can be converted into monotonic CP-ABE with large universe (and bounds on the size of attribute sets and span programs). In this section, we extend their result to show that non-monotonic CP-ABE with large universe and the same bounds can be constructed from DSE. We note that our transformation is very different from that of [25] even if we only consider monotonic CP-ABE because of expositional reasons. We also note that by simply swapping key and ciphertext attributes, we immediately obtain DSE-to-non-monotonic-KP-ABE conversion. Again, we first describe the conversion, provide some intuition later below.

4.1   The Conversion

Mapping Parameters. We map $f_{\mathsf{p}}^{\mathsf{CP}\rightarrow\mathsf{DSE}} : (\bar{k}, \bar{\ell}, \bar{m}, \bar{\ell}) \mapsto n = 4\bar{\ell} + \bar{m} + 2\bar{k}\bar{\ell}$. We assume that the universe of attributes is $\mathbb{Z}_p$. This restriction can be easily removed by using collision resistant hash.

Mapping Ciphertext Attributes. For a span program $(\mathbf{L}, \rho)$, we map $f_{\mathsf{e}}^{\mathsf{CP}\rightarrow\mathsf{DSE}} : (\mathbf{L}, \rho) \mapsto (\mathbf{x}_0, \mathbf{X})$ as follows. Let $\ell \times \bar{m}$ be the dimension of $\mathbf{L}$, where $\ell \leq \bar{\ell}$. (If the number of columns is smaller, we can adjust the size by padding zeroes.) Let ℓ_0, ℓ_1 be such that $\ell = \ell_0 + \ell_1$, and without loss of generality, we assume that the first ℓ_0 rows of $\mathbf{L}$ are associated with positive attributes and the last ℓ_1 rows with negative attributes by the map ρ. We denote $\mathbf{L}$ as $\mathbf{L} = [\mathbf{L}_0; \mathbf{L}_1]$ using matrices $\mathbf{L}_0 \in \mathbb{Z}_p^{\ell_0 \times \bar{m}}$ and $\mathbf{L}_1 \in \mathbb{Z}_p^{\ell_1 \times \bar{m}}$. We then define $f_{\mathsf{e}}^{\mathsf{CP}\rightarrow\mathsf{DSE}}(\mathbf{L}, \rho) = (\mathbf{x}_0, \mathbf{X})$ with

$$
\mathbf{x}_0 = -\mathbf{e}_1 \in \mathbb{Z}_p^n, \qquad \mathbf{X}^\top = \begin{pmatrix} \mathbf{L}_0 \overset{\bar{\ell}}{\frown} \mathbf{G}_0 & & \\ \mathbf{L}_1 & \mathbf{I}_{\ell_1} \overset{\bar{\ell}-\ell_1}{\frown} & \\ & & \mathbf{G}_1 \end{pmatrix} \in \mathbb{Z}_p^{(\ell_0 + 2\ell_1) \times n},
$$

where $\mathbf{G}_b \in \mathbb{Z}_p^{\ell_b \times \bar{\ell}(\bar{k}+1)}$ for each $b \in \{0,1\}$ is defined as

$$
\mathbf{G}_b = \begin{pmatrix} \mathbf{p}\big(\rho(b\ell_0 + 1)\big)^\top & & & \overset{(\bar{\ell}-\ell_b)(\bar{k}+1)}{\frown} \\ & \mathbf{p}\big(\rho(b\ell_0 + 2)\big)^\top & & \\ & & \ddots & \\ & & & \mathbf{p}\big(\rho(b\ell_0 + \ell_b)\big)^\top \end{pmatrix}
$$

where $\mathbf{p}()$ is a function that takes an element of $\mathbb{Z}_p$ or its negation ($\{\neg x | x \in \mathbb{Z}_p\}$) as an input and outputs a vector $\mathbf{p}(x) = (1, x, x^2, \ldots, x^{\bar{k}})^\top \in \mathbb{Z}_p^{\bar{k}+1}$.

Mapping Key Attributes. For a set $S = (S_1, \ldots, S_k)$ such that $k \leq \bar{k}$, we map $f_{\mathsf{k}}^{\mathsf{CP}\rightarrow\mathsf{DSE}} : S \mapsto (\mathbf{y}_0, \mathbf{Y})$ where

$$
\mathbf{y}_0 = \mathbf{0}_n \in \mathbb{Z}_p^n, \qquad \mathbf{Y}^\top = \begin{pmatrix} \overset{\bar{m}}{\frown} & \mathbf{H}\mathbf{I}_{(\bar{k}+1)\bar{\ell}} & \& \\ & & \mathbf{H}\mathbf{I}_{(\bar{k}+1)\bar{\ell}} \end{pmatrix} \in \mathbb{Z}_p^{2(\bar{k}+1)\bar{\ell} \times n},
$$

of which $\mathbf{H}$ is defined as

$$\mathbf{H} = \mathbf{I}_{\bar{\ell}} \otimes \mathbf{q}_S = \begin{pmatrix} \mathbf{q}_S & & & \\ & \mathbf{q}_S & & \\ & & \ddots & \\ & & & \mathbf{q}_S \end{pmatrix} \in \mathbb{Z}_p^{\left((\bar{k}+1)\bar{\ell}\right) \times \bar{\ell}},$$

where $\mathbf{q}_S = (\mathbf{q}_S[1], \ldots, \mathbf{q}_S[\bar{k}+1])^\top \in \mathbb{Z}_p^{\bar{k}+1}$ is defined as a coefficient vector from

$$Q_S[Z] = \sum_{i=1}^{k+1} \mathbf{q}_S[i] \cdot Z^{i-1} = \prod_{i=1}^{k} (Z - S_i).$$

If $k < \bar{k}$, the coordinates $\mathbf{q}_S[k+2], \ldots, \mathbf{q}_S[\bar{k}+1]$ are all set to 0.

Intuition. The matrices $\mathbf{X}$ and $\mathbf{Y}$ constructed above can be divided into two parts. The first ℓ_0 rows of $\mathbf{X}^\top$ and the first $(\bar{k}+1)\bar{\ell}$ rows of $\mathbf{Y}^\top$ deal with positive attributes. The lower parts of $\mathbf{X}^\top$ and $\mathbf{Y}^\top$ deal with negation of attributes. Here, we explain how we handle negated attributes. Positive attributes are handled by a similar mechanism. $\mathbf{I}_{(\bar{k}+1)\bar{\ell}}$ in $\mathbf{Y}^\top$ and $\mathbf{G}_1$ in $\mathbf{X}^\top$ restricts the linear combination of the rows of $\mathbf{X}^\top$ and $\mathbf{Y}^\top$ to a certain form in order to two affine spaces to have a intersection. As a result, we can argue that the coefficient of the i-th row of $\mathbf{L}_1$ in the linear combination should be multiple of $Q_S(\rho(\ell_0 + i))$[5]. Since we have that $Q_S(x) = 0$ iff $x \in S$ for any $x \in \mathbb{Z}_p$, this means that the coefficient of the vector in the linear combination should be 0 if $\rho(\ell_0 + i) = \neg\mathsf{Att}$ and $\mathsf{Att} \in S$. This restriction is exactly what we need to emulate predicate of non-monotonic CP-ABE.

4.2 Correctness of the Conversion

We show the following theorem. The implication from DSE to non-monotonic CP-ABE with large universe would then follow from the embedding lemma.

Theorem 2. *For any span program* $(\mathbf{L} \in \mathbb{Z}_p^{\ell \times m}, \rho)$ *such that* $\ell \leq \bar{\ell}$ *and* $m \leq \bar{m}$ *and* S *such that* $|S| \leq \bar{k}$, *let* $N = (\bar{k}, \bar{\ell}, \bar{m}, \bar{\ell})$, *we have that*

$$R_n^{\mathsf{DSE}}((\mathbf{x}_0, \mathbf{X}), (\mathbf{y}_0, \mathbf{Y})) = 1 \Leftrightarrow R_N^{\mathsf{CP}}(S, (\mathbf{L}, \rho)) = 1$$

where $n = f_p^{\mathsf{CP} \to \mathsf{DSE}}(N)$, $(\mathbf{x}_0, \mathbf{X}) = f_e^{\mathsf{CP} \to \mathsf{DSE}}(\mathbf{L}, \rho)$, *and* $(\mathbf{y}_0, \mathbf{Y}) = f_k^{\mathsf{CP} \to \mathsf{DSE}}(S)$.

Proof. Let $I \subset [1, \ell]$ be $I = \{i | (\rho(i) = \mathsf{Att} \wedge \mathsf{Att} \in S) \vee (\rho(i) = \neg\mathsf{Att} \wedge \mathsf{Att} \notin S)\}$. We also let $\mathbf{L}_I$ be the sub-matrix of $\mathbf{L}$ formed by rows whose index is in I.

To prove the theorem statement is equivalent to prove that

$$\left(\mathbf{x}_0 + \mathrm{span}(\mathbf{X})\right) \cap \left(\mathbf{y}_0 + \mathrm{span}(\mathbf{Y})\right) \neq \emptyset \quad \Leftrightarrow \quad \mathbf{e}_1 \in \mathrm{span}(\mathbf{L}_I^\top).$$

---

[5] Here, We treat negated attributes $(\{\neg x | x \in \mathbb{Z}_p\})$ as elements of $\mathbb{Z}_p$. Namely, if $\rho(\ell_0 + i) = \neg\mathsf{Att}$ for some $\mathsf{Att} \in \mathbb{Z}_p$, $Q_S(\rho(\ell_0 + i)) := Q_S(\mathsf{Att})$..

Forward Direction ($\Rightarrow$). Suppose that there exist $\mathbf{u} \in \mathbb{Z}_p^{\ell_0 + 2\ell_1}$ and $\mathbf{v} \in \mathbb{Z}_p^{2(\bar{k}+1)\bar{\ell}}$ such that $\mathbf{x}_0^\top + \mathbf{u}^\top \mathbf{X}^\top = \mathbf{y}_0^\top + \mathbf{v}^\top \mathbf{Y}^\top = \mathbf{v}^\top \mathbf{Y}^\top$. We denote these vectors as

$$\mathbf{u}^\top = (\underbrace{\mathbf{u}_0^\top}_{\ell_0}, \underbrace{\mathbf{u}_1^\top}_{\ell_1}, \underbrace{\mathbf{u}_2^\top}_{\ell_1}), \qquad \mathbf{v}^\top = (\underbrace{\mathbf{v}_1^\top, \ldots, \mathbf{v}_{\bar{\ell}}^\top}_{\bar{k}+1}, \underbrace{\mathbf{w}_1^\top, \ldots, \mathbf{w}_{\bar{\ell}}^\top}_{\bar{k}+1} \underbrace{}_{\bar{k}+1} \underbrace{}_{\bar{k}+1}.)$$

Hence, $\mathbf{x}_0^\top + \mathbf{u}^\top \mathbf{X}$ and $\mathbf{v}^\top \mathbf{Y}$ can be written as

$$\mathbf{x}_0^\top + \mathbf{u}^\top \mathbf{X} = \Big(\underbrace{-\mathbf{e}_1^\top + \mathbf{u}_0^\top \mathbf{L}_0 + \mathbf{u}_1^\top \mathbf{L}_1}_{\bar{m}}, \mathbf{0}_{\bar{\ell}}^\top, \underbrace{\mathbf{u}_0[1] \cdot \mathbf{p}(\rho(1))^\top, \ldots, \mathbf{u}_0[\ell_0] \cdot \mathbf{p}(\rho(\ell_0))^\top}_{(\bar{k}+1)\ell_0},$$

$$\mathbf{0}_{(\bar{\ell}-\ell_0)(\bar{k}+1)}^\top, \underbrace{\mathbf{u}_1^\top}_{\ell_1}, \mathbf{0}_{\bar{\ell}-\ell_1}^\top,$$

$$\underbrace{\mathbf{u}_2[1] \cdot \mathbf{p}(\rho(\ell_0 + 1))^\top, \ldots, \mathbf{u}_2[\ell_1] \cdot \mathbf{p}(\rho(\ell_0 + \ell_1))^\top}_{(\bar{k}+1)\ell_1}, \mathbf{0}_{(\bar{\ell}-\ell_1)(\bar{k}+1)}^\top \Big) \quad (4)$$

and

$$\mathbf{v}^\top \mathbf{Y} = (\mathbf{0}_{\bar{m}}^\top, \underbrace{\langle \mathbf{v}_1, \mathbf{q}_S \rangle, \ldots, \langle \mathbf{v}_{\bar{\ell}}, \mathbf{q}_S \rangle}_{\bar{\ell}}, \underbrace{\mathbf{v}_1^\top, \ldots, \mathbf{v}_{\bar{\ell}}^\top}_{(\bar{k}+1)\bar{\ell}},$$

$$\underbrace{\langle \mathbf{w}_1, \mathbf{q}_S \rangle, \ldots, \langle \mathbf{w}_{\bar{\ell}}, \mathbf{q}_S \rangle}_{\bar{\ell}}, \underbrace{\mathbf{w}_1^\top, \ldots, \mathbf{w}_{\bar{\ell}}^\top}_{(\bar{k}+1)\bar{\ell}}). \quad (5)$$

First, by comparing the $\bar{m}+\bar{\ell}+1$-th to $\bar{m}+(\bar{k}+2)\bar{\ell}$-th elements of the vector, we obtain that $\mathbf{v}_i = \mathbf{u}_0[i] \cdot \mathbf{p}(\rho(i))$ for $i \in [1, \ell_0]$ and $\mathbf{v}_i = \mathbf{0}_{\bar{k}+1}$ for $i \in [\ell_0 + 1, \bar{\ell}]$. Furthermore, by comparing $\bar{m}+1$-th to $\bar{m}+\bar{\ell}$-th elements of the vector, we have

$$\langle \mathbf{v}_i, \mathbf{q}_S \rangle = \mathbf{u}_0[i] \cdot \langle \mathbf{p}(\rho(i)), \mathbf{q}_S \rangle = \mathbf{u}_0[i] \cdot Q_S(\rho(i)) = 0$$

for $i \in [1, \ell_0]$. The second equation above follows from the definition of $\mathbf{p}()$ and $\mathbf{q}_S$. Since $Q_S(\rho(i)) = \prod_{\omega \in S}(\rho(i) - \omega) \neq 0$ if $\rho(i) \notin S$, we have that $\mathbf{u}_0[i] = 0$ if $\rho(i) \notin S$. That is, $\mathbf{u}_0[i] = 0$ for $i \in [1, \ell_0] \backslash I$.

Next, by comparing the last $(\bar{k}+1)\bar{\ell}$ elements in the vector, we obtain that $\mathbf{w}_i = \mathbf{u}_2[i] \cdot \mathbf{p}(\rho(\ell_0 + i))$ for $i \in [1, \ell_1]$ and $\mathbf{w}_i = \mathbf{0}_{\bar{k}+1}$ for $i \in [\ell_1 + 1, \bar{\ell}]$. By comparing the $\bar{m} + (\bar{k}+2)\bar{\ell} + 1$-th to $\bar{m} + (\bar{k}+3)\bar{\ell}$-th elements in the vector, we have that $(\mathbf{u}_1^\top, \mathbf{0}_{\bar{\ell}-\ell_1}^\top) = (\langle \mathbf{w}_1, \mathbf{q}_S \rangle, \ldots, , \langle \mathbf{w}_{\bar{\ell}}, \mathbf{q}_S \rangle)$ and thus

$$\mathbf{u}_1[i] = \langle \mathbf{w}_i, \mathbf{q}_S \rangle = \mathbf{u}_2[i] \cdot \langle \mathbf{p}(\rho(\ell_0 + i)), \mathbf{q}_S \rangle = \mathbf{u}_2[i] \cdot Q_S(\rho(\ell_0 + i))$$

holds for $i \in [1, \ell_1]$. From the above, we have that $\mathbf{u}_1[i] = 0$ if $\rho(\ell_0 + i) = \neg \mathsf{Att}$ and $\mathsf{Att} \in S$ for some Att. This implies that $\mathbf{u}_1[i] = 0$ if $(\ell_0 + i) \notin I$ for $i \in [1, \ell_1]$.

Finally, by comparing the first $\bar{m}$ elements in the vector, we obtain that $-\mathbf{e}_1^\top + \mathbf{u}_0^\top \mathbf{L}_0 + \mathbf{u}_1^\top \mathbf{L}_1 = \mathbf{0}^\top$. Let $\mathbf{u}_{0,I}$ be a subvector of $\mathbf{u}_0$ which is obtained by deleting all elements $\mathbf{u}_0[i]$ for $i \notin I$. Similarly, we define $\mathbf{u}_{1,I}$ as a vector obtained by deleting all elements $\mathbf{u}_1[i]$ for i such that $(\ell_0 + i) \notin I$ from $\mathbf{u}_1$. Since $\mathbf{u}_0[i] = 0$ for $i \in [1, \ell_0] \backslash I$ and $\mathbf{u}_1[i] = 0$ for $i \in [1, \ell_1]$ such that $(\ell_0 + i) \notin I$, it follows that $(\mathbf{u}_{0,I}^\top, \mathbf{u}_{1,I}^\top)\mathbf{L}_I = \mathbf{u}_0^\top \mathbf{L}_0 + \mathbf{u}_1^\top \mathbf{L}_1 = \mathbf{e}_1^\top$ and thus $\mathbf{e}_1 \in \mathrm{span}(\mathbf{L}_I^\top)$ as desired.

Converse Direction ($\Leftarrow$). The converse direction can be shown by repeating the above discussion in reverse order. Assume that $\mathbf{e}_1 \in \mathrm{span}(\mathbf{L}_I^\top)$. Then there exists $\mathbf{u}' \in \mathbb{Z}_p^{|I|}$ such that $\mathbf{u}'^\top \mathbf{L}_I = \mathbf{e}_1^\top$. We extend $\mathbf{u}'$ to define $\mathbf{u}'' \in \mathbb{Z}_p^{\ell_0 + \ell_1}$ so that $\mathbf{u}''_I = \mathbf{u}'$ and $\mathbf{u}''[i] = 0$ for $i \notin I$ hold. Here, $\mathbf{u}''_I \in \mathbb{Z}_p^{|I|}$ is a subvector of $\mathbf{u}''$ which is obtained by deleting all elements $\mathbf{u}''[i]$ for $i \notin I$. These conditions completely determine $\mathbf{u}''$. We denote this $\mathbf{u}''$ as $\mathbf{u}''^\top = (\mathbf{u}_0^\top, \mathbf{u}_1^\top)$ using $\mathbf{u}_0 \in \mathbb{Z}_p^{\ell_0}$ and $\mathbf{u}_1 \in \mathbb{Z}_p^{\ell_1}$. We note that $\mathbf{u}_0^\top \mathbf{L}_0 + \mathbf{u}_1^\top \mathbf{L}_1 = \mathbf{e}_1^\top$ holds by the definition.

Next we define v_i for $i \in [\bar{\ell}]$ as $v_i = \mathbf{u}_0[i] \cdot \mathbf{p}(\rho(i))$ if $i \in [\ell_0]$ and $v_i = \mathbf{0}_{\bar{k}+1}$ if $i \in [\ell_0 + 1, \bar{\ell}]$. We claim that $\langle v_i, \mathbf{q}_S \rangle = 0$ holds for $i \in [\bar{\ell}]$. Here, we prove this. The case for $i \in [\ell_0 + 1, \bar{\ell}]$ is trivial. For the case of $i \in [1, \ell_0]$, we have

$$\langle v_i, \mathbf{q}_S \rangle = \mathbf{u}_0[i] \cdot \langle \mathbf{p}(\rho(i)), \mathbf{q}_S \rangle = \mathbf{u}_0[i] \cdot Q_S(\rho(i)) = 0.$$

The last equation above holds because we have $Q_S(\rho(i)) = 0$ if $i \in I$ and $\mathbf{u}_0[i] = 0$ otherwise, by the definition of $\mathbf{u}_0[i]$.

We define $\mathbf{u}_2[i] \in \mathbb{Z}_p$ for $i \in [1, \ell_1]$ as $\mathbf{u}_2[i] = \mathbf{u}_1[i]/Q_S(\rho(\ell_0 + i))$ if $\mathbf{u}_1[i] \neq 0$ and $\mathbf{u}_2[i] = 0$ if $\mathbf{u}_1[i] = 0$. We have to show that $\mathbf{u}_2[i]$ are well defined by showing that $Q_S(\rho(\ell_0 + i)) \neq 0$ if $\mathbf{u}_1[i] \neq 0$ (i.e., division by 0 does not occur). If $\mathbf{u}_1[i] \neq 0$, then $(\ell_0 + i) \in I$ by the definition of $\mathbf{u}_1$. It implies that $(\rho(\ell_0 + i) = \neg \mathsf{Att}) \wedge (\mathsf{Att} \notin S)$ for some $\mathsf{Att} \in \mathbb{Z}_p$ and thus $Q_S(\rho(\ell_0 + i)) = \prod_{\omega \in S}(\mathsf{Att} - \omega) \neq 0$ holds as desired.

We also define $\mathbf{w}_i$ as $\mathbf{w}_i = \mathbf{u}_2[i] \cdot \mathbf{p}(\rho(\ell_0 + i))$ for $i \in [1, \ell_1]$ and $\mathbf{w}_i = \mathbf{0}_{\bar{k}+1}$ for $i \in [\ell_1 + 1, \bar{\ell}]$. Then, we have

$$\langle \mathbf{w}_i, \mathbf{q}_S \rangle = \mathbf{u}_2[i] \cdot \langle \mathbf{p}(\rho(\ell_0 + i)), \mathbf{q}_S \rangle = \mathbf{u}_2[i] \cdot Q_S(\rho(\ell_0 + i)) = \mathbf{u}_1[i]$$

for $i \in [1, \ell_1]$ and $\langle \mathbf{w}_i, \mathbf{q}_S \rangle = 0$ for $i \in [\ell_1 + 1, \bar{\ell}]$.

Finally, we define $\mathbf{u}$ and v as $\mathbf{u}^\top = (\mathbf{u}_0^\top, \mathbf{u}_1^\top, \mathbf{u}_2^\top)$ and $v^\top = (v_1^\top, \ldots, v_{\bar{\ell}}^\top, \mathbf{w}_1^\top, \ldots, \mathbf{w}_{\bar{\ell}}^\top)$. Then, Eqs. (4) and (5) hold. By the properties of $\mathbf{u}$ and v we investigated so far, it is straightforward to see that $x_0^\top + \mathbf{u}^\top \mathbf{X}^\top = y_0^\top + v^\top \mathbf{Y}$ holds. This concludes the proof of the theorem.

5   From KP(CP)-ABE to KASP(CASP)

In this section, we show that monotonic KP-ABE with small universe (without bounds on the size of span programs) can be converted into KASP. We note that we can also obtain CP-ABE-to-CASP conversion by simply swapping key and ciphertext attribute.

5.1 The Conversion

Mapping Parameters. We show how to construct KASP for dimension n from monotonic KP-ABE for parameter $N = (n\kappa + 1, -, -, -)$ and the size of attribute universe is $|\mathcal{U}| = 2n\kappa + 1$. Here, $\kappa = \lceil \log_2 p \rceil$. That is, we define $f_{\mathsf{p}}^{\mathsf{KASP} \to \mathsf{KP}}(n) = N$. We set the universe of attributes as

$$\mathcal{U} = \Big\{ \, \mathsf{Att}[i][j][b] \, \Big| \, (i, j, b) \in [1, n] \times [1, \kappa] \times \{0, 1\} \, \Big\} \cup \{\mathsf{D}\}.$$

Intuitively, $\mathsf{Att}[i][j][b]$ represents an indicator for the condition "the j-th least significant bit of the binary representation of the i-th element of the vector x is $b \in \{0, 1\}$". D is a dummy attribute which will be assigned for all ciphertexts.

Mapping Ciphertext Attributes. For $\mathrm{x} \in \mathbb{Z}_p^n$, we map $f_{\mathsf{e}}^{\mathsf{KASP} \to \mathsf{KP}} : \mathrm{x} \mapsto S$ where

$$S = \Big\{ \, \mathsf{Att}[i][j][b] \, \Big| \, (i, j) \in [1, n] \times [1, \kappa], \ b = \mathrm{x}[i][j] \, \Big\} \cup \{\mathsf{D}\},$$

where we define $\mathrm{x}[i][j] \in \{0, 1\}$ in such a way that $\mathrm{x}[i] = \sum_{j=1}^{\kappa} 2^{j-1} \cdot \mathrm{x}[i][j]$. In other words, $\mathrm{x}[i][j]$ is the j-th least significant bit of the binary representation of $\mathrm{x}[i] \in \mathbb{Z}_p$.

Mapping Key Attributes. For an arithmetic span program $(\mathbf{Y} = (\mathrm{y}_1, \ldots, \mathrm{y}_\ell) \in \mathbb{Z}_p^{m \times \ell}, \mathbf{Z} = (\mathrm{z}_1, \ldots, \mathrm{z}_\ell) \in \mathbb{Z}_p^{m \times \ell}, \rho)$ such that $\mathbf{Y}, \mathbf{Z} \in \mathbb{Z}_p^{m \times \ell}$, we define the map $f_{\mathsf{k}}^{\mathsf{KASP} \to \mathsf{KP}} : (\mathbf{Y}, \mathbf{Z}, \rho) \mapsto (\mathbf{L}, \rho')$ as follows. First, we define

$$\mathbf{L} = \begin{pmatrix} \mathbf{G}_1 & \mathbf{J} & & \\ \mathbf{G}_2 & & \mathbf{J} & \\ \vdots & & & \ddots \\ \mathbf{G}_\ell & & & & \mathbf{J} \end{pmatrix} \in \mathbb{Z}_p^{\big((2\kappa+1)\ell\big) \times \big(\kappa\ell + m\big)}, \tag{6}$$

where the matrix $\mathbf{J} \in \mathbb{Z}_p^{(2\kappa+1) \times \kappa}$ is defined as in Equation (3) (by setting $n = 1$) while $\mathbf{G}_i$ is defined as

$$\mathbf{G}_i = [\mathbf{g} \cdot \mathrm{y}_i^\top ; \mathrm{z}_i^\top] = (\mathbf{0}_m, \mathrm{y}_i, \mathbf{0}_m, 2\mathrm{y}_i, \cdots, \mathbf{0}_m, 2^{\kappa-1}\mathrm{y}_i, \mathrm{z}_i)^\top \in \mathbb{Z}_p^{(2\kappa+1) \times m}$$

where $\mathbf{g} = (0, 1, 0, 2, \ldots, 0, 2^i, \ldots, 0, 2^{\kappa-1})^\top \in \mathbb{Z}_p^{2\kappa}$.

Next, we define the map $\rho' : [(2\kappa + 1)\ell] \to \mathcal{U}$ as follows.

- If $i = 0 \mod (2\kappa + 1)$, we set $\rho(i) := \mathsf{D}$.
- Else, we write

$$i = (2\kappa + 1)i' + 2j' + b' + 1$$

with unique $i' \in [0, \ell - 1]$, $j' \in [0, \kappa - 1]$, and $b' \in \{0, 1\}$. We finally set $\rho'(i) = \mathsf{Att}[\rho(i' + 1)][j' + 1][b']$.

Intuition. S can be seen as a binary representation of the information of x. In the span program $(\mathbf{L}, \rho')$, $\mathbf{J}$ is used to constrain the form of linear combination among rows to a certain form. $\mathbf{G}_i$ as well as ρ', along with the above restriction, are designed so that linear combination of rows of $\mathbf{G}_i$ only can be a scalar multiple of the vector $(\mathbf{x}[\rho(i)]\mathbf{y}_i + \mathbf{z}_i)^\top$. Therefore, $(\mathbf{L}, \rho')$ essentially works as an arithmetic span program.

5.2   Correctness of the Conversion

We show the following theorem. The implication from KP-ABE with parameter $N = (n\kappa + 1, -, -, -)$ to KASP with dimension n would then follow from the embedding lemma.

Theorem 3. *For any* $\mathbf{x} \in \mathbb{Z}_p^n$, $\mathbf{Y} \in \mathbb{Z}_p^{m \times \ell}$, $\mathbf{Z} \in \mathbb{Z}_p^{m \times \ell}$, *and* $\rho : [\ell] \to [n]$, *it holds that*

$$R_N^{\mathsf{KP}}(S, (\mathbf{L}, \rho')) = 1 \Leftrightarrow R_n^{\mathsf{KASP}}(\mathbf{x}, (\mathbf{Y}, \mathbf{Z}, \rho)) = 1$$

where $N = f_{\mathsf{p}}^{\mathsf{KASP} \to \mathsf{KP}}(n)$, $S = f_{\mathsf{e}}^{\mathsf{KASP} \to \mathsf{KP}}(\mathbf{x})$, *and* $(\mathbf{L}, \rho') = f_{\mathsf{k}}^{\mathsf{KASP} \to \mathsf{KP}}(\mathbf{Y}, \mathbf{Z}, \rho)$.

Proof. Define $I \subset [1, (2\kappa + 1)\ell]$ as $I = \{i | \rho'(i) \in S\}$. We define $\mathbf{L}_I$ as the submatrix of $\mathbf{L}$ formed by rows whose index is in I. From the definition of $f_{\mathsf{e}}^{\mathsf{KASP} \to \mathsf{KP}}$, we have that $\mathbf{L}_I$ is in the form of

$$\mathbf{L}_I = \begin{pmatrix} \mathbf{G}'_1 & \mathbf{J}' & & \\ \mathbf{G}'_2 & & \mathbf{J}' & \\ \vdots & & & \ddots \\ \mathbf{G}'_\ell & & & & \mathbf{J}' \end{pmatrix} \in \mathbb{Z}_p^{((\kappa+1)\ell) \times (\kappa\ell + m)},$$

where

$$\mathbf{G}'_i = [\mathbf{g}_i \cdot \mathbf{y}_i^\top ; \mathbf{z}_i^\top] \in \mathbb{Z}_p^{(\kappa+1) \times m}, \qquad \mathbf{J}' = \begin{pmatrix} -1 & & & \\ & -1 & & \\ & & \ddots & \\ & & & -1 \\ 1 & 1 & \cdots & 1 \end{pmatrix} \in \mathbb{Z}_p^{(\kappa+1) \times \kappa},$$

and where $\mathbf{g}_i = (\mathbf{x}[\rho(i)][1], 2\mathbf{x}[\rho(i)][2], \ldots, 2^{\kappa-1}\mathbf{x}[\rho(i)][\kappa])^\top \in \mathbb{Z}_p^\kappa$. We note that we have $\langle \mathbf{1}_\kappa, \mathbf{g}_i \rangle = \mathbf{x}[\rho(i)]$ by the definition of $\mathbf{x}[\rho(i)][j]$ and thus $\mathbf{G}'^\top_i \cdot \mathbf{1}_{\kappa+1} = \mathbf{x}[\rho(i)]\mathbf{y}_i + \mathbf{z}_i$ holds. We also remark that if $\mathbf{v}^\top \mathbf{J}' = \mathbf{0}$ holds for some $\mathbf{v} \in \mathbb{Z}_p^{\kappa+1}$, then there exists $v \in \mathbb{Z}_p$ such that $\mathbf{v} = v\mathbf{1}_{\kappa+1}$. These properties will be used later below.

To prove the theorem statement is equivalent to prove that

$$\mathbf{e}_1 \in \mathrm{span}(\mathbf{L}_I^\top) \quad \Leftrightarrow \quad \mathbf{e}_1 \in \mathrm{span}(\{\mathbf{x}[\rho(i)]\mathbf{y}_i + \mathbf{z}_i\}_{i \in [\ell]}).$$

Forward Direction ($\Rightarrow$). We assume that $\mathrm{e}_1 \in \mathrm{span}(\mathbf{L}_I^\top)$. From this, there exists $\mathbf{u} \in \mathbb{Z}_p^{(\kappa+1)\ell}$ such that $\mathbf{u}^\top \mathbf{L}_I = \mathrm{e}_1^\top$. We write this $\mathbf{u}$ as

$$\mathbf{u}^\top = (\underbrace{\mathbf{u}_1^\top}_{\kappa+1}, \underbrace{\mathbf{u}_2^\top}_{\kappa+1}, \dots, \underbrace{\mathbf{u}_\ell^\top}_{\kappa+1}).$$

Therefore, we have that

$$\mathrm{e}_1^\top = \mathbf{u}^\top \cdot \mathbf{L}_I = \left(\sum_{i \in [\ell]} \mathbf{u}_i^\top \mathbf{G}_i', \mathbf{u}_1^\top \mathbf{J}', \dots, \mathbf{u}_\ell^\top \mathbf{J}'\right).$$

Since $\mathbf{u}_i^\top \cdot \mathbf{J}' = \mathbf{0}$ for $i \in [\ell]$, there exist $\{u_i \in \mathbb{Z}_p\}_{i \in [\ell]}$ such that $\mathbf{u}_i = u_i \mathbf{1}_{\kappa+1}$. Then, we have

$$\mathrm{e}_1^\top = \sum_{i \in [\ell]} \mathbf{u}_i^\top \mathbf{G}_i' = \sum_{i \in [\ell]} u_i \mathbf{1}_{\kappa+1}^\top \mathbf{G}_i' = \sum_{i \in [\ell]} u_i (\mathrm{x}[\rho(i)] \cdot \mathrm{y}_i + \mathbf{z}_i)^\top.$$

This implies $\mathrm{e}_1 \in \mathrm{span}(\{\mathrm{x}[\rho(i)]\mathrm{y}_i + \mathbf{z}_i\}_{i \in [\ell]})$, as desired.

Converse Direction ($\Leftarrow$). We assume that $\mathrm{e}_1 \in \mathrm{span}(\{\mathrm{x}[\rho(i)]\mathrm{y}_i + \mathbf{z}_i\}_{i \in [\ell]})$. Then, there exist $\{u_i \in \mathbb{Z}_p\}_{i \in [\ell]}$ such that $\sum_{i \in [\ell]} u_i(\mathrm{x}[\rho(i)] \cdot \mathrm{y}_i + \mathbf{z}_i) = \mathrm{e}_1$. We set a vector $\mathbf{u} \in \mathbb{Z}_p^{(\kappa+1)\ell}$ as $\mathbf{u}^\top = (u_1 \mathbf{1}_{\kappa+1}^\top, \dots, u_\ell \mathbf{1}_{\kappa+1}^\top)$. Then, we have that

$$\mathbf{u}^\top \cdot \mathbf{L}_I = \left(\sum_{i \in [\ell]} u_i \mathbf{1}_{\kappa+1}^\top \mathbf{G}_i', u_1 \mathbf{1}_{\kappa+1}^\top \mathbf{J}', \dots, u_\ell \mathbf{1}_{\kappa+1}^\top \mathbf{J}'\right)$$

$$= \left(\sum_{i \in [\ell]} u_i (\mathrm{x}[\rho(i)]\mathrm{y}_i + \mathbf{z}_i)^\top, \mathbf{0}_\kappa^\top, \dots, \mathbf{0}_\kappa^\top\right) = \mathrm{e}_1^\top.$$

This implies $\mathrm{e}_1 \in \mathrm{span}(\mathbf{L}_I^\top)$, as desired. This concludes the proof of the theorem.

6   Implications of Our Result

In this section, we discuss consequences of our results.

Equivalence between (bounded) ABE and DSE. We have shown that monotonic KP/CP-ABE for $(\bar{k}, \bar{\ell}, \bar{m}, \varphi)$ implies DSE (without delegation) in Sect. 3 and DSE implies non-monotonic KP/CP-ABE with large universe for $(\bar{k}, \bar{\ell}, \bar{m}, \varphi)$ in Sect. 4. Since non-monotonic KP/CP-ABE with large universe for $(\bar{k}, \bar{\ell}, \bar{m}, \varphi)$ trivially implies monotonic KP/CP-ABE with small universe for $(\bar{k}, \bar{\ell}, \bar{m}, \varphi)$, our results indicate that these PE schemes are essentially equivalent in the sense that they imply each other.

Equivalence between K(C)ASP and KP(CP)-ABE. Next, we consider the case where there is no restriction on the size of span programs. In Sect. 5,

Table 1. Comparison among DSE Schemes

Schemes	$|\mathsf{mpk}|$	$|C|$	$|\mathsf{sk}|$	Delegation	Security	Assumption
Hamburg11 [25]	$O(n)$	$O(d_1)$	$O(d_2)$	✓	Selective	Parameterized
CW14 [17]	$O(n^2)$	$O(nd_1)$	$O(n)$	✓	Selective	Static
CZF12 [14]	$O(n)$	$O(d_1)$	$O(d_2)$	✓	Adaptive	Static
Section 3 + RW13 [37]	$O(1)$	$O(nd_1\kappa)$	$O(n^2\kappa)$	✓	Selective	Parameterized
Section 3 + ALP11 [5]	$O(n^2\kappa)$	$O(1)$	$O(n^4\kappa^2)$	✓	Selective	Parameterized
Section 3 + OT12 [35]	$O(1)$	$O(n^2 d_1\kappa)$	$O(n^2\kappa)$	?	Adaptive	Static
Section 3 + A15 [3]	$O(1)$	$O(nd_1\kappa)$	$O(n^2\kappa)$	?	Adaptive	Parameterized
Section 3 + A15 [3]	$O(n^2\kappa)$	$O(1)$	$O(n^4\kappa^2)$	?	Adaptive	Parameterized
Section 3 + A15 [3]	$O(n^2\kappa)$	$O(n^4\kappa^2)$	$O(1)$	?	Adaptive	Parameterized

[a] n is the dimension of the scheme; d_1 and d_2 denote the dimension of the space associated with the ciphertext and private key, respectively; $\kappa = \lceil \log_2 p \rceil$.

[b] "Delegation" shows if key delegation is supported. "?" means unknown.

we showed that monotonic KP-ABE for $((\bar{k}+1)\kappa, -, -, -)$ implies KASP for $(\bar{k}, -, -, -)$. In the full version [4], we also show the converse direction. That is, we show that KASP for $(\bar{k}+1, -, -, -)$ implies non-monotonic KP-ABE for $(\bar{k}, -, -, -)$ with large universe. Since non-monotonic KP-ABE for $(\bar{k}, -, -, -)$ trivially implies monotonic KP-ABE for $(\bar{k}, -, -, -)$, our results indicate that these PE schemes are essentially equivalent similarly to the above case. Similar implications hold for CP-ABE. See Fig. 1 for the overview.

By applying the conversions to existing schemes, we obtain various new schemes. The overviews of properties of resulting schemes and comparison with existing schemes are provided in Tables 1, 2, 3 and 4. All schemes in the tables are constructed in pairing groups. In the tables, we count the number of group elements to measure the size of master public keys ($|\mathsf{mpk}|$), ciphertexts ($|C|$), and private keys ($|\mathsf{sk}|$). Note that our conversions only can be applied to ABE schemes supporting span programs over $\mathbb{Z}_p$. Therefore, for ABE schemes constructed on composite order groups [2,30], our conversions are not applicable since they support span programs over $\mathbb{Z}_N$ where N is a product of several large primes. Similar restrictions are posed on DSE and K(C)ASP. Though it is quite plausible that our conversions work even in such cases assuming hardness of factoring N, we do not prove this in this paper.

New DSE Schemes. By applying our KP(CP)-ABE-to-DSE conversion to existing KP(CP)-ABE schemes, we obtain many new DSE schemes. Table 1 shows overview of obtained schemes.[6] Specifically,

- From the unbounded KP-ABE schemes [3,35,37], we obtain the first DSE scheme with constant-size master public key (without delegation). Note that all previous schemes [14,17,25] require at least $O(n)$ group elements in master public key where n is the dimension of the scheme.

---

[6] In the table, parameterized assumptions refer to q-type assumptions, which are non-interactive and falsifiable but parameterized by some parameters of the scheme such as $k, \bar{k}$.

Table 2. Comparison among CP-ABE Schemes

Schemes	Expressiveness		Efficiency			Security	Assumption
	Universe	Policy	$|\mathsf{mpk}|$	$|C|$	$|\mathsf{sk}|$		
OT12 [35]	Large	Non-mono. Span	$O(1)$	$O(\ell)$	$O(k\varphi)$	Adaptive	Static
AY15 [6], A15 [3]	Large	Mono. Span	$O(1)$	$O(\ell)$	$O(k)$	Adaptive	Parametrized
AY15 [6], A15 [3]	Large	Mono. Span	$O(\bar{k})$	$O(\bar{k}\ell)$	$O(1)$	Adaptive	Parametrized
EMN+09 [18]	Small	AND-only	$O(\bar{k})$	$O(1)$	$O(\bar{k})$	Selective	Static
CZF11 [13]	Small	AND-only	$O(\bar{k})$	$O(1)$	$O(\bar{k}^2)$	Selective	Static
CCL+13 [12]	Small	Threshold	$O(\bar{k})$	$O(1)$	$O(\bar{k}^2)$	Adaptive	Static
Sections 3,4 + ALP11 [5]	Large	Non-mono. Span	$O((\bar{k}\bar{\ell})^2\kappa)$	$O(1)$	$O((\bar{k}\bar{\ell})^4\kappa^2)$	Selective	Parametrized
Sections 3,4 + T14 [39]	Large	Non-mono. Span.	$O((\bar{k}\bar{\ell})^2\kappa)$	$O(1)$	$O((\bar{k}\bar{\ell})^4\kappa^2)$	Semi-adapt	Static
Section 3,4 + A15 [3]	Large	Non-mono. Span	$O((\bar{k}\bar{\ell})^2\kappa)$	$O(1)$	$O((\bar{k}\bar{\ell})^4\kappa^2)$	Adaptive	Parametrized

[a] k is the size of an attribute set associated with a key, ℓ is the number of rows of a span program matrix associated with a ciphertext; $\bar{k}, \bar{\ell}$ are the maximums of k, ℓ (if bounded); φ is the maximum number of allowed attribute multi-use in one policy (if bounded); $\kappa = \lceil \log_2 p \rceil$.

- From KP-ABE scheme with constant-size ciphertexts [3,5,27,39], we obtain the first DSE scheme with constant-size ciphertexts. All previous schemes [14,17,25] require at least $O(d_1)$ group elements in ciphertexts where d_1 is the dimension of the affine space associated to a ciphertext.
- From CP-ABE scheme with constant-size keys [6], we obtain the first DSE scheme with constant-size private keys. All previous schemes require at least $O(d_2)$ group elements in private keys where d_2 is the dimension of the affine space associated to a private key.

The schemes obtained from [3,35] achieves adaptive security. Furthermore, for schemes obtained from [5,27,37], we can define key delegation algorithm. The details of the key delegation algorithm will be given in the full version [4].

CP-ABE with Constant-Size Ciphertexts. By applying our DSE-to-non-monotonic-CP-ABE conversion in Sect. 4 to the DSE scheme with constant-size ciphertexts obtained above, we obtain the first non-monotonic CP-ABE with constant-size ciphertexts. Previous CP-ABE schemes with constant-size ciphertexts [12,13,18] only support threshold or more limited predicates[7]. See Table 2 for comparison (we list only relevant schemes).

KP-ABE with Constant-Size Keys. By applying our DSE-to-non-monotonic-KP-ABE conversion in Sect. 4 to the DSE scheme with constant-size keys obtained above, we obtain the first non-monotonic KP-ABE with constant-size keys. See Table 3 for comparison (we list only relevant schemes).

---

[7] One would be able to obtain CP-ABE with constant-size ciphertexts supporting threshold formulae by applying the generic conversion in [22] to a KP-ABE scheme proposed in [5]. However, the resulting scheme supports more limited predicate compared to ours. To the best of our knowledge, this observation has not appeared elsewhere.

Table 3. Comparison among KP-ABE Schemes

Schemes	Expressiveness		Efficiency			Security	Assumption
	Universe	Policy	$|\mathsf{mpk}|$	$|C|$	$|\mathsf{sk}|$		
OT12 [35]	Large	Non-mono. Span.	$O(1)$	$O(k\varphi)$	$O(\ell)$	Adaptive	Static
AY15 [6], A15 [3]	Large	Mono. Span	$O(1)$	$O(k)$	$O(\ell)$	Adaptive	Parameterized
AY15 [6], A15 [3]	Large	Mono. Span	$O(\bar{k})$	$O(1)$	$O(\bar{k}\ell)$	Adaptive	Parameterized
Sections 3,4 + A15 [3]	Large	Non-mono. Span.	$O((\bar{k}\bar{\ell})^2\kappa)$	$O((\bar{k}\bar{\ell})^4\kappa^2)$	$O(1)$	Adaptive	Parameterized

[a] k is the size of an attribute set associated with a ciphertext, ℓ is the number of rows of a span program matrix associated with a key; $\bar{k}, \bar{\ell}$ are the maximums of k, ℓ (if bounded); φ is the maximum number of allowed attribute multi-use in one policy (if bounded); $\kappa = \lceil \log_2 p \rceil$.

New KASP and CASP Schemes. By applying the KP(CP)-ABE-to-K(C)ASP conversion in Sect. 5, we obtain many new K(C)ASP schemes. See Table 4 for the overview. Specifically,

- From the unbounded KP-ABE, CP-ABE schemes of [3,37], we obtain the first KASP, CASP schemes with constant-size master public key.
- From adaptively secure KP-ABE, CP-ABE schemes of [3,32], we obtain the first adaptively secure KASP, CASP schemes with unbounded attribute multi-use.
- From KP-ABE schemes with constant-size ciphertexts [3,5,27,39], we obtain the first KASP schemes with constant-size ciphertexts.
- From CP-ABE schemes with constant-size keys [3], we obtain the first CASP schemes with constant-size keys.

Until recently, the only (K)ASP scheme in the literature was proposed by [28], which is selectively secure and the master public key and ciphertext size are linear in the dimension of the scheme. Very recently, adaptively secure KASP and CASP were given in [16], albeit with the restriction of one-time use (of the same attribute in one policy).

We remark that the conversion is not applicable for schemes in [34,35] since these schemes are KP-ABE for $(*,*,*,\varphi)$ where φ is polynomially bounded, whereas our conversion requires the last parameter to be unbounded.

7   Application to Attribute-Based Signature

Here, we discuss that our techniques developed in previous sections are also applicable to construct attribute-based signatures (ABS) [33,34]. ABS is an advanced form of signature and can be considered as a signature analogue of ABE. In particular, it resembles CP-ABE in the sense that a private key is associated with a set of attributes while a signature is associated with a policy and a message. A user can sign on a message with a policy if and only if she has a private key associated with a set satisfying the policy. Roughly speaking, this property corresponds to the correctness and unforgeability. For ABS, we

Table 4. Comparison among KASP and CASP Schemes

Schemes	Type	Efficiency			Security	Attribute multi-use	Assumption
		$|mpk|$	$|C|$	$|sk|$			
IW14 [28]	KASP	$O(n)$	$O(n)$	$O(\ell)$	Selective	yes	Static
CGW15 [16]	KASP	$O(n)$	$O(n)$	$O(\ell)$	Adaptive	no	Static
CGW15 [16]	CASP	$O(n)$	$O(\ell)$	$O(n)$	Adaptive	no	Static
Section 5 + LW12 [32]	KASP	$O(n\kappa)$	$O(n\kappa)$	$O(\ell\kappa)$	Adaptive	yes	Parameterized
Section 5 + ALP11 [5]	KASP	$O(n\kappa)$	$O(1)$	$O(\ell n\kappa^2)$	Selective	yes	Parameterized
Section 5 + RW13 [37]	KASP	$O(1)$	$O(n\kappa)$	$O(\ell\kappa)$	Selective	yes	Parameterized
Section 5 + A15 [3]	KASP	$O(n\kappa)$	$O(1)$	$O(\ell n\kappa^2)$	Adaptive	yes	Parameterized
Section 5 + A15 [3]	KASP	$O(1)$	$O(n\kappa)$	$O(\ell\kappa)$	Adaptive	yes	Parameterized
Section 5 + LW12 [32]	CASP	$O(n\kappa)$	$O(\ell\kappa)$	$O(n\kappa)$	Adaptive	yes	Parameterized
Section 5 + RW13 [37]	CASP	$O(1)$	$O(\ell\kappa)$	$O(n\kappa)$	Selective	yes	Parameterized
Section 5 + A15 [3]	CASP	$O(1)$	$O(\ell\kappa)$	$O(n\kappa)$	Adaptive	yes	Parameterized
Section 5 + A15 [3]	CASP	$O(n\kappa)$	$O(\ell n\kappa^2)$	$O(1)$	Adaptive	yes	Parameterized

[a] n is the dimension of the scheme; ℓ is the number of the columns of the matrices that define an arithmetic span program (ℓ reflects the size of an arithmetic span program); $\kappa = \lceil \log_2 p \rceil$.

also require privacy. That is, we require that one cannot obtain any information about the attribute of the signer from a signature.

The construction of expressive ABS scheme with constant-size signatures has been open. All previous ABS schemes with constant-size signatures [12,26] only supports threshold predicates. The difficulty of constructing ABS with constant-size signatures seems to be related to the difficulty of construction of CP-ABE with constant-size ciphertexts. That is, it is hard to set constant number of group elements so that they include very complex information such as span programs.

To solve the problem, we first define the notion of predicate signature (PS) that is a signature analogue of PE. Then we construct a PS scheme that is dual of ABS: a private key is associated with a policy and a signature with a set. The scheme achieves constant-size signatures. This is not difficult to achieve because the signature is associated with a set which is a simpler object compared to a policy. The scheme is based on PS scheme for threshold predicate with constant-size signatures by [26]. We change the scheme mainly in two ways. At first, instead of using Shamir's secret sharing scheme, we use linear secret sharing scheme so that they support more general predicate. We also add some modification so that the signature size be even shorter. The signatures of the resulting scheme only consist of two group elements.

Since signature analogue of Lemma 1 holds, we can apply KP-ABE-to-non-monotonic-CP-ABE conversion (combination of the results in Sects. 3 and 4) to obtain the first ABS scheme with constant-size signatures supporting non-monotone span programs. We refer to the full version [4] for the details.

Acknowledgements. We would like to thank anonymous reviewers and members of Shin-Akarui-Angou-Benkyou-Kai for their helpful comments.

References

1. Agrawal, S., Chase, M.: A study of pair encodings: predicate encryption in prime order groups. IACR Cryptology ePrint Archive 2015, p. 413 (2015)
2. Attrapadung, N.: Dual system encryption via doubly selective security: framework, fully secure functional encryption for regular languages, and more. In: Nguyen, P.Q., Oswald, E. (eds.) EUROCRYPT 2014. LNCS, vol. 8441, pp. 557–577. Springer, Heidelberg (2014)
3. Attrapadung, N.: Dual system encryption framework in prime-order groups. IACR Cryptology ePrint Archive 2015, p. 390 (2015)
4. Attrapadung, N., Hanaoka, G., Yamada, S.: Conversions among Several Classes ofPredicate Encryption and Applications to ABE with Various Compactness Tradeoffs.IACR Cryptology ePrint Archive 2015:431 (2015)
5. Attrapadung, N., Libert, B., de Panafieu, E.: Expressive key-policy attribute-based encryption with constant-size ciphertexts. In: Catalano, D., Fazio, N., Gennaro, R., Nicolosi, A. (eds.) PKC 2011. LNCS, vol. 6571, pp. 90–108. Springer, Heidelberg (2011)
6. Attrapadung, N., Yamada, S.: Duality in ABE: converting attribute based encryption for dual predicate and dual policy via computational encodings. In: Nyberg, K. (ed.) CT-RSA 2015. LNCS, vol. 9048, pp. 87–105. Springer, Heidelberg (2015)
7. Belovs, A.: Span-program-based quantum algorithm for the rank problem. Technical Report arXiv:1103.0842, arXiv.org, 2011. Available from arXiv:1103.0842
8. Beimel, A., Ishai, Y.: On the power of nonlinear secret-sharing. In: IEEE Conference on Computational Complexity, pp. 188–202 (2001)
9. Boneh, D., Gentry, C., Gorbunov, S., Halevi, S., Nikolaenko, V., Segev, G., Vaikuntanathan, V., Vinayagamurthy, D.: Fully key-homomorphic encryption, arithmetic circuit ABE and compact garbled circuits. In: Nguyen, P.Q., Oswald, E. (eds.) EUROCRYPT 2014. LNCS, vol. 8441, pp. 533–556. Springer, Heidelberg (2014)
10. Boneh, D., Hamburg, M.: Generalized identity based and broadcast encryption schemes. In: Pieprzyk, J. (ed.) ASIACRYPT 2008. LNCS, vol. 5350, pp. 455–470. Springer, Heidelberg (2008)
11. Boneh, D., Sahai, A., Waters, B.: Functional encryption: definitions and challenges. In: Ishai, Y. (ed.) TCC 2011. LNCS, vol. 6597, pp. 253–273. Springer, Heidelberg (2011)
12. Chen, C., Chen, J., Lim, H.W., Zhang, Z., Feng, D., Ling, S., Wang, H.: Fully secure attribute-based systems with short ciphertexts/signatures and threshold access structures. In: Dawson, E. (ed.) CT-RSA 2013. LNCS, vol. 7779, pp. 50–67. Springer, Heidelberg (2013)
13. Chen, C., Zhang, Z., Feng, D.: Efficient ciphertext policy attribute-based encryption with constant-size ciphertext and constant computation-cost. In: Boyen, X., Chen, X. (eds.) ProvSec 2011. LNCS, vol. 6980, pp. 84–101. Springer, Heidelberg (2011)
14. Chen, C., Zhang, Z., Feng, D.: Fully secure doubly-spatial encryption under simple assumptions. In: Takagi, T., Wang, G., Qin, Z., Jiang, S., Yu, Y. (eds.) ProvSec 2012. LNCS, vol. 7496, pp. 253–263. Springer, Heidelberg (2012)
15. Chen, J., Lim, H., Ling, S., Wang, H.: The relation and transformation between hierarchical inner product encryption and spatial encryption. Des. Codes Crypt. 71(2), 347–364 (2014)
16. Chen, J., Gay, R., Wee, H.: Improved dual system ABE in prime-order groups via predicate encodings. In: Oswald, E., Fischlin, M. (eds.) EUROCRYPT 2015. LNCS, vol. 9057, pp. 595–624. Springer, Heidelberg (2015)

17. Chen, J., Wee, H.: Doubly spatial encryption from DBDH. Theor. Comput. Sci. **543**, 79–89 (2014)

18. Emura, K., Miyaji, A., Nomura, A., Omote, K., Soshi, M.: A ciphertext-policy attribute-based encryption scheme with constant ciphertext length. In: Bao, F., Li, H., Wang, G. (eds.) ISPEC 2009. LNCS, vol. 5451, pp. 13–23. Springer, Heidelberg (2009)

19. Garg, S., Gentry, C., Halevi, S., Sahai, A., Waters, B.: Attribute-based encryption for circuits from multilinear maps. In: Canetti, R., Garay, J.A. (eds.) CRYPTO 2013, Part II. LNCS, vol. 8043, pp. 479–499. Springer, Heidelberg (2013)

20. Ge, A., Zhang, R., Chen, C., Ma, C., Zhang, Z.: Threshold ciphertext policy attribute-based encryption with constant size ciphertexts. In: Susilo, W., Mu, Y., Seberry, J. (eds.) ACISP 2012. LNCS, vol. 7372, pp. 336–349. Springer, Heidelberg (2012)

21. Gorbunov, S., Vaikuntanathan, V., Wee, H.: Attribute-based encryption for circuits. In: STOC, pp. 545–554 (2013)

22. Goyal, V., Jain, A., Pandey, O., Sahai, A.: Bounded ciphertext policy attribute based encryption. In: Aceto, L., Damgård, I., Goldberg, L.A., Halldórsson, M.M., Ingólfsdóttir, A., Walukiewicz, I. (eds.) ICALP 2008, Part II. LNCS, vol. 5126, pp. 579–591. Springer, Heidelberg (2008)

23. Goyal, V., Kumar, V., Lokam, S., Mahmoody, M.: On black-box reductions between predicate encryption schemes. In: Cramer, R. (ed.) TCC 2012. LNCS, vol. 7194, pp. 440–457. Springer, Heidelberg (2012)

24. Goyal, V., Pandey, O., Sahai, A., Waters B.: Attribute-based encryption for fine-grained access control of encrypted data. In: ACM Conference on Computer and Communications Security, pp. 89–98 (2006)

25. Hamburg, M.: Spatial encryption. IACR Cryptology ePrint Archive 2011, p. 389 (2011)

26. Herranz, J., Laguillaumie, F., Libert, B., Ràfols, C.: Short attribute-based signatures for threshold predicates. In: Dunkelman, O. (ed.) CT-RSA 2012. LNCS, vol. 7178, pp. 51–67. Springer, Heidelberg (2012)

27. Hohenberger, S., Waters, B.: Attribute-based encryption with fast decryption. In: Kurosawa, K., Hanaoka, G. (eds.) PKC 2013. LNCS, vol. 7778, pp. 162–179. Springer, Heidelberg (2013)

28. Ishai, Y., Wee, H.: Partial garbling and their applications. ICALP **1**, 650–662 (2014)

29. Katz, J., Sahai, A., Waters, B.: Predicate encryption supporting disjunctions, polynomial equations, and inner products. In: Smart, N.P. (ed.) EUROCRYPT 2008. LNCS, vol. 4965, pp. 146–162. Springer, Heidelberg (2008)

30. Lewko, A., Okamoto, T., Sahai, A., Takashima, K., Waters, B.: Fully secure functional encryption: attribute-based encryption and (hierarchical) inner product encryption. In: Gilbert, H. (ed.) EUROCRYPT 2010. LNCS, vol. 6110, pp. 62–91. Springer, Heidelberg (2010)

31. Lewko, A., Waters, B.: Unbounded HIBE and attribute-based encryption. In: Paterson, K.G. (ed.) EUROCRYPT 2011. LNCS, vol. 6632, pp. 547–567. Springer, Heidelberg (2011)

32. Lewko, A., Waters, B.: New proof methods for attribute-based encryption: achieving full security through selective techniques. In: Safavi-Naini, R., Canetti, R. (eds.) CRYPTO 2012. LNCS, vol. 7417, pp. 180–198. Springer, Heidelberg (2012)

33. Maji, H.K., Prabhakaran, M., Rosulek, M.: Attribute-based signatures. In: Kiayias, A. (ed.) CT-RSA 2011. LNCS, vol. 6558, pp. 376–392. Springer, Heidelberg (2011)

34. Okamoto, T., Takashima, K.: Efficient attribute-based signatures for non-monotone predicates in the standard model. In: PKC, pp. 35–52 (2011)

35. Okamoto, T., Takashima, K.: Fully secure unbounded inner-product and attribute-based encryption. In: Wang, X., Sako, K. (eds.) ASIACRYPT 2012. LNCS, vol. 7658, pp. 349–366. Springer, Heidelberg (2012)

36. Parno, B., Raykova, M., Vaikuntanathan, V.: How to delegate and verify in public: verifiable computation from attribute-based encryption. In: Cramer, R. (ed.) TCC 2012. LNCS, vol. 7194, pp. 422–439. Springer, Heidelberg (2012)

37. Rouselakis, Y., Waters, B.: Practical constructions and new proof methods for large universe attribute-based encryption. In: ACM Conference on Computer and Communications Security, pp. 463–474 (2013)

38. Sahai, A., Waters, B.: Fuzzy identity-based encryption. In: Cramer, R. (ed.) EUROCRYPT 2005. LNCS, vol. 3494, pp. 457–473. Springer, Heidelberg (2005)

39. Takashima, K.: Expressive attribute-based encryption with constant-size ciphertexts from the decisional linear assumption. In: Abdalla, M., De Prisco, R. (eds.) SCN 2014. LNCS, vol. 8642, pp. 298–317. Springer, Heidelberg (2014)

40. Waters, B.: Functional encryption for regular languages. In: Safavi-Naini, R., Canetti, R. (eds.) CRYPTO 2012. LNCS, vol. 7417, pp. 218–235. Springer, Heidelberg (2012)

Zero-Knowledge

QA-NIZK Arguments in Asymmetric Groups: New Tools and New Constructions

Alonso González[1]([⊠]), Alejandro Hevia[1], and Carla Ràfols[2]

[1] Departamento de Ciencias de la Computación,
Universidad de Chile, Santiago, Chile
alonso.gon@gmail.com
[2] Faculty of Mathematics, Horst-Görtz Institute for IT Security,
Ruhr-Universität Bochum, Bochum, Germany

Abstract. A sequence of recent works have constructed constant-size quasi-adaptive (QA) NIZK arguments of membership in linear subspaces of $\hat{\mathbb{G}}^m$, where $\hat{\mathbb{G}}$ is a group equipped with a bilinear map $e : \hat{\mathbb{G}} \times \check{\mathbb{H}} \to \mathbb{T}$. Although applicable to any bilinear group, these techniques are less useful in the asymmetric case. For example, Jutla and Roy (Crypto 2014) show how to do QA aggregation of Groth-Sahai proofs, but the types of equations which can be aggregated are more restricted in the asymmetric setting. Furthermore, there are natural statements which cannot be expressed as membership in linear subspaces, for example the satisfiability of quadratic equations.

In this paper we develop specific techniques for asymmetric groups. We introduce a new computational assumption, under which we can recover all the aggregation results of Groth-Sahai proofs known in the symmetric setting. We adapt the arguments of membership in linear spaces of $\hat{\mathbb{G}}^m$ to linear subspaces of $\hat{\mathbb{G}}^m \times \check{\mathbb{H}}^n$. In particular, we give a constant-size argument that two sets of Groth-Sahai commitments, defined over different groups $\hat{\mathbb{G}}, \check{\mathbb{H}}$, open to the same scalars in $\mathbb{Z}_q$, a useful tool to prove satisfiability of quadratic equations in $\mathbb{Z}_q$. We then use one of the arguments for subspaces in $\hat{\mathbb{G}}^m \times \check{\mathbb{H}}^n$ and develop new techniques to give constant-size QA-NIZK proofs that a commitment opens to a bit-string. To the best of our knowledge, these are the first constant-size proofs for quadratic equations in $\mathbb{Z}_q$ under standard and falsifiable assumptions. As a result, we obtain improved threshold Groth-Sahai proofs for pairing product equations, ring signatures, proofs of membership in a list, and various types of signature schemes.

1 Introduction

Ideally, a NIZK proof system should be both expressive and efficient, meaning that it should allow to prove statements which are general enough to be useful in

A. González—Funded by CONICYT, CONICYT-PCHA/Doctorado Nacional/2013-21130937.

C. Ràfols—Part of this work was done while visiting Centro de Modelamiento Matemático, U. Chile. Gratefully acknowledges the support of CONICYT via Basal in Applied Mathematics.

T. Iwata and J.H. Cheon (Eds.): ASIACRYPT 2015, Part I, LNCS 9452, pp. 605–629, 2015.
DOI: 10.1007/978-3-662-48797-6_25

practice using a small amount of resources. Furthermore, it should be constructed under mild security assumptions. As it is usually the case for most cryptographic primitives, there is a trade off between these three design goals. For instance, there exist constant-size proofs for any language in NP (e.g. [12]) but based on very strong and controversial assumptions, namely knowledge-of-exponent type of assumptions (which are non-falsifiable, according to Naor's classification [25]) or the random oracle model.

The Groth-Sahai proof system (GS proofs) [16] is an outstanding example of how these three goals (expressivity, efficiency, and mild assumptions) can be combined successfully. It provides a proof system for satisfiability of quadratic equations over bilinear groups. This language suffices to capture almost all of the statements which appear in practice when designing public-key cryptographic schemes over bilinear groups. Although GS proofs are quite efficient, proving satisfiability of m equations in n variables requires sending some commitments of size $\Theta(n)$ and some proofs of size $\Theta(m)$ and they easily get expensive unless the statement is very simple. For this reason, several recent works have focused on further improving proof efficiency (e.g. [7,8,26])

Among those, a recent line of work [19–22] has succeeded in constructing constant-size arguments for very specific statements, namely, for membership in subspaces of $\hat{\mathbb{G}}^m$, where $\hat{\mathbb{G}}$ is some group equipped with a bilinear map where the discrete logarithm is hard. The soundness of the schemes is based on standard, falsifiable assumptions and the proof size is independent of both m and the witness size. These improvements are in a *quasi-adaptive* model (QA-NIZK, [19]). This means that the common reference string of these proof systems is specialized to the linear space where one wants to prove membership.

Interestingly, Jutla and Roy [20] also showed that their techniques to construct constant-size NIZK in linear spaces can be used to aggregate the GS proofs of m equations in n variables, that is, the total proof size can be reduced to $\Theta(n)$. Aggregation is also quasi-adaptive, which means that the common reference string depends on the set of equations one wants to aggregate. Further, it is only possible if the equations meet some restrictions. The first one is that only linear equations can be aggregated. The second one is that, in asymmetric bilinear groups, the equations must be one-sided linear, i.e. linear equations which have variables in only one of the $\mathbb{Z}_q$ modules $\hat{\mathbb{G}}, \check{\mathbb{H}}$, or $\mathbb{Z}_q$.[1]

Thus, it is worth to investigate if we can develop new techniques to aggregate other types of equations, for example, quadratic equations in $\mathbb{Z}_q$ and also recover all the aggregation results of [20] (in particular, for two-sided linear equations) in asymmetric bilinear groups. The latter (Type III bilinear groups, according to the classification of [11]) are the most attractive from the perspective of a performance and security trade off, specially since the recent attacks on discrete logarithms in finite fields by Joux [18] and subsequent improvements. Considerable research

---

[1] Jutla and Roy show how to aggregate two-sided linear equations in symmetric bilinear groups. The asymmetric case is not discussed, yet for one-sided linear equations it can be easily derived from their results. This is not the case for two-sided ones, see Sect. 4.

effort (e.g. [1,10]) has been put into translating pairing-based cryptosystems from a setting with more structure in which design is simpler (e.g. composite-order or symmetric bilinear groups) to a more efficient setting (e.g. prime order or asymmetric bilinear groups). In this line, we aim not only at obtaining new results in the asymmetric setting but also to translate known results and develop new tools specifically designed for it which might be of independent interest.

1.1   Our Results

In Sect. 3, we give constructions of constant-size QA-NIZK arguments of membership in linear spaces of $\hat{\mathbb{G}}^m \times \check{\mathbb{H}}^n$. Denote the elements of $\hat{\mathbb{G}}$ (respectively of $\check{\mathbb{H}}$) with a hat (resp. with an inverted hat) , as $\hat{x} \in \hat{\mathbb{G}}$ (respectively, as $\check{y} \in \check{\mathbb{H}}$). Given $\hat{\mathbf{M}} \in \hat{\mathbb{G}}^{m \times t}$ and $\check{\mathbf{N}} \in \check{\mathbb{H}}^{n \times t}$, we construct QA-NIZK arguments of membership in the language

$$\mathcal{L}_{\hat{\mathbf{M}},\check{\mathbf{N}}} := \{(\hat{\mathbf{x}}, \check{\mathbf{y}}) \in \hat{\mathbb{G}}^m \times \check{\mathbb{H}}^n : \exists \mathbf{w} \in \mathbb{Z}_q^t, \ \hat{\mathbf{x}} = \hat{\mathbf{M}}\mathbf{w}, \ \check{\mathbf{y}} := \check{\mathbf{N}}\mathbf{w}\},$$

which is the subspace of $\hat{\mathbb{G}}^m \times \check{\mathbb{H}}^n$ spanned by $\begin{pmatrix} \hat{\mathbf{M}} \\ \check{\mathbf{N}} \end{pmatrix}$. This construction is based on the recent constructions of [21]. When $m = n$, we construct QA-NIZK arguments of membership in

$$\mathcal{L}_{\hat{\mathbf{M}},\check{\mathbf{N}},+} := \{(\hat{\mathbf{x}}, \check{\mathbf{y}}) \in \hat{\mathbb{G}}^m \times \check{\mathbb{H}}^m : \exists \mathbf{w} \in \mathbb{Z}_q^t, \ \mathbf{x} + \mathbf{y} = (\mathbf{M} + \mathbf{N})\mathbf{w}\},$$

which is the linear subspace of $\hat{\mathbb{G}}^m \times \check{\mathbb{H}}^m$ of vectors $(\hat{\mathbf{x}}, \check{\mathbf{y}})$ such that the sum of their discrete logarithms is in the image of $\mathbf{M}+\mathbf{N}$ (the sum of discrete logarithms of $\hat{\mathbf{M}}$ and $\check{\mathbf{N}}$).

From the argument for $\mathcal{L}_{\hat{\mathbf{M}},\check{\mathbf{N}}}$, we easily derive another constant-size QA-NIZK argument in the space

$$\mathcal{L}_{\mathsf{com},\hat{\mathbf{U}},\check{\mathbf{V}},\nu} := \left\{ (\hat{\mathbf{c}}, \check{\mathbf{d}}) \in \hat{\mathbb{G}}^m \times \check{\mathbb{H}}^n : \exists (\mathbf{w}, \mathbf{r}, \mathbf{s}), \ \hat{\mathbf{c}} = \hat{\mathbf{U}} \begin{pmatrix} \mathbf{w} \\ \mathbf{r} \end{pmatrix}, \ \check{\mathbf{d}} = \check{\mathbf{V}} \begin{pmatrix} \mathbf{w} \\ \mathbf{s} \end{pmatrix} \right\},$$

where $\hat{\mathbf{U}} \in \hat{\mathbb{G}}^{m \times \tilde{m}}$, $\check{\mathbf{V}} \in \check{\mathbb{H}}^{n \times \tilde{n}}$ and $\mathbf{w} \in \mathbb{Z}_q^\nu$. Membership in this space captures the fact that two commitments (or sets of commitments) in $\hat{\mathbb{G}}, \check{\mathbb{H}}$ open to the same vector $\mathbf{w} \in \mathbb{Z}_q^\nu$. This is significant for the efficiency of quadratic GS proofs in asymmetric groups since, because of the way the proofs are constructed, one can only prove satisfiability of equations of degree one in each variable. Therefore, to prove a quadratic statement one needs to add auxiliary variables with commitments in the other group. For instance, to prove that $\hat{c}$ opens to $b \in \{0, 1\}$, one proves that some commitment $\check{\mathbf{d}}$ opens to $\bar{b}$ such that $\{b(\bar{b} - 1) = 0, b - \bar{b} = 0\}$. Our result allows us to aggregate the n proofs of the second statement.

To construct these arguments we introduce a new assumption, the *Split Kernel Matrix Diffie-Hellman Assumption* (SKerMDH). This assumption is derived from the recently introduced Kernel Matrix Diffie-Hellman Assumption (KerMDH, [24]), which says that it is hard to find a vector in the co-kernel of $\hat{\mathbf{A}} \in \hat{\mathbb{G}}^{\ell \times k}$ when $\mathbf{A}$ is such that it is hard to decide membership in $\mathbf{Im}(\hat{\mathbf{A}})$ (i.e. when $\mathbf{A}$ is an instance of a Matrix DH Assumption [8]). Our SKerMDH Assumption says that one cannot find a solution to the KerMDH problem which is "split" between the groups $\hat{\mathbb{G}}$ and $\check{\mathbb{H}}$. We think this assumption can be useful in other protocols in asymmetric bilinear groups. A particular case of Kernel MDH Assumption is the *Simultaneous Double Pairing Assumption* (SDP, [2]), which is a well established assumption in symmetric bilinear maps, and its "split" variant is the SSDP Assumption (see Sect. 2.1) [2].

In Sect. 4 we use the SKerMDH Assumption to lift the known aggregation results in symmetric groups to asymmetric ones. More specifically, we show how to extend the results of [20] to aggregate proofs of two-sided linear equations in asymmetric groups. While the original aggregation results of [20] were based on decisional assumptions, our proof shows that they are implied by computational assumptions.

Next, in Sect. 5, we address the problem of aggregating the proof of quadratic equations in $\mathbb{Z}_q$. For concreteness, we study the problem of proving that a commitment in $\hat{\mathbb{G}}$ opens to a bit-string of length n. Such a construction was unknown even in symmetric bilinear groups (yet, it can be easily generalized to this setting, see the full version). More specifically, we prove membership in

$$\mathcal{L}_{\hat{\mathbf{U}},\text{bits}} := \{\hat{\mathbf{c}} \in \hat{\mathbb{G}}^{n+m} : \hat{\mathbf{c}} := \hat{\mathbf{U}}_1 \mathbf{b} + \hat{\mathbf{U}}_2 \mathbf{w}, (\mathbf{b}, \mathbf{w}) \in \{0,1\}^n \times \mathbb{Z}_q^m\},$$

where $(\hat{\mathbf{U}}_1, \hat{\mathbf{U}}_2) \in \hat{\mathbb{G}}^{(n+m) \times n} \times \hat{\mathbb{G}}^{(n+m) \times m}$ are matrices which define a perfectly binding and computationally hiding commitment to $\mathbf{b}$. Specifically, we give instantiations for $m = 1$ (when $\hat{\mathbf{c}}$ is a single commitment to $\mathbf{b}$), and $m = n$ (when $\hat{\mathbf{c}}$ is the concatenation of n Groth-Sahai commitments to a bit).

We stress that although our proof is constant-size, we need the commitment to be perfectly binding, thus the size of the commitment is linear in n. The common reference string which we need for this construction is quadratic in the size of the bit-string. Our proof is compatible with proving linear statements about the bit-string, for instance, that $\sum_{i \in [n]} b_i = t$ by adding a linear number (in n) of elements to the CRS (see the full version). We observe that in the special case where $t = 1$ the common reference string can be linear in n. The costs of our constructions and the cost of GS proofs are summarized in Table 1.

We stress that our results rely solely on falsifiable assumptions. More specifically, in the asymmetric case we need some assumptions which are weaker than the Symmetric External DH Assumption plus the SSDP Assumption. Interestingly, our construction in the symmetric setting relies on assumptions which are all weaker than the 2-Lin Assumption (see the full version).

We think that our techniques for constructing QA-NIZK arguments for bit-strings might be of independent interest. In the asymmetric case, we combine our QA-NIZK argument for $\mathcal{L}_{\hat{\mathbf{M}},\check{\mathbf{N}},+}$ with decisional assumptions in $\hat{\mathbb{G}}$ and $\check{\mathbb{H}}$. We do this with the purpose of using QA-NIZK arguments even when $\mathbf{M} + \mathbf{N}$ has full rank. In this case, strictly speaking "proving membership in the space" looses all meaning, as every vector in $\hat{\mathbb{G}}^m \times \check{\mathbb{H}}^m$ is in the space. However, using decisional assumptions, we can argue that the generating matrix of the space is indistinguishable from a lower rank matrix which spans a subspace in which it is meaningful to prove membership.

Finally, in Sect. 6 we discuss some applications of our results. In particular, our results provide shorter signature size of several schemes, more efficient ring signatures, more efficient proofs of membership in a list, and improved threshold GS proofs for pairing product equations.

Table 1. Comparison for proofs of $b_i \in \{0, 1\}$, for $i \in [n]$, between GS proofs and our different constructions. Our NIZK construction for bit-strings is denoted by Π_{bit} and the construction for proving that two sets of commitments open to the same value $\Psi_{\overline{\mathcal{D}}_k, \text{com}}$. Row "$\Pi_{\text{bit}}$ $m = 1$" is for our construction for a single commitment of size $n + 1$ to a bit-string of size n, and "Π_{bit} $m = n$" is for our construction for n concatenated GS commitments. Row "Π_{bit} weight 1" is for our construction for bit-strings of weight 1 with $m = 1$. Column "Comms" contains the size of the commitments, "CK" the size of the commitment keys in the CRS, and "CRS(ρ)" the size of the language dependent part of the CRS. The size of elements in $\hat{\mathbb{G}}$ and $\check{\mathbb{H}}$ is $\mathfrak{g}$ and $\mathfrak{h}$, respectively. The table is computed for $\mathcal{D}_k = \mathcal{L}_2$, the 2-Linear matrix distribution.

	Comms	Proof	CK	CRS(ρ)	#Pairings
GS [15]	$2n(\mathfrak{g} + \mathfrak{h})$	$4n(\mathfrak{g} + \mathfrak{h})$	$4(\mathfrak{g} + \mathfrak{h})$	0	$28n$
GS + $\Psi_{\overline{\mathcal{D}}_k, \text{com}}$	$2n(\mathfrak{g} + \mathfrak{h})$	$(2n + 2)(\mathfrak{g} + \mathfrak{h})$	$4(\mathfrak{g} + \mathfrak{h})$	$(10n + 4)(\mathfrak{g} + \mathfrak{h})$	$20n + 8$
Π_{bit} $m = 1$	$(n + 1)\mathfrak{g}$	$10(\mathfrak{g} + \mathfrak{h})$	$(n + 1)\mathfrak{g}$	$(6n^2 + 10n + 32)(\mathfrak{g} + \mathfrak{h})$	$n + 55$
Π_{bit} $m = n$	$2n\mathfrak{g}$	$10(\mathfrak{g} + \mathfrak{h})$	$4\mathfrak{g}$	$(12n^2 + 14n + 22)\mathfrak{g} + (12n^2 + 12n + 24)\mathfrak{h}$	$2n + 54$
Π_{bit} weight 1, $m = 1$	$(n + 1)\mathfrak{g}$	$10(\mathfrak{g} + \mathfrak{h})$	$(n + 1)\mathfrak{g}$	$(20n + 26)(\mathfrak{g} + \mathfrak{h})$	$n + 55$

2 Preliminaries

Let Gen_a be some probabilistic polynomial time algorithm which on input 1^λ, where λ is the security parameter, returns the description of an asymmetric bilinear group $(q, \hat{\mathbb{G}}, \check{\mathbb{H}}, \mathbb{T}, e, \hat{g}, \check{h})$, where $\hat{\mathbb{G}}, \check{\mathbb{H}}$ and $\mathbb{T}$ are groups of prime order q, the elements $\hat{g}, \check{h}$ are generators of $\hat{\mathbb{G}}, \check{\mathbb{H}}$ respectively, and $e : \hat{\mathbb{G}} \times \check{\mathbb{H}} \to \mathbb{T}$ is an efficiently computable, non-degenerate bilinear map.

We denote by $\mathfrak{g}$ and $\mathfrak{h}$ the bit-size of the elements of $\hat{\mathbb{G}}$ and $\check{\mathbb{H}}$, respectively. Elements $\hat{x} \in \hat{\mathbb{G}}$ (resp. $\check{y} \in \check{\mathbb{H}}$, $z_{\mathbb{T}} \in \mathbb{T}$) are written with a hat (resp, with inverted hat, sub-index $\mathbb{T}$) and $\hat{0}, \check{0}$ and $0_{\mathbb{T}}$ denote the neutral elements. Given $\hat{x} \in \hat{\mathbb{G}}, \check{y} \in \check{\mathbb{H}}$, $\hat{x}\check{y}$ refers to the pairing operation, i.e. $\hat{x}\check{y} = e(\hat{x}, \check{y})$. Vectors and matrices are denoted in boldface and any product of vectors/matrices of elements in $\hat{\mathbb{G}}$ and $\check{\mathbb{H}}$ is defined in the natural way via the pairing operation. That is, given $\hat{\mathbf{X}} \in \hat{\mathbb{G}}^{n \times m}$ and $\check{\mathbf{Y}} \in \check{\mathbb{H}}^{m \times \ell}$, $\hat{\mathbf{X}}\check{\mathbf{Y}} \in \mathbb{T}^{n \times \ell}$. The product $\check{\mathbf{X}}\hat{\mathbf{Y}} \in \mathbb{T}^{n \times \ell}$ is defined similarly by switching the arguments of the pairing. Given a matrix $\mathbf{T} = (t_{i,j}) \in \mathbb{Z}_q^{m \times n}$, $\hat{\mathbf{T}}$ (resp. $\check{\mathbf{T}}$) is the natural embedding of $\mathbf{T}$ in $\hat{\mathbb{G}}$ (resp. in $\check{\mathbb{H}}$), that is, the matrix whose (i, j)th entry is $t_{i,j}\hat{g}$ (resp. $t_{i,j}\check{h}$). Conversely, given $\hat{\mathbf{T}}$ or $\check{\mathbf{T}}$, we use $\mathbf{T} \in \mathbb{Z}_q^{n \times m}$ for the matrix of discrete logarithms of $\hat{\mathbf{T}}$ (resp. $\check{\mathbf{T}}$). We denote by $\mathbf{I}_{n \times n}$ the identity matrix in $\mathbb{Z}_q^{n \times n}$ and $\mathbf{e}_i^n$ the ith element of the canonical basis of $\mathbb{Z}_q^n$ (simply $\mathbf{e}_i$ if n is clear from the context). We make extensive use of the set $[n+k] \times [n+k] \setminus \{(i,i) : i \in [n]\}$ and for brevity we denote it by $\mathcal{I}_{N,K}$.

2.1 Computational Assumptions

Definition 1. *Let $\ell, k \in \mathbb{N}$ with $\ell > k$. We call $\mathcal{D}_{\ell,k}$ a matrix distribution if it outputs (in poly time, with overwhelming probability) matrices in $\mathbb{Z}_q^{\ell \times k}$. We define $\mathcal{D}_k := \mathcal{D}_{k+1,k}$ and $\overline{\mathcal{D}_k}$ the distribution of the first k rows when $\mathbf{A} \leftarrow \mathcal{D}_k$.*

Definition 2 (Matrix Diffie-Hellman Assumption [8]). *Let $\mathcal{D}_{\ell,k}$ be a matrix distribution and $\Gamma := (q, \hat{\mathbb{G}}, \check{\mathbb{H}}, \mathbb{T}, e, \hat{g}, \check{h}) \leftarrow \mathsf{Gen}_a(1^\lambda)$. We say that the $\mathcal{D}_{\ell,k}$-Matrix Diffie-Hellman ($\mathcal{D}_{\ell,k}$-MDDH$_{\hat{\mathbb{G}}}$) Assumption holds relative to Gen_a if for all PPT adversaries D,*

$$\mathbf{Adv}_{\mathcal{D}_{\ell,k},\mathsf{Gen}_a}(\mathsf{D}) := \Big| \Pr[\mathsf{D}(\Gamma, \hat{\mathbf{A}}, \hat{\mathbf{A}}\mathbf{w}) = 1] - \Pr[\mathsf{D}(\Gamma, \hat{\mathbf{A}}, \hat{\mathbf{u}}) = 1] \Big| = \mathsf{negl}(\lambda),$$

where the probability is taken over $\Gamma \leftarrow \mathsf{Gen}_a(1^\lambda)$, $\mathbf{A} \leftarrow \mathcal{D}_{\ell,k}, \mathbf{w} \leftarrow \mathbb{Z}_q^k, \hat{\mathbf{u}} \leftarrow \hat{\mathbb{G}}^\ell$ and the coin tosses of adversary D.

The $\mathcal{D}_{\ell,k}$-MDDH$_{\check{\mathbb{H}}}$ problem is defined similarly. In this paper we will refer to the following matrix distributions:

$$\mathcal{L}_k : \mathbf{A} = \begin{pmatrix} a_1 & 0 & \dots & 0 \\ 0 & a_2 & \dots & 0 \\ \vdots & \vdots & \ddots & \vdots \\ 0 & 0 & \dots & a_k \\ 1 & 1 & \dots & 1 \end{pmatrix}, \mathcal{L}_{\ell,k} : \mathbf{A} = \begin{pmatrix} \mathbf{B} \\ \mathbf{C} \end{pmatrix}, \mathcal{U}_{\ell,k} : \mathbf{A} = \begin{pmatrix} a_{1,1} & \dots & a_{1,k} \\ \vdots & \ddots & \vdots \\ a_{\ell,1} & \dots & a_{\ell,k} \end{pmatrix},$$

where $a_i, a_{i,j} \leftarrow \mathbb{Z}_q$, for each $i, j \in [k]$, $\mathbf{B} \leftarrow \overline{\mathcal{L}_k}$, $\mathbf{C} \leftarrow \mathbb{Z}_q^{\ell-k,k}$.

The $\mathcal{L}_k$-MDDH Assumption is the k-linear family of Decisional Assumptions [17,27]. The $\mathcal{L}_1$-MDDH$_X$, $X \in \{\hat{\mathbb{G}}, \check{\mathbb{H}}\}$, is the Decisional Diffie-Hellman (DDH) Assumption in X, and the assumption that it holds in both groups is the Symmetric External DH Assumption (SXDH). The $\mathcal{L}_{\ell,k}$-MDDH Assumption is used in our construction to commit to multiple elements simultaneously. It can be shown tightly equivalent to the $\mathcal{L}_k$-MDDH Assumption. The $\mathcal{U}_{\ell,k}$ Assumption is the *Uniform* Assumption and is weaker than the $\mathcal{L}_k$-MDDH. Additionally, we will be using the following family of computational assumptions:

Definition 3 (Kernel Diffie-Hellman Assumption [24]). *Let $\Gamma \leftarrow \mathsf{Gen}_a(1^\lambda)$. The Kernel Diffie-Hellman Assumption in $\check{\mathbb{H}}$ ($\mathcal{D}_{\ell,k}$-KerMDH$_{\check{\mathbb{H}}}$) says that every PPT Algorithm has negligible advantage in the following game: given $\check{\mathbf{A}}$, where $\mathbf{A} \leftarrow \mathcal{D}_{\ell,k}$, find $\hat{\mathbf{x}} \in \hat{\mathbb{G}}^\ell \setminus \{\hat{\mathbf{0}}\}$, such that $\hat{\mathbf{x}}^\top \check{\mathbf{A}} = \mathbf{0}_{\mathbb{T}}$.*

The Simultaneous Pairing Assumption in $\check{\mathbb{H}}$ (SP$_{\check{\mathbb{H}}}$) is the $\mathcal{U}_1$-KerMDH$_{\check{\mathbb{H}}}$ Assumption and the Simultaneous Double Pairing Assumption (SDP$_{\check{\mathbb{H}}}$) is the $\mathcal{L}_{2,3}$-KerMDH$_{\check{\mathbb{H}}}$ Assumption. The Kernel Diffie-Hellman assumption is a generalization and abstraction of these two assumptions to other matrix distributions. The

$\mathcal{D}_{\ell,k}$-KerMDH$_{\breve{\mathbb{H}}}$ Assumption is weaker than the $\mathcal{D}_{\ell,k}$-MDDH$_{\breve{\mathbb{H}}}$ Assumption, since a solution allows to decide membership in $\mathsf{Im}(\breve{\mathbf{A}})$.

For our construction, we need to introduce a new family of computational assumptions.

Definition 4 (Split Kernel Diffie-Hellman Assumption). *Let* $\Gamma \leftarrow \mathsf{Gen}_a(1^\lambda)$. *The Split Kernel Diffie-Hellman Assumption in* $\hat{\mathbb{G}}, \breve{\mathbb{H}}$ *($\mathcal{D}_{\ell,k}$-SKerMDH) says that every PPT Algorithm has negligible advantage in the following game: given* $(\hat{\mathbf{A}}, \breve{\mathbf{A}})$, $\mathbf{A} \leftarrow \mathcal{D}_{\ell,k}$, *find a pair of vectors* $(\hat{\mathbf{r}}, \breve{\mathbf{s}}) \in \hat{\mathbb{G}}^\ell \times \breve{\mathbb{H}}^\ell$, $\mathbf{r} \neq \mathbf{s}$, *such that* $\hat{\mathbf{r}}^\top \mathbf{A} = \breve{\mathbf{s}}^\top \hat{\mathbf{A}}$.

As a particular case we consider the *Split Simultaneous Double Pairing Assumption in* $\hat{\mathbb{G}}, \breve{\mathbb{H}}$ (SSDP) which is the $\mathcal{L}_2$-SKerMDH Assumption. Intuitively, the Kernel Diffie-Hellman Assumption says one cannot find a non-zero vector in $\hat{\mathbb{G}}^\ell$ which is in the co-kernel of $\breve{\mathbf{A}}$, while the new assumption says one cannot find a pair of vectors in $\hat{\mathbb{G}}^\ell \times \breve{\mathbb{H}}^\ell$ such that the difference of the vector of their discrete logarithms is in the co-kernel of $\breve{\mathbf{A}}$. The name "split" comes from the idea that the output of a successful adversary would break the Kernel Diffie-Hellman Assumption, but this instance is "split" between the groups $\hat{\mathbb{G}}$ and $\breve{\mathbb{H}}$. When $k = 1$, the $\mathcal{D}_{\ell,k}$-SKerMDH Assumption does not hold. The assumption is generically as least as hard as the standard, "non-split" assumption in symmetric bilinear groups. This means, in particular, that in Type III bilinear groups, one can use the SSDP Assumption with the same security guarantees as the SDP Assumption, which is a well established assumption (used for instance in [2,23]).

Lemma 1. *If* $\mathcal{D}_{\ell,k}$-KerMDH *holds in generic symmetric bilinear groups, then* $\mathcal{D}_{\ell,k}$-SKerMDH *holds in generic asymmetric bilinear groups.*

Suppose there is a generic algorithm which breaks the $\mathcal{D}_{\ell,k}$-SKerMDH Assumption. Intuitively, given two different encodings of $\mathbf{A} \leftarrow \mathcal{D}_{\ell,k}$, $(\hat{\mathbf{A}}, \breve{\mathbf{A}})$, this algorithm finds $\hat{\mathbf{r}}$ and $\breve{\mathbf{s}}$, $\mathbf{r} \neq \mathbf{s}$ such that $\hat{\mathbf{r}}^\top \mathbf{A} = \breve{\mathbf{s}}^\top \hat{\mathbf{A}}$. But since the algorithm is generic, it also works when $\hat{\mathbb{G}} = \breve{\mathbb{H}}$, and then $\hat{\mathbf{r}} - \hat{\mathbf{s}}$ is a solution to $\mathcal{D}_{\ell,k}$-KerMDH. We provide a formal proof in the full version.

2.2 Groth-Sahai NIZK Proofs

The GS proof system allows to prove satisfiability of a set of quadratic equations in a bilinear group. The admissible equation types must be in the following form:

$$\sum_{j=1}^{m_y} f(\alpha_j, \mathsf{y}_j) + \sum_{i=1}^{m_x} f(\mathsf{x}_i, \beta_i) + \sum_{i=1}^{m_x}\sum_{j=1}^{m_y} f(\mathsf{x}_i, \gamma_{i,j}\mathsf{y}_j) = t, \tag{1}$$

where A_1, A_2, A_T are $\mathbb{Z}_q$-vector spaces equipped with some bilinear map $f : A_1 \times A_2 \rightarrow A_T$, $\boldsymbol{\alpha} \in A_1^{m_y}$, $\boldsymbol{\beta} \in A_2^{m_x}$, $\boldsymbol{\Gamma} = (\gamma_{i,j}) \in \mathbb{Z}_q^{m_x \times m_y}$, $t \in A_T$. The modules and the map f can be defined in different ways as: (a) in pairing-product equations (PPEs), $A_1 = \hat{\mathbb{G}}$, $A_2 = \breve{\mathbb{H}}$, $A_T = \mathbb{T}$, $f(\hat{x}, \breve{y}) = \hat{x}\breve{y} \in \mathbb{T}$, in which case $t = 0_\mathbb{T}$, (b1) in multi-scalar multiplication equations in $\hat{\mathbb{G}}$ (MMEs),

$A_1 = \hat{\mathbb{G}}$, $A_2 = \mathbb{Z}_q$, $A_T = \hat{\mathbb{G}}$, $f(\hat{x}, y) = y\hat{x} \in \hat{\mathbb{G}}$, (b2) MMEs in $\check{\mathbb{H}}$ (MMEs), $A_1 = \mathbb{Z}_q$, $A_2 = \check{\mathbb{H}}$, $A_T = \check{\mathbb{H}}$, $f(x, \check{y}) = x\check{y} \in \check{\mathbb{H}}$, and (c) in quadratic equations in $\mathbb{Z}_q$ (QEs), $A_1 = A_2 = A_T = \mathbb{Z}_q$, $f(x, y) = xy \in \mathbb{Z}_q$. An equation is linear if $\boldsymbol{\Gamma} = \mathbf{0}$, it is *two-sided linear* if both $\boldsymbol{\alpha} \neq \mathbf{0}$ and $\boldsymbol{\beta} \neq \mathbf{0}$, and *one-sided* otherwise.

We briefly recall some facts about GS proofs in the SXDH instantiation used in the rest of the paper. Let $\Gamma \leftarrow \mathsf{Gen}_a(1^\lambda)$, $\mathbf{u}_2, \mathbf{v}_2 \leftarrow \mathcal{L}_1$, $\mathbf{u}_1 := \mathbf{e}_1 + \mu\mathbf{u}_2$, $\mathbf{v}_1 := \mathbf{e}_1 + \epsilon\mathbf{v}_2$, $\mu, \epsilon \leftarrow \mathbb{Z}_q$. The common reference string is $\mathsf{crs}_{\mathsf{GS}} := (\Gamma, \hat{\mathbf{u}}_1, \hat{\mathbf{u}}_2, \check{\mathbf{v}}_1, \check{\mathbf{v}}_2)$ and is known as the *perfectly sound CRS*. There is also a *perfectly witness-indistinguishable CRS*, with the only difference being that $\mathbf{u}_1 := \mu\mathbf{u}_2$ and $\mathbf{v}_1 := \epsilon\mathbf{v}_2$ and the simulation trapdoor is (μ, ϵ). These two CRS distributions are computationally indistinguishable. Implicitly, $\mathsf{crs}_{\mathsf{GS}}$ defines the maps:

$$\iota_1 : \hat{\mathbb{G}} \cup \mathbb{Z}_q \to \hat{\mathbb{G}}^2, \qquad \iota_1(\hat{x}) := (\hat{x}, \hat{0})^\top, \qquad \iota_1(x) := x\hat{\mathbf{u}}_1.$$
$$\iota_2 : \check{\mathbb{H}} \cup \mathbb{Z}_q \to \check{\mathbb{H}}^2, \qquad \iota_2(\check{y}) := (\check{y}, \check{0})^\top, \qquad \iota_2(y) := y\check{\mathbf{v}}_1.$$

The maps ι_X, $X \in \{1, 2\}$ can be naturally extended to vectors of arbitrary length $\boldsymbol{\delta} \in A_X^m$ and we write $\iota_X(\boldsymbol{\delta})$ for $(\iota_X(\delta_1) || \ldots || \iota_X(\delta_m))$.

The perfectly sound CRS defines perfectly binding commitments for any variable in A_1 or A_2. Specifically, the commitment to $x \in A_1$ is $\hat{\mathbf{c}} := \iota_1(x) + r_1(\hat{\mathbf{u}}_1 - \hat{\mathbf{e}}_1) + r_2\hat{\mathbf{u}}_2 \in \hat{\mathbb{G}}^2$, and to $y \in A_2$ is $\check{\mathbf{d}} := \iota_2(y) + s_1(\check{\mathbf{v}}_1 - \check{\mathbf{e}}_1) + s_2\check{\mathbf{v}}_2$, where $r_1, r_2, s_1, s_2 \leftarrow \mathbb{Z}_q$, except if $A_1 = \mathbb{Z}_q$ (resp. $A_2 = \mathbb{Z}_q$) in which case $r_1 = 0$ (resp. $s_1 = 0$).

2.3 Quasi-Adaptive NIZK Arguments

We recall the definition of Quasi Adaptive NIZK (QA-NIZK) Arguments of Jutla et al. [19]. A QA-NIZK proof system enables to prove membership in a language defined by a relation $\mathcal{R}_\rho$, which in turn is completely determined by some parameter ρ sampled from a distribution $\mathcal{D}_\Gamma$. We say that $\mathcal{D}_\Gamma$ is *witness samplable* if there exist an efficient algorithm that samples (ρ, ω) such that ρ is distributed according to $\mathcal{D}_\Gamma$, and membership of ρ in the *parameter language* $\mathcal{L}_{\mathsf{par}}$ can be efficiently verified with ω. While the Common Reference String can be set based on ρ, the zero-knowledge simulator is required to be a single probabilistic polynomial time algorithm that works for the whole collection of relations $\mathcal{R}_\Gamma$.

A tuple of algorithms $(\mathsf{K}_0, \mathsf{K}_1, \mathsf{P}, \mathsf{V})$ is called a QA-NIZK proof system for witness-relations $\mathcal{R}_\Gamma = \{\mathcal{R}_\rho\}_{\rho \in \mathrm{sup}(\mathcal{D}_\Gamma)}$ with parameters sampled from a distribution $\mathcal{D}_\Gamma$ over associated parameter language $\mathcal{L}_{\mathsf{par}}$, if there exists a probabilistic polynomial time simulator $(\mathsf{S}_1, \mathsf{S}_2)$, such that for all non-uniform PPT adversaries $\mathsf{A}_1, \mathsf{A}_2, \mathsf{A}_3$ we have:

Quasi-Adaptive Completeness:

$$\Pr\begin{bmatrix} \Gamma \leftarrow \mathsf{K}_0(1^\lambda); \rho \leftarrow \mathcal{D}_\Gamma; \psi \leftarrow \mathsf{K}_1(\Gamma, \rho); (x, w) \leftarrow \mathsf{A}_1(\Gamma, \psi); \\ \pi \leftarrow \mathsf{P}(\psi, x, w) : \mathsf{V}(\psi, x, \pi) = 1 \text{ if } \mathcal{R}_\rho(x, w) \end{bmatrix} = 1.$$

Computational Quasi-Adaptive Soundness:

$$\Pr\left[\begin{array}{l} \Gamma \leftarrow \mathsf{K}_0(1^\lambda); \rho \leftarrow \mathcal{D}_\Gamma; \psi \leftarrow \mathsf{K}_1(\Gamma, \rho); \\ (x, \pi) \leftarrow \mathsf{A}_2(\Gamma, \psi) : \mathsf{V}(\psi, x, \pi) = 1 \text{ and } \neg(\exists w : \mathcal{R}_\rho(x, w)) \end{array}\right] \approx 0.$$

Perfect Quasi-Adaptive Zero-Knowledge:

$$\Pr[\Gamma \leftarrow \mathsf{K}_0(1^\lambda); \rho \leftarrow \mathcal{D}_\Gamma; \psi \leftarrow \mathsf{K}_1(\Gamma, \rho) : \mathsf{A}_3^{\mathsf{P}(\psi, \cdot, \cdot)}(\Gamma, \psi) = 1] =$$
$$\Pr[\Gamma \leftarrow \mathsf{K}_0(1^\lambda); \rho \leftarrow \mathcal{D}_\Gamma; (\psi, \tau) \leftarrow \mathsf{S}_1(\Gamma, \rho) : \mathsf{A}_3^{\mathsf{S}(\psi, \tau, \cdot, \cdot)}(\Gamma, \psi) = 1]$$

where

- $\mathsf{P}(\psi, \cdot, \cdot)$ emulates the actual prover. It takes input (x, w) and outputs a proof π if $(x, w) \in \mathcal{R}_\rho$. Otherwise, it outputs $\bot$.
- $\mathsf{S}(\psi, \tau, \cdot, \cdot)$ is an oracle that takes input (x, w). It outputs a simulated proof $\mathsf{S}_2(\psi, \tau, x)$ if $(x, w) \in \mathcal{R}_\rho$ and $\bot$ if $(x, w) \notin \mathcal{R}_\rho$.

Note that ψ is the CRS in the above definitions. We assume that ψ contains an encoding of ρ, which is thus available to V.

2.4 QA-NIZK Argument for Linear Spaces

In this section we recall the two constructions of QA-NIZK arguments of membership in linear spaces given by Kiltz and Wee [21], for the language:

$$\mathcal{L}_{\hat{\mathbf{M}}} := \{\hat{\mathbf{x}} \in \hat{\mathbb{G}}^n : \exists \mathbf{w} \in \mathbb{Z}_q^t, \ \hat{\mathbf{x}} = \hat{\mathbf{M}}\mathbf{w}\}.$$

Algorithm $\mathsf{K}_0(1^\lambda)$ just outputs $\Gamma := (q, \hat{\mathbb{G}}, \check{\mathbb{H}}, \mathbb{T}, e, \hat{g}, \check{h}) \leftarrow \mathsf{Gen}_a(1^\lambda)$, the rest of the algorithms are described in Fig. 1.

$\mathsf{K}_1(\Gamma, \hat{\mathbf{M}}, n)$ $(\mathsf{S}_1(\Gamma, \hat{\mathbf{M}}, n))$	$\mathsf{P}(\mathrm{crs}, \hat{\mathbf{x}}, \mathbf{w}) \ \backslash\backslash \hat{\mathbf{x}} = \hat{\mathbf{M}}\mathbf{w}$	$\mathsf{S}_2(\mathrm{crs}, \hat{\mathbf{x}}, \tau_{sim})$
$\mathbf{A} \leftarrow \widetilde{\mathcal{D}_k}, \mathbf{\Delta} \leftarrow \mathbb{Z}_q^{\tilde{k} \times n}$	Return $\hat{\boldsymbol{\sigma}} := \hat{\mathbf{M}}_\Delta \mathbf{w}$.	Return $\hat{\boldsymbol{\sigma}} := \mathbf{\Delta}\hat{\mathbf{x}}$
$\check{\mathbf{A}}_\Delta := \mathbf{\Delta}^\top \check{\mathbf{A}}, \hat{\mathbf{M}}_\Delta := \mathbf{\Delta}\hat{\mathbf{M}}$		
Return $\mathrm{crs} := (\hat{\mathbf{M}}_\Delta, \check{\mathbf{A}}_\Delta, \check{\mathbf{A}})$	$\mathsf{V}(\mathrm{crs}, \hat{\mathbf{x}}, \hat{\boldsymbol{\sigma}})$	
$(\tau_{sim} := \mathbf{\Delta})$	Return $(\hat{\mathbf{x}}^\top \check{\mathbf{A}}_\Delta = \hat{\boldsymbol{\sigma}}^\top \check{\mathbf{A}})$	

Fig. 1. The figure describes $\Psi_{\mathcal{D}_k}$ when $\widetilde{\mathcal{D}_k} = \mathcal{D}_k$ and $\tilde{k} = k+1$ and $\Psi_{\overline{\mathcal{D}}_k}$ when $\widetilde{\mathcal{D}_k} = \overline{\mathcal{D}}_k$ and $\tilde{k} = k$. Both are QA-NIZK arguments for $\mathcal{L}_{\hat{\mathbf{M}}}$. $\Psi_{\mathcal{D}_k}$ is the construction of [21, Sect. 3.1], which is a generalization of Libert *et al*'s QA-NIZK [22] to any $\mathcal{D}_k$-KerMDH$_{\check{\mathbb{H}}}$ Assumption. $\Psi_{\overline{\mathcal{D}}_k}$ is the construction of [21, Sect. 3.2.].

Theorem 1 (Theorem 1 of [21]). *If $\widetilde{\mathcal{D}_k} = \mathcal{D}_k$ and $\tilde{k} = k+1$, Fig. 1 describes a QA-NIZK proof system with perfect completeness, computational adaptive soundness based on the $\mathcal{D}_k$-KerMDH$_{\check{\mathbb{H}}}$ Assumption, perfect zero-knowledge, and proof size $k+1$.*

A. González et al.

Theorem 2 (Theorem 2 of [21]). *If $\widetilde{\mathcal{D}}_k = \overline{\mathcal{D}}_k$ and $\tilde{k} = k$, and $\mathcal{D}_\Gamma$ is a witness samplable distribution, Fig. 1 describes a QA-NIZK proof system with perfect completeness, computational adaptive soundness based on the $\mathcal{D}_k$-KerMDH$_{\check{\mathbb{H}}}$ Assumption, perfect zero-knowledge, and proof size k.*

3 New QA-NIZK Arguments in Asymmetric Groups

In this section we construct three QA-NIZK arguments of membership in different subspaces of $\hat{\mathbb{G}}^m \times \check{\mathbb{H}}^n$. Their soundness relies on the Split Kernel Assumption.

3.1 Argument of Membership in Subspace Concatenation

Figure 2 describes a QA-NIZK Argument of Membership in the language

$$\mathcal{L}_{\hat{\mathbf{M}},\check{\mathbf{N}}} := \{(\hat{\mathbf{x}}, \check{\mathbf{y}}) : \exists \mathbf{w} \in \mathbb{Z}_q^t, \ \hat{\mathbf{x}} = \hat{\mathbf{M}}\mathbf{w}, \check{\mathbf{y}} = \check{\mathbf{N}}\mathbf{w}\} \subseteq \hat{\mathbb{G}}^m \times \check{\mathbb{H}}^n.$$

We refer to this as the *Concatenation Language*, because if we define $\mathbf{P}$ as the concatenation of $\hat{\mathbf{M}}, \check{\mathbf{N}}$, that is $\mathbf{P} := \begin{pmatrix} \hat{\mathbf{M}} \\ \check{\mathbf{N}} \end{pmatrix}$, then $(\hat{\mathbf{x}}, \check{\mathbf{y}}) \in \mathcal{L}_{\hat{\mathbf{M}},\check{\mathbf{N}}}$ iff $\begin{pmatrix} \hat{\mathbf{x}} \\ \check{\mathbf{y}} \end{pmatrix}$ is in the span of $\mathbf{P}$.

$\underline{\mathsf{K}_1(\Gamma, \hat{\mathbf{M}}, \check{\mathbf{N}}, m, n) \quad (\mathsf{S}_1(\Gamma, \hat{\mathbf{M}}, \check{\mathbf{N}}, m, n))}$

$\mathbf{A} \leftarrow \widetilde{\mathcal{D}}_k$

$\mathbf{\Lambda} \leftarrow \mathbb{Z}_q^{\tilde{k} \times m}, \mathbf{\Xi} \leftarrow \mathbb{Z}_q^{\tilde{k} \times n}, \mathbf{Z} \leftarrow \mathbb{Z}_q^{\tilde{k} \times t}$

$\check{\mathbf{A}}_\Lambda := \mathbf{\Lambda}^\top \check{\mathbf{A}}$

$\hat{\mathbf{A}}_\Xi := \mathbf{\Xi}^\top \hat{\mathbf{A}}$

$\hat{\mathbf{M}}_\Lambda := \mathbf{\Lambda}\hat{\mathbf{M}} + \hat{\mathbf{Z}}$

$\check{\mathbf{N}}_\Xi := \mathbf{\Xi}\check{\mathbf{N}} - \check{\mathbf{Z}}$

Return $\mathsf{crs} := (\hat{\mathbf{M}}_\Lambda, \check{\mathbf{A}}_\Lambda, \check{\mathbf{A}}, \check{\mathbf{N}}_\Xi,$

$\hat{\mathbf{A}}_\Xi, \hat{\mathbf{A}}).$

$(\tau_{sim} := (\mathbf{\Lambda}, \mathbf{\Xi}).)$

$\underline{\mathsf{P}(\mathsf{crs}, \hat{\mathbf{x}}, \check{\mathbf{y}}, \mathbf{w})}$

$\backslash\backslash(\hat{\mathbf{x}} = \hat{\mathbf{M}}\mathbf{w}, \check{\mathbf{y}} = \check{\mathbf{N}}\mathbf{w})$

$\mathbf{z} \leftarrow \mathbb{Z}_q^{\tilde{k}}$

$\hat{\boldsymbol{\rho}} := \hat{\mathbf{M}}_\Lambda \mathbf{w} + \hat{\mathbf{z}}$

$\check{\boldsymbol{\sigma}} := \check{\mathbf{N}}_\Xi \mathbf{w} - \check{\mathbf{z}}$

Return $(\hat{\boldsymbol{\rho}}, \check{\boldsymbol{\sigma}}).$

$\underline{\mathsf{V}(\mathsf{crs}, (\hat{\mathbf{x}}, \check{\mathbf{y}}), (\hat{\boldsymbol{\rho}}, \check{\boldsymbol{\sigma}}))}$

Return $(\hat{\mathbf{x}}^\top \check{\mathbf{A}}_\Lambda - \hat{\boldsymbol{\rho}}^\top \check{\mathbf{A}}$

$= \check{\boldsymbol{\sigma}}^\top \hat{\mathbf{A}} - \check{\mathbf{y}}^\top \hat{\mathbf{A}}_\Xi).$

$\underline{\mathsf{S}_2(\mathsf{crs}, (\hat{\mathbf{x}}, \check{\mathbf{y}}), \tau_{sim})}$

$\mathbf{z} \leftarrow \mathbb{Z}_q^{\tilde{k}}$

$\hat{\boldsymbol{\rho}} := \mathbf{\Lambda}\hat{\mathbf{x}} + \hat{\mathbf{z}}$

$\check{\boldsymbol{\sigma}} := \mathbf{\Xi}\check{\mathbf{y}} - \check{\mathbf{z}}$

Return $(\hat{\boldsymbol{\rho}}, \check{\boldsymbol{\sigma}}).$

Fig. 2. Two QA-NIZK Arguments for $\mathcal{L}_{\hat{\mathbf{M}},\check{\mathbf{N}}}$. $\Psi_{\mathcal{D}_k,\mathsf{spl}}$ is defined for $\widetilde{\mathcal{D}}_k = \mathcal{D}_k$ and $\tilde{k} = k + 1$, and is a generalization of [21] Sect. 3.1 in two groups. The second construction $\Psi_{\overline{\mathcal{D}}_k,\mathsf{spl}}$ corresponds to $\widetilde{\mathcal{D}}_k = \overline{\mathcal{D}}_k$ and $\tilde{k} = k$, and is a generalization of [21] Sect. 3.2 in two groups. Computational soundness is based on the $\mathcal{D}_k$-SKerMDH Assumption. The CRS size is $(\tilde{k}k + \tilde{k}t + mk)\mathfrak{g} + (\tilde{k}k + \tilde{k}t + nk)\mathfrak{h}$ and the proof size $\tilde{k}(\mathfrak{g} + \mathfrak{h})$. Verification requires $2\tilde{k}k + (m + n)k$ pairing computations.

Soundness Intuition. If we ignore for a moment that $\hat{\mathbb{G}}, \check{\mathbb{H}}$ are different groups, $\Psi_{\mathcal{D}_k,\mathsf{spl}}$ (resp. $\Psi_{\overline{\mathcal{D}}_k,\mathsf{spl}}$) is almost identical to $\Psi_{\mathcal{D}_k}$ (resp. to $\Psi_{\overline{\mathcal{D}}_k}$) for the language $\mathcal{L}_{\hat{\mathbf{P}}}$, and $\mathbf{\Delta} := (\mathbf{\Lambda} \| \mathbf{\Xi})$, where $\mathbf{\Lambda} \in \mathbb{Z}_q^{\tilde{k} \times m}, \mathbf{\Xi} \in \mathbb{Z}_q^{\tilde{k} \times n}$. Further, the information that an unbounded adversary can extract from the CRS about $\mathbf{\Delta}$ is:

1. $\left\{ \mathbf{P}_{\Delta} = \mathbf{\Lambda M} + \mathbf{\Xi N}, \mathbf{A}_{\Delta} = \mathbf{\Delta}^{\top} \mathbf{A} = \begin{pmatrix} \mathbf{\Lambda}^{\top} \mathbf{A} \\ \mathbf{\Xi}^{\top} \mathbf{A} \end{pmatrix} \right\}$ from $\mathsf{crs}_{\Psi_{\mathcal{D}_k}}$,

2. $\left\{ \mathbf{M}_{\Lambda} = \mathbf{\Lambda M} + \mathbf{Z}, \mathbf{N}_{\Xi} = \mathbf{\Xi N} - \mathbf{Z}, \begin{pmatrix} \mathbf{A}_{\Lambda} \\ \mathbf{A}_{\Xi} \end{pmatrix} = \begin{pmatrix} \mathbf{\Lambda}^{\top} \mathbf{A} \\ \mathbf{\Xi}^{\top} \mathbf{A} \end{pmatrix} \right\}$ from $\mathsf{crs}_{\Psi_{\mathcal{D}_k},\mathsf{spl}}$.

Given that the matrix $\mathbf{Z}$ is uniformly random, $\mathsf{crs}_{\Psi_{\mathcal{D}_k}}$ and $\mathsf{crs}_{\Psi_{\mathcal{D}_k},\mathsf{spl}}$ reveal the same information about $\mathbf{\Delta}$ to an unbounded adversary. Therefore, as the proof of soundness is essentially based on the fact that parts of $\mathbf{\Delta}$ are information theoretically hidden to the adversary, the original proof of [21] can be easily adapted for the new arguments. The proofs can be found in the full version.

Theorem 3. *If $\widetilde{\mathcal{D}_k} = \mathcal{D}_k$ and $\tilde{k} = k + 1$, Fig. 2 describes a QA-NIZK proof system with perfect completeness, computational adaptive soundness based on the $\mathcal{D}_k$-SKerMDH Assumption, and perfect zero-knowledge.*

Theorem 4. *If $\widetilde{\mathcal{D}_k} = \overline{\mathcal{D}}_k$ and $\tilde{k} = k$, and $\mathcal{D}_{\Gamma}$ is a witness samplable distribution, Fig. 2 describes a QA-NIZK proof system with perfect completeness, computational adaptive soundness based on the $\mathcal{D}_k$-SKerMDH Assumption, and perfect zero-knowledge.*

3.2 Argument of Sum in Subspace

We can adapt the previous construction to the *Sum in Subspace* Language,

$$\mathcal{L}_{\hat{\mathbf{M}},\check{\mathbf{N}},+} := \{(\hat{\mathbf{x}}, \check{\mathbf{y}}) \in \hat{\mathbb{G}}^m \times \check{\mathbb{H}}^m : \exists \mathbf{w} \in \mathbb{Z}_q^t, \ \mathbf{x} + \mathbf{y} = (\mathbf{M} + \mathbf{N})\mathbf{w}\}.$$

We define two proof systems $\Psi_{\mathcal{D}_k,+}$, $\Psi_{\overline{\mathcal{D}}_k,+}$ as in Fig. 2, but now with $\mathbf{\Lambda} = \mathbf{\Xi}$. Intuitively, soundness follows from the same argument because the information about $\mathbf{\Lambda}$ in the CRS is now $\mathbf{\Lambda}^{\top} \mathbf{A}, \mathbf{\Lambda}(\mathbf{M} + \mathbf{N})$.

3.3 Argument of Equal Opening in Different Groups

Given the results for Subspace Concatenation of Sect. 3.1, it is direct to construct constant-size NIZK Arguments of membership in:

$$\mathcal{L}_{\mathsf{com},\hat{\mathbf{U}},\check{\mathbf{V}},\nu} := \left\{ (\hat{\mathbf{c}}, \check{\mathbf{d}}) \in \hat{\mathbb{G}}^m \times \check{\mathbb{H}}^n : \exists (\mathbf{w}, \mathbf{r}, \mathbf{s}), \hat{\mathbf{c}} = \hat{\mathbf{U}} \begin{pmatrix} \mathbf{w} \\ \mathbf{r} \end{pmatrix}, \check{\mathbf{d}} = \check{\mathbf{V}} \begin{pmatrix} \mathbf{w} \\ \mathbf{s} \end{pmatrix} \right\},$$

where $\hat{\mathbf{U}} \in \hat{\mathbb{G}}^{m \times \tilde{m}}$, $\check{\mathbf{V}} \in \check{\mathbb{H}}^{n \times \tilde{n}}$ and $\mathbf{w} \in \mathbb{Z}_q^{\nu}$. The witness is $(\mathbf{w}, \mathbf{r}, \mathbf{s}) \in \mathbb{Z}_q^{\nu} \times \mathbb{Z}_q^{\tilde{m}-\nu} \times \mathbb{Z}_q^{\tilde{n}-\nu}$. This language is interesting because it can express the fact that $(\hat{\mathbf{c}}, \check{\mathbf{d}})$ are commitments to the same vector $\mathbf{w} \in \mathbb{Z}_q^{\nu}$ in different groups.

The construction is an immediate consequence of the observation that $\mathcal{L}_{\mathsf{com},\hat{\mathbf{U}},\check{\mathbf{V}},\nu}$ can be rewritten as some concatenation language $\mathcal{L}_{\hat{\mathbf{M}},\check{\mathbf{N}}}$. Denote by $\hat{\mathbf{U}}_1$ the first ν columns of $\hat{\mathbf{U}}$ and $\hat{\mathbf{U}}_2$ the remaining ones, and $\check{\mathbf{V}}_1$ the first ν columns of $\check{\mathbf{V}}$ and $\check{\mathbf{V}}_2$ the remaining ones. If we define:

$$\hat{\mathbf{M}} := (\hat{\mathbf{U}}_1 || \hat{\mathbf{U}}_2 || \hat{\mathbf{0}}_{m \times (\tilde{n}-\nu)}) \qquad \check{\mathbf{N}} := (\check{\mathbf{V}}_1 || \check{\mathbf{0}}_{n \times (\tilde{m}-\nu)} || \check{\mathbf{V}}_2).$$

then it is immediate to verify that $\mathcal{L}_{\mathsf{com},\hat{\mathbf{U}},\check{\mathbf{V}},\nu} = \mathcal{L}_{\hat{\mathbf{M}},\check{\mathbf{N}}}$.

In the rest of the paper, we denote as $\Psi_{\overline{\mathcal{D}}_k,\mathsf{com}}$ the proof system for $\mathcal{L}_{\mathsf{com},\hat{\mathbf{U}},\check{\mathbf{V}},\nu}$ which corresponds to $\Psi_{\overline{\mathcal{D}}_k,\mathsf{spl}}$ for $\mathcal{L}_{\hat{\mathbf{M}},\check{\mathbf{N}}}$, where $\hat{\mathbf{M}}, \check{\mathbf{N}}$ are the matrices defined above. Note that for commitment schemes we can generally assume $\hat{\mathbf{U}}, \check{\mathbf{V}}$ to be drawn from some witness samplable distribution.

4   Aggregating Groth-Sahai Proofs in Asymmetric Groups

Jutla and Roy [20] show how to aggregate GS proofs of two-sided linear equations in symmetric bilinear groups. In the original construction of [20] soundness is based on a decisional assumption (a weaker variant of the 2-Lin Assumption). Its natural generalization in asymmetric groups (where soundness is based on the SXDH Assumption) only enables to aggregate the proofs of one-sided linear equations.

In this section, we revisit their construction. We give an alternative, simpler, proof of soundness under a computational assumption which avoids altogether the "Switching Lemma" of [20]. Further, we extend it to two-sided equations in the asymmetric setting. For one-sided linear equations we can prove soundness under any kernel assumption and for two-sided linear equations, under any split kernel assumption.[2]

Let A_1, A_2, A_T be $\mathbb{Z}_q$-vector spaces compatible with some Groth-Sahai equation as detailed in Sect. 2.2. Let $\mathcal{D}_\Gamma$ be a witness samplable distribution which outputs n pairs of vectors $(\boldsymbol{\alpha}_\ell, \boldsymbol{\beta}_\ell) \in A_1^{m_y} \times A_2^{m_x}$, $\ell \in [n]$, for some $m_x, m_y \in \mathbb{N}$. Given some fixed pairs $(\boldsymbol{\alpha}_\ell, \boldsymbol{\beta}_\ell)$, we define, for each $\tilde{\mathbf{t}} \in A_T^n$, the set of equations $\mathcal{S}_{\tilde{\mathbf{t}}}$ as:

$$\mathcal{S}_{\tilde{\mathbf{t}}} = \left\{ E_\ell(\mathbf{x}, \mathbf{y}) = \tilde{t}_\ell : \ell \in [n] \right\}, \quad E_\ell(\mathbf{x}, \mathbf{y}) := \sum_{j \in [m_y]} f(\alpha_{\ell,j}, y_j) + \sum_{i \in [m_x]} f(x_i, \beta_{\ell,i}).$$

We note that, as in [20], we only achieve *quasi-adaptive aggregation*, that is, the common reference string is specific to a particular set of equations. More specifically, it depends on the constants $\boldsymbol{\alpha}_\ell, \boldsymbol{\beta}_\ell$ (but not on $\tilde{t}_\ell$, which can be chosen by the prover) and it can be used to aggregate the proofs of $\mathcal{S}_{\tilde{\mathbf{t}}}$, for any $\tilde{\mathbf{t}}$.

Given the equation types for which we can construct NIZK GS proofs, there always exists (1) $t_\ell \in A_1$, such that $\tilde{t}_\ell = f(t_\ell, \mathsf{base}_2)$ or (2) $\tilde{t}_\ell \in A_2$, such that $\tilde{t}_\ell = f(\mathsf{base}_1, t_\ell)$, where $\mathsf{base}_i = 1$ if $A_i = \mathbb{Z}_q$, $\mathsf{base}_1 = \hat{g}$ if $A_1 = \hat{\mathbb{G}}$ and $\mathsf{base}_2 = \check{h}$ if $A_2 = \check{\mathbb{H}}$. This is because $\tilde{t}_\ell = 0_T$ for PPEs, and $A_T = A_i$, for some $i \in [2]$, for other types of equations. For simplicity, in the construction we assume that (1) is the case, otherwise change $\iota_2(a_{\ell,i}), \iota_1(t_\ell)$ for $\iota_1(a_{\ell,i}), \iota_2(t_\ell)$ in the construction below.

---

[2] The results of [20] are based on what they call the "Switching Lemma". As noted in [24], it is implicit in the proof of this lemma that the same results can be obtained under computational assumptions.

$\mathsf{K}_0(1^\Lambda)$: Return $\Gamma := (q, \hat{\mathbb{G}}, \check{\mathbb{H}}, \mathbb{T}, e, \hat{g}, \check{h}) \leftarrow \mathsf{Gen}_a(1^\Lambda)$.

$\mathcal{D}_\Gamma$: $\mathcal{D}_\Gamma$ is some distribution over n pairs of vectors $(\boldsymbol{\alpha}_\ell, \boldsymbol{\beta}_\ell) \in A_1^{m_x} \times A_2^{m_y}$.

$\mathsf{K}_1(\Gamma, \mathcal{S}_{\check{\mathfrak{t}}})$: Let $\mathbf{A} = (a_{i,j}) \leftarrow \mathcal{D}_{n,k}$. Define

$$\mathsf{crs} := \left(\mathsf{crs}_{\mathsf{GS}}, \left\{ \sum_{\ell \in [n]} \iota_1(a_{\ell,i}\boldsymbol{\alpha}_\ell), \sum_{\ell \in [n]} \iota_2(a_{\ell,i}\boldsymbol{\beta}_\ell), \left\{ \iota_2(a_{\ell,i}) : \ell \in [n] \right\} : i \in [k] \right\} \right)$$

$\mathsf{P}(\Gamma, \mathcal{S}_{\check{\mathfrak{t}}}, \mathbf{x}, \mathbf{y})$: Given a solution $\mathbf{x} = x$, $\mathbf{y} = y$ to $\mathcal{S}_{\check{\mathfrak{t}}}$, the prover proceeds as follows:

  – Commit to all $x_j \in A_1$ as $\hat{\mathbf{c}}_j \leftarrow \mathsf{Comm}_{\mathsf{GS}}(x_j)$, and to all $y_j \in A_2$ as $\check{\mathbf{d}}_j \leftarrow \mathsf{Comm}_{\mathsf{GS}}(y_j)$.
  – For each $i \in [k]$, run the GS prover for the equation $\sum_{\ell \in [n]} a_{\ell,i} E_\ell(\mathbf{x}, \mathbf{y}) = \sum_{\ell \in [n]} f(t_\ell, a_{\ell,i})$ to obtain the proof, which is a pair $(\hat{\boldsymbol{\Theta}}_i, \check{\boldsymbol{\Pi}}_i)$.

Output $(\{\hat{\mathbf{c}}_j : j \in [m_x]\}, \{\check{\mathbf{d}}_j : j \in [m_y]\}, \{(\check{\boldsymbol{\Pi}}_i, \hat{\boldsymbol{\Theta}}_i) : i \in [k]\})$.

$\mathsf{V}(\mathsf{crs}, \mathcal{S}_{\check{\mathfrak{t}}}, \{\hat{\mathbf{c}}_j\}_{j \in [m_x]}, \{\check{\mathbf{d}}_j\}_{j \in [m_y]}, \{\hat{\boldsymbol{\Theta}}_i, \check{\boldsymbol{\Pi}}_i\}_{i \in [k]})$: For each $i \in [k]$, run the GS verifier for equation

$$\sum_{\ell \in [n]} a_{\ell,i} E_\ell(\mathbf{x}, \mathbf{y}) = \sum_{\ell \in [n]} f(t_\ell, a_{\ell,i}).$$

Theorem 5. *The above protocol is a QA-NIZK proof system for two-sided linear equations.*

Proof. Completeness. Observe that

$$\sum_{\ell \in [n]} a_{\ell,i} E_\ell(\mathbf{x}, \mathbf{y}) = \sum_{j \in [m_y]} f(a_{\ell,i}\alpha_{\ell,j}, \mathsf{y}_j) + \sum_{j \in [m_x]} f(\mathsf{x}_j, a_{\ell,i}\beta_{\ell,j}). \qquad (2)$$

Completeness follows from the observation that to efficiently compute the proof, the GS Prover [16] only needs, a part from a satisfying assignment to the equation, the randomness used in the commitments, plus a way to compute the inclusion map of all involved constants, in this case $\iota_1(a_{\ell,i}\alpha_{\ell,j})$, $\iota_2(a_{\ell,i}\beta_{\ell,j})$ and the latter is part of the CRS.

Soundness. We change to a game Game_1 where we know the discrete logarithm of the GS commitment key, as well as the discrete logarithms of $(\boldsymbol{\alpha}_\ell, \boldsymbol{\beta}_\ell)$, $\ell \in [n]$. This is possible because they are both chosen from a witness samplable distribution.

We now prove that an adversary against the soundness in Game_1 can be used to construct an adversary B against the $\mathcal{D}_{n,k}$-SKerMDH Assumption, where $\mathcal{D}_{n,k}$ is the matrix distribution used in the CRS generation.

B receives a challenge $(\hat{\mathbf{A}}, \check{\mathbf{A}}) \in \hat{\mathbb{G}}^{n \times k} \times \check{\mathbb{H}}^{n \times k}$. Given all the discrete logarithms that B knows, it can compute a properly distributed CRS even without knowledge of the discrete logarithm of $\hat{\mathbf{A}}$. The soundness adversary outputs commitments $\{\hat{\mathbf{c}}_j\}_{j \in [m_x]}, \{\check{\mathbf{d}}_j\}_{j \in [m_y]}$ together with proofs $\{\hat{\boldsymbol{\Theta}}_i, \check{\boldsymbol{\Pi}}_i\}_{i \in [k]}$, which are accepted by the verifier.

Let $\mathbf{x}$ (resp. $\hat{\mathbf{x}}$) be the vector of openings of $\{\hat{\mathbf{c}}_j\}_{j \in [m_x]}$ in A_1 (resp. in the group $\hat{\mathbb{G}}$) and $\mathbf{y}$ (resp. $\check{\mathbf{y}}$) the vector of openings of $\{\check{\mathbf{d}}_j\}_{j \in [m_y]}$ in A_2 (resp. in the group $\check{\mathbb{H}}$). If $A_1 = \hat{\mathbb{G}}$ (resp. $A_2 = \check{\mathbb{H}}$) then $\mathbf{x} = \hat{\mathbf{x}}$ (resp. $\mathbf{y} = \check{\mathbf{y}}$). The vectors $\hat{\mathbf{x}}$ and $\check{\mathbf{y}}$ are efficiently computable by B who knows the discrete logarithm of the commitment keys. We claim that the pair $(\hat{\boldsymbol{\rho}}, \check{\boldsymbol{\sigma}}) \in \hat{\mathbb{G}}^n \times \check{\mathbb{H}}^n$, $\hat{\boldsymbol{\rho}} := (\boldsymbol{\beta}_1^\top \hat{\mathbf{x}} - \hat{t}_1, \ldots, \boldsymbol{\beta}_n^\top \hat{\mathbf{x}} - \hat{t}_n), \check{\boldsymbol{\sigma}} := (\boldsymbol{\alpha}_1^\top \check{\mathbf{y}}, \ldots, \boldsymbol{\alpha}_n^\top \check{\mathbf{y}})$, solves the $\mathcal{D}_{n,k}$-SKerMDH challenge.

First, observe that if the adversary is successful in breaking the soundness property, then $\boldsymbol{\rho} \neq \boldsymbol{\sigma}$. Indeed, if this is the case there is some index $\ell \in [n]$ such that $E_\ell(\mathbf{x}, \mathbf{y}) \neq \tilde{t}_\ell$, which means that $\sum_{j \in [m_y]} f(\alpha_{\ell,j}, \mathsf{y}_j) \neq \sum_{j \in [m_x]} f(\mathsf{x}_j, \beta_{\ell,j}) - f(t_\ell, \mathsf{base}_2)$. If we take discrete logarithms in each side of the equation, this inequality is exactly equivalent to $\boldsymbol{\rho} \neq \boldsymbol{\sigma}$.

Further, because GS proofs have perfect soundness, $\mathbf{x}$ and $\mathbf{y}$ satisfy the equation $\sum_{\ell \in [n]} a_{\ell,i} E_\ell(\mathbf{x}, \mathbf{y}) = \sum_{\ell \in [n]} f(t_\ell, a_{\ell,i})$, for all $i \in [k]$, Thus, for all $i \in [k]$,

$$\sum_{\ell \in [n]} \check{a}_{\ell,i} \left(\boldsymbol{\beta}_\ell^\top \hat{\mathbf{x}} - \hat{t}_\ell \right) = \sum_{\ell \in [n]} \hat{a}_{\ell,i} \left(\boldsymbol{\alpha}_\ell^\top \check{\mathbf{y}} \right), \tag{3}$$

which implies that $\hat{\boldsymbol{\rho}} \check{\mathbf{A}} = \check{\boldsymbol{\sigma}} \hat{\mathbf{A}}$.

Zero-Knowledge. The same simulator of GS proofs can be used. Specifically the simulated proof corresponds to k simulated GS proofs.

One-Sided Equations. In the case when $\boldsymbol{\alpha}_\ell = \mathbf{0}$ and $\tilde{t}_\ell = f(t_\ell, \mathsf{base}_2)$ for some $t_\ell \in A_1$, for all $\ell \in [n]$, proofs can be aggregated under a standard Kernel Assumption (and thus, in asymmetric bilinear groups we can choose $k = 1$). Indeed, in this case, in the soundness proof, the adversary B receives $\check{\mathbf{A}} \in \check{\mathbb{H}}^{n \times k}$, an instance of the $\mathcal{D}_{n,k} - \mathsf{KerMDH}_{\check{\mathbb{H}}}$ problem. The adversary B outputs $\hat{\boldsymbol{\rho}} := (\boldsymbol{\beta}_1^\top \hat{\mathbf{x}} - \hat{t}_1, \ldots, \boldsymbol{\beta}_n^\top \hat{\mathbf{x}} - \hat{t}_n)$ as a solution to the challenge. To see why this works, note that, when $\boldsymbol{\alpha}_\ell = \mathbf{0}$ for all $\ell \in [n]$, equation (3) reads $\sum_{\ell \in [n]} \check{a}_{\ell,i} \left(\boldsymbol{\beta}_\ell^\top \hat{\mathbf{x}} - \hat{t}_\ell \right) = \mathbf{0}_\mathbb{T}$ and thus $\hat{\boldsymbol{\rho}} \check{\mathbf{A}} = \mathbf{0}_\mathbb{T}$. The case when $\boldsymbol{\beta}_\ell = \mathbf{0}$ and $\tilde{t}_\ell = f(\mathsf{base}_1, t_\ell)$ for some $t_\ell \in A_2$, for all $\ell \in [n]$, is analogous.

5   QA-NIZK Arguments for Bit-Strings

We construct a constant-size QA-NIZK for proving that a perfectly binding commitment opens to a bit-string. That is, we prove membership in the language:

$$\mathcal{L}_{\hat{\mathbf{U}}, \mathsf{bits}} := \{\hat{\mathbf{c}} \in \hat{\mathbb{G}}^{n+m} : \hat{\mathbf{c}} := \hat{\mathbf{U}}_1 \mathbf{b} + \hat{\mathbf{U}}_2 \mathbf{w}, (\mathbf{b}, \mathbf{w}) \in \{0,1\}^n \times \mathbb{Z}_q^m\},$$

where $\hat{\mathbf{U}} := (\hat{\mathbf{U}}_1, \hat{\mathbf{U}}_2) \in \hat{\mathbb{G}}^{(n+m) \times n} \times \hat{\mathbb{G}}^{(n+m) \times m}$ defines perfectly binding and computationally hiding commitment keys. The witness for membership is $(\mathbf{b}, \mathbf{w})$ and $\hat{\mathbf{U}} \leftarrow \mathcal{D}_\Gamma$, where $\mathcal{D}_\Gamma$ is some witness samplable distribution.

To prove that a commitment in $\hat{\mathbb{G}}$ opens to a vector of bits $\mathbf{b}$, the usual strategy is to compute another commitment $\check{\mathbf{d}} \in \check{\mathbb{H}}^{\tilde{n}}$ to a vector $\bar{\mathbf{b}} \in \mathbb{Z}_q^n$ and prove (1) $b_i(\bar{b}_i - 1) = 0$, for all $i \in [n]$, and (2) $b_i - \bar{b}_i = 0$, for all $i \in [n]$.

For statement (2), since $\hat{\mathbf{U}}$ is witness samplable, we can use our most efficient QA-NIZK from Sect. 3.3 for equal opening in different groups. Under the SSDP Assumption, which is the SKerMDH Assumption of minimal size conjectured to hold in asymmetric groups, the proof is of size $2(\mathfrak{g} + \mathfrak{h})$. Thus, the challenge is to aggregate n equations of the form $b_i(\overline{b}_i - 1) = 0$. We note that this is a particular case of the problem of aggregating proofs of quadratic equations, which was left open in [20].

We finally remark that the proof must include $\check{\mathbf{d}}$ and thus it may be not of size independent of n. However, it turns out that $\check{\mathbf{d}}$ needs not be perfectly binding, in fact $\bar{n} = 2$ suffices.

Intuition. A prover wanting to show satisfiability of the equation $\mathsf{x}(\mathsf{y} - 1) = 0$ using GS proofs, will commit to a solution $\mathsf{x} = b$ and $\mathsf{y} = \overline{b}$ as $\hat{\mathbf{c}} = b\hat{\mathbf{u}}_1 + r\hat{\mathbf{u}}_2$ and $\check{\mathbf{d}} = \overline{b}\check{\mathbf{v}}_1 + s\check{\mathbf{v}}_2$, for $r, s \leftarrow \mathbb{Z}_q$, and then give a pair $(\hat{\boldsymbol{\theta}}, \check{\boldsymbol{\pi}}) \in \hat{\mathbb{G}}^2 \times \check{\mathbb{H}}^2$ which satisfies the following verification equation[3]:

$$\hat{\mathbf{c}} \left(\check{\mathbf{d}} - \check{\mathbf{v}}_1\right)^\top = \hat{\mathbf{u}}_2 \check{\boldsymbol{\pi}}^\top + \hat{\boldsymbol{\theta}} \check{\mathbf{v}}_2^\top. \tag{4}$$

The reason why this works is that, if we express both sides of the equation in the basis of $\mathbb{T}^{2 \times 2}$ given by $\{\hat{\mathbf{u}}_1 \check{\mathbf{v}}_1^\top, \hat{\mathbf{u}}_2 \check{\mathbf{v}}_1^\top, \hat{\mathbf{u}}_1 \check{\mathbf{v}}_2^\top, \hat{\mathbf{u}}_2 \check{\mathbf{v}}_2^\top\}$, the coefficient of $\hat{\mathbf{u}}_1 \check{\mathbf{v}}_1^\top$ is $b(\overline{b} - 1)$ on the left side and 0 on the right side (regardless of $(\hat{\boldsymbol{\theta}}, \check{\boldsymbol{\pi}})$). Our observation is that the verification equation can be abstracted as saying:

$$\hat{\mathbf{c}} \left(\check{\mathbf{d}} - \check{\mathbf{v}}_1\right)^\top \in \mathsf{Span}(\hat{\mathbf{u}}_2 \check{\mathbf{v}}_1^\top, \hat{\mathbf{u}}_1 \check{\mathbf{v}}_2^\top, \hat{\mathbf{u}}_2 \check{\mathbf{v}}_2^\top) \subset \mathbb{T}^{2 \times 2}. \tag{5}$$

Now consider commitments to $(b_1, \ldots, b_n)$ and $(\overline{b}_1, \ldots, \overline{b}_n)$ constructed with some commitment key $\{(\hat{\mathbf{g}}_i, \check{\mathbf{h}}_i) : i \in [n+1]\} \subset \hat{\mathbb{G}}^{\bar{n}} \times \check{\mathbb{H}}^{\bar{n}}$, for some $\bar{n} \in \mathbb{N}$, to be determined later, and defined as $\hat{\mathbf{c}} := \sum_{i \in [n]} b_i \hat{\mathbf{g}}_i + r\hat{\mathbf{g}}_{n+1}$, $\check{\mathbf{d}} := \sum_{i \in [n]} \overline{b}_i \check{\mathbf{h}}_i + s\check{\mathbf{h}}_{n+1}$, $r, s \leftarrow \mathbb{Z}_q$. Suppose for a moment that $\{\hat{\mathbf{g}}_i \check{\mathbf{h}}_j^\top : i, j \in [n+1]\}$ is a set of linearly independent vectors. Then,

$$\hat{\mathbf{c}} \left(\check{\mathbf{d}}^\top - \sum_{j \in [n]} \check{\mathbf{h}}_j^\top\right) \in \mathsf{Span}\{\hat{\mathbf{g}}_i \check{\mathbf{h}}_j^\top : (i, j) \in \mathcal{I}_{N,1}\} \tag{6}$$

if and only if $b_i(\overline{b}_i - 1) = 0$ for all $i \in [n]$, because $b_i(\overline{b}_i - 1)$ is the coordinate of $\hat{\mathbf{g}}_i \check{\mathbf{h}}_i^\top$ in the left side of the equation.

Equation 6 suggests to use one of the constant-size QA-NIZK Arguments for linear spaces to get a constant-size proof that $b_i(\overline{b}_i - 1) = 0$ for all $i \in [n]$. Unfortunately, these arguments are only defined for membership in subspaces in $\hat{\mathbb{G}}^m$ or $\check{\mathbb{H}}^m$ but not in $\mathbb{T}^m$. Our solution is to include information in the CRS to "bring back" this statement from $\mathbb{T}$ to $\hat{\mathbb{G}}$, i.e. the matrices $\hat{\mathbf{C}}_{i,j} := \hat{\mathbf{g}}_i \check{\mathbf{h}}_j^\top$, for each $(i, j) \in \mathcal{I}_{N,1}$. Then, to prove that $b_i(\overline{b}_i - 1) = 0$ for all $i \in [n]$, the prover

---

[3] For readers familiar with the Groth-Sahai notation, Eq. (4) corresponds to $\mathbf{c} \bullet (\mathbf{d} - \iota_2(1)) = \mathbf{u}_2 \bullet \boldsymbol{\pi} + \boldsymbol{\theta} \bullet \mathbf{v}_2$.

computes $\hat{\Theta}_{b(\bar{b}-1)}$ as a linear combination of $\mathcal{C} := \{\hat{\mathbf{C}}_{i,j} : (i,j) \in \mathcal{I}_{N,1}\}$ (with coefficients which depend on $\mathbf{b}, \bar{\mathbf{b}}, r, s$) such that

$$\hat{\mathbf{c}}\left(\check{\mathbf{d}} - \sum_{j\in[n]} \check{\mathbf{h}}_j\right)^{\top} = \hat{\Theta}_{b(\bar{b}-1)}\check{\mathbf{I}}_{\bar{n}\times\bar{n}}, \tag{7}$$

and gives a QA-NIZK proof of $\hat{\Theta}_{b(\bar{b}-1)} \in \mathsf{Span}(\mathcal{C})$.

This reasoning assumes that $\{\hat{\mathbf{g}}_i\check{\mathbf{h}}_j^{\top}\}$ (or equivalently, $\{\hat{\mathbf{C}}_{i,j}\}$) are linearly independent, which can only happen if $\bar{n} \geq n+1$. If that is the case, the proof cannot be constant because $\hat{\Theta}_{b(\bar{b}-1)} \in \hat{\mathbb{G}}^{\bar{n}\times\bar{n}}$ and this matrix is part of the proof. Instead, we choose $\hat{\mathbf{g}}_1, \ldots, \hat{\mathbf{g}}_{n+1} \in \hat{\mathbb{G}}^2$ and $\check{\mathbf{h}}_1, \ldots, \check{\mathbf{h}}_{n+1} \in \check{\mathbb{H}}^2$, so that $\{\hat{\mathbf{C}}_{i,j}\} \subseteq \hat{\mathbb{G}}^{2\times2}$. Intuitively, this should still work because the prover receives these vectors as part of the CRS and he does not know their discrete logarithms, so to him, they behave as linearly independent vectors.

With this change, the statement $\hat{\Theta}_{b(\bar{b}-1)} \in \mathsf{Span}(\mathcal{C})$ seems no longer meaningful, as $\mathsf{Span}(\mathcal{C})$ is all of $\hat{\mathbb{G}}^{2\times2}$ with overwhelming probability. But this is not the case, because by means of decisional assumptions in $\hat{\mathbb{G}}^2$ and in $\check{\mathbb{H}}^2$, we switch to a game where the matrices $\hat{\mathbf{C}}_{i,j}$ span a non-trivial space of $\hat{\mathbb{G}}^{2\times2}$. Specifically, to a game where $\hat{\mathbf{C}}_{i^*,i^*} \notin \mathsf{Span}(\mathcal{C})$ and $i^* \leftarrow [n]$ remains hidden to the adversary. Once we are in such a game, perfect soundness is guaranteed for equation $b_{i^*}(\bar{b}_{i^*} - 1) = 0$ and a cheating adversary is caught with probability at least $1/n$. We think this technique might be of independent interest.

The last obstacle is that, using decisional assumptions on the set of vectors $\{\check{\mathbf{h}}_j\}_{j\in[n+1]}$ is incompatible with using the discrete logarithms of $\check{\mathbf{h}}_j$ to compute the matrices $\hat{\mathbf{C}}_{i,j} := \hat{\mathbf{g}}_i\check{\mathbf{h}}_j^{\top}$ given in the CRS. To account for the fact that, in some games, we only know $\mathbf{g}_i \in \mathbb{Z}_q$ and, in some others, only $\mathbf{h}_j \in \mathbb{Z}_q$, we replace each matrix $\hat{\mathbf{C}}_{i,j}$ by a pair $(\hat{\mathbf{C}}_{i,j}, \check{\mathbf{D}}_{i,j})$ which is uniformly distributed conditioned on $\mathbf{C}_{i,j} + \mathbf{D}_{i,j} = \mathbf{g}_i\mathbf{h}_j^{\top}$. This randomization completely hides the group in which we can compute $\mathbf{g}_i\mathbf{h}_j^{\top}$. Finally, we use our QA-NIZK Argument for sum in a subspace (Sect. 3.2) to prove membership in this space.

Instantiations. We discuss in detail two particular cases of languages $\mathcal{L}_{\hat{\mathbf{U}},\mathsf{bits}}$. First, in Sect. 5.1 we discuss the case when

(a) $\hat{\mathbf{c}}$ is a vector in $\hat{\mathbb{G}}^{n+1}$, $\hat{\mathbf{u}}_{n+1} \leftarrow \mathcal{L}_{n+1,1}$ and $\hat{\mathbf{U}}_1 := \begin{pmatrix} \hat{\mathbf{I}}_{n\times n} \\ \hat{\mathbf{0}}_{1\times n} \end{pmatrix} \in \hat{\mathbb{G}}^{(n+1)\times n}, \hat{\mathbf{U}}_2 :=$

$\hat{\mathbf{u}}_{n+1} \in \hat{\mathbb{G}}^{n+1}, \hat{\mathbf{U}} = (\hat{\mathbf{U}}_1\|\hat{\mathbf{U}}_2)$.

In this case, the vectors $\hat{\mathbf{g}}_i$ in the intuition are defined as $\hat{\mathbf{g}}_i = \Delta\hat{\mathbf{u}}_i$, where $\Delta \leftarrow \mathbb{Z}_q^{2\times(n+1)}$, and the commitment to $\mathbf{b}$ is computed as $\hat{\mathbf{c}} := \sum_{i\in[n]} b_i\hat{\mathbf{u}}_i + w\hat{\mathbf{u}}_{n+1}$. Then in Sect. 5.3 we discuss how to generalize the construction for a) to

(b) $\hat{\mathbf{c}}$ is the concatenation of n GS commitments. That is, given the GS CRS $\mathsf{crs}_{\mathrm{GS}} = (\Gamma, \hat{\mathbf{u}}_1, \hat{\mathbf{u}}_2, \check{\mathbf{v}}_1, \check{\mathbf{v}}_2)$, we define,

$$\hat{\mathbf{U}}_1 := \begin{pmatrix} \hat{\mathbf{u}}_1 & \dots & \hat{\mathbf{0}} \\ \vdots & \ddots & \vdots \\ \hat{\mathbf{0}} & \dots & \hat{\mathbf{u}}_1 \end{pmatrix} \in \hat{\mathbb{G}}^{2n \times n}, \hat{\mathbf{U}}_2 := \begin{pmatrix} \hat{\mathbf{u}}_2 & \dots & \hat{\mathbf{0}} \\ \vdots & \ddots & \vdots \\ \hat{\mathbf{0}} & \dots & \hat{\mathbf{u}}_2 \end{pmatrix} \in \hat{\mathbb{G}}^{2n \times n}.$$

Although the proof size is constant, in both of our instantiations the commitment size is $\Theta(n)$. Specifically, $(n+1)\mathfrak{g}$ for case a) and $2n\mathfrak{g}$ for case b).

5.1   The Scheme

$\mathsf{K}_0(1^\lambda)$: Return $\Gamma := (q, \hat{\mathbb{G}}, \check{\mathbb{H}}, \mathbb{T}, e, \hat{g}, \check{h}) \leftarrow \mathsf{Gen}_a(1^\lambda)$.

$\mathcal{D}_\Gamma$: The distribution $\mathcal{D}_\Gamma$ over $\hat{\mathbb{G}}^{(n+1)\times(n+1)}$ is some witness samplable distribution which defines the relation $\mathcal{R}_\Gamma = \{\mathcal{R}_{\hat{\mathbf{U}}}\} \subseteq \hat{\mathbb{G}}^{n+1} \times (\{0,1\}^n \times \mathbb{Z}_q)$, where $\hat{\mathbf{U}} \leftarrow \mathcal{D}_\Gamma$, such that $(\hat{\mathbf{c}}, \langle \mathbf{b}, w \rangle) \in \mathcal{R}_{\hat{\mathbf{U}}}$ iff $\hat{\mathbf{c}} = \hat{\mathbf{U}}\binom{\mathbf{b}}{w}$. The relation $\mathcal{R}_{par}$ consists of pairs $(\hat{\mathbf{U}}, \mathbf{U})$ where $\hat{\mathbf{U}} \leftarrow \mathcal{D}_\Gamma$.

$\mathsf{K}_1(\Gamma, \hat{\mathbf{U}})$: Let $\mathbf{h}_{n+1} \leftarrow \mathbb{Z}_q^2$ and for all $i \in [n]$, $\mathbf{h}_i := \epsilon_i \mathbf{h}_{n+1}$, where $\epsilon_i \leftarrow \mathbb{Z}_q$. Define $\check{\mathbf{H}} := (\check{\mathbf{h}}_1 || \dots || \check{\mathbf{h}}_{n+1})$. Choose $\mathbf{\Delta} \leftarrow \mathbb{Z}_q^{2\times(n+1)}$, define $\hat{\mathbf{G}} := \mathbf{\Delta}\hat{\mathbf{U}}$ and $\hat{\mathbf{g}}_i := \mathbf{\Delta}\hat{\mathbf{u}}_i \in \hat{\mathbb{G}}^2$, for all $i \in [n+1]$. Let $\mathbf{a} \leftarrow \mathcal{L}_1$ and define $\check{\mathbf{a}}_\Delta := \mathbf{\Delta}^\top \check{\mathbf{a}} \in \check{\mathbb{H}}^{n+1}$. For any pair $(i,j) \in \mathcal{I}_{N,1}$, let $\mathbf{T}_{i,j} \leftarrow \mathbb{Z}_q^{2\times2}$ and set:

$$\hat{\mathbf{C}}_{i,j} := \hat{\mathbf{g}}_i \mathbf{h}_j^\top - \hat{\mathbf{T}}_{i,j} \in \hat{\mathbb{G}}^{2\times2}, \qquad \check{\mathbf{D}}_{i,j} := \check{\mathbf{T}}_{i,j} \in \check{\mathbb{H}}^{2\times2}.$$

Note that $\hat{\mathbf{C}}_{i,j}$ can be efficiently computed as $\mathbf{h}_j \in \mathbb{Z}_q^2$ is the vector of discrete logarithms of $\check{\mathbf{h}}_j$.

Let $\Psi_{\overline{\mathcal{D}}_k,+}$ be the proof system for Sum in Subspace (Sect. 3.2) and $\Psi_{\overline{\mathcal{D}}_k,\mathsf{com}}$ be an instance of our proof system for Equal Opening (Sect. 3.3).

Let $\mathsf{crs}_{\Psi_{\overline{\mathcal{D}}_k,+}} \leftarrow \mathsf{K}_1(\Gamma, \{\hat{\mathbf{C}}_{i,j}, \check{\mathbf{D}}_{i,j}\}_{(i,j)\in\mathcal{I}_{N,1}})$ and[4] $\mathsf{crs}_{\Psi_{\overline{\mathcal{D}}_k,\mathsf{com}}} \leftarrow \mathsf{K}_1(\Gamma, \hat{\mathbf{G}}, \check{\mathbf{H}}, n)$.

The common reference string is given by:

$$\mathsf{crs}_P := \left(\hat{\mathbf{U}}, \hat{\mathbf{G}}, \check{\mathbf{H}}, \{\hat{\mathbf{C}}_{i,j}, \check{\mathbf{D}}_{i,j}\}_{(i,j)\in\mathcal{I}_{N,1}}, \mathsf{crs}_{\Psi_{\overline{\mathcal{D}}_k,+}}, \mathsf{crs}_{\Psi_{\overline{\mathcal{D}}_k,\mathsf{com}}} \right),$$

$$\mathsf{crs}_V := \left(\check{\mathbf{a}}, \check{\mathbf{a}}_\Delta, \mathsf{crs}_{\Psi_{\overline{\mathcal{D}}_k,+}}, \mathsf{crs}_{\Psi_{\overline{\mathcal{D}}_k,\mathsf{com}}} \right).$$

$\mathsf{P}(\mathsf{crs}_P, \hat{\mathbf{c}}, \langle \mathbf{b}, w_g \rangle)$: Pick $w_h \leftarrow \mathbb{Z}_q$, $\mathbf{R} \leftarrow \mathbb{Z}_q^{2\times2}$ and then:
1. Define

$$\hat{\mathbf{c}}_\Delta := \hat{\mathbf{G}}\begin{pmatrix} \mathbf{b} \\ w_g \end{pmatrix}, \qquad \check{\mathbf{d}} := \check{\mathbf{H}}\begin{pmatrix} \mathbf{b} \\ w_h \end{pmatrix}.$$

---

[4] We identify matrices in $\hat{\mathbb{G}}^{2\times2}$ (resp. in $\check{\mathbb{H}}^{2\times2}$) with vectors in $\hat{\mathbb{G}}^4$ (resp. in $\check{\mathbb{H}}^4$).

2. Compute $(\hat{\boldsymbol{\Theta}}_{b(\bar{b}-1)}, \check{\boldsymbol{\Pi}}_{b(\bar{b}-1)}) :=$

$$\sum_{i \in [n]} \left(b_i w_h (\hat{\mathbf{C}}_{i,n+1}, \check{\mathbf{D}}_{i,n+1}) + w_g(b_i - 1)(\hat{\mathbf{C}}_{n+1,i}, \check{\mathbf{D}}_{n+1,i}) \right)$$

$$+ \sum_{i \in [n]} \sum_{\substack{j \in [n] \\ j \neq i}} b_i(b_j - 1)(\hat{\mathbf{C}}_{i,j}, \check{\mathbf{D}}_{i,j})$$

$$+ w_g w_h (\hat{\mathbf{C}}_{n+1,n+1}, \check{\mathbf{D}}_{n+1,n+1}) + (\hat{\mathbf{R}}, -\check{\mathbf{R}}). \tag{8}$$

3. Compute a proof $(\hat{\boldsymbol{\rho}}_{b(\bar{b}-1)}, \check{\boldsymbol{\sigma}}_{b(\bar{b}-1)})$ that $\boldsymbol{\Theta}_{b(\bar{b}-1)} + \boldsymbol{\Pi}_{b(\bar{b}-1)}$ belongs to the space spanned by $\{\mathbf{C}_{i,j} + \mathbf{D}_{i,j}\}_{(i,j) \in \mathcal{I}_{N,1}}$, and a proof $(\hat{\boldsymbol{\rho}}_{b-\bar{b}}, \check{\boldsymbol{\sigma}}_{b-\bar{b}})$ that $(\hat{\mathbf{c}}_\Delta, \check{\mathbf{d}})$ open to the same value, using $\mathbf{b}, w_g$, and w_h.

$\mathsf{V}(\mathsf{crs}_V, \hat{\mathbf{c}}, \langle \hat{\mathbf{c}}_\Delta, \check{\mathbf{d}}, (\hat{\boldsymbol{\Theta}}_{b(\bar{b}-1)}, \check{\boldsymbol{\Pi}}_{b(\bar{b}-1)}), \{(\hat{\boldsymbol{\rho}}_X, \check{\boldsymbol{\sigma}}_X)\}_{X \in \{b(\bar{b}-1), b-\bar{b}\}}\rangle)$:

1. Check if $\hat{\mathbf{c}}^\top \check{\mathbf{a}}_\Delta = \hat{\mathbf{c}}_\Delta^\top \check{\mathbf{a}}$.
2. Check if

$$\hat{\mathbf{c}}_\Delta \left(\check{\mathbf{d}} - \sum_{j \in [n]} \check{\mathbf{h}}_j \right)^\top = \hat{\boldsymbol{\Theta}}_{b(\bar{b}-1)} \check{\mathbf{I}}_{2 \times 2} + \hat{\mathbf{I}}_{2 \times 2} \check{\boldsymbol{\Pi}}_{b(\bar{b}-1)}. \tag{9}$$

3. Verify that $(\hat{\boldsymbol{\rho}}_{b(\bar{b}-1)}, \check{\boldsymbol{\sigma}}_{b(\bar{b}-1)}), (\hat{\boldsymbol{\rho}}_{b-\bar{b}}, \check{\boldsymbol{\sigma}}_{b-\bar{b}})$ are valid proofs for $(\hat{\boldsymbol{\Theta}}_{b(\bar{b}-1)}, \check{\boldsymbol{\Pi}}_{b(\bar{b}-1)})$ and $(\hat{\mathbf{c}}_\Delta, \check{\mathbf{d}})$ using $\mathsf{crs}_{\Psi_{\overline{\mathcal{D}}_k,+}}$ and $\mathsf{crs}_{\Psi_{\overline{\mathcal{D}}_k,\mathsf{com}}}$ respectively.

If any of these checks fails, the verifier outputs 0, else it outputs 1.

$\mathsf{S}_1(\Gamma, \hat{\mathbf{U}})$: The simulator receives as input a description of an asymmetric bilinear group Γ and a matrix $\hat{\mathbf{U}} \in \hat{\mathbb{G}}^{(n+1) \times (n+1)}$ sampled according to distribution $\mathcal{D}_\Gamma$. It generates and outputs the CRS in the same way as K_1, but additionally it also outputs the simulation trapdoor

$$\tau = \left(\mathbf{H}, \boldsymbol{\Delta}, \tau_{\Psi_{\overline{\mathcal{D}}_k,+}}, \tau_{\Psi_{\overline{\mathcal{D}}_k,\mathsf{com}}} \right),$$

where $\tau_{\Psi_{\overline{\mathcal{D}}_k,+}}$ and $\tau_{\Psi_{\overline{\mathcal{D}}_k,\mathsf{com}}}$ are, respectively, $\Psi_{\overline{\mathcal{D}}_k,+}$'s and $\Psi_{\overline{\mathcal{D}}_k,\mathsf{com}}$'s simulation trapdoors.

$\mathsf{S}_2(\mathsf{crs}_P, \hat{\mathbf{c}}, \tau)$: Compute $\hat{\mathbf{c}}_\Delta := \boldsymbol{\Delta}\hat{\mathbf{c}}$. Then pick random $\overline{w}_h \leftarrow \mathbb{Z}_q, \mathbf{R} \leftarrow \mathbb{Z}_q^{2 \times 2}$ and define $\mathbf{d} := \overline{w}_h \mathbf{h}_{n+1}$. Then set:

$$\hat{\boldsymbol{\Theta}}_{b(\bar{b}-1)} := \hat{\mathbf{c}}_\Delta \left(\mathbf{d} - \sum_{i \in [n]} \mathbf{h}_i \right)^\top + \hat{\mathbf{R}}, \quad \check{\boldsymbol{\Pi}}_{b(\bar{b}-1)} := -\check{\mathbf{R}}.$$

Finally, simulate proofs $(\hat{\boldsymbol{\rho}}_X, \check{\boldsymbol{\sigma}}_X)$ for $X \in \{b(\bar{b}-1), b - \bar{b}\}$ using $\tau_{\Psi_{\overline{\mathcal{D}}_k,+}}$ and $\tau_{\Psi_{\overline{\mathcal{D}}_k,\mathsf{com}}}$.

5.2   Proof of Security

Completeness is proven in the full version. The following theorem guarantees Soundness.

Theorem 6. *Let* $\mathsf{Adv}_{\mathcal{PS}}(\mathsf{A})$ *be the advantage of an adversary* A *against the soundness of the proof system described above. There exist PPT adversaries* $\mathsf{B}_1, \mathsf{B}_2, \mathsf{B}_3, \mathsf{P}_1^*, \mathsf{P}_2^*$ *such that*

$$\mathsf{Adv}_{\mathcal{PS}}(\mathsf{A}) \leq n \left(6/q + \mathsf{Adv}_{\mathcal{U}_1, \hat{\mathbb{G}}}(\mathsf{B}_1) + \mathsf{Adv}_{\mathcal{U}_1, \check{\mathbb{H}}}(\mathsf{B}_2) + \mathsf{Adv}_{\mathsf{SP}_{\check{\mathbb{H}}}}(\mathsf{B}_3) \right.$$
$$\left. + \mathsf{Adv}_{\Psi_{\overline{\mathcal{D}}_k, +}}(\mathsf{P}_1^*) + \mathsf{Adv}_{\Psi_{\overline{\mathcal{D}}_k, \mathrm{com}}}(\mathsf{P}_2^*) \right).$$

The proof follows from the indistinguishability of the following games:

Real This is the real soundness game. The output is 1 if the adversary breaks the soundness, i.e. the adversary submits some $\hat{\mathbf{c}} = \hat{\mathbf{U}} \left(\begin{smallmatrix} \mathbf{b} \\ w_g \end{smallmatrix} \right)$, for some $\mathbf{b} \notin \{0,1\}^n$ and $w \in \mathbb{Z}_q$, and the corresponding proof which is accepted by the verifier.

Game_0 This game is identical to Real except that algorithm K_1 does not receive $\hat{\mathbf{U}}$ as a input but it samples $(\hat{\mathbf{U}}, \mathbf{U}) \in \mathcal{R}_{par}$ itself according to $\mathcal{D}_\Gamma$.

Game_1 This game is identical to Game_0 except that the simulator picks a random $i^* \in [n]$, and uses $\mathbf{U}$ to check if the output of the adversary A is such that $b_{i^*} \in \{0,1\}$. It aborts if $b_{i^*} \in \{0,1\}$.

Game_2 This game is identical to Game_1 except that now the vectors $\hat{\mathbf{g}}_i$, $i \in [n]$ and $i \neq i^*$, are uniform vectors in the space spanned by $\hat{\mathbf{g}}_{n+1}$.

Game_3 This game is identical to Game_2 except that now the vector $\check{\mathbf{h}}_{i^*}$ is a uniform vector in $\check{\mathbb{H}}^2$, sampled independently of $\check{\mathbf{h}}_{n+1}$.

It is obvious that the first two games are indistinguishable. The rest of the argument goes as follows (the remaining proofs are in the full version).

Lemma 2. $\Pr\left[\mathsf{Game}_1(\mathsf{A}) = 1\right] \geq \dfrac{1}{n} \Pr\left[\mathsf{Game}_0(\mathsf{A}) = 1\right].$

Lemma 3. *There exists a* $\mathcal{U}_1$-*MDDH*$_{\hat{\mathbb{G}}}$ *adversary* B *such that* $|\Pr\left[\mathsf{Game}_1(\mathsf{A}) = 1\right] - \Pr\left[\mathsf{Game}_2(\mathsf{A}) = 1\right]| \leq \mathsf{Adv}_{\mathcal{U}_1, \hat{\mathbb{G}}}(\mathsf{B}) + 2/q.$

Proof. The adversary B receives $(\hat{\mathbf{s}}, \hat{\mathbf{t}})$ an instance of the $\mathcal{U}_1$-MDDH$_{\hat{\mathbb{G}}}$ problem. B defines all the parameters honestly except that it embeds the $\mathcal{U}_1$-MDDH$_{\hat{\mathbb{G}}}$ challenge in the matrix $\hat{\mathbf{G}}$.

Let $\hat{\mathbf{E}} := (\hat{\mathbf{s}}||\hat{\mathbf{t}})$. B picks $i^* \leftarrow [n]$, $\mathbf{W}_0 \leftarrow \mathbb{Z}_q^{2 \times (i^*-1)}$, $\mathbf{W}_1 \leftarrow \mathbb{Z}_q^{2 \times (n-i^*)}$, $\hat{\mathbf{g}}_{i^*} \leftarrow \hat{\mathbb{G}}^2$, and defines $\hat{\mathbf{G}} := (\hat{\mathbf{E}}\mathbf{W}_0||\hat{\mathbf{g}}_{i^*}||\hat{\mathbf{E}}\mathbf{W}_1||\hat{\mathbf{s}})$. In the real algorithm K_1, the generator picks the matrix $\mathbf{\Delta} \in \mathbb{Z}_q^{2 \times (n+1)}$. Although B does not know $\mathbf{\Delta}$, it can compute $\hat{\mathbf{\Delta}}$ as $\hat{\mathbf{\Delta}} = \hat{\mathbf{G}}\mathbf{U}^{-1}$, given that $\mathbf{U}$ is full rank and was sampled by B, so it can compute the rest of the elements of the common reference string using the discrete logarithms of $\hat{\mathbf{U}}$, $\check{\mathbf{H}}$ and $\check{\mathbf{a}}$.

In case $\hat{\mathbf{t}}$ is uniform over $\hat{\mathbb{G}}^2$, by the Schwartz-Zippel lemma $\det(\hat{\mathbf{E}}) = 0$ with probability at most $2/q$. Thus, with probability at least $1 - 2/q$, the matrix $\hat{\mathbf{E}}$ is full-rank and $\hat{\mathbf{G}}$ is uniform over $\hat{\mathbb{G}}^{2 \times (n+1)}$ as in Game_1. On the other hand, in case $\hat{\mathbf{t}} = \gamma \hat{\mathbf{s}}$, all of $\hat{\mathbf{g}}_i$, $i \neq i^*$, are in the space spanned by $\hat{\mathbf{g}}_{n+1}$ as in Game_2.

Lemma 4. *There exists a* $\mathcal{U}_1$*-MDDH*$_{\hat{\mathbb{H}}}$ *adversary* B *such that* $|\Pr[\mathsf{Game}_2(\mathsf{A}) = 1]$ $- \Pr[\mathsf{Game}_3(\mathsf{A}) = 1]| \leq \mathsf{Adv}_{\mathcal{U}_1,\hat{\mathbb{H}}}(\mathsf{B})$.

Lemma 5. *There exists a* $\mathsf{SP}_{\hat{\mathbb{H}}}$ *adversary* B *and soundness adversaries* $\mathsf{P}_1^*, \mathsf{P}_2^*$ *for* $\Psi_{\overline{\mathcal{D}}_k,+}$ *and* $\Psi_{\overline{\mathcal{D}}_k,\mathsf{com}}$ *such that*

$$\Pr[\mathsf{Game}_3(\mathsf{A}) = 1] \leq 4/q + \mathbf{Adv}_{\mathsf{SP}_{\hat{\mathbb{H}}}}(\mathsf{B}) + \mathbf{Adv}_{\Psi_{\overline{\mathcal{D}}_k,+}}(\mathsf{P}_1^*) + \mathbf{Adv}_{\Psi_{\overline{\mathcal{D}}_k,\mathsf{com}}}(\mathsf{P}_2^*).$$

Proof. $\Pr[\det((\mathbf{g}_{i^*}\|\mathbf{g}_{n+1})) = 0] = \Pr[\det((\mathbf{h}_{i^*}\|\mathbf{h}_{n+1})) = 0] \leq 2/q$, by the Schwartz-Zippel lemma. Then, with probability at least $1 - 4/q$, $\mathbf{g}_{i^*}\mathbf{h}_{i^*}^{\top}$ is linearly independent from $\{\mathbf{g}_i\mathbf{h}_j^{\top} : (i,j) \in [n+1]^2 \setminus \{(i^*,i^*)\}\}$ which implies that $\mathbf{g}_{i^*}\mathbf{h}_{i^*}^{\top} \notin \mathsf{Span}(\{\mathbf{C}_{i,j} + \mathbf{D}_{i,j} : (i,j) \in \mathcal{I}_{N,1}\})$. Additionally $\mathsf{Game}_3(\mathsf{A}) = 1$ implies that $b_{i^*} \notin \{0,1\}$ while the verifier accepts the proof produced by A, which is $(\hat{\mathbf{c}}_{\Delta}, \check{\mathbf{d}}, (\hat{\mathbf{\Theta}}_{b(\overline{b}-1)}, \check{\mathbf{\Pi}}_{b(\overline{b}-1)}), \{(\hat{\rho}_X, \check{\sigma}_X)\}_{X \in \{b(\overline{b}-1),b-\overline{b}\}})$. Since $\{\check{\mathbf{h}}_{i^*}, \check{\mathbf{h}}_{n+1}\}$ is a basis of $\check{\mathbb{H}}^2$, we can define $\overline{w}_h, \overline{b}_{i^*}$ as the unique coefficients in $\mathbb{Z}_q$ such that $\check{\mathbf{d}} = \overline{b}_{i^*}\check{\mathbf{h}}_{i^*} + \overline{w}_h\check{\mathbf{h}}_{n+1}$. We distinguish three cases:

(1) If $\hat{\mathbf{c}}_{\Delta} \neq \mathbf{\Delta}\hat{\mathbf{c}}$, we can construct an adversary B against the $\mathsf{SP}_{\hat{\mathbb{H}}}$ Assumption that outputs $\hat{\mathbf{c}}_{\Delta} - \mathbf{\Delta}\hat{\mathbf{c}} \in \ker(\check{\mathbf{a}}^{\top})$.

(2) If $\hat{\mathbf{c}}_{\Delta} = \mathbf{\Delta}\hat{\mathbf{c}}$ but $b_{i^*} \neq \overline{b}_{i^*}$. Given that $(b_i\mathbf{g}_{i^*}, \overline{b}_{i^*}\mathbf{h}_{i^*})$ is linearly independent from $\{(\mathbf{g}_{i^*}, \mathbf{h}_{i^*}), (\mathbf{g}_{n+1}, \mathbf{h}_{n+1})\}$ whenever $b_{i^*} \neq \overline{b}_{i^*}$, an adversary P_2^* against $\Psi_{\overline{\mathcal{D}}_k,\mathsf{com}}$ outputs the pair $(\hat{\rho}_{b-\overline{b}}, \check{\sigma}_{b-\overline{b}})$ which is a fake proof for $(\hat{\mathbf{c}}_{\Delta}, \check{\mathbf{d}})$.

(3) If $\hat{\mathbf{c}}_{\Delta} = \mathbf{\Delta}\hat{\mathbf{c}}$ and $b_{i^*} = \overline{b}_{i^*}$, then $b_{i^*}(\overline{b}_{i^*} - 1) \neq 0$. If we express $\mathbf{\Theta}_{b(\overline{b}-1)} + \mathbf{\Pi}_{b(\overline{b}-1)}$ as a linear combination of $\mathbf{g}_i\mathbf{h}_j^{\top}$, the coordinate of $\mathbf{g}_{i^*}\mathbf{h}_{i^*}^{\top}$ is $b_{i^*}(\overline{b}_{i^*} - 1) \neq 0$ and thus $\mathbf{\Theta}_{b(\overline{b}-1)} + \mathbf{\Pi}_{b(\overline{b}-1)} \notin \mathsf{Span}(\{\mathbf{C}_{i,j} + \mathbf{D}_{i,j} : (i,j) \in \mathcal{I}_{N,1}\})$. The adversary P_1^* against $\Psi_{\overline{\mathcal{D}}_k,+}$ outputs the pair $(\hat{\rho}_{b(\overline{b}-1)}, \check{\sigma}_{b(\overline{b}-1)})$ which is a fake proof for $(\hat{\mathbf{\Theta}}_{b(\overline{b}-1)}, \check{\mathbf{\Pi}}_{b(\overline{b}-1)})$.

This concludes the proof of soundness. Now we prove Zero-Knowledge.

Theorem 7. *The proof system is perfect quasi-adaptive zero-knowledge.*

Proof. First, note that the vector $\check{\mathbf{d}} \in \check{\mathbb{H}}^2$ output by the prover and the vector output by S_2 follow exactly the same distribution. This is because the rank of $\check{\mathbf{H}}$ is 1. In particular, although the simulator S_2 does not know the opening of $\hat{\mathbf{c}}$, which is some $\mathbf{b} \in \{0,1\}^n$, there exists $w_h \in \mathbb{Z}_q$ such that $\check{\mathbf{d}} = \check{\mathbf{H}}\left(\begin{smallmatrix} \mathbf{b} \\ w_h \end{smallmatrix}\right)$. Since $\mathbf{R}$ is chosen uniformly at random in $\mathbb{Z}_q^{2 \times 2}$, the proof $(\hat{\mathbf{\Theta}}_{b(\overline{b}-1)}, \check{\mathbf{\Pi}}_{b(\overline{b}-1)})$ is uniformly distributed conditioned on satisfying check 2) of algorithm V. Therefore, these elements of the simulated proof have the same distribution as in a real proof. This fact combined with the perfect zero-knowledge property of $\Psi_{\overline{\mathcal{D}}_k,+}$ and $\Psi_{\overline{\mathcal{D}}_k,\mathsf{com}}$ concludes the proof.

5.3   Extensions

CRS Generation for Individual Commitments. When using individual commitments (distribution b) from Sect. 5), the only change is that Δ is sampled uniformly from $\mathbb{Z}_q^{2 \times 2n}$ (the distribution of $\check{\mathbf{H}}$ is not changed). Thus, the matrix $\hat{\mathbf{G}} := \Delta \hat{\mathbf{U}}$ has $2n$ columns instead of $n+1$ and $\hat{\mathbf{c}}_\Delta := \hat{\mathbf{G}} \left(\begin{smallmatrix} \mathbf{b} \\ \mathbf{w}_g \end{smallmatrix} \right)$ for some $\mathbf{w}_g \in \mathbb{Z}_q^n$. In the soundness proof, the only change is that in Game_2, the extra columns are also changed to span a one-dimensional space, *i.e.* in this game $\hat{\mathbf{g}}_i$, $i \in [2n - 1]$ and $i \neq i^*$, are uniform vectors in the space spanned by $\hat{\mathbf{g}}_{2n}$.

Bit-Strings of Weight 1. In the special case when the bit-string has only one 1 (this case is useful in some applications, see Sect. 6), the size of the CRS can be made linear in n, instead of quadratic. To prove this statement we would combine our proof system for bit-strings of Sect. 5.1 and a proof that $\sum_{i \in [n]} b_i = 1$ as described above. In the definition of $(\hat{\mathbf{\Theta}}_{b(\overline{b}-1)}, \check{\mathbf{\Pi}}_{b(\overline{b}-1)})$ in Eq. 8, one sees that for all pairs $(i, j) \in [n] \times [n]$, the coefficient of $(\hat{\mathbf{C}}_{i,j}, \check{\mathbf{D}}_{i,j})$ is $b_i(b_j - 1)$. If i^* is the only index such that $b_{i^*} = 1$, then we have:

$$\sum_{i \in [n]} \sum_{j \in [n]} b_i(b_j - 1)(\hat{\mathbf{C}}_{i,j}, \check{\mathbf{D}}_{i,j}) = \sum_{j \neq i^*} (\hat{\mathbf{C}}_{i^*,j}, \check{\mathbf{D}}_{i^*,j}) =: (\hat{\mathbf{C}}_{i^*,\neq}, \check{\mathbf{D}}_{i^*,\neq}).$$

Therefore, one can replace in the CRS the pairs of matrices $(\hat{\mathbf{C}}_{i,j}, \check{\mathbf{D}}_{i,j})$ by $(\hat{\mathbf{C}}_{i,\neq}, \check{\mathbf{D}}_{i,\neq})$, $i \in [n]$. The resulting CRS is linear in n.

6   Applications

Many protocols use proofs that a commitment opens to a bit-string as a building block. Since our commitments are still of size $\Theta(n)$, our results may not apply to some of these protocols (*e.g.* range proofs). Yet, there are several applications where bits need to be used independently and our results provide significant improvements. Table 2 summarizes them.

Signatures. Some application examples are the signature schemes of [3–5,9]. For example, in the revocable attribute-based signature scheme of Escala *et. al* [9], every signature includes a proof that a set of GS commitments, whose size is the number of attributes, opens to a bit-string. Further, the proof of membership in a list which is discussed below can also be used to reduce the size of Ring Signature scheme of [6], which is the most efficient ring signature in the standard model. To sign a message m, among other things, the signer picks a one-time signature key and certifies the one-time verification key by signing it with a Boneh-Boyen signature under vk_α. Then, the signer commits to vk_α and shows that vk_α belongs to the list of Boneh-Boyen verification keys $(vk_1, \ldots, vk_n)$ of the parties in the ring R.

Table 2. Comparison of the application of our techniques and results from the literature. In rows labeled as "Threshold GS" we give the size of the proof of satisfiability of t-out-of-n sets $\mathcal{S}_i$, where m_x is the sum of the number of variables in $\hat{\mathbb{G}}$ in each set $\mathcal{S}_i$, and $\bar{n}$ is the total number of two-sided and quadratic equations in some $\bigcup_{i\in[n]}\mathcal{S}_i$. For all rows, we must add to the proof size the cost of a GS proof of each equation in one of the sets $\mathcal{S}_i$. In the other rows n is the size of the list.

Proof System	Author	Proof Size
Threshold GS	Ràfols [26] (1)	$(m_x + 3(n-t) + 2\bar{n})\mathfrak{g}$
	Ràfols [26] (2)	$2(n-t+1)\mathfrak{h} + 2n(\mathfrak{g}+\mathfrak{h})$
	This work	$2(n+1)\mathfrak{g} + 10(\mathfrak{g}+\mathfrak{h})$
Dynamic list (ring signature)	Chandran et al. [6]	$(16\sqrt{n}+4)(\mathfrak{g}+\mathfrak{h})$
	Ràfols [26]	$(8\sqrt{n}+6)\mathfrak{g} + 12\sqrt{n}\mathfrak{h}$
	This work	$(4\sqrt{n}+14)\mathfrak{g} + (8\sqrt{n}+14)\mathfrak{h}$

Threshold GS Proofs for PPEs. There are two approaches to construct threshold GS proofs for PPEs, i.e. proofs of satisfiability of t-out-of-n equations. One is due to [13] and consists of compiling the n equations into a single equation which is satisfied only if t of the original equations are satisfied. For the case of PPEs, this method adds new variables and proves that each of them opens to a bit. Our result reduces the cost of this approach, but we omit any further discussion as it is quite inefficient because the number of additional variables is $\Theta(m_{var} + n)$, where m_{var} is the total number of variables in the original n equations.

The second approach is due to Ràfols [26]. The basic idea behind [26], which extends [14], follows from the observation that for each GS equation type tp, the CRS space $\mathcal{K}$ is partitioned into a perfectly sound CRS space $\mathcal{K}_{\mathsf{tp}}^b$ and a perfectly witness indistinguishable CRS space $\mathcal{K}_{\mathsf{tp}}^h$.

In particular, to prove satisfiability of t-out-of-n sets of equations from $\{\mathcal{S}_i : i \in [n]\}$ of type tp, it suffices to construct an algorithm $\mathsf{K}_{\mathsf{corr}}$ which on input $\mathsf{crs}_{\mathsf{GS}}$ and some set of indexes $A \subset [n]$, $|A| = t$, generates n GS common reference strings $\{\mathsf{crs}_i, i \in [n]\}$ and simulation trapdoors $\tau_{i,sim}$, $i \in A^c$, in a such a way that[5]:

(a) it can be publicly verified the set of perfectly sound keys, $\{\mathsf{crs}_i : \mathsf{crs}_i \in \mathcal{K}_{\mathsf{tp}}^b\}$ is of size at least t,

(b) there exists a simulator $\mathsf{S}_{\mathsf{corr}}$ who outputs $(\mathsf{crs}_i, \tau_{i,sim})$ for all $i \in [n]$, and the distribution of $\{\mathsf{crs}_i : i \in [n]\}$ is the same as the one of the keys output by $\mathsf{K}_{\mathsf{corr}}$ when $\mathsf{crs}_{\mathsf{GS}}$ is the perfectly witness-indistinguishable CRS.

---

[5] More technically, this is the notion of *Simulatable Verifiable Correlated Key Generation* in [26], which extends the definition of Verifiable Correlated Key Generation of [14].

The prover of t-out-of-n satisfiability can run $\mathsf{K_{corr}}$ and, for all $i \in [n]$, compute a real (resp. simulated) proof for $\mathcal{S}_i$ with respect to crs_i when $i \in A$ (resp. when $i \in A^c$).

Ràfols gives two constructions for PPEs, the first one can be found in [26], App. C and the other follows from [26, Sect. 7][6]. Our algorithm $\mathsf{K_{corr}}$ for PPEs[7] goes as follows:

- Define $(b_1, \ldots, b_n)$ as $b_i = 1$ if $i \in A$ and $b_i = 0$ if $i \in A^c$. For all $i \in [n]$, let $\hat{\mathbf{z}}_i := \mathsf{Comm}(b_i) = b_i \hat{\mathbf{u}}_1 + r_i \hat{\mathbf{u}}_2$, $r_i \in \mathbb{Z}_q$, and define $\tau_{sim,i} = r_i$, for all $i \in A^c$. Define $\mathsf{crs}_i := (\Gamma, \hat{\mathbf{z}}_i, \hat{\mathbf{u}}_2, \check{\mathbf{v}}_1, \check{\mathbf{v}}_2)$.
- Prove that $\{\hat{\mathbf{c}}_i\}$ opens to $\mathbf{b} \in \{0,1\}^n$ and that $\sum_{i \in [n]} b_i = t$.

The simulator just defines $\mathbf{b} = \mathbf{0}$. The reason why this works is that when $b_i = 1$, $(\hat{\mathbf{z}}_i - \hat{\mathbf{u}}_1) \in \mathsf{Span}(\hat{\mathbf{u}}_2)$, therefore $\mathsf{crs}_i \in \mathcal{K}^b_{PPE}$ and when $b_i = 0$, $(\hat{\mathbf{z}}_i - \hat{\mathbf{u}}_1) \notin \mathsf{Span}(\hat{\mathbf{u}}_2)$ so $\mathsf{crs}_i \in \mathcal{K}^h_{PPE}$.

More Efficient Proof of Membership in a List. Chandran *et al.* construct a ring signature of size $\Theta(\sqrt{n})$ [6], which is the most efficient ring signature in the standard model. Their construction uses as a subroutine a non-interactive proof of membership in some list $L = (\hat{l}_1, \ldots, \hat{l}_n)$ which is of size $\Theta(\sqrt{n})$. The trick of Chandran *et al.* to achieve this asymptotic complexity is to view L as a matrix $\hat{\mathbf{L}} \in \hat{\mathbb{G}}^{m \times m}$, for $m = \sqrt{n}$, where the i, j th element of $\hat{\mathbf{L}}$ is $\hat{l}_{i,j} := \hat{l}_{(i,j)}$ and $(i,j) := (i-1)m + j$. Given a commitment $\hat{c}$ to some element $\hat{l}_\alpha$, where $\alpha = (i_\alpha, j_\alpha)$, their construction in asymmetric bilinear groups works as follows :

1. Compute GS commitments in $\check{\mathbb{H}}$ to $b_1 \ldots, b_m$ and $b'_1, \ldots, b'_m$, where $b_i = 1$ if $i = i_\alpha$ and 0 otherwise, and $b'_j = 1$ if $j = j_\alpha$, and 0 otherwise.
2. Compute a GS proof that $b_i \in \{0,1\}$ and $b'_j \in \{0,1\}$ for all $i, j \in [m]$, and that $\sum_{i \in [m]} b_i = 1$, and $\sum_{j \in [m]} b'_j = 1$.
3. Compute GS commitments to $\hat{x}_1 := \hat{l}_{(i_\alpha,1)}, \ldots, \hat{x}_m := \hat{l}_{(i_\alpha,m)}$.
4. Compute a GS proof that $\hat{x}_j = \sum_{i \in [m]} b_i \hat{l}_{(i,j)}$, for all $j \in [m]$, is satisfied.
5. Compute a GS proof that $\hat{l}_\alpha = \sum_{j \in [m]} b'_j \hat{x}_j$ is satisfied.

With respect to the naive use of GS proofs, Step 2 was improved by Ràfols [26]. Using our proofs for bit-strings of weight 1 from Sect. 5.3, we can further reduce the size of the proof in step 2, see table.

We note that although in step 4 the equations are all two-sided linear equations, proofs can only be aggregated if the list comes from a witness samplable distribution and the CRS is set to depend on that specific list. This is not useful for the application to ring signatures, since the CRS should be independent of the ring R (which defines the list). If aggregation is possible then the size of the

---

[6] The construction in [26, Sect. 7] is for other equation types but can be used to prove that t-out-of-n of $\mathsf{crs}_1, \ldots, \mathsf{crs}_n$ are perfectly binding for PPEs.

[7] Properly speaking the construction is for PPEs which are left-simulatable in the terminology of [26].

proof in step 4 is reduced from $(2\mathfrak{g} + 4\mathfrak{h})\sqrt{n}$ to $4\mathfrak{g} + 8\mathfrak{h}$. A complete description of the proof can be found in the full version, where we also show that when the CRS depends on the list and the list is witness samplable, the proof can be further reduced to $\Theta(\sqrt[3]{n})$.

References

1. Abe, M., Groth, J., Ohkubo, M., Tango, T.: Converting cryptographic schemes from symmetric to asymmetric bilinear groups. In: Garay, J.A., Gennaro, R. (eds.) CRYPTO 2014, Part I. LNCS, vol. 8616, pp. 241–260. Springer, Heidelberg (2014)
2. Abe, M., Haralambiev, K., Ohkubo, M.: Signing on elements in bilinear groups for modular protocol design. Cryptology ePrint Archive, Report 2010/133 (2010). http://eprint.iacr.org/2010/133
3. Blazy, O., Fuchsbauer, G., Pointcheval, D., Vergnaud, D.: Signatures on randomizable ciphertexts. In: Catalano, D., Fazio, N., Gennaro, R., Nicolosi, A. (eds.) PKC 2011. LNCS, vol. 6571, pp. 403–422. Springer, Heidelberg (2011)
4. Blazy, O., Pointcheval, D., Vergnaud, D.: Compact round-optimal partially-blind signatures. In: Visconti, I., De Prisco, R. (eds.) SCN 2012. LNCS, vol. 7485, pp. 95–112. Springer, Heidelberg (2012)
5. Camacho, P.: Fair exchange of short signatures without trusted third party. In: Dawson, E. (ed.) CT-RSA 2013. LNCS, vol. 7779, pp. 34–49. Springer, Heidelberg (2013)
6. Chandran, N., Groth, J., Sahai, A.: Ring signatures of sub-linear size without random oracles. In: Arge, L., Cachin, C., Jurdziński, T., Tarlecki, A. (eds.) ICALP 2007. LNCS, vol. 4596, pp. 423–434. Springer, Heidelberg (2007)
7. Escala, A., Groth, J.: Fine-tuning Groth-Sahai proofs. In: Krawczyk, H. (ed.) PKC 2014. LNCS, vol. 8383, pp. 630–649. Springer, Heidelberg (2014)
8. Escala, A., Herold, G., Kiltz, E., Ràfols, C., Villar, J.: An algebraic framework for Diffie-Hellman assumptions. In: Canetti, R., Garay, J.A. (eds.) CRYPTO 2013, Part II. LNCS, vol. 8043, pp. 129–147. Springer, Heidelberg (2013)
9. Escala, A., Herranz, J., Morillo, P.: Revocable attribute-based signatures with adaptive security in the standard model. In: Nitaj, A., Pointcheval, D. (eds.) AFRICACRYPT 2011. LNCS, vol. 6737, pp. 224–241. Springer, Heidelberg (2011)
10. Freeman, D.M.: Converting pairing-based cryptosystems from composite-order groups to prime-order groups. In: Gilbert, H. (ed.) EUROCRYPT 2010. LNCS, vol. 6110, pp. 44–61. Springer, Heidelberg (2010)
11. Galbraith, S.D., Paterson, K.G., Smart, N.P.: Pairings for cryptographers. Discrete Appl. Math. 156(16), 3113–3121 (2008)
12. Gennaro, R., Gentry, C., Parno, B., Raykova, M.: Quadratic span programs and succinct NIZKs without PCPs. In: Johansson, T., Nguyen, P.Q. (eds.) EUROCRYPT 2013. LNCS, vol. 7881, pp. 626–645. Springer, Heidelberg (2013)
13. Groth, J.: Simulation-sound NIZK proofs for a practical language and constant size group signatures. In: Lai, X., Chen, K. (eds.) ASIACRYPT 2006. LNCS, vol. 4284, pp. 444–459. Springer, Heidelberg (2006)
14. Groth, J., Ostrovsky, R., Sahai, A.: Non-interactive zaps and new techniques for NIZK. In: Dwork, C. (ed.) CRYPTO 2006. LNCS, vol. 4117, pp. 97–111. Springer, Heidelberg (2006)
15. Groth, J., Sahai, A.: Efficient non-interactive proof systems for bilinear groups. In: Smart, N.P. (ed.) EUROCRYPT 2008. LNCS, vol. 4965, pp. 415–432. Springer, Heidelberg (2008)

16. Groth, J., Sahai, A.: Efficient noninteractive proof systems for bilinear groups. SIAM J. Comput. **41**(5), 1193–1232 (2012)

17. Hofheinz, D., Kiltz, E.: Secure hybrid encryption from weakened key encapsulation. In: Menezes, A. (ed.) CRYPTO 2007. LNCS, vol. 4622, pp. 553–571. Springer, Heidelberg (2007)

18. Joux, A.: A new index calculus algorithm with complexity $L(1/4 + o(1))$ in small characteristic. In: Lange, T., Lauter, K., Lisoněk, P. (eds.) SAC 2013. LNCS, vol. 8282, pp. 355–380. Springer, Heidelberg (2014)

19. Jutla, C.S., Roy, A.: Shorter quasi-adaptive NIZK proofs for linear subspaces. In: Sako, K., Sarkar, P. (eds.) ASIACRYPT 2013, Part I. LNCS, vol. 8269, pp. 1–20. Springer, Heidelberg (2013)

20. Jutla, C.S., Roy, A.: Switching lemma for bilinear tests and constant-size NIZK proofs for linear subspaces. In: Garay, J.A., Gennaro, R. (eds.) CRYPTO 2014, Part II. LNCS, vol. 8617, pp. 295–312. Springer, Heidelberg (2014)

21. Kiltz, E., Wee, H.: Quasi-adaptive NIZK for linear subspaces revisited. In: Oswald, E., Fischlin, M. (eds.) EUROCRYPT 2015. LNCS, vol. 9057, pp. 101–128. Springer, Heidelberg (2015)

22. Libert, B., Peters, T., Joye, M., Yung, M.: Non-malleability from malleability: simulation-sound quasi-adaptive NIZK proofs and CCA2-secure encryption from homomorphic signatures. In: Nguyen, P.Q., Oswald, E. (eds.) EUROCRYPT 2014. LNCS, vol. 8441, pp. 514–532. Springer, Heidelberg (2014)

23. Libert, B., Peters, T., Yung, M.: Group signatures with almost-for-free revocation. In: Safavi-Naini, R., Canetti, R. (eds.) CRYPTO 2012. LNCS, vol. 7417, pp. 571–589. Springer, Heidelberg (2012)

24. Morillo, P., Ràfols, C., Villar, J.L.: Matrix computational assumptions in multilinear groups. Cryptology ePrint Archive, Report 2015/353 (2015). http://eprint.iacr.org/2015/353

25. Naor, M.: On cryptographic assumptions and challenges. In: Boneh, D. (ed.) CRYPTO 2003. LNCS, vol. 2729, pp. 96–109. Springer, Heidelberg (2003)

26. Ràfols, C.: Stretching Groth-Sahai: NIZK proofs of partial satisfiability. In: Dodis, Y., Nielsen, J.B. (eds.) TCC 2015, Part II. LNCS, vol. 9015, pp. 247–276. Springer, Heidelberg (2015)

27. Shacham, H.: A cramer-shoup encryption scheme from the linear assumption and from progressively weaker linear variants. Cryptology ePrint Archive, Report 2007/074 (2007). http://eprint.iacr.org/

Dual-System Simulation-Soundness with Applications to UC-PAKE and More

Charanjit S. Jutla[1]([⊠]) and Arnab Roy[2]

[1] IBM T. J. Watson Research Center, Yorktown Heights, NY, USA
csjutla@us.ibm.com
[2] Fujitsu Laboratories of America, Sunnyvale, CA, USA
aroy@us.fujitsu.com

Abstract. We introduce a novel concept of dual-system simulation-sound non-interactive zero-knowledge (NIZK) proofs. Dual-system NIZK proof system can be seen as a two-tier proof system. As opposed to the usual notion of zero-knowledge proofs, dual-system defines an intermediate partial-simulation world, where the proof simulator may have access to additional auxiliary information about the word, for example a membership bit, and simulation of proofs is only guaranteed if the membership bit is correct. Further, dual-system NIZK proofs allow a quasi-adaptive setting where the CRS can be generated based on language parameters. This allows for the further possibility that the partial-world CRS simulator may have access to additional trapdoors related to the language parameters. We show that for important hard languages like the Diffie-Hellman language, such dual-system proof systems can be given which allow unbounded partial simulation soundness, and which further allow transition between partial simulation world and single-theorem full simulation world even when proofs are sought on non-members. The construction is surprisingly simple, involving only two additional group elements for general linear-subspace languages in asymmetric bilinear pairing groups.

As a direct application we give a short keyed-homomorphic CCA-secure encryption scheme. The ciphertext in this scheme consists of only six group elements (under the SXDH assumption) and the security reduction is tight. An earlier scheme of Libert et al. based on their efficient unbounded simulation-sound QA-NIZK proofs only provided a loose security reduction, and further had ciphertexts almost twice as long as ours.

We also show a single-round universally-composable password authenticated key-exchange (UC-PAKE) protocol which is secure under adaptive corruption in the erasure model. The single message flow only requires **four** group elements under the SXDH assumption.

This is the *shortest known* UC-PAKE even without considering adaptive corruption. The latest published scheme which considered adaptive corruption, by Abdalla et al [ABB+13], required non-constant (more than 10 times the bit-size of the password) number of group elements.

Keywords: NIZK · Bilinear pairings · UC-PAKE · Keyed-homomorphic encryption · SXDH

© International Association for Cryptologic Research 2015
T. Iwata and J.H. Cheon (Eds.): ASIACRYPT 2015, Part I, LNCS 9452, pp. 630–655, 2015.
DOI: 10.1007/978-3-662-48797-6_26

1   Introduction

Since the introduction of simulation-sound non-interactive zero-knowledge proofs (NIZK) in [Sah99] (based on the concept of non-malleability [DDN91]), simulation-soundness has become an essential cryptographic tool. While the idea of zero-knowledge simulation [GMR89] brought rigor to the concept of semantic security, simulation-soundness of some form is usually implicit in most cryptographic applications. While the original construction of [Sah99] was rather inefficient, the advent of pairing based cryptography, and in particular Groth-Sahai NIZK proofs [GS08], has led to much more efficient simulation-sound NIZK constructions. Pairing-based cryptography has also led to efficient construction of powerful primitives where simulation-soundness is not very explicit.

It has been shown that different forms of simulation-soundness suffice for many applications. Indeed, the original application (CCA2-secure encryption) considered in [Sah99] only required what is known as single-theorem simulation-soundness (also known as one-time simulation-soundness). However, many other cryptographic constructions are known only using unbounded simulation-sound NIZK proofs. In this paper, we introduce the concept of **dual-system simulation-sound** NIZK proofs, which lie somewhere in between one-time and unbounded simulation-sound NIZK proofs. The aim is to show that this weaker concept suffices for constructions where unbounded simulation-soundness was being used till now. We also show that in many applications this new concept of dual-system simulation soundness is implicit, in the sense that although we cannot get a generic construction from a NIZK proof, we can use the underlying ideas of the dual-system simulation-sound NIZK proofs.

Indeed, our novel definition is inspired by the dual-system identity-based encryption (IBE) scheme of Waters [Wat09], where such a concept was implicit, and led to the first IBE scheme which was fully-secure under static and standard assumptions. So without further ado, we jump straight into the main idea of the new concept. In dual-system simulation-sound NIZK proof systems we will consider three worlds: the real-world, the partial-simulation world, and the one-time full-simulation world. The real world consists of a common-reference string (CRS), an efficient prover P, and an efficient verifier V. The concept of completeness and soundness of P and V with respect to a witness-relation R is well-understood. The full-simulation world is also standard, and it includes two simulators: a CRS simulator and a proof simulator. The proof simulator is a zero-knowledge simulator in the sense that it can simulate proofs even without access to the witness. In order to achieve this, the CRS simulator generates the CRS in a potentially different way and produces a trapdoor for the proof simulator. The **partial-simulation** world we consider also has a CRS simulator, and a proof simulator, but this proof simulator is allowed partial access to the witness (or some other auxiliary information) about the member on which the proof is sought.

At this point, we also bring in the possibility of the CRS being generated as a function of the language or witness-relation under consideration. The recent quasi-adaptive NIZK (QA-NIZK) proofs of [JR13] allow this possibility for distributions of witness-relations. The CRS in the real and the full-simulation world is generated based on a language parameter generated according to some distribution. Now we consider the possibility that in the partial-simulation world, the CRS simulator actually generates the language parameter itself. In other words, the CRS simulator has access to the "witness" of the language parameter. For example, the CRS simulator may know the discrete-logs of the language parameters. This leads to the possibility that in the partial simulation world the proof simulator may have access to additional trapdoors which makes simulation and/or simulation soundness easier to achieve.

In this paper, we will only define and consider dual-system simulation sound QA-NIZK proofs (called DSS-QA-NIZK), where the only auxiliary information that the partial proof simulator gets is a single bit which is called the **membership bit**. The membership bit indicates whether the word on which the proof is sought is in the language or not. We show that we can achieve unbounded partial-simulation soundness for important languages like the Diffie-Hellman language by relatively simple constructions. The constructions also allow one-time full-ZK simulation, and hence form a DSS-QA-NIZK for the Diffie-Hellman language. We actually give a general construction for arbitrary languages which allow smooth and universal$_2$ projective hash proofs [CS02] and have QA-NIZKs for the language augmented with such a hash proof. We show that for linear subspace languages (over bilinear groups), like the Diffie-Hellman and decisional-linear (DLIN) languages, the requirements for the general construction are easy to obtain. Thus, for all such languages, under the standard and static SXDH assumption in bilinear pairing groups, we get a DSS-QA-NIZK proof of only two group elements.

Table 1 summarizes comparison among existing schemes and ours. DSS is weaker than unbounded simulation soundness, and although incomparable with one time simulation soundness, it seems to enjoy better properties. Consistent with this, we observe that the proof sizes also place in the middle of the shortest known OTSS-NIZKs [ABP15,KW15] and the shortest known USS-NIZKs [KW15] for linear subspaces.

Applications. We now give the main idea as to why such a construction is useful. The security of most applications is shown by reduction to a hard language. However, a particular application may have a more complex language for which the NIZK proofs are required, and the security proof may require soundness of the NIZK system while proofs of many elements (real or fake) of such a complex language are being simulated. The idea is that multiple simulations of such elements can be performed in a partial-simulation manner (i.e. it is always possible to supply the correct membership-bit), and full simulation is only required of one member at a time, on which the hardness assumption can then be invoked.

Keyed-Homomorphic CCA-secure Encryption. As a first application we consider the keyed-homomorphic CCA-secure encryption scheme notion of [EHO+13]. In such an encryption scheme, a further functionality called Eval is available which

Table 1. Comparison with existing NIZK schemes for linear subspaces with table adapted from [KW15]. The language of interest is a t dimensional subspace of an n dimensional ambient space. m is the bit-size of the tag. AS is adaptive-soundness. OTSS is one-time simulation-soundness and USS is unbounded simulation-soundness.

	Soundness	Assumption	Proof	CRS	#pairings
[GS08]	AS	DLIN	$2n + 3t$	6	$3n(t+3)$
[LPJY14]	AS	DLIN	3	$2n + 3t + 3$	$2n + 4$
[JR13]	AS	k-Linear	$k(n-t)$	$2kt(n-t) + k + 1$	$k(n-t)(t+2)$
[JR14a]	AS	k-Linear	k	$kn + kt + k^2$	$kn + k^2$
[ABP15]	AS	k-Linear	k	$kn + kt + k$	$kn + k$
[KW15]	AS	k-Linear	k	$kn + kt + k - 1$	$kn + k - 1$
[ABP15]	OTSS	k-Linear	k	$2m(kn + (k+1)t) + k$	$mkn + k$
[KW15]	OTSS	k-Linear	k	$2m(kn + (k+1)t)+k-1$	$mkn+k-1$
This paper	DSS	k-Linear	$k + 1$	$k(n+1) + kt + k^2$	$k(n+1) + k^2$
[CCS09]	USS	DLIN	$2n + 6t + 52$	18	$O(tn)$
[LPJY14]	USS	DLIN	20	$2n + 3t + 3m + 10$	$2n + 30$
[KW15]	USS	k-Linear	$2k + 2$	$kn + 4(k+t+1)k + 2k$	$k(n+k+1) + k$

using a key can homomorphically combine valid ciphertexts. The scheme should provide IND-CCA2 security when this Eval key is unavailable to the adversary, and should continue to enjoy IND-CCA1 security when the Eval key is exposed to the adversary. Emura et al. also gave constructions for such a scheme, albeit schemes which are not publicly verifiable, and further satisfying a weaker notion than CCA1-security when Eval key is revealed. Recently, Libert et al. gave a publicly-verifiable construction which is more efficient and also CCA1-secure when Eval key is revealed. Their construction is based on a new and improved unbounded simulation-sound QA-NIZK for linear subspace languages. We show in this paper that a DSS-QA-NIZK for the Diffie-Hellman language suffice, and leads to a much improved construction. While the construction in [LPJY14], under the SXDH assumption, requires nine group elements in one group, and two more in the other plus a one-time signature key pair, our construction *only requires six group elements* in any one of the bilinear groups. Further, while the earlier construction was loose (i.e. loses a factor quadratic in number of Eval calls), our reduction is tight.

UC Password-Authenticated Key Exchange (UC-PAKE). The UC-PAKE ideal functionality was introduced in [CHK+05] where they also gave a three-round construction. In [KV11] a single-round construction for UC-PAKE was given using Groth-Sahai NIZK proofs along with unbounded simulation-soundness construction of [CCS09] (also see [JR12]). Later [BBC+13] gave a UC-PAKE construction based on novel trapdoor smooth projective hash functions, but secure only under static corruption; each message consisted of six group elements in one group, and another five elements in the other group (under the SXDH assumption).

Table 2. Comparison with existing UC-PAKE schemes. m is the password size in bits and λ is the security parameter. AC stands for Adaptive Corruption. For one-round schemes, message size is per flow.

	AC	One-round	Assumption	Message size
[ACP09]	yes	no	DDH	$O(m\lambda)$
[KV11]	no	yes	DLIN	$> 65 \times \mathbb{G}$
[JR12]	no	yes	SXDH	> 30 total group elements
[BBC+13]	no	yes	SXDH	$6 \times \mathbb{G}_1 + 5 \times \mathbb{G}_2$
[ABB+13]	yes	yes	SXDH	$10 * m \times \mathbb{G}_1 + m \times \mathbb{G}_2$
This paper	yes	yes	SXDH	$3 \times \mathbb{G}_1 + 1 \times \mathbb{G}_2$

In this paper, we construct a a single-round construction based on dual-system simulation-soundness which is UC-secure under adaptive corruption (in the erasure model), and which has only a total of **four** group elements in each message. The key is generated in the target group. The construction is not a black-box application of the DSS-QA-NIZK for the Diffie-Hellman language, but uses its underlying idea as well as the various component algorithms of the DSS-QA-NIZK. The main idea of the construction is given in more detail in Sect. 6.2.

To the best of our knowledge, this is the *shortest known UC-PAKE*, even without considering adaptive corruption. The first UC-PAKE to consider adaptive corruption was by Abdalla, Chevalier and Pointcheval [ACP09], which was a two round construction. Recently, Abdalla et al [ABB+13] also constructed a single round protocol, which required a non-constant (more than 10 times the bit-size of the password) number of group elements in each flow. Comparison with existing UC-PAKEs is given in Table 2.

Identity-Based Encryption (IBE). In the full version of this paper [JR14b], we show that the recent efficient dual-system IBE [JR13] (inspired by the original dual-system IBE of Waters [Wat09]) can also be obtained using the ideas of DSS-QA-NIZK. While the construction is not black-box and utilizes additional "smoothness" and "single-pairing-product test" properties of the verifier, it along with the other two applications clearly demonstrate the power and utility of the new notion, which we expect will find many more applications.

2   Preliminaries: Quasi-Adaptive NIZK Proofs

A witness relation is a binary relation on pairs of inputs, the first called a word and the second called a witness. Note that each witness relation R defines a corresponding language L which is the set of all x for which there exists a witness w, such that $R(x, w)$ holds.

We will consider Quasi-Adaptive NIZK proofs [JR13] for a probability distribution $\mathcal{D}$ on a collection of (witness-) relations $\mathcal{R} = \{R_\rho\}$ (with corresponding languages L_ρ). Recall that in a quasi-adaptive NIZK, the CRS can be set after the language parameter has been chosen according to $\mathcal{D}$. Please refer to [JR13] for detailed definitions.

Definition 1. *([JR13]) We call* (pargen, crsgen, prover, ver) *a* (labeled) *quasi-adaptive non-interactive zero-knowledge (QA-NIZK) proof system for witness-relations* $\mathcal{R}_\lambda = \{R_\rho\}$ *with parameters sampled from a distribution $\mathcal{D}$ over associated parameter language* Lpar, *if there exist simulators* crs $-$ sim, sim *such that for all non-uniform PPT adversaries* $\mathcal{A}_1, \mathcal{A}_2, \mathcal{A}_3$ *we have (in all of the following probabilistic experiments, the experiment starts by setting λ as* $\lambda \leftarrow$ pargen(1^m)*, and choosing ρ as $\rho \leftarrow \mathcal{D}_\lambda$):*

Quasi-Adaptive Completeness:
$$\Pr\left[\begin{array}{c} \text{CRS} \leftarrow \text{crsgen}(\lambda, \rho); (x, w, l) \leftarrow \mathcal{A}_1(\text{CRS}, \rho); \\ \pi \leftarrow \text{prover}(\text{CRS}, x, w, l) : \text{ver}(\text{CRS}, x, l, \pi) = 1 \text{ if } R_\rho(x, w) \end{array} \right] = 1$$

Quasi-Adaptive Soundness:
$\Pr[\text{CRS} \leftarrow \text{crsgen}(\lambda, \rho); (x, l, \pi) \leftarrow \mathcal{A}_2(\text{CRS}, \rho) : x \notin L_\rho \wedge \text{ver}(\text{CRS}, x, l, \pi) = 1] \approx 0$

Quasi-Adaptive Zero-Knowledge:
$\Pr[\text{CRS} \leftarrow \text{crsgen}(\lambda, \rho) : \mathcal{A}_3^{\text{prover}(\text{CRS}, \cdot, \cdot, \cdot)}(\text{CRS}, \rho) = 1] \approx$
$\Pr[(\text{CRS}, \text{trap}) \leftarrow \text{crs} - \text{sim}(\lambda, \rho) : \mathcal{A}_3^{\text{sim}^*(\text{CRS}, \text{trap}, \cdot, \cdot, \cdot)}(\text{CRS}, \rho) = 1],$
where sim$^*(\text{CRS}, \text{trap}, x, w, l) = $ sim$(\text{CRS}, \text{trap}, x, l)$ *for $(x, w) \in R_\rho$ and both oracles (i.e.* prover *and* sim**) output failure if $(x, w) \notin R_\rho$.*

The QA-NIZK is called a **statistical zero-knowledge** QA-NIZK if the view of adversary $\mathcal{A}_3$ above in the two experiments is statistically indistinguishable.

3   Dual-System Simulation-Soundness

To define dual-system simulation soundness of QA-NIZK proofs, we will consider three worlds: the real-world, the partial-simulation world, and the one-time (or single theorem) full-simulation world. While the real-world and the full-simulation world should be familiar from earlier definitions of NIZK proof systems, the partial-simulation world leads to interesting possibilities. To start with, in the partial simulation world, one would like the proof simulator to have access to partial or complete witness of the word[1]. Finally, in the quasi-adaptive setting, the language parameters may actually be generated by the CRS simulator and hence the simulator may have access to, say, the discrete logs of the language parameters, which can serve as further trapdoors.

Rather than considering these general settings, we focus on a simple partial-simulation setting, where (a) the CRS simulator can generate the language parameters itself and (b) the proof simulator when invoked with a word x is given

---

[1] In case the proof simulator is being invoked on a non-language word, it is not immediately clear what this witness can be, unless we also define a language and a distribution for a super-language which includes the language under consideration as a subset.

an additional bit β, which we call the **membership bit**, that represents the information whether x is indeed a member or not.

The partial simulation world is required to be unbounded simulation-sound, and hopefully this should be easier to prove than usual unbounded simulation-soundness (given that its simulators have additional information). We also allow the partial simulation world to be sound with respect to a private verifier (this concept has been considered earlier in [JR12]), and this further leads to the possibility of easier and/or simpler constructions. A surprising property achievable under such a definition is that one can go back and forth between the partial-simulation world and the one-time full-simulation world even when simulating fake tuples.

Definition 2 (Dual – System Non – Interactive Proofs). *A Dual-system non-interactive proof system consists of PPT algorithms defined in three worlds as follows:*

Real World *consisting of:*
- *A pair of* **CRS generators** (K_0, K_1), *where* K_0 *takes a unary string and produces an ensemble parameter* λ. *(The ensemble parameter* λ *is used to sample a witness-relation parameter* ρ *using* $\mathcal{D}_\lambda$ *in the security definition.) PPT algorithm* K_1 *uses* ρ *(and* λ*) to produce the real-world CRS* ψ.
- *A* **prover** P *that takes as input a CRS, a language member and its witness, a label, and produces a proof.*
- *A* **verifier** V *that takes as input a CRS, a word, a label, and a proof, and outputs a single bit.*

Partial-Simulation World *consisting of:*
- *A* **semi-functional CRS simulator** sfK_1 *that takes ensemble parameter* λ *as input and produces a witness relation parameter* ρ, *a semi-functional CRS* σ, *as well as two trapdoors* τ *and* η. *The first trapdoor is used by the proof simulator, and the second by the private verifier.*
- *A* **semi-functional simulator** sfSim *that takes a CRS, a trapdoor* τ, *a word, a membership-bit* β, *and a label, to produce a proof.*
- *A* **private verifier** pV *that takes a CRS, a trapdoor* η, *a word, a label, and a proof and outputs a single bit.*

One-time Full Simulation World *consisting of:*
- *A* **one-time full-simulation CRS generator** otfK_1, *that takes as input the ensemble parameter* λ, *the witness relation parameter* ρ *to produce a CRS and three trapdoors* τ, τ_1 *and* η.
- *A* **one-time full simulator** otfSim *that takes as input a CRS, a trapdoor* τ_1, *a word, a label, and produces a proof[2].*

---

[2] We remark here that the One-time Full Simulation World also uses a semi-functional simulator as can be seen in Fig. 1. It has the same black-box properties as in the Partial-Simulation World, but could potentially have a different internal construction. In this paper it turns out that the same construction suffices for both the worlds, so for the sake of simplicity we forgo making this explicit in the definition.

– A **semi-functional verifier** sfV *that takes as input a CRS, a trapdoor η, a word, a label, a proof and outputs a bit. The adversaries also have access to the semi-functional simulator.*

Definition 3 (DSS-QA-NIZK). *The definition of the real-world components of a dual-system non-interactive proof to be complete and (computationally) sound are same as in QA-NIZK Definition 1. Such a proof system is called a* **dual-system simulation-sound quasi-adaptive NIZK (DSS-QA-NIZK)** *for a collection of witness relations $\mathcal{R}_\lambda = \{R_\rho\}$, with parameters sampled from a distribution $\mathcal{D}$, if its real-world components are complete and (computationally) sound, and if for all non-uniform PPT adversaries $\mathcal{A} = (\mathcal{A}_0, \mathcal{A}_1, \mathcal{A}_2, \mathcal{A}_3, \mathcal{A}_4)$ all of the following properties are satisfied (in all of the following probabilistic experiments, the experiment starts by setting λ as $\lambda \leftarrow \mathsf{K}_0(1^m)$):*

- **(Composable) Partial-ZK:**

$$\Pr[\rho \leftarrow \mathcal{D}_\lambda; \sigma \leftarrow \mathsf{K}_1(\lambda, \rho) : \mathcal{A}_0(\sigma, \rho) = 1] \approx$$
$$\Pr[(\rho, \sigma, \tau, \eta) \leftarrow \mathsf{sfK}_1(\lambda) : \mathcal{A}_0(\sigma, \rho) = 1],$$

and

$$\Pr[(\rho, \sigma, \tau, \eta) \leftarrow \mathsf{sfK}_1(\lambda) : \mathcal{A}_1^{\mathsf{P}(\sigma, \cdot, \cdot, \cdot),\ \mathsf{sfSim}(\sigma, \tau, \cdot, \cdot, \cdot),\ \mathsf{V}(\sigma, \cdot, \cdot, \cdot)}(\sigma, \rho) = 1] \approx$$
$$\Pr[(\rho, \sigma, \tau, \eta) \leftarrow \mathsf{sfK}_1(\lambda) : \mathcal{A}_1^{\mathsf{sfSim}^*(\sigma, \tau, \cdot, \cdot, \cdot),\ \mathsf{sfSim}(\sigma, \tau, \cdot, \cdot, \cdot),\ \mathsf{pV}(\sigma, \eta, \cdot, \cdot, \cdot)}(\sigma, \rho) = 1],$$

where $\mathsf{sfSim}^*(\sigma, \tau, x, w, l)$ is defined to be $\mathsf{sfSim}(\sigma, \tau, x, \beta = 1, l)$ (i.e. witness is dropped, and membership-bit $\beta = 1$), and the experiment aborts if either a call to the first oracle (i.e. P and sfSim^*) is with (x, w, l) s.t. $\neg R_\rho(x, w)$, or call to the second oracle is with an (x, β, l) s.t. $x \notin L_\rho$ or $\beta = 0$.

- **Unbounded Partial-Simulation Soundness:**

$$\Pr\left[\begin{array}{l} (\rho, \sigma, \tau, \eta) \leftarrow \mathsf{sfK}_1(\lambda); (x, l, \pi) \leftarrow \mathcal{A}_2^{\mathsf{sfSim}(\sigma, \tau, \cdot, \cdot, \cdot),\ \mathsf{pV}(\sigma, \eta, \cdot, \cdot, \cdot)}(\sigma, \rho) : \\ ((x \notin L_\rho) \vee \mathsf{V}(\sigma, x, l, \pi) = 0) \wedge \mathsf{pV}(\sigma, \eta, x, l, \pi) = 1 \end{array}\right] \approx 0.$$

- **One-time Full-ZK:**

$$\Pr\left[\begin{array}{l} (\rho, \sigma, \tau, \eta) \leftarrow \mathsf{sfK}_1(\lambda); (x^*, l^*, \beta^*, s) \leftarrow \mathcal{A}_3^{\mathsf{sfSim}(\sigma, \tau, \cdot, \cdot, \cdot),\ \mathsf{pV}(\sigma, \eta, \cdot, \cdot, \cdot)}(\sigma, \rho); \\ \pi^* \leftarrow \mathsf{sfSim}(\sigma, \tau, x^*, \beta^*, l^*) : \mathcal{A}_4^{\mathsf{sfSim}(\sigma, \tau, \cdot, \cdot, \cdot),\ \mathsf{pV}(\sigma, \eta, \cdot, \cdot, \cdot)}(\pi^*, s) = 1 \end{array}\right]$$
$$\approx \Pr\left[\begin{array}{l} \rho \leftarrow \mathcal{D}_\lambda; (\sigma, \tau, \tau_1, \eta) \leftarrow \mathsf{otfK}_1(\lambda, \rho); \\ (x^*, l^*, \beta^*, s) \leftarrow \mathcal{A}_3^{\mathsf{sfSim}(\sigma, \tau, \cdot, \cdot, \cdot),\ \mathsf{sfV}(\sigma, \eta, \cdot, \cdot, \cdot)}(\sigma, \rho); \\ \pi^* \leftarrow \mathsf{otfSim}(\sigma, \tau_1, x^*, l^*) : \mathcal{A}_4^{\mathsf{sfSim}(\sigma, \tau, \cdot, \cdot, \cdot),\ \mathsf{sfV}(\sigma, \eta, \cdot, \cdot, \cdot)}(\pi^*, s) = 1 \end{array}\right],$$

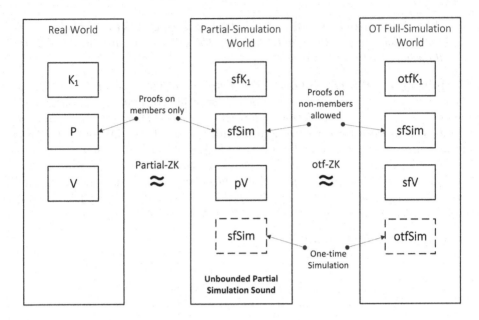

Fig. 1. The three worlds of a DSS-QA-NIZK

*where the experiment aborts if either in the call to the first oracle, or in the (x^*, β^*) produced by $\mathcal{A}_3$, the membership-bit provided is not correct for L_ρ, **or** if $\langle x^*, l^*, \pi^* \rangle$ is queried to sfV/pV. Here s is a state variable.*

The three worlds and the properties of a DSS-QA-NIZK are depicted in Fig. 1.

Remark 1. In the partial-simulation soundness definition, there is *no restriction* of x, l, π being *not* the same as that obtained from a call to the first oracle sfSim.

Remark 2. Note that in the partial-ZK definition, the calls to the prover are restricted to ones satisfying the relation. However, the calls to the simulator sfSim in the one-time full-ZK definition are only restricted to having the correct membership bit β.

Remark 3. It can be shown that sfSim generated proofs on words (whether members or not) are accepted by real-world verifier V (with semi-functional CRS). Of course, the private verifier pV will *even* reject proofs generated by sfSim on non-language words. This justifies the name "semi-functional simulator". See [JR14b] for a precise claim and proof.

It can also be shown that the semi-functional verifier sfV is still complete, i.e. it accepts language members and proofs generated on them by $P(\sigma, \cdot, \cdot, \cdot)$ (with σ generated by $otfK_1$). As opposed to P and pV, it may no longer be sound. This justifies the name "semi-functional verifier" a la Waters' dual-system IBE construction. However, if the one-time full-ZK property holds statistically, it can be shown that the semi-functional verifier is sound in the one-time full-simulation world. See [JR14b] for a precise statement.

Remark 4. The composable partial-ZK and unbounded partial-simulation soundness imply that that the system is true-simulation-sound (cf. true-simulation extractable [Har11]) w.r.t. the semi-functional simulator, as stated below.

Lemma 1. (true-simulation-soundness) For a DSS-QA-NIZK, for all PPT $\mathcal{A}$,

$$\Pr\left[\begin{array}{c}(\rho,\sigma,\tau,\eta) \leftarrow \mathsf{sfK}_1(\lambda);\ (x,l,\pi) \leftarrow \mathcal{A}^{\mathsf{sfSim}(\sigma,\tau,\cdot,\cdot,\cdot)}(\sigma,\rho):\\ (x \notin L_\rho) \wedge \mathsf{V}(\sigma,x,l,\pi) = 1\end{array}\right] \approx 0, \text{ where the}$$

experiment aborts if $\mathcal{A}$ calls the oracle with some (y, β, l), s.t. $y \notin L_\rho$ or $\beta = 0$.

4 DSS-QA-NIZK for Linear Subspaces

In this section we show that languages that are linear subspaces of vector spaces of hard bilinear groups have very short dual-system simulation sound QA-NIZK. In fact, under the Symmetric-eXternal Diffie-Hellman (SXDH) assumption, such proofs only require two group elements, regardless of the subspace. It was shown in [JR14a] that such subspaces have a QA-NIZK proof of just one group element (under the SXDH assumption). Our construction essentially shows that with one additional group element, one can make the QA-NIZK dual-system simulation-sound. We will actually show a more general construction which is more widely applicable, and does not even refer to bilinear groups or linear subspaces. Informally speaking, the requirement for such a general construction for parameterized languages is that each language has a 2-universal projective hash proof system and the augmented language with this hash proof attached has a QA-NIZK proof system with statistical zero-knowledge. A few other properties of the QA-NIZK are required for this construction, and we show that such properties already hold for the construction of [JR14a]. Since for linear subspaces, 2-universal projective hash proofs are rather easy to obtain, the general construction along with the QA-NIZK of [JR14a] allows us to obtain a short DSS-QA-NIZK for linear subspaces. Apart from abstracting the main ideas involved in the DSS-QA-NIZK construction for linear subspaces, the general construction's wider applicability also allows us to extend our results to linear subspaces with tags.

We start this section by briefly reviewing projective hash proofs [CS02], and their extensions to distributions of languages, as they are extensively used in the rest of the section.

Projective Hash Proof System. For a language L, let X be a superset of L and let $H = (H_k)_{k \in K}$ be a collection of (hash) functions indexed by K with domain X and range another set Π. The hash function family is generalized to a notion of *projective hash function family* if there is a set S of projection keys, and a projection map $\alpha : K \rightarrow S$, and further the action of H_k on subset L of X is completely determined by the projection key $\alpha(k)$. Finally, the projective hash function family is defined to be ϵ-**universal$_2$** is for all $s \in S$, $x, x^* \in X$, and $\pi, \pi^* \in \Pi$ with $x \notin L \cup \{x^*\}$, the following holds:

$$\Pr[H_k(x) = \pi \mid H_k(x^*) = \pi^* \wedge \alpha(k) = s] \leq \epsilon.$$

A projective hash function family is called ϵ-**smooth** if for all $x \in X \setminus L$, the statistical difference between the following two distributions is ϵ: sample k uniformly from K and π' uniformly from Π; the first distribution is given by the pair $(\alpha(k), H_k(x))$ and the second by the pair $(\alpha(k), \pi')$. For languages defined by a witness-relation R, the projective hash proof family constitutes a *projective hash proof system* (PHPS) if α, H_k, and another *public evaluation function* $\hat{H}$ that computes H_k on $x \in L$, given a witness of x and *only* the projection key $\alpha(k)$, are all efficiently computable. An efficient algorithm for sampling the key $k \in K$ is also assumed.

The above notions can also incorporate labels. In an *extended PHPS*, the hash functions take an additional input called *label*. The public evaluation algorithm also takes this additional input called label. All the above notions are now required to hold for each possible value of label. The extended PHPS is now defined to be ϵ-**universal$_2$** is for all $s \in S$, $x, x^* \in X$, all labels l and l^*, and $\pi, \pi^* \in \Pi$ with $x \notin L$ and $(x, l) \neq (x^*, l^*)$, the following holds: $\Pr[H_k(x, l) = \pi \mid H_k(x^*, l^*) = \pi^* \wedge \alpha(k) = s] \leq \epsilon$.

Since, we are interested in distributions of languages, we extend the above definition to distribution of languages. So consider a parametrized class of languages $\{L_\rho\}_{\rho \in \mathsf{Lpar}}$ with the parameters coming from an associated parameter language Lpar. Assume that all the languages in this collection are subsets of X. Let H as above be a collection of hash functions from X to Π. We say that the hash family is a projective hash family if for all L_ρ, the action of H_k on L_ρ is determined by $\alpha(k)$. Similarly, the hash family is ϵ-universal$_2$ (ϵ-smooth) for $\{L_\rho\}_{\rho \in \mathsf{Lpar}}$ if for all languages L_ρ the ϵ-universal$_2$ (resp. ϵ-smooth) property holds.

Intuition for the Construction. The main idea of the construction is to first attach (as a proof component) a universal$_2$ and smooth projective hash proof T. The DSS-QA-NIZK is then just (T, π), where π is a QA-NIZK proof of the original language augmented with hash proof T. So, why should this work? First note that the smooth projective hash function is a designated-verifier NIZK, and hence this component T is used in private verification. Secondly, since it is universal$_2$, its soundness will hold even when the Adversary gets to see the projection key $\alpha(k)$ plus one possibly fake hash proof (i.e. $H_k(x)$, where x not in the language).

We will assume in our general construction that the parameterized language is such that the simulator can sample the language parameters along with auxiliary information that allows it to easily verify a language member. For example, this auxiliary information can be discrete logs of the language parameters. The idea of obtaining partial-ZK and unbounded partial-simulation soundness is then pretty simple. The proof simulation of T is easy to accomplish given the hash keys and, crucially, the correct membership-bit. In fact, if the membership-bit is false, T can just be set randomly (by smoothness). The simulation of π part of the proof is done using the QA-NIZK simulation trapdoor. The private verification is done as conjunction of three separate checks: (a) using the auxiliary information, (b) using the hash proof and (c) using the real-world verifier.

Now, in the one-time full simulation, the auxiliary information is not available, but the semi-functional verifier can still use hash keys. Further, we can have one bad use of keys (in full simulation of one proof. Since the oracle calls to semi-functional simulator sfSim are restricted to having correct membership-bit, they do not yield any additional information about the hash keys.

Requirements of the General Construction. Consider a parameterized class of languages $\{L_\rho\}_{\rho \in \mathsf{Lpar}}$, and a probability distribution $\mathcal{D}$ on Lpar. Assume that this class has a projective hash proof system as above. Let R_ρ be the corresponding witness relation of L_ρ. Now consider the augmented witness-relation $R^*_{\rho,s}$ defined as follows (for $\rho \in \mathsf{Lpar}$ and $s \in S$):

$$R_{\rho,s}(\langle x, T, l \rangle, w) \equiv (R_\rho(x, w) \wedge T \stackrel{?}{=} \hat{H}(s, \langle x, l \rangle, w)).$$

Note, the witness remains the same for the augmented relation. Since H is a projective hash function, it follows that for $s = \alpha(k)$, the corresponding augmented language is $L^*_{\rho,s} = \{(x, T, l) \mid x \in L_\rho \wedge T \stackrel{?}{=} H_k(x, l)\}$. Let the distribution $\mathcal{D}'$ on pairs (ρ, s) be defined by sampling ρ according to $\mathcal{D}$ and sampling k uniformly from K, and setting $s = \alpha(k)$. We remark that the language parameters of the augmented language include projection keys s (instead of keys k) because it is crucial that the CRS simulator in the quasi-adaptive NIZK gets only the projection key s (and not k).

We will also assume that the distribution $\mathcal{D}$ on Lpar is efficiently *witness samplable* which is defined by requiring that there are two efficient (probabilistic) algorithms E_1, E_2 such that E_1 can sample ρ from $\mathcal{D}$ along with auxiliary information ψ (which can be thought of as witness of ρ in the language Lpar), and E_2 can decide w.h.p. if a word x is in L_ρ given ρ and ψ, where the probability is defined over choice of ρ according to $\mathcal{D}$ and the internal coins of E_2.

Finally, we need a few additional properties of QA-NIZK proofs (Sect. 2) that we now define. We will later show that the single group element QA-NIZK construction for linear-subspaces of [JR14a] already satisfies these properties.

Definition 4. *There are various specializations of QA-NIZK of interest:*

- *The QA-NIZK (Sect. 2) is said to have* **composable zero-knowledge** *[GS08] if the CRS are indistinguishable in the real and simulation worlds, and the simulation is indistinguishable even if the adversary is given the trapdoor.*

More precisely, for all PPT adversary $\mathcal{A}_1, \mathcal{A}_2$,

$$\Pr[\text{CRS} \leftarrow \text{crsgen}(\lambda, \rho) : \mathcal{A}_1(\text{CRS}, \rho) = 1] \approx$$
$$\Pr[(\text{CRS}, \text{trap}) \leftarrow \text{crs} - \text{sim}(\lambda, \rho) : \mathcal{A}_1(\text{CRS}, \rho) = 1],$$

and

$$\Pr[(\text{CRS}, \text{trap}) \leftarrow \text{crs} - \text{sim}(\lambda, \rho) : \mathcal{A}_2^{\text{prover}(\text{CRS}, \cdot, \cdot, \cdot)}(\text{CRS}, \rho, \text{trap}) = 1] \approx$$
$$\Pr[(\text{CRS}, \text{trap}) \leftarrow \text{crs} - \text{sim}(\lambda, \rho) : \mathcal{A}_2^{\text{sim}^*(\text{CRS}, \text{trap}, \cdot, \cdot, \cdot)}(\text{CRS}, \rho, \text{trap}) = 1],$$

where $\mathcal{A}_2$ is restricted to calling the oracle only on (x, w, l) with $(x, w) \in R_\rho$.

- *The QA-NIZK is called* **true-simulation-sound** *[Har11] if the verifier is sound even when an adaptive adversary has access to simulated proofs on language members. More precisely, for all PPT $\mathcal{A}$,*

$$\Pr\left[\begin{array}{l} (\text{CRS}, \text{trap}) \leftarrow \text{crs} - \text{sim}(\lambda, \rho) \\ (x, l, \pi) \leftarrow \mathcal{A}^{\text{sim}(\text{CRS}, \text{trap}, \cdot, \cdot)}(\text{CRS}, \rho) \end{array} : x \notin L_\rho \wedge \text{ver}(\text{CRS}, x, l, \pi) = 1 \right] \approx 0,$$

where the experiment aborts if the oracle is called with some $y \notin L_\rho$.

- *The simulator is said to generate* **unique acceptable proofs** *if for all x, all labels l, and all proofs π^*,*

$$\Pr\left[\begin{array}{l} (\text{CRS}, \text{trap}) \leftarrow \text{crs} - \text{sim}(\lambda, \rho) \\ \pi \leftarrow \text{sim}(\text{CRS}, \text{trap}, x, l) \end{array} : (\pi^* \neq \pi) \wedge \text{ver}(\text{CRS}, x, l, \pi^*) = 1 \right] \approx 0.$$

General Construction. We now show that given:

1. An ϵ-smooth and ϵ-universal$_2$ (labeled) projective hash proof system for the collection $\{L_\rho\}_{\rho \in \text{Lpar}}$, and
2. A composable zero-knowledge, true-simulation-sound QA-NIZK $Q=$ (pargen, crsgen, prover, ver, crs-sim, sim) for the augmented parameterized language $L_{\rho,s}^*$ with probability distribution $\mathcal{D}'$, such that the simulator *generates unique acceptable proofs*, and
3. Efficient algorithms (E_1, E_2) s.t. $\mathcal{D}$ is efficiently witness-samplable using (E_1, E_2), and
4. An efficient algorithm E_3 to sample uniformly from Π,

one can construct a DSS-QA-NIZK for $\{L_\rho\}_{\rho \in \text{Lpar}}$ with probability distribution $\mathcal{D}$. We first give the construction, and then prove the required properties. The QA-NIZK Q *need not take any labels as input.* The various components of the dual-system non-interactive proof system Σ are as follows.

Real World consisting of:

- The algorithm K_0 takes a unary string 1^m as input and generates parameters λ using pargen of Q on 1^m. The CRS generation algorithm K_1 uses crsgen of Q and produces the CRS as follows: it takes λ and the language parameter ρ, and first samples k uniformly from K_λ (recalling that the hash function families are ensembles, one for each λ). It then outputs the CRS to be the pair $(\text{crsgen}(\lambda, \langle \rho, \alpha(k) \rangle), \alpha(k))$.

- The **prover** P takes a CRS (σ, s), input x, witness w, and label l and outputs the proof to be (T, W) where T is computed using the public evaluation algorithm $\hat{H}$ as $\hat{H}(s, \langle x, l \rangle, w)$ and $W = \mathsf{prover}(\sigma, \langle x, T, l \rangle, w)$.
- The **verifier** V on input CRS $= (\sigma', s)$, x, l, and proof (T, W), returns the value $\mathsf{ver}(\sigma', \langle x, T, l \rangle, W)$ (using ver of Q).

Partial-Simulation World consisting of:
- The **semi-functional CRS simulator** sfK_1 takes λ as input and samples (ρ, ψ) using E_1, and also samples k uniformly from K_λ. It then uses $\mathsf{crs} - \mathsf{sim}$ of Q, and key projection algorithm α to generate the CRS σ as follows: Let $(\sigma', \mathsf{trap}) = \mathsf{crs} - \mathsf{sim}(\lambda, \langle \rho, \alpha(k) \rangle)$. The CRS σ is then the pair $(\sigma', \alpha(k))$. sfK_1 also outputs k, trap as proof simulator trapdoors τ, and ρ, ψ, k as private verifier trapdoors η.
- The **semi-functional simulator** sfSim uses trapdoors k, trap to produce a (partially-simulated) proof for a word x, a label l and a binary bit β using sim of Q as follows: if $\beta = 1$, output

$$T = H_k(x, l), \ W = \mathsf{sim}(\sigma, \mathsf{trap}, \langle x, T, l \rangle),$$

else sample π' at random from Π (using E_3) and output

$$T = \pi', \ W = \mathsf{sim}(\sigma, \mathsf{trap}, \langle x, T, l \rangle).$$

This proof is partially simulated as it uses the bit β.
- The **private verifier** pV uses trapdoors (ρ, ψ, k) to check a word x, label l and a proof T, W as follows: it outputs 1 iff (a) E_2 using ρ and ψ confirms that x is in L_ρ, and (b) $H_k(x, l) = T$, and (c) verifier of Q accepts, i.e. $\mathsf{ver}(\sigma, \langle x, T, l \rangle, W) = 1$.

One-time Full Simulation World consisting of:
- The **one-time full-simulation CRS generator** otfK_1 takes as input λ and language parameter ρ, and using $\mathsf{crs} - \mathsf{sim}$ of Q outputs σ as follows: first it samples k uniformly from K_λ. Let $(\sigma', \mathsf{trap}) = \mathsf{crs} - \mathsf{sim}(\lambda, \langle \rho, \alpha(k) \rangle)$. Then $\sigma = (\sigma', \alpha(k))$. otfK_1 also outputs k, trap as proof simulator trapdoors τ and τ_1, and outputs k as private verifier trapdoor η.
- The **one-time full simulator** otfSim takes as input the trapdoors k, trap and a word x and a label l to produce a proof as follows:

$$T = H_k(x, l), \ \ W = \mathsf{sim}(\sigma, \mathsf{trap}, \langle x, T, l \rangle).$$

- The **semi-functional verifier** sfV uses trapdoors k to verify a word x, a label l and a proof T, W as follows: output 1 iff (a) $H_k(x, l) = T$, and (b) $\mathsf{ver}(\sigma, \langle x, T, l \rangle, W) = 1$.

Theorem 1. *For a parameterized class of languages $\{L_\rho\}_{\rho \in \mathsf{Lpar}}$ with probability distribution $\mathcal{D}$, if the above four conditions hold for projective hash family H, QA-NIZK Q, and efficient algorithms E_1, E_2, E_3, then the above dual-system non-interactive proof system Σ is a DSS-QA-NIZK for $\{L_\rho\}_{\rho \in \mathsf{Lpar}}$ with probability distribution $\mathcal{D}$.*

Remark. In [JR14b] we instantiate the general construction for linear subspaces of vector spaces of hard bilinear groups. As a corollary, it follows that under the SXDH assumption the Diffie-Hellman (DH) language has a DSS-QA-NIZK with only two group elements.

Due to space limitations, we will focus on only the proof of one-time zero-knowledge (otzk) property, as that is the most non-trivial proof. Indeed, this property is a significant generalization of the usual dual-system technique employed in IBE constructions because although in otzk only one proof needs to be fully simulated (i.e. without its membership bit being available), all the private verifier calls in the partial-simulation world need to be simulated in the otzk world without the quasi-adaptive trapdoors (i.e. trapdoor obtained by witness-sampling the language parameters). Recall, in the IBE construction the ciphertext is the counterpart of our verifier, and the IBE private keys are the QA-NIZK proofs. Thus, in IBE only a single ciphertext needs to be simulated when the different private keys are being "fixed" one-by-one by otzk simulation.

The detailed proof of all other properties is given in [JR14b]. The main idea of the proof of these properties is already sketched earlier in this section.

Lemma 2. *In the context of Theorem 1, let the maximum probability that the simulator of Q does not generate unique acceptable proofs be δ. Let H be an ϵ-smooth and ϵ-universal$_2$ (labeled) projective hash proof system for the collection $\{L_\rho\}_{\rho \in \mathsf{Lpar}}$. Let M be the number of calls to the second oracle (verifier) by $\mathcal{A}_3$ and $\mathcal{A}_4$ combined in the two experiments of the one-time full-ZK property of DSS-QA-NIZK Σ. Then the maximum statistical distance (over all PPT Adversaries $\mathcal{A}_3$ and $\mathcal{A}_4$) between the views of the adversaries $(\mathcal{A}_3, \mathcal{A}_4)$ in these two experiments, denoted $dist^{otzk}(\Sigma)$, is at most $(\epsilon + \delta) * (1 + M)$.*

Proof. We will show that the one-time full-ZK property holds statistically. We will define a sequence of experiments and show that the view of the PPT adversary is statistically indistinguishable in every two consecutive experiments. The first experiment $\mathbf{H}_0$ is identical to the partial-simulation world. First, note that ρ is identically generated using $\mathcal{D}$ in both worlds. Next, note that the CRS σ and trapdoors τ generated by sfK_1 is identically distributed to the CRS σ and both the trapdoors τ and τ_1 generated by otfK_1.

The next experiment $\mathbf{H}_1$ is identical to $\mathbf{H}_0$ except that on $\mathcal{A}_3$ supplied input (x^*, l^*, β^*) the proof π^* generated by sfSim is replaced by proof generated by otfSim. If β^* provided by $\mathcal{A}_3$ is not the valid membership bit for x^* then both experiments abort. So, assume that β^* is the correct membership bit. In case $\beta^* = 1$, both sfSim and otfSim behave identically. When $\beta^* = 0$, the random T^* produced by sfSim is identically distributed to the T^* generated by $H_k(x^*, l^*)$ since H is assumed to be smooth.

The next experiment $\mathbf{H}_2$ is identical to $\mathbf{H}_1$ except that the second oracle is replaced by sfV (from being pV). In order to show that the view of the adversary is indistinguishable in experiments $\mathbf{H}_2$ and $\mathbf{H}_1$, we define several hybrid experiments $\mathbf{H}_{1,i}$ (for $0 \le i \le N$, where N is the total number of calls to the second-oracle by $\mathcal{A}_3$ and $\mathcal{A}_4$ combined). Experiment $\mathbf{H}_{1,0}$ is identical to $\mathbf{H}_1$, and the intermediate experiments are defined inductively, by modifying the response of one additional second-oracle call starting with the last (N-th) second-oracle call, and ending with the changed response of the first second-oracle call. The last hybrid experiment $\mathbf{H}_{1,N}$ will then be same as $\mathbf{H}_2$. The second-oracle call response in experiment $\mathbf{H}_{1,i+1}$ differs only in the ($N-i$)-th second-oracle call response in $\mathbf{H}_{1,i}$. In the latter experiment, this call is still served as in $\mathbf{H}_1$ (i.e. using pV). In the former experiment $\mathbf{H}_{1,i+1}$, the ($N-i$)-th call is responded to as defined in $\mathbf{H}_2$ above (i.e. using sfV).

To show that the view of the adversary is statistically indistinguishable in $\mathbf{H}_{1,i}$ and $\mathbf{H}_{1,i+1}$, first note that the view of the adversary ($\mathcal{A}_3$ and A_4 combined) till it's ($N-i$)-th call in both experiments is identical. Moreover, as we next show, the dependence on k of this partial view (i.e. till the ($N-i$)-th call) is limited to $\alpha(k)$ and at most one evaluation of H_k (by otfSim) on an input that is not in L_ρ. To start with, the CRS generated by sfK$_1$ depends only on $\alpha(k)$. Next, the first oracle sfSim produces T using H_k on its input only if the membership bit β is 1 and correct, and since H is projective this hash value is then completely determined by $\alpha(k)$. Finally, all calls to the second oracle till the ($N-i$)-th call are still served using pV, and again using the projective property of H, it is clear that the conjunct (b) in pV can be computed using only $\alpha(k)$, because for non L_ρ members, the conjunct (a) is already false, and hence (b) is redundant.

Now, the difference in the ($N-i$)-th call is that the conjunct (a) of pV is missing in sfV. Let x, l, T, W be the input supplied by the PPT Adversary to this call. If $H_k(x, l)$ is not equal to the supplied T, then both pV and sfV return 0. So, suppose $H_k(x, l)$ is equal to T, and yet x is not in L_ρ, i.e. conjunct (a) of pV is false. First, if this input x, l, T, W is same as (x^*, l^*, T^*, W^*) associated with the one-time call to otfSim, then the experiment aborts. Thus, we can assume that this is a different input. If (x, l) is same as (x^*, l^*), then $(T, W) \neq (T^*, W^*)$. Now, by construction (i.e. by definition of otfSim) $T^* = H_k(x^*, l^*)$, and hence either $T \neq H_k(x, l)$ which is not possible by hypothesis, or $(x, l, T) = (x^*, l^*, T^*)$ and $W \neq W^*$. But, W^* is proof generated by the simulator of Q, and since the simulator of Q generates unique acceptable proofs (by assumption), the verifier ver of Q rejects (x, l, T, W), and thus both pV and sfV return 0.

On the other hand, if $(x, l) \neq (x^*, l^*)$ then by the ϵ-universal$_2$ property of H, the probability of T being same as $H_k(x, l)$ is at most ϵ. Thus, both pV and sfV return 0. That completes the induction step, and thus the view of the adversary in experiments $\mathbf{H}_1$ and $\mathbf{H}_2$ is statistically indistinguishable.

The next experiment $\mathbf{H}_3$ is identical to $\mathbf{H}_2$ except that the CRS is generated using otfK$_1$. The only difference is that the (verifier) trapdoor does not include ρ, ψ. But, since the second oracle is served by sfV and it does not need ρ, ψ, the experiment $\mathbf{H}_3$ is well-defined and statistically indistinguishable from $\mathbf{H}_2$, Further, $\mathbf{H}_3$ is identical to the one-time simulation world, and that completes the proof.

The statistical distance between the views of the adversaries ($\mathcal{A}_3, \mathcal{A}_4$) in $\mathbf{H}_0$ and $\mathbf{H}_3$ is at most $(\epsilon + \delta) * (1 + M)$. $\qquad\square$

5   Keyed-Homomorphic CCA Encryption

Keyed-Homomorphic Encryption is a primitive, first developed in [EHO+13], which allows homomorphic operations with a restricted evaluation key, while preserving different flavors of semantic security depending on whether access to the evaluation key is provided or not. For an adversary not having access to the evaluation key, the homomorphic operation should not be available and this is ensured by requiring CCA security. However, if an adversary comes into possession of the evaluation key, CCA security can no longer be preserved and thus weaker forms of security, such as CCA1, are required. In [LPJY14], the authors gave improved constructions for multiplicative homomorphism with better security guarantees.

A **KH-PKE** scheme consists of algorithms $(KeyGen, Enc, Dec, Eval)$, where the first three are familiar from public-key encryption, and KeyGen generates a public key pk, a decryption key sk_d and an Eval key sk_h. Algorithm Eval takes two ciphertexts and returns a ciphertext or $\perp$. Detailed definitions can be found in [JR14b]. The scheme is said to be correct if (i) for Enc we have $Dec(sk_d, Enc(pk, M)) = M$, where sk_d is the secret decryption key, and (ii) for Eval we have $Dec(sk_d, Eval(sk_h, C_1, C_2)) = Dec(sk_d, C_1) \odot Dec(sk_d, C_2)$, where $\odot$ is a binary operation on plaintexts, and if any operand of $\odot$ is $\perp$ then the result is $\perp$. The KH-PKE scheme is defined to be **KH-CCA** secure by a usual public-key CCA experiment with the following twists: the challenger maintains a set D of ciphertexts dependent on the challenge ciphertext (via Eval); decryption queries are not allowed on ciphertexts in D. Further, an adversary $\mathcal{A}$ can adaptively ask for sk_h, which we call the *reveal event*. After the reveal event, the Eval oracle is not available. Similarly, decryption is not available after $\mathcal{A}$ has both requested sk_h and obtained the challenge ciphertext, in any order. Again, detailed definitions can be found in [JR14b].

Construction. We present a construction of a KH-CCA secure KH-PKE encryption scheme with multiplicative homomorphism which utilizes our general DSS-QA-NIZK construction for the Diffie-Hellman (DH) language. In fact, if we assume that the adversary never invokes RevHK, we can prove security generically assuming any DSS-QA-NIZK (with statistical one-time full-ZK) for the DH language. When the adversary invokes RevHK, the partial-simulation trapdoor is revealed to the Adversary, and hence the one-time full-ZK property of DSS-QA-NIZK may not hold. Thus, we a need a stronger notion of DSS-QA-NIZK that incorporates the reveal event, and includes an additional requirement that the semi-functional verifier remains sound as before. Using this stronger notion, we can prove generic security of the KH-PKE scheme even with RevHK, and we further show that our general construction of Sect. 4 continues to satisfy this stronger property.

We start with the observation that a standard ElGamal encryption scheme $(\mathbf{g}^x, m \cdot \mathbf{f}^x)$ is multiplicatively homomorphic, but is not CCA secure due to the exact same reason. The main idea of our construction is as follows. The ciphertexts include an ElGamal encryption of the message M, say $\mathbf{g}^r, M \cdot \mathbf{g}^{kr}$ for a public key $\mathbf{g}^k$. The public key also consists of a member $\mathbf{g}^a$, and the ciphertext also include $\mathbf{g}^{ar}$ (we refer to this triple in the ciphertext as *augmented ElGamal encryption*). It is well-known [JR12] that if a one-time simulation-sound NIZK proof of $\mathbf{g}^r$ and $\mathbf{g}^{ar}$ being of the correct form is also included in the ciphertext then it becomes a publicly-verifiable CCA2-secure encryption scheme. In our keyed-homomorphic construction, we include a DSS-QA-NIZK for $\mathbf{g}^r$ and $\mathbf{g}^{ar}$ being of the correct form (i.e. being a DH tuple). Although the DSS-QA-NIZK itself is not homomorphic, we can take advantage of the corresponding Semi-Functional Simulator sfSim and simulate the proof of a multiplicatively generated (augmented) ElGamal encryption when computing a homomorphic evaluation.

So, given a dual-system non-interactive proof Σ, consider the following algorithms for a KH-PKE scheme $\mathcal{P}$:

KeyGen: Generate $\mathbf{g}, a, k$ randomly. Use sfK_1 of Σ to get CRS σ and trapdoors τ and η, and language parameters $\rho = (\mathbf{g}, \mathbf{g}^a)$. Set $pk = (\mathbf{g}, \mathbf{g}^a, \mathbf{g}^k, \sigma)$, $sk_h = \tau$, $sk_d = k$.

Enc: Given plaintext m, generate $w \leftarrow \mathbb{Z}_q$ and compute (using P of Σ) $c := (\mathbf{g}^w, \mathbf{g}^{aw}, \gamma, \mathsf{P}(\sigma, (\mathbf{g}^w, \mathbf{g}^{aw}), w, l = \gamma))$, where $\gamma := m \cdot \mathbf{g}^{kw}$.

Dec: Given ciphertext $c = (\rho, \hat{\rho}, \gamma, \pi)$, first check if $\mathsf{V}(\sigma, \pi, (\rho, \hat{\rho}), \gamma)$ of Σ holds, then compute $m := \gamma / \rho^k$.

Eval (Multiplicative): Given ciphertexts $c_1 = (\rho_1, \hat{\rho}_1, \gamma_1, \pi_1)$ and $c_2 = (\rho_2, \hat{\rho}_2, \gamma_2, \pi_2)$, first check if $\mathsf{V}(\sigma, \pi_i, (\rho_i, \hat{\rho}_i), \gamma_i)$ of Σ holds for all $i \in \{1, 2\}$. Then compute: $\rho = \rho_1 \rho_2 \rho_3$, $\hat{\rho} = \hat{\rho}_1 \hat{\rho}_2 \hat{\rho}_3$, $\gamma = \gamma_1 \gamma_2 \gamma_3$, where $\langle \rho_3, \hat{\rho}_3, \gamma_3 \rangle$ is a fresh random tuple obtained by picking r at random and setting the tuple to be $\langle \mathbf{g}^r, (\mathbf{g}^a)^r, (\mathbf{g}^k)^r \rangle$. Then compute $\pi := \mathsf{sfSim}(\sigma, \tau, (\rho, \hat{\rho}), \beta = 1, l = \gamma)$ using sfSim of Σ. Output ciphertext $c := (\rho, \hat{\rho}, \gamma, \pi)$.

Theorem 2. (Security of Construction). *The above algorithms* $\mathcal{P} =$ *(Key-Gen, Enc, Dec, Eval) constitute a KH-CCA secure Keyed-Homomorphic Public Key Encryption scheme with multiplicative homomorphism, if Σ is a DSS-QA-NIZK for the parameterized Diffie-Hellman language (with language parameters distributed randomly) and RevHK is not available.*

The main idea of the proof of the above theorem is similar to proofs of CCA2-secure public key encryption schemes using alternate decryption. In other words, the ciphertext can be decrypted as $m := \gamma / \rho^k$, or as $m := \gamma / (\rho^{k_0} \hat{\rho}^{k_1})$, where $k = k_0 + a k_1$. But, this requires that the ciphertext has correct $\hat{\rho}$ component, i.e. $\hat{\rho} = \rho^a$. The ciphertexts include a NIZK for this purpose, but the NIZK needs to be simulation-sound. Additional complication arises because of dependent ciphertexts. To handle this, we first build an intermediate experiment where all dependent ciphertexts are generated using fresh random ElGamal tuples. Indistinguishability of such an intermediate experiment from the KH-CCA experiment is shown inductively, by carefully employing one-time full-ZK and partial-simulation unbounded simulation soundness. The theorem is proved in detail in [JR14b]. The Adversary's advantage in the KH-CCA security game is at most $(8L + 1) \cdot \mathrm{ADV}_{\mathrm{DDH}} + O(L/q)$, where L is the total number of calls to Eval.

Functionality $\mathcal{F}_{\mathbf{pake}}$

The functionality $\mathcal{F}_{\text{PAKE}}$ is parameterized by a security parameter k. It interacts with an adversary S and a set of parties via the following queries:

Upon receiving a query (NewSession, $sid, P_i, P_j, pw, role$) **from party P_i:**
Send (NewSession, $sid, P_i, P_j, role$) to S. In addition, if this is the first NewSession query, or if this is the second NewSession query and there is a record (P_j, P_i, pw'), then record (P_i, P_j, pw) and mark this record fresh.

Upon receiving a query (TestPwd, sid, P_i, pw') **from the adversary S:**
If there is a record of the form (P_i, P_j, pw) which is fresh, then do: If $pw = pw'$, mark the record compromised and reply to S with "correct guess". If $pw \neq pw'$, mark the record interrupted and reply with "wrong guess".

Upon receiving a query (NewKey, sid, P_i, sk) **from S, where $|sk| = k$:**
If there is a record of the form (P_i, P_j, pw), and this is the first NewKey query for P_i, then:
- If this record is compromised, or either P_i or P_j is corrupted, then output (sid, sk) to player P_i.
- If this record is fresh, and there is a record (P_j, P_i, pw') with $pw' = pw$, and a key sk' was sent to P_j, and (P_j, P_i, pw) was fresh at the time, then output (sid, sk') to P_i.
- In any other case, pick a new random key sk' of length k and send (sid, sk') to P_i. Either way, mark the record (P_i, P_j, pw) as completed.

Upon receiving (Corrupt, sid, P_i) **from S:** if there is a (P_i, P_j, pw) recorded, return pw to S, and mark P_i corrupted.

Fig. 2. The password-based key-exchange functionality $\mathcal{F}_{\text{PAKE}}$

The more general theorem (with RevHK) is stated and proved in [JR14b]. Under the SXDH assumption, the above construction leads to ciphertexts of size only five group elements. Further, using an *augmented Diffie Hellman language* (augmented with a smooth hash proof of DH tuple) and its DSS-QA-NIZK, we also extend our result to get CCA1-security despite the key being revealed (see [JR14b]). The resulting scheme has KH-PKE ciphertexts of size six group elements.

6   Single-Round UC Password-Based Key Exchange

The essential elements of the Universal Composability framework can be found in [Can01]. In the following, we adopt the definition for password-based key exchange (UC-PAKE) from Canetti et al [CHK+05].

6.1   UC-PAKE Definition

Just as in the normal key-exchange functionality, if both participating parties are not corrupted, then they receive the same uniformly distributed session key and the adversary learns nothing of the key except that it was generated. However, if one of the parties is corrupted, then the adversary determines the session key. This power to the adversary is *also* given in case it succeeds in guessing the parties' shared password. Participants also detect when the adversary makes an unsuccessful attempt. If the adversary makes a wrong password guess in a given

Generate $\mathbf{g}_1 \leftarrow \mathbb{G}_1, \mathbf{g}_2 \leftarrow \mathbb{G}_2$ and $a,b,c,d,e,u_1,u_2 \leftarrow \mathbb{Z}_q$, and let $\mathcal{H}$ be a CRHF. Compute $\mathbf{a} = \mathbf{g}_1^a$, $\mathbf{d} = \mathbf{g}_1^d$, $\mathbf{e} = \mathbf{g}_1^e$, $\mathbf{w}_1 = \mathbf{g}_1^{u_1}$, $\mathbf{w}_2 = \mathbf{g}_1^{u_2}$ $\quad\mathbf{b} = \mathbf{g}_2^b$, $\mathbf{c} = \mathbf{g}_2^c$, $\mathbf{v}_1 = \mathbf{g}_2^{u_1 b - d - ca}$, $\mathbf{v}_2 = \mathbf{g}_2^{u_2 b - e}$. $$\text{CRS} := (\mathbf{g}_1, \mathbf{g}_2, \mathbf{a}, \mathbf{b}, \mathbf{c}, \mathbf{d}, \mathbf{e}, \mathbf{w}_1, \mathbf{w}_2, \mathbf{v}_1, \mathbf{v}_2, \mathcal{H}).$$	
Party P_i	Network
Input $(\mathsf{NewSession}, sid, \mathsf{ssid}, P_i, P_j, \mathsf{pwd}, initiator/responder)$ Choose $r_1, s_1 \xleftarrow{\$} \mathbb{Z}_q$. Set $R_1 = \mathbf{g}_1^{r_1}$, $S_1 = \mathsf{pwd} \cdot \mathbf{a}^{r_1}$, $T_1 = (\mathbf{d} \cdot \mathbf{e}^{i_1})^{r_1}$, $\hat{\rho}_1 = \mathbf{b}^{s_1}$, $W_1 = (\mathbf{w}_1 \mathbf{w}_2^{i_1})^{r_1}$, where $i_1 = \mathcal{H}(sid, \mathsf{ssid}, P_i, P_j, R_1, S_1, \hat{\rho}_1)$ and erase r_1. Send R_1, S_1, T_1 and $\hat{\rho}_1$, and retain W_1.	$\xrightarrow{R_1, S_1, T_1, \hat{\rho}_1} P_j$
Receive $R_2', S_2', T_2', \hat{\rho}_2'$. If any of $R_2', S_2', T_2', \hat{\rho}_2'$ is not in their respective group or is 1, set $\mathsf{sk}_1 \xleftarrow{\$} \mathbb{G}_T$, else compute $i_2' = \mathcal{H}(sid, \mathsf{ssid}, P_j, P_i, R_2', S_2', \hat{\rho}_2')$, $\rho_1 = \mathbf{g}_2^{s_1}$, $\theta_1 = \mathbf{c}^{s_1}$, $\gamma_1 = (\mathbf{v}_1 \mathbf{v}_2^{i_2'})^{s_1}$. Compute $\mathsf{sk}_1 = e(T_2', \rho_1) \cdot e(S_2'/\mathsf{pwd}, \theta_1) \cdot e(R_2', \gamma_1) \cdot e(W_1, \hat{\rho}_2')$ Output $(sid, \mathsf{ssid}, \mathsf{sk}_1)$.	$\xleftarrow{R_2', S_2', T_2', \hat{\rho}_2'} P_j$

Fig. 3. Single round UC-secure Password-authenticated KE under SXDH Assumption.

session, then the session is marked interrupted and the parties are provided random and independent session keys. If however the adversary makes a successful guess, then the session is marked compromised, and the adversary is allowed to set the session key. If a session remains marked fresh, meaning that it is neither interrupted nor compromised. uncorrupted parties conclude with both parties receiving the same, uniformly distributed session key. The formal description of the UC-PAKE functionality $\mathcal{F}_{\mathrm{PAKE}}$ is given in Fig. 2.

The real-world protocol we provide is also shown to be secure when different sessions use the same common reference string (CRS). To achieve this goal, we consider the *universal Composability with joint state* (JUC) formalism of Canetti and Rabin [CR03]. This formalism provides a "wrapper layer" that deals with "joint state" among different copies of the protocol. In particular, defining a functionality $\mathcal{F}$ also implicitly defines the multi-session extension of $\mathcal{F}$ (denoted by $\hat{\mathcal{F}}$): $\hat{\mathcal{F}}$ runs multiple independent copies of $\mathcal{F}$, where the copies are distinguished via sub-session IDs ssid. The JUC theorem [CR03] asserts that for any protocol π that uses multiple independent copies of $\mathcal{F}$, composing π instead with a single copy of a protocol that realizes $\hat{\mathcal{F}}$, preserves the security of π.

6.2 Main Idea of the UC Protocol Using DSS-QA-NIZK

For the sake of exposition, let's call one party in the session the server and the other the client. (There is no such distinction in the actual protocol, and in fact each party will run two parallel protocols, one as a client and another as a server, and output the product of the two keys generated). The common reference string (CRS) defines a Diffie-Hellman language, i.e. $\rho = \mathbf{g}_1, \mathbf{g}_1^a$. The client picks a fresh Diffie-Hellman tuple by picking a witness r and computing $\langle \mathbf{x}_1 = \mathbf{g}_1^r, \mathbf{x}_2 = \mathbf{g}_1^{a \cdot r} \rangle$. It also computes a DSS-QA-NIZK proof on this tuple, which is a hash proof T and a QA-NIZK proof W of the augmented Diffie-Hellman tuple. Note, the QA-NIZK proof W is just a single group element [JR14a] (see [JR14b] for details). It next modifies the Diffie-Hellman tuple using the password pwd it possesses. Essentially, it multiplies $\mathbf{x}_2$ by pwd to get a modified group element which we will denote by S - in fact $(\mathbf{x}_1, S)$ is an ElGamal encryption of pwd. It next sends this ElGamal encryption $\mathbf{x}_1, S$ and the T component of the proof to the server. It retains W for later use. At this point it can erase the witness r.

As a first step, we intend to utilize an interesting property of the real-world verifier V of the DSS-QA-NIZK: the verifier is just the verifier of the QA-NIZK for the DH language augmented with the hash proof, and the QA-NIZK verifiers for linear subspaces are just a single bi-linear product test. Specifically (see [JR14b]), V on input $\mathbf{x}_1, \mathbf{x}_2$ and proof T, W, computes $\iota = \mathcal{H}(\mathbf{x}_1, \mathbf{x}_2)$, and outputs true iff

$$e(\mathbf{x}_1, (\mathbf{v}_1 \mathbf{v}_2^\iota)) \cdot e(\mathbf{x}_2, \mathbf{c}) \cdot e(T, \mathbf{g}_2) = e(W, \mathbf{b}).$$

Thus, it outputs true iff the left-hand-size (LHS) equals the right-hand-side (RHS) of the above equation. Note that the client sent $\mathbf{x}_1, S$ (i.e. $\mathbf{x}_2$ linearly modified by pwd) and T to the server. Assuming the server has the same password pwd, it can un-modify the received message and get $\mathbf{x}_2 = S/\text{pwd}$, and hence can compute this LHS (using the CRS). The client retained W, and can compute the RHS (using the CRS).

The intuition is that unless an adversary out-right guesses the password, it cannot produce a different $\mathbf{x}_1', S', T'$, such that $\mathbf{x}_1', S'/pwd, T'$ used to compute the LHS will match the RHS above. While we make this intuition rigorous later by showing a UC simulator, to complete the description of the protocol, and using this intuition, the client and server actually compute the LHS and RHS respectively of the following equation (for a fresh random $s \in \mathbb{Z}_q$ picked by the server):

$$e(\mathbf{x}_1, (\mathbf{v}_1 \mathbf{v}_2^\iota)^s) \cdot e(\mathbf{x}_2, \mathbf{c}^s) \cdot e(T, \mathbf{g}_2^s) = e(W, \mathbf{b}^s). \tag{1}$$

Now note that for the client to be able to compute the RHS, it must have $\mathbf{b}^s$, since s was picked by the server afresh. For this purpose, the protocol requires that the server send $\mathbf{b}^s$ to the client (note this can be done independently and asynchronously of the message coming from the client). It is not difficult to see, from completeness of the prover and verifier of the DSS-QA-NIZK, that both parties compute the same quantity.

As mentioned earlier, each pair of parties actually run two versions of the above protocol, where-in each party plays the part of client in one version, and the part of server in the other version. Each party then outputs the product of the LHS of (1) computation (in the server version) and the RHS of (1) computation (in the client version) as the session-key. We will refer to these two factors in the session-key computation as the *server factor* and the client factor resp. This is the final UC-PAKE protocol described in Fig. 3 (with the parties identities, session identifiers and $\mathbf{b}^s$ from its server version, used as label). The quantity $\mathbf{x}_1$ is called R in the protocol, as subscripts will be used for other purposes.

Theorem 3. *Assuming the existence of SXDH-hard groups, the protocol given in Fig. 3 securely realizes the $\widehat{\mathcal{F}}_{\text{PAKE}}$ functionality in the $\mathcal{F}_{\text{CRS}}$ hybrid model, in the presence of adaptive corruption adversaries.*

The theorem is proved in [JR14b]. We provide the intuition below.

6.3 Main Idea of the UC Simulator

We first *re-define* the various verifiers in the DSS-QA-NIZK for the DH language described in [JR14b], to bring them in line with the above description. In particular, the real-world verifier V is defined equivalently to be: the verifier V takes as input CRS_v, a word $\langle \mathbf{x}_1, \mathbf{x}_2 \rangle$, and a proof $\pi = (T, W)$, computes $\iota = \mathcal{H}(\mathbf{x}_1, \mathbf{x}_2, l)$, picks a fresh random $s \in \mathbb{Z}_q$, and outputs true iff

$$e(\mathbf{x}_1, (\mathbf{v}_1 \mathbf{v}_2^\iota))^s \cdot e(\mathbf{x}_2, \mathbf{c})^s \cdot e(T, \mathbf{g}_2)^s \ = \ e(W, \mathbf{b}^s).$$

This is equivalent as long as $s \neq 0$.

The partial-simulation world private-verifier pV is now defined as: it checks a word $\langle \mathbf{x}_1, \mathbf{x}_2 \rangle$ and a proof T, W as follows: compute $\iota = \mathcal{H}(\mathbf{x}_1, \mathbf{x}_2, l)$; pick s and s' randomly and independently from $\mathbb{Z}_q$, and if $\mathbf{x}_2 = \mathbf{x}_1^a$ and $T = \mathbf{x}_1^{d+\iota e}$ then set $\xi = 1_T$ else set $\xi = e(\mathbf{g}_1, \mathbf{g}_2)^{s'}$ and output true iff

$$e(\mathbf{x}_1, (\mathbf{v}_1 \mathbf{v}_2^\iota))^s \cdot e(\mathbf{x}_2, \mathbf{c})^s \cdot e(T, \mathbf{g}_2)^s \cdot \xi \ = \ e(W, \mathbf{b}^s). \qquad (2)$$

This is equivalent to the earlier definition of pV with high probability by an information-theoretic argument, if the trapdoors used were generated by the semi-functional CRS generator sfK$_1$.

The UC simulator $\mathcal{S}$ works as follows: It will generate the CRS for $\widehat{\mathcal{F}}_{\text{PAKE}}$ using the semi-functional CRS generator sfK$_1$ for the Diffie-Hellman language. The next main difference is in the simulation of the outgoing message of the real world parties: $\mathcal{S}$ uses a dummy message μ instead of the real password which it does not have access to. Further, it postpones computation of W till the session-key generation time. Finally, another difference is in the processing of the incoming message, where $\mathcal{S}$ decrypts the incoming message R_2', S_2', T_2' to compute a pwd$'$, which it uses to call the ideal functionality's test function. It next generates a sk similar to how it is generated in the real-world (recall the computation of server factor and client factor by LHS and RHS of (1)) except that it uses the Eq. (2) corresponding to the private verifier. It sends sk to the ideal functionality to be output to the party concerned.

Note, $\mathcal{S}$ simulating the server factor computation can compute the LHS of Eq. (2), except $\mathcal{S}$ does not have direct access to pwd and hence cannot get $\mathbf{x}_2$ from the modified $\hat{S}$ that it receives. However, it can do the following: Use the TESTPWD functionality of the ideal functionality $\widehat{\mathcal{F}}_{\text{PAKE}}$ with a pwd$'$ computed as

$\hat{S}/\mathbf{x}_1^a$. If this pwd' does not match the pwd recorded in $\widehat{\mathcal{F}}_{\text{PAKE}}$ for this session and party, then $\widehat{\mathcal{F}}_{\text{PAKE}}$ anyway outputs a fresh random session key, which will then turn out to be correct simulation (note, this case is same as $\mathbf{x}_2 (= S/pwd) \neq \mathbf{x}_1^a$, which would also have resulted in the same computation on the LHS). If the pwd' matched the pwd, the simulator is notified the same, and hence it can now do the following: if $T = \mathbf{x}_1^{d+\iota e}$ then set $\xi = \mathbf{1}_T$ else set $\xi = e(\mathbf{g}_1, \mathbf{g}_2)^{s'}$. Next, it calls $\widehat{\mathcal{F}}_{\text{PAKE}}$'s NewKey with session key $e(\mathbf{x}_1, (\mathbf{v}_1\mathbf{v}_2^\iota))^s \cdot e(\mathbf{x}_1^a, \mathbf{c})^s \cdot e(T, \mathbf{g}_2)^s \cdot \xi$ (multiplied by a RHS computation of (2) in simulation of the client factor, which we will discuss later).

The UC Simulator $\mathcal{S}$ must also simulate $\mathbf{g}_1^r$, $pwd \cdot (\mathbf{g}_1^a)^r$ and the T component of the DSS-QA-NIZK, as that is the message sent out to the adversary by the real party ("client" part of the protocol). However, $\mathcal{S}$ does not have access to pwd. It can just generate a fake tuple $\mathbf{g}_1^r, \mu \cdot (\mathbf{g}_1^a)^r \cdot \mathbf{g}_1^{r'}$ (for some constant or randomly chosen group element μ, and some random and independent $r' \in \mathbb{Z}_q$). Now, the semi-functional (proof) simulator sfSim of the DSS-QA-NIZK of [JR14b] has an interesting property that when the tuple $\langle \mathbf{x}_1, \mathbf{x}_2 \rangle$ does not belong to the language (language membership-bit zero), the T component of the simulated proof can just be generated randomly.

The simulator also needs W to compute the client factor, and we had postponed it till the session-key computation phase. As mentioned above, if the password pwd' "decrypted" from the incoming message is not correct then the key is anyway set to be random, and hence a proper W is not even required. However, if the pwd' is correct, the simulator is notified of same, and hence it can compute W component of the proof by passing $\mathbf{x}_2 = \mu \cdot (\mathbf{g}_1^a)^r \cdot \mathbf{g}_1^{r'}/pwd'$ along with $\mathbf{x}_1 (= \mathbf{g}_1^r)$ to sfSim.

Of course, fixing the above fake tuples employs one-time full-simulation property of the DSS-QA-NIZK (and the DDH assumption).

6.4 Main Idea of the Proof of UC Realization

The proof that the simulator $\mathcal{S}$ described above simulates the Adversary in the real-world protocol, follows essentially from the properties of the DSS-QA-NIZK, although not generically since the real-world protocol and the simulator use the verifiers V and pV (resp.) in a split fashion. However, as described above the proof is very similar and we give a broad outline here. The proof will describe various experiments between a challenger $\mathcal{C}$ and the adversary, which we will just assume to be the environment $\mathcal{Z}$ (as the adversary $\mathcal{A}$ can be assumed to be just dummy and following $\mathcal{Z}$'s commands). In the first experiment the challenger $\mathcal{C}$ will just be the combination of the code of the simulator $\mathcal{S}$ above and $\widehat{\mathcal{F}}_{\text{PAKE}}$. In particular, after the environment issues a NewSession request with a password pwd, the challenger gets that password. So, while in the first experiment, the challenger (copying $\mathcal{S}$) does not use pwd directly, from the next experiment onwards, it can use pwd. Thus, the main goal of the ensuing experiments is to modify the fake tuples $\mathbf{g}_1^r, \mu \cdot (\mathbf{g}_1^a)^r \cdot \mathbf{g}_1^{r'}$ by real tuples (as in real-world) $\mathbf{g}_1^r, pwd \cdot (\mathbf{g}_1^a)^r$, since the challenger has access to pwd. This is accomplished by a hybrid argument, modifying one instance at a time using DDH assumption in group $\mathbb{G}_1$ and using one-time full-ZK property (and using the otfSim proof simulator for that instance). A variant of the one-time full-ZK semi-functional verifier sfV (just as the variants for pV and V described above) is easily obtained. Note that in each experiment, whenever the simulator invokes partial proof simulation it can provide the correct membership bit (with high probability) as in each experiment it knows exactly which tuples are real and which are fake.

Once all the instances are corrected, i.e. R, S generated as $\mathbf{g}_1^r, \mathrm{pwd} \cdot (\mathbf{g}_1^a)^r$, the challenger can switch to the real-world because the tuples $R, S/\mathrm{pwd}$ are now Diffie-Hellman tuples. This implies that the session keys are generated using the V variant described above, which is exactly as in the real-world.

6.5 Adaptive Corruption

The UC protocol described above is also UC-secure against adaptive corruption of parties by the Adversary in the erasure model. In the real-world when the adversary corrupts a party (with a CORRUPT command), it gets the internal state of the party. Clearly, if the party has already been invoked with a NEWSESSION command then the password pwd is leaked at the minimum, and hence the ideal functionality $\mathcal{F}_{\mathrm{PAKE}}$ leaks the password to the Adversary in the ideal world. In the protocol described above, the Adversary also gets W and s, as this is the only state maintained by each party between sending $R, S, T, \hat{\rho}$, and the final issuance of session-key. Simulation of s is easy for the simulator $\mathcal{S}$ since $\mathcal{S}$ generates s exactly as in the real world. For generating W, which $\mathcal{S}$ had postponed to computing till it received an incoming message from the adversary, it can now use the pwd which it gets from $\widehat{\mathcal{F}}_{\mathrm{PAKE}}$ by issuing a CORRUPT call to $\widehat{\mathcal{F}}_{\mathrm{PAKE}}$. More precisely, it issues the CORRUPT call, and gets pwd, and then calls the semi-functional simulator with $\mathbf{x}_2 = \mu \cdot (\mathbf{g}_1^a)^r \cdot \mathbf{g}_1^{r'}/\mathrm{pwd}$ along with $\mathbf{x}_1 \, (= \mathbf{g}_1^r)$ to get W. Note that this computation of W is identical to the postponed computation of W in the computation of client factor of sk_1 (which is really used in the output to the environment when $\mathrm{pwd}' = \mathrm{pwd}$).

References

[ABB+13] Abdalla, M., Benhamouda, F., Blazy, O., Chevalier, C., Pointcheval, D.: SPHF-friendly non-interactive commitments. In: Sako, K., Sarkar, P. (eds.) ASIACRYPT 2013, Part I. LNCS, vol. 8269, pp. 214–234. Springer, Heidelberg (2013)

[ABP15] Abdalla, M., Benhamouda, F., Pointcheval, D.: Disjunctions for hash proof systems: new constructions and applications. In: Oswald, E., Fischlin, M. (eds.) EUROCRYPT 2015. LNCS, vol. 9057, pp. 69–100. Springer, Heidelberg (2015)

[ACP09] Abdalla, M., Chevalier, C., Pointcheval, D.: Smooth projective hashing for conditionally extractable commitments. In: Halevi, S. (ed.) CRYPTO 2009. LNCS, vol. 5677, pp. 671–689. Springer, Heidelberg (2009)

654   C.S. Jutla and A. Roy

[BBC+13] Benhamouda, F., Blazy, O., Chevalier, C., Pointcheval, D., Vergnaud, D.: New techniques for SPHFs and efficient one-round PAKE protocols. In: Canetti, R., Garay, J.A. (eds.) CRYPTO 2013, Part I. LNCS, vol. 8042, pp. 449–475. Springer, Heidelberg (2013)

[Can01] Canetti, R.: Universally composable security: a new paradigm for cryptographic protocols. In: 42nd FOCS, pp. 136–145. IEEE Computer Society Press, October 2001

[CCS09] Camenisch, J., Chandran, N., Shoup, V.: A public key encryption scheme secure against key dependent chosen plaintext and adaptive chosen ciphertext attacks. In: Joux, A. (ed.) EUROCRYPT 2009. LNCS, vol. 5479, pp. 351–368. Springer, Heidelberg (2009)

[CHK+05] Canetti, R., Halevi, S., Katz, J., Lindell, Y., MacKenzie, P.: Universally composable password-based key exchange. In: Cramer, R. (ed.) EURO-CRYPT 2005. LNCS, vol. 3494, pp. 404–421. Springer, Heidelberg (2005)

[CR03] Canetti, R., Rabin, T.: Universal composition with joint state. In: Boneh, D. (ed.) CRYPTO 2003. LNCS, vol. 2729, pp. 265–281. Springer, Heidelberg (2003)

[CS02] Cramer, R., Shoup, V.: Universal hash proofs and a paradigm for adaptive chosen ciphertext secure public-key encryption. In: Knudsen, L.R. (ed.) EUROCRYPT 2002. LNCS, vol. 2332, pp. 45–64. Springer, Heidelberg (2002)

[DDN91] Dolev, D., Dwork, C., Naor, M.: Non-malleable cryptography (extended abstract). In: Proceedings of 23rd ACM STOC, pp. 542–552 (1991)

[EHO+13] Emura, K., Hanaoka, G., Ohtake, G., Matsuda, T., Yamada, S.: Chosen ciphertext secure keyed-homomorphic public-key encryption. In: Kurosawa, K., Hanaoka, G. (eds.) PKC 2013. LNCS, vol. 7778, pp. 32–50. Springer, Heidelberg (2013)

[GMR89] Goldwasser, S., Micali, S., Rackoff, C.: The knowledge complexity of interactive proof systems. SIAM J. Comput. **18**(1), 186–208 (1989)

[GS08] Groth, J., Sahai, A.: Efficient non-interactive proof systems for bilinear groups. In: Smart, N.P. (ed.) EUROCRYPT 2008. LNCS, vol. 4965, pp. 415–432. springer, Heidelberg (2008)

[Har11] Haralambiev, K.:Efficient cryptographic primitives for non-interactive zero-knowledge proofs and applications. Ph.D. dissertation (2011)

[JR12] Jutla, C., Roy, A.: Relatively-sound NIZKs and password-based key-exchange. In: Fischlin, M., Buchmann, J., Manulis, M. (eds.) PKC 2012. LNCS, vol. 7293, pp. 485–503. Springer, Heidelberg (2012)

[JR13] Jutla, C.S., Roy, A.: Shorter quasi-adaptive NIZK proofs for linear subspaces. In: Sako, K., Sarkar, P. (eds.) ASIACRYPT 2013, Part I. LNCS, vol. 8269, pp. 1–20. Springer, Heidelberg (2013)

[JR14a] Jutla, C.S., Roy, A.: Switching lemma for bilinear tests and constant-size NIZK proofs for linear subspaces. In: Garay, J.A., Gennaro, R. (eds.) CRYPTO 2014, Part II. LNCS, vol. 8617, pp. 295–312. Springer, Heidelberg (2014)

[JR14b] Jutla, C.S., Roy, A.: Dual-system simulation-soundness with applications to UC-PAKE and more. Cryptology ePrint Archive, Report 2014/805. https://eprint.iacr.org/2014/805

[KV11] Katz, J., Vaikuntanathan, V.: Round-optimal password-based authenticated key exchange. In: Ishai, Y. (ed.) TCC 2011. LNCS, vol. 6597, pp. 293–310. Springer, Heidelberg (2011)

[KW15]   Kiltz, E., Wee, H.: Quasi-adaptive NIZK for linear subspaces revisited. In: Oswald, E., Fischlin, M. (eds.) EUROCRYPT 2015. LNCS, vol. 9057, pp. 101–128. Springer, Heidelberg (2015)

[LPJY14] Libert, B., Peters, T., Joye, M., Yung, M.: Non-malleability from malleability: simulation-sound quasi-adaptive NIZK proofs and CCA2-secure encryption from homomorphic signatures. In: Nguyen, P.Q., Oswald, E. (eds.) EUROCRYPT 2014. LNCS, vol. 8441, pp. 514–532. Springer, Heidelberg (2014)

[Sah99]   Sahai, A.: Non-malleable non-interactive zero knowledge and adaptive chosen-ciphertext security. In: 40th FOCS, pp. 543–553. IEEE Computer Society Press, October 1999

[Wat09]   Waters, B.: Dual system encryption: realizing fully secure IBE and HIBE under simple assumptions. In: Halevi, S. (ed.) CRYPTO 2009. LNCS, vol. 5677, pp. 619–636. Springer, Heidelberg (2009)

Secret Sharing and Statistical Zero Knowledge

Vinod Vaikuntanathan$^{(\boxtimes)}$ and Prashant Nalini Vasudevan

MIT CSAIL, Cambridge, USA
vinodv@csail.mit.edu, prashvas@mit.edu

Abstract. We show a general connection between various types of statistical zero-knowledge (SZK) proof systems and (unconditionally secure) secret sharing schemes. Viewed through the SZK lens, we obtain several new results on secret-sharing:

- *Characterizations:* We obtain an almost-characterization of access structures for which there are secret-sharing schemes with an efficient sharing algorithm (but not necessarily efficient reconstruction). In particular, we show that for every language $L \in \mathbf{SZK_L}$ (the class of languages that have statistical zero knowledge proofs with log-space verifiers and simulators), a (monotonized) access structure associated with L has such a secret-sharing scheme. Conversely, we show that such secret-sharing schemes can only exist for languages in $\mathbf{SZK}$.

- *Constructions:* We show new constructions of secret-sharing schemes with both efficient sharing and efficient reconstruction for access structures associated with languages that are in $\mathbf{P}$, but are not known to be in $\mathbf{NC}$, namely Bounded-Degree Graph Isomorphism and constant-dimensional lattice problems. In particular, this gives us the first combinatorial access structure that is conjectured to be outside $\mathbf{NC}$ but has an efficient secret-sharing scheme. Previous such constructions (Beimel and Ishai; CCC 2001) were algebraic and number-theoretic in nature.

- *Limitations:* We also show that *universally-efficient* secret-sharing schemes, where the complexity of computing the shares is a polynomial independent of the complexity of deciding the access structure, cannot exist for all (monotone languages in) $\mathbf{P}$, unless there is a polynomial q such that $\mathbf{P} \subseteq \mathbf{DSPACE}(q(n))$.

Keywords: Statistical zero knowledge · Secret sharing

1 Introduction

Secret-sharing [8, 29], a foundational primitive in information-theoretic cryptography, enables a dealer to distribute shares of a secret to n parties such that

V. Vaikuntanathan—Supported in part by NSF CNS-1350619, CNS-1414119, the Qatar Computing Research Institute, NEC Corporation, Alfred P. Sloan Research Fellowship, Microsoft Faculty Fellowship, and a Steven and Renee Finn Career Development Chair.

P.N. Vasudevan—Supported by the Qatar Computing Research Institute.

© International Association for Cryptologic Research 2015
T. Iwata and J.H. Cheon (Eds.): ASIACRYPT 2015, Part I, LNCS 9452, pp. 656–680, 2015.
DOI: 10.1007/978-3-662-48797-6_27

only some predefined authorized sets of parties will be able to reconstruct the secret from their shares. Moreover, the shares of any unauthorized set of parties should reveal *no information* about the secret, even if the parties are computationally unbounded. The (monotone) collection of authorized sets is called an *access structure*.

We call a secret-sharing scheme *efficient* if both the sharing algorithm (executed by the dealer) and reconstruction algorithm (executed by the parties) run in time polynomial in n. Associating sets $S \subseteq [n]$ with their characteristic vectors $x_S \in \{0,1\}^n$, we can define a language $L_{\mathcal{A}}$ associated with an access structure $\mathcal{A}$.[1] Namely, $L_{\mathcal{A}}$ is simply the set of all x_S such that $S \in \mathcal{A}$. For an access structure $\mathcal{A}$ to have an efficient secret sharing scheme, it must be the case that the language $L_{\mathcal{A}}$ is computable in polynomial time.

A major open question in information-theoretic cryptography is:

Q1: Characterize access structures with efficient secret-sharing schemes

Indeed, this question has been widely studied [6,8,19,20,29], culminating with the result of Karchmer and Wigderson [20] who showed efficient secret sharing schemes for various log-space classes.[2] We refer the reader to Beimel's excellent survey [4] for more details. In any event, it is wide open whether all of **mP**, the class of languages recognized by monotone polynomial-size circuits, has efficient secret sharing schemes.

Restricting the reconstruction algorithm to be a linear function of the shares gives us a special kind of secret-sharing scheme called a *linear secret-sharing scheme*. The Karchmer-Wigderson secret sharing scheme [20] for log-space classes is a linear secret-sharing scheme. We also know that linear and even the slightly more general quasi-linear schemes [5,20] cannot exist for access structures outside **NC**, the class of languages computable by boolean circuits of polylogarithmic depth. Finally, Beimel and Ishai [5] showed *non-linear* secret-sharing schemes for two specific access structures associated to algebraic problems (related to computing quadratic residuosity and co-primality) which are in **P** but are believed not to be in **NC**.

We will also study secret-sharing schemes (which we call *semi-efficient*) where the dealer is efficient, namely runs in time polynomial in n, however the reconstruction algorithm need not be efficient. Aside from their theoretical interest, such secret-sharing schemes may find use in scenarios where sharing happens in the present (and thus has to be efficient) but reconstruction happens in a future where computational resources might be cheaper. This also justifies our desire to achieve information-theoretic (unconditional) security since not only the honest parties, but also the adversary gains more computational resources with time.

Beimel and Ishai [5] show a semi-efficient secret-sharing scheme for the language of quadratic residuosity modulo a composite, which is believed not to be

---

[1] More formally, we have to speak of a family of access structures $\{\mathcal{A}_n\}_{n \in \mathbb{N}}$, one for every n. We abuse notation slightly and denote $\mathcal{A}$, consisting of subsets of n parties, as the access structure.

[2] We use this as a short-hand to say "secret sharing schemes for access structures $\mathcal{A}$ whose associated language $L_{\mathcal{A}}$ can be recognized in log-space".

in **P**. However, quite surprisingly, a characterization of access structures with semi-efficient secret-sharing schemes also appears to be open:

Q2: Characterize access structures with semi-efficient secret-sharing schemes

As a parenthetical remark, we note that a different interpretation of *efficiency* is sometimes used in the secret-sharing literature. Namely, a secret-sharing scheme is termed *efficient* [9,11,21] if the total length of the n shares is polynomial in n. Let us call this notion *size efficiency*. This makes no reference to the complexity of either the sharing or the reconstruction algorithms. In this work, we use the strong interpretation of *efficient*, namely where both the sharing and reconstruction algorithms run in time $\mathsf{poly}(n)$ and that of *semi-efficient* where only the sharing algorithm needs to run in time $\mathsf{poly}(n)$. We note that either of these two notions is stronger than size efficiency.

It is against this backdrop that we begin our study. Our main contribution is to develop an *interactive proof lens* to study these questions. As concrete results of this connection, we obtain an almost-characterization of access structures with semi-efficent secret-sharing schemes (almost solving $Q2$), new combinatorial access structures conjectured to lie outside **NC** which have efficient secret-sharing schemes (extending [5]), and limitations on an ambitious notion of universally efficient secret-sharing. We describe our results in detail below.

1.1   Our Results

Our central tool is a special type of two-message interactive proof system (that we call *Special Interactive Proofs*). Roughly speaking, the restriction on the proof system for a language L (aside from the fact that it has two messages) is that the verifier uses a special procedure to accept or reject. In particular, the verifier V on input x and a uniformly random bit b, comes up with a message m to send to the prover. The prover wins (the verifier accepts) if he can guess the bit b, given m. If $x \in L$, the prover should have a distinguishing (and therefore an accepting) strategy. However, if $x \notin L$, the verifier messages for bits 0 and 1 should be statistically indistinguishable.

Before we proceed, we must clarify what it means to have a secret sharing scheme for a language L which is not necessarily monotone. We follow the approach of Beimel and Ishai [5] and define a (monotonized) access structure on $2n$ parties $\{P_{i,0}, P_{i,1}\}_{i \in [n]}$ associated with L (more precisely, $L \cap \{0,1\}^n$): for every i, the pair of parties $\{P_{i,0}, P_{i,1}\}$ is in the access structure, as is every set of parties $\{P_{1,x_1}, P_{2,x_2}, \ldots, P_{n,x_n}\}$ for all $x \in L$. These are the minimal sets that make up the access structure $\mathcal{A}_L$. Note that the complexity of deciding whether a set $S \in \mathcal{A}_L$ is precisely the complexity of deciding the language L.

Our research in this direction was motivated by the fact that if, for some language L, $\mathcal{A}_L$ has a semi-efficient secret sharing scheme, then L has a special interactive proof: the verifier simply shares a random bit b according to the sharing algorithm and sends the prover the shares corresponding to the input, and the prover has to guess b. The honest prover runs the reconstruction algorithm, and completeness and soundness are guaranteed by correctness and privacy of the secret sharing scheme, respectively. We then investigated the circumstances under which the converse might also hold. We were able to show the following:

Theorem 1 (Informal). *Let L be a language and let $\mathcal{A}_L$ be the associated access structure. If L has a special interactive proof with a log-space verifier, then $\mathcal{A}_L$ has a semi-efficient secret-sharing scheme. Conversely, if $\mathcal{A}_L$ has a semi-efficient secret-sharing scheme, then L has a special interactive proof.*

Our proof goes through the notion of partial garbling schemes, defined and studied in the work of Ishai and Wee [18].

Characterizing Semi-Efficient Secret-Sharing. Using Theorem 1, we characterize access structures that have semi-efficient secret-sharing schemes: we show that all languages in $\mathbf{SZK_L}$, the class of languages with statistical zero knowledge proof systems [28] where the verifier and simulator run in log-space, have semi-efficient secret-sharing schemes. This follows from the observation, using a result of Sahai and Vadhan [28], that L has a special interactive proof with a log-space verifier if and only if $L \in \mathbf{SZK_L}$. Conversely, it is easy to see that if a language L has a semi-efficient secret-sharing scheme, then $L \in \mathbf{SZK}$, the class of languages with statistical zero knowledge proof systems with polynomial-time verifier and simulator. Together, this almost characterizes languages with semi-efficient secret-sharing schemes.

The class $\mathbf{SZK_L}$, which is contained in $\mathbf{SZK}$, and hence in $\mathbf{AM} \cap \mathbf{coAM}$, contains several problems of both historical and contemporary significance to cryptography, such as *Quadratic Residosity*, *Discrete Logarithm*, and the *Approximate Closest Vector Problem*, as well as other well-studied problems like *Graph Isomorphism*. For further details, including those about complete problems and about prospects of basing cryptography on the worst-case hardness of $\mathbf{SZK_L}$, see [12]. As a result of these containments, our characterization captures as a special case the Beimel-Ishai secret-sharing scheme for the language of quadratic residuosity modulo composites [5].

We also show a version of this theorem for efficient (as opposed to semi-efficient) secret-sharing schemes. In particular:

Theorem 2 (Informal). *Let L be a language and let $\mathcal{A}_L$ be the associated access structure. If L has a special interactive proof with a log-space verifier and a polynomial-time prover, then $\mathcal{A}_L$ has an efficient secret-sharing scheme. Conversely, if $\mathcal{A}_L$ has an efficient secret-sharing scheme, then L has a special interactive proof with a polynomial-time prover.*

Constructions of Efficient Secret-Sharing Schemes. We show new constructions of efficient secret-sharing schemes for languages that are in $\mathbf{P}$ but are not known to be in $\mathbf{NC}$, namely Bounded-Degree Graph Isomorphism [2,26], and lattice Shortest and Closest Vector problems in constant dimensions [16,24]. Our constructions arise from special interactive proofs for these languages together with an application of Theorem 2. In particular, our construction for Bounded-Degree Graph Isomorphism gives us *the first* efficient secret-sharing scheme for a *combinatorial* access structure conjectured to be in $\mathbf{P} \backslash \mathbf{NC}$ (The results of Beimel and Ishai were for algebraic access structures associated to quadratic residuosity modulo primes and co-primality). Moreover, our interactive proofs and secret-sharing schemes are simple, natural and easy to describe.

Limitations on Universally Efficient Secret-Sharing Schemes. Consider secret sharing schemes that are defined not for a given access structure, but uniformly for some class of access structures. The sharing algorithm in such a case gets a description of the access structure, in the form of a circuit or a Turing machine that decides membership in the access structure. Typically, the sharing algorithm runs for as much time as the Turing machine (and therefore as much time as required to decide membership). However, there is no a-priori reason why this should be the case. Indeed, one can reasonably require that the sharing algorithm runs in some fixed polynomial time $t(n)$, even though the access structure may take arbitrary polynomial time to decide. (We allow the reconstruction algorithm to run in arbitrary polynomial time to make up for the deficiency of the sharing algorithm). Can such *universally efficient* secret-sharing schemes exist?

Our definition is inspired by the recent progress on (computationally secure) succinct randomized encodings [7,10,23,25]. Indeed, these works show, assuming indistinguishability obfuscation [3,14], that $\mathbf{P}$ has *computationally secure* succinct randomized encoding schemes. One could also reasonably ask: Can such *succinct* randomized encodings exist unconditionally for all of $\mathbf{P}$? It was observed in [7] that this cannot be the case under certain complexity-theoretic assumptions about speeding up non-deterministic algorithms.

Using our interactive proof characterization, we show that unconditionally secure universally efficient secret-sharing schemes (and succinct randomized encodings) cannot exist for all languages in $\mathbf{P}$, unless there is a fixed polynomial q such that $\mathbf{P} \subseteq \mathbf{DSPACE}(q(n))$ (the class of languages computable by a deterministic single-tape Turing machine with $q(n)$ space). We remind the reader that $\mathbf{P} \neq \mathbf{DSPACE}(q(n))$ for any fixed q, although non-containment either way is not known.

1.2 Related Work and Open Problems

In this work, we insist on statistical (or unconditional) security from our secret-sharing schemes. A number of works relax this to computational security and achieve stronger positive results. Settling for computational security and assuming the existence of one-way functions, Yao [32,34] showed an efficient secret-sharing scheme for all monotone languages in $\mathbf{P}$ recognized by polynomial-sized monotone circuits. We mention that even here, we are far from a characterization as there are monotone languages in $\mathbf{P}$ that cannot be recognized by polynomial-sized monotone circuits [27,30].

Komargodski, Naor and Yogev [22] also exploit the relaxation to computational security, and show secret-sharing schemes for all of monotone **NP**, where the sharing algorithm is polynomial-time, and the reconstruction algorithm is polynomial-time given the **NP** witness. Their result relies on strong computational assumptions related to indistinguishability obfuscation [3,14].

While we show semi-efficient secret-sharing schemes for *monotonized* access structures corresponding to all languages in **SZK_L**, it remains open to characterize *which monotone languages in* **SZK** *have semi-efficient secret-sharing schemes*. The central difficulty is that even if a language is monotone, there is no reason why the verifier in the SZK proof for the language should inherit monotonicity-like properties (and indeed, this is hard to even define).

2   Preliminaries and Definitions

Notation. Given a set S, we denote by 2^S the set of all subsets of S. Let $T = (t_1, \ldots, t_n)$ and $B = \{i_1, \ldots, i_m\} \subseteq [n]$; T_B is used to denote the tuple $(t_{i_1}, \ldots, t_{i_m})$.

We use languages and Boolean functions interchangeably. Given a language L, we overload L to also denote the corresponding Boolean function, namely, $L(x) = 0$ if $x \notin L$ and $L(x) = 1$ otherwise. Given a randomized algorithm A, we denote by $A(x)$ the random variable arising from running A on x, and by $A(x; r)$ the output when A is run on x with randomness r.

Given a distribution D over a finite set X and an $x \in X$, we denote by $D(x)$ the probability mass D places on x, and for a subset $S \subseteq X$, $D(S) = \sum_{x \in S} D(x)$. $x \leftarrow D$ indicates that x is a sample drawn according to the distribution D. For a set S, $x \leftarrow S$ indicates that x is drawn uniformly at random from S.

We use the notion of statistical distance (also called total variation distance or ℓ_1 distance) between distributions, defined as follows.

Definition 3 (Statistical Distance). *The* statistical distance *between two distributions D_1 and D_2 over the domain X is defined as*

$$d(D_1, D_2) = \frac{1}{2} \sum_{x \in X} |D_1(x) - D_2(x)| = \max_{S \subseteq X} (D_1(S) - D_2(S))$$

Of particular interest to us is the following relationship of statistical distance to the advantage of any unbounded procedure in distinguishing between two distributions given a uniform prior.

Fact 4. *Given distributions D_1, D_2 over a domain X, for functions $f : X \rightarrow \{0, 1\}$, we have:*

$$\max_f Pr[f(x) = b : b \leftarrow \{0, 1\}, x \leftarrow D_b] = \frac{1}{2} + \frac{d(D_1, D_2)}{2}$$

2.1   Complexity Classes

We briefly define the following complexity classes that are referred to frequently in the rest of the paper. To start with, **P** (resp. **BPP**) is the class of languages decidable in deterministic (resp. randomized) polynomial time and **L** is the class of languages decidable in deterministic logarithmic space. $\mathbf{NC}^k$ is the class of languages decidable by circuits of depth $O((\log n)^k)$ (here, n denotes the input length). A language is in **NC** if it is in $\mathbf{NC}^k$ for some k.

Definition 5 ($\mathbf{NC}^k$). *For any $k \in \mathbb{N} \cup \{0\}$, $\mathbf{NC}^k$ is the class of languages L for which there exists a family of boolean circuits $\{C_n\}_{n \in \mathbb{N}}$ such that:*

- *There is a constant c such that for all n, C_n has depth at most $c(\log n)^k$.*
- *For any input x of length n, $x \in L \Leftrightarrow C_n(x) = 1$*

DSPACE($p(n)$) is the class of languages decidable by deterministic uring machines running with space $p(n)$. Thus, **L** is the union of **DSPACE**($c \log n$) over all constants c.

Definition 6 (DSPACE). *For any function $p : \mathbb{N} \to \mathbb{N}$, **DSPACE**($p(n)$) is the class of languages L for which there exists a deterministic Turing machine L such that for any input x:*

- *$x \in L \Leftrightarrow M(x) = 1$*
- *M uses at most $p(|x|)$ cells on its work tape.*

And finally, **SZK** consists of languages that have Statistical Zero Knowledge (SZK) proofs, which are interactive proofs with some additional properties, as described below.

Definition 7 (SZK). *A language L is in **SZK** if there exist a tuple of Turing machines (P, V, S), where the verifier V and simulator S run in probabilistic polynomial time, satisfying the following:*

- *(P, V) is an interactive proof for L with negligible completeness and soundness errors.*
- *Let $(P, V)(x)$ denote the distribution of transcripts of the interaction between P and V on input x. For any $x \in L$ of large enough size,*

$$d(S(x), (P, V)(x)) \leq \mathrm{negl}(|x|)$$

The above is actually a definition of honest-verifier Statistical Zero Knowledge, but we know from [28] that any language with an honest-verifier SZK proof also has an SZK proof against cheating verifiers. So this follows as a definition of **SZK** as well. We refer the reader to [31] for extensive definitions and explanations.

$\mathbf{SZK_L}$ is the same as **SZK**, but with the verifier and simulator running with logarithmic space. In this case too, the above definition is only for honest verifiers, but as this would only define a larger class, and we show positive results for this class, we will work with this definition.

2.2   Secret Sharing

Definition 8 (Access Structure). *Given a set of parties* $P = \{P_1, \ldots, P_n\}$, *an access structure* $\mathcal{A}$ *is a monotone collection of subsets of* P. *That is, if* $S \in \mathcal{A}$ *and* $T \supseteq S$, *then* $T \in \mathcal{A}$.

In the context of a secret-sharing scheme, the access structure consists of all subsets of parties that are allowed to reconstruct a secret shared among them. Of course, as the access structure is monotone, it suffices to specify its minimal elements. Along the lines of [5], we associate with every language L an family of access structures $\{\mathcal{A}_{L,n}\}_{n \in \mathbb{N}}$ where $\mathcal{A}_{L,n}$ is defined for $2n$ parties. We will then study the efficiency of secret sharing schemes for access structures in such families as a function of n. As will be evident from the definition below, the complexity of deciding whether a set $S \in \mathcal{A}_{L,n}$ is exactly the hardness of deciding the language.

Definition 9 (Access Structure associated with Language L). *For a language* L, *its associated access structure, denoted by* $\mathcal{A}_{L,n}$, *for a set of* $2n$ *parties* $\mathcal{P}_n = \{P_{i,b}\}_{i \in [n], b \in \{0,1\}}$ *is defined by the following minimal elements:*

- $\forall i : \{P_{i,0}, P_{i,1}\} \in \mathcal{A}_{L,n}$
- $\forall x \in L \cap \{0,1\}^n : \{P_{1,x_1}, \ldots, P_{n,x_n}\} \in \mathcal{A}_{L,n}$

We use the following definition of secret sharing schemes.

Definition 10 (Statistical Secret Sharing). *An* (ϵ, δ)-*Secret Sharing Scheme for* n *parties* $\mathcal{P} = \{P_1, \ldots, P_n\}$ *and a domain of secrets* D *under access structure* $\mathcal{A} \subseteq 2^{\mathcal{P}}$ *is a pair of algorithms* (S, R), *where*

- *S is the randomized sharing algorithm that takes as input a secret* $s \in D$ *and outputs a sequence of shares* $(s_1, s_2, \ldots, s_n)$; *and*
- *R is the deterministic reconstruction algorithm that takes as input a subset of parties* $B \subseteq [n]$ *and the corresponding subset of shares* $(s_i)_{i \in B}$ *and outputs either a secret* s *or a special symbol* $\perp$.

We require (S, R) *to satisfy the following conditions:*

1. *Correctness: For any* $B \in \mathcal{A}$ *and any* $s \in D$, *the reconstruction algorithm* R *works:* $Pr[R(B, S(s)_B) = s] \geq 1 - \epsilon(n)$
2. *Privacy: For any* $B \notin \mathcal{A}$ *and any* $s, s' \in D$: $d(S(s)_B, S(s')_B) \leq \delta(n)$.

The scheme is said to be semi-efficient *if S is computable in* $poly(n)$ *time, and it is said to be* efficient *if both S and R are computable in* $poly(n)$ *time.*

Unless otherwise specified, the domain of secrets for all schemes we talk about in this work shall be $\{0, 1\}$, which is without loss of generality.

Remark 11. When we talk about access structures associated with promise problems, we require no guarantees from a secret sharing scheme for sets corresponding to inputs that do not satisfy the promise (even though technically they are not part of the associated access structure, and so privacy would otherwise be expected to hold).

While much of the literature on secret sharing schemes studies the size of the shares (and call schemes that produce shares of size poly(n) efficient), we use a stronger interpretation of efficiency. Namely, in all our exposition, the sharing algorithm S is required to run in time polynomial in n. Thus, we will not discuss the sizes of the shares produced by the schemes, which is always poly(n).

2.3   Partial Randomized Encodings

We use the notion of partial randomized encodings (defined as *partial garbling schemes* in [18]). They are essentially randomized encodings [17] where part of the input is allowed to be public.

Definition 12 (Partial Randomized Encodings). *An (ϵ, δ)-partial randomized encoding (PRE) of a (bi-variate) function $f : \{0,1\}^* \times \{0,1\}^* \to \{0,1\}^*$ is a pair of (randomized) functions (E_f, D_f), called the encoding and decoding functions, respectively, that satisfy the following conditions for all n, n':*

1. *Correctness:* $\forall (x, z) \in \{0,1\}^n \times \{0,1\}^{n'}$:

$$Pr[D_f(x, E_f(x, z)) = f(x, z)] \geq 1 - \epsilon(n)$$

   Note that the decoder gets the first half of the input, namely the public part x, in addition to the randomized encoding $E_f(x, z)$.

2. *Privacy:* $\forall x \in \{0,1\}^n$ *and* $\forall z_1, z_2 \in \{0,1\}^{n'}$:

$$f(x, z_1) = f(x, z_2) \Rightarrow d(E_f(x, z_1), E_f(x, z_2)) \leq \delta(n)$$

Furthermore:

- *(E_f, D_f) is* local *(or* locally computable*) if E_f can be decomposed into a set of functions $\{E_f^{(i)}(x_i, z)\}_{i \in [|x|]}$, where $E_f^{(i)}$ depends only on the ith bit of x and on z.*
- *(E_f, D_f) is* perfect *if $\epsilon(n) = \delta(n) = 0$.*
- *(E_f, D_f) is said to be* semi-efficient *if E_f is computable in poly($|x|, |z|$) time, and it is said to be* efficient *if both E_f and D_f are computable in poly($|x|, |z|$) time.*

We can extend the above definition to PREs of randomized functions in a natural way. Namely, to construct an (ϵ, δ)-PRE for a randomized function $A(x, z; r)$, simply construct an (ϵ, δ)-PRE $(E_{A'}, D_{A'})$ for the deterministic function $A'(x, (z, r)) = A(x, z; r)$, and take $E_A(x, z)$ to be the random variable $E_{A'}(x, (z, r))$ when r is chosen uniformly at random, and have D_A be the same as $D_{A'}$. Note that in $E_{A'}$, the randomness r used by A is part of the private input. This is crucial, as revealing r along with x and $A(x, b; r)$ could end up revealing b.

We then have the following lemma, whose proof is in Appendix A.

Lemma 13. *Let $A(x, z)$ be a randomized function, and (E_A, D_A) be an (ϵ, δ)-PRE of A as described above. Then, for any x and any z_1, z_2:*

$$d(A(x, z_1), A(x, z_2)) \leq \delta' \Rightarrow d(E_A(x, z_1), E_A(x, z_2)) \leq \delta(|x|) + \delta'$$

We also use the following lemma.

Lemma 14 ([1,18]). *Every function $f : \{0,1\}^* \times \{0,1\}^* \rightarrow \{0,1\}^*$ that can be computed in $\mathbf{L}/\mathrm{poly}$ has efficient perfect locally computable PREs, with encoding in $\mathbf{NC}^0$ and decoding in $\mathbf{NC}^2$.*

Finally, we abuse notation slightly and define partial randomized encodings for languages (boolean functions) a bit differently, for somewhat technical reasons (instead of calling this object something different).

Definition 15 (PREs for Languages). *An (ϵ, δ)-partial randomized encoding (PRE) of a language $L \subseteq \{0,1\}^*$ is a pair of (randomized) functions (E_L, D_L), called the encoding and decoding functions, respectively, that satisfy the following conditions:*

1. *Correctness:* $\forall x \in L$ *and* $b \in \{0,1\}$: $Pr[D_L(x, E_L(x, b)) = b] \geq 1 - \epsilon(|x|)$.
2. *Privacy:* $\forall x \notin L$, $d(E_L(x, 0), E_L(x, 1)) \leq \delta(|x|)$.

Semi-efficiency, efficiency and locality are defined as for general partial randomized encodings.

In other words, a PRE for a language L is a PRE for the following function:

$$f_L(x, b) = \begin{cases} b & \text{if } x \in L \\ \bot & \text{otherwise} \end{cases}$$

Using the above equivalence and Lemma 14, we have the following:

Lemma 16 ([18]). *Every language in $\mathbf{L}/\mathrm{poly}$ has efficient perfect locally computable PREs, with encoding in $\mathbf{NC}^0$ and decoding in $\mathbf{NC}^2$.*

2.4 Special Interactive Proofs

We define a special type of interactive proof system with two messages. Roughly speaking, the restriction on the proof system (aside from the fact that it has two messages) is that the verifier uses a special procedure to accept or reject. In particular, the verifier V on input x and a uniformly random bit b, comes up with a message m to send to the prover. The prover wins if he can guess the bit b, given m.

Definition 17 (SIP). *An (ϵ, δ)-Special Interactive Proof (SIP) for a language L is a pair (P, V), where:*

1. *V is a PPT algorithm that takes as input an instance x and a bit b, and outputs a message m; and*

2. *P takes as input the instance x and the verifier message m, and outputs a bit b'.*

We require (P, V) to satisfy the following conditions, when $b \leftarrow \{0, 1\}$:

1. Completeness: $\forall x \in L, \ Pr[P(x, V(x, b)) = b] \geq 1 - \epsilon(|x|)$.
2. Soundness: $\forall x \notin L$, and for any P^*, $Pr[P^*(x, V(x, b)) = b] \leq 1/2 + \delta(|x|)$.

While the restrictions imposed on these proofs seem rather severe, they turn out to be quite general. In fact, it follows from the work of Sahai and Vadhan [28] that the set of languages with such proofs is exactly the class **SZK**. See Theorem 20.

2.5   Statistical Zero Knowledge

Recall that the class **SZK** is the set of languages that have statistical zero-knowledge proofs, and the class **SZK_L** is set of languages that have statistical zero-knowledge proofs where the verifier and the simulator (for a statistically close simulation) both run in log-space.

Definition 18 (Promise Problems SD, SD_L). *The promise problem (ϵ, δ)-Statistical Difference (SD) is defined by the following YES and NO instances:*

$$SD^{YES} = \{(M_1, M_2, 1^n) : d(M_1^n, M_2^n) > 1 - \epsilon(n)\}$$
$$SD^{NO} = \{(M_1, M_2, 1^n) : d(M_1^n, M_2^n) < \delta(n)\}$$

where M_1, M_2 are deterministic Turing machines, and M_1^n, M_2^n represent the random variables corresponding to their outputs when the input is distributed uniformly at random in $\{0, 1\}^n$.

If M_1 and M_2 are log-space machines, then the language is called (ϵ, δ)-Statistical Difference for Log-space Machines, or simply SD_L.

Theorem 19 ([28]). *For every $\epsilon(n), \delta(n) = 2^{-n^{O(1)}}$ such that $\delta(n) < (1 - \epsilon(n))^2$, the (ϵ, δ)-SD problem is complete for **SZK**, and the (ϵ, δ)-SD_L problem is complete for **SZK_L**.*

We will use the following theorem which is a slightly stronger version of Theorem 19. We describe the proof (which follows from the proof of Theorem 19 in [28]) in Appendix B for completeness.

Theorem 20 ([28]). *There exist negligible functions $\epsilon(n), \delta(n) = n^{-\omega(1)}$ such that for any language $L \in$ **SZK**, L has an (ϵ, δ)-special interactive proof system (P, V). Furthermore, if $L \in$ **SZK_L**, then the verifier V can be computed in log-space.*

Sketch of Proof. For the main statement, we observe that the complete problem for **SZK**, namely (ϵ, δ)-SD, has a simple $(\epsilon/2, \delta/2)$-special interactive proof which works as follows.

- The verifier V, on input an instance $(M_0, M_1, 1^n)$ of the SD problem chooses a uniformly random bit b, and outputs a sample from M_b^n; and
- The prover's goal is to guess the bit b.

By Fact 4, it follows that the best success probability of any prover in this game is $\frac{1 + d(M_0^n, M_1^n)}{2}$. By the completeness of SD (Theorem 19), we get that **SZK** has (ϵ, δ)-special interactive proofs for some $\epsilon(n), \delta(n) = n^{-\omega(1)}$.

The proof for $\mathbf{SZK_L}$ works in exactly the same way, except it is now a concern that the verifier has to first run the SZK-completeness reduction to obtain an instance of the statistical distance problem SD_L, since it is not guaranteed that the reduction runs in log-space. However, we show that the Sahai-Vadhan reduction indeed does. We refer the reader to Appendix B for more details.

In fact, the connection between languages with special interactive proofs and **SZK** goes both ways. Namely,

Fact 21. *Let* $(1 - 2\epsilon(n))^2 > 2\delta(n)$. *If a language L has an (ϵ, δ)-SIP, then $L \in \mathbf{SZK}$.*

This is because deciding a language L that has an (ϵ, δ)-SIP (P, V) is the same as deciding whether $(V_{(x,0)}, V_{(x,1)}, 1^{|r(|x|)|}) \in (2\epsilon, 2\delta)$-$SD$, where $V_{(x,b)}(r) = V(x, b; r)$, and $(2\epsilon, 2\delta)$-SD is in **SZK** for ϵ and δ satisfying the above property.

3 From Zero Knowledge to Secret Sharing and Back

In this section, we show tight connections between languages with special interactive proofs, partial randomized encodings (PRE), and secret sharing schemes. In particular, we show:

Theorem 22 (Main Theorem). *For any language L and parameters $\epsilon(n)$ and $\delta(n)$, the following three statements are equivalent:*

1. *There are parameters $\epsilon_1 = O(\epsilon)$ and $\delta_1 = O(\delta)$ such that L has an (ϵ_1, δ_1)-special interactive proof (P, V), where the verifier V has a semi-efficient, locally computable, (ϵ_1, δ_1)-PRE.*
2. *There are parameters $\epsilon_2 = O(\epsilon)$ and $\delta_2 = O(\delta)$ such that L has a semi-efficient, locally computable, (ϵ_2, δ_2)-PRE.*
3. *There are parameters $\epsilon_3 = O(\epsilon)$ and $\delta_3 = O(\delta)$ such that for all n, there is a semi-efficient (ϵ_3, δ_3)-secret sharing scheme under the access structure $\mathcal{A}_{L,n}$.*

We will prove Theorem 22 in Sect. 3.1, and here we state a number of interesting corollaries. The first two corollaries "almost" characterize the languages L whose associated access structure $\mathcal{A}_{L,n}$ (as defined in Definition 9) has a semi-efficient secret-sharing scheme. Corollary 23 shows that any language in $\mathbf{SZK_L}$ has a semi-efficient secret-sharing scheme. Corollary 24 shows that furthermore, if $\mathbf{P}/\text{poly}$ has semi-efficient, locally computable PREs, then any language in the entire class **SZK** has a semi-efficient secret-sharing scheme. Moreover, it also says that no language outside **SZK** has semi-efficient secret-sharing schemes, implying that our characterization is almost tight.

Corollary 23. *Let $\epsilon(n), \delta(n) = n^{-\omega(1)}$ be negligible functions. For any language $L \in \mathbf{SZK_L}$, and for every n, there is a semi-efficient (ϵ, δ)-secret sharing scheme under the associated access structure $\mathcal{A}_{L,n}$.*

Proof. Theorem 20 asserts that for any $L \in \mathbf{SZK_L}$, there is an (ϵ, δ)-special interactive proof (P, V) for some $\epsilon(n), \delta(n) = n^{-\omega(1)}$, where the verifier algorithm V can be computed in log-space. Therefore, by Lemma 14, V has an efficient (and not just semi-efficient) perfect, locally computable PRE. Applying Theorem 22 (in particular, that $(1) \Rightarrow (3)$), there is a semi-efficient $(O(\epsilon), O(\delta))$-secret sharing scheme for $\mathcal{A}_{L,n}$.

Corollary 24. *Let $\epsilon(n), \delta(n) = n^{-\omega(1)}$ be negligible functions.*

- *Assume that $\mathbf{P}/\mathrm{poly}$ has semi-efficient (ϵ, δ)-locally computable PREs. Then, for any language $L \in \mathbf{SZK}$, and for every n, there is a semi-efficient (ϵ, δ)-secret sharing scheme under the associated access structure $\mathcal{A}_{L,n}$.*
- *Conversely, if $\mathcal{A}_{L,n}$ has a semi-efficient (ϵ, δ)-secret sharing scheme, then $L \in \mathbf{SZK}$.*

This follows from the same arguments as Corollary 23, but with the absence of something like Lemma 14 to complete the argument. In fact, one may replace $\mathbf{P}/\mathrm{poly}$ in Corollary 24 with any complexity class $\mathbf{C}$ that is closed under the operations involved in the reduction used in the proof of Theorem 36 (while replacing $\mathbf{SZK}$ with the appropriate $\mathbf{SZK_C}$). The converse is true because of Theorem 22 and Fact 21

We also have the following theorem about *efficient* secret sharing schemes, where both the sharing and reconstruction algorithms run in time polynomial in n. The difference from Theorem 22 is that here, we require the prover in the special interactive proof to be *efficient*, namely run in time polynomial in n. We view this theorem as an avenue to constructing efficient secret sharing schemes for languages L outside $\mathbf{L}$: namely, to construct a secret-sharing scheme for $\mathcal{A}_{L,n}$, it suffices to construct special interactive proofs for L wherein the verifier algorithm can be computed in $\mathbf{L}$.

The proof of Theorem 25 follows directly from that of Theorem 22.

Theorem 25. *For any language L and parameters $\epsilon(n)$ and $\delta(n)$, the following three statements are equivalent:*

1. *There are parameters $\epsilon_1 = O(\epsilon)$ and $\delta_1 = O(\delta)$ such that L has an (ϵ_1, δ_1)-special interactive proof (P, V), where the prover algorithm is computable in polynomial time, and the verifier V has an efficient, locally computable, (ϵ_1, δ_1)-PRE.*
2. *There are parameters $\epsilon_2 = O(\epsilon)$ and $\delta_2 = O(\delta)$ such that L has an efficient, locally computable, (ϵ_2, δ_2)-PRE.*
3. *There are parameters $\epsilon_3 = O(\epsilon)$ and $\delta_3 = O(\delta)$ such that for all n, there is an efficient (ϵ_3, δ_3)-secret sharing scheme under the access structure $\mathcal{A}_{L,n}$.*

3.1   Proof of the Main Theorem

We prove Theorem 22 by showing that $(1) \Rightarrow (2) \Rightarrow (3) \Rightarrow (1)$.

$(1) \Rightarrow (2)$. Let (P, V) be an (ϵ, δ)-special interactive proof for the language L, and let (E_V, D_V) be the hypothesized semi-efficient, locally computable (ϵ, δ)-PRE for V. The PRE for the language L works as follows:

- $E_L(x, b) = E_V(x, b)$
- $D_L(x, y) = P(x, D_V(x, y))$

We first show correctness. Let $x \in L$ and $b \in \{0, 1\}$. From the correctness of the PRE for the verifier algorithm V, we know that:

$$D_V(x, E_V(x, b)) = V(x, b)$$

with probability at least $1 - \epsilon$. Now, by the completeness of the special interactive proof, we know that:

$$P(x, V(x, b)) = b$$

with probability at least $1 - 2\epsilon$ (because this probability is at least $1 - \epsilon$ when b is chosen at random). Putting these together, we have:

$$D_L(x, E_L(x, b)) = P\big(x, D_V(x, E_V(x, b))\big) = P\big(x, V(x, b)\big) = b$$

with probability at least $1 - 3\epsilon$.

Next, we turn to privacy. Let $x \notin L$. We will show that $E_L(x, 0)$ and $E_L(x, 1)$ are statistically close. First, note that by the δ-soundness of the special interactive proof, we know that the distributions $V(x, 0)$ and $V(x, 1)$ are $O(\delta)$-close. Now, by Lemma 13 and using the δ-privacy of the PRE scheme for V, this means that $E_V(x, 0)$ and $E_V(x, 1)$ are also $O(\delta)$-close. This demonstrates privacy of our PRE scheme for L.

Since E_L is the same as E_V, it is clear that if the PRE scheme (E_V, D_V) is locally computable, so is (E_L, D_L). Moreover, if (E_V, D_V) is semi-efficient, so is (E_L, D_L). Finally, if (E_V, D_V) is efficient and the prover P in the special interactive proof is computable in polynomial time, then (E_L, D_L) is also efficient.

$(2) \Rightarrow (3)$. This implication follows from the work of Ishai and Wee [18]. We provide a proof here for completeness.

Given a locally computable (ϵ, δ)-PRE (E_L, D_L) for a language L, let the set of functions $\{E_L^{(i)}(x_i, b)\}_{i \in [n]}$ be the local decomposition of $E_L(x, b)$. The following is the secret sharing scheme (S, R) for the access structure $\mathcal{A}_{L,n}$:

- *Sharing*: Let $s \in \{0, 1\}$ be the secret bit to be shared. $S(s)$ works as follows:
    1. For each i, pick $s_{i,0}, s_{i,1} \in \{0, 1\}$ at random such that $s_{i,0} \oplus s_{i,1} = s$, and give $s_{i,b}$ to the party $P_{i,b}$.
    2. Select bits $\{s_0, \ldots, s_n\}$ at random such that $\bigoplus_{i=0}^{n} s_i = s$. For each $i \in [n]$, give s_i to both $P_{i,0}$ and $P_{i,1}$.

670   V. Vaikuntanathan and P.N. Vasudevan

3. Choose a random string r, compute $\psi_{i,b} \leftarrow E_L^{(i)}(b, s_0; r)$ for every $i \in [n]$ and $b \in \{0, 1\}$, and give $\psi_{i,b}$ to party $P_{i,b}$.

- *Reconstruction*: Any authorized set $B \in \mathcal{A}_{L,n}$ reconstructs the secret as follows:
  - If B contains $P_{i,0}$ and $P_{i,1}$ for some i, the secret s can be retrieved as $s = s_{i,0} \oplus s_{i,1}$.
  - If not, then $B = \{P_{i,x_i}\}$ for some $x \in L$. This means that between them, the parties contain $E_L(x, s_0; r) = \{E_L^{(i)}(x_i, s_0; r)\}_{i \in [n]}$. Output

$$D_L(x, E_L(x, s_0; r)) \oplus \bigoplus_{i \in [n]} s_i$$

  as the secret.

For correctness, note that there are two possible types of authorized sets B in $\mathcal{A}_{L,n}$. If the set B contains parties $P_{i,0}$ and $P_{i,1}$ for some i, they recover the secret as $s_{i,0} \oplus s_{i,1}$. If not, the authorized set contains the parties $P_{1,x_1}, \ldots, P_{n,x_n}$ for some $x = (x_1, x_2, \ldots, x_n) \in L$. By the correctness of the PRE scheme for L, we know that $D_L(x, E_L(x, s_0; r)) = s_0$ with probability at least $1 - \epsilon$. Thus, the recovered secret is

$$D_L(x, E_L(x, s_0; r)) \oplus \bigoplus_{i \in [n]} s_i = \bigoplus_{i \in \{0,1,\ldots,n\}} s_i = s$$

with probability at least $1 - \epsilon$.

For privacy, there are again two types of sets B that are not present in $\mathcal{A}_{L,n}$. If there is an i such that the set of parties B does not contain either of $P_{i,0}$ and $P_{i,1}$, then B's shares look completely random due to the absence of any information about s_i. The other case is when $B = \{P_{i,x_i}\}$ for some $x \notin L$. In this case, $d(S(0)_B, S(1)_B)$ is exactly the distance between $E_L(x, 0)$ and $E_L(x, 1)$ due to how the s_i's are picked, which is at most δ by the privacy of the randomized encoding of L.

It is also easy to see from the definition of S and R that if (E_L, D_L) is semi-efficient, then so is (S, R); and the same if it is efficient.

$(3) \Rightarrow (1)$. Given an (ϵ, δ)-secret sharing scheme (S, R) for the access structure $\mathcal{A}_{L,n}$, we construct a special interactive proof (P, V) for L, as follows:

- The verifier V, on input x and a bit b, outputs $S(b)_{B_x}$, where $B_x = \{P_{i,x_i}\}$.
- The prover P on input x and the verifier message m, outputs $R(B_x, m)$, where $B_x = \{P_{i,x_i}\}$.

For completeness, we have that for any $x \in L$, when $b \leftarrow \{0, 1\}$,

$$\Pr[P(x, V(x, b)) = b] = \Pr[R(B_x, (S(b)_{B_x}) = b] \geq 1 - \epsilon$$

by the correctness of secret sharing scheme, as $B_x \in \mathcal{A}_{(L.n)}$.

For privacy, we have that for any $x \notin L$, when $b \leftarrow \{0,1\}$, for any P^*,

$$\Pr\left[P^*(x, V(x, b)) = b\right] \leq \frac{1 + d(V(x, 0), V(x, 1))}{2} \leq \frac{1}{2} + \frac{\delta}{2}$$

by privacy of the secret sharing scheme, as $B_x \notin \mathcal{A}_{L,n}$.

V is a PPT algorithm if (S, R) is semi-efficient, and P is computable in polynomial time if (S, R) is efficient. Also, V is local because it can be split into the collection $\{V^{(i)}(x_i, b) = S(b)_{\{P_{i,x_i}\}}\}$, so it serves as its own semi-efficient locally computable PRE.

4   Positive Results on Efficient Secret Sharing

In this section we present efficient secret sharing schemes for access structures associated with Bounded-Degree Graph Non-Isomorphism, Lattice Closest Vector in small dimensions, and Co-Primality. These are obtained by the application of Theorem 25 (in particular, the implication $(1) \Rightarrow (2)$ in the theorem).

Useful throughout this section is the fact that arithmetic over integers (and rational numbers) may be performed in $\mathbf{NC}^1$ (see [33] for details).

4.1   Bounded-Degree Graph Non-Isomorphism

Notation. Given an upper triangular matrix $M \in \{0,1\}^{n \times n}$, denote by $G(M)$ the undirected graph whose adjacency matrix is $(M + M^T)$, and for a symmetric matrix M, the undirected graph whose adjacency matrix is M. The degree of a graph, $deg(G)$, is the maximum degree of any vertex in the graph. If G_1 and G_2 are isomorphic, we denote this as $G_1 \equiv G_2$.

Definition 26 (dBDGNI). *d-Bounded Degree Graph Non-Isomorphism is the promise problem given by the following sets of YES and NO instances over pairs of upper triangular matrices:*

$$\mathsf{dBDGNI}^{YES} = \{(M_0, M_1) | G(M_0) \not\equiv G(M_1); deg(G(M_0)), deg(G(M_1)) \leq d\}$$
$$\mathsf{dBDGNI}^{NO} = \{(M_0, M_1) | G(M_0) \equiv G(M_1); deg(G(M_0)), deg(G(M_1)) \leq d\}$$

While Graph (Non-)Isomorphism is not known to be in $\mathbf{P}$, there is a classical polynomial time algorithm known for dBDGNI due to Luks [26]. However, it appears to be a long open question whether dBDGNI is in $\mathbf{NC}$ (or even in $\mathbf{RNC}$) [2].

Theorem 27. *For every constant d and every n, there is an efficient (perfect) secret sharing scheme for the access structure $\mathcal{A}_{\mathsf{dBDGNI},n}$. The complexity of the reconstruction algorithm grows as $n^{O(d)}$, whereas sharing runs in time polynomial in n.*

Proof. We prove this by showing a special interactive proof for dBDGNI where the verifier runs in log-space (and therefore, has efficient perfect locally computable PREs) and the prover runs in polynomial time. This satisfies statement (1) in Theorem 25, and hence implies the existence of the required secret sharing scheme.

The SIP proof (P,V) works along the lines of the classical SZK proof for Graph Non-Isomorphism [15], as follows:

- The verifier $V((M_0, M_1), b)$, on input upper triangular matrices $M_0, M_1 \in \{0,1\}^{n \times n}$ and bit b, selects a random permutation matrix $P \in S_n$, and outputs $P(M_b + M_b^T)P^T$.
- The prover $P((M_0, M_1), M)$, checks whether $G(M) \equiv G(M_0)$. If so, it outputs 0, else 1.

Note that the operation $P(M + M^T)P^T$ is equivalent to permuting the vertices of the graph $G(M)$ by the permutation P.

Perfect completeness of this protocol follows from the fact that if $M_0 \not\equiv M_1$, then the verifier's output M will be such that $G(M)$ is isomorphic to exactly one of $G(M_0)$ and $G(M_1)$, and P can identify which by running the algorithm for dBDGNI [26].

The protocol is perfectly sound because if $M_0 \equiv M_1$, then the distribution of the verifier's output is the same whether $b = 0$ or 1, and P has probability exactly $1/2$ of guessing b correctly.

The complexity of the verifier V in the above protocol is that of selecting a random permutation and performing two matrix multiplications, both of which can be done in log-space. Hence by Lemma 14, V has efficient perfect locally computable PREs. The prover P is computable in polynomial time because all the prover does is run the (polynomial time) algorithm for dBDGNI.

(That the running time of reconstruction algorithm of the resulting secret sharing scheme is $n^{O(d)}$ can be seen by tracing its dependence on the running time of the algorithm for dBDGNI - the one in [26] runs in time $n^{O(d)}$ - in the proof of Theorem 22.)

4.2   Lattice Closest Vectors

Notation. For a full-rank (over $\mathbb{Q}$) matrix $B \in \mathbb{Z}^{d \times d}$, let $\Lambda(B)$ denote the integer lattice (of dimension d) whose basis is B, and $\mathcal{P}(B)$ denote the fundamental parallelepiped of the same lattice (the parallelepiped formed by the column vectors of B and the origin). We denote by $\mathcal{B}(y, \delta)$ the set of points in the ball of radius δ centered at the point y (note that as we work with discretised space and not with $\mathbb{R}^d$, the number of points in this set is finite).

Given full-rank matrix $B \in \mathbb{Z}^{d \times d}$, a vector $y \in \mathbb{Z}^d$, $\delta \in \mathbb{Z}^+$ and $\gamma \in [0,1]$, the (decision version of the) gap closest vector problem in d dimensions ($\mathsf{GapCVP}_{\gamma,d}$) asks whether the Euclidean distance of y from (any point in) $\Lambda(B)$ is at most $(\gamma\delta)$ or at least δ.

While classical algorithms due to Gauss, and Lenstra, Lenstra and Lovasz (from [24]) show that for any d, $\mathsf{GapCVP}_{\gamma,d}$ is in **P** for any γ, it is not known to be (and conjectured not to be) in **NC**. We are interested in the complement of this problem, as defined below.

Definition 28 (coGapCVP$_{\gamma,d}$). *For any* $d \in \mathbb{Z}^+$ *and* $\gamma \in [0,1]$, **coGapCVP$_{\gamma,d}$** *is the promise problem defined by the following YES and NO instances over triples* (B, y, δ), *where* $B \in \mathbb{Z}^{d \times d}$ *is full-rank over* $\mathbb{Q}$, $y \in \mathbb{Z}^d$ *and* $\delta \in \mathbb{Z}^+$:

$$\mathsf{coGapCVP}_{\gamma,d}^{YES} = \{(B, y, \delta) \mid \forall x \in \Lambda(B) : ||y - x|| > \delta\}$$

$$\mathsf{coGapCVP}_{\gamma,d}^{NO} = \{(B, y, \delta) \mid \exists x \in \Lambda(B) : ||y - x|| \leq \gamma\delta\}$$

The following theorem asserts the existence of efficient secret sharing schemes under access structures associated with the above problem.

Theorem 29. *For every* c, d, n, *and any* $\gamma = \left(1 - \Omega(\frac{1}{n^c})\right)$, *there is an efficient* $(o(1), o(1))$-*secret sharing scheme under the access structure* $\mathcal{A}_{\mathsf{coGapCVP}_{\gamma,d},n}$.

We prove this theorem by constructing a $(o(1), o(1))$-Special Interactive Proof for coGapCVP$_{\gamma,d}$ with a log-space verifier and a poly time prover. As the verifier is computable in log-space, it has efficient perfect locally computable PREs, by Lemma 14. The existence of such an SIP, along with Theorem 25, implies the efficient secret sharing schemes we need. In interest of space, we defer details of the proof to the full version of this paper.

4.3 Co-primality

Efficient secret sharing schemes for non-co-primality and semi-efficient ones for quadratic non-residuosity were shown by Beimel and Ishai [5] as an illustration of the power of non-linear secret sharing schemes over linear ones. We note that these follow as implications of our Theorem 22 given the existence of SZK proofs for these languages with logspace verifiers (which are indeed known to exist).

We demonstrate here, as an example, the case of Non-Co-Primality, which is in **P**, but again, as noted in [5], not known to be in **NC**.

Definition 30 (NCoP). *The language* Non-Co-Primality (NCoP) *consists of pairs of positive integers that are not co-prime, represented as strings, that is,*

$$\mathsf{NCoP} = \{(u, v) \mid u, v \in \mathbb{Z}^+, gcd(u, v) > 1\}$$

Theorem 31 asserts the existence of statistically correct, statistically private efficient secret sharing schemes under the access structure associated with NCoP.

Theorem 31. *For every* n, *there is an efficient* (ϵ, δ)-*secret sharing scheme under the access structure* $\mathcal{A}_{\mathsf{NCoP},n}$ *from some* $\epsilon(n), \delta(n) = o(1)$.

Proof. Again, we prove this by demonstrating a $(o(1), o(1))$-SIP for Non-co-primality where the prover is efficient and the verifier has efficient perfect locally computable PREs. This implies what we need, by Theorem 25.

We denote by $|u|$ the length of the representation of u as a boolean string. Below, we assume $|u| \geq |v|$. The SIP proof (P, V) is roughly as follows, for some $m = \Theta(|u|)$:

- The verifier V takes as input (u, v) and a bit b.
  - If $b = 1$, it outputs m random multiples of u modulo v; that is, it picks m random numbers $\{r_i\}_{i \in [m]} \leftarrow \{0,1\}^{|u|}$ and outputs $\{(r_i u)(mod\ v)\}$.
  - If $b = 0$, it outputs m random numbers in $[v]$.
- The prover P takes as input (u, v) and the verifiers message, which is a set of m numbers $\{a_i\}_{i \in [m]}$. If $gcd(\{a_i\}) = 1$, the prover outptus 0, else 1.

The above SIP is complete because if $gcd(u, v) > 1$, then if $b = 1$, all multiples of u modulo v will be divisible by $gcd(u, v)$, and the prover will always output 1, and if $b = 0$, with high probability the gcd of m random numbers in $[v]$ will be 1 and the prover will output 0. It is sound because when $gcd(u, v) = 1$, the distribution of multiples of u (drawn from a large enough range) modulo v is negligibly close to uniform, and the cases $b = 0$ and $b = 1$ are indistinguishable.

The verifier V is computable in $\mathbf{L}$, as all it does is multiply n-bit numbers, and so has efficient perfect locally computable PREs, by Lemma 14. The prover is efficient, as all it has to do is compute the gcd of some numbers.

5   Negative Results on Universally Efficient Secret Sharing

In this section, we show that a natural strengthening of efficient secret-sharing, that we call *universally efficient secret-sharing*, cannot exist for all of $\mathbf{P}$, if for every polynomial t, $\mathbf{P} \not\subseteq \mathbf{DSPACE}(t(n))$.

Notation. Below, by L we denote both a language in a class $\mathcal{C}$, and its standard representation as a member of this class, say, for example, as a Turing machine that decides the language in case $\mathcal{C} = \mathbf{P}$. For a function f that takes two arguments (as $f(x, y)$), by $f(x, \cdot)$, we denote f curried with x, that is, the function $g(y) = f(x, y)$; this extends naturally to the case where f takes more than two arguments.

Definition 32 (Universal Secret Sharing). *An (ϵ, δ)-Universally Efficient Secret Sharing Scheme (USS), or simply a universal secret sharing scheme, for a class of languages $\mathcal{C}$ over a domain D is a pair of (randomized) algorithms (S, R) such that for any $L \in \mathcal{C}$ and any n, $(S(L, 1^n, \cdot), R(L, 1^n, \cdot, \cdot))$ is an (ϵ, δ)-secret sharing scheme under the access structure $A_{L,n}$ over the domain D.*

For any polynomial t, a universal secret sharing scheme is said to be t-semi-efficient if for any $L \in \mathcal{C}$, $S(L, 1^n, \cdot)$ is computable in time $t(n)$. The scheme is said to be t-efficient if both $S(L, 1^n, \cdot)$ and $R(L, 1^n, \cdot, \cdot)$ are computable in time $t(n)$.

Theorem 33. *Let, for all n, $1 - \epsilon(n) > \delta(n)$. If a class of languages $\mathcal{C}$ has t-semi-efficient (ϵ, δ)-universal secret sharing (USS) schemes, then there exists t' such that $t'(n) = O(t(n))$ and $\mathcal{C} \subseteq \mathbf{DSPACE}(t'(n))$.*

Sketch of Proof. Suppose (S, R) is a t-semi-efficient (ϵ, δ) USS scheme for the class $\mathcal{C}$. Theorem 33 follows from applying Lemma 34 to each language $L \in \mathcal{C}$, using the fact that by definition, $(S(L, 1^n, \cdot), R(L, 1^n, \cdot, \cdot))$ is an (ϵ, δ)-secret sharing scheme for $A_{L,n}$ where the sharing algorithm runs in time $t(n)$.

In particular, Theorem 33 implies that if $\mathbf{P}$ had a t-semi-efficient USS scheme, then it would be contained in $\mathbf{DSPACE}(t(n))$ for some polynomial $t(n)$.

Lemma 34. *Let, for all n, $1 - 3\epsilon(n) > 3\delta(n)$. If, for some language L, there is an (ϵ, δ)-secret sharing scheme (S, R) for $A_{L,n}$ for all n, where S runs in time $t(n)$, then $L \in \mathbf{DSPACE}(t'(n))$, where $t'(n) = O(t(n))$.*

The proof below is adapted from that of a more general statement from [13].

Proof. We start by using Theorem 22 to recognize the existence of an (ϵ', δ')-SIP (P, V) for L where V runs in time $t(n)$, where $\epsilon' = 3\epsilon$ and $\delta' = 3\delta$ (the constant 3 comes out of the proof of Theorem 22), and we have $1 - \epsilon'(n) > \delta'(n)$.

In order to decide whether $x \in L$, it is sufficient to determine whether any P' can guess b given $V(x, b)$ with probability $\geq (1 - \epsilon'(|x|))$ or only $\leq (1/2 + \delta'(|x|)/2)$. This is equivalent to whether $d(V(x, 0), V(x, 1))$ is $\geq (1 - \epsilon(|x|))$ or $\leq \delta(|x|)$. But $d(V(x, 0), V(x, 1))$ itself can be computed in space $O(t(|x|))$ as follows.

First, for any v of length at most $t(|x|)$, $\Pr_r [V(x, b; r) = v]$ can be computed by iterating over the possible values of r – note that $|r| \leq t(|x|)$– and simulating V to see if it outputs v, and counting the number of r's for which it does. This requires only $O(t(|x|))$ space because V can be simulated in this much space, and the count of r's is at most $2^{t(|x|)}$.

So for each v, we can also compute

$$p(v) := |\Pr_r [V(x, 0; r) = v] - \Pr_r [V(x, 1; r) = v]|$$

in $O(t(|x|))$ space. What we need is the sum $\left(\sum_{v: |v| \leq t(|x|)} p(v)\right)$. To compute this, we simply iterate over all the v's, storing at the end of each iteration only the sum $\left(\sum_{v': v' \leq v} p(v)\right)$. As each $p(v) \geq 2^{-t(|x|)}$, and the cumulative sum is at most 1, this adds at most $O(t(|x|))$ space to what is needed for each iteration. Hence, the entire computation of $d(V(x, 0), V(x, 1))$ can be done in space $t'(|x|) = O(t(|x|))$, and hence $L \in \mathbf{DSPACE}(t'(n))$.

Acknowledgments. We thank an anonymous ASIACRYPT reviewer for comments that helped improve the presentation of this paper.

A   Proof of Lemma 13

In this section, we restate and prove Lemma 13. This essential lemma extends the privacy properties of PREs to the case of PREs of randomized functions - while the original definition of PREs (for deterministic functions) states that if for some x, $f(x, z_1) = f(x, z_2)$, then $E_f(x, z_1)$ and $E_f(x, z_2)$ are statistically close, Lemma 13 states that even for a randomized function g, if $g(x, z_1)$ and $g(x, z_2)$ are statistically close, then so are $E_g(x, z_1)$ and $E_g(x, z_2)$.

Note that PREs for randomized functions are defined as described in Sect. 2: To construct an (ϵ, δ)-PRE for a randomized function $A(x, z; r)$, simply construct an (ϵ, δ)-PRE $(E_{A'}, D_{A'})$ for the deterministic function $A'(x, (z, r)) = A(x, z; r)$, and let $E_A(x, z)$ be the random variable $E_{A'}(x, (z, r))$ when r is chosen uniformly at random, and have D_A be the same as $D_{A'}$.

Lemma 35. *Let $A(x, z)$ be a randomized function, and (E_A, D_A) be an (ϵ, δ)-PRE of A as described above. Then, for any x and any z_1, z_2:*

$$d(A(x, z_1), A(x, z_2)) \leq \delta' \Rightarrow d(E_A(x, z_1), E_A(x, z_2)) \leq \delta(|x|) + \delta'$$

Proof. As above, consider the deterministic function $A'(x, (z, r)) = A(x, z; r)$. By definition, $d(E_A(x, z_1), E_A(x, z_2)) = d(E_{A'}(x, (z_1, r)), E_{A'}(x, (z_2, r)))$, which is given by:

$$\sum_{\hat{v}} |\Pr[E_{A'}(x, (z_1, r)) = \hat{v}] - \Pr[E_{A'}(x, (z_2, r)) = \hat{v}]|$$

where r is distributed uniformly over its domain. We wish to prove that this expression is small. From the privacy of PREs, we have promises on the behaviour of $E_{A'}$ on inputs for which A' has the same output value. Towards exploiting this, we expand the above expression, conditioning on possible values of A' to get:

$$\sum_{\hat{v}} \left| \sum_v \Pr[A'(x, (z_1, r)) = v] \Pr\left[E_{A'}(x, (z_1, r)) = \hat{v} \middle| A'(x, (z_1, r)) = v\right] \right.$$

$$\left. - \sum_v \Pr[A'(x, (z_2, r)) = v] \Pr\left[E_{A'}(x, (z_2, r)) = \hat{v} \middle| A'(x, (z_2, r)) = v\right] \right|$$

For the same reason - so that we may compare $E_{A'}$ on points where A' has the same output value - we add and subtract $(\sum_v \Pr[A'(x, (z_1, r)) = v])$ to the factor in the second term above and use the triangle inequality to say that what we have is at most:

$$\sum_v \Pr[A'(x, (z_1, r)) = v] \left(\sum_{\hat{v}} \left| \Pr\left[E_{A'}(x, (z_1, r)) = \hat{v} \middle| A'(x, (z_1, r)) = v\right] \right. \right.$$

$$\left. \left. - \Pr\left[E_{A'}(x, (z_2, r)) = \hat{v} \middle| A'(x, (z_2, r)) = v\right] \right| \right)$$

$$+ \sum_v \sum_{\hat{v}} \Pr\left[E_A(x, (z_2, r)) = \hat{v} \middle| A(x, (z_2, r)) = v\right] \cdot$$

$$|\Pr[A(x, (z_1, r)) = v] - \Pr[A(x, (z_2, r)) = v]|$$

The first summand above is a convex combination of several terms, each of which is at most $\delta(|x|)$ by the privacy guarantee of $E_{A'}$ (as each of these terms is some convex combination of the distance between $E_{A'}$ on input values for which A' produces the same output). The second summand is simply equal to $d(A'(x, z_1), A'(x, z_2)) = \delta'$. Hence the whole thing is at most $(\delta(|x|) + \delta')$, which is what we wanted to prove.

B   A Refined Completeness Theorem for SZK$_\mathbf{L}$

In this section, we complete the proof sketch of Theorem 20. In order to do so, we shall first demonstrate Lemma 36.

Lemma 36 ([28]). *There exist negligible functions $\epsilon(n), \delta(n) = n^{-\omega(1)}$ such that every language L in* **SZK$_\mathbf{L}$** *reduces to (ϵ, δ)-SD$_L$. Furthermore, there is a logspace program D_L such that, if an instance x is mapped to the instance (C_0, C_1) by the above reduction, $D_L(b, x, r) = C_b(r)$.*

Given D_L from Lemma 36 for a language $L \in$ **SZK$_\mathbf{L}$**, we can prove Theorem 20 by constructing a special interactive proof (P, V) for L as follows:

- $V(x, b; r) = D_L(b, x, r)$
- $P(x, m)$ outputs 0 if $\Pr[D_L(0, x, r) = m] > \Pr[D(1, x, r) = m]$, and 1 otherwise.

Note that the above is an $(\epsilon/2, \delta/2)$-SIP proof for L where the verifier can be computed in logspace.

We shall now sketch a proof of Lemma 36, for which we shall need the following amplification lemma for statistical distance of distributions.

Lemma 37 (Polarization Lemma, [28]). *Let $\alpha, \beta \in [0,1]$ be constants such that $\alpha^2 > \beta$. Given two logspace machines $X_0, X_1 : \{0,1\}^n \to \{0,1\}^m$, there are logspace machines $Y_0, Y_1 : \{0,1\}^{n'} \to \{0,1\}^{m'}$ (where n', m' grow polynomially with n, m) that use X_0, X_1 only as blackboxes such that:*

$$d(X_0, X_1) \geq \alpha \Rightarrow d(Y_0, Y_1) \geq 1 - 2^{-n'}$$
$$d(X_0, X_1) \leq \beta \Rightarrow d(Y_0, Y_1) \leq 2^{-n'}$$

Both the above lemmas are not stated in precisely this manner in either [28] or [31], but these extensions follow easily from the proofs of statements that are indeed made in these works.

Sketch of Proof. of Lemma 36 (The lemma follows directly from the proof of completeness of SD for SZK presented in [31], noticing that the reduction from any $L \in SZK$ to SD, outlined below, leads to logspace machines if one starts with an $L \in SZKL$, as L has a logspace simulator.)

Suppose L has an SZK proof (P, V) in which, on inputs of length n, the total communication is $t(n)$ over $v(n)$ messages, V uses $r(n)$ bits of randomness, and there is a logspace simulator S that achieves deviation $\mu(n) \leq 1/(Ct(n)^2)$, for some constant C to be determined. Let S_i denote the distribution of the output of S (on a given input) truncated to the first i rounds. We assume, without loss of generality, that the prover speaks first, messages alternate, and that the last message of the verifier consists of all its randomness. We shall describe now distributions that witness the reduction of L to SD_L. Proofs and further details may be found in [31], Chap. 3.

Define the following distributions:

$$X : S_2 \otimes S_4 \otimes \cdots \otimes S_{2v}$$
$$Y_1 : S_1 \otimes S_3 \otimes \cdots \otimes S_{2v-1} \otimes U_{r-7}$$
$$Y_2 : \text{Run } S \ 8ln(tv + 2) \text{ times, and if the transcript is rejecting in}$$
$$\text{a majority of these, output } U_{tv+2}, \text{ else output nothing.}$$
$$Y : Y_1 \otimes Y_2$$

We may arrange, again without loss of generality, for a given input length n of L, for both X and Y to use at most m' bits of randomness and have output length n'. Let $q = 9\,km'^2$ for some constant k to be determined later.

Let $X'' = \otimes^q X$ and $Y'' = \otimes^q Y$, and m'' and n'' be the (upper bound on) number of bits of randomness used and output length of X'' and Y''. Let $H = H_{m''+n'',m''}$ be a family of 2-universal hash functions from $\{0,1\}^{m''+n''} \to \{0,1\}^{m''}$. Define now the following distributions:

$A :$ Choose $r \leftarrow \{0,1\}^{m''}, h \leftarrow H, y \leftarrow Y$, let $x = X''(r)$. Output $(x, h, h(r, y))$.

$B :$ Choose $x \leftarrow X'', h \leftarrow H, z \leftarrow \{0,1\}^{m''}$. Output (x, h, z).

As proven in [31], if $x \in L$, then $d(A, B) \geq 1 - O(2^{-k})$, and if $x \notin L$, $d(A, B) \leq 2^{-\Omega(k)}$. Note that all steps involved so far, including evaluating the hash function, may be done in logspace, meaning that there is a randomised logspace program that on input x can sample A (or B).

This lets us apply Lemma 37 to (A, B) to get distributions (A', B') which are still sampleable in logspace given x (as they are logspace programs that only use the samplers for A and B as blackboxes), and are such that if $x \in L$, $d(A', B') \geq 1 - 2^{-r}$ and if $x \notin L$, $d(A', B') \leq 2^{-r}$, where r (a polynomial in $|x|$ and $\Omega(|x|)$) is the amount of randomness used by the sampler for A' (or B'). This gives us the reduction to SD_L.

We now define D_L to simply emulate the above steps. On input (b, x, r), where $|r|$ is a function of $|x|$ resulting from above operations (D_L is undefined on input lengths that do not obey this relation between $|x|$ and $|r|$), if $b = 0$, D_L runs the logspace sampler for A' with input x and randomness r, and similiarly the sampler for B' if $b = 1$. Note that D_L is still in logspace, and that if $x \in L$, $d(D_L(0, x, r), D_L(1, x, r)) \geq 1 - 2^{-|x|}$, and if $x \notin L$, $d(D_L(0, x, r), D_L(1, x, r)) \leq 2^{-|x|}$. $\square$

References

1. Applebaum, B., Ishai, Y., Kushilevitz, E.: Cryptography in nc^0. SIAM J. Comput. **36**(4), 845–888 (2006)
2. Arvind, V., Torán, J.: Isomorphism testing: perspective and open problems. Bull. EATCS **86**, 66–84 (2005)
3. Barak, B., Goldreich, O., Impagliazzo, R., Rudich, S., Sahai, A., Vadhan, S.P., Yang, K.: On the (im)possibility of obfuscating programs. J. ACM **59**(2), 6 (2012)
4. Beimel, A.: Secret-sharing schemes: a survey. In: Chee, Y.M., Guo, Z., Ling, S., Shao, F., Tang, Y., Wang, H., Xing, C. (eds.) IWCC 2011. LNCS, vol. 6639, pp. 11–46. Springer, Heidelberg (2011)
5. Beimel, A., Ishai, Y.: On the power of nonlinear secret-sharing. IACR Cryptol. ePrint Arch. **2001**, 30 (2001)
6. Benaloh, J.C., Leichter, J.: Generalized secret sharing and monotone functions. In: Goldwasser, S. (ed.) CRYPTO 1988. LNCS, vol. 403, pp. 27–35. Springer, Heidelberg (1990)
7. Bitansky, N., Garg, S., Telang, S.: Succinct randomized encodings and their applications. IACR Cryptol. ePrint Arch. **2014**, 771 (2014)
8. Blakley, G.: Safeguarding cryptographic keys. In: Proceedings of the National Computer Conference, vol. 48, pp. 313–317 (1979)
9. Blundo, C., De Santis, A., De Simone, R., Vaccaro, U.: Tight bounds on the information rate of secret sharing schemes. Des. Codes Cryptography **11**(2), 107–122 (1997)
10. Canetti, R., Holmgren, J., Jain, A., Vaikuntanathan, V.: Indistinguishability obfuscation of iterated circuits and RAM programs. IACR Cryptology ePrint Archive **2014**, 769 (2014)
11. Csirmaz, L.: The size of a share must be large. J. Cryptology **10**(4), 223–231 (1997)
12. Dvir, Z., Gutfreund, D., Rothblum, G. N., Vadhan, S.: On approximating the entropy of polynomial mappings. In: Proceedings of the 2nd Innovations in Computer Science Conference, pp. 460–475 (2011)
13. Fortnow, L., Lund, C.: Interactive proof systems and alternating time-space complexity. Theor. Comput. Sci. **113**(1), 55–73 (1993)
14. Garg, S., Gentry, C., Halevi, S., Raykova, M., Sahai, A., Waters, B.: Candidate indistinguishability obfuscation and functional encryption for all circuits. In: FOCS, pp. 40–49 (2013)
15. Goldreich, O., Micali, S., Wigderson, A.: Proofs that yield nothing but their validity and a methodology of cryptographic protocol design (extended abstract). In: 27th Annual Symposium on Foundations of Computer Science, Toronto, Canada, 27–29 October 1986, pp. 174–187 (1986)
16. Greenlaw, R., Hoover, H.J., Ruzzo, W.L.: Limits to Parallel Computation: P-completeness Theory. Oxford University Press Inc, New York (1995)
17. Ishai, Y., Kushilevitz, E.: Randomizing polynomials: A new representation with applications to round-efficient secure computation. In: FOCS, pp. 294–304 (2000)
18. Ishai, Y., Wee, H.: Partial garbling schemes and their applications. In: Esparza, J., Fraigniaud, P., Husfeldt, T., Koutsoupias, E. (eds.) ICALP 2014. LNCS, vol. 8572, pp. 650–662. Springer, Heidelberg (2014)
19. Ito, M., Saio, A., Nishizeki, T.: Multiple assignment scheme for sharing secret. J. Cryptology **6**(1), 15–20 (1993)
20. Karchmer, M., Wigderson, A.: On span programs. In: Proceedings of the Eigth Annual Structure in Complexity Theory Conference, San Diego, CA, USA, May 18–21, 1993, pp. 102–111. IEEE Computer Society (1993)

21. Karnin, E.D., Greene, J.W., Hellman, M.E.: On secret sharing systems. IEEE Trans. Inf. Theor. **29**(1), 35–41 (1983)
22. Komargodski, I., Naor, M., Yogev, E.: Secret-sharing for NP. In: Sarkar, P., Iwata, T. (eds.) ASIACRYPT 2014, Part II. LNCS, vol. 8874, pp. 254–273. Springer, Heidelberg (2014)
23. Koppula, V., Lewko, A. B., Waters, B.: Indistinguishability obfuscation for turing machines with unbounded memory. IACR Cryptology ePrint Archive, 2014/925 (2014)
24. Lenstra Jr., A.K., Lenstra, H.W., Lovász, L.: Factoring polynomials with rational coefficients. Math. Ann. **261**(4), 515–534 (1982)
25. Lin, H., Pass, R.: Succinct garbling schemes and applications. IACR Cryptology ePrint Archive **2014**, 766 (2014)
26. Luks, E.M.: Isomorphism of graphs of bounded valence can be tested in polynomial time. In: FOCS, pp. 42–49 (1980)
27. Razborov, A.A.: Lower bounds on the monotone complexity of some Boolean functions. Doklady Akademii Nauk SSSR **285**, 798–801 (1985)
28. Sahai, A., Vadhan, S.P.: A complete problem for statistical zero knowledge. Electron. Colloquium on Comput. Complex. (ECCC) 7(84) (2000)
29. Shamir, A.: How to share a secret. Commun. ACM **22**(11), 612–613 (1979)
30. Tardos, É.: The gap between monotone and non-monotone circuit complexity is exponential. Combinatorica **8**(1), 141–142 (1988)
31. Vadhan, S.: A study of statistical zero-knowledge proofs. Ph.D. thesis, Massachusetts Institute of Technology (1999)
32. Vinod, V., Narayanan, A., Srinathan, K., Pandu Rangan, C., Kim, K.: On the power of computational secret sharing. In: Johansson, T., Maitra, S. (eds.) INDOCRYPT 2003. LNCS, vol. 2904, pp. 162–176. Springer, Heidelberg (2003)
33. Wegener, I.: The Complexity of Boolean Functions. Wiley, New York (1987)
34. Yao, A.: Unpublished manuscript (1989). Presented at Oberwolfach and DIMACS Workshops

Compactly Hiding Linear Spans
Tightly Secure Constant-Size Simulation-Sound QA-NIZK Proofs and Applications

Benoît Libert[1]([✉]), Thomas Peters[2], Marc Joye[3], and Moti Yung[4,5]

[1] Ecole Normale Supérieure de Lyon, Lyon, France
benoit.libert@ens-lyon.fr
[2] Ecole Normale Supérieure, Paris, France
thomas.peters@ens.fr
[3] Technicolor, Los Altos, USA
marc.joye@technicolor.com
[4] Google Inc., New York, NY, USA
moti@cs.columbia.edu
[5] Columbia University, New York, NY, USA

Abstract. Quasi-adaptive non-interactive zero-knowledge (QA-NIZK) proofs is a recent paradigm, suggested by Jutla and Roy (ASIACRYPT '13), which is motivated by the Groth-Sahai seminal techniques for efficient non-interactive zero-knowledge (NIZK) proofs. In this paradigm, the common reference string may depend on specific language parameters, a fact that allows much shorter proofs in important cases. It even makes certain standard model applications competitive with the Fiat-Shamir heuristic in the Random Oracle idealization. Such QA-NIZK proofs were recently optimized to constant size by Jutla and Roy (CRYPTO '14) and Libert *et al.* (EUROCRYPT '14) for the important case of proving that a vector of group elements belongs to a linear subspace. While the QA-NIZK arguments of Libert *et al.* provide unbounded simulation-soundness and constant proof length, their simulation-soundness is only loosely related to the underlying assumption (with a gap proportional to the number of adversarial queries) and it is unknown how to alleviate this limitation without sacrificing efficiency. In this paper, we deal with the question of whether we can simultaneously optimize the proof size and the tightness of security reductions, allowing for important applications with tight security (which are typically quite lengthy) to be of shorter size. We resolve this question by designing a novel simulation-sound QA-NIZK argument showing that a vector $v \in \mathbb{G}^n$ belongs to a subspace of rank $t < n$ using a constant number of group elements. Unlike previous short QA-NIZK proofs of such statements, the unbounded simulation-soundness of our system is nearly tightly related (i.e., the reduction only loses a factor proportional to the security parameter) to the standard Decision Linear assumption. To show simulation-soundness in the constrained context of tight reductions, we explicitly point at a technique— which may be of independent interest—of hiding the linear span of a vector defined by a signature (which is part of an OR proof). As an application, we design a public-key cryptosystem with almost tight CCA2-security in the multi-challenge, multi-user setting with improved length

© International Association for Cryptologic Research 2015
T. Iwata and J.H. Cheon (Eds.): ASIACRYPT 2015, Part I, LNCS 9452, pp. 681–707, 2015.
DOI: 10.1007/978-3-662-48797-6_28

(asymptotically optimal for long messages). We also adapt our scheme to provide CCA security in the key-dependent message scenario (KDM-CCA2) with ciphertext length reduced by 75 % when compared to the best known tightly secure KDM-CCA2 system so far.

Keywords: Security tightness · Constant-size QA-NIZK proofs · Simulation soundness · CCA2 security

1 Introduction

In this paper, we consider the problem of achieving (almost) tight security in short simulation-sound non-interactive zero-knowledge proofs and chosen-ciphertext-secure encryption. While tight security results are known in both cases [35,38], they incur quite long proofs and ciphertexts. A natural question is to develop tools and techniques to make them short and, in the process, develop deeper understanding of this highly constrained setting. As an answer in this direction, we describe space-efficient methods and constructions with almost tight security. For the specific problem of proving that a vector of group elements belongs to a linear subspace, our main result is the first constant-size NIZK arguments whose simulation-soundness tightly relates to a standard assumption.

TIGHT AND ALMOST TIGHT SECURITY. Any public-key system must rely on some hardness assumption. To provide concrete guarantees, the security proof should preferably give a tight reduction from a well-established assumption. Namely, a successful adversary should imply a probabilistic polynomial time (PPT) algorithm breaking the assumption with nearly the same advantage. Tightness matters because the loss in the reduction may necessitate the use of a larger (at times prohibitively larger) security parameter to counteract the loss. The importance of tightness was first advocated by Bellare and Rogaway [10] in the context of digital signatures 18 years ago. Since then, it received a continuous attention with a flurry of positive and negative results in the random oracle model [2,11,24–26,44,46,57] and in the standard model [6,14,39,40,57].

A highly challenging problem has been to obtain tight security under standard assumptions in the standard model. For many primitives, satisfactory solutions have remained elusive until very recently. Bellare, Boldyreva and Micali [7] raised the problem of constructing a chosen-ciphertext-secure public-key cryptosystem based on a standard assumption and whose exact security does not degrade with the number of users or the number of challenge ciphertexts. The first answer to this question was only given more than a decade later by Hofheinz and Jager [38] and it was more a feasibility result than a practical solution. In the context of identity-based encryption (IBE), Chen and Wee [23] designed the first "almost tightly" secure system —meaning that the degradation factor only depends on the security parameter λ, and not on the number q of adversarial queries— based on a simple assumption in the standard model,[1] which resolved an 8-year-old open problem [58].

NIZK PROOFS AND SIMULATION-SOUNDNESS. Non-interactive zero-knowledge proofs [15] are crucial tools used in the design of countless cryptographic protocols. In the standard model, truly efficient constructions remained lacking until the last decade, when Groth and Sahai [36] gave nearly practical non-interactive witness indistinguishable (NIWI) and zero-knowledge (NIZK) proof systems for a wide class of languages in groups endowed with a bilinear map. While quite powerful, their methods remain significantly more costly than the non-interactive proof heuristics enabled by the Fiat-Shamir paradigm [30] in the idealized random oracle model [9]. recently, Jutla and Roy [42] showed that important efficiency improvements are possible for *quasi-adaptive* NIZK (QA-NIZK) proofs, i.e., where the common reference string (CRS) may depend on the specific language for which proofs are being generated but a single CRS simulator works for the entire class of languages. For the specific task of proving that a vector of n group elements belongs to a linear subspace of rank t, Jutla and Roy [42] gave computationally sound QA-NIZK proofs of length $\Theta(n - t)$ where the Groth-Sahai (GS) techniques entail $\Theta(n+t)$ group elements per proof. They subsequently refined their techniques, reducing the proof's length to a constant [43], regardless of the number of equations or the number of variables. Libert *et al.* [49] independently obtained similar improvements using different techniques. Other constructions were recently given by Abdalla *et al.* [1] and Kiltz and Wee [47] who gave a general methodology for building short QA-NIZK arguments.

The design of non-malleable protocols, primarily IND-CCA2-secure encryption schemes, at times appeals to NIZK proofs endowed with a property named *simulation-soundness* by Sahai [56]: informally, an adversary should remain unable to prove a false statement by itself, even with the help of an oracle generating simulated proofs for (possibly false) adversarially-chosen statements. Groth [35] and Camenisch *et al.* [19] extended the Groth-Sahai techniques so as to obtain simulation-sound NIZK proofs. Their techniques incur a substantial overhead due to the use of quadratic pairing product equations, OR proofs or IND-CCA2-secure encryption schemes. It was shown [41,45,51] that one-time simulation-soundness —where the adversary obtains only one simulated proof— is much cheaper to achieve than unbounded simulation-soundness (USS). When it comes to proving membership of linear subspaces, Libert, Peters, Joye and Yung [49] gave very efficient unbounded simulation-sound quasi-adaptive NIZK proofs which do not require quadratic pairing product equations or IND-CCA2-secure encryption. As in the improved solution of Kiltz and Wee [47], their USS QA-NIZK arguments have constant size, regardless of the dimensions of the considered subspace. Unfortunately, the simulation-soundness of their proof system does not tightly reduce to the underlying assumption. The multiplicative gap between the reduction's probability of success and the adversary's advantage

---

[1] Using random oracles, Katz and Wang [46] previously gave a tightly secure variant of the Boneh-Franklin IBE [17].

depends on the number q of simulated proofs observed by the adversary. As a consequence, the results of [47,49] do not imply tight chosen-ciphertext security [38] in a scenario —first envisioned by Bellare, Boldyreva and Micali [7]— where the adversary obtains polynomially many challenge ciphertexts. As of now, USS proof systems based on OR proofs [35,38] are the only ones to enable tight multi-challenge security and it is unclear how to render them as efficient as [49] for linear equations.

TIGHTNESS AND CHOSEN-CIPHERTEXT SECURITY. Bellare, Boldyreva and Micali [7] showed that, if a public-key cryptosystem is secure in the sense of the one-user, one-challenge security definition [55], it remains secure in a more realistic multi-user setting where the adversary obtains polynomially many challenge ciphertexts. Their reduction involves a loss of exact security which is proportional to the number of users *and* the number of challenge ciphertexts. They also showed that, in the Cramer-Shoup encryption scheme [28], the degradation factor only depends on the number of challenges per user. Hofheinz and Jager [38] used a tightly secure simulation-sound proof system to build the first encryption system whose IND-CCA2 security tightly reduces to a standard assumption in the multi-user, multi-challenge setting. Due to very large ciphertexts, their scheme was mostly a feasibility result and the same holds for the improved constructions of Abe *et al.* [5]. Until recently, the only known CCA2-secure encryption schemes with tight security in the multi-challenge, multi-user setting either relied on non-standard q-type assumptions [37] —where the number of input elements depends on the number of adversarial queries— or incurred long ciphertexts [5,38] comprised of hundreds of group elements (or both). One of the reasons is that solutions based on standard assumptions [5,38,50] build on simulation-sound proof systems relying on OR proofs. Libert *et al.* [50] gave an almost tightly IND-CCA2 system in the multi-challenge setting where, despite their use of OR proofs, ciphertexts only require 69 group elements under the Decision Linear assumption. Unfortunately, their result falls short of implying constant-size simulation-sound QA-NIZK proofs of linear subspace membership since each vector coordinate would require its own proof elements. In particular, the technique of [50] would result in long proofs made of $O(\lambda)$ group elements in the setting of key-dependent message CCA2 security, where $O(1)$ group elements per proof suffices [43, Section6] if we accept a loose reduction.

Very recently, Hofheinz *et al.* [40] put forth an almost tightly secure IBE scheme in the multi-challenge, multi-instance scenario. While their result implies an almost tightly CCA2 secure public-key encryption scheme via the Canetti-Halevi-Katz paradigm [21], it relies on composite order groups. In [40], it was left as an open problem to apply the same technique under standard assumptions in the (notoriously much more efficient) prime order setting.

OUR CONTRIBUTIONS. As a core technical innovation, this paper presents short QA-NIZK proofs of linear subspace membership (motivated by those in [43,49]) where the unbounded simulation-soundness property can be *almost tightly* —in the terminology of Chen and Wee [23]— related to the standard Decision Linear

(DLIN) assumption [16]. As in [23], the loss of concrete security only depends on the security parameter, and not on the number of simulated proofs obtained by the adversary, which solves a problem left open in [49]. Our construction only lengthens the QA-NIZK proofs of Libert *et al.* [49] by a factor of 2 and thus retains the constant proof length of [49], independently of the dimensions of the subspace. In particular, it does not rely on an IND-CCA2-secure encryption scheme —which, in this context, would require a tightly secure CCA2 cryptosystem to begin with— and it does not even require quadratic equations.

Building on our QA-NIZK proofs and the Naor-Yung paradigm [54], we obtain a new public-key encryption scheme which is proved IND-CCA2-secure in the multi-challenge, multi-user setting under the Decision Linear assumption via an almost tight reduction. While the reduction is slightly looser than those of [5,38], our security bound does not depend on the number of users or the number of challenges, so that our scheme is as secure in the multi-challenge, multi-user scenario as in the single-challenge, single-user setting. Like [5,38], our construction features publicly recognizable well-formed ciphertexts, which makes it suitable for non-interactive threshold decryption. Moreover, our ciphertexts are much shorter than those of [5,38] as they only consist of 48 group elements under the DLIN assumption, whereas the most efficient construction based on the same assumption [50] entails 69 group elements per ciphertext.

Our constant-size proofs offer more dramatic savings when it comes to encrypting long messages without affecting the compatibility with zero-knowledge proofs. We can encrypt N group elements at once while retaining short proofs, which only takes $2N+46$ group elements per ciphertext. The asymptotic expansion ratio of 2 —which is inherent to the Naor-Yung technique— is thus optimal. To our knowledge, all prior results on tight CCA2 security would incur $\Theta(N)$ elements per proof and thus a higher expansion rate in this situation. In turn, our encryption schemes imply tightly secure non-interactive universally composable (UC) commitments [20,27] with adaptive security in the erasure model. In particular, using the same design principle as previous UC commitments [31,42,52] based on CCA2-secure cryptosystems, our scheme for long messages allows committing to N group elements at once with a two-fold expansion rate.

Using our QA-NIZK proof system, we also construct an almost tightly secure encryption scheme with key-dependent message chosen-ciphertext security (KDM-CCA2) [12,18] —in the sense of [19]— with shorter ciphertexts. Analogously to the Jutla-Roy construction [43, Section6], our system offers substantial savings w.r.t. [19] as it allows for constant-size proofs even though, due to the use of the Boneh *et al.* approach [18] to KDM security, the dimension of underlying vectors of group elements depends on the security parameter. Like the Jutla-Roy construction [43], our KDM-CCA2 system only lengthens the ciphertexts of its underlying KDM-CPA counterpart by a constant number of group elements. Unlike [43], however, the KDM-CCA2 security of our scheme is almost tightly related to the DLIN assumption. So far, the most efficient tightly KDM-CCA2 system was implied by the results of Hofheinz-Jager [38] and Abe *et al.* [5], which incur rather long proofs. Our QA-NIZK proofs yield ciphertexts that are about 75 % shorter, as we show in the full version of the paper.

OUR TECHNIQUES. Our QA-NIZK arguments (as the construction in [49]) build on linearly homomorphic structure-preserving signatures (LHSPS) [48]. In [49], each proof of subspace membership is a Groth-Sahai NIWI proof of knowledge of a homomorphic signature on the vector v whose membership is being proved. The security analysis relies on the fact that, with some probability, all simulated proofs take place on a perfectly NIWI Groth-Sahai CRS while the adversary's fake proof pertains to a perfectly binding CRS. Here, in order to do this without applying Waters' partitioning method [58] to the CRS space as in [53], we let the prover generate a Groth-Sahai CRS $\mathbf{F} = (\boldsymbol{f_1}, \boldsymbol{f_2}, \boldsymbol{F})$ of its choice (a similar technique was used by Escala and Groth [29] in a different context), for vectors of group elements $\boldsymbol{f_1}, \boldsymbol{f_2}, \boldsymbol{F} \in \mathbb{G}^3$, and first prove that this CRS is perfectly binding (i.e., $\boldsymbol{F}$ lives in $\mathsf{span}\langle \boldsymbol{f_1}, \boldsymbol{f_2} \rangle$). This seemingly additional "freedom" that we give the prover ends up allowing a stronger simulator (tight simulation-soundness).

Simulation-soundness is, in fact, obtained by having the prover demonstrate that either: (i) The prover's CRS $\mathbf{F}$ is perfectly binding; or (ii) The prover knows a signature which only the NIZK simulator would be able to compute using some simulation trapdoor. One key idea is that, since the latter OR proof involves a relatively short statement (namely, the membership of a two-dimensional subspace) which the adversary has no control on, it can be generated using a constant number of group elements and using *only* linear pairing product equations.

In order to efficiently prove the above OR statement, we leverage the algebraic properties of a variant of the Chen-Wee signature scheme [23], which was proved almost tightly secure under the DLIN assumption, recently proposed by Libert *et al.* [50]. In short, the real prover computes a pseudo-signature σ (without knowing the signing key) on the verification key of a one-time signature and uses the real witnesses to prove that $\mathbf{F}$ is a perfectly binding CRS. In contrast, the simulator computes a real signature σ using the private key instead of the real witnesses. In order to make sure that simulated proofs will be indistinguishable from real proofs, we apply a technique —implicitly used in [50]— consisting of hiding the linear subspace from where a partially committed vector of group elements defined by the signature σ is chosen: while a pseudo-signature fits within a proper subspace of a linear space specified by the public key, real signatures live in the full linear space. A difference between our approach and the one of [50] is our non-modular and more involved use of the signature scheme, yet the technique we point at above may be useful elsewhere. Our QA-NIZK CRS actually contains the description of a linear subspace which mixes the public key components of the signature and vectors used to build the prover's Groth-Sahai CRS $\mathbf{F}$. In order to implement the OR proof, our idea is to make sure that the only way to prove a non-perfectly-binding CRS $\mathbf{F}$ is to compute the committed σ as a real signature for a legally modified public key. By "legally modified key," we mean that some of its underlying private components may be scaled by an adversarially-chosen factor $x \in \mathbb{Z}_p$ as long as the adversary also outputs g^x. While we rely on an unusual security property of the signature which allows the adversary to tamper with the public key, this property can be proved under the standard DLIN assumption in the scheme of [50]. This unusual property is a crucial technique allowing us to prove the OR statement about the ephemeral CRS **F** without using quadratic equations.

In turn, the simulation-soundness relies on the fact that, unless some security property of the signature of [50] is broken, the adversary still has to generate its fake proof on a perfectly binding CRS. If this condition is satisfied, we can employ the arguments as in [49] to show that the reduction is able to extract a non-trivial homomorphic signature, thus breaking the DLIN assumption.

FULL VERSION. The full version of this paper is available as Cryptology ePrint Archive, Report 2015/242 at URL http://eprint.iacr.org/2015/242.

2   Background and Definitions

2.1   Hardness Assumptions

We consider groups $(\mathbb{G}, \mathbb{G}_T)$ of prime-order p endowed with a bilinear map $e : \mathbb{G} \times \mathbb{G} \to \mathbb{G}_T$. In this setting, we rely on the standard Decision Linear assumption.

Definition 1. [16] The *Decision Linear Problem* (DLIN) in $\mathbb{G}$, is to distinguish the distributions $(g^a, g^b, g^{ac}, g^{bd}, g^{c+d})$ and $(g^a, g^b, g^{ac}, g^{bd}, g^z)$, with $a, b, c, d \xleftarrow{R} \mathbb{Z}_p$, $z \xleftarrow{R} \mathbb{Z}_p$. The DLIN assumption asserts the intractability of DLIN for any PPT distinguisher.

We also use the following problem, which is at least as hard as DLIN [22].

Definition 2. The *Simultaneous Double Pairing problem* (SDP) in $(\mathbb{G}, \mathbb{G}_T)$ is, given group elements $(g_z, g_r, h_z, h_u) \in \mathbb{G}^4$, to find a non-trivial triple $(z, r, u) \in \mathbb{G}^3 \setminus \{(1_\mathbb{G}, 1_\mathbb{G}, 1_\mathbb{G})\}$ such that $e(z, g_z) \cdot e(r, g_r) = 1_{\mathbb{G}_T}$ and $e(z, h_z) \cdot e(u, h_u) = 1_{\mathbb{G}_T}$.

2.2   Quasi-Adaptive NIZK Proofs and Simulation-Soundness

Quasi-Adaptive NIZK (QA-NIZK) proofs are NIZK proofs where the CRS is allowed to depend on the specific language for which proofs have to be generated. The CRS is divided into a fixed part Γ, produced by an algorithm K_0, and a language-dependent part ψ. However, there should be a single simulator for the entire class of languages.

Let λ be a security parameter. For public parameters $\Gamma \leftarrow \mathsf{K}_0(\lambda)$, let $\mathcal{D}_\Gamma$ be a probability distribution over a collection of relations $\mathcal{R} = \{R_\rho\}$ parametrized by a string ρ with an associated language $\mathcal{L}_\rho = \{x \mid \exists w : R_\rho(x, w) = 1\}$.

We consider proof systems where the prover and the verifier both take a label lbl as additional input. For example, this label can be the message-carrying part of an ElGamal-like encryption. Formally, a tuple of algorithms $(\mathsf{K}_0, \mathsf{K}_1, \mathsf{P}, \mathsf{V})$ is a QA-NIZK proof system for $\mathcal{R}$ if there exists a PPT simulator $(\mathsf{S}\_1, \mathsf{S}\_2)$ such that, for any PPT adversaries $\mathcal{A}_1, \mathcal{A}_2$ and $\mathcal{A}_3$, we have the following properties:

Quasi-Adaptive Completeness:

$$\Pr[\Gamma \leftarrow \mathsf{K}_0(\lambda);\ \rho \leftarrow \mathcal{D}_\Gamma;\ \psi \leftarrow \mathsf{K}_1(\Gamma, \rho);$$
$$(x, w, \mathsf{lbl}) \leftarrow \mathcal{A}_1(\Gamma, \psi, \rho);\ \pi \leftarrow \mathsf{P}(\psi, x, w, \mathsf{lbl}) :$$
$$\mathsf{V}(\psi, x, \pi, \mathsf{lbl}) = 1\ \text{if}\ R_\rho(x, w) = 1] = 1.$$

Quasi-Adaptive Soundness:

$$\Pr[\Gamma \leftarrow \mathsf{K}_0(\lambda);\ \rho \leftarrow \mathcal{D}_\Gamma;\ \psi \leftarrow \mathsf{K}_1(\Gamma, \rho);\ (x, \pi, \mathsf{lbl}) \leftarrow \mathcal{A}_2(\Gamma, \psi, \rho) :$$
$$\mathsf{V}(\psi, x, \pi, \mathsf{lbl}) = 1\ \wedge\ \neg(\exists w : R_\rho(x, w) = 1)] \in \mathsf{negl}(\lambda).$$

Quasi-Adaptive Zero-Knowledge:

$$\Pr[\Gamma \leftarrow \mathsf{K}_0(\lambda);\ \rho \leftarrow \mathcal{D}_\Gamma;\ \psi \leftarrow \mathsf{K}_1(\Gamma, \rho)\ :\ \mathcal{A}_3^{\mathsf{P}(\psi, \cdot, \cdot)}(\Gamma, \psi, \rho) = 1]$$
$$\approx \Pr[\Gamma \leftarrow \mathsf{K}_0(\lambda);\ \rho \leftarrow \mathcal{D}_\Gamma;\ (\psi, \tau_{sim}) \leftarrow \mathsf{S}_1(\Gamma, \rho)\ :$$
$$\mathcal{A}_3^{\mathsf{S}(\psi, \tau_{sim}, \cdot, \cdot)}(\Gamma, \psi, \rho) = 1],$$

where

$\mathsf{P}(\psi, ., ., .)$ emulates the actual prover. It takes as input (x, w) and lbl and outputs a proof π if $(x, w) \in R_\rho$. Otherwise, it outputs $\perp$.

$\mathsf{S}(\psi, \tau_{sim}, ., ., .)$ is an oracle that takes as input (x, w) and lbl. It outputs a simulated proof $\mathsf{S}_2(\psi, \tau_{sim}, x, \mathsf{lbl})$ if $(x, w) \in R_\rho$ and $\perp$ if $(x, w) \notin R_\rho$.

We assume that the CRS ψ contains an encoding of ρ, which is thus available to V. The definition of Quasi-Adaptive Zero-Knowledge requires a single simulator for the entire family of relations $\mathcal{R}$.

The property called *simulation-soundness* [56] requires that the adversary remain unable to prove false statements even after having seen simulated proofs for potentially false statements. We consider the strongest form, called *unbounded simulation-soundness* (USS) as opposed to one-time simulation-soundness, where the adversary is allowed to see polynomially many simulated proofs.

In order to use QA-NIZK proofs in a modular manner without degrading the exact security of our constructions, we will require simulation-soundness to hold *even* if the adversary $\mathcal{A}_4$ has a trapdoor τ_m that allows deciding membership in the language $\mathcal{L}_\rho$. We thus assume that the algorithm $\mathcal{D}_\Gamma$ outputs a language parameter ρ and a trapdoor τ_m that allows recognizing elements of $\mathcal{L}_\rho$. This trapdoor τ_m is revealed to $\mathcal{A}_4$ and should not help prove false statements.

Enhanced Unbounded Simulation-Soundness: For any PPT adversary $\mathcal{A}_4$,

$$\Pr[\Gamma \leftarrow \mathsf{K}_0(\lambda);\ (\rho, \tau_m) \leftarrow \mathcal{D}_\Gamma;\ (\psi, \tau_{sim}) \leftarrow \mathsf{S}_1(\Gamma, \rho);$$
$$(x, \pi, \mathsf{lbl}) \leftarrow \mathcal{A}_4^{\mathsf{S}_2(\psi, \tau_{sim}, \cdot, \cdot)}(\Gamma, \psi, \rho, \tau_m) :$$
$$\mathsf{V}(\psi, x, \pi, \mathsf{lbl}) = 1\ \wedge\ \neg(\exists w : R_\rho(x, w) = 1)\ \wedge\ (x, \pi, \mathsf{lbl}) \notin Q] \in \mathsf{negl}(\lambda),$$

where the adversary is allowed unbounded access to an oracle $\mathsf{S}_2(\psi, \tau, ., .)$ that takes as input statement-label pairs (x, lbl) (where x may be outside $\mathcal{L}_\rho$) and outputs simulated proofs $\pi \leftarrow \mathsf{S}_2(\psi, \tau_{sim}, x, \mathsf{lbl})$ before updating the set $Q = Q \cup \{(x, \pi, \mathsf{lbl})\}$, which is initially empty.

The standard notion of soundness can be enhanced in a similar way, by handing the membership testing trapdoor τ_m to $\mathcal{A}_2$. In the weaker notion of one-time simulation-soundness, only one query to the S_2 oracle is allowed.

In order to achieve tight security in the multi-user setting, we also consider a notion of unbounded simulation-soundness in the multi-CRS setting. Namely, the adversary is given a set of μ reference strings $\{\psi_\kappa\}_{\kappa=1}^\mu$ for language parameters $\{\rho_\kappa\}_{\kappa=1}^\mu$ and should remain unable to break the soundness of one these after having seen multiple simulated proofs for each CRS ψ_κ. A standard argument shows that (enhanced) unbounded simulation-soundness in the multi CRS setting is implied by the same notion in the single CRS setting. However, the reduction is far from being tight as it loses a factor μ. In our construction, the random self-reducibility of the underlying hard problems fortunately allows avoiding this security loss in a simple and natural way.

Enhanced Unbounded Simulation-Soundness in the multi-CRS setting: For any PPT adversary $\mathcal{A}_4$, we have

$$\Pr[\Gamma \leftarrow \mathsf{K}_0(\lambda); \{\rho_\kappa, \tau_{m,\kappa}\}_{\kappa=1}^\mu \leftarrow \mathcal{D}_\Gamma; (\{\psi_\kappa, \tau_{sim,\kappa}\}_{\kappa=1}^\mu) \leftarrow \mathsf{S}_1(\Gamma, \{\rho_\kappa\}_{\kappa=1}^\mu);$$
$$(\kappa^*, x, \pi, \mathsf{lbl}) \leftarrow \mathcal{A}_4^{\mathsf{S}_2(\{\psi_\kappa\}_{\kappa=1}^\mu, \{\tau_{sim,\kappa}\}_{\kappa=1}^\mu, \cdot, \cdot, \cdot)}(\Gamma, \{\psi_\kappa, \rho_\kappa, \tau_{m,\kappa}\}_{\kappa=1}^\mu):$$
$$\mathsf{V}(\psi_{\kappa^*}, x, \pi, \mathsf{lbl}) = 1 \ \wedge \ \neg(\exists w : R_{\rho_{\kappa^*}}(x, w) = 1) \ \wedge \ (\kappa^*, x, \pi, \mathsf{lbl}) \notin Q] \in \mathsf{negl}(\lambda).$$

Here, $\mathcal{A}_4$ has access to an oracle $\mathsf{S}_2(\{\psi_\kappa\}_{\kappa=1}^\mu, \{\tau_{sim,\kappa}\}_{\kappa=1}^\mu, ., ., .)$ that takes as input tuples (j, x, lbl) (where x may be outside $\mathcal{L}_{\rho_j}$) and outputs simulated proofs $\pi \leftarrow \mathsf{S}_2(\{\psi_\kappa\}_{\kappa=1}^\mu, \{\tau_{sim,\kappa}\}_{\kappa=1}^\mu, j, x, \mathsf{lbl})$ for $\mathcal{L}_{\rho_j}$ before updating the set $Q = Q \cup \{(j, x, \pi, \mathsf{lbl})\}$, which is initially empty.

The standard notion of soundness extends to the multi-CRS setting in a similar way and it can be enhanced by giving $\{\psi_\kappa\}_{\kappa=1}^\mu$ and the membership trapdoors $\{\tau_{m,\kappa}\}_{\kappa=1}^\mu$ to the adversary. The definition of quasi-adaptive zero-knowledge readily extends as well, by having S_1 output $\{\psi_\kappa, \tau_{sim,\kappa}\}_{\kappa=1}^\mu$ while the oracle S and the simulator S_2 both take an additional index $j \in \{1, \ldots, \mu\}$ as input.

2.3 Linearly Homomorphic Structure-Preserving Signatures

Structure-preserving signatures [3,4] are signature schemes where messages and public keys consist of elements in the group $\mathbb{G}$ of a bilinear configuration $(\mathbb{G}, \mathbb{G}_T)$.

Libert *et al.* [48] considered structure-preserving with linear homomorphic properties (see the full version of the paper for formal definitions). This section reviews the one-time linearly homomorphic structure-preserving signature (LHSPS) of [48].

Keygen(λ, n): given a security parameter λ and the subspace dimension $n \in \mathbb{N}$, choose bilinear group $(\mathbb{G}, \mathbb{G}_T)$ of prime order $p > 2^\lambda$. Then, choose $g_z, g_r, h_z, h_u \xleftarrow{R} \mathbb{G}$. For $i = 1$ to n, choose $\chi_i, \gamma_i, \delta_i \xleftarrow{R} \mathbb{Z}_p$ and compute $g_i = g_z{}^{\chi_i} g_r{}^{\gamma_i}$, $h_i = h_z{}^{\chi_i} h_u{}^{\delta_i}$. The private key is $\mathsf{sk} = \{(\chi_i, \gamma_i, \delta_i)\}_{i=1}^n$ and the public key is $\mathsf{pk} = \big(g_z, \ g_r, \ h_z, \ h_u, \{(g_i, h_i)\}_{i=1}^n\big) \in \mathbb{G}^{2n+4}$.

Sign$(\mathsf{sk}, (M_1, \ldots, M_n))$: to sign a vector $(M_1, \ldots, M_n) \in \mathbb{G}^n$ using $\mathsf{sk} = \{(\chi_i, \gamma_i, \delta_i)\}_{i=1}^n$, output $\sigma = (z, r, u) = \big(\prod_{i=1}^n M_i^{-\chi_i}, \prod_{i=1}^n M_i^{-\gamma_i}, \prod_{i=1}^n M_i^{-\delta_i}\big)$.

SignDerive$(\mathsf{pk}, \{(\omega_i, \sigma^{(i)})\}_{i=1}^\ell)$: given pk as well as ℓ tuples $(\omega_i, \sigma^{(i)})$, parse $\sigma^{(i)}$ as $\sigma^{(i)} = (z_i, r_i, u_i)$ for $i = 1$ to ℓ. Return the triple $\sigma = (z, r, u) \in \mathbb{G}^3$, where $z = \prod_{i=1}^\ell z_i^{\omega_i}$, $r = \prod_{i=1}^\ell r_i^{\omega_i}$, $u = \prod_{i=1}^\ell u_i^{\omega_i}$.

Verify$(\mathsf{pk}, \sigma, (M_1, \ldots, M_n))$: given $\sigma = (z, r, u) \in \mathbb{G}^3$ and $(M_1, \ldots, M_n)$, return 1 if and only if $(M_1, \ldots, M_n) \neq (1_{\mathbb{G}}, \ldots, 1_{\mathbb{G}})$ and (z, r, u) satisfy

$$1_{\mathbb{G}_T} = e(g_z, z) \cdot e(g_r, r) \cdot \prod_{i=1}^n e(g_i, M_i) = e(h_z, z) \cdot e(h_u, u) \cdot \prod_{i=1}^n e(h_i, M_i) \ . \quad (1)$$

Our simulation-sound proof system will rely on the fact that the above scheme provides tight security under the DLIN assumption, as implicitly shown in [48].

3   Constant-Size QA-NIZK Proofs of Linear Subspace Membership with Tight Simulation-Soundness

At a high level, our proof system can be seen as a variant of the construction of Libert et al. [49] with several modifications allowing to tightly relate the simulation-soundness property to the DLIN assumption. The construction also uses the tightly signature scheme of [50].

3.1   Intuition

Like [49], we combine linearly homomorphic signatures and Groth-Sahai proofs for pairing product equations. Each QA-NIZK proof consists of a Groth-Sahai NIWI proof of knowledge of a homomorphic signature on the candidate vector[2] $\boldsymbol{v}$. By making sure that all simulated proofs take place on a perfectly WI CRS, the simulator is guaranteed to leak little information about its simulation trapdoor,

---

[2] At first, tight simulation-soundness may seem achievable via an OR proof showing the knowledge of either a homomorphic signature on $\boldsymbol{v}$ or a digital signature on the verification key of a one-time signature. However, proving that a disjunction of pairing product equations [35] is satisfiable requires a proof length proportional to the number of pairings (which is linear in the dimension n here) in pairing product equations.

which is the private key of the homomorphic signature. At the same time, if the adversary's proof involves a perfectly binding CRS, the reduction can extract a homomorphic signature that it would have been unable to compute and solve a DLIN instance. To implement this approach, the system of [49] uses Waters' partitioning technique [58] in the fashion of [53], which inevitably [39] affects the concrete security by a factor proportional to the number q of queries.

Our first main modification is that we let the prover compute the Groth-Sahai NIWI proof on a CRS $\mathbf{F}$ of his own and append a proof π_F that the chosen CRS is perfectly binding, which amounts to proving the membership of a two-dimensional linear subspace $\mathsf{span}\langle f_1, f_2 \rangle$. At first, it appears that π_F has to be simulation-sound itself since, in all simulated proofs, the reduction must trick the adversary into believing that the ephemeral CRS $\mathbf{F}$ is perfectly sound. Fortunately, the reduction only needs to do this for vectors of its choice —rather than adversarially chosen vectors— and this scenario can be accommodated by appropriately mixing the subspace of Groth-Sahai vectors $f_1, f_2 \in \mathbb{G}^3$ with the one in the public key of the signature scheme of [50].

The NIWI proof of knowledge is thus generated for a Groth-Sahai CRS $\mathbf{F} = (f_1, f_2, F)$ where f_1 and f_2 are part of the global CRS but $F \in \mathbb{G}^3$ is chosen by the prover and included in the proof. To prove that $\mathbf{F}$ is a perfectly sound CRS, honest provers derive a homomorphic signature (Z, R, U) from the first $4L + 2$ rows of a matrix $\mathbf{M} \in \mathbb{G}^{(4L+5) \times (4L+6)}$ defined by the public key of the signature scheme and fixed vectors $f_1, f_2, f_0 \in \mathbb{G}^3$. The first two rows allow deriving a signature on the honestly generated $F = f_1^{\mu_1} \cdot f_2^{\mu_2}$ from publicly available homomorphic signatures on f_1 and f_2. The next $4L$ rows are used to demonstrate the validity of a pseudo-signature $(\sigma_1, \sigma_2, \sigma_3) = (H(V, \mathsf{VK})^r \cdot H(W, \mathsf{VK})^s, f^r, h^s)$ on the verification key VK of a one-time signature. This allows the prover to derive a homomorphic signature (Z, R, U) that authenticates a specific vector $\sigma \in \mathbb{G}^{(4L+6)}$ determined by F and the pseudo-signature $(\sigma_1, \sigma_2, \sigma_3)$.

The proof of simulation-soundness uses a strategy where, with high probability, all simulated proofs will take place on a perfectly NIWI CRS $\mathbf{F} = (f_1, f_2, F)$ —where $F \in \mathbb{G}^3$ is linearly independent of (f_1, f_2)— whereas the adversary's fake proof $\pi^\star$ will contain a vector $F^\star \in \mathbb{G}^3$ such that $\mathbf{F} = (f_1, f_2, F^\star)$ is an extractable CRS (namely, $F^\star \in \mathsf{span}\langle f_1, f_2 \rangle$). In order to satisfy the above conditions, the key idea is to have each QA-NIZK proof demonstrate that either: (i) The vector F contained in π satisfies $F \in \mathsf{span}\langle f_1, f_2 \rangle$; (ii) $(\sigma_1, \sigma_2, \sigma_3)$ is a real signature rather than a pseudo-signature. Since $F \in \mathbb{G}^3$ is chosen by the simulator, we can prove this compound statement *without* resorting to quadratic equations, by appropriately mixing linear subspaces. In more details, using a perfectly NIWI CRS in all simulated proofs requires the reduction to introduce a dependency on the fixed $f_0 \in \mathbb{G}^3$ in the vector F which is included in the proof π. In turn, in order to obtain a valid homomorphic signature on the vector $\sigma \in \mathbb{G}^{(4L+6)}$ determined by F and $(\sigma_1, \sigma_2, \sigma_3)$, this forces the simulator to use the last row of the matrix $\mathbf{M}$ which contains the vector $f_0 \in \mathbb{G}^3$ and the public key components Ω_1, Ω_2 of the signature scheme in [50]. To satisfy the verification algorithm, the vector σ must contain $1_\mathbb{G}$ in the coordinates where

Ω_1, Ω_2 are located in the last row of $\mathbf{M}$. In order to retain these $1_{\mathbb{G}}$'s at these places, the simulator must use two other rows of $\mathbf{M}$ to cancel out the introduction of Ω_1, Ω_2 in $\boldsymbol{\sigma}$. Applying such a "correction" implies the capability of replacing the pseudo-signature $(\sigma_1, \sigma_2, \sigma_3, Z, R, U)$ by a pair $(\sigma, X = g^x)$, where $\sigma = (\sigma_1, \sigma_2, \sigma_3, Z, R, U)$ is a real signature for a possibly modified key.

In order to obtain a perfectly NIZK proof system, we need to unconditionally hide the actual subspace where $\boldsymbol{\sigma} \in \mathbb{G}^{(4L+6)}$ lives as well as the fact that $(\sigma_1, \sigma_2, \sigma_3)$ is a real signature in simulated proofs. To this end, we refrain from letting (σ_1, Z, R, U) appear in the clear and replace them by perfectly hiding commitments $\boldsymbol{C}_{\sigma_1}, \boldsymbol{C}_Z, \boldsymbol{C}_R, \boldsymbol{C}_U$ to the same values and a NIWI proof that (Z, R, U) is a valid homomorphic signature on the partially committed vector $\boldsymbol{\sigma}$. Using our technique, we only need to prove *linear* pairing product equations.

In a construction of nearly tightly CCA2-secure cryptosystem, Libert et al. [50] used a somewhat similar approach based on pseudo-signatures and consisting of hiding the subspace where a partially committed vector is chosen. However, besides falling short of providing constant-size QA-NIZK proofs of subspace membership, the approach of [50] requires quadratic equations. In contrast, while we also relying on pseudo-signatures, our technique for compactly hiding the underlying linear span completely avoids quadratic equations. It further yields simulation-sound QA-NIZK arguments that is constant size fitting within 42 group elements, regardless of the dimensions of the subspace.

3.2   Construction

For simplicity, the description below assumes symmetric pairings $e : \mathbb{G} \times \mathbb{G} \to \mathbb{G}_T$ but instantiations in asymmetric pairings $e : \mathbb{G} \times \hat{\mathbb{G}} \to \mathbb{G}_T$, with $\mathbb{G} \neq \hat{\mathbb{G}}$, are possible, as explained in the full version of the paper.

As in [42], we assume that the language parameter ρ is a matrix in $\mathbb{G}^{t \times n}$, for some integers $t, n \in \mathsf{poly}(\lambda)$ such that $t < n$, with an underlying witness relation R_{par} such that, for any $\mathbf{A} \in \mathbb{Z}_p^{t \times n}$ and $\rho \in \mathbb{G}^{t \times n}$, $R_{\mathrm{par}}(\mathbf{A}, \rho) = 1$ if and only if $\rho = g^{\mathbf{A}}$. We consider distributions $\mathcal{D}_\Gamma \subset \mathbb{G}^{t \times n}$ that are efficiently witness-samplable: namely, there is a PPT algorithm which outputs a pair $(\rho, \mathbf{A})$ such that $R_{\mathrm{par}}(\mathbf{A}, \rho) = 1$ and describing a relation R_ρ with its associated language $\mathcal{L}_\rho$ according to $\mathcal{D}_\Gamma$. For example, the sampling algorithm could pick a random matrix $\mathbf{A} \xleftarrow{R} \mathbb{Z}_p^{t \times n}$ and define $\rho = g^{\mathbf{A}}$.

$\mathsf{K}_0(\lambda)$: choose symmetric bilinear groups $(\mathbb{G}, \mathbb{G}_T)$ of prime order $p > 2^\lambda$ with $f, g, h \xleftarrow{R} \mathbb{G}$. Choose a strongly unforgeable one-time signature $\Sigma = (\mathcal{G}, \mathcal{S}, \mathcal{V})$ with verification keys consisting of L-bit strings, for a suitable $L \in \mathsf{poly}(\lambda)$. Then, output $\Gamma = (\mathbb{G}, \mathbb{G}_T, f, g, h, \Sigma)$.

The dimensions (t, n) of the matrix $\mathbf{A} \in \mathbb{Z}_p^{t \times n}$ such that $\rho = g^{\mathbf{A}}$ can be part of the language, so that t, n can be given as input to algorithm K_1.

$\mathsf{K}_1(\boldsymbol{\Gamma}, \boldsymbol{\rho})$: parse Γ as $(\mathbb{G}, \mathbb{G}_T, f, g, h, \Sigma)$ and ρ as $\rho = \big(G_{i,j}\big)_{1 \leq i \leq t, \, 1 \leq j \leq n} \in \mathbb{G}^{t \times n}$.

1. Generate key pairs $\{(\mathsf{sk}_b, \mathsf{pk}_b)\}_{b=0}^{1}$ for the one-time homomorphic signature of Sect. 2.3 in order to sign vectors of $\mathbb{G}^n$ and $\mathbb{G}^{4L+6}$, respectively. Namely, choose $g_z, g_r, h_z, h_u \xleftarrow{R} \mathbb{G}$, $G_z, G_r, H_z, H_u \xleftarrow{R} \mathbb{G}$. Then, for $i = 1$ to n, pick $\chi_i, \gamma_i, \delta_i \xleftarrow{R} \mathbb{Z}_p$ and compute $g_i = g_z{}^{\chi_i} g_r{}^{\gamma_i}$ and $h_i = h_z{}^{\chi_i} h_u{}^{\delta_i}$. Let $\mathsf{sk}_0 = \{\chi_i, \gamma_i, \delta_i\}_{i=1}^{n}$ be the private key and let $\mathsf{pk}_0 = (g_z, g_r, h_z, h_u, \{g_i, h_i\}_{i=1}^{n})$ be the public key. The second LHSPS key pair $(\mathsf{sk}_1, \mathsf{pk}_1)$ is generated analogously as $\mathsf{sk}_1 = \{\varphi_i, \phi_i, \vartheta_i\}_{i=1}^{4L+6}$ and

$$\mathsf{pk}_1 = \left(G_z,\ G_r,\ H_z,\ H_u,\ \{G_i = G_z^{\varphi_i} G_r^{\phi_i},\ H_i = H_z^{\varphi_i} H_u^{\vartheta_i}\}_{i=1}^{4L+6}\right).$$

2. Choose $y_1, y_2, \xi_1, \xi_2, \xi_3 \xleftarrow{R} \mathbb{Z}_p$ and compute $f_1 = g^{y_1}$, $f_2 = g^{y_2}$. Define vectors $\boldsymbol{f_1} = (f_1, 1_{\mathbb{G}}, g)$, $\boldsymbol{f_2} = (1_{\mathbb{G}}, f_2, g)$ and $\boldsymbol{f_3} = \boldsymbol{f_1}^{\xi_1} \cdot \boldsymbol{f_2}^{\xi_2} \cdot \iota(g)^{\xi_3}$, where $\iota(g) = (1_{\mathbb{G}}, 1_{\mathbb{G}}, g)$. Define the Groth-Sahai CRS $\mathbf{f} = (\boldsymbol{f_1}, \boldsymbol{f_2}, \boldsymbol{f_3})$. Then, define yet another vector $\boldsymbol{f_0} = \boldsymbol{f_1}^{\nu_1} \cdot \boldsymbol{f_2}^{\nu_2}$, with $\nu_1, \nu_2 \xleftarrow{R} \mathbb{Z}_p$.

3. For $\ell = 1$ to L, choose $V_{\ell,0}, V_{\ell,1}, W_{\ell,0}, W_{\ell,1} \xleftarrow{R} \mathbb{G}$ and define row vectors $\boldsymbol{V} = (V_{1,0}, V_{1,1}, \ldots, V_{L,0}, V_{L,1})$, $\boldsymbol{W} = (W_{1,0}, W_{1,1}, \ldots, W_{L,0}, W_{L,1})$.

4. Choose random exponents $\omega_1, \omega_2 \xleftarrow{R} \mathbb{Z}_p$ and group elements $u_1, u_2 \xleftarrow{R} \mathbb{G}$, and compute $\Omega_1 = u_1^{\omega_1} \in \mathbb{G}$, $\Omega_2 = u_2^{\omega_2} \in \mathbb{G}$.

5. Define the matrix $\mathbf{M} = \left(M_{i,j}\right)_{i,j} \in \mathbb{G}^{(4L+5)\times(4L+6)}$ as

$$\left(M_{i,j}\right)_{i,j} = \begin{pmatrix} 1 & \mathbf{1}^{1\times 2L} & \mathbf{1}^{1\times 2L} & 1 & 1 & \boldsymbol{f_1} \\ 1 & \mathbf{1}^{1\times 2L} & \mathbf{1}^{1\times 2L} & 1 & 1 & \boldsymbol{f_2} \\ \boldsymbol{V}^{\top} & \mathbf{Id}_{f,2L} & \mathbf{1}^{2L\times 2L} & \mathbf{1}^{2L\times 1} & \mathbf{1}^{2L\times 1} & \mathbf{1}^{2L\times 3} \\ \boldsymbol{W}^{\top} & \mathbf{1}^{2L\times 2L} & \mathbf{Id}_{h,2L} & \mathbf{1}^{2L\times 1} & \mathbf{1}^{2L\times 1} & \mathbf{1}^{2L\times 3} \\ g & \mathbf{1}^{1\times 2L} & \mathbf{1}^{1\times 2L} & u_1 & 1 & \mathbf{1}^{1\times 3} \\ g & \mathbf{1}^{1\times 2L} & \mathbf{1}^{1\times 2L} & 1 & u_2 & \mathbf{1}^{1\times 3} \\ 1 & \mathbf{1}^{1\times 2L} & \mathbf{1}^{1\times 2L} & \Omega_1^{-1} & \Omega_2^{-1} & \boldsymbol{f_0} \end{pmatrix} \quad (2)$$

with $\mathbf{Id}_{f,2L} = f^{\mathbf{I}_{2L}} \in \mathbb{G}^{2L\times 2L}$, $\mathbf{Id}_{h,2L} = h^{\mathbf{I}_{2L}} \in \mathbb{G}^{2L\times 2L}$, and where $\mathbf{I}_{2L} \in \mathbb{Z}_p^{2L\times 2L}$ stands for the identity matrix. Note that the last row allows linking $\boldsymbol{f_0}$ and Ω_1, Ω_2.

6. Use sk_0 to generate one-time homomorphic signatures $\{(z_i, r_i, u_i)\}_{i=1}^{t}$ on the vectors $(G_{i1}, \ldots, G_{in}) \in \mathbb{G}^n$ that form the rows of $\boldsymbol{\rho} \in \mathbb{G}^{t\times n}$. These are given by $(z_i, r_i, u_i) = \left(\prod_{j=1}^{n} G_{i,j}^{-\chi_j}, \prod_{j=1}^{n} G_{i,j}^{-\gamma_j}, \prod_{j=1}^{n} G_{i,j}^{-\delta_j}\right)$ for each $i \in \{1, \ldots, t\}$. Likewise, use sk_1 to sign the rows $\mathbf{M}_j = (M_{j,1}, \ldots, M_{j,4L+6})$ of the matrix (2) and obtain signatures

$$(Z_j, R_j, U_j) = \left(\prod_{k=1}^{4L+6} M_{j,k}^{-\varphi_k},\ \prod_{k=1}^{4L+6} M_{j,k}^{-\phi_k},\ \prod_{k=1}^{4L+6} M_{j,k}^{-\vartheta_k} \right)$$

for each $j \in \{1, \ldots, 4L+5\}$.

7. The CRS $\psi = (\mathbf{CRS}_1, \mathbf{CRS}_2)$ consists of two parts which are defined as

$$\mathbf{CRS}_1 = \Big(\rho, \ \mathbf{f}, \ \boldsymbol{f_0}, \ u_1, \ u_2, \ \Omega_1, \ \Omega_2, \ \boldsymbol{V}, \ \boldsymbol{W}, \ \mathsf{pk}_0, \ \mathsf{pk}_1, $$
$$\{(z_i, r_i, u_i)\}_{i=1}^t, \ \{(Z_j, R_j, U_j)\}_{j=1}^{4L+5} \Big),$$

$$\mathbf{CRS}_2 = \Big(\mathbf{f}, \ \boldsymbol{f_0}, \ \mathsf{pk}_0, \ \mathsf{pk}_1, \ \Omega_1, \ \Omega_2, \ \boldsymbol{V}, \ \boldsymbol{W} \Big),$$

while the simulation trapdoor is $\tau_{sim} = \big(\omega_1, \omega_2, \{\chi_i, \gamma_i, \delta_i\}_{i=1}^n\big)$.

$P(\boldsymbol{\Gamma}, \psi, \boldsymbol{v}, \boldsymbol{x}, \mathsf{lbl})$: given $\boldsymbol{v} \in \mathbb{G}^n$ and a witness $\boldsymbol{x} = (x_1, \ldots, x_t) \in \mathbb{Z}_p^t$ such that $\boldsymbol{v} = g^{\boldsymbol{x} \cdot \mathbf{A}}$, generate a one-time signature key pair $(\mathsf{VK}, \mathsf{SK}) \leftarrow \mathcal{G}(\lambda)$.

1. Using $\{(z_j, r_j, u_j)\}_{j=1}^t$, derive a one-time linearly homomorphic signature (z, r, u) on the vector $\boldsymbol{v}$ with respect to pk_0. Namely, compute $z = \prod_{i=1}^t z_i^{x_i}$, $r = \prod_{i=1}^t r_i^{x_i}$ and $u = \prod_{i=1}^t u_i^{x_i}$.
2. Choose a vector $\boldsymbol{F} = (F_1, F_2, F_3) = \boldsymbol{f_1}^{\mu_1} \cdot \boldsymbol{f_2}^{\mu_2}$, for random $\mu_1, \mu_2 \xleftarrow{R} \mathbb{Z}_p$.
3. Pick $r, s \xleftarrow{R} \mathbb{Z}_p$ and compute a pseudo-signature on $\mathsf{VK} = \mathsf{VK}[1] \ldots \mathsf{VK}[L]$, which is obtained as $(\sigma_1, \sigma_2, \sigma_3) = (H(\boldsymbol{V}, \mathsf{VK})^r \cdot H(\boldsymbol{W}, \mathsf{VK})^s, f^r, h^s)$, where $H(\boldsymbol{V}, \mathsf{VK}) = \prod_{\ell=1}^L V_{\ell, \mathsf{VK}[\ell]}$ and $H(\boldsymbol{W}, \mathsf{VK}) = \prod_{\ell=1}^L W_{\ell, \mathsf{VK}[\ell]}$.
4. Derive a one-time linearly homomorphic signature $(Z, R, U) \in \mathbb{G}^3$ for pk_1 on the vector

$$\boldsymbol{\sigma} = (\sigma_1, \sigma_2^{1-\mathsf{VK}[1]}, \sigma_2^{\mathsf{VK}[1]}, \ldots, \sigma_2^{1-\mathsf{VK}[L]}, \sigma_2^{\mathsf{VK}[L]}, \sigma_3^{1-\mathsf{VK}[1]},$$
$$\sigma_3^{\mathsf{VK}[1]}, \ldots, \sigma_3^{1-\mathsf{VK}[L]}, \sigma_3^{\mathsf{VK}[L]}, 1_{\mathbb{G}}, 1_{\mathbb{G}}, F_1, F_2, F_3) \in \mathbb{G}^{4L+6} \quad (3)$$

which belongs to subspace spanned by the first $4L + 2$ rows of the matrix $\mathbf{M} \in \mathbb{G}^{(4L+5) \times (4L+6)}$. Hence, the coefficients $r, s, \mu_1, \mu_2 \in \mathbb{Z}_p$ allow deriving a homomorphic signature (Z, R, U) on $\boldsymbol{\sigma}$ in (3). Note that the $(4L+2)$-th and the $(4L+3)$-th coordinates of $\boldsymbol{\sigma}$ must both equal $1_{\mathbb{G}}$.
5. Using the CRS $\mathbf{f} = (\boldsymbol{f_1}, \boldsymbol{f_2}, \boldsymbol{f_3})$, generate Groth-Sahai commitments $C_{\sigma_1}, C_Z, C_R, C_U \in \mathbb{G}^3$. Then, compute NIWI proofs $\boldsymbol{\pi}_{\sigma,1}, \boldsymbol{\pi}_{\sigma,2} \in \mathbb{G}^3$ that committed variables (σ_1, Z, R, U) satisfy

$$e(Z, G_z) \cdot e(R, G_r) \cdot e(\sigma_1, G_1) = t_G,$$
$$e(Z, H_z) \cdot e(U, H_u) \cdot e(\sigma_1, H_1) = t_H, \quad (4)$$

where

$$t_G = e(\sigma_2, \prod_{i=1}^L G_{2i+\mathsf{VK}[i]})^{-1} \cdot e(\sigma_3, \prod_{i=1}^L G_{2L+2i+\mathsf{VK}[i]})^{-1} \cdot \prod_{i=1}^3 e(F_i, G_{4L+3+i})^{-1}$$

and

$$t_H = e(\sigma_2, \prod_{i=1}^L H_{2i+\mathsf{VK}[i]})^{-1} \cdot e(\sigma_3, \prod_{i=1}^L H_{2L+2i+\mathsf{VK}[i]})^{-1}$$
$$\cdot \prod_{i=1}^3 e(F_i, H_{4L+3+i})^{-1}.$$

6. Using the vector $\boldsymbol{F} = (F_1, F_2, F_3)$ of Step 2, define a new Groth-Sahai CRS $\mathbf{F} = (\boldsymbol{f}_1, \boldsymbol{f}_2, \boldsymbol{F})$ and use it to compute commitments

$$\boldsymbol{C}_z = \iota(z) \cdot \boldsymbol{f}_1^{\theta_{z,1}} \cdot \boldsymbol{f}_2^{\theta_{z,2}} \cdot \boldsymbol{F}^{\theta_{z,3}}, \quad \boldsymbol{C}_r = \iota(r) \cdot \boldsymbol{f}_1^{\theta_{r,1}} \cdot \boldsymbol{f}_2^{\theta_{r,2}} \cdot \boldsymbol{F}^{\theta_{r,3}},$$
$$\boldsymbol{C}_u = \iota(u) \cdot \boldsymbol{f}_1^{\theta_{u,1}} \cdot \boldsymbol{f}_2^{\theta_{u,2}} \cdot \boldsymbol{F}^{\theta_{u,3}}$$

to the components of (z, r, u) along with NIWI proofs $(\boldsymbol{\pi}_1, \boldsymbol{\pi}_2) \in \mathbb{G}^6$ that $\boldsymbol{v}$ and (z, r, u) satisfy (1). Let $(\boldsymbol{C}_z, \boldsymbol{C}_r, \boldsymbol{C}_u, \boldsymbol{\pi}_1, \boldsymbol{\pi}_2) \in \mathbb{G}^{15}$ be the resulting commitments and proofs.

7. Set $\sigma = \mathcal{S}(\mathsf{SK}, (\boldsymbol{v}, \boldsymbol{F}, \boldsymbol{C}_{\sigma_1}, \sigma_2, \sigma_3, \boldsymbol{C}_Z, \boldsymbol{C}_R, \boldsymbol{C}_U, \boldsymbol{C}_z, \boldsymbol{C}_r, \boldsymbol{C}_u, \boldsymbol{\pi}_{\sigma,1}, \boldsymbol{\pi}_{\sigma,2}, \boldsymbol{\pi}_1, \boldsymbol{\pi}_2, \mathsf{lbl}))$ and output

$$\begin{aligned} \pi = \big(&\mathsf{VK}, \boldsymbol{F}, \boldsymbol{C}_{\sigma_1}, \sigma_2, \sigma_3, \boldsymbol{C}_Z, \boldsymbol{C}_R, \boldsymbol{C}_U, \boldsymbol{C}_z, \boldsymbol{C}_r, \boldsymbol{C}_u, \\ &\boldsymbol{\pi}_{\sigma,1}, \boldsymbol{\pi}_{\sigma,2}, \boldsymbol{\pi}_1, \boldsymbol{\pi}_2, \sigma \big). \end{aligned} \tag{5}$$

$\mathsf{V}(\boldsymbol{\Gamma}, \psi, \boldsymbol{v}, \pi, \mathsf{lbl})$: parse π as in (5) and $\boldsymbol{v}$ as $(v_1, \dots, v_n) \in \mathbb{G}^n$. Return 1 if the conditions hereunder all hold. Otherwise, return 0.

(i) $\mathcal{V}(\mathsf{VK}, (\boldsymbol{v}, \boldsymbol{F}, \boldsymbol{C}_{\sigma_1}, \sigma_2, \sigma_3, \boldsymbol{C}_Z, \boldsymbol{C}_R, \boldsymbol{C}_U, \boldsymbol{C}_z, \boldsymbol{C}_r, \boldsymbol{C}_u, \boldsymbol{\pi}_{\sigma,1}, \boldsymbol{\pi}_{\sigma,2}, \boldsymbol{\pi}_1, \boldsymbol{\pi}_2, \mathsf{lbl}), \sigma) = 1$;

(ii) $\boldsymbol{\pi}_{\sigma,1}, \boldsymbol{\pi}_{\sigma,2}$ are valid proofs that the variables (σ_1, Z, R, U), which are contained in commitments $\boldsymbol{C}_{\sigma_1}, \boldsymbol{C}_Z, \boldsymbol{C}_R, \boldsymbol{C}_U$, satisfy equations (4).

(iii) The tuple $(\boldsymbol{C}_z, \boldsymbol{C}_r, \boldsymbol{C}_u, \boldsymbol{\pi}_1, \boldsymbol{\pi}_2)$ forms a valid a valid NIWI proof for the Groth-Sahai CRS $\mathbf{F} = (\boldsymbol{f}_1, \boldsymbol{f}_2, \boldsymbol{F})$. Namely, $\boldsymbol{\pi}_1 = (\pi_{1,1}, \pi_{1,2}, \pi_{1,3})$ and $\boldsymbol{\pi}_2 = (\pi_{2,1}, \pi_{2,2}, \pi_{2,3})$ satisfy

$$\prod_{i=1}^{n} E\big(g_i, \iota(v_i)\big)^{-1} = E\big(g_z, \boldsymbol{C}_z\big) \cdot E\big(g_r, \boldsymbol{C}_r\big) \cdot E(\pi_{1,1}, \boldsymbol{f}_1) \cdot$$

$$E(\pi_{1,2}, \boldsymbol{f}_2) \cdot E(\pi_{1,3}, \boldsymbol{F}) \tag{6}$$

$$\prod_{i=1}^{n} E\big(h_i, \iota(v_i)\big)^{-1} = E\big(h_z, \boldsymbol{C}_z\big) \cdot E\big(h_u, \boldsymbol{C}_u\big) \cdot E(\pi_{2,1}, \boldsymbol{f}_1) \cdot$$

$$E(\pi_{2,2}, \boldsymbol{f}_2) \cdot E(\pi_{2,3}, \boldsymbol{F}).$$

The proof only requires 38 elements of $\mathbb{G}$ and a pair (VK, σ). In instantiations using the one-time signature of [38], its total size amounts to 42 group elements, which only lengthens the QA-NIZK proofs of [49] by a factor of 2.

4 Security

To avoid unnecessarily overloading notations, we will prove our results in the single CRS setting. At the main steps, we will explain how the proof can be adapted to the multi-CRS setting without affecting the tightness of reductions.

Theorem 1. *The above proof system is perfectly quasi-adaptive zero-knowledge.*

Proof (sketch). We describe the QA-NIZK simulator here but we refer to the full paper for a detailed proof that the simulation is perfect. This simulator (S_1, S_2) is defined by having S_1 generate the CRS ψ as in the real K_0 algorithm but retain the simulation trapdoor $\tau_{sim} = (\omega_1, \omega_2, \{\chi_i, \gamma_i, \delta_i\}_{i=1}^n)$ for later use. As for S_2, it generates a simulated proof for $v = (v_1, \ldots, v_n) \in \mathbb{G}^n$ by using $\{(\chi_i, \gamma_i, \delta_i)\}_{i=1}^n$ to compute $(z, r, u) = \left(\prod_{j=1}^n v_j^{-\chi_j}, \prod_{j=1}^n v_j^{-\gamma_j}, \prod_{j=1}^n v_j^{-\delta_j}\right)$ at step 1 of the simulation instead of using the witness $x \in \mathbb{Z}_p^t$ as in the real proving algorithm P. At step 2, it defines $(F_1, F_2, F_3) = f_0 \cdot f_1^{\mu_1} \cdot f_2^{\mu_2}$ with $\mu_1, \mu_2 \xleftarrow{R} \mathbb{Z}_p$. At step 3, it picks $r, s \xleftarrow{R} \mathbb{Z}_p$ to compute $(\sigma_1, \sigma_2, \sigma_3) = (g^{\omega_1 + \omega_2} \cdot H(V, \mathsf{VK})^r \cdot H(W, \mathsf{VK})^s, f^r, h^s)$ before using the coefficients $\mu_1, \mu_2, r, s, \omega_1, \omega_2, 1 \in \mathbb{Z}_p$ to derive a homomorphic signature (Z, R, U) from $\{(Z_j, R_j, U_j)\}_{j=1}^{4L+5}$ at step 4. Steps 5 to 7 are conducted as in the real P. In the full paper, we prove that the simulation is perfect in that the simulated CRS ψ is distributed as a real CRS and, for all $v \in \mathbb{G}^n$ such that $v = g^{x \cdot A}$ for some $x \in \mathbb{Z}_p^t$, simulated proofs are distributed as real proofs. $\qquad\square$

We now prove that the system remains computationally sound and simulation-sound, even when the adversary is given the matrix $\mathbf{A} = \log_g(\boldsymbol{\rho}) \in \mathbb{Z}_p^{t \times n}$, which allows recognizing elements of $\mathcal{L}_\rho$. Although the enhanced soundness property is implied by that of enhanced simulation-soundness, we prove it separately (see the full paper for the proof) in Theorem 2 since the reduction is optimal.

Theorem 2. *The system provides quasi-adaptive soundness under the DLIN assumption. Any enhanced soundness adversary $\mathcal{A}$ with running time $t_{\mathcal{A}}$ implies a DLIN distinguisher $\mathcal{B}$ with running time $t_{\mathcal{B}} \leq t_{\mathcal{A}} + q \cdot \mathsf{poly}(\lambda, L, t, n)$ and such that $\mathbf{Adv}_{\mathcal{A}}^{e\text{-sound}}(\lambda) \leq 2 \cdot \mathbf{Adv}_{\mathcal{B}}^{\mathrm{DLIN}}(\lambda) + 2/p$.*

Theorem 3. *The above system provides quasi-adaptive unbounded simulation-soundness if: (i) Σ is a strongly unforgeable one-time signature; (ii) The DLIN assumption holds. For any enhanced unbounded simulation-soundness adversary $\mathcal{A}$, there exist a one-time signature forger $\mathcal{B}'$ in the multi-key setting and a DLIN distinguisher $\mathcal{B}$ with running times $t_{\mathcal{B}}, t_{\mathcal{B}'} \leq t_{\mathcal{A}} + q \cdot \mathsf{poly}(\lambda, L, t, n)$ such that*

$$\mathbf{Adv}_{\mathcal{A}}^{e\text{-uss}}(\lambda) \leq \mathbf{Adv}_{\mathcal{B}'}^{q\text{-suf-ots}}(\lambda) + 3 \cdot (L+2) \cdot \mathbf{Adv}_{\mathcal{B}}^{\mathrm{DLIN}}(\lambda) + 4/p, \qquad (7)$$

where L is the verification key length of Σ and q is the number of simulations.

Proof. To prove the result, we consider a sequence of games. In Game_i, we denote by S_i the event that the challenger outputs 1.

Game_1: This game is the actual attack. Namely, the adversary $\mathcal{A}$ receives as input the description of the language $\mathcal{L}_\rho$ and has access to a simulated CRS ψ and the simulated prover $S_2(\psi, \tau_{sim}, ., .)$ which is described in the proof of Theorem 1. At each invocation, $S_2(\psi, \tau_{sim}, ., .)$ inputs a vector-label pair (v, lbl) and outputs a simulated proof π that $v \in \mathcal{L}_\rho$. In order a generate the matrix $\boldsymbol{\rho} \in \mathbb{G}^{t \times n}$ with the appropriate distribution D_Γ, the challenger chooses a matrix $\mathbf{A} \in \mathbb{Z}_p^{t \times n}$ with the suitable distribution (which is possible since D_Γ is efficiently witness-samplable) and computes $\boldsymbol{\rho} = g^{\mathbf{A}}$. Also, the challenger $\mathcal{B}$ computes a basis $\mathbf{W} \in \mathbb{Z}_p^{n \times (n-t)}$ of the nullspace of $\mathbf{A}$. The adversary receives as input the simulated CRS ψ and the matrix $\mathbf{A} \in \mathbb{Z}_p^{t \times n}$, which serves as a membership testing trapdoor τ_m, and queries the simulator $S_2(\psi, \tau_{sim}, ., .)$ on a polynomial number of occasions. When the adversary $\mathcal{A}$ halts, it outputs an element $\boldsymbol{v}^\star$, a proof $\pi^\star$ and a label $\mathsf{lbl}^\star$. The adversary is declared successful and the challenger outputs 1 if and only if $(\pi^\star, \mathsf{lbl}^\star)$ is a verifying proof but $\boldsymbol{v}^\star \notin \mathcal{L}_\rho$ (i.e., $\boldsymbol{v}^\star$ is linearly independent of the rows of $\rho \in \mathbb{G}^{t \times n}$) and $(\pi^\star, \mathsf{lbl}^\star)$ was not trivially obtained from the simulator. We call S_1 the latter event, which is easily recognizable by the challenger $\mathcal{B}$ since the latter knows a basis $\mathbf{W} \in \mathbb{Z}_p^{n \times (n-t)}$ of the right kernel of $\mathbf{A}$. Indeed, $\mathbf{W}$ allows testing if $\boldsymbol{v} = (v_1, \ldots, v_n) \in \mathbb{G}^n$ satisfies $\prod_{j=1}^n v_j^{w_{ji}} = 1_{\mathbb{G}}$ for each column $\boldsymbol{w}_i^\top = (w_{1i}, \ldots, w_{ni})^\top$ of $\mathbf{W}$. By definition, the adversary's advantage is $\mathbf{Adv}(\mathcal{A}) := \Pr[S_1]$.

Game$_2$: We modify the generation of the CRS $\psi = (\mathbf{CRS}_1, \mathbf{CRS}_2)$. Instead of choosing $\boldsymbol{f}_3 \in_R \mathbb{G}^3$ as a uniformly random vector, S_1 sets $\boldsymbol{f}_3 = \boldsymbol{f}_1^{\xi_1} \cdot \boldsymbol{f}_2^{\xi_2}$, for random $\xi_1, \xi_2 \xleftarrow{R} \mathbb{Z}_p$. Hence, $\boldsymbol{f}_1, \boldsymbol{f}_2$ and $\boldsymbol{f}_3$ now underlie a subspace of dimension 2 and $\mathbf{f} = (\boldsymbol{f}_1, \boldsymbol{f}_2, \boldsymbol{f}_3)$ thus becomes a perfectly binding CRS. Under the DLIN assumption, this modification should have no noticeable impact on $\mathcal{A}$'s probability of success. We have $|\Pr[S_2] - \Pr[S_1]| \leq \mathbf{Adv}^{\mathrm{DLIN}}(\mathcal{B})$.

Game$_3$: We modify again the generation of ψ. Now, instead of choosing $\boldsymbol{f}_0$ in $\mathrm{span}\langle \boldsymbol{f}_1, \boldsymbol{f}_2 \rangle$, S_1 sets $\boldsymbol{f}_0 = \boldsymbol{f}_1^{\nu_1} \cdot \boldsymbol{f}_2^{\nu_2} \cdot \iota(g)$, for random $\nu_1, \nu_2 \xleftarrow{R} \mathbb{Z}_p^*$. The vector $\boldsymbol{f}_0$ is now linearly independent of $(\boldsymbol{f}_1, \boldsymbol{f}_2)$. Under the DLIN assumption, this modification will remain unnoticed to the adversary. In particular, $\mathcal{A}$'s winning probability should only change by a negligible amount. A two-step reduction from DLIN shows that $|\Pr[S_3] - \Pr[S_2]| \leq 2 \cdot \mathbf{Adv}^{\mathrm{DLIN}}(\mathcal{B})$.

Game$_4$: This game is like Game$_3$ but $\mathcal{B}$ halts and outputs a random bit if $\mathcal{A}$ outputs a proof $\pi^\star$ containing a one-time verification key $\mathsf{VK}^\star$ that is recycled from an output of the $S_2(\psi, \tau_{sim}, ., .)$ oracle. Game$_4$ and Game$_3$ proceed identically until the latter event occurs. This event further contradicts the strong unforgeability of Σ. If Σ has tight multi-key security[3] (in the sense of [38]), the probability of this event can be bounded independently of the number q of queries to $S_2(\psi, \tau_{sim}, ., .)$. We have $|\Pr[S_4] - \Pr[S_3]| \leq \mathbf{Adv}_{\mathcal{B}}^{q\text{-suf-ots}}(\lambda)$.

Game$_5$: This game is identical to Game$_4$ but we raise a failure event E_5. When $\mathcal{A}$ outputs its fake proof $\pi^\star = (\mathsf{VK}^\star, \boldsymbol{F}^\star, \boldsymbol{C}_{\sigma_1}^\star, \sigma_2^\star, \sigma_3^\star, \boldsymbol{C}_Z^\star, \boldsymbol{C}_R^\star, \boldsymbol{C}_U^\star, \boldsymbol{C}_z^\star, \boldsymbol{C}_r^\star, \boldsymbol{C}_u^\star, \pi_{\sigma,1}^\star, \pi_{\sigma,2}^\star, \pi_1^\star, \pi_2^\star, \sigma^\star)$, $\mathcal{B}$ parses the vector $\boldsymbol{F}^\star$ as $(F_1^\star, F_2^\star, F_3^\star) \in \mathbb{G}^3$ and uses the extraction trapdoor $(y_1, y_2) = (\log_g(f_1), \log_g(f_2))$ of the Groth-Sahai CRS $\mathbf{f} = (\boldsymbol{f}_1, \boldsymbol{f}_2, \boldsymbol{f}_3)$ to test if $F_3^\star \neq F_1^{\star 1/y_1} \cdot F_2^{\star 1/y_2}$, meaning that $\boldsymbol{F}^\star = (\boldsymbol{f}_1, \boldsymbol{f}_2, \boldsymbol{F}^\star)$ is not a perfectly binding Groth-Sahai CRS. We denote by E_5 the latter event, which causes $\mathcal{B}$ to abort and output a random bit

---

[3] This notion (see Definition 4 in [38]) is defined via a game where the adversary is given q verification keys $\{\mathsf{VK}_i\}_{i=1}^q$ and an oracle that returns exactly one signature for each key. The adversary's tasks is to output a triple $(i^\star, M^\star, \sigma^\star)$, where $i^\star \in \{1, \ldots, q\}$ and $(M^\star, \sigma^\star)$ was not produced by the signing oracle for $\mathsf{VK}_{i^\star}$. Hofheinz and Jager [38, Section 4.2] gave a discrete-log-based one-time signature with tight security in the multi-key setting.

if it occurs. Clearly, Game_5 is identical to Game_4 unless E_5 occurs, so that $|\Pr[S_5] - \Pr[S_4]| \le \Pr[E_5]$. Lemma 1 demonstrates that event E_5 occurs with negligible probability if the DLIN assumption holds. More precisely, the probability $\Pr[E_5]$ is at most $\Pr[E_5] \le (2 \cdot L + 1) \cdot \mathbf{Adv}_{\mathcal{B}}^{\mathrm{DLIN}}(\lambda) + 2/p$, where $\mathcal{B}$ is a DLIN distinguisher whose computational complexity only exceeds that of $\mathcal{A}$ by the cost of a polynomial number of exponentiations in $\mathbb{G}$ and a constant number of pairing evaluations.

In Game_5, we have $\Pr[S_5] = \Pr[S_5 \wedge E_5] + \Pr[S_5 \wedge \neg E_5] = \frac{1}{2} \cdot \Pr[E_5] + \Pr[S_5 \wedge \neg E_5]$, so that $\Pr[S_5] \le (L+1) \cdot \mathbf{Adv}_{\mathcal{B}}^{\mathrm{DLIN}}(\lambda) + \frac{1}{p} + \Pr[S_5 \wedge \neg E_5]$.

In Game_5, we show that event $S_5 \wedge \neg E_5$ implies an algorithm $\mathcal{B}$ solving a given SDP instance (g_z, g_r, h_z, h_u), which also contradicts the DLIN assumption.

Assuming that event $S_5 \wedge \neg E_5$ indeed occurs, we know that the adversary $\mathcal{A}$ manages to output a correct proof $\pi^\star = (\mathsf{VK}^\star, \boldsymbol{F}^\star, \boldsymbol{C}_{\sigma_1}^\star, \sigma_2^\star, \sigma_3^\star, \boldsymbol{C}_Z^\star, \boldsymbol{C}_R^\star, \boldsymbol{C}_U^\star, \boldsymbol{C}_z^\star,$ $\boldsymbol{C}_r^\star, \boldsymbol{C}_u^\star, \boldsymbol{\pi}_{\sigma,1}^\star, \boldsymbol{\pi}_{\sigma,2}^\star, \boldsymbol{\pi}_1^\star, \boldsymbol{\pi}_2^\star, \sigma^\star)$ for a vector $\boldsymbol{v}^\star = (v_1^\star, \ldots, v_n^\star)$ outside the row space of $\boldsymbol{\rho} = g^{\mathbf{A}}$ and such that $\boldsymbol{F}^\star = (F_1^\star, F_2^\star, F_3^\star)$ is a BBS encryption of $1_{\mathbb{G}}$ (namely, $F_3^\star = F_1^{\star 1/y_1} \cdot F_2^{\star 1/y_1}$). This means that, although the simulated proofs produced by $\mathsf{S}_2(\psi, \tau_{sim}, ., .)$ were all generated for a perfectly NIWI Groth-Sahai CRS $\boldsymbol{F} = (\boldsymbol{f}_1, \boldsymbol{f}_2, \boldsymbol{F})$, the last part $(\boldsymbol{C}_z^\star, \boldsymbol{C}_r^\star, \boldsymbol{C}_u^\star, \boldsymbol{\pi}_1^\star, \boldsymbol{\pi}_2^\star)$ of $\mathcal{A}$'s proof $\pi^\star$ takes place on a perfectly binding CRS $\boldsymbol{F}^\star = (\boldsymbol{f}_1, \boldsymbol{f}_2, \boldsymbol{F}^\star)$. Moreover, although $\mathcal{B}$ does not know $\mu_1^\star, \mu_2^\star \in \mathbb{Z}_p$ such that $\boldsymbol{F}^\star = \boldsymbol{f}_1^{\mu_1^\star} \cdot \boldsymbol{f}_2^{\mu_2^\star}$, $\mathcal{B}$ can still use the extraction trapdoor $(y_1, y_2) = (\log_g(f_1), \log_g(f_2))$ to recover $(z^\star, r^\star, u^\star)$ from $(\boldsymbol{C}_z^\star, \boldsymbol{C}_r^\star, \boldsymbol{C}_u^\star)$ by performing BBS decryptions. Indeed, $\boldsymbol{C}_z^\star = \iota(z^\star) \cdot \boldsymbol{f}_1^{\theta_{z,1}} \cdot \boldsymbol{f}_2^{\theta_{z,2}} \cdot \boldsymbol{F}^{\star \theta_{z,3}}$ is of the form $\boldsymbol{C}_z^\star = \iota(z^\star) \cdot \boldsymbol{f}_1^{\theta_{z,1} + \mu_1^\star \cdot \theta_{z,3}} \cdot \boldsymbol{f}_2^{\theta_{z,2} + \mu_2^\star \cdot \theta_{z,3}}$, which decrypts to $z^\star$.

The perfect soundness of the Groth-Sahai CRS $\boldsymbol{F}^\star = (\boldsymbol{f}_1, \boldsymbol{f}_2, \boldsymbol{F}^\star)$ ensures that extracted group elements $(z^\star, r^\star, u^\star)$ satisfy the pairing product equations

$$e(g_z, z^\star) \cdot e(g_r, r^\star) \cdot \prod_{i=1}^{n} e(g_i, v_i^\star) = e(h_z, z^\star) \cdot e(h_u, u^\star) \cdot \prod_{i=1}^{n} e(h_i, v_i^\star) = 1_{\mathbb{G}_T}. \quad (8)$$

In addition, $\mathcal{B}$ computes $(z^\dagger, r^\dagger, u^\dagger) = \left(\prod_{i=1}^{n} v_i^{\star -\chi_i}, \prod_{i=1}^{n} v_i^{\star -\gamma_i}, \prod_{i=1}^{n} v_i^{\star -\delta_i} \right)$, which also satisfies the equations (8). Since $(z^\dagger, r^\dagger, u^\dagger)$ and $(z^\star, r^\star, u^\star)$ both satisfy (8), the triple $(z^\ddagger, r^\ddagger, u^\ddagger) = \left(\frac{z^\star}{z^\dagger}, \frac{r^\star}{r^\dagger}, \frac{u^\star}{u^\dagger} \right)$ necessarily satisfies the equalities $e(g_z, z^\ddagger) \cdot e(g_r, r^\ddagger) = e(h_z, z^\ddagger) \cdot e(h_u, u^\ddagger) = 1_{\mathbb{G}_T}$. We argue that $z^\ddagger \neq 1_{\mathbb{G}}$ with probability $1 - 1/p$, so that $(z^\ddagger, r^\ddagger, u^\ddagger)$ breaks the SDP assumption.

To see this, we remark that, if event $S_5 \wedge \neg E_5$ actually happens, $\mathcal{B}$ never reveals any information about $(\chi_1, \ldots, \chi_n)$ when it emulates $\mathsf{S}_2(\psi, \tau_{sim}, ., .)$. Indeed, in simulated proofs, the only components that depend on $(\chi_1, \ldots, \chi_n)$ are $(\boldsymbol{C}_z, \boldsymbol{C}_r, \boldsymbol{C}_u, \boldsymbol{\pi}_1, \boldsymbol{\pi}_2)$, which are generated for a perfectly NIWI Groth-Sahai CRS $(\boldsymbol{f}_1, \boldsymbol{f}_2, \boldsymbol{F})$. Consequently, the same arguments as in [48, Theorem 1] show that $z^\dagger \neq z^\star$ with probability $1 - 1/p$. In the CRS, $\{(g_i, h_i)\}_{i=1}^{n}$ and $\{(z_i, r_i, u_i)\}_{i=1}^{t}$ provide $\mathcal{A}$ with a linear system of $2n + t < 3n$ equations in $3n$ unknowns $\{(\chi_i, \gamma_i, \delta_i)\}_{i=1}^{n}$, which leaves $z^\dagger$ completely undetermined in $\mathcal{A}$'s view if $\boldsymbol{v}^\star$ is linearly independent of the rows of $\boldsymbol{\rho} = (G_{i,j})_{i,j}$. We thus find $\Pr[S_5 \wedge \neg E_5] \le \mathbf{Adv}_{\mathcal{B}}^{\mathrm{SDP}}(\lambda) + 1/p$, which yields the bound (7) since $\mathbf{Adv}_{\mathcal{B}}^{\mathrm{SDP}}(\lambda) \le \frac{1}{2} \cdot \mathbf{Adv}_{\mathcal{B}}^{\mathrm{DLIN}}(\lambda)$ if we translate the SDP solver $\mathcal{B}$ into a DLIN distinguisher. $\qquad \square$

The result easily extends to the multi-CRS setting via the following changes. In the transitions from Game_1 to Game_2 and Game_2 to Game_3, we can simultaneously modify all CRSes $\{\psi^{(\kappa)}\}_{\kappa=1}^{\mu}$ by using the random self-reducibility of DLIN to build μ instances of the DLIN assumption from a given instance. In Game_5, the probability $\Pr[E_5]$ can be bounded by implicitly relying on the multi-user security (in the sense of [33]) of the signature scheme of [50], which remains almost tight in the multi-key setting. In the proof of the following lemma, we will explain at each step how the proof can be adapted to the multi-CRS setting. Finally, the probability of event $S_5 \wedge \neg E_5$ in Game_5 can be proved by applying the same arguments as in the proof (see [50, AppendixG]) that the signature of [50] provides tight security in the multi-user setting.

Lemma 1. *In* Game_5*, there is a DLIN distinguisher* $\mathcal{B}$ *such that the probability* $\Pr[E_5]$ *is at most* $\Pr[E_5] \leq (2{\cdot}L{+}1){\cdot}\mathbf{Adv}_{\mathcal{B}}^{\mathrm{DLIN}}(\lambda){+}2/p$*. Moreover,* $\mathcal{B}$*'s complexity only exceeds that of* $\mathcal{A}$ *by a polynomial number of exponentiations and a constant number of pairing computations.* (The proof is given in the full version).

5   Applications to Tightly Secure Primitives

As an application of our QA-NIZK proof system, we present a new encryption scheme whose IND-CCA2 security in the multi-challenge-multi-user setting (almost) tightly relates to the DLIN assumption. We show that the resulting construction allows improving the expansion rate of non-interactive universally composable commitments based on IND-CCA2-secure public-key encryption.

5.1   CCA2-Secure (Threshold) Encryption with Shorter Ciphertexts

Like [38,50], our scheme builds on the Naor-Yung paradigm [54] and the encryption scheme of Boneh, Boyen and Shacham (BBS) [16].

The encryption phase computes $(C_0, C_1, C_2) = (M \cdot g^{\theta_1 + \theta_2}, X_1^{\theta_1}, Y_1^{\theta_2})$ and $(D_0, D_1, D_3) = (M \cdot g^{\theta_3 + \theta_4}, X_2^{\theta_3}, Y_2^{\theta_4})$, where (X_1, Y_1, X_2, Y_2) are part of the public key, and generates a QA-NIZK proof π that the vector

$$\boldsymbol{v} = \left(C_1/D_1, C_2/D_2, C_0/D_0, C_1 \cdot C_2, D_1^{-1} \cdot D_2^{-1}\right) \in \mathbb{G}^5$$
$$= \left(X_1^{\theta_1} \cdot X_2^{-\theta_3}, \; Y_1^{\theta_2} \cdot Y_2^{-\theta_4}, \; g^{(\theta_1+\theta_2)-(\theta_3+\theta_4)}, \; X_1^{\theta_1} \cdot Y_1^{\theta_2}, X_2^{-\theta_3} \cdot Y_2^{-\theta_4}\right)$$

is in the subspace spanned by $\boldsymbol{X}_1 = (X_1, 1, g, X_1, 1)$, $\boldsymbol{Y}_1 = (1, Y_1, g, Y_1, 1)$, $\boldsymbol{X}_2 = (X_2, 1, g, 1, X_2)$ and $\boldsymbol{Y}_2 = (1, X_2, g, 1, X_2)$. As in [50], our reduction is not quite as tight as in [5,38] since a factor $\Theta(\lambda)$ is lost. On the other hand, our scheme becomes nearly practical as the ciphertext overhead now decreases to 48 group elements. In comparison, the solution of Libert *et al.* [50] incurs 69 group elements per ciphertext. Our technique thus improves upon [50] by 30 % and also outperforms the most efficient perfectly tight solution [5], which entails over 300 group elements per ciphertext.

The CRS of the proof system is included in the user's public key rather than in the common public parameters since, in the QA-NIZK setting, it depends on the considered language which is defined by certain public key components.

Par-Gen(λ): Run the K_0 algorithm of Sect. 3 in order to obtain common public parameters $\Gamma = ((\mathbb{G}, \mathbb{G}_T), f, g, h, \Sigma)$.

Keygen(Γ): Parse Γ as $((\mathbb{G}, \mathbb{G}_T), f, g, h, \Sigma)$ and conduct the following steps.

1. Choose random exponents $x_1, x_2, y_1, y_2 \overset{R}{\leftarrow} \mathbb{Z}_p$ and define $X_1 = g^{x_1}$, $X_2 = g^{x_2}$, $Y_1 = g^{y_1}$, $Y_2 = g^{y_2}$. Then, define the independent vectors $\boldsymbol{X}_1 = (X_1, 1, g, X_1, 1)$, $\boldsymbol{Y}_1 = (1, Y_1, g, Y_1, 1)$, $\boldsymbol{X}_2 = (X_2, 1, g, 1, X_2)$ and $\boldsymbol{Y}_2 = (1, X_2, g, 1, X_2)$.

2. Run algorithm $\mathsf{K}_1(\Gamma, \rho)$ of Sect. 3 to generate the language-dependent part of the CRS for the proof system, where the rows of the matrix $\rho \in \mathbb{G}^{4 \times 5}$ consist of $\boldsymbol{X}_1$, $\boldsymbol{Y}_1$, $\boldsymbol{X}_2$ and $\boldsymbol{Y}_2$. Let $\psi = (\mathbf{CRS}_1, \mathbf{CRS}_2)$ be the obtained CRS, where

$$\mathbf{CRS}_1 = \left(\rho, \boldsymbol{f}, \boldsymbol{f}_0, \{u_i\}_{i=1}^2, \{\Omega_i\}_{i=1}^2, \boldsymbol{V}, \boldsymbol{W}, \right.$$
$$\left. \{\mathsf{pk}_i\}_{i=1}^2, \{(z_i, r_i, u_i)\}_{i=1}^4, \{(Z_j, R_j, U_j)\}_{j=1}^{4L+5} \right),$$

$$\mathbf{CRS}_2 = \left(\boldsymbol{f}, \boldsymbol{f}_0, \{\mathsf{pk}_i\}_{i=1}^2, \{\Omega_i\}_{i=1}^2, \boldsymbol{V}, \boldsymbol{W} \right).$$

3. Define the private key as the pair $SK = (x_1, y_1) \in \mathbb{Z}_p^4$. The public key is

$$PK = \left(g, \boldsymbol{X}_1, \boldsymbol{Y}_1, \boldsymbol{X}_2, \boldsymbol{Y}_2, \psi = (\mathbf{CRS}_1, \mathbf{CRS}_2) \right).$$

Encrypt(M, PK): to encrypt $M \in \mathbb{G}$, conduct the following steps.

1. Pick random exponents $\theta_1, \theta_2, \theta_3, \theta_4 \overset{R}{\leftarrow} \mathbb{Z}_p$ and compute

$$(C_0, C_1, C_2) = (M \cdot g^{\theta_1 + \theta_2}, X_1^{\theta_1}, Y_1^{\theta_2})$$
$$(D_0, D_1, D_3) = (M \cdot g^{\theta_3 + \theta_4}, X_2^{\theta_3}, Y_2^{\theta_4}).$$

2. Define $\mathsf{lbl} = (C_0, C_1, C_2, D_0, D_1, D_2)$. Using the witness $\boldsymbol{x} = (\theta_1, \theta_2, -\theta_3, -\theta_4) \in \mathbb{Z}_p^4$ and the label lbl, run Steps 1–7 of Algorithm P in Sect. 3 to generate a proof π that the vector

$$\boldsymbol{v} = (C_1/D_1, C_2/D_2, C_0/D_0, C_1 \cdot C_2, D_1^{-1} \cdot D_2^{-1}) \in \mathbb{G}^5$$
$$= (X_1^{\theta_1} \cdot X_2^{-\theta_3}, Y_1^{\theta_2} \cdot Y_2^{-\theta_4}, g^{(\theta_1+\theta_2)-(\theta_3+\theta_4)}, X_1^{\theta_1} \cdot Y_1^{\theta_2}, X_2^{-\theta_3} \cdot Y_2^{-\theta_4})$$

belongs to $\mathsf{span}\langle \boldsymbol{X}_1, \boldsymbol{Y}_1, \boldsymbol{X}_2, \boldsymbol{Y}_2 \rangle$. The QA-NIZK proof is

$$\pi = \left(\mathsf{VK}, \boldsymbol{F}, \boldsymbol{C}_{\sigma_1}, \sigma_2, \sigma_3, \boldsymbol{C}_Z, \boldsymbol{C}_R, \boldsymbol{C}_U, \boldsymbol{C}_z, \boldsymbol{C}_r, \boldsymbol{C}_u, \boldsymbol{\pi}_{\sigma,1}, \boldsymbol{\pi}_{\sigma,2}, \boldsymbol{\pi}_1, \boldsymbol{\pi}_2, \sigma \right).$$

3. Output the ciphertext $C = (C_0, C_1, C_2, D_0, D_1, D_2, \pi)$.

Decrypt(SK, C): given $C = (C_0, C_1, C_2, D_0, D_1, D_2, \pi)$, do the following.

1. Run the verification algorithm V of Sect. 3 on input of $\mathsf{lbl} = (C_0, C_1, C_2, D_0, D_1, D_2)$, the vector $\boldsymbol{v} = (C_1/D_1, C_2/D_2, C_0/D_0, C_1 \cdot C_2, D_1^{-1} \cdot D_2^{-1})$ and π. Return $\perp$ if π is not a valid proof for the label lbl that $\boldsymbol{v}$ is in $\mathsf{span}\langle \boldsymbol{X}_1, \boldsymbol{Y}_1, \boldsymbol{X}_2, \boldsymbol{Y}_2 \rangle$.

2. Using $SK = (x_1, y_1) \in \mathbb{Z}_p^2$, compute and return $M = C_0 \cdot C_1^{-1/x_1} \cdot C_2^{-1/y_1}$.

Using our proof system of Sect. 3 and the one-time signature of [38], the ciphertext size amounts to that of 48 group elements, instead of 69 in [50].

While our construction is described in terms of symmetric pairings in order to lighten notations as much as possible, it readily extends to asymmetric pairings.

Theorem 4. *The scheme is* $(1, q_e)$*-IND-CCA secure provided: (i)* Σ *is a strongly unforgeable one-time signature; (ii) The DLIN assumption holds in* $\mathbb{G}$*. For any adversary* $\mathcal{A}$*, there exist a one-time signature forger* $\mathcal{B}'$ *and a DLIN distinguisher* $\mathcal{B}$ *with running times* $t_\mathcal{B}, t_{\mathcal{B}'} \leq t_\mathcal{A} + q_e \cdot \mathsf{poly}(\lambda, L)$ *such that*

$$\mathbf{Adv}_\mathcal{A}^{(1,q_e)\text{-}cca}(\lambda) \leq \mathbf{Adv}_{\mathcal{B}'}^{q_e\text{-}\mathsf{suf}\text{-}\mathsf{ots}}(\lambda) + (3L + 10) \cdot \mathbf{Adv}_\mathcal{B}^{\mathrm{DLIN}}(\lambda) + 8/p,$$

where L *is the length of one-time verification keys and* q_e *is the number of encryption queries.* (The proof is given in the full version of the paper.)

The result of Theorem 4 carries over to a scenario involving $\mu > 1$ public keys modulo an additional negligible term μ/p in the bound which is inherited from [38, Theorem 6]. This is achieved by relying on the enhanced USS property of the QA-NIZK proof system in the multi-CRS setting.

Similarly to previous IND-CCA2-secure encryption schemes based on the Naor-Yung paradigm (e.g., [32]), the public verifiability of ciphertexts makes our scheme amenable for non-interactive threshold decryption in a static corruption model.

By instantiating the construction of Camenisch *et al.* [19] with our QA-NIZK proofs, we similarly obtain more efficient KDM-CCA2-secure systems with tight security, as explained in the full version of the paper.

5.2   Encrypting Long Messages

In some applications, it is useful to encrypt long messages while preserving the feasibility of efficiently proving statements about encrypted values using Groth-Sahai proofs. In this case, the amortized efficiency of our system can be significantly improved. Suppose that we want to encrypt messages $(M_1, \ldots, M_N) \in \mathbb{G}^N$. The technique of Bellare *et al.* [8] allows doing so while making optimal use of encryption exponents. In more details, the public key consists of group elements $\left(g, h, \{(X_{i,1}, Y_{i,1}, X_{i,2}, Y_{i,2})\}_{i=1}^N\right)$, with $(X_{i,1}, Y_{i,1}, X_{i,2}, Y_{i,2}) = (g^{x_{i,1}}, h^{y_{i,1}}, g^{x_{i,2}}, h^{y_{i,2}})$ and the secret key is $\{(x_{i,1}, y_{i,1})\}_{i=1}^N$. The vector is encrypted by choosing $\theta_1, \theta_2, \theta_3, \theta_4 \xleftarrow{R} \mathbb{Z}_p$ and computing

$$C_0 = f^{\theta_1}, \qquad C_0' = h^{\theta_2}, \qquad \{C_i = M_i \cdot X_{i,1}^{\theta_1} \cdot Y_{i,1}^{\theta_2}\}_{i=1}^N,$$
$$D_0 = f^{\theta_3}, \qquad D_0' = h^{\theta_4}, \qquad \{D_i = M_i \cdot X_{i,2}^{\theta_3} \cdot Y_{i,2}^{\theta_4}\}_{i=1}^N,$$

while appending a simulation-sound QA-NIZK argument that the vector

$$
(C_1/D_1, \ldots, C_N/D_N, \overbrace{C_0, \ldots, C_0}^{N \text{ times}},
$$

$$
\overbrace{D_0^{-1}, \ldots, D_0^{-1}}^{N \text{ times}}, \overbrace{C_0', \ldots, C_0'}^{N \text{ times}}, \overbrace{D_0'^{-1}, \ldots, D_0'^{-1}}^{N \text{ times}}) \in \mathbb{G}^{5N}
$$

lives in the $4N$-dimensional linear subspace $\mathrm{span}\langle \boldsymbol{X}_{i,1}, \boldsymbol{X}_{i,2}, \boldsymbol{Y}_{i,1}, \boldsymbol{Y}_{i,2} \rangle_{i=1}^N$, with

$$
\boldsymbol{X}_{i,1} = (\mathbf{1}^{i-1}, X_{i,1}, \mathbf{1}^{N-i}, \mathbf{1}^{i-1}, f, \mathbf{1}^{N-i}, \mathbf{1}^{3N}),
$$
$$
\boldsymbol{X}_{i,2} = (\mathbf{1}^{i-1}, X_{i,2}, \mathbf{1}^{N-i}, \mathbf{1}^{N}, \mathbf{1}^{i-1}, f, \mathbf{1}^{N-i}, \mathbf{1}^{2N}),
$$
$$
\boldsymbol{Y}_{i,1} = (\mathbf{1}^{i-1}, Y_{i,1}, \mathbf{1}^{N-i}, \mathbf{1}^{2N}, \mathbf{1}^{i-1}, h, \mathbf{1}^{N-i}, \mathbf{1}^{N}),
$$
$$
\boldsymbol{Y}_{i,2} = (\mathbf{1}^{i-1}, Y_{i,2}, \mathbf{1}^{N-i}, \mathbf{1}^{3N}, \mathbf{1}^{i-1}, h, \mathbf{1}^{N-i}),
$$

where, for each $i \in \mathbb{N}$, $\mathbf{1}^i$ stands for the i-dimensional vector $(1_{\mathbb{G}}, \ldots, 1_{\mathbb{G}}) \in \mathbb{G}^i$. The entire ciphertext fits within $2N + 46$ group elements, of which only 42 elements are consumed by the QA-NIZK proof.

The tight IND-CCA2 security can be proved in the same way as in Theorem 4. In particular, we rely on the tight IND-CPA security in the multi-challenge setting of a variant of the BBS encryption scheme where messages M are encrypted[4] as $(f^{\theta_1}, h^{\theta_2}, M \cdot X^{\theta_1} \cdot Y^{\theta_2})$.

In Sect. 5.3, we explain how the compatibility of this construction with zero-knowledge proofs comes in handy to build non-interactive and adaptively secure universally composable commitments based on CCA2-secure encryption.

5.3 Application to UC Commitments

Universally composable commitments [20,27] are commitment schemes that provably remain secure when composed with arbitrary other protocols. They are known [20] to require some setup assumption like a common reference string. In some constructions, the CRS can only be used in a single commitment. Back in 2001, Canetti and Fischlin [20] gave re-usable bit commitments based on chosen-ciphertext-secure public-key encryption. In [52], Lindell described a simple and practical re-usable construction which allows committing to strings rather than individual bits. In short, each commitment consists of an IND-CCA2-secure encryption. In order to open a commitment later on, the sender generates an interactive zero-knowledge proof that the ciphertext encrypts the underlying plaintext. In its basic variant, Lindell's commitment only provides security against static adversaries that have to choose whom to corrupt upfront[5]. Subsequently, Fischlin et al. [31] showed that Lindell's commitment can be made

---

[4] The reduction from the DLIN assumption is straightforward and sets up $X = f^\alpha \cdot g^\gamma$, $Y = h^\beta \cdot g^\gamma$. From a given DLIN instance $(f, g, h, f^a, h^b, \eta)$, where $\eta = g^{a+b}$ or $\eta \in_R \mathbb{G}$, the challenge ciphertext is computed as $(C_1, C_2, C_3) = (f^a, h^b, M_\beta \cdot (f^a)^\alpha \cdot (h^b)^\beta \cdot \eta^\gamma)$.

[5] Lindell's commitment can actually be made adaptively secure (modulo a patch [13]), but even its optimized variant [13] remains interactive with 3 rounds of communication during the commitment phase.

adaptively secure in the erasure model by the simple expedient of opening commitments via a NIZK proof (rather than an interactive one) which the sender generates at commitment time before erasing his encryption coins. Jutla and Roy [42] gave an optimization of the latter approach where the use of QA-NIZK proofs allows reducing the size of commitments and openings.

Using our CCA2-secure encryption scheme for long messages, we can build a tightly secure non-interactive universally composable commitment [20,27] that allows committing to long messages with expansion rate 2. In constructions of UC commitments from IND-CCA2-secure encryption (e.g., [20,31,42]), a multi-challenge definition of IND-CCA2 security is usually considered in proofs of UC security. In the erasure model, the non-interactive and adaptively secure variants of Lindell's commitment [31,42] can be optimized using the techniques of [43,49] to achieve a two-fold expansion rate. However, these solutions are not known to provide tight security. At the cost of a CRS of size $\Theta(N)$, the labeled version of our encryption scheme for long messages (where the label L of the ciphertext is simply included in |b|) allows eliminating this limitation. As in [42], the sender can encrypt the message $(M_1, \ldots, M_N)$ he wants to commit to and open the commitment via a QA-NIZK proof that

$$\left(C_1/M_1, \ldots, C_N/M_N, \overbrace{C_0, \ldots, C_0}^{N \text{ times}}, \overbrace{1, \ldots, 1}^{N \text{ times}}, \overbrace{C_0', \ldots, C_0'}^{N \text{ times}}, \overbrace{1, \ldots, 1}^{N \text{ times}}\right) \in \mathbb{G}^{5N}$$

is in $\mathsf{span}\langle \boldsymbol{X}_{i,1}, \boldsymbol{X}_{i,2}, \boldsymbol{Y}_{i,1}, \boldsymbol{Y}_{i,2} \rangle_{i=1}^{N}$. For long messages, this construction thus achieves a two-fold expansion rate. While not as efficient as the recent rate-1 commitments of Garay et al. [34], it retains adaptive security assuming reliable erasures while [34] is only known to be secure against static adversaries.

Acknowledgments. The first author's work was supported by the "Programme Avenir Lyon Saint-Etienne de l'Université de Lyon" in the framework of the programme "Investissements d'Avenir" (ANR-11-IDEX-0007). The second author was supported by the European Research Council (FP7/2007-2013 Grant Agreement no. 339563 CryptoCloud). Part of this work of the fourth author was done while visiting the Simons Institute for Theory of Computing, U.C. Berkeley.

References

1. Abdalla, M., Benhamouda, F., Pointcheval, D.: Disjunctions for hash proof systems: new constructions and applications. In: Oswald, E., Fischlin, M. (eds.) EUROCRYPT 2015. LNCS, vol. 9057, pp. 69–100. Springer, Heidelberg (2015)
2. Abdalla, M., Fouque, P.-A., Lyubashevsky, V., Tibouchi, M.: Tightly-secure signatures from lossy identification schemes. In: Pointcheval, D., Johansson, T. (eds.) EUROCRYPT 2012. LNCS, vol. 7237, pp. 572–590. Springer, Heidelberg (2012)
3. Abe, M., Fuchsbauer, G., Groth, J., Haralambiev, K., Ohkubo, M.: Structure-preserving signatures and commitments to group elements. In: Rabin, T. (ed.) CRYPTO 2010. LNCS, vol. 6223, pp. 209–236. Springer, Heidelberg (2010)
4. Abe, M., Haralambiev, K., Ohkubo, M.: Signing on elements in bilinear groups for modular protocol design. In: Cryptology ePrint Archive: Report 2010/133 (2010)

5. Abe, M., David, B., Kohlweiss, M., Nishimaki, R., Ohkubo, M.: Tagged one-time signatures: tight security and optimal tag size. In: Kurosawa, K., Hanaoka, G. (eds.) PKC 2013. LNCS, vol. 7778, pp. 312–331. Springer, Heidelberg (2013)
6. Bader, C., Hofheinz, D., Jager, T., Kiltz, E., Li, Y.: Tightly-secure authenticated key exchange. In: Dodis, Y., Nielsen, J.B. (eds.) TCC 2015, Part I. LNCS, vol. 9014, pp. 629–658. Springer, Heidelberg (2015)
7. Bellare, M., Boldyreva, A., Micali, S.: Public-key encryption in a multi-user setting: security proofs and improvements. In: Preneel, B. (ed.) EUROCRYPT 2000. LNCS, vol. 1807, p. 259. Springer, Heidelberg (2000)
8. Bellare, M., Boldyreva, A., Kurosawa, K., Staddon, J.: Multi-recipient encryption schemes: how to save on bandwidth and computation without sacrificing security. IEEE Trans. Inf. Theor. 53(11), 3927–3943 (2007)
9. Bellare, M., Rogaway, P.: Random oracles are practical: a paradigm for designing efficient protocols. In: ACM CCS 1993, pp. 62–73. ACM Press (1993)
10. Bellare, M., Rogaway, P.: The exact security of digital signatures - how to sign with RSA and Rabin. In: Maurer, U.M. (ed.) EUROCRYPT 1996. LNCS, vol. 1070, pp. 399–416. Springer, Heidelberg (1996)
11. Bernstein, D.J.: Proving tight security for Rabin-Williams signatures. In: Smart, N.P. (ed.) EUROCRYPT 2008. LNCS, vol. 4965, pp. 70–87. Springer, Heidelberg (2008)
12. Black, J., Rogaway, P., Shrimpton, T.: Encryption scheme security in the presence of key-dependent messages. In: Nyberg, K., Heys, H.M. (eds.) SAC 2002. LNCS, vol. 2595, pp. 62–75. Springer, Heidelberg (2002)
13. Blazy, O., Chevalier, C., Pointcheval, D., Vergnaud, D.: Analysis and improvement of Lindell's UC-secure commitment schemes. In: Jacobson, M., Locasto, M., Mohassel, P., Safavi-Naini, R. (eds.) ACNS 2013. LNCS, vol. 7954, pp. 534–551. Springer, Heidelberg (2013)
14. Blazy, O., Kiltz, E., Pan, J.: (Hierarchical) identity-based encryption from affine message authentication. In: Garay, J.A., Gennaro, R. (eds.) CRYPTO 2014, Part I. LNCS, vol. 8616, pp. 408–425. Springer, Heidelberg (2014)
15. Blum, M., Feldman, P., Micali, S.: Non-interactive zero-knowledge and its applications. In: STOC 1988, pp. 103–112. ACM Press (1988)
16. Boneh, D., Boyen, X., Shacham, H.: Short group signatures. In: Franklin, M. (ed.) CRYPTO 2004. LNCS, vol. 3152, pp. 41–55. Springer, Heidelberg (2004)
17. Boneh, D., Franklin, M.: Identity-based encryption from the Weil pairing. SIAM J. Comput. 32(3), 586–615 (2003). Earlier version in Crypto 2001
18. Boneh, D., Halevi, S., Hamburg, M., Ostrovsky, R.: Circular-secure encryption from decision Diffie-Hellman. In: Wagner, D. (ed.) CRYPTO 2008. LNCS, vol. 5157, pp. 108–125. Springer, Heidelberg (2008)
19. Camenisch, J., Chandran, N., Shoup, V.: A public key encryption scheme secure against key dependent chosen plaintext and adaptive chosen ciphertext attacks. In: Joux, A. (ed.) EUROCRYPT 2009. LNCS, vol. 5479, pp. 351–368. Springer, Heidelberg (2009)
20. Canetti, R., Fischlin, M.: Universally composable commitments. In: Kilian, J. (ed.) CRYPTO 2001. LNCS, vol. 2139, p. 19. Springer, Heidelberg (2001)
21. Canetti, R., Halevi, S., Katz, J.: Chosen-ciphertext security from identity-based encryption. In: Cachin, C., Camenisch, J.L. (eds.) EUROCRYPT 2004. LNCS, vol. 3027, pp. 207–222. Springer, Heidelberg (2004)
22. Cathalo, J., Libert, B., Yung, M.: Group encryption: non-interactive realization in the standard model. In: Matsui, M. (ed.) ASIACRYPT 2009. LNCS, vol. 5912, pp. 179–196. Springer, Heidelberg (2009)

23. Chen, J., Wee, H.: Fully, (almost) tightly secure IBE and dual system groups. In: Canetti, R., Garay, J.A. (eds.) CRYPTO 2013, Part II. LNCS, vol. 8043, pp. 435–460. Springer, Heidelberg (2013)

24. Chevallier-Mames, B.: An efficient CDH-based signature scheme with a tight security reduction. In: Shoup, V. (ed.) CRYPTO 2005. LNCS, vol. 3621, pp. 511–526. Springer, Heidelberg (2005)

25. Coron, J.-S.: On the exact security of full domain hash. In: Bellare, M. (ed.) CRYPTO 2000. LNCS, vol. 1880, p. 229. Springer, Heidelberg (2000)

26. Coron, J.-S.: Optimal security proofs for PSS and other signature schemes. In: Knudsen, L.R. (ed.) EUROCRYPT 2002. LNCS, vol. 2332, p. 272. Springer, Heidelberg (2002)

27. Canetti, R.: Universally composable security: a new paradigm for cryptographic protocols. In: FOCS 2001, pp. 136–145 2001

28. Cramer, R., Shoup, V.: A practical public key cryptosystem provably secure against adaptive chosen ciphertext attack. In: Krawczyk, H. (ed.) CRYPTO 1998. LNCS, vol. 1462, p. 13. Springer, Heidelberg (1998)

29. Escala, A., Groth, J.: Fine-tuning Groth-Sahai proofs. In: Krawczyk, H. (ed.) PKC 2014. LNCS, vol. 8383, pp. 630–649. Springer, Heidelberg (2014)

30. Fiat, A., Shamir, A.: How to prove yourself: practical solutions to identification and signature problems. In: Odlyzko, A.M. (ed.) CRYPTO 1986. LNCS, vol. 263, pp. 186–194. Springer, Heidelberg (1987)

31. Fischlin, M., Libert, B., Manulis, M.: Non-interactive and re-usable universally composable string commitments with adaptive security. In: Lee, D.H., Wang, X. (eds.) ASIACRYPT 2011. LNCS, vol. 7073, pp. 468–485. Springer, Heidelberg (2011)

32. Fouque, P.-A., Pointcheval, D.: Threshold cryptosystems secure against chosen-ciphertext attacks. In: Boyd, C. (ed.) ASIACRYPT 2001. LNCS, vol. 2248, p. 351. Springer, Heidelberg (2001)

33. Galbraith, S., Malone-Lee, J., Smart, N.: Public-key signatures in the multi-user setting. Inf. Process. Lett. **83**(5), 263–266 (2002)

34. Garay, J.A., Ishai, Y., Kumaresan, R., Wee, H.: On the complexity of UC commitments. In: Nguyen, P.Q., Oswald, E. (eds.) EUROCRYPT 2014. LNCS, vol. 8441, pp. 677–694. Springer, Heidelberg (2014)

35. Groth, J.: Simulation-sound NIZK proofs for a practical language and constant size group signatures. In: Lai, X., Chen, K. (eds.) ASIACRYPT 2006. LNCS, vol. 4284, pp. 444–459. Springer, Heidelberg (2006)

36. Groth, J., Sahai, A.: Efficient non-interactive proof systems for bilinear groups. In: Smart, N.P. (ed.) EUROCRYPT 2008. LNCS, vol. 4965, pp. 415–432. Springer, Heidelberg (2008)

37. Hofheinz, D.: All-but-many lossy trapdoor functions. In: Pointcheval, D., Johansson, T. (eds.) EUROCRYPT 2012. LNCS, vol. 7237, pp. 209–227. Springer, Heidelberg (2012)

38. Hofheinz, D., Jager, T.: Tightly secure signatures and public-key encryption. In: Safavi-Naini, R., Canetti, R. (eds.) CRYPTO 2012. LNCS, vol. 7417, pp. 590–607. Springer, Heidelberg (2012)

39. Hofheinz, D., Jager, T., Knapp, E.: Waters signatures with optimal security reduction. In: Fischlin, M., Buchmann, J., Manulis, M. (eds.) PKC 2012. LNCS, vol. 7293, pp. 66–83. Springer, Heidelberg (2012)

40. Hofheinz, D., Koch, J., Striecks, C.: Identity-based encryption with (almost) tight security in the multi-instance, multi-ciphertext setting. In: Katz, J. (ed.) PKC 2015. LNCS, vol. 9020, pp. 799–822. Springer, Heidelberg (2015)

41. Jutla, C., Roy, A.: Relatively-sound NIZKs and password-based key-exchange. In: Fischlin, M., Buchmann, J., Manulis, M. (eds.) PKC 2012. LNCS, vol. 7293, pp. 485–503. Springer, Heidelberg (2012)

42. Jutla, C.S., Roy, A.: Shorter quasi-adaptive NIZK proofs for linear subspaces. In: Sako, K., Sarkar, P. (eds.) ASIACRYPT 2013, Part I. LNCS, vol. 8269, pp. 1–20. Springer, Heidelberg (2013)

43. Jutla, C.S., Roy, A.: Switching lemma for bilinear tests and constant-size NIZK proofs for linear subspaces. In: Garay, J.A., Gennaro, R. (eds.) CRYPTO 2014, Part II. LNCS, vol. 8617, pp. 295–312. Springer, Heidelberg (2014)

44. Kakvi, S.A., Kiltz, E.: Optimal security proofs for full domain hash, revisited. In: Pointcheval, D., Johansson, T. (eds.) EUROCRYPT 2012. LNCS, vol. 7237, pp. 537–553. Springer, Heidelberg (2012)

45. Katz, J., Vaikuntanathan, V.: Round-optimal password-based authenticated key exchange. In: Ishai, Y. (ed.) TCC 2011. LNCS, vol. 6597, pp. 293–310. Springer, Heidelberg (2011)

46. Katz, J., Wang, N.: Efficiency improvements for signature schemes with tight security reductions. In: ACM-CCS 2003, pp. 155–164. ACM Press (2003)

47. Kiltz, E., Wee, H.: Quasi-adaptive NIZK for linear subspaces revisited. In: Oswald, E., Fischlin, M. (eds.) EUROCRYPT 2015. LNCS, vol. 9057, pp. 101–128. Springer, Heidelberg (2015)

48. Libert, B., Peters, T., Joye, M., Yung, M.: Linearly homomorphic structure-preserving signatures and their applications. In: Canetti, R., Garay, J.A. (eds.) CRYPTO 2013, Part II. LNCS, vol. 8043, pp. 289–307. Springer, Heidelberg (2013)

49. Libert, B., Peters, T., Joye, M., Yung, M.: Non-malleability from malleability: simulation-sound quasi-adaptive NIZK proofs and CCA2-secure encryption from homomorphic signatures. In: Nguyen, P.Q., Oswald, E. (eds.) EUROCRYPT 2014. LNCS, vol. 8441, pp. 514–532. Springer, Heidelberg (2014)

50. Libert, B., Joye, M., Yung, M., Peters, T.: Concise multi-challenge CCA-secure encryption and signatures with almost tight security. In: Sarkar, P., Iwata, T. (eds.) ASIACRYPT 2014, Part II. LNCS, vol. 8874, pp. 1–21. Springer, Heidelberg (2014)

51. Libert, B., Yung, M.: Non-interactive CCA-secure threshold cryptosystems with adaptive security: new framework and constructions. In: Cramer, R. (ed.) TCC 2012. LNCS, vol. 7194, pp. 75–93. Springer, Heidelberg (2012)

52. Lindell, Y.: Highly-efficient universally-composable commitments based on the DDH assumption. In: Paterson, K.G. (ed.) EUROCRYPT 2011. LNCS, vol. 6632, pp. 446–466. Springer, Heidelberg (2011)

53. Malkin, T., Teranishi, I., Vahlis, Y., Yung, M.: Signatures resilient to continual leakage on memory and computation. In: Ishai, Y. (ed.) TCC 2011. LNCS, vol. 6597, pp. 89–106. Springer, Heidelberg (2011)

54. Naor, M., Yung, M.: Public-key cryptosystems provably secure against chosen ciphertext attacks. In: STOC 1990. ACM Press (1990)

55. Rackoff, C., Simon, D.R.: Non-interactive zero-knowledge proof of knowledge and chosen ciphertext attack. In: Feigenbaum, J. (ed.) CRYPTO 1991. LNCS, vol. 576, pp. 433–444. Springer, Heidelberg (1992)

56. Sahai, A.: Non-malleable non-interactive zero-knowledge and adaptive chosen-ciphertext security. In: FOCS 1999, pp. 543–553, IEEE Press (1999)
57. Schäge, S.: Tight proofs for signature schemes without random oracles. In: Paterson, K.G. (ed.) EUROCRYPT 2011. LNCS, vol. 6632, pp. 189–206. Springer, Heidelberg (2011)
58. Waters, B.: Efficient Identity-Based Encryption Without Random Oracles. In: Cramer, R. (ed.) EUROCRYPT 2005. LNCS, vol. 3494, pp. 114–127. Springer, Heidelberg (2005)

Multiparty Computation II

A Unified Approach to MPC
with Preprocessing Using OT

Tore Kasper Frederiksen[1], Marcel Keller[2]([✉]),
Emmanuela Orsini[2], and Peter Scholl[2]

[1] Department of Computer Science, Aarhus University, Aarhus, Denmark
jot2re@cs.au.dk
[2] Department of Computer Science, University of Bristol, Bristol, UK
{m.keller,emmanuela.orsini,peter.scholl}@bristol.ac.uk

Abstract. SPDZ, TinyOT and MiniMAC are a family of MPC protocols based on secret sharing with MACs, where a preprocessing stage
produces multiplication triples in a finite field. This work describes new
protocols for generating multiplication triples in fields of characteristic two using OT extensions. Before this work, TinyOT, which works on
binary circuits, was the only protocol in this family using OT extensions.
Previous SPDZ protocols for triples in large finite fields require somewhat homomorphic encryption, which leads to very inefficient runtimes in
practice, while no dedicated preprocessing protocol for MiniMAC (which
operates on vectors of small field elements) was previously known. Since
actively secure OT extensions can be performed very efficiently using
only symmetric primitives, it is highly desirable to base MPC protocols
on these rather than expensive public key primitives. We analyze the
practical efficiency of our protocols, showing that they should all perform favorably compared with previous works; we estimate our protocol
for SPDZ triples in $\mathbb{F}_{2^{40}}$ will perform around 2 orders of magnitude faster
than the best known previous protocol.

Keywords: MPC · SPDZ · TinyOT · MiniMAC · Preprocessing · OT
extension

1 Introduction

Secure multi-party computation (MPC) allows parties to perform computations
on their private inputs, without revealing their inputs to each other. Recently,
there has been much progress in the design of practical MPC protocols that can
be efficiently implemented in the real world. These protocols are based on secret
sharing over a finite field, and they provide security against an active, static
adversary who can corrupt up to $n - 1$ of n parties (dishonest majority).

In the *preprocessing model*, an MPC protocol is divided into two phases: a
preprocessing (or *offline*) phase, which is independent of the parties' inputs and
hence can be performed in advance, and an *online* phase. The preprocessing stage

© International Association for Cryptologic Research 2015
T. Iwata and J.H. Cheon (Eds.): ASIACRYPT 2015, Part I, LNCS 9452, pp. 711–735, 2015.
DOI: 10.1007/978-3-662-48797-6_29

only generates random, correlated data, often in the form of secret shared multiplication triples [2]. The online phase then uses this correlated randomness to perform the actual computation; the reason for this separation is that the online phase can usually be much more efficient than the preprocessing, which results in a lower latency during execution than if the whole computation was done together. This paper builds on the so-called 'MPC with MACs' family of protocols, which use information-theoretic MACs to authenticate secret-shared data, efficiently providing active security in the online phase, starting with the work of Bendlin et al. [4]. We focus on the SPDZ [10], MiniMAC [11] and TinyOT [19] protocols, which we now describe.

The 'SPDZ' protocol of Damgård et al. [8,10] evaluates arithmetic circuits over a finite field of size at least 2^k, where k is a statistical security parameter. All values in the computation are represented using additive secret sharing and with an additive secret sharing of a MAC that is the product of the value and a secret key. The online phase can be essentially performed with only information theoretic techniques and thus is extremely efficient, with throughputs of almost 1 million multiplications per second as reported by Keller et al. [16]. The preprocessing of the triples uses somewhat homomorphic encryption (SHE) to create an initial set of triples, which may have errors due to the faulty distributed decryption procedure used. These are then paired up and a 'sacrificing' procedure is done: one triple is wasted to check the correctness of another. Using SHE requires either expensive zero knowledge proofs or cut-and-choose techniques to achieve active security, which are *much* slower than the online phase – producing a triple in $\mathbb{F}_p$ (for 64-bit prime p) takes around 0.03 s [8], whilst $\mathbb{F}_{2^{40}}$ triples are even more costly due to the algebra of the homomorphic encryption scheme, taking roughly 0.27 s [7].

TinyOT [19] is a two-party protocol for binary circuits based on OT extensions. It has similar efficiency to SPDZ in the online phase but has faster preprocessing, producing around 10000 $\mathbb{F}_2$ triples per second. Larraia et al. [17] extended TinyOT to the multi-party setting and adapted it to fit with the SPDZ online phase. The multi-party TinyOT protocol also checks correctness of triples using sacrificing, and two-party TinyOT uses a similar procedure called combining to remove possible leakage from a triple, but when working in small fields simple pairwise checks are not enough. Instead an expensive 'bucketing' method is used, which gives an overhead of around 3–8 times for each check, depending on the number of triples required and the statistical security parameter.

MiniMAC [11] is another protocol in the SPDZ family, which reduces the size of MACs in the online phase for the case of binary circuits (or arithmetic circuits over small fields). Using SPDZ or multi-party TinyOT requires the MAC on every secret shared value to be at least as big as the statistical security parameter, whereas MiniMAC can authenticate vectors of bits at once combining them into a codeword, allowing the MAC size to be constant. Damgård et al. [9] implemented the online phase of MiniMAC and found it to be faster than TinyOT for performing many operations in parallel, however no dedicated preprocessing protocol for MiniMAC has been published.

1.1 Our Contributions

In this paper we present new, improved protocols for the preprocessing stages of the 'MPC with MACs' family of protocols based on OT extensions, focusing on finite fields of characteristic two. Our main contribution is a new method of creating SPDZ triples in $\mathbb{F}_{2^k}$ using only symmetric primitives, so it is much more efficient than previous protocols using SHE. Our protocol is based on a novel correlated OT extension protocol that increases efficiency by allowing an adversary to introduce errors of a specific form, which may be of independent interest. Additionally, we revisit the multi-party TinyOT protocol by Larraia et al. from CRYPTO 2014 [17], and identify a crucial security flaw that results in a selective failure attack. A standard fix has an efficiency cost of at least 9x, which we show how to reduce to just 3x with a modified protocol. Finally, we give the first dedicated preprocessing protocol for MiniMAC, by building on the same correlated OT that lies at the heart of our SPDZ triple generation protocol.

Table 1 gives the main costs of our protocols in terms of the number of correlated and random OTs required, as well as an estimate of the total time per triple, based on OT extension implementation figures. We include the SPDZ protocol timings based on SHE to give a rough comparison with our new protocol for $\mathbb{F}_{2^{40}}$ triples. For a full explanation of the derivation of our time estimates, see Sect. 7. Our protocol for $\mathbb{F}_{2^{40}}$ triples has the biggest advantage over previous protocols, with an estimated 200x speed-up over the SPDZ implementation. For binary circuits, our multi-party protocol is comparable with the two-party TinyOT protocol and around 3x faster than the fixed protocol of Larraia et al. [17]. For MiniMAC, we give figures for the amortized cost of a single multiplication in $\mathbb{F}_{2^8}$. This seems to incur a slight cost penalty compared with using SPDZ triples and embedding the circuit in $\mathbb{F}_{2^{40}}$, however this is traded off by the more efficient online phase of MiniMAC when computing highly parallel circuits [9].

We now highlight our contributions in detail.

$\mathbb{F}_{2^k}$ **Triples.** We show how to use a new variant of correlated OT extension to create multiplication triples in the field $\mathbb{F}_{2^k}$, where k is at least the statistical security parameter. Note that this finite field allows much more efficient evaluation of AES in MPC than using binary circuits [7], and is also more efficient than $\mathbb{F}_p$ for computing ORAM functionalities for secure computation on RAM programs [15]. Previously, creating big field triples for the SPDZ protocol required using somewhat homomorphic encryption and therefore was very slow (particularly for the binary field case, due to limitations of the underlying SHE plaintext algebra [7]). It seems likely that our OT based protocol can improve the performance of SPDZ triples by 2 orders of magnitude, since OT extensions can be performed very efficiently using just symmetric primitives.

The naive approach to achieving this is to create k^2 triples in $\mathbb{F}_2$, and use these to evaluate the $\mathbb{F}_{2^k}$ multiplication circuit. Each of these $\mathbb{F}_2$ triples would need sacrificing and combining, in total requiring many more than k^2 OT extensions. Instead, our protocol in Sect. 5.1 creates a $\mathbb{F}_{2^k}$ triple using only $O(k)$ OTs. The key insight into our technique lies in the way we look at OT: instead of taking

Table 1. Number of OTs and estimates of time required to create a multiplication triple using our protocols and previous protocols, for n parties. See Sect. 7 for details.

Finite field	Protocol	# Correlated OTs	# Random OTs	Time estimate, ms ($n = 2$)
$\mathbb{F}_2$	2-party TinyOT [5,19]	0	54	0.07
	n-party TinyOT [5,17]	$81n(n-1)$	$27n(n-1)$	0.24
	This work Sect. 5.2	$27n(n-1)$	$9n(n-1)$	0.08
$\mathbb{F}_{2^{40}}$	SPDZ [7]	N/A	N/A	272
	This work Sect. 5.1	$240n(n-1)$	$240n(n-1)$	1.13
$\mathbb{F}_{2^8}$ (MiniMAC)	**This work** Sect. 6	$1020n(n-1)$	$175n(n-1)$	2.63

the traditional view of a sender and a receiver, we use a linear algebra approach with matrices, vectors and tensor products, which pinpoints the precise role of OT in secure computation. A correlated OT is a set of OTs where the sender's messages are all $(x, x + \Delta)$ for some fixed string Δ. We represent a set of k correlated OTs between two parties, with inputs $\mathbf{x}, \mathbf{y} \in \mathbb{F}_2^k$, as:

$$Q + T = \mathbf{x} \otimes \mathbf{y}$$

where $Q, T \in \mathbb{F}_2^{k \times k}$ are the respective outputs to each party. Thus, correlated OT gives precisely a secret sharing of the tensor product of two vectors. From the tensor product it is then straightforward to obtain a $\mathbb{F}_{2^k}$ multiplication of the corresponding field elements by taking the appropriate linear combination of the components.

An actively secure protocol for correlated OT was presented by Nielsen et al. [19], with an overhead of ≈ 7.3 calls to the base OT protocol due to the need for consistency checks and privacy amplification, to avoid any leakage on the secret correlation. In our protocol, we choose to miss out the consistency check, allowing the party creating correlation to input different correlations to each OT. We show that if this party attempts to cheat then the error introduced will be amplified by the privacy amplification step so much that it can always be detected in the pairwise sacrificing check we later perform on the triples. Allowing these errors significantly complicates the analysis and security proofs, but reduces the overhead of the correlated OT protocol down to just 3 times that of a basic OT extension.

$\mathbb{F}_2$ **Triples.** The triple production protocol by Larraia et al. [17] has two main stages: first, unauthenticated shares of triples are created (using the aBit protocol by Nielsen et al. [19] as a black box) and secondly the shares are authenticated, again using aBit, and checked for correctness with a sacrificing procedure. The main problem with this approach is that given shares of an unauthenticated triple for $a, b, c \in \mathbb{F}_2$ where $c = a \cdot b$, the parties may not input their correct shares of this triple into the authentication step. A corrupt party can change their share such that $a + 1$ is authenticated instead of a; if $b = 0$ (with probability $1/2$) then $(a + 1) \cdot b = a \cdot b$, the sacrificing check still passes, and the corrupt party hence learns the value of b.[1]

To combat this problem, an additional *combining* procedure can be done: similarly to sacrificing, a batch of triples are randomly grouped together into buckets and combined, such that as long as one of them is secure, the resulting triple remains secure, as was done by Nielsen et al. [19]. However, combining only removes leakage on either a or b. To remove leakage on both a and b, combining must be done twice, which results in an overhead of at least 9x, depending on the batch size. Note that this fix is described in full in a recent preprint [5], which is a merged and extended version of the two TinyOT papers [17,19].

In Sect. 5.2 we modify the triple generation procedure so that combining only needs to be done once, reducing the overhead on top of the original (insecure) protocol to just 3x (for a large enough batch of triples). Our technique exploits the structure of the OT extension protocol to allow a triple to be created, whilst simultaneously authenticating one of the values a or b, preventing the selective failure attack on the other value. Combining still needs to be performed once to prevent leakage, however.

MiniMAC Triples. The MiniMAC protocol [11] uses multiplication triples of the form $C^*(\mathbf{c}) = C(\mathbf{a}) * C(\mathbf{b})$, where $\mathbf{a}, \mathbf{b} \in \mathbb{F}_{2^u}^k$ and C is a systematic, linear code over $\mathbb{F}_{2^u}$, for 'small' u (e.g. $\mathbb{F}_2$ or $\mathbb{F}_{2^8}$), $*$ denotes the component-wise vector product and C^* is the product code given by the span of all products of codewords in C. Based on the protocol for correlated OT used for the $\mathbb{F}_{2^k}$ multiplication triples, we present the first dedicated construction of MiniMAC multiplication triples. The major obstacles to overcome are that we must somehow guarantee that the triples produced form valid codewords. This must be ensured both during the triple generation stage and the authentication stage, otherwise another subtle selective failure attack can arise. To do this, we see $\mathbf{a}$ and $\mathbf{b}$ as vectors over $\mathbb{F}_2^{u \cdot k}$ and input these to the same secure correlated OT procedure as used for the $\mathbb{F}_{2^k}$ multiplication triples. From the resulting shared tensor product, we can compute shares of all of the required products in $C(\mathbf{a}) * C(\mathbf{b})$, due to the linearity of the code. For authentication we use the same correlated OT as used for authentication of the $\mathbb{F}_{2^k}$ triples. However, this only allows us to authenticate components in $\mathbb{F}_{2^u}$ one at a time, so we also add a "compression" step to combine individual authentications of each component in $C(\mathbf{x})$ into a single MAC. Finally, the construction is ended with a pairwise sacrificing step.

Furthermore, since the result of multiplication of two codewords results in an element in the Schur transform, we need some more preprocessed material,

---

[1] We stress that this attack only applies to the multi-party protocol from CRYPTO 2014 [17], and not the original two-party protocol of Nielsen et al. [19].

in order to move such an element back down to an "ordinary" codeword. This is done using an authenticated pair of equal elements; one being an ordinary codeword and one in the Schur transform of the code. We also construct these pairs by authenticating the k components in $\mathbb{F}_{2^u}$ and then, using the linearity of the code, computing authenticated shares of the entire codeword. Since this again results in a MAC for each component of the codeword we execute a compression step to combine the MAC's into a single MAC.

Efficient Authentication from Passively Secure OT. All of our protocols are unified by a common method of authenticating shared values using correlated OT extension. Instead of using an actively secure correlated OT extension protocol as was previously done [17,19], we use just a passively secure protocol, which is simply the passive OT extension of Ishai et al. [13], without the hashing at the end of the protocol (which removes the correlation).

This allows corrupt parties to introduce errors on MACs that depend on the secret MAC key, which could result in a few bits of the MAC key being leaked if the MAC check protocol still passes. Essentially, this means that corrupt parties can try to guess subsets of the field in which the MAC key shares lie, but if their guess is incorrect the protocol aborts. We model this ability in all the relevant functionalities,

Overview of our Protocols

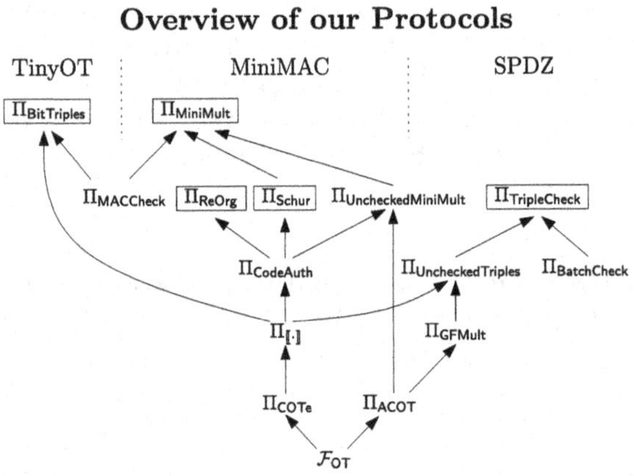

Fig. 1. Illustration of the relationship between our protocols. Protocols in boxes indicate final elements for use in online execution.

showing that the resulting protocols are actively secure, even when this leakage is present.

Security. The security of our protocols is proven in the standard UC framework of Canetti [6] (see the full version for details). We consider security against malicious, static adversaries, i.e. corruption may only take place before the protocols start, corrupting up to $n-1$ of n parties.

Setup Assumption. The security of our protocols is in the $\mathcal{F}_{OT}$-hybrid model, i.e. all parties have access to an ideal 1-out-of-2 OT functionality. Moreover we assume *authenticated* communication between parties, in the form of a functionality $\mathcal{F}_{AT}$ which, on input (m, i, j) from P_i, gives m to P_j and also leaks m

to the adversary. Our security proof for $\mathbb{F}_2$ triples also uses the random oracle (RO) model [3] to model the hash function used in an OT extension protocol. This means that the parties and the adversaries have access to a uniformly random $H : \{0,1\}^* \rightarrow \{0,1\}^\kappa$, such that if it is queried on the same input twice, it returns the same output. We also use a standard coin flipping functionality, $\mathcal{F}_{\mathsf{Rand}}$, which can be efficiently implemented using hash-based commitments in the random oracle model as done previously [8].

Overview. The rest of this paper is organized as follows: In Sect. 2 we go through our general notation, variable naming and how we represent shared values. We continue in Sect. 3 with a description of the passively secure OT extensions we use as building block for our triple generation and authentication. We then go into more details on our authentication procedure in Sect. 4. This is followed by a description of how we generate TinyOT ($\mathbb{F}_2$) and SPDZ ($\mathbb{F}_{2^k}$) triples in Sect. 5 and MiniMAC triples in Sect. 6. We end with a complexity analysis in Sect. 7. Many protocols and proofs are omitted due to space reasons; we refer the reader to the full version for details [12].

We illustrate the relationship between all of our protocols in Fig. 1. In the top we have the protocol producing final triples used in online execution and on the bottom the protocols for correlated OT extension and authentication.

2   Notation

We denote by κ the computational security parameter and s the statistical security parameter. We let $\mathsf{negl}(\kappa)$ denote some unspecified function $f(\kappa)$, such that $f = o(\kappa^{-c})$ for every fixed constant c, saying that such a function is *negligible* in κ. We say that a probability is *overwhelming* in κ if it is $1 - \mathsf{negl}(\kappa)$. We denote by $a \xleftarrow{\$} A$ the random sampling of a from a distribution A, and by $[d]$ the set of integers $\{1, \dots d\}$.

We consider the sets $\{0,1\}$ and $\mathbb{F}_2^\kappa$ endowed with the structure of the fields $\mathbb{F}_2$ and $\mathbb{F}_{2^\kappa}$, respectively. We denote by $\mathbb{F}$ any finite field of characteristic two, and use roman lower case letters to denote elements in $\mathbb{F}$, and bold lower case letters for vectors. We will use the notation $\mathbf{v}[i]$ to denote the i-th entry of $\mathbf{v}$. Sometimes we will use $\mathbf{v}[i; j]$ to denote the range of bits from i to j when viewing $\mathbf{v}$ as a bit vector. Given matrix A, we denote its rows by subindices $\mathbf{a}_i$ and its columns by superindices $\mathbf{a}^j$. If we need to denote a particular entry we use the notation $A[i, j]$. We will use $\mathcal{O}$ to denote the matrix full of ones and $D_\mathbf{x}$ for some vector $\mathbf{x}$ to denote the square matrix whose diagonal is $\mathbf{x}$ and where every other positions is 0.

We use $\cdot$ to denote multiplication of elements in a finite field; note that in this case we often switch between elements in the field $\mathbb{F}_{2^\kappa}$, vectors in $\mathbb{F}_2^\kappa$ and vectors in $\mathbb{F}_{2^u}^{\kappa/u}$ (where $u|\kappa$), but when multiplication is involved we always imply multiplication over the field, or and entry-wise multiplication if the first operand is a scalar. If $\mathbf{a}, \mathbf{b}$ are vectors over $\mathbb{F}$ then $\mathbf{a} * \mathbf{b}$ denotes the component-wise product of the vectors, and $\mathbf{a} \otimes \mathbf{b}$ to denote the matrix containing the tensor (or outer) product of the two vectors.

We consider a systematic linear error correcting code C over finite field $\mathbb{F}_{2^u}$ of length m, dimension k and distance d. So if $\mathbf{a} \in \mathbb{F}_{2^u}^k$, we denote by $C(\mathbf{a}) \in \mathbb{F}_{2^u}^m$ the encoding of $\mathbf{a}$ in C, which contains $\mathbf{a}$ in its first k positions, due to the systematic property of the code. We let C^* denote the product code (or Schur transform) of C, which consists of the linear span of $C(\mathbf{a}) * C(\mathbf{b})$, for all vectors $\mathbf{a}, \mathbf{b} \in \mathbb{F}_{2^u}^k$. If C is a $[m, k, d]$ linear error correcting code then C^* is a $[m, k^*, d^*]$ linear error correcting code for which it holds that $k^* \geq k$ and $d^* \leq d$.

2.1   Authenticating Secret-Shared Values

Let $\mathbb{F}$ be a finite field, we additively secret share bits and elements in $\mathbb{F}$ among a set of parties $\mathcal{P} = \{P_1, \ldots, P_n\}$, and sometimes abuse notation identifying subsets $\mathcal{I} \subseteq \{1, \ldots, n\}$ with the subset of parties indexed by $i \in \mathcal{I}$. We write $\langle a \rangle$ if a is additively secret shared amongst the set of parties, with party P_i holding a value $a^{(i)}$, such that $\sum_{i \in \mathcal{P}} a^{(i)} = a$. We adopt the convention that, if $a \in \mathbb{F}$ then the shares also lie in the same field, i.e. $a^{(i)} \in \mathbb{F}$.

Our main technique for authentication of secret shared values is similar to the one by Larraia et al. [17] and Damgård et al. [10], i.e. we authenticate a secret globally held by a system of parties, by placing an *information theoretic tag* (MAC) on the secret shared value. We will use a fixed global key $\Delta \in \mathbb{F}_{2^M}$, $M \geq \kappa$, which is additively secret shared amongst parties, and we represent an authenticated value $x \in \mathbb{F}$, where $\mathbb{F} = \mathbb{F}_{2^u}$ and $u | M$, as follows:

$$[\![x]\!] = (\langle x \rangle, \langle \mathbf{m}_x \rangle, \langle \Delta \rangle),$$

where $\mathbf{m}_x = x \cdot \Delta$ is the MAC authenticating x under Δ. We drop the dependence on x in $\mathbf{m}_x$ when it is clear from the context. In particular this notation indicates that each party P_i has a share $x^{(i)}$ of $x \in \mathbb{F}$, a share $\mathbf{m}^{(i)} \in \mathbb{F}_2^M$ of the MAC, and a uniform share $\Delta^{(i)}$ of Δ; hence a $[\![\cdot]\!]$-representation of x implies that x is both *authenticated* with the global key Δ and $\langle \cdot \rangle$-shared, i.e. its value is actually unknown to the parties. Looking ahead, we say that $[\![x]\!]$ is *partially open* if $\langle x \rangle$ is opened, i.e. the parties reveal x, but not the shares of the MAC value $\mathbf{m}$. It is straightforward to see that all the linear operations on $[\![\cdot]\!]$ can be performed locally on the $[\![\cdot]\!]$-sharings. We describe the ideal functionality for generating elements in the $[\![\cdot]\!]$-representation in Fig. 4.

In Sect. 6 we will see a generalization of this representation for codewords, i.e. we denote an authenticated codeword $C(\mathbf{x})$ by $[\![C(\mathbf{x})]\!]^* = (\langle C(\mathbf{x}) \rangle, \langle \mathbf{m} \rangle, \langle \Delta \rangle)$, where the $*$ is used to denote that the MAC will be "component-wise" on the codeword $C(\mathbf{x})$, i.e. that $\mathbf{m} = C(\mathbf{x}) * \Delta$.

3   OT Extension Protocols

In this section we describe the OT extensions that we use as building blocks for our triple generation protocols. Two of these are standard – a 1-out-of-2 OT functionality and a passively secure correlated OT functionality – whilst

---

Functionality $\mathcal{F}_{\mathsf{COTe}}^{\kappa,\ell}$

The **Initialize** step is independent of inputs and only needs to be called once. After this, **Extend** can be called multiple times. The functionality is parametrized by the number ℓ of resulting OTs and by the bit length κ.

Running with parties P_S, P_R and an ideal adversary denoted by $\mathcal{S}$, it operates as follows.

Initialize: Upon receiving $\Delta \in \mathbb{F}_2^\kappa$ from P_S, the functionality stores Δ.

Extend(R,S): Upon receiving $(P_R, (\mathbf{x}_1, \ldots, \mathbf{x}_\ell))$ from P_R, where $\mathbf{x}_h \in \mathbb{F}_2^\kappa$, it does the following:

- It samples $\mathbf{t}_h \in \mathbb{F}_2^\kappa$, $h = 1, \ldots, \ell$, for P_R. If P_R is corrupted then it waits for $\mathcal{S}$ to input $\mathbf{t}_h$.
- It computes $\mathbf{q}_h = \mathbf{t}_h + \mathbf{x}_h * \Delta$, $h = 1, \ldots, \ell$, and sends them to P_S. If P_S is corrupted, the functionality waits for $\mathcal{S}$ to input $\mathbf{q}_h$, and then it outputs to P_R values of $\mathbf{t}_h$ consistent with the adversarial inputs.

Fig. 2. IKNP extension functionality $\mathcal{F}_{\mathsf{COTe}}^{\kappa,\ell}$

the third protocol is our variant on passively secure correlated OT with privacy amplification, which may be of independent interest for other uses.

We denote by $\mathcal{F}_{\mathsf{OT}}$ the standard $\binom{2}{1}$ OT functionality, where the *sender* P_S inputs two messages $\mathbf{v}_0, \mathbf{v}_1 \in \mathbb{F}_2^\kappa$ and the receiver inputs a choice bit b, and at the end of the protocol the *receiver* P_R learns only the selected message $\mathbf{v}_b$. We use the notation $\mathcal{F}_{\mathsf{OT}}^{\kappa,\ell}$ to denote the functionality that provides ℓ $\binom{2}{1}$ OTs in $\mathbb{F}_2^\kappa$. Note that $\mathcal{F}_{\mathsf{OT}}^{\kappa,\ell}$ can be implemented very efficiently for any $\ell = \mathsf{poly}(\kappa)$ using just one call to $\mathcal{F}_{\mathsf{OT}}^{\kappa,\kappa}$ and symmetric primitives, for example with actively secure OT extensions [1,14,19].

A slightly different variant of $\mathcal{F}_{\mathsf{OT}}$ is correlated OT, which is a batch of OTs where the sender's messages are correlated, i.e. $\mathbf{v}_0^i + \mathbf{v}_1^i = \Delta$ for some constant Δ, for every pair of messages. We do not use an actively secure correlated OT protocol but a *passively* secure protocol, which is essentially the OT extension of Ishai et al. [13] without the hashing that removes correlation at the end of the protocol. We model this protocol with a functionality that accounts for the deviations an active adversary could make, introducing errors into the output, and call this *correlated OT with errors* (Fig. 2). The implementation of this is exactly the same as the first stage of the IKNP protocol, but for completeness we include the description in the full version. The security was proven e.g. by Nielsen [18], where it was referred to as the ABM box.

3.1 Amplified Correlated OT with Errors

Our main new OT extension protocol is a variant of correlated OT that we call *amplified correlated OT with errors*. To best illustrate our use of the protocol, we find it useful to use the concept of a tensor product to describe it. We observe that performing k correlated OTs on k-bit strings between two parties P_R and P_S gives a symmetric protocol: if the input strings of the two parties are $\mathbf{x}$ and $\mathbf{y}$ then the output is given by

$$Q + T = \mathbf{x} \otimes \mathbf{y}$$

where Q and T are the $k \times k$ matrices over $\mathbb{F}_2$ output to each respective party. Thus we view correlated OT as producing a secret sharing of the tensor product of two input vectors. The matrix $\mathbf{x} \otimes \mathbf{y}$ consists of every possible bit product between bits in $\mathbf{x}$ held by P_R and bits in $\mathbf{y}$ held by P_S. We will later use this to compute a secret sharing of the product in an extension field of $\mathbb{F}_2$.

The main difficulty in implementing this with active security is ensuring that a corrupt P_R inputs the same correlation into each OT: if they cheat in just one OT, for example, they can guess P_S's corresponding input bit, resulting in a selective failure attack in a wider protocol. The previous construction used in the TinyOT protocol [19] first employed a consistency check to ensure that P_R used the same correlation on most of the inputs. Since the consistency check cannot completely eliminate cheating, a *privacy amplification* step is then used, which multiplies all of the OTs by a random binary matrix to remove any potential leakage on the sender's input from the few, possibly incorrect OTs.

In our protocol, we choose to omit the consistency check, since the correctness of SPDZ multiplication triples is later checked in the sacrificing procedure. This means that an adversary is able to break the correlation, but the output will be distorted in a way such that sacrificing will fail for all but one possible $\mathbf{x}$ input by P_R. Without amplification, the adversary could craft a situation where the latter check succeeds if, for example, first bit is zero, allowing the selective failure attack. On the other hand, if the success of the adversary depends on guessing k random bits, the probability of a privacy breach is 2^{-k}, which is negligible in k. In the functionality $\mathcal{F}_{\mathsf{ACOT}}^{k,s}$ (see the full version), the amplification manifests itself in the fact that the environment does not learn $\mathbf{x}'$ which amplifies the error Y'.

The protocol $\Pi_{\mathsf{ACOT}}^{k,s}$ (Fig. 3) requires parties to create the initial correlated OTs on strings of length $\ell' = 2k+s$, where s is the statistical security parameter. The sender P_S is then allowed to input a $\ell' \times k$ matrix Y instead of a vector $\mathbf{y}$, whilst the receiver chooses a random string $\mathbf{x}' \in \mathbb{F}_2^{\ell'}$. $\mathcal{F}_{\mathsf{OT}}$ then produces a sharing of $D_{\mathbf{x}'}Y$, instead of $\mathbf{x}' \otimes \mathbf{y}$ in the honest case. For the privacy amplification, a random $k \times \ell'$ binary matrix M is chosen, and everything is multiplied by this to give outputs of length k as required. Finally, P_R sends $M\mathbf{x}' + \mathbf{x}$ to switch to their real input $\mathbf{x}$. Multiplying by M ensures that even if P_S learns a few bits of $\mathbf{x}'$, all of $\mathbf{x}$ remains secure as every bit of $\mathbf{x}'$ is combined into every bit of the output.

Lemma 1. *The protocol $\Pi_{\mathsf{ACOT}}^{k,s}$ (Fig. 3) implements the functionality $\mathcal{F}_{\mathsf{ACOT}}^{k,s}$ (see the full version) in the $\mathcal{F}_{\mathsf{OT}}^{k,\ell'}$-hybrid model with statistical security s.*

Proof. The proof essentially involves checking that $Q + T = \mathbf{x} \otimes \mathbf{y}$ for honest parties, that at most k deviations by P_S are canceled by M with overwhelming

---

Protocol $\Pi_{\mathsf{ACOT}}^{k,s}$

Let $\mathbf{x} \in \mathbb{F}_2^k$ and $\mathbf{y} \in \mathbb{F}_2^k$ denote the inputs of P_R and P_S, respectively. Let $\ell' := 2k+s$.

1. Parties run $\mathcal{F}_{\mathsf{OT}}^{k,\ell'}$:
   (a) P_S samples $Q' \xleftarrow{\$} \mathbb{F}_2^{\ell' \times k}$, sets $Y = \mathcal{O}D_{\mathbf{y}}$ where $\mathcal{O} \in \mathbb{F}_2^{\ell' \times k}$ is the matrix full of ones and inputs $(Q', Q' + Y)$.
   (b) P_R samples and inputs $\mathbf{x}' \xleftarrow{\$} \mathbb{F}_2^{\ell'}$.
   (c) P_R receives $T' = Q' + D_{\mathbf{x}'}Y$.
2. Parties sample a random matrix $M \in \mathbb{F}_2^{k \times \ell'}$ using $\mathcal{F}_{\mathsf{Rand}}$ (see full version).
3. P_R sends $\delta = M\mathbf{x}' + \mathbf{x}$ to P_S and outputs $T = MT'$.
4. P_S outputs $Q = MQ' + \delta \otimes \mathbf{y}$.

Fig. 3. Amplified correlated OT

probability, and that more than k deviations cause the desired entropy in the output. The two cases are modeled by two different possible adversarial inputs to the functionality. See the full version for further details.

4 Authentication Protocol

In this section we describe our protocol to authenticate secret shared values over characteristic two finite fields, using correlated OT extension. The resulting MACs, and the relative MAC keys, are always elements of a finite field $\mathbb{F} := \mathbb{F}_{2^M}$, where $M \geq \kappa$ and κ is a computational security parameter, whilst the secret values may lie in $\mathbb{F}_{2^u}$ for any $u|M$. We then view the global MAC key as an element of $\mathbb{F}_{2^u}^{M/u}$ and the MAC multiplicative relation as componentwise multiplication in this ring. Our authentication method is similar to that by Larraia et al. [17] (with modifications to avoid the selective failure attack) but here we only use a passively secure correlated OT functionality ($\mathcal{F}_{\mathsf{COTe}}$), allowing an adversary to introduce errors in the MACs that depend on arbitrary bits of other parties' MAC key shares. When combined with the MAC check protocol by Damgård et al. [8] (see full version), this turns out to be sufficient for our purposes, avoiding the need for additional consistency checks in the OTs.

Our authentication protocol $\Pi_{[\![\cdot]\!]}$ (see the full version) begins with an Initialize stage, which initializes a $\mathcal{F}_{\mathsf{COTe}}$ instance between every pair of parties (P_i, P_j), where P_j inputs their MAC key share $\Delta^{(j)}$. This introduces the subtle issue that a corrupt P_j may initialize $\mathcal{F}_{\mathsf{COTe}}$ with two different MAC shares for P_{i_1} and P_{i_2}, say $\Delta^{(j)}$ and $\hat{\Delta}^{(j)}$, which allows for the selective failure attack mentioned earlier – if P_{i_2} authenticates a bit b, the MAC check will still pass if $b = 0$, despite being authenticated under the wrong key. However, since $\mathcal{F}_{\mathsf{COTe}}$.Initialize is only called once, the MAC key shares are fixed for the entire protocol, so it is clear that P_j could not remain undetected if enough random values are authenticated and checked. To ensure this in our protocol we add a consistency check to the Initialize

---

Functionality $\mathcal{F}^{\mathbb{F}}_{[\![\cdot]\!]}$

Let $\mathbb{F} = \mathbb{F}_{2^M}$, with $M \geq \kappa$. Let A be the indices of corrupt parties. Running with parties $P_1, \ldots, P_n$ and an ideal adversary $\mathcal{S}$, the functionality authenticates values in $\mathbb{F}_{2^u}$ for $u | M$.

Initialize: On input (Init) the functionality activates and waits for the adversary to input a set of shares $\{\Delta^{(j)}\}_{j \in A}$ in $\mathbb{F}$. It samples random $\{\Delta^{(i)}\}_{i \notin A}$ in $\mathbb{F}$ for the honest parties, defining $\Delta := \sum_{i \in [n]} \Delta^{(i)}$. If any $j \in A$ outputs Abort then the functionality aborts.

n-**Share:** On input (Authenticate, $\mathbf{x}_1^{(i)}, \ldots, \mathbf{x}_\ell^{(i)}$) from the honest parties and the adversary where $\mathbf{x}_h^{(i)} \in \mathbb{F}_{2^u}$, the functionality proceeds as follows.

Honest parties: $\forall h \in [\ell]$, it computes $\mathbf{x}_h = \sum_{i \in \mathcal{P}} \mathbf{x}_h^{(i)}$ and $\mathbf{m}_h = \mathbf{x}_h \cdot \Delta$. [a] Then it creates a sharing $\langle \mathbf{m}_h \rangle = \{\mathbf{m}_h^{(1)}, \ldots, \mathbf{m}_h^{(n)}\}$ and outputs $\mathbf{m}_h^{(i)}$ to P_i for each $i \in \mathcal{P}, h \in [\ell]$.

Corrupted parties: The functionality waits for the adversary $\mathcal{S}$ to input the set A of corrupted parties. Then it proceeds as follows:

  - $\forall h \in [\ell]$, the functionality waits for $\mathcal{S}$ to input shares $\{\mathbf{m}_h^{(j)}\}_{j \in A}$ and it generates $\langle \mathbf{m}_h \rangle$, with honest shares $\{\mathbf{m}_h^{(i)}\}_{i \notin A, h \in [\ell]}$, consistent with adversarial shares but otherwise random.
  - If the adversary inputs (Error, $\{e_{h,j}^{(k)}\}_{k \notin A, h \in [\ell], j \in [M]}$) with elements in $\mathbb{F}_{2^M}$, the functionality sets $\mathbf{m}_h^{(k)} = \mathbf{m}_h^{(k)} + \sum_{j=1}^{M} e_{h,j}^{(k)} \cdot \Delta_j^{(k)} \cdot X^{j-1}$ where $\Delta_j^{(k)}$ denotes the j-th bit of $\Delta^{(k)}$.
  - For each $k \notin A$, the functionality outputs $\{\mathbf{m}_h^{(k)}\}$ to P_k.

Key queries: On input of a description of an affine subspace $S \subset (\mathbb{F}_2^M)^n$, return Success if $(\Delta^{(1)}, \ldots, \Delta^{(n)}) \in S$. Otherwise return Abort.

---

[a] If $u \neq M$ we view Δ as an element of $\mathbb{F}_{2^u}^{M/u}$ and perform the multiplication by $\mathbf{x}_h$ componentwise.

Fig. 4. Ideal Generation of $[\![\cdot]\!]$-representations

stage, where κ dummy values are authenticated, then opened and checked. If the check passes then every party's MAC key has been initialized correctly, except with probability $2^{-\kappa}$. Although in practice this overhead is not needed when authenticating $\ell \geq \kappa$ values, modeling this would introduce additional errors into the functionality and make the analysis of the triple generation protocols more complex.

Now we present the protocol $\Pi_{[\![\cdot]\!]}$, realizing the ideal functionality of Fig. 4, more in detail. We describe the authentication procedure for bits first and then the extension to $\mathbb{F}_{2^u}$.

Suppose parties need to authenticate an additively secret shared random bit $x = x^{(1)} + \cdots + x^{(n)}$. Once the global key Δ is initialized, the parties call the subprotocol $\Pi_{[\cdot]}$ (see the full version) n times. Output of each of these calls is a value $\mathbf{u}^{(i)}$ for P_i and values $\mathbf{q}^{(j,i)}$ for each P_j, $j \neq i$, such that

$$\mathbf{u}^{(i)} + \mathbf{q}^{(j,i)} = \sum_{j \neq i} \mathbf{t}^{(i,j)} + x^{(i)} \cdot \Delta^{(i)} + \sum_{j \neq i} \mathbf{q}^{(j,i)} = x^{(i)} \cdot \Delta. \qquad (1)$$

To create a complete authentication $[\![x]\!]$, each party sets $\mathbf{m}^{(i)} = \mathbf{u}^{(i)} + \sum_{j \neq i} \mathbf{q}^{(i,j)}$. Notice that if we add up all the MAC shares, we obtain:

$$\mathbf{m} = \sum_{i \in \mathcal{P}} \mathbf{m}^{(i)} = \sum_{i \in \mathcal{P}} \left(\mathbf{u}^{(i)} + \sum_{j \neq i} \mathbf{q}^{(i,j)} \right) = \sum_{i \in \mathcal{P}} \left(\mathbf{u}^{(i)} + \sum_{j \neq i} \mathbf{q}^{(j,i)} \right) = \sum_{i \in \mathcal{P}} x^{(i)} \cdot \Delta = x \cdot \Delta,$$

where the second equality holds for the symmetry of the notation $\mathbf{q}^{(i,j)}$ and the third follows from (1).

Finally, if P_i wants to authenticate a bit $x^{(i)}$, it is enough, from Equation (1), setting $\mathbf{m}^{(i)} = \mathbf{u}^{(i)}$ and $\mathbf{m}^{(j)} = \mathbf{q}^{(j,i)}$, $\forall j \neq i$. Clearly, from (1), we have $\sum_{i \in \mathcal{P}} \mathbf{m}^{(i)} = x^{(i)} \cdot \Delta$.

Consider now the case where parties need to authenticate elements in $\mathbb{F}_{2^u}$. We can represent any element $\mathbf{x} \in \mathbb{F}_{2^u}$ as a binary vector $(x_1, \ldots, x_u) \in \mathbb{F}_2^u$. In order to obtain a representation $[\![\mathbf{x}]\!]$ it is sufficient to repeat the previous procedure u times to get $[\![x_i]\!]$ and then compute $[\![\mathbf{x}]\!]$ as $\sum_{k=1}^{u} [\![x_k]\!] \cdot X^{k-1}$ (see the full version for details). Here we let X denote the variable in polynomial representation of $\mathbb{F}_{2^u}$ and $[\![x_k]\!]$ the k'th coefficient.

We now describe what happens to the MAC representation in presence of corrupted parties. As we have already pointed out before, a corrupt party could input different MAC key shares when initializing $\mathcal{F}_{\mathsf{COTe}}$ with different parties. Moreover a corrupt P_i could input vectors $\mathbf{x}_1^{(i)}, \ldots \mathbf{x}_\ell^{(i)}$ instead of bits to n-**Share**(i) (i.e. to $\mathcal{F}_{\mathsf{COTe}}$). This will produce an error in the authentication depending on the MAC key. Putting things together we obtain the following faulty representation:

$$\mathbf{m} = x \cdot \Delta + \sum_{k \notin A} x^{(k)} \cdot \delta^{(i)} + \sum_{k \notin A} \mathbf{e}^{(i,k)} * \Delta^{(k)}, \quad \text{for some } i \in A$$

where A is the set of corrupt parties, $\delta^{(i)}$ is an offset vector known to the adversary which represents the possibility that corrupted parties input different MAC key shares, whilst $\mathbf{e}^{(i,k)}$ depends on the adversary inputting vectors and not just bits to $\mathcal{F}_{\mathsf{COTe}}$. More precisely, if P_i inputs a vector $\mathbf{x}^{(i)}$ to n-**Share**(i), we can rewrite it as $\mathbf{x}^{(i)} = x^{(i)} \cdot \mathbf{1} + \mathbf{e}^{(i,k)}$, where $\mathbf{e}^{(i,k)} \in \mathbb{F}_2^M$ is an error vector known to the adversary. While we prevent the first type of errors by adding a MACCheck step in the Initialize phase, we allow the second type of corruption. This faulty authentication suffices for our purposes due to the MAC checking procedure used later on.

Lemma 2. *In the $\mathcal{F}_{\mathsf{COTe}}^{\kappa,\ell}$-hybrid model, the protocol $\Pi_{[\![\cdot]\!]}$ implements $\mathcal{F}_{[\![\cdot]\!]}$ against any static adversary corrupting up to $n-1$ parties.*

Proof. See the full version.

Functionality $\mathcal{F}_{\mathsf{Triples}}^{\mathbb{F}}$

Let A be the indices of corrupt parties. Running with parties $P_1, \ldots, P_n$ and an adversary $\mathcal{S}$, the functionality operates as follows.

Initialize: On input (Init) the functionality activates and waits for $\mathcal{S}$ to input a set of shares $\{\Delta^{(j)}\}_{j \in A}$. It samples random $\{\Delta^{(i)}\}_{i \notin A}$ in $\mathbb{F}_2^{\kappa}$ for the honest parties, defining $\Delta := \sum_{i \in [n]} \Delta^{(i)}$. If any $j \in A$ outputs Abort then the functionality aborts.

Honest Parties: On input (Triples), the functionality outputs random $[\![x_h]\!]_\Delta, [\![y_h]\!]_\Delta, [\![z_h]\!]_\Delta$, such that
$\langle z_h \rangle = \langle x_h \rangle \cdot \langle y_h \rangle$ and $z_h, y_h, x_h \in \mathbb{F}$.

Corrupted Parties: The functionality samples $x_h, y_h \xleftarrow{\$} \mathbb{F}$ and computes $z_h = x_h \cdot y_h$. To produce $[\![a]\!]_\Delta = (\langle a \rangle, \langle \mathbf{m} \rangle, \langle \Delta \rangle)$, where $a \in \{x_h, y_h, z_h\}_{h \in [\ell]}$ it does the following:

    - It waits the adversary to input shares $\{a^{(i)}\}_{i \in A}$ and $\{\mathbf{m}^{(i)}\}_{i \in A}$.
    - It waits for the adversary to input (ValueError, e) and (MacError, $\mathbf{e}$).
    - It selects the shares of honest parties at random, but consistent with adversarial shares and with $a + e$ and $a \cdot \Delta + \mathbf{e}$, that is, such that $\sum_{i=1}^{n} a^{(i)} = a + e$ and $\sum_{i=1}^{n} \mathbf{m}^{(i)} = a \cdot \Delta + \mathbf{e}$.

Key queries: On input of a description of an affine subspace $S \subset (\mathbb{F}_2^\kappa)^n$, return Success if $(\Delta^{(1)}, \ldots, \Delta^{(n)}) \in S$. Otherwise return Abort.

Fig. 5. Ideal functionality for triples generation

5 Triple Generation in $\mathbb{F}_2$ and $\mathbb{F}_{2^k}$

In this section we describe our protocols generating triples in finite fields. First we describe the protocols for multiplication triples in $\mathbb{F}_{2^\kappa}$ (Figs. 7 and 8), and then the protocol for bit triples (Fig. 9). Both approaches implement the functionality $\mathcal{F}_{\mathsf{Triples}}^{\mathbb{F}}$, given in Fig. 5. Note that the functionality allows an adversary to try and guess an affine subspace containing the parties' MAC key shares, which is required because of our faulty authentication procedure described in the previous section.

5.1 $\mathbb{F}_{2^k}$ Triples

In this section, we show how to generate $\mathbb{F}_{2^k}$ authenticated triples using two functionalities $\mathcal{F}_{\mathsf{GFMult}}^{k,s}$ and $\mathcal{F}_{[\cdot]}^{\mathbb{F}_{2^k}}$ (see the full version). We realize the functionality $\mathcal{F}_{\mathsf{GFMult}}^{k,s}$ with protocol $\Pi_{\mathsf{GFMult}}^{k,s}$ (Fig. 6). This protocol is a simple extension of $\mathcal{F}_{\mathsf{ACOT}}$ that converts the sharing of a tensor product matrix in $\mathbb{F}_2^{k \times k}$ to the sharing of a product in $\mathbb{F}_{2^k}$. Taking this modular approach simplifies the proof for triple generation, as we can deal with the complex errors from $\mathcal{F}_{\mathsf{ACOT}}$ separately. Our first triple generation protocol ($\Pi_{\mathsf{UncheckedTriples}}$) will not reveal any information about the values or the authentication key, but an active adversary

---

Protocol $\Pi_{\mathsf{GFMult}}^{k,s}$

Let $\mathbf{x}$ and $\mathbf{y}$ denote the inputs of P_R and P_S respectively, in $\mathbb{F}_{2^k}$, and let s be a statistical security parameter. Furthermore, let $\mathbf{e} = (1, X, \ldots, X^{k-1})$ and $\ell' = 2k + s$.

1. The parties run $\mathcal{F}_{\mathsf{ACOT}}^{k,s}$:
   (a) P_R inputs $\mathbf{x}$ and P_S inputs $\mathbf{y}$.
   (b) P_R receives T and P_S receives Q such that $T + Q = \mathbf{x} \otimes \mathbf{y}$.
2. P_R outputs $\mathbf{t} = \mathbf{e}T\mathbf{e}^\top$ and P_S outputs $\mathbf{q} = \mathbf{e}Q\mathbf{e}^\top$.

Fig. 6. $\mathbb{F}_{2^k}$ multiplication

Protocol $\Pi_{\mathsf{UncheckedTriples}}$

Initialize: The parties initialize $\mathcal{F}_{[\![\cdot]\!]}^{\mathbb{F}_{2^k}}$, which outputs $\Delta^{(i)}$ to party i.
Triple generation:
1. Every party i samples random $\mathbf{a}^{(i)} \xleftarrow{\$} \mathbb{F}_{2^k}$ and $\mathbf{b}^{(i)} \xleftarrow{\$} \mathbb{F}_{2^k}$.
2. Every tuple of parties $(i,j) \in [n]^2, i \neq j$ call $\mathcal{F}_{\mathsf{GFMult}}^{k,s}$ with P_i inputting $\mathbf{a}^{(i)}$ and P_j inputting $\mathbf{b}^{(j)}$ to generate a random secret sharing $\mathbf{c}_{i,j}^{(i,j)} + \mathbf{c}_{i,j}^{(j,i)} = \mathbf{a}^{(i)} \cdot \mathbf{b}^{(j)}$.
3. Every party i computes $\mathbf{c}^{(i)} = \mathbf{a}^{(i)} \cdot \mathbf{b}^{(i)} + \sum_{j \neq i}(\mathbf{c}_{i,j}^{(i,j)} + \mathbf{c}_{j,i}^{(i,j)})$.
4. Party i calls $\mathcal{F}_{[\![\cdot]\!]}^{\mathbb{F}_{2^k}}$ with inputs $\mathbf{a}^{(i)}$, $\mathbf{b}^{(i)}$, and $\mathbf{c}^{(i)}$, and receives $\mathbf{m}_{\mathbf{a}}^{(i)}$, $\mathbf{m}_{\mathbf{b}}^{(i)}$, and $\mathbf{m}_{\mathbf{c}}^{(i)}$.

Fig. 7. Protocol for generation of unchecked $\mathbb{F}_{2^k}$ triples.

can distort the output in various ways. We then present a protocol ($\Pi_{\mathsf{TripleCheck}}$) to check the generated triples from $\Pi_{\mathsf{UncheckedTriples}}$, similarly to the sacrificing step of the SPDZ protocol [8], to ensure that an adversary has not distorted them.

The protocol is somewhat similar to the one in the previous section. Instead of using $n(n-1)$ instances of $\mathcal{F}_{\mathsf{COTe}}$, it uses $n(n-1)$ instances of $\mathcal{F}_{\mathsf{GFMult}}^{k,s}$, which is necessary to compute a secret sharing of $\mathbf{x} \cdot \mathbf{y}$, where $\mathbf{x}$ and $\mathbf{y}$ are known to different parties.

Lemma 3. *The protocol $\Pi_{\mathsf{UncheckedTriples}}$ (see the full version) implements the functionality $\mathcal{F}_{\mathsf{UncheckedTriples}}$ in the $(\mathcal{F}_{\mathsf{GFMult}}^{k,s}, \mathcal{F}_{[\![\cdot]\!]}^{\mathbb{F}_{2^k}})$-hybrid model with perfect security.*

Proof. The proof is straightforward using an appropriate simulator. See the full version for further details.

The protocol $\Pi_{\mathsf{TripleCheck}}$ produces N triples using $2N$ unchecked triples similar to the sacrificing step of the SPDZ protocol. However, corrupted parties

---

Protocol $\Pi_{\mathsf{TripleCheck}}$

Initialize: Each party receives $\Delta^{(i)}$ from $\mathcal{F}_{\mathsf{UncheckedTriples}}$.

Triple Generation:

1. Generate $2N$ $\{[\![\mathbf{a}_j]\!], [\![\mathbf{b}_j]\!], [\![\mathbf{c}_j]\!]\}_{j \in [2N]}$ unchecked triples using $\mathcal{F}_{\mathsf{UncheckedTriples}}$.
2. Sample $\mathbf{t}, \mathbf{t}', \mathbf{t}'' \xleftarrow{\$} \mathbb{F}_{2^k}$ using $\mathcal{F}_{\mathsf{Rand}}$.
3. For all $j \in [N]$, open $\mathbf{t} \cdot \langle \mathbf{b}_j \rangle + \mathbf{t}' \cdot \langle \mathbf{b}_{j+N} \rangle$ as $\mathbf{r}_j$ and $\mathbf{t}' \cdot \langle \mathbf{a}_j \rangle + \mathbf{t}'' \cdot \langle \mathbf{a}_{j+N} \rangle$ as $\mathbf{s}_j$.
4. Use $\mathcal{F}_{\mathsf{BatchCheck}}$ with $\{\mathbf{r}_j \cdot \langle \Delta \rangle + \mathbf{t} \cdot \langle \mathbf{m}_{\mathbf{b}_j} \rangle + \mathbf{t}' \cdot \langle \mathbf{m}_{\mathbf{b}_{j+N}} \rangle)\}_{j \in [N]}$ and $\{\mathbf{s}_j \cdot \langle \Delta \rangle + \mathbf{t}' \cdot \langle \mathbf{m}_{\mathbf{a}_j} \rangle + \mathbf{t}'' \cdot \langle \mathbf{m}_{\mathbf{a}_{j+N}} \rangle\}_{j \in [N]}$, and abort if it returns $\bot$.
5. Use $\mathcal{F}_{\mathsf{BatchCheck}}$ with $\{\mathbf{t} \cdot \langle \mathbf{m}_{\mathbf{c}_j} \rangle + \mathbf{t}'' \cdot \langle \mathbf{m}_{\mathbf{c}_{j+N}} \rangle + \mathbf{r}_j \cdot \langle \mathbf{m}_{\mathbf{a}_j} \rangle + \mathbf{s}_j \cdot \langle \mathbf{m}_{\mathbf{b}_{j+N}} \rangle\}_{j \in [N]}$, and abort if it returns $\bot$.
6. Output $\{[\![\mathbf{a}_j]\!], [\![\mathbf{b}_j]\!], [\![\mathbf{c}_j]\!]\}_{j \in [N]}$.

Fig. 8. Triple checking protocol.

have more options to deviate here, which we counter by using more random coefficients for checking. Recall that, in the SPDZ protocol, parties input their random shares by broadcasting a homomorphic encryption thereof. Here, the parties have to input such a share by using an instance of $\mathcal{F}_{\mathsf{GFMult}}^{k,s}$ and $\mathcal{F}_{[\![\cdot]\!]}^{\mathbb{F}_{2^k}}$ with every other party, which opens up the possibility of using a different value in every instance. We will prove that, if the check passes, the parties have used consistent inputs to $\mathcal{F}_{\mathsf{GFMult}}^{k,s}$. On the other hand, $\mathcal{F}_{[\![\cdot]\!]}^{\mathbb{F}_{2^k}}$ provides less security guarantees. However, we will also prove that the more deviation there is with $\mathcal{F}_{[\![\cdot]\!]}^{\mathbb{F}_{2^k}}$, the more likely the check is to fail. This is modeled using the key query access of $\mathcal{F}_{\mathsf{Triples}}$. Note that, while this reveals some information about the MAC key Δ, this does not contradict the security of the resulting MPC protocol because Δ does not protect any private information. Furthermore, breaking correctness corresponds to guessing Δ, which will only succeed with probability negligible in k because incorrect guesses lead to an abort.

We use a supplemental functionality $\mathcal{F}_{\mathsf{BatchCheck}}$, which checks that a batch of shared values are equal to zero, and can be easily implemented using commitment and $\mathcal{F}_{\mathsf{Rand}}$ (see the full version for details). The first use of $\mathcal{F}_{\mathsf{BatchCheck}}$ corresponds to using the SPDZ MAC check protocol for $\mathbf{r}_j$ and $\mathbf{s}_j$ for all $j \in [N]$, and the second use corresponds to the sacrificing step, which checks whether $\mathbf{t} \cdot \mathbf{c}_j + \mathbf{t}'' \cdot \mathbf{c}_{j+N} + \mathbf{r}_j \mathbf{a}_j + \mathbf{s}_j \cdot \mathbf{b}_{j+N} = 0$ for all $j \in [N]$.

Theorem 1. *The protocol $\Pi_{\mathsf{TripleCheck}}$, described in Fig. 8, implements $\mathcal{F}_{\mathsf{Triples}}$ in the $(\mathcal{F}_{\mathsf{UncheckedTriples}}, \mathcal{F}_{\mathsf{Rand}})$-hybrid model with statistical security $(k - 4)$.*

Proof. The proof mainly consists of proving that, if $\mathbf{c}_j \neq \mathbf{a}_j \cdot \mathbf{b}_j$ or the MAC values are incorrect for some j, and the check passes, then the adversary can compute the offset of $\mathbf{c}_j$ or the MAC values. See the full version.

5.2 $\mathbb{F}_2$ Triples

This section shows how to produce a large number ℓ of random, authenticated bit triples using the correlated OT with errors functionality $\mathcal{F}_{\mathsf{COTe}}$ from Sect. 3. We describe the main steps of the protocol in Fig. 9. The main difference with respect to the protocol by Larraia et al. [17] is that here we use the outputs of $\mathcal{F}_{\mathsf{COTe}}$ to *simultaneously* generate triples, $\langle z_h \rangle = \langle x_h \rangle \cdot \langle y_h \rangle$, and authenticate the random bits x_h, for $h = 1, \ldots, \ell$, under the fixed global key Δ, giving $[\![x_h]\!] = (\langle x_h \rangle, \langle \mathbf{m}_h \rangle, \langle \Delta \rangle)$. To do this, we need to double the length of the correlation used in $\mathcal{F}_{\mathsf{COTe}}$, so that half of the output is used to authenticate x_h, and the other half is hashed to produce shares of the random triple.[2]

The shares $\langle y_h \rangle, \langle z_h \rangle$ are then authenticated with additional calls to $\mathcal{F}_{\mathsf{COTe}}$ to obtain $[\![y_h]\!], [\![z_h]\!]$. We then use a random bucketing technique to combine the x_h values in several triples, removing any potential leakage due to incorrect authentication of y_h (avoiding the selective failure attack present in the previous protocol [17]) and then sacrifice to check for correctness (as in the previous protocol).

The **Initialize** stage consists of initializing the functionality $\mathcal{F}_{\mathsf{COTe}}^{2\kappa,\ell}$ with $\hat{\Delta} \in \mathbb{F}_2^{2\kappa}$. Note that $\hat{\Delta}$ is the concatenation of a random $\tilde{\Delta} \in \mathbb{F}_2^{\kappa}$ and the MAC key Δ. We add a consistency check to ensure that each party initialize $\hat{\Delta}$ correctly, as we did in $\Pi_{[\![\cdot]\!]}$.

Then, in **COTe.Extend**, each party P_i runs a $\mathsf{COTe}^{2\kappa,\ell}$ with all other parties on input $\mathbf{x}^{(i)} = (x_1^{(i)}, \ldots, x_\ell^{(i)}) \in \mathbb{F}_2^\ell$. For each $i \in \mathcal{P}$, we obtain $\hat{\mathbf{q}}_h^{(j,i)} = \hat{\mathbf{t}}_h^{(i,j)} + x_h^{(i)} \cdot \hat{\Delta}^{(j)}$, $h \in [\ell]$, where

$$\hat{\mathbf{q}}_h^{(j,i)} = (\tilde{\mathbf{q}}_h^{(j,i)} \| \mathbf{q}_h^{(j,i)}) \in \mathbb{F}_2^{2\kappa} \quad \text{and} \quad \hat{\mathbf{t}}_h^{(j,i)} = (\tilde{\mathbf{t}}_h^{(j,i)} \| \mathbf{t}_h^{(j,i)}) \in \mathbb{F}_2^{2\kappa}.$$

Note that we allow corrupt parties to input vectors $\mathbf{x}_h^{(i)}$ instead of bits.

Parties use the first κ components of their shares during the **Triple Generation** phase. More precisely, each party P_i samples ℓ random bits $y_h^{(i)}$ and then uses the first κ components of the output of $\mathsf{COTe}^{2\kappa,\ell}$ to generate shares $z_h^{(i)}$. The idea (as previously [17]) is that of using OT-relations to produce multiplicative triples. In step 2, in order to generate ℓ random and independent triples, we need to break the correlation generated by COTe. For this purpose we use a hash function H, but after that, as we need to "bootstrap" to an n-parties representation, we must create new correlations for each $h \in [\ell]$. P_i sums all the values $n_h^{(i,j)}, j \neq i$, and $x_h^{(i)} \cdot y_h^{(i)}$ to get $u_h^{(i,j)} = \sum_{j \neq i} n_h^{(j,i)} + x_h^{(i)} \cdot y_h^{(i)}$. Notice that adding up the share $u_h^{(i,j)}$ held by P_i and all the shares of other parties, after step 2 we have:

$$u_h^{(i,j)} + \sum_{j \neq i} v_{0,h}^{(j,i)} = x_h^{(i)} \cdot y_h.$$

---

[2] If the correlation length is not doubled, and the same output is used both for authentication *and* as input to the hash function, we cannot prove UC security as the values and MACs of a triple are no longer independent.

Protocol $\Pi_{\mathsf{BitTriples}}$

The goal of the protocol is to generate ℓ $\mathbb{F}_2$ triples $\langle x_h \rangle$, $\langle y_h \rangle$, $\langle z_h \rangle$, $h = 1, \ldots, \ell$, such that $z_h = x_h \cdot y_h$, together with $[\![x_h]\!]$, $[\![y_h]\!]$, $[\![z_h]\!]$. The protocol is parametrized by the number ℓ of authenticated triples, and it assumes access to a random oracle $H : \{0,1\}^* \to \{0,1\}$.

Initialize:
1. Each party P_i samples a random MAC key share $\Delta^{(i)}$, a second value $\tilde{\Delta}^{(i)} \in \mathbb{F}_2^{\kappa}$ and sets $\hat{\Delta}^{(i)} = (\tilde{\Delta}^{(i)} \| \Delta^{(i)}) \in \mathbb{F}_2^{2\kappa}$.
2. Each pair of parties (P_i, P_j) (for $i \neq j$) calls $\mathcal{F}_{\mathsf{COTe}}$.Initialize, where P_j inputs $\hat{\Delta}^{(j)}$, and $\mathcal{F}_{[\![\cdot]\!]}$.Init, where P_j inputs $\Delta^{(j)}$.
3. Parties check consistency of the $\mathcal{F}_{\mathsf{COTe}}$ inputs $\hat{\Delta} = \hat{\Delta}^{(1)} + \cdots + \hat{\Delta}^{(n)}$ as in the Initialize step of $\Pi_{[\![\cdot]\!]}$, using κ random values. If Π_{MACCheck} fails, output Abort.

COTe.Extend: Each P_i, $i \in \mathcal{P}$, runs $\mathcal{F}_{\mathsf{COTe}}$.Extend with P_j, $\forall j \neq i$: P_i inputs $\mathbf{x}^{(i)} = (x_1^{(i)}, \ldots, x_\ell^{(i)}) \in \mathbb{F}_2^\ell$, and then it receives $\{\hat{\mathbf{t}}_h^{(i,j)}\}_{h \in [\ell]}$ and P_j receives $\hat{\mathbf{q}}_h^{(j,i)} = \hat{\mathbf{t}}_h^{(i,j)} + x_h^{(i)} \cdot \hat{\Delta}^{(j)}, h \in [\ell]$.

Triple generation: Each party P_i uses only the first κ components of its shares. We denote them by $\tilde{\mathbf{q}}_h^{(i,j)}$, $\tilde{\Delta}^{(i)}$ and $\tilde{\mathbf{t}}_h^{(i,j)}$.
1. Each party P_i generates ℓ random $y_h^{(i)} \in \mathbb{F}_2$.
2. For each $i \in \mathcal{P}$ do:
   (a) Using a random oracle $H : \{0,1\}^* \to \{0,1\}$, break the correlation from the previous step. P_i locally computes $H(\tilde{\mathbf{t}}_h^{(i,j)}) = w_h^{(i,j)}$, and P_j locally computes $H(\tilde{\mathbf{q}}_h^{(j,i)}) = v_{0,h}^{(j,i)}, H(\tilde{\mathbf{q}}_h^{(j,i)} + \tilde{\Delta}^{(j)}) = v_{1,h}^{(j,i)}, \forall j \neq i, \forall h \in [\ell]$.
   (b) Parties need to create new correlations corresponding to y_h:
      - Each P_j, $j \neq i$, sends a vector $\mathbf{s}^{(j,i)} \in \mathbb{F}_2^\ell$ to P_i such that each component is $s_h^{(j,i)} = v_{0,h}^{(j,i)} + v_{1,h}^{(j,i)} + y_h^{(j)}$.
      - $\forall j \neq i$, P_i computes $n_h^{(i,j)} = w_h^{(i,j)} + x_h^{(i)} \cdot s_h^{(j,i)} = v_{0,h}^{(j,i)} + x_h^{(i)} \cdot y_h^{(j)}$.
3. Each P_i computes

$$z_h^{(i)} = \sum_{j \neq i} n_h^{(i,j)} + x_h^{(i)} \cdot y_h^{(i)} + \sum_{j \neq i} v_{0,h}^{(i,j)}.$$

Authentication: 1. Authenticate x_h by summing up the last κ components of the outputs from the COTe step to obtain $[\![x_h]\!]$, for $h = 1, \ldots, \ell$.
2. Call $\mathcal{F}_{[\![\cdot]\!]}^{\mathbb{F}_2^\kappa}$ with input Authenticate to authenticate $y_h^{(j)}, z_h^{(j)}$ for $j = 1, \ldots, n$ and $h = 1, \ldots, \ell$, obtaining $[\![y_h]\!]$, $[\![z_h]\!]$

Check triples: This step performs sacrificing and combining, to check that the triples are correctly generated and to prevent any leakage on x_h in case y_h was authenticated incorrectly. The parties call the subprotocol $\Pi_{\mathsf{CheckTriples}}$ (see the full version).

Fig. 9. $\mathbb{F}_2$-triples generation

Repeating this procedure for each $i \in \mathcal{P}$ and adding up, we get $z_h = x_h \cdot y_h$.

Once the multiplication triples are generated the parties **Authenticate** z_h and y_h using $\mathcal{F}_{[\cdot]}$, while to authenticate x_h they use the remaining κ components of the outputs of the COTe.Extend step.

Checking Triples. In the last step we want to check that the authenticated triples are correctly generated. For this we use the bucket-based cut-and-choose technique by Larraia et al. [17]. In the full version we generalize and optimize the parameters for this method.

The bucket-cut-and-choose step ensures that the generated triples are correct. Privacy on x is then guaranteed by the combine step, whereas privacy on y follows from the use of the original COTe for both creating triples and authenticating x.

Note also that if a corrupt party inputs an inconsistent bit $x_h^{(i)}$ in $n_h^{(i,k)}$, for some $k \notin A$ in step 2.b, then the resulting triples $z_h = x_h \cdot y_h + s_h^{(k,i)} \cdot y_h$ will pass the checks if and only if $s_h^{(k,i)} = 0$, revealing nothing about y_h.

We conclude by stating the main result of this section.

Theorem 2. *For every static adversary $\mathcal{A}$ corrupting up to $n - 1$ parties, the protocol $\Pi_{\mathsf{BitTriples}}$ κ-securely implements $\mathcal{F}_{\mathsf{Triples}}$ (Fig. 5) in the $(\mathcal{F}_{\mathsf{COTe}}^{\kappa,\ell}, \mathcal{F}_{[\cdot]})$-hybrid model.*

Proof. Correctness easily follows from the above discussion. For more details see the full version.

6 Triple Generation for MiniMACs

In this section we describe how to construct the preprocessing data needed for the online execution of the MiniMAC protocol [9,11]. The complete protocols and security proofs are in the full version. Here we briefly outline the protocols and give some intuition of security.

6.1 Raw Material

The raw material used for MiniMAC is very similar to the raw material in both TinyOT and SPDZ. In particular this includes random multiplication triples. These are used in the same manner as $\mathbb{F}_2$ and $\mathbb{F}_{2^k}$ triples to allow for multiplication during an online phase. However, remember that we work on elements which are codewords of some systematic linear error correcting code, C. Thus an authenticated element is defined as $[\![C(\mathbf{x})]\!]^* = \{ \langle C(\mathbf{x}) \rangle, \langle \mathbf{m} \rangle, \langle \Delta \rangle \}$ where $\mathbf{m} = C(\mathbf{x}) * \Delta$ with $C(\mathbf{x})$, $\mathbf{m}$ and Δ elements of $\mathbb{F}_{2^u}^m$ and $\mathbf{x} \in \mathbb{F}_{2^u}^k$. Similarly a triple is a set of three authenticated elements, $\{ [\![C(\mathbf{a})]\!]^*, [\![C(\mathbf{b})]\!]^*, [\![C^*(\mathbf{c})]\!]^* \}$ under the constraint that $C^*(\mathbf{c}) = C(\mathbf{a}) * C(\mathbf{b})$, where $*$ denotes component-wise multiplication. We notice that the multiplication of two codewords results in an element in the Schur transform. Since we might often be doing multiplication involving the result of another multiplication, that thus lives in C^*, we need some way of bringing elements from C^* back down to C. To do this we need another piece of raw material: the Schur pair. Such a pair is simply two authenticated elements of the same message, one in the codespace and one in the Schur transform. That is, the pair $\{[\![C(\mathbf{r})]\!]^*, [\![C^*(\mathbf{s})]\!]^*\}$ with $\mathbf{r} = \mathbf{s}$. After doing a multiplication using a preprocessed random triple in the online phase, we use the $[\![C^*(\mathbf{s})]\!]^*$ element to onetime pad the result, which can then be partially opened. This opened value is re-encoded using C and then added to $[\![C(\mathbf{r})]\!]^*$. This gives a shared codeword element in C, that is the correct output of the multiplication.

Finally, to avoid being restricted to just parallel computation within each codeword vector, we also need a way to reorganize these components within a codeword. To do so we need to construct "reorganization pairs". Like the Schur pairs, these will simply be two elements with a certain relation on the values they authenticate. Specifically, one will encode a random element and the other a linear function applied to the random element encoded by the first. Thus the pair will be $\{[\![C(\mathbf{r})]\!]^*, [\![C(f(\mathbf{r}))]\!]^*\}$ for some linear function $f : \mathbb{F}_{2^u}^k \to \mathbb{F}_{2^u}^k$. We use these by subtracting $[\![C(\mathbf{r})]\!]^*$ from the shared element we will be working on. We then partially open and decode the result. This is then re-encoded and added to $[\![C(f(\mathbf{r}))]\!]^*$, resulting in the linear computation defined by $f(\cdot)$ on each of the components.

6.2   Authentication

For the MiniMAC protocol to be secure, we need a way of ensuring that authenticated vectors always form valid codewords. We do this based on the functionality $\mathcal{F}_{\mathsf{CodeAuth}}$ in two steps, first a 'BigMAC' authentication, which is then compressed to give a 'MiniMAC' authentication. For the BigMAC authentication, we simply use the $\mathcal{F}_{[\![\cdot]\!]}$ functionality to authenticate each component of $\mathbf{x}$ (living in $\mathbb{F}_{2^u}$) separately under the whole of $\Delta \in \mathbb{F}_{2^u}^m$. Because every component of $\mathbf{x}$ is then under the same MAC key, we can compute MACs for the rest of the codeword $C(\mathbf{x})$ by simply linearly combining the MACs on $\mathbf{x}$, due to the linearity of C. We use the notation $[\![C(\mathbf{x})]\!] = \left\{ \langle C(\mathbf{x}) \rangle, \{\langle \mathbf{m}_{\mathbf{x}_i} \rangle\}_{i \in [m]}, \langle \Delta \rangle \right\}$ to denote the BigMAC share. To go from BigMAC to MiniMAC authentication, we just extract the relevant $\mathbb{F}_{2^u}$ element from each MAC. We then use $[\![C(\mathbf{x})]\!] = \{\langle C(\mathbf{x}) \rangle, \langle \mathbf{m}_{\mathbf{x}} \rangle, \langle \Delta \rangle\}$ to denote a MiniMAC element, where $\mathbf{m}_{\mathbf{x}}$ is made up of one component of each of the m BigMACs. The steps are described in detail in the full version.

6.3   Multiplication Triples

To generate a raw, unauthenticated MiniMAC triple, we need to be able to create vectors of shares $\langle C(\mathbf{a}) \rangle, \langle C(\mathbf{b}) \rangle, \langle C^*(\mathbf{c}) \rangle$ where $C^*(\mathbf{c}) = C(\mathbf{a}) * C(\mathbf{b})$ and $\mathbf{a}, \mathbf{b} \in \mathbb{F}_{2^u}^k$. These can then be authenticated using the $\mathcal{F}_{\mathsf{CodeAuth}}$ functionality described above.

Since the authentication procedure only allows shares of valid codewords to be authenticated, it might be tempting to directly use the SPDZ triple generation protocol from Sect. 5.1 in $\mathbb{F}_{2^u}$ for each component of the codewords $C(\mathbf{a})$ and

$C(\mathbf{b})$. In this case, it is possible that parties do not input valid codewords, but this would be detected in the authentication stage. However, it turns out this approach is vulnerable to a subtle selective failure attack – a party could input to the triple protocol a share for $C(\mathbf{a})$ that differs from a codeword in just one component, and then change their share to the correct codeword before submitting it for authentication. If the corresponding component of $C(\mathbf{b})$ is zero then this would go undetected, leaking that fact to the adversary.

To counter this, we must ensure that shares output by the triple generation procedure are guaranteed to be codewords. To do this, we only generate shares of the $\mathbb{F}_{2^u}^k$ vectors $\mathbf{a}$ and $\mathbf{b}$ – since C is a linear $[m, k, d]$ code, the shares for the parity components of $C(\mathbf{a})$ and $C(\mathbf{b})$ can be computed locally. For the product $C^*(\mathbf{c})$, we need to ensure that the first $k^* \geq k$ components can be computed, since C^* is a $[m, k^*, d^*]$ code. Note that the first k components are just $(\mathbf{a}_1, \ldots, \mathbf{a}_k) * (\mathbf{b}_1, \ldots, \mathbf{b}_k)$, which could be computed similarly to the SPDZ triples. However, for the next $k^* - k$ components, we also need the cross terms $\mathbf{a}_i \cdot \mathbf{b}_j$, for every $i, j \in [k]$. To ensure that these are computed correctly, we input vectors containing all the bits of $\mathbf{a}, \mathbf{b}$ to $\mathcal{F}_{\mathsf{ACOT}}$, which outputs the tensor product $\mathbf{a} \otimes \mathbf{b}$, from which all the required codeword shares can be computed locally. Similarly to the BigMAC authentication technique, this results in an overhead of $O(k \cdot u) = O(\kappa \log \kappa)$ for every multiplication triple when using Reed-Solomon codes.

Taking our departure in the above description we generate the multiplication triples in two steps: First unauthenticated multiplication triples are generated by using the CodeOT subprotocol, which calls $\mathcal{F}_{\mathsf{ACOT}}$ and takes the diagonal of the resulting shared matrices. The codewords of these diagonals are then used as inputs to $\mathcal{F}_{\mathsf{CodeAuth}}$, which authenticates them. This is described by protocol $\Pi_{\mathsf{UncheckedMiniTriples}}$ in (see the full version). Then a random pairwise sacrificing is done to ensure that it was in fact shares of multiplication being authenticated. This is done using protocol $\Pi_{\mathsf{MiniTriples}}$ (see the full version). One minor issue that arises during this stage is that we also need to use a Schur pair to perform the sacrifice, to change one of the multiplication triple outputs back down to the code C, before it is multiplied by a challenge codeword and checked.

Security intuition. Since the CodeOT procedure is guaranteed to produces shares of valid codewords, and the authentication procedure can only be used to authenticate valid codewords, if an adversary changes their share before authenticating it, they must change it in at least d positions, where d is the minimum distance of the code. For the pairwise sacrifice check to pass, the adversary then has to essentially guess d components of the random challenge codeword to win, which only happens with probability $2^{-u \cdot d}$.

6.4   Schur and Reorganization Pairs

The protocols Π_{Schur} and Π_{Reorg} (see the full version for more details) describe how to create the Schur and reorganization pairs. We now give a brief intuition of how these work.

Schur Pairs. We require random authenticated codewords $[\![C(\mathbf{r})]\!]^*, [\![C^*(\mathbf{s})]\!]^*$ such that the first k components of $\mathbf{r}$ and $\mathbf{s}$ are equal. Note that since $C \subset C^*$, it might be tempting to use the same codeword (in C) for both elements. However, this will be insecure – during the online phase, parties reveal elements of the form $[\![C^*(\mathbf{x} * \mathbf{y})]\!]^* - [\![C^*(\mathbf{s})]\!]^*$. If $C^*(\mathbf{s})$ is actually in the code C then it is uniquely determined by its first k components, which means $C^*(\mathbf{x} * \mathbf{y})$ will not be masked properly and could leak information on $\mathbf{x}, \mathbf{y}$.

Instead, we have parties authenticate a random codeword in C^* that is zero in the first k positions, reveal the MACs at these positions to check that this was honestly generated, and then add this to $[\![C(\mathbf{r})]\!]^*$ to obtain $[\![C^*(\mathbf{s})]\!]^*$. This results in a pair where the parties' shares are identical in the first k positions, however we prove in the full version that this does not introduce any security issues for the online phase.

Reorganizing Pairs. To produce the pairs $[\![C(\mathbf{r})]\!]^*, [\![C(f(\mathbf{r}))]\!]^*$, we take advantage of the fact that during BigMAC authentication, every component of a codeword vector has the same MAC key. This means linear functions can be applied across the components, which makes creating the required data very straightforward. Note that with MiniMAC shares, this would not be possible, since you cannot add two elements with different MAC keys.

7   Complexity Analysis

We now turn to analyzing the complexity of our triple generation protocols, in terms of the required number of correlated and random OTs (on κ-bit strings) and the number of parties n.

Two-Party TinyOT. The appendix of TinyOT [19] states that 54 aBits are required to compute an AND gate, when using a bucket size of 4. An aBit is essentially a passive correlated OT combined with a consistency check and some hashes, so we choose to model this as roughly the cost of an actively secure random OT.

Multi-party TinyOT. Note that although the original protocol of Larraia et al. [17] and the fixed protocol of Burra et al. [5] construct secret-shared OT quadruples, these are locally equivalent to multiplication triples, which turn out to be simpler to produce as one less authentication is required. Producing a triple requires one random OT per pair of parties, and the 3 correlated OTs per pair of parties to authenticate the 3 components of each triple. Combining twice, and sacrificing gives an additional overhead of B^3, where B is the bucket size. When creating a batch of at least 1 million triples with statistical security parameter 40, the proofs in the full version show that we can use bucket size 3, giving $81n(n-1)$ calls to $\mathcal{F}_{\mathsf{COTe}}$ and $27n(n-1)$ to $\mathcal{F}_{\mathsf{OT}}$.

Authentication. To authenticate a single bit, the $\Pi_{[\cdot]}$ protocol requires $n(n-1)$ calls to $\mathcal{F}_{\mathsf{COTe}}$. For full field elements in $\mathbb{F}_{2^k}$ this is simply performed k times, taking $kn(n-1)$ calls.

$\mathbb{F}_2$ **Triples.** The protocol starts with $n(n-1)$ calls to $\mathcal{F}_{\mathsf{COTe}}$ to create the initial triple and authenticate x; however, these are on strings of length 2κ rather than κ and also require a call to H, so we choose to count this as $n(n-1)$ calls to both $\mathcal{F}_{\mathsf{OT}}$ and $\mathcal{F}_{\mathsf{COTe}}$ to give a conservative estimate. Next, y and z are authenticated using $\mathcal{F}_{[\cdot]}$, needing a further $2n(n-1) \times \mathcal{F}_{\mathsf{COTe}}$.

We need to sacrifice once and combine once, and if we again use buckets of size 3 this gives a total overhead of 9x. So the total cost of an $\mathbb{F}_2$ triple with our protocol is $27n(n-1)$ $\mathcal{F}_{\mathsf{COTe}}$ calls and $9n(n-1)$ $\mathcal{F}_{\mathsf{OT}}$ calls.

$\mathbb{F}_{2^k}$ **Triples.** We start with $n(n-1)$ calls to $\mathcal{F}_{\mathsf{ACOT}}^{k,s}$, each of which requires $3k$ $\mathcal{F}_{\mathsf{OT}}$ calls, assuming that k is equal to the statistical security parameter. We then need to authenticate the resulting triple (three field elements) for a cost of $3kn(n-1)$ calls to $\mathcal{F}_{\mathsf{COTe}}$. The sacrificing step in the checked triple protocol wastes one triple to check one, so doubling these numbers gives $6kn(n-1)$ for each of $\mathcal{F}_{\mathsf{OT}}$ and $\mathcal{F}_{\mathsf{COTe}}$.

MiniMAC Triples. Each MiniMAC triple also requires one Schur pair for the sacrificing step and one Schur pair for the online phase multiplication protocol.

Codeword Authentication. Authenticating a codeword with Π_{CodeAuth} takes k calls to $\mathcal{F}_{[\cdot]}$ on u-bit field elements, giving $kun(n-1)$ COTe's on a $u \cdot m$-bit MAC key. Since COTe is usually performed with a κ-bit MAC key and scales linearly, we choose to scale by $u \cdot m/\kappa$ and model this as $ku^2mn(n-1)/\kappa$ calls to $\mathcal{F}_{\mathsf{COTe}}$.

Schur and Reorganization Pairs. These both just perform 1 call to $\mathcal{F}_{\mathsf{CodeAuth}}$, so have the same cost as above.

Multiplication Triples. Creating an unchecked triple first uses $n(n-1)$ calls to CodeOT on $k \cdot u$-bit strings, each of which calls $\mathcal{F}_{\mathsf{ACOT}}$, for a total of $(2ku+s)n(n-1)$ $\mathcal{F}_{\mathsf{OT}}$'s. The resulting shares are then authenticated with 3 calls to $\mathcal{F}_{\mathsf{CodeAuth}}$. Pairwise sacrificing doubles all of these costs, to give $2kun(n-1)(2ku+s)/\kappa$ $\mathcal{F}_{\mathsf{OT}}$'s and 6 calls to $\mathcal{F}_{\mathsf{CodeAuth}}$, which becomes $8ku^2mn(n-1)/\kappa$ $\mathcal{F}_{\mathsf{COTe}}$'s when adding on the requirement for two Schur pairs.

Parameters. [9] implemented the online phase using Reed-Solomon codes over $\mathbb{F}_{2^8}$, with $(m,k) = (256,120)$ and $(255,85)$, for a 128-bit statistical security level. The choice $(255,85)$ allowed for efficient FFT encoding, resulting in a much faster implementation, so we choose to follow this and use $u = 8, k = 85$. This means the cost of a single (vector) multiplication triple is $86700n(n-1)$ calls to $\mathcal{F}_{\mathsf{COTe}}$ and $14875(n-1)$ calls to $\mathcal{F}_{\mathsf{OT}}$. Scaling this down by k, the amortized cost of a single $\mathbb{F}_{2^u}$ multiplication becomes $1020(n-1)$ and $175(n-1)$ calls. Note that this is around twice the cost of $\mathbb{F}_{2^{40}}$ triples, which were used to embed the AES circuit by Damgård et al. [7], so it seems that although the MiniMAC online phase was reported by Damgård et al. [9] to be more efficient than other protocols for certain applications, there is some extra cost when it comes to the preprocessing using our protocol.

7.1   Estimating Runtimes

To provide rough estimates of the runtimes for generating triples, we use the OT extension implementation of Asharov et al. [1] to provide estimates for $\mathcal{F}_{\mathsf{COTe}}$ and $\mathcal{F}_{\mathsf{OT}}$. For $\mathcal{F}_{\mathsf{COTe}}$, we simply use the time required for a passively secure extended OT ($1.07\,\mu\mathrm{s}$), and for $\mathcal{F}_{\mathsf{OT}}$ the time for an actively secure extended OT ($1.29\,\mu\mathrm{s}$) (both running over a LAN). Note that these estimates will be too high, since $\mathcal{F}_{\mathsf{COTe}}$ does not require hashing, unlike a passively secure random OT. However, there will be additional overheads due to communication etc., so the figures given in Table 1 are only supposed to be a rough guide.

Acknowledgements. We would like to thank Nigel Smart, Rasmus Zakarias and the anonymous reviewers, whose comments helped to improve the paper. The first author has been supported by the Danish National Research Foundation and The National Science Foundation of China (under the grant 61361136003) for the Sino-Danish Center for the Theory of Interactive Computation and from the Center for Research in Foundations of Electronic Markets (CFEM), supported by the Danish Strategic Research Council. Furthermore, partially supported by Danish Council for Independent Research via DFF Starting Grant 10-081612 and the European Research Commission Starting Grant 279447. The second, third and fourth authors have been supported in part by EPSRC via grant EP/I03126X.

References

1. Asharov, G., Lindell, Y., Schneider, T., Zohner, M.: More efficient oblivious transfer extensions with security for malicious adversaries. In: Oswald, E., Fischlin, M. (eds.) EUROCRYPT 2015. LNCS, vol. 9056, pp. 673–701. Springer, Heidelberg (2015)
2. Beaver, D.: Efficient multiparty protocols using circuit randomization. In: Feigenbaum, J. (ed.) CRYPTO 1991. LNCS, vol. 576, pp. 420–432. Springer, Heidelberg (1992)
3. Bellare, M., Rogaway, P.: Random oracles are practical: a paradigm for designing efficient protocols. In: CCS 1993, Proceedings of the 1st ACM Conference on Computer and Communications Security, Fairfax, Virginia, USA, 3–5 November 1993, pp. 62–73 (1993)
4. Bendlin, R., Damgård, I., Orlandi, C., Zakarias, S.: Semi-homomorphic encryption and multiparty computation. In: Paterson, K.G. (ed.) EUROCRYPT 2011. LNCS, vol. 6632, pp. 169–188. Springer, Heidelberg (2011)

5. Burra, S.S., Larraia, E., Nielsen, J.B., Nordholt, P.S., Orlandi, C., Orsini, E., Scholl, P., Smart, N.P.: High performance multi-party computation for binary circuits based on oblivious transfer. Cryptology ePrint Archive, Report 2015/472 (2015). https://eprint.iacr.org/

6. Canetti, R.: Universally composable security: a new paradigm for cryptographic protocols. In: 42nd Annual Symposium on Foundations of Computer Science, FOCS 2001, Las Vegas, Nevada, USA, 14–17 October 2001, pp. 136–145 (2001)

7. Damgård, I., Keller, M., Larraia, E., Miles, C., Smart, N.P.: Implementing AES via an actively/covertly secure dishonest-majority MPC protocol. In: Visconti, I., De Prisco, R. (eds.) SCN 2012. LNCS, vol. 7485, pp. 241–263. Springer, Heidelberg (2012)

8. Damgård, I., Keller, M., Larraia, E., Pastro, V., Scholl, P., Smart, N.P.: Practical covertly secure MPC for dishonest majority – or: breaking the SPDZ limits. In: Crampton, J., Jajodia, S., Mayes, K. (eds.) ESORICS 2013. LNCS, vol. 8134, pp. 1–18. Springer, Heidelberg (2013)

9. Damgård, I., Lauritsen, R., Toft, T.: An empirical study and some improvements of the minimac protocol for secure computation. In: Abdalla, M., De Prisco, R. (eds.) SCN 2014. LNCS, vol. 8642, pp. 398–415. Springer, Heidelberg (2014)

10. Damgård, I., Pastro, V., Smart, N., Zakarias, S.: Multiparty computation from somewhat homomorphic encryption. In: Safavi-Naini, R., Canetti, R. (eds.) CRYPTO 2012. LNCS, vol. 7417, pp. 643–662. Springer, Heidelberg (2012)

11. Damgård, I., Zakarias, S.: Constant-overhead secure computation of Boolean circuits using preprocessing. In: Sahai, A. (ed.) TCC 2013. LNCS, vol. 7785, pp. 621–641. Springer, Heidelberg (2013)

12. Frederiksen, T.K., Keller, M., Orsini, E., Scholl, P.: A unified approach to MPC with preprocessing using OT. Cryptology ePrint Archive (2015, to appear). https://eprint.iacr.org/

13. Ishai, Y., Kilian, J., Nissim, K., Petrank, E.: Extending oblivious transfers efficiently. In: Boneh, D. (ed.) CRYPTO 2003. LNCS, vol. 2729, pp. 145–161. Springer, Heidelberg (2003)

14. Keller, M., Orsini, E., Scholl, P.: Actively secure OT extension with optimal overhead. In: Advances in Cryptology - CRYPTO 2015–35th Annual Cryptology Conference, Santa Barbara, CA, USA, 16–20 August 2015, Proceedings, Part I, pp. 724–741 (2015)

15. Keller, M., Scholl, P.: Efficient, oblivious data structures for MPC. In: Sarkar, P., Iwata, T. (eds.) ASIACRYPT 2014, Part II. LNCS, vol. 8874, pp. 506–525. Springer, Heidelberg (2014)

16. Keller, M., Scholl, P., Smart, N.P.: An architecture for practical actively secure MPC with dishonest majority. In: ACM Conference on Computer and Communications Security, pp. 549–560 (2013)

17. Larraia, E., Orsini, E., Smart, N.P.: Dishonest majority multi-party computation for binary circuits. In: Garay, J.A., Gennaro, R. (eds.) CRYPTO 2014, Part II. LNCS, vol. 8617, pp. 495–512. Springer, Heidelberg (2014)

18. Nielsen, J.B.: Extending oblivious transfers efficiently - how to get robustness almost for free. IACR Cryptology ePrint Archive 2007:215 (2007)

19. Nielsen, J.B., Nordholt, P.S., Orlandi, C., Burra, S.S.: A new approach to practical active-secure two-party computation. In: Safavi-Naini, R., Canetti, R. (eds.) CRYPTO 2012. LNCS, vol. 7417, pp. 681–700. Springer, Heidelberg (2012)

Secure Computation from Millionaire

Abhi Shelat[1]($\boxtimes$) and Muthuramakrishnan Venkitasubramaniam[2]

[1] University of Virginia, Charlottesville, VA, USA
abhi@virginia.edu
[2] University of Rochester, Rochester, NY, USA
muthuv@cs.rochester.edu

Abstract. The standard method for designing a secure computation protocol for function f first transforms f into either a circuit or a RAM program and then applies a generic secure computation protocol that either handles boolean gates or translates the RAM program into oblivious RAM instructions.

In this paper, we show a large class of functions for which a different *iterative* approach to secure computation results in more efficient protocols. The first such examples of this technique was presented by Aggarwal, Mishra, and Pinkas (J. of Cryptology, 2010) for computing the median; later, Brickell and Shmatikov (Asiacrypt 2005) showed a similar technique for shortest path problems.

We generalize the technique in both of those works and show that it applies to a large class of problems including certain matroid optimizations, sub-modular optimization, convex hulls, and other scheduling problems. The crux of our technique is to *securely reduce* these problems to secure comparison operations and to employ the idea of *gradually releasing* part of the output. We then identify conditions under which both of these techniques for protocol design are compatible with achieving simulation-based security in the honest-but-curious and covert adversary models. In special cases such as median, we also show how to achieve malicious security.

Keywords: Secure computation · Semi-honest · Covert security · Greedy algorithms

1 Introduction

Secure two-party computation allows Alice with private input x and Bob, with input y, to jointly compute $f(x, y)$ without revealing any information other than the output $f(x, y)$.

A. Shelat — Research supported by Google Faculty Research Grant, Microsoft Faculty Fellowship, SAIC Scholars Research Award, and NSF Awards TC-1111781, 0939718, 0845811.

M. Venkitasubramaniam — Research supported by Google Faculty Research Grant and NSF Award CNS-1526377.

© International Association for Cryptologic Research 2015
T. Iwata and J.H. Cheon (Eds.): ASIACRYPT 2015, Part I, LNCS 9452, pp. 736–757, 2015.
DOI: 10.1007/978-3-662-48797-6_30

Building on Yao's celebrated garbled circuits construction [25], many recent works [4,10,11,14–18,21] construct such protocols by first translating f into a boolean circuit and then executing a protocol to securely evaluate *each gate* of that circuit. Alternatively, Ostrovsky and Shoup [20] demonstrated a way to construct two-party secure computation protocols for RAM programs by first translating the RAM program into a sequence of oblivious RAM (ORAM) instructions and then applying a secure computation protocol to implement each ORAM operation. Further refinements of this idea and state of the art approaches to ORAM design [7,22,23] limit the overhead in terms of bandwidth, client storage and total storage to roughly $\tilde{O}(\log^3(n))$ for each operation on a memory of size n resulting in protocols [9,12,16,24] that are efficient enough for some problems in practice.

Reduction-based techniques. In both of the above approaches, the secure evaluation of f is reduced to the secure evaluation of either a boolean gate or an ORAM instruction.

Instead of reducing function f into such low-level primitives and securely evaluating *each* primitives, one can also consider reducing f into a program that only makes secure evaluations of a higher-level primitive. A natural candidate for this secure primitive is the comparison function, or the *millionaires* problem.

Aggarwal, Mishra, and Pinkas [1] begin to investigate this approach by studying the problem of securely computing the kth-ranked element of dataset $D_A \cup D_B$ where Alice privately holds dataset $D_A \subset F$ and Bob privately holds dataset $D_B \subset F$. They reduce the computation of the kth-ranked element to $O(\log k)$ secure comparisons of $(\log M)$-bit inputs where $\log M$ is the number of bits needed to describe the elements in F; this protocol outperforms the naive method for the same problem since a circuit for computing ranked elements has size at least $|D_A \cup D_B|$.

Their algorithm follows the classic communication-optimal protocol for this problem: each party computes the median of its own dataset, the parties then jointly compare their medians, and depending on whose median is larger, each party eliminates half of its input values and then recurses on the smaller datasets. Aggarwal, Mishra and Pinkas observe that by replacing each comparison between the parties' medians with a secure protocol for comparing two elements, they can argue that the overall protocol is secure in the honest-but-curious setting. In particular, for the case of median, they observe that the sequence of answers from each secure comparison operation can be simulated using only the output k^{th}-ranked element.

Brickell and Shmatikov [6] use a similar approach to construct semi-honest secure computation protocols for the all pairs shortest distance (APSD) and single source shortest distance (SSSD) problems. In both cases, their protocols are more efficient than circuit-based secure computation protocols. While the work of [6] considers only the two-party setting the work of [1] additionally considers the multiparty setting.

1.1 Our Results

We continue the study of reducing the secure evaluation of function f to secure evaluation of comparisons in the *two-party* setting. Our first contribution is to generalize the approach of Aggarwal, Mishra, and Pinkas and that of Brickell and Shmatikov as the parameterized protocol in Fig. 1. The parameters to this protocol are the comparison function $\mathbf{LT}_f$, and the method F. The comparison function takes two input elements with their corresponding key values and returns the element with the smaller key value; F is the local update function that determines how each party determines its input for the next iteration based on the answers from the previous iteration $(c_1 \ldots, c_j)$ and its local input U (or V).

We show that this parameterized protocol can be used to construct more efficient secure computation protocols for a much larger class of optimization problems than the three specific instances they considered. In Sect. 4, we construct choices for $\mathbf{LT}, F$ that securely compute several combinatorial optimization problems, matroid optimization problems, sub-modular optimizations, and computation of the convex hull.

A key reason for the improved efficiency of this approach over both circuits and ORAM techniques is the fact that the output is *gradually released* to both parties. The result of one iteration of the loop is used to select inputs to the next iteration of the loop; more generally, the output of the secure computation can be thought to be released *bit-by-bit*, *node-by-node*, or *edge-by-edge*. Thus it is not immediately clear that such an approach can be secure, even against honest-but-curious adversaries.

Our next contribution is to show that an instantiation of the above generic protocol can be made secure in the honest-but-curious model when the functions f is *greedy-compatible*, i.e. it satisfies a few simple properties. First, the problem must have a unique solution. One can often guarantee this property by specifying simple rules to break ties among comparisons. Second, the order in which the output is revealed must be unique, and finally, we require a local updatability property for the function F which essentially states that F has a very weak homomorphism with the comparison function $\mathbf{LT}$. (See Definition 1). When these

---

GENERIC ITERATIVE SECURE COMPUTATION

Alice Input: Distinct elements $U = \{u_1, \ldots, u_n\}$

Bob Input: Distinct elements $V = \{v_1, \ldots, v_n\}$

Output: The final output is $c_1, \ldots, c_\ell$

1. Alice initializes $(u_a, k_a) \leftarrow F(\perp, U)$ and Bob initializes $(v_b, k_b) \leftarrow F(\perp, V)$.
2. Repeat for $\ell(|U|, |V|)$ times:
   (a) Alice and Bob execute the secure protocol $c_j \leftarrow \mathbf{LT}_f((u_a, k_a), (v_b, k_b))$.
   (b) Alice updates $(u_a, k_a) \leftarrow F((c_1, \ldots, c_j), U)$ and Bob updates $(v_b, k_b) \leftarrow F((c_1, \ldots, c_j), V)$.

Fig. 1. The generic structure of a secure iterative protocol.

Algorithm	This Work	Circuit	ORAM
Convex Hull	$O(|Z|l_M)$	$\Omega(|I|^2 l_M)$	$\Omega(|I|\log^3|I|l_M)$
MST	$O(Vl_M)$	$\Omega((V\alpha(V))^2 l_M)$	$\Omega(V\alpha(V)\log^3 Vl_M)$
Unit Job Sched	$O(|Z|l_M)$	$\Omega(|I|^2 l_M)$	$\Omega(|I|\log^3|I|l_M)$
Single-Src ADSP	$O(Vl_M)$	$\Omega(E^2 l_M)$	$\Omega(E\log^3 El_M)$
Submodular Opt	$O(|Z|l_M)$	$\Omega(|I_S|^2 l_M)$	$\Omega(|I_S|\log|I_S|l_M)$

Fig. 2. Communication costs for secure protocols in the semi-honest case. $I = U \cup V$ the union of Alice and Bob's inputs and Z is the output. V and E are the number of vertices and edges in graph problems. $\alpha(\cdot)$ is the Inverse Ackermann function. For problems where each element is a set, I_S represents the sum of the set sizes. $l_M = \log M$ where M typically represents the maximum integer values the inputs can take. For each case, the complexity for the generic Circuit-based approach was obtained by relating it to the number of (dependent) memory accesses made by the best algorithm and the ORAM complexity was obtained by relating it to the time complexity of best known algorithm. In many cases, our communication complexity is related to the output-size, which can be much smaller than the input size for many problems.

three conditions are met, and when the $\mathbf{LT}_f$ function can be securely evaluated efficiently, then the instantiated protocol can be asymptotically (and practically) superior to other approaches. See Fig. 2 for several examples.

1.2 Malicious and Covert Security

We also consider stronger notions of security for our protocols, namely security against fully malicious adversaries, and security against covert adversaries (which can be achieved much more efficiently) in the two-party setting.

Recall from the previous section that efficiency gain of our approach owes in part to the gradual release of output during each iteration. Herein lies a difficulty: A malicious adversary can manipulate its input used at each iteration of the protocol based on the results from previous iterations. This cheating ability complicates the construction of a proper Ideal-simulator.

As a first step, we can require the adversary to commit to its input before the protocol starts and force the adversary to only use committed data. To perform a simulation, we will first attempt to extract the adversaries' input, and then use this input to simulate the rest of the computation. We can use standard ideas with extractable commitments to perform this extraction. However, this is not enough. The main technical problem arises in the case that the adversary *selectively aborts* or *selectively uses* his committed inputs in the iterative protocol based on the intermediate results of the computation.

The prior work of Aggarwal, Mishra and Pinkas [1] claim that their simulation works for malicious adversaries; however, their simulation fails to account for the case when the adversary *selectively aborts*. Our second technical contributions is to present a *hardened* version of the protocol by Aggarwal, Mishra and Pinkas [1] for securely computing the median and a simulation argument which proves that it achieves malicious security.

As we discuss in Sect. 6.1, the techniques we use to show full malicious security rely on two specific properties that holds for the median problem. At a high level, for any input A of Alice and any element $a \in A$, we need that only one sequence of outputs from the previous iterations of our general protocol framework lead to Alice using a as an input. Furthermore, there are at most polynomially-many execution "traces" for any set of inputs from Alice and Bob (in contrast, graph problems have exponentially many traces). If there were multiple traces that lead to the use of element $a \in A$, then the adversary can selectively decide to abort on one of the traces and such an adversary cannot be simulated since its view depends on the trace and therefore the honest party's input. If the second property fails to hold then it can be argued that it would be hard for the simulator to extract the "right" input of the adversary.

Indeed, the selective abort issue seems to be a fundamental bottleneck to overcome. When these two properties fail to hold, e.g. in the case of the convex hull, or submodular optimization problems, we augment our basic protocol into one that achieves *covert security* as introduced Aumann and Lindell [2]. Covert security guarantees that if an adversary deviates in a way that would enable it to "cheat", then the honest party is guaranteed to detect the cheating with reasonable probability. The covert model handles situations in which a malicious party has a strong incentive "not to be caught cheating," while offering substantial improvements in efficiency versus the fully malicious model.

To achieve covert security, we must handle both selective aborts, and also ensure that an adversary does not cheat by using only a subset of its committed input during the protocol (perhaps based on a predicate of intermediate output). To handle this last issue, we require the adversary to prove at the end of the protocol that all of the committed inputs are either part of the output or used properly during the protocol. The adversary will provide one proof per "element" of the input, and thus, we need to design proofs that are sub-linear in the input size n, preferably logarithmic or even constant-sized.

For each of our selected problems, we provide these novel consistency checks. In cases such as the convex-hull, single-source shortest paths and job-scheduling, these checks are simple and have constant size (modulo the security parameter). For the case of the Minimum Spanning Tree, however, we required an elegant application of the Union-Find data structure to achieve communication efficient consistency checks for this problem. We summarize our performance for many problems in Fig. 3.

Although one might be able to use either Universal arguments [3,19] or SNARKS [5,8] to achieve malicious security with low communication, both of those techniques dramatically increase the computational overhead of the protocol. In particular, when proving an NP-relation of size t on an input statement x, the prover's computational complexity is proportional to $\tilde{O}(t)$ and the

Algorithm	This Work (covert)	Circuit (malicious)
Convex Hull	$O(|Z|l_M + |I|l_M^2)$	$\Omega(|I|^2 l_M)$
MST	$O(V \log V l_M)$	$\Omega((V\alpha(V))^2 l_M)$
Unit Job Scheduling	$O((|Z| + |I|)l_M)$	$\Omega(|I|^2 l_M)$
Single-Source ADSP	$O((V + E)l_M)$	$\Omega(E^2 l_M)$

Fig. 3. Comparison of the communication costs of covert security with the malicious security using circuits ignoring $poly(k)$ factors. $l_M = \log M$. $I = U \cup V$ the union of Alice and Bob's inputs. V and E are the number of vertices and edges in graph problems. We remark that since we ignore $poly(k)$ factors, the complexity of the Circuit-based approach would be the same as above even if we considered only achieving covert security. We were unable to estimate costs for achieving malicious security with ORAM.

verifier's computational complexity is proportional to $\tilde{O}(|x|)$. In our context, since such a proof will be required in each iteration, the computational complexity for both parties would be $\tilde{O}(|Z| \times |I|) + c(f)$ where Z and I are the inputs and outputs of the computation and $c(f)$ is the complexity for computing f itself. In contrast, our covert security and semi-honest protocol computation complexity is $O(|Z| + |I|) + c(f)$.

2 Preliminaries

We denote $(c_1, \ldots, c_j)$ by $c_{\leq j}$. Two sequences of distributions $\{C_n\}_{n \in \mathbb{N}}$ and $\{D_n\}_{n \in \mathbb{N}}$ are said to be computationally indistinguishable if for any probabilistic polynomial time algorithm A, $|\Pr[A(C_n) = 1] - \Pr[A(D_n) = 1]|$ is a negligible function in n. We formally describe a generalized millionaire (comparison) function in Fig. 4.

GENERALIZED COMPARE

Alice Input: Tuple (u, x) with k-bit integer key x

Bob Input: Tuple (v, y) k-bit integer key y

LT$_f$ Output: Return u if $x < y$ and v otherwise

Fig. 4. Generic comparison protocol

3 Honest-But-Curious Protocols

For many well-known greedy algorithms we show how to securely compute them using our Generic Protocol specified in Fig. 1. On a high-level, in this protocol abstraction, Alice and Bob have a set of inputs U and V. In every iteration, each of them provide an input element e from their local inputs and an associated key k_e to the comparison functionality $\mathbf{LT}_f$ which returns the element with smaller key value. More precisely, in iteration i, Alice supplies input (u_a, k_a) and Bob supplies input (v_b, k_b) where $u_a \in U$ and $v_b \in V$ to the $\mathbf{LT}_f$-functionality. The functionality returns as output $c_i = k_a < k_b ? u_a : v_b$. At the end of each iteration there is a local update rule that determines the next input and key for the next iteration. Finally, Alice and Bob output $c_1, \ldots, c_\ell$ as their outputs.

We make the following requirements on the function f that we wish to compute using a greedy-algorithm. For each instantiation, we show that the requirements are satisfied.

Definition 1. *We say that a two-party function f is greedy compatible if there exists functions $\mathbf{LT}, F$ such that the following holds:*

1. **Unique Solution:** *Given the inputs U and V of Alice and Bob, there is a unique solution.*
2. **Unique Order:** *There is a unique order in which the greedy-strategy outputs the solution. More precisely,*

$$f(U, V) = (c_1, \ldots, c_\ell)$$

   where $c_1 = F(\perp, U \cup V)$ and $c_{i+1} = F(c_{\leq i}, U \cup V)$ for every $i = 1, \ldots, \ell - 1$.
3. **Local Updatability:** *Informally, we require that F on the union of Alice and Bob's inputs can be obtained by applying F locally to U and V and then computing a comparison. More precisely, we require that*

$$F_1(c_{\leq j}, U \cup V) = \mathbf{LT}_f(F(c_{\leq j}, U), F(c_{\leq j}, V))$$

   where F_1 represents the first member in the tuple output by F.

3.1   Honest-but-Curious Security

Theorem 1. *For any function f that is greedy compatible, the Generic Iterative Secure Computation algorithm from Fig. 1 securely computes f on the union of the inputs held by Alice and Bob, for the case of semi-honest adversaries.*

We argue correctness and privacy of our protocols. Our analysis of the protocol will be in the $\mathbf{LT}_f$ hybrid, where the parties are assumed to have access to a trusted party computing the $\mathbf{LT}_f$.

Correctness: First we observe that if U and V are Alice and Bob's inputs, then from the Unique Order property it holds that $f(U, V) = (c_1, \ldots, c_\ell)$ where $c_1 = F(\perp, U \cup V)$ and $c_{i+1} = F(c_{\leq i}, U \cup V)$ for $i = 1, \ldots, \ell - 1$. The output computed by Alice and Bob by executing the Generic Iterative Secure Computation algorithm is $\tilde{c}_1, \ldots, \tilde{c}_\ell$ where

$$\tilde{c}_1 = \mathbf{LT}_f(F(\perp, U), F(\perp, V))$$
$$\tilde{c}_{i+1} = \mathbf{LT}_f(F(\tilde{c}_{\leq i}, U), F(\tilde{c}_{\leq i}, V)) \text{ for } i \text{ in } \{1, \ldots, \ell - 1\}$$

Correctness now follows from the Local Updatability property of f.

Privacy: Next to prove security in the honest-but-curious case, we construct a simulator that given the parties input and output can simulate the interaction indistinguishably.

Recall that, our analysis of the security of the protocol is in the $\mathbf{LT}_f$ hybrid. Thus the simulator that we describe will play the trusted party implementing $\mathbf{LT}_f$, when simulating the adversary. Below we prove security when one of the parties are corrupted. We argue for the case when Alice is corrupted and the case for Bob follows symmetrically since the protocol is symmetric in both parties.

Alice is corrupted. The simulator needs to produce a transcript indistinguishable to the honest adversary A_h in the $\mathbf{LT}_f$ hybrid.

- The simulator upon corrupting Alice receives her input U. It feeds U to the ideal functionality computing f to receive the output $c_{\leq l}$.
- Next run the honest Alice's code for the Generic Algorithm. Alice in iteration i for $i = 1, \ldots, \ell$, submits an input (u_a, k_a) to the $\mathbf{LT}_f$ functionality. S simulates the output by feeding c_i to Alice.
- Finally, at the end of ℓ-iterations, S outputs the view of Alice.

From the Unique Order property and the fact that Alice is honest, the view generated by S in the $\mathbf{LT}_f$-hybrid is identical to the view of Alice in the real experiment. More precisely,

$$\mathrm{IDEAL}^{\mathbf{LT}_f}_{f,S(z),I}(U, V, k) \equiv \mathrm{REAL}^{\mathbf{LT}_f}_{f,A_h(z),I}(U, V, k)$$

Security against semi-honest adversaries follows from a standard composition theorem (omitted) which concludes this proof sketch.

4 Instantiations of Our Protocol

4.1 Convex Hull

In this problem, Alice and Bob have as input sets of points U and V in a plane and the goal is to securely compute the convex hull of the union of points. Each element $u = (x, y)$ consists of two $\log M$-bit integers that represent the X and Y coordinate of the point. We assume that the union of points are such that no two points share the same X-coordinate and no three of them are collinear. The function F for the convex hull is defined as $F(c_{\leq j}, U) = (u_a, k_a)$ where:

- If $j = 0$, then u_a is point with the least value for the x-coordinate (i.e. the leftmost point) and k_a is set to be the x-coordinate of u_a.
- If $j > 0$, u_a is the point in U that attains the minimum value for $\mathsf{angle}(c_j, c)$ where $\mathsf{angle}(\mathsf{pt}_1, \mathsf{pt}_2)$ is the (clockwise) angle made by the line joining pt_1 and pt_2 with the vertical drawn through pt_1 and $k_a = \mathsf{angle}(c_j, a)$.

The correctness of the Convex-hull instantiation follows from the Gift-Wrapping (or Jarvis march) algorithm. Furthermore, it is easy to verify that Convex Hull is **greedy compatible** with F if no two-points have the same x or y coordinate and no three-points are collinear. Hence, we have the following theorem.

Theorem 2. *The* GENERIC ITERATIVE SECURE COMPUTATION *protocol instantiated with the F described above securely computes the convex hull of the union of inputs of Alice and Bob, for the case of semi-honest adversaries, assuming all inputs of Alice and Bob are distinct, no two of which share the same x-coordinate and no three points are collinear.*

Overhead: The total number of rounds of communication is $|Z|$, the size of the convex-hull Z which is at most $|I|$ where $I = U \cup V$. In each round, the protocol performs at most one secure comparison of $\log M$-bit integers. A circuit for performing the comparison has $O(\log M)$ gates and $\log M$ inputs. The overhead of the protocol for computing this circuit, secure against semi-honest adversaries, is $\log M$ oblivious-transfers. This can be thought of as $O(\log M)$ public-key operations, $O(\log M)$ symmetric key operations and communication of $O(\log M)$. The overall communication complexity is $O(|Z| \log M)$.

In comparison, the naive circuit implementation will have at least $|I|$ (dependent) memory accesses which will result in a circuit size of $\Omega(|I|^2 \log M)$. If we considered an ORAM implementation it would result in total communication of $\Omega(|I| \log^3 |I| \log M)$ since the best algorithm would require $O(|I| \log |I|)$ steps and the overhead for each step is $\log^2 |I|$ since we need to maintain a memory of size $O(|I|)$.

In the full version, we provide more examples: Job Interval Scheduling problem; general Matroid optimization problems for which membership in set $\mathcal{I}$ can be tested locally including minimum spanning tree problems and unit cost scheduling problems; the single-source shortest distance problem; and sub modular optimization problems such as set-cover and max cover approximations.

5 Covert Security

We describe the main issues to overcome with the current protocol:

Adaptively chosen inputs. As our iterative protocol gradually releases the answer, it is possible for the adversary to modify its input as the protocol proceeds. To defend, we include an input commitment phase. Then in the secure computation phase, the adversary provides decommitments with every input it uses in the computation of the LT_f-functionality.

Missing inputs. Consider an adversary that commits to its inputs but fails to follow the greedy strategy, namely, does not perform the local update rule using F honestly. This is an issue even if the adversary is restricted to only use inputs that it committed to because it can adaptively decide to use only a subset of them. Consider the minimum spanning tree problem in which the adversary can decide to drop a certain edge based on the partial output released before an iteration. To prevent this attack, we will rely on digital signatures.

Alice and Bob will first pick signature keys and share their verification keys. Next, in every computation using LT_f, Alice and Bob will obtain signatures of the output along with some specific *auxiliary* information that will later be used by each party to demonstrate honest behavior. More precisely, after the secure computation phase and the output is obtained, for every input $u \in U$ of Alice, it does the following:

- If u is part of the output, then we require Alice to prove to Bob that it has a signature on u under Bob's key and modify $\mathbf{LT}_f$ to reveal the Commitment of u to Bob in that iteration. This will allow Bob to determine which of the Commitments made by Alice in the input commitment phase is not part of the output.
- If u is not part of the output, Alice proves to Bob that u is not part of the solution. We prove in many of our examples how we can achieve this efficiently. In essence, Alice will show that in the iteration after which u was eliminated, a better element was chosen. For instance, in the minimum spanning tree problem, we demonstrate that an edge $e = (a, b)$ was eliminated because a cheaper edge e' got added to the output that connected the components containing vertices a and b.

Input Commitment Phase: To resolve, we add an Input Commitment Phase at the beginning of the protocol and a Consistency-Check Phase at the end of the protocol described in the Sect. 5.1. In an Input Commitment Phase executed at the beginning of the protocol, both parties commit to their input using an extractable commitment scheme Π_{Ext}.

Modifications to $\mathbf{LT}_f$ functionality: Besides the inputs (u_a, k_a) and (v_b, k_b) that Alice and Bob submit, they also submit $(\mathsf{aux}_a, \mathsf{sk}_A)$ and $(\mathsf{aux}_b, \mathsf{sk}_B)$ which are the auxiliary information corresponding to their inputs and their signing keys respectively. The function besides outputting the answer as in Fig. 1 additionally signs (u_a, aux_a) if u_a is the output and (u_b, aux_b) if u_b is the output using both keys sk_A and sk_B. We remark here that for modularity we describe that the signatures are computed by $\mathbf{LT}$ functionality. However, in all our instantiations the message to be signed (u, aux) in the ith iteration can be computed directly from the outputs of the current and previous calls to the $\mathbf{LT}_f$ functionality, namely, $c_1, \ldots, c_i$ and signature of these messages under the keys of Alice and Bob can be computed and sent directly to the other party. In particular, these signatures need not be computed securely.

Consistency-Check Phase: Alice and Bob need to prove they followed the greedy strategy at every iteration. Recall that, c_i for each i belongs to Alice or Bob. Alice proves that corresponding to every commitment C in the Input Commitment Phase, there exists an input u such that either

- u is one of the c_i's and it has a signature on c_i using sk_B, or
- u could not have been selected by the greedy strategy.

We achieve this by securely evaluating this consistency check procedure where in the first case, u is revealed to both parties and in the second case, only the result of the check is revealed.

5.1   Generic Algorithm for Covert Security

We make the following requirements on the function f we compute. For each instantiation, we show that the requirements are satisfied. We say that a function f is covert-greedy compatible if it is greedy compatible and additionally the following holds:

GENERALIZED COMPARE WITH COVERT SECURITY

Alice Input: Tuple $(u_a, x, \mathsf{aux}_a, \mathsf{sk}_A)$ with k-bit integer key x

Bob Input: Tuple $(v_b, y, \mathsf{aux}_b, \mathsf{sk}_B)$ k-bit integer key y

$\mathbf{LT}_f$ **Output:** $(u_a, \mathsf{aux}_a, \sigma_A, \sigma_B)$ if $x < y$ and $(v_b, \mathsf{aux}_b, \sigma_A, \sigma_B)$ otherwise where σ_A and σ_B are signatures on message m under keys sk_A and sk_B respectively and $m = (u_a, \mathsf{aux}_a)$ if $x < y$ and $m = (v_b, \mathsf{aux}_b)$ otherwise.

Fig. 5. Generic Millionaire's protocol with Covert security

- **Consistency Check:** There exists a consistency-check procedure CC, functions key, wit and aux which satisfies the following property: Given inputs U and V for Alice and Bob and any output $\tilde{c}_1, \ldots, \tilde{c}_\ell$, it holds that, for every input u of Alice (respectively, v of Bob), such that: $CC(u, \mathsf{key}(u), \{\tilde{c}_i, \mathsf{aux}(i, \tilde{c}_i)\}_{i \in I})$
  - Returns TRUE: if u is not part of the solution and $I = \mathsf{wit}(U)$ or $u = u_i$ for some $i \in I$
  - Returns FALSE: if u is in the solution and $u \neq u_i$ for any $i \in I$.
  Furthermore, we require that $\mathsf{aux}(u_a)$ for an input u_a in iteration i can be determined by $c_{<i}$.

GENERIC CONSISTENCY CHECK

Prover Input: Tuple $(u, \mathsf{key}(u), C, D, \{u_i, \mathsf{aux}_i, \sigma^i\}_{i \in I})$

CC_{vk} **Output:** It outputs 1 to Verifier and additionally outputs u when $u = u_i$ for some i, if all the following hold:

1. *Correct Input:* D is a valid decommitment information for C to $(u, \mathsf{key}(u))$.
2. *Consistency Check:* Either $CC(u, \mathsf{key}(u), \{u_i, \mathsf{aux}_i\}_{i \in I})$ returns true or $u_i = u$ for some $i \in I$.
3. *Signature Check:* $\mathsf{Ver}_{\mathsf{vk}}((u_i, \mathsf{aux}_i), \sigma^i) = 1$ for $i \in I$

Fig. 6. Generic Consistency Check Procedure GCC_{vk}

Let $\Pi_{\mathsf{Ext}} = \langle C, R \rangle$ be an extractable commitment scheme. In Fig. 7, we give our general protocol to achieve covert security. Then for each of our problems, we

---

GENERIC ITERATIVE SECURE COMPUTATION WITH COVERT SECURITY

Alice Input: A set of distinct elements $U = \{u_1, \ldots, u_n\}$

Bob Input: A set of distinct elements $V = \{v_1, \ldots, v_n\}$

Output: $c_1, \ldots, c_\ell$.

Input Commitment Phase:

1. For every $i \in [n]$, Alice acting as the Sender with input $m = (u_i, \mathsf{key}(u_i))$ interacts with Bob as the Receiver using the protocol Π_{Ext}.

2. For every $i \in [n]$, Bob acting as the Sender with input $m = (v_i, \mathsf{key}(v_i))$ interacts with Alice as the Receiver using the protocol Π_{Ext}.

3. Alice and Bob run $\mathsf{Gen}(1^k)$ to obtain the key-pairs $(\mathsf{sk}_A, \mathsf{vk}_A)$ and $(\mathsf{sk}_B, \mathsf{vk}_B)$ respectively. Alice sends vk_A to Bob and Bob sends vk_B to Alice.

Secure Computation Phase:

1. Alice initializes $(u_a, k_a) \leftarrow F(\bot, U)$ and Bob initializes $(v_b, k_b) \leftarrow F(\bot, V)$.

2. Repeat for $\ell(|U|, |V|)$ times:

   (a) Alice and Bob execute the protocol computing $\mathbf{LT}_{f, \mathsf{vk}_A, \mathsf{vk}_B}$ on inputs $(u_a, k_a, \mathsf{aux}(i, u_a), \mathsf{sk}_A)$ and $(v_b, k_b, \mathsf{aux}(i, u_b), \mathsf{sk}_B))$ and receives as output $(c_j, \sigma_A, \sigma_B)$ where i is the iteration number.

   (b) Alice updates $(u_a, k_a) \leftarrow F(c_{\leq j}, U)$ and Bob updates $(v_b, k_b) \leftarrow F(c_{\leq j}, V)$. They store σ_B and σ_A respectively.

3. Alice outputs $c_1, \ldots, c_\ell$. Bob outputs $c_1, \ldots, c_\ell$.

Consistency Check Phase:

1. For every commitment C made by Alice in the Input Commitment Phase to an element u_a, Alice and Bob execute the protocol

$$CC_{\mathsf{vk}_B}((u_a, key(u_a), C, D, \sigma, \{(c_i, \mathsf{aux}_i, \sigma_B^i)\}_{i \in I}), \bot)$$

where (a) $I = \mathsf{wit}(u_a)$, if u_a is not part of the output, (b) σ is a signature on a message of the form $(u_a, \cdot)$, otherwise. If CC outputs 0, then Bob outputs $corrupt_A$. If CC returns u_a, Bob stores u_a in store $OutCheck_A$.

2. For every commitment C made by Bob in the Input Commitment Phase to an element v_b, Alice and Bob execute the protocol

$$CC_{\mathsf{vk}_A}((v_b, key(v_b), C, D, \sigma, \{(c_i, \mathsf{aux}_i, \sigma_A^i)\}_{i \in I}), \bot)$$

where (a) $I = \mathsf{wit}(v_b)$, if v_b is not part of the output, (b)σ is a signature on a message of the form $(v_b, \cdot)$, otherwise. If CC outputs 0, then Alice outputs $corrupt_B$. If CC returns v_b, Bob stores v_b in store $OutCheck_B$.

3. If $OutCheck_A$ is not equal to $c_{\leq \ell}$ minus the elements that are part of Bob's input, then Bob outputs $corrupt_A$.

4. If $OutCheck_B$ is not equal to $c_{\leq \ell}$ minus the elements that are part of Alice's input, then Alice outputs $corrupt_A$.

Fig. 7. The generic structure of a secure iterative protocol with covert security

specify how we modify the $\mathbf{LT}_f$ functionality and provide the Consistency Check procedure. Let com be a statistically-binding commitment scheme. In Fig. 6, we give the generic structure of the consistency-check procedure.

Theorem 3. *Let f be a functionality that is* **covert-greedy compatible.** *Then the Generic Covert Security protocol described in Fig. 7 securely computes f in the presence of covert adversaries with 1-deterrence.*

Proof. Our analysis of the security of the protocol is in the $\mathbf{LT}_f, CC_{vk}$ hybrid, where the parties are assumed to have access to a trusted party computing the respective functionalities. Thus the simulator that we describe will play the trusted party implementing the two functionalities, when simulating the adversary. We consider the different corruption cases: (1) When no party is corrupted (2) When one of the parties are corrupted. In the first case, the security reduces to the semi-honest case and follows the proof presented in Sect. 3.1. Below we prove security when one of the parties are corrupted. We argue for the case when Alice is corrupted and the case for Bob follows symmetrically since the protocol is symmetric in both parties.

Alice is corrupted. On a high-level, by our assumptions there is a unique solution and a unique order in which the output is revealed in every iteration. More precisely, given the optimal solution, the output of each iteration using $\mathbf{LT}_f$ is determined. Furthermore, this output is either an element in Alice's input or Bob's input. The simulator S fixes A's random tape unfiromly at random and proceeds as follows:

1. S executes the Input Commitment Phase playing the role of Bob. For all commitments made by Alice, S runs the extractor algorithm E provided by the Π_{Ext} protocol to create a transcript and extract all of Alice's input. For all of Bob's, S commits to the all 0 string.
2. Now S has Alice's input which it feeds to the ideal functionality computing f and receives the output $c_1, \ldots, c_\ell$. Next S interacts with Alice in the Secure Computation Phase. In iteration i, S receives Alice's input $(u_a, x, \mathsf{aux}_a, \mathsf{sk}_A)$ for $\mathbf{LT}_f$. S can check if Alice's input is correct, by performing the computation of $\mathbf{LT}_f$ with Alice's input as (u_a, x). If $u_a = c_i$ then S simply outputs c_i as the output of the computation. Otherwise, c_i must be Bob's input and the simulator checks if the computation with Bob's input as c_i results in c_i. If it is not, S outputs BAD$_i$ and halts.
3. If S successfully completes the Secure Computation Phase, it proceeds to simulate the Consistency Check Phase. In this phase, Alice first proves consistency for every commitment C it made in the Input Commitment Phase by providing input to the CC_{vk_B} functionality. The simulator evaluates the input using the procedure honestly and sends $corrupt_A$ if the procedure returns 0.
4. Finally, for every Commitment made by Bob that is not part of the input, S simply sends what the CC_{vk} functionality should send if Bob is honest, namely, it sends 1 to Alice.

This concludes the description of the simulator S. We now proceed to prove covert-security. First, we prove the following claim.

Consider an adversarial Alice A^*. We prove indistinguishability in a hybrid experiment H where we construct another simulator S' that knows Bob's input.

In this experiment S' proceeds identically to S with the exception that in the Input Commitment Phase it commits to the real inputs of Bob instead of the all 0 string as S would. Indistinguishability of the output of S and S' follows directly from the hiding property of the commitment scheme and the fact that the views are in the $\mathbf{LT}, CC$-hybrid. More precisely,

$$\text{IDEAL}^{\mathbf{LT},CC}_{f,S(z),I}(x_1,x_2,k) \approx \text{IDEAL}^{\mathbf{LT},CC}_{f,S'(z),I}(x_1,x_2,k)$$

Next, we argue indistinguishability of the hybrid experiment H and the real experiment. First, we observe that both these experiment proceed identically in the Input Commitment Phase. Fix a partial transcript τ of the view at the end of Input Commitment Phase. Let $c_1,\ldots,c_\ell$ be the output obtained by S'. It now follows that, conditioned on the Secure Computation Phase outputting $c_1,\ldots,c_\ell$, the views in H and the real experiment are identically distributed. This is because the simulator honestly computes the $\mathbf{LT}$-functionality and the output is completely determined by $c_{\leq i}$ (c_i is the output of the iteration and the other previous outputs are required to determine aux_b). For any view v of the experiment, let $\mathsf{Success}(v)$ denote this event. Let BAD denote the union of all events BAD_i. It follows that

$$\Pr[v \leftarrow \text{IDEAL}^{\mathbf{LT},CC}_{f,S'(z),I}(x_1,x_2,k) : D(v)=1 \wedge \neg\text{BAD}]$$
$$= \Pr[v \leftarrow \text{REAL}^{\mathbf{LT},CC}_{f,A^*(z),I}(x_1,x_2,k)) : D(v)=1 \wedge \mathsf{Success}(v)] \quad (1)$$

Assume for contradiction, the simulation did not satisfy covert security with 1-deterrence. Then there exists an adversary A, distinguisher D and polynomial $p(\cdot)$ such that

$$\left| \Pr[D(\text{IDEAL}^{\mathbf{LT},CC}_{f,S'(z),I}(x_1,x_2,k)) = 1] - \Pr[D(\text{REAL}^{\mathbf{LT},CC}_{f,A^*(z),I}(x_1,x_2,k))) = 1] \right|$$
$$\geq \Pr[out_B(\text{REAL}_{f,A^*(z),I}(x_1,x_2,k)) = \mathsf{corrupt}_A] + \frac{1}{p(k)}$$

Using Eq. 1, we rewrite the above equation as follows:

$$\left| \Pr[v \leftarrow \text{IDEAL}^{\mathbf{LT},CC}_{f,S'(z),I}(x_1,x_2,k) : D(v)=1 \wedge \text{BAD}] \right.$$
$$\left. - \Pr[v \leftarrow \text{REAL}^{\mathbf{LT},CC}_{f,A^*(z),I}(x_1,x_2,k)) : D(v)=1 \wedge \neg\mathsf{Success}(v)] \right|$$
$$\geq \Pr[out_B(\text{REAL}_{f,A^*(z),I}(x_1,x_2,k)) = \mathsf{corrupt}_A] + \frac{1}{p(k)} \quad (2)$$

Below in Claim 1, we show that $\Pr[\text{BAD}]$ is negligible close to $\Pr[\neg\mathsf{Success}]$. Furthemore, if $\neg\mathsf{Success}$ occurs, then it must be the case that Bob outputs $corrupt_A$. Therefore,

$$\left| \Pr[v \leftarrow \text{IDEAL}^{\mathbf{LT},CC}_{f,S'(z),I}(x_1,x_2,k) : D(v)=1 \wedge \text{BAD}] \right.$$
$$\left. - \Pr[v \leftarrow \text{REAL}^{\mathbf{LT},CC}_{f,A^*(z),I}(x_1,x_2,k)) : D(v)=1 \wedge \neg\mathsf{Success}(v)] \right|$$
$$\leq \Pr[\neg\mathsf{Success}] - \mu_1(n)$$
$$= \Pr[out_B(\text{REAL}_{f,A^*(z),I}(x_1,x_2,k)) = \mathsf{corrupt}_A] - \mu_1(k)$$

This is a contradiction to Eq. 2.

Claim 1. $\left| \Pr[\text{BAD}] - \Pr[\neg\text{Success}] \right| < \mu_1(k)$

where the first probability is over the experiment in H and the second is over the real experiment.

Proof. Observe that if BAD occurs, then S' outputs BAD_i for some i. This means that in iteration i, the result of the computation using **LT** was not c_i. Since, we consider unique inputs, it must be the case that the output was something different from c_i. There are two cases:

c_i **was part of Alice's input** In this case, Alice must not have used the input corresponding to c_i in iteration i. Hence, the output of the ith iteration must have been different. Since there is a unique order in which the outputs are revealed, the computation between Alice and Bob must have resulted in an output different from $c_{\leq \ell}$. Then by the consistency-check property of f, it will follow that Alice cannot convince Bob in the Consistency-Check Phase for the commitment corresponding to C_i on the same transcript output by S'. This means that Bob would output $corrupt_A$ in Step 1 on such a transcript.

c_i **was part of Bob's input** Suppose that Alice used an input u in iteration i that was chosen by the greedy procedure. In this case, it cannot be that Alice committed to u in the Input Commitment Phase. This is because, S' extracted all the inputs from Alice and the output is unique given Alice's input. In this case, we show that Alice will fail in the Consistency-Check Procedure. First, we observe that Alice cannot produce an input to CC such that the output is u. Recall that it would have to produce both a signature using Bob's key for u and a commitment C from the input commitment phase containing the input u, since there is no such commitment, it cannot achieve this. Hence, $OutCheck_A$ computed by Bob will not contain u but $c_{\leq \ell}$ does. Therefore, Bob will output $corrupt_A$ in Step 3.

We recall that the view in H and the real experiments are identically distributed up until the iteration where BAD_i occurs. Therefore, from the above argument, it follows that if BAD_i was output by S_i on any partial transcript τ in hybrid experiment H up until iteration i, then Bob must have output $corrupt_A$ on a continuation from τ in the real experiment except with negligible probability. This concludes the proof of the Claim and the Theorem. $\square$

5.2 Convex Hull

We present a consistency-check for the convex-hull problem that will provide covert-security with 1-deterrence. Recall from the general covert-secure protocol that an adversary can fail to consider some of its input committed in the Commitment Phase. In the Consistency-Check Phase, the adversary needs to show that a particular point p committed to is not part of the convex hull. Towards this, it will choose three points p_1, p_2 and p_3 on the convex-hull that was output and prove that p lies strictly in the interior of the triangle formed by the p_1, p_2

and p_3. As before we assume that no two points share the same x or y coordinate and no three points are collinear.

We describe below how to instantiate this problem in our framework. First we show that convex-hull is covert-greedy compatible:

greedy compatible From Sect. 4.1 we know this is greedy compatible.

Consistency Check: We define the key$(u) = \bot$ and aux$(i, u) = \bot$. The function wit on input u is defined to be the index of three points in the output $c_1, \ldots, c_\ell$ such that u resides within the triangle formed by the three points. Observe that if a particular point u is not on the convex-hull, it lies inside and there must be three points on the hull for which u is contained in the triangle formed by the three points. Moreover, for any point outside the hull, there exists no set of three points for which this conditions will be true. The function CC on input $(u, C, D, (u_1, u_2, u_3))$ outputs 1 only if u is contained in the triangle formed by u_1, u_2 and u_3.

Theorem 4. *The* Generic Iterative Secure Computation with Covert Security *protocol instantiated with the Consistency Check Procedure CC, functions aux, key and wit described above securely computes the convex hull of the union of inputs of Alice and Bob, in the presence of covert-adversaries with 1-deterrence, assuming all inputs of Alice and Bob are distinct, no two of which share the same x-coordinate and no three points are collinear.*

Overhead: The total number of rounds of communication is $O(|Z| + |I|)$. This is because the secure computation phase take $O(|Z|)$-rounds and the consistency check phase is $O(|I|)$-rounds. In each round of the secure computation phase, the protocol performs at most one secure comparison of $\log M$-bit integers and one signature computation on a $\log M$ bit string. As mentioned before, the signatures need not be securely computed and can be computed locally and sent to the other party. In particular, for the case of convex-hull, the message to be signed is the point output in the current iteration. Therefore, the communication complexity of each round of iteration in this phase is $O(\log M) + O(k)$. In each round of the consistency check phase, the protocol performs (a) One decommitment verification that will cost $poly(k)$ (b) $O(1)$ signature verifications that will cost $poly(k)$ (c) $O(1)$ subtractions of $\log M$-bit integers (this will require $O(\log M)$ gates) and $O(1)$ multiplications of $\log M$-bit integers (this will require $O(\log^2 M)$ gates). This is for checking if a point is in a triangle. Since all the circuits need to be securely computed against malicious adversaries there will be a $poly(k)$ overhead. The overall communication complexity will be $O(|Z| \log M + |I| \log^2 M)$ times $poly(k)$. In comparison, the naive circuit implementation will have at least n memory accesses which will result in a circuit size of $\Omega(|I|^2 \log M)$ times $poly(k)$.

5.3   Matroids

We begin with a simple consistency-check for matroids that will yield covert-security with 1-deterrence. The communication complexity of implementing this check would be $O(|S|)$ where the matroid is $(S, \mathcal{I})$.

We recall some basic notions regarding matroids (see [13]). All sets in $\mathcal{I}$ are referred to as *independent sets* and any set not in $\mathcal{I}$ is called a *dependent set*. A *cycle* of a matroid $(S, \mathcal{I})$ is a setwise minimal dependent set. A *cut* in $(S, \mathcal{I})$ is a setwise minimal subset of S intersecting all maximal independent sets. The names have been so chosen to maintain the intuitive connection with the special case of MST. [1] Suppose B is an independent set and $B \cup \{x\}$ is dependent. Then $B \cup \{x\}$ contains a cycle. We refer to this cycle as the *fundamental cycle* of x and B. The cycle must contain x, since $C - \{x\} \subseteq B$ and hence independent. The following proposition follows directly form the properties of a matroid.

Proposition 1. *An element $x \in S$ is in no minimum weight maximal independent set iff it has the largest weight on some cycle.*

This proposition will form the basis of the consistency-check as for every element x not in the minimum weight maximal independent set B, there is a unique cycle, i.e. the fundamental cycle of x and B which can be computed by both parties given B and x. Then this cycle can be used to demonstrate that x is not part of the solution. In the full version, we consider other examples such as the (a) minimum spanning tree,[2] (b) unit job scheduling, and (c) single source shortest distance examples.

6   Computing the Median: Revisiting the AMP Protocol

To incorporate the median protocol to our framework we need to make a few modifications; the output of each iteration is not part of the final output. The output of the final iteration is the output of the protocol.

We define $\mathbf{LT}_f(x, y)$ function simply returns either 0 or 1 depending on whether $x > y$. The definition of F is slightly more complicated and subsumes the *pruning* of the input sets that is explicit in the AMP protocol. Specifically, we define

$$
F_A(i, m, x, S) = \begin{cases} n/2 & \text{if } i = 0 \\ x/2 & \text{if } m = 0 \ \wedge \ i > 0 \\ x + x/2 & \text{if } m = 1 \ \wedge \ i > 0 \end{cases}
$$
$$
F_B(i, m, x, S) = F_A(i, \overline{m}, x, S)
$$

In words, Alice begins the protocol using her $n/2^{\text{th}}$ (i.e. her median) element in comparison with Bob's median element. If Alice's element is larger, then she updates her index to the median of the smaller half of her array (i.e., the $n/4$ rank element) and conversely Bob updates his index to the median of the larger half of his array (i.e., his $n/2 + n/4$ rank element). The loop is repeated $\ell(|U|, |V|) = \lceil \log(|U|) \rceil = \lceil \log(n) \rceil$ times and at the end of the looping, the output O.

---

[1] Note however, the notion of cuts in graphs are a union of cuts as defined here since the notion of a cut in a graph need not necessarily be setwise minimal.

[2] It is possible to achieve better efficiency then the general matroid approach for MST.

Malicious Security for aborting adversaries. The security proof in [1] for the median protocol does not handle the case when the adversary aborts during extraction. We improve their simulation to handle this case.

As in [1], we construct a simulator in a hybrid model where all parties have access to an ideal functionality that computes comparisons. In the malicious protocol, at every step a lower bound l and an upper bound u is maintained by secret sharing the values between the two parties. This enforces the input provided by the parties to lie between l and u. Consider an adversary A for the protocol. We can visualize the actions of A in each iteration as either going left or right in a binary tree depending on the output of the comparison at that iteration. In the malicious setting, the simulator is required to extract the adversary's input to feed it to the ideal functionality. Since the actual inputs are revealed at each node, the simulator needs a mechanism to visit every node. Since every node can be arrived by a sequence of left or right traversal to a child, the simulator can reach every node by providing the appropriate sequence of outputs for the comparison operation and then allowing the adversary to reveal the value at that node. This value is given as input to the next iteration. Since we are in the hybrid where the simulator can see every input that the adversary sends, the simulator can extract the entire tree. Recall that the adversary is required to provide inputs that respect the lower and upper bound. Hence after extraction, an *in-order* traversal of the binary tree gives a sorted set of values. Using this set as the adversary's input, the simulator feeds it to the ideal functionality and obtains the output, and then traverses the tree again giving the actual inputs as dictated by the median.

We show how to construct a simulator that also allows the adversary to abort. If the adversary aborts when the execution reaches some particular node, then the simulator will not be able to extract the values in the entire tree rooted at that node. Our simulation proceeds as follows. It explores the tree just as in [1] and marks the nodes on which the adversary aborted. At the end of the extraction, the simulator has a partial binary tree. However, the simulator needs to send some input the ideal functionality on behalf of the adversary to obtain the output. Towards that, the simulator extends the partial tree to a full binary tree by adding dummy nodes. Next, the simulator assigns values for the dummy nodes. To do that every dummy node is given a unique label and then following [1], we perform an in-order traversal of the tree (excluding leaves) to obtain the adversary's input. Then each label is assigned a value equal to the first value in the sequence before the label. Then the sequence is sanitized to handle duplicate items as in [1].

The following observations follows from [1]. When the computation reaches a leaf, the adversary provides a single value to the comparison. For the rightmost leaf, the value is the largest value among all nodes. For the other nodes, the value is the same value on the lowermost internal node of the path from the root to the leaf, for which the comparison result returned true. Each item in the input of Alice appears exactly once in an internal node and exactly once in a leaf node. Finally, the values for the labels have so been chosen so that if for the actual (some hidden) inputs of the adversary the median is revealed in a particular leaf node, the median calculated by the input constructed by simulator will have the property that the leaf node corresponding to the median on the simulator's reconstructed input and the actual node corresponding to the adversary's input will be part of the same subtree of all dummy nodes. Thus, any traversal to either of these nodes from the root will result on the same node on which the adversary aborted.

Proof sketch. When S receives the output from the trusted party, it simulates the route that the execution takes in the tree, and performs any additional operation that Alice might apply to its view of the protocol. There are two cases depending on the median: Either the median is the value revealed in one of the comparisons in a leaf where the adversary aborted in some ancestor of that node, or it is the value revealed in a leaf that is not a dummy node. In the latter case, the simulation traverses the route the execution would take using the median and reach the leaf node. In the former case, the simulation will result in a traversal to the leaf node corresponding to the input computed by the simulator. However, by construction, the actual median lies in the same subtree as this leaf node and the simulator will abort at the same place as it would have with the actual inputs of the adversary. Hence the simulation proceeds identical to real experiment.

6.1   On Achieving Malicious Security in Our General Framework

Recall that in our general iterative framework the outputs are gradually released and this can allow a malicious adversary to alter its input to an intermediate iteration of the protocol based on the results from previous iterations. As in the covert security protocol, this can be circumvented by requiring the adversary to commit to their input before the protocol starts and the simulation extract these inputs at the beginning to be fed to the ideal functionality. However, this is not enough, as an adversary can selectively abort or selectively use his committed inputs in the iterative protocol based on the intermediate results of the computation. One approach would be to simply require each party to prove in zero-knowledge that it used the correct input. While this can be made communication efficient (by relying on universal arguments or SNARKs) it will blow up the computational complexity. This might be efficient if a short witness can establish correctness (analogous to our consistency checks for covert security) but seems unlikely for the applications discussed in this work. For the specific case of the median, as explained above, we are able to obtain malicious security without significant blow up in the communication or computation complexity. Towards obtaining efficient simulation of malicious adversaries, we next try to identify what property of the median protocol enables such efficient simulation.

Towards this, define $trace_i$ in our general framework to be the trace of the outputs in the first i iterations, namely, $c_1, \ldots, c_i$ which are the outputs of $\mathbf{LT}_f$ in each iteration. Let Alice's input be $A = \{a_1, \ldots, a_n\}$. For any particular data element a_j held by Alice (analogously for Bob) define $T(a_j)$ to be the set containing all traces $trace_i$ such that there is some input set B for Bob such that a_j is Alice's input in the $i+1$'st iteration when Alice and Bob interact with inputs A and B and $trace_i$ is the trace for the first i iterations.

When we consider the secure median protocol the first property we observe is that $T(a_j)$ is exactly 1. To see this consider two traces $trace_i = (c_1, \ldots, c_i)$ and $\widehat{trace}_i = \hat{c}_1, \ldots, \hat{c}_i$ such that the maximum prefix length that they share is k, namely, for $t = 1, \ldots, k$, $c_t = \hat{c}_t$ and $c_{k+1} \neq \hat{c}_{k+1}$. Since the outputs of each iteration are results of comparisons, let us assume without loss of generality that $c_{k+1} = 0$ (meaning Alice's input was less than Bob's input in that iteration) and $\hat{c}_{k+1} = 1$. Since they share the same prefix until iteration k, the input fed by Alice in the $k + 1^{st}$ iteration in both the traces must be identical. Call this element a_{j_1}. Let a_{j_2} and a_{j_3} be the input used by Alice in iteration $i + 1$ at the end of traces $trace_i$ and $\widehat{trace}_i$. It follows from the AMP protocol that $a_{j_2} < a_{j_1} < a_{j_3}$. This is because at iteration $k+1$ the result of the comparisons either prunes the data of elements less than a_{j_1} or greater than a_{j_1}. Hence, each input a_j cannot be in two different traces. A second property of the median protocol is that given an adversary every possible trace can be simulated in polynomial time. This is because the number of iterations is $O(\log n)$ and in each iteration there are at most two outcomes. We next give a rough intuition as to why these two properties are necessary.

If the first property fails to hold, then for a particular input element there are two different traces. This means the adversary can decide to selectively abort in one of the traces and such an adversary cannot be simulated without knowing the honest parties input. If the second property fails to hold, the adversary can selectively use its data. Since the all traces cannot enumerated in polynomial time, the adversary's input needs to be extracted by other means. If we relied on extractable commitments as in the case of covert security, the simulator will not know what to feed to the ideal functionality as some of the committed inputs may never be used by the adversary. Another example besides the median that satisfies this would be the bisection method to find roots of a polynomial where Alice and Bob hold parts of a polynomial and wish to find a root in a prescribed interval. In future work, we plan to explore more examples that admit malicious security in our framework.

References

1. Aggarwal, G., Mishra, N., Pinkas, B.: Secure computation of the median (and other elements of specified ranks). J. Cryptology **23**(3), 373–401 (2010)
2. Aumann, Y., Lindell, Y.: Security against covert adversaries: efficient protocols for realistic adversaries. J. Cryptology **23**, 281–343 (2010)
3. Barak, B., Goldreich, O.: Universal arguments and their applications. In: IEEE Conference on Computational Complexity, pp. 194–203 (2002)
4. Ben-David, A., Nisan, N., Pinkas, B.: FairplayMP: a system for secure multi-party computation. In: ACM Conference on Computer and Communications Security (2008)
5. Bitansky, N., Canetti, R., Chiesa, A., Tromer, E.: From extractable collision resistance to succinct non-interactive arguments of knowledge, and back again. In: Innovations in Theoretical Computer Science 2012, Cambridge, MA, USA, 8–10 January 2012, pp. 326–349 (2012)
6. Brickell, J., Shmatikov, V.: Privacy-preserving graph algorithms in the semi-honest model. In: Roy, B. (ed.) ASIACRYPT 2005. LNCS, vol. 3788, pp. 236–252. Springer, Heidelberg (2005)
7. Chung, K.-M., Liu, Z., Pass, R.: Statistically-secure oram with $\tilde{O}(\log^2 n)$ overhead (2013). arXiv preprint arXiv:1307.3699
8. Di Crescenzo, G., Lipmaa, H.: Succinct NP proofs from an extractability assumption. In: Beckmann, A., Dimitracopoulos, C., Löwe, B. (eds.) CiE 2008. LNCS, vol. 5028, pp. 175–185. Springer, Heidelberg (2008)
9. Gordon, S.D., Katz, J., Kolesnikov, V., Krell, F., Malkin, T., Raykova, M., Vahlis, Y.: Secure two-party computation in sublinear (amortized) time. In: CCS, pp. 513–524 (2012)
10. Shen, C.H., Shelat, A.: Fast two-party secure computation with minimal assumptions. In: ACM CCS 2013 (2012)
11. Huang, Y., Evans, D., Katz, J., Malka, L.: Faster secure two-party computation using garbled circuits. In: USENIX Security Symposium (2011)
12. Keller, M. Scholl, P.: Efficient, oblivious data structures for MPC. Cryptology ePrint Archive, Report 2014/137 (2014). http://eprint.iacr.org/
13. Kozen, D.C.: Design and Analysis of Algorithms. Texts and Monographs in Computer Science. Springer, New York (1992)
14. Kreuter, B., Mood, B., Shelat, A., Butler, K.: PCF: a portable circuit format for scalable two-party secure computation. In: USENIX Security Symposium (2013)
15. Kreuter, B., Shelat, A., Shen, C.H.: Billion-gate secure computation with malicious adversaries. In: USENIX Security Symposium (2012)
16. Liu, C., Huang, Y., Shi, E., Katz, J., Hicks, M.: Automating efficient ram-model secure computation. In: IEEE S & P (2014)
17. MacKenzie, P., Oprea, A., Reiter, M.: Automatic generation of two-party computations. In: ACM Conference on Computer and Communications Security (2003)
18. Malkhi, D., Nisan, N., Pinkas, B., Sella, Y.: Fairplay: a secure two-party computation system. In: USENIX Security (2004)
19. Micali, S.: Computationally sound proofs. SIAM J. Comput. **30**(4), 1253–1298 (2000)
20. Ostrovsky, R., Shoup, V.: Private information storage. In: STOC 1997, pp. 294–303 (1997)
21. Rastogi, A., Hammer, M.A., Hicks, M.: Wysteria: a programming language for generic, mixed-mode multiparty computations. In: IEEE S & P (2014)

22. Shi, E., Chan, T.-H.H., Stefanov, E., Li, M.: Oblivious RAM with $O((\log N)^3)$ worst-case cost. In: Lee, D.H., Wang, X. (eds.) ASIACRYPT 2011. LNCS, vol. 7073, pp. 197–214. Springer, Heidelberg (2011)
23. Stefanov, E., van Dijk, M., Shi, E., Fletcher, C., Ren, L., Yu, X., Devadas, S.: Path ORAM: an extremely simple oblivious ram protocol. In: CCS (2013)
24. Wang, X.S., Huang, Y., Hubert Chan, T-H., Shelat, A., Shi, E.: Scoram: Oblivious ram for secure computation. In: CCS 2014 (2014)
25. Yao, A.C.-C.: How to generate and exchange secrets. In: FOCS (1986)

Garbling Scheme for Formulas with Constant Size of Garbled Gates

Carmen Kempka, Ryo Kikuchi, Susumu Kiyoshima, and Koutarou Suzuki$^{(\boxtimes)}$

NTT Secure Platform Laboratories, Tokyo, Japan
{kempka.carmen,kikuchi.ryo,kiyoshima.susumu,
suzuki.koutarou}@lab.ntt.co.jp

Abstract. We provide a garbling scheme which creates garbled circuits of a very small constant size (four bits per gate) for circuits with fan-out one (formulas). For arbitrary fan-out, we additionally need only two ciphertexts per additional connection of each gate output wire. We make use of a trapdoor permutation for which we define a generalized notion of correlation robustness. We show that our notion is implied by PRIV-security, a notion for deterministic (searchable) encryption. We prove our scheme secure in the programmable random oracle model.

Keywords: Garbled circuits · Constant size of garbled gates · Correlation robustness · PRIV-security

1 Introduction

Yao's *garbled circuit* technique [33] is one of the most important techniques on secure computation. Very roughly speaking, this technique allows a party (the *garbler*) to create an encrypted form of a circuit—a "garbled" circuit—and an encoding of input with which the other party (the *evaluator*) can evaluate the circuit on the input but cannot compute anything other than the output. Compared with other techniques on secure computation (e.g., the technique by Goldreich et al. [12]), the garbled circuit technique has a big advantage on efficiency since we can construct *constant-round* protocols by using it.

Traditionally, the garbled circuit technique was considered to be a theoretical feasibility result; however, recently many works have demonstrated that the garbled circuit technique can also be used to construct two-party computation protocols with *practical efficiency*. The first implementation of the garbled circuit technique was shown by Malkhi et al. [28]. Since then, significant efforts have been devoted toward making the technique more practical.

A major line of research on the garbled circuit technique is the reduction of the size of garbled circuits. Since the main efficiency bottleneck of garbled-circuit-based two-party computation protocols is usually network bandwidth,

© International Association for Cryptologic Research 2015
T. Iwata and J.H. Cheon (Eds.): ASIACRYPT 2015, Part I, LNCS 9452, pp. 758–782, 2015.
DOI: 10.1007/978-3-662-48797-6_31

reducing the size of garbled circuits typically leads to a big improvement of efficiency in practice.[1]

Reduction of Garbled Circuit Size. Originally, the garbled circuit technique uses four ciphertexts for each gate to create "garbled truth tables", and thus, the size of a garbled circuit is $O(k)$ bits per gate (where k is the security parameter). In [22], Kolesnikov and Schneider proposed a technique, called *free-XOR technique*, with which we can construct a garbled circuit that contains no ciphertexts for XOR gates. In [29], Naor et al. proposed a technique that reduces the number of ciphertexts from four to three for each gate. In [30], Pinkas et al. proposed a technique that reduces the number of ciphertexts to two for each gate. Recently, Kolesnikov et al. [21] introduced the *fleXOR technique*, which requires zero, one, or two ciphertexts to garble an XOR gate—thus, garbling XOR gates is not "free" in general—but is compatible with the garbled row-reduction technique of [30]. Very recently, Zahur et al. [34] introduced a technique that is compatible with the free-XOR technique and can garble each AND gate by using only two ciphertexts; thus, this technique requires two ciphertexts for each AND gate and no ciphertexts for each XOR gate. We remark that although all of these techniques do not achieve an asymptotic reduction of the size of garbled circuits—it remains to be $O(k)$ bits per gates—they offer a significant reduction of communication cost in practice.

A different approach for reducing the size of garbled circuits is the technique of Kolesnikov [19], which is an information-theoretic variant of Yao's garbled circuit technique. In this technique, a circuit is garbled by using secret sharing (instead of encryption), and a garbled circuit is evaluated by recovering a share assigned to the output wire of each gate from shares assigned to the input wires of that gate. The size of the garbled circuit is zero (since there is no garbled truth table) and the size of the encoded input grows with the depth of the circuit; specifically, the size of the shares is quadratic in the depth of the gate for formulas, and exponential for circuits. For shallow circuits, the technique of Kolesnikov [19] is more efficient than other techniques.

Our Contribution. In this paper, we propose a garbling technique for formulas (i.e., circuits with fan-out 1) such that the size of the garbled circuit is *four bits* per gate. Unlike the optimization techniques of [21,22,29,34], our technique achieves asymptotic reduction of the size of the garbled circuits. Also, unlike the information-theoretic garbled circuit technique of [19], our technique encodes input in such a way that the size of the encoded input is independent of the depth of the circuit. (For detailed comparisons, see Sect. 3.3.)

In our technique, ciphertexts include trapdoor permutations (instead of hash functions as in most of the previous techniques). To prove the security, we extend

---

[1] In the case of security against malicious adversaries, the number of circuits that are generated in the cut-and-choose technique also has an impact on efficiency [1,10,14, 15,23,26,27].

the definition of correlation robustness (which is originally defined for hash functions [11]) to the case of trapdoor permutations, and assume that the underlying trapdoor permutation is correlation robust. We also show that our notion of correlation robustness is implied by PRIV-security as defined in [7].

Idea of Our Garbling Scheme. The idea of our construction is as follows: Unlike most existing techniques, our construction garbles circuits *backwards*, starting from the output gate. This allows us to reduce communication cost drastically: We can compute the ciphertexts needed for each gate as a hash of the gate-ID, so the evaluator can re-compute them by himself and they need not be included in the garbled circuit. These ciphertexts are then interpreted by the garbler as the XOR of the output key K_i and an image of a trapdoor one-way permutation of a function of the input keys, which he inverts to compute appropriate input keys corresponding to a given output key. Altogether we will have a system of four equations of the form

$$c_0 = E(f_0(K, L)) \oplus K_C^0$$
$$c_1 = E(f_1(K, L')) \oplus K_C^0$$
$$c_2 = E(f_2(K', L)) \oplus K_C^0$$
$$c_3 = E(f_3(K', L')) \oplus K_C^1$$

in permuted order, where K, K', L, L' are the input keys, K_C^0 and K_C^1 the output keys, and f_i are linear functions. The garbler can solve this system of equations to compute the input keys by inverting the trapdoor permutation E. The evaluator can only go through the circuit *forward*, using the one-way permutation to obtain output keys corresponding to his input. However, lacking the trapdoor, he cannot go backwards to compute any of the other keys.

One caveat of our backwards garbling technique is that the input keys for each gate are uniquely determined given the output keys and the ciphertexts assigned to this gate. Thus, we have no freedom in choosing any keys but the circuit output keys. Therefore, our garbling scheme only allows fan-out one, i. e., formulas. Moreover, to communicate wire choice bits, we cannot use the usual technique of defining the least significant bit of the keys as choice bit, since we have no freedom in choosing the input keys or their LSBs. Therefore, we use a hash function H' with one bit output, and publish $H'(K_a, K_b) \oplus l_i$ for input keys K_a, K_b and corresponding choice bit l_i of the output key, giving a garbled circuit with l gates an overall size of $4l$ bits, plus the number of bits needed to communicate the key of a keyed hash function and the index of a trapdoor one-way permutation.

Somewhat surprisingly, we can use the free-XOR technique to garble XOR gates at no additional cost, by using a "local" difference per XOR-sub-tree rather than a global difference. In the case of formulas, this will only safe us the 4 bit per gate for the choice bits. However in the case of general circuits, the freedom an XOR-gate gives us in choosing input keys can safe us additional ciphertexts needed for dealing with arbitrary fan-out in some cases.

Since our basic construction only allows fan-out one, a problem occurs when we garble circuits which use the same input variable multiple times, such as $(a \wedge b) \vee (a \wedge c)$. In such cases, we can duplicate the input wire for this variable, and assign a different input key pair to each occurrence of the variable. In the semi-honest setting, this does not affect security. In the malicious case, additional care needs to be taken to ensure that the garbler provides the same input for each occurrence of a variable. We discuss this in Sect. 5.3.

Related Works. The garbled circuit technique was introduced in the seminal paper of Yao [33]. A formal analysis of the garbled circuit technique (or, more precisely, the two-party computation protocol based on it) was presented by Lindell and Pinkas [25]. Bellare et al. [8] introduced an abstraction of the garbled circuit technique, which they call *garbling schemes*.

There are a lot of works that studied the size of garbled circuits. Other than the works we mentioned above [21,22,29,30,34], Choi et al. [11] and Applebaum [2] studied what assumptions are needed by the free-XOR technique. Choi et al. showed that a circular security assumption on the underlying hash function is sufficient. Applebaum showed that the LPN assumption is sufficient. Also, Boneh et al. [9] showed that an asymptotic reduction of garbled circuit size is possible under the learning-with-errors (LWE) assumption.

Other than the technique of [19], there are several information-theoretic variants of the garbled circuit technique, e.g., [17,18,20,32].

The correlation security of trapdoor permutations has also been studied by previous work in other contexts. For example, Rosen and Segev [31] introduced *correlated product security* of trapdoor permutations and used it to construct a CCA-secure encryption scheme. Also, Hemenway et al. [13] studied the relation between the decisional variant of correlated product security and the security of deterministic encryption schemes. We remark however that these notions of correlation security are different from the one we consider in this work. Roughly speaking, in correlated product security [31], correlated inputs are applied to k functions $f_1, \ldots, f_k$ that are independently chosen from a family of functions, whereas in our notion of correlation robustness, correlated inputs are applied to a *single* function f.

The size of inputs of garbled circuits has been studied in the context of randomized encoding [3,4,16]. Applebaum et al. [5] proposed a garbling scheme with constant online rate, i.e., they improve the online communication complexity for input keys from nk to $n + k$, where n is the number of inputs and k is security parameter. In contrast, our proposed scheme improves, only for formulas, communication complexity for garbled circuits from lk to $l + k$ and has communication complexity nk for input keys, where l is the number of gates, n is the number of inputs, and k a security parameter. We can combine the scheme of [5] and our proposed scheme to realize a randomized encoding for formulas with online communication complexity $n + k$ for input keys and communication complexity $l + k$ for garbled circuit.

Outline of This Work. The rest of this paper is organized as follows. We explain preliminaries and notation in Sect. 2, where we also recap the formal definition of garbling schemes, and introduce our notion of correlation robustness. We describe our basic garbling scheme for formulas in the semi-honest setting in Sect. 3, and prove its security in Sect. 4. In Sect. 5, we discuss possible extensions like arbitrary fan-out, incorporation of the free-XOR technique as well as extending our construction to the case of active adversaries. We discuss the instantiation of our correlation robust trapdoor one-way permutation with a PRIV-secure deterministic encryption scheme in Sect. 6.

2   Preliminaries

2.1   Notation

We use the following notations. By $x \xleftarrow{U} X$, we denote that x is randomly selected from set X according to the uniform distribution. By $x \leftarrow \mathsf{Algo}$, we denote that probabilistic algorithm Algo outputs x. By $A := B$, we denote that A is defined by B. By $[S]_x$, we denote the x-th bit of bitstring S.

2.2   Garbling Scheme

In this section, we recall the definition of a garbling scheme and the notion of *simulation-based privacy* of Bellare et al. [8].

A circuit is described as $f = (n, m, l, A, B, G)$. Here, $n \geq 2$ is the number of circuit input wires, $m \geq 1$ is the number of circuit output wires, and $l \geq 1$ is the number of gates (and their output wires). Let $W = \{1, ..., n + l\}$ be the set of all wires, $W_{input} = \{1, ..., n\}$ the set of circuit input wires, $W_{output} = \{n+l-m+1, ..., n+l\}$ the set of circuit output wires, and $W_{gate} = \{n+1, ..., n+l\}$ the set of gates (and their output wires). $A : W_{gate} \to W \setminus W_{output}$ is a function to specify the first input wire $A(i)$ of each gate i. $B : W_{gate} \to W \setminus W_{output}$ is a function to specify the second input wire $B(i)$ of each gate i. We require $A(i) < B(i) < i$ for all $i \in W_{gate}$. $G : W_{gate} \times \{0,1\}^2 \to \{0,1\}$ is a function to specify the gate function $G(i, \cdot, \cdot)$ of each gate i. We will later in our garbling scheme assign to each wire i two keys $K_{i,0}$ and $K_{i,1}$, representing the truth values 0 and 1 on this wire. To each wire i, we assign a permute bit λ_i, and to each key $K_{i,a}$ representing truth value $a \in \{0,1\}$, we assign a choice bit $l_i^a = \lambda_i \oplus a$.

We define the notion of garbling schemes as follows.

Definition 1 (Garbling Scheme). *A garbling scheme for a family of circuits* $\mathcal{F} = \{\mathcal{F}_n\}_{n \in \mathbb{N}}$, *where n is a polynomial in a security parameter k, consists of probabilistic polynomial-time algorithms* $\mathsf{GC} = (\mathsf{Garble}, \mathsf{Encode}, \mathsf{Eval}, \mathsf{Decode})$ *defined as follows.*

– Garble *takes as input security parameter 1^k and circuit $f \in \mathcal{F}_n$, and outputs garbled circuit F, encoding information e, and decoding information d, i.e.,* $(F, e, d) \leftarrow \mathsf{Garble}(1^k, f)$.

- Encode *takes as input encoding information e and circuit input $x \in \{0,1\}^n$, and outputs garbled input X, i.e., $X \leftarrow \mathsf{Encode}(e,x)$.*
- Eval *takes as input garbled circuit F and garbled input X, and outputs garbled output Y, i.e., $Y \leftarrow \mathsf{Eval}(F,X)$*
- Decode *takes as input decoding information d and garbled output Y, and outputs circuit output y, i.e., $y \leftarrow \mathsf{Decode}(d,Y)$.*

A garbling scheme should have the following correctness property: for all security parameters k, circuits $f \in \mathcal{F}_n$, and input values $x \in \{0,1\}^n$, $(F,e,d) \leftarrow \mathsf{Garble}(1^k, f)$, $X \leftarrow \mathsf{Encode}(e,x)$, $Y \leftarrow \mathsf{Eval}(F,X)$, $y \leftarrow \mathsf{Decode}(d,Y)$, it holds that $y = f(x)$.

We then define the security notion of garbling schemes called *simulation-based privacy* as follows. We adapt the notion of Bellare et al. [8] slightly to allow the adversary access to a random oracle H. We denote by $\Phi(f)$ the information about circuit f that is allowed to be leaked by the garbling scheme, e.g., size $\Phi_{size}(f) = (n,m,l)$, topology $\Phi_{topo}(f) = (n,m,l,A,B)$, or the entire information $\Phi_{circ}(f) = (n,m,l,A,B,G)$ of circuit $f = (n,m,l,A,B,G)$.

Definition 2 (Simulation-based Privacy). *For a garbling scheme GC = (Garble, Encode, Eval, Decode), function $f \in \mathcal{F}_n$, input values $x \in \{0,1\}^n$, simulator Sim, adversary $\mathcal{A}$, and random oracle H, we define the advantage*

$$Adv^{\mathrm{prv.sim}}_{\mathsf{GC,Sim},\Phi,\mathcal{A}}(k) :=$$
$$\left| \Pr\left[\begin{matrix} st \leftarrow \mathcal{A}^H(1^k), (F,e,d) \leftarrow \mathsf{Garble}(1^k,f), \\ X \leftarrow \mathsf{Encode}(e,x) \end{matrix} : \mathcal{A}^H(st,F,X,d) = 1 \right] \right.$$
$$\left. - \Pr\left[\begin{matrix} st \leftarrow \mathcal{A}^H(1^k), \\ (F,X,d) \leftarrow \mathsf{Sim}(1^k, f(x), \Phi(f)) \end{matrix} : \mathcal{A}^H(st,F,X,d) = 1 \right] \right|.$$

A garbling scheme GC = (Garble, Encode, Eval, Decode) is private, if there exists a probabilistic polynomial-time simulator Sim, such that for any function $f \in \mathcal{F}_n$, input values $x \in \{0,1\}^n$, and probabilistic polynomial-time adversary $\mathcal{A}$, the advantage $Adv^{\mathrm{prv.sim}}_{\mathsf{GC,Sim},\Phi,\mathcal{A}}(k)$ is negligible.

2.3 Generalized Correlation Robustness

We define a generalized notion of correlation robustness for trapdoor one-way permutations, in which we extend correlation robustness as defined by Choi et al. [11]. Choi et al. considered ciphertexts of the form $H(K \oplus a\Delta || L \oplus b\Delta || i) \oplus m$, where i is a gate-ID, H is a hash function, K and L are input keys, $a,b \in \{0,1\}$ and Δ is a global difference as needed for the free-XOR technique, meaning $K_{i,1} = K_{i,0} \oplus \Delta$ for each wire i. Given four such ciphertexts and two input keys $K_{A,\alpha} = K \oplus \alpha\Delta$ and $K_{B,\beta} = L \oplus \beta\Delta$ for $\alpha, \beta \in \{0,1\}$, the evaluator should only be able to decrypt one of them. Our ciphertexts have a similar form. However, we need to extend the definition of Choi et al. in two aspects. We do not have a global difference Δ. Instead, our definition considers general correlations defined by arbitrary functions of input keys, rather than correlations given by a global difference. Since we garble gates backwards, the garbler needs to be able to invert H. Therefore, instead of a hash function, we use a trapdoor one-way permutation E_ι. Thus, our notion of correlation robustness allows for a trapdoor ι.

Before we define correlation robustness, we recall the syntax of trapdoor one-way permutations.

Definition 3 (Family of Trapdoor one-way Permutations). *A family of trapdoor one-way permutations $E = \{E_\iota : D_\iota \to D_\iota\}_{\iota \in I}$ for finite index set I is defined by a tuple of PPT algorithms $E = (\mathsf{Gen}_E, \mathsf{Samp}_E, \mathsf{Eval}_E, \mathsf{Inv}_E)$ such that:*

- $\mathsf{Gen}_E(1^k)$ *is a probabilistic algorithm that outputs a pair (ι, t_ι) of index $\iota \in I$ and trapdoor t_ι.*
- $\mathsf{Samp}_E(1^k, \iota)$ *is a probabilistic algorithm that outputs a uniformly random element $x \in D_\iota$.*
- $\mathsf{Eval}_E(1^k, \iota, x)$ *is a deterministic algorithm that outputs $y = E_\iota(x)$ (assuming that ι is output by Gen_E and it holds that $x \in D_\iota$).*
- $\mathsf{Inv}_E(\iota, t_\iota, y)$ *is a deterministic algorithm that outputs an element $x \in D_\iota$ such that $y = E_\iota(x)$ (assuming that (ι, t_ι) is output by Gen_E and it holds that $y \in D_\iota$).*

In abuse of notation, we write $x \xleftarrow{U} D_\iota$ to denote $x \leftarrow \mathsf{Samp}_E(1^k, \iota)$, $y = E_\iota(x)$ to denote $y = \mathsf{Eval}_E(1^k, \iota, x)$, and $x = E_\iota^{-1}(t_\iota, y)$ to denote $x = \mathsf{Inv}_E(\iota, t_\iota, y)$.

We define generalized correlation robustness of trapdoor one-way permutations as follows.

Definition 4 (Generalized Correlation Robustness). *Let f_0, f_1, f_2, f_3 be any two-input functions. For a family of trapdoor one-way permutations $E = \{E_\iota : D_\iota \to D_\iota\}_{\iota \in I}$ and a probabilistic polynomial-time adversary $\mathcal{A}$, let us consider the following probabilistic experiment $\mathsf{Exp}^{\mathrm{corr}}_{E, f_a, f_b, f_c, \mathcal{A}}(k)$ for $a < b < c \in \{0, 1, 2, 3\}$.*

Experiment $\mathsf{Exp}^{\mathrm{corr}}_{E, f_a, f_b, f_c, \mathcal{A}}(k)$:
---

1. $\beta \xleftarrow{U} \{0, 1\}$.
2. $(K, L) \leftarrow \mathcal{A}(1^k)$.
3. $(\iota, t_\iota) \leftarrow \mathsf{Gen}_E(1^k)$ and $K' \xleftarrow{U} D_\iota$, $L' \xleftarrow{U} D_\iota$.
4. If $\beta = 0$, $Z_a := E_\iota(f_a(K, L))$, $Z_b := E_\iota(f_b(K', L))$, $Z_c := E_\iota(f_c(K', L'))$, *otherwise*, $Z_a \xleftarrow{U} D_\iota$, $Z_b \xleftarrow{U} D_\iota$, $Z_c \xleftarrow{U} D_\iota$.
5. $\beta' \leftarrow \mathcal{A}(\iota, Z_a, Z_b, Z_c)$.
6. *Output 1 if and only if $\beta = \beta'$.*

Let $Adv^{\mathrm{corr}}_{E, f_0, f_1, f_2, f_3, \mathcal{A}}(k) := \max_{a < b < c \in \{0, 1, 2, 3\}} \{\Pr\left[\mathsf{Exp}^{\mathrm{corr}}_{E, f_a, f_b, f_c, \mathcal{A}}(k) = 1\right] - 1/2\}$. Then, a family of trapdoor one-way permutations E is correlation robust w.r.t. f_0, f_1, f_2, f_3, if for any probabilistic polynomial-time adversary $\mathcal{A}$, the advantage $Adv^{\mathrm{corr}}_{E, f_0, f_1, f_2, f_3, \mathcal{A}}(k)$ is negligible.

In the proposed garbling scheme, we use the following invertible linear function $f = (f_0, f_1, f_2, f_3)$, with

$$\begin{aligned} f_0 : \quad &(x, y) \mapsto x + 2y, \\ f_1, f_2, f_3 : \; &(x, y) \mapsto x + y. \end{aligned}$$

3   Garbling Scheme for Formulas

We describe our basic garbling scheme for circuits with fan-out one. An extension to general circuits is given in Sect. 5.1. Our garbling scheme is designed for the semi-honest case, but can be extended to the malicious case using standard techniques. A brief discussion about this can be found in Sect. 5.3. As mentioned in the introduction, we can use each input wire only once. Multiple occurrences of a variable are handled by duplicating the corresponding input wire and assigning a new key pair for each occurrence of the variable, i. e. we treat multiple occurrences of the same input variable as different variables. This only affects security in the malicious case (see Sect. 5.3).

3.1   Garbling

We describe our garbling scheme informally. A formal description of our garbling algorithm and encoding function is given in Figs. 1 and 2, and the evaluation algorithm is given in Fig. 3. Our decoding function is defined as

$$Decode : (Y, d) \mapsto d \oplus Y.$$

Let q be a prime number such that $(2^k - q)/2^k$ is negligible, e.g., we can use a Mersenne prime $q = 2^k - 1$ for appropriate k. k and q are public. We can regard a random element $a \in \{0,1\}^k$ as an element $a \in \mathbb{F}_q \subset \{0,1\}^k$ with negligible error probability. Let $H : \{0,1\}^* \to \{0,1\}^k$ be a keyed hash function, modeled as a programmable random oracle. Let $H' : \{0,1\}^* \to \{0,1\}$ be a hash function which outputs one bit, modeled as a (non-programmable) random oracle. Let $E_\iota : \{0,1\}^k \to \{0,1\}^k$ be a trapdoor one-way permutation on $\{0,1\}^k$ which is correlation robust with respect to functions (f_0, f_1, f_2, f_3) on $\{0,1\}^k$, with

$$f_0 : (x, y) \mapsto x + 2y \in \mathbb{F}_q,$$

$$f_1, f_2, f_3 : (x, y) \mapsto x + y \in \mathbb{F}_q.$$

Let l denote the number of gates in circuit f; since the number of input wires is n, the circuit output wire is wire $l + n$. We assume that the evaluator knows the circuit topology.

During the garbling process, we assign a permute bit λ_i to each wire i, a key $K_{i,0}$ with choice bit $l_{i,0} = \lambda_i$, and a key $K_{i,1}$ with choice bit $l_{i,1} = 1 - \lambda_i$. $K_{i,0}$ corresponds to truth value 0, and $K_{i,1}$ to truth value 1 on this wire. To garble a circuit, the garbler first chooses a key R for the hash function H uniformly at random, and includes it in the garbled circuit. Then, he chooses a pair of output keys $K_{l+n,0}$ and $K_{l+n,1}$ for the circuit output wire $l+n$ uniformly at random. He assigns to the circuit output wire the permute bit $\lambda_{l+n} := 0$, and sets the output key choice bits $l_{l+n,0} := 0$ and $l_{l+n,1} := 1$. Then, starting from gate $l + n$, the garbler iteratively computes the remaining keys by computing the input keys of each gate i depending on its output key pair. Since for the input wires $A(i)$ and $B(i)$ of gate i we have $A(i) < B(i) < i$ for all i, we can simply iterate over i

backwards and be sure that output keys are defined before their corresponding input keys. Before the input keys of a gate i are computed, a uniformly random permute bit $\lambda_{A(i)}$ and, respectively, $\lambda_{B(i)}$ is chosen for its two input wires, which defines the input key choice bits $l^0_{A(i)} = \lambda_{A(i)}$ and $l^1_{A(i)} = 1 - \lambda_{A(i)}$ for wire $A(i)$, and analog for $B(i)$. To compute the input keys for each gate, the garbler computes four ciphertexts c_0, c_1, c_2, c_3 by computing $c_x := H(R, i||x)$ for $x = 0, 1, 2, 3$. The choice bits of the (yet undefined) input keys map each possible input combination $(a, b) \in \{0, 1\}^2$ to a ciphertext $c_{2l^a_{A(i)} + l^b_{B(i)}}$. This way, the evaluator can infer which ciphertext to use when processing gate i, without knowing the actual truth values on the input wires.

Using his trapdoor t_ι, for each of the four possible inputs $(a, b) \in \{0, 1\}^2$, the garbler computes $P_{2l^a_A + l^b_B} := E_\iota^{-1}(t_\iota, c_{2l^a_A + l^b_B} \oplus K_{i,G(a,b)})$, and solves the equation system

$$P_0 = K_{A,\lambda_A} + 2K_{B,\lambda_B}$$
$$P_1 = K_{A,\lambda_A} + K_{B,1-\lambda_B}$$
$$P_2 = K_{A,1-\lambda_A} + K_{B,\lambda_B}$$
$$P_3 = K_{A,1-\lambda_A} + K_{B,1-\lambda_B}$$

to compute the input keys $K_{A(i),0}$, $K_{A(i),1}$, $K_{B(i),0}$ and $K_{B(i),1}$. To enable the evaluator to compute the choice bit of output key $K_{i,G(a,b)}$, for all four possible inputs $(a, b) \in \{0, 1\}^2$, the garbler includes in the garbled circuit i the bit $b_{2l^a_A + l^b_B} := H'(K_{A,a}||K_{B,b}) \oplus l^{G_i(a,b)}_i$. Since H' is a random oracle, each bit $H'(K_{A,a}||K_{B,b})$ is random, and therefore, each bit $b_{2l^a_A + l^b_B}$ is also random and independent of $l^{G_i(a,b)}_i$, so the four published bits give no information about the permute bits or the choice bits. These four bits are sorted according to the choice bits of the input keys, so the evaluator knows which one to use. The choice bits assigned to the circuit input keys are directly provided with these keys by extending them by one bit.

We set the permute bit of the circuit output wire to 0, so the choice bits of its keys correspond to the actual truth value on this wire. Apart from the keys for the circuit input wires and their choice bits, the only values communicated to the evaluator are the key R for the hash function H, the index ι of the trapdoor one-way function, and the four bits $b_{2l^a_A + l^b_B}$ for each gate. Altogether, our garbled circuit has size $4l + |R| + |\iota|$.

3.2   Evaluation

Evaluation (see Fig. 3) is then straightforward: after obtaining the garbled circuit F and the input keys $K_1, \ldots, K_n$, the evaluator processes the circuit *forward*. For each gate i, he computes the ciphertext $c_{2l_A + l_B} := H(R, i||2l_A + l_B)$. Then he computes $K_i := E_\iota(K_A + 2K_B) \oplus c_{2l_A + l_B}$, if $2l_A + l_B = 0$, and $K_i := E_\iota(K_A + K_B) \oplus c_{2l_A + l_B}$, otherwise.

Garbling algorithm $\mathsf{Garble}(1^k, f)$

Input: Security parameter k, Circuit $f = (n, m, l, A, B, G)$ computing a formula

Output: Garbled circuit F, encoding e, decoding d

Algorithm: 1. Initialize:

Choose a trapdoor permutation $E_\iota : \{0,1\}^k \to \{0,1\}^k$ with trapdoor t_ι.

Choose a key $R \in \{0,1\}^k$ for hash function H uniformly at random

Initialize empty arrays $L[]$, $e[]$ with $|L| = l$ and $|e| = n$.

Choose circuit output keys $K_{l+n,0}, K_{l+n,1} \in \{0,1\}^k$ uniformly at random.

Set permute bit $\lambda_{l+n} := 0$ and choice bits $l^0_{l+n} := 0$, $l^1_{l+n} := 1$.

2. Garbling the gates:

For $i := l + n$ to $n + 1$ do (count i downwards):

(a) Set $A := A(i)$ and $B := B(i)$

(b) Choose permute bits $\lambda_A, \lambda_B \in \{0,1\}$ for wires A and B at random.

For all $(a, b) \in \{0,1\}^2$, set input key choice bits
$l^a_A := \lambda_A \oplus a$, $l^b_B := \lambda_B \oplus b \in \{0,1\}$.

(c) Deriving the input keys:

– Compute $c_x := H(R, i\|x) \in \{0,1\}^k$ for $x = 0, 1, 2, 3$.

– For all $(a, b) \in \{0,1\}^2$, use trapdoor t_ι to compute
$P_{2l^a_A + l^b_B} := E_\iota^{-1}(t_\iota, c_{2l^a_A + l^b_B} \oplus K_{i,G(a,b)}) \in \{0,1\}^k$

– Solve the equation system in $\mathbb{F}_q \subset \{0,1\}^k$

$$P_0 = K_{A,\lambda_A}\quad + 2K_{B,\lambda_B} \in \mathbb{F}_q$$
$$P_1 = K_{A,\lambda_A}\quad + K_{B,1-\lambda_B} \in \mathbb{F}_q$$
$$P_2 = K_{A,1-\lambda_A} + K_{B,\lambda_B} \in \mathbb{F}_q$$
$$P_3 = K_{A,1-\lambda_A} + K_{B,1-\lambda_B} \in \mathbb{F}_q$$

to obtain the input keys $K_{A,0}, K_{A,1}, K_{B,0}, K_{B,1} \in \mathbb{F}_q$. Abort if $P_i \notin \mathbb{F}_q$ for some i. This occurs with negligible probability.

If A is a circuit input wire, set $e[A] := (K^0_A\|l^0_A, K^1_A\|l^1_A)$.

If B is a circuit input wire, set $e[B] := (K^0_B\|l^0_B, K^1_B\|l^1_B)$.

(d) Indicate choice bits:

For all $(a, b) \in \{0,1\}^2$, compute
$b_{2l^a_A + l^b_B} := H'(K_{A,a}\|K_{B,b}) \oplus l^{G_i(a,b)}_i \in \{0,1\}$.

(e) Set $L[i] := (b_0, b_1, b_2, b_3)$

3. Output $F := (R, L, \iota)$, e and $d := \lambda_{l+n}$.

Fig. 1. The proposed garbling algorithm.

To obtain the choice bit l_i of the output key K_i, the evaluator computes $l_i := H'(K_{A(i)}\|K_{B(i)})) \oplus b_{2l_A + l_B}$. The evaluator proceeds until he finally obtains the choice bit l_{l+n} of the circuit output key K_{l+n}, which equals the output $f(x)$.

Encoding algorithm Encode(e, x)

Inputs: Garbled input keys e, input x
Algorithm: Parse x to $x = x_1 \ldots x_n$
    For $i = 1$ to n do:
        Parse $e[i] = (e_0, e_1)$
        $X[i] := e_{x_i}$
    Return X

Fig. 2. The function Encode.

Evaluation algorithm Eval(F, X)

Inputs: Garbled circuit F, garbled input X
Algorithm: 1. Parse F to $F = (R, L)$
      2. For $j = 1$ to n do
          $K_j \| l_j := X[j]$
      3. Compute gate output keys and choice bits:
          For $i := n + 1$ to $l + n$ do
              Set $A := A(i)$ and $B := B(i)$.
              Compute $c_{2l_A + l_B} := H(R, i \| 2l_A + l_B) \in \{0, 1\}^k$.
              If $x = 0$, set $K_i := E_\iota(K_A + 2K_B) \oplus c_x \in \{0, 1\}^k$,
              else set $K_i := E_\iota(K_A + K_B) \oplus c_x \in \{0, 1\}^k$.
              Parse $L[i]$ to (b_0, b_1, b_2, b_3).
              Set choice bit $l_i := H'(K_A \| K_B) \oplus b_x \in \{0, 1\}$.
      4. Return $Y := l_{l+n}$.

Fig. 3. The evaluation algorithm.

3.3 Efficiency Comparison with Previous Schemes

Comparison with the Half-Gates Construction. We compare our garbling scheme with the best known result (the half-gate construction) proposed by Zahur et al. [34] on efficiency. It is difficult to compare them directly, since ours uses a public-key primitive, while the half-gate construction uses a symmetric-key one. Therefore, we evaluate the cost in an abstract way and later discuss the concrete efficiency in the circuit size.

Let L_E be the length of the domain of E, L_I the length of ι, T_{Eval} the computation cost of Eval_E, T_{Inv} the computation cost of Inv_E, and $|R|$ the length of a hash key. Also let L_H be the length of the range of the correlation robust hash function used in [34], T_H the computation cost of hashing, and l_{AND} the number of AND gates. Table 1 shows the communication and computation cost of the two garbling schemes[2].

---

[2] We estimate the cost of our basic scheme here although our garbling scheme can be combined with the free-XOR technique as discussed in Subsect. 5.2.

Table 1. Comparison with the half-gates scheme [34].

	Communication cost		Computation cost	
	Circuit	Input	Garbling	Evaluation
Ours	$4l + \|R\| + L_\mathrm{I}$	$n \cdot L_\mathrm{E}$	$4T_\mathrm{Inv} \cdot l$	$T_\mathrm{Eval} \cdot l$
[34]	$L_\mathrm{H} \cdot l_\mathrm{AND}$	$n \cdot L_\mathrm{H}$	$2T_\mathrm{H} \cdot l_\mathrm{AND}$	$T_\mathrm{H} \cdot l_\mathrm{AND}$

Regarding communication cost, the size of the garbled circuit is a constant multiple of l in our garbling scheme. Therefore, communication cost is asymptotically small when a formula is large. Regarding computation cost, our garbling scheme requires executions of Eval_E and Inv_E, which are computationally expensive compared to the computation of a hash function.

We additionally evaluate the circuit size for concrete example parameters. We assume that the trapdoor one-way permutation is instantiated by an RSA-based primitive such as RSA-DOAEP [6], the correlation robust hash function is instantiated by fixed-key AES, and $l_\mathrm{AND} = 0.14l$, since this is the case of the most significant difference among the examples in [34]. AES is regarded with 128-bit security, so we set $L_\mathrm{E} = 4096$, $L_\mathrm{I} = 8192$, $|R| = 128$ and $L_\mathrm{H} = 128$. In this case, the circuit size in our garbling scheme is smaller than the one of the half-gates scheme if $l \geq 709$.

Regarding total communication cost, however, our scheme cannot beat the half-gate construction in its current state. We consider our scheme a proof of concept of a new way of circuit construction. Finding more efficient instantiations for the trapdoor permutation or getting rid of the public key primitive altogether are interesting open problems.

Comparison with Information-Theoretic Garbling Scheme of [19]. We compare the efficiency of our scheme with that of the information-theoretic garbling scheme of [19], which garbles formulas more efficiently than other techniques. For simplicity, we consider the case of garbling a "balanced" formula such that all the gates connecting to the input wires have the same depth.

Regarding the communication cost, the scheme of [19] garbles a balanced formula with depth d in a way that the size of the garbled circuit is zero and the size of the encoded input is approximately $2^{d+1} \cdot d^2$ (each of the 2^{d+1} wires has a share with size approximately d^2), and our scheme garbles such a formula in a way that the size of the garbled circuit is $4l + |R| + L_\mathrm{I} = 4 \cdot 2^d + |R| + L_\mathrm{I}$ and the size of the encoded input is $n \cdot L_\mathrm{E} = 2^{d+1} \cdot L_\mathrm{E}$. Hence, in total, the communication cost of the scheme of [19] is $2^{d+1} \cdot d^2$ whereas that of ours is $2^{d+1}(L_\mathrm{E} + 2) + |R| + L_\mathrm{I} \approx 2^{d+1} \cdot L_\mathrm{E}$ (see Table 2), and thus our scheme has smaller communication cost when $d > \sqrt{L_\mathrm{E}}$.

Regarding the computation cost, we note that since the scheme of [19] requires no cryptographic primitive whereas ours uses a public-key primitive, the computation cost of our scheme is likely to be much bigger than that of the scheme of [19].

Table 2. Comparison with information-theoretic garbling scheme of [19].

	Communication cost		
	Circuit	Input	Total
Ours	$4 \cdot 2^d + |R| + L_I$	$2^{d+1} \cdot L_E$	$2^{d+1}(L_E + 2) + |R| + L_I$
[19]	0	$2^{d+1} \cdot d^2$	$2^{d+1} \cdot d^2$

4   Security of the Proposed Scheme

The proposed garbling scheme is simulation-based private in the (programmable) random oracle model if the trapdoor permutation E is correlation robust.

Theorem 1 (Simulation-based Privacy). *The proposed garbling scheme described in Sect. 3 satisfies simulation-based privacy of Definition 2 if we assume that E satisfies correlation robustness as defined in Definition 4, H is a programmable random oracle and H' is a non-programmable random oracle. More precisely, for any adversary $\mathcal{A}$ there exists an adversary $\mathcal{B}$ such that*

$$Adv_{\mathsf{GC},\mathsf{Sim},\Phi,\mathcal{A}}^{\mathrm{prv.sim}}(k) \leq l \cdot Adv_{E,f_0,f_1,f_2,f_3,\mathcal{B}}^{\mathrm{corr}}(k) + q_H \cdot 2^{-k},$$

where l is the number of gates and q_H is the number of queries $\mathcal{A}$ makes to H.

Proof. We consider the following hybrid games $H_{real}, H_0, H_1, ..., H_l, H_{sim}$, where H_{real} is identical to the real experiment ($\beta = 0$), and H_{sim} is identical to the simulated experiment ($\beta = 1$) of Definition 2. The simulator $\mathsf{Sim}(1^k, f(x), \Phi_{topo}(f))$ of Definition 2 is provided in Fig. 6.

Game H_{real} : This game is identical to the real experiment of Definition 2. The garbled circuit (F, e, d) is generated by the garbling algorithm $\mathsf{Garble}(1^k, f)$ of the real garbling scheme, as given in Fig. 1, and garbled input X is generated by the encoding algorithm $\mathsf{Encode}(e, x)$ (Fig. 2) of the real scheme.

Game H_0 : In this game, the garbled circuit is generated forward, from input gates to output gate, which is in contrast to the real scheme. The garbled circuit is generated by algorithm $\mathsf{Sim}_0(1^k, f)$ as described in Fig. 4, and garbled input X is generated by encoding algorithm $\mathsf{Encode}(e, x)$ (Fig. 2) of the real scheme. We note that in step 2-(c), we need to program the random oracle H to keep consistency.

Game H_s : With the Games H_s, we move from Game H_0 to the full simulation with a gate-wise replacement. We incrementally replace variables not touched in an evaluation with input x with random values gate by gate, such that finally, in the full simulation, the "unused" variables of each gate are replaced by randomness. For $s = 1, ..., l$, in each Game H_s, the garbled circuit is generated by algorithm $\mathsf{Sim}_s(1^k, x, f)$, given in Fig. 5, and garbled input X is generated by the encoding algorithm $\mathsf{Encode}(e, x)$ (Fig. 2) of the real scheme. We say that key $K_{i,\beta}$ is active if and only if the bit obtained on wire i is β when circuit f is evaluated with input x. Only in these intermediate games

---

Simulator $\mathsf{Sim}_0(1^k, f)$

Input: Security parameter k, circuit $f = (n, m, l, A, B, G)$ computing a formula.
Output: Garbled circuit F, encoding e, decoding d.
Algorithm: 1. Initialize:
   Choose a trapdoor permutation $E_\iota : \mathbb{F}_q \rightarrow \mathbb{F}_q$ with trapdoor t_ι.
   Choose a key R for hash function H uniformly at random
   Initialize empty arrays $L[]$, $e[]$ with $|L| = l$ and $|e| = n$.
   For $i := 1$ to n do:
   (a) Choose circuit input keys $K_{i,0}, K_{i,1} \in \mathbb{F}_q$ uniformly at random.
   (b) Choose permute bit $\lambda_i \in \{0, 1\}$ uniformly at random and set choice bits $l_i^0 := \lambda_i \oplus 0$, $l_i^1 := \lambda_i \oplus 1$.
   (c) Set $e[i] := (K_i^0 || l_i^0, K_i^1 || l_i^1)$.
   2. Garbling the gates:
   For $i := n + 1$ to $n + l$ do (count i upwards):
   (a) Set $A := A(i)$ and $B := B(i)$.
   (b) Choose permute bit λ_i for output wire i uniformly at random.
       Set output key choice bit $l_i^0 := \lambda_i \oplus 0$, $l_i^1 := \lambda_i \oplus 1$.
   (c) 'Deriving' the output keys:
       – For all $(a, b) \in \{0, 1\}^2$, compute
       $Y_{2l_A^a + l_B^b} := E_\iota(K_{A,a} + 2K_{B,b})$, if $2l_A^a + l_B^b = 0$ or
       $Y_{2l_A^a + l_B^b} := E_\iota(K_{A,a} + K_{B,b})$, otherwise.
       – Choose output keys $K_{i,0}, K_{i,1}$ uniformly random
       – Program random oracle such that
       $c_{2l_A^a + l_B^b} := K_{i, G(a,b)} \oplus Y_{2l_A^a + l_B^b}$
   (d) Indicate choice bits:
       For all $(a, b) \in \{0, 1\}^2$, compute
       $b_{2l_A^a + l_B^b} := H'(K_{A,a} || K_{B,b}) \oplus l_i^{G(a,b)}$.
   (e) Set $L[i] := (b_0, b_1, b_2, b_3)$
   3. Output $F := (R, L, \iota)$, e and $d := \lambda_{n+l}$.

Fig. 4. The algorithm Sim_0 for game H_0.

H_s, we give the simulator knowledge of input x and circuit f, so he can label each key $K_{i,\beta}$ as active or inactive. We say that value $E_\iota(K_{A,a} + K_{B,b})$ is active if and only if both $K_{A,a}$ and $K_{B,b}$ are active. We replace inactive values with random values in step 2-(c) of algorithm $\mathsf{Sim}_s(1^k, x, f)$ for gates $i = n + 1, ..., n + s$; gates $i = n + s + 1, ..., n + l$ are generated as in game H_0. We also replace the inactive output key by the active output key in this step.

Game H_{sim}: This game is identical to the simulated experiment in Definition 2. The garbled circuit and garbled input (F, X, d) are generated by algorithm $\mathsf{Sim}(1^k, f(x), \Phi_{topo}(f))$, given in Fig. 6, without knowledge of x.

The difference between the advantage of adversary $\mathcal{A}$ in H_{real} and his advantage in H_0 is bound by $q_H \cdot 2^{-k}$ as follows, where q_H is the number of queries

---

Simulator $\mathsf{Sim}_s(1^k, x, f)$

Input: Security parameter k, circuit input x, circuit $f = (n, m, l, A, B, G)$
computing a formula.
Output: Garbled circuit F, encoding e, decoding d.
Algorithm: 1. Initialize:
   Same as Sim_0.
   2. Garbling the gates:
   For $i := n + 1$ to $n + s$ do (count i upwards):
   (a) Set $A := A(i)$ and $B := B(i)$.
   (b) Choose permute bit λ_i for output wire i uniformly at random.
      Set output key choice bit $l_i^0 := \lambda_i \oplus 0$, $l_i^1 := \lambda_i \oplus 1$.
   (c) Deriving the output keys:
      – For all $(a, b) \in \{0, 1\}^2$,
         if $K_{A,a}$, $K_{B,b}$ are active keys, compute
         $Y_{2l_A^a + l_B^b} := E_\iota(K_{A,a} + 2K_{B,b})$, if $2l_A^a + l_B^b = 0$ or
         $Y_{2l_A^a + l_B^b} := E_\iota(K_{A,a} + K_{B,b})$ if $2l_A^a + l_B^b \neq 0$,
         otherwise choose $Y_{2l_A^a + l_B^b} \in \mathbb{F}_q$ uniformly at random.
      – Choose output keys $K_{i,0}, K_{i,1}$ uniformly random.
      – Let $(a^*, b^*) \in \{0, 1\}^2$ be the values such that K_{A,a^*} and K_{B,b^*}
         are active keys. Then, for all $(a, b) \in \{0, 1\}^2$, program random
         oracle such that $c_{2l_A^a + l_B^b} := K_{i,G(a^*,b^*)} \oplus Y_{2l_A^a + l_B^b}$.
   (d) Indicate choice bits:
      For all $(a, b) \in \{0, 1\}^2$, compute
      $b_{2l_A^a + l_B^b} := H'(K_{A,a}||K_{B,b}) \oplus l_i^{G_i(a,b)}$.
   (e) Set $L[i] := (b_0, b_1, b_2, b_3)$
   3. Garbling the gates:
   For $i := n + s + 1$ to $n + l$ do (count i upwards):
   Same as Sim_0.
   4. Output $F := (R, L, \iota)$, e and $d := \lambda_{n+l}$.

Fig. 5. The algorithm Sim_s for game H_s.

that $\mathcal{A}$ makes to H. Since E_ι is a bijective map and $f = (f_0, f_1, f_2, f_3)$ is an invertible linear map, the system of four linear equations

$$c_{2l_A^a + l_B^b} = K_{i,G(a,b)} \oplus E_\iota(f_{2l_A^a + l_B^b}(K_{A,a}, K_{B,b})), \quad (a, b) \in \{0, 1\}^2$$

is uniquely solvable with the input keys $K_{A,0}$, $K_{A,1}$, $K_{B,0}$ and $K_{B,1}$ as variables. Therefore, it maps uniformly random output keys $K_{i,0}, K_{i,1}$ to uniformly random input keys $K_{A,0}$, $K_{A,1}$, $K_{B,0}$, $K_{B,1}$ in H_{real}. In H_0, both the input keys $K_{A,0}$, $K_{A,1}$, $K_{B,0}$, $K_{B,1}$ and the output keys $K_{i,0}, K_{i,1}$ are chosen uniformly at random. However, the relation between the values $c_{2l_A^a + l_B^b}$, $K_{i,G(a,b)}$ and $(K_{A,a}, K_{B,b})$, which is given by these four equations, is preserved for all $(a, b) \in \{0, 1\}^2$ by programming the random oracle H to output consistent values for c_0, c_1, c_2 and c_3. So, all keys are uniformly random in both H_{real} and H_0, and the distributions of the garbled circuits created in H_{real} and H_0 are indistinguishable to the

---

Simulator $\mathsf{Sim}(1^k, f(x), \Phi_{topo}(f))$

Input: Security parameter k, output value $f(x)$, topology of circuit $\Phi_{topo}(f) = (n, m, l, A, B)$ of formula f.

Output: Garbled circuit F, garbled input X, decoding d.

Algorithm: 1. Initialize:

Choose a trapdoor permutation $E_\iota : \mathbb{F}_q \to \mathbb{F}_q$ with trapdoor t_ι.

Choose a key R for hash function H uniformly at random

Initialize empty arrays $L[]$, $X[]$ with $|L| = l$ and $|X| = n$.

Set permute bit $\lambda_{n+l} := 0$

For $i := 1$ to n do:

(a) Choose $K_i \in \mathbb{F}_q$ uniformly at random.

(b) Choose $l_i \in \{0, 1\}$ uniformly at random.

(c) Set $X[i] := K_i || l_i$.

2. Garbling the gates:

For $i := n + 1$ to $n + l$ do (count i upwards):

(a) Set $A := A(i)$ and $B := B(i)$.

(b) If $i = n + l$, set $l_i := f(x)$.

Otherwise, choose $l_i \in \{0, 1\}$ uniformly at random.

(c) 'Deriving' the output key of gate i:

   – If $2l_A + l_B = 0$ compute $Y_{2l_A + l_B} := E_\iota(K_A + 2K_B)$.

     Otherwise, compute $Y_{2l_A + l_B} := E_\iota(K_A + K_B)$.

     Choose Y_x uniformly random, for $x = 0, 1, 2, 3, x \neq 2l_A + l_B$.

   – Choose K_i uniformly at random.

   – Program random oracle such that

     $c_x := K_i \oplus Y_x$ for all $x \in \{0, 1, 2, 3\}$.

(d) Indicate choice bits:

   – Set $b_{2l_A + l_B} := H'(K_A || K_B) \oplus l_i$.

   – For each $x \in \{0, 1, 2, 3\} \setminus \{2l_A + l_B\}$, choose $b_x \in \{0, 1\}$ uniformly at random.

(e) Set $L[i] := (b_0, b_1, b_2, b_3)$

3. Output $F := (R, L, \iota)$, X and $d := \lambda_{n+l}$.

Fig. 6. The algorithm Sim to create the simulated garbled circuit.

adversary, except for the case where the adversary asks $H(R, i||x)$ for some $x \in \{0, 1, 2, 3\}$ before the simulator garbles gate i, where he defines this value. However, the hash key R is given to the adversary only after garbling the whole circuit. Before that, $\mathcal{A}$ can guess R with probability 2^{-k} in each query to H. Thus, from the union bound, the probability that $\mathcal{A}$ makes "a bad query" to H is bound by $q_H \cdot 2^{-k}$.

The advantage of adversary $\mathcal{A}$ in distinguishing H_s and H_{s+1} for $s = 0, ..., l-1$ is bound by the advantage $Adv^{corr}_{E, f_0, f_1, f_2, f_3, \mathcal{B}}(k)$ of the following adversary $\mathcal{B}$ of the correlation robustness game as follows. Let A and B denote the input wires of gate $n+s+1$. For simplicity, we assume that $K_{A,0}$ and $K_{B,1}$ are the active input keys of gate $n+s+1$; the three other cases are analogous. Remember that gates are labeled after their output wires; since the first n wires are circuit input wires, in Game H_{s+1}, Gate $n + s + 1$ is the next to be replaced by a "simulated" gate. At the beginning of the correlation robustness game, the adversary $\mathcal{B}$ selects $K_{A,0}$ and $K_{B,1}$ uniformly at random, and outputs $K := K_{A,0}$ and $L := K_{B,1}$ as the target keys of the correlation robustness experiment. Then, $\mathcal{B}$ receives index ι of E and challenge Z_1, Z_2, Z_3 of the correlation robustness experiment.

To provide adversary $\mathcal{A}$ with a "challenge circuit", adversary $\mathcal{B}$ simulates gates $n + 1$ to $n + s$ as in Step 2 of algorithm $\mathsf{Sim}_s(1^k, x, f)$, while setting the active keys of wire A and B to $K_{A,0}$ and $K_{B,1}$, respectively. Since they were chosen uniformly at random, this does not contradict the construction in Game H_s or H_{s+1}.

The adversary $\mathcal{B}$ then creates Gate $n + s + 1$ as follows:

1. Compute active value $Y_{2l_A^0+l_B^1} := E_\iota(f_{2l_A^0+l_B^1}(K, L))$, and set the inactive values to the random values from the challenge as follows: $Y_{2l_A^0+(1-l_B^1)} := Z_1$, $Y_{2(1-l_A^0)+l_B^1} := Z_2$ and $Y_{2(1-l_A^0)+(1-l_B^1)} := Z_3$.
2. Choose $K_{n+s+1,0}$ and $K_{n+s+1,1}$ uniformly at random. Program random oracle such that $c_{2l_A^a+l_B^b} := K_{n+s+1,G(a,b)} \oplus Y_{2l_A^a+l_B^b}$ for all $(a,b) \in \{0,1\}^2$.

All remaining gates (Gate $n + s + 2$ to Gate $n + l$), if there are any, are created as in algorithm $\mathsf{Sim}_s(1^k, x, f)$. Adversary $\mathcal{B}$ gives the created garbled circuit to $\mathcal{A}$, and outputs whatever $\mathcal{A}$ outputs.

If Z_1, Z_2 and Z_3 are correctly formed, i.e., $Z_1 = E_\iota(f_{2l_A^0+(1-l_B^1)}(K_{A,0}, K_{B,1}))$, $Z_2 = E_\iota(f_{2(1-l_A^0)+l_B^1}(K_{A,1}, K_{B,0}))$, and $Z_3 = E_\iota(f_{2(1-l_A^0)+(1-l_B^1)}(K_{A,1}, K_{B,1}))$, the simulated garbled circuit is generated as in H_s. If Z_1, Z_2 and Z_3 are uniformly random, $c_{2l_A^0+l_B^1}$ is $K_{n+s+1,G(a,b)} \oplus E_\iota(f_{2l_A^0+l_B^1}(K, L))$, and $c_{2l_A^0+(1-l_B^1)}$, $c_{2(1-l_A^0)+l_B^1}$ and $c_{2(1-l_A^0)+(1-l_B^1)}$ are uniformly random; thus, they are distributed identically to those in H_{s+1}. Thus, Adversary $\mathcal{B}$ is successful in the correlation robustness game whenever Adversary $\mathcal{A}$ is successful in distinguishing H_s and H_{s+1}. Therefore, the advantage of adversary $\mathcal{A}$ in distinguishing H_s and H_{s+1} is bound by advantage $Adv_{E,f_0,f_1,f_2,f_3,\mathcal{B}}^{corr}(k)$ of the correlation robustness experiment.

The distributions of the simulated garbled circuits in H_l and H_{sim} are indistinguishable as follows. In H_{sim}, the simulator "loses knowledge" of input x, i.e., he does not know which values are active, and therefore, which inputs should result in the active key. He does not need this knowledge anymore, since the inactive output keys have been replaced by the active ones in the previous games. The bits b_x for $x = 0, 1, 2, 3$ in H_{sim} are chosen uniformly at random. Since H' is a random oracle, this is indistinguishable from setting $b_{2l_A^a+l_B^b} := H'(K_{A,a}\|K_{B,b}) \oplus l_i^{G_i(a,b)}$.

Thus, by summing up the differences between the games, we have $Adv_{GC,Sim,\Phi,\mathcal{A}}^{prv.sim}(k) \leq l \cdot Adv_{E,f_0,f_1,f_2,f_3,\mathcal{A}}^{corr}(k) + q_H \cdot 2^{-k}$. $\square$

5   Extensions

In this section, we generalize our garbling scheme to allow circuits with arbitrary fan-out. Then we show how to combine our garbling scheme with the free-XOR technique to further reduce communication cost. We also discuss the malicious case.

5.1 Arbitrary Fan-Out

Our scheme can handle arbitrary fan-out with a slight modification, which comes at the cost of going back to linear size of the garbled circuit in the worst case. Note that if a circuit has more than one circuit output wire, the decoding function needs to be adjusted in the obvious way. The reason our basic scheme can only be used for formulas is that our garbling algorithm leaves no degree of freedom when computing the input keys of a gate. This implies a conflict when a wire is the input wire of two different gates i and j: W.l.o.g., consider the case of a shared wire $s := B(i) = A(j)$ and $i < j$. When garbling gate i, the garbler computes corresponding input keys $K_{s,0}$ and $K_{s,1}$ for the shared wire s. However, when garbling gate j, he obtains different input keys $K'_{s,0} \neq K_{s,0}$ and $K'_{s,1} \neq K_{s,1}$ for the same wire. We can solve this conflict by providing two ciphertexts for this wire by encrypting $K'_{s,0}$ with key $K_{s,0}$, and $K'_{s,1}$ with $K_{s,1}$. Then we can use $K_{s,0}$ and $K_{s,1}$ for gate i, and $K'_{s,0}$ and $K'_{s,1}$ for gate j. We sort the two ciphertexts according to the permute bit λ_s of wire s. Since λ_s applies to both key pairs, $K'_{s,b}$ and $K_{s,b}$ have the same choice bit for $b = 0, 1$. This way, whenever a wire is input for a second gate, the evaluator can compute the "second version" of his obtained key by decrypting the corresponding ciphertext. Our garbled circuit now has the size

$$4l + |R| + |\iota| + \sum_{d=1}^{d_{max}} (2(d-1) \cdot k \cdot l_d),$$

where l_d is the number of gates with fan-out d, and d_{max} is the maximal number of output wires of a single gate.

In the case that a gate has two conflicting input wires, this gate might need four ciphertexts, making our scheme seemingly inefficient. However, depending on the number of circuit input and circuit output wires, we still have an average of considerably less than four ciphertexts per gate. Consider a circuit with k gates. According to the model we use, each gate has two input wires and one output wire. So altogether we have $2k$ input wires, of which n are circuit input wires (i.e., cannot origin from other gates), and k output wires, of which m are circuit output wires (i.e., those m wires cannot induce a conflict). So we have $k - m$ wires which need to "go somewhere", and $2k - n$ places where they can go. Since the two ciphertexts are only needed for each *additional* connection of an output wire, each wire can connect to one input slot "for free". So in the worst case, we end up with $(2k-n)-(k-m) = k+m-n$ conflicts, meaning $2(k-n+m)$ ciphertexts. In the case of $m \leq n$, we have at most two ciphertexts per gate in the worst case. If $m > n$, we might have more than two ciphertexts per gate in the case of an unfortunate layout, but still less than four since $m < k$. So in the case of arbitrary fan-out, whether or not our scheme provides an efficiency gain compared to other optimizations like the half-gate or fleXOR constructions, strongly depends on the circuit layout. As we can see in the following section, our basic scheme is compatible to the free-XOR technique. This compatibility translates to the case of arbitrary fan-out with similar conflict issues arising from conflicting wire offsets as well as conflicting keys. However, in some cases where XOR gates are involved in a conflict, only one ciphertext is needed to solve it. This is elaborated in more detail in Sect. 5.2.

5.2 Incorporating Free-XOR

Going back to the world of formulas, our garbling scheme is compatible with a slightly adapted version of the free-XOR technique introduced in [22]. The free-XOR technique allows us to garble XOR gates "for free", by choosing each key pair with a constant offset Δ, such that output keys can be obtained by XORing corresponding input keys. To be compatible with our garbling scheme, XOR gates, too, have to be garbled backwards, but induce no communication cost.

For each gate i, let $K_{i,0}$ and $K_{i,1}$ be the already given output keys for gate i. (If i is a circuit output wire, choose the two keys at random.) If i is a non-XOR gate, the garbler does exactly what he does in our basic scheme. If i is an XOR gate, the garbler sets $\Delta := K_{i,0} \oplus K_{i,1}$. Then for the input wires $A(i)$ and $B(i)$, the garbler chooses random keys $K_{A(i),0}$ and $K_{B(i),0}$ such that $K_{A(i),0} \oplus K_{B(i),0} = K_{i,0}$ and sets $K_{A(i),1} := K_{A(i),0} \oplus \Delta$ and $K_{B(i),1} := K_{B(i),0} \oplus \Delta$. The permute bits $\lambda_{A(i)}$ and $\lambda_{B(i)}$ of the input wires are set such that $\lambda_{A(i)} \oplus \lambda_{B(i)} = \lambda_i$. This way, the evaluator can obtain the choice bit of the output key by applying the XOR function to the choice bits of the input keys. Using this technique, our XOR gates are free: they induce no communication cost. This works as long as the circuit is cycle-free, which is required by the property $A(i) < B(i) < i$ for each gate.

When incorporating free-XOR, we need our trapdoor permutation E_ι to additionally achieve a circular security property similar to the one introduced by Choi et al. [11]. Our incorporation of free-XOR seems similar to the combination of the fleXOR technique with two-row-reduction in [21]. However, we have different dependencies: In the fleXOR technique, the offset Δ is determined by the input keys of a gate, such that if there is a sub-tree in the circuit consisting only of XOR gates, there might be different offsets Δ_i within this sub-tree. In our scheme, Δ is determined by the output key, so there is only one Δ within each sub-tree. Since input keys depend on output keys, we can define the input keys of an XOR sub-tree to have the same offset Δ.

Free-XOR and Arbitrary Fan-Out. Incorporating the free-XOR technique allows us some freedom in choosing the input keys of XOR gates. This can reduce the number of ciphertexts needed when dealing with conflicting wires in some cases. Consider gates i and j with shared input wire $B(i) = A(j)$ (other shared wires are analogous). There are essentially three types of conflicting wires:

- **Case 1:** The keys of the input wires of both gates i and j cannot be chosen freely:

This occurs for example when one of the gates has more than one conflicting input wire, or when both i and j are non-XOR gates. In this case, we solve the conflict exactly as described in Sect. 5.1, by encrypting the keys determined by garbling gate j with those determined for gate i. Solving this conflict induces a communication cost of two ciphertexts, as before.

- **Case 2:** i and j are both XOR gates with no additional conflict:
  The problem here is that for the input wires of i and j we might have to use different values Δ_i and Δ_j. In this case, let λ_B be the permute bit of wire $B(i)$ (i.e., $K_{B(i),\lambda_B}$ is the key with choice bit zero) and choose $K_{B(i),\lambda_B} = K_{A(j),\lambda_B}$ at random (the other input wires of i and j then have to be chosen accordingly such that the correct output key is obtained). Then set $K_{B(i),1-\lambda_B} := K_{B(i),\lambda_B} \oplus \Delta_i$ and $K_{A(j),1-\lambda_B} := K_{A(j),\lambda_B} \oplus \Delta_j$, and include the ciphertext $Enc_{K_{B(i),1-\lambda_B}}(K_{A(j),1-\lambda_B})$ in the garbled circuit. No additional ciphertext is needed, so this conflict can be solved with only one ciphertext.

- **Case 3:** i is an non-XOR gate or an XOR gate with additional conflicts, and j is an XOR gate with no additional conflict:
  This can be solved similar to Case 2: let λ_B be the permute bit of wire $B(i)$. Set $K_{A(j),\lambda_B} := K_{B(i),\lambda_B}$. Set $K_{A(j),1-\lambda_B} := K_{A(j),\lambda_B} \oplus \Delta_j$ and include the ciphertext $Enc_{K_{B(i),1-\lambda_B}}(K_{A(j),1-\lambda_B})$ in the garbled circuit. Like Case 2, this conflict can be solved with only one ciphertext.

As we can see, if an XOR gate is involved in the conflict, and this is the only conflict of this XOR gate, we need to include only one ciphertext for the conflicting wire.

5.3 Security Against Malicious Adversaries

Our scheme extents to the malicious case by using the cut-and-choose-based techniques described by Lindell and Pinkas [24].

One additional issue we need to address is that our scheme requires us to duplicate input wires if the same input variable is used multiple times. In the malicious case, we need to prove that consistent input keys are sent for each wire representing the same variable.

The case of a malicious evaluator is simply handled by sending all input keys for one variable in a single OT. The case of a malicious garbler can be solved by adapting the consistency check of the garbler's input used by Lindell and Pinkas (see Fig. 2 in [24]) in a straightforward manner, by grouping the input wires representing the same variable.

6 Instantiation with PRIV-secure Deterministic Encryption

In this section, we show that we can instantiate our trapdoor one-way permutation E_ι with a deterministic encryption function which achieves the notion of PRIV-security, introduced by Bellare et al. in [6]. Though not limited to deterministic encryption, PRIV-security has been introduced as the strongest security notion one can achieve for deterministic encryption. It was proven equivalent to various other notions of deterministic encryption by Bellare et al. in [7].

In this section, we show that PRIV-security implies correlation robustness with respect to addition, as needed in our garbling scheme. At the end of this section, we discuss a concrete instantiation.

We briefly recall the PRIV-notion here. Though we use the term PRIV, we actually use one of its equivalents called *IND* in [7]. Let $E(pk, m)$ denote a deterministic encryption function with key space $\mathcal{K}(1^k)$, domain space $Dom(E)$ and image space $Im(E)$. Let $x \leftarrow_R S$ denote the assignment of a random element in set S to variable x. As a shortcut, we write $E(pk, \boldsymbol{x})$ for componentwise encryption of a vector $\boldsymbol{x}$. PRIV-security is defined via the experiment $Exp_{\mathcal{A}}^{PRIV}$ (see Fig. 7). It uses an adversary algorithm $\mathcal{A} = (A_t, A_m, A_g)$, which is split into three adversaries A_t, A_m and A_g. The first one, A_t, chooses a string st which is readable (but not writable) by the other two adversaries. It does not exist in the original PRIV-definition of [6], and only used to show equivalence in [7], where st is then required to be the empty string in the actual security definition, while the most fit st is assumed hard-wired into the other two adversaries A_m and A_g. We use A_t and st for our reduction to correlation robustness in a similar way. The adversary A_m chooses vectors $\boldsymbol{x}_0$ and $\boldsymbol{x}_1$ of plaintexts knowing choice bit $\tilde{b}$, but without knowledge of the public encryption key pk. The vector $\boldsymbol{x}_{\tilde{b}}$ is then encrypted componentwise. The third adversary, A_g, then tries to guess $\tilde{b}$, getting the public key pk, the ciphertext $E(pk, \boldsymbol{x}_{\tilde{b}})$ as well as st as input.

Experiment $Exp_{\mathcal{A}}^{PRIV}(k)$

Input: Security parameter k

$\quad st \leftarrow A_t(1^k)$

$\quad \tilde{b} \leftarrow_R \{0, 1\}$

$\quad (\boldsymbol{x}_0, \boldsymbol{x}_1) \leftarrow A_m(1^k, \tilde{b}, st)$

$\quad (pk, sk) \leftarrow_R \mathcal{K}(1^k)$

$\quad \boldsymbol{c} := E(pk, \boldsymbol{x}_{\tilde{b}})$

$\quad b' \leftarrow A_g(1^k, pk, \boldsymbol{c}, st)$

$\quad Return(b' = \tilde{b})$

Fig. 7. The experiment Exp^{PRIV}.

Definition 5 (PRIV-security). *An encryption function E is PRIV-secure, if the advantage* $\mathsf{Adv}_{\mathcal{A}}^{\mathsf{PRIV}}(k) := \Pr[\mathsf{Exp}_{\mathcal{A}}^{\mathsf{PRIV}}(k) = 1] - \frac{1}{2}$ *is negligible in k for all security parameters k and all adversaries $\mathcal{A} = (A_t, A_m, A_g)$, under the following requirements:*

1. $|\boldsymbol{x}_0| = |\boldsymbol{x}_1|$
2. $\boldsymbol{x}_0[i] = \boldsymbol{x}_0[j]$ *iff* $\boldsymbol{x}_1[i] = \boldsymbol{x}_1[j]$
3. *High min-entropy:* $Pr[\boldsymbol{x}_\beta[i] = x : \boldsymbol{x} \leftarrow A_m(1^k, \tilde{b}, st)]$ *is negligible in k for all* $\beta \in \{0, 1\}$, *all* $1 \le i \le |\boldsymbol{x}|$, *all* $x \in Dom(E)$, *and all* $st \in \{0, 1\}^*$.

4. st is the empty string.

The first requirement prevents the adversary from trivially distinguishing the two ciphertexts by their length, while the second requirement addresses the fact that a deterministic encryption maps equal plaintexts to equal ciphertexts, and prevents the adversary from distinguishing ciphertexts using their equality pattern.

PRIV-security is sufficient to achieve our notion of correlation robustness with respect to the linear function we use in our garbling scheme. Therefore, we can instantiate our trapdoor permutation with a PRIV-secure deterministic encryption function. We prove this with the following theorem:

Theorem 2 (PRIV-security implies generalized correlation robustness). *Let E be a PRIV-secure deterministic encryption function with key space $\mathcal{K} = \mathcal{PK} \times \mathcal{SK}$, plaintext space $\mathcal{P}$ and ciphertext space $\mathcal{C}$. Let $f = (f_0, f_1, f_2, f_3)$ be an invertible linear function, consisting of the four functions $f_0, f_1, f_2, f_3 : Dom(E) \to Dom(E)$.*

Then the trapdoor permutation $E_\iota : \mathcal{P} \to \mathcal{C}; \ m \mapsto E(\iota, m)$ is correlation robust with respect to (f_0, f_1, f_2, f_3) for all $\iota \in \mathcal{PK}$.

Proof. We show that for each A^{corr} who breaks correlation robustness with non-negligible advantage p, there exists a PRIV-Adversary $\mathcal{A}$ who breaks PRIV-security with advantage $\frac{1}{2}p$.

We first construct an adversary with non-empty state st. Let us consider an adversary A^{corr}, who can break correlation robustness of E_ι with non-negligible advantage p. The adversary $\mathcal{A} = (A_t, A_m, A_g)$ uses A^{corr} to break PRIV-security as follows: First, adversary A_t asks A^{corr} to output a tuple (K, L), and sets $st := (K, L)$. Then, A_m chooses K', L', R_1, R_2 and R_3 uniformly at random. In the correlation robustness experiment, there are four ways for the challenger to pick a, b, c out of $\{0, 1, 2, 3\}$ with $a < b < c$. Adversary A_m chooses such (a, b, c) uniformly at random, and sets $x_0 := (f_a(K, L'), f_b(K', L), f_c(K', L'))$. He sets $x_1 := (R_1, R_2, R_3)$. Since K' and L' are chosen uniformly at random, the plaintexts chosen by A_m all have negligible probability of occurring, and therefore the desired min-entropy. The plaintext vector elements $K + L'$, $K' + L$ and $K' + L'$ of x_0 are all mutually different with overwhelming probability, so x_0 and x_1 have the same equality pattern.

A_g then obtains a public key pk and a ciphertext $c = (c_1, c_2, c_3)$.

If $\tilde{b} = 0$, c has the form of a challenge in $Exp^{corr}_{E, f_0, f_1, f_2, f_3, \mathcal{A}}(k)$ if $\beta = 0$ in this experiment, while in the case $\tilde{b} = 1$, c has the form of a challenge for $\beta = 1$. Therefore, A_g can feed the adversary A^{corr} with an appropriate challenge by handing him (pk, c). Adversary A_g then outputs whatever A^{corr} outputs.

Now we need to modify $\mathcal{A}$ such that st is the empty string. Without loss of generality, we can divide the random coins of A^{corr} into (r_1, r_2), such that r_1 is the randomness used to choose (K, L). Then for each choice of r_1, there is an adversary $A^{corr}_{r_1}$ which has r_1 hard-wired and always chooses a specific (K_{r_1}, L_{r_1}). From an average argument, if A^{corr} breaks correlation-robustness with advantage p, then there is random coin r_p for which the adversary $A_{r_p}^{corr}$ with hard-wired r_p breaks correlation-robustness with advantage p.

Consider the PRIV-Adversary $\mathcal{A}^{r_p} = (A_t^{r_p}, A_m^{r_p}, A_g^{r_p})$ which has his random coins hard-wired, such that (1) A_t always chooses st to be the empty string, (2) A_m always chooses $K = K_{r_p}$ and $L = L_{r_p}$, randomly chooses $a, b, c \in \{1, 2, 3, 4\}$ with $a < b < c$, and sets $x_0 := (f_a(K, L'), f_b(K', L), f_c(K', L'))$, and (3) Adversary $A_g^{r_p}$ internally executes $A_{r_p}^{corr}$ and outputs whatever $A_{r_p}^{corr}$ outputs. Since $A_m^{r_p}$ has K and L hard wired, $A_g^{r_p}$ is the only adversary who communicates with $A_{r_p}^{corr}$.

If $A_{r_p}^{corr}$ breaks correlation robustness with advantage p, then $\mathcal{A}^{r_p}$ breaks PRIV-security with advantage at least $\frac{1}{4}p$.   □

We can instantiate our correlation robust trapdoor permutation $E : \{0,1\}^k \to \{0,1\}^k$ with a PRIV-secure deterministic encryption scheme. In [6], Bellare et al. introduce a PRIV-secure length-preserving scheme *RSA-DOAEP*, which works on bitstrings. This scheme can be used to instantiate our correlation robust trapdoor permutation.

Acknowledgements. We thank the anonymous reviewers for helpful comments.

References

1. Shelat, A., Shen, C.: Two-output secure computation with malicious adversaries. In: Paterson, K.G. (ed.) EUROCRYPT 2011. LNCS, vol. 6632, pp. 386–405. Springer, Heidelberg (2011)
2. Applebaum, B.: Garbling XOR gates "for free" in the standard model. In: Sahai, A. (ed.) TCC 2013. LNCS, vol. 7785, pp. 162–181. Springer, Heidelberg (2013)
3. Applebaum, B., Ishai, Y., Kushilevitz, E.: Cryptography in NC⁰. FOCS **2004**, 166–175 (2004)
4. Applebaum, B., Ishai, Y., Kushilevitz, E.: Computationally private randomizing polynomials and their applications. Comput. Complex. **15**(2), 115–162 (2006)
5. Applebaum, B., Ishai, Y., Kushilevitz, E., Waters, B.: Encoding functions with constant online rate or how to compress garbled circuits keys. In: Canetti, R., Garay, J.A. (eds.) CRYPTO 2013, Part II. LNCS, vol. 8043, pp. 166–184. Springer, Heidelberg (2013)
6. Bellare, M., Boldyreva, A., O'Neill, A.: Deterministic and efficiently searchable encryption. In: Menezes, A. (ed.) CRYPTO 2007. LNCS, vol. 4622, pp. 535–552. Springer, Heidelberg (2007)
7. Bellare, M., Fischlin, M. O'Neill, A., Ristenpart, T.: Deterministic encryption: definitional equivalences and constructions without random oracles. Cryptology ePrint Archive, Report 2008/267 (2008). http://eprint.iacr.org/
8. Bellare, M., Hoang, V.T., Rogaway, P.: Foundations of garbled circuits. Cryptology ePrint Archive, Report 2012/265 (2012). http://eprint.iacr.org/
9. Boneh, D., Gentry, C., Gorbunov, S., Halevi, S., Nikolaenko, V., Segev, G., Vaikuntanathan, V., Vinayagamurthy, D.: Fully key-homomorphic encryption, arithmetic circuit ABE and compact garbled circuits. In: Nguyen, P.Q., Oswald, E. (eds.) EUROCRYPT 2014. LNCS, vol. 8441, pp. 533–556. Springer, Heidelberg (2014)

10. Brandão, L.T.A.N.: Secure two-party computation with reusable bit-commitments, via a cut-and-choose with forge-and-lose technique. In: Sako, K., Sarkar, P. (eds.) ASIACRYPT 2013, Part II. LNCS, vol. 8270, pp. 441–463. Springer, Heidelberg (2013)
11. Choi, S.G., Katz, J., Kumaresan, R., Zhou, H.-S.: On the security of the "Free-XOR" technique. In: Cramer, R. (ed.) TCC 2012. LNCS, vol. 7194, pp. 39–53. Springer, Heidelberg (2012)
12. Goldreich, O., Micali, S., Wigderson, A.: How to play any mental game or a completeness theorem for protocols with honest majority. In: STOC, pp. 218–229 (1987)
13. Hemenway, B., Lu, S., Ostrovsky, R.: Correlated product security from any one-way function. In: Fischlin, M., Buchmann, J., Manulis, M. (eds.) PKC 2012. LNCS, vol. 7293, pp. 558–575. Springer, Heidelberg (2012)
14. Huang, Y., Katz, J., Evans, D.: Efficient secure two-party computation using symmetric cut-and-choose. In: Canetti, R., Garay, J.A. (eds.) CRYPTO 2013, Part II. LNCS, vol. 8043, pp. 18–35. Springer, Heidelberg (2013)
15. Huang, Y., Katz, J., Kolesnikov, V., Kumaresan, R., Malozemoff, A.J.: Amortizing garbled circuits. In: Garay, J.A., Gennaro, R. (eds.) CRYPTO 2014, Part II. LNCS, vol. 8617, pp. 458–475. Springer, Heidelberg (2014)
16. Ishai, Y., Kushilevitz, E.: Randomizing polynomials: a new representation with applications to round-efficient secure computation. In: FOCS, pp. 294–304 (2000)
17. Ishai, Y., Kushilevitz, E.: Perfect constant-round secure computation via perfect randomizing polynomials. In: Widmayer, P., Triguero, F., Morales, R., Hennessy, M., Eidenbenz, S., Conejo, R. (eds.) ICALP 2002. LNCS, vol. 2380, p. 244. Springer, Heidelberg (2002)
18. Kilian, J.: Founding cryptography on oblivious transfer. In: STOC, pp. 20–31 (1988)
19. Kolesnikov, V.: Gate evaluation secret sharing and secure one-round two-party computation. In: Roy, B. (ed.) ASIACRYPT 2005. LNCS, vol. 3788, pp. 136–155. Springer, Heidelberg (2005)
20. Kolesnikov, V., Kumaresan, R.: Improved secure two-party computation via information-theoretic garbled circuits. In: Visconti, I., De Prisco, R. (eds.) SCN 2012. LNCS, vol. 7485, pp. 205–221. Springer, Heidelberg (2012)
21. Kolesnikov, V., Mohassel, P., Rosulek, M.: FleXOR: flexible garbling for XOR gates that beats free-XOR. In: Garay, J.A., Gennaro, R. (eds.) CRYPTO 2014, Part II. LNCS, vol. 8617, pp. 440–457. Springer, Heidelberg (2014)
22. Kolesnikov, V., Schneider, T.: Improved garbled circuit: free XOR gates and applications. In: Aceto, L., Damgård, I., Goldberg, L.A., Halldórsson, M.M., Ingólfsdóttir, A., Walukiewicz, I. (eds.) ICALP 2008, Part II. LNCS, vol. 5126, pp. 486–498. Springer, Heidelberg (2008)
23. Lindell, Y.: Fast cut-and-choose based protocols for malicious and covert adversaries. In: Canetti, R., Garay, J.A. (eds.) CRYPTO 2013, Part II. LNCS, vol. 8043, pp. 1–17. Springer, Heidelberg (2013)
24. Lindell, Y., Pinkas, B.: An efficient protocol for secure two-party computation in the presence of malicious adversaries. In: Naor, M. (ed.) EUROCRYPT 2007. LNCS, vol. 4515, pp. 52–78. Springer, Heidelberg (2007)
25. Lindell, Y., Pinkas, B.: A proof of security of Yao's protocol for two-party computation. J. Cryptology 22(2), 161–188 (2009)
26. Lindell, Y., Pinkas, B.: Secure two-party computation via cut-and-choose oblivious transfer. In: Ishai, Y. (ed.) TCC 2011. LNCS, vol. 6597, pp. 329–346. Springer, Heidelberg (2011)

27. Lindell, Y., Riva, B.: Cut-and-choose Yao-based secure computation in the online/offline and batch settings. In: Garay, J.A., Gennaro, R. (eds.) CRYPTO 2014, Part II. LNCS, vol. 8617, pp. 476–494. Springer, Heidelberg (2014)
28. Malkhi, D., Nisan, N., Pinkas, B., Sella, Y.: Fairplay - secure two-party computation system. In: USENIX Security Symposium, pp. 287–302 (2004)
29. Naor, M., Pinkas, B., Sumner, R.: Privacy preserving auctions and mechanism design. In: EC, pp. 129–139 (1999)
30. Pinkas, B., Schneider, T., Smart, N.P., Williams, S.C.: Secure two-party computation is practical. In: Matsui, M. (ed.) ASIACRYPT 2009. LNCS, vol. 5912, pp. 250–267. Springer, Heidelberg (2009)
31. Rosen, A., Segev, G.: Chosen-ciphertext security via correlated products. In: Reingold, O. (ed.) TCC 2009. LNCS, vol. 5444, pp. 419–436. Springer, Heidelberg (2009)
32. Sander, T., Young, A.L., Yung, M.: Non-interactive cryptocomputing for NC^1. In: FOCS, pp. 554–567 (1999)
33. Yao, A.C.-C.: How to generate and exchange secrets (extended abstract). In: FOCS, pp. 162–167 (1986)
34. Zahur, S., Rosulek, M., Evans, D.: Two halves make a whole. In: Oswald, E., Fischlin, M. (eds.) EUROCRYPT 2015. LNCS, vol. 9057, pp. 220–250. Springer, Heidelberg (2015)

Card-Based Cryptographic Protocols Using a Minimal Number of Cards

Alexander Koch$^{(\boxtimes)}$, Stefan Walzer, and Kevin Härtel

Karlsruhe Institute of Technology (KIT), Karlsruhe, Germany
alexander.koch@kit.edu, {stefan.walzer,kevin.haertel}@student.kit.edu

Abstract. Secure multiparty computation can be done with a deck of playing cards. For example, den Boer (EUROCRYPT '89) devised his famous "five-card trick", which is a secure two-party AND protocol using five cards. However, the output of the protocol is revealed in the process and it is therefore not suitable for general circuits with hidden intermediate results. To overcome this limitation, protocols in committed format, i.e., with concealed output, have been introduced, among them the six-card AND protocol of (Mizuki and Sone, FAW 2009). In their paper, the authors ask whether six cards are minimal for committed format AND protocols.

We give a comprehensive answer to this problem: there is a four-card AND protocol with a runtime that is finite in expectation (i.e., a Las Vegas protocol), but no protocol with finite runtime. Moreover, we show that five cards are sufficient for finite runtime. In other words, improving on (Mizuki, Kumamoto and Sone, ASIACRYPT 2012) "The Five-Card Trick can be done with four cards", our results can be stated as "The Five-Card Trick can be done in committed format" and furthermore it "can be done with four cards in Las Vegas committed format".

By devising a Las Vegas protocol for any k-ary boolean function using $2k$ cards, we address the open question posed by (Nishida et al., TAMC 2015) on whether $2k + 6$ cards are necessary for computing any k-ary boolean function. For this we use the shuffle abstraction as introduced in the computational model of card-based protocols in (Mizuki and Shizuya, Int. J. Inf. Secur., 2014). We augment this result by a discussion on implementing such general shuffle operations.

Keywords: Card-based protocols · Committed format · Boolean AND · Secure computation · Cryptography without computers

1 Introduction

The most well known card-based cryptographic protocol uses five cards showing two different types of symbols, $\heartsuit$ and $\clubsuit$, which are otherwise assumed to be physically indistinguishable. Let us quickly describe the elegant "five-card trick" of den Boer [B89] for computing a logical AND operation on the bits of two players. For this, the players input their bits as a commitment, which is two

© International Association for Cryptologic Research 2015
T. Iwata and J.H. Cheon (Eds.): ASIACRYPT 2015, Part I, LNCS 9452, pp. 783–807, 2015.
DOI: 10.1007/978-3-662-48797-6_32

face-down cards either as $\heartsuit\clubsuit$ or $\clubsuit\heartsuit$, encoding 1 or 0, respectively, with a separating $\clubsuit$ card in between, so that the possible input sequences look like this:

Now, the second player inverts his bit by swapping his cards, leading to the following situation:

Observe that only in the case of $a = b = 1$, the three $\clubsuit$s are consecutive. The following cyclic arrangement of cards as seen from below a "glass table" makes it obvious that this property is preserved under cyclic shifts of the cards:

By applying a cyclic shift by a random offset, the correspondence of the positions to the players is obscured. This "shuffling" of the cards can be done by the players taking turns in applying a cyclic shift of a random offset, without letting the other players observe the permutation that has been applied to the cards. By revealing all cards afterwards, the players can check whether the three $\clubsuit$s are consecutive and deduce that the output is 1 if this is the case, and 0 otherwise.

This example illustrates that a deck of cards can be used to securely evaluate functions, without the players giving away anything about their inputs that cannot be deduced from the result of the execution of such a card-based cryptographic protocol. The utility of these protocols is evident from their use in classrooms and lectures to illustrate secure multiparty computation to non-experts to the field of cryptography, or in an introductory course. Moreover, the possibility of performing these protocols without the use of computers is an interesting distinctive feature.

In their ASIACRYPT 2012 paper, Mizuki, Kumamoto, and Sone [MKS12] were able to reduce the number of cards to the best possible of 4, which is already necessary to encode the inputs. However, both protocols have an important caveat: They unavoidably reveal the final result during the computation. This makes them inadequate for use in larger protocols, for instance when evaluating complex logical circuits.

Therefore, starting with [NR98, S01, CK93], several researchers came up with so-called committed format protocols, which output a commitment encoding the result by two cards, as described above. This allows for using the output commitment of the protocol as an input to another protocol and for having a fine-grained control on who learns what about the result.

So far, the protocols using the least number of cards for computing AND in committed format are

- the six-card protocol of Mizuki and Sone [MS09], which has a deterministic runtime (cf. Fig. 2), and
- the five-card Las Vegas protocol of [CHL13], as described in Example 1. Note that this protocol may end in a configuration which needs restarting with probability $1/2$ and utilizes a rather complex shuffle operation. (These operations will be discussed in Sect. 8).

This leads to the natural question on the minimality of cards needed for a secure committed format AND, which has been posed in several places in the literature, see, e.g., [MS09, MS14a, MKS12]. Moreover in [CHL13], the authors ask whether there is a "deterministic" five-card variant of their protocol. In this paper, we answer these questions comprehensively.

To cope with these questions, [MS14a] defined a formal computational model stating the possible operations that a card-based protocol can make. To allow for strong impossibility results, the authors give a rather wide palette of possible operations that can be applied to the cards, e.g., shuffling with an arbitrary probability distribution on the set of permutations. Our paper shows that this yields rather strong possibility results by utilizing "non-closed" shuffles, as defined in Sect. 8.

Note that all protocols are in the *honest-but-curious* setting (although some analysis of malicious behavior has been done in [MS14b]), i.e., the players execute the protocol according to its description, but gather any information they can possibly obtain.

Contribution. In this paper, we

- introduce a four-card Las Vegas protocol for the AND of two players' bits,
- give a five-card variant, which has an a priori bound on the number of execution steps, i.e., a finite-runtime protocol,
- show that this is optimal, as four-card finite-runtime protocols computing AND in committed format are impossible,
- define a method of enriching the description of a protocol, that makes correctness and security transparent and gives a good understanding of how these protocols work, which can be used as a leverage to devise impossibility results. We therefore believe that this method is of general interest for research in card-based cryptography,
- state a general $2k$-card protocol for any k-ary boolean function, which can be seen as a touchstone for the practicability of the underlying computational model,
- discuss the computational model of [MS14a] briefly.

For comparison with other protocols, we refer the reader to Tables 1 and 2. For the former, we have three key parameters in describing the properties of protocols: whether it is committed format, whether it is a finite-runtime or a

Table 1. Minimal number of cards required by protocols computing AND of two bits, subject to the requirements specified in the first three columns.

Format	Runtime	Shuffles		#cards	Reference
Committed	Exp. finite	Non-uniform	Closed	4	Theorem 1
Committed	Exp. finite	Uniform	Non-closed	4	Theorem 4 (k=2)
Non-committed	Finite	Uniform	Closed	4	[MKS12]
Committed	Finite	Non-uniform	Non-closed	5	Theorems 2 and 3
Committed	Finite	Uniform	Closed	≤ 6	[MS09]

Table 2. Comparison of protocols for k-ary boolean functions.

#cards	Success probability	Shuffles		#steps	Reference
$2k$	2^{-k}	Uniform	Non-closed	Constant	Theorem 4
$2k + 6$	1	Uniform	Closed	Large	[N+15]

Las Vegas algorithm, and whether "non-closed" or "non-uniform" shuffles are used in the protocols, for which it is not yet apparent how they can be run in practice, cf. Section 8 for a discussion. Table 1 states the minimal number of cards for protocols with the given parameters and gives the corresponding references.

In Table 2 we compare our $2k$-card protocol of Sect. 7 with the best protocol for general boolean functions in the literature, with respect to the number of cards, namely [N+15]. While our protocol reduces the number of cards by six, it is a Las Vegas protocol with a substantial probability to end in a state which requires to restart the protocol. Moreover, it uses the non-closed shuffles mentioned above. Even though the expected number of restarts until a successful run is of order $O(2^k)$, each run of our protocol requires only a constant number of steps. This result can also be interpreted as a touchstone of the plausibility of the computational model for card-based protocols.

Outline. In Sect. 2 we introduce the basic computational model of card-based protocols and a strong information-theoretic security definition. We describe a method for the analysis of protocols in Sect. 3. We give a description of our four- and five-card protocols in Sects. 4 and 5, respectively. In the subsequent Sect. 6 we show that five cards are necessary for finite-runtime protocols. In Sect. 7 we state a Las Vegas protocol for general boolean functions using a strong shuffle operation that the computational model allows. We discuss these shuffle operations in Sect. 8. Finally, we conclude the paper in Sect. 9.

Notation. In the paper we use the following notation.

- *Cycle Decomposition.* For $n \in \mathbb{N}$ and numbers $a_1, a_2, \ldots, a_k \leq n$ we write $\pi = (a_1 \ a_2 \ \ldots \ a_k)$ for the permutation $\pi \in S_n$ that maps a_i to a_{i+1} for

$1 \leq i \leq k - 1$ and a_k to a_1 and all other $x \leq n$ to themselves. We call this a *cycle*. Cycles are maps, so they can be composed with $\circ$, which we will omit in the following, e.g. $(1\ 3\ 5)(2\ 4)$ maps $1 \mapsto 3$, $3 \mapsto 5$, $5 \mapsto 1$, $2 \mapsto 4$ and $4 \mapsto 2$.

- *Drawing from a Probability Distribution.* If $\mathcal{F}$ is a probability distribution on a set X, we write $x \leftarrow \mathcal{F}$ to indicate that $x \in X$ should be randomly chosen from X according to $\mathcal{F}$.
- *Sequence Indices.* Given a sequence $x = (\alpha_1, \ldots, \alpha_l)$ and an index i with $1 \leq i \leq l$, we denote by $x[i]$, the ith entry of the sequence, namely α_i.

2   Machine Model and Security of Card-Based Protocols

Mizuki and Shizuya [MS14a] came up with an elegant framework to model a computation with card-based cryptographic protocols. We adopt their setting to our needs and quickly review the important definitions in the following.

A *deck* $\mathcal{D}$ is a finite multiset of symbols, its elements are *cards*. We will restrict ourselves to the case where $\mathcal{D}$ contains two types of symbols, depicted by $\heartsuit$ and $\clubsuit$. For a symbol $c \in \mathcal{D}$, $\frac{c}{?}$ denotes a *face-up card* and $\frac{?}{c}$ a *face-down card* with symbol c, respectively. Here, '?' is a special backside symbol, not contained in $\mathcal{D}$. For a face-up or face-down card α, $\mathsf{top}\,(\alpha)$ and $\mathsf{atom}\,(\alpha)$ denote the symbol in the "numerator" and the symbol distinct from '?', respectively.

Cards are lying on the table in a *sequence*. A sequence is obtained by permuting $\mathcal{D}$ and choosing face-up or face-down for each card. For example, $(\frac{?}{\clubsuit}, \frac{\clubsuit}{?}, \frac{?}{\heartsuit}, \frac{?}{\heartsuit}, \frac{?}{\clubsuit})$ is a sequence of $\mathcal{D} = [\clubsuit, \clubsuit, \clubsuit, \heartsuit, \heartsuit]$. We extend $\mathsf{top}\,(\cdot)$ and $\mathsf{atom}\,(\cdot)$ from single cards to sequences of cards in the canonical way. For a sequence Γ, $\mathsf{top}\,(\Gamma)$ is the *visible sequence* of Γ. For example, $\mathsf{top}\,(\frac{?}{\clubsuit}, \frac{\clubsuit}{?}, \frac{?}{\heartsuit}, \frac{?}{\heartsuit}, \frac{?}{\clubsuit}) = (?, \clubsuit, ?, ?, ?)$. We denote the set of all visible sequences of $\mathcal{D}$ by $\mathsf{Vis}^{\mathcal{D}}$, or Vis for short. Furthermore, we define the set of *atomic sequences* $\mathsf{AtSeq}^{\mathcal{D}}$, or AtSeq for short, as the set of all permutations of $\mathcal{D}$.

A *protocol* $\mathcal{P}$ is a quadruple $(\mathcal{D}, U, Q, A)$, where $\mathcal{D}$ is a deck, U is a set of input sequences, Q is a set of states with two distinguished states q_0 and q_f, being the initial and the final state. Moreover, we have a (partial) action function

$$A \colon (Q \setminus \{q_f\}) \times \mathsf{Vis} \to Q \times \mathsf{Action},$$

depending only on the current state and visible sequence, specifying the next state and an operation on the sequence from Action that contains the following actions:

- (perm, π) for a permutation $\pi \in S_{|\mathcal{D}|}$ from the symmetric group $S_{|\mathcal{D}|}$ on elements $\{1, \ldots, |\mathcal{D}|\}$. This transforms a sequence $\Gamma = (\alpha_1, \ldots, \alpha_{|\mathcal{D}|})$ into

$$\mathsf{perm}_\pi(\Gamma) := (\alpha_{\pi^{-1}(1)}, \ldots, \alpha_{\pi^{-1}(|\mathcal{D}|)}),$$

i.e., it permutes the cards according to π.

- (turn, T) for $T \subseteq \{1, \dots, |\mathcal{D}|\}$. This transforms a sequence $\Gamma = (\alpha_1, \dots, \alpha_{|\mathcal{D}|})$ into

$$\mathsf{turn}_T(\Gamma) := (\beta_1, \dots, \beta_{|\mathcal{D}|}), \text{ where } \beta_i = \begin{cases} \mathsf{swap}(\alpha_i), & \text{if } i \in T, \\ \alpha_i, & \text{otherwise,} \end{cases}$$

  i.e., it turns over all cards from a *turn set* T. Here $\mathsf{swap}(\frac{c}{?}) := \frac{?}{c}$ and $\mathsf{swap}(\frac{?}{c}) := \frac{c}{?}$, for $c \in \mathcal{D}$.
- (shuffle, $\Pi, \mathcal{F}$) for a probability distribution $\mathcal{F}$ on $S_{|\mathcal{D}|}$ with support Π. This transforms a sequence Γ into the random sequence

$$\mathsf{shuffle}_{\Pi, \mathcal{F}}(\Gamma) := \mathsf{perm}_\pi(\Gamma), \text{ for } \pi \leftarrow \mathcal{F},$$

  i.e., $\pi \in \Pi$ is drawn according to $\mathcal{F}$ and then applied to Γ. Note that the players do not learn the chosen permutation when executing the protocol (unless they can derive it from $\mathcal{F}$ and the visible sequence after the operation). If $\mathcal{F}$ is the uniform distribution on Π, we may omit it and write (shuffle, Π).
- (rflip, $\Phi, \mathcal{G}$) for a probability distribution $\mathcal{G}$ on $2^{\{1, \dots, |\mathcal{D}|\}}$ with support Φ. This transforms a sequence Γ into

$$\mathsf{rflip}_{\Phi, \mathcal{G}}(\Gamma) := \mathsf{turn}_T(\Gamma), \text{ for } T \leftarrow \mathcal{G},$$

  i.e., $T \subseteq \{1, \dots, |\mathcal{D}|\}$ is drawn according to $\mathcal{G}$ and then the corresponding cards of Γ are turned.
- (restart). This transforms a sequence into the start sequence. This special operation requires that the first component of A's output, i.e., the next state, is q_0. This allows for Las Vegas protocols that "fail" and start over with a certain probability. Protocols with a (deterministic) finite runtime do not need this operation.
- (result, $p_1, \dots, p_l$) for a list of positions $p_1, \dots, p_l \in \{1, \dots, |\mathcal{D}|\}$. This special operation occurs if and only if the first component of A's output is q_f. This halts the protocol and specifies that $(\alpha_{p_1}, \dots, \alpha_{p_l})$ is the *output*, where $\Gamma = (\alpha_1, \dots, \alpha_{|\mathcal{D}|})$ is the current sequence.

A tuple $(\Gamma_0, \Gamma_1, \dots, \Gamma_t)$ of sequences such that $\Gamma_0 \in U$ and Γ_{i+1} arises from Γ_i by an operation as specified by the action function in a protocol run is a *sequence trace*; in that case $(\mathsf{top}(\Gamma_0), \mathsf{top}(\Gamma_1), \dots, \mathsf{top}(\Gamma_t))$ is a *visible sequence trace*.[1]

A protocol *terminates* when entering the final state q_f. A protocol is called *finite-runtime*[2] if there is a fixed bound on the number of steps, and in contrast *Las Vegas*, if it terminates almost surely (i.e., with probability 1) and in a number of steps that is *only expectedly* finite.

Next we describe a canonical form for protocols computing boolean functions. For this we interpret two cards with distinct symbols as 1, if their symbols are arranged $\heartsuit\clubsuit$, and 0, if they are arranged as $\clubsuit\heartsuit$.

---

[1] Note that traces in our sense also capture prefixes of complete protocol runs.

[2] We avoid the term "deterministic" here, as, for their security, card-based protocols use randomness as an intrinsic property, albeit not necessarily as a speedup of the protocol.

Definition 1. *Let $f\colon \{0,1\}^k \to \{0,1\}$ be a boolean function. Then we say a protocol $\mathcal{P} = (\mathcal{D}, U, Q, A)$ computes f, if the following holds:*

- *the deck $\mathcal{D}$ contains at least k cards of each symbol,*
- *there is a one-to-one correspondence between inputs and input sequences, with the convention that for $b \in \{0,1\}^k$ we have that U contains $\Gamma^b = (\alpha_1, \ldots, \alpha_{|\mathcal{D}|})$, where*

$$
(\alpha_{2i-1}, \alpha_{2i}) = \begin{cases} (\frac{?}{\heartsuit}, \frac{?}{\clubsuit}), & \text{if } b[i] = 1, \\ (\frac{?}{\clubsuit}, \frac{?}{\heartsuit}), & \text{if } b[i] = 0, \end{cases}
$$

for $1 \le i \le k$. The remaining $|\mathcal{D}| - 2k$ "helping" cards are arranged in some canonical way (their arrangement does not depend on b). In this paper we assume that the helping $\clubsuit$s are to the left of the helping $\heartsuit$s.
- *it terminates almost surely,*
- *for an execution starting with Γ^b for $b \in \{0,1\}^k$ the protocol ends with the action $(\mathsf{result}, p_1, p_2)$, such that*

$$
\mathsf{atom}\,(\beta_{p_1}, \beta_{p_2}) = \begin{cases} (\heartsuit, \clubsuit), & \text{if } f(b) = 1, \\ (\clubsuit, \heartsuit), & \text{otherwise,} \end{cases}
$$

where $\Gamma = (\beta_1, \ldots, \beta_{|\mathcal{D}|})$ is the final sequence.

Example 1. Let us describe, as an example, the Las Vegas five-card AND protocol of Hawthorne, and Lee [CHL13]. Here, the deck is $\mathcal{D} = [\heartsuit, \heartsuit, \clubsuit, \clubsuit, \clubsuit]$ and the set of inputs is given by $U = \{\Gamma^{11}, \Gamma^{10}, \Gamma^{01}, \Gamma^{00}\}$, where $\Gamma^{11} = (\frac{?}{\heartsuit}, \frac{?}{\clubsuit}, \frac{?}{\heartsuit}, \frac{?}{\clubsuit}, \frac{?}{\clubsuit})$, $\Gamma^{10} = (\frac{?}{\heartsuit}, \frac{?}{\clubsuit}, \frac{?}{\clubsuit}, \frac{?}{\heartsuit}, \frac{?}{\clubsuit})$, $\Gamma^{01} = (\frac{?}{\clubsuit}, \frac{?}{\heartsuit}, \frac{?}{\heartsuit}, \frac{?}{\clubsuit}, \frac{?}{\clubsuit})$, and $\Gamma^{00} = (\frac{?}{\clubsuit}, \frac{?}{\heartsuit}, \frac{?}{\clubsuit}, \frac{?}{\heartsuit}, \frac{?}{\clubsuit})$. The protocol $\mathcal{P} = (\mathcal{D}, U, \{q_0, q_1, q_2, q_3, q_f\}, A)$ is then described by A as follows:

1. $A(q_0, v) = (q_1, (\mathsf{perm}, (2\ 3\ 4\ 5)))$, i.e., insert the helping card at position 2.[3]
2. $A(q_1, v) = (q_2, (\mathsf{shuffle}, \Pi))$, where $\Pi = \{\mathrm{id}, (1\ 4\ 2\ 5\ 3)\}$.
3. $A(q_2, v) = (q_3, (\mathsf{turn}, \{1\}))$, i.e., turn the first card.
4. $A(q_3, v) = \begin{cases} (q_f, (\mathsf{result}, 2, 3)), & \text{if } v[1] = \clubsuit, \\ (q_0, (\mathsf{restart})), & \text{otherwise.} \end{cases}$

Here, v denotes the current visible sequence in each step. Note that there is no obvious way to implement the shuffle in step 2 efficiently, as it is non-closed. See Sect. 8 for discussion.

Definition 2 (secure, committed format). *Let $\mathcal{P} = (\mathcal{D}, U, Q, A)$ be a protocol. Let Γ_0 be a random variable with values in the set of input sequences U and a distribution $\mathcal{M}$ on U. Let V be a random variable for the visible sequence trace of the protocol execution.*

$\mathcal{P}$ is secure or private if Γ_0 and V are stochastically independent.

---

[3] Note that this step is only needed because our input convention from Definition 1 differs from the input convention of [CHL13].

Moreover, let R be a random variable that encodes the output of the protocol. Then $\mathcal{P}$ is said to be in committed format, *if* atom (R) *and V are stochastically independent. (In particular, this implies that an index occurring in the* result *action points to a face-down card, unless this part of the output is constant.)*

From this definition it is apparent that if there is a functional dependency between the inputs and the output, then security implies committed format. Note that it is stronger than other security definitions in the literature that were defined to also capture non-committed format protocols, such as the five-card trick of [B89].

When the input is provided by players, each of them have a partial knowledge on Γ_0. The definition then implies that, even given this partial knowledge, Γ_0 and V are still independent. Therefore the players cannot learn anything about the inputs of the other players, as the result is not part of V.

3   A Calculus of States

From a specification of a protocol it is not immediately obvious whether it is correct and private. We describe a new method to obtain a rich description of possible protocol runs, from which correctness and privacy can be more easily recognized. We use this method in later sections to describe our constructions and prove the impossibility of finite-runtime four-card AND in Sect. 6. We believe this method is of general interest for researchers in the field of card-based cryptography.

When describing all possible executions of a protocol we obtain a tree which branches when the visible sequence differs. The nodes of this tree correspond to the visible sequence traces that can occur during the run of the protocol. Each node has an action associated to it, namely the action that the protocol prescribes for that situation. In the following, this action is a label on the outgoing edges.

Take for instance the six-card AND protocol of [MS09], as shown in Fig. 1. We hope that it will soon become clear why the protocol works.

Until the fourth step (the turn step) there is no observable difference, i.e., all visible sequences contain only '?'. After the turn, there are two types of executions that can be distinguished by players. If security was violated, i.e., players can deduce information about the input, then this is because some inputs are more likely to lead to a specific visible sequence than other inputs.

While the actual sequence on the table and the actual input of the players is typically unknown, knowledge about the former implies knowledge about the latter and vice versa. To facilitate the privacy analysis, we annotate the nodes of the tree with this dependent knowledge. A state in our sense captures the probability distribution of atomic sequences conditioned on the input sequence.

Definition 3. *Let $\mathcal{P}$ be a secure protocol computing $f\colon \{0,1\}^k \to \{0,1\}$ and V be a visible sequence trace of $\mathcal{P}$. The* state S *of $\mathcal{P}$ belonging to V is the map $S\colon \mathsf{AtSeq} \to \mathbb{X}_k$, with $s \mapsto \Pr[s|V]$, where:*

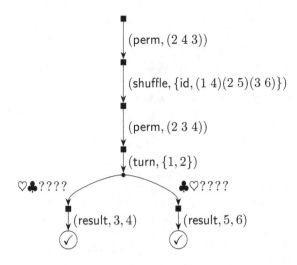

Fig. 1. Six-card AND protocol in committed format of [MS09].

- $\mathbb{X}_k$ denotes the polynomials over the variables X_b for $b \in \{0,1\}^k$ of the form $\sum_{b \in \{0,1\}^k} \beta_b X_b$, for $\beta_b \in [0,1] \subseteq \mathbb{R}$. We interpret these polynomials as probabilities which depend on the probabilities of the inputs b, symbolized by the variables X_b for $b \in \{0,1\}^k$.
- for $s \in$ AtSeq, $\Pr[s|V]$ denotes the (symbolic) probability that the current atomic sequence is s given that current visible sequence trace is V. (It will later be apparent that the probability $\Pr[s|V]$ is indeed in $\mathbb{X}_k$.)

We say a state S contains an atomic sequence s (or s is in S for short) if $S(s)$ is not the zero polynomial. For $k \geq 2$, we introduce the additional shorthands $X_0 := \sum_{f(b)=0} X_b$ and $X_1 := \sum_{f(b)=1} X_b$.

Let S be a state. Given a probability distribution $\mathcal{M}$ on the inputs, then substituting each variable X_b with the probability of the input b, yields a probability distribution on the atomic sequences in S. In particular, if s is an atomic sequence in S and $S(s)$ the corresponding polynomial, substituting 1 for the variable X_b and 0 for the other variables in $S(s)$, yields the probability that s is the current atomic sequence, given the input b and any information observed so far. Accordingly, we can use our notions to analyze player knowledge in multiparty computations where an agent has partial information about the input.

As an illustration of our method, consider the states of the six-card AND protocol from above, see Fig. 2 on page 10, where states are represented by a box with atomic sequences on the left and the associated polynomials on the right. In such a 2-ary protocol, a state maps each atomic sequence to a polynomial of the form $\beta_{11}X_{11} + \beta_{10}X_{10} + \beta_{01}X_{01} + \beta_{00}X_{00}$, where $\beta_{11}, \beta_{10}, \beta_{01}, \beta_{00} \in [0,1]$.

- In the start state, each input $b \in \{00, 01, 10, 11\}$ is associated with a unique input sequence $\Gamma^b \in U$, which, by our conventions in Definition 1, are

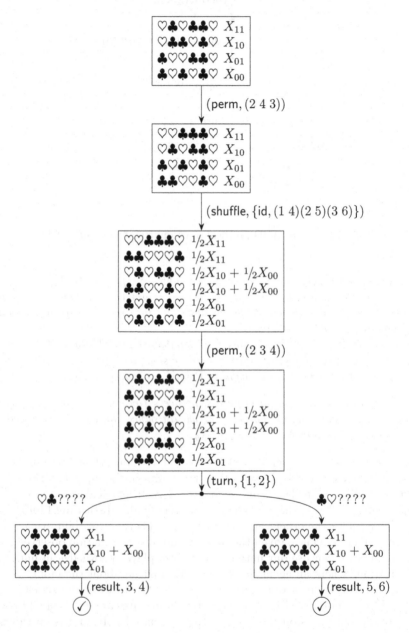

Fig. 2. Six-card AND protocol in committed format of [MS09] augmented with state information as in Definition 3.

$\Gamma^{11} = (\heartsuit, \clubsuit, \heartsuit, \clubsuit, \clubsuit, \heartsuit)$, $\Gamma^{10} = (\heartsuit, \clubsuit, \clubsuit, \heartsuit, \clubsuit, \heartsuit)$, $\Gamma^{01} = (\clubsuit, \heartsuit, \heartsuit, \clubsuit, \clubsuit, \heartsuit)$ and $\Gamma^{00} = (\clubsuit, \heartsuit, \clubsuit, \heartsuit, \clubsuit, \heartsuit)$. The probability of atom (Γ^b) being the current atomic sequence is therefore exactly X_b, i.e., the probability that b is the input. The remaining $\binom{6}{3} - 4$ atomic sequences are mapped to zero and omitted in the presentation.

- The first (and third) action is a permutation. Mathematically, nothing interesting happens here: If an atomic sequence s had its probability captured by $S(s)$, then after permuting with a permutation π, these probabilities are then assigned to the atomic sequence $\pi(s)$.
- The shuffle introduces uncertainty. Consider for instance the case that the input was "10". Then, before the shuffle, we must have had the atomic sequence $s = (\heartsuit, \clubsuit, \heartsuit, \clubsuit, \clubsuit, \heartsuit)$. It was either permuted by id or by $\pi = (1\ 4)(2\ 5)(3\ 6)$, yielding either s itself or $s' = (\clubsuit, \clubsuit, \heartsuit, \heartsuit, \clubsuit, \heartsuit)$, both with probability $1/2$. This explains the coefficients of X_{10} in the polynomials for s and s'.
- The turn step can yield two possible visible sequences: $(\heartsuit, \clubsuit, ?, ?, ?, ?)$ and $(\clubsuit, \heartsuit, ?, ?, ?, ?)$. Crucially, the probability of observing $(\clubsuit, \heartsuit, ?, ?, ?, ?)$ is the same for each possible input, so no information about the actual sequence is leaked: If $(\clubsuit, \heartsuit, ?, ?, ?, ?)$ would be observed slightly more frequently for, say, the input "01" than for the input "10", then observing $(\clubsuit, \heartsuit, ?, ?, ?, ?)$ would be weak evidence that the input was "01". In the case at hand, however, the probability for the right branch is $1/2$ for each input, as the sum of the polynomials of the atomic sequences branching right is $1/2(X_{11} + X_{10} + X_{01} + X_{11})$.
  After the turn our knowledge has changed, for instance, if we have observed $(\heartsuit, \clubsuit, ?, ?, ?, ?)$ and know that the input was "11" then we know beyond doubt that the atomic sequence must then be $(\heartsuit, \clubsuit, \heartsuit, \clubsuit, \clubsuit, \heartsuit)$, explaining the coefficient 1 of X_{11}.
- The output given by the result actions is correct: For all polynomials containing X_{11} with non-zero coefficient, the corresponding atomic sequence has $(\heartsuit, \clubsuit)$ at the specified positions and for all polynomials containing one of the other variables with non-zero coefficient, the corresponding atomic sequence has $(\clubsuit, \heartsuit)$ there.
  Note that "mixed" polynomials with non-zero coefficients of both types cannot occur in a final state of a protocol.

Derivation Rules for States. To compute the states we first identify the start state and then specify how subsequent states arise from a given state when performing an action. The rules of our calculus can also be seen as an inductive proof that our definition of a state is sound in secure protocols, as the probabilities are in $\mathbb{X}_k$ as claimed.

The *start state* S_0 with initial visible sequence trace V_0 contains exactly the input sequences in U. Each $\Gamma_b \in U$ of input $b \in \{0, 1\}^k$ is mapped to the probability $\Pr[\text{atom}(\Gamma_b)|V_0] = X_b$.

An action $\text{act} \in \text{Action}$ on a state S belonging to a visible sequence trace V can result in visible sequences $v_1, \ldots, v_n$. In the following, we state the rules for

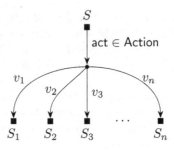

Fig. 3. Performing an action on a state can result in different visible sequences corresponding to a state each.

the derivation of these subsequent states $S_1, \ldots, S_n$ belonging to the extended visible sequences traces $V \parallel v_i$, obtained by appending the new visible sequence v_i to the trace, for $1 \leq i \leq n$. We restrict the presentation to shuffle and randomized flip operations, as the permutation and turn operations are special cases. For an illustration, we refer to Fig. 3.

Shuffle Action. Let $\mathsf{act} = (\mathsf{shuffle}, \Pi, \mathcal{F})$. If all cards are face-down before the shuffle, act can result in only one visible sequence, but in general let Π_v be the subset of Π that leads to some visible sequence v with corresponding state S'. If $\mathcal{F}_{|v}$ denotes the probability distribution on Π_v conditioned on the fact that v is observed, we have that

$$S'(s) = \sum_{\pi \in \Pi_v} \mathcal{F}_{|v}(\pi) \cdot S(\pi^{-1}(s)).$$

In other words, the probability for the atomic sequence s in the new state S' is obtained by considering all atomic sequences $\pi^{-1}(s)$ from which s may have originated through some $\pi \in \Pi_v$ and summing the probability of those atomic sequences in the old state, weighted with the probabilities that the corresponding π is chosen.

Randomized Flip Action. Let $\mathsf{act} = (\mathsf{rflip}, \Phi, \mathcal{G})$. Consider the state S' belonging to the visible sequence trace $V' := V \parallel v$ for the new visible sequence v, resulting from a flip of some turn set $T \in \Phi$. We say that v is *compatible* with an atomic sequence s from S if v and s agree in all positions that are not '?' in v. The set of all atomic sequences compatible with v is denoted by C_v.

Let $P_v := \sum_{s \in C_v} S(s)$. This polynomial represents the probability of observing v if T is turned in state S. Let β_b be the coefficients of P_v, i.e., $P_v = \sum_{b \in \{0,1\}^k} \beta_b X_b$. If the coefficients differ, i.e., $\beta_{b_1} \neq \beta_{b_2}$ for two inputs b_1 and b_2, then the probability of observing v when turning T in state S depends on the input. This must not be the case in secure protocols where visible sequences and inputs are independent.

In secure protocols, we therefore know that

$$P_v = \sum_{b \in \{0,1\}^k} \beta_v X_b = \beta_v \sum_{b \in \{0,1\}^k} X_b,$$

for some $\beta_v \in \mathbb{R}$. In our interpretation as probabilities, we have $\sum_{b \in \{0,1\}^k} X_b = 1$, i.e., the sum over all input probabilities is 1. From this, we obtain $P_v = \beta_v$.

Then, using Bayes' formula yields

$$S'(s) = \Pr[s|V'] = \Pr[s|(V \parallel v)] = \Pr[v|V, s] \cdot \frac{\Pr[s|V]}{\Pr[v|V]}$$

$$= \Pr[v|V, s] \cdot \frac{S(s)}{P_v} = \begin{cases} S(s)/\beta_v, & \text{if } s \in C_v, \\ 0, & \text{otherwise,} \end{cases}$$

where $\Pr[v|V, s]$ denotes the probability that v occurs, given that the visible sequence trace is V and the actual atomic sequence is s, and $\Pr[v|V]$ denotes the probability that v occurs, given that the visible sequence trace is V. Note that the actual atomic sequence s determines the visible sequence of the turn action, so $\Pr[v|V, s]$ is either 0 or 1.

Checking Correctness and Security. Since we keep track of the set of possible atomic sequences for any state of the protocol, we can decide for any result action whether it yields the correct output in all cases.

To check privacy, first note that shuffle actions never reveal new critical information: When shuffling with face-up cards, the shuffle may reveal information about which permutation was used to shuffle, but this information is a fresh random variable independent of all previous information. Considering turns or randomized flips, we already identified the condition before: A turn does not violate privacy if for every visible sequence v that may result from the turn, the set C_v of atomic sequences that are compatible with v must fulfill $\sum_{s \in C_v} S(s) = \beta_v \in [0, 1]$ since this exactly means that the probability to observe a visible sequence does not depend on the inputs. As this was a precondition for the derivation rule of randomized flips, being able to construct a diagram by the rules above is a witness to the security of the protocol. (In this sense, Fig. 2 is an alternative proof for the security of the six-card AND protocol of [MS09].)

Las Vegas vs Finite-Runtime. In our formalism, the states of a finite-runtime protocol form a finite tree without restart actions. A Las Vegas protocol, in contrast, makes use of restart actions, or its states form a cyclic or infinite diagram.

4   A Four-Card Las Vegas AND Protocol

We present a secure protocol to compute AND on two bits in committed format and without restarts. An algorithmic description is given in Protocol 1 and a representation in the state calculus of Sect. 3, from which correctness and privacy can be deduced, is given in Fig. 4.

Note that the state diagram contains a cycle, i.e., it is possible to return to a state that was encountered before. This implies that the protocol is not finite-runtime. However, on the cycle there are two turn operations each of which have a chance of $1/3$ to yield a final state and therefore leave the cycle. The probability to return to a state on the cycle is therefore $(\frac{2}{3})^2 = \frac{4}{9}$ and the probability to take the cycle k times is $(\frac{4}{9})^k$. The expected number of times the cycle is taken is therefore $\sum_{k \geq 0}(\frac{4}{9})^k = (1 - \frac{4}{9})^{-1} = \frac{9}{5}$. In particular, the expected runtime of the protocol is bounded. We summarize our result in the following theorem.

Theorem 1. *There is a secure Las Vegas protocol to compute AND on two bits in committed format and without restarts.*

In contrast to the protocol for general boolean functions presented in Sect. 7 the shuffle operations are "closed", a circumstance we discuss more closely in Sect. 8.

5   A Five-Card Finite-Runtime AND Protocol

In the presentation of our five-card finite-runtime AND protocol in committed format, we reuse part of our four-card protocol from Sect. 4. We just have to show that we can "break out" of the cycle of the four card protocol by using the fifth card. This yields a finite-runtime protocol with at most 12 steps in every execution. Here, the fifth card is chosen to have symbol $\heartsuit$.

An algorithmic description is given in Protocol 2 and a representation of the crucial component in the state calculus of Sect. 3, from which correctness and privacy can be deduced, is given in Fig. 5. We summarize our result in the following theorem.

Theorem 2. *There is a secure five-card finite-runtime protocol to compute AND on two bits in committed format.*

6   Finite-Runtime AND Requires Five Cards

There are secure protocols with four cards computing AND in committed format using either the restart operation (see Sect. 7) or running in cycles for a number of iterations that is finite only in expectation (see Sect. 4). However, it would be nice to have a protocol that is finite-runtime, i.e., is guaranteed to terminate after a finite number of steps. In the following we show that this is impossible.

To this end, we distinguish several different types of states and later analyze which state transitions are possible. We need the following definitions and observations only for the deck $\mathcal{D} = [\heartsuit, \heartsuit, \clubsuit, \clubsuit]$, but choose to state some of them in a more general form to better convey the underlying ideas.

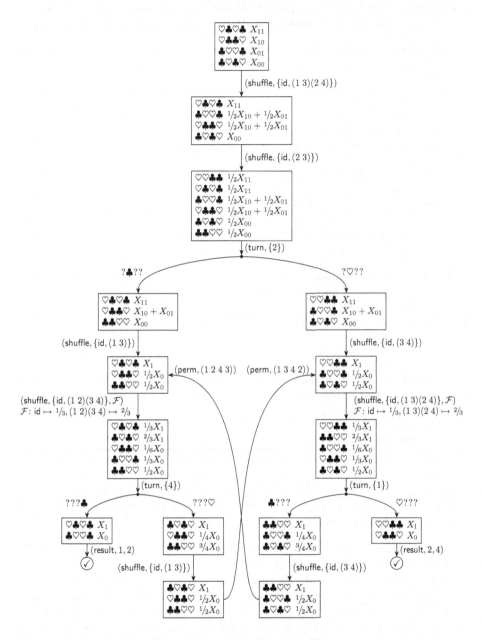

Fig. 4. The four-card Las Vegas AND protocol without restart operations from Protocol 1. Note that we make use of the shorthands $X_1 := X_{11}$ and $X_0 := X_{00} + X_{10} + X_{01}$ and omit the turn actions that merely turn cards back to face-down. Starting at certain points the tree becomes self-similar, which we represent by drawing backwards edges.

Protocol 1. Protocol to compute AND in committed format using four cards. Note that, because of the **goto** operations, no bound on the number of steps can be given.

(shuffle, {id, (1 3)(2 4)})
(shuffle, {id, (2 3)})
(turn, {2})
if $v = (?, \clubsuit, ?, ?)$ **then**

    (turn, {2}) // turn back
    (shuffle, {id, (1 3)})

1   (shuffle, {id, (1 2)(3 4)}, $\mathcal{F}$: id $\mapsto$ ¹/₃, (1 2)(3 4) $\mapsto$ ²/₃)
    (turn, {4})
    if $v = (?, ?, ?, \clubsuit)$ **then**
       | (result, 1, 2)
    else if $v = (?, ?, ?, \heartsuit)$ **then**
       (turn, {4}) // turn back
       (shuffle, {id, (1 3)})
       (perm, (1 3 4 2))
       goto 2
else if $v = (?, \heartsuit, ?, ?)$ **then**

    (turn, {2}) // turn back
    (shuffle, {id, (3 4)})

2   (shuffle, {id, (1 3)(2 4)}, $\mathcal{F}$: id $\mapsto$ ¹/₃, (1 3)(2 4) $\mapsto$ ²/₃)
    (turn, {1})
    if $v = (\heartsuit, ?, ?, ?)$ **then**
       | (result, 2, 4)
    else if $v = (\clubsuit, ?, ?, ?)$ **then**
       (turn, {1}) // turn back
       (shuffle, {id, (3 4)})
       (perm, (1 2 4 3))
       goto 1

Definition 4. *Let $\mathcal{P}$ be a protocol with deck $\mathcal{D}$ computing a boolean function f. Let s be an atomic sequence, S a state of $\mathcal{P}$ and $P = S(s)$ the polynomial representing the probability of s in S.*

1. *If P contains only variables X_b with $f(b) = 1$ or $f(b) = 0$, then s is called a 1-sequence or 0-sequence, respectively.*
2. *If P contains variables of both types, then s is called a $\bot$-sequence.*
3. *We say that S is of type i/j, or an i/j-state, if its number of 0-sequences and 1-sequences is i and j, respectively, and it does not contain any $\bot$-sequences.*
4. *We call a state S final if it does not contain a $\bot$-sequence and there are indices $m, n \in \{1, \ldots, |\mathcal{D}|\}$, such that all 1-sequences have $\heartsuit$ at position m, all 0-sequences have $\clubsuit$ at position m, and the other way round at position n. In that case (result, m, n) is a correct output operation.*

Note that a protocol that produces a $\bot$-sequence cannot be finite-runtime: once the $\bot$-sequence is lying on the table, it is impossible to decide whether the

Protocol 2. A five-card finite-runtime AND protocol. It proceeds as in Protocol 1 (ignoring card 5) until reaching the line marked as 1, when instead of executing the line, an alternative path is taken using the fifth card.

(shuffle, {id, (1 3)(2 4)})
(shuffle, {id, (2 3)})
(turn, {2})
if $v = (?, \clubsuit, ?, ?, ?)$ then
    (turn, {2}) // turn back
    (shuffle, {id, (1 3)})
$\star$ (perm, (1 5 2 4)) // sort in the fifth card
    (shuffle, {id, (5 4 3 2 1)}, $\mathcal{F}$: id $\mapsto$ 1/3, (5 4 3 2 1) $\mapsto$ 2/3)
    (turn, {5})
    if $v = (?, ?, ?, ?, \clubsuit)$ then
        (result, 4, 3)
    else if $v = (?, ?, ?, ?, \heartsuit)$ then
        (result, 3, 1)
else if $v = (?, \heartsuit, ?, ?, ?)$ then
    (turn, {2}) // turn back
    (shuffle, {id, (3 4)})
    (shuffle, {id, (1 3)(2 4)}, $\mathcal{F}$: id $\mapsto$ 1/3, (1 3)(2 4) $\mapsto$ 2/3)
    (turn, {1})
    if $v = (\heartsuit, ?, ?, ?, ?)$ then
        (result, 2, 4)
    else if $v = (\clubsuit, ?, ?, ?, ?)$ then
        (turn, {1}) // turn back
        (shuffle, {id, (3 4)})
        (perm, (1 2 4 3))
        goto $\star$

---

output should be 0 or 1. Thus, any protocol that proceeds to output something without restarting in between produces an incorrect result with positive probability; and any protocol that may use a restart, may take this execution path an unbounded number of times.

Since we are interested in the existence of finite-runtime protocols, we restrict our attention to protocols that never produce $\perp$-sequences. We now bundle a few simple properties about i/j-states in the following lemma.

Lemma 1. Given a secure protocol computing a non-constant boolean function with deck $\mathcal{D}$, consisting of n $\heartsuit$s and m $\clubsuit$s where $n, m \geq 1$, the following holds.

1. In a state of type i/j, we have $i, j \geq 1$, otherwise players could derive the the result, contradicting the committed format property.
2. If a turn in a state S of type i/j can result in two different successor states S_1 and S_2 of type i_1/j_1 and i_2/j_2, respectively, then $i = i_1 + i_2$ and $j = j_1 + j_2$. In particular, $i \geq 2$ and $j \geq 2$.
3. In a state of type i/j resulting from a turn that revealed a $\heartsuit$ or $\clubsuit$ we have $i + j \leq \binom{n+m-1}{n-1}$ or $i + j \leq \binom{n+m-1}{m-1}$, respectively.

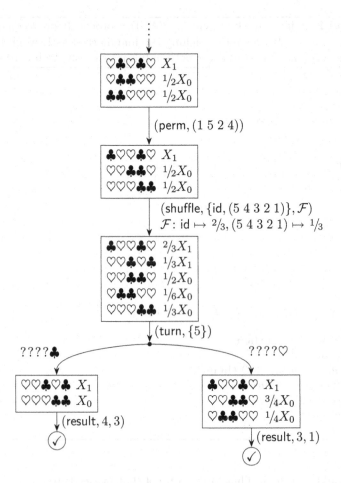

Fig. 5. The crucial part of a five-card finite-runtime AND protocol that allows to "break out" of the cycle in the four-card Las Vegas AND protocol.

4. Let S be a state of type i/j and S' a state of type i'/j' resulting from S via a shuffle operation. Then we have $i' \geq i$, $j' \geq j$.
5. If S is a final state of type i/j, then $i, j \leq \binom{n+m-2}{n-1}$.
6. Two atomic sequences differ in an even number of positions, i.e., have even *distance*.
7. Given an atomic sequence $s \in \mathsf{AtSeq}$, there are

$$\binom{n}{\frac{d}{2}}\binom{m}{\frac{d}{2}}$$

atomic sequences of (even) distance d to s.
8. Any two sequences have distance at most $\min\{2m, 2n\}$.
9. After a single-card turn revealing $\heartsuit$ or $\clubsuit$, any two sequences of the state have distance at most $2n - 2$ or $2m - 2$, respectively.

Theorem 3. *There is no secure finite-runtime four-card* AND *protocol in committed format.*

Proof. Let P be a secure protocol computing AND with four cards in committed format.

We will define a set of *good states*, denoted by $\mathcal{G}$, that contain all final states but not the starting state and show that any operation on a non-good state will produce at least one non-good state as a successor. From this it is then clear by induction that P is not finite-runtime.

A state S is *good* iff it fulfills one of the following properties:

- S is a 1/1-state,
- S is a 2/2-state,
- S is a 1/2- or 2/1-state containing two atomic sequences of distance 4.

We first observe which state types i/j can occur with our deck: Since there are $6 = \binom{4}{2}$ atomic sequences in total, we need $i + j \leq 6$. By Lemma 1, item 1, states with $i = 0$ or $j = 0$ cannot occur.

Final States are Good. From item 5 in Lemma 1 we know that final states fulfil $i, j \leq 2$ so the only candidate for final states are 1/1, 2/2, 1/2 and 2/1. We need to show that they are good which is true by definition for 1/1 and 2/2. Consider a final 1/2-state (the argument for the 2/1-state is symmetric). Its 0-sequence differs from both 1-sequences in the two positions used for the output. Since the two 1-sequences are distinct, at least one of them must differ from the 0-sequence in another position, meaning they must have distance at least 3 and therefore distance 4 (item 6 in Lemma 1).

Therefore, all final sequences are good, but the start state, which is a 3/1-state, is non-good. Consider an action $\mathsf{act} \in \mathsf{Action}$ that acts on a non-good state. We show that act has a non-good successor state by considering all cases for the type of act:

Non-trivial Single-card Turns. Let S be a non-good state of type i/j, and $S_\heartsuit$ and $S_\clubsuit$ the two possible states after a turn of a single card. From item 2 in Lemma 1, we know that S has to be of type i/j, with $i, j \geq 2$, excluding the case of 2/2, as S is non-good. This leaves the following possible types for S: 2/3, 3/3, 2/4 where we assume without loss of generality that $i \leq j$. The turn partitions the sequences onto the two branches in one of the following ways:

```
    2/3         3/3         3/3         2/4         2/4
    / \         / \         / \         / \         / \
 1/1   1/2   1/1   2/2   1/2   2/1   1/2   1/2   1/3   1/1
```

From item 3 in Lemma 1, we know that a state resulting directly from a turn contains at most 3 atomic sequences, thereby ruling out turn-transitions that lead to a 2/2- or 1/3-state. Moreover, any 2/1- or 1/2-state occurring after a turn has the property that all atomic sequences have pairwise distance 2 by item 9 in Lemma 1. By definition, such 2/1-states are non-good. Note that a turn action on a 2/3-state – while producing a good and even final 1/1-state – produces a non-good 1/2-state on the other branch.[4]

Non-branching Shuffles. Now consider a shuffle that produces a unique subsequent state S' of type i'/j'. We want to show that S' is non-good. Using item 4 in Lemma 1 and the fact that a good S' would require $i', j' \leq 2$, we only need to consider the case that S is a non-good state with $i, j \leq 2$, i.e., S is of type 1/2 or 2/1 with pairwise distance 2 – without loss of generality of type 1/2 and with a 0-sequence s_0 and two 1-sequences s_1 and s'_1. We argue that without loss of generality S is of the form *6pt

This is because

- S contains a constant column: Let k and l be the positions where s_0 differs from s_1, and m, n the positions where s_0 differs from s'_1. If $\{k, l\}$ and $\{m, n\}$ are disjoint, then s_1 and s'_1 have distance 4 – a contradiction. Otherwise $\{k, l, m, n\}$ has size at most 3 so there is one position where all atomic sequences agree.
- The constant column can be assumed to be in position 1 and to contain $\heartsuit$s. This completely determines the atomic sequences occurring in S. Our choice to pick the 0-sequence is arbitrary, but inconsequential.

If all permutations in the shuffle map 1 to the same $i \in \{1, 2, 3, 4\}$, then S' will have a constant column in position i. Then S' is still of type 1/2 with sequences of pairwise distance 2, so non-good. If there are two permutations in the shuffle that map 1 to different positions $i \neq j$, then S' will contain all three atomic sequences with $\heartsuit$ in position i and all three atomic sequences with $\heartsuit$ in position j. There is only one atomic sequence with $\heartsuit$ in both positions. So S' contains at least $3 + 3 - 1 = 5$ atomic sequences and is therefore non-good.

Other Actions. The hard work is done, but for completeness, we need to consider the remaining actions as well:

> *Restart.* This action is not allowed in our finite-runtime setting.
> *Result.* Since non-good states are non-final this action cannot be applied.
> *Permutation.* This is just a special case of a non-branching shuffle.

---

[4] Moreover, this is the only way to produce a good state from a non-good state via a turn action. We make use of such a turn in our four-card protocol in Sect. 4, which did not require finite-runtime. (In contrast to our protocol in Sect. 7 this allows us to avoid restart actions.).

Trivial turn. If act is a turn operation that can only result in a single visible sequence (the turn is *trivial*), then the outcome of the turn was known in advance and the state does not change.

Multi-card turn. If act turns more than one card, then act can be decomposed into single-card turn actions, turning the cards one after the other. We already know that a single-card turn from a non-good state yields a non-good subsequent state, so following a "trail" of non-good states shows act produces a non-good state as well.

Randomized flip. If act is a randomized flip then consider any turn set T that act might be picked. We already know that turning T yields a non-good subsequent state and this is also a subsequent state of act.

Branching shuffle. If act is a shuffle that produces several subsequent states (this requires shuffling with a face-up card), then restricting the set of allowed permutations to those corresponding to one of the visible sequences yields an ordinary shuffle that therefore yields a single subsequent non-good state. This state is also a subsequent state of act.

This concludes the proof.   □

7   A 2k-Card Protocol for any k-ary Boolean Function

The following protocol will compute a k-ary boolean function with $2k$ cards and success probability 2^{-k} in three steps: One shuffle, one turn and one result or restart action. The "hard work" is done in an "irregularly complex" shuffle operation, which may pose practical problems we expand upon in Sect. 8.

Theorem 4. *For any boolean function* $f\colon \{0,1\}^k \to \{0,1\}$ *there is a secure Las Vegas protocol in committed format using* $2k$ *cards. The expected number of* restart *actions in a run is* $2^k - 1$.

Proof. Note first that all unary boolean functions can easily be implemented: The identity and not-function is simple (just output the input or the inversed input) and for the constant functions we may shuffle the two cards (to obscure the input), then turn the cards over, arrange them to represent the constant and then return the positions of the corresponding cards, via result.

We now assume $k \geq 2$. For each input $b = (b_1, b_2, \ldots, b_k) \in \{0,1\}^k$ we define the permutation:

$$\pi_b := (2\ 3)^{1-f(b)} \circ (1\ 2)^{b_1}(3\ 4)^{b_2} \cdots (2k-1\ 2k)^{b_k}.$$

In other words, when applied to an input sequence, π_b first swaps the i-th input bit for each i such that $b_i = 1$. Afterwards, it swaps the second and third card if $f(b) = 0$.

We can now describe the steps of our protocol:

1. (shuffle, $\{\pi_b : b \in \{0,1\}^k\}$), i.e., pick $b \in \{0,1\}^k$ uniformly at random and permute the cards with π_b.
2. (turn, $\{1, 4, 6, 8, \ldots, 2k\}$), i.e., turn over the first card and all cards with even indices except 2.
3. If the turn revealed ♣ in position 1 and ♡ everywhere else, i.e., the visible sequence is $(\clubsuit, ?, ?, \heartsuit, ?, \heartsuit, \ldots, ?, \heartsuit)$, then perform (result, 2, 3). Otherwise, (restart).

For a deeper understanding of what is actually going on, we suggest contemplating on Fig. 6 (which is, admittedly, somewhat intimidating), but correctness and privacy are surprisingly easy to show directly:

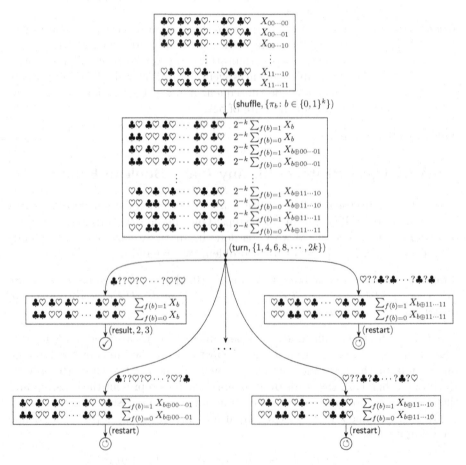

Fig. 6. The $2k$-card protocol for an arbitrary boolean function f of Theorem 4. We use the notation $b_1 \oplus b_2$ to denote the bitwise exclusive-or operation, e.g. $0011 \oplus 0101 = 0110$.

Correctness. Assume the input is $b \in \{0, 1\}^k$ and a result action is performed. Then the visible sequence after the turn was $(\clubsuit, ?, ?, \heartsuit, ?, \heartsuit, \ldots, ?, \heartsuit)$. This means the permutation π done by the shuffle must have first transformed the input sequence to $(\clubsuit, \heartsuit, \clubsuit, \heartsuit, \clubsuit, \heartsuit, \ldots, \clubsuit, \heartsuit)$ (before potentially flipping the cards in position 2 and 3). This can be interpreted as the sequence encoding only 0s, therefore π has flipped exactly the card pairs, where the input sequence had $(\heartsuit, \clubsuit)$ encoding 1. This implies $\pi = \pi_b$. From the definition of π_b it is now clear that the output is $(\heartsuit, \clubsuit)$ if $f(b) = 1$ and $(\clubsuit, \heartsuit)$ if $f(b) = 0$.

Privacy. Let v be a visible sequence after the turn step. Consider an input sequence Γ_b belonging to the input $b \in \{0, 1\}^k$. The probability that Γ_b yields the visible sequence v in the turn is exactly 2^{-k} since exactly one of the 2^k permutations in the shuffle action swaps the appropriate set of pairs of positions. This means the probability to observe v is 2^{-k} – and thus independent of the input sequence.

Runtime. The probability to observe $(\clubsuit, ?, ?, \heartsuit, \ldots, ?, \heartsuit)$ in the turn step is 2^{-k}, the probability to restart is therefore $1 - 2^{-k}$. This yields a runtime that is finite in expectation – of order $O(2^k)$. □

8   On the Implementation of Shuffle Operations

The shuffle used in the protocol in Sect. 7, while allowed in the formalism by [MS14a], is of questionable practicality: in general there is no obvious way to perform it in a real world situation with actual people and actual cards such that the players do not learn anything about the permutation that was done in the shuffle. In a weaker form this also applies to the protocols in Sects. 4 and 5.

Other shuffle operations, such as (shuffle, $\{\mathrm{id}, (1\ 2)\}$) that either perform a swap or do nothing, both with probability $\frac{1}{2}$, are unproblematic to implement with two players Alice and Bob: first let Alice perform the shuffle while Bob is looking away and then have Bob perform the shuffle while Alice is looking away. Provided they do not tell each other what they did, to both of them the cards seem to be swapped with probability 1/2. Here, it is crucial that performing the swap twice yields the identity: one of the allowed permutations.

In general, a shuffle action act = (shuffle, $\Pi, \mathcal{F}$) can be implemented in this way if act is *closed*, i.e., $\Pi^2 := \{\pi_1 \circ \pi_2 \mid \pi_1, \pi_2 \in \Pi\} = \Pi$ and *uniform*, i.e., $\mathcal{F}$ is the uniform distribution on Π. Note that our protocols in Sects. 4, 5 and 7 use shuffles that are not uniform and/or not closed, see Tables 1 and 2. Therefore, it may be worthwhile to continue studying shuffles in several directions:

- Restrict the computational model to uniform closed shuffles and examine the properties of the new model.
- Replace the action shuffle of the computational model by an alternative action playerPerm executed by a single player, while other players are not allowed to look on the table. Here, (playerPerm, $p, \Pi, \mathcal{F}$) is like (shuffle, $\Pi, \mathcal{F}$), with the difference that the executing player p learns which permutation has been

chosen. As argued above, this at least as powerful as allowing uniform closed shuffles.
- Search for a more clever way to implement shuffles with everyday objects.
- Weaken the honest-but-curious assumption and discuss implementations of shuffles with respect to, e.g., robustness against active attacks.

9   Conclusion

To summarize our results, we have extensively considered the question on tight lower bound on the number of cards for AND protocols, which has been open for several years. We believe that our answer to this question is satisfactory, as we do not only give two concrete AND protocols with different properties, we also show an impossibility result. Apart from the impossibility for perfect copy of a single card in [MS14a], we are the first to give such a type of result. This may be because of the sparsity of good ways to speak about card-based protocols. We believe to have overcome this problem by introducing an elegant "calculus of protocol states" in Sect. 3. Finally, we give a protocol for evaluating a k-ary boolean function with the theoretical minimum of cards, i.e., the $2k$ cards which are already necessary for encoding the input.

Open Problems. Our paper identifies a number of open problems in the field of card-based cryptographic protocols. This is, for example, how to implement non-closed or non-uniform shuffles and in consequence back up the current computational model with more evidence that its definition is rooted in reality. In the same way, we ask whether there is a finite-runtime five-card protocol using only closed and/or uniform shuffles.

The same set of questions which have been answered in Table 1 can also be asked for general boolean functions: What is the minimal number of cards for finite-runtime protocols with and without closed shuffles. Analogously, a tight lower bound on the number of cards in Las Vegas protocols using only uniform closed shuffles would be interesting.

Acknowledgments. We would like to thank the anonymous reviewers and Gunnar Hartung for helpful comments.

References

[B89] den Boer, B.: More efficient match-making and satisfiability. In: Quisquater, J.-J., Vandewalle, J. (eds.) EUROCRYPT 1989. LNCS, vol. 434, pp. 208–217. Springer, Heidelberg (1990)

[CHL13] Cheung, E., Hawthorne, C., Lee, P.: CS 758 Project: Secure Computation with Playing Cards (2013). https://cs.uwaterloo.ca/~p3lee/projects/cs758.pdf. Accessed on 02 October 2015

[CK93] Crépeau, C., Kilian, J.: Discreet solitary games. In: Stinson, D.R. (ed.) CRYPTO 1993. LNCS, vol. 773, pp. 319–330. Springer, Heidelberg (1994)

[MKS12] Mizuki, T., Kumamoto, M., Sone, H.: The five-card trick can be done with four cards. In: Wang, X., Sako, K. (eds.) ASIACRYPT 2012. LNCS, vol. 7658, pp. 598–606. Springer, Heidelberg (2012)

[MS09] Mizuki, T., Sone, H.: Six-card secure AND and four-card secure XOR. In: Deng, X., Hopcroft, J.E., Xue, J. (eds.) FAW 2009. LNCS, vol. 5598, pp. 358–369. Springer, Heidelberg (2009)

[MS14a] Mizuki, T., Shizuya, H.: A formalization of card-based cryptographic protocols via abstract machine. Int. J. Inf. Secur. **13**(1), 15–23 (2014). doi:10. 1007/s10207-013-0219-4

[MS14b] Mizuki, T., Shizuya, H.: Practical card-based cryptography. In: Ferro, A., Luccio, F., Widmayer, P. (eds.) FUN 2014. LNCS, vol. 8496, pp. 313–324. Springer, Heidelberg (2014)

[N+15] Nishida, T., Hayashi, Y., Mizuki, T., Sone, H.: Card-based protocols for any boolean function. In: Jain, R., Jain, S., Stephan, F. (eds.) TAMC 2015. LNCS, vol. 9076, pp. 110–121. Springer, Heidelberg (2015)

[NR98] Niemi, V., Renvall, A.: Secure multiparty computations without computers. Theor. Comput. Sci. **191**(1–2), 173–183 (1998). doi:10.1016/ S0304-3975(97)00107-2

[S01] Stiglic, A.: Computations with a deck of cards. Theor. Comput. Sci. **259**(1–2), 671–678 (2001). doi:10.1016/S0304-3975(00)00409-6

Author Index

Printed in the United States
By Bookmasters